Your Guide to Success in Math

Complete **Step 0** as soon as you begin your math course.

STEP 0: PLAN YOUR SEMESTER

☐ Register for the online part of the course (if there is one) ~~as possible.~~

☐ Fill in your Course and Contact information on this pull-out card.

☐ Write important dates from your syllabus on the Semester Organizer on this pull-out card.

Follow **Steps 1–4** during your course. Your instructor will tell you which resources to use—and when—in the textbook or eText, *MyMathGuide* workbook, videos, and **MyMathLab. Use these resources for extra help and practice.**

STEP 1: LEARN THE SKILLS AND CONCEPTS

☐ Read the **textbook** or **eText,** listen to your instructor's lecture, and/or watch the **videos.** You can work in *MyMathGuide* as you do this. As you are learning:

  ☐ Take notes, write down your questions, and save all your work (including homework solutions, quizzes, and tests) to review throughout the course.

  ☐ Work the *Skill to Review* exercises at the beginning of each section.

  ☐ Stop and do the *Margin* and *Guided Solution Exercises* as directed.

  ☐ Watch the videos. Answer the *Interactive Your Turn* questions in the videos and in *MyMathGuide.*

STEP 2: CHECK YOUR UNDERSTANDING

☐ Answer the *Reading Checks* in the Section Exercise sets or in MyMathLab.

☐ Explore the concepts using the *Active Learning Figures* in MyMathLab.

STEP 3: DO YOUR HOMEWORK

☐ Plan to spend 2 hours studying and doing homework for every hour of class.

☐ Complete your assigned homework from the textbook and/or in MyMathLab.

  ☐ When doing homework from the textbook, use the answer section to check your work.

  ☐ When doing homework in MyMathLab, use the Learning Aids, such as Help Me Solve This and View an Example, as needed, working toward being able to complete exercises without the aids.

STEP 4: REVIEW AND TEST YOUR UNDERSTANDING

☐ Work the exercises in the *Mid-Chapter Review.*

☐ Make your own chapter study sheet by doing the *Chapter Summary and Review.*

☐ Take the *Chapter Test* as a practice exam. To watch an instructor solve each problem, go to the Chapter Test Prep Videos in MyMathLab or on YouTube (search "BittingerIntro" and click on "Channels").

Use the *Studying for Success* tips in the text and the MyMathLab **Study Skills modules** (with videos, tips, and activities) to help you develop effective time-management, note-taking, test-prep, and other skills.

Student Organizer

Course Information

Course Number: _____ Name: _____

Location: _____ Days/Time: _____

Contact Information

Contact	Name	Email	Phone	Office Hours	Location
Instructor					
Tutor					
Math Lab					
Classmate					
Classmate					

Semester Organizer

Week	Homework	Quizzes and Tests	Other

INTRODUCTORY ALGEBRA

TWELFTH EDITION

MARVIN L. BITTINGER

Indiana University Purdue University Indianapolis

JUDITH A. BEECHER

BARBARA L. JOHNSON

Indiana University Purdue University Indianapolis

PEARSON

Boston   Columbus   Indianapolis   New York   San Francisco   Upper Saddle River
Amsterdam   Cape Town   Dubai   London   Madrid   Milan   Munich   Paris   Montréal   Toronto
Delhi   Mexico City   São Paulo   Sydney   Hong Kong   Seoul   Singapore   Taipei   Tokyo

Editorial Director	Christine Hoag
Editor in Chief	Maureen O'Connor
Executive Editor	Cathy Cantin
Content Editor	Katherine Minton
Editorial Assistant	Chase Hammond
Senior Managing Editor	Karen Wernholm
Senior Production Supervisor	Ron Hampton
Composition	PreMediaGlobal
Editorial and Production Services	Martha K. Morong/Quadrata, Inc.
Art Editor and Photo Researcher	The Davis Group, Inc.
Manager, Multimedia Production	Christine Stavrou
Associate Producer	Jonathan Wooding
Executive Content Manager	Rebecca Williams (MathXL)
Senior Content Developer	John Flanagan (TestGen)
Marketing Manager	Rachel Ross
Marketing Assistant	Kelly Cross
Senior Manufacturing Buyer	Debbie Rossi
Text Designer	The Davis Group, Inc.
Associate Design Director	Andrea Nix
Senior Designer/Cover Design	Barbara Atkinson
Cover Photograph	Frank Krahmer/Getty Images Inc.

Photo and Text Credits
Photo and text credits appear on page vi.

Library of Congress Cataloging-in-Publication Data
Bittinger, Marvin L.
    Introductory algebra / Marvin L. Bittinger,
Judith A. Beecher, Barbara L. Johnson.—Twelfth edition.
        pages cm
    Includes bibliographical references and index.
    ISBN: Student: 0-321-86796-3
  1. Arithmetic—Textbooks.   I. Beecher, Judith A.
II. Johnson, Barbara L. (Barbara Loreen),   III. Title.
    QA152.3.B57 2014
    512—dc23   2013026663

1 2 3 4 5 6 7 8 9 10—V011—16 15 14 13

www.pearsonhighered.com

ISBN-13: 978-0-321-86796-4
ISBN-10: 0-321-86796-3

Contents

CREDITS

8, Dr. Paulus Gerdes **30 (left),** Dorset Media Service/Alamy **30 (right),** Juice Images/Glow Images **32 (left),** ArenaCreative/Fotolia **32 (right),** IIIadam36III/Shutterstock **40,** Kokhanchikov/Shutterstock **46,** Veniamin Kraskov/Shutterstock **48,** Carlos Santa Maria/Fotolia **52,** Ed Metz/Shutterstock **59,** Dave King/Dorling Kindersley, Ltd. **66,** Comstock/Getty Images **72,** Mellowbox/Fotolia **82,** Ivan Alvarado/Reuters **85,** David Peart/Dorling Kindersley, Ltd. **121,** Brian Snyder/Reuters **149,** Leonid Tit/Fotolia **150,** London Photos/Alamy **153,** SoCalBatGal/Fotolia **154 (left),** Sky Bonillo/PhotoEdit **154 (right),** Anthony Berenyi/Shutterstock **161,** Jennifer Pritchard/MCT KRT/Newscom **162,** Gjeerawut/Fotolia **165 (left),** Iofoto/Shutterstock **165 (right),** Stew Milne/Associated Press **166 (left),** Wilson Araujo/Shutterstock **166 (right),** Michael Jung/Shutterstock **175,** Magic Mountain/Associated Press **176,** Stephen VanHorn/Shutterstock **179 (left),** Rodney Todt/Alamy **179 (right),** Corbis/SuperStock **180 (left),** courtesy of Indianapolis Motor Speedway **180 (right),** Lars Lindblad/Shutterstock **181 (left),** Barbara Johnson **181 (right),** Studio 8/Pearson Education Ltd. **182,** Elena Yakusheva/Shutterstock **197 (photo),** Kike Calvo VWPics/SuperStock **197 (figure),** U.S. Department of Health and Human Services and U.S. Department of Agriculture **200,** Andrey N. Bannov/Shutterstock **203 (left),** Reggie Lavoie/Shutterstock **203 (right),** Monkey Business/Fotolia **207,** pearlguy/Fotolia **208,** Outdoorsman/Fotolia **222,** Denis_pc/Fotolia **246,** David Pearson/Alamy **247,** Newspaper Association of America **248,** *Editor and Publisher* **251 (photo, left),** Alex Robinson/Dorling Kindersley, Ltd. **251 (photo, right),** Evan Meyer/Shutterstock **251 (figure, left),** *China Daily* **251 (figure, right),** Centers for Disease Control and Prevention **252 (left),** Maria Dryfhout/Shutterstock **252 (right),** Onizu3d/Fotolia **284,** Arinahabich/Fotolia **292,** Wei Ming/Shutterstock **295,** NASA **306 (left),** Anna Omelchenko/Fotolia **306 (right),** Nathan Devery/Science Source/Photo Researchers, Inc. **309,** Lorraine Swanson/Fotolia **310,** OJO Images, Ltd./Alamy **312 (left),** Terry Leung/Pearson Education Asia Ltd **312 (right),** Lana Rastro/Alamy **313 (top),** Engine Images/Fotolia **313 (bottom),** NASA **314 (left),** SeanPavonePhoto/Fotolia **314 (right),** Science Source/Photo Researchers, Inc. **323,** Brian Buckland **324,** Joern Sackermann/Alamy **357,** Linda Whitwam/Dorling Kindersley, Ltd. **361,** Coneyl Jay/Science Source/Photo Researchers, Inc. **415,** NASA **446,** Hans Neleman/Corbis/Glow Images **451,** Stephen Barnes/Hobbies and Crafts/Alamy **459,** Pattie Steib/Shutterstock **464,** Pedrosala/Shutterstock **484,** Igor Normann/Shutterstock **519,** Brocreative/Shutterstock **522,** Ortodox/Shutterstock **527 (left),** Elenathewise/Fotolia **527 (right),** Simon Johnsen/Shutterstock **528,** HandmadePictures/Shutterstock **529 (left),** Lightpoet/Shutterstock **529 (right),** Drazen/Shutterstock **530,** Catmando/Shutterstock **533,** Minadezhda/Fotolia **535,** Heather Charles/*The Indianapolis Star* **536,** EduardSV/Fotolia **537 (left and right),** NASA **539,** Ibooo7/Shutterstock **545 (left),** Dmitry Kalinovsky/Shutterstock **545 (right),** James McConnachie © Rough Guides/DK Images **549,** Clive Chilvers/Shutterstock **564,** Kyod/Landov **576,** Steve Pepple/Shutterstock **577,** Ron Brown/AGE Fotostock **581,** Lightpoet/Shutterstock **584,** Lev Kropotov/Shutterstock **602,** moodboard/Corbis **606,** Oskar Eyb/Glow Images **609 (left),** Glow Images **609 (right),** Caro/Alamy **610,** Action Plus Sports Images/Alamy **619,** Fotokostic/Shutterstock **626,** Renaud Visage/AGE Fotostock **636,** Steve Smith/SuperStock **640,** Holger Leue/AGE Fotostock **644,** Andrey Yurlov/Shutterstock **647 (top),** Photo courtesy LandWave Products, Inc., Pioneer, OH **647 (bottom),** Circle of Lights presented by the Contractors of Quality Connection and Electrical Workers of IBEW 481 **675,** National Geographic Image Collection/Alamy **689,** Wavebreakmedia/Shutterstock **698,** EWAStock/Glow Images **699,** Vladimir Sazonov/Fotolia **715,** EditorialByDarrellYoung/Alamy **717 (left),** AfricaMediaOnline/Glow Images **717 (right),** Spotmatik/Shutterstock **730,** Barbara Johnson **731,** National Geographic Image Collection/Glow Images **750,** Khuong Hoang/Getty Images **752 (left),** Aspen Rock/Shutterstock **752,** fStop/Alamy

Index of Applications

Preface

The Bittinger Program

Math hasn't changed, but students—and the way they learn it—have.

Introductory Algebra, Twelfth Edition, continues the Bittinger tradition of objective-based, guided learning, while integrating timely updates to the proven pedagogy. In this edition, there is a greater emphasis on guided learning and helping students get the most out of all of the course resources available with the Bittinger program, including new opportunities for mobile learning.

The program has expanded to include these comprehensive new teaching and learning resources: ***MyMathGuide* workbook**, **To-the-Point Objective Videos**, and enhanced, media-rich **MyMathLab** courses. Feedback from instructors and students motivated these and several other significant improvements: a new design to support guided learning, new figures and photos to help students visualize both concepts and applications, and many new and updated real-data applications to bring the math to life.

With so many resources available in so many formats, the trusted guidance of the Bittinger team on *what to do* and *when* will help today's math students stay on task. Students are encouraged to use ***Your Guide to Success in Math***, a four-step learning path and checklist available on the handy reference card in the front of this text and in MyMathLab. The guide will help students identify the resources in the textbook, supplements, and MyMathLab that support *their* learning style, as they develop and retain the skills and conceptual understanding they need to succeed in this and future courses.

In this preface, a look at the key new *and* hallmark resources and features of the *Introductory Algebra* program—including the textbook/eText, video program, *MyMathGuide* workbook, and MyMathLab—is organized around ***Your Guide to Success in Math***. This will help instructors direct students to the tools and resources that will help them most in a traditional lecture, hybrid, lab-based, or online environment.

NEW AND HALLMARK FEATURES IN RELATION TO
Your Guide to Success in Math

STEP 1 Learn the Skills and Concepts

Students have several options for learning, reviewing, and practicing the math concepts and skills.

Textbook/eText

☐ **Skill to Review.** At the beginning of nearly every text section, *Skill to Review* offers a just-in-time review of a previously presented skill that relates to the new material in the section. Section and objective references are included for the student's convenience, and two practice exercises are provided for review and reinforcement.

☐ **Margin Exercises.** For each objective, problems labeled "Do Exercise . . ." give students frequent opportunities to solve exercises while they learn.

☐ *New!* **Guided Solutions.** Nearly every section has *Guided Solution* margin exercises with fill-in blanks at key steps in the problem-solving process.

☐ *Enhanced!* **MyMathLab.** MyMathLab now includes *Active Learning Figures* for directed exploration of concepts; more problem types, including *Reading Checks* and *Guided Solutions*; and new, objective-based videos. (See pp. xiv–xvii for a detailed description of the features of MyMathLab.)

   ☐ *New!* **Skills Checks.** In the Learning Path for Ready-to-Go MyMathLab, each chapter begins with a brief assessment of students' mastery of the prerequisite skills needed to learn the new material in the chapter. Based on the results of this pretest, a personalized homework set is designed to help each student prepare for the chapter.

☐ *New!* **To-the-Point Objective Videos.** This is a comprehensive new program of objective-based, interactive videos that are incorporated into the Learning Path in MyMathLab and can be used hand-in-hand with the *MyMathGuide* workbook.

   ☐ *New!* **Interactive Your Turn Exercises.** For each objective in the videos, students solve exercises and receive instant feedback on their work.

☐ *New!* **MyMathGuide: Notes, Practice, and Video Path.** This is an objective-based workbook (available printed and in MyMathLab) for guided, hands-on learning. It offers vocabulary, skill, and concept review—along with problem-solving practice—with space to show work and write notes. Incorporated in the Learning Path in MyMathLab, it can be used together with the To-the-Point Objective Video program, instructor lectures, and the textbook.

STEP 2   Check Your Understanding

Throughout the program, students have frequent opportunities to check their work and confirm that they understand each skill and concept before moving on to the next topic.

☐ *New!* **Reading Checks.** At the beginning of each set of section exercises in the text, students demonstrate their grasp of the skills and concepts.

☐ *New!* **Active Learning Figures.** In MyMathLab, Active Learning Figures guide students in exploring math concepts and reinforcing their understanding.

☐ **Translating/Visualizing for Success.** In the text and in MyMathLab, these activities offer students extra practice with the important first step of the process for solving applied problems.

STEP 3   Do Your Homework

Introductory Algebra, Twelfth Edition, has a wealth of proven and updated exercises. Prebuilt assignments are available for instructors in MyMathLab, and they are preassigned and incorporated into the Learning Path in the Ready-to-Go course.

☐ **Skill Maintenance.** In each section, these exercises offer a thorough review of the math in the preceding text.

☐ **Synthesis Exercises.** To help build critical-thinking skills, these section exercises require students to use what they know and combine learning objectives from the current section with those from previous sections.

STEP 4 Review and Test Your Understanding

Students have a variety of resources to check their skills and understanding along the way and to help them prepare for tests.

- ☐ **Mid-Chapter Review.** Midway through each chapter, students work a set of exercises (*Concept Reinforcement, Guided Solutions, Mixed Review,* and *Understanding Through Discussion and Writing*) to confirm that they have grasped the skills and concepts covered in the first half before moving on to new material.

- ☐ **Summary and Review.** This resource provides an in-text opportunity for active learning and review for each chapter. *Vocabulary Reinforcement, Concept Reinforcement,* objective-based *Study Guide* (examples paired with similar exercises), *Review Exercises* (including *Synthesis* problems), and *Understanding Through Discussion and Writing* are included in these comprehensive chapter reviews.

- ☐ **Chapter Test.** Chapter Tests offer students the opportunity for comprehensive review and reinforcement prior to taking their instructor's exam. **Chapter Test-Prep Videos** (in MyMathLab and on YouTube) show step-by-step solutions to the Chapter Tests.

- ☐ **Cumulative Review.** Following every chapter beginning with Chapter 2, a Cumulative Review revisits skills and concepts from all preceding chapters to help students retain previously learned material.

Study Skills

Developing solid time-management, note-taking, test-taking, and other study skills is key to student success in math courses (as well as professionally and personally). Instructors can direct students to related study skills resources as needed.

- ☐ *New!* **Student Study Reference.** This pull-out card at the front of the text is perforated, three-hole-punched, and binder-ready for convenient reference. It includes **Your Guide to Success in Math** course checklist, **Student Organizer**, and **At a Glance**, a list of key information and expressions for quick reference as students work exercises and review for tests.

- ☐ *New!* **Studying for Success.** Checklists of study skills—designed to ensure that students develop the skills they need to succeed in math, school, and life—are integrated throughout the text at the beginning of selected sections.

- ☐ *New!* **Study Skills Modules.** In MyMathLab, interactive modules address common areas of weakness, including time-management, test-taking, and note-taking skills. Additional modules support career-readiness.

Learning Math in Context

- ☐ *New!* **Applications.** Throughout the text in examples and exercises, real-data applications encourage students to see and interpret the mathematics that appears every day in the world around them. Applications that use real data are drawn from business and economics, life and physical sciences, medicine, technology, and areas of general interest such as sports and daily life. New applications include "Cycling in Vietnam" (p. 167), "Gold Leaf on Georgia's State Capitol Dome" (p. 314), "Butterfly Wings" (p. 445), "Speed of Sea Animals" (p. 517), and "Produce Prices" (p. 576). For a complete list of applications, please refer to the Index of Applications (p. vii).

BREAK THROUGH
To improving results

MyMathLab
Ties the Complete Learning Program Together

MyMathLab® Online Course (access code required)

MyMathLab from Pearson is the world's leading online resource in mathematics, integrating interactive homework, assessment, and media in a flexible, easy-to-use format. MyMathLab delivers **proven results** in helping individual students succeed. It provides **engaging experiences** that personalize, stimulate, and measure learning for each student. And it comes from an **experienced partner** with educational expertise and an eye on the future.

MyMathLab for Developmental Mathematics

Prepared to go wherever you want to take your students.

Personalized Support for Students

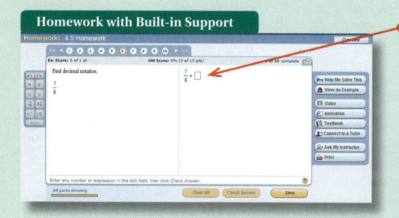

Exercises: The homework and practice exercises in MyMathLab are correlated to the exercises in the textbook, and they regenerate algorithmically to give students unlimited opportunities for practice and mastery. The software offers immediate, helpful feedback when students enter incorrect answers.

Multimedia Learning Aids: Exercises include guided solutions, sample problems, animations, videos, and eText access for extra help at point of use.

Expert Tutoring: Although many students describe the whole of MyMathLab as "like having your own personal tutor," students using MyMathLab do have access to live tutoring from qualified math instructors.

To help students achieve mastery, MyMathLab can generate **personalized homework** based on individual performance on tests or quizzes. Personalized homework allows students to focus on topics they have not yet mastered.

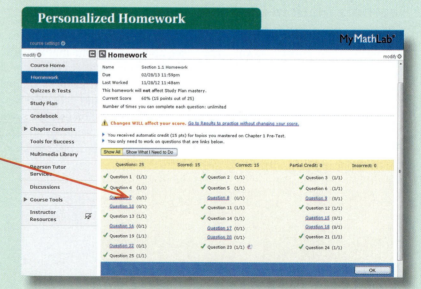

The **Adaptive Study Plan** makes studying more efficient and effective for every student. Performance and activity are assessed continually in real time. The data and analytics are used to provide personalized content—reinforcing concepts that target each student's strengths and weaknesses.

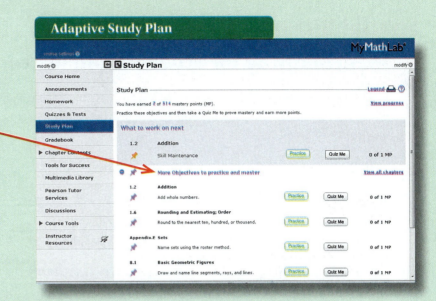

Flexible Design, Easy Start-Up, and Results for Instructors

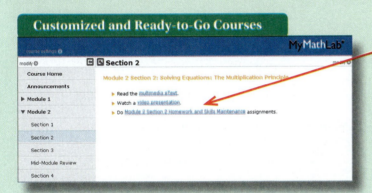

Instructors can modify the site navigation and insert their own directions on course-level landing pages; also, a **custom MyMathLab** course can be built that reorganizes and structures the course material by chapters, modules, units— whatever the need may be.

Ready-to-Go courses include preassigned homework, quizzes, and tests to make it even easier to get started. The Bittinger Ready-to-Go courses include new *Mid-Chapter Reviews* and *Reading Check Assignments*, plus a four-step Learning Path on each section-level landing page to help instructors direct students where to go and what resources to use.

The **comprehensive online gradebook** automatically tracks students' results on tests, quizzes, and homework and in the study plan. Instructors can use the gradebook to quickly intervene if students have trouble, or to provide positive feedback on a job well done. The data within MyMathLab are easily exported to a variety of spreadsheet programs, such as Microsoft Excel.® Instructors can determine which points of data to export and then analyze the results to determine success.

New features, such as **Search/Email by criteria**, make the gradebook a powerful tool for instructors. With this feature, instructors can easily communicate with both at-risk and successful students. They can search by score on specific assignments, noncompletion of assignments within a given time frame, last login date, or overall score.

Special Bittinger Resources
in MyMathLab for Students and Instructors

In addition to robust course delivery, MyMathLab offers the full Bittinger eText, additional Bittinger Program features, and the entire set of instructor and student resources in one easy-to-access online location.

New! Active Learning Figures

In MyMathLab, Active Learning Figures guide students in exploring math concepts and reinforcing their understanding. Instructors can use Active Learning Figures in class or as media assignments in MyMathLab.

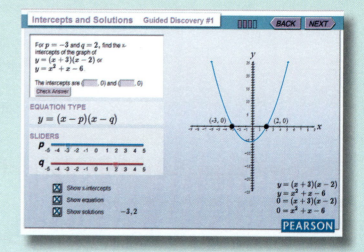

New! Four-Step Learning Path

Each of the section-level landing pages in the Ready-to-Go MyMathLab course includes a Learning Path that aligns with *Your Guide to Success in Math* to link students directly to the resources they should use when they need them. This also allows instructors to point students to the best resources to use at particular times.

New! Integrated Bittinger Video Program and *MyMathGuide* workbook
Bittinger Video Program*
(DVD ISBN: 978-0-321-92289-2)

The Video Program is available in MyMathLab and on DVD and includes closed captioning and the following video types:

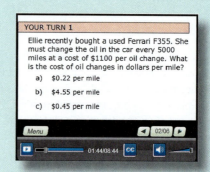

New! To-the-Point Objective Videos. These objective-based, interactive videos are incorporated into the Learning Path in MyMathLab and can be used along with the *MyMathGuide* workbook.

Chapter Test Prep Videos. The Chapter Test Prep Videos let students watch instructors work through step-by-step solutions to all the Chapter Test exercises from the textbook. Chapter Test Prep Videos are also available on YouTube (search using author name and book title).

New! *MyMathGuide: Notes, Practice, and Video Path* workbook*

(Printed Workbook ISBN: 978-0-321-92069-0)

This objective-based workbook for guided, hands-on learning offers vocabulary, skill, and concept review—along with problem-solving practice—with space to show work and write notes. Incorporated in the Learning Path in MyMathLab, *MyMathGuide* can be used together with the To-the-Point Objective Video program, instructor lectures, and the textbook. Instructors can assign To-the-Point Objective Videos in MyMathLab in conjunction with the *MyMathGuide* workbook.

Equations and Solutions

ESSENTIALS

An **equation** is a number sentence that says that the expressions on either side of the equals sign, =, represent the same number.

Any replacement for the variable that makes an equation true is called a **solution** of the equation. To solve an equation means to find *all* of its solutions.

Examples

- $2 + 5 = 7$ The equation is *true*.
- $9 - 3 = 3$ The equation is *false*.
- $x - 8 = 11$ The equation is *neither* true nor false, because we do not know what number x represents.

GUIDED LEARNING	📖 Textbook 👤 Instructor ▶ Video
EXAMPLE 1	YOUR TURN 1
Determine whether the equation is true, false, or neither. $4 - 6 = 2$ The equation is false.	Determine whether the equation is true, false, or neither. $5 - 9 = -4$
EXAMPLE 2	YOUR TURN 2
Determine whether the equation is true, false, or neither. $13 + 7 = 5 + 15$ The equation is true.	Determine whether the equation is true, false, or neither. $12 + 4 = 7 + 7$
EXAMPLE 3	YOUR TURN 3
Determine whether the equation is true, false, or neither. $x + 5 = 14$ The equation is neither true nor false, because we do not know what number x represents.	Determine whether the equation is true, false, or neither. $7 + 3 = x$

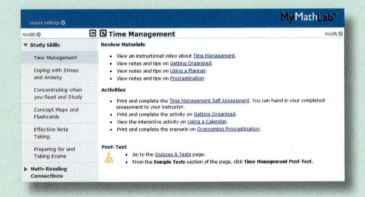

Study Skills Modules

In MyMathLab, interactive modules address common areas of weakness, including time-management, test-taking, and note-taking skills. Additional modules support career-readiness. Instructors can assign module material with a post-quiz.

Additional Resources in MyMathLab

For Students

Student's Solutions Manual*
(ISBN: 978-0-321-92299-1)

By Judy Penna

Contains completely worked-out annotated solutions for all the odd-numbered exercises in the text. Also includes fully worked-out annotated solutions for all the exercises (odd- and even-numbered) in the Mid-Chapter Reviews, the Summary and Reviews, the Chapter Tests, and the Cumulative Reviews.

For Instructors

Annotated Instructor's Edition**
(ISBN: 978-0-321-92295-3)

This version of the text includes answers to all exercises presented in the book, as well as helpful teaching tips.

Instructor's Resource Manual with Tests and Mini Lectures**
(download only)

By Laurie Hurley

This manual includes resources designed to help both new and experienced instructors with course preparation and classroom management. This includes chapter-by-chapter teaching tips and support for media supplements. Contains two multiple-choice tests per chapter, six free-response tests per chapter, and eight final exams.

Instructor's Solutions Manual**
(download only)

By Judy Penna

This manual contains detailed, worked-out solutions to all odd-numbered exercises and brief solutions to the even-numbered exercises in the exercise sets.

PowerPoint® Lecture Slides**
(download only)

Present key concepts and definitions from the text.

To learn more about how MyMathLab combines proven learning applications with powerful assessment, visit www.mymathlab.com or contact your Pearson representative.

---

*Printed supplements or DVDs are also available for separate purchase through MyMathLab, MyPearsonStore.com, or other retail outlets. They can also be value-packed with a textbook or MyMathLab code at a discount.

**Also available in print or for download from the Instructor Resource Center (IRC) on www.pearsonhighered.com.

Acknowledgments

Our deepest appreciation to all of you who helped to shape this edition by reviewing and spending time with us on your campuses. In particular, we would like to thank the following reviewers:

Afsheen Akbar, *Bergen Community College*
Erin Cooke, *Gwinnett Technical College*
Kay Davis, *Del Mar College*
Beverlee Drucker, *Northern Virginia Community College*
Sabine Eggleston, *Edison State College*
Dylan Faullin, *Dodge City Community College*
Rebecca Gubitti, *Edison State College*
Exie Hall, *Del Mar College*
Stephanie Houdek, *St. Cloud Technical Institute*
Linda Kass, *Bergen Community College*
Jamie Kleine, *Southwestern Illinois College*

Dorothy Marshall, *Edison State College*
Kimberley McHale, *Heartland Community College*
Arda Melkonian, *Victor Valley College*
Christian Miller, *Glendale Community College*
Joan Monaghan, *County College of Morris*
Thomas Pulver, *Waubonsee Community College*
Renee Quick, *Wallace State Community College–Hanceville*
Richard Semmler, *Northern Virginia Community College*
Jane Serbousek, *Northern Virginia Community College*
Alissa Sustarsic, *Lake Sumter Community College*
Melanie Walker, *Bergen Community College*

The endless hours of hard work by Martha Morong and Geri Davis have led to products of which we are immensely proud. We also want to thank Judy Penna for writing the Student's and Instructor's Solutions Manuals. Other strong support has come from Laurie Hurley for the *Instructor's Resource Manual* and for accuracy checking, along with checkers Judy Penna, Laurie Hurley, and Mike Rosenborg, and from proofreader David Johnson. Michelle Lanosga assisted with applications research. We also wish to recognize Nelson Carter, Tom Atwater, Judy Penna, and Laurie Hurley who prepared the videos.

In addition, a number of people at Pearson have contributed in special ways to the development and production of this textbook, including the Developmental Math team: Senior Production Supervisor Ron Hampton, Senior Designer Barbara Atkinson, Content Editor Katherine Minton, Editorial Assistant Chase Hammond, and Associate Media Producer Jonathan Wooding. Executive Editor Cathy Cantin and Marketing Manager Rachel Ross encouraged our vision and provided marketing insight.

CHAPTER
R

Prealgebra Review

R.1   Factoring, GCF, and LCM

OBJECTIVES

a Find all the factors of numbers and find prime factorizations of numbers.

b Find the GCF, the greatest common factor, of two or more numbers using prime factorizations.

c Find the LCM, the least common multiple, of two or more numbers using prime factorizations.

Find all the factors of each number.

1. 16

2. 63

Find some factorizations.

$$63 = 1 \cdot \boxed{},$$
$$63 = 3 \cdot \boxed{},$$
$$63 = 7 \cdot \boxed{}$$

The factors of 63 are
1, 3, $\boxed{}$, 9, 21, and 63.

3. 180

Answers

1. 1, 2, 4, 8, 16   **2.** 1, 3, 7, 9, 21, 63   **3.** 1, 2, 3, 4, 5, 6, 9, 10, 12, 15, 18, 20, 30, 36, 45, 60, 90, 180

Guided Solution:
2. 63, 21, 9, 7

a FACTORS AND PRIME FACTORIZATIONS

We begin our review with *factoring*, a necessary skill for addition and subtraction with fraction notation. Factoring is also an important skill in algebra.

The numbers we will be factoring are **natural numbers**:

1, 2, 3, 4, 5,   and so on.

To **factor** a number means to express the number as a product. Consider the product $12 = 3 \cdot 4$. We say that 3 and 4 are **factors** of 12 and that $3 \cdot 4$ is a **factorization** of 12. Since $12 = 1 \cdot 12$ and $12 = 2 \cdot 6$, we also know that 1, 12, 2, and 6 are factors of 12 and that $1 \cdot 12$ and $2 \cdot 6$ are factorizations of 12.

EXAMPLE 1   Find all the factors of 150.

We first find some factorizations:

$$150 = 1 \cdot 150, \qquad 150 = 2 \cdot 75, \qquad 150 = 3 \cdot 50,$$
$$150 = 5 \cdot 30, \qquad 150 = 6 \cdot 25, \qquad 150 = 10 \cdot 15.$$

The factors of 150 are 1, 2, 3, 5, 6, 10, 15, 25, 30, 50, 75, and 150.

Note that the word "factor" is used both as a noun and as a verb. We **factor** when we express a number as a product. The numbers we multiply together to get the product are **factors**.

◀ **Do Margin Exercises 1–3.**

PRIME NUMBER

A natural number that has *exactly two different factors*, itself and 1, is called a **prime number**.

EXAMPLE 2   Which of these numbers are prime?   7, 4, 11, 18, 1

7 is prime. It has exactly two different factors, 1 and 7.

4 is *not* prime. It has three different factors, 1, 2, and 4.

11 is prime. It has exactly two different factors, 1 and 11.

18 is *not* prime. It has factors 1, 2, 3, 6, 9, and 18.

1 is *not* prime. It does not have two *different* factors.

In the margin at right is a table of the prime numbers from 2 to 157. There are more extensive tables, but these prime numbers will be the most helpful to you in this text.

Do Exercise 4. ▶

If a natural number, other than 1, is not prime, we call it **composite**. Every composite number can be factored into a product of prime numbers. Such a factorization is called a **prime factorization**.

EXAMPLE 3 Find the prime factorization of 36.

We begin by factoring 36 any way we can. One way is like this:

$$36 = 4 \cdot 9.$$

The factors 4 and 9 are not prime, so we factor them:

$$36 = \quad 4 \quad \cdot \quad 9$$
$$= 2 \cdot 2 \cdot 3 \cdot 3.$$

The factors in the last factorization are all prime, so we now have the *prime factorization* of 36. Note that 1 is *not* part of this factorization because it is not prime.

Another way to find the prime factorization of 36 is like this:

$$36 = 2 \cdot 18 = 2 \cdot 3 \cdot 6 = 2 \cdot 3 \cdot 2 \cdot 3.$$

In effect, we begin factoring any way we can think of and keep factoring until all factors are prime. Using a **factor tree** might also be helpful.

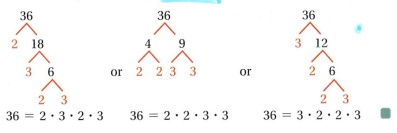

No matter which way we begin, the result is the same: The prime factorization of 36 contains two factors of 2 and two factors of 3. Every composite number has a *unique* prime factorization.

EXAMPLE 4 Find the prime factorization of 60.

This time, we use the list of primes from the table. We go through the table until we find a prime that is a factor of 60. The first such prime is 2.

$$60 = 2 \cdot 30$$

We keep dividing by 2 until it is not possible to do so.

$$60 = 2 \cdot 2 \cdot 15$$

Now we go to the next prime in the table that is a factor of 60. It is 3.

$$60 = 2 \cdot 2 \cdot 3 \cdot 5$$

Each factor in $2 \cdot 2 \cdot 3 \cdot 5$ is a prime. Thus this is the prime factorization.

Do Exercises 5–7. ▶

4. Which of these numbers are prime?

8, 6, 13, 14, 1

Find the prime factorization.

5. 48   **6.** 50

GS **7.** 770

Using a factor tree gives

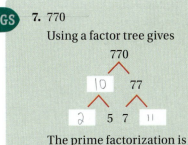

The prime factorization is $2 \cdot 5 \cdot \boxed{7} \cdot 11.$

Answers

4. 13   **5.** $2 \cdot 2 \cdot 2 \cdot 2 \cdot 3$ **6.** $2 \cdot 5 \cdot 5$
7. $2 \cdot 5 \cdot 7 \cdot 11$

Guided Solution:
7. 10, 2, 11, 7

b FINDING THE GREATEST COMMON FACTOR

The numbers 20 and 30 have several factors in common, among them 2 and 5. The greatest of the common factors is called the **greatest common factor, GCF**. One way to find the GCF is by making a list of factors of each number.

List all the factors of 20: 1, 2, 4, 5, 10, and 20.

List all the factors of 30: 1, 2, 3, 5, 6, 10, 15, and 30.

We now list the numbers common to both lists, the common factors:

1, 2, 5, and 10.

The greatest common factor, the GCF, is 10, the largest number in the common list.

The preceding procedure gives meaning to the notion of a GCF, but the following method, using prime factorizations, is generally faster.

EXAMPLE 5 Find the GCF of 20 and 30.

We find the prime factorization of each number. Then we draw lines between the common factors.

$$20 = 2 \cdot 2 \cdot 5$$
$$30 = 2 \cdot 3 \cdot 5$$

The GCF $= 2 \cdot 5 = 10$.

EXAMPLE 6 Find the GCF of 54, 90, and 252.

We find the prime factorization of each number. Then we draw lines between the common factors.

$$54 = 2 \cdot 3 \cdot 3 \cdot 3$$
$$90 = 2 \cdot 3 \cdot 3 \cdot 5$$
$$252 = 2 \cdot 2 \cdot 3 \cdot 3 \cdot 7$$

The GCF $= 2 \cdot 3 \cdot 3 = 18$.

EXAMPLE 7 Find the GCF of 30 and 77.

We find the prime factorization of each number. Then we draw lines between the common factors, if any exist.

$$30 = 2 \cdot 3 \cdot 5$$
$$77 = 7 \cdot 11$$

Since there is no common prime factor, the GCF is 1.

◀ **Do Exercises 8–11.**

Find the GCF.

8. 40, 100

GS
Find the prime factorization of each number and draw lines between common factors.

$$40 = 2 \cdot \boxed{} \cdot 2 \cdot 5$$
$$100 = 2 \cdot 2 \cdot 5 \cdot \boxed{}$$
The GCF $= 2 \cdot \boxed{} \cdot 5 = \boxed{}$.

9. 7, 21

10. 72, 360, 432

11. 3, 5, 22

Answers

8. 20 **9.** 7 **10.** 72 **11.** 1

Guided Solution:
8. 2, 5, 2, 20

c LEAST COMMON MULTIPLES

Least common multiples are used to add and subtract with fraction notation.
The **multiples** of a number all have that number as a factor. For example, the multiples of 2 are

2, 4, 6, 8, 10, 12, 14, 16,. . . .

We could name each of them in such a way as to show 2 as a factor. For example, $14 = 2 \cdot 7$.

The multiples of 3 all have 3 as a factor:

3, 6, 9, 12, 15, 18,. . . .

Two or more numbers always have many multiples in common. From lists of multiples, we can find common multiples.

EXAMPLE 8   Find the common multiples of 2 and 3.

We make lists of their multiples and circle the multiples that appear in both lists.

2, 4, ⑥, 8, 10, ⑫, 14, 16, ⑱, 20, 22, ㉔, 26, 28, ㉚, 32, 34, ㊱,. . . ;

3, ⑥, 9, ⑫, 15, ⑱, 21, ㉔, 27, ㉚, 33, ㊱,. . . .

The common multiples of 2 and 3 are

6, 12, 18, 24, 30, 36,. . . .

Do Exercises 12 and 13. ▶

In Example 8, we found common multiples of 2 and 3. The *least*, or smallest, of those common multiples is 6. We abbreviate **least common multiple** as **LCM**.

There are several methods that work well for finding the LCM of several numbers. Some of these do not work well when we consider expressions with variables such as $4ab$ and $12abc$. We now review a method that will work in arithmetic *and in algebra as well*. To see how it works, let's look at the prime factorizations of 9 and 15 in order to find the LCM:

$$9 = 3 \cdot 3, \qquad 15 = 3 \cdot 5.$$

Any multiple of 9 must have *two* 3's as factors. Any multiple of 15 must have *one* 3 and *one* 5 as factors. The smallest multiple of 9 and 15 is

  ┌─── Two 3's; 9 is a factor
$3 \cdot 3 \cdot 5 = 45.$
  └─── One 3, one 5; 15 is a factor

The LCM must have all the factors of 9 and all the factors of 15, *but the factors are not repeated when they are common to both numbers.*

To find the LCM of several numbers using prime factorizations:

a) Write the prime factorization of each number.

b) Form the LCM by writing the product of the different factors from step (a), using each factor the greatest number of times that it occurs in any *one* of the factorizations.

---

12. Find the common multiples of 3 and 5 by making lists of multiples.

13. Find the common multiples of 9 and 15 by making lists of multiples.

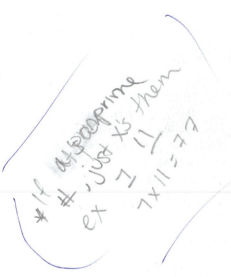

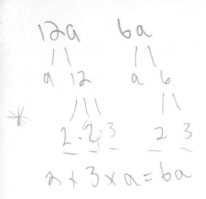

Find each LCM by factoring.

14. 8 and 10   **15.** 18 and 27

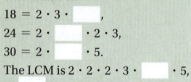

16. Find the LCM of 18, 24, and 30. (GS)
   Find the prime factorization of each number:
   $18 = 2 \cdot 3 \cdot \boxed{}$,
   $24 = 2 \cdot \boxed{} \cdot 2 \cdot 3$,
   $30 = 2 \cdot \boxed{} \cdot 5$.
   The LCM is $2 \cdot 2 \cdot 2 \cdot 3 \cdot \boxed{} \cdot 5$, or $\boxed{}$.

Find each LCM.

17. 3, 18   **18.** 12, 24

Find each LCM.

19. 4, 9   **20.** 5, 6, 7

Answers

14. 40   **15.** 54   **16.** 360   **17.** 18   **18.** 24
19. 36   **20.** 210

Guided Solution:
16. 3, 2, 3, 3, 360

EXAMPLE 9   Find the LCM of 40 and 100.

a) We find the prime factorizations:
$$40 = 2 \cdot 2 \cdot 2 \cdot 5,$$
$$100 = 2 \cdot 2 \cdot 5 \cdot 5.$$

b) The different prime factors are 2 and 5. We write 2 as a factor three times (the greatest number of times that it occurs in any *one* factorization). We write 5 as a factor two times (the greatest number of times that it occurs in any *one* factorization).

The LCM is $2 \cdot 2 \cdot 2 \cdot 5 \cdot 5$, or 200.

◀ **Do Exercises 14 and 15.**

EXAMPLE 10   Find the LCM of 27, 90, and 84.

a) We factor:
$$27 = 3 \cdot 3 \cdot 3,$$
$$90 = 2 \cdot 3 \cdot 3 \cdot 5,$$
$$84 = 2 \cdot 2 \cdot 3 \cdot 7.$$

b) We write 2 as a factor two times, 3 three times, 5 one time, and 7 one time.

The LCM is $2 \cdot 2 \cdot 3 \cdot 3 \cdot 3 \cdot 5 \cdot 7$, or 3780.

◀ **Do Exercise 16.**

EXAMPLE 11   Find the LCM of 7 and 21.

Since 7 is prime, it has no prime factorization. It still, however, must be a factor of the LCM:
$$7 = 7,$$
$$21 = 3 \cdot 7.$$

The LCM is $7 \cdot 3$, or 21.

> If one number is a factor of another, then the LCM is the larger of the two numbers.

◀ **Do Exercises 17 and 18.**

EXAMPLE 12   Find the LCM of 8 and 9.

We have
$$8 = 2 \cdot 2 \cdot 2,$$
$$9 = 3 \cdot 3.$$

The LCM is $2 \cdot 2 \cdot 2 \cdot 3 \cdot 3$, or 72.

> If two or more numbers have no common prime factor, then the LCM is the product of the numbers.

◀ **Do Exercises 19 and 20.**

✓ Reading Check

Complete each statement by choosing the correct term from the words under the blank.

RC1. If one number is a factor of another, then the GCF is the _____ of the two numbers.
smaller/larger

RC2. A number that has more than two factors is called a _____ number.
prime/composite

RC3. If one number is a factor of another, then the LCM is the _____ of the two numbers.
smaller/larger

RC4. A number that has exactly two different factors, itself and 1, is called a _____ number.
prime/composite

Always review the objectives before doing an exercise set. See page 2. Note how the objectives are keyed to the exercises.

a   Find all the factors of each number.

1. 20   **2.** 36   **3.** 72   **4.** 81

Find the prime factorization of each number.

5. 15   **6.** 33   **7.** 49   **8.** 121

9. 18   **10.** 24   **11.** 40   **12.** 56

13. 90   **14.** 120   **15.** 210   **16.** 330

17. 91   **18.** 143

b   Find the GCF of each number.

19. 36, 48   **20.** 40, 88   **21.** 13, 52   **22.** 14, 42

23. 15, 28   **24.** 27, 40   **25.** 54, 180   **26.** 40, 220

27. 18, 66   **28.** 30, 135   **29.** 15, 40, 60   **30.** 25, 70, 125

31. 70, 105, 350   **32.** 99, 110, 825

c   Find the prime factorization of the numbers. Then find the LCM.

33. 24, 36   **34.** 24, 27   **35.** 3, 15   **36.** 20, 40

37. 30, 40   **38.** 50, 60   **39.** 13, 23   **40.** 17, 29

41. 18, 30   **42.** 45, 72   **43.** 30, 36   **44.** 30, 50

45. 12, 28
46. 35, 45
47. 24, 36, 12
48. 8, 16, 22

49. 5, 12, 15
50. 12, 18, 40
51. 6, 12, 18
52. 24, 35, 45

Planet Orbits. The earth, Jupiter, Saturn, and Uranus all revolve around the sun. The earth takes 1 year, Jupiter 12 years, Saturn 30 years, and Uranus 84 years to make a complete revolution. On a certain night, you look at those three distant planets and wonder how many years it will be before they have the same position. (*Hint*: To find out, determine the LCM of 12, 30, and 84. It will be that number of years.)

Source: *The Handy Science Answer Book.*

53. How often will Jupiter and Saturn have the same position in the night sky as seen from the earth?

54. How often will Jupiter and Uranus have the same position in the night sky as seen from the earth?

55. How often will Saturn and Uranus have the same position in the night sky as seen from the earth?

56. How often will Jupiter, Saturn, and Uranus have the same position in the night sky as seen from the earth?

African Artistry. In Africa, the design of every woven handbag, or *gipatsi* (plural, *sipatsi*), is created by repeating two or more geometric patterns. Each pattern encircles the bag, sharing the strands of fabric with any pattern above or below. The length, or period, of each pattern is the number of strands required to construct the pattern. For a gipatsi to be considered beautiful, each individual pattern must fit a whole number of times around the bag.

Source: Gerdes, Paulus. *Women, Art and Geometry in Southern Africa.* Asmara, Eritrea: Africa World Press, Inc., p. 5.

57. A weaver is using two patterns to create a gipatsi. Pattern A is 10 strands long, and pattern B is 3 strands long. What is the smallest number of strands that can be used to complete the gipatsi?

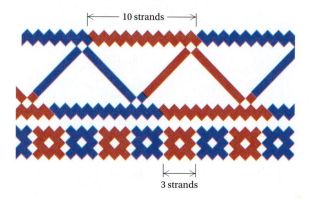

58. A weaver is using a four-strand pattern, a six-strand pattern, and an eight-strand pattern. What is the smallest number of strands that can be used to complete the gipatsi?

Synthesis

To the student and the instructor: The Synthesis exercises found at the end of every exercise set challenge students to combine concepts or skills studied in that section or in preceding parts of the text. Exercises marked with a ▦ symbol are meant to be solved using a calculator.

59. Consider the numbers 8 and 12. Determine whether each of the following is the LCM of 8 and 12. Tell why or why not.

a) $2 \cdot 2 \cdot 3 \cdot 3$
b) $2 \cdot 2 \cdot 2 \cdot 3 \cdot 5$
c) $2 \cdot 3 \cdot 3$
d) $2 \cdot 2 \cdot 2 \cdot 3$

Fraction Notation

We now review fraction notation and its use with addition, subtraction, multiplication, and division of *arithmetic numbers*.

a EQUIVALENT EXPRESSIONS AND FRACTION NOTATION

An example of **fraction notation** for a number is

$$\frac{2}{3} \begin{array}{l} \leftarrow \text{Numerator} \\ \leftarrow \text{Denominator} \end{array}$$

The top number is called the **numerator**, and the bottom number is called the **denominator**.

The **whole numbers** consist of the natural numbers and 0:

$$0, \quad 1, \quad 2, \quad 3, \quad 4, \quad 5, \ldots.$$

The **arithmetic numbers**, also called the **nonnegative rational numbers**, consist of the whole numbers and the fractions, such as $\frac{2}{3}$ and $\frac{9}{5}$.

ARITHMETIC NUMBERS (NONNEGATIVE RATIONAL NUMBERS)

The **arithmetic numbers** are the whole numbers and the fractions, such as 8, $\frac{3}{4}$, and $\frac{6}{5}$. All these numbers can be named with fraction notation $\frac{a}{b}$, where a and b are whole numbers and $b \neq 0$.

Note that all whole numbers can be named with fraction notation. For example, we can name the whole number 8 as $\frac{8}{1}$. We call 8 and $\frac{8}{1}$ **equivalent expressions**.

Being able to find an equivalent expression is critical to a study of algebra. Two simple but powerful properties of numbers that allow us to find equivalent expressions are the identity properties of 0 and 1.

THE IDENTITY PROPERTY OF 0 (ADDITIVE IDENTITY)

For any number a,

$$a + 0 = a.$$

(Adding 0 to any number gives that same number—for example, $12 + 0 = 12$.)

THE IDENTITY PROPERTY OF 1 (MULTIPLICATIVE IDENTITY)

For any number a,

$$a \cdot 1 = a.$$

(Multiplying any number by 1 gives that same number—for example, $\frac{3}{5} \cdot 1 = \frac{3}{5}$.)

OBJECTIVES

a Find equivalent fraction expressions by multiplying by 1.

b Simplify fraction notation.

c Add, subtract, multiply, and divide using fraction notation.

SKILL TO REVIEW

Objective R.1a: Find prime factorizations of numbers.

Find the prime factorization.

1. 20
2. 84

Answers

Skill to Review:
1. $2 \cdot 2 \cdot 5$ 2. $2 \cdot 2 \cdot 3 \cdot 7$

Here are some ways to name the number 1:

$$\frac{5}{5}, \quad \frac{3}{3}, \quad \text{and} \quad \frac{26}{26}.$$

The following property allows us to find equivalent fraction expressions.

EQUIVALENT EXPRESSIONS FOR 1

For any number a, $a \neq 0$,

$$\frac{a}{a} = 1.$$

We can use the identity property of 1 and the preceding result to find equivalent fraction expressions.

EXAMPLE 1 Write a fraction expression equivalent to $\frac{2}{3}$ with a denominator of 15.

Note that $15 = 3 \cdot 5$. We want fraction notation for $\frac{2}{3}$ that has a denominator of 15, but the denominator 3 is missing a factor of 5. We multiply by 1, using $\frac{5}{5}$ as an equivalent expression for 1. Recall from arithmetic that to multiply with fraction notation, we multiply numerators and we multiply denominators:

$$\frac{2}{3} = \frac{2}{3} \cdot 1 \qquad \text{Using the identity property of 1}$$

$$= \frac{2}{3} \cdot \frac{5}{5} \qquad \text{Using } \frac{5}{5} \text{ for 1}$$

$$= \frac{10}{15}. \qquad \text{Multiplying numerators and denominators}$$

◀ **Do Exercises 1–3.**

b SIMPLIFYING FRACTION NOTATION

We know that $\frac{1}{2}, \frac{2}{4}, \frac{4}{8}$, and so on, all name the same number. Any arithmetic number can be named in many ways. The **simplest fraction notation** is the notation that has the smallest numerator and denominator. We call the process of finding the simplest fraction notation **simplifying**. We reverse the process of Example 1 by first factoring the numerator and the denominator. Then we factor the fraction expression and remove a factor of 1 using the identity property of 1.

EXAMPLE 2 Simplify: $\dfrac{10}{15}$.

$$\frac{10}{15} = \frac{2 \cdot 5}{3 \cdot 5} \qquad \text{Factoring the numerator and the denominator.}$$
$$\qquad\qquad \text{In this case, each is the prime factorization.}$$

$$= \frac{2}{3} \cdot \frac{5}{5} \qquad \text{Factoring the fraction expression}$$

$$= \frac{2}{3} \cdot 1$$

$$= \frac{2}{3} \qquad \text{Using the identity property of 1 (removing a factor of 1)}$$

---

1. Write a fraction expression equivalent to $\frac{2}{3}$ with a denominator of 12.

2. Write a fraction expression equivalent to $\frac{3}{4}$ with a denominator of 28.

3. Multiply by 1 to find three different fraction expressions for $\frac{7}{8}$.

Answers

1. $\frac{8}{12}$ 2. $\frac{21}{28}$ 3. $\frac{14}{16}, \frac{21}{24}, \frac{28}{32};$
answers may vary

EXAMPLE 3 Simplify: $\frac{36}{24}$.

$$\frac{36}{24} = \frac{2 \cdot 3 \cdot 2 \cdot 3}{2 \cdot 2 \cdot 3 \cdot 2}$$ Factoring the numerator and the denominator

$$= \frac{2 \cdot 3 \cdot 2}{2 \cdot 3 \cdot 2} \cdot \frac{3}{2}$$ Factoring the fraction expression

$$= 1 \cdot \frac{3}{2}$$

$$= \frac{3}{2}$$ Removing a factor of 1

It is always a good idea to check at the end to see if you have indeed factored out all the common factors of the numerator and the denominator.

Canceling

Canceling is a shortcut that you may have used to remove a factor of 1 when working with fraction notation. With *great* concern, we mention it as a possible way to speed up your work. You should use canceling only when removing common factors in numerators and denominators. Each common factor allows us to remove a factor of 1 in a product. **Canceling *cannot* be done when adding.** Our concern is that "canceling" be performed with care and understanding. Example 3 might have been done faster as follows:

$$\frac{36}{24} = \frac{\cancel{2} \cdot \cancel{3} \cdot 2 \cdot 3}{\cancel{2} \cdot 2 \cdot \cancel{3} \cdot 2} = \frac{3}{2}, \quad \text{or} \quad \frac{36}{24} = \frac{3 \cdot \cancel{12}}{2 \cdot \cancel{12}} = \frac{3}{2}, \quad \text{or} \quad \frac{\overset{\overset{3}{\cancel{18}}}{\cancel{36}}}{\underset{\underset{2}{\cancel{12}}}{\cancel{24}}} = \frac{3}{2}.$$

Do Exercises 4–7. ▶

We can always insert the number 1 as a factor. The identity property of 1 allows us to do that.

EXAMPLE 4 Simplify: $\frac{18}{72}$.

$$\frac{18}{72} = \frac{2 \cdot \cancel{9}}{8 \cdot \cancel{9}} = \frac{2}{8} = \frac{2 \cdot 1}{2 \cdot 4} = \frac{1}{4}, \quad \text{or} \quad \frac{18}{72} = \frac{1 \cdot \cancel{18}}{4 \cdot \cancel{18}} = \frac{1}{4}$$

EXAMPLE 5 Simplify: $\frac{72}{9}$.

$$\frac{72}{9} = \frac{8 \cdot 9}{1 \cdot 9}$$ Factoring and inserting a factor of 1 in the denominator

$$= \frac{8 \cdot \cancel{9}}{1 \cdot \cancel{9}}$$ Removing a factor of 1: $\frac{9}{9} = 1$

$$= \frac{8}{1} = 8$$ Simplifying

Do Exercises 8 and 9. ▶

Simplify.

GS **4.** $\frac{18}{45} = \frac{2 \cdot 3 \cdot 3}{3 \cdot 3 \cdot 5}$

$$= \frac{3 \cdot 3}{3 \cdot 3} \cdot \frac{\boxed{}}{5} = \frac{2}{\boxed{}}$$

5. $\frac{38}{18}$ **6.** $\frac{72}{27}$

GS **7.** $\frac{32}{56} = \frac{8 \cdot \boxed{}}{8 \cdot 7} = \frac{\boxed{}}{7}$

Simplify.

8. $\frac{27}{54}$ **9.** $\frac{48}{12}$

Answers

4. $\frac{2}{5}$ **5.** $\frac{19}{9}$ **6.** $\frac{8}{3}$ **7.** $\frac{4}{7}$ **8.** $\frac{1}{2}$ **9.** 4

Guided Solutions:
4. 2, 5 **7.** 4, 4

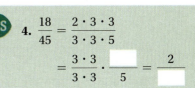

C MULTIPLICATION, ADDITION, SUBTRACTION, AND DIVISION

After we have performed an operation of multiplication, addition, subtraction, or division, the answer may not be in simplified form. We simplify, if at all possible.

Multiplication

To multiply using fraction notation, we multiply the numerators to get the new numerator, and we multiply the denominators to get the new denominator.

> **MULTIPLYING FRACTIONS**
>
> To multiply fractions, multiply the numerators and multiply the denominators:
>
> $$\frac{a}{b} \cdot \frac{c}{d} = \frac{a \cdot c}{b \cdot d}.$$

EXAMPLE 6 Multiply and simplify: $\frac{5}{6} \cdot \frac{9}{25}$.

$$\frac{5}{6} \cdot \frac{9}{25} = \frac{5 \cdot 9}{6 \cdot 25} \qquad \text{Multiplying numerators and denominators}$$

$$= \frac{5 \cdot 3 \cdot 3}{2 \cdot 3 \cdot 5 \cdot 5} \qquad \text{Factoring the numerator and the denominator}$$

$$= \frac{\cancel{5} \cdot \cancel{3} \cdot 3}{2 \cdot \cancel{3} \cdot \cancel{5} \cdot 5} \qquad \text{Removing a factor of 1: } \frac{3 \cdot 5}{3 \cdot 5} = 1$$

$$= \frac{3}{10} \qquad \text{Simplifying}$$

◀ **Do Exercises 10 and 11.**

Multiply and simplify.

10. $\frac{6}{5} \cdot \frac{25}{12}$ **11.** $\frac{3}{8} \cdot \frac{5}{3} \cdot \frac{7}{2}$

Addition

When denominators are the same, we can add by adding the numerators and keeping the same denominator.

> **ADDING FRACTIONS WITH LIKE DENOMINATORS**
>
> To add fractions when denominators are the same, add the numerators and keep the same denominator:
>
> $$\frac{a}{c} + \frac{b}{c} = \frac{a + b}{c}.$$

EXAMPLE 7 Add: $\frac{4}{8} + \frac{5}{8}$.

The common denominator is 8. We add the numerators and keep the common denominator:

$$\frac{4}{8} + \frac{5}{8} = \frac{4 + 5}{8} = \frac{9}{8}.$$

Answers

10. $\frac{5}{2}$ **11.** $\frac{35}{16}$

In arithmetic, we generally write $\frac{9}{8}$ as $1\frac{1}{8}$. (See a review of converting from a mixed numeral to fraction notation at right.) In algebra, you will find that *improper fraction* symbols such as $\frac{9}{8}$ are more useful and are quite *proper* for our purposes.

What do we do when denominators are different? We find a common denominator. We can do this by multiplying by 1. Consider adding $\frac{1}{6}$ and $\frac{3}{4}$. There are several common denominators that can be obtained. Let's look at two possibilities.

A. $\dfrac{1}{6} + \dfrac{3}{4} = \dfrac{1}{6} \cdot 1 + \dfrac{3}{4} \cdot 1$

$\qquad = \dfrac{1}{6} \cdot \dfrac{4}{4} + \dfrac{3}{4} \cdot \dfrac{6}{6}$

$\qquad = \dfrac{4}{24} + \dfrac{18}{24}$

$\qquad = \dfrac{22}{24}$

$\qquad = \dfrac{11}{12} \qquad$ Simplifying

B. $\dfrac{1}{6} + \dfrac{3}{4} = \dfrac{1}{6} \cdot 1 + \dfrac{3}{4} \cdot 1$

$\qquad = \dfrac{1}{6} \cdot \dfrac{2}{2} + \dfrac{3}{4} \cdot \dfrac{3}{3}$

$\qquad = \dfrac{2}{12} + \dfrac{9}{12}$

$\qquad = \dfrac{11}{12}$

We had to simplify in **A**. We didn't have to simplify in **B**. In **B**, we used the least common multiple of the denominators, 12. That number is called the **least common denominator**, or **LCD**. Using the LCD allows us to add fractions using the smallest numbers possible.

> To convert from a mixed numeral to fraction notation:
>
> $\overset{\text{(b)}}{\underset{\text{(a)}}{\curvearrowright}}\ 3\frac{5}{8} = \frac{29}{8} \leftarrow \text{(c)}$
>
> (a) Multiply the whole number by the denominator:
>
> $\qquad 3 \cdot 8 = 24.$
>
> (b) Add the result to the numerator:
>
> $\qquad 24 + 5 = 29.$
>
> (c) Keep the denominator.

ADDING FRACTIONS WITH DIFFERENT DENOMINATORS

To add fractions when denominators are different:

a) Find the least common multiple of the denominators. That number is the least common denominator, LCD.

b) Multiply by 1, using the appropriate notation n/n for each fraction to express fractions in terms of the LCD.

c) Add the numerators, keeping the same denominator.

d) Simplify, if possible.

EXAMPLE 8 Add and simplify: $\dfrac{3}{8} + \dfrac{5}{12}$.

The LCM of the denominators, 8 and 12, is 24. Thus the LCD is 24. We multiply each fraction by 1 to obtain the LCD:

$\dfrac{3}{8} + \dfrac{5}{12} = \dfrac{3}{8} \cdot \dfrac{3}{3} + \dfrac{5}{12} \cdot \dfrac{2}{2}$ Multiplying by 1. Since $3 \cdot 8 = 24$, we multiply the first number by $\frac{3}{3}$. Since $2 \cdot 12 = 24$, we multiply the second number by $\frac{2}{2}$.

$\qquad = \dfrac{9}{24} + \dfrac{10}{24}$

$\qquad = \dfrac{9 + 10}{24}$ Adding the numerators and keeping the same denominator

$\qquad = \dfrac{19}{24}.$

Add and simplify.

12. $\dfrac{4}{5} + \dfrac{3}{5}$ **13.** $\dfrac{5}{6} + \dfrac{7}{6}$

14. $\dfrac{5}{6} + \dfrac{7}{10}$ **GS**

$$= \dfrac{5}{2 \cdot 3} + \dfrac{7}{\boxed{} \cdot 5}$$

The LCD is $2 \cdot 3 \cdot \boxed{}$, or 30.

$$= \dfrac{5}{2 \cdot 3} \cdot \dfrac{\boxed{}}{5} + \dfrac{7}{2 \cdot 5} \cdot \dfrac{3}{3}$$

$$= \dfrac{5 \cdot 5 + 7 \cdot 3}{2 \cdot 3 \cdot 5}$$

$$= \dfrac{25 + \boxed{}}{2 \cdot 3 \cdot 5} = \dfrac{46}{2 \cdot 3 \cdot 5}$$

$$= \dfrac{2 \cdot \boxed{}}{2 \cdot 3 \cdot 5} = \dfrac{23}{\boxed{}}$$

15. $\dfrac{13}{24} + \dfrac{7}{40}$

EXAMPLE 9 Add and simplify: $\dfrac{11}{30} + \dfrac{5}{18}$.

We first look for the LCM of 30 and 18. That number is then the LCD. We find the prime factorization of each denominator:

$$\dfrac{11}{30} + \dfrac{5}{18} = \dfrac{11}{5 \cdot 2 \cdot 3} + \dfrac{5}{2 \cdot 3 \cdot 3}.$$

The LCD is $5 \cdot 2 \cdot 3 \cdot 3$, or 90. To get the LCD in the first denominator, we need a factor of 3. To get the LCD in the second denominator, we need a factor of 5. We get these numbers by multiplying by 1:

$$\dfrac{11}{30} + \dfrac{5}{18} = \dfrac{11}{5 \cdot 2 \cdot 3} \cdot \dfrac{3}{3} + \dfrac{5}{2 \cdot 3 \cdot 3} \cdot \dfrac{5}{5} \qquad \text{Multiplying by 1}$$

$$= \dfrac{33}{5 \cdot 2 \cdot 3 \cdot 3} + \dfrac{25}{2 \cdot 3 \cdot 3 \cdot 5} \qquad \text{The denominators are now the LCD.}$$

$$= \dfrac{58}{5 \cdot 2 \cdot 3 \cdot 3} \qquad \text{Adding the numerators and keeping the LCD}$$

$$= \dfrac{2 \cdot 29}{5 \cdot 2 \cdot 3 \cdot 3} \qquad \text{Factoring the numerator and removing a factor of 1}$$

$$= \dfrac{29}{45}. \qquad \text{Simplifying}$$

◀ **Do Exercises 12–15.**

Subtraction

When subtracting, we also multiply by 1 to obtain the LCD. After we have made the denominators the same, we can subtract by subtracting the numerators and keeping the same denominator.

EXAMPLE 10 Subtract and simplify: $\dfrac{9}{8} - \dfrac{4}{5}$.

$$\dfrac{9}{8} - \dfrac{4}{5} = \dfrac{9}{8} \cdot \dfrac{5}{5} - \dfrac{4}{5} \cdot \dfrac{8}{8} \qquad \text{The LCD is 40.}$$

$$= \dfrac{45}{40} - \dfrac{32}{40}$$

$$= \dfrac{45 - 32}{40} = \dfrac{13}{40} \qquad \text{Subtracting the numerators and keeping the same denominator}$$

EXAMPLE 11 Subtract and simplify: $\dfrac{7}{10} - \dfrac{1}{5}$.

$$\dfrac{7}{10} - \dfrac{1}{5} = \dfrac{7}{10} - \dfrac{1}{5} \cdot \dfrac{2}{2} \qquad \text{The LCD is 10; } \dfrac{7}{10} \text{ already has the LCD.}$$

$$= \dfrac{7}{10} - \dfrac{2}{10} = \dfrac{7 - 2}{10}$$

$$= \dfrac{5}{10}$$

$$= \dfrac{1 \cdot 5}{2 \cdot 5} = \dfrac{1}{2} \qquad \text{Removing a factor of 1: } \dfrac{5}{5} = 1$$

Subtract and simplify.

16. $\dfrac{7}{8} - \dfrac{2}{5}$ **17.** $\dfrac{5}{12} - \dfrac{2}{9}$

◀ **Do Exercises 16 and 17.**

Answers

12. $\dfrac{7}{5}$ **13.** 2 **14.** $\dfrac{23}{15}$ **15.** $\dfrac{43}{60}$ **16.** $\dfrac{19}{40}$

17. $\dfrac{7}{36}$

Guided Solution:

14. 2, 5, 5, 21, 23, 15

Reciprocals

Two numbers whose product is 1 are called **reciprocals**, or **multiplicative inverses**, of each other. All the arithmetic numbers, except zero, have reciprocals.

EXAMPLES

12. The reciprocal of $\frac{2}{3}$ is $\frac{3}{2}$ because $\frac{2}{3} \cdot \frac{3}{2} = \frac{6}{6} = 1$.

13. The reciprocal of 9 is $\frac{1}{9}$ because $9 \cdot \frac{1}{9} = \frac{9}{9} = 1$.

14. The reciprocal of $\frac{1}{4}$ is 4 because $\frac{1}{4} \cdot 4 = \frac{4}{4} = 1$.

Do Exercises 18–21.

Find each reciprocal.

18. $\dfrac{4}{11}$ **19.** $\dfrac{15}{7}$

20. 5   **21.** $\dfrac{1}{3}$

Reciprocals and Division

Reciprocals and the number 1 can be used to justify a fast way to divide arithmetic numbers. We multiply by 1, carefully choosing the expression for 1.

EXAMPLE 15 Divide $\dfrac{2}{3}$ by $\dfrac{7}{5}$.

This is a symbol for 1.

$$\frac{2}{3} \div \frac{7}{5} = \frac{\frac{2}{3}}{\frac{7}{5}} = \frac{\frac{2}{3}}{\frac{7}{5}} \cdot \frac{\frac{5}{7}}{\frac{5}{7}} \qquad \text{Multiplying by } \tfrac{5}{7}. \text{ We use } \tfrac{5}{7} \text{ because it is the reciprocal of the divisor, } \tfrac{7}{5}.$$

$$= \frac{\frac{2}{3} \cdot \frac{5}{7}}{\frac{7}{5} \cdot \frac{5}{7}} \qquad \text{Multiplying numerators and denominators}$$

$$= \frac{\frac{10}{21}}{\frac{35}{35}} = \frac{\frac{10}{21}}{1} \qquad \tfrac{35}{35} = 1$$

$$= \frac{10}{21} \qquad \text{Simplifying}$$

Do Exercise 22.

After multiplying in Example 15, we had a denominator of $\frac{35}{35}$, or 1. In the numerator, we had $\frac{2}{3} \cdot \frac{5}{7}$, which is the product of $\frac{2}{3}$, the dividend, and $\frac{5}{7}$, the reciprocal of the divisor. This gives us a procedure for dividing fractions.

DIVIDING FRACTIONS

To divide fractions, multiply by the reciprocal of the divisor:

$$\frac{a}{b} \div \frac{c}{d} = \frac{a}{b} \cdot \frac{d}{c}.$$

EXAMPLE 16 Divide by multiplying by the reciprocal of the divisor: $\frac{1}{2} \div \frac{3}{5}$.

$$\frac{1}{2} \div \frac{3}{5} = \frac{1}{2} \cdot \frac{5}{3} \qquad \tfrac{5}{3} \text{ is the reciprocal of } \tfrac{3}{5}$$

$$= \frac{5}{6} \qquad \text{Multiplying}$$

22. Divide by multiplying by 1:

$$\frac{\frac{3}{5}}{\frac{4}{7}}.$$

CALCULATOR CORNER

Operations on Fractions We can perform operations on fractions on a graphing calculator. Selecting the ▶FRAC option from the MATH menu causes the result to be expressed in fraction form. The calculator display is shown below.

3/4 + 1/2 ▶ Frac $\frac{5}{4}$

EXERCISES: Perform each calculation. Give the answer in fraction notation.

1. $\dfrac{5}{6} + \dfrac{7}{8}$ **2.** $\dfrac{13}{16} - \dfrac{4}{7}$

3. $\dfrac{15}{4} \cdot \dfrac{7}{12}$ **4.** $\dfrac{1}{5} \div \dfrac{3}{10}$

Answers

18. $\frac{11}{4}$ **19.** $\frac{7}{15}$ **20.** $\frac{1}{5}$ **21.** 3   **22.** $\frac{21}{20}$

Divide by multiplying by the reciprocal of the divisor. Then simplify.

23. $\dfrac{4}{3} \div \dfrac{7}{2}$

24. $\dfrac{5}{4} \div \dfrac{3}{2} = \dfrac{5}{4} \cdot \dfrac{\boxed{}}{3} = \dfrac{5 \cdot 2}{\boxed{} \cdot 3}$ **GS**

$= \dfrac{5 \cdot 2}{\boxed{} \cdot 2 \cdot 3} = \dfrac{\boxed{}}{6}$

25. $\dfrac{\frac{2}{9}}{\frac{5}{12}}$ **26.** $\dfrac{\frac{5}{6}}{\frac{45}{22}}$

Divide and simplify.

27. $\dfrac{7}{8} \div 56$ **28.** $36 \div \dfrac{4}{9}$

Answers

23. $\dfrac{8}{21}$ 24. $\dfrac{5}{6}$ 25. $\dfrac{8}{15}$ 26. $\dfrac{11}{27}$

27. $\dfrac{1}{64}$ 28. 81

Guided Solution:
24. 2, 4, 2, 5

EXAMPLE 17 Divide and simplify: $\dfrac{2}{3} \div \dfrac{4}{9}$.

$\dfrac{2}{3} \div \dfrac{4}{9} = \dfrac{2}{3} \cdot \dfrac{9}{4}$ $\dfrac{9}{4}$ is the reciprocal of $\dfrac{4}{9}$

$= \dfrac{2 \cdot 9}{3 \cdot 4}$ Multiplying numerators and denominators

$= \dfrac{2 \cdot 3 \cdot 3}{3 \cdot 2 \cdot 2}$ Removing a factor of 1: $\dfrac{2 \cdot 3}{2 \cdot 3} = 1$

$= \dfrac{3}{2}$

◀ **Do Exercises 23–26.**

EXAMPLE 18 Divide and simplify: $\dfrac{5}{6} \div 30$.

$\dfrac{5}{6} \div 30 = \dfrac{5}{6} \div \dfrac{30}{1} = \dfrac{5}{6} \cdot \dfrac{1}{30} = \dfrac{5 \cdot 1}{6 \cdot 30} = \dfrac{5 \cdot 1}{6 \cdot 5 \cdot 6} = \dfrac{1}{6 \cdot 6} = \dfrac{1}{36}$

Removing a factor of 1: $\dfrac{5}{5} = 1$

EXAMPLE 19 Divide and simplify: $24 \div \dfrac{3}{8}$.

$24 \div \dfrac{3}{8} = \dfrac{24}{1} \div \dfrac{3}{8} = \dfrac{24}{1} \cdot \dfrac{8}{3} = \dfrac{24 \cdot 8}{1 \cdot 3} = \dfrac{3 \cdot 8 \cdot 8}{1 \cdot 3} = \dfrac{8 \cdot 8}{1} = 64$

Removing a factor of 1: $\dfrac{3}{3} = 1$

◀ **Do Exercises 27 and 28.**

R.2 Exercise Set

✓ Reading Check

Choose from the column on the right the description that best matches each expression.

RC1. $9 + 0 = 9$

RC2. $\dfrac{6}{15}$ and $\dfrac{2}{5}$

RC3. $\dfrac{5}{12} + \dfrac{1}{15}$

RC4. $\dfrac{11}{10}$ and $\dfrac{10}{11}$

RC5. $\dfrac{31}{31}$

RC6. $24 \cdot 1 = 24$

a) Equivalent expressions
b) Identity property of 1
c) The LCD is $2 \cdot 2 \cdot 3 \cdot 5$.
d) Identity property of 0
e) Equivalent expression for 1
f) Reciprocals

a   Write an equivalent expression for each of the following. Use the indicated name for 1.

1. $\dfrac{3}{4}\left(\text{Use } \dfrac{3}{3} \text{ for 1.}\right)$

2. $\dfrac{5}{6}\left(\text{Use } \dfrac{10}{10} \text{ for 1.}\right)$

3. $\dfrac{3}{5}\left(\text{Use } \dfrac{20}{20} \text{ for 1.}\right)$

4. $\dfrac{8}{9}\left(\text{Use } \dfrac{4}{4} \text{ for 1.}\right)$

5. $\dfrac{13}{20}\left(\text{Use } \dfrac{8}{8} \text{ for 1.}\right)$

6. $\dfrac{13}{32}\left(\text{Use } \dfrac{40}{40} \text{ for 1.}\right)$

Write an equivalent expression with the given denominator.

7. $\dfrac{7}{8}$ (Denominator: 24)

8. $\dfrac{5}{6}$ (Denominator: 48)

9. $\dfrac{5}{4}$ (Denominator: 16)

10. $\dfrac{2}{9}$ (Denominator: 54)

11. $\dfrac{17}{19}$ (Denominator: 437)

12. $\dfrac{15}{23}$ (Denominator: 437)

b   Simplify.

13. $\dfrac{18}{27}$

14. $\dfrac{49}{56}$

15. $\dfrac{56}{14}$

16. $\dfrac{48}{27}$

17. $\dfrac{6}{42}$

18. $\dfrac{13}{104}$

19. $\dfrac{56}{7}$

20. $\dfrac{132}{11}$

21. $\dfrac{19}{76}$

22. $\dfrac{17}{51}$

23. $\dfrac{100}{20}$

24. $\dfrac{150}{25}$

25. $\dfrac{425}{525}$

26. $\dfrac{625}{325}$

27. $\dfrac{2600}{1400}$

28. $\dfrac{4800}{1600}$

29. $\dfrac{8 \cdot x}{6 \cdot x}$

30. $\dfrac{13 \cdot v}{39 \cdot v}$

c   Compute and simplify.

31. $\dfrac{1}{3} \cdot \dfrac{1}{4}$

32. $\dfrac{15}{16} \cdot \dfrac{8}{5}$

33. $\dfrac{15}{4} \cdot \dfrac{3}{4}$

34. $\dfrac{10}{11} \cdot \dfrac{11}{10}$

35. $\dfrac{1}{3} + \dfrac{1}{3}$

36. $\dfrac{1}{4} + \dfrac{1}{3}$

37. $\dfrac{4}{9} + \dfrac{13}{18}$

38. $\dfrac{4}{5} + \dfrac{8}{15}$

39. $\dfrac{3}{10} + \dfrac{8}{15}$

40. $\dfrac{9}{8} + \dfrac{7}{12}$

41. $\dfrac{7}{30} + \dfrac{5}{12}$

42. $\dfrac{3}{16} - \dfrac{1}{18}$

43. $\dfrac{5}{4} - \dfrac{3}{4}$

44. $\dfrac{12}{5} - \dfrac{2}{5}$

45. $\dfrac{11}{12} - \dfrac{3}{8}$

46. $\dfrac{15}{16} - \dfrac{5}{12}$

47. $\dfrac{11}{12} - \dfrac{2}{5}$

48. $\dfrac{15}{16} - \dfrac{2}{3}$

49. $5 \cdot \dfrac{2}{3}$

50. $\dfrac{4}{11} \cdot 9$

51. $\dfrac{8}{9} \div \dfrac{4}{15}$

52. $\dfrac{3}{4} \div \dfrac{3}{7}$

53. $\dfrac{1}{8} \div \dfrac{1}{4}$

54. $\dfrac{1}{20} \div \dfrac{1}{5}$

55. $\dfrac{\frac{13}{12}}{\frac{39}{5}}$

56. $\dfrac{\frac{17}{6}}{\frac{3}{8}}$

57. $100 \div \dfrac{1}{5}$

58. $78 \div \dfrac{1}{6}$

59. $\dfrac{3}{4} \div 10$

60. $\dfrac{5}{6} \div 15$

61. $1000 - \dfrac{1}{100}$

62. $\dfrac{147}{50} - 2$

63. $30 \div \dfrac{1}{30}$

64. $\dfrac{1}{30} \div 30$

Skill Maintenance

This heading indicates that the exercises that follow are *Skill Maintenance exercises*, which review any skill previously studied in the text. You can expect such exercises in every exercise set. Answers to *all* skill maintenance exercises are found at the back of the book. If you miss an exercise, restudy the objective shown in red.

Find the prime factorization.   [R.1a]

65. 28

66. 56

67. 1000

68. 192

Find the GCF.   [R.1b]

69. 28, 42

70. 35, 55

71. 248, 400

72. 90, 162

Find each LCM.   [R.1c]

73. 16, 24

74. 28, 49, 56

75. 48, 64, 96

76. 25, 75, 150

Synthesis

Simplify.

77. $\dfrac{192}{256}$

78. $\dfrac{p \cdot q}{r \cdot q}$

79. $\dfrac{64 \cdot a \cdot b}{16 \cdot a \cdot b}$

80. $\dfrac{4 \cdot 9 \cdot 24}{2 \cdot 8 \cdot 15}$

81. $\dfrac{36 \cdot (2 \cdot h)}{8 \cdot (9 \cdot h)}$

82. $\dfrac{256 \cdot a \cdot b \cdot c \cdot d}{192 \cdot b \cdot c \cdot d}$

Decimal Notation

Let's say that the cost of a sound system is

$1768.95.

This amount is given in **decimal notation**. The following place-value chart shows the place value of each digit in 1768.95.

PLACE-VALUE CHART								
Ten Thousands	Thousands	Hundreds	Tens	Ones	Ten*ths*	Hundred*ths*	Thousand*ths*	Ten-Thousand*ths*
10,000	1000	100	10	1	$\frac{1}{10}$	$\frac{1}{100}$	$\frac{1}{1000}$	$\frac{1}{10,000}$

$$1 \quad 7 \quad 6 \quad 8 \ . \ 9 \quad 5$$

a CONVERTING FROM DECIMAL NOTATION TO FRACTION NOTATION

When we multiply by 1, a number is not changed. If we choose the notation $\frac{10}{10}, \frac{100}{100}, \frac{1000}{1000}$, and so on for 1, we can move a decimal point in a numerator to the right to convert from decimal notation to fraction notation.

Look for a pattern in the following products:

$$0.1 = 0.1 \times 1 = 0.1 \times \frac{10}{10} = \frac{0.1 \times 10}{10} = \frac{1}{10};$$

$$0.6875 = 0.6875 \times 1 = 0.6875 \times \frac{10,000}{10,000} = \frac{0.6875 \times 10,000}{10,000} = \frac{6875}{10,000};$$

$$53.47 = 53.47 \times 1 = 53.47 \times \frac{100}{100} = \frac{53.47 \times 100}{100} = \frac{5347}{100}.$$

To convert from decimal notation to fraction notation:

a) Count the number of decimal places.

$$4.98$$

2 places

b) Move the decimal point that many places to the right.

$$4.98.$$

Move 2 places.

c) Write the result over a denominator with that number of zeros.

$$\frac{498}{100}$$

2 zeros

EXAMPLE 1 Convert 0.876 to fraction notation. Do not simplify.

$$0.876 \qquad 0.876. \qquad 0.876 = \frac{876}{1000}$$

3 places   3 places   3 zeros

OBJECTIVES

a Convert from decimal notation to fraction notation.

b Add, subtract, multiply, and divide using decimal notation.

c Round numbers to a specified decimal place.

SKILL TO REVIEW

Objective R.2c: Multiply using fraction notation.

Multiply and simplify.

1. $\frac{3}{5} \cdot \frac{2}{15}$ **2.** $6 \cdot \frac{4}{11}$

Answers

Skill to Review:

1. $\frac{2}{25}$ 2. $\frac{24}{11}$

Convert to fraction notation. Do not simplify.

1. 0.568

0.568. 0.568 = $\dfrac{568}{\boxed{}}$

$\boxed{}$ places

GS

2. 2.3

3. 0.009

Convert to decimal notation.

4. $\dfrac{4131}{1000}$

5. $\dfrac{4131}{10,000}$

6. $\dfrac{573}{100}$

7. $\dfrac{49}{10}$

Add.

8. 69 + 1.785 + 213.67

$$\begin{array}{r}
\overset{1}{}\ \overset{1}{}\ \overset{1}{}\ \\
6\ 9.\ 0\ \boxed{}\ 0 \\
1.\ 7\ 8\ 5 \\
+\ 2\ 1\ 3.\ 6\ 7\ \boxed{} \\
\hline
2\ \boxed{}\ 4.\ \boxed{}\ 5\ 5
\end{array}$$

GS

9. 17.95 + 14.68 + 236

Answers

1. $\dfrac{568}{1000}$ **2.** $\dfrac{23}{10}$ **3.** $\dfrac{9}{1000}$ **4.** 4.131

5. 0.4131 **6.** 5.73 **7.** 4.9 **8.** 284.455

9. 268.63

Guided Solutions:

1. 3, 1000 **8.** 0, 0, 8, 4

EXAMPLE 2 Convert 1.5018 to fraction notation. Do not simplify.

1.5018   1.5018. $1.5018 = \dfrac{15,018}{10,000}$

4 places   4 zeros

◀ **Do Exercises 1–3.**

To convert from fraction notation to decimal notation when the denominator is a number like 10, 100, or 1000:

a) Count the number of zeros.

$$\dfrac{8679}{1000}$$

3 zeros

b) Move the decimal point that number of places to the left. Leave off the denominator.

8.679.

Move 3 places.

EXAMPLE 3 Convert to decimal notation: $\dfrac{123,067}{10,000}$.

$\dfrac{123,067}{10,000}$ 12.3067. $\dfrac{123,067}{10,000} = 12.3067$

4 zeros   4 places

◀ **Do Exercises 4–7.**

b ADDITION, SUBTRACTION, MULTIPLICATION, AND DIVISION

ADDITION WITH DECIMAL NOTATION

Adding with decimal notation is similar to adding whole numbers. First, line up the decimal points. Then add the thousandths, then the hundredths, and so on, carrying if necessary.

EXAMPLE 4 Add: 74 + 26.46 + 0.998.

$$\begin{array}{r}
\overset{1}{}\ \overset{1}{}\ \overset{1}{}\ \\
7\ 4. \\
2\ 6.4\ 6 \\
+\ \ \ \ 0.9\ 9\ 8 \\
\hline
1\ 0\ 1.4\ 5\ 8
\end{array}$$

You can place extra zeros to the right of any decimal point so that there are the same number of decimal places in all the addends, but this is not necessary. If you did so, the preceding problem would look like this:

$$\begin{array}{r}
\overset{1}{}\ \overset{1}{}\ \overset{1}{}\ \\
7\ 4.0\ 0\ 0 \\
2\ 6.4\ 6\ 0 \\
+\ \ \ \ 0.9\ 9\ 8 \\
\hline
1\ 0\ 1.4\ 5\ 8
\end{array}$$ Adding zeros to 74
Adding zeros to 26.46

◀ **Do Exercises 8 and 9.**

SUBTRACTION WITH DECIMAL NOTATION

Subtracting with decimal notation is similar to subtracting whole numbers. First, line up the decimal points. Then subtract the thousandths, then the hundredths, the tenths, and so on, borrowing if necessary. Extra zeros can be added if needed.

EXAMPLES

5. Subtract: $76.14 - 18.953$.

$$
\begin{array}{r}
\overset{15\ \ 10\ \ 13}{\overset{6\ \ \cancel5\ \ \cancel0\ \ \cancel3\ \ 10}{7\ 6 . \cancel1\ 4\ \cancel0}} \\
-\ 1\ 8 . 9\ 5\ 3 \\
\hline
5\ 7 . 1\ 8\ 7
\end{array}
$$

6. Subtract: $200 - 0.68$.

$$
\begin{array}{r}
\overset{1\ \ 9\ \ 9\ \ 9\ 10}{\cancel2\ \cancel0\ \cancel0 . \cancel0\ \cancel0} \\
-\ \ \ \ \ 0 . 6\ 8 \\
\hline
1\ 9\ 9 . 3\ 2
\end{array}
$$

Do Exercises 10–13. ▶

Subtract.

10. $29.35 - 1.674$

11. $92.375 - 27.692$

12. $100 - 0.41$

13. $240 - 0.117$

Look at this product.

$$
\underset{\text{2 places}}{5.\underset{\uparrow}{14}} \quad \times \quad \underset{\text{1 place}}{0.\underset{\uparrow}{8}} \quad = \frac{514}{100} \times \frac{8}{10} = \frac{514 \times 8}{100 \times 10} = \frac{4112}{1000} = \underset{\text{3 places}}{4.\underset{\uparrow}{112}}
$$

We can also do this calculation more quickly by multiplying the whole numbers 8 and 514 and then determining the position of the decimal point.

MULTIPLICATION WITH DECIMAL NOTATION

a) Ignore the decimal points and multiply as whole numbers.

b) Place the decimal point in the result of step (a) by adding the number of decimal places in the original factors.

EXAMPLE 7 Multiply: 5.14×0.8.

a) We ignore the decimal points and multiply as whole numbers.

$$
\begin{array}{r}
\overset{1\ \ 3}{5.1\ 4} \\
\times\ \ \ \ 0.8 \\
\hline
4\ 1\ 1\ 2
\end{array}
$$

b) We then place the decimal point in the result of step (a) by adding the number of decimal places in the original factors.

$$
\begin{array}{r}
5.1\ 4 \leftarrow \text{2 decimal places} \\
\times\ \ \ \ 0.8 \leftarrow \text{1 decimal place} \\
\hline
4.1\ 1\ 2 \\
\overset{\uparrow}{} \text{3 decimal places}
\end{array}
$$

Do Exercises 14–17. ▶

Multiply.

14.
$$
\begin{array}{r}
6.5\ 2 \\
\times\ \ \ 0.9 \\
\hline
\end{array}
$$

15.
$$
\begin{array}{r}
6.5\ 2 \\
\times\ 0.0\ 9 \\
\hline
\end{array}
$$

16.
$$
\begin{array}{r}
5\ 6.7\ 6 \\
\times\ 0.9\ 0\ 8 \\
\hline
\end{array}
$$

17.
$$
\begin{array}{r}
0.0\ 3 \\
\times\ 0.0\ 0\ 1 \\
\hline
\end{array}
$$

Answers

10. 27.676 **11.** 64.683 **12.** 99.59
13. 239.883 **14.** 5.868 **15.** 0.5868
16. 51.53808 **17.** 0.00003

Divide.

18. $7 \overline{)\ 3\ 4\ 2.3}$

19. $1\ 6 \overline{)\ 2\ 5\ 3.1\ 2}$

Divide.

20. $2\ 5 \overline{)\ 3\ 2}$

21. $3\ 8 \overline{)\ 6\ 8\ 2.1}$

Note that $37.6 \div 8 = 4.7$ because $8 \times 4.7 = 37.6$. If we write this as shown at right, we see how the following method can be used to divide by a whole number.

$$
\begin{array}{r}
4.7 \\
8 \overline{)\ 3\ 7.6} \\
3\ 2 \\
\hline
5\ 6 \\
5\ 6 \\
\hline
0
\end{array}
$$

DIVIDING WHEN THE DIVISOR IS A WHOLE NUMBER

a) Place the decimal point in the quotient directly above the decimal point in the dividend.

b) Divide as whole numbers.

EXAMPLE 8 Divide: $216.75 \div 25$.

a)

$$
25 \overline{)\ 2\ 1\ 6.7\ 5}
$$

↑ Place the decimal point.

b)

$$
\begin{array}{r}
8.6\ 7 \\
25 \overline{)\ 2\ 1\ 6.7\ 5} \\
2\ 0\ 0 \\
\hline
1\ 6\ 7 \\
1\ 5\ 0 \\
\hline
1\ 7\ 5 \\
1\ 7\ 5 \\
\hline
0
\end{array}
$$

Divide as though dividing whole numbers.

◀ **Do Exercises 18 and 19.**

Sometimes it is helpful to write extra zeros to the right of the decimal point. Doing so does not change the answer. Remember that the decimal point for a whole number, though not normally written, is to the right of the number.

EXAMPLE 9 Divide: $54 \div 8$.

a)

$$
8 \overline{)\ 5\ 4.}
$$

b)

$$
\begin{array}{r}
6.7\ 5 \\
8 \overline{)\ 5\ 4.0\ 0} \\
4\ 8 \\
\hline
6\ 0 \\
5\ 6 \\
\hline
4\ 0 \\
4\ 0 \\
\hline
0
\end{array}
$$

Extra zeros are written to the right of the decimal point as needed.

◀ **Do Exercises 20 and 21.**

DIVIDING WHEN THE DIVISOR IS NOT A WHOLE NUMBER

a) Move the decimal point in the divisor as many places to the right as it takes to make it a whole number. Move the decimal point in the dividend the same number of places to the right and place the decimal point in the quotient.

b) Divide as whole numbers, inserting zeros if necessary.

EXAMPLE 10 Divide: $83.79 \div 0.098$.

a)

$$0.0\,9\,8.\,\overline{)\,8\,3.7\,9\,0.}$$

b)

$$\begin{array}{r} 8\,5\,5. \\ 0.0\,9\,8\,_{\wedge}\overline{)\,8\,3.7\,9\,0\,_{\wedge}} \\ \underline{7\,8\,4} \\ 5\,3\,9 \\ \underline{4\,9\,0} \\ 4\,9\,0 \\ \underline{4\,9\,0} \\ 0 \end{array}$$

Do Exercises 22 and 23. ▶

Divide.

22. $0.0\,2\,4\,\overline{)\,2\,0.5\,4\,4}$

23. $4.6\,\overline{)\,3.9\,1}$

Converting from Fraction Notation to Decimal Notation

To convert from fraction notation to decimal notation when the denominator is not a number like 10, 100, or 1000, we divide the numerator by the denominator.

EXAMPLE 11 Convert to decimal notation: $\dfrac{5}{16}$.

$$\begin{array}{r} 0.3\,1\,2\,5 \\ 1\,6\,\overline{)\,5.0\,0\,0\,0} \\ \underline{4\,8} \\ 2\,0 \\ \underline{1\,6} \\ 4\,0 \\ \underline{3\,2} \\ 8\,0 \\ \underline{8\,0} \\ 0 \end{array}$$

If we get a remainder of 0, the decimal *terminates*. Thus, $\frac{5}{16} = 0.3125$.

EXAMPLE 12 Convert to decimal notation: $\dfrac{7}{12}$.

$$\begin{array}{r} 0.5\,8\,3\,3 \\ 1\,2\,\overline{)\,7.0\,0\,0\,0} \\ \underline{6\,0} \\ 1\,0\,0 \\ \underline{9\,6} \\ 4\,0 \\ \underline{3\,6} \\ 4\,0 \\ \underline{3\,6} \\ 4 \end{array}$$

The number 4 repeats as a remainder, so the digit 3 will repeat in the quotient. Therefore,

$$\frac{7}{12} = 0.583333\ldots.$$

Instead of dots, we often put a bar over the repeating part—in this case, only the 3. Thus,

$$\frac{7}{12} = 0.58\overline{3}.$$

Convert to decimal notation.

24. $\dfrac{5}{8}$ **25.** $\dfrac{2}{3}$

26. $\dfrac{84}{11}$ **27.** $\dfrac{7}{40}$

Do Exercises 24–27. ▶

Answers

22. 856 **23.** 0.85 **24.** 0.625
25. $0.\overline{6}$ **26.** $7.\overline{63}$ **27.** 0.175

ROUNDING DECIMAL NOTATION

To round to a certain place:

a) Locate the digit in that place.

b) Consider the digit to its right.

c) If the digit to the right is 5 or higher, round up. If the digit to the right is less than 5, round down.

Round to the nearest tenth.

28. 13.85

29. 7.009

Round to the nearest hundredth.

30. 34.675

31. 100.9748

Round to the nearest thousandth.

32. 0.9434

33. 37.4005

Round 7459.8549 to the nearest:

34. Thousandth.

35. Hundredth.

36. Tenth.

37. One.

38. Ten.

Answers

28. 13.9 **29.** 7.0 **30.** 34.68 **31.** 100.97
32. 0.943 **33.** 37.401 **34.** 7459.855
35. 7459.85 **36.** 7459.9 **37.** 7460 **38.** 7460

C ROUNDING

When working with decimal notation in real-life situations, we often shorten notation by **rounding**. Although there are many rules for rounding, we will use the rules listed at left.

EXAMPLE 13 Round 3872.2459 to the nearest tenth.

a) We locate the digit in the tenths place, 2.

3 8 7 2.2 4 5 9

b) Then we consider the next digit to the right, 4.

3 8 7 2.2 4 5 9

c) Since that digit, 4, is less than 5, we round down.

3 8 7 2.2 ⟵ This is the answer.

EXAMPLE 14 Round 3872.2459 to the nearest thousandth, hundredth, tenth, one, ten, hundred, and thousand.

thousandth:	3872.246
hundredth:	3872.25
tenth:	3872.2
one:	3872
ten:	3870
hundred:	3900
thousand:	4000

Caution!

Each time you round, use the original number.

◀ **Do Exercises 28–38.**

In rounding, we sometimes use the symbol ≈, which means "is approximately equal to." Thus, 46.124 ≈ 46.1.

R.3 Exercise Set

For Extra Help MyMathLab MathXL PRACTICE WATCH READ REVIEW

✓ Reading Check

Name the place value of the underlined digit in each exercise.

RC1. 0.30<u>5</u>7

RC2. 15.4<u>2</u>

RC3. 49<u>0</u>.4

RC4. 0.0002<u>7</u>

RC5. 0.6<u>7</u>8

RC6. <u>5</u>041.283

a Convert to fraction notation. Do not simplify.

1. 5.3

2. 2.7

3. 0.67

4. 0.93

5. 2.0007

6. 4.0008

7. 7889.8

8. 1122.3

Convert to decimal notation.

9. $\dfrac{1}{10}$

10. $\dfrac{1}{100}$

11. $\dfrac{1}{10,000}$

12. $\dfrac{1}{1000}$

13. $\dfrac{9999}{1000}$

14. $\dfrac{39}{10,000}$

15. $\dfrac{4578}{10,000}$

16. $\dfrac{94}{100,000}$

b Add.

17.
```
   4 1 5.7 8
+   2 9.1 6
```

18.
```
   7 0 8.9 9
+   7 5.4 8
```

19.
```
   2 3 4.0 0 0
+ 1 5 6.6 1 7
```

20.
```
   1 3 4 5.1 2
+    5 6 6.9 8
```

21. $85 + 67.95 + 2.774$

22. $119 + 43.74 + 18.876$

23. $17.95 + 16.99 + 28.85$

24. $14.59 + 16.79 + 19.95$

Subtract.

25.
```
   7 8.1 1 0
-  4 5.8 7 6
```

26.
```
   1 4.0 8 0
-    9.1 9 9
```

27.
```
   3 8.7
- 1 1.8 6 5
```

28.
```
   3 0 0.
-  2 4.6 7 7
```

29. $57.86 - 9.95$

30. $2.6 - 1.08$

31. $3 - 1.0807$

32. $5 - 3.4051$

Multiply.

33.
```
   7.3 4
×   1.8
```

34.
```
   6.5 5
×   3.2
```

35.
```
   0.8 6
× 0.9 3
```

36.
```
   0.0 2 8
× 7.4 0 9
```

37.
```
   1 7.9 5
×     1 0
```

38.
```
   1 8.9 4
×     0.1
```

39.
```
   0.4 5 7
× 3.0 8
```

40.
```
   0.0 0 2 4
× 0.0 1 5
```

41.
```
   3.6 4 2
× 0.9 9
```

42.
```
   2 8 7.4
×   1.0 8
```

Divide.

43. $7\,2\,\overline{)\,1\,6\,5.6}$ **44.** $5.2\,\overline{)\,4\,4.2}$ **45.** $8.5\,\overline{)\,4\,4.2}$ **46.** $7.8\,\overline{)\,7\,2.5\,4}$

47. $9.9\,\overline{)\,0.2\,2\,7\,7}$ **48.** $1\,0\,0\,\overline{)\,9\,5}$ **49.** $0.6\,4\,\overline{)\,1\,2}$ **50.** $1.6\,\overline{)\,7\,5}$

51. $1.0\,5\,\overline{)\,6\,9\,3}$ **52.** $2\,5\,\overline{)\,4}$ **53.** $8.6\,\overline{)\,5.8\,4\,8}$ **54.** $0.4\,7\,\overline{)\,0.1\,2\,2\,2}$

Convert to decimal notation.

55. $\dfrac{11}{32}$ **56.** $\dfrac{17}{32}$ **57.** $\dfrac{13}{11}$ **58.** $\dfrac{17}{12}$

59. $\dfrac{5}{9}$ **60.** $\dfrac{5}{6}$ **61.** $\dfrac{19}{9}$ **62.** $\dfrac{9}{11}$

c   Round to the nearest hundredth, tenth, one, ten, and hundred.

63. 745.06534 **64.** 317.18565 **65.** 6780.50568 **66.** 840.15493

Round to the nearest cent (nearest hundredth) and to the nearest dollar (nearest one).

67. $17.988 **68.** $20.492 **69.** $346.075 **70.** $4.718

Round to the nearest dollar.

71. $16.95 **72.** $17.50 **73.** $189.50 **74.** $567.24

Divide and round to the nearest ten-thousandth, thousandth, hundredth, tenth, and one.

75. $\dfrac{2}{7}$ **76.** $\dfrac{23}{17}$ **77.** $\dfrac{23}{39}$ **78.** $\dfrac{1000}{81}$

Skill Maintenance

Calculate.   [R.2c]

79. $\dfrac{7}{8} + \dfrac{5}{32}$ **80.** $\dfrac{15}{16} - \dfrac{11}{12}$ **81.** $\dfrac{15}{16} \cdot \dfrac{11}{12}$ **82.** $\dfrac{15}{32} \div \dfrac{3}{8}$

83. $\dfrac{9}{70} + \dfrac{8}{15}$ **84.** $\dfrac{11}{21} + \dfrac{13}{16}$ **85.** $\dfrac{9}{10} + \dfrac{1}{100} + \dfrac{113}{1000}$ **86.** $\dfrac{1}{7} + \dfrac{4}{21} + \dfrac{9}{10}$

Find the prime factorization.   [R.1a]

87. 208 **88.** 2560

Find the GCF.   [R.1b]

89. 72, 150 **90.** 100, 1000

Synthesis

Convert to decimal notation.

91. $\dfrac{5}{7}$ **92.** $\dfrac{8}{13}$ **93.** $\dfrac{9}{14}$ **94.** $\dfrac{6}{17}$

Percent Notation

a CONVERTING TO DECIMAL NOTATION

On average, 43% of residential energy use is for heating and cooling. This means that of every 100 units of energy used, 43 units are used for heating and cooling. Thus, 43% is a ratio of 43 to 100.

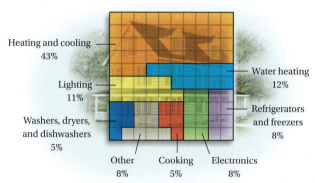

Heating and cooling
43%

Water heating
12%

Lighting
11%

Refrigerators
and freezers
8%

Washers, dryers,
and dishwashers
5%

Other
8%

Cooking
5%

Electronics
8%

SOURCE: U. S. Department of Energy

The percent symbol % means "per hundred." We can regard the percent symbol as a part of a name for a number. For example,

$$28\% \quad \text{is defined to mean} \quad 28 \times 0.01, \quad \text{or} \quad 28 \times \frac{1}{100}, \quad \text{or} \quad \frac{28}{100}.$$

NOTATION FOR $n\%$

$n\%$ means $n \times 0.01$, or $n \times \dfrac{1}{100}$, or $\dfrac{n}{100}$.

EXAMPLE 1 *Energy Use.* The U.S. Department of Energy has determined that, on average, 8% of residential energy use is for electronics. Convert 8% to decimal notation.

$$8\% = 8 \times 0.01 \qquad \text{Replacing \% with } \times 0.01$$
$$= 0.08$$

FROM PERCENT NOTATION TO DECIMAL NOTATION

To convert from percent notation to decimal notation, move the decimal point *two* places to the *left* and drop the percent symbol.

EXAMPLE 2 Convert 43.67% to decimal notation.

43.67%   0.43.67   43.67% = 0.4367

Move the decimal point two places to the left.

Do Margin Exercises 1–3. ▶

OBJECTIVES

a Convert from percent notation to decimal notation.

b Convert from percent notation to fraction notation.

c Convert from decimal notation to percent notation.

d Convert from fraction notation to percent notation.

SKILL TO REVIEW

Objective R.3b: Multiply using decimal notation.

Multiply.
1. 4.8 × 0.02
2. 329 × 0.15

1. *Texting and Driving.* A 2011 survey of high school seniors found that 58% say they text while driving. Convert 58% to decimal notation.

   Source: *The Indianapolis Star*, 6/18/12, Mike Stobbe, Associated Press

Convert to decimal notation.

2. 100%   3. 66.67%

Answers

Skill to Review:
1. 0.096   2. 49.35

Margin Exercises:
1. 0.58   2. 1   3. 0.6667

4. *Number of Charge Cards.* In 2012, 11% of adults ages 18–49 had four or five credit cards. Convert 11% to fraction notation.

Source: Study conducted by ORC International, February 23–26, 2012

Convert to fraction notation.

5. 53%　　**6.** 45.9%　　**7.** 4.375%

8. 0.23%

$$0.23\% = 0.23 \cdot \frac{1}{\boxed{}}$$

$$= \frac{0.23}{100} \cdot \frac{100}{\boxed{}}$$

$$= \frac{\boxed{}}{10{,}000}$$

GS

b　CONVERTING TO FRACTION NOTATION

EXAMPLE 3　Convert 88% to fraction notation.

$$88\% = 88 \times \frac{1}{100} \qquad \text{Replacing } \% \text{ with } \times \frac{1}{100}$$

$$= \frac{88}{100} \qquad \text{Multiplying. You need not simplify.}$$

EXAMPLE 4　Convert 34.8% to fraction notation.

$$34.8\% = 34.8 \times \frac{1}{100} \qquad \text{Replacing } \% \text{ with } \times \frac{1}{100}$$

$$= \frac{34.8}{100}$$

$$= \frac{34.8}{100} \cdot \frac{10}{10} \qquad \text{Multiplying by 1 to get a whole number in the numerator}$$

$$= \frac{348}{1000} \qquad \text{You need not simplify.}$$

◀ Do Exercises 4–8.

c　CONVERTING FROM DECIMAL NOTATION

By applying the definition of percent in reverse, we can convert from decimal notation to percent notation. We multiply by 1, expressing it as 100×0.01 and replacing $\times 0.01$ with %.

EXAMPLE 5　*Foreign-Born Residents.*　In 2010, 0.272 of the residents of California were foreign-born. Convert 0.272 to percent notation.

Source: Pew Hispanic Center Tabulations of 2010 American Community Survey

$$0.272 = 0.272 \times 1 \qquad \text{Identity property of 1}$$

$$= 0.272 \times (100 \times 0.01) \qquad \text{Expressing 1 as } 100 \times 0.01$$

$$= (0.272 \times 100) \times 0.01$$

$$= 27.2 \times 0.01$$

$$= 27.2\% \qquad \text{Replacing } \times 0.01 \text{ with } \%$$

9. *Foreign-Born Residents.* In 2010, 0.188 of the residents of Nevada were foreign-born. In West Virginia, 0.012 of the residents were foreign-born. Convert 0.188 and 0.012 to percent notation.

Source: Pew Hispanic Center Tabulations of 2010 American Community Survey

> ### FROM DECIMAL NOTATION TO PERCENT NOTATION
>
> To convert from decimal notation to percent notation, move the decimal point *two* places to the *right* and write the percent symbol.

EXAMPLE 6　Convert 0.082 to percent notation.

$$0.082 \qquad 0.08.2 \qquad 0.082 = 8.2\%$$

Move the decimal point two places to the right.

Convert to percent notation.

10. 6.77　　　　　**11.** 0.9944

◀ Do Exercises 9–11.

Answers

4. $\frac{11}{100}$　**5.** $\frac{53}{100}$　**6.** $\frac{459}{1000}$　**7.** $\frac{4375}{100{,}000}$

8. $\frac{23}{10{,}000}$　**9.** 18.8%; 1.2%　**10.** 677%

11. 99.44%

Guided Solution:
8. 100, 100, 23

d CONVERTING FROM FRACTION NOTATION

We can convert from fraction notation to percent notation by converting first to decimal notation. Then we move the decimal point two places to the *right* and write a percent symbol.

EXAMPLE 7 Convert $\frac{5}{8}$ to percent notation.

a) We first find decimal notation for $\frac{5}{8}$ using long division.

$$
\begin{array}{r}
0.6\ 2\ 5 \\
8\ )\overline{\ 5.0\ 0\ 0} \\
\underline{4\ 8} \\
2\ 0 \\
\underline{1\ 6} \\
4\ 0 \\
\underline{4\ 0} \\
0
\end{array}
$$

Thus, $\dfrac{5}{8} = 0.625$.

b) We then convert the decimal notation to percent notation by moving the decimal point two places to the right and writing the percent symbol.

$$0.62.5 \qquad \frac{5}{8} = 62.5\%, \quad \text{or} \quad 62\frac{1}{2}\% \qquad 0.5 = \frac{5}{10} = \frac{1}{2}$$

EXAMPLE 8 Convert $\frac{227}{150}$ to percent notation.

a) We first find decimal notation for $\frac{227}{150}$ using long division.

$$
\begin{array}{r}
1.5\ 1\ 3\ 3\ \ldots \\
1\ 5\ 0\ )\overline{\ 2\ 2\ 7.0\ 0\ 0\ 0} \\
\underline{1\ 5\ 0} \\
7\ 7\ 0 \\
\underline{7\ 5\ 0} \\
2\ 0\ 0 \\
\underline{1\ 5\ 0} \\
5\ 0\ 0 \\
\underline{4\ 5\ 0} \\
5\ 0
\end{array}
$$

We get a repeating decimal: $1.51\overline{3}$.

b) Next, we convert the decimal notation to percent notation by moving the decimal point two places to the right and writing the percent symbol.

$$1.51.\overline{3} \qquad \frac{227}{150} = 151.\overline{3}\%, \quad \text{or} \quad 151\frac{1}{3}\% \qquad 0.\overline{3} = \frac{1}{3}$$

Do Exercises 12–15. ▶

EXAMPLE 9 *Ages 0–14 in Kenya.* As of July 2011, $\frac{21}{50}$ of the population of Kenya was 0–14 years old. Convert $\frac{21}{50}$ to percent notation.

Source: The CIA World Factbook 2012

We can use long division. Or, since $2 \cdot 50 = 100$, we can multiply by a form of 1 to obtain 100 in the denominator:

$$\frac{21}{50} = \frac{21}{50} \cdot \frac{2}{2} = \frac{42}{100} = 42 \times \frac{1}{100} = 42\%.$$

Do Exercise 16. ▶

Convert to percent notation.

12. $\dfrac{1}{8}$

13. $\dfrac{436}{75}$

14. $\dfrac{7}{16}$

15. $\dfrac{5}{12}$

16. *Renewable Energy.*
Renewable energy provides $\frac{2}{25}$ of the total energy in the United States. Convert $\frac{2}{25}$ to percent notation.

Source: U.S. Department of Energy

Answers

12. 12.5%, or $12\frac{1}{2}\%$ **13.** 581.$\overline{3}$%, or $581\frac{1}{3}\%$

14. 43.75%, or $43\frac{3}{4}\%$ **15.** 41.$\overline{6}$%, or $41\frac{2}{3}\%$

16. 8%

✓ Reading Check

Choose from the columns on the right an equivalent expression for each exercise.

RC1. 500%

RC2. 5%

a) $\dfrac{1}{1000}$

b) 0.05%

RC3. $\dfrac{1}{400}$

RC4. 0.0005

c) 0.25%

d) 10%

RC5. 50%

RC6. 0.5%

e) $\dfrac{1}{200}$

f) $\dfrac{1}{20}$

RC7. $\dfrac{1}{10}$

RC8. 0.001

g) 5

h) $\dfrac{1}{2}$

a Convert the percent notation in each sentence to decimal notation.

1. *Walking to Work.*   In Boston, 13% of commuters walk to work.

Source: SustainLane

2. *Hospital Patients.*   The average age of hospital patients in the United States is 52.5 years old. Those ages 75 and older make up 24% of the patient population.

Source: U.S. Centers for Disease Control and Prevention

3. *Under 15 Years Old.*   In 2011, 35.4% of the population in Pakistan was under 15 years old. In the United States, 20.1% was under 15 years old.

Source: U.S. Census Bureau

4. *65 Years and Older.*   In 2011, 4.2% of the population of Pakistan was ages 65 and older. In the United States, 13.1% was ages 65 and older.

Source: U.S. Census Bureau

Convert to decimal notation.

5. 63%

6. 64%

7. 94.1%

8. 34.6%

9. 1%

10. 100%

11. 0.61%

12. 125%

13. 240%

14. 0.73%

15. 3.25%

16. 2.3%

b Convert the percent notation in each sentence to fraction notation.

17. *Olympic Gold Medals.*   Each gold medal awarded at the 2012 Summer Olympics contains 93% silver, 6% copper, and 1% gold.

Sources: economist.com; howmuchisgoldworth.net

18. *Ice Cream Flavors.*   Vanilla and chocolate ice cream are each selected by 23% of adults as their favorite flavor. Butter pecan was chosen by 9%, and mint chocolate chip came in at only 6%.

Source: Rasmussen survey, 2011

19. Lunch Break. Of all corporate executives, 39% have 30 min or less for a lunch break.

Source: OfficeTeam Survey

20. College Applications. About 71% of high school graduates who apply to colleges submit three or more college applications.

Source: Higher Education Research Institute, UCLA

21. Fluent in English. Of all second-generation Latinos in the United States, 88% are fluent in English.

Source: Pew Research Center

22. Melanoma. The number of melanoma cancer cases rose 31% from 2000 to 2008.

Source: American Cancer Society

Convert to fraction notation. Do not simplify.

23. 60%

24. 40%

25. 28.9%

26. 37.5%

27. 110%

28. 120%

29. 0.042%

30. 0.68%

31. 250%

32. 3.2%

33. 3.47%

34. 12.557%

C Convert the decimal notation in each sentence to percent notation.

35. Junk Mail. The number of pieces of junk mail increased by 0.214 from 2010 to 2011.

Source: U.S. Postal Service

36. Level of Education. In Utah, 0.912 of the population ages 25 and older are high school graduates, and 0.27 have a bachelor's degree or more.

Source: U.S. Census Bureau

37. Private Schools. In the United States, 0.14 of school-age children attend private schools.

Source: National Center for Education Statistics

38. Eating and Drinking in Vehicles. In a recent survey, 0.21 of adult drivers said that they prohibit eating and drinking in their vehicles.

Source: Kelton Research for YES Essentials

Convert to percent notation.

39. 0.99

40. 0.83

41. 1

42. 8.56

43. 0.0047

44. 2

45. 0.072

46. 1.34

47. 9.2

48. 0.013

49. 0.0068

50. 0.675

d Convert to percent notation.

51. $\dfrac{1}{6}$ **52.** $\dfrac{1}{5}$ **53.** $\dfrac{13}{20}$ **54.** $\dfrac{14}{25}$

55. $\dfrac{29}{100}$ **56.** $\dfrac{123}{100}$ **57.** $\dfrac{8}{10}$ **58.** $\dfrac{7}{10}$

59. $\dfrac{3}{5}$ **60.** $\dfrac{17}{50}$ **61.** $\dfrac{2}{3}$ **62.** $\dfrac{7}{8}$

63. $\dfrac{7}{4}$ **64.** $\dfrac{3}{8}$ **65.** $\dfrac{3}{4}$ **66.** $\dfrac{99.4}{100}$

Convert the fraction notation in each sentence to percent notation.

67. *Checking e-Mail.* The longest that approximately $\frac{2}{5}$ of workers go without checking their e-mail is less than 30 min.

Source: Studylogic for Starwood Hotels and Resorts/Sheraton

68. *Truck Drivers.* Of all truck drivers, $\frac{11}{50}$ spend 40 or more weeks per year on the road.

Source: Atlas Van Lines Survey

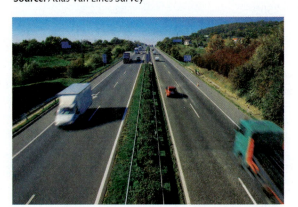

Skill Maintenance

Convert to decimal notation. [R.3b]

69. $\dfrac{9}{4}$ **70.** $\dfrac{17}{11}$

Convert to fraction notation. [R.3a]

73. 0.0123 **74.** 5.07

Round to the nearest hundredth. [R.3c]

77. 18.234 **78.** 0.0989

Calculate. [R.3b]

71. 23.458 × 7.03 **72.** $7.8\overline{)440.154}$

Simplify. [R.2b]

75. $\dfrac{87}{24}$ **76.** $\dfrac{35}{420}$

Convert to fraction notation. [R.4b]

79. 34.3% **80.** 80%

Synthesis

Simplify. Express the answer in percent notation.

81. 18% + 14% **82.** 84% − 12% **83.** 1 − 30% **84.** 50% − 0.5% **85.** 1 + 5%

86. 42% − (1 − 58%) **87.** 3(1 + 15%) **88.** 7(1% + 13%) **89.** $\dfrac{100\%}{40}$ **90.** $\dfrac{3}{4}$ + 20%

Exponential Notation and Order of Operations

a EXPONENTIAL NOTATION

Exponents provide a shorter way of writing products. An abbreviation for a product in which the factors are the same is called a **power**. An expression for a power is called **exponential notation**. For

$$10 \cdot 10 \cdot 10, \quad \text{we write} \quad 10^3.$$

3 factors

This is read "ten to the third power." We call the number 3 an **exponent** and we say that 10 is the **base**. For example,

$$a \cdot a \cdot a \cdot a = a^4.$$

This is the exponent.

This is the base.

An exponent of 2 or greater tells how many times the base is used as a factor.

EXPONENTIAL NOTATION

For any natural number n greater than or equal to 2,

n factors

$$b^n = \overbrace{b \cdot b \cdot b \cdot b \cdots b}.$$

EXAMPLE 1 Write exponential notation for $10 \cdot 10 \cdot 10 \cdot 10 \cdot 10$.

$$10 \cdot 10 \cdot 10 \cdot 10 \cdot 10 = 10^5$$

Do Margin Exercises 1–3. ▶

b EVALUATING EXPONENTIAL EXPRESSIONS

EXAMPLE 2 Evaluate: 3^4.

We have

$$3^4 = 3 \cdot 3 \cdot 3 \cdot 3 = 9 \cdot 9 = 81.$$

Do Exercises 4–7. ▶

c ORDER OF OPERATIONS

What does $4 + 5 \times 2$ mean? If we add 4 and 5 and multiply the result by 2, we get 18. If we multiply 5 and 2 and add 4 to the result, we get 14. Since the results are different, we see that the order in which we carry out operations is important. To indicate which operation is to be done first, we use grouping symbols such as parentheses ( ), or brackets [ ], or braces { }. For example, $(3 \times 5) + 6 = 15 + 6 = 21$, but $3 \times (5 + 6) = 3 \times 11 = 33$.

OBJECTIVES

a Write exponential notation for a product.

b Evaluate exponential expressions.

c Simplify expressions using the rules for order of operations.

SKILL TO REVIEW

Objective R.2c: Multiply using fraction notation.

Multiply.

1. $\dfrac{2}{15} \cdot \dfrac{1}{4}$ **2.** $\dfrac{3}{10} \cdot \dfrac{3}{10}$

Write exponential notation.

1. $4 \cdot 4 \cdot 4$

2. $6 \cdot 6 \cdot 6 \cdot 6 \cdot 6$

3. 1.08×1.08

Evaluate.

4. 10^4 **5.** 8^3

6. $(1.1)^3$ **7.** $\left(\dfrac{2}{9}\right)^2$

Answers

Skill to Review:

1. $\dfrac{1}{30}$ **2.** $\dfrac{9}{100}$

Margin Exercises:

1. 4^3 **2.** 6^5 **3.** 1.08^2 **4.** 10,000

5. 512 **6.** 1.331 **7.** $\dfrac{4}{81}$

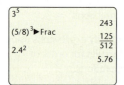
Grouping symbols tell us what to do first. If there are no grouping symbols, there is a set of rules for the order in which operations should be done.

✳ RULES FOR ORDER OF OPERATIONS

1. Do all calculations within grouping symbols before operations outside.

2. Evaluate all exponential expressions.

3. Do all multiplications and divisions in order from left to right.

4. Do all additions and subtractions in order from left to right.

EXAMPLE 3   Calculate: $15 - 2 \times 5 + 3$.

$$15 - 2 \times 5 + 3 = 15 - 10 + 3 \qquad \text{Multiplying}$$
$$= 5 + 3 \qquad \text{Subtracting}$$
$$= 8 \qquad \text{Adding}$$

◀ **Do Exercises 8 and 9.**

Always calculate within parentheses first. When there are exponents and no parentheses, simplify powers first.

EXAMPLE 4   Calculate: $(3 \times 4)^2$.

$$(3 \times 4)^2 = (12)^2 \qquad \text{Working within parentheses first}$$
$$= 144 \qquad \text{Evaluating the exponential expression}$$

EXAMPLE 5   Calculate: 3×4^2.

$$3 \times 4^2 = 3 \times 16 \qquad \text{Evaluating the exponential expression}$$
$$= 48 \qquad \text{Multiplying}$$

Note that Examples 4 and 5 show that $(3 \times 4)^2 \neq 3 \times 4^2$.

EXAMPLE 6   Calculate: $7 + 3 \times 29 - 4^2$.

$$7 + 3 \times 29 - 4^2 = 7 + 3 \times 29 - 16 \qquad \text{There are no parentheses, so we find } 4^2 \text{ first.}$$
$$= 7 + 87 - 16 \qquad \text{Multiplying}$$
$$= 94 - 16 \qquad \text{Adding}$$
$$= 78 \qquad \text{Subtracting}$$

◀ **Do Exercises 10–13.**

EXAMPLE 7   Calculate: $100 \div 20 \div 2$.

$$100 \div 20 \div 2 = 5 \div 2 \qquad \text{Doing the divisions in order from left to right}$$
$$= \frac{5}{2}, \text{ or } 2.5 \qquad \text{Doing the second division}$$

Calculate.

8. $16 - 3 \times 5 + 4$

9. $4 + 5 \times 2$

Calculate.

10. $18 - 4 \times 3 + 7$

11. $(2 \times 5)^3$

12. 2×5^3

13. $8 + 2 \times 5^3 - 4 \cdot 20$ **GS**
$= 8 + 2 \times \boxed{} - 4 \cdot 20$
$= 8 + \boxed{} - 4 \cdot 20$
$= 8 + 250 - \boxed{}$
$= \boxed{} - 80$
$= 178$

EXAMPLE 8 Calculate: $1000 \div \frac{1}{10} \cdot \frac{4}{5}$.

$$1000 \div \frac{1}{10} \cdot \frac{4}{5} = (1000 \cdot 10) \cdot \frac{4}{5}$$ Doing the division first, multiplying by the reciprocal of the divisor

$$= 10,000 \cdot \frac{4}{5}$$ Multiplying inside the parentheses

$$= 8000$$ Multiplying

Do Exercises 14 and 15. ▶

Calculate.

14. $51.2 \div 0.64 \div 40$

15. $1000 \cdot \frac{1}{10} \div \frac{4}{5}$

Sometimes combinations of grouping symbols are used. The rules still apply. We begin with the innermost grouping symbols and work to the outside.

EXAMPLE 9 Calculate: $5[4 + (8 - 2)]$.

$$5[4 + (8 - 2)] = 5[4 + 6]$$ Subtracting within the parentheses first

$$= 5[10]$$ Adding inside the brackets

$$= 50$$ Multiplying

A fraction bar can play the role of a grouping symbol.

EXAMPLE 10 Calculate: $\dfrac{12(9 - 7) + 4 \cdot 5}{3^3 - 2^4}$.

We do the calculations separately in the numerator and in the denominator, and then divide the results:

$$\frac{12(9 - 7) + 4 \cdot 5}{3^3 - 2^4} = \frac{12(2) + 4 \cdot 5}{27 - 16} = \frac{24 + 20}{11} = \frac{44}{11} = 4.$$

Do Exercises 16 and 17. ▶

Calculate.

16. $4[(8 - 3) + 7]$

17. $\dfrac{13(10 - 6) + 4 \cdot 9}{5^2 - 3^2}$

Answers

14. 2   **15.** 125   **16.** 48   **17.** $\dfrac{11}{2}$

CALCULATOR CORNER

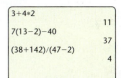

Order of Operations Computations are generally entered on a graphing calculator as they are written. We enter grouping symbols (parentheses, brackets, and braces) using the ⬛ and ⬛ keys. We indicate that a fraction bar acts as a grouping symbol by enclosing both the numerator and the denominator in parentheses. To calculate $\dfrac{38 + 142}{47 - 2}$, for example, we rewrite it with grouping symbols as $(38 + 142) \div (47 - 2)$.

EXERCISES: Calculate.

1. $68 - 8 \div 4 + 3 \cdot 5$

2. $\dfrac{311 - 17^2}{13 - 2}$

3. $(15 + 3)^3 + 4(12 - 7)^2$

4. $3.2 + 4.7[159.3 - 2.1(60.3 - 59.4)]$

5. $785 - \dfrac{5^4 - 285}{17 + 3 \cdot 51}$

6. $12^5 - 12^4 + 11^5 \div 11^3 - 10.2^2$

✓ Reading Check

In each exercise, name the operation that should be performed first. Do not perform the calculations.

RC1. $58 - 2(4 + 3)$

RC2. $9 + 6 \div 2 \cdot 4$

RC3. $\dfrac{100 - 2 \cdot 3}{2}$

RC4. $13 - 11 + 7$

a   Write exponential notation.

1. $5 \times 5 \times 5 \times 5$

2. $3 \times 3 \times 3 \times 3 \times 3$

3. $10 \cdot 10 \cdot 10$

4. $1 \cdot 1 \cdot 1$

5. $10 \times 10 \times 10 \times 10 \times 10 \times 10$

6. $18 \cdot 18$

b   Evaluate.

7. 7^2

8. 4^3

9. 9^5

10. 12^4

11. 1^4

12. 1^5

13. $(2.3)^2$

14. $(1.8)^2$

15. $(0.2)^3$

16. $(0.1)^3$

17. $(20.4)^2$

18. $(14.8)^2$

19. $\left(\dfrac{3}{8}\right)^2$

20. $\left(\dfrac{4}{5}\right)^2$

21. 5^3

22. 2^4

23. $\left(\dfrac{3}{2}\right)^3$

24. $\left(\dfrac{5}{3}\right)^3$

25. $1000 \times (1.02)^3$

26. $2000 \times (1.06)^2$

c   Calculate.

27. $9 + 2 \times 8$

28. $14 + 6 \times 6$

29. $9(8) + 7(6)$

30. $30(5) + 2(2)$

31. $39 - 4 \times 2 + 2$

32. $14 - 2 \times 6 + 7$

33. $9 \div 3 + 16 \div 8$

34. $32 - 8 \div 4 - 2$

35. $(6 \cdot 3)^2$

36. $(5 \cdot 4)^2$

37. $4 \cdot 5^2$

38. $3 \cdot 2^3$

39. $(8 + 2)^3$

40. $(5 - 2)^2$

41. $6 + 4^2$

42. $10 - 3^2$

43. $4^3 \div 8 - 4$

44. $20 + 4^3 \div 8 - 4$

45. $120 - 3^3 \cdot 4 \div 6$

46. $7 \times 3^4 + 18$

47. $6[9 + (3 + 4)]$

48. $8[(13 + 6) - 11]$

49. $8 + (7 + 9)$

50. $(8 + 7) + 9$

51. $15(4 + 2)$

52. $15 \cdot 4 + 15 \cdot 2$

53. $12 - (8 - 4)$

54. $(12 - 8) - 4$

55. $1000 \div 100 \div 10$

56. $256 \div 32 \div 4$

57. $2000 \div \dfrac{3}{50} \cdot \dfrac{3}{2}$

58. $400 \times 0.64 \div 3.2$

59. $75 \div 15 \cdot 4 \cdot 8 \div 32$

60. $84 \div 12 \cdot 10 \div 35 \cdot 8 \cdot 2 \div 16$

61. $16 \cdot 5 \div 80 \div 12 \cdot 36 \cdot 9$

62. $20 \cdot 45 \div 15 \div 15 \cdot 60 \div 12$

63. $\dfrac{80 - 6^2}{9^2 + 3^2}$

64. $\dfrac{5^2 + 4^3 - 3}{9^2 - 2^2 + 1^5}$

65. $\dfrac{3(6 + 7) - 5 \cdot 4}{6 \cdot 7 + 8(4 - 1)}$

66. $\dfrac{20(8 - 3) - 4(10 - 3)}{10(6 + 2) + 2(5 + 2)}$

67. $8 \cdot 2 - (12 - 0) \div 3 - (5 - 2)$

68. $95 - 2^3 \cdot 5 \div (24 - 4)$

Skill Maintenance

Find percent notation. [R.4d]

69. $\dfrac{5}{16}$

70. $\dfrac{11}{6}$

Simplify. [R.2b]

71. $\dfrac{9}{2001}$

72. $\dfrac{2005}{3640}$

73. Find the prime factorization of 48. [R.1a]

74. Find the LCM of 12, 24, and 56. [R.1c]

Synthesis

Write each of the following with a single exponent.

75. $\dfrac{10^5}{10^3}$

76. $\dfrac{10^7}{10^2}$

77. $5^4 \cdot 5^2$

78. $\dfrac{2^8}{8^2}$

Vocabulary Reinforcement

Complete each statement with the correct word or words from the column on the right. Some of the choices may not be used.

1. In the expression 8^2, 8 is called the _____. [R.5a]	equivalent
	base
2. The _____ of 15 and 40 is 120. [R.1c]	exponent
	less than
3. To _____ a number means to express the number as a product. [R.1a]	GCF
	LCM
4. For the fraction $\frac{4}{13}$, 4 is called the _____. [R.2a]	denominator
	numerator
5. $417 \cdot 1 = 417$ illustrates the _____. [R.2a]	prime
	composite
6. If a natural number, other than 1, is not prime, it is called a(n) _____ number. [R.1a]	reciprocal
	opposite
7. The fraction expression $\frac{2}{13}$ is _____ to the fraction	identity property of 0
expression $\frac{14}{91}$. [R.2a]	identity property of 1
	factor

8. $24 + 0 = 24$ illustrates the _____. [R.2a]

9. A natural number that has exactly two different factors, itself and 1, is called a(n) _____ number. [R.1a]

10. The _____ of 15 and 40 is 5. [R.1b]

11. $\frac{11}{3}$ is the _____ of $\frac{3}{11}$. [R.2c]

12. In the expression 7^5, 5 is called the _____. [R.5a]

Concept Reinforcement

Determine whether each statement is true or false.

_____ **1.** The least common multiple of two numbers is always larger than or equal to the larger number. [R.1c]

_____ **2.** To convert from decimal notation to percent notation, move the decimal point two places to the left and write the percent symbol. [R.4c]

_____ **3.** The number 1 is not prime. [R.1a]

_____ **4.** The greatest common factor of two numbers is always smaller than the smaller number. [R.1b]

_____ **5.** A percent that is greater than 100% is greater than 1. [R.4a]

_____ **6.** The prime factorization of a composite number contains that number. [R.1a]

Review Exercises

The review exercises that follow are for practice. Answers are at the back of the book. If you miss an exercise, restudy the objective indicated in red next to the exercise or direction line that precedes it.

Find the prime factorization.   [R.1a]

1. 92

2. 1400

Find the GCF.   [R.1b]

3. 20, 50

4. 18, 45, 900

Find the LCM.   [R.1c]

5. 13, 32

6. 5, 18, 45

Write an equivalent expression using the indicated number for 1.   [R.2a]

7. $\dfrac{2}{5}$ $\left(\text{Use } \dfrac{6}{6} \text{ for 1.} \right)$

8. $\dfrac{12}{23}$ $\left(\text{Use } \dfrac{8}{8} \text{ for 1.} \right)$

Write an equivalent expression with the given denominator.   [R.2a]

9. $\dfrac{5}{8}$ (Denominator: 64)

10. $\dfrac{13}{12}$ (Denominator: 84)

Simplify.   [R.2b]

11. $\dfrac{20}{48}$

12. $\dfrac{1020}{1820}$

Compute and simplify.   [R.2c]

13. $\dfrac{4}{9} + \dfrac{5}{12}$

14. $\dfrac{3}{4} \div 3$

15. $\dfrac{2}{3} - \dfrac{1}{15}$

16. $\dfrac{9}{10} \cdot \dfrac{16}{5}$

17. $\dfrac{11}{18} + \dfrac{13}{16}$

18. $\dfrac{35}{36} + \dfrac{23}{24}$

19. $\dfrac{25}{27} + \dfrac{17}{18}$

20. $\dfrac{29}{42} + \dfrac{17}{28}$

21. $\dfrac{35}{36} - \dfrac{19}{24}$

22. $\dfrac{13}{16} - \dfrac{11}{18}$

23. $\dfrac{29}{42} - \dfrac{17}{28}$

24. $\dfrac{11}{36} - \dfrac{1}{20}$

25. Convert to fraction notation: 17.97.   [R.3a]

26. Convert to decimal notation: $\dfrac{2337}{10,000}$.   [R.3a]

Add.   [R.3b]

27.
$$\begin{array}{r} 2\ 3\ 4\ 4.5\ 6 \\ +\quad 9\ 8.3\ 4\ 5 \\ \hline \end{array}$$

28. $6.04 + 78 + 1.9898$

Subtract.   [R.3b]

29. $20.4 - 11.058$

30.
$$\begin{array}{r} 7\ 8\ 9.0\ 3\ 2 \\ -\ 6\ 5\ 5.7\ 6\ 8 \\ \hline \end{array}$$

Multiply.   [R.3b]

31.
$$\begin{array}{r} 1\ 7.9\ 5 \\ \times\quad 2\ 4 \\ \hline \end{array}$$

32.
$$\begin{array}{r} 5\ 6.9\ 5 \\ \times\quad 1.9\ 4 \\ \hline \end{array}$$

Divide.   [R.3b]

33. $2.8\,\overline{)\,1\ 5\ 5.6\ 8}$

34. $5\ 2\,\overline{)\,2\ 3.4}$

35. Convert to decimal notation: $\frac{19}{12}$.   [R.3b]

36. Round to the nearest tenth: 34.067.   [R.3c]

37. *Population of Europe.*   In 2012, the population of the world was 7,047,000,000. About 10.5% of the world population is in Europe. Convert 10.5% to decimal notation.   [R.4a]

Source: U.S. Census Bureau

38. *Blood Type.*   In the United States, 35.7% of the population has blood type A, Rh positive. Convert 35.7% to fraction notation.   [R.4b]

Source: Stanford School of Medicine, Blood Center

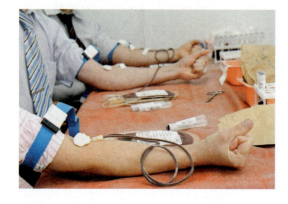

39. *Farms in Missouri.*   Of all the farms in the United States, 0.0502 of them are in Missouri. Convert 0.0502 to percent notation.   [R.4c]

Source: U.S. Department of Agriculture, National Agricultural Statistics Service

40. *Soccer.*   Of Americans ages 12–17, 396 of every 1000 have played soccer. Convert $\frac{396}{1000}$ to percent notation.   [R.4d]

Source: ESPN Sports Poll, a service of TNS Sport

Convert to percent notation.   [R.4d]

41. $\frac{5}{8}$

42. $\frac{29}{25}$

43. Write exponential notation: $6 \cdot 6 \cdot 6$.   [R.5a]

44. Evaluate: $(1.06)^2$.   [R.5b]

Calculate.   [R.5c]

45. $120 - 6^2 \div 4 + 8$

46. $64 \div 16 \cdot 32 \div 48 \div 12 \cdot 18$

47. $(120 - 6^2) \div 4 + 8$

48. $64 \cdot 16 \div 32 \div 48 \div 12 \cdot 18$

49. $(120 - 6^2) \div (4 + 8)$

50. $8^2 \cdot 2^4 \div 2^2 \cdot 8 \div 48 \div 12 \cdot 18$

51. Calculate: $\dfrac{4(18 - 8) + 7 \cdot 9}{9^2 - 8^2}$.   [R.5c]

Understanding Through Discussion and Writing

To the student and the instructor: The *Understanding Through Discussion and Writing* exercises are meant to be answered with one or more sentences. They can be discussed and answered collaboratively by the entire class or by small groups.

1. Explain in your own words when it *is* possible to cancel and when it is *not* possible to cancel.

2. The expression $9 - (4 \cdot 2)$ contains parentheses. Are they necessary? Why or why not?

3. A student insists that $5.367 \div 0.1$ is 0.5367. How could you convince this student that a mistake has been made?

CHAPTER

R **Test**

For Extra Help For step-by-step test solutions, access the Chapter Test Prep Videos in MyMathLab® or on YouTube (search "BittingerIntro" and click on "Channels").

1. Find the prime factorization of 300.

2. Find the GCF of 42, 56, and 140.

3. Find the LCM of 15, 24, and 60.

4. Write an expression equivalent to $\frac{3}{7}$ using $\frac{7}{7}$ as a name for 1.

5. Write an expression equivalent to $\frac{11}{16}$ with a denominator of 48.

Simplify.

6. $\dfrac{16}{24}$

7. $\dfrac{925}{1525}$

Compute and simplify.

8. $\dfrac{10}{27} \div \dfrac{8}{3}$

9. $\dfrac{9}{10} - \dfrac{5}{8}$

10. $\dfrac{11}{12} + \dfrac{17}{18}$

11. $\dfrac{10}{27} \cdot \dfrac{3}{8}$

12. Convert to fraction notation (do not simplify): 6.78.

13. Convert to decimal notation: $\dfrac{1895}{1000}$.

14. Add: $7.14 + 89 + 2.8787$.

15. Subtract: $1800 - 3.42$.

16. Multiply: 1 2 3.6
                    $\times$ 3.5 2

17. Divide: $7.2\,\overline{)\,1\,1\,1.5\,2}$

18. Convert to decimal notation: $\dfrac{23}{11}$.

19. Round 234.7284 to the nearest tenth.

20. Round 234.7284 to the nearest thousandth.

21. Convert to decimal notation: 0.7%.

22. Convert to fraction notation: 91%.

23. Convert to percent notation: $\dfrac{11}{25}$.

24. Evaluate: 5^4.

25. Evaluate: $(1.2)^2$.

26. Calculate: $200 - 2^3 + 5 \times 10$.

27. Calculate: $8000 \div 0.16 \div 2.5$.

28. *Home Equity Loan Rate.* A recent bank promotion offered a home equity loan at a low interest rate of 5.4%. Convert 5.4% to decimal notation.

29. *Dermatologists.* Of the 902,100 doctors in the United States, 12 of every 1000 are dermatologists. Convert $\frac{12}{1000}$ to percent notation.

Source: American Medical Association

Synthesis

30. Simplify: $\dfrac{38xy}{2x}$.

Introduction to Real Numbers and Algebraic Expressions

1.1 Introduction to Algebra

OBJECTIVES

a Evaluate algebraic expressions by substitution.

b Translate phrases to algebraic expressions.

SKILL TO REVIEW

Objective R.2b: Simplify fraction notation.

Simplify.

1. $\dfrac{100}{20}$ **2.** $\dfrac{78}{3}$

The study of algebra involves the use of equations to solve problems. Equations are constructed from algebraic expressions.

a EVALUATING ALGEBRAIC EXPRESSIONS

In arithmetic, you have worked with expressions such as

$$49 + 75, \quad 8 \times 6.07, \quad 29 - 14, \quad \text{and} \quad \frac{5}{6}.$$

In algebra, we can use letters to represent numbers and work with *algebraic expressions* such as

$$x + 75, \quad 8 \times y, \quad 29 - t, \quad \text{and} \quad \frac{a}{b}.$$

Sometimes a letter can represent various numbers. In that case, we call the letter a **variable**. Let a = your age. Then a is a variable since a changes from year to year. Sometimes a letter can stand for just one number. In that case, we call the letter a **constant**. Let b = your date of birth. Then b is a constant.

Where do algebraic expressions occur? Most often we encounter them when we are solving applied problems. For example, consider the bar graph shown at left, one that we might find in a book or a magazine. Suppose we want to know how much greater the average population density per square mile is in New Jersey than in Illinois. Using arithmetic, we might simply subtract. But let's see how we can determine this using algebra. We translate the problem into a statement of equality, an equation. It could be done as follows:

Population Density

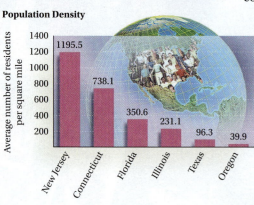

Average number of residents per square mile

- New Jersey: 1195.5
- Connecticut: 738.1
- Florida: 350.6
- Illinois: 231.1
- Texas: 96.3
- Oregon: 39.9

SOURCE: 2010 U.S. Census

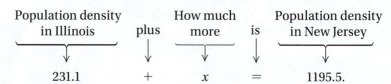

Population density in Illinois	plus	How much more	is	Population density in New Jersey
↓	↓	↓	↓	↓
231.1	+	x	=	1195.5.

Note that we have an algebraic expression, $231.1 + x$, on the left of the equals sign. To find the number x, we can subtract 231.1 on both sides of the equation:

$$231.1 + x = 1195.5$$
$$231.1 + x - 231.1 = 1195.5 - 231.1$$
$$x = 964.4.$$

This value of x gives the answer, 964.4 residents per square mile.

Answers

Skill to Review:
1. 5   2. 26

We call 231.1 + x an *algebraic expression* and 231.1 + x = 1195.5 an *algebraic equation*. Note that there is no equals sign, =, in an algebraic expression.

Do Margin Exercise 1. ▶

An **algebraic expression** consists of variables, constants, numerals, operation signs, and/or grouping symbols. When we replace a variable with a number, we say that we are **substituting** for the variable. When we replace all of the variables in an expression with numbers and carry out the operations in the expression, we are **evaluating the expression**.

EXAMPLE 1 Evaluate $x + y$ when $x = 37$ and $y = 29$.

We substitute 37 for x and 29 for y and carry out the addition:

$$x + y = 37 + 29 = 66.$$

The number 66 is called the **value** of the expression when $x = 37$ and $y = 29$.

Algebraic expressions involving multiplication can be written in several ways. For example, "8 times a" can be written as

$$8 \times a, \quad 8 \cdot a, \quad 8(a), \quad \text{or simply} \quad 8a.$$

Two letters written together without an operation symbol, such as ab, also indicate a multiplication.

EXAMPLE 2 Evaluate $3y$ when $y = 14$.

$$3y = 3(14) = 42$$

Do Exercises 2–4. ▶

EXAMPLE 3 *Area of a Rectangle.* The area A of a rectangle of length l and width w is given by the formula $A = lw$. Find the area when l is 24.5 in. and w is 16 in.

We substitute 24.5 in. for l and 16 in. for w and carry out the multiplication:

$$
\begin{aligned}
A = lw &= (24.5 \text{ in.})(16 \text{ in.}) \\
&= (24.5)(16)(\text{in.})(\text{in.}) \\
&= 392 \text{ in}^2, \text{ or } 392 \text{ square inches.}
\end{aligned}
$$

Do Exercise 5. ▶

Algebraic expressions involving division can also be written in several ways. For example, "8 divided by t" can be written as

$$8 \div t, \quad \frac{8}{t}, \quad 8/t, \quad \text{or} \quad 8 \cdot \frac{1}{t},$$

where the fraction bar is a division symbol.

EXAMPLE 4 Evaluate $\dfrac{a}{b}$ when $a = 63$ and $b = 9$.

We substitute 63 for a and 9 for b and carry out the division:

$$\frac{a}{b} = \frac{63}{9} = 7.$$

---

1. Translate this problem to an equation. Then solve the equation.

Population Density. The average number of residents per square mile in six U.S. states is shown in the bar graph on the preceding page. How much greater is the population density in Connecticut than in Oregon?

2. Evaluate $a + b$ when $a = 38$ and $b = 26$.

3. Evaluate $x - y$ when $x = 57$ and $y = 29$.

4. Evaluate $4t$ when $t = 15$.

GS **5.** Find the area of a rectangle when l is 24 ft and w is 8 ft.

$$
\begin{aligned}
A &= lw \\
A &= (24 \text{ ft})(\quad) \\
&= (24)(\quad)(\text{ft})(\text{ft}) \\
&= 192 \quad, \text{ or} \\
&\quad 192 \text{ square feet}
\end{aligned}
$$

Answers

1. $39.9 + x = 738.1$; 698.2 residents per square mile   **2.** 64   **3.** 28   **4.** 60
5. 192 ft^2

Guided Solution:
5. 8 ft, 8, ft^2

6. Evaluate a/b when $a = 200$ and $b = 8$.

7. Evaluate $10p/q$ when $p = 40$ and $q = 25$.

8. *Commuting via Bicycle.* Find the time it takes to bike 22 mi if the speed is 16 mph.

EXAMPLE 5 Evaluate $\dfrac{12m}{n}$ when $m = 8$ and $n = 16$.

$$\frac{12m}{n} = \frac{12 \cdot 8}{16} = \frac{96}{16} = 6$$

◄ **Do Exercises 6 and 7.**

EXAMPLE 6 *Commuting Via Bicycle.* Commuting to work via bicycle has increased in popularity with the emerging concept of sharing bicycles. Bikes are picked up and returned at docking stations. The payment is approximately $1.50 per 30 min. Richard bicycles 18 mi to work. The time t, in hours, that it takes to bike 18 mi is given by

$$t = \frac{18}{r},$$

where r is the speed. Find the time for Richard to commute to work if his speed is 15 mph.

We substitute 15 for r and carry out the division:

$$t = \frac{18}{r} = \frac{18}{15} = 1.2 \text{ hr.}$$

◄ **Do Exercise 8.**

b TRANSLATING TO ALGEBRAIC EXPRESSIONS

We translate problems to equations. The different parts of an equation are translations of word phrases to algebraic expressions. It is easier to translate if we know that certain words often translate to certain operation symbols.

Key Words, Phrases, and Concepts

ADDITION (+)	SUBTRACTION (−)	MULTIPLICATION (·)	DIVISION (÷)
add	subtract	multiply	divide
added to	subtracted from	multiplied by	divided by
sum	difference	product	quotient
total	minus	times	
plus	less than	of	
more than	decreased by		
increased by	take away		

EXAMPLE 7 Translate to an algebraic expression:

Twice (or two times) some number.

Think of some number, say, 8. We can write 2 times 8 as 2×8, or $2 \cdot 8$. We multiplied by 2. Do the same thing using a variable. We can use any variable we wish, such as x, y, m, or n. Let's use y to represent some number. If we multiply by 2, we get an expression

$$y \times 2, \quad 2 \times y, \quad 2 \cdot y, \quad \text{or} \quad 2y.$$

Answers

6. 25   **7.** 16   **8.** 1.375 hr

EXAMPLE 8 Translate to an algebraic expression:

Thirty-eight percent of some number.

Let $n = $ the number. The word "of" translates to a multiplication symbol, so we could write any of the following expressions as a translation:

$38\% \cdot n, \quad 0.38 \times n, \quad \text{or} \quad 0.38n.$

EXAMPLE 9 Translate to an algebraic expression:

Seven less than some number.

We let x represent the number. If the number were 10, then 7 less than 10 is $10 - 7$, or 3. If we knew the number to be 34, then 7 less than the number would be $34 - 7$. Thus if the number is x, then the translation is

$x - 7.$

EXAMPLE 10 Translate to an algebraic expression:

Eighteen more than a number.

We let $t = $ the number. If the number were 6, then the translation would be $6 + 18$, or $18 + 6$. If we knew the number to be 17, then the translation would be $17 + 18$, or $18 + 17$. Thus if the number is t, then the translation is

$t + 18, \quad \text{or} \quad 18 + t.$

EXAMPLE 11 Translate to an algebraic expression:

A number divided by 5.

We let $m = $ the number. If the number were 7, then the translation would be $7 \div 5$, or $7/5$, or $\frac{7}{5}$. If the number were 21, then the translation would be $21 \div 5$, or $21/5$, or $\frac{21}{5}$. If the number is m, then the translation is

$m \div 5, \quad m/5, \quad \text{or} \quad \dfrac{m}{5}.$

> **Caution!**
>
> Note that $7 - x$ is *not* a correct translation of the expression in Example 9. The expression $7 - x$ is a translation of "seven minus some number" or "some number less than seven."

EXAMPLE 12 Translate each phrase to an algebraic expression.

PHRASE	ALGEBRAIC EXPRESSION
Five more than some number	$n + 5$, or $5 + n$
Half of a number	$\frac{1}{2}t, \frac{t}{2}$, or $t/2$
Five more than three times some number	$3p + 5$, or $5 + 3p$
The difference of two numbers	$x - y$
Six less than the product of two numbers	$mn - 6$
Seventy-six percent of some number	$76\%z$, or $0.76z$
Four less than twice some number	$2x - 4$

Do Exercises 9–17.

Translate each phrase to an algebraic expression.

9. Eight less than some number

10. Eight more than some number

11. Four less than some number

12. One-third of some number

13. Six more than eight times some number

14. The difference of two numbers

15. Fifty-nine percent of some number

16. Two hundred less than the product of two numbers

17. The sum of two numbers

Answers

9. $x - 8$ 10. $y + 8$, or $8 + y$ 11. $m - 4$
12. $\frac{1}{3} \cdot p$, or $\frac{p}{3}$ 13. $8x + 6$, or $6 + 8x$
14. $a - b$ 15. $59\%x$, or $0.59x$ 16. $xy - 200$
17. $p + q$

✓ Reading Check

Classify each expression as an algebraic expression involving either multiplication or division.

RC1. $3/q$
RC2. $3q$
RC3. $3 \cdot q$
RC4. $\dfrac{3}{q}$

 Substitute to find values of the expressions in each of the following applied problems.

1. *Commuting Time.* It takes Abigail 24 min less time to commute to work than it does Jayden. Suppose that the variable x stands for the time it takes Jayden to get to work. Then $x - 24$ stands for the time it takes Abigail to get to work. How long does it take Abigail to get to work if it takes Jayden 56 min? 93 min? 105 min?

2. *Enrollment Costs.* At Mountain View Community College, it costs $600 to enroll in the 8 A.M. section of Elementary Algebra. Suppose that the variable n stands for the number of students who enroll. Then $600n$ stands for the total amount of tuition collected for this course. How much is collected if 34 students enroll? 78 students? 250 students?

3. *Distance Traveled.* A driver who drives at a constant speed of r miles per hour for t hours will travel a distance of d miles given by $d = rt$ miles. How far will a driver travel at a speed of 65 mph for 4 hr?

4. *Simple Interest.* The simple interest I on a principal of P dollars at interest rate r for time t, in years, is given by $I = Prt$. Find the simple interest on a principal of $4800 at 3% for 2 years.

5. *Wireless Internet Sign.* The U.S. Department of Transportation has designed a new sign that indicates the availability of wireless internet. The square sign measures 24 in. on each side. Find its area.

Source: *Manual on Uniform Traffic Control Devices*, U.S. Department of Transportation, Federal Highway Administration

6. *Yield Sign.* The U.S. Department of Transportation has designed a new yield sign. Each side of the triangular sign measures 30 in., and the height of the triangle is 26 in. Find its area. The area of a triangle with base b and height h is given by $A = \frac{1}{2}bh$.

Source: *Manual on Uniform Traffic Control Devices*, U.S. Department of Transportation, Federal Highway Administration

7. *Area of a Triangle.* The area A of a triangle with base b and height h is given by $A = \frac{1}{2}bh$. Find the area when $b = 45$ m (meters) and $h = 86$ m.

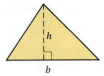

8. *Area of a Parallelogram.* The area A of a parallelogram with base b and height h is given by $A = bh$. Find the area of the parallelogram when the height is 15.4 cm (centimeters) and the base is 6.5 cm.

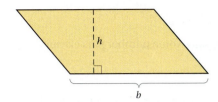

Evaluate.

9. $8x$, when $x = 7$

10. $6y$, when $y = 7$

11. $\dfrac{c}{d}$, when $c = 24$ and $d = 3$

12. $\dfrac{p}{q}$, when $p = 16$ and $q = 2$

13. $\dfrac{3p}{q}$, when $p = 2$ and $q = 6$

14. $\dfrac{5y}{z}$, when $y = 15$ and $z = 25$

15. $\dfrac{x + y}{5}$, when $x = 10$ and $y = 20$

16. $\dfrac{p + q}{2}$, when $p = 2$ and $q = 16$

17. $\dfrac{x - y}{8}$, when $x = 20$ and $y = 4$

18. $\dfrac{m - n}{5}$, when $m = 16$ and $n = 6$

b   Translate each phrase to an algebraic expression. Use any letter for the variable(s) unless directed otherwise.

19. Seven more than some number

20. Some number increased by thirteen

21. Twelve less than some number

22. Fourteen less than some number

23. b more than a

24. c more than d

25. x divided by y

26. c divided by h

27. x plus w

28. s added to t

29. m subtracted from n

30. p subtracted from q

31. Twice some number

32. Three times some number

33. Three multiplied by some number

34. The product of eight and some number

35. Six more than four times some number

36. Two more than six times some number

37. Eight less than the product of two numbers

38. The product of two numbers minus seven

39. Five less than twice some number

40. Six less than seven times some number

41. Three times some number plus eleven

42. Some number times 8 plus 5

43. The sum of four times a number plus three times another number

44. Five times a number minus eight times another number

45. Your salary after a 5% salary increase if your salary before the increase was s

46. The price of a chain saw after a 30% reduction if the price before the reduction was P

47. Aubrey drove at a speed of 65 mph for t hours. How far did she travel? (See Exercise 3.)

48. Liam drove his pickup truck at 55 mph for t hours. How far did he travel? (See Exercise 3.)

49. Lisa had $50 before spending x dollars on pizza. How much money remains?

50. Juan has d dollars before spending $820 on four new tires for his truck. How much did Juan have after the purchase?

51. Sid's part-time job pays $12.50 per hour. How much does he earn for working n hours?

52. Meredith pays her babysitter $10 per hour. What does it cost her to hire the sitter for m hours?

Skill Maintenance

This heading indicates that the exercises that follow are Skill Maintenance exercises, which review any skill previously studied in the text. You can expect such exercises in every exercise set. Answers to *all* skill maintenance exercises are found at the back of the book. If you miss an exercise, restudy the objective shown in red.

Find the prime factorization.   [R.1a]

53. 108

54. 192

Add.   [R.2c]

55. $\dfrac{3}{8} + \dfrac{5}{14}$

56. $\dfrac{11}{27} + \dfrac{1}{6}$

Multiply.   [R.3b]

57. 0.05×1.03

58. 43.5×1000

Find the LCM.   [R.1c]

59. 16, 24, 32

60. 18, 36, 44

Synthesis

To the student and the instructor: The Synthesis exercises found at the end of most exercise sets challenge students to combine concepts or skills studied in that section or in preceding parts of the text.

Evaluate.

61. $\dfrac{a - 2b + c}{4b - a}$, when $a = 20$, $b = 10$, and $c = 5$

62. $\dfrac{x}{y} - \dfrac{5}{x} + \dfrac{2}{y}$, when $x = 30$ and $y = 6$

63. $\dfrac{12 - c}{c + 12b}$, when $b = 1$ and $c = 12$

64. $\dfrac{2w - 3z}{7y}$, when $w = 5$, $y = 6$, and $z = 1$

The Real Numbers

A **set** is a collection of objects. For our purposes, we will most often be considering sets of numbers. One way to name a set uses what is called **roster notation**. For example, roster notation for the set containing the numbers 0, 2, and 5 is $\{0, 2, 5\}$.

Sets that are part of other sets are called **subsets**. In this section, we become acquainted with the set of *real numbers* and its various subsets.

Two important subsets of the real numbers are listed below using roster notation.

NATURAL NUMBERS

The set of **natural numbers** = $\{1, 2, 3, \ldots\}$. These are the numbers used for counting.

WHOLE NUMBERS

The set of **whole numbers** = $\{0, 1, 2, 3, \ldots\}$. This is the set of natural numbers and 0.

We can represent these sets on the number line. The natural numbers are to the right of zero. The whole numbers are the natural numbers and zero.

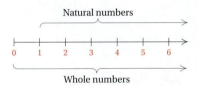

We create a new set, called the *integers*, by starting with the whole numbers, 0, 1, 2, 3, and so on. For each natural number 1, 2, 3, and so on, we obtain a new number to the left of zero on the number line:

For the number 1, there will be an *opposite* number -1 (negative 1).

For the number 2, there will be an *opposite* number -2 (negative 2).

For the number 3, there will be an *opposite* number -3 (negative 3), and so on.

The **integers** consist of the whole numbers and these new numbers.

INTEGERS

The set of **integers** = $\{\ldots, -5, -4, -3, -2, -1, 0, 1, 2, 3, 4, 5, \ldots\}$.

OBJECTIVES

a State the integer that corresponds to a real-world situation.

b Graph rational numbers on the number line.

c Convert from fraction notation for a rational number to decimal notation.

d Determine which of two real numbers is greater and indicate which, using $<$ or $>$. Given an inequality like $a > b$, write another inequality with the same meaning. Determine whether an inequality like $-3 \leq 5$ is true or false.

e Find the absolute value of a real number.

SKILL TO REVIEW

Objective R.3b: Convert fraction notation to decimal notation.

Convert to decimal notation.

1. $\dfrac{5}{8}$ 2. $\dfrac{7}{9}$

Answers

Skill to Review:
1. 0.625 2. $0.\overline{77}$

We picture the integers on the number line as follows.

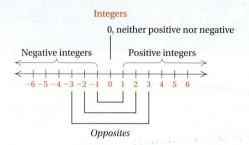

We call the integers to the left of zero **negative integers**. The natural numbers are also called **positive integers**. Zero is neither positive nor negative. We call −1 and 1 **opposites** of each other. Similarly, −2 and 2 are opposites, −3 and 3 are opposites, −100 and 100 are opposites, and 0 is its own opposite. Pairs of opposite numbers like −3 and 3 are the same distance from zero. The integers extend infinitely on the number line to the left and right of zero.

a INTEGERS AND THE REAL WORLD

Integers correspond to many real-world problems and situations. The following examples will help you get ready to translate problem situations that involve integers to mathematical language.

EXAMPLE 1 Tell which integer corresponds to this situation: The temperature is 4 degrees below zero.

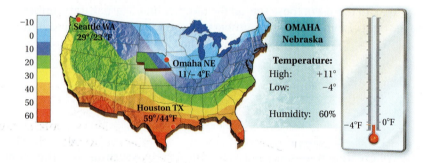

The integer −4 corresponds to the situation. The temperature is −4°.

EXAMPLE 2 *Water Level.* Tell which integer corresponds to this situation: As the water level of the Mississippi River fell during the drought of 2012, barge traffic was restricted, causing a severe decline in shipping volumes. On August 24, the river level at Greenville, Mississippi, was 10 ft below normal.

Source: Rick Jervis, *USA TODAY*, August 24, 2012

The integer −10 corresponds to the drop in water level.

EXAMPLE 3 *Stock Price Change.* Tell which integers correspond to this situation: Hal owns a stock whose price decreased $16 per share over a recent period. He owns another stock whose price increased $2 per share over the same period.

The integer -16 corresponds to the decrease in the value of the first stock. The integer 2 represents the increase in the value of the second stock.

Do Exercises 1–5. ▶

b THE RATIONAL NUMBERS

We created the set of integers by obtaining a negative number for each natural number and also including 0. To create a larger number system, called the set of **rational numbers**, we consider quotients of integers with nonzero divisors. The following are some examples of rational numbers:

$$\frac{2}{3}, \quad -\frac{2}{3}, \quad \frac{7}{1}, \quad 4, \quad -3, \quad 0, \quad \frac{23}{-8}, \quad 2.4, \quad -0.17, \quad 10\frac{1}{2}.$$

The number $-\frac{2}{3}$ (read "negative two-thirds") can also be named $\frac{-2}{3}$ or $\frac{2}{-3}$; that is,

$$-\frac{a}{b} = \frac{-a}{b} = \frac{a}{-b}.$$

The number 2.4 can be named $\frac{24}{10}$ or $\frac{12}{5}$, and -0.17 can be named $-\frac{17}{100}$. We can describe the set of rational numbers as follows.

RATIONAL NUMBERS

The set of **rational numbers** = the set of numbers $\dfrac{a}{b}$,

where a and b are integers and b is not equal to 0 ($b \neq 0$).

Note that this new set of numbers, the rational numbers, contains the whole numbers, the integers, the arithmetic numbers (also called the non-negative rational numbers), and the negative rational numbers.

We picture the rational numbers on the number line as follows.

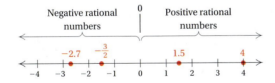

To **graph** a number means to find and mark its point on the number line. Some rational numbers are graphed in the preceding figure.

Tell which integers correspond to each situation.

1. *Temperature High and Low.* The highest recorded temperature in Illinois is $117°$F on July 14, 1954, in East St. Louis. The lowest recorded temperature in Illinois is $36°$F below zero on January 5, 1999, in Congerville.

   Source: National Climate Data Center, NESDIS, NOAA, U.S. Department of Commerce (through 2010)

2. *Stock Decrease.* The price of a stock decreased $3 per share over a recent period.

3. At 10 sec before liftoff, ignition occurs. At 148 sec after liftoff, the first stage is detached from the rocket.

4. The halfback gained 8 yd on first down. The quarterback was sacked for a 5-yd loss on second down.

5. A submarine dove 120 ft, rose 50 ft, and then dove 80 ft.

Answers

1. $117; -36$ **2.** -3 **3.** $-10; 148$
4. $8; -5$ **5.** $-120; 50; -80$

Graph each number on the number line.

6. $-\dfrac{7}{2}$

7. 1.4

8. $-\dfrac{11}{4}$

EXAMPLES Graph each number on the number line.

4. -3.2 The graph of -3.2 is $\dfrac{2}{10}$ of the way from -3 to -4.

5. $\dfrac{13}{8}$ The number $\dfrac{13}{8}$ can also be named $1\dfrac{5}{8}$, or 1.625. The graph is $\dfrac{5}{8}$ of the way from 1 to 2.

◀ **Do Exercises 6–8.**

C NOTATION FOR RATIONAL NUMBERS

Each rational number can be named using fraction notation or decimal notation.

EXAMPLE 6 Convert to decimal notation: $-\dfrac{5}{8}$.

We first find decimal notation for $\dfrac{5}{8}$. Since $\dfrac{5}{8}$ means $5 \div 8$, we divide.

$$
\begin{array}{r}
0.6\,2\,5 \\
8\,\overline{)\,5.0\,0\,0} \\
\underline{4\,8} \\
2\,0 \\
\underline{1\,6} \\
4\,0 \\
\underline{4\,0} \\
0
\end{array}
$$

Thus, $\dfrac{5}{8} = 0.625$, so $-\dfrac{5}{8} = -0.625$.

Decimal notation for $-\dfrac{5}{8}$ is -0.625. We consider -0.625 to be a **terminating decimal**. Decimal notation for some numbers repeats.

EXAMPLE 7 Convert to decimal notation: $\dfrac{7}{11}$.

$$
\begin{array}{r}
0.6\,3\,6\,3\ldots \\
1\,1\,\overline{)\,7.0\,0\,0\,0} \\
\underline{6\,6} \\
4\,0 \\
\underline{3\,3} \\
7\,0 \\
\underline{6\,6} \\
4\,0 \\
\underline{3\,3} \\
7
\end{array}
$$

Dividing

We can abbreviate **repeating decimal** notation by writing a bar over the repeating part—in this case, we write $0.\overline{63}$. Thus, $\dfrac{7}{11} = 0.\overline{63}$.

Answers

6. $-\dfrac{7}{2}$

7. 1.4

8. $-\dfrac{11}{4}$

Each rational number can be expressed in either terminating decimal notation or repeating decimal notation.

The following are other examples showing how rational numbers can be named using fraction notation or decimal notation:

$$0 = \frac{0}{8}, \qquad \frac{27}{100} = 0.27, \qquad -8\frac{3}{4} = -8.75, \qquad -\frac{13}{6} = -2.1\overline{6}.$$

Do Exercises 9–11. ▶

Find decimal notation.

9. $-\dfrac{3}{8}$

10. $-\dfrac{6}{11}$

11. $\dfrac{4}{3}$

d   THE REAL NUMBERS AND ORDER

Every rational number has a point on the number line. However, there are some points on the line for which there is no rational number. These points correspond to what are called **irrational numbers**.

What kinds of numbers are irrational? One example is the number π, which is used in finding the area and the circumference of a circle: $A = \pi r^2$ and $C = 2\pi r$.

Another example of an irrational number is the square root of 2, named $\sqrt{2}$. It is the length of the diagonal of a square with sides of length 1. It is also the number that when multiplied by itself gives 2—that is, $\sqrt{2} \cdot \sqrt{2} = 2$. There is no rational number that can be multiplied by itself to get 2. But the following are rational *approximations*:

1.4 is an approximation of $\sqrt{2}$ because $(1.4)^2 = 1.96$;

1.41 is a better approximation because $(1.41)^2 = 1.9881$;

1.4142 is an even better approximation because $(1.4142)^2 = 1.99996164$.

We can find rational approximations for square roots using a calculator.

Decimal notation for rational numbers *either* terminates *or* repeats.

Decimal notation for irrational numbers *neither* terminates *nor* repeats.

Some other examples of irrational numbers are $\sqrt{3}$, $-\sqrt{8}$, $\sqrt{11}$, and $0.121221222122221.\ldots$ Whenever we take the square root of a number that is not a perfect square, we will get an irrational number.

The rational numbers and the irrational numbers together correspond to all the points on the number line and make up what is called the **real-number system**.

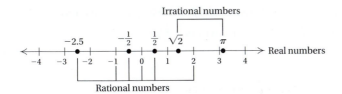

CALCULATOR CORNER

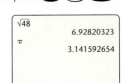

Approximating Square Roots and π   Square roots are found by pressing **2ND** ✓. ($\sqrt{\ }$ is the second operation associated with the **x²** key.)

To find an approximation for $\sqrt{48}$, we press **2ND** ✓ **4** **8** **ENTER**.

The number π is used widely enough to have its own key. (π is the second operation associated with the ⌢ key.) To approximate π, we press **2ND** **π** **ENTER**.

$\sqrt{48}$	
	6.92820323
π	
	3.141592654

EXERCISES: Approximate.

1. $\sqrt{76}$

2. $\sqrt{317}$

3. $15 \cdot \sqrt{20}$

4. $29 + \sqrt{42}$

5. π

6. $29 \cdot \pi$

7. $\pi \cdot 13^2$

8. $5 \cdot \pi + 8 \cdot \sqrt{237}$

Answers

9. -0.375 10. $-0.\overline{54}$ 11. $1.\overline{3}$

CALCULATOR CORNER

Negative Numbers on a Calculator; Converting to Decimal Notation We use the opposite key ⊖ to enter negative numbers on a graphing calculator. Note that this is different from the subtraction key, ⊖.

To convert $-\frac{5}{8}$ to decimal notation, we press ⊖ ⑤ ÷ ⑧ **ENTER**. The result is -0.625.

EXERCISES: Convert to decimal notation.

1. $-\dfrac{3}{4}$

2. $-\dfrac{9}{20}$

3. $-\dfrac{1}{8}$

4. $-\dfrac{9}{5}$

5. $-\dfrac{27}{40}$

6. $-\dfrac{11}{16}$

7. $-\dfrac{7}{2}$

8. $-\dfrac{19}{25}$

The real numbers consist of the rational numbers and the irrational numbers. The following figure shows the relationships among various kinds of numbers.

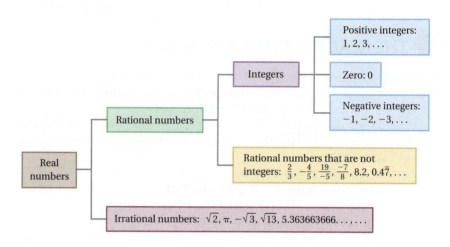

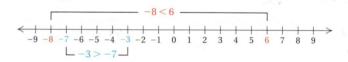

Order

Real numbers are named in order on the number line, increasing as we move from left to right. For any two numbers on the line, the one on the left is less than the one on the right.

We use the symbol **<** to mean "**is less than.**" The sentence $-8 < 6$ means "-8 is less than 6." The symbol **>** means "**is greater than.**" The sentence $-3 > -7$ means "-3 is greater than -7." The sentences $-8 < 6$ and $-3 > -7$ are **inequalities**.

EXAMPLES Use either $<$ or $>$ for ☐ to write a true sentence.

8. 2 ☐ 9 Since 2 is to the left of 9, 2 is less than 9, so $2 < 9$.

9. -7 ☐ 3 Since -7 is to the left of 3, we have $-7 < 3$.

10. 6 ☐ -12 Since 6 is to the right of -12, then $6 > -12$.

11. -18 ☐ -5 Since -18 is to the left of -5, we have $-18 < -5$.

12. -2.7 ☐ $-\frac{3}{2}$ The answer is $-2.7 < -\frac{3}{2}$.

13. 1.5 ☐ -2.7 The answer is $1.5 > -2.7$.

14. 1.38 ☐ 1.83 The answer is $1.38 < 1.83$.

15. $-3.45 \;\square\; 1.32$ The answer is $-3.45 < 1.32$.

16. $-4 \;\square\; 0$ The answer is $-4 < 0$.

17. $5.8 \;\square\; 0$ The answer is $5.8 > 0$.

18. $\frac{5}{8} \;\square\; \frac{7}{11}$ We convert to decimal notation: $\frac{5}{8} = 0.625$ and $\frac{7}{11} = 0.6363\ldots$. Thus, $\frac{5}{8} < \frac{7}{11}$.

19. $-\frac{1}{2} \;\square\; -\frac{1}{3}$ The answer is $-\frac{1}{2} < -\frac{1}{3}$.

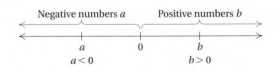

20. $-2\frac{3}{5} \;\square\; -\frac{11}{4}$ The answer is $-2\frac{3}{5} > -\frac{11}{4}$.

Do Exercises 12–19. ▶

Note that both $-8 < 6$ and $6 > -8$ are true. Every true inequality yields another true inequality when we interchange the numbers or the variables and reverse the direction of the inequality sign.

ORDER; >, <

$a < b$ also has the meaning $b > a$.

EXAMPLES   Write another inequality with the same meaning.

21. $-3 > -8$ The inequality $-8 < -3$ has the same meaning.

22. $a < -5$ The inequality $-5 > a$ has the same meaning.

A helpful mental device is to think of an inequality sign as an "arrow" with the arrowhead pointing to the smaller number.

Do Exercises 20 and 21. ▶

Note that all positive real numbers are greater than zero and all negative real numbers are less than zero.

```
        Negative numbers a    Positive numbers b
      ←————————————————————┤———————————————————→
      ←—————————+—————————————+——————————+————————→
                a             0          b
              a < 0                    b > 0
```

If b is a positive real number, then $b > 0$.

If a is a negative real number, then $a < 0$.

Use either $<$ or $>$ for $\square$ to write a true sentence.

12. $-3 \;\square\; 7$

13. $-8 \;\square\; -5$

14. $7 \;\square\; -10$

15. $3.1 \;\square\; -9.5$

16. $-4.78 \;\square\; -5.01$

17. $-\dfrac{2}{3} \;\square\; -\dfrac{5}{9}$

18. $-\dfrac{11}{8} \;\square\; \dfrac{23}{15}$

19. $0 \;\square\; -9.9$

Write another inequality with the same meaning.

20. $-5 < 7$

21. $x > 4$

Expressions like $a \leq b$ and $b \geq a$ are also inequalities. We read $a \leq b$ as "***a* is less than or equal to *b*.**" We read $a \geq b$ as "***a* is greater than or equal to *b*.**"

Write true or false for each statement.

22. $-4 \leq -6$

23. $7.8 \geq 7.8$

24. $-2 \leq \dfrac{3}{8}$

EXAMPLES Write true or false for each statement.

23. $-3 \leq 5.4$ True since $-3 < 5.4$ is true

24. $-3 \leq -3$ True since $-3 = -3$ is true

25. $-5 \geq 1\frac{2}{3}$ False since neither $-5 > 1\frac{2}{3}$ nor $-5 = 1\frac{2}{3}$ is true

◀ **Do Exercises 22–24.**

CALCULATOR CORNER

Absolute Value Finding absolute value is the first item in the Catalog on the 11-84 Plus graphing calculator. To find $|-7|$, we first press **2ND** **CATALOG** **ENTER** to copy "abs(" to the home screen. (**CATALOG** is the second operation associated with the **0** numeric key.) Then we press **(−)** **7** **)** **ENTER**. The result is 7.

    To find $\left|-\frac{1}{2}\right|$ and express the result as a fraction, we press **2ND** **CATALOG** **ENTER** **(−)** **1** **÷** **2** **)** **MATH** **1** **ENTER**. The result is $\frac{1}{2}$.

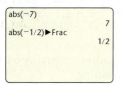

```
abs(−7)
                    7
abs(−1/2)▶Frac
                  1/2
```

EXERCISES: Find the absolute value.

1. $|-5|$ **2.** $|17|$

3. $|0|$ **4.** $|6.48|$

5. $|-12.7|$ **6.** $|-0.9|$

7. $\left|-\dfrac{5}{7}\right|$ **8.** $\left|\dfrac{4}{3}\right|$

e   ABSOLUTE VALUE

From the number line, we see that numbers like 4 and -4 are the same distance from zero. Distance is always a nonnegative number. We call the distance of a number from zero on the number line the **absolute value** of the number.

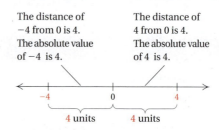

The distance of -4 from 0 is 4. The absolute value of -4 is 4.

The distance of 4 from 0 is 4. The absolute value of 4 is 4.

4 units   4 units

ABSOLUTE VALUE

The **absolute value** of a number is its distance from zero on the number line. We use the symbol $|x|$ to represent the absolute value of a number x.

FINDING ABSOLUTE VALUE

a) If a number is negative, its absolute value is its opposite.

b) If a number is positive or zero, its absolute value is the same as the number.

Find the absolute value.

25. $|8|$ **26.** $|-9|$

27. $\left|-\dfrac{2}{3}\right|$ **28.** $|5.6|$

EXAMPLES Find the absolute value.

26. $|-7|$ The distance of -7 from 0 is 7, so $|-7| = 7$.

27. $|12|$ The distance of 12 from 0 is 12, so $|12| = 12$.

28. $|0|$ The distance of 0 from 0 is 0, so $|0| = 0$.

29. $\left|\frac{3}{2}\right| = \frac{3}{2}$

30. $|-2.73| = 2.73$

◀ **Do Exercises 25–28.**

Answers

22. False   **23.** True   **24.** True   **25.** 8

26. 9   **27.** $\dfrac{2}{3}$ **28.** 5.6

1.2 Exercise Set

For Extra Help
MyMathLab®
MathXL®
PRACTICE   WATCH   READ   REVIEW

✓ Reading Check

Use the number line below for Exercises RC1–RC10.

```
        A          K  H J C      E        G B   D      F
←───┼──┼──┼──●──┼──┼──┼●●┼●┼──┼──●──┼──┼●┼●─┼●─┼──┼●─┼──┼───→
   −10          −5          0          5          10
```

Match each number with its graph.

RC1. $-2\dfrac{5}{7}$ **RC2.** $\left|\dfrac{0}{-8}\right|$ **RC3.** -2.25 **RC4.** $\dfrac{17}{3}$ **RC5.** $\left|-4\right|$ **RC6.** $3.\overline{4}$

Write true or false. The letters name numbers on the number line shown above.

RC7. $K < B$ **RC8.** $H < B$ **RC9.** $E < C$ **RC10.** $J > D$

---

a State the integers that correspond to each situation.

1. On Wednesday, the temperature was 24° above zero. On Thursday, it was 2° below zero.

2. A student deposited her tax refund of $750 in a savings account. Two weeks later, she withdrew $125 to pay technology fees.

3. *Temperature Extremes.* The highest temperature ever created in a lab is 7,200,000,000,000°F. The lowest temperature ever created is approximately 460°F below zero.

Sources: *Live Science; Guinness Book of World Records*

4. *Extreme Climate.* Verkhoyansk, a river port in northeast Siberia, has the most extreme climate on the planet. Its average monthly winter temperature is 58.5°F below zero, and its average monthly summer temperature is 56.5°F.

Source: *Guinness Book of World Records*

5. *Empire State Building.* The Empire State Building has a total height, including the lightning rod at the top, of 1454 ft. The foundation depth is 55 ft below ground level.

Source: www.empirestatebuildingfacts.com

6. *Shipwreck.* There are numerous shipwrecks to explore near Bermuda. One of the most frequently visited sites is L'Herminie, a French warship that sank in 1838. This ship is 35 ft below the surface.

Source: www.10best.com/interest/adventures/scuba-diving-in-pirate-territory

Graph the number on the number line.

7. $\dfrac{10}{3}$ ←—+——+——+——+——+——+——+——+——+——+——+——+——→
      −6 −5 −4 −3 −2 −1   0   1   2   3   4   5   6

8. $-\dfrac{17}{4}$ ←—+——+——+——+——+——+——+——+——+——+——+——+——→
      −6 −5 −4 −3 −2 −1   0   1   2   3   4   5   6

9. -5.2 ←—+——+——+——+——+——+——+——+——+——+——+——+——→
      −6 −5 −4 −3 −2 −1   0   1   2   3   4   5   6

10. 4.78 ←—+——+——+——+——+——+——+——+——+——+——+——+——→
      −6 −5 −4 −3 −2 −1   0   1   2   3   4   5   6

11. $-4\dfrac{2}{5}$ ←—+——+——+——+——+——+——+——+——+——+——+——+——→
      −6 −5 −4 −3 −2 −1   0   1   2   3   4   5   6

12. $2\dfrac{6}{11}$ ←—+——+——+——+——+——+——+——+——+——+——+——+——→
      −6 −5 −4 −3 −2 −1   0   1   2   3   4   5   6

c   Convert to decimal notation.

13. $-\dfrac{7}{8}$ **14.** $-\dfrac{3}{16}$ **15.** $\dfrac{5}{6}$ **16.** $\dfrac{5}{3}$ **17.** $-\dfrac{7}{6}$

18. $-\dfrac{5}{12}$ **19.** $\dfrac{2}{3}$ **20.** $-\dfrac{11}{9}$ **21.** $\dfrac{1}{10}$ **22.** $\dfrac{1}{4}$

23. $-\dfrac{1}{2}$ **24.** $\dfrac{9}{8}$ **25.** $\dfrac{4}{25}$ **26.** $-\dfrac{7}{20}$

d   Use either $<$ or $>$ for $\square$ to write a true sentence.

27. $8 \,\square\, 0$ **28.** $3 \,\square\, 0$ **29.** $-8 \,\square\, 3$ **30.** $6 \,\square\, -6$

31. $-8 \,\square\, 8$ **32.** $0 \,\square\, -9$ **33.** $-8 \,\square\, -5$ **34.** $-4 \,\square\, -3$

35. $-5 \,\square\, -11$ **36.** $-3 \,\square\, -4$ **37.** $2.14 \,\square\, 1.24$ **38.** $-3.3 \,\square\, -2.2$

39. $-12.88 \,\square\, -6.45$

40. $17.2 \,\square\, -1.67$

41. $-\dfrac{1}{2} \,\square\, -\dfrac{2}{3}$

42. $-\dfrac{5}{4} \,\square\, -\dfrac{3}{4}$

43. $-\dfrac{2}{3} \,\square\, \dfrac{1}{3}$

44. $\dfrac{3}{4} \,\square\, -\dfrac{5}{4}$

45. $\dfrac{5}{12} \,\square\, \dfrac{11}{25}$

46. $-\dfrac{13}{16} \,\square\, -\dfrac{5}{9}$

Write an inequality with the same meaning.

47. $-6 > x$

48. $x < 8$

49. $-10 \le y$

50. $12 \ge t$

Write true or false.

51. $-5 \le -6$

52. $-7 \ge -10$

53. $4 \ge 4$

54. $7 \le 7$

55. $-3 \ge -11$

56. $-1 \le -5$

57. $0 \ge 8$

58. $-5 \le 7$

e Find the absolute value.

59. $|-3|$

60. $|-6|$

61. $|11|$

62. $|0|$

63. $\left|-\dfrac{2}{3}\right|$

64. $|325|$

65. $\left|\dfrac{0}{4}\right|$

66. $|14.8|$

67. $|-2.65|$

68. $\left|-3\dfrac{5}{8}\right|$

Skill Maintenance

Convert to decimal notation. [R.4a]

69. 110%

70. $23\dfrac{4}{5}\%$

Convert to percent notation. [R.4d]

71. $\dfrac{13}{25}$

72. $\dfrac{19}{32}$

Evaluate. [R.5b]

73. 3^4

74. 5^0

Simplify. [R.5c]

75. $3(7 + 2^3)$

76. $48 \div 8 - 6$

Synthesis

List in order from the least to the greatest.

77. $\dfrac{2}{3}, -\dfrac{1}{7}, \dfrac{1}{3}, -\dfrac{2}{7}, -\dfrac{2}{3}, \dfrac{2}{5}, -\dfrac{1}{3}, -\dfrac{2}{5}, \dfrac{9}{8}$

78. $-8\dfrac{7}{8}, 7^1, -5, |-6|, 4, |3|, -8\dfrac{5}{8}, -100, 0, 1^7, \dfrac{7}{2}, -\dfrac{67}{8}$

Given that $0.\overline{3} = \frac{1}{3}$ and $0.\overline{6} = \frac{2}{3}$, express each of the following as a quotient or a ratio of two integers.

79. $0.\overline{9}$

80. $0.\overline{1}$

81. $5.\overline{5}$

OBJECTIVES

a Add real numbers without using the number line.

b Find the opposite, or additive inverse, of a real number.

c Solve applied problems involving addition of real numbers.

SKILL TO REVIEW

Objective R.2c: Add using fraction notation.

Add.

1. $\dfrac{1}{6} + \dfrac{2}{9}$ **2.** $\dfrac{5}{8} + \dfrac{7}{30}$

In this section, we consider addition of real numbers. First, to gain an understanding, we add using the number line. Then we consider rules for addition.

ADDITION ON THE NUMBER LINE

To do the addition $a + b$ on the number line, start at 0, move to a, and then move according to b.

a) If b is positive, move from a to the right.

b) If b is negative, move from a to the left.

c) If b is 0, stay at a.

EXAMPLE 1   Add: $3 + (-5)$.

We start at 0 and move to 3. Then we move 5 units left since -5 is negative.

$$3 + (-5) = -2$$

EXAMPLE 2   Add: $-4 + (-3)$.

We start at 0 and move to -4. Then we move 3 units left since -3 is negative.

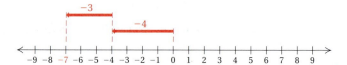

$$-4 + (-3) = -7$$

EXAMPLE 3   Add: $-4 + 9$.

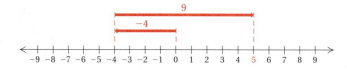

$$-4 + 9 = 5$$

Answers

Skill to Review:

1. $\dfrac{7}{18}$ **2.** $\dfrac{103}{120}$

EXAMPLE 4   Add: $-5.2 + 0$.

Stay at -5.2.

-5.2

-5.2

$-5.2 + 0 = -5.2$

Do Exercises 1–6. ▶

a ADDING WITHOUT THE NUMBER LINE

You may have noticed some patterns in the preceding examples. These lead us to rules for adding without using the number line that are more efficient for adding larger numbers.

RULES FOR ADDITION OF REAL NUMBERS

1. *Positive numbers:* Add the same as arithmetic numbers. The answer is positive.
2. *Negative numbers:* Add absolute values. The answer is negative.
3. *A positive number and a negative number:*
   - If the numbers have the same absolute value, the answer is 0.
   - If the numbers have different absolute values, subtract the smaller absolute value from the larger. Then:
      a) If the positive number has the greater absolute value, the answer is positive.
      b) If the negative number has the greater absolute value, the answer is negative.
4. *One number is zero:* The sum is the other number.

Rule 4 is known as the **identity property of 0.** It says that for any real number a, $a + 0 = a$.

EXAMPLES   Add without using the number line.

5. $-12 + (-7) = -19$ Two negatives. Add the absolute values: $|-12| + |-7| = 12 + 7 = 19$. Make the answer *negative*: -19.

6. $-1.4 + 8.5 = 7.1$ One negative, one positive. Find the absolute values: $|-1.4| = 1.4$; $|8.5| = 8.5$. Subtract the smaller absolute value from the larger: $8.5 - 1.4 = 7.1$. The *positive* number, 8.5, has the larger absolute value, so the answer is *positive*: 7.1.

7. $-36 + 21 = -15$ One negative, one positive. Find the absolute values: $|-36| = 36$; $|21| = 21$. Subtract the smaller absolute value from the larger: $36 - 21 = 15$. The *negative* number, -36, has the larger absolute value, so the answer is *negative*: -15.

Add using the number line.

1. $0 + (-3)$

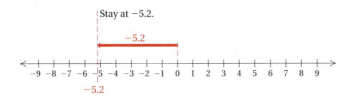

2. $1 + (-4)$

3. $-3 + (-2)$

4. $-3 + 7$

5. $-2.4 + 2.4$

6. $-\dfrac{5}{2} + \dfrac{1}{2}$

Answers

1. -3 2. -3 3. -5
4. 4 5. 0 6. -2

Add without using the number line.

7. $-5 + (-6)$ **8.** $-9 + (-3)$

9. $-4 + 6$ **10.** $-7 + 3$

11. $5 + (-7)$ **12.** $-20 + 20$

13. $-11 + (-11)$ **14.** $10 + (-7)$

15. $-0.17 + 0.7$ **16.** $-6.4 + 8.7$

17. $-4.5 + (-3.2)$

18. $-8.6 + 2.4$

19. $\dfrac{5}{9} + \left(-\dfrac{7}{9}\right)$

20. $-\dfrac{1}{5} + \left(-\dfrac{3}{4}\right)$ **GS**

$= -\dfrac{4}{20} + \left(-\dfrac{\boxed{}}{20}\right)$

$= -\dfrac{19}{\boxed{}}$

Add.

21. $(-15) + (-37) + 25 + 42 +$
$(-59) + (-14)$

22. $42 + (-81) + (-28) + 24 +$
$18 + (-31)$

23. $-2.5 + (-10) + 6 + (-7.5)$

24. $-35 + 17 + 14 + (-27) +$
$31 + (-12)$

8. $1.5 + (-1.5) = 0$ The numbers have the same absolute value. The sum is 0.

9. $-\dfrac{7}{8} + 0 = -\dfrac{7}{8}$ One number is zero. The sum is $-\dfrac{7}{8}$.

10. $-9.2 + 3.1 = -6.1$

11. $-\dfrac{3}{2} + \dfrac{9}{2} = \dfrac{6}{2} = 3$

12. $-\dfrac{2}{3} + \dfrac{5}{8} = -\dfrac{16}{24} + \dfrac{15}{24} = -\dfrac{1}{24}$

◀ **Do Exercises 7–20.**

Suppose we want to add several numbers, some positive and some negative, as follows. How can we proceed?

$$15 + (-2) + 7 + 14 + (-5) + (-12)$$

We can change grouping and order as we please when adding. For instance, we can group the positive numbers together and the negative numbers together and add them separately. Then we add the two results.

EXAMPLE 13   Add: $15 + (-2) + 7 + 14 + (-5) + (-12)$.

a) $15 + 7 + 14 = 36$ Adding the positive numbers
b) $-2 + (-5) + (-12) = -19$ Adding the negative numbers
    $36 + (-19) = 17$ Adding the results in (a) and (b)

We can also add the numbers in any other order we wish—say, from left to right—as follows:

$$
\begin{aligned}
15 + (-2) + 7 + 14 + (-5) + (-12) &= 13 + 7 + 14 + (-5) + (-12) \\
&= 20 + 14 + (-5) + (-12) \\
&= 34 + (-5) + (-12) \\
&= 29 + (-12) \\
&= 17
\end{aligned}
$$

◀ **Do Exercises 21–24.**

b   OPPOSITES, OR ADDITIVE INVERSES

Suppose we add two numbers that are **opposites**, such as 6 and -6. The result is 0. When opposites are added, the result is always 0. Opposites are also called **additive inverses**. Every real number has an opposite, or additive inverse.

> **OPPOSITES, OR ADDITIVE INVERSES**
>
> Two numbers whose sum is 0 are called **opposites**, or **additive inverses**, of each other.

Answers

7. -11 **8.** -12 **9.** 2   **10.** -4
11. -2 **12.** 0   **13.** -22 **14.** 3
15. 0.53   **16.** 2.3   **17.** -7.7 **18.** -6.2
19. $-\dfrac{2}{9}$ **20.** $-\dfrac{19}{20}$ **21.** -58 **22.** -56
23. -14 **24.** -12

Guided Solution:
20. 15, 20

EXAMPLES Find the opposite, or additive inverse, of each number.

14. 34   The opposite of 34 is -34 because $34 + (-34) = 0$.

15. -8 The opposite of -8 is 8 because $-8 + 8 = 0$.

16. 0   The opposite of 0 is 0 because $0 + 0 = 0$.

17. $-\dfrac{7}{8}$ The opposite of $-\dfrac{7}{8}$ is $\dfrac{7}{8}$ because $-\dfrac{7}{8} + \dfrac{7}{8} = 0$.

Do Exercises 25–30. ▶

Find the opposite, or additive inverse, of each number.

25. -4 **26.** 8.7

27. -7.74 **28.** $-\dfrac{8}{9}$

29. 0   **30.** 12

To name the opposite, we use the symbol $-$, as follows.

SYMBOLIZING OPPOSITES

The opposite, or additive inverse, of a number a can be named $-a$ (read "the opposite of a," or "the additive inverse of a").

Note that if we take a number, say, 8, and find its opposite, -8, and then find the opposite of the result, we will have the original number, 8, again.

THE OPPOSITE OF AN OPPOSITE

The **opposite of the opposite** of a number is the number itself. (The additive inverse of the additive inverse of a number is the number itself.) That is, for any number a,

$$-(-a) = a.$$

EXAMPLE 18   Evaluate $-x$ and $-(-x)$ when $x = 16$.

If $x = 16$, then $-x = -16$.   The opposite of 16 is -16.

If $x = 16$, then $-(-x) = -(-16) = 16$.   The opposite of the opposite of 16 is 16.

EXAMPLE 19   Evaluate $-x$ and $-(-x)$ when $x = -3$.

If $x = -3$, then $-x = -(-3) = 3$.

If $x = -3$, then $-(-x) = -(-(-3)) = -(3) = -3$.

Note that in Example 19 we used a second set of parentheses to show that we are substituting the negative number -3 for x. Symbolism like $--x$ is not considered meaningful.

Do Exercises 31–34. ▶

Evaluate $-x$ and $-(-x)$ when:

31. $x = 14$.

GS **32.** $x = -1.6$.

$-x = -(\boxed{}) = 1.6;$

$-(-x) = -(-(\boxed{}))$

$= -(\boxed{}) = -1.6$

33. $x = \dfrac{2}{3}$.   **34.** $x = -\dfrac{9}{8}$.

A symbol such as -8 is usually read "negative 8." It could be read "the additive inverse of 8," because the additive inverse of 8 is negative 8. It could also be read "the opposite of 8," because the opposite of 8 is -8. Thus a symbol like -8 can be read in more than one way. It is never correct to read -8 as "minus 8."

·   **Caution!**   ·

A symbol like $-x$, which has a variable, should be read "the opposite of x" or "the additive inverse of x" and *not* "negative x," because we do not know whether x represents a positive number, a negative number, or 0. You can check this in Examples 18 and 19.

· ·

Answers

25. 4   **26.** -8.7 **27.** 7.74   **28.** $\dfrac{8}{9}$

29. 0   **30.** -12 **31.** -14; 14

32. 1.6; -1.6 **33.** $-\dfrac{2}{3}; \dfrac{2}{3}$ **34.** $\dfrac{9}{8}; -\dfrac{9}{8}$

Guided Solution:

32. -1.6; -1.6, 1.6

We can use the symbolism $-a$ to restate the definition of opposite, or additive inverse.

> **OPPOSITES, OR ADDITIVE INVERSES**
>
> For any real number a, the **opposite**, or **additive inverse**, of a, denoted $-a$, is such that
> $$a + (-a) = (-a) + a = 0.$$

Signs of Numbers

A negative number is sometimes said to have a "negative sign." A positive number is said to have a "positive sign." When we replace a number with its opposite, we can say that we have "changed its sign."

EXAMPLES   Find the opposite. (Change the sign.)

20. -3 $-(-3) = 3$

21. $-\dfrac{2}{13}$ $-\left(-\dfrac{2}{13}\right) = \dfrac{2}{13}$

22. 0 $-(0) = 0$

23. 14 $-(14) = -14$

◀ **Do Exercises 35–38.**

Find the opposite. (Change the sign.)

35. -4

36. -13.4

37. 0

38. $\dfrac{1}{4}$

C   APPLICATIONS AND PROBLEM SOLVING

Addition of real numbers occurs in many real-world situations.

EXAMPLE 24   *Banking Transactions.*   On August 1st, Martias checks his bank account balance on his phone and sees that it is $54. During the next week, the following transactions were recorded: a debit-card purchase of $71, an overdraft fee of $29, a direct deposit of $160, and an ATM withdrawal of $80. What is Martias's balance at the end of the week?

We let $B =$ the ending balance of the bank account. Then the problem translates to the following:

Ending balance	is	Beginning balance	plus	Debit-card purchase	plus	Overdraft fee	plus	Direct deposit	plus	ATM withdrawal
B	$=$	54	$+$	(-71)	$+$	(-29)	$+$	160	$+$	$(-80).$

Adding, we have

$B = 54 + (-71) + (-29) + 160 + (-80)$

$= 214 + (-180)$ Adding the positive numbers and adding the negative numbers

$= 34.$

Martias's balance at the end of the week was $34.

◀ **Do Exercise 39.**

39. *Change in Class Size.*   During the first two weeks of the semester in Jim's algebra class, 4 students withdrew, 8 students enrolled late, and 6 students were dropped as "no shows." By how many students had the class size changed at the end of the first two weeks?

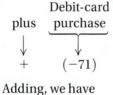

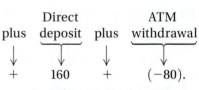

Answers

35. 4   **36.** 13.4   **37.** 0

38. $-\dfrac{1}{4}$ **39.** -2 students

✓ Reading Check

Fill in each blank with either "left" or "right" so that the statements describe the steps when adding numbers with the number line.

RC1. To add $7 + 2$, start at 0, move _____ to 7, and then move 2 units _____. The sum is 9.

RC2. To add $-3 + (-5)$, start at 0, move _____ to -3, and then move 5 units _____. The sum is -8.

RC3. To add $4 + (-6)$, start at 0, move _____ to 4, and then move 6 units _____. The sum is -2.

RC4. To add $-8 + 3$, start at 0, move _____ to -8, and then move 3 units _____. The sum is -5.

a   Add. Do not use the number line except as a check.

1. $2 + (-9)$ **2.** $-5 + 2$ **3.** $-11 + 5$ **4.** $4 + (-3)$ **5.** $-6 + 6$

6. $8 + (-8)$ **7.** $-3 + (-5)$ **8.** $-4 + (-6)$ **9.** $-7 + 0$ **10.** $-13 + 0$

11. $0 + (-27)$ **12.** $0 + (-35)$ **13.** $17 + (-17)$ **14.** $-15 + 15$ **15.** $-17 + (-25)$

16. $-24 + (-17)$ **17.** $18 + (-18)$ **18.** $-13 + 13$ **19.** $-28 + 28$ **20.** $11 + (-11)$

21. $8 + (-5)$ **22.** $-7 + 8$ **23.** $-4 + (-5)$ **24.** $10 + (-12)$ **25.** $13 + (-6)$

26. $-3 + 14$ **27.** $-25 + 25$ **28.** $50 + (-50)$ **29.** $53 + (-18)$ **30.** $75 + (-45)$

31. $-8.5 + 4.7$ **32.** $-4.6 + 1.9$ **33.** $-2.8 + (-5.3)$ **34.** $-7.9 + (-6.5)$ **35.** $-\dfrac{3}{5} + \dfrac{2}{5}$

36. $-\dfrac{4}{3} + \dfrac{2}{3}$ **37.** $-\dfrac{2}{9} + \left(-\dfrac{5}{9}\right)$ **38.** $-\dfrac{4}{7} + \left(-\dfrac{6}{7}\right)$ **39.** $-\dfrac{5}{8} + \dfrac{1}{4}$ **40.** $-\dfrac{5}{6} + \dfrac{2}{3}$

41. $-\dfrac{5}{8} + \left(-\dfrac{1}{6}\right)$ **42.** $-\dfrac{5}{6} + \left(-\dfrac{2}{9}\right)$ **43.** $-\dfrac{3}{8} + \dfrac{5}{12}$ **44.** $-\dfrac{7}{16} + \dfrac{7}{8}$

45. $-\dfrac{1}{6} + \dfrac{7}{10}$ **46.** $-\dfrac{11}{18} + \left(-\dfrac{3}{4}\right)$ **47.** $\dfrac{7}{15} + \left(-\dfrac{1}{9}\right)$ **48.** $-\dfrac{4}{21} + \dfrac{3}{14}$

49. $76 + (-15) + (-18) + (-6)$ **50.** $29 + (-45) + 18 + 32 + (-96)$

51. $-44 + \left(-\dfrac{3}{8}\right) + 95 + \left(-\dfrac{5}{8}\right)$ **52.** $24 + 3.1 + (-44) + (-8.2) + 63$

 Find the opposite, or additive inverse.

53. 24   **54.** −64   **55.** −26.9   **56.** 48.2

Evaluate $-x$ when:

57. $x = 8.$ **58.** $x = -27.$ **59.** $x = -\dfrac{13}{8}.$ **60.** $x = \dfrac{1}{236}.$

Evaluate $-(-x)$ when:

61. $x = -43.$ **62.** $x = 39.$ **63.** $x = \dfrac{4}{3}.$ **64.** $x = -7.1.$

Find the opposite. (Change the sign.)

65. −24   **66.** −12.3   **67.** $-\dfrac{3}{8}$ **68.** 10

c Solve.

69. *Tallest Mountain.* The tallest mountain in the world, when measured from base to peak, is Mauna Kea (White Mountain) in Hawaii. From its base 19,684 ft below sea level in the Hawaiian Trough, it rises 33,480 ft. What is the elevation of the peak above sea level?

Source: *The Guinness Book of Records*

70. *Copy Center Account.* Rachel's copy-center bill for July was $327. She made a payment of $200 and then made $48 worth of copies in August. How much did she then owe on her account?

71. Temperature Changes. One day, the temperature in Lawrence, Kansas, is 32°F at 6:00 A.M. It rises 15° by noon, but falls 50° by midnight when a cold front moves in. What is the final temperature?

72. Stock Changes. On a recent day, the price of a stock opened at a value of $61.38. During the day, it rose $4.75, dropped $7.38, and rose $5.13. Find the value of the stock at the end of the day.

73. "Flipping" Houses. Buying run-down houses, fixing them up, and reselling them is referred to as "flipping" houses. Charlie and Sophia bought and sold four houses in a recent year. The profits and losses are shown in the following bar graph. Find the sum of the profits and losses.

Flipping Houses

74. Football Yardage. In a college football game, the quarterback attempted passes with the following results. Find the total gain or loss.

ATTEMPT	GAIN OR LOSS
1st	13-yd gain
2nd	12-yd loss
3rd	21-yd gain

75. Credit-Card Bills. On August 1, Lyle's credit-card bill shows that he owes $470. During the month of August, Lyle makes a payment of $45 to the credit-card company, charges another $160 in merchandise, and then pays off another $500 of his bill. What is the new amount that Lyle owes at the end of August?

76. Account Balance. Emma has $460 in a checking account. She uses her debit card for a purchase of $530, makes a deposit of $75, and then writes a check for $90. What is the balance in her account?

Skill Maintenance

77. Evaluate $5a - 2b$ when $a = 9$ and $b = 3$. [1.1a]

78. Write an inequality with the same meaning as the inequality $-3 < y$. [1.2d]

Convert to decimal notation. [1.2c]

79. $-\dfrac{1}{12}$

80. $\dfrac{5}{8}$

Find the absolute value. [1.2e]

81. $|0|$

82. $|-21.4|$

Synthesis

83. For what numbers x is $-x$ negative?

84. For what numbers x is $-x$ positive?

85. If a is positive and b is negative, then $-a + b$ is which of the following?
  A. Positive
  C. 0
  B. Negative
  D. Cannot be determined without more information

86. If $a = b$ and a and b are negative, then $-a + (-b)$ is which of the following?
  A. Positive
  C. 0
  B. Negative
  D. Cannot be determined without more information

a SUBTRACTION

We now consider subtraction of real numbers.

SUBTRACTION

The difference $a - b$ is the number c for which $a = b + c$.

Consider, for example, $45 - 17$. *Think*: What number can we add to 17 to get 45? Since $45 = 17 + 28$, we know that $45 - 17 = 28$. Let's consider an example whose answer is a negative number.

EXAMPLE 1 Subtract: $3 - 7$.

Think: What number can we add to 7 to get 3? The number must be negative. Since $7 + (-4) = 3$, we know the number is -4: $3 - 7 = -4$. That is, $3 - 7 = -4$ because $7 + (-4) = 3$.

◀ **Do Exercises 1–3.**

The definition above does not provide the most efficient way to do subtraction. We can develop a faster way to subtract. As a rationale for the faster way, let's compare $3 + 7$ and $3 - 7$ on the number line.

To find $3 + 7$ on the number line, we start at 0, move to 3, and then move 7 units farther to the right since 7 is positive.

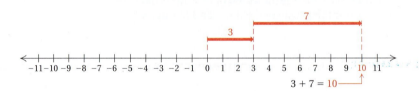

To find $3 - 7$, we do the "opposite" of adding 7: We move 7 units to the *left* to do the subtracting. This is the same as *adding* the opposite of 7, -7, to 3.

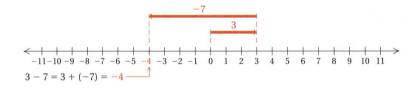

◀ **Do Exercises 4–6.**

Look for a pattern in the examples shown at right.

SUBTRACTING	ADDING AN OPPOSITE
$5 - 8 = -3$	$5 + (-8) = -3$
$-6 - 4 = -10$	$-6 + (-4) = -10$
$-7 - (-2) = -5$	$-7 + 2 = -5$

Subtract.

1. $-6 - 4$

Think: What number can be added to 4 to get -6:
$$\square + 4 = -6?$$

2. $-7 - (-10)$

Think: What number can be added to -10 to get -7:
$$\square + (-10) = -7?$$

3. $-7 - (-2)$

Think: What number can be added to -2 to get -7:
$$\square + (-2) = -7?$$

Subtract. Use the number line, doing the "opposite" of addition.

4. $5 - 9$

5. $-3 - 2$

6. $-4 - (-3)$

Answers

1. -10 **2.** 3   **3.** -5 **4.** -4
5. -5 **6.** -1

Do Exercises 7–10. ▶

Perhaps you have noticed that we can subtract by adding the opposite of the number being subtracted. This can always be done.

SUBTRACTING BY ADDING THE OPPOSITE

For any real numbers a and b,

$$a - b = a + (-b).$$

(To subtract, add the opposite, or additive inverse, of the number being subtracted.)

This is the method generally used for quick subtraction of real numbers.

EXAMPLES   Subtract.

2. $2 - 6 = 2 + (-6) = -4$

The opposite of 6 is -6. We change the subtraction to addition and add the opposite. *Check*: $-4 + 6 = 2.$

3. $4 - (-9) = 4 + 9 = 13$

The opposite of -9 is 9. We change the subtraction to addition and add the opposite. *Check*: $13 + (-9) = 4.$

4. $-4.2 - (-3.6) = -4.2 + 3.6 = -0.6$

Adding the opposite. *Check*: $-0.6 + (-3.6) = -4.2.$

5. $-\dfrac{1}{2} - \left(-\dfrac{3}{4}\right) = -\dfrac{1}{2} + \dfrac{3}{4}$

$\qquad = -\dfrac{2}{4} + \dfrac{3}{4} = \dfrac{1}{4}$

Adding the opposite. *Check*: $\dfrac{1}{4} + \left(-\dfrac{3}{4}\right) = -\dfrac{1}{2}.$

Do Exercises 11–16. ▶

EXAMPLES   Subtract by adding the opposite of the number being subtracted.

6. $3 - 5$ *Think*: "Three minus five is three plus the opposite of five"

$3 - 5 = 3 + (-5) = -2$

7. $\dfrac{1}{8} - \dfrac{7}{8}$ *Think*: "One-eighth minus seven-eighths is one-eighth plus the opposite of seven-eighths"

$\dfrac{1}{8} - \dfrac{7}{8} = \dfrac{1}{8} + \left(-\dfrac{7}{8}\right) = -\dfrac{6}{8}, \text{ or } -\dfrac{3}{4}$

8. $-4.6 - (-9.8)$ *Think*: "Negative four point six minus negative nine point eight is negative four point six plus the opposite of negative nine point eight"

$-4.6 - (-9.8) = -4.6 + 9.8 = 5.2$

9. $-\dfrac{3}{4} - \dfrac{7}{5}$ *Think*: "Negative three-fourths minus seven-fifths is negative three-fourths plus the opposite of seven-fifths"

$-\dfrac{3}{4} - \dfrac{7}{5} = -\dfrac{3}{4} + \left(-\dfrac{7}{5}\right) = -\dfrac{15}{20} + \left(-\dfrac{28}{20}\right) = -\dfrac{43}{20}$

Do Exercises 17–21. ▶

Complete the addition and compare with the subtraction.

7. $4 - 6 = -2;$
$\quad 4 + (-6) = $ _____

8. $-3 - 8 = -11;$
$\quad -3 + (-8) = $ _____

9. $-5 - (-9) = 4;$
$\quad -5 + 9 = $ _____

10. $-5 - (-3) = -2;$
$\quad -5 + 3 = $ _____

Subtract.

GS 11. $2 - 8 = 2 + ($ ☐ $) = $ ☐

12. $-6 - 10$

13. $12.4 - 5.3$

14. $-8 - (-11)$

15. $-8 - (-8)$

16. $\dfrac{2}{3} - \left(-\dfrac{5}{6}\right)$

Subtract by adding the opposite of the number being subtracted.

17. $3 - 11$

18. $12 - 5$

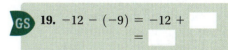

GS 19. $-12 - (-9) = -12 + $ ☐
$\qquad\qquad\qquad = $ ☐

20. $-12.4 - 10.9$

21. $-\dfrac{4}{5} - \left(-\dfrac{4}{5}\right)$

Answers

7. -2 **8.** -11 **9.** 4   **10.** -2 **11.** -6
12. -16 **13.** 7.1   **14.** 3   **15.** 0   **16.** $\dfrac{3}{2}$
17. -8 **18.** 7   **19.** -3 **20.** -23.3 **21.** 0
Guided Solutions:
11. $-8, -6$ **19.** $9, -3$

When several additions and subtractions occur together, we can make them all additions.

EXAMPLES Simplify.

10. $8 - (-4) - 2 - (-4) + 2 = 8 + 4 + (-2) + 4 + 2$ Adding the opposite
$$= 16$$

11. $8.2 - (-6.1) + 2.3 - (-4) = 8.2 + 6.1 + 2.3 + 4 = 20.6$

12. $\dfrac{3}{4} - \left(-\dfrac{1}{12}\right) - \dfrac{5}{6} - \dfrac{2}{3} = \dfrac{9}{12} + \dfrac{1}{12} + \left(-\dfrac{10}{12}\right) + \left(-\dfrac{8}{12}\right)$

$$= \dfrac{9 + 1 + (-10) + (-8)}{12}$$

$$= \dfrac{-8}{12} = -\dfrac{8}{12} = -\dfrac{2}{3}$$

Simplify.

22. $-6 - (-2) - (-4) - 12 + 3$

23. $\dfrac{2}{3} - \dfrac{4}{5} - \left(-\dfrac{11}{15}\right) + \dfrac{7}{10} - \dfrac{5}{2}$

24. $-9.6 + 7.4 - (-3.9) - (-11)$

◀ **Do Exercises 22–24.**

b APPLICATIONS AND PROBLEM SOLVING

Let's now see how we can use subtraction of real numbers to solve applied problems.

EXAMPLE 13 *Surface Temperatures on Mars.* Surface temperatures on Mars vary from $-128°$C during polar night to $27°$C at the equator during midday at the closest point in orbit to the sun. Find the difference between the highest value and the lowest value in this temperature range.

Source: Mars Institute

25. *Temperature Extremes.* The highest temperature ever recorded in the United States is 134°F in Greenland Ranch, California, on July 10, 1913. The lowest temperature ever recorded is −80°F in Prospect Creek, Alaska, on January 23, 1971. How much higher was the temperature in Greenland Ranch than the temperature in Prospect Creek?

Source: National Oceanographic and Atmospheric Administration

We let D = the difference in the temperatures. Then the problem translates to the following subtraction:

Difference in temperature	is	Highest temperature	minus	Lowest temperature
↓	↓	↓	↓	↓
D	$=$	27	$-$	(-128)
D	$= 27 + 128 = 155.$			

The difference in the temperatures is $155°$C.

◀ **Do Exercise 25.**

Answers

22. -9 **23.** $-\dfrac{6}{5}$ **24.** 12.7 **25.** 214°F

✓ Reading Check

Match the expression with an expression from the column on the right that names the same number.

RC1. $18 - 6$

RC2. $-18 - (-6)$

RC3. $-18 - 6$

RC4. $18 - (-6)$

a) $18 + 6$

b) $-18 + 6$

c) $18 + (-6)$

d) $-18 + (-6)$

a Subtract.

1. $2 - 9$

2. $3 - 8$

3. $-8 - (-2)$

4. $-6 - (-8)$

5. $-11 - (-11)$

6. $-6 - (-6)$

7. $12 - 16$

8. $14 - 19$

9. $20 - 27$

10. $30 - 4$

11. $-9 - (-3)$

12. $-7 - (-9)$

13. $-40 - (-40)$

14. $-9 - (-9)$

15. $7 - (-7)$

16. $4 - (-4)$

17. $8 - (-3)$

18. $-7 - 4$

19. $-6 - 8$

20. $6 - (-10)$

21. $-4 - (-9)$

22. $-14 - 2$

23. $-6 - (-5)$

24. $-4 - (-3)$

25. $8 - (-10)$

26. $5 - (-6)$

27. $-5 - (-2)$

28. $-3 - (-1)$

29. $-7 - 14$

30. $-9 - 16$

31. $0 - (-5)$

32. $0 - (-1)$

33. $-8 - 0$ **34.** $-9 - 0$ **35.** $7 - (-5)$ **36.** $7 - (-4)$

37. $2 - 25$ **38.** $18 - 63$ **39.** $-42 - 26$ **40.** $-18 - 63$

41. $-71 - 2$ **42.** $-49 - 3$ **43.** $24 - (-92)$ **44.** $48 - (-73)$

45. $-50 - (-50)$ **46.** $-70 - (-70)$ **47.** $-\dfrac{3}{8} - \dfrac{5}{8}$ **48.** $\dfrac{3}{9} - \dfrac{9}{9}$

49. $\dfrac{3}{4} - \dfrac{2}{3}$ **50.** $\dfrac{5}{8} - \dfrac{3}{4}$ **51.** $-\dfrac{3}{4} - \dfrac{2}{3}$ **52.** $-\dfrac{5}{8} - \dfrac{3}{4}$

53. $-\dfrac{5}{8} - \left(-\dfrac{3}{4}\right)$ **54.** $-\dfrac{3}{4} - \left(-\dfrac{2}{3}\right)$ **55.** $6.1 - (-13.8)$ **56.** $1.5 - (-3.5)$

57. $-2.7 - 5.9$ **58.** $-3.2 - 5.8$ **59.** $0.99 - 1$ **60.** $0.87 - 1$

61. $-79 - 114$ **62.** $-197 - 216$ **63.** $0 - (-500)$ **64.** $500 - (-1000)$

65. $-2.8 - 0$ **66.** $6.04 - 1.1$ **67.** $7 - 10.53$ **68.** $8 - (-9.3)$

69. $\dfrac{1}{6} - \dfrac{2}{3}$ **70.** $-\dfrac{3}{8} - \left(-\dfrac{1}{2}\right)$ **71.** $-\dfrac{4}{7} - \left(-\dfrac{10}{7}\right)$ **72.** $\dfrac{12}{5} - \dfrac{12}{5}$

73. $-\dfrac{7}{10} - \dfrac{10}{15}$

74. $-\dfrac{4}{18} - \left(-\dfrac{2}{9}\right)$

75. $\dfrac{1}{5} - \dfrac{1}{3}$

76. $-\dfrac{1}{7} - \left(-\dfrac{1}{6}\right)$

77. $\dfrac{5}{12} - \dfrac{7}{16}$

78. $-\dfrac{1}{35} - \left(-\dfrac{9}{40}\right)$

79. $-\dfrac{2}{15} - \dfrac{7}{12}$

80. $\dfrac{2}{21} - \dfrac{9}{14}$

Simplify.

81. $18 - (-15) - 3 - (-5) + 2$

82. $22 - (-18) + 7 + (-42) - 27$

83. $-31 + (-28) - (-14) - 17$

84. $-43 - (-19) - (-21) + 25$

85. $-34 - 28 + (-33) - 44$

86. $39 + (-88) - 29 - (-83)$

87. $-93 - (-84) - 41 - (-56)$

88. $84 + (-99) + 44 - (-18) - 43$

89. $-5.4 - (-30.9) + 30.8 + 40.2 - (-12)$

90. $14.9 - (-50.7) + 20 - (-32.8)$

91. $-\dfrac{7}{12} + \dfrac{3}{4} - \left(-\dfrac{5}{8}\right) - \dfrac{13}{24}$

92. $-\dfrac{11}{16} + \dfrac{5}{32} - \left(-\dfrac{1}{4}\right) + \dfrac{7}{8}$

b Solve.

93. *Elevations in Asia.* The elevation of the highest point in Asia, Mt. Everest, Nepal–Tibet, is 29,035 ft. The lowest elevation, at the Dead Sea, Israel–Jordan, is −1348 ft. What is the difference in the elevations of the two locations?

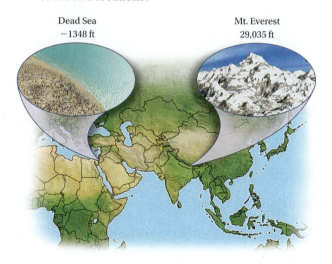

Dead Sea
−1348 ft

Mt. Everest
29,035 ft

94. *Ocean Depth.* The deepest point in the Pacific Ocean is the Marianas Trench, with a depth of 10,924 m. The deepest point in the Atlantic Ocean is the Puerto Rico Trench, with a depth of 8605 m. What is the difference in the elevation of the two trenches?

Source: *The World Almanac and Book of Facts*

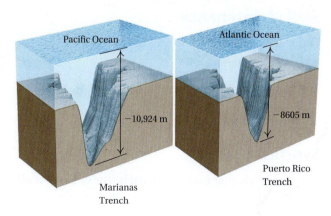

Pacific Ocean

Atlantic Ocean

−10,924 m

−8605 m

Marianas
Trench

Puerto Rico
Trench

95. Francisca has a charge of $476.89 on her credit card, but she then returns a sweater that cost $128.95. How much does she now owe on her credit card?

96. Jacob has $825 in a checking account. What is the balance in his account after he has written a check for $920 to pay for a laptop?

97. *Difference in Elevation.*   The highest elevation in Japan is 3776 m above sea level at Fujiyama. The lowest elevation in Japan is 4 m below sea level at Hachirogata. Find the difference in the elevations.

Source: *Information Please Almanac*

98. *Difference in Elevation.*   The lowest elevation in North America—Death Valley, California—is 282 ft below sea level. The highest elevation in North America—Mount McKinley, Alaska—is 20,320 ft above sea level. Find the difference in elevation between the highest point and the lowest point.

Source: National Geographic Society

99. *Low Points on Continents.*   The lowest point in Africa is Lake Assal, which is 512 ft below sea level. The lowest point in South America is the Valdes Peninsula, which is 131 ft below sea level. How much lower is Lake Assal than the Valdes Peninsula?

Source: National Geographic Society

100. *Temperature Records.*   The greatest recorded temperature change in one 24-hr period occurred between January 14 and January 15, 1972, in Loma, Montana, where the temperature rose from $-54°F$ to $49°F$. By how much did the temperature rise?

Source: *The Guinness Book of Records*

101. *Surface Temperature on Mercury.*   Surface temperatures on Mercury vary from $840°F$ on the equator when the planet is closest to the sun to $-290°F$ at night. Find the difference between these two temperatures.

102. *Run Differential.*   In baseball, the difference between the number of runs that a team scores and the number of runs that it allows its opponents to score is called the *run differential*. That is,

$$\text{Run differential} = \begin{array}{c}\text{Number of}\\\text{runs scored}\end{array} - \begin{array}{c}\text{Number of}\\\text{runs allowed}.\end{array}$$

Teams strive for a positive run differential.

Source: Major League Baseball

a) In a recent season, the Kansas City Royals scored 676 runs and allowed 746 runs to be scored on them. Find the run differential.
b) In a recent season, the Atlanta Braves scored 700 runs and allowed 600 runs to be scored on them. Find the run differential.

Skill Maintenance

Translate to an algebraic expression.   [1.1b]

103. 7 more than y

104. 41 less than t

105. h subtracted from a

106. The product of 6 and c

107. r more than s

108. x less than y

Synthesis

Determine whether each statement is true or false for all integers a and b. If false, give an example to show why. Examples may vary.

109. $a - 0 = 0 - a$

110. $0 - a = a$

111. If $a \neq b$, then $a - b \neq 0$.

112. If $a = -b$, then $a + b = 0$.

113. If $a + b = 0$, then a and b are opposites.

114. If $a - b = 0$, then $a = -b$.

Mid-Chapter Review

Concept Reinforcement

Determine whether each statement is true or false.

_____ **1.** All rational numbers can be named using fraction notation. [1.2c]

_____ **2.** If $a > b$, then a lies to the left of b on the number line. [1.2d]

_____ **3.** The absolute value of a number is always nonnegative. [1.2e]

_____ **4.** We can translate "7 less than y" as $7 - y$. [1.1b]

Guided Solutions

 Fill in each blank with the number that creates a correct statement or solution.

5. Evaluate $-x$ and $-(-x)$ when $x = -4$. [1.3b]

$$-x = -(\boxed{}) = \boxed{} \; ;$$
$$-(-x) = -(-(\boxed{})) = -(\boxed{}) = \boxed{}$$

Subtract. [1.4a]

6. $5 - 13 = 5 + (\boxed{}) = \boxed{}$

7. $-6 - 7 = -6 + (\boxed{}) = \boxed{}$

Mixed Review

Evaluate. [1.1a]

8. $\dfrac{3m}{n}$, when $m = 8$ and $n = 6$

9. $\dfrac{a + b}{2}$, when $a = 5$ and $b = 17$

Translate each phrase to an algebraic expression. Use any letter for the variable. [1.1b]

10. Three times some number

11. Five less than some number

12. State the integers that correspond to this situation: Jerilyn deposited $450 in her checking account. Later that week, she wrote a check for $79. [1.2a]

13. Graph -3.5 on the number line. [1.2b]

$$\xleftarrow{\quad}\overset{\displaystyle|\;|\;|\;|\;|\;|\;|\;|\;|\;|\;|\;|\;|}{\underset{-6\;-5\;-4\;-3\;-2\;-1\;\;0\;\;1\;\;2\;\;3\;\;4\;\;5\;\;6}{\quad}}\xrightarrow{\quad}$$

Convert to decimal notation. [1.2c]

14. $-\dfrac{4}{5}$

15. $\dfrac{7}{3}$

Use either $<$ or $>$ for $\boxed{}$ to write a true sentence. [1.2d]

16. $-5 \boxed{} -3$

17. $-9.9 \boxed{} -10.1$

Write true or false. [1.2d]

18. $-8 \geq -5$ **19.** $-4 \leq -4$

true

Write an inequality with the same meaning. [1.2d]

20. $y < 5$ **21.** $-3 \geq t$

Find the absolute value. [1.2e]

22. $|15.6|$ **23.** $|-18|$ **24.** $|0|$ **25.** $\left| -\dfrac{12}{5} \right|$

Find the opposite, or additive inverse, of each number. [1.3b]

26. -5.6 **27.** $\dfrac{7}{4}$ **28.** 0 **29.** -49

30. Evaluate $-x$ when x is -19. [1.3b]

31. Evaluate $-(-x)$ when x is 2.3. [1.3b]

Compute and simplify. [1.3a], [1.4a]

32. $7 + (-9)$ **33.** $-\dfrac{3}{8} + \dfrac{1}{4}$ **34.** $3.6 + (-3.6)$ **35.** $-8 + (-9)$

36. $\dfrac{2}{3} + \left(-\dfrac{9}{8} \right)$ **37.** $-4.2 + (-3.9)$ **38.** $-14 + 5$ **39.** $19 + (-21)$

40. $-4.1 - 6.3$ **41.** $5 - (-11)$ **42.** $-\dfrac{1}{4} - \left(-\dfrac{3}{5} \right)$ **43.** $12 - 24$

44. $-8 - (-4)$ **45.** $-\dfrac{1}{2} - \dfrac{5}{6}$ **46.** $12.3 - 14.1$ **47.** $6 - (-7)$

48. $16 - (-9) - 20 - (-4)$ **49.** $-4 + (-10) - (-3) - 12$

50. $17 - (-25) + 15 - (-18)$ **51.** $-9 + (-3) + 16 - (-10)$

Solve. [1.3c], [1.4b]

52. *Temperature Change.* In chemistry lab, Ben works with a substance whose initial temperature is 25°C. During an experiment, the temperature falls to -8°C. Find the difference between the two temperatures.

53. *Stock Price Change.* The price of a stock opened at $56.12. During the day, it dropped $1.18, then rose $1.22, and then dropped $1.36. Find the value of the stock at the end of the day.

Understanding Through Discussion and Writing

54. Give three examples of rational numbers that are not integers. Explain. [1.2b]

55. Give three examples of irrational numbers. Explain the difference between an irrational number and a rational number. [1.2b, d]

56. Explain in your own words why the sum of two negative numbers is always negative. [1.3a]

57. If a negative number is subtracted from a positive number, will the result always be positive? Why or why not? [1.4a]

Multiplication of Real Numbers

1.5

a MULTIPLICATION

Multiplication of real numbers is very much like multiplication of arithmetic numbers. The only difference is that we must determine whether the product is positive or negative.

Multiplication of a Positive Number and a Negative Number

To see how to multiply a positive number and a negative number, consider the pattern of the following.

This number decreases by 1 each time.

$$
\begin{array}{rl}
4 \cdot 5 = & 20 \\
3 \cdot 5 = & 15 \\
2 \cdot 5 = & 10 \\
1 \cdot 5 = & 5 \\
0 \cdot 5 = & 0 \\
-1 \cdot 5 = & -5 \\
-2 \cdot 5 = & -10 \\
-3 \cdot 5 = & -15
\end{array}
$$

This number decreases by 5 each time.

Do Exercise 1. ▶

According to this pattern, it looks as though the product of a negative number and a positive number is negative. That is the case, and we have the first part of the rule for multiplying real numbers.

> ### THE PRODUCT OF A POSITIVE NUMBER AND A NEGATIVE NUMBER
>
> To multiply a positive number and a negative number, multiply their absolute values. The product is negative.

EXAMPLES Multiply.

1. $8(-5) = -40$ **2.** $-\dfrac{1}{3} \cdot \dfrac{5}{7} = -\dfrac{5}{21}$ **3.** $(-7.2)5 = -36$

Do Exercises 2–5 on the following page. ▶

OBJECTIVES

a Multiply real numbers.

b Solve applied problems involving multiplication of real numbers.

SKILL TO REVIEW

Objective R.2c: Multiply using fraction notation.

Multiply.

1. $\dfrac{5}{3} \cdot \dfrac{4}{7}$ **2.** $\dfrac{1}{8} \cdot \dfrac{3}{11}$

---

1. Complete, as in the example.

$$
\begin{array}{l}
4 \cdot 10 = 40 \\
3 \cdot 10 = 30 \\
2 \cdot 10 = \\
1 \cdot 10 = \\
0 \cdot 10 = \\
-1 \cdot 10 = \\
-2 \cdot 10 = \\
-3 \cdot 10 =
\end{array}
$$

Answers

Skill to Review:
1. $\dfrac{20}{21}$ 2. $\dfrac{3}{88}$

Margin Exercise:
1. $20; 10; 0; -10; -20; -30$

Multiply.

2. $-3 \cdot 6$

3. $20 \cdot (-5)$

4. $-\dfrac{2}{3} \cdot \dfrac{5}{6}$

5. $-4.23(7.1)$

6. Complete, as in the example.

$$3 \cdot (-10) = -30$$
$$2 \cdot (-10) = -20$$
$$1 \cdot (-10) =$$
$$0 \cdot (-10) =$$
$$-1 \cdot (-10) =$$
$$-2 \cdot (-10) =$$
$$-3 \cdot (-10) =$$

Multiply.

7. $-9 \cdot (-3)$

8. $-16 \cdot (-2)$

9. $-7 \cdot (-5)$

10. $-\dfrac{4}{7}\left(-\dfrac{5}{9}\right)$

11. $-\dfrac{3}{2}\left(-\dfrac{4}{9}\right)$

12. $-3.25(-4.14)$

Multiplication of Two Negative Numbers

How do we multiply two negative numbers? Again, we look for a pattern.

This number decreases by 1 each time.

This number increases by 5 each time.

$$4 \cdot (-5) = -20$$
$$3 \cdot (-5) = -15$$
$$2 \cdot (-5) = -10$$
$$1 \cdot (-5) = -5$$
$$0 \cdot (-5) = 0$$
$$-1 \cdot (-5) = 5$$
$$-2 \cdot (-5) = 10$$
$$-3 \cdot (-5) = 15$$

◀ **Do Exercise 6.**

According to the pattern, it appears that the product of two negative numbers is positive. That is actually so, and we have the second part of the rule for multiplying real numbers.

> ## THE PRODUCT OF TWO NEGATIVE NUMBERS
>
> To multiply two negative numbers, multiply their absolute values. The product is positive.

◀ **Do Exercises 7–12.**

The following is another way to consider the rules that we have for multiplication.

> To multiply two nonzero real numbers:
>
> **a)** Multiply the absolute values.
> **b)** If the signs are the same, the product is positive.
> **c)** If the signs are different, the product is negative.

Multiplication by Zero

The only case that we have not considered is multiplying by zero. As with nonnegative numbers, the product of any real number and 0 is 0.

> ## THE MULTIPLICATION PROPERTY OF ZERO
>
> For any real number a,
> $$a \cdot 0 = 0 \cdot a = 0.$$
> (The product of 0 and any real number is 0.)

EXAMPLES Multiply.

4. $(-3)(-4) = 12$

5. $-1.6(2) = -3.2$

6. $-19 \cdot 0 = 0$

7. $\left(-\dfrac{5}{6}\right)\left(-\dfrac{1}{9}\right) = \dfrac{5}{54}$

8. $0 \cdot (-452) = 0$

9. $23 \cdot 0 \cdot \left(-8\dfrac{2}{3}\right) = 0$

Do Exercises 13–18. ▶

Multiplying More Than Two Numbers

When multiplying more than two real numbers, we can choose order and grouping as we please.

EXAMPLES Multiply.

10. $-8 \cdot 2(-3) = -16(-3)$ Multiplying the first two numbers
$$= 48$$

11. $-8 \cdot 2(-3) = 24 \cdot 2$ Multiplying the negative numbers. Every pair of negative numbers gives a positive product.
$$= 48$$

12. $-3(-2)(-5)(4) = 6(-5)(4)$ Multiplying the first two numbers
$$= (-30)4$$
$$= -120$$

13. $\left(-\dfrac{1}{2}\right)(8)\left(-\dfrac{2}{3}\right)(-6) = (-4)4$ Multiplying the first two numbers and the last two numbers
$$= -16$$

14. $-5 \cdot (-2) \cdot (-3) \cdot (-6) = 10 \cdot 18 = 180$

15. $(-3)(-5)(-2)(-3)(-6) = (-30)(18) = -540$

Considering that the product of a pair of negative numbers is positive, we see the following pattern.

> The product of an even number of negative numbers is positive.
> The product of an odd number of negative numbers is negative.

Do Exercises 19–24. ▶

EXAMPLE 16 Evaluate $2x^2$ when $x = 3$ and when $x = -3$.

$$2x^2 = 2(3)^2 = 2(9) = 18;$$
$$2x^2 = 2(-3)^2 = 2(9) = 18$$

Let's compare the expressions $(-x)^2$ and $-x^2$.

EXAMPLE 17 Evaluate $(-x)^2$ and $-x^2$ when $x = 5$.

$$(-x)^2 = (-5)^2 = (-5)(-5) = 25;$$ Substitute 5 for x. Then evaluate the power.

$$-x^2 = -(5)^2 = -(25) = -25$$ Substitute 5 for x. Evaluate the power. Then find the opposite.

In Example 17, we see that the expressions $(-x)^2$ and $-x^2$ are *not* equivalent. That is, they do not have the same value for every allowable replacement of the variable by a real number. To find $(-x)^2$, we take the opposite and then square. To find $-x^2$, we find the square and then take the opposite.

Multiply.

13. $5(-6)$

14. $(-5)(-6)$

15. $(-3.2) \cdot 10$

16. $\left(-\dfrac{4}{5}\right)\left(\dfrac{10}{3}\right)$

17. $0 \cdot (-34.2)$

18. $-\dfrac{5}{7} \cdot 0 \cdot \left(-4\dfrac{2}{3}\right)$

Multiply.

19. $5 \cdot (-3) \cdot 2$

20. $-3 \times (-4.1) \times (-2.5)$

21. $-\dfrac{1}{2} \cdot \left(-\dfrac{4}{3}\right) \cdot \left(-\dfrac{5}{2}\right)$

22. $-2 \cdot (-5) \cdot (-4) \cdot (-3)$

23. $(-4)(-5)(-2)(-3)(-1)$

24. $(-1)(-1)(-2)(-3)(-1)(-1)$

Answers

13. -30 **14.** 30 **15.** -32 **16.** $-\dfrac{8}{3}$
17. 0 **18.** 0 **19.** -30 **20.** -30.75
21. $-\dfrac{5}{3}$ **22.** 120 **23.** -120 **24.** 6

25. Evaluate $3x^2$ when $x = 4$ and when $x = -4$.

26. Evaluate $(-x)^2$ and $-x^2$ when $x = 2$.

27. Evaluate $(-x)^2$ and $-x^2$ when $x = -3$.

EXAMPLE 18 Evaluate $(-a)^2$ and $-a^2$ when $a = -4$.

To make sense of the substitutions and computations, we introduce extra grouping symbols into the expressions.

$$(-a)^2 = [-(-4)]^2 = [4]^2 = 16;$$
$$-a^2 = -(-4)^2 = -(16) = -16$$

◄ **Do Exercises 25–27.**

b **APPLICATIONS AND PROBLEM SOLVING**

We now consider multiplication of real numbers in real-world applications.

EXAMPLE 19 *Mine Rescue.* The San Jose copper and gold mine near Copiapó, Chile, collapsed on August 5, 2010, trapping 33 miners. Each miner was safely brought out of the mine with a specially designed capsule that could be lowered into the mine at -137 feet per minute. It took approximately 15 minutes to lower the capsule to the miners' location. Determine how far below the surface of the earth the miners were trapped.

Source: Reuters News

28. *Chemical Reaction.* During a chemical reaction, the temperature in a beaker increased by 3°C every minute until 1:34 P.M. If the temperature was -17°C at 1:10 P.M., when the reaction began, what was the temperature at 1:34 P.M.?

Since the capsule moved -137 feet per minute and it took 15 minutes to reach the miners, we have the depth d given by

$$d = 15 \cdot (-137) = -2055.$$

Thus the miners were trapped at -2055 ft.

◄ **Do Exercise 28.**

Answers

25. 48; 48 **26.** 4; -4 **27.** 9; -9 **28.** 55°C

1.5 **Exercise Set**

✓ **Reading Check**

Fill in the blank with either "positive" or "negative."

RC1. To multiply a positive number and a negative number, multiply their absolute values. The answer is _____.

RC2. To multiply two negative numbers, multiply their absolute values. The answer is _____.

RC3. The product of an even number of negative numbers is _____.

RC4. The product of an odd number of negative numbers is _____.

Evaluate.

RC5. -3^2

RC6. $(-3)^2$

RC7. $-\left(\dfrac{1}{2}\right)^2$

RC8. $-\left(-\dfrac{1}{2}\right)^2$

Multiply.

1. $-4 \cdot 2$ 2. $-3 \cdot 5$ 3. $8 \cdot (-3)$ 4. $9 \cdot (-5)$ 5. $-9 \cdot 8$

6. $-10 \cdot 3$ 7. $-8 \cdot (-2)$ 8. $-2 \cdot (-5)$ 9. $-7 \cdot (-6)$ 10. $-9 \cdot (-2)$

11. $15 \cdot (-8)$ 12. $-12 \cdot (-10)$ 13. $-14 \cdot 17$ 14. $-13 \cdot (-15)$ 15. $-25 \cdot (-48)$

16. $39 \cdot (-43)$ 17. $-3.5 \cdot (-28)$ 18. $97 \cdot (-2.1)$ 19. $9 \cdot (-8)$ 20. $7 \cdot (-9)$

21. $4 \cdot (-3.1)$ 22. $3 \cdot (-2.2)$ 23. $-5 \cdot (-6)$ 24. $-6 \cdot (-4)$ 25. $-7 \cdot (-3.1)$

26. $-4 \cdot (-3.2)$ 27. $\frac{2}{3} \cdot \left(-\frac{3}{5}\right)$ 28. $\frac{5}{7} \cdot \left(-\frac{2}{3}\right)$ 29. $-\frac{3}{8} \cdot \left(-\frac{2}{9}\right)$

30. $-\frac{5}{8} \cdot \left(-\frac{2}{5}\right)$ 31. -6.3×2.7 32. -4.1×9.5 33. $7 \cdot (-4) \cdot (-3) \cdot 5$

34. $9 \cdot (-2) \cdot (-6) \cdot 7$ 35. $-\frac{2}{3} \cdot \frac{1}{2} \cdot \left(-\frac{6}{7}\right)$ 36. $-\frac{1}{8} \cdot \left(-\frac{1}{4}\right) \cdot \left(-\frac{3}{5}\right)$ 37. $-3 \cdot (-4) \cdot (-5)$

38. $-2 \cdot (-5) \cdot (-7)$ 39. $-2 \cdot (-5) \cdot (-3) \cdot (-5)$ 40. $-3 \cdot (-5) \cdot (-2) \cdot (-1)$

41. $-4 \cdot (-1.8) \cdot 7$ 42. $-8 \cdot (-1.3) \cdot (-5)$ 43. $-\frac{1}{9}\left(-\frac{2}{3}\right)\left(\frac{5}{7}\right)$

44. $-\frac{7}{2}\left(-\frac{5}{7}\right)\left(-\frac{2}{5}\right)$ 45. $4 \cdot (-4) \cdot (-5) \cdot (-12)$ 46. $-2 \cdot (-3) \cdot (-4) \cdot (-5)$

47. $0.07 \cdot (-7) \cdot 6 \cdot (-6)$

48. $80 \cdot (-0.8) \cdot (-90) \cdot (-0.09)$

49. $\left(-\dfrac{5}{6}\right)\left(\dfrac{1}{8}\right)\left(-\dfrac{3}{7}\right)\left(-\dfrac{1}{7}\right)$

50. $\left(\dfrac{4}{5}\right)\left(-\dfrac{2}{3}\right)\left(-\dfrac{15}{7}\right)\left(\dfrac{1}{2}\right)$

51. $(-14) \cdot (-27) \cdot 0$

52. $7 \cdot (-6) \cdot 5 \cdot (-4) \cdot 3 \cdot (-2) \cdot 1 \cdot 0$

53. $(-8)(-9)(-10)$

54. $(-7)(-8)(-9)(-10)$

55. $(-6)(-7)(-8)(-9)(-10)$

56. $(-5)(-6)(-7)(-8)(-9)(-10)$

57. $(-1)^{12}$

58. $(-1)^{9}$

59. Evaluate $(-x)^2$ and $-x^2$ when $x = 4$ and when $x = -4$.

60. Evaluate $(-x)^2$ and $-x^2$ when $x = 10$ and when $x = -10$.

61. Evaluate $(-y)^2$ and $-y^2$ when $y = \frac{2}{5}$ and when $y = -\frac{2}{5}$.

62. Evaluate $(-w)^2$ and $-w^2$ when $w = \frac{1}{10}$ and when $w = -\frac{1}{10}$.

63. Evaluate $-(-t)^2$ and $-t^2$ when $t = 3$ and when $t = -3$.

64. Evaluate $-(-s)^2$ and $-s^2$ when $s = 1$ and when $s = -1$.

65. Evaluate $(-3x)^2$ and $-3x^2$ when $x = 7$ and when $x = -7$.

66. Evaluate $(-2x)^2$ and $-2x^2$ when $x = 3$ and when $x = -3$.

67. Evaluate $5x^2$ when $x = 2$ and when $x = -2$.

68. Evaluate $2x^2$ when $x = 5$ and when $x = -5$.

69. Evaluate $-2x^3$ when $x = 1$ and when $x = -1$.

70. Evaluate $-3x^3$ when $x = 2$ and when $x = -2$.

b Solve.

71. *Chemical Reaction.* The temperature of a chemical compound was 0°C at 11:00 A.M. During a reaction, it dropped 3°C per minute until 11:08 A.M. What was the temperature at 11:08 A.M.?

72. *Chemical Reaction.* The temperature of a chemical compound was −5°C at 3:20 P.M. During a reaction, it increased 2°C per minute until 3:52 P.M. What was the temperature at 3:52 P.M.?

73. *Weight Loss.* Dave lost 2 lb each week for a period of 10 weeks. Express his total weight change as an integer.

74. *Stock Loss.* Each day for a period of 5 days, the value of a stock that Lily owned dropped $3. Express Lily's total loss as an integer.

75. *Stock Price.* The price of a stock began the day at $23.75 per share and dropped $1.38 per hour for 8 hr. What was the price of the stock after 8 hr?

76. *Population Decrease.* The population of Bloomtown was 12,500. It decreased 380 each year for 4 years. What was the population of the town after 4 years?

77. *Diver's Position.* After diving 95 m below sea level, a diver rises at a rate of 7 m/min for 9 min. Where is the diver in relation to the surface at the end of the 9-min period?

78. *Bank Account Balance.* Karen had $68 in her bank account. After she used her debit card to make seven purchases at $13 each, what was the balance in her bank account?

79. *Drop in Temperature.* The temperature in Osgood was 62°F at 2:00 P.M. It dropped 6°F per hour for the next 4 hr. What was the temperature at the end of the 4-hr period?

80. *Juice Consumption.* Oliver bought a 64-oz container of cranberry juice and drank 8 oz per day for a week. How much juice was left in the container at the end of the week?

Skill Maintenance

81. Evaluate $\dfrac{x - 2y}{3}$ when $x = 20$ and $y = 7$.   [1.1a]

82. Evaluate $\dfrac{d - e}{3d}$ when $d = 5$ and $e = 1$.   [1.1a]

Subtract.   [1.4a]

83. $-\dfrac{1}{2} - \left(-\dfrac{1}{6}\right)$

84. $8 - 12.3$

85. $31 - (-13)$

86. $-\dfrac{5}{12} - \left(-\dfrac{1}{3}\right)$

Write true or false.   [1.2d]

87. $-10 > -12$

88. $0 \le -1$

89. $4 < -8$

90. $-7 \ge -6$

Synthesis

91. If a is positive and b is negative, then $-ab$ is which of the following?

  A. Positive
  B. Negative
  C. 0
  D. Cannot be determined without more information

92. If a is positive and b is negative, then $(-a)(-b)$ is which of the following?

  A. Positive
  B. Negative
  C. 0
  D. Cannot be determined without more information

93. Of all possible quotients of the numbers 10, $-\frac{1}{2}$, -5, and $\frac{1}{5}$, which two produce the largest quotient? Which two produce the smallest quotient?

1.6 Division of Real Numbers

OBJECTIVES

a Divide integers.

b Find the reciprocal of a real number.

c Divide real numbers.

d Solve applied problems involving division of real numbers.

SKILL TO REVIEW

Objective 1.5a: Multiply real numbers.

Multiply.

1. $-\dfrac{2}{9} \cdot \dfrac{4}{11}$ **2.** $\dfrac{13}{2} \cdot \left(-\dfrac{2}{25}\right)$

Divide.

1. $6 \div (-3)$

Think: What number multiplied by -3 gives 6?

2. $\dfrac{-15}{-3}$

Think: What number multiplied by -3 gives -15?

3. $-24 \div 8$

Think: What number multiplied by 8 gives -24?

4. $\dfrac{-48}{-6}$ **5.** $\dfrac{30}{-5}$

6. $\dfrac{30}{-7}$

Answers

Skill to Review:

1. $-\dfrac{8}{99}$ **2.** $-\dfrac{13}{25}$

Margin Exercises:

1. -2 **2.** 5 **3.** -3 **4.** 8

5. -6 **6.** $-\dfrac{30}{7}$

We now consider division of real numbers. The definition of division results in rules for division that are the same as those for multiplication.

a DIVISION OF INTEGERS

> **DIVISION**
>
> The quotient $a \div b$, or $\dfrac{a}{b}$, where $b \neq 0$, is that unique real number c for which $a = b \cdot c$.

Let's use the definition to divide integers.

EXAMPLES Divide, if possible. Check your answer.

1. $14 \div (-7) = -2$ *Think*: What number multiplied by -7 gives 14? That number is 2. *Check*: $(-2)(-7) = 14$.

2. $\dfrac{-32}{-4} = 8$ *Think*: What number multiplied by -4 gives -32? That number is 8. *Check*: $8(-4) = -32$.

3. $\dfrac{-10}{7} = -\dfrac{10}{7}$ *Think*: What number multiplied by 7 gives -10? That number is $-\frac{10}{7}$. *Check*: $-\frac{10}{7} \cdot 7 = -10$.

4. $\dfrac{-17}{0}$ is **not defined**.   *Think*: What number multiplied by 0 gives -17? There is no such number because the product of 0 and *any* number is 0.

The rules for division are the same as those for multiplication.

> To multiply or divide two real numbers (where the divisor is nonzero):
>
> **a)** Multiply or divide the absolute values.
>
> **b)** If the signs are the same, the answer is positive.
>
> **c)** If the signs are different, the answer is negative.

◀ **Do Margin Exercises 1–6.**

Excluding Division by 0

Example 4 shows why we cannot divide -17 by 0. We can use the same argument to show why we cannot divide any nonzero number b by 0. Consider $b \div 0$. We look for a number that when multiplied by 0 gives b. There is no such number because the product of 0 and any number is 0. Thus we cannot divide a nonzero number b by 0.

On the other hand, if we divide 0 by 0, we look for a number c such that $0 \cdot c = 0$. But $0 \cdot c = 0$ for any number c. Thus it appears that $0 \div 0$ could be any number we choose. Getting any answer we want when we divide 0 by 0 would be very confusing. Thus we agree that division by 0 is not defined.

Dividing 0 by Other Numbers

Note that

$$0 \div 8 = 0 \text{ because } 0 = 0 \cdot 8; \qquad \frac{0}{-5} = 0 \text{ because } 0 = 0 \cdot (-5).$$

EXAMPLES Divide.

5. $0 \div (-6) = 0$ **6.** $\frac{0}{12} = 0$ **7.** $\frac{-3}{0}$ is not defined.

Do Exercises 7 and 8. ▶

Divide, if possible.

7. $\frac{-5}{0}$ **8.** $\frac{0}{-3}$

b   RECIPROCALS

When two numbers like $\frac{1}{2}$ and 2 are multiplied, the result is 1. Such numbers are called **reciprocals** of each other. Every nonzero real number has a reciprocal, also called a **multiplicative inverse**.

EXAMPLES Find the reciprocal.

8. $\frac{7}{8}$ The reciprocal of $\frac{7}{8}$ is $\frac{8}{7}$ because $\frac{7}{8} \cdot \frac{8}{7} = 1$.

9. -5 The reciprocal of -5 is $-\frac{1}{5}$ because $-5\left(-\frac{1}{5}\right) = 1$.

10. 3.9 The reciprocal of 3.9 is $\frac{1}{3.9}$ because $3.9\left(\frac{1}{3.9}\right) = 1$.

11. $-\frac{1}{2}$ The reciprocal of $-\frac{1}{2}$ is -2 because $\left(-\frac{1}{2}\right)(-2) = 1$.

12. $-\frac{2}{3}$ The reciprocal of $-\frac{2}{3}$ is $-\frac{3}{2}$ because $\left(-\frac{2}{3}\right)\left(-\frac{3}{2}\right) = 1$.

13. $\frac{3y}{8x}$ The reciprocal of $\frac{3y}{8x}$ is $\frac{8x}{3y}$ because $\left(\frac{3y}{8x}\right)\left(\frac{8x}{3y}\right) = 1$.

Answers

7. Not defined   **8.** 0

Find the reciprocal.

9. $\dfrac{2}{3}$ **10.** $-\dfrac{5}{4}$

11. -3 **12.** $-\dfrac{1}{5}$

13. 1.3 **14.** $\dfrac{a}{6b}$

RECIPROCAL PROPERTIES

For $a \neq 0$, the reciprocal of a can be named $\dfrac{1}{a}$ and the reciprocal of $\dfrac{1}{a}$ is a.

The reciprocal of a nonzero number $\dfrac{a}{b}$ can be named $\dfrac{b}{a}$.

The number 0 has no reciprocal.

◀ **Do Exercises 9–14.**

The reciprocal of a positive number is also a positive number, because the product of the two numbers must be the positive number 1. The reciprocal of a negative number is also a negative number, because the product of the two numbers must be the positive number 1.

THE SIGN OF A RECIPROCAL

The reciprocal of a number has the same sign as the number itself.

························· **Caution!** ·························

It is important *not* to confuse *opposite* with *reciprocal.* Keep in mind that the opposite, or additive inverse, of a number is what we add to the number to get 0. The reciprocal, or multiplicative inverse, is what we multiply the number by to get 1.

15. Complete the following table.

NUMBER	OPPOSITE	RECIPROCAL
$\dfrac{2}{9}$		
$-\dfrac{7}{4}$		
0		
1		
-8		
-4.7		

Compare the following.

NUMBER	OPPOSITE (Change the sign.)	RECIPROCAL (Invert but do not change the sign.)
$-\dfrac{3}{8}$	$\dfrac{3}{8}$	$-\dfrac{8}{3}$
19	-19	$\dfrac{1}{19}$
$\dfrac{18}{7}$	$-\dfrac{18}{7}$	$\dfrac{7}{18}$
-7.9	7.9	$-\dfrac{1}{7.9}$, or $-\dfrac{10}{79}$
0	0	None

$\left(-\dfrac{3}{8}\right)\left(-\dfrac{8}{3}\right) = 1$

$-\dfrac{3}{8} + \dfrac{3}{8} = 0$

◀ **Do Exercise 15.**

Answers

9. $\dfrac{3}{2}$ **10.** $-\dfrac{4}{5}$ **11.** $-\dfrac{1}{3}$ **12.** -5 **13.** $\dfrac{1}{1.3}$, or $\dfrac{10}{13}$ **14.** $\dfrac{6b}{a}$ **15.** $-\dfrac{2}{9}$ and $\dfrac{9}{2}$; $\dfrac{7}{4}$ and $-\dfrac{4}{7}$; 0 and none; -1 and 1; 8 and $-\dfrac{1}{8}$; 4.7 and $-\dfrac{1}{4.7}$, or $-\dfrac{10}{47}$

c DIVISION OF REAL NUMBERS

We know that we can subtract by adding an opposite. Similarly, we can divide by multiplying by a reciprocal.

RECIPROCALS AND DIVISION

For any real numbers a and b, $b \neq 0$,

$$a \div b = \frac{a}{b} = a \cdot \frac{1}{b}.$$

(To divide, multiply by the reciprocal of the divisor.)

EXAMPLES Rewrite each division as a multiplication.

14. $-4 \div 3$ $\qquad$ $-4 \div 3$ is the same as $-4 \cdot \dfrac{1}{3}$

15. $\dfrac{6}{-7}$ $\qquad$ $\dfrac{6}{-7} = 6\left(-\dfrac{1}{7}\right)$

16. $\dfrac{3}{5} \div \left(-\dfrac{9}{7}\right)$ $\qquad$ $\dfrac{3}{5} \div \left(-\dfrac{9}{7}\right) = \dfrac{3}{5}\left(-\dfrac{7}{9}\right)$

17. $\dfrac{x+2}{5}$ $\qquad$ $\dfrac{x+2}{5} = (x+2)\dfrac{1}{5}$ $\qquad$ Parentheses are necessary here.

18. $\dfrac{-17}{1/b}$ $\qquad$ $\dfrac{-17}{1/b} = -17 \cdot b$

Do Exercises 16–20. ▶

When actually doing division calculations, we sometimes multiply by a reciprocal and we sometimes divide directly. With fraction notation, it is usually better to multiply by a reciprocal. With decimal notation, it is usually better to divide directly.

EXAMPLES Divide by multiplying by the reciprocal of the divisor.

19. $\dfrac{2}{3} \div \left(-\dfrac{5}{4}\right) = \dfrac{2}{3} \cdot \left(-\dfrac{4}{5}\right) = -\dfrac{8}{15}$

20. $-\dfrac{5}{6} \div \left(-\dfrac{3}{4}\right) = -\dfrac{5}{6} \cdot \left(-\dfrac{4}{3}\right) = \dfrac{20}{18} = \dfrac{10 \cdot 2}{9 \cdot 2} = \dfrac{10}{9} \cdot \dfrac{2}{2} = \dfrac{10}{9}$

⋯⋯⋯⋯⋯⋯⋯⋯ **Caution!** ⋯⋯⋯⋯⋯⋯⋯⋯

Be careful *not* to change the sign when taking a reciprocal!

⋯⋯⋯⋯⋯⋯⋯⋯⋯⋯⋯⋯⋯⋯⋯⋯⋯⋯⋯⋯

21. $-\dfrac{3}{4} \div \dfrac{3}{10} = -\dfrac{3}{4} \cdot \left(\dfrac{10}{3}\right) = -\dfrac{30}{12} = -\dfrac{5 \cdot 6}{2 \cdot 6} = -\dfrac{5}{2} \cdot \dfrac{6}{6} = -\dfrac{5}{2}$

Do Exercises 21 and 22. ▶

Rewrite each division as a multiplication.

16. $\dfrac{4}{7} \div \left(-\dfrac{3}{5}\right)$

17. $\dfrac{5}{-8}$

18. $\dfrac{a-b}{7}$

19. $\dfrac{-23}{1/a}$

20. $-5 \div 7$

Divide by multiplying by the reciprocal of the divisor.

GS **21.** $\dfrac{4}{7} \div \left(-\dfrac{3}{5}\right)$

$\qquad = \dfrac{4}{7} \cdot \left(-\dfrac{5}{\boxed{}}\right) = \boxed{}$

22. $-\dfrac{12}{7} \div \left(-\dfrac{3}{4}\right)$

Answers

16. $\dfrac{4}{7} \cdot \left(-\dfrac{5}{3}\right)$ **17.** $5 \cdot \left(-\dfrac{1}{8}\right)$ **18.** $(a-b) \cdot \dfrac{1}{7}$

19. $-23 \cdot a$ **20.** $-5 \cdot \left(\dfrac{1}{7}\right)$ **21.** $-\dfrac{20}{21}$

22. $\dfrac{16}{7}$

Guided Solution:

21. $3, -\dfrac{20}{21}$

With decimal notation, it is easier to carry out long division than to multiply by the reciprocal.

EXAMPLES Divide.

22. $-27.9 \div (-3) = \dfrac{-27.9}{-3} = 9.3$

Do the long division $3\overline{)27.9}$.
The answer is positive.

23. $-6.3 \div 2.1 = -3$

Do the long division $2.1\overline{)6.3}$.
The answer is negative.

◀ **Do Exercises 23 and 24.**

Divide.
23. $21.7 \div (-3.1)$

24. $-20.4 \div (-4)$

Consider the following:

1. $\dfrac{2}{3} = \dfrac{2}{3} \cdot 1 = \dfrac{2}{3} \cdot \dfrac{-1}{-1} = \dfrac{2(-1)}{3(-1)} = \dfrac{-2}{-3}$. Thus, $\dfrac{2}{3} = \dfrac{-2}{-3}$.

(A negative number divided by a negative number is positive.)

2. $-\dfrac{2}{3} = -1 \cdot \dfrac{2}{3} = \dfrac{-1}{1} \cdot \dfrac{2}{3} = \dfrac{-1 \cdot 2}{1 \cdot 3} = \dfrac{-2}{3}$. Thus, $-\dfrac{2}{3} = \dfrac{-2}{3}$.

(A negative number divided by a positive number is negative.)

3. $\dfrac{-2}{3} = \dfrac{-2}{3} \cdot 1 = \dfrac{-2}{3} \cdot \dfrac{-1}{-1} = \dfrac{-2(-1)}{3(-1)} = \dfrac{2}{-3}$. Thus, $-\dfrac{2}{3} = \dfrac{2}{-3}$.

(A positive number divided by a negative number is negative.)

We can use the following properties to make sign changes in fraction notation.

SIGN CHANGES IN FRACTION NOTATION

For any numbers a and b, $b \neq 0$:

1. $\dfrac{-a}{-b} = \dfrac{a}{b}$

(The opposite of a number a divided by the opposite of another number b is the same as the quotient of the two numbers a and b.)

2. $\dfrac{-a}{b} = \dfrac{a}{-b} = -\dfrac{a}{b}$

(The opposite of a number a divided by another number b is the same as the number a divided by the opposite of the number b, and both are the same as the opposite of a *divided by* b.)

Find two equal expressions for each number with negative signs in different places.

25. $\dfrac{-5}{6} = \dfrac{5}{\boxed{}} = -\dfrac{\boxed{}}{6}$ (GS)

26. $-\dfrac{8}{7}$

27. $\dfrac{10}{-3}$

◀ **Do Exercises 25–27.**

d APPLICATIONS AND PROBLEM SOLVING

EXAMPLE 24 *Chemical Reaction.* During a chemical reaction, the temperature in a beaker decreased every minute by the same number of degrees. The temperature was 56°F at 10:10 A.M. By 10:42 A.M., the temperature had dropped to -12°F. By how many degrees did it change each minute?

Answers

23. -7 **24.** 5.1 **25.** $\dfrac{5}{-6}; -\dfrac{5}{6}$ **26.** $\dfrac{8}{-7}; \dfrac{-8}{7}$

27. $\dfrac{-10}{3}; -\dfrac{10}{3}$

Guided Solution:
25. $-6, 5$

We first determine by how many degrees d the temperature changed altogether. We subtract -12 from 56:

$$d = 56 - (-12) = 56 + 12 = 68.$$

The temperature changed a total of $68°$. We can express this as $-68°$ since the temperature dropped.

The amount of time t that passed was $42 - 10$, or 32 min. Thus the number of degrees T that the temperature dropped each minute is given by

$$T = \frac{d}{t} = \frac{-68}{32} = -2.125.$$

The change was $-2.125°$F per minute.

Do Exercise 28. ▶

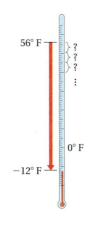

28. *Chemical Reaction.* During a chemical reaction, the temperature in a beaker decreased every minute by the same number of degrees. The temperature was $71°$F at 2:12 P.M. By 2:37 P.M., the temperature had changed to $-14°$F. By how many degrees did it change each minute?

Answer

28. $-3.4°$F per minute

CALCULATOR CORNER

Operations on the Real Numbers To perform operations on the real numbers on a graphing calculator, recall that negative numbers are entered using the opposite key, (-), rather than the subtraction operation key, ⊖. Consider the sum $-5 + (-3.8)$. On a graphing calculator, the parentheses are not necessary. The result is -8.8. Note that it is not incorrect to enter the parentheses. The result will be the same if this is done. We can also subtract, multiply, and divide real numbers. At right, we see $10 - (-17)$, $-5 \cdot (-7)$, and $45 \div (-9)$. Again, it is not necessary to use parentheses.

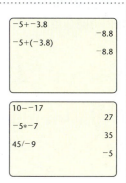

EXERCISES: Use a calculator to perform each operation.

1. $-8 + 4$
2. $-7 - (-5)$
3. $-8 \cdot 4$
4. $-7 \div (-5)$
5. $1.2 - (-1.5)$
6. $-7.6 + (-1.9)$
7. $1.2 \div (-1.5)$
8. $-7.6 \cdot (-1.9)$

1.6 Exercise Set

For Extra Help  MathXL® PRACTICE WATCH READ REVIEW

✓ Reading Check

Choose the word or the number below the blank that will make the sentence true.

RC1. The numbers 4 and -4 are called _____ of each other.
　　　　　　　　　　　　　　　　　opposites/reciprocals

RC2. The multiplicative inverse, or reciprocal, of a number is what we multiply the number by to get ____.
　　　　　　　　　　　　　　　　　　　　　　　　　　　　　　　　　　　　　0/1

RC3. The additive inverse, or opposite, of a number is what we add to the number to get ____.
　　　　　　　　　　　　　　　　　　　　　　　　　　　　　　　　　　　　0/1

RC4. The numbers $-\dfrac{9}{4}$ and $-\dfrac{4}{9}$ are called _____ of each other.
　　　　　　　　　　　　　　　　　　　　　　　opposites/reciprocals

a Divide, if possible. Check each answer.

1. $48 \div (-6)$

2. $\dfrac{42}{-7}$

3. $\dfrac{28}{-2}$

4. $24 \div (-12)$

5. $\dfrac{-24}{8}$

6. $-18 \div (-2)$

7. $\dfrac{-36}{-12}$

8. $-72 \div (-9)$

9. $\dfrac{-72}{9}$

10. $\dfrac{-50}{25}$

11. $-100 \div (-50)$

12. $\dfrac{-200}{8}$

13. $-108 \div 9$

14. $\dfrac{-63}{-7}$

15. $\dfrac{200}{-25}$

16. $-300 \div (-16)$

17. $\dfrac{75}{0}$

18. $\dfrac{0}{-5}$

19. $\dfrac{0}{-2.6}$

20. $\dfrac{-23}{0}$

b Find the reciprocal.

21. $\dfrac{15}{7}$

22. $\dfrac{3}{8}$

23. $-\dfrac{47}{13}$

24. $-\dfrac{31}{12}$

25. 13

26. -10

27. -32

28. 15

29. $\dfrac{1}{-7.1}$

30. $\dfrac{1}{-4.9}$

31. $\dfrac{1}{9}$

32. $\dfrac{1}{16}$

33. $\dfrac{1}{4y}$

34. $\dfrac{-1}{8a}$

35. $\dfrac{2a}{3b}$

36. $\dfrac{-4y}{3x}$

C Rewrite each division as a multiplication.

37. $4 \div 17$

38. $5 \div (-8)$

39. $\dfrac{8}{-13}$

40. $-\dfrac{13}{47}$

41. $\dfrac{13.9}{-1.5}$

42. $-\dfrac{47.3}{21.4}$

43. $\dfrac{2}{3} \div \left(-\dfrac{4}{5}\right)$

44. $\dfrac{3}{4} \div \left(-\dfrac{7}{10}\right)$

45. $\dfrac{\dfrac{x}{1}}{\dfrac{1}{y}}$

46. $\dfrac{\dfrac{13}{1}}{\dfrac{1}{x}}$

47. $\dfrac{3x + 4}{5}$

48. $\dfrac{4y - 8}{-7}$

Divide.

49. $\dfrac{3}{4} \div \left(-\dfrac{2}{3}\right)$

50. $\dfrac{7}{8} \div \left(-\dfrac{1}{2}\right)$

51. $-\dfrac{5}{4} \div \left(-\dfrac{3}{4}\right)$

52. $-\dfrac{5}{9} \div \left(-\dfrac{5}{6}\right)$

53. $-\dfrac{2}{7} \div \left(-\dfrac{4}{9}\right)$

54. $-\dfrac{3}{5} \div \left(-\dfrac{5}{8}\right)$

55. $-\dfrac{3}{8} \div \left(-\dfrac{8}{3}\right)$

56. $-\dfrac{5}{8} \div \left(-\dfrac{6}{5}\right)$

57. $-\dfrac{5}{6} \div \dfrac{2}{3}$

58. $-\dfrac{7}{16} \div \dfrac{3}{8}$

59. $-\dfrac{9}{4} \div \dfrac{5}{12}$

60. $-\dfrac{3}{5} \div \dfrac{7}{10}$

61. $\dfrac{-11}{-13}$

62. $\dfrac{-21}{-25}$

63. $-6.6 \div 3.3$

64. $-44.1 \div (-6.3)$

65. $\dfrac{48.6}{-3}$

66. $\dfrac{-1.9}{20}$

67. $\dfrac{-12.5}{5}$

68. $\dfrac{-17.8}{3.2}$

69. $11.25 \div (-9)$

70. $-9.6 \div (-6.4)$

71. $\dfrac{-9}{17 - 17}$

72. $\dfrac{-8}{-5 + 5}$

d To determine percent increase or decrease, we first determine the change by subtracting the new amount from the original amount. Then we divide the change, which is *positive* if an increase, or *negative* if a decrease, by the original amount. Finally, we convert the decimal answer to percent notation.

73. *Passports.* In 2006, approximately 71 million valid passports were in circulation in the United States. This number increased to approximately 102 million in 2011. What is the percent increase?

Source: U.S. State Department

74. *Super Bowl Spending.* The average amount spent during the Super Bowl per television viewer increased from $59.33 in 2011 to $63.87 in 2012. What is the percent increase?

Source: Retail Advertising and Marketing Association, Super Bowl Consumer Intentions and Actions Survey

75. *Pieces of Mail.* The number of pieces of mail handled by the U.S. Postal Service decreased from 212 billion in 2007 to 168 billion in 2011. What is the percent decrease?

Source: U.S. Postal Service

76. *Beef Consumption.* Beef consumption per capita in the United States decreased from 67.5 lb in 1990 to 57.4 in 2011. What is the percent decrease?

Source: U.S. Department of Agriculture

Skill Maintenance

Simplify.

77. $\dfrac{1}{4} - \dfrac{1}{2}$ [1.4a]

78. $-9 - 3 + 17$ [1.4a]

79. $35 \cdot (-1.2)$ [1.5a]

80. $4 \cdot (-6) \cdot (-2) \cdot (-1)$ [1.5a]

81. $13.4 + (-4.9)$ [1.3a]

82. $-\dfrac{3}{8} - \left(-\dfrac{1}{4}\right)$ [1.4a]

Convert to decimal notation. [1.2c]

83. $-\dfrac{1}{11}$

84. $\dfrac{11}{12}$

85. $\dfrac{15}{4}$

86. $-\dfrac{10}{3}$

Synthesis

87. Find the reciprocal of -10.5. What happens if you take the reciprocal of the result?

88. Determine those real numbers a for which the opposite of a is the same as the reciprocal of a.

Determine whether each expression represents a positive number or a negative number when a and b are negative.

89. $\dfrac{-a}{b}$

90. $\dfrac{-a}{-b}$

91. $-\left(\dfrac{a}{-b}\right)$

92. $-\left(\dfrac{-a}{b}\right)$

93. $-\left(\dfrac{-a}{-b}\right)$

94   CHAPTER 1   Introduction to Real Numbers and Algebraic Expressions

Properties of Real Numbers

a EQUIVALENT EXPRESSIONS

In solving equations and doing other kinds of work in algebra, we manipulate expressions in various ways. For example, instead of $x + x$, we might write $2x$, knowing that the two expressions represent the same number for any allowable replacement of x. In that sense, the expressions $x + x$ and $2x$ are **equivalent**, as are $\frac{3}{x}$ and $\frac{3x}{x^2}$, even though 0 is not an allowable replacement because division by 0 is not defined.

> ### EQUIVALENT EXPRESSIONS
>
> Two expressions that have the same value for all allowable replacements are called **equivalent**.

The expressions $x + 3x$ and $5x$ are *not* equivalent, as we see in Margin Exercise 2.

Do Margin Exercises 1 and 2. ▶

In this section, we will consider several laws of real numbers that will allow us to find equivalent expressions. The first two laws are the *identity properties of 0 and 1*.

> ### THE IDENTITY PROPERTY OF 0
>
> For any real number a,
>
> $$a + 0 = 0 + a = a.$$
>
> (The number 0 is the *additive identity*.)

> ### THE IDENTITY PROPERTY OF 1
>
> For any real number a,
>
> $$a \cdot 1 = 1 \cdot a = a.$$
>
> (The number 1 is the *multiplicative identity*.)

We often refer to the use of the identity property of 1 as "multiplying by 1." We can use this method to find equivalent fraction expressions. Recall from arithmetic that to multiply with fraction notation, we multiply the numerators and multiply the denominators.

EXAMPLE 1 Write a fraction expression equivalent to $\frac{2}{3}$ with a denominator of $3x$:

$$\frac{2}{3} = \frac{\square}{3x}.$$

OBJECTIVES

a Find equivalent fraction expressions and simplify fraction expressions.

b Use the commutative laws and the associative laws to find equivalent expressions.

c Use the distributive laws to multiply expressions like 8 and $x - y$.

d Use the distributive laws to factor expressions like $4x - 12 + 24y$.

e Collect like terms.

SKILL TO REVIEW

Objective 1.3a: Add real numbers.

Add.

1. $-16 + 5$

2. $29 + (-23)$

Complete the table by evaluating each expression for the given values.

1.

VALUE	$x + x$	$2x$
$x = 3$		
$x = -6$		
$x = 4.8$		

2.

VALUE	$x + 3x$	$5x$
$x = 2$		
$x = -6$		
$x = 4.8$		

Answers

Skill to Review:
1. -11 **2.** 6

Margin Exercises:
1. 6, 6; -12, -12; 9.6, 9.6   **2.** 8, 10; -24, -30; 19.2, 24

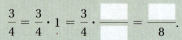

3. Write a fraction expression equivalent to $\frac{3}{4}$ with a denominator of 8:

$$\frac{3}{4} = \frac{3}{4} \cdot 1 = \frac{3}{4} \cdot \frac{\boxed{}}{\boxed{}} = \frac{\boxed{}}{8}.$$

4. Write a fraction expression equivalent to $\frac{3}{4}$ with a denominator of $4t$:

$$\frac{3}{4} = \frac{3}{4} \cdot 1 = \frac{3}{4} \cdot \frac{\boxed{}}{\boxed{}} = \frac{\boxed{}}{4t}.$$

Simplify.

5. $\dfrac{3y}{4y}$

6. $-\dfrac{16m}{12m}$

7. $\dfrac{5xy}{40y}$

8. $\dfrac{18p}{24pq} = \dfrac{6p \cdot 3}{6p \cdot \boxed{}}$

$$= \dfrac{6p}{6p} \cdot \dfrac{\boxed{}}{4q}$$

$$= 1 \cdot \dfrac{3}{4q} = \dfrac{3}{4q}$$

9. Evaluate $x + y$ and $y + x$ when $x = -2$ and $y = 3$.

10. Evaluate xy and yx when $x = -2$ and $y = 5$.

Answers

3. $\dfrac{6}{8}$ **4.** $\dfrac{3t}{4t}$ **5.** $\dfrac{3}{4}$ **6.** $-\dfrac{4}{3}$ **7.** $\dfrac{x}{8}$

8. $\dfrac{3}{4q}$ **9.** 1; 1 **10.** $-10; -10$

Guided Solutions:

3. $\dfrac{2}{2}, 6$ **4.** $\dfrac{t}{t}, 3t$ **8.** $4q, 3$

Note that $3x = 3 \cdot x$. We want fraction notation for $\frac{2}{3}$ that has a denominator of $3x$, but the denominator 3 is missing a factor of x. Thus we multiply by 1, using x/x as an equivalent expression for 1:

$$\frac{2}{3} = \frac{2}{3} \cdot 1 = \frac{2}{3} \cdot \frac{x}{x} = \frac{2x}{3x}.$$

The expressions $2/3$ and $2x/(3x)$ are equivalent. They have the same value for any allowable replacement. Note that $2x/3x$ is not defined for a replacement of 0, but for all nonzero real numbers, the expressions $2/3$ and $2x/(3x)$ have the same value.

◀ **Do Exercises 3 and 4.**

In algebra, we consider an expression like $2/3$ to be "simplified" from $2x/(3x)$. To find such simplified expressions, we use the identity property of 1 to remove a factor of 1.

EXAMPLE 2 Simplify: $-\dfrac{20x}{12x}$.

$$-\frac{20x}{12x} = -\frac{5 \cdot 4x}{3 \cdot 4x} \qquad \text{We look for the largest factor common to both the numerator and the denominator and factor each.}$$

$$= -\frac{5}{3} \cdot \frac{4x}{4x} \qquad \text{Factoring the fraction expression}$$

$$= -\frac{5}{3} \cdot 1 \qquad \frac{4x}{4x} = 1$$

$$= -\frac{5}{3} \qquad \text{Removing a factor of 1 using the identity property of 1}$$

EXAMPLE 3 Simplify: $\dfrac{14ab}{56a}$.

$$\frac{14ab}{56a} = \frac{14a \cdot b}{14a \cdot 4} = \frac{14a}{14a} \cdot \frac{b}{4} = 1 \cdot \frac{b}{4} = \frac{b}{4}$$

◀ **Do Exercises 5–8.**

b THE COMMUTATIVE LAWS AND THE ASSOCIATIVE LAWS

The Commutative Laws

Let's examine the expressions $x + y$ and $y + x$, as well as xy and yx.

EXAMPLE 4 Evaluate $x + y$ and $y + x$ when $x = 4$ and $y = 3$.

We substitute 4 for x and 3 for y in both expressions:

$$x + y = 4 + 3 = 7; \qquad y + x = 3 + 4 = 7.$$

EXAMPLE 5 Evaluate xy and yx when $x = 3$ and $y = -12$.

We substitute 3 for x and -12 for y in both expressions:

$$xy = 3 \cdot (-12) = -36; \qquad yx = (-12) \cdot 3 = -36.$$

◀ **Do Exercises 9 and 10.**

The expressions $x + y$ and $y + x$ have the same values no matter what the variables stand for. Thus they are equivalent. Therefore, when we add two numbers, the order in which we add does not matter. Similarly, the expressions xy and yx are equivalent. They also have the same values, no matter what the variables stand for. Therefore, when we multiply two numbers, the order in which we multiply does not matter.

The following are examples of general patterns or laws.

THE COMMUTATIVE LAWS

Addition. For any numbers a and b,

$$a + b = b + a.$$

(We can change the order when adding without affecting the answer.)

Multiplication. For any numbers a and b,

$$ab = ba.$$

(We can change the order when multiplying without affecting the answer.)

Using a commutative law, we know that $x + 2$ and $2 + x$ are equivalent. Similarly, $3x$ and $x(3)$ are equivalent. Thus, in an algebraic expression, we can replace one with the other and the result will be equivalent to the original expression.

EXAMPLE 6 Use the commutative laws to write an equivalent expression: **(a)** $y + 5$; **(b)** mn; **(c)** $7 + xy$.

a) An expression equivalent to $y + 5$ is $5 + y$ by the commutative law of addition.

b) An expression equivalent to mn is nm by the commutative law of multiplication.

c) An expression equivalent to $7 + xy$ is $xy + 7$ by the commutative law of addition. Another expression equivalent to $7 + xy$ is $7 + yx$ by the commutative law of multiplication. Another equivalent expression is $yx + 7$.

Do Exercises 11–13. ▶

Use a commutative law to write an equivalent expression.

11. $x + 9$

12. pq

13. $xy + t$

The Associative Laws

Now let's examine the expressions $a + (b + c)$ and $(a + b) + c$. Note that these expressions involve the use of parentheses as *grouping* symbols, and they also involve three numbers. Calculations within parentheses are to be done first.

EXAMPLE 7 Calculate and compare: $3 + (8 + 5)$ and $(3 + 8) + 5$.

$$3 + (8 + 5) = 3 + 13 \quad \text{Calculating within parentheses first; adding 8 and 5}$$
$$= 16;$$
$$(3 + 8) + 5 = 11 + 5 \quad \text{Calculating within parentheses first; adding 3 and 8}$$
$$= 16$$

14. Calculate and compare:

$8 + (9 + 2)$ and $(8 + 9) + 2$.

15. Calculate and compare:

$10 \cdot (5 \cdot 3)$ and $(10 \cdot 5) \cdot 3$.

The two expressions in Example 7 name the same number. Moving the parentheses to group the additions differently does not affect the value of the expression.

EXAMPLE 8 Calculate and compare: $3 \cdot (4 \cdot 2)$ and $(3 \cdot 4) \cdot 2$.

$$3 \cdot (4 \cdot 2) = 3 \cdot 8 = 24; \quad (3 \cdot 4) \cdot 2 = 12 \cdot 2 = 24$$

◀ **Do Exercises 14 and 15.**

You may have noted that when only addition is involved, numbers can be grouped in any way we please without affecting the answer. When only multiplication is involved, numbers can also be grouped in any way we please without affecting the answer.

THE ASSOCIATIVE LAWS

Addition. For any numbers a, b, and c,

$$a + (b + c) = (a + b) + c.$$

(Numbers can be grouped in any manner for addition.)

Multiplication. For any numbers a, b, and c,

$$a \cdot (b \cdot c) = (a \cdot b) \cdot c.$$

(Numbers can be grouped in any manner for multiplication.)

Use an associative law to write an equivalent expression.

16. $r + (s + 7)$

17. $9(ab)$

EXAMPLE 9 Use an associative law to write an equivalent expression: **(a)** $(y + z) + 3$; **(b)** $8(xy)$.

a) An expression equivalent to $(y + z) + 3$ is $y + (z + 3)$ by the associative law of addition.

b) An expression equivalent to $8(xy)$ is $(8x)y$ by the associative law of multiplication.

◀ **Do Exercises 16 and 17.**

The associative laws say that numbers can be grouped in any way we please when only additions or only multiplications are involved. Thus we often omit the parentheses. For example,

$$x + (y + 2) \quad \text{means} \quad x + y + 2, \quad \text{and} \quad (lw)h \quad \text{means} \quad lwh.$$

Using the Commutative Laws and the Associative Laws Together

EXAMPLE 10 Use the commutative laws and the associative laws to write at least three expressions equivalent to $(x + 5) + y$.

a) $(x + 5) + y = x + (5 + y)$ Using the associative law first and then
$ = x + (y + 5)$ using the commutative law

b) $(x + 5) + y = y + (x + 5)$ Using the commutative law twice
$ = y + (5 + x)$

c) $(x + 5) + y = (5 + x) + y$ Using the commutative law first and
$ = 5 + (x + y)$ then the associative law

EXAMPLE 11 Use the commutative laws and the associative laws to write at least three expressions equivalent to $(3x)y$.

a) $(3x)y = 3(xy)$ Using the associative law first and then using the commutative law
$ = 3(yx)$

b) $(3x)y = y(3x)$ Using the commutative law twice
$ = y(x \cdot 3)$

c) $(3x)y = (x \cdot 3)y$ Using the commutative law, and then the associative law, and then the commutative law again
$ = x(3y)$
$ = x(y \cdot 3)$

Do Exercises 18 and 19. ▶

Use the commutative laws and the associative laws to write at least three equivalent expressions.

18. $4(tu)$

19. $r + (2 + s)$

c THE DISTRIBUTIVE LAWS

The *distributive laws* are the basis of many procedures in both arithmetic and algebra. They are probably the most important laws that we use to manipulate algebraic expressions. The distributive law of multiplication over addition involves two operations: addition and multiplication.

Let's begin by considering a multiplication problem from arithmetic:

$$
\begin{array}{r}
4\,5 \\
\times \quad 7 \\
\hline
3\,5 \\
2\,8\,0 \\
\hline
3\,1\,5
\end{array}
$$
←This is $7 \cdot 5$.
←This is $7 \cdot 40$.
←This is the sum $7 \cdot 5 + 7 \cdot 40$.

To carry out the multiplication, we actually added two products. That is,

$$7 \cdot 45 = 7(5 + 40) = 7 \cdot 5 + 7 \cdot 40.$$

Let's examine this further. If we wish to multiply a sum of several numbers by a factor, we can either add and then multiply, or multiply and then add.

EXAMPLE 12 Compute in two ways: $5 \cdot (4 + 8)$.

a) $5 \cdot (4 + 8)$ Adding within parentheses first, and then multiplying

$= 5 \cdot \quad 12$
$= 60$

b) $5 \cdot (4 + 8) = (5 \cdot 4) + (5 \cdot 8)$ Distributing the multiplication to terms within parentheses first and then adding

$ = \quad 20 \quad + \quad 40$
$ = \quad 60$

Do Exercises 20–22. ▶

Compute.

20. a) $7 \cdot (3 + 6)$
    b) $(7 \cdot 3) + (7 \cdot 6)$

21. a) $2 \cdot (10 + 30)$
    b) $(2 \cdot 10) + (2 \cdot 30)$

22. a) $(2 + 5) \cdot 4$
    b) $(2 \cdot 4) + (5 \cdot 4)$

THE DISTRIBUTIVE LAW OF MULTIPLICATION OVER ADDITION

For any numbers a, b, and c,

$$a(b + c) = ab + ac.$$

Answers

18. $(4t)u,\ (tu)4,\ t(4u)$; answers may vary
19. $(2 + r) + s,\ (r + s) + 2,\ s + (r + 2)$; answers may vary   **20. (a)** $7 \cdot 9 = 63$;
(b) $21 + 42 = 63$ **21. (a)** $2 \cdot 40 = 80$;
(b) $20 + 60 = 80$ **22. (a)** $7 \cdot 4 = 28$;
(b) $8 + 20 = 28$

In the statement of the distributive law, we know that in an expression such as $ab + ac$, the multiplications are to be done first according to the rules for order of operations. So, instead of writing $(4 \cdot 5) + (4 \cdot 7)$, we can write $4 \cdot 5 + 4 \cdot 7$. However, in $a(b + c)$, we cannot omit the parentheses. If we did, we would have $ab + c$, which means $(ab) + c$. For example, $3(4 + 2) = 3(6) = 18$, but $3 \cdot 4 + 2 = 12 + 2 = 14$.

Another distributive law relates multiplication and subtraction. This law says that to multiply by a difference, we can either subtract and then multiply, or multiply and then subtract.

THE DISTRIBUTIVE LAW OF MULTIPLICATION OVER SUBTRACTION

For any numbers a, b, and c,
$$a(b - c) = ab - ac.$$

We often refer to "*the* distributive law" when we mean *either* or *both* of these laws.

◀ **Do Exercises 23–25.**

What do we mean by the *terms* of an expression? **Terms** are separated by addition signs. If there are subtraction signs, we can find an equivalent expression that uses addition signs.

EXAMPLE 13   What are the terms of $3x - 4y + 2z$?

We have

$$3x - 4y + 2z = 3x + (-4y) + 2z. \qquad \text{Separating parts with } + \text{ signs}$$

The terms are $3x$, $-4y$, and $2z$.

◀ **Do Exercises 26 and 27.**

The distributive laws are a basis for **multiplying** algebraic expressions. In an expression like $8(a + 2b - 7)$, we multiply each term inside the parentheses by 8:

$$8(a + 2b - 7) = 8 \cdot a + 8 \cdot 2b - 8 \cdot 7 = 8a + 16b - 56.$$

EXAMPLES   Multiply.

14. $9(x - 5) = 9 \cdot x - 9 \cdot 5$ Using the distributive law of multiplication over subtraction

$\quad = 9x - 45$

15. $\frac{2}{3}(w + 1) = \frac{2}{3} \cdot w + \frac{2}{3} \cdot 1$ Using the distributive law of multiplication over addition

$\quad = \frac{2}{3}w + \frac{2}{3}$

16. $\frac{4}{3}(s - t + w) = \frac{4}{3}s - \frac{4}{3}t + \frac{4}{3}w$ Using both distributive laws

◀ **Do Exercises 28–30.**

Calculate.

23. a) $4(5 - 3)$
  b) $4 \cdot 5 - 4 \cdot 3$

24. a) $-2 \cdot (5 - 3)$
  b) $-2 \cdot 5 - (-2) \cdot 3$

25. a) $5 \cdot (2 - 7)$
  b) $5 \cdot 2 - 5 \cdot 7$

What are the terms of each expression?

26. $5x - 8y + 3$

27. $-4y - 2x + 3z$

Multiply.

28. $3(x - 5)$

29. $5(x + 1)$

30. $\frac{3}{5}(p + q - t)$

Answers

23. (a) $4 \cdot 2 = 8$; (b) $20 - 12 = 8$
24. (a) $-2 \cdot 2 = -4$; (b) $-10 + 6 = -4$
25. (a) $5(-5) = -25$; (b) $10 - 35 = -25$
26. $5x, -8y, 3$ **27.** $-4y, -2x, 3z$ **28.** $3x - 15$
29. $5x + 5$ **30.** $\frac{3}{5}p + \frac{3}{5}q - \frac{3}{5}t$

EXAMPLE 17 Multiply: $-4(x - 2y + 3z)$.

$$-4(x - 2y + 3z) = -4 \cdot x - (-4)(2y) + (-4)(3z)$$
　Using both distributive laws

$$= -4x - (-8y) + (-12z)$$
　Multiplying

$$= -4x + 8y - 12z$$

We can also do this problem by first finding an equivalent expression with all plus signs and then multiplying:

$$-4(x - 2y + 3z) = -4[x + (-2y) + 3z]$$
$$= -4 \cdot x + (-4)(-2y) + (-4)(3z)$$
$$= -4x + 8y - 12z.$$

Do Exercises 31–33. ▶

Multiply.

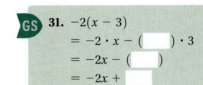
31. $-2(x - 3)$
$$= -2 \cdot x - (\boxed{}) \cdot 3$$
$$= -2x - (\boxed{})$$
$$= -2x + \boxed{}$$

32. $5(x - 2y + 4z)$

33. $-5(x - 2y + 4z)$

EXAMPLES Name the property or the law illustrated by each equation.

Equation	*Property*
18. $5x = x(5)$	Commutative law of multiplication
19. $a + (8.5 + b) = (a + 8.5) + b$	Associative law of addition
20. $0 + 11 = 11$	Identity property of 0
21. $(-5s)t = -5(st)$	Associative law of multiplication
22. $\dfrac{3}{4} \cdot 1 = \dfrac{3}{4}$	Identity property of 1
23. $12.5(w - 3) = 12.5w - 12.5(3)$	Distributive law of multiplication over subtraction
24. $y + \dfrac{1}{2} = \dfrac{1}{2} + y$	Commutative law of addition

Do Exercises 34–40. ▶

Name the property or the law illustrated by each equation.

34. $(-8a)b = -8(ab)$

35. $p \cdot 1 = p$

36. $m + 34 = 34 + m$

37. $2(t + 5) = 2t + 2(5)$

38. $0 + k = k$

39. $-8x = x(-8)$

40. $x + (4.3 + b) = (x + 4.3) + b$

d FACTORING

Factoring is the reverse of multiplying. To factor, we can use the distributive laws in reverse:

$$ab + ac = a(b + c) \quad \text{and} \quad ab - ac = a(b - c).$$

FACTORING

To **factor** an expression is to find an equivalent expression that is a product.

To factor $9x - 45$, for example, we find an equivalent expression that is a product: $9(x - 5)$. This reverses the multiplication that we did in Example 14. When all the terms of an expression have a factor in common, we can "factor it out" using the distributive laws. Note the following.

$9x$ has the factors $9, -9, 3, -3, 1, -1, x, -x, 3x, -3x, 9x, -9x$;

-45 has the factors $1, -1, 3, -3, 5, -5, 9, -9, 15, -15, 45, -45$

We generally remove the largest common factor. In this case, that factor is 9. Thus,

$$9x - 45 = 9 \cdot x - 9 \cdot 5$$
$$= 9(x - 5).$$

Remember that an expression has been factored when we have found an equivalent expression that is a product. Above, we note that $9x - 45$ and $9(x - 5)$ are equivalent expressions. The expression $9x - 45$ is the difference of $9x$ and 45; the expression $9(x - 5)$ is the product of 9 and $(x - 5)$.

EXAMPLES   Factor.

25. $5x - 10 = 5 \cdot x - 5 \cdot 2$ Try to do this step mentally.
$\qquad\quad = 5(x - 2)$ You can check by multiplying.

26. $ax - ay + az = a(x - y + z)$

27. $9x + 27y - 9 = 9 \cdot x + 9 \cdot 3y - 9 \cdot 1 = 9(x + 3y - 1)$

Note in Example 27 that you might, at first, just factor out a 3, as follows:

$$9x + 27y - 9 = 3 \cdot 3x + 3 \cdot 9y - 3 \cdot 3$$
$$= 3(3x + 9y - 3).$$

At this point, the mathematics is correct, but the answer is not because there is another factor of 3 that can be factored out, as follows:

$$3 \cdot 3x + 3 \cdot 9y - 3 \cdot 3 = 3(3x + 9y - 3)$$
$$= 3(3 \cdot x + 3 \cdot 3y - 3 \cdot 1)$$
$$= 3 \cdot 3(x + 3y - 1)$$
$$= 9(x + 3y - 1).$$

We now have a correct answer, but it took more work than we did in Example 27. Thus it is better to look for the *greatest common factor* at the outset.

EXAMPLES   Factor. Try to write just the answer, if you can.

28. $5x - 5y = 5(x - y)$

29. $-3x + 6y - 9z = -3(x - 2y + 3z)$

We generally factor out a negative factor when the first term is negative. The way we factor can depend on the situation in which we are working. We might also factor the expression in Example 29 as follows:

$$-3x + 6y - 9z = 3(-x + 2y - 3z).$$

30. $18z - 12x - 24 = 6(3z - 2x - 4)$

31. $\frac{1}{2}x + \frac{3}{2}y - \frac{1}{2} = \frac{1}{2}(x + 3y - 1)$

Remember that you can always check factoring by multiplying. Keep in mind that an expression is factored when it is written as a product.

◀ **Do Exercises 41–46.**

Factor.

41. $6x - 12$

42. $3x - 6y + 9$

43. $bx + by - bz$

44. $16a - 36b + 42$ GS
$\quad = 2 \cdot 8a - \boxed{} \cdot 18b + 2 \cdot 21$
$\quad = \boxed{}(8a - 18b + 21)$

45. $\dfrac{3}{8}x - \dfrac{5}{8}y + \dfrac{7}{8}$

46. $-12x + 32y - 16z$

e COLLECTING LIKE TERMS

Terms such as $5x$ and $-4x$, whose variable factors are exactly the same, are called **like terms**. Similarly, numbers, such as -7 and 13, are like terms. Also, $3y^2$ and $9y^2$ are like terms because the variables are raised to the same power. Terms such as $4y$ and $5y^2$ are not like terms, and $7x$ and $2y$ are not like terms.

The process of **collecting like terms** is also based on the distributive laws. We can apply a distributive law when a factor is on the right-hand side because of the commutative law of multiplication.

Later in this text, terminology like "collecting like terms" and "combining like terms" will also be referred to as "simplifying."

EXAMPLES Collect like terms. Try to write just the answer, if you can.

32. $4x + 2x = (4 + 2)x = 6x$ Factoring out x using a distributive law

33. $2x + 3y - 5x - 2y = 2x - 5x + 3y - 2y$
$$= (2 - 5)x + (3 - 2)y = -3x + 1y = -3x + y$$

34. $3x - x = 3x - 1x = (3 - 1)x = 2x$

35. $x - 0.24x = 1 \cdot x - 0.24x = (1 - 0.24)x = 0.76x$

36. $x - 6x = 1 \cdot x - 6 \cdot x = (1 - 6)x = -5x$

37. $4x - 7y + 9x - 5 + 3y - 8 = 13x - 4y - 13$

38. $\frac{2}{3}a - b + \frac{4}{5}a + \frac{1}{4}b - 10 = \frac{2}{3}a - 1 \cdot b + \frac{4}{5}a + \frac{1}{4}b - 10$
$$= (\tfrac{2}{3} + \tfrac{4}{5})a + (-1 + \tfrac{1}{4})b - 10$$
$$= (\tfrac{10}{15} + \tfrac{12}{15})a + (-\tfrac{4}{4} + \tfrac{1}{4})b - 10$$
$$= \tfrac{22}{15}a - \tfrac{3}{4}b - 10$$

Do Exercises 47–53. ▶

Collect like terms.

47. $6x - 3x$

48. $7x - x$

49. $x - 9x$

50. $x - 0.41x$

51. $5x + 4y - 2x - y$

GS **52.** $3x - 7x - 11 + 8y + 4 - 13y$
$$= (3 - \boxed{})x + (8 - 13)y +$$
$$(\boxed{} + 4)$$
$$= \boxed{}\, x + (\boxed{})y + (\boxed{})$$
$$= -4x - 5y - 7$$

53. $-\dfrac{2}{3} - \dfrac{3}{5}x + y + \dfrac{7}{10}x - \dfrac{2}{9}y$

Answers

47. $3x$ **48.** $6x$ **49.** $-8x$ **50.** $0.59x$
51. $3x + 3y$ **52.** $-4x - 5y - 7$
53. $\dfrac{1}{10}x + \dfrac{7}{9}y - \dfrac{2}{3}$

Guided Solution:
52. $7, -11, -4, -5, -7$

1.7 Exercise Set

✓ Reading Check

Choose from the column on the right an equation that illustrates the property or law.

RC1. Associative law of multiplication

RC2. Identity property of 1

RC3. Distributive law of multiplication over subtraction

RC4. Commutative law of addition

RC5. Identity property of 0

RC6. Commutative law of multiplication

RC7. Associative law of addition

a) $3 \cdot 5 = 5 \cdot 3$

b) $8 + (\tfrac{1}{2} + 9) = (8 + \tfrac{1}{2}) + 9$

c) $5(6 + 3) = 5 \cdot 6 + 5 \cdot 3$

d) $3 + 0 = 3$

e) $3 + 5 = 5 + 3$

f) $5(6 - 3) = 5 \cdot 6 - 5 \cdot 3$

g) $8 \cdot \left(\dfrac{1}{2} \cdot 9\right) = \left(8 \cdot \dfrac{1}{2}\right) \cdot 9$

h) $\dfrac{6}{5} \cdot 1 = \dfrac{6}{5}$

a Find an equivalent expression with the given denominator.

1. $\dfrac{3}{5} = \dfrac{\square}{5y}$

2. $\dfrac{5}{8} = \dfrac{\square}{8t}$

3. $\dfrac{2}{3} = \dfrac{\square}{15x}$

4. $\dfrac{6}{7} = \dfrac{\square}{14y}$

5. $\dfrac{2}{x} = \dfrac{\square}{x^2}$

6. $\dfrac{4}{9x} = \dfrac{\square}{9xy}$

Simplify.

7. $-\dfrac{24a}{16a}$

8. $-\dfrac{42t}{18t}$

9. $-\dfrac{42ab}{36ab}$

10. $-\dfrac{64pq}{48pq}$

11. $\dfrac{20st}{15t}$

12. $\dfrac{21w}{7wz}$

b Write an equivalent expression. Use a commutative law.

13. $y + 8$

14. $x + 3$

15. mn

16. yz

17. $9 + xy$

18. $11 + ab$

19. $ab + c$

20. $rs + t$

Write an equivalent expression. Use an associative law.

21. $a + (b + 2)$

22. $3(vw)$

23. $(8x)y$

24. $(y + z) + 7$

25. $(a + b) + 3$

26. $(5 + x) + y$

27. $3(ab)$

28. $(6x)y$

Use the commutative laws and the associative laws to write three equivalent expressions.

29. $(a + b) + 2$

30. $(3 + x) + y$

31. $5 + (v + w)$

32. $6 + (x + y)$

33. $(xy)3$

34. $(ab)5$

35. $7(ab)$

36. $5(xy)$

c Multiply.

37. $2(b + 5)$

38. $4(x + 3)$

39. $7(1 + t)$

40. $4(1 + y)$

41. $6(5x + 2)$

42. $9(6m + 7)$

43. $7(x + 4 + 6y)$

44. $4(5x + 8 + 3p)$

45. $7(x - 3)$ **46.** $15(y - 6)$ **47.** $-3(x - 7)$ **48.** $1.2(x - 2.1)$

49. $\frac{2}{3}(b - 6)$ **50.** $\frac{5}{8}(y + 16)$ **51.** $7.3(x - 2)$ **52.** $5.6(x - 8)$

53. $-\frac{3}{5}(x - y + 10)$ **54.** $-\frac{2}{3}(a + b - 12)$ **55.** $-9(-5x - 6y + 8)$ **56.** $-7(-2x - 5y + 9)$

57. $-4(x - 3y - 2z)$ **58.** $8(2x - 5y - 8z)$

59. $3.1(-1.2x + 3.2y - 1.1)$ **60.** $-2.1(-4.2x - 4.3y - 2.2)$

List the terms of each expression.

61. $4x + 3z$ **62.** $8x - 1.4y$ **63.** $7x + 8y - 9z$ **64.** $8a + 10b - 18c$

d   Factor. Check by multiplying.

65. $2x + 4$ **66.** $5y + 20$ **67.** $30 + 5y$ **68.** $7x + 28$

69. $14x + 21y$ **70.** $18a + 24b$ **71.** $14t - 7$ **72.** $25m - 5$

73. $8x - 24$ **74.** $10x - 50$ **75.** $18a - 24b$ **76.** $32x - 20y$

77. $-4y + 32$ **78.** $-6m + 24$ **79.** $5x + 10 + 15y$ **80.** $9a + 27b + 81$

81. $16m - 32n + 8$ **82.** $6x + 10y - 2$ **83.** $12a + 4b - 24$ **84.** $8m - 4n + 12$

85. $8x + 10y - 22$ **86.** $9a + 6b - 15$ **87.** $ax - a$ **88.** $by - 9b$

89. $ax - ay - az$ **90.** $cx + cy - cz$ **91.** $-18x + 12y + 6$ **92.** $-14x + 21y + 7$

93. $\dfrac{2}{3}x - \dfrac{5}{3}y + \dfrac{1}{3}$ **94.** $\dfrac{3}{5}a + \dfrac{4}{5}b - \dfrac{1}{5}$ **95.** $36x - 6y + 18z$ **96.** $8a - 4b + 20c$

e   Collect like terms.

97. $9a + 10a$ **98.** $12x + 2x$ **99.** $10a - a$

100. $-16x + x$ **101.** $2x + 9z + 6x$ **102.** $3a - 5b + 7a$

103. $7x + 6y^2 + 9y^2$ **104.** $12m^2 + 6q + 9m^2$ **105.** $41a + 90 - 60a - 2$

106. $42x - 6 - 4x + 2$ **107.** $23 + 5t + 7y - t - y - 27$ **108.** $45 - 90d - 87 - 9d + 3 + 7d$

109. $\dfrac{1}{2}b + \dfrac{1}{2}b$ **110.** $\dfrac{2}{3}x + \dfrac{1}{3}x$ **111.** $2y + \dfrac{1}{4}y + y$

112. $\dfrac{1}{2}a + a + 5a$ **113.** $11x - 3x$ **114.** $9t - 17t$

115. $6n - n$

116. $100t - t$

117. $y - 17y$

118. $3m - 9m + 4$

119. $-8 + 11a - 5b + 6a - 7b + 7$

120. $8x - 5x + 6 + 3y - 2y - 4$

121. $9x + 2y - 5x$

122. $8y - 3z + 4y$

123. $11x + 2y - 4x - y$

124. $13a + 9b - 2a - 4b$

125. $2.7x + 2.3y - 1.9x - 1.8y$

126. $6.7a + 4.3b - 4.1a - 2.9b$

127. $\dfrac{13}{2}a + \dfrac{9}{5}b - \dfrac{2}{3}a - \dfrac{3}{10}b - 42$

128. $\dfrac{11}{4}x + \dfrac{2}{3}y - \dfrac{4}{5}x - \dfrac{1}{6}y + 12$

Skill Maintenance

Compute and simplify.   [1.3a], [1.4a], [1.5a], [1.6a, c]

129. $18 - (-20)$

130. $-3.8 + (-1.1)$

131. $-\dfrac{4}{15} \cdot (-15)$

132. $-500 \div (-50)$

133. $\dfrac{2}{7} \div \left(-\dfrac{7}{2}\right)$

134. $2 \cdot (-53)$

135. $-\dfrac{1}{2} + \dfrac{3}{2}$

136. $-6 - 28$

137. Evaluate $9w$ when $w = 20$.   [1.1a]

138. Find the absolute value: $\left| -\dfrac{4}{13} \right|$.   [1.2e]

Write true or false.   [1.2d]

139. $-43 < -40$

140. $-3 \geq 0$

141. $-6 \leq -6$

142. $0 > -4$

Synthesis

Determine whether the expressions are equivalent. Explain why if they are. Give an example if they are not. Examples may vary.

143. $3t + 5$ and $3 \cdot 5 + t$

144. $4x$ and $x + 4$

145. $5m + 6$ and $6 + 5m$

146. $(x + y) + z$ and $z + (x + y)$

147. Factor: $q + qr + qrs + qrst$.

148. Collect like terms:
$21x + 44xy + 15y - 16x - 8y - 38xy + 2y + xy.$

1.8 Simplifying Expressions; Order of Operations

OBJECTIVES

a Find an equivalent expression for an opposite without parentheses, where an expression has several terms.

b Simplify expressions by removing parentheses and collecting like terms.

c Simplify expressions with parentheses inside parentheses.

d Simplify expressions using the rules for order of operations.

SKILL TO REVIEW

Objective 1.7c: Use the distributive laws to multiply expressions like 8 and $x - y$.

Multiply.

1. $4(x + 5)$

2. $-7(a + b)$

We now expand our ability to manipulate expressions by first considering opposites of sums and differences. Then we simplify expressions involving parentheses.

a OPPOSITES OF SUMS

What happens when we multiply a real number by -1? Consider the following products:

$$-1(7) = -7, \qquad -1(-5) = 5, \qquad -1(0) = 0.$$

From these examples, it appears that when we multiply a number by -1, we get the opposite, or additive inverse, of that number.

THE PROPERTY OF -1

For any real number a,

$$-1 \cdot a = -a.$$

(Negative one times a is the opposite, or additive inverse, of a.)

The property of -1 enables us to find expressions equivalent to opposites of sums.

EXAMPLES Find an equivalent expression without parentheses.

1.
$$\begin{aligned}
-(3 + x) &= -1(3 + x) && \text{Using the property of } -1 \\
&= -1 \cdot 3 + (-1)x && \text{Using a distributive law, multiplying each term by } -1 \\
&= -3 + (-x) && \text{Using the property of } -1 \\
&= -3 - x
\end{aligned}$$

2.
$$\begin{aligned}
-(3x + 2y + 4) &= -1(3x + 2y + 4) && \text{Using the property of } -1 \\
&= -1(3x) + (-1)(2y) + (-1)4 && \text{Using a distributive law} \\
&= -3x - 2y - 4 && \text{Using the property of } -1
\end{aligned}$$

◀ **Do Margin Exercises 1 and 2.**

Find an equivalent expression without parentheses.

1. $-(x + 2)$

2. $-(5x + 2y + 8)$

Suppose we want to remove parentheses in an expression like

$$-(x - 2y + 5).$$

We can first rewrite any subtractions inside the parentheses as additions. Then we take the opposite of each term:

$$\begin{aligned}
-(x - 2y + 5) &= -[x + (-2y) + 5] \\
&= -x + 2y + (-5) = -x + 2y - 5.
\end{aligned}$$

The most efficient method for removing parentheses is to replace each term in the parentheses with its opposite ("change the sign of every term"). Doing so for $-(x - 2y + 5)$, we obtain $-x + 2y - 5$ as an equivalent expression.

Answers

Skill to Review:
1. $4x + 20$ 2. $-7a - 7b$

Margin Exercises:
1. $-x - 2$ 2. $-5x - 2y - 8$

EXAMPLES Find an equivalent expression without parentheses.

3. $-(5 - y) = -5 + y$ Changing the sign of each term

4. $-(2a - 7b - 6) = -2a + 7b + 6$

5. $-(-3x + 4y + z - 7w - 23) = 3x - 4y - z + 7w + 23$

Do Exercises 3–6. ▶

b **REMOVING PARENTHESES AND SIMPLIFYING**

When a sum is added to another expression, as in $5x + (2x + 3)$, we can simply remove, or drop, the parentheses and collect like terms because of the associative law of addition:

$$5x + (2x + 3) = 5x + 2x + 3 = 7x + 3.$$

On the other hand, when a sum is subtracted from another expression, as in $3x - (4x + 2)$, we cannot simply drop the parentheses. However, we can subtract by adding an opposite. We then remove parentheses by changing the sign of each term inside the parentheses and collecting like terms.

EXAMPLE 6 Remove parentheses and simplify.

$$
\begin{aligned}
3x - (4x + 2) &= 3x + [-(4x + 2)] &&\text{Adding the opposite of } (4x + 2)\\
&= 3x + (-4x - 2) &&\text{Changing the sign of each term inside the parentheses}\\
&= 3x - 4x - 2\\
&= -x - 2 &&\text{Collecting like terms}
\end{aligned}
$$

·········· **Caution!** ··········

Note that $3x - (4x + 2) \neq 3x - 4x + 2$. You cannot simply drop the parentheses.

Do Exercises 7 and 8. ▶

In practice, the first three steps of Example 6 are usually combined by changing the sign of each term in parentheses and then collecting like terms.

EXAMPLES Remove parentheses and simplify.

7. $5y - (3y + 4) = 5y - 3y - 4$ Removing parentheses by changing the sign of every term inside the parentheses

$$= 2y - 4 \quad\text{Collecting like terms}$$

8. $3x - 2 - (5x - 8) = 3x - 2 - 5x + 8$

$$= -2x + 6$$

9. $(3a + 4b - 5) - (2a - 7b + 4c - 8)$

$$= 3a + 4b - 5 - 2a + 7b - 4c + 8$$

$$= a + 11b - 4c + 3$$

Do Exercises 9–11. ▶

Find an equivalent expression without parentheses. Try to do this in one step.

3. $-(6 - t)$

4. $-(x - y)$

5. $-(-4a + 3t - 10)$

6. $-(18 - m - 2n + 4z)$

Remove parentheses and simplify.

7. $5x - (3x + 9)$

8. $5y - 2 - (2y - 4)$

Remove parentheses and simplify.

9. $6x - (4x + 7)$

10. $8y - 3 - (5y - 6)$

11. $(2a + 3b - c) - (4a - 5b + 2c)$

Answers

3. $-6 + t$ **4.** $-x + y$ **5.** $4a - 3t + 10$
6. $-18 + m + 2n - 4z$ **7.** $2x - 9$
8. $3y + 2$ **9.** $2x - 7$ **10.** $3y + 3$
11. $-2a + 8b - 3c$

Next, consider subtracting an expression consisting of several terms multiplied by a number other than 1 or −1.

EXAMPLE 10 Remove parentheses and simplify.

$$x - 3(x + y) = x + [-3(x + y)] \qquad \text{Adding the opposite of } 3(x + y)$$
$$= x + [-3x - 3y] \qquad \text{Multiplying } x + y \text{ by } -3$$
$$= x - 3x - 3y$$
$$= -2x - 3y \qquad \text{Collecting like terms}$$

Remove parentheses and simplify.

12. $y - 9(x + y)$

13. $5a - 3(7a - 6)$
$$= 5a - \boxed{} + \boxed{}$$
$$= \boxed{} + 18$$

14. $4a - b - 6(5a - 7b + 8c)$

15. $5x - \dfrac{1}{4}(8x + 28)$

16. $4.6(5x - 3y) - 5.2(8x + y)$

EXAMPLES Remove parentheses and simplify.

11. $3y - 2(4y - 5) = 3y - 8y + 10 \qquad$ Multiplying each term in the parentheses by -2
$$= -5y + 10$$

12. $(2a + 3b - 7) - 4(-5a - 6b + 12)$
$$= 2a + 3b - 7 + 20a + 24b - 48 = 22a + 27b - 55$$

13. $2y - \frac{1}{3}(9y - 12) = 2y - 3y + 4 = -y + 4$

14. $6(5x - 3y) - 2(8x + y) = 30x - 18y - 16x - 2y = 14x - 20y$

◀ **Do Exercises 12–16.**

C **PARENTHESES WITHIN PARENTHESES**

In addition to parentheses, some expressions contain other grouping symbols such as brackets $[\ ]$ and braces $\{\ \}$.

> When more than one kind of grouping symbol occurs, do the computations in the innermost ones first. Then work from the inside out.

Simplify.

17. $12 - (8 + 2)$

18. $9 - [10 - (13 + 6)]$
$$= 9 - [10 - (\boxed{})]$$
$$= 9 - [\boxed{}]$$
$$= 9 + \boxed{}$$
$$= 18$$

19. $[24 \div (-2)] \div (-2)$

20. $5(3 + 4) - \{8 - [5 - (9 + 6)]\}$

EXAMPLES Simplify.

15. $2[3 - (7 + 3)] = 2[3 - 10] = 2[-7] = -14$

16. $8 - [9 - (12 + 5)] = 8 - [9 - 17] \qquad$ Computing $12 + 5$
$$= 8 - [-8] \qquad \text{Computing } 9 - 17$$
$$= 8 + 8 = 16$$

17. $\left[-4 - 2\left(-\frac{1}{2}\right)\right] \div \frac{1}{4} = [-4 + 1] \div \frac{1}{4} \qquad$ Working within parentheses
$$= -3 \div \frac{1}{4} \qquad \text{Computing } -4 + 1$$
$$= -3 \cdot 4 = -12$$

18. $4(2 + 3) - \{7 - [4 - (8 + 5)]\}$
$$= 4(5) - \{7 - [4 - 13]\} \qquad \text{Working with the innermost parentheses first}$$
$$= 4(5) - \{7 - [-9]\} \qquad \text{Computing } 4 - 13$$
$$= 4(5) - 16 \qquad \text{Computing } 7 - [-9]$$
$$= 20 - 16 = 4$$

◀ **Do Exercises 17–20.**

Answers

12. $-9x - 8y$ 13. $-16a + 18$
14. $-26a + 41b - 48c$ 15. $3x - 7$
16. $-18.6x - 19y$ 17. 2   18. 18   19. 6
20. 17

Guided Solutions:
13. $21a, 18, -16a$ 18. $19, -9, 9$

EXAMPLE 19 Simplify.

$$[5(x + 2) - 3x] - [3(y + 2) - 7(y - 3)]$$
$$= [5x + 10 - 3x] - [3y + 6 - 7y + 21]$$ Working with the innermost parentheses first

$$= [2x + 10] - [-4y + 27]$$ Collecting like terms within brackets

$$= 2x + 10 + 4y - 27$$ Removing brackets
$$= 2x + 4y - 17$$ Collecting like terms

Do Exercise 21. ▶

21. Simplify:
$$[3(x + 2) + 2x] - [4(y + 2) - 3(y - 2)].$$

d   ORDER OF OPERATIONS

When several operations are to be done in a calculation or a problem, we apply the following.

RULES FOR ORDER OF OPERATIONS

1. Do all calculations within grouping symbols before operations outside.
2. Evaluate all exponential expressions.
3. Do all multiplications and divisions in order from left to right.
4. Do all additions and subtractions in order from left to right.

These rules are consistent with the way in which most computers and scientific calculators perform calculations.

EXAMPLE 20 Simplify: $-34 \cdot 56 - 17$.

There are no parentheses or powers, so we start with the third step.

$$-34 \cdot 56 - 17 = -1904 - 17$$ Doing all multiplications and divisions in order from left to right

$$= -1921$$ Doing all additions and subtractions in order from left to right

EXAMPLE 21 Simplify: $25 \div (-5) + 50 \div (-2)$.

There are no calculations inside parentheses and no powers. The parentheses with (-5) and (-2) are used only to represent the negative numbers. We begin by doing all multiplications and divisions.

$$\underbrace{25 \div (-5)} + \underbrace{50 \div (-2)}$$

$$= -5 + (-25)$$ Doing all multiplications and divisions in order from left to right

$$= -30$$ Doing all additions and subtractions in order from left to right

Do Exercises 22–24. ▶

Simplify.

22. $23 - 42 \cdot 30$

23. $32 \div 8 \cdot 2$

24. $-24 \div 3 - 48 \div (-4)$

Answers

21. $5x - y - 8$ **22.** -1237 **23.** 8 **24.** 4

EXAMPLE 22 Simplify: $-2^4 + 51 \cdot 4 - (37 + 23 \cdot 2)$.

$$-2^4 + 51 \cdot 4 - (37 + 23 \cdot 2)$$
$$= -2^4 + 51 \cdot 4 - (37 + 46)$$ Following the rules for order of operations within the parentheses first
$$= -2^4 + 51 \cdot 4 - 83$$ Completing the addition inside parentheses
$$= -16 + 51 \cdot 4 - 83$$ Evaluating exponential expressions. Note that $-2^4 \neq (-2)^4$.
$$= -16 + 204 - 83$$ Doing all multiplications
$$= 188 - 83$$ Doing all additions and subtractions in order from left to right
$$= 105$$

A fraction bar can play the role of a grouping symbol.

EXAMPLE 23 Simplify: $\dfrac{-64 \div (-16) \div (-2)}{2^3 - 3^2}$.

An equivalent expression with brackets as grouping symbols is

$$[-64 \div (-16) \div (-2)] \div [2^3 - 3^2].$$

This shows, in effect, that we do the calculations in the numerator and then in the denominator, and divide the results:

$$\frac{-64 \div (-16) \div (-2)}{2^3 - 3^2} = \frac{4 \div (-2)}{8 - 9} = \frac{-2}{-1} = 2.$$

◀ **Do Exercises 25 and 26.**

Simplify.

25. $-4^3 + 52 \cdot 5 + 5^3 -$ **GS**
$(4^2 - 48 \div 4)$
$= \boxed{} + 52 \cdot 5 + 125 -$
$(\boxed{} - 48 \div 4)$
$= -64 + 52 \cdot 5 + 125 -$
$(16 - \boxed{})$
$= -64 + 52 \cdot 5 + 125 - 4$
$= -64 + \boxed{} + 125 - 4$
$= \boxed{} + 125 - 4$
$= 321 - 4$
$= 317$

26. $\dfrac{5 - 10 - 5 \cdot 23}{2^3 + 3^2 \quad 7}$

Answers
25. 317   **26.** -12

Guided Solution:
25. $-64, 16, 12, 260, 196$

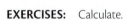

☑ Reading Check

In each of Exercises 1–6, name the operation that should be performed first. Do not calculate.

RC1. $10 - 4 \cdot 2 + 5$ **RC2.** $10 - 4(2 + 5)$ **RC3.** $(10 - 4) \cdot 2 + 5$

RC4. $5[2(10 \div 5) - 3]$ **RC5.** $5(10 \div 2 + 5 - 3)$ **RC6.** $5 \cdot 2 - 4 \cdot 8 \div 2$

a   Find an equivalent expression without parentheses.

1. $-(2x + 7)$ **2.** $-(8x + 4)$ **3.** $-(8 - x)$ **4.** $-(a - b)$

5. $-(4a - 3b + 7c)$ **6.** $-(x - 4y - 3z)$ **7.** $-(6x - 8y + 5)$ **8.** $-(4x + 9y + 7)$

9. $-(3x - 5y - 6)$ **10.** $-(6a - 4b - 7)$ **11.** $-(-8x - 6y - 43)$ **12.** $-(-2a + 9b - 5c)$

b   Remove parentheses and simplify.

13. $9x - (4x + 3)$ **14.** $4y - (2y + 5)$ **15.** $2a - (5a - 9)$

16. $12m - (4m - 6)$ **17.** $2x + 7x - (4x + 6)$ **18.** $3a + 2a - (4a + 7)$

19. $2x - 4y - 3(7x - 2y)$ **20.** $3a - 9b - 1(4a - 8b)$ **21.** $15x - y - 5(3x - 2y + 5z)$

22. $4a - b - 4(5a - 7b + 8c)$ **23.** $(3x + 2y) - 2(5x - 4y)$ **24.** $(-6a - b) - 5(2b + a)$

25. $(12a - 3b + 5c) - 5(-5a + 4b - 6c)$ **26.** $(-8x + 5y - 12) - 6(2x - 4y - 10)$

c   Simplify.

27. $9 - 2(5 - 4)$ **28.** $6 - 5(8 - 4)$ **29.** $8[7 - 6(4 - 2)]$ **30.** $10[7 - 4(7 - 5)]$

31. $[4(9 - 6) + 11] - [14 - (6 + 4)]$ **32.** $[7(8 - 4) + 16] - [15 - (7 + 8)]$

33. $[10(x + 3) - 4] + [2(x - 1) + 6]$ **34.** $[9(x + 5) - 7] + [4(x - 12) + 9]$

35. $[7(x + 5) - 19] - [4(x - 6) + 10]$ **36.** $[6(x + 4) - 12] - [5(x - 8) + 14]$

37. $3\{[7(x - 2) + 4] - [2(2x - 5) + 6]\}$ **38.** $4\{[8(x - 3) + 9] - [4(3x - 2) + 6]\}$

39. $4\{[5(x - 3) + 2] - 3[2(x + 5) - 9]\}$ **40.** $3\{[6(x - 4) + 5] - 2[5(x + 8) - 3]\}$

d   Simplify.

41. $8 - 2 \cdot 3 - 9$ **42.** $8 - (2 \cdot 3 - 9)$ **43.** $(8 - 2) \div (3 - 9)$

44. $(8 - 2) \div 3 - 9$ **45.** $[(-24) \div (-3)] \div \left(-\frac{1}{2}\right)$ **46.** $[32 \div (-2)] \div \left(-\frac{1}{4}\right)$

47. $16 \cdot (-24) + 50$

48. $10 \cdot 20 - 15 \cdot 24$

49. $2^4 + 2^3 - 10$

50. $40 - 3^2 - 2^3$

51. $5^3 + 26 \cdot 71 - (16 + 25 \cdot 3)$

52. $4^3 + 10 \cdot 20 + 8^2 - 23$

53. $4 \cdot 5 - 2 \cdot 6 + 4$

54. $4 \cdot (6 + 8)/(4 + 3)$

55. $4^3/8$

56. $5^3 - 7^2$

57. $8(-7) + 6(-5)$

58. $10(-5) + 1(-1)$

59. $19 - 5(-3) + 3$

60. $14 - 2(-6) + 7$

61. $9 \div (-3) + 16 \div 8$

62. $-32 - 8 \div 4 - (-2)$

63. $-4^2 + 6$

64. $-5^2 + 7$

65. $-8^2 - 3$

66. $-9^2 - 11$

67. $12 - 20^3$

68. $20 + 4^3 \div (-8)$

69. $2 \cdot 10^3 - 5000$

70. $-7(3^4) + 18$

71. $6[9 - (3 - 4)]$

72. $8[3(6 - 13) - 11]$

73. $-1000 \div (-100) \div 10$

74. $256 \div (-32) \div (-4)$

75. $8 - (7 - 9)$

76. $(16 - 6) \cdot \dfrac{1}{2} + 9$

77. $\dfrac{10 - 6^2}{9^2 + 3^2}$

78. $\dfrac{5^2 - 4^3 - 3}{9^2 - 2^2 - 1^5}$

79. $\dfrac{3(6 - 7) - 5 \cdot 4}{6 \cdot 7 - 8(4 - 1)}$

80. $\dfrac{20(8 - 3) - 4(10 - 3)}{10(2 - 6) - 2(5 + 2)}$

81. $\dfrac{\left| 2^3 - 3^2 \right| + \left| 12 \cdot 5 \right|}{-32 \div (-16) \div (-4)}$

82. $\dfrac{\left| 3 - 5 \right|^2 - \left| 7 - 13 \right|}{\left| 12 - 9 \right| + \left| 11 - 14 \right|}$

Skill Maintenance

Evaluate.   [1.1a]

83. $\dfrac{x - y}{y}$, when $x = 38$ and $y = 2$

84. $a - 3b$, when $a = 50$ and $b = 5$

Find the absolute value.   [1.2e]

85. $\left| -0.4 \right|$

86. $\left| \dfrac{15}{2} \right|$

Find the reciprocal.   [1.6b]

87. -9

88. $\dfrac{7}{3}$

Subtract.   [1.4a]

89. $5 - 30$

90. $-5 - 30$

91. $-5 - (-30)$

92. $5 - (-30)$

Synthesis

Simplify.

93. $x - [f - (f - x)] + [x - f] - 3x$

94. $x - \{x - 1 - [x - 2 - (x - 3 - \{x - 4 - [x - 5 - (x - 6)]\})]\}$

95. ▦ Use your calculator to do the following.

    a) Evaluate $x^2 + 3$ when $x = 7$, when $x = -7$, and when $x = -5.013$.

    b) Evaluate $1 - x^2$ when $x = 5$, when $x = -5$, and when $x = -10.455$.

96. Express $3^3 + 3^3 + 3^3$ as a power of 3.

Find the average.

97. $-15, 20, 50, -82, -7, -2$

98. $-1, 1, 2, -2, 3, -8, -10$

Vocabulary Reinforcement

Complete each statement with the correct term from the column on the right. Some of the choices may not be used.

natural numbers

whole numbers

integers

real numbers

multiplicative inverses

additive inverses

commutative law

associative law

distributive law

identity property of 0

identity property of 1

property of -1

1. The set of _____ is $\{\ldots, -5, -4, -3, -2, -1, 0, 1, 2, 3, 4, 5, \ldots\}$. [1.2a]

2. Two numbers whose sum is 0 are called _____ of each other. [1.3b]

3. The _____ of addition says that $a + b = b + a$ for any real numbers a and b. [1.7b]

4. The _____ states that for any real number a, $a \cdot 1 = 1 \cdot a = a$. [1.7a]

5. The _____ of multiplication says that $a(bc) = (ab)c$ for any real numbers a, b, and c. [1.7b]

6. Two numbers whose product is 1 are called _____ of each other. [1.6b]

7. The equation $y + 0 = y$ illustrates the _____. [1.7a]

Concept Reinforcement

Determine whether each statement is true or false.

_____ 1. Every whole number is also an integer. [1.2a]

_____ 2. The product of an even number of negative numbers is positive. [1.5a]

_____ 3. The product of a number and its multiplicative inverse is -1. [1.6b]

_____ 4. $a < b$ also has the meaning $b \geq a$. [1.2d]

Study Guide

Objective 1.1a Evaluate algebraic expressions by substitution.

Example Evaluate $y - z$ when $y = 5$ and $z = -7$.
$y - z = 5 - (-7) = 5 + 7 = 12$

Practice Exercise

1. Evaluate $2a + b$ when $a = -1$ and $b = 16$.

Objective 1.2d Determine which of two real numbers is greater and indicate which, using $<$ or $>$.

Example Use $<$ or $>$ for $\square$ to write a true sentence:
$-5 \;\square\; -12$.
Since -5 is to the right of -12 on the number line, we have $-5 > -12$.

Practice Exercise

2. Use $<$ or $>$ for $\square$ to write a true sentence:
$-6 \;\square\; -3$.

Objective 1.2e Find the absolute value of a real number.

Example Find the absolute value: **(a)** $|21|$; **(b)** $|-3.2|$; **(c)** $|0|$.

a) The number is positive, so the absolute value is the same as the number.

$$|21| = 21$$

b) The number is negative, so we make it positive.

$$|-3.2| = 3.2$$

c) The number is 0, so the absolute value is the same as the number.

$$|0| = 0$$

Practice Exercise

3. Find the absolute value: $\left| -\dfrac{5}{4} \right|$.

Objective 1.3a Add real numbers without using the number line.

Example Add without using the number line: **(a)** $-13 + 4$; **(b)** $-2 + (-3)$.

a) We have a negative number and a positive number. The absolute values are 13 and 4. The difference is 9. The negative number has the larger absolute value, so the answer is negative.

$$-13 + 4 = -9$$

b) We have two negative numbers. The sum of the absolute values is $2 + 3$, or 5. The answer is negative.

$$-2 + (-3) = -5$$

Practice Exercise

4. Add without using the number line:

$$-5.6 + (-2.9).$$

Objective 1.4a Subtract real numbers.

Example Subtract: $-4 - (-6)$.

$$-4 - (-6) = -4 + 6 = 2$$

Practice Exercise

5. Subtract: $7 - 9$.

Objective 1.5a Multiply real numbers.

Example Multiply: **(a)** $-1.9(4)$; **(b)** $-7(-6)$.

a) The signs are different, so the answer is negative.

$$-1.9(4) = -7.6$$

b) The signs are the same, so the answer is positive.

$$-7(-6) = 42$$

Practice Exercise

6. Multiply: $-8(-7)$.

Objective 1.6a Divide integers.

Example Divide: **(a)** $15 \div (-3)$; **(b)** $-72 \div (-9)$.

a) The signs are different, so the answer is negative.

$$15 \div (-3) = -5$$

b) The signs are the same, so the answer is positive.

$$-72 \div (-9) = 8$$

Practice Exercise

7. Divide: $-48 \div 6$.

Objective 1.6c Divide real numbers.

Example Divide: **(a)** $-\dfrac{1}{4} \div \dfrac{3}{5}$; **(b)** $-22.4 \div (-4)$.

a) We multiply by the reciprocal of the divisor:

$$-\frac{1}{4} \div \frac{3}{5} = -\frac{1}{4} \cdot \frac{5}{3} = -\frac{5}{12}.$$

b) We carry out the long division. The answer is positive.

$$
\begin{array}{r}
5.6 \\
4\overline{)22.4} \\
\underline{20} \\
2\,4 \\
\underline{2\,4} \\
0
\end{array}
$$

Practice Exercise

8. Divide: $-\dfrac{3}{4} \div \left(-\dfrac{5}{3} \right)$.

Objective 1.7a Simplify fraction expressions.

Example Simplify: $-\dfrac{18x}{15x}$.

$$-\frac{18x}{15x} = -\frac{6 \cdot 3x}{5 \cdot 3x} \quad \text{Factoring the numerator and the denominator}$$

$$= -\frac{6}{5} \cdot \frac{3x}{3x} \quad \text{Factoring the fraction expression}$$

$$= -\frac{6}{5} \cdot 1 \quad \frac{3x}{3x} = 1$$

$$= -\frac{6}{5} \quad \text{Removing a factor of 1}$$

Practice Exercise

9. Simplify: $\dfrac{45y}{27y}$.

Objective 1.7c Use the distributive laws to multiply expressions like 8 and $x - y$.

Example Multiply: $3(4x - y + 2z)$.

$$3(4x - y + 2z) = 3 \cdot 4x - 3 \cdot y + 3 \cdot 2z$$
$$= 12x - 3y + 6z$$

Practice Exercise

10. Multiply: $5(x + 3y - 4z)$.

Objective 1.7d Use the distributive laws to factor expressions like $4x - 12 + 24y$.

Example Factor: $12a - 8b + 4c$.

$$12a - 8b + 4c = 4 \cdot 3a - 4 \cdot 2b + 4 \cdot c$$
$$= 4(3a - 2b + c)$$

Practice Exercise

11. Factor: $27x + 9y - 36z$.

Objective 1.7e Collect like terms.

Example Collect like terms: $3x - 5y + 8x + y$.

$$3x - 5y + 8x + y = 3x + 8x - 5y + y$$
$$= 3x + 8x - 5y + 1 \cdot y$$
$$= (3 + 8)x + (-5 + 1)y$$
$$= 11x - 4y$$

Practice Exercise

12. Collect like terms: $6a - 4b - a + 2b$.

Objective 1.8b Simplify expressions by removing parentheses and collecting like terms.

Example Remove parentheses and simplify:
$$5x - 2(3x - y).$$
$$5x - 2(3x - y) = 5x - 6x + 2y = -x + 2y$$

Practice Exercise

13. Remove parentheses and simplify:
$$8a - b - (4a + 3b).$$

Objective 1.8d Simplify expressions using the rules for order of operations.

Example Simplify: $12 - (7 - 3 \cdot 6)$.
$$
\begin{aligned}
12 - (7 - 3 \cdot 6) &= 12 - (7 - 18) \\
&= 12 - (-11) \\
&= 12 + 11 \\
&= 23
\end{aligned}
$$

Practice Exercise

14. Simplify: $75 \div (-15) + 24 \div 8$.

Review Exercises

The review exercises that follow are for practice. Answers are at the back of the book. If you miss an exercise, restudy the objective indicated in red after the exercise or the direction line that precedes it.

1. Evaluate $\dfrac{x - y}{3}$ when $x = 17$ and $y = 5$. [1.1a]

2. Translate to an algebraic expression: [1.1b]

Nineteen percent of some number.

3. Tell which integers correspond to this situation: [1.2a]

Josh earned \$620 for one week's work. While driving to work one day, he received a speeding ticket for \$125.

Find the absolute value. [1.2e]

4. $|-38|$ **5.** $|126|$

Graph the number on the number line. [1.2b]

6. -2.5 **7.** $\dfrac{8}{9}$

Use either $<$ or $>$ for $\square$ to write a true sentence. [1.2d]

8. $-3 \ \square \ 10$ **9.** $-1 \ \square \ -6$

10. $0.126 \ \square \ -12.6$ **11.** $-\dfrac{2}{3} \ \square \ -\dfrac{1}{10}$

12. Write another inequality with the same meaning as $-3 < x$. [1.2d]

Write true or false. [1.2d]

13. $-9 \le 9$ **14.** $-11 \ge -3$

Find the opposite. [1.3b]

15. 3.8 **16.** $-\dfrac{3}{4}$

Find the reciprocal. [1.6b]

17. $\dfrac{3}{8}$ **18.** -7

19. Evaluate $-x$ when $x = -34$. [1.3b]

20. Evaluate $-(-x)$ when $x = 5$. [1.3b]

Compute and simplify.

21. $4 + (-7)$ [1.3a]

22. $6 + (-9) + (-8) + 7$ [1.3a]

23. $-3.8 + 5.1 + (-12) + (-4.3) + 10$ [1.3a]

24. $-3 - (-7) + 7 - 10$ [1.4a]

25. $-\dfrac{9}{10} - \dfrac{1}{2}$ [1.4a]

26. $-3.8 - 4.1$ [1.4a]

27. $-9 \cdot (-6)$ [1.5a]

28. $-2.7(3.4)$ [1.5a]

29. $\dfrac{2}{3} \cdot \left(-\dfrac{3}{7}\right)$ [1.5a]

30. $3 \cdot (-7) \cdot (-2) \cdot (-5)$ [1.5a]

31. $35 \div (-5)$ [1.6a]

32. $-5.1 \div 1.7$ [1.6c]

33. $-\dfrac{3}{11} \div \left(-\dfrac{4}{11}\right)$ [1.6c]

Simplify. [1.8d]

34. $2(-3.4 - 12.2) - 8(-7)$

35. $\dfrac{-12(-3) - 2^3 - (-9)(-10)}{3 \cdot 10 + 1}$

36. $-16 \div 4 - 30 \div (-5)$

37. $\dfrac{-4[7 - (10 - 13)]}{|-2(8) - 4|}$

Solve.

38. On the first, second, and third downs, a football team had these gains and losses: 5-yd gain, 12-yd loss, and 15-yd gain, respectively. Find the total gain (or loss). [1.3c]

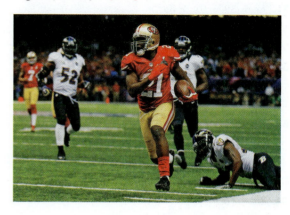

39. Chang's total assets are $2140. He borrows $2500. What are his total assets now? [1.4b]

40. *Stock Price.* The value of EFX Corp. stock began the day at $17.68 per share and dropped $1.63 per hour for 8 hr. What was the price of the stock after 8 hr? [1.5b]

41. *Bank Account Balance.* Yuri had $68 in his bank account. After using his debit card to buy seven equally priced tee shirts, the balance in his account was $-$64.65. What was the price of each shirt? [1.6d]

Multiply. [1.7c]

42. $5(3x - 7)$

43. $-2(4x - 5)$

44. $10(0.4x + 1.5)$

45. $-8(3 - 6x)$

Factor. [1.7d]

46. $2x - 14$

47. $-6x + 6$

48. $5x + 10$

49. $-3x + 12y - 12$

Collect like terms.   [1.7e]

50. $11a + 2b - 4a - 5b$

51. $7x - 3y - 9x + 8y$

52. $6x + 3y - x - 4y$

53. $-3a + 9b + 2a - b$

Remove parentheses and simplify.

54. $2a - (5a - 9)$ [1.8b]

55. $3(b + 7) - 5b$ [1.8b]

56. $3[11 - 3(4 - 1)]$ [1.8c]

57. $2[6(y - 4) + 7]$ [1.8c]

58. $[8(x + 4) - 10] - [3(x - 2) + 4]$ [1.8c]

59. $5\{[6(x - 1) + 7] - [3(3x - 4) + 8]\}$ [1.8c]

60. Factor out the greatest common factor:
$18x - 6y + 30$.   [1.7d]

A. $2(9x - 2y + 15)$ **B.** $3(6x - 2y + 10)$
C. $6(3x + 5)$ **D.** $6(3x - y + 5)$

61. Which expression is *not* equivalent to $mn + 5$?
[1.7b]

A. $nm + 5$ **B.** $5n + m$
C. $5 + mn$ **D.** $5 + nm$

Synthesis

Simplify.   [1.2e], [1.4a], [1.6a], [1.8d]

62. $-\left| \dfrac{7}{8} - \left(-\dfrac{1}{2}\right) - \dfrac{3}{4} \right|$

63. $(|2.7 - 3| + 3^2 - |-3|) \div (-3)$

64. $2000 - 1990 + 1980 - 1970 + \cdots + 20 - 10$

65. Find a formula for the perimeter of the figure
below.   [1.7e]

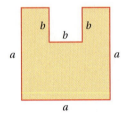

Understanding Through Discussion and Writing

1. Without actually performing the addition, explain why the sum of all integers from -50 to 50 is 0. [1.3b]

2. What rule have we developed that would tell you the sign of $(-7)^8$ and of $(-7)^{11}$ without doing the computations? Explain.   [1.5a]

3. Explain how multiplication can be used to justify why a negative number divided by a negative number is positive.   [1.6c]

4. Explain how multiplication can be used to justify why a negative number divided by a positive number is negative.   [1.6c]

5. The distributive law was introduced before the discussion on collecting like terms. Why do you think this was done?   [1.7c, e]

6. ▦ Jake keys in $18/2 \cdot 3$ on his calculator and expects the result to be 3. What mistake is he making?   [1.8d]

CHAPTER

1 **Test**

For Extra Help   For step-by-step test solutions, access the Chapter Test Prep Videos in MyMathLab® or on YouTube (search "BittingerIntro" and click on "Channels").

1. Evaluate $\dfrac{3x}{y}$ when $x = 10$ and $y = 5$.

2. Translate to an algebraic expression: Nine less than some number.

Use either $<$ or $>$ for $\square$ to write a true sentence.

3. $-3 \;\square\; -8$

4. $-\dfrac{1}{2} \;\square\; -\dfrac{1}{8}$

5. $-0.78 \;\square\; -0.87$

6. Write an inequality with the same meaning as $x < -2$.

7. Write true or false: $-13 \le -3$.

Simplify.

8. $|-7|$

9. $\left|\dfrac{9}{4}\right|$

10. $|-2.7|$

Find the opposite.

11. $\dfrac{2}{3}$

12. -1.4

Find the reciprocal.

13. -2

14. $\dfrac{4}{7}$

15. Evaluate $-x$ when $x = -8$.

Compute and simplify.

16. $3.1 - (-4.7)$

17. $-8 + 4 + (-7) + 3$

18. $-\dfrac{1}{5} + \dfrac{3}{8}$

19. $2 - (-8)$

20. $3.2 - 5.7$

21. $\dfrac{1}{8} - \left(-\dfrac{3}{4}\right)$

22. $4 \cdot (-12)$

23. $-\dfrac{1}{2} \cdot \left(-\dfrac{3}{8}\right)$

24. $-45 \div 5$

25. $-\dfrac{3}{5} \div \left(-\dfrac{4}{5}\right)$

26. $4.864 \div (-0.5)$

27. $-2(16) - |2(-8) - 5^3|$

28. $-20 \div (-5) + 36 \div (-4)$

29. Isabella kept track of the changes in the stock market over a period of 5 weeks. By how many points had the market risen or fallen over this time?

WEEK 1	WEEK 2	WEEK 3	WEEK 4	WEEK 5
Down 13 pts	Down 16 pts	Up 36 pts	Down 11 pts	Up 19 pts

30. *Difference in Elevation.* The lowest elevation in Australia, Lake Eyre, is 15 m below sea level. The highest elevation in Australia, Mount Kosciuszko, is 2229 m. Find the difference in elevation between the highest point and the lowest point.

Source: *The CIA World Factbook*, 2012

Lake Eyre
−15 m

Sydney

Mt. Kosciuszko
2229 m

31. *Population Decrease.* The population of Stone City was 18,600. It dropped 420 each year for 6 years. What was the population of the city after 6 years?

32. *Chemical Experiment.* During a chemical reaction, the temperature in a beaker decreased every minute by the same number of degrees. The temperature was 16°C at 11:08 A.M. By 11:52 A.M., the temperature had dropped to −17°C. By how many degrees did it change each minute?

Multiply.

33. $3(6 - x)$

34. $-5(y - 1)$

Factor.

35. $12 - 22x$

36. $7x + 21 + 14y$

Simplify.

37. $6 + 7 - 4 - (-3)$

38. $5x - (3x - 7)$

39. $4(2a - 3b) + a - 7$

40. $4\{3[5(y - 3) + 9] + 2(y + 8)\}$

41. $256 \div (-16) \div 4$

42. $2^3 - 10[4 - 3(-2 + 18)]$

43. Which of the following is *not* a true statement?
   A. $-5 \leq -5$
   B. $-5 < -5$
   C. $-5 \geq -5$
   D. $-5 = -5$

Synthesis

Simplify.

44. $|-27 - 3(4)| - |-36| + |-12|$

45. $a - \{3a - [4a - (2a - 4a)]\}$

46. Find a formula for the perimeter of the figure shown here.

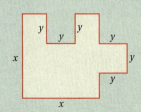

CHAPTER

2

Solving Equations and Inequalities

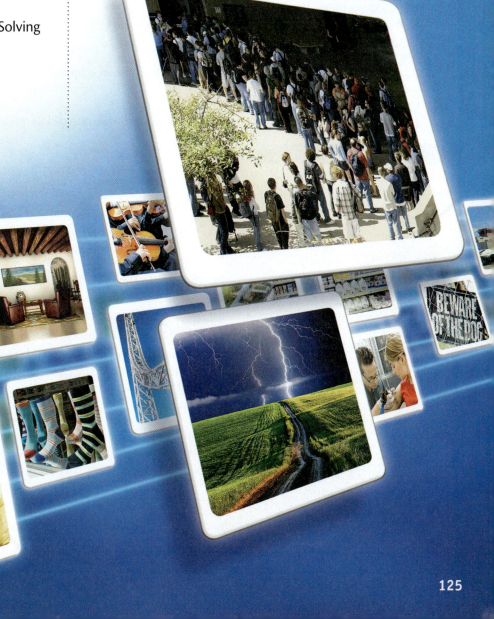

2.1 Solving Equations: The Addition Principle

OBJECTIVES

a Determine whether a given number is a solution of a given equation.

b Solve equations using the addition principle.

SKILL TO REVIEW

Objective 1.1a: Evaluate algebraic expressions by substitution.

1. Evaluate $x - 7$ when $x = 5$.
2. Evaluate $2x + 3$ when $x = -1$.

Determine whether each equation is true, false, or neither.

1. $5 - 8 = -4$

2. $12 + 6 = 18$

3. $x + 6 = 7 - x$

a   EQUATIONS AND SOLUTIONS

In order to solve problems, we must learn to solve equations.

EQUATION

An **equation** is a number sentence that says that the expressions on either side of the equals sign, $=$, represent the same number.

Here are some examples of equations:

$$3 + 2 = 5, \quad 14 - 10 = 1 + 3, \quad x + 6 = 13, \quad 3x - 2 = 7 - x.$$

Equations have expressions on each side of the equals sign. The sentence "$14 - 10 = 1 + 3$" asserts that the expressions $14 - 10$ and $1 + 3$ name the same number.

Some equations are true. Some are false. Some are neither true nor false.

EXAMPLES   Determine whether each equation is true, false, or neither.

1. $3 + 2 = 5$ The equation is *true*.
2. $7 - 2 = 4$ The equation is *false*.
3. $x + 6 = 13$ The equation is *neither* true nor false, because we do not know what number x represents.

◀ Do Margin Exercises 1–3.

SOLUTION OF AN EQUATION

Any replacement for the variable that makes an equation true is called a **solution** of the equation. To solve an equation means to find *all* of its solutions.

One way to determine whether a number is a solution of an equation is to evaluate the expression on each side of the equals sign by substitution. If the values are the same, then the number is a solution.

Answers

Skill to Review:
1. -2 2. 1

Margin Exercises:
1. False   2. True   3. Neither

EXAMPLE 4   Determine whether 7 is a solution of $x + 6 = 13$.

We have

$$x + 6 = 13 \qquad \text{Writing the equation}$$
$$7 + 6 \ ?\ 13 \qquad \text{Substituting 7 for } x$$
$$13 \mid \qquad \text{TRUE}$$

Since the left-hand side and the right-hand side are the same, 7 is a solution. No other number makes the equation true, so the only solution is the number 7.

EXAMPLE 5   Determine whether 19 is a solution of $7x = 141$.

We have

$$7x = 141 \qquad \text{Writing the equation}$$
$$7(19) \ ?\ 141 \qquad \text{Substituting 19 for } x$$
$$133 \mid \qquad \text{FALSE}$$

Since the left-hand side and the right-hand side are not the same, 19 is not a solution of the equation.

Do Exercises 4–7. ▶

Determine whether the given number is a solution of the given equation.

4. 8; $x + 4 = 12$

5. 0; $x + 4 = 12$

6. -3; $7 + x = -4$

7. $-\dfrac{3}{5}$; $-5x = 3$

b USING THE ADDITION PRINCIPLE

Consider the equation

$$x = 7.$$

We can easily see that the solution of this equation is 7. If we replace x with 7, we get

$$7 = 7, \quad \text{which is true.}$$

Now consider the equation of Example 4: $x + 6 = 13$. In Example 4, we discovered that the solution of this equation is also 7, but the fact that 7 is the solution is not as obvious. We now begin to consider principles that allow us to start with an equation like $x + 6 = 13$ and end up with an *equivalent equation,* like $x = 7$, in which the variable is alone on one side and for which the solution is easier to find.

EQUIVALENT EQUATIONS

Equations with the same solutions are called **equivalent equations**.

One of the principles that we use in solving equations involves addition. An equation $a = b$ says that a and b stand for the same number. Suppose this is true, and we add a number c to the number a. We get the same answer if we add c to b, because a and b are the same number.

THE ADDITION PRINCIPLE FOR EQUATIONS

For any real numbers a, b, and c,

$$a = b \quad \text{is equivalent to} \quad a + c = b + c.$$

Let's solve the equation $x + 6 = 13$ using the addition principle. We want to get x alone on one side. To do so, we use the addition principle, choosing to add -6 because $6 + (-6) = 0$:

$$x + 6 = 13$$
$$x + 6 + (-6) = 13 + (-6) \qquad \text{Using the addition principle:}$$
$$\qquad\qquad\qquad\qquad\qquad\qquad \text{adding } -6 \text{ on both sides}$$
$$x + 0 = 7 \qquad \text{Simplifying}$$
$$x = 7. \qquad \text{Identity property of 0: } x + 0 = x$$

The solution of $x + 6 = 13$ is 7.

◀ **Do Exercise 8.**

8. Solve $x + 2 = 11$ using the addition principle.
$$x + 2 = 11$$
$$x + 2 + (-2) = 11 + (\boxed{})$$
$$x + \boxed{} = 9$$
$$x = \boxed{}$$

When we use the addition principle, we sometimes say that we "add the same number on both sides of the equation." This is also true for subtraction, since we can express every subtraction as an addition. That is, since

$$a - c = b - c \quad \text{is equivalent to} \quad a + (-c) = b + (-c),$$

the addition principle tells us that we can "subtract the same number on both sides of the equation."

EXAMPLE 6 Solve: $x + 5 = -7$.

We have

$$x + 5 = -7$$
$$x + 5 - 5 = -7 - 5 \qquad \text{Using the addition principle: adding } -5 \text{ on}$$
$$\qquad\qquad\qquad\qquad\qquad \text{both sides or subtracting 5 on both sides}$$
$$x + 0 = -12 \qquad \text{Simplifying}$$
$$x = -12. \qquad \text{Identity property of 0}$$

The solution of the original equation is -12. The equations $x + 5 = -7$ and $x = -12$ are *equivalent*.

◀ **Do Exercise 9.**

9. Solve using the addition principle, subtracting 5 on both sides:
$$x + 5 = -8.$$

Now we use the addition principle to solve an equation that involves a subtraction.

EXAMPLE 7 Solve: $a - 4 = 10$.

We have

$$a - 4 = 10$$
$$a - 4 + 4 = 10 + 4 \qquad \text{Using the addition principle:}$$
$$\qquad\qquad\qquad\qquad\qquad \text{adding 4 on both sides}$$
$$a + 0 = 14 \qquad \text{Simplifying}$$
$$a = 14. \qquad \text{Identity property of 0}$$

Check:
$$\begin{array}{c} a - 4 = 10 \\ \hline 14 - 4 \; ? \; 10 \\ 10 \mid \qquad \text{TRUE} \end{array}$$

The solution is 14.

◀ **Do Exercise 10.**

10. Solve: $t - 3 = 19$.

Answers

8. 9   **9.** -13 **10.** 22

Guided Solution:
8. $-2, 0, 9$

EXAMPLE 8 Solve: $-6.5 = y - 8.4$.

We have

$$-6.5 = y - 8.4$$

$$-6.5 + 8.4 = y - 8.4 + 8.4$$ Using the addition principle: adding 8.4 on both sides to eliminate -8.4 on the right

$$1.9 = y.$$

Check:
$$\frac{-6.5 = y - 8.4}{-6.5 \;\overset{?}{|}\; 1.9 - 8.4}$$
$$\Big|\; -6.5 \qquad \text{TRUE}$$

The solution is 1.9.

Note that equations are reversible. That is, if $a = b$ is true, then $b = a$ is true. Thus when we solve $-6.5 = y - 8.4$, we can reverse it and solve $y - 8.4 = -6.5$ if we wish.

Do Exercises 11 and 12. ▶

Solve.

11. $8.7 = n - 4.5$

12. $y + 17.4 = 10.9$

EXAMPLE 9 Solve: $-\dfrac{2}{3} + x = \dfrac{5}{2}$.

We have

$$-\frac{2}{3} + x = \frac{5}{2}$$

$$\frac{2}{3} - \frac{2}{3} + x = \frac{2}{3} + \frac{5}{2}$$ Adding $\frac{2}{3}$ on both sides

$$x = \frac{2}{3} + \frac{5}{2}$$

$$x = \frac{2}{3} \cdot \frac{2}{2} + \frac{5}{2} \cdot \frac{3}{3}$$ Multiplying by 1 to obtain equivalent fraction expressions with the least common denominator 6

$$x = \frac{4}{6} + \frac{15}{6}$$

$$x = \frac{19}{6}.$$

Check:
$$-\frac{2}{3} + x = \frac{5}{2}$$
$$-\frac{2}{3} + \frac{19}{6} \;\overset{?}{|}\; \frac{5}{2}$$
$$-\frac{4}{6} + \frac{19}{6}$$
$$\frac{15}{6}$$
$$\frac{5}{2} \;\Big|\; \qquad \text{TRUE}$$

The solution is $\dfrac{19}{6}$.

Solve.

13. $x + \dfrac{1}{2} = -\dfrac{3}{2}$

14. $t - \dfrac{13}{4} = \dfrac{5}{8}$

Answers

11. 13.2 **12.** -6.5 **13.** -2 **14.** $\frac{31}{8}$

Do Exercises 13 and 14. ▶

✓ Reading Check

Choose from the column on the right the most appropriate first step in solving each equation.

RC1. $9 = x - 4$

RC2. $3 + x = -15$

RC3. $x - 3 = 9$

RC4. $x + 4 = 3$

  a) Add -4 on both sides.
  b) Add 15 on both sides.
  c) Subtract 3 on both sides.
  d) Subtract 9 on both sides.
  e) Add 3 on both sides.
  f) Add 4 on both sides.

a   Determine whether the given number is a solution of the given equation.

1. 15; $x + 17 = 32$

2. 35; $t + 17 = 53$

3. 21; $x - 7 = 12$

4. 36; $a - 19 = 17$

5. -7; $6x = 54$

6. -9; $8y = -72$

7. 30; $\dfrac{x}{6} = 5$

8. 49; $\dfrac{y}{8} = 6$

9. 20; $5x + 7 = 107$

10. 9; $9x + 5 = 86$

11. -10; $7(y - 1) = 63$

12. -5; $6(y - 2) = 18$

b   Solve using the addition principle. Don't forget to check!

13. $x + 2 = 6$

Check: $x + 2 = 6$

14. $y + 4 = 11$

Check: $y + 4 = 11$

15. $x + 15 = -5$

Check: $x + 15 = -5$

16. $t + 10 = 44$

Check: $t + 10 = 44$

17. $x + 6 = -8$

Check: $x + 6 = -8$

18. $z + 9 = -14$

19. $x + 16 = -2$

20. $m + 18 = -13$

21. $x - 9 = 6$

22. $x - 11 = 12$

23. $x - 7 = -21$

24. $x - 3 = -14$

25. $5 + t = 7$

26. $8 + y = 12$

27. $-7 + y = 13$

28. $-8 + y = 17$ **29.** $-3 + t = -9$ **30.** $-8 + t = -24$ **31.** $x + \dfrac{1}{2} = 7$ **32.** $24 = -\dfrac{7}{10} + r$

33. $12 = a - 7.9$ **34.** $2.8 + y = 11$ **35.** $r + \dfrac{1}{3} = \dfrac{8}{3}$ **36.** $t + \dfrac{3}{8} = \dfrac{5}{8}$

37. $m + \dfrac{5}{6} = -\dfrac{11}{12}$ **38.** $x + \dfrac{2}{3} = -\dfrac{5}{6}$ **39.** $x - \dfrac{5}{6} = \dfrac{7}{8}$ **40.** $y - \dfrac{3}{4} = \dfrac{5}{6}$

41. $-\dfrac{1}{5} + z = -\dfrac{1}{4}$ **42.** $-\dfrac{1}{8} + y = -\dfrac{3}{4}$ **43.** $7.4 = x + 2.3$ **44.** $8.4 = 5.7 + y$

45. $7.6 = x - 4.8$ **46.** $8.6 = x - 7.4$ **47.** $-9.7 = -4.7 + y$ **48.** $-7.8 = 2.8 + x$

49. $5\dfrac{1}{6} + x = 7$ **50.** $5\dfrac{1}{4} = 4\dfrac{2}{3} + x$ **51.** $q + \dfrac{1}{3} = -\dfrac{1}{7}$ **52.** $52\dfrac{3}{8} = -84 + x$

Skill Maintenance

53. Divide: $\dfrac{2}{3} \div \left(-\dfrac{4}{9}\right)$. [1.6c]

54. Add: $-8.6 + 3.4$. [1.3a]

55. Subtract: $-\dfrac{2}{3} - \left(-\dfrac{5}{8}\right)$. [1.4a]

56. Multiply: $(-25.4)(-6.8)$. [1.5a]

Translate to an algebraic expression. [1.1b]

57. Jane had \$83 before paying x dollars for a pair of tennis shoes. How much does she have left?

58. Justin drove his S-10 pickup truck 65 mph for t hours. How far did he drive?

Synthesis

Solve.

59. $x + \dfrac{4}{5} = -\dfrac{2}{3} - \dfrac{4}{15}$

60. $x + x = x$

61. $16 + x - 22 = -16$

62. $x + 4 = 5 + x$

63. $x + 3 = 3 + x$

64. $|x| + 6 = 19$

OBJECTIVE

a Solve equations using the multiplication principle.

SKILL TO REVIEW

Objective 1.6b: Find the reciprocal of a real number.

Find the reciprocal.

1. 5

2. $-\dfrac{5}{4}$

a USING THE MULTIPLICATION PRINCIPLE

Suppose that $a = b$ is true, and we multiply a by some number c. We get the same number if we multiply b by c, because a and b are the same number.

THE MULTIPLICATION PRINCIPLE FOR EQUATIONS

For any real numbers a, b, and c, $c \neq 0$,

$$a = b \quad \text{is equivalent to} \quad a \cdot c = b \cdot c.$$

When using the multiplication principle, we sometimes say that we "multiply on both sides of the equation by the same number."

EXAMPLE 1 Solve: $5x = 70$.

To get x alone on one side, we multiply by the *multiplicative inverse*, or *reciprocal*, of 5. Then we get the *multiplicative identity* 1 times x, or $1 \cdot x$, which simplifies to x. This allows us to eliminate 5 on the left.

$$5x = 70 \qquad \text{The reciprocal of 5 is } \tfrac{1}{5}.$$

$$\frac{1}{5} \cdot 5x = \frac{1}{5} \cdot 70 \qquad \begin{array}{l}\text{Multiplying by } \tfrac{1}{5} \text{ to get } 1 \cdot x \text{ and}\\ \text{eliminate 5 on the left}\end{array}$$

$$1 \cdot x = 14 \qquad \text{Simplifying}$$

$$x = 14 \qquad \text{Identity property of 1: } 1 \cdot x = x$$

Check:
$$\begin{array}{c} 5x = 70 \\ \hline 5 \cdot 14 \;?\; 70 \\ 70 \;\bigm|\; \quad \textcolor{red}{\text{TRUE}}\end{array}$$

The solution is 14.

The multiplication principle also tells us that we can "divide on both sides of the equation by the same nonzero number." This is because dividing is the same as multiplying by a reciprocal. That is,

$$\frac{a}{c} = \frac{b}{c} \quad \text{is equivalent to} \quad a \cdot \frac{1}{c} = b \cdot \frac{1}{c}, \quad \text{when } c \neq 0.$$

In an expression like $5x$ in Example 1, the number 5 is called the **coefficient**. Example 1 could be done as follows, dividing by 5 on both sides, the coefficient of x.

EXAMPLE 2 Solve: $5x = 70$.

$$5x = 70$$

$$\frac{5x}{5} = \frac{70}{5} \qquad \text{Dividing by 5 on both sides}$$

$$1 \cdot x = 14 \qquad \text{Simplifying}$$

$$x = 14 \qquad \text{Identity property of 1. The solution is 14.}$$

Do Exercises 1 and 2. ▶

EXAMPLE 3 Solve: $-4x = 92$.

We have

$$-4x = 92$$

$$\frac{-4x}{-4} = \frac{92}{-4}$$ Using the multiplication principle. Dividing by -4 on both sides is the same as multiplying by $-\frac{1}{4}$.

$$1 \cdot x = -23$$ Simplifying

$$x = -23.$$ Identity property of 1

Check: $\dfrac{-4x = 92}{-4(-23) \overset{?}{\vert} 92}$

$\qquad\qquad\quad 92 \,\vert\quad$ TRUE

The solution is -23.

Do Exercise 3. ▶

EXAMPLE 4 Solve: $-x = 9$.

We have

$$-x = 9$$

$$-1 \cdot x = 9$$ Using the property of -1: $-x = -1 \cdot x$

$$\frac{-1 \cdot x}{-1} = \frac{9}{-1}$$ Dividing by -1 on both sides: $-1/(-1) = 1$

$$1 \cdot x = -9$$

$$x = -9.$$

Check: $\dfrac{-x = 9}{-(-9) \overset{?}{\vert} 9}$

$\qquad\qquad\quad 9 \,\vert\quad$ TRUE

The solution is -9.

Do Exercise 4. ▶

We can also solve the equation $-x = 9$ by multiplying as follows.

EXAMPLE 5 Solve: $-x = 9$.

We have

$$-x = 9$$

$$-1 \cdot (-x) = -1 \cdot 9$$ Multiplying by -1 on both sides

$$-1 \cdot (-1) \cdot x = -9$$ $\qquad -x = (-1) \cdot x$

$$1 \cdot x = -9$$ $\qquad -1 \cdot (-1) = 1$

$$x = -9.$$

The solution is -9.

Do Exercise 5. ▶

GS **1.** Solve $6x = 90$ by multiplying on both sides.

$$6x = 90$$

$$\frac{1}{6} \cdot 6x = \boxed{} \cdot 90$$

$$1 \cdot x = 15$$

$$\boxed{} = 15$$

Check: $\dfrac{6x = 90}{6 \cdot \boxed{} \overset{?}{\vert} 90}$

$\qquad\qquad\quad 90 \,\vert\quad$ TRUE

2. Solve $4x = -7$ by dividing on both sides.

$$4x = -7$$

$$\frac{4x}{4} = \frac{-7}{\boxed{}}$$

$$1 \cdot x = -\frac{7}{4}$$

$$\boxed{} = -\frac{7}{4}$$

Don't forget to check.

3. Solve: $-6x = 108$.

4. Solve. Divide on both sides.

$$-x = -10$$

5. Solve. Multiply on both sides.

$$-x = -10$$

Answers

1. 15 **2.** $-\dfrac{7}{4}$ **3.** -18 **4.** 10 **5.** 10

Guided Solutions:

1. $\dfrac{1}{6}$, x, 15 **2.** 4, x

In practice, it is generally more convenient to divide on both sides of the equation if the coefficient of the variable is in decimal notation or is an integer. If the coefficient is in fraction notation, it is usually more convenient to multiply by a reciprocal.

EXAMPLE 6 Solve: $\dfrac{3}{8} = -\dfrac{5}{4}x$.

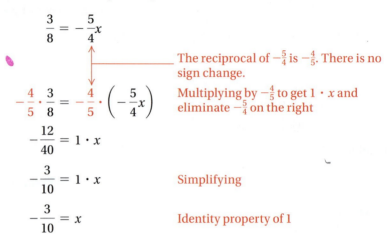

$$\dfrac{3}{8} = -\dfrac{5}{4}x$$

The reciprocal of $-\frac{5}{4}$ is $-\frac{4}{5}$. There is no sign change.

$$-\dfrac{4}{5} \cdot \dfrac{3}{8} = -\dfrac{4}{5} \cdot \left(-\dfrac{5}{4}x\right)$$

Multiplying by $-\frac{4}{5}$ to get $1 \cdot x$ and eliminate $-\frac{5}{4}$ on the right

$$-\dfrac{12}{40} = 1 \cdot x$$

$$-\dfrac{3}{10} = 1 \cdot x \qquad \text{Simplifying}$$

$$-\dfrac{3}{10} = x \qquad \text{Identity property of 1}$$

Check:

$$\dfrac{3}{8} = -\dfrac{5}{4}x$$

$$\dfrac{3}{8} \;?\; -\dfrac{5}{4}\left(-\dfrac{3}{10}\right)$$

$$\left.\dfrac{3}{8}\right| \qquad \text{TRUE}$$

The solution is $-\dfrac{3}{10}$.

As noted in Section 2.1, if $a = b$ is true, then $b = a$ is true. Thus we can reverse the equation $\frac{3}{8} = -\frac{5}{4}x$ and solve $-\frac{5}{4}x = \frac{3}{8}$ if we wish.

◀ **Do Exercise 6.**

EXAMPLE 7 Solve: $1.16y = 9744$.

$$1.16y = 9744$$

$$\dfrac{1.16y}{1.16} = \dfrac{9744}{1.16} \qquad \text{Dividing by 1.16 on both sides}$$

$$y = \dfrac{9744}{1.16}$$

$$y = 8400 \qquad \text{Simplifying}$$

Check:

$$1.16y = 9744$$

$$1.16(8400) \;?\; 9744$$

$$\left.9744\right| \qquad \text{TRUE}$$

The solution is 8400.

◀ **Do Exercises 7 and 8.**

---

6. Solve: $\dfrac{2}{3} = -\dfrac{5}{6}y$.

$$\dfrac{2}{3} = -\dfrac{5}{6}y$$

$$\boxed{} \cdot \dfrac{2}{3} = -\dfrac{6}{5} \cdot \left(-\dfrac{5}{6}y\right)$$

$$-\dfrac{\boxed{}}{15} = 1 \cdot y$$

$$-\dfrac{\boxed{}}{5} = y$$

Solve.

7. $1.12x = 8736$

8. $6.3 = -2.1y$

Answers

6. $-\dfrac{4}{5}$ 7. 7800   8. -3

Guided Solution:

6. $-\dfrac{6}{5},\, 12,\, 4$

Now we use the multiplication principle to solve an equation that involves division.

EXAMPLE 8 Solve: $\dfrac{-y}{9} = 14$.

$$\frac{-y}{9} = 14$$

$$9 \cdot \frac{-y}{9} = 9 \cdot 14 \qquad \text{Multiplying by 9 on both sides}$$

$$-y = 126$$

$$-1 \cdot (-y) = -1 \cdot 126 \qquad \text{Multiplying by } -1 \text{ on both sides}$$

$$y = -126$$

Check:

$$\frac{-y}{9} = 14$$

$$\frac{-(-126)}{9} \;?\; 14$$

$$\frac{126}{9}$$

$$14 \qquad \text{TRUE}$$

The solution is -126.

There are other ways to solve the equation in Example 8. One is by multiplying by -9 on both sides as follows:

$$-9 \cdot \frac{-y}{9} = -9 \cdot 14$$

$$\frac{9y}{9} = -126$$

$$y = -126.$$

Do Exercise 9. ▶

9. Solve: $-14 = \dfrac{-y}{2}$.

Answer

9. 28

 Reading Check

Choose from the column on the right the most appropriate first step in solving each equation.

RC1. $3 = -\dfrac{1}{12}x$

RC2. $-6x = 12$

RC3. $12x = -6$

RC4. $\dfrac{1}{6}x = 12$

a) Divide by 12 on both sides.
b) Multiply by 6 on both sides.
c) Multiply by 12 on both sides.
d) Divide by -6 on both sides.
e) Divide by 6 on both sides.
f) Multiply by -12 on both sides.

Solve using the multiplication principle. Don't forget to check!

1. $6x = 36$

Check: $6x = 36$
$$\overline{} \atop ?$$

2. $3x = 51$

Check: $3x = 51$
$$\overline{} \atop ?$$

3. $5y = 45$

Check: $5y = 45$
$$\overline{} \atop ?$$

4. $8y = 72$

Check: $8y = 72$
$$\overline{} \atop ?$$

5. $84 = 7x$

6. $63 = 9x$

7. $-x = 40$

8. $-x = 53$

9. $-1 = -z$

10. $-47 = -t$

11. $7x = -49$

12. $8x = -56$

13. $-12x = 72$

14. $-15x = 105$

15. $-21w = -126$

16. $-13w = -104$

17. $\dfrac{t}{7} = -9$

18. $\dfrac{y}{5} = -6$

19. $\dfrac{n}{-6} = 8$

20. $\dfrac{y}{-8} = 11$

21. $\dfrac{3}{4}x = 27$

22. $\dfrac{4}{5}x = 16$

23. $-\dfrac{2}{3}x = 6$

24. $-\dfrac{3}{8}x = 12$

25. $\dfrac{-t}{3} = 7$

26. $\dfrac{-x}{6} = 9$

27. $-\dfrac{m}{3} = \dfrac{1}{5}$

28. $\dfrac{1}{8} = -\dfrac{y}{5}$

29. $-\dfrac{3}{5}r = \dfrac{9}{10}$

30. $-\dfrac{2}{5}y = \dfrac{4}{15}$

31. $-\dfrac{3}{2}r = -\dfrac{27}{4}$

32. $-\dfrac{3}{8}x = -\dfrac{15}{16}$

33. $6.3x = 44.1$

34. $2.7y = 54$

35. $-3.1y = 21.7$

36. $-3.3y = 6.6$

37. $38.7m = 309.6$

38. $29.4m = 235.2$

39. $-\dfrac{2}{3}y = -10.6$

40. $-\dfrac{9}{7}y = 12.06$

41. $\dfrac{-x}{5} = 10$

42. $\dfrac{-x}{8} = -16$

43. $-\dfrac{t}{2} = 7$

44. $\dfrac{m}{-3} = 10$

Skill Maintenance

Collect like terms.   [1.7e]

45. $3x + 4x$

46. $6x + 5 - 7x$

47. $-4x + 11 - 6x + 18x$

48. $8y - 16y - 24y$

Remove parentheses and simplify.   [1.8b]

49. $3x - (4 + 2x)$

50. $2 - 5(x + 5)$

51. $8y - 6(3y + 7)$

52. $-2a - 4(5a - 1)$

Translate to an algebraic expression.   [1.1b]

53. Patty drives her van for 8 hr at a speed of r miles per hour. How far does she drive?

54. A triangle has a height of 10 meters and a base of b meters. What is the area of the triangle?

Synthesis

Solve.

55. $-0.2344m = 2028.732$

56. $0 \cdot x = 0$

57. $0 \cdot x = 9$

58. $4|x| = 48$

59. $2|x| = -12$

Solve for x.

60. $ax = 5a$

61. $3x = \dfrac{b}{a}$

62. $cx = a^2 + 1$

63. $\dfrac{a}{b}x = 4$

64. A student makes a calculation and gets an answer of 22.5. On the last step, she multiplies by 0.3 when she should have divided by 0.3. What is the correct answer?

SKILL TO REVIEW

Objective 1.7e: Collect like terms.

Collect like terms.

1. $q + 5t - 1 + 5q - t$

2. $7d + 16 - 11w - 2 - 10d$

a **APPLYING BOTH PRINCIPLES**

Consider the equation $3x + 4 = 13$. It is more complicated than those we discussed in the preceding two sections. In order to solve such an equation, we first isolate the x-term, $3x$, using the addition principle. Then we apply the multiplication principle to get x by itself.

EXAMPLE 1 Solve: $3x + 4 = 13$.

$$3x + 4 = 13$$

$$3x + 4 - 4 = 13 - 4 \quad \text{Using the addition principle: subtracting 4 on both sides}$$

First isolate the x-term. $\longrightarrow \quad 3x = 9 \quad \text{Simplifying}$

$$\frac{3x}{3} = \frac{9}{3} \quad \text{Using the multiplication principle: dividing by 3 on both sides}$$

Then isolate x. $\longrightarrow \quad x = 3 \quad \text{Simplifying}$

Check:

$$\frac{3x + 4 = 13}{3 \cdot 3 + 4 \; ? \; 13}$$
$$9 + 4$$
$$13 \quad | \quad \text{TRUE}$$

We use the rules for order of operations to carry out the check. We find the product $3 \cdot 3$. Then we add 4.

The solution is 3.

◀ **Do Margin Exercise 1.**

1. Solve: $9x + 6 = 51$.

EXAMPLE 2 Solve: $-5x - 6 = 16$.

$$-5x - 6 = 16$$

$$-5x - 6 + 6 = 16 + 6 \quad \text{Adding 6 on both sides}$$

$$-5x = 22$$

$$\frac{-5x}{-5} = \frac{22}{-5} \quad \text{Dividing by } -5 \text{ on both sides}$$

$$x = -\frac{22}{5}, \text{ or } -4\frac{2}{5} \quad \text{Simplifying}$$

Solve.

2. $8x - 4 = 28$

3. $-\frac{1}{2}x + 3 = 1$

Check:

$$\frac{-5x - 6 = 16}{-5\left(-\frac{22}{5}\right) - 6 \; ? \; 16}$$
$$22 - 6$$
$$16 \quad | \quad \text{TRUE}$$

The solution is $-\frac{22}{5}$.

◀ **Do Margin Exercises 2 and 3.**

Answers

Skill to Review:
1. $6q + 4t - 1$
2. $-3d + 14 - 11w$

Margin Exercises:
1. 5 **2.** 4 **3.** 4

EXAMPLE 3 Solve: $45 - t = 13$.

$$45 - t = 13$$
$$-45 + 45 - t = -45 + 13 \qquad \text{Adding } -45 \text{ on both sides}$$
$$-t = -32$$
$$-1(-t) = -1(-32) \qquad \text{Multiplying by } -1 \text{ on both sides}$$
$$t = 32$$

The number 32 checks and is the solution.

Do Exercise 4. ▶

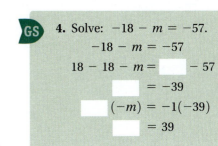

4. Solve: $-18 - m = -57$.
$$-18 - m = -57$$
$$18 - 18 - m = \boxed{} - 57$$
$$\boxed{} = -39$$
$$\boxed{} (-m) = -1(-39)$$
$$\boxed{} = 39$$

EXAMPLE 4 Solve: $16.3 - 7.2y = -8.18$.

$$16.3 - 7.2y = -8.18$$
$$-16.3 + 16.3 - 7.2y = -16.3 + (-8.18) \qquad \text{Adding } -16.3 \text{ on both sides}$$
$$-7.2y = -24.48$$
$$\frac{-7.2y}{-7.2} = \frac{-24.48}{-7.2} \qquad \text{Dividing by } -7.2 \text{ on both sides}$$
$$y = 3.4$$

Check:
$$\begin{array}{c|c} 16.3 - 7.2y = -8.18 \\ \hline 16.3 - 7.2(3.4) \,?\, -8.18 \\ 16.3 - 24.48 \\ -8.18 \end{array} \quad \text{TRUE}$$

The solution is 3.4.

Do Exercises 5 and 6. ▶

Solve.

5. $-4 - 8x = 8$

6. $41.68 = 4.7 - 8.6y$

b COLLECTING LIKE TERMS

If there are like terms on one side of the equation, we collect them before using the addition principle or the multiplication principle.

EXAMPLE 5 Solve: $3x + 4x = -14$.

$$3x + 4x = -14$$
$$7x = -14 \qquad \text{Collecting like terms}$$
$$\frac{7x}{7} = \frac{-14}{7} \qquad \text{Dividing by 7 on both sides}$$
$$x = -2$$

The number -2 checks, so the solution is -2.

Do Exercises 7 and 8. ▶

Solve.

7. $4x + 3x = -21$

8. $x - 0.09x = 728$

If there are like terms on opposite sides of the equation, we get them on the same side by using the addition principle. Then we collect them. In other words, we get all the terms with a variable on one side of the equation and all the terms without a variable on the other side.

Answers

4. 39 **5.** $-\dfrac{3}{2}$ **6.** -4.3 **7.** -3 **8.** 800

Guided Solution:
4. $18, -m, -1, m$

EXAMPLE 6 Solve: $2x - 2 = -3x + 3$.

$$2x - 2 = -3x + 3$$
$$2x - 2 + 2 = -3x + 3 + 2 \qquad \text{Adding 2}$$
$$2x = -3x + 5 \qquad \text{Collecting like terms}$$
$$2x + 3x = -3x + 3x + 5 \qquad \text{Adding } 3x$$
$$5x = 5 \qquad \text{Simplifying}$$
$$\frac{5x}{5} = \frac{5}{5} \qquad \text{Dividing by 5}$$
$$x = 1 \qquad \text{Simplifying}$$

Check:

$$\begin{array}{c|c} \multicolumn{2}{c}{2x - 2 = -3x + 3} \\ \hline 2 \cdot 1 - 2 \ ? \ -3 \cdot 1 + 3 & \text{Substituting in the} \\ 2 - 2 \ \big| \ -3 + 3 & \text{original equation} \\ 0 \ \big| \ 0 & \text{TRUE} \end{array}$$

The solution is 1.

◀ **Do Exercises 9 and 10.**

In Example 6, we used the addition principle to get all the terms with an x on one side of the equation and all the terms without an x on the other side. Then we collected like terms and proceeded as before. If there are like terms on one side at the outset, they should be collected first.

EXAMPLE 7 Solve: $6x + 5 - 7x = 10 - 4x + 3$.

$$6x + 5 - 7x = 10 - 4x + 3$$
$$-x + 5 = 13 - 4x \qquad \text{Collecting like terms}$$
$$4x - x + 5 = 13 - 4x + 4x \qquad \text{Adding } 4x \text{ to get all terms with a variable on one side}$$
$$3x + 5 = 13 \qquad \text{Simplifying; that is, collecting like terms}$$
$$3x + 5 - 5 = 13 - 5 \qquad \text{Subtracting 5}$$
$$3x = 8 \qquad \text{Simplifying}$$
$$\frac{3x}{3} = \frac{8}{3} \qquad \text{Dividing by 3}$$
$$x = \frac{8}{3} \qquad \text{Simplifying}$$

The number $\frac{8}{3}$ checks, so it is the solution.

◀ **Do Exercises 11 and 12.**

Clearing Fractions and Decimals

In general, equations are easier to solve if they do not contain fractions or decimals. Consider, for example, the equations

$$\frac{1}{2}x + 5 = \frac{3}{4} \quad \text{and} \quad 2.3x + 7 = 5.4.$$

Solve.

9. $7y + 5 = 2y + 10$

10. $5 - 2y = 3y - 5$

Solve.

11. $7x - 17 + 2x = 2 - 8x + 15$ GS

$$\boxed{} \cdot x - 17 = 17 - 8x$$
$$8x + 9x - 17 = 17 - 8x + \boxed{}$$
$$\boxed{} \cdot x - 17 = 17$$
$$17x - 17 + 17 = 17 + \boxed{}$$
$$17x = 34$$
$$\frac{17x}{17} = \frac{34}{\boxed{}}$$
$$\boxed{} = 2$$

12. $3x - 15 = 5x + 2 - 4x$

Answers

9. 1   **10.** 2   **11.** 2   **12.** $\dfrac{17}{2}$

Guided Solution:
11. 9, 8x, 17, 17, 17, x

If we multiply by 4 on both sides of the first equation and by 10 on both sides of the second equation, we have

$$4\left(\frac{1}{2}x + 5\right) = 4 \cdot \frac{3}{4} \quad \text{and} \qquad 10(2.3x + 7) = 10 \cdot 5.4$$

$$4 \cdot \frac{1}{2}x + 4 \cdot 5 = 4 \cdot \frac{3}{4} \quad \text{and} \quad 10 \cdot 2.3x + 10 \cdot 7 = 10 \cdot 5.4$$

$$2x + 20 = 3 \qquad \text{and} \qquad 23x + 70 = 54.$$

The first equation has been "cleared of fractions" and the second equation has been "cleared of decimals." Both resulting equations are equivalent to the original equations and are easier to solve. *It is your choice* whether to clear fractions or decimals, but doing so often eases computations.

The easiest way to clear an equation of fractions is to multiply *every term on both sides* by the **least common multiple of all the denominators**.

EXAMPLE 8   Solve: $\frac{2}{3}x - \frac{1}{6} + \frac{1}{2}x = \frac{7}{6} + 2x.$

The denominators are 3, 6, and 2. The number 6 is the least common multiple of all the denominators. We multiply by 6 on both sides of the equation.

$$6\left(\frac{2}{3}x - \frac{1}{6} + \frac{1}{2}x\right) = 6\left(\frac{7}{6} + 2x\right) \qquad \text{Multiplying by 6 on both sides}$$

$$6 \cdot \frac{2}{3}x - 6 \cdot \frac{1}{6} + 6 \cdot \frac{1}{2}x = 6 \cdot \frac{7}{6} + 6 \cdot 2x \qquad \text{Using the distributive law (\textit{Caution!} Be sure to multiply \textit{all} the terms by 6.)}$$

$$4x - 1 + 3x = 7 + 12x \qquad \text{Simplifying. Note that the fractions are cleared.}$$

$$7x - 1 = 7 + 12x \qquad \text{Collecting like terms}$$

$$7x - 1 - 12x = 7 + 12x - 12x \qquad \text{Subtracting } 12x$$

$$-5x - 1 = 7 \qquad \text{Collecting like terms}$$

$$-5x - 1 + 1 = 7 + 1 \qquad \text{Adding 1}$$

$$-5x = 8 \qquad \text{Collecting like terms}$$

$$\frac{-5x}{-5} = \frac{8}{-5} \qquad \text{Dividing by } -5$$

$$x = -\frac{8}{5}$$

Check:

$$\frac{2}{3}x - \frac{1}{6} + \frac{1}{2}x = \frac{7}{6} + 2x$$

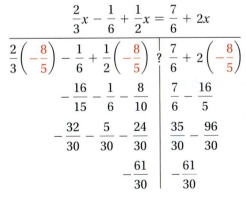

················ **Caution!** ············

Check the possible solution in the *original* equation rather than in the equation that has been cleared of fractions.

The solution is $-\frac{8}{5}$.

13. Solve: $\frac{7}{8}x - \frac{1}{4} + \frac{1}{2}x = \frac{3}{4} + x.$

$$\frac{7}{8}x - \frac{1}{4} + \frac{1}{2}x = \frac{3}{4} + x$$

$$8 \cdot \left(\frac{7}{8}x - \frac{1}{4} + \frac{1}{2}x\right) = \boxed{} \cdot \left(\frac{3}{4} + x\right)$$

$$8 \cdot \frac{7}{8}x - \boxed{} \cdot \frac{1}{4} + 8 \cdot \frac{1}{2}x$$

$$= 8 \cdot \frac{3}{4} + \boxed{} \cdot x$$

$$\boxed{}\,x - \boxed{} + 4x = 6 + 8x$$

$$\boxed{}\,x - 2 = 6 + 8x$$

$$11x - 2 - 8x = 6 + 8x - \boxed{}$$

$$3x - 2 = \boxed{}$$

$$3x - 2 + \boxed{} = 6 + 2$$

$$3x = \boxed{}$$

$$\frac{3x}{3} = \frac{8}{\boxed{}}$$

$$x = \frac{8}{3}$$

14. Solve: $41.68 = 4.7 - 8.6y.$

Solve.

15. $2(2y + 3) = 14$

16. $5(3x - 2) = 35$

◄ **Do Exercise 13.**

To illustrate clearing decimals, we repeat Example 4, but this time we clear the equation of decimals first. Compare the methods.

To clear an equation of decimals, we count the greatest number of decimal places in any one number. If the greatest number of decimal places is 1, we multiply every term on both sides by 10; if it is 2, we multiply by 100; and so on.

EXAMPLE 9 Solve: $16.3 - 7.2y = -8.18.$

The greatest number of decimal places in any one number is *two*. Multiplying by 100, which has *two* 0's, will clear all decimals.

$$100(16.3 - 7.2y) = 100(-8.18) \qquad \text{Multiplying by 100 on both sides}$$

$$100(16.3) - 100(7.2y) = 100(-8.18) \qquad \text{Using the distributive law}$$

$$1630 - 720y = -818 \qquad \text{Simplifying}$$

$$1630 - 720y - 1630 = -818 - 1630 \qquad \text{Subtracting 1630}$$

$$-720y = -2448 \qquad \text{Collecting like terms}$$

$$\frac{-720y}{-720} = \frac{-2448}{-720} \qquad \text{Dividing by } -720$$

$$y = \frac{17}{5}, \text{ or } 3.4$$

The number $\frac{17}{5}$, or 3.4, checks, as shown in Example 4, so it is the solution.

◄ **Do Exercise 14.**

c EQUATIONS CONTAINING PARENTHESES

To solve certain kinds of equations that contain parentheses, we first use the distributive laws to remove the parentheses. Then we proceed as before.

EXAMPLE 10 Solve: $8x = 2(12 - 2x).$

$$8x = 2(12 - 2x)$$

$$8x = 24 - 4x \qquad \text{Using the distributive laws to multiply and remove parentheses}$$

$$8x + 4x = 24 - 4x + 4x \qquad \text{Adding } 4x \text{ to get all the } x\text{-terms on one side}$$

$$12x = 24 \qquad \text{Collecting like terms}$$

$$\frac{12x}{12} = \frac{24}{12} \qquad \text{Dividing by 12}$$

$$x = 2$$

The number 2 checks, so the solution is 2.

◄ **Do Exercises 15 and 16.**

Answers

13. $\frac{8}{3}$ **14.** $-\frac{43}{10}$, or -4.3 **15.** 2   **16.** 3

Guided Solution:
13. 8, 8, 8, 7, 2, 11, 8x, 6, 2, 8, 3

Here is a procedure for solving the types of equation discussed in this section.

AN EQUATION-SOLVING PROCEDURE

....

1. Multiply on both sides to clear the equation of fractions or decimals. (This is optional, but it can ease computations.)
2. If parentheses occur, multiply to remove them using the *distributive laws*.
3. Collect like terms on each side, if necessary.
4. Get all terms with variables on one side and all numbers (constant terms) on the other side, using the *addition principle*.
5. Collect like terms again, if necessary.
6. Multiply or divide to solve for the variable, using the *multiplication principle*.
7. Check all possible solutions in the original equation.

EXAMPLE 11 Solve: $2 - 5(x + 5) = 3(x - 2) - 1$.

$$2 - 5(x + 5) = 3(x - 2) - 1$$

$2 - 5x - 25 = 3x - 6 - 1$ Using the distributive laws to multiply and remove parentheses

$-5x - 23 = 3x - 7$ Collecting like terms

$-5x - 23 + 5x = 3x - 7 + 5x$ Adding $5x$

$-23 = 8x - 7$ Collecting like terms

$-23 + 7 = 8x - 7 + 7$ Adding 7

$-16 = 8x$ Collecting like terms

$\dfrac{-16}{8} = \dfrac{8x}{8}$ Dividing by 8

$-2 = x$

Check:

$$\begin{array}{c|c} 2 - 5(x + 5) = 3(x - 2) - 1 \\ \hline 2 - 5(-2 + 5) \;?\; 3(-2 - 2) - 1 \\ 2 - 5(3) \;\big|\; 3(-4) - 1 \\ 2 - 15 \;\big|\; -12 - 1 \\ -13 \;\big|\; -13 \end{array}$$ TRUE

The solution is -2.

Do Exercises 17 and 18. ▶

Equations with Infinitely Many Solutions

The types of equations we have considered thus far in Sections 2.1–2.3 have all had exactly one solution. We now look at two other possibilities.

  Consider

$$3 + x = x + 3.$$

Let's explore the equation and possible solutions in Margin Exercises 19–22.

Do Exercises 19–22. ▶

Solve.

17. $3(7 + 2x) = 30 + 7(x - 1)$

18. $4(3 + 5x) - 4 = 3 + 2(x - 2)$

Determine whether the given number is a solution of the given equation.

19. $10;\ 3 + x = x + 3$

20. $-7;\ 3 + x = x + 3$

21. $\dfrac{1}{2};\ 3 + x = x + 3$

22. $0;\ 3 + x = x + 3$

Answers

17. -2 **18.** $-\dfrac{1}{2}$ **19.** Yes **20.** Yes
21. Yes **22.** Yes

We know by the commutative law of addition that the equation $3 + x = x + 3$ holds for any replacement of x with a real number. (See Section 1.7.) We have confirmed some of these solutions in Margin Exercises 19–22. Suppose we try to solve this equation using the addition principle:

$$3 + x = x + 3$$
$$-x + 3 + x = -x + x + 3 \qquad \text{Adding } -x$$
$$3 = 3. \qquad \text{True}$$

We end with a true equation. The original equation holds for all real-number replacements. Every real number is a solution. Thus the number of solutions is **infinite**.

EXAMPLE 12 Solve: $7x - 17 = 4 + 7(x - 3)$.

$$7x - 17 = 4 + 7(x - 3)$$
$$7x - 17 = 4 + 7x - 21 \qquad \text{Using the distributive law to multiply and remove parentheses}$$
$$7x - 17 = 7x - 17 \qquad \text{Collecting like terms}$$
$$-7x + 7x - 17 = -7x + 7x - 17 \qquad \text{Adding } -7x$$
$$-17 = -17 \qquad \text{True for all real numbers}$$

Every real number is a solution. There are infinitely many solutions. ●

Equations with No Solution

Now consider

$$3 + x = x + 8.$$

Let's explore the equation and possible solutions in Margin Exercises 23–26.

◀ **Do Exercises 23–26.**

None of the replacements in Margin Exercises 23–26 is a solution of the given equation. In fact, there are no solutions. Let's try to solve this equation using the addition principle:

$$3 + x = x + 8$$
$$-x + 3 + x = -x + x + 8 \qquad \text{Adding } -x$$
$$3 = 8. \qquad \text{False}$$

We end with a false equation. The original equation is false for all real-number replacements. Thus it has **no** solution.

EXAMPLE 13 Solve: $3x + 4(x + 2) = 11 + 7x$.

$$3x + 4(x + 2) = 11 + 7x$$
$$3x + 4x + 8 = 11 + 7x \qquad \text{Using the distributive law to multiply and remove parentheses}$$
$$7x + 8 = 11 + 7x \qquad \text{Collecting like terms}$$
$$7x + 8 - 7x = 11 + 7x - 7x \qquad \text{Subtracting } 7x$$
$$8 = 11 \qquad \text{False}$$

There are no solutions.

◀ **Do Exercises 27 and 28.**

Determine whether the given number is a solution of the given equation.

23. $10;\ 3 + x = x + 8$

24. $-7;\ 3 + x = x + 8$

25. $\dfrac{1}{2};\ 3 + x = x + 8$

26. $0;\ 3 + x = x + 8$

Solve.

27. $30 + 5(x + 3) = -3 + 5x + 48$

28. $2x + 7(x - 4) = 13 + 9x$

When solving an equation, if the result is:

- an equation of the form $x = a$, where a is a real number, then there is one solution, the number a;
- a true equation like $3 = 3$ or $-1 = -1$, then every real number is a solution;
- a false equation like $3 = 8$ or $-4 = 5$, then there is no solution.

Answers

23. No **24.** No **25.** No **26.** No
27. All real numbers **28.** No solution

For Extra Help

MyMathLab® MathXL®
PRACTICE   WATCH   READ   REVIEW

✓ Reading Check

Choose from the column on the right the operation that will clear each equation of fractions or decimals.

RC1. $\frac{2}{5}x - 5 + \frac{1}{2}x = \frac{3}{10} + x$

RC2. $0.003y - 0.1 = 0.03 + y$

RC3. $\frac{1}{4} - 8t + \frac{5}{6} = t - \frac{1}{12}$

RC4. $0.5 + 2.15y = 1.5y - 10$

RC5. $\frac{3}{5} - x = \frac{2}{7}x + 4$

a) Multiply by 1000 on both sides.
b) Multiply by 35 on both sides.
c) Multiply by 12 on both sides.
d) Multiply by 10 on both sides.
e) Multiply by 100 on both sides.

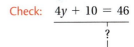

a Solve. Don't forget to check!

1. $5x + 6 = 31$
Check: $5x + 6 = 31$
?

2. $7x + 6 = 13$
Check: $7x + 6 = 13$
?

3. $8x + 4 = 68$
Check: $8x + 4 = 68$
?

4. $4y + 10 = 46$
Check: $4y + 10 = 46$
?

5. $4x - 6 = 34$

6. $5y - 2 = 53$

7. $3x - 9 = 33$

8. $4x - 19 = 5$

9. $7x + 2 = -54$

10. $5x + 4 = -41$

11. $-45 = 3 + 6y$

12. $-91 = 9t + 8$

13. $-4x + 7 = 35$

14. $-5x - 7 = 108$

15. $\frac{5}{4}x - 18 = -3$

16. $\frac{3}{2}x - 24 = -36$

b Solve.

17. $5x + 7x = 72$
Check: $5x + 7x = 72$
?

18. $8x + 3x = 55$
Check: $8x + 3x = 55$
?

19. $8x + 7x = 60$
Check: $8x + 7x = 60$
?

20. $8x + 5x = 104$
Check: $8x + 5x = 104$
?

21. $4x + 3x = 42$ **22.** $7x + 18x = 125$ **23.** $-6y - 3y = 27$ **24.** $-5y - 7y = 144$

25. $-7y - 8y = -15$ **26.** $-10y - 3y = -39$ **27.** $x + \dfrac{1}{3}x = 8$ **28.** $x + \dfrac{1}{4}x = 10$

29. $10.2y - 7.3y = -58$ **30.** $6.8y - 2.4y = -88$ **31.** $8y - 35 = 3y$ **32.** $4x - 6 = 6x$

33. $8x - 1 = 23 - 4x$ **34.** $5y - 2 = 28 - y$ **35.** $2x - 1 = 4 + x$ **36.** $4 - 3x = 6 - 7x$

37. $6x + 3 = 2x + 11$ **38.** $14 - 6a = -2a + 3$ **39.** $5 - 2x = 3x - 7x + 25$

40. $-7z + 2z - 3z - 7 = 17$ **41.** $4 + 3x - 6 = 3x + 2 - x$ **42.** $5 + 4x - 7 = 4x - 2 - x$

43. $4y - 4 + y + 24 = 6y + 20 - 4y$ **44.** $5y - 7 + y = 7y + 21 - 5y$

Solve. Clear fractions or decimals first.

45. $\dfrac{7}{2}x + \dfrac{1}{2}x = 3x + \dfrac{3}{2} + \dfrac{5}{2}x$ **46.** $\dfrac{7}{8}x - \dfrac{1}{4} + \dfrac{3}{4}x = \dfrac{1}{16} + x$

47. $\dfrac{2}{3} + \dfrac{1}{4}t = \dfrac{1}{3}$ **48.** $-\dfrac{3}{2} + x = -\dfrac{5}{6} - \dfrac{4}{3}$

49. $\dfrac{2}{3} + 3y = 5y - \dfrac{2}{15}$ **50.** $\dfrac{1}{2} + 4m = 3m - \dfrac{5}{2}$

51. $\dfrac{5}{3} + \dfrac{2}{3}x = \dfrac{25}{12} + \dfrac{5}{4}x + \dfrac{3}{4}$ **52.** $1 - \dfrac{2}{3}y = \dfrac{9}{5} - \dfrac{y}{5} + \dfrac{3}{5}$

53. $2.1x + 45.2 = 3.2 - 8.4x$

54. $0.96y - 0.79 = 0.21y + 0.46$

55. $1.03 - 0.62x = 0.71 - 0.22x$

56. $1.7t + 8 - 1.62t = 0.4t - 0.32 + 8$

57. $\dfrac{2}{7}x - \dfrac{1}{2}x = \dfrac{3}{4}x + 1$

58. $\dfrac{5}{16}y + \dfrac{3}{8}y = 2 + \dfrac{1}{4}y$

C Solve.

59. $3(2y - 3) = 27$

60. $8(3x + 2) = 30$

61. $40 = 5(3x + 2)$

62. $9 = 3(5x - 2)$

63. $-23 + y = y + 25$

64. $17 - t = -t + 68$

65. $-23 + x = x - 23$

66. $y - \dfrac{2}{3} = -\dfrac{2}{3} + y$

67. $2(3 + 4m) - 9 = 45$

68. $5x + 5(4x - 1) = 20$

69. $5r - (2r + 8) = 16$

70. $6b - (3b + 8) = 16$

71. $6 - 2(3x - 1) = 2$

72. $10 - 3(2x - 1) = 1$

73. $5(d + 4) = 7(d - 2)$

74. $3(t - 2) = 9(t + 2)$

75. $8(2t + 1) = 4(7t + 7)$

76. $7(5x - 2) = 6(6x - 1)$

77. $5x + 5 - 7x = 15 - 12x + 10x - 10$

78. $3 - 7x + 10x - 14 = 9 - 6x + 9x - 20$

79. $22x - 5 - 15x + 3 = 10x - 4 - 3x + 11$

80. $11x - 6 - 4x + 1 = 9x - 8 - 2x + 12$

81. $3(r - 6) + 2 = 4(r + 2) - 21$

82. $5(t + 3) + 9 = 3(t - 2) + 6$

83. $19 - (2x + 3) = 2(x + 3) + x$

84. $13 - (2c + 2) = 2(c + 2) + 3c$

85. $2[4 - 2(3 - x)] - 1 = 4[2(4x - 3) + 7] - 25$

86. $5[3(7 - t) - 4(8 + 2t)] - 20 = -6[2(6 + 3t) - 4]$

87. $11 - 4(x + 1) - 3 = 11 + 2(4 - 2x) - 16$

88. $6(2x - 1) - 12 = 7 + 12(x - 1)$

89. $22x - 1 - 12x = 5(2x - 1) + 4$

90. $2 + 14x - 9 = 7(2x + 1) - 14$

91. $0.7(3x + 6) = 1.1 - (x + 2)$

92. $0.9(2x + 8) = 20 - (x + 5)$

Skill Maintenance

93. Divide: $-22.1 \div 3.4$. [1.6c]

94. Multiply: $-22.1(3.4)$. [1.5a]

95. Factor: $7x - 21 - 14y$. [1.7d]

96. Factor: $8y - 88x + 8$. [1.7d]

Simplify.

97. $-3 + 2(-5)^2(-3) - 7$ [1.8d]

98. $3x + 2[4 - 5(2x - 1)]$ [1.8c]

99. $23(2x - 4) - 15(10 - 3x)$ [1.8b]

100. $256 \div 64 \div 4^2$ [1.8d]

Synthesis

Solve.

101. $\dfrac{2}{3}\left(\dfrac{7}{8} - 4x\right) - \dfrac{5}{8} = \dfrac{3}{8}$

102. $\dfrac{1}{4}(8y + 4) - 17 = -\dfrac{1}{2}(4y - 8)$

103. $\dfrac{4 - 3x}{7} = \dfrac{2 + 5x}{49} - \dfrac{x}{14}$

104. The width of a rectangle is 5 ft, its length is $(3x + 2)$ ft, and its area is 75 ft^2. Find x.

Formulas

a EVALUATING FORMULAS

A **formula** is a "recipe" for doing a certain type of calculation. Formulas are often given as equations. When we replace the variables in an equation with numbers and calculate the result, we are **evaluating** the formula. Evaluating was introduced in Section 1.1.

Let's consider a formula that has to do with weather. Suppose you see a flash of lightning during a storm. Then a few seconds later, you hear thunder. Your distance from the place where the lightning struck is given by the formula $M = \frac{1}{5}t$, where t is the number of seconds from the lightning flash to the sound of the thunder and M is in miles.

EXAMPLE 1 *Distance from Lightning.* Consider the formula $M = \frac{1}{5}t$. Suppose it takes 10 sec for the sound of thunder to reach you after you have seen a flash of lightning. How far away did the lightning strike?

We substitute 10 for t and calculate M:

$$M = \tfrac{1}{5}t = \tfrac{1}{5}(10) = 2.$$

The lightning struck 2 mi away.

Do Margin Exercise 1. ▶

EXAMPLE 2 *Cost of Operating a Microwave Oven.* The cost C of operating a microwave oven for 1 year is given by the formula

$$C = \frac{W \times h \times 365}{1000} \cdot e,$$

where $W =$ the wattage, $h =$ the number of hours used per day, and $e =$ the energy cost per kilowatt-hour. Find the cost of operating a 1500-W microwave oven for 0.25 hr per day if the energy cost is $0.13 per kilowatt-hour.

Substituting, we have

$$C = \frac{W \times h \times 365}{1000} \cdot e = \frac{1500 \times 0.25 \times 365}{1000} \cdot \$0.13 \approx \$17.79.$$

The cost for operating a 1500-W microwave oven for 0.25 hr per day for 1 year is about $17.79.

Do Margin Exercise 2. ▶

OBJECTIVES

a Evaluate a formula.

b Solve a formula for a specified letter.

SKILL TO REVIEW

Objective 2.3a: Solve equations using both the addition principle and the multiplication principle.

Solve.

1. $28 = 7 - 3a$

2. $\frac{1}{2}x - 22 = -20$

1. *Storm Distance.* Refer to Example 1. Suppose that it takes the sound of thunder 14 sec to reach you. How far away is the storm?

2. *Microwave Oven.* Refer to Example 2. Determine the cost of operating an 1100-W microwave oven for 0.5 hr per day for 1 year if the energy cost is $0.16 per kilowatt-hour.

Answers

Skill to Review:
1. -7 **2.** 4

Margin Exercises:
1. 2.8 mi **2.** $32.12

3. Socks from Cotton. Refer to Example 3. Determine the number of socks that can be made from 65 bales of cotton.

EXAMPLE 3 *Socks from Cotton.* Consider the formula $S = 4321x$, where S is the number of socks of average size that can be produced from x bales of cotton. You see a shipment of 300 bales of cotton taken off a ship. How many socks can be made from the cotton?

Source: *Country Woman Magazine*

We substitute 300 for x and calculate S:

$$S = 4321x = 4321(300) = 1{,}296{,}300.$$

Thus, 1,296,300 socks can be made from 300 bales of cotton.

◀ **Do Exercise 3.**

b SOLVING FORMULAS

Refer to Example 3. Suppose a clothing company wants to produce S socks and needs to know how many bales of cotton to order. If this calculation is to be repeated many times, it might be helpful to first solve the formula for x:

$$S = 4321x$$

$$\frac{S}{4321} = x. \qquad \text{Dividing by 4321}$$

Then we can substitute a number for S and calculate x. For example, if the number of socks S to be produced is 432,100, then

$$x = \frac{S}{4321} = \frac{432{,}100}{4321} = 100.$$

The company would need to order 100 bales of cotton.

EXAMPLE 4 Solve for z: $H = \frac{1}{4}z$.

$$H = \frac{1}{4}z \qquad \text{We want this letter alone.}$$
$$4 \cdot H = 4 \cdot \frac{1}{4}z \qquad \text{Multiplying by 4 on both sides}$$
$$4H = z$$

For $H = 2$ in Example 4, $z = 4H = 4(2)$, or 8.

EXAMPLE 5 *Distance, Rate, and Time.* Solve for t: $d = rt$.

$$d = rt \qquad \text{We want this letter alone.}$$
$$\frac{d}{r} = \frac{rt}{r} \qquad \text{Dividing by } r$$
$$\frac{d}{r} = \frac{r}{r} \cdot t$$
$$\frac{d}{r} = t \qquad \text{Simplifying}$$

4. Solve for q: $B = \frac{1}{3}q$.

5. Solve for m: $n = mz$.

6. *Electricity.* Solve for I: $V = IR$. (This formula relates voltage V, current I, and resistance R.)

◀ **Do Exercises 4–6.**

Answers

3. 280,865 socks   **4.** $q = 3B$

5. $m = \dfrac{n}{z}$ **6.** $I = \dfrac{V}{R}$

EXAMPLE 6 Solve for x: $y = x + 3$.

$$y = x + 3 \qquad \text{We want this letter alone.}$$
$$y - 3 = x + 3 - 3 \qquad \text{Subtracting 3}$$
$$y - 3 = x \qquad \text{Simplifying}$$

EXAMPLE 7 Solve for x: $y = x - a$.

$$y = x - a \qquad \text{We want this letter alone.}$$
$$y + a = x - a + a \qquad \text{Adding } a$$
$$y + a = x \qquad \text{Simplifying}$$

Do Exercises 7–9. ▶

Solve for x.

7. $y = x + 5$

8. $y = x - 7$

9. $y = x - b$

EXAMPLE 8 Solve for y: $6y = 3x$.

$$6y = 3x \qquad \text{We want this letter alone.}$$
$$\frac{6y}{6} = \frac{3x}{6} \qquad \text{Dividing by 6}$$
$$y = \frac{x}{2}, \text{ or } \frac{1}{2}x \qquad \text{Simplifying}$$

EXAMPLE 9 Solve for y: $by = ax$.

$$by = ax \qquad \text{We want this letter alone.}$$
$$\frac{by}{b} = \frac{ax}{b} \qquad \text{Dividing by } b$$
$$y = \frac{ax}{b} \qquad \text{Simplifying}$$

10. Solve for y: $9y = 5x$.

11. Solve for p: $ap = bt$.

Do Exercises 10 and 11. ▶

EXAMPLE 10 Solve for x: $ax + b = c$.

$$ax + b = c \qquad \text{We want this letter alone.}$$
$$ax + b - b = c - b \qquad \text{Subtracting } b$$
$$ax = c - b \qquad \text{Simplifying}$$
$$\frac{ax}{a} = \frac{c - b}{a} \qquad \text{Dividing by } a$$
$$x = \frac{c - b}{a} \qquad \text{Simplifying}$$

GS **12.** Solve for x: $y = mx + b$.

$$y = mx + b$$
$$y - \boxed{} = mx + b - b$$
$$y - b = \boxed{}$$
$$\frac{y - b}{m} = \frac{mx}{\boxed{}}$$
$$\frac{y - b}{m} = \boxed{}$$

13. Solve for Q: $tQ - p = a$.

Do Exercises 12 and 13. ▶

Answers

7. $x = y - 5$ **8.** $x = y + 7$

9. $x = y + b$ **10.** $y = \dfrac{5x}{9}$, or $\dfrac{5}{9}x$

11. $p = \dfrac{bt}{a}$ **12.** $x = \dfrac{y - b}{m}$

13. $Q = \dfrac{a + p}{t}$

Guided Solution:
12. b, mx, m, x

A FORMULA-SOLVING PROCEDURE

To solve a formula for a given letter, identify the letter and:

1. Multiply on both sides to clear fractions or decimals, if that is needed.

2. Collect like terms on each side, if necessary.

3. Get all terms with the letter to be solved for on one side of the equation and all other terms on the other side.

4. Collect like terms again, if necessary.

5. Solve for the letter in question.

EXAMPLE 11 *Circumference.* Solve for r: $C = 2\pi r$. This is a formula for the circumference C of a circle of radius r.

$C = 2\pi r$ We want this letter alone.

$\dfrac{C}{2\pi} = \dfrac{2\pi r}{2\pi}$ Dividing by 2π

$\dfrac{C}{2\pi} = r$

EXAMPLE 12 *Averages.* Solve for a: $A = \dfrac{a + b + c}{3}$. This is a formula for the average A of three numbers a, b, and c.

$A = \dfrac{a + b + c}{3}$ We want the letter a alone.

$3 \cdot A = 3 \cdot \dfrac{a + b + c}{3}$ Multiplying by 3 on both sides

$3A = a + b + c$ Simplifying

$3A - b - c = a$ Subtracting b and c

◀ **Do Exercises 14 and 15.**

14. *Circumference.* Solve for D:
$$C = \pi D.$$
This is a formula for the circumference C of a circle of diameter D.

15. *Averages.* Solve for c:
$$A = \dfrac{a + b + c + d}{4}.$$

Answers

14. $D = \dfrac{C}{\pi}$ **15.** $c = 4A - a - b - d$

☑ Reading Check

Solve each equation for the indicated letter and choose the correct solution from the column on the right.

RC1. Solve $4x = 7y$ for x.

RC2. Solve $y = \dfrac{1}{4}x - w$ for w.

RC3. Solve $xz + 4y = w$ for x.

RC4. Solve $z = w + 4$ for w.

a) $w = z - 4$

b) $x = \dfrac{7y}{4}$, or $\dfrac{7}{4}y$

c) $w = \dfrac{1}{4}x - y$

d) $x = \dfrac{w - 4y}{z}$

 a   Solve.

1. *Wavelength of a Musical Note.* The wavelength w, in meters per cycle, of a musical note is given by

$$w = \frac{r}{f},$$

where r is the speed of the sound, in meters per second, and f is the frequency, in cycles per second. The speed of sound in air is 344 m/sec. What is the wavelength of a note whose frequency in air is 24 cycles per second?

2. *Furnace Output.* Contractors in the Northeast use the formula $B = 30a$ to determine the minimum furnace output B, in British thermal units (Btu's), for a well-insulated house with a square feet of flooring. Determine the minimum furnace output for an 1800-ft^2 house that is well insulated.

Source: U.S. Department of Energy

3. *Calorie Density.* The calorie density D, in calories per ounce, of a food that contains c calories and weighs w ounces is given by

$$D = \frac{c}{w}.$$

Eight ounces of fat-free milk contains 84 calories. Find the calorie density of fat-free milk.

Source: *Nutrition Action Healthletter*, March 2000, p. 9.

4. *Size of a League Schedule.* When all n teams in a league play every other team twice, a total of N games are played, where

$$N = n^2 - n.$$

A soccer league has 7 teams and all teams play each other twice. How many games are played?

5. *Distance, Rate, and Time.* The distance d that a car will travel at a rate, or speed, r in time t is given by

$$d = rt.$$

a) A car travels at 75 miles per hour (mph) for 4.5 hr. How far will it travel?

b) Solve the formula for t.

6. *Surface Area of a Cube.* The surface area A of a cube with side s is given by

$$A = 6s^2.$$

a) Find the surface area of a cube with sides of 3 in.

b) Solve the formula for s^2.

7. College Enrollment. At many colleges, the number of "full-time-equivalent" students f is given by

$$f = \frac{n}{15},$$

where n is the total number of credits for which students have enrolled in a given semester.

a) Determine the number of full-time-equivalent students on a campus in which students registered for a total of 21,345 credits.

b) Solve the formula for n.

8. Electrical Power. The power rating P, in watts, of an electrical appliance is determined by

$$P = I \cdot V,$$

where I is the current, in amperes, and V is measured in volts.

a) A microwave oven requires 12 amps of current and the voltage in the house is 115 volts. What is the wattage of the microwave?

b) Solve the formula for I; for V.

---

b Solve for the indicated letter.

9. $y = 5x$, for x

10. $d = 55t$, for t

11. $a = bc$, for c

12. $y = mx$, for x

13. $n = m + 11$, for m

14. $z = t + 21$, for t

15. $y = x - \dfrac{3}{5}$, for x

16. $y = x - \dfrac{2}{3}$, for x

17. $y = 13 + x$, for x

18. $t = 6 + s$, for s

19. $y = x + b$, for x

20. $y = x + A$, for x

21. $y = 5 - x$, for x

22. $y = 10 - x$, for x

23. $y = a - x$, for x

24. $y = q - x$, for x

25. $8y = 5x$, for y

26. $10y = -5x$, for y

27. $By = Ax$, for x

28. $By = Ax$, for y

29. $W = mt + b$, for t

30. $W = mt - b$, for t

31. $y = bx + c$, for x

32. $y = bx - c$, for x

33. *Area of a Parallelogram:*
$A = bh$, for h
(Area A, base b, height h)

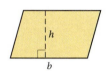

34. *Distance, Rate, Time:*
$d = rt$, for r
(Distance d, speed r, time t)

Speed, r Time, t

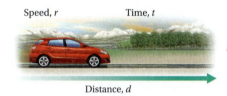

Distance, d

35. *Perimeter of a Rectangle:*
$P = 2l + 2w$, for w
(Perimeter P, length l, width w)

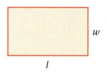

36. *Area of a Circle:*
$A = \pi r^2$, for r^2
(Area A, radius r)

37. *Average of Two Numbers:*
$A = \dfrac{a + b}{2}$, for a

a $A = \dfrac{a + b}{2}$ b

38. *Area of a Triangle:*
$A = \dfrac{1}{2}bh$, for b

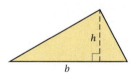

39. $A = \dfrac{a + b + c}{3}$, for b

40. $A = \dfrac{a + b + c}{3}$, for c

41. $A = at + b$, for t

42. $S = rx + s$, for x

43. $Ax + By = c$, for x

44. $Q = \dfrac{p - q}{2}$, for p

45. *Force:*

$$F = ma, \text{ for } a$$

(Force F, mass m, acceleration a)

46. *Simple Interest:*

$$I = Prt, \text{ for } P$$

(Interest I, principal P, interest rate r, time t)

47. *Relativity:*

$$E = mc^2, \text{ for } c^2$$

(Energy E, mass m, speed of light c)

48. $Ax + By = c$, for y

49. $v = \dfrac{3k}{t}$, for t

50. $P = \dfrac{ab}{c}$, for c

Skill Maintenance

51. Evaluate $\dfrac{3x - 2y}{y}$ when $x = 6$ and $y = 2$.   [1.1a]

52. Remove parentheses and simplify:
$$4a - 8b - 5(5a - 4b). \quad [1.8b]$$

Subtract.   [1.4a]

53. $-45.8 - (-32.6)$

54. $-\dfrac{2}{3} - \dfrac{5}{6}$

55. $87\dfrac{1}{2} - 123$

Add.   [1.3a]

56. $-\dfrac{5}{12} + \dfrac{1}{4}$

57. $0.082 + (-9.407)$

58. $-2\dfrac{1}{2} + 6\dfrac{1}{4}$

Solve.

59. $2y - 3 + y = 8 - 5y$ [2.3b]

60. $10x + 4 = 3x - 2 + x$ [2.3b]

61. $2(5x + 6) = x - 15$ [2.3c]

62. $5a = 3(6 - 3a)$ [2.3c]

Synthesis

Solve.

63. $H = \dfrac{2}{a - b}$, for b; for a

64. $P = 4m + 7mn$, for m

65. In $A = lw$, if l and w both double, what is the effect on A?

66. In $P = 2a + 2b$, if P doubles, do a and b necessarily both double?

67. In $A = \frac{1}{2}bh$, if b increases by 4 units and h does not change, what happens to A?

68. Solve for F: $D = \dfrac{1}{E + F}$.

Mid-Chapter Review

Concept Reinforcement

Determine whether each statement is true or false.

_____ **1.** $3 - x = 4x$ and $5x = -3$ are equivalent equations.   [2.1b]

_____ **2.** For any real numbers a, b, and c, $a = b$ is equivalent to $a + c = b + c$.   [2.1b]

_____ **3.** We can use the multiplication principle to divide on both sides of an equation by the same nonzero number.   [2.2a]

_____ **4.** Every equation has at least one solution.   [2.3c]

Guided Solutions

 Fill in each blank with the number, variable, or expression that creates a correct statement or solution.

Solve.   [2.1b], [2.2a]

5.
$$x + 5 = -3$$
$$x + 5 - 5 = -3 - \boxed{}$$
$$x + \boxed{} = -8$$
$$x = \boxed{}$$

6.
$$-6x = 42$$
$$\frac{-6x}{-6} = \frac{42}{\boxed{}}$$
$$\boxed{} \cdot x = -7$$
$$x = \boxed{}$$

7. Solve for y: $5y + z = t$.   [2.4b]
$$5y + z = t$$
$$5y + z - z = t - \boxed{}$$
$$5y = \boxed{}$$
$$\frac{5y}{5} = \frac{t - z}{\boxed{}}$$
$$y = \frac{\boxed{}}{5}$$

Mixed Review

Solve.   [2.1b], [2.2a], [2.3a, b, c]

8. $x + 5 = 11$

9. $x + 9 = -3$

10. $8 = t + 1$

11. $-7 = y + 3$

12. $x - 6 = 14$

13. $y - 7 = -2$

14. $-\dfrac{3}{2} + z = -\dfrac{3}{4}$

15. $-3.3 = -1.9 + t$

16. $7x = 42$

17. $17 = -t$

18. $6x = -54$

19. $-5y = -85$

20. $\dfrac{x}{7} = 3$

21. $\dfrac{2}{3}x = 12$

22. $-\dfrac{t}{5} = 3$

23. $\dfrac{3}{4}x = -\dfrac{9}{8}$

24. $3x + 2 = 5$

25. $5x + 4 = -11$

26. $6x - 7 = 2$

27. $-4x - 9 = -5$

28. $6x + 5x = 33$ **29.** $-3y - 4y = 49$ **30.** $3x - 4 = 12 - x$ **31.** $5 - 6x = 9 - 8x$

32. $4y - \dfrac{3}{2} = \dfrac{3}{4} + 2y$ **33.** $\dfrac{4}{5} + \dfrac{1}{6}t = \dfrac{1}{10}$ **34.** $0.21n - 1.05 = 2.1 - 0.14n$

35. $5(3y - 1) = -35$ **36.** $7 - 2(5x + 3) = 1$ **37.** $-8 + t = t - 8$

38. $z + 12 = -12 + z$ **39.** $4(3x + 2) = 5(2x - 1)$ **40.** $8x - 6 - 2x = 3(2x - 4) + 6$

Solve for the indicated letter.   [2.4b]

41. $A = 4b$, for b **42.** $y = x - 1.5$, for x **43.** $n = s - m$, for m

44. $4t = 9w$, for t **45.** $B = at - c$, for t **46.** $M = \dfrac{x + y + z}{2}$, for y

Understanding Through Discussion and Writing

47. Explain the difference between equivalent expressions and equivalent equations.   [1.7a], [2.1b]

48. Are the equations $x = 5$ and $x^2 = 25$ equivalent? Why or why not?   [2.1b]

49. When solving an equation using the addition principle, how do you determine which number to add or subtract on both sides of the equation?   [2.1b]

50. Explain the following mistake made by a fellow student.   [2.1b]

$$x + \frac{1}{3} = -\frac{5}{3}$$
$$x = -\frac{4}{3}$$

51. When solving an equation using the multiplication principle, how do you determine by what number to multiply or divide on both sides of the equation?   [2.2a]

52. Devise an application in which it would be useful to solve the equation $d = rt$ for r.   [2.4b]

Applications of Percent

2.5

a   TRANSLATING AND SOLVING

Many applied problems involve percent. Here we begin to see how equation solving can enhance our problem-solving skills.

In solving percent problems, we first *translate* the problem to an equation. Then we *solve* the equation using the techniques discussed in Sections 2.1–2.3. The key words in the translation are as follows.

> **KEY WORDS IN PERCENT TRANSLATIONS**
>
> "**Of**" translates to "·" or "×".
>
> "**Is**" translates to "=".
>
> "**What number**" or "**what percent**" translates to any letter.
>
> "**%**" translates to "× $\frac{1}{100}$" or "× 0.01".

EXAMPLE 1   Translate:

28% of 5 is what number?

28% · 5 = a This is a percent equation.

EXAMPLE 2   Translate:

45% of what number is 28?

45% × b = 28

EXAMPLE 3   Translate:

What percent of 90 is 7?

n · 90 = 7

Do Margin Exercises 1–6. ▶

Percent problems are actually of three different types. Although the method we present does *not* require that you be able to identify which type we are studying, it is helpful to know them. Let's begin by using a specific example to find a standard form for a percent problem.

OBJECTIVE

a Solve applied problems involving percent.

SKILL TO REVIEW

Objective 2.2a: Solve equations using the multiplication principle.

Solve.

1. $20 = 0.05a$

2. $0.3z = 327$

Translate to an equation. Do not solve.

1. 13% of 80 is what number?

2. What number is 60% of 70?

3. 43 is 20% of what number?

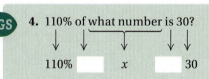

GS 4. 110% of what number is 30?

110% ☐ x ☐ 30

5. 16 is what percent of 80?

6. What percent of 94 is 10.5?

Answers

Skill to Review:
1. 400   2. 1090

Answers to Margin Exercises 1–6 and Guided Solution are on p. 160.

We know that

$$15 \text{ is } 25\% \text{ of } 60, \quad \text{or} \quad 15 = 25\% \times 60.$$

We can think of this as:

Amount = Percent number × Base.

Each of the three types of percent problem depends on which of the three pieces of information is missing in the statement

Amount = Percent number × Base.

1. Finding the ***amount*** (the result of taking the percent)

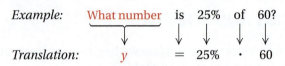

Example: What number is 25% of 60?

Translation: $y = 25\% \cdot 60$

2. Finding the ***base*** (the number you are taking the percent of)

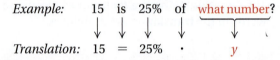

Example: 15 is 25% of what number?

Translation: $15 = 25\% \cdot y$

3. Finding the ***percent number*** (the percent itself)

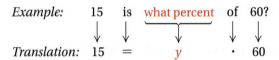

Example: 15 is what percent of 60?

Translation: $15 = y \cdot 60$

Finding the Amount

EXAMPLE 4 What number is 11% of 49?

What number is 11% of 49?

Translate: $a = 11\% \times 49$

Solve: The letter is by itself. To solve the equation, we need only convert 11% to decimal notation and multiply:

$$a = 11\% \times 49 = 0.11 \times 49 = 5.39.$$

Thus, 5.39 is 11% of 49. The answer is 5.39.

7. What number is 2.4% of 80?

◀ **Do Exercise 7.**

Finding the Base

EXAMPLE 5 3 is 16% of what number?

3 is 16% of what number?

Translate: $3 = 16\% \times b$

$3 = 0.16 \times b$ Converting 16% to decimal notation

Solve: In this case, the letter is not by itself. To solve the equation, we divide by 0.16 on both sides:

$$3 = 0.16 \times b$$

$$\frac{3}{0.16} = \frac{0.16 \times b}{0.16} \qquad \text{Dividing by 0.16}$$

$$18.75 = b. \qquad \text{Simplifying}$$

The answer is 18.75.

Do Exercise 8. ▶

Finding the Percent Number

In solving these problems, you *must* remember to convert to percent notation after you have solved the equation.

EXAMPLE 6 $32 is what percent of $50?

$$\underset{32}{\$32} \quad \underset{=}{\text{is}} \quad \underset{p}{\text{what percent}} \quad \underset{\times}{\text{of}} \quad \underset{50}{\$50?}$$

Translate: $\quad 32 \quad = \quad p \quad \times \quad 50$

Solve: To solve the equation, we divide by 50 on both sides and convert the answer to percent notation:

$$32 = p \times 50$$

$$\frac{32}{50} = \frac{p \times 50}{50} \qquad \text{Dividing by 50}$$

$$0.64 = p$$

$$64\% = p. \qquad \text{Converting to percent notation}$$

Thus, $32 is 64% of $50. The answer is 64%.

Do Exercise 9. ▶

EXAMPLE 7   *Donated Girl Scout Cookies.*   Through Operation Cookie Drop, Girl Scout cookies can be donated to all branches of the U.S. military. In 2012, the Girl Scouts of Western Washington sold 3,093,834 boxes of cookies. Of this number, 4.11% were donated to military personnel. How many boxes of cookies did the Girl Scouts of Western Washington donate to Operation Cookie Drop?

Source: Girl Scouts of the USA

To solve this problem, we first reword and then translate. We let $c =$ the number of boxes of cookies donated.

Rewording: $\quad \underset{c}{\underbrace{\text{What number}}} \quad \underset{=}{\text{is}} \quad \underset{4.11\%}{4.11\%} \quad \underset{\times}{\text{of}} \quad \underset{3{,}093{,}834}{3{,}093{,}834?}$

Translating: $\quad\quad c \quad\quad = \quad 4.11\% \quad \times \quad 3{,}093{,}834$

Solve: The letter is by itself. To solve the equation, we need only convert 4.11% to decimal notation and multiply:

$$c = 4.11\% \times 3{,}093{,}834 = 0.0411 \times 3{,}093{,}834 \approx 127{,}157.$$

Thus, 127,157 is about 4.11% of 3,093,834, so 127,157 boxes of Girl Scout cookies were donated to Operation Cookie Drop in 2012.

Do Exercise 10. ▶

GS **8.** 25.3 is 22% of what number?

$$25.3 \quad \boxed{} \quad 22\% \quad \boxed{} \quad x$$

$$25.3 = \boxed{} \cdot x$$

$$\frac{25.3}{\boxed{}} = \frac{0.22x}{0.22}$$

$$\boxed{} = x$$

9. What percent of $50 is $18?

10. *Haitian Population Ages 0–14.* The population of Haiti is approximately 9,720,000. Of this number, 35.9% are ages 0–14. How many Haitians are ages 14 and younger? Round to the nearest 1000.

Source: Central Intelligence Agency

MCT KidNews 05/05

Answers

8. 115   **9.** 36%
10. About 3,489,000

Guided Solution:
8. = , · , 0.22, 0.22, 115

11. Areas of Texas and Alaska.
The area of the second largest state, Texas, is 268,581 mi^2. This is about 40.5% of the area of the largest state, Alaska. What is the area of Alaska?

12. Median Income. The U.S. median family income in 2004 was $49,800. This number decreased to $45,800 in 2010. What is the percent decrease?

Source: Federal Reserve's Survey of Consumer Finances, June 2012

Answers

11. About 663,163 mi^2
12. About 8.0% decrease

EXAMPLE 8 *Motor Vehicle Production.* In 2010, 7.632 million motor vehicles were produced in the United States. This was 10.4% of the world production of motor vehicles. How many motor vehicles were produced worldwide in 2010?

Sources: R. L. Polk, Automotive News Data Center

To solve this problem, we first reword and then translate. We let $P =$ the total worldwide production, in millions, of motor vehicles in 2010.

Rewording: 7.632   is   10.4%   of   what number?

Translating: 7.632   =   10.4%   ×   P

Solve: To solve the equation, we convert 10.4% to decimal notation and divide by 0.104 on both sides:

$$7.632 = 10.4\% \times P$$
$$7.632 = 0.104 \times P \qquad \text{Converting to decimal notation}$$
$$\frac{7.632}{0.104} = \frac{0.104 \times P}{0.104} \qquad \text{Dividing by 0.104}$$
$$73.4 \approx P. \qquad \text{Simplifying and rounding to the nearest tenth}$$

About 73.4 million motor vehicles were produced worldwide in 2010.

◀ **Do Exercise 11.**

EXAMPLE 9 *Employment Outlook.* Jobs at United States auto plants and parts factories (including domestic and foreign-owned) totaled approximately 650,000 in 2012. This number is expected to grow to 756,800 in 2015. What is the percent increase?

Sources: Center for Automotive Research & IHS Global Insight

To solve the problem, we must first determine the amount of the increase:

Jobs in 2015   minus   Jobs in 2012   =   Increase

756,800   −   650,000   =   106,800.

Using the job increase of 106,800, we reword and then translate. We let $p =$ the percent increase. We want to know, "what percent of the number of jobs in 2012 is 106,800?"

Rewording: 106,800   is   what percent   of   650,000

Translating: 106,800   = p ×   650,000

Solve: To solve the equation, we divide by 650,000 on both sides and convert the answer to percent notation:

$$106,800 = p \times 650,000$$
$$\frac{106,800}{650,000} = \frac{p \times 650,000}{650,000} \qquad \text{Dividing by 650,000}$$
$$0.164 \approx p \qquad \text{Simplifying}$$
$$16.4\% \approx p. \qquad \text{Converting to percent notation}$$

The percent increase is about 16.4%.

◀ **Do Exercise 12.**

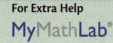
✓ Reading Check

Match each question with the most appropriate translation from the column on the right.

RC1. 13 is 82% of what number?

RC2. What number is 13% of 82?

RC3. 82 is what percent of 13?

RC4. 82 is 13% of what number?

RC5. 13 is what percent of 82?

RC6. What number is 82% of 13?

a) $82 = 13\% \cdot b$

b) $a = 13\% \cdot 82$

c) $a = 82\% \cdot 13$

d) $13 = 82\% \cdot b$

e) $82 = p \cdot 13$

f) $13 = p \cdot 82$

a Solve.

1. What percent of 180 is 36?

2. What percent of 76 is 19?

3. 45 is 30% of what number?

4. 20.4 is 24% of what number?

5. What number is 65% of 840?

6. What number is 50% of 50?

7. 30 is what percent of 125?

8. 57 is what percent of 300?

9. 12% of what number is 0.3?

10. 7 is 175% of what number?

11. 2 is what percent of 40?

12. 16 is what percent of 40?

13. What percent of 68 is 17?

14. What percent of 150 is 39?

15. What number is 35% of 240?

16. What number is 1% of one million?

17. What percent of 575 is 138?

18. What percent of 60 is 75?

19. What percent of 300 is 48?

20. What percent of 70 is 70?

21. 14 is 30% of what number?

22. 54 is 24% of what number?

23. What number is 2% of 40?

24. What number is 40% of 2?

25. 0.8 is 16% of what number?

26. 40 is 2% of what number?

27. 54 is 135% of what number?

28. 8 is 2% of what number?

World Population by Continent.   It has been projected that in 2050, the world population will be 8909 million, or 8.909 billion. The following circle graph shows the breakdown of this total population by continent.

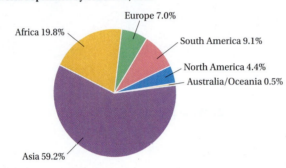

World Population by Continent, 2050

Europe 7.0%
Africa 19.8%
South America 9.1%
North America 4.4%
Australia/Oceania 0.5%
Asia 59.2%

SOURCE: Central Intelligence Agency

Using the data in the figure, complete the following table of projected populations in 2050. Round to the nearest million.

	Continent	Population		Continent	Population
29.	South America		**30.**	Europe	
31.	Asia		**32.**	North America	
33.	Africa		**34.**	Australia/Oceania	

35. *eBook Revenue.*   Net revenue from 2011 adult book sales totaled approximately $2360 million. eBook revenue accounted for 41% of this amount. What was the net revenue from adult eBook sales in 2011? Round to the nearest million.

Source: Association for American Publishers

36. *Hardcover Book Revenue.*   Net revenue from 2011 adult book sales totaled approximately $2360 million. Revenue from hardcover books accounted for 54.8% of this amount. What was the net revenue from adult hardcover sales in 2011? Round to the nearest million.

Source: Association for American Publishers

37. *Student Loans.*   To finance her community college education, Sarah takes out a Stafford loan for $6500. After a year, Sarah decides to pay off the interest, which is 3.4% of $6500. How much will she pay?

38. *Student Loans.*   Paul takes out a PLUS loan for $5400. After a year, Paul decides to pay off the interest, which is 7.9% of $5000. How much will he pay?

39. Tattoos. Of the 237,400,000 adults ages 18 and older in the United States, approximately 49,854,000 have at least one tattoo. What percent of adults ages 18 and older have at least one tattoo?

Sources: U.S. Census Bureau; Harris Poll of 2016 adults; UPI.com

40. Boston Marathon. The 2012 Boston Marathon was the 116th running of the race. Since its first race, the United States has won the men's open division 43 times. What percent of the years did the United States win the men's open?

Source: Boston Athletic Association

41. Tipping. William left a $1.50 tip for a meal that cost $12.
   a) What percent of the cost of the meal was the tip?
   b) What was the total cost of the meal including the tip?

42. Tipping. Sam, Selena, Rachel, and Clement left a 20% tip for a meal that cost $75.
   a) How much was the tip?
   b) What was the total cost of the meal including the tip?

43. Tipping. David left a 15% tip of $4.65 for a meal.
   a) What was the cost of the meal before the tip?
   b) What was the total cost of the meal including the tip?

44. Tipping. Addison left an 18% tip of $6.75 for a meal.
   a) What was the cost of the meal before the tip?
   b) What was the total cost of the meal including the tip?

45. City Park Space. Portland, Oregon, has 12,959 acres of park space. This is 15.1% of the acreage of the entire city. What is the total acreage of Portland?

Source: Indy Parks and Recreation Master Plan

46. Junk Mail. About 46.2 billion pieces of unopened junk mail end up in landfills each year. This is about 44% of all the junk mail that is sent annually. How many pieces of junk mail are sent annually?

Source: Globaljunkmailcrisis.org

47. Employment Growth. In 1980, there were 1.7 million licensed registered nurses in the United States. This number increased to 3.1 million in 2012. What is the percent increase?

Source: Maria Sonnenberg, *Florida Today*, May 14, 2012

48. Artificial-Tree Sales. From 2008 to 2012, sales of artificial Christmas trees grew from $950 million to approximately $1070 million. What is the percent increase?

Sources: BalsamHill.com and USATradeonline.gov

49. *Newspaper Advertiser Spending.* In 2000, advertisers spent $48.6 billion in newspapers. This number dropped to $22.8 billion in 2010. What is the percent decrease?

Source: Paul Glader, "Black and White, Read No More", *The American Legion Magazine*, August 2012

50. *New Magazines.* In 2010, 301 magazines were launched in North America. In 2011, only 273 were launched. What is the percent decrease?

Source: Media Finder

51. *Dog Bites.* Over one-third of all homeowner insurance liability claims paid in 2011 were for dog bites. The average cost of dog-bite claims in the United States increased from $19,162 in 2004 to $29,396 in 2011. What is the percent increase?

Sources: Adam Belz, Insurance Information Institute, *USA TODAY*

52. *International Students.* In 2001–2002, 582,996 international students were enrolled in U.S. colleges. By 2011–2012, this number had increased to 764,495. What is the percent increase?

Source: Institute of International Education, Open Doors 2012

53. *Little League Baseball.* The number of Little League baseball players worldwide declined from 2.6 million in 1997 to only 2.1 million in 2011. What is the percent decrease?

Source: Little League International

54. *Adoptions from Russia.* The number of Russian child adoptions by Americans has declined from 4381 in 1999 to only 962 in 2011. What is the percent decrease?

Source: U.S. State Department

Skill Maintenance

Multiply. [1.7c]

55. $3(4 + q)$

56. $-\dfrac{1}{2}(-10x + 42)$

Simplify. [1.7a]

57. $\dfrac{75yw}{40y}$

58. $-\dfrac{18b}{12b}$

Simplify. [1.8c]

59. $-2[3 - 5(7 - 2)]$

60. $[3(x + 4) - 6] - [8 + 2(x - 5)]$

Synthesis

61. It has been determined that at the age of 15, a boy has reached 96.1% of his final adult height. Jaraan is 6 ft 4 in. at the age of 15. What will his final adult height be?

62. It has been determined that at the age of 10, a girl has reached 84.4% of her final adult height. Dana is 4 ft 8 in. at the age of 10. What will her final adult height be?

Applications and Problem Solving

a FIVE STEPS FOR SOLVING PROBLEMS

We have discussed many new equation-solving tools in this chapter and used them for applications and problem solving. Here we consider a five-step strategy that can be very helpful in solving problems.

FIVE STEPS FOR PROBLEM SOLVING IN ALGEBRA

1. *Familiarize* yourself with the problem situation.
2. *Translate* the problem to an equation.
3. *Solve* the equation.
4. *Check* the answer in the original problem.
5. *State* the answer to the problem clearly.

Of the five steps, the most important is probably the first one: becoming familiar with the problem situation. The box below lists some hints for familiarization.

FAMILIARIZING YOURSELF WITH A PROBLEM

- If a problem is given in words, read it carefully. Reread the problem, perhaps aloud. Try to verbalize the problem as though you were explaining it to someone else.
- Choose a variable (or variables) to represent the unknown and clearly state what the variable represents. Be descriptive! For example, let L = the length, d = the distance, and so on.
- Make a drawing and label it with known information, using specific units if given. Also, indicate unknown information.
- Find further information. Look up formulas or definitions with which you are not familiar. (Geometric formulas appear on the inside back cover of this text.) Consult the Internet or a reference librarian.
- Create a table that lists all the information you have available. Look for patterns that may help in the translation to an equation.
- Think of a possible answer and check the guess. Note the manner in which the guess is checked.

EXAMPLE 1   *Cycling in Vietnam.*   National Highway 1, which runs along the coast of Vietnam, is considered one of the top routes for avid bicyclists. While on sabbatical, a history professor spent six weeks biking 1720 km on National Highway 1 from Hanoi through Ha Tinh to Ho Chi Minh City (commonly known as Saigon). At Ha Tinh, he was four times as far from Ho Chi Minh City as he was from Hanoi. How far had he biked and how far did he still need to bike in order to reach the end?

Sources: www.smh.com; *Lonely Planet's Best in 2010*

1. Familiarize. Let's look at a map.

To become familiar with the problem, let's guess a possible distance that the professor is from Hanoi—say, 400 km. Four times 400 km is 1600 km. Since 400 km + 1600 km = 2000 km and 2000 km is greater than 1720 km, we see that our guess is too large. Rather than guess again, let's use the equation-solving skills that we learned in this chapter. We let

d = the distance, in kilometers, to Hanoi, and

$4d$ = the distance, in kilometers, to Ho Chi Minh City.

(We also could let d = the distance to Ho Chi Minh City and $\frac{1}{4}d$ = the distance to Hanoi.)

2. Translate. From the map, we see that the lengths of the two parts of the trip must add up to 1720 km. This leads to our translation.

$$
\underbrace{\text{Distance to Hanoi}}_{d} \quad \underset{+}{\text{plus}} \quad \underbrace{\text{Distance to Ho Chi Minh}}_{4d} \quad \underset{=}{\text{is}} \quad \underset{1720}{1720 \text{ km.}}
$$

3. Solve. We solve the equation:

$$d + 4d = 1720$$
$$5d = 1720 \qquad \text{Collecting like terms}$$
$$\frac{5d}{5} = \frac{1720}{5} \qquad \text{Dividing by 5}$$
$$d = 344.$$

4. Check. As we expected, d is less than 400 km. If $d = 344$ km, then $4d = 1376$ km. Since 344 km + 1376 km = 1720 km, the answer checks.

5. State. At Ha Tinh, the professor had biked 344 km from Hanoi and had 1376 km to go to reach Ho Chi Minh City.

◀ **Do Exercise 1.**

1. *Running.* Yiannis Kouros of Australia holds the record for the greatest distance run in 24 hr by running 188 mi. After 8 hr, he was approximately twice as far from the finish line as he was from the start line. How far had he run?

Source: Australian Ultra Runners Association

Answer

1. $62\frac{2}{3}$ mi

EXAMPLE 2 *Knitted Scarf.* Lily knitted a scarf with shades of orange and red yarn, starting with an orange section, then a medium-red section, and finally a dark-red section. The medium-red section is one-half the length of the orange section. The dark-red section is one-fourth the length of the orange section. The scarf is 7 ft long. Find the length of each section of the scarf.

1. **Familiarize.** Because the lengths of the medium-red section and the dark-red section are expressed in terms of the length of the orange section, we let

$$x = \text{the length of the orange section.}$$

Then $\frac{1}{2}x =$ the length of the medium-red section

and $\frac{1}{4}x =$ the length of the dark-red section.

We make a drawing and label it.

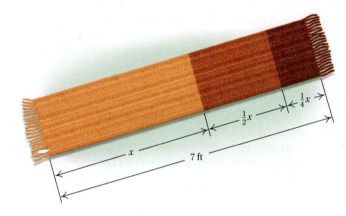

2. **Translate.** From the statement of the problem and the drawing, we know that the lengths add up to 7 ft. This gives us our translation:

Length of orange section	plus	Length of medium-red section	plus	Length of dark-red section	is	Total length
↓	↓	↓	↓	↓	↓	↓
x	$+$	$\frac{1}{2}x$	$+$	$\frac{1}{4}x$	$=$	$7.$

3. **Solve.** First, we clear fractions and then carry out the solution as follows:

$$x + \frac{1}{2}x + \frac{1}{4}x = 7 \qquad \text{\color{red}{The LCM of the denominators is 4.}}$$

$$4\left(x + \frac{1}{2}x + \frac{1}{4}x\right) = 4 \cdot 7 \qquad \text{\color{red}{Multiplying by the LCM, 4}}$$

$$4 \cdot x + 4 \cdot \frac{1}{2}x + 4 \cdot \frac{1}{4}x = 4 \cdot 7 \qquad \text{\color{red}{Using the distributive law}}$$

$$4x + 2x + x = 28 \qquad \text{\color{red}{Simplifying}}$$

$$7x = 28 \qquad \text{\color{red}{Collecting like terms}}$$

$$\frac{7x}{7} = \frac{28}{7} \qquad \text{\color{red}{Dividing by 7}}$$

$$x = 4.$$

2. *Gourmet Sandwiches.* A sandwich shop specializes in sandwiches prepared in buns of length 18 in. Jenny, Emma, and Sarah buy one of these sandwiches and take it back to their apartment. Since they have different appetites, Jenny cuts the sandwich in such a way that Emma gets one-half of what Jenny gets and Sarah gets three-fourths of what Jenny gets. Find the length of each person's sandwich.

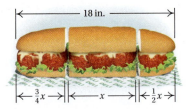

18 in.

$\frac{3}{4}x$ x $\frac{1}{2}x$

4. Check. Do we have an answer to the *original problem*? If the length of the orange section is 4 ft, then the length of the medium-red section is $\frac{1}{2} \cdot$ 4 ft, or 2 ft, and the length of the dark-red section is $\frac{1}{4} \cdot$ 4 ft, or 1 ft. The sum of these lengths is 7 ft, so the answer checks.

5. State. The length of the orange section is 4 ft, the length of the medium-red section is 2 ft, and the length of the dark-red section is 1 ft. (Note that we must include the unit, feet, in the answer.)

◀ **Do Exercise 2.**

Recall that the set of integers = {..., −5, −4, −3, −2, −1, 0, 1, 2, 3, 4, 5,...}. Before we solve the next problem, we need to learn some additional terminology regarding integers.

The following are examples of **consecutive integers:** 16, 17, 18, 19, 20; and −31, −30, −29, −28. Note that consecutive integers can be represented in the form $x, x + 1, x + 2$, and so on.

The following are examples of **consecutive even integers:** 16, 18, 20, 22, 24; and −52, −50, −48, −46. Note that consecutive even integers can be represented in the form $x, x + 2, x + 4$, and so on.

The following are examples of **consecutive odd integers:** 21, 23, 25, 27, 29; and −71, −69, −67, −65. Note that consecutive odd integers can be represented in the form $x, x + 2, x + 4$, and so on.

EXAMPLE 3 *Limited-Edition Prints.* A limited-edition print is usually signed and numbered by the artist. For example, a limited edition with only 50 prints would be numbered 1/50, 2/50, 3/50, and so on. An estate donates two prints numbered consecutively from a limited edition with 150 prints. The sum of the two numbers is 263. Find the numbers of the prints.

1. Familiarize. The numbers of the prints are consecutive integers. If we let x = the smaller number, then $x + 1$ = the larger number. Since there are 150 prints in the edition, the first number must be 149 or less. If we guess that $x = 138$, then $x + 1 = 139$. The sum of the numbers is 277. We see that the numbers need to be smaller. We could continue guessing and solve the problem this way, but let's work on developing algebra skills.

Answer

2. Jenny: 8 in.; Emma: 4 in.; Sarah: 6 in.

2. Translate. We reword the problem and translate as follows:

Rewording: First integer plus Second integer is 263

Translating: x + $(x + 1)$ = 263.

3. Solve. We solve the equation:

$x + (x + 1) = 263$

$2x + 1 = 263$ Collecting like terms

$2x + 1 - 1 = 263 - 1$ Subtracting 1

$2x = 262$

$\dfrac{2x}{2} = \dfrac{262}{2}$ Dividing by 2

$x = 131.$

If $x = 131$, then $x + 1 = 132$.

4. Check. Our possible answers are 131 and 132. These are consecutive positive integers and $131 + 132 = 263$, so the answers check.

5. State. The print numbers are 131/150 and 132/150.

Do Exercise 3. ▶

EXAMPLE 4 *Delivery Truck Rental.* An appliance business needs to rent a delivery truck for 6 days while one of its trucks is being repaired. The cost of renting a 16-ft truck is $29.95 per day plus $0.29 per mile. If $550 is budgeted for the rental, how many miles can be driven and stay within budget?

1. Familiarize. Suppose the van is driven 1100 mi. The cost is given by the daily charge plus the mileage charge, so we have

6($29.95) + Cost per mile times Number of miles

$179.70 + $0.29 · 1100,

which is $498.70. We see that the van can be driven more than 1100 mi on the business' budget of $550. This process familiarizes us with the way in which a calculation is made.

 3. *Interstate Mile Markers.* The sum of two consecutive mile markers on I-90 in upstate New York is 627. (On I-90 in New York, the marker numbers increase from east to west.) Find the numbers on the markers.

Source: New York State Department of Transportation

Let $x =$ the first marker and $x + 1 =$ the second marker.

Translate and *Solve:*

First marker + Second marker = 627

☐ + (☐) = 627

☐ + 1 = 627

$2x + 1 - 1 = 627 -$ ☐

$2x =$ ☐

$\dfrac{2x}{☐} = \dfrac{626}{2}$

$x = 313.$

If $x = 313$, then $x + 1 =$ ☐. The mile markers are ☐ and 314.

We let m = the number of miles that can be driven on the budget of $550.

2. **Translate.** We reword the problem and translate as follows:

Daily cost	plus	Cost per mile	times	Number of miles	is	Budget
$\downarrow$	$\downarrow$	$\downarrow$	$\downarrow$	$\downarrow$	$\downarrow$	$\downarrow$
6($29.95)	+	$0.29	$\cdot$	m	=	$550.

3. **Solve.** We solve the equation:

$$6(29.95) + 0.29m = 550$$
$$179.70 + 0.29m = 550$$
$$0.29m = 370.30 \qquad \text{\textcolor{red}{Subtracting 179.70}}$$
$$\frac{0.29m}{0.29} = \frac{370.30}{0.29} \qquad \text{\textcolor{red}{Dividing by 0.29}}$$
$$m \approx 1277. \qquad \text{\textcolor{red}{Rounding to the nearest one}}$$

4. **Check.** We check our answer in the original problem. The cost for driving 1277 mi is 1277($0.29) = $370.33. The rental for 6 days is 6($29.95) = $179.70. The total cost is then

$$\$370.33 + \$179.70 \approx \$550,$$

which is the $550 budget that was allowed.

5. **State.** The truck can be driven 1277 mi on the truck-rental allotment.

◀ **Do Exercise 4.**

4. *Delivery Truck Rental.* Refer to Example 4. The business decides to increase its 6-day rental budget to $625. How many miles can be driven for $625?

EXAMPLE 5 *Perimeter of a Lacrosse Field.* The perimeter of a lacrosse field is 340 yd. The length is 50 yd longer than the width. Find the dimensions of the field.

Source: www.sportsknowhow.com

1. **Familiarize.** We first make a drawing.

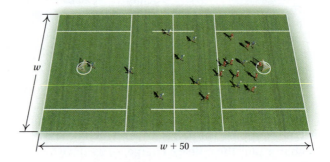

We let w = the width of the rectangle. Then $w + 50$ = the length. The perimeter P of a rectangle is the distance around the rectangle and is given by the formula $2l + 2w = P$, where

$$l = \text{the length} \quad \text{and} \quad w = \text{the width.}$$

Answer

4. 1536 mi

2. **Translate.** To translate the problem, we substitute $w + 50$ for l and 340 for P:

$$2l + 2w = P$$
$$2(w + 50) + 2w = 340.$$

···· **Caution!** ····

Parentheses are necessary here.

3. **Solve.** We solve the equation:

$$2(w + 50) + 2w = 340$$
$$2w + 100 + 2w = 340 \qquad \text{Using the distributive law}$$
$$4w + 100 = 340 \qquad \text{Collecting like terms}$$
$$4w + 100 - 100 = 340 - 100 \qquad \text{Subtracting 100}$$
$$4w = 240$$
$$\frac{4w}{4} = \frac{240}{4} \qquad \text{Dividing by 4}$$
$$w = 60.$$

Thus the possible dimensions are

$$w = 60 \text{ yd} \quad \text{and} \quad l = w + 50 = 60 + 50, \text{ or } 110 \text{ yd}.$$

4. **Check.** If the width is 60 yd and the length is 110 yd, then the perimeter is $2(60 \text{ yd}) + 2(110 \text{ yd})$, or 340 yd. This checks.

5. **State.** The width is 60 ft and the length is 110 yd.

Do Exercise 5. ▶

········· **Caution!** ·········

Always be sure to answer the original problem completely. For instance, in Example 1, we need to find *two* numbers: the distances from *each* city to the biker. Similarly, in Example 3, we need to find two print numbers, and in Example 5, we need to find two dimensions, not just the width.

EXAMPLE 6 *Roof Gable.* In a triangular gable end of a roof, the angle of the peak is twice as large as the angle of the back side of the house. The measure of the angle on the front side is 20° greater than the angle on the back side. How large are the angles?

Peak angle
$2x$
$x + 20$
Front angle
x
Back angle

1. **Familiarize.** We first make a drawing as shown above. We let

measure of back angle $= x$.

Then measure of peak angle $= 2x$

and measure of front angle $= x + 20$.

5. *Perimeter of High School Basketball Court.* The perimeter of a standard high school basketball court is 268 ft. The length is 34 ft longer than the width. Find the dimensions of the court.

Source: Indiana High School Athletic Association

Answer

5. Length: 84 ft; width: 50 ft

2. Translate. To translate, we need to know that the sum of the measures of the angles of a triangle is 180°. You might recall this fact from geometry or you can look it up in a geometry book or in the list of formulas inside the back cover of this book. We translate as follows:

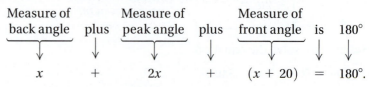

Measure of back angle	plus	Measure of peak angle	plus	Measure of front angle	is	180°
x	$+$	$2x$	$+$	$(x + 20)$	$=$	180°.

3. Solve. We solve the equation:

$$x + 2x + (x + 20) = 180$$
$$4x + 20 = 180$$
$$4x + 20 - 20 = 180 - 20$$
$$4x = 160$$
$$\frac{4x}{4} = \frac{160}{4}$$
$$x = 40.$$

The possible measures for the angles are as follows:

Back angle: $x = 40°$;

Peak angle: $2x = 2(40) = 80°$;

Front angle: $x + 20 = 40 + 20 = 60°$.

4. Check. Consider our answers: 40°, 80°, and 60°. The peak is twice the back, and the front is 20° greater than the back. The sum is 180°. The angles check.

5. State. The measures of the angles are 40°, 80°, and 60°.

.... **Caution!** ....

Units are important in answers. Remember to include them, where appropriate.

....

6. The second angle of a triangle is three times as large as the first. The third angle measures 30° more than the first angle. Find the measures of the angles.

◀ **Do Exercise 6.**

EXAMPLE 7 *Fastest Roller Coasters.* The average top speed of the three fastest steel roller coasters in the United States is 116 mph. The third-fastest roller coaster, Superman: The Escape (located at Six Flags Magic Mountain, Valencia, California), reaches a top speed of 28 mph less than the fastest roller coaster, Kingda Ka (located at Six Flags Great Adventure, Jackson, New Jersey). The second-fastest roller coaster, Top Thrill Dragster (located at Cedar Point, Sandusky, Ohio), has a top speed of 120 mph. What is the top speed of the fastest steel roller coaster?

Source: Coaster Grotto

Answer

6. First: 30°; second: 90°; third: 60°

1. **Familiarize.** The **average** of a set of numbers is the sum of the numbers divided by the number of addends.

   We are given that the second-fastest speed is 120 mph. Suppose the three top speeds are 131, 120, and 103. The average is then

$$\frac{131 + 120 + 103}{3} = \frac{354}{3} = 118,$$

which is too high. Instead of continuing to guess, let's use the equation-solving skills we have learned in this chapter. We let $x =$ the top speed of the fastest roller coaster. Then $x - 28 =$ the top speed of the third-fastest roller coaster.

2. **Translate.** We reword the problem and translate as follows:

$$\frac{\text{Speed of fastest coaster} + \text{Speed of second-fastest coaster} + \text{Speed of third-fastest coaster}}{\text{Number of roller coasters}} = \text{Average speed of three fastest roller coasters}$$

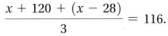

$$\frac{x + 120 + (x - 28)}{3} = 116.$$

3. **Solve.** We solve as follows:

$$\frac{x + 120 + (x - 28)}{3} = 116$$

$$3 \cdot \frac{x + 120 + (x - 28)}{3} = 3 \cdot 116 \qquad \text{Multiplying by 3 on both sides to clear the fraction}$$

$$x + 120 + (x - 28) = 348$$

$$2x + 92 = 348 \qquad \text{Collecting like terms}$$

$$2x = 256 \qquad \text{Subtracting 92}$$

$$x = 128. \qquad \text{Dividing by 2}$$

4. **Check.** If the top speed of the fastest roller coaster is 128 mph, then the top speed of the third-fastest is $128 - 28$, or 100 mph. The average of the top speeds of the three fastest is

$$\frac{128 + 120 + 100}{3} = \frac{348}{3} = 116 \text{ mph.}$$

   The answer checks.

5. **State.** The top speed of the fastest steel roller coaster in the United States is 128 mph.

Do Exercise 7. ▶

7. *Average Test Score.* Sam's average score on his first three math tests is 77. He scored 62 on the first test. On the third test, he scored 9 more than he scored on his second test. What did he score on the second and third tests?

Answer

7. Second: 80; third: 89

EXAMPLE 8 *Simple Interest.* An investment is made at 3% simple interest for 1 year. It grows to $746.75. How much was originally invested (the principal)?

1. **Familiarize.** Suppose that $100 was invested. Recalling the formula for simple interest, $I = Prt$, we know that the interest for 1 year on $100 at 3% simple interest is given by $I = \$100 \cdot 0.03 \cdot 1 = \$3$. Then, at the end of the year, the amount in the account is found by adding the principal and the interest:

$$\begin{array}{ccccc} \text{Principal} & + & \text{Interest} & = & \text{Amount} \\ \downarrow & & \downarrow & & \downarrow \\ \$100 & + & \$3 & = & \$103. \end{array}$$

In this problem, we are working backward. We are trying to find the principal, which is the original investment. We let $x =$ the principal. Then the interest earned is 3%x.

2. **Translate.** We reword the problem and then translate:

$$\begin{array}{ccccc} \text{Principal} & + & \text{Interest} & = & \text{Amount} \\ \downarrow & & \downarrow & & \downarrow \\ x & + & 3\%x & = & 746.75. \end{array}$$ Interest is 3% of the principal.

3. **Solve.** We solve the equation:

$$x + 3\%x = 746.75$$
$$x + 0.03x = 746.75 \qquad \text{\color{red}Converting to decimal notation}$$
$$1x + 0.03x = 746.75 \qquad \text{\color{red}Identity property of 1}$$
$$(1 + 0.03)x = 746.75$$
$$1.03x = 746.75 \qquad \text{\color{red}Collecting like terms}$$
$$\frac{1.03x}{1.03} = \frac{746.75}{1.03} \qquad \text{\color{red}Dividing by 1.03}$$
$$x = 725.$$

4. **Check.** We check by taking 3% of $725 and adding it to $725:

$$3\% \times \$725 = 0.03 \times 725 = \$21.75.$$

Then $725 + $21.75 = $746.75, so $725 checks.

5. **State.** The original investment was $725.

◀ **Do Exercise 8.**

---

8. *Simple Interest.* An investment is made at 5% simple interest for 1 year. It grows to $2520. How much was originally invested (the principal)? GS

Let $x =$ the principal. Then the interest earned is 5%x.

Translate and *Solve*:

$$\begin{array}{ccccc} \text{Principal} & + & \text{Interest} & = & \text{Amount} \\ \downarrow & & \downarrow & & \downarrow \\ x & + & \boxed{} & = & 2520 \end{array}$$
$$x + 0.05x = 2520$$
$$(1 + \boxed{})x = 2520$$
$$\boxed{}\,x = 2520$$
$$\frac{1.05x}{1.05} = \frac{2520}{\boxed{}}$$
$$x = 2400.$$

EXAMPLE 9 *Selling a House.* The Patels are planning to sell their house. If they want to be left with $130,200 after paying 7% of the selling price to a realtor as a commission, for how much must they sell the house?

1. **Familiarize.** Suppose the Patels sell the house for $138,000. A 7% commission can be determined by finding 7% of $138,000:

$$7\% \text{ of } \$138,000 = 0.07(\$138,000) = \$9660.$$

Subtracting this commission from $138,000 would leave the Patels with

$$\$138,000 - \$9660 = \$128,340.$$

This shows that in order for the Patels to clear $130,200, the house must sell for more than $138,000. Our guess shows us how to translate to an equation. We let $x =$ the selling price, in dollars. With a 7% commission, the realtor would receive $0.07x$.

Answer

8. $2400

Guided Solution:

8. 5%x, 0.05, 1.05, 1.05

176 CHAPTER 2 Solving Equations and Inequalities

2. Translate. We reword the problem and translate as follows:

$$\underbrace{\text{Selling price}}_{x} \quad \underbrace{\text{less}}_{-} \quad \underbrace{\text{Commission}}_{0.07x} \quad \underbrace{\text{is}}_{=} \quad \underbrace{\text{Amount remaining}}_{130{,}200.}$$

3. Solve. We solve the equation:

$$x - 0.07x = 130{,}200$$
$$1x - 0.07x = 130{,}200$$
$$(1 - 0.07)x = 130{,}200$$
$$0.93x = 130{,}200 \qquad \textcolor{red}{\text{Collecting like terms. Had we noted that after}}$$
$$\textcolor{red}{\text{the commission has been paid, 93\% remains,}}$$
$$\textcolor{red}{\text{we could have begun with this equation.}}$$

$$\frac{0.93x}{0.93} = \frac{130{,}200}{0.93} \qquad \textcolor{red}{\text{Dividing by 0.93}}$$

$$x = 140{,}000.$$

4. Check. To check, we first find 7% of $140,000:

$$7\% \text{ of } \$140{,}000 = 0.07(\$140{,}000) = \$9800. \qquad \textcolor{red}{\text{This is the}}$$
$$\textcolor{red}{\text{commission.}}$$

Next, we subtract the commission to find the remaining amount:

$$\$140{,}000 - \$9800 = \$130{,}200.$$

Since, after the commission, the Patels are left with $130,200, our answer checks. Note that the $140,000 selling price is greater than $138,000, as predicted in the *Familiarize* step.

5. State. To be left with $130,200, the Patels must sell the house for $140,000.

Do Exercise 9. ▶

.... **Caution!** ....

The problem in Example 9 is easy to solve with algebra. Without algebra, it is not. A common error in such a problem is to take 7% of the price after commission and then subtract or add. Note that 7% of the selling price $(7\% \cdot \$140{,}000 = \$9800)$ is not equal to 7% of the amount that the Patels want to be left with $(7\% \cdot \$130{,}200 = \$9114)$.

....

9. *Selling a Condominium.*
An investor needs to sell a condominium in New York City. If she wants to be left with $761,400 after paying a 6% commission, for how much must she sell the condominium?

Answer

8. $810,000

Translating for Success

1. **Angle Measures.** The measure of the second angle of a triangle is 51° more than that of the first angle. The measure of the third angle is 3° less than twice the first angle. Find the measures of the angles.

2. **Sales Tax.** Tina paid $3976 for a used car. This amount included 5% for sales tax. How much did the car cost before tax?

3. **Perimeter.** The perimeter of a rectangle is 2347 ft. The length is 28 ft greater than the width. Find the length and the width.

4. **Fraternity or Sorority Membership.** At Arches Tech University, 3976 students belong to a fraternity or a sorority. This is 35% of the total enrollment. What is the total enrollment at Arches Tech?

5. **Fraternity or Sorority Membership.** At Moab Tech University, thirty-five percent of the students belong to a fraternity or a sorority. The total enrollment of the university is 11,360 students. How many students belong to either a fraternity or a sorority?

The goal of these matching questions is to practice step (2), Translate, of the five-step problem-solving process. Translate each word problem to an equation and select a correct translation from equations A–O.

A. $x + (x - 3) + \frac{4}{5}x = 384$

B. $x + (x + 51) + (2x - 3) = 180$

C. $x + (x + 96) = 180$

D. $2 \cdot 96 + 2x = 3976$

E. $x + (x + 1) + (x + 2) = 384$

F. $3976 = x \cdot 11,360$

G. $2x + 2(x + 28) = 2347$

H. $3976 = x + 5\%x$

I. $x + (x + 28) = 2347$

J. $x = 35\% \cdot 11,360$

K. $x + 96 = 3976$

L. $x + (x + 3) + \frac{4}{5}x = 384$

M. $x + (x + 2) + (x + 4) = 384$

N. $35\% \cdot x = 3976$

O. $2x + (x + 28) = 2347$

Answers on page A-5

6. **Island Population.** There are 180 thousand people living on a small Caribbean island. The women outnumber the men by 96 thousand. How many men live on the island?

7. **Wire Cutting.** A 384-m wire is cut into three pieces. The second piece is 3 m longer than the first. The third is four-fifths as long as the first. How long is each piece?

8. **Locker Numbers.** The numbers on three adjoining lockers are consecutive integers whose sum is 384. Find the integers.

9. **Fraternity or Sorority Membership.** The total enrollment at Canyonlands Tech University is 11,360 students. Of these, 3976 students belong to a fraternity or a sorority. What percent of the students belong to a fraternity or a sorority?

10. **Width of a Rectangle.** The length of a rectangle is 96 ft. The perimeter of the rectangle is 3976 ft. Find the width.

✓ Reading Check

Choose from the column on the right the word that completes each step in the five steps for problem solving.

RC1. _____ yourself with the problem situation.

RC2. _____ the problem to an equation.

RC3. _____ the equation.

RC4. _____ the answer in the original problem.

RC5. _____ the answer to the problem clearly.

Solve
Familiarize
State
Translate
Check

a   Solve. *Although you might find the answer quickly in some other way, practice using the five-step problem-solving strategy.*

1. *Medals of Honor.*   In 1863, the U.S. Secretary of War presented the first Medals of Honor. The two wars with the most Medals of Honor awarded are the Civil War and World War II. There were 464 recipients of this medal for World War II. This number is 1058 fewer than the number of recipients for the Civil War. How many Medals of Honor were awarded for valor in the Civil War?

Sources: U.S. Army Center of Military History; U.S. Department of Defense

2. *Milk Alternatives.*   Milk alternatives such as rice, soy, almond, and flax are becoming more available and increasingly popular. A cup of almond milk contains only 60 calories. This number is 89 calories less than the number of calories in a cup of whole milk. How many calories are in a cup of whole milk?

Source: Janet Kinosian, "Nutrition Udder Chaos," *AARP Magazine,* August/September, 2012

3. *Pipe Cutting.*   A 240-in. pipe is cut into two pieces. One piece is three times the length of the other. Find the lengths of the pieces.

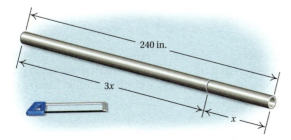

240 in.

$3x$

x

4. *Board Cutting.*   A 72-in. board is cut into two pieces. One piece is 2 in. longer than the other. Find the lengths of the pieces.

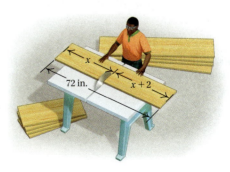

x

72 in.

$x + 2$

5. Public Transit Systems. In the first quarter of 2012, the ridership on the public transit system in Boston was 99.2 million. This number is 77.4 million more than the ridership in San Diego over the same period of time. What was the ridership in San Diego during the first quarter of 2012?

Source: American Public Transportation Association

6. Home Listing Price. In 2011, the average listing price of a home in Hawaii was $72,000 more than three times the average listing price of a home in Arizona. The average listing price of a home in Hawaii was $876,000. What was the average listing price of a home in Arizona?

Source: Trulia

7. 500 Festival Mini-Marathon. On May 4, 2013, 35,000 runners participated in the 13.1-mi One America 500 Festival Mini-Marathon. If a runner stopped at a water station that is twice as far from the start as from the finish, how far is the runner from the finish? Round the answer to the nearest hundredth of a mile.

Source: www.500festival.com

8. Airport Control Tower. At a height of 385 ft, the FAA airport traffic control tower in Atlanta is the tallest traffic control tower in the United States. Its height is 59 ft greater than the height of the tower at the Memphis airport. How tall is the traffic control tower at the Memphis airport?

Source: Federal Aviation Administration

9. Consecutive Apartment Numbers. The apartments in Vincent's apartment house are numbered consecutively on each floor. The sum of his number and his next-door neighbor's number is 2409. What are the two numbers?

10. Consecutive Post Office Box Numbers. The sum of the numbers on two consecutive post office boxes is 547. What are the numbers?

11. Consecutive Ticket Numbers. The numbers on Sam's three raffle tickets are consecutive integers. The sum of the numbers is 126. What are the numbers?

12. Consecutive Ages. The ages of Whitney, Wesley, and Wanda are consecutive integers. The sum of their ages is 108. What are their ages?

13. Consecutive Odd Integers. The sum of three consecutive odd integers is 189. What are the integers?

14. Consecutive Integers. Three consecutive integers are such that the first plus one-half the second plus seven less than twice the third is 2101. What are the integers?

15. *Photo Size.* A hotel orders a large photo for its newly renovated lobby. The perimeter of the photo is 292 in. The width is 2 in. more than three times the height. Find the dimensions of the photo.

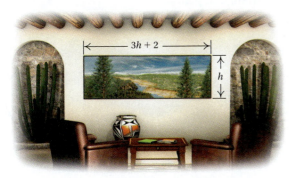

16. *Two-by-Four.* The perimeter of a cross section or end of a "two-by-four" piece of lumber is 10 in. The length is 2 in. more than the width. Find the actual dimensions of the cross section of a two-by-four.

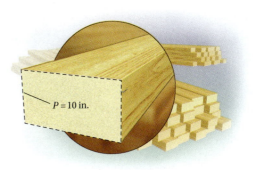

$P = 10$ in.

17. *Price of Coffee Beans.* A student-owned and -operated coffee shop near a campus purchases gourmet coffee beans from Costa Rica. During a recent 30%-off sale, a 3-lb bag could be purchased for $44.10. What is the regular price of a 3-lb bag?

18. *Price of an iPad Case.* Makayla paid $33.15 for an iPad case during a 15%-off sale. What was the regular price?

19. *Price of a Security Wallet.* Caleb paid $26.70, including a 7% sales tax, for a security wallet. How much did the wallet itself cost?

20. *Price of a Car Battery.* Tyler paid $117.15, including a 6.5% sales tax, for a car battery. How much did the battery itself cost?

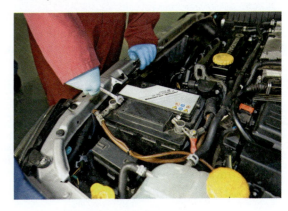

21. *Parking Costs.* A hospital parking lot charges $1.50 for the first hour or part thereof, and $1.00 for each additional hour or part thereof. A weekly pass costs $27.00 and allows unlimited parking for 7 days. Suppose that each visit Hailey makes to the hospital lasts $1\frac{1}{2}$ hr. What is the minimum number of times that Hailey would have to visit per week to make it worthwhile for her to buy the pass?

22. *Van Rental.* Value Rent-A-Car rents vans at a daily rate of $84.45 plus 55¢ per mile. Molly rents a van to deliver electrical parts to her customers. She is allotted a daily budget of $250. How many miles can she drive for $250? (*Hint:* 60¢ = $0.60.)

23. *Triangular Field.* The second angle of a triangular field is three times as large as the first angle. The third angle is 40° greater than the first angle. How large are the angles?

24. *Triangular Parking Lot.* The second angle of a triangular parking lot is four times as large as the first angle. The third angle is 45° less than the sum of the other two angles. How large are the angles?

25. *Triangular Backyard.* A home has a triangular backyard. The second angle of the triangle is 5° more than the first angle. The third angle is 10° more than three times the first angle. Find the angles of the triangular yard.

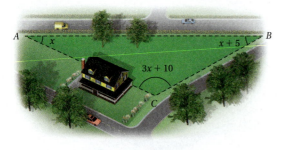

26. *Boarding Stable.* A rancher needs to form a triangular horse pen using ropes next to a stable. The second angle is three times the first angle. The third angle is 15° less than the first angle. Find the angles of the triangular pen.

27. *Stock Prices.* Diego's investment in a technology stock grew 28% to $448. How much did he invest?

28. *Savings Interest.* Ella invested money in a savings account at a rate of 6% simple interest. After 1 year, she has $6996 in the account. How much did Ella originally invest?

29. *Credit Cards.* The balance on Will's credit card grew 2%, to $870, in one month. What was his balance at the beginning of the month?

30. *Loan Interest.* Alvin borrowed money from a cousin at a rate of 10% simple interest. After 1 year, $7194 paid off the loan. How much did Alvin borrow?

31. *Taxi Fares.* In New Orleans, Louisiana, taxis charge an initial charge of $3.50 plus $2.00 per mile. How far can one travel for $39.50?

Source: www.taxifarefinders.com

32. *Taxi Fares.* In Baltimore, Maryland, taxis charge an initial charge of $1.80 plus $2.20 per mile. How far can one travel for $26?

Source: www.taxifarefinders.com

33. *Tipping.* Isabella left a 15% tip for a meal. The total cost of the meal, including the tip, was $44.39. What was the cost of the meal before the tip was added?

34. *Tipping.* Nicolas left a 20% tip for a meal. The total cost of the meal, including the tip, was $24.90. What was the cost of the meal before the tip was added?

35. *Average Test Score.* Mariana averaged 84 on her first three history exams. The first score was 67. The second score was 7 less than the third score. What did she score on the second and third exams?

36. *Average Price.* David paid an average of $34 per shirt for a recent purchase of three shirts. The price of one shirt was twice as much as another, and the remaining shirt cost $27. What were the prices of the other two shirts?

37. If you double a number and then add 16, you get $\frac{2}{3}$ of the original number. What is the original number?

38. If you double a number and then add 85, you get $\frac{3}{4}$ of the original number. What is the original number?

Skill Maintenance

Calculate.

39. $-\frac{4}{5} - \frac{3}{8}$ [1.4a]

40. $-\frac{4}{5} + \frac{3}{8}$ [1.3a]

41. $-\frac{4}{5} \cdot \frac{3}{8}$ [1.5a]

42. $-\frac{4}{5} \div \frac{3}{8}$ [1.6c]

43. $\frac{1}{10} \div \left(-\frac{1}{100}\right)$ [1.6c]

44. $-25.6 \div (-16)$ [1.6c]

45. $-25.6\,(-16)$ [1.5a]

46. $-25.6 - (-16)$ [1.4a]

47. $-25.6 + (-16)$ [1.3a]

48. $(-0.02) \div (-0.2)$ [1.6c]

49. Use a commutative law to write an equivalent expression for $12 + yz$.   [1.7b]

50. Use an associative law to write an equivalent expression for $(c + 4) + d$.   [1.7b]

Synthesis

51. Apples are collected in a basket for six people. One-third, one-fourth, one-eighth, and one-fifth are given to four people, respectively. The fifth person gets ten apples, leaving one apple for the sixth person. Find the original number of apples in the basket.

52. *Test Questions.* A student scored 78 on a test that had 4 seven-point fill-in questions and 24 three-point multiple-choice questions. The student answered one fill-in question incorrectly. How many multiple-choice questions did the student answer correctly?

53. The area of this triangle is 2.9047 in². Find x.

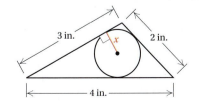

54. Susanne goes to the bank to get $20 in quarters, dimes, and nickels to use to make change at her yard sale. She gets twice as many quarters as dimes and 10 more nickels than dimes. How many of each type of coin does she get?

2.7

Solving Inequalities

OBJECTIVES

a Determine whether a given number is a solution of an inequality.

b Graph an inequality on the number line.

c Solve inequalities using the addition principle.

d Solve inequalities using the multiplication principle.

e Solve inequalities using the addition principle and the multiplication principle together.

SKILL TO REVIEW

Objective 1.2d: Determine whether an inequality like $-3 \leq 5$ is true or false.

Write true or false.

1. $-6 \leq -8$ **2.** $1 \geq 1$

We now extend our equation-solving principles to the solving of inequalities.

a SOLUTIONS OF INEQUALITIES

In Section 1.2, we defined the symbols $>$ (is greater than), $<$ (is less than), $\geq$ (is greater than or equal to), and $\leq$ (is less than or equal to).

An **inequality** is a number sentence with $>, <, \geq$, or $\leq$ as its verb—for example,

$$-4 > t, \quad x < 3, \quad 2x + 5 \geq 0, \quad \text{and} \quad -3y + 7 \leq -8.$$

Some replacements for a variable in an inequality make it true and some make it false. (There are some exceptions to this statement, but we will not consider them here.)

SOLUTION OF AN INEQUALITY

A replacement that makes an inequality true is called a **solution**. The set of all solutions is called the **solution set**. When we have found the set of all solutions of an inequality, we say that we have **solved** the inequality.

EXAMPLES Determine whether each number is a solution of $x < 2$.

1. -2.7 Since $-2.7 < 2$ is true, -2.7 is a solution.

2. 2 Since $2 < 2$ is false, 2 is not a solution.

EXAMPLES Determine whether each number is a solution of $y \geq 6$.

3. 6 Since $6 \geq 6$ is true, 6 is a solution.

4. $-\dfrac{4}{3}$ Since $-\dfrac{4}{3} \geq 6$ is false, $-\dfrac{4}{3}$ is not a solution.

◀ **Do Margin Exercises 1 and 2.**

b GRAPHS OF INEQUALITIES

Some solutions of $x < 2$ are $-3, 0, 1, 0.45, -8.9, -\pi, \frac{5}{8}$, and so on. In fact, there are infinitely many real numbers that are solutions. Because we cannot list them all individually, it is helpful to make a drawing that represents all the solutions.

A **graph** of an inequality is a drawing that represents its solutions. An inequality in one variable can be graphed on the number line. An inequality in two variables can be graphed on the coordinate plane. We will study such graphs in Chapter 3.

Determine whether each number is a solution of the inequality.

1. $x > 3$

   a) 2   **b)** 0

   c) -5 **d)** 15.4

   e) 3   **f)** $-\dfrac{2}{5}$

2. $x \leq 6$

   a) 6   **b)** 0

   c) -4.3 **d)** 25

   e) -6 **f)** $\dfrac{5}{8}$

EXAMPLE 5 Graph: $x < 2$.

The solutions of $x < 2$ are all those numbers less than 2. They are shown on the number line by shading all points to the left of 2. The parenthesis at 2 indicates that 2 *is not* part of the graph.

EXAMPLE 6 Graph: $x \geq -3$.

The solutions of $x \geq -3$ are shown on the number line by shading the point for -3 and all points to the right of -3. The bracket at -3 indicates that -3 *is* part of the graph.

EXAMPLE 7 Graph: $-3 \leq x < 2$.

The inequality $-3 \leq x < 2$ is read "-3 is less than or equal to x *and* x is less than 2," or "x is greater than or equal to -3 *and* x is less than 2." In order to be a solution of this inequality, a number must be a solution of both $-3 \leq x$ and $x < 2$. The number 1 is a solution, as are -1.7, 0, 1.5, and $\frac{3}{8}$. We can see from the following graphs that the solution set consists of the numbers that overlap in the two solution sets in Examples 5 and 6.

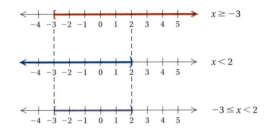

The parenthesis at 2 means that 2 *is not* part of the graph. The bracket at -3 means that -3 *is* part of the graph. The other solutions are shaded.

Do Exercises 3–5. ▶

C SOLVING INEQUALITIES USING THE ADDITION PRINCIPLE

Consider the true inequality $3 < 7$. If we add 2 on both sides, we get another true inequality:

$$3 + 2 < 7 + 2, \quad \text{or} \quad 5 < 9.$$

Similarly, if we add -4 on both sides of $x + 4 < 10$, we get an *equivalent* inequality:

$$x + 4 + (-4) < 10 + (-4),$$

or

$$x < 6.$$

To say that $x + 4 < 10$ and $x < 6$ are **equivalent** is to say that they have the same solution set. For example, the number 3 is a solution of $x + 4 < 10$. It is also a solution of $x < 6$. The number -2 is a solution of $x < 6$. It is also a solution of $x + 4 < 10$. Any solution of one inequality is a solution of the other—they are equivalent.

Graph.

3. $x \leq 4$

4. $x > -2$

5. $-2 < x \leq 4$

Answers

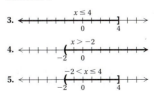

As with equation solving, when solving inequalities, our goal is to isolate the variable on one side. Then it is easier to determine the solution set.

EXAMPLE 8 Solve: $x + 2 > 8$. Then graph.

We use the addition principle, subtracting 2 on both sides:

$$x + 2 - 2 > 8 - 2$$
$$x > 6.$$

From the inequality $x > 6$, we can determine the solutions directly. Any number greater than 6 makes the last sentence true and is a solution of that sentence. Any such number is also a solution of the original sentence. Thus the inequality is solved. The graph is as follows:

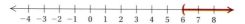

We cannot check all the solutions of an inequality by substitution, as we usually can for an equation, because there are too many of them. A partial check can be done by substituting a number greater than 6—say, 7—into the original inequality:

$$\frac{x + 2 > 8}{7 + 2 \ ? \ 8}$$
$$9 \ \bigm| \quad \text{TRUE}$$

Since $9 > 8$ is true, 7 is a solution. This is a partial check that any number greater than 6 is a solution.

EXAMPLE 9 Solve: $3x + 1 \leq 2x - 3$. Then graph.

We have

$$3x + 1 \leq 2x - 3$$
$$3x + 1 - 1 \leq 2x - 3 - 1 \qquad \text{Subtracting 1}$$
$$3x \leq 2x - 4 \qquad \text{Simplifying}$$
$$3x - 2x \leq 2x - 4 - 2x \qquad \text{Subtracting } 2x$$
$$x \leq -4. \qquad \text{Simplifying}$$

Any number less than or equal to -4 is a solution. The graph is as follows:

In Example 9, any number less than or equal to -4 is a solution. The following are some solutions:

$$-4, \quad -5, \quad -6, \quad -\frac{13}{3}, \quad -204.5, \quad \text{and} \quad -18\pi.$$

Besides drawing a graph, we can also describe all the solutions of an inequality using **set notation**. We could just begin to list them in a set using roster notation (see p. 51), as follows:

$$\left\{ -4, -5, -6, -\frac{13}{3}, -204.5, -18\pi, \ldots \right\}.$$

We can never list them all this way, however. Seeing this set without knowing the inequality makes it difficult for us to know what real numbers we are considering. There is, however, another kind of notation that we can use. It is

$$\{x \mid x \le -4\},$$

which is read

"The set of all x such that x is less than or equal to -4."

This shorter notation for sets is called **set-builder notation**.
From now on, we will use this notation when solving inequalities.

Do Exercises 6–8. ▶

EXAMPLE 10 Solve: $x + \frac{1}{3} > \frac{5}{4}$.

We have

$$x + \frac{1}{3} > \frac{5}{4}$$

$$x + \frac{1}{3} - \frac{1}{3} > \frac{5}{4} - \frac{1}{3} \qquad \text{Subtracting } \frac{1}{3}$$

$$x > \frac{5}{4} \cdot \frac{3}{3} - \frac{1}{3} \cdot \frac{4}{4} \qquad \begin{array}{l}\text{Multiplying by 1 to obtain} \\ \text{a common denominator}\end{array}$$

$$x > \frac{15}{12} - \frac{4}{12}$$

$$x > \frac{11}{12}.$$

Any number greater than $\frac{11}{12}$ is a solution. The solution set is

$$\left\{ x \mid x > \frac{11}{12} \right\},$$

which is read

"The set of all x such that x is greater than $\frac{11}{12}$."

When solving inequalities, you may obtain an answer like $\frac{11}{12} < x$. Recall from Chapter 1 that this has the same meaning as $x > \frac{11}{12}$. Thus the solution set in Example 10 can be described as $\left\{ x \mid \frac{11}{12} < x \right\}$ or as $\left\{ x \mid x > \frac{11}{12} \right\}$. The latter is used most often.

Do Exercises 9 and 10. ▶

d SOLVING INEQUALITIES USING THE MULTIPLICATION PRINCIPLE

There is a multiplication principle for inequalities that is similar to that for equations, but it must be modified. When we are multiplying on both sides by a negative number, the direction of the inequality symbol must be changed.

Solve. Then graph.

6. $x + 3 > 5$

7. $x - 1 \le 2$

8. $5x + 1 < 4x - 2$

Solve.

9. $x + \dfrac{2}{3} \ge \dfrac{4}{5}$

GS **10.** $5y + 2 \le -1 + 4y$

$$5y + 2 - \boxed{} \le -1 + 4y - 4y$$

$$y + 2 \le -1$$

$$y + 2 - 2 \le -1 - \boxed{}$$

$$y \le \boxed{}$$

The solution set is $\{y \mid y \boxed{} -3\}$.

Consider the true inequality $3 < 7$. If we multiply on both sides by a *positive* number, like 2, we get another true inequality:

$$3 \cdot 2 < 7 \cdot 2, \quad \text{or} \quad 6 < 14. \qquad \text{True}$$

If we multiply on both sides by a *negative* number, like -2, and we do not change the direction of the inequality symbol, we get a *false* inequality:

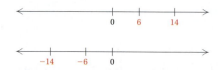

$$3 \cdot (-2) < 7 \cdot (-2), \quad \text{or} \quad -6 < -14. \qquad \text{False}$$

The fact that $6 < 14$ is true but $-6 < -14$ is false stems from the fact that the negative numbers, in a sense, mirror the positive numbers. That is, whereas 14 is to the *right* of 6 on the number line, the number -14 is to the *left* of -6. Thus, if we reverse (change the direction of) the inequality symbol, we get a *true* inequality: $-6 > -14$.

THE MULTIPLICATION PRINCIPLE FOR INEQUALITIES

For any real numbers a and b, and any *positive* number c:

$a < b$ is equivalent to $ac < bc$;

$a > b$ is equivalent to $ac > bc$.

For any real numbers a and b, and any *negative* number c:

$a < b$ is equivalent to $ac > bc$;

$a > b$ is equivalent to $ac < bc$.

Similar statements hold for $\leq$ and $\geq$.

In other words, when we multiply or divide by a positive number on both sides of an inequality, the direction of the inequality symbol stays the same. When we multiply or divide by a negative number on both sides of an inequality, the direction of the inequality symbol is reversed.

EXAMPLE 11 Solve: $4x < 28$. Then graph.

We have

$$4x < 28$$

$$\frac{4x}{4} < \frac{28}{4} \qquad \text{Dividing by 4}$$

The symbol stays the same.

$$x < 7. \qquad \text{Simplifying}$$

The solution set is $\{x \mid x < 7\}$. The graph is as follows:

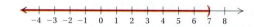

◀ **Do Exercises 11 and 12.**

Solve. Then graph.

11. $8x < 64$

12. $5y \geq 160$

Answers

11. $\{x \mid x < 8\}$;

12. $\{y \mid y \geq 32\}$;

EXAMPLE 12 Solve: $-2y < 18$. Then graph.

$$-2y < 18$$

$$\frac{-2y}{-2} > \frac{18}{-2} \qquad \text{Dividing by } -2$$

The symbol must be reversed!

$$y > -9. \qquad \text{Simplifying}$$

The solution set is $\{y \mid y > -9\}$. The graph is as follows:

Do Exercises 13 and 14. ▶

Solve.

13. $-4x \le 24$

14. $-5y > 13$

e USING THE PRINCIPLES TOGETHER

All of the equation-solving techniques used in Sections 2.1–2.3 can be used with inequalities, provided we remember to reverse the inequality symbol when multiplying or dividing on both sides by a negative number.

EXAMPLE 13 Solve: $6 - 5x > 7$.

$$6 - 5x > 7$$

$$-6 + 6 - 5x > -6 + 7 \qquad \text{Adding } -6. \text{ The symbol stays the same.}$$

$$-5x > 1 \qquad \text{Simplifying}$$

$$\frac{-5x}{-5} < \frac{1}{-5} \qquad \text{Dividing by } -5$$

The symbol must be reversed because we are dividing by a *negative* number, -5.

$$x < -\frac{1}{5}. \qquad \text{Simplifying}$$

The solution set is $\left\{ x \mid x < -\frac{1}{5} \right\}$.

Do Exercise 15. ▶

15. Solve: $7 - 4x < 8$.

EXAMPLE 14 Solve: $17 - 5y > 8y - 9$.

$$-17 + 17 - 5y > -17 + 8y - 9 \qquad \text{Adding } -17. \text{ The symbol stays the same.}$$

$$-5y > 8y - 26 \qquad \text{Simplifying}$$

$$-8y - 5y > -8y + 8y - 26 \qquad \text{Adding } -8y$$

$$-13y > -26 \qquad \text{Simplifying}$$

$$\frac{-13y}{-13} < \frac{-26}{-13} \qquad \text{Dividing by } -13$$

The symbol must be reversed because we are dividing by a *negative* number, -13.

$$y < 2$$

The solution set is $\{y \mid y < 2\}$.

16. Solve. Begin by subtracting 24 on both sides.

$$24 - 7y \le 11y - 14$$

Do Exercise 16. ▶

Answers

13. $\{x \mid x \ge -6\}$ **14.** $\left\{ y \mid y < -\frac{13}{5} \right\}$

15. $\left\{ x \mid x > -\frac{1}{4} \right\}$ **16.** $\left\{ y \mid y \ge \frac{19}{9} \right\}$

Typically, we solve an equation or an inequality by isolating the variable on the left side. When we are solving an inequality, however, there are situations in which isolating the variable on the right side will eliminate the need to reverse the inequality symbol. Let's solve the inequality in Example 14 again, but this time we will isolate the variable on the right side.

EXAMPLE 15 Solve: $17 - 5y > 8y - 9$.

Note that if we add $5y$ on both sides, the coefficient of the y-term will be positive after like terms have been collected.

$$17 - 5y + 5y > 8y - 9 + 5y \qquad \text{Adding } 5y$$
$$17 > 13y - 9 \qquad \text{Simplifying}$$
$$17 + 9 > 13y - 9 + 9 \qquad \text{Adding } 9$$
$$26 > 13y \qquad \text{Simplifying}$$
$$\frac{26}{13} > \frac{13y}{13} \qquad \begin{array}{l}\text{Dividing by 13. We leave the} \\ \text{inequality symbol the same} \\ \text{because we are dividing by a} \\ \text{positive number.}\end{array}$$
$$2 > y$$

The solution set is $\{y \,|\, 2 > y\}$, or $\{y \,|\, y < 2\}$.

◀ **Do Exercise 17.**

17. Solve. Begin by adding $7y$ on both sides.
$$24 - 7y \le 11y - 14$$

EXAMPLE 16 Solve: $3(x - 2) - 1 < 2 - 5(x + 6)$.

First, we use the distributive law to remove parentheses. Next, we collect like terms and then use the addition and multiplication principles for inequalities to get an equivalent inequality with x alone on one side.

$$3(x - 2) - 1 < 2 - 5(x + 6)$$
$$3x - 6 - 1 < 2 - 5x - 30 \qquad \begin{array}{l}\text{Using the distributive law to} \\ \text{multiply and remove parentheses}\end{array}$$
$$3x - 7 < -5x - 28 \qquad \text{Collecting like terms}$$
$$3x + 5x < -28 + 7 \qquad \begin{array}{l}\text{Adding } 5x \text{ and 7 to get all } x\text{-terms} \\ \text{on one side and all other terms} \\ \text{on the other side}\end{array}$$
$$8x < -21 \qquad \text{Simplifying}$$
$$x < \frac{-21}{8}, \text{ or } -\frac{21}{8}. \qquad \text{Dividing by 8}$$

The solution set is $\left\{x \,|\, x < -\frac{21}{8}\right\}$.

◀ **Do Exercise 18.**

18. Solve:
$$3(7 + 2x) \le 30 + 7(x - 1)$$
$$\boxed{} + 6x \le 30 + 7x - \boxed{}$$
$$21 + 6x \le \boxed{} + 7x$$
$$21 + 6x - 6x \le 23 + 7x - \boxed{}$$
$$21 \le 23 + \boxed{}$$
$$21 - \boxed{} \le 23 + x - 23$$
$$-2 \le x, \text{ or}$$
$$x \boxed{} -2$$
The solution set is $\{x \,|\, x \ge \boxed{}\}$.

GS

EXAMPLE 17 Solve: $16.3 - 7.2p \le -8.18$.

The greatest number of decimal places in any one number is *two*. Multiplying by 100, which has two 0's, will clear decimals. Then we proceed as before.

$$16.3 - 7.2p \le -8.18$$
$$100(16.3 - 7.2p) \le 100(-8.18) \quad \text{Multiplying by 100}$$
$$100(16.3) - 100(7.2p) \le 100(-8.18) \quad \text{Using the distributive law}$$
$$1630 - 720p \le -818 \quad \text{Simplifying}$$
$$1630 - 720p - 1630 \le -818 - 1630 \quad \text{Subtracting 1630}$$
$$-720p \le -2448 \quad \text{Simplifying}$$
$$\frac{-720p}{-720} \ge \frac{-2448}{-720} \quad \text{Dividing by } -720$$

The symbol must be reversed.

$$p \ge 3.4$$

The solution set is $\{p \mid p \ge 3.4\}$.

Do Exercise 19. ▶

19. Solve:
$$2.1x + 43.2 \ge 1.2 - 8.4x.$$

EXAMPLE 18 Solve: $\dfrac{2}{3}x - \dfrac{1}{6} + \dfrac{1}{2}x > \dfrac{7}{6} + 2x$.

The number 6 is the least common multiple of all the denominators. Thus we first multiply by 6 on both sides to clear fractions.

$$\frac{2}{3}x - \frac{1}{6} + \frac{1}{2}x > \frac{7}{6} + 2x$$
$$6\left(\frac{2}{3}x - \frac{1}{6} + \frac{1}{2}x\right) > 6\left(\frac{7}{6} + 2x\right) \quad \text{Multiplying by 6 on both sides}$$
$$6 \cdot \frac{2}{3}x - 6 \cdot \frac{1}{6} + 6 \cdot \frac{1}{2}x > 6 \cdot \frac{7}{6} + 6 \cdot 2x \quad \text{Using the distributive law}$$
$$4x - 1 + 3x > 7 + 12x \quad \text{Simplifying}$$
$$7x - 1 > 7 + 12x \quad \text{Collecting like terms}$$
$$7x - 1 - 7x > 7 + 12x - 7x \quad \text{Subtracting } 7x.\text{ The coefficient of the } x\text{-term will be positive.}$$
$$-1 > 7 + 5x \quad \text{Simplifying}$$
$$-1 - 7 > 7 + 5x - 7 \quad \text{Subtracting 7}$$
$$-8 > 5x \quad \text{Simplifying}$$
$$\frac{-8}{5} > \frac{5x}{5} \quad \text{Dividing by 5}$$
$$-\frac{8}{5} > x$$

The solution set is $\left\{x \mid -\dfrac{8}{5} > x\right\}$, or $\left\{x \mid x < -\dfrac{8}{5}\right\}$.

Do Exercise 20. ▶

20. Solve:
$$\frac{3}{4} + x < \frac{7}{8}x - \frac{1}{4} + \frac{1}{2}x.$$

Answers

19. $\{x \mid x \ge -4\}$ **20.** $\left\{x \mid x > \dfrac{8}{3}\right\}$

For Extra Help
MyMathLab®   MathXL®   PRACTICE   WATCH   READ   REVIEW

✓ Reading Check

Classify each pair of inequalities as "equivalent" or "not equivalent."

RC1. $x + 10 \geq 12$; $x \leq 2$

RC2. $3x - 5 \leq -x + 1$; $2x \leq 6$

RC3. $-\dfrac{3}{4}y < 6$; $y > -8$

RC4. $2 - t > -3t + 4$; $2t > 2$

a Determine whether each number is a solution of the given inequality.

1. $x > -4$
  a) 4
  b) 0
  c) −4
  d) 6
  e) 5.6

2. $x \leq 5$
  a) 0
  b) 5
  c) −1
  d) −5
  e) $7\dfrac{1}{4}$

3. $x \geq 6.8$
  a) −6
  b) 0
  c) 6
  d) 8
  e) $-3\dfrac{1}{2}$

4. $x < 8$
  a) 8
  b) −10
  c) 0
  d) 11
  e) −4.7

b Graph on the number line.

5. $x > 4$

6. $x < 0$

7. $t < -3$

8. $y > 5$

9. $m \geq -1$

10. $x \leq -2$

11. $-3 < x \leq 4$

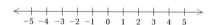

12. $-5 \leq x < 2$

13. $0 < x < 3$

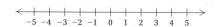

14. $-5 \leq x \leq 0$

c Solve using the addition principle. Then graph.

15. $x + 7 > 2$

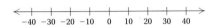

16. $x + 5 > 2$

17. $x + 8 \leq -10$

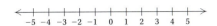

18. $x + 8 \leq -11$

Solve using the addition principle.

19. $y - 7 > -12$

20. $y - 9 > -15$

21. $2x + 3 > x + 5$

22. $2x + 4 > x + 7$

23. $3x + 9 \leq 2x + 6$

24. $3x + 18 \leq 2x + 16$

25. $5x - 6 < 4x - 2$

26. $9x - 8 < 8x - 9$

27. $-9 + t > 5$

28. $-8 + p > 10$

29. $y + \dfrac{1}{4} \leq \dfrac{1}{2}$

30. $x - \dfrac{1}{3} \leq \dfrac{5}{6}$

31. $x - \dfrac{1}{3} > \dfrac{1}{4}$

32. $x + \dfrac{1}{8} > \dfrac{1}{2}$

d Solve using the multiplication principle. Then graph.

33. $5x < 35$

34. $8x \geq 32$

35. $-12x > -36$

36. $-16x > -64$

Solve using the multiplication principle.

37. $5y \geq -2$

38. $3x < -4$

39. $-2x \leq 12$

40. $-3x \leq 15$

41. $-4y \geq -16$

42. $-7x < -21$

43. $-3x < -17$

44. $-5y > -23$

45. $-2y > \dfrac{1}{7}$

46. $-4x \leq \dfrac{1}{9}$

47. $-\dfrac{6}{5} \leq -4x$

48. $-\dfrac{7}{9} > 63x$

e Solve using the addition principle and the multiplication principle.

49. $4 + 3x < 28$

50. $3 + 4y < 35$

51. $3x - 5 \leq 13$

52. $5y - 9 \leq 21$

53. $13x - 7 < -46$

54. $8y - 6 < -54$

55. $30 > 3 - 9x$

56. $48 > 13 - 7y$

57. $4x + 2 - 3x \leq 9$

58. $15x + 5 - 14x \leq 9$

59. $-3 < 8x + 7 - 7x$

60. $-8 < 9x + 8 - 8x - 3$

61. $6 - 4y > 4 - 3y$

62. $9 - 8y > 5 - 7y + 2$

63. $5 - 9y \leq 2 - 8y$

64. $6 - 18x \leq 4 - 12x - 5x$

65. $19 - 7y - 3y < 39$

66. $18 - 6y - 4y < 63 + 5y$

67. $0.9x + 19.3 > 5.3 - 2.6x$

68. $0.96y - 0.79 \leq 0.21y + 0.46$

69. $\dfrac{x}{3} - 2 \leq 1$

70. $\dfrac{2}{3} + \dfrac{x}{5} < \dfrac{4}{15}$

71. $\dfrac{y}{5} + 1 \leq \dfrac{2}{5}$

72. $\dfrac{3x}{4} - \dfrac{7}{8} \geq -15$

73. $3(2y - 3) < 27$

74. $4(2y - 3) > 28$

75. $2(3 + 4m) - 9 \geq 45$

76. $3(5 + 3m) - 8 \leq 88$

77. $8(2t + 1) > 4(7t + 7)$

78. $7(5y - 2) > 6(6y - 1)$

79. $3(r - 6) + 2 < 4(r + 2) - 21$

80. $5(x + 3) + 9 \leq 3(x - 2) + 6$

81. $0.8(3x + 6) \geq 1.1 - (x + 2)$

82. $0.4(2x + 8) \geq 20 - (x + 5)$

83. $\dfrac{5}{3} + \dfrac{2}{3}x < \dfrac{25}{12} + \dfrac{5}{4}x + \dfrac{3}{4}$

84. $1 - \dfrac{2}{3}y \geq \dfrac{9}{5} - \dfrac{y}{5} + \dfrac{3}{5}$

Skill Maintenance

Add or subtract. [1.3a], [1.4a]

85. $-\dfrac{3}{4} + \dfrac{1}{8}$

86. $8.12 - 9.23$

87. $-2.3 - 7.1$

88. $-\dfrac{3}{4} - \dfrac{1}{8}$

Simplify.

89. $5 - 3^2 + (8 - 2)^2 \cdot 4$ [1.8d]

90. $10 \div 2 \cdot 5 - 3^2 + (-5)^2$ [1.8d]

91. $5(2x - 4) - 3(4x + 1)$ [1.8b]

92. $9(3 + 5x) - 4(7 + 2x)$ [1.8b]

Synthesis

93. Determine whether each number is a solution of the inequality $|x| < 3$.

  a) 0
  b) -2
  c) -3
  d) 4
  e) 3
  f) 1.7
  g) -2.8

94. Graph $|x| < 3$ on the number line.

$$\xleftarrow{\hspace{0.3cm}} \overset{\hspace{0.2cm}}{\underset{-5 \;\; -4 \;\; -3 \;\; -2 \;\; -1 \;\;\; 0 \;\;\; 1 \;\;\; 2 \;\;\; 3 \;\;\; 4 \;\;\; 5}{|\;\;|\;\;|\;\;|\;\;|\;\;|\;\;|\;\;|\;\;|\;\;|\;\;|}} \xrightarrow{\hspace{0.3cm}}$$

Solve.

95. $x + 3 < 3 + x$

96. $x + 4 > 3 + x$

OBJECTIVES

a Translate number sentences to inequalities.

b Solve applied problems using inequalities.

SKILL TO REVIEW

Objective 2.7d: Solve inequalities using the multiplication principle.

Solve.

1. $-8x \leq -512$

2. $-300 > 15y$

Translate.

1. Sara worked no fewer than 15 hr last week.

2. The price of that Volkswagen Beetle convertible is at most $31,210.

3. The time of the test was between 45 min and 55 min.

4. Camila's weight is less than 110 lb.

5. That number is more than -2.

6. The costs of production of that marketing video cannot exceed $12,500.

7. At most 1250 people attended the concert.

8. Yesterday, at least 23 people got tickets for speeding.

The five steps for problem solving can be used for problems involving inequalities.

a TRANSLATING TO INEQUALITIES

Before solving problems that involve inequalities, we list some important phrases to look for. Sample translations are listed as well.

IMPORTANT WORDS	SAMPLE SENTENCE	TRANSLATION
is at least	Bill is at least 21 years old.	$b \geq 21$
is at most	At most 5 students dropped the course.	$n \leq 5$
cannot exceed	To qualify, earnings cannot exceed $12,000.	$r \leq 12{,}000$
must exceed	The speed must exceed 15 mph.	$s > 15$
is less than	Tucker's weight is less than 50 lb.	$w < 50$
is more than	Nashville is more than 200 mi away.	$d > 200$
is between	The film is between 90 min and 100 min long.	$90 < t < 100$
no more than	Cooper weighs no more than 90 lb.	$w \leq 90$
no less than	Sofia scored no less than 8.3.	$s \geq 8.3$

The following phrases deserve special attention.

TRANSLATING "AT LEAST" AND "AT MOST"

A quantity x is at least some amount q: $x \geq q$.

   (If x is at least q, it cannot be less than q.)

A quantity x is at most some amount q: $x \leq q$.

   (If x is at most q, it cannot be more than q.)

◀ **Do Margin Exercises 1–8.**

b SOLVING PROBLEMS

EXAMPLE 1 *Catering Costs.* To cater a company's annual lobster-bake cookout, Jayla's Catering charges a $325 setup fee plus $18.50 per person. The cost cannot exceed $3200. How many people can attend the cookout?

1. **Familiarize.** Suppose that 130 people were to attend the cookout. The cost would then be $325 + $18.50(130)$, or $2730. This shows that more than 130 people could attend the picnic without exceeding $3200. Instead of making another guess, we let $n =$ the number of people in attendance.

Answers

Skill to Review:

1. $\{x \,|\, x \geq 64\}$ 2. $\{y \,|\, y < -20\}$

Margin Exercises:

1. $h \geq 15$ 2. $p \leq 31{,}210$ 3. $45 < t < 55$
4. $w < 110$ 5. $n > -2$ 6. $c \leq 12{,}500$
7. $p \leq 1250$ 8. $s \geq 23$

2. **Translate.** Our guess shows us how to translate. The cost of the cookout will be the $325 setup fee plus $18.50 times the number of people attending. We translate to an inequality:

Rewording: The setup fee   plus   the cost of the meals   cannot exceed $3200.

Translating: 325   +   18.50n   ≤   3200.

3. **Solve.** We solve the inequality for n:

$$325 + 18.50n \le 3200$$

$$325 + 18.50n - 325 \le 3200 - 325 \qquad \text{Subtracting 325}$$

$$18.50n \le 2875 \qquad \text{Simplifying}$$

$$\frac{18.50n}{18.50} \le \frac{2875}{18.50} \qquad \text{Dividing by 18.50}$$

$$n \le 155.4. \qquad \text{Rounding to the nearest tenth}$$

4. **Check.** Although the solution set of the inequality is all numbers less than or equal to about 155.4, since n = the number of people in attendance, we round *down* to 155 people. If 155 people attend, the cost will be $325 + $18.50(155)$, or $3192.50. If 156 attend, the cost will exceed $3200.

5. **State.** At most, 155 people can attend the lobster-bake cookout.

Do Exercise 9. ▶

·· **Caution!** ··

Solutions of problems should always be checked using the original wording of the problem. In some cases, answers might need to be whole numbers or integers or rounded off in a particular direction.

··

EXAMPLE 2 *Nutrition.* The U.S. Department of Agriculture recommends that for a typical 2000-calorie daily diet, no more than 20 g of saturated fat be consumed. In the first three days of a four-day vacation, Ethan consumed 26 g, 17 g, and 22 g of saturated fat. Determine (in terms of an inequality) how many grams of saturated fat Ethan can consume on the fourth day if he is to average no more than 20 g of saturated fat per day.

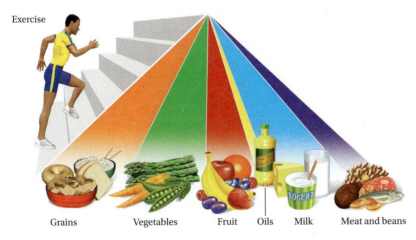

Exercise

Grains   Vegetables   Fruit   Oils   Milk   Meat and beans

SOURCES: U.S. Department of Health and Human Services; U.S. Department of Agriculture

1. **Familiarize.** Suppose Ethan consumed 19 g of saturated fat on the fourth day. His daily average for the vacation would then be

$$\frac{26\,g + 17\,g + 22\,g + 19\,g}{4} = \frac{84\,g}{4} = 21\,g.$$

This shows that Ethan cannot consume 19 g of saturated fat on the fourth day, if he is to average no more than 20 g of fat per day. We let $x =$ the number of grams of fat that Ethan can consume on the fourth day.

2. **Translate.** We reword the problem and translate to an inequality as follows:

Rewording: The average consumption of saturated fat   should be no more than   20 g.

Translating: $\dfrac{26 + 17 + 22 + x}{4}$ $\leq$ 20.

3. **Solve.** Because of the fraction expression, it is convenient to use the multiplication principle first to clear the fraction:

$$\frac{26 + 17 + 22 + x}{4} \leq 20$$

$$4\left(\frac{26 + 17 + 22 + x}{4}\right) \leq 4 \cdot 20 \qquad \text{Multiplying by 4}$$

$$26 + 17 + 22 + x \leq 80$$

$$65 + x \leq 80 \qquad \text{Simplifying}$$

$$x \leq 15. \qquad \text{Subtracting 65}$$

4. **Check.** As a partial check, we show that Ethan can consume 15 g of saturated fat on the fourth day and not exceed a 20-g average for the four days:

$$\frac{26 + 17 + 22 + 15}{4} = \frac{80}{4} = 20.$$

5. **State.** Ethan's average intake of saturated fat for the vacation will not exceed 20 g per day if he consumes no more than 15 g of saturated fat on the fourth day.

◀ **Do Exercise 10.**

Translate to an inequality and solve.

10. *Test Scores.* A pre-med student is taking a chemistry course in which four tests are given. To get an A, she must average at least 90 on the four tests. The student got scores of 91, 86, and 89 on the first three tests. Determine (in terms of an inequality) what scores on the last test will allow her to get an A.

Answer

10. $\dfrac{91 + 86 + 89 + s}{4} \geq 90; \{s \mid s \geq 94\}$

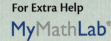
✓ Reading Check

Match each sentence with one of the following.

$$q < r \qquad q \leq r \qquad r < q \qquad r \leq q$$

RC1. r is at most q.

RC2. q is no more than r.

RC3. r is less than q.

RC4. r is at least q.

RC5. q exceeds r.

RC6. q is no less than r.

a   Translate to an inequality.

1. A number is at least 7.

2. A number is greater than or equal to 5.

3. The baby weighs more than 2 kilograms (kg).

4. Between 75 and 100 people attended the concert.

5. The speed of the train was between 90 mph and 110 mph.

6. The attendance was no more than 180.

7. Brianna works no more than 20 hr per week.

8. The amount of acid must exceed 40 liters (L).

9. The cost of gasoline is no less than $3.20 per gallon.

10. The temperature is at most $-2°$.

11. A number is greater than 8.

12. A number is less than 5.

13. A number is less than or equal to -4.

14. A number is greater than or equal to 18.

15. The number of people is at least 1300.

16. The cost is at most $4857.95.

17. The amount of water is not to exceed 500 liters.

18. The cost of ground beef is no less than $3.19 per pound.

19. Two more than three times a number is less than 13.

20. Five less than one-half a number is greater than 17.

b   Solve.

21. *Test Scores.*   Xavier is taking a geology course in which four tests are given. To get a B, he must average at least 80 on the four tests. He got scores of 82, 76, and 78 on the first three tests. Determine (in terms of an inequality) what scores on the last test will allow him to get at least a B.

22. *Test Scores.*   Chloe is taking a French class in which five quizzes are given. Her first four quiz grades are 73, 75, 89, and 91. Determine (in terms of an inequality) what scores on the last quiz will allow her to get an average quiz grade of at least 85.

23. *Gold Temperatures.*   Gold stays solid at Fahrenheit temperatures below 1945.4°. Determine (in terms of an inequality) those Celsius temperatures for which gold stays solid. Use the formula given in Margin Exercise 9.

24. *Body Temperatures.*   The human body is considered to be fevered when its temperature is higher than 98.6°F. Using the formula given in Margin Exercise 9, determine (in terms of an inequality) those Celsius temperatures for which the body is fevered.

25. *World Records in the 1500-m Run.*   The formula
$$R = -0.075t + 3.85$$
can be used to predict the world record in the 1500-m run t years after 1930. Determine (in terms of an inequality) those years for which the world record will be less than 3.5 min.

26. *World Records in the 200-m Dash.*   The formula
$$R = -0.028t + 20.8$$
can be used to predict the world record in the 200-m dash t years after 1920. Determine (in terms of an inequality) those years for which the world record will be less than 19.0 sec.

27. *Blueprints.*   To make copies of blueprints, Vantage Reprographics charges a $5 setup fee plus $4 per copy. Myra can spend no more than $65 for copying her blueprints. What numbers of copies will allow her to stay within budget?

28. *Banquet Costs.*   The Shepard College women's volleyball team can spend at most $750 for its awards banquet at a local restaurant. If the restaurant charges an $80 setup fee plus $16 per person, at most how many can attend?

29. *Envelope Size.* For a direct-mail campaign, Hollcraft Advertising determines that any envelope with a fixed width of $3\frac{1}{2}$ in. and an area of at least $17\frac{1}{2}$ in^2 can be used. Determine (in terms of an inequality) those lengths that will satisfy the company constraints.

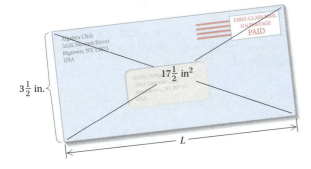

30. *Package Sizes.* Logan Delivery Service accepts packages of up to 165 in. in length and girth combined. (Girth is the distance around the package.) A package has a fixed girth of 53 in. Determine (in terms of an inequality) those lengths for which a package is acceptable.

31. *Phone Costs.* Simon claims that it costs him at least $3.00 every time he calls an overseas customer. If his typical call costs 75¢ plus 45¢ for each minute, how long do his calls typically last? (*Hint*: 75¢ = $0.75.)

32. *Parking Costs.* Laura is certain that every time she parks in the municipal garage it costs her at least $6.75. If the garage charges $1.50 plus 75¢ for each half hour, for how long is Laura's car generally parked?

33. *College Tuition.* Angelica's financial aid stipulates that her tuition cannot exceed $1000. If her local community college charges a $35 registration fee plus $375 per course, what is the greatest number of courses for which Angelica can register?

34. *Furnace Repairs.* RJ's Plumbing and Heating charges $45 for a service call plus $30 per hour for emergency service. Gary remembers being billed over $150 for an emergency call. How long was RJ's there?

35. *Nutrition.* Following the guidelines of the Food and Drug Administration, Dale tries to eat at least 5 servings of fruits or vegetables each day. For the first six days of one week, he had 4, 6, 7, 4, 6, and 4 servings. How many servings of fruits or vegetables should Dale eat on Saturday in order to average at least 5 servings per day for the week?

36. *College Course Load.* To remain on financial aid, Millie needs to complete an average of at least 7 credits per quarter each year. In the first three quarters of 2013, Millie completed 5, 7, and 8 credits. How many credits of course work must Millie complete in the fourth quarter if she is to remain on financial aid?

37. Perimeter of a Rectangle. The width of a rectangle is fixed at 8 ft. What lengths will make the perimeter at least 200 ft? at most 200 ft?

38. Perimeter of a Triangle. One side of a triangle is 2 cm shorter than the base. The other side is 3 cm longer than the base. What lengths of the base will allow the perimeter to be greater than 19 cm?

39. Area of a Rectangle. The width of a rectangle is fixed at 4 cm. For what lengths will the area be less than 86 cm^2?

$$L$$

4 cm $A < 86 \text{ cm}^2$ 4 cm

$$L$$

40. Area of a Rectangle. The width of a rectangle is fixed at 16 yd. For what lengths will the area be at least 264 yd^2?

41. Insurance-Covered Repairs. Most insurance companies will replace a vehicle if an estimated repair exceeds 80% of the "blue-book" value of the vehicle. Rachel's insurance company paid $8500 for repairs to her Toyota after an accident. What can be concluded about the blue-book value of the car?

42. Insurance-Covered Repairs. Following an accident, Jeff's Ford pickup was replaced by his insurance company because the damage was so extensive. Before the damage, the blue-book value of the truck was $21,000. How much would it have cost to repair the truck? (See Exercise 41.)

43. Reduced-Fat Foods. In order for a food to be labeled "reduced fat," it must have at least 25% less fat than the regular item. One brand of reduced-fat peanut butter contains 12 g of fat per serving. What can you conclude about the fat content in a serving of the brand's regular peanut butter?

44. Reduced-Fat Foods. One brand of reduced-fat chocolate chip cookies contains 5 g of fat per serving. What can you conclude about the fat content of the brand's regular chocolate chip cookies? (See Exercise 43.)

45. Area of a Triangular Flag. As part of an outdoor education course at Baxter YMCA, Wendy needs to make a bright-colored triangular flag with an area of at least 3 ft^2. What heights can the triangle be if the base is $1\frac{1}{2}$ ft?

46. Area of a Triangular Sign. Zoning laws in Harrington prohibit displaying signs with areas exceeding 12 ft^2. If Flo's Marina is ordering a triangular sign with an 8-ft base, how tall can the sign be?

47. Pond Depth. On July 1, Garrett's Pond was 25 ft deep. Since that date, the water level has dropped $\frac{2}{3}$ ft per week. For what dates will the water level not exceed 21 ft?

48. Weight Gain. A 3-lb puppy is gaining weight at a rate of $\frac{3}{4}$ lb per week. When will the puppy's weight exceed $22\frac{1}{2}$ lb?

49. Electrician Visits. Dot's Electric made 17 customer calls last week and 22 calls this week. How many calls must be made next week in order to maintain a weekly average of at least 20 calls for the three-week period?

50. Volunteer Work. George and Joan do volunteer work at a hospital. Joan worked 3 more hr than George, and together they worked more than 27 hr. What possible numbers of hours did each work?

Skill Maintenance

Solve.

51. $-13 + x = 27$ [2.1b]

52. $-6y = 132$ [2.2a]

53. $4a - 3 = 45$ [2.3a]

54. $8x + 3x = 66$ [2.3b]

55. $-\frac{1}{2} + x = x - \frac{1}{2}$ [2.3c]

56. $9x - 1 + 11x - 18 = 3x - 15 + 4 + 17x$ [2.3c]

Solve.   [2.5a]

57. What percent of 200 is 15?

58. What is 10% of 310?

59. 25 is 2% of what number?

60. 80 is what percent of 96?

Synthesis

Solve.

61. Ski Wax. Green ski wax works best between 5° and 15° Fahrenheit. Determine those Celsius temperatures for which green ski wax works best. Use the formula given in Margin Exercise 9.

62. Parking Fees. Mack's Parking Garage charges $4.00 for the first hour and $2.50 for each additional hour. For how long has a car been parked when the charge exceeds $16.50?

63. Low-Fat Foods. In order for a food to be labeled "low fat," it must have fewer than 3 g of fat per serving. One brand of reduced-fat tortilla chips contains 60% less fat than regular nacho cheese tortilla chips, but still cannot be labeled low fat. What can you conclude about the fat content of a serving of nacho cheese tortilla chips?

64. Parking Fees. When asked how much the parking charge is for a certain car, Mack replies "between 14 and 24 dollars." For how long has the car been parked? (See Exercise 62.)

Vocabulary Reinforcement

Complete each statement with the correct word or words from the column on the right. Some of the choices may not be used.

addition principle
multiplication principle
solution
equivalent
equation
inequality

1. Any replacement for the variable that makes an equation true is called a(n) _____ of the equation.   [2.1a]

2. The _____ for equations states that for any real numbers $a, b,$ and $c, a = b$ is equivalent to $a + c = b + c$.   [2.1b]

3. The _____ for equations states that for any real numbers $a, b,$ and $c, a = b$ is equivalent to $a \cdot c = b \cdot c$.   [2.2a]

4. An _____ is a number sentence with $<, \leq, >,$ or $\geq$ as its verb.   [2.7a]

5. Equations with the same solutions are called _____ equations.   [2.1b]

Concept Reinforcement

Determine whether each statement is true or false.

_____ 1. Some equations have no solution.   [2.3c]

_____ 2. For any number $n, n \geq n$.   [2.7a]

_____ 3. $2x - 7 < 11$ and $x < 2$ are equivalent inequalities.   [2.7c]

_____ 4. If $x > y$, then $-x < -y$.   [2.7d]

Study Guide

Objective 2.3a   Solve equations using both the addition principle and the multiplication principle.

Objective 2.3b   Solve equations in which like terms may need to be collected.

Objective 2.3c   Solve equations by first removing parentheses and collecting like terms.

Example   Solve: $6y - 2(2y - 3) = 12$.

$$6y - 2(2y - 3) = 12$$

$$6y - 4y + 6 = 12 \qquad \text{Removing parentheses}$$

$$2y + 6 = 12 \qquad \text{Collecting like terms}$$

$$2y + 6 - 6 = 12 - 6 \qquad \text{Subtracting 6}$$

$$2y = 6$$

$$\frac{2y}{2} = \frac{6}{2} \qquad \text{Dividing by 2}$$

$$y = 3$$

The solution is 3.

Practice Exercise

1. Solve: $4(x - 3) = 6(x + 2)$.

Objective 2.3c Solve equations with an infinite number of solutions and equations with no solutions.

Example Solve: $8 + 2x - 4 = 6 + 2(x - 1)$.

$$8 + 2x - 4 = 6 + 2(x - 1)$$
$$8 + 2x - 4 = 6 + 2x - 2$$
$$2x + 4 = 2x + 4$$
$$2x + 4 - 2x = 2x + 4 - 2x$$
$$4 = 4$$

Every real number is a solution of the equation $4 = 4$, so all real numbers are solutions of the original equation. The equation has infinitely many solutions.

Example Solve: $2 + 5(x - 1) = -6 + 5x + 7$.

$$2 + 5(x - 1) = -6 + 5x + 7$$
$$2 + 5x - 5 = -6 + 5x + 7$$
$$5x - 3 = 5x + 1$$
$$5x - 3 - 5x = 5x + 1 - 5x$$
$$-3 = 1$$

This is a false equation, so the original equation has no solution.

Practice Exercises

2. Solve: $4 + 3y - 7 = 3 + 3(y - 2)$.

3. Solve: $4(x - 3) + 7 = -5 + 4x + 10$.

Objective 2.4b Solve a formula for a specified letter.

Example Solve for n: $M = \dfrac{m + n}{5}$.

$$M = \frac{m + n}{5}$$
$$5 \cdot M = 5\left(\frac{m + n}{5}\right)$$
$$5M = m + n$$
$$5M - m = m + n - m$$
$$5M - m = n$$

Practice Exercise

4. Solve for b: $A = \dfrac{1}{2}bh$.

Objective 2.7b Graph an inequality on the number line.

Example Graph each inequality: **(a)** $x < 2$; **(b)** $x \geq -3$.

a) The solutions of $x < 2$ are all numbers less than 2. We shade all points to the left of 2, and we use a parenthesis at 2 to indicate that 2 *is not* part of the graph.

b) The solutions of $x \geq -3$ are all numbers greater than -3 and the number -3 as well. We shade all points to the right of -3, and we use a bracket at -3 to indicate that -3 *is* part of the graph.

Practice Exercises

5. Graph: $x > 1$.

6. Graph: $x \leq -1$.

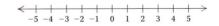

Objective 2.7e Solve inequalities using the addition principle and the multiplication principle together.

Example Solve: $8y - 7 \leq 5y + 2$.

$$8y - 7 \leq 5y + 2$$
$$8y - 7 - 8y \leq 5y + 2 - 8y$$
$$-7 \leq -3y + 2$$
$$-7 - 2 \leq -3y + 2 - 2$$
$$-9 \leq -3y$$
$$\frac{-9}{-3} \geq \frac{-3y}{-3} \qquad \textcolor{red}{\text{Reversing the symbol}}$$
$$3 \geq y$$

The solution set is $\{y \mid 3 \geq y\}$, or $\{y \mid y \leq 3\}$.

Practice Exercise

7. Solve: $6y + 5 > 3y - 7$.

Review Exercises

Solve. [2.1b]

1. $x + 5 = -17$

2. $n - 7 = -6$

3. $x - 11 = 14$

4. $y - 0.9 = 9.09$

Solve. [2.2a]

5. $-\dfrac{2}{3}x = -\dfrac{1}{6}$

6. $-8x = -56$

7. $-\dfrac{x}{4} = 48$

8. $15x = -35$

9. $\dfrac{4}{5}y = -\dfrac{3}{16}$

Solve. [2.3a]

10. $5 - x = 13$

11. $\dfrac{1}{4}x - \dfrac{5}{8} = \dfrac{3}{8}$

Solve. [2.3b, c]

12. $5t + 9 = 3t - 1$

13. $7x - 6 = 25x$

14. $14y = 23y - 17 - 10$

15. $0.22y - 0.6 = 0.12y + 3 - 0.8y$

16. $\dfrac{1}{4}x - \dfrac{1}{8}x = 3 - \dfrac{1}{16}x$

17. $14y + 17 + 7y = 9 + 21y + 8$

18. $4(x + 3) = 36$

19. $3(5x - 7) = -66$

20. $8(x - 2) - 5(x + 4) = 20 + x$

21. $-5x + 3(x + 8) = 16$

22. $6(x - 2) - 16 = 3(2x - 5) + 11$

Determine whether the given number is a solution of the inequality $x \leq 4$. [2.7a]

23. -3 **24.** 7   **25.** 4

Solve. Write set notation for the answers. [2.7c, d, e]

26. $y + \dfrac{2}{3} \geq \dfrac{1}{6}$

27. $9x \geq 63$

28. $2 + 6y > 14$

29. $7 - 3y \geq 27 + 2y$

30. $3x + 5 < 2x - 6$

31. $-4y < 28$

32. $4 - 8x < 13 + 3x$

33. $-4x \leq \dfrac{1}{3}$

Graph on the number line. [2.7b, e]

34. $4x - 6 < x + 3$

35. $-2 < x \leq 5$

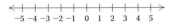

36. $y > 0$

Solve. [2.4b]

37. $C = \pi d$, for d

38. $V = \dfrac{1}{3}Bh$, for B

39. $A = \dfrac{a + b}{2}$, for a

40. $y = mx + b$, for x

Solve. [2.6a]

41. *Dimensions of Wyoming.* The state of Wyoming is roughly in the shape of a rectangle whose perimeter is 1280 mi. The length is 90 mi more than the width. Find the dimensions.

42. *Interstate Mile Markers.* The sum of two consecutive mile markers on I-5 in California is 691. Find the numbers on the markers.

43. An entertainment center sold for $2449 in June. This was $332 more than the cost in February. What was the cost in February?

44. Ty is paid a commission of $4 for each magazine subscription he sells. One week, he received $108 in commissions. How many subscriptions did he sell?

45. The measure of the second angle of a triangle is 50° more than that of the first angle. The measure of the third angle is 10° less than twice the first angle. Find the measures of the angles.

Solve. [2.5a]

46. What number is 20% of 75?

47. Fifteen is what percent of 80?

48. 18 is 3% of what number?

49. Black Bears. The population of black bears in the 26 U.S. states that border on or are east of the Mississippi River increased from 87,872 in 1988 to 164,440 in 2008. What is the percent increase?

Source: Hank Hristienko, Manitoba Conservation Wildlife and Ecosystem Protection Branch

50. First-Class Mail. The volume of first-class mail decreased from 102.4 billion pieces in 2002 to only 73.5 billion pieces in 2011. What is the percent decrease?

Source: United States Postal Service

Solve. [2.6a]

51. After a 30% reduction, a bread maker is on sale for $154. What was the marked price (the price before the reduction)?

52. A restaurant manager's salary is $78,300, which is an 8% increase over the previous year's salary. What was the previous salary?

53. A tax-exempt organization received a bill of $145.90 for janitorial supplies. The bill incorrectly included sales tax of 5%. How much does the organization actually owe?

Solve. [2.8b]

54. Test Scores. Noah's test grades are 71, 75, 82, and 86. What is the lowest grade that he can get on the next test and still have an average test score of at least 80?

55. The length of a rectangle is 43 cm. What widths will make the perimeter greater than 120 cm?

56. The solution of the equation
$$4(3x - 5) + 6 = 8 + x$$
is which of the following? [2.3c]
  A. Less than -1 **B.** Between -1 and 1
  C. Between 1 and 5   **D.** Greater than 5

57. Solve for y: $3x + 4y = P$. [2.4b]
  A. $y = \dfrac{P - 3x}{4}$ **B.** $y = \dfrac{P + 3x}{4}$

  C. $y = P - \dfrac{3x}{4}$ **D.** $y = \dfrac{P}{4} - 3x$

Synthesis

Solve.

58. $2|x| + 4 = 50$ [1.2e], [2.3a]

59. $|3x| = 60$ [1.2e], [2.2a]

60. $y = 2a - ab + 3$, for a [2.4b]

Understanding Through Discussion and Writing

1. Would it be better to receive a 5% raise and then an 8% raise or the other way around? Why? [2.5a]

2. Erin returns a tent that she bought during a storewide 25%-off sale that has ended. She is offered store credit for 125% of what she paid (not to be used on sale items). Is this fair to Erin? Why or why not? [2.5a]

3. Are the inequalities $x > -5$ and $-x < 5$ equivalent? Why or why not? [2.7d]

4. Explain in your own words why it is necessary to reverse the inequality symbol when multiplying on both sides of an inequality by a negative number. [2.7d]

5. If f represents Fran's age and t represents Todd's age, write a sentence that would translate to $t + 3 < f$. [2.8a]

6. Explain how the meanings of "Five more than a number" and "Five is more than a number" differ. [2.8a]

CHAPTER

2 Test

For Extra Help For step-by-step test solutions, access the Chapter Test Prep Videos in MyMathLab® or on YouTube (search "BittingerIntro" and click on "Channels").

Solve.

1. $x + 7 = 15$

2. $t - 9 = 17$

3. $3x = -18$

4. $-\dfrac{4}{7}x = -28$

5. $3t + 7 = 2t - 5$

6. $\dfrac{1}{2}x - \dfrac{3}{5} = \dfrac{2}{5}$

7. $8 - y = 16$

8. $-\dfrac{2}{5} + x = -\dfrac{3}{4}$

9. $3(x + 2) = 27$

10. $-3x - 6(x - 4) = 9$

11. $0.4p + 0.2 = 4.2p - 7.8 - 0.6p$

12. $4(3x - 1) + 11 = 2(6x + 5) - 8$

13. $-2 + 7x + 6 = 5x + 4 + 2x$

Solve. Write set notation for the answers.

14. $x + 6 \le 2$

15. $14x + 9 > 13x - 4$

16. $12x \le 60$

17. $-2y \ge 26$

18. $-4y \le -32$

19. $-5x \ge \dfrac{1}{4}$

20. $4 - 6x > 40$

21. $5 - 9x \ge 19 + 5x$

Graph on the number line.

22. $y \le 9$

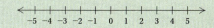

23. $6x - 3 < x + 2$

24. $-2 \le x \le 2$

Solve.

25. What number is 24% of 75?

26. 15.84 is what percent of 96?

27. 800 is 2% of what number?

28. *Bottled Water.* Annual bottled water consumption per person in the United States increased from 18.2 gal in 2001 to 29.2 gal in 2011. What is the percent increase?

Source: Beverage Marketing Corporation

29. Perimeter of a Photograph. The perimeter of a rectangular photograph is 36 cm. The length is 4 cm greater than the width. Find the width and the length.

30. Cost of Raising a Child. It is estimated that $41,500 will be spent for child care and K–12 education for a child to age 17. This number represents approximately 18% of the total cost of raising a child to age 17. What is the total cost of raising a child to age 17?

Sources: U.S. Department of Agriculture

31. Raffle Tickets. The numbers on three raffle tickets are consecutive integers whose sum is 7530. Find the integers.

32. Savings Account. Money is invested in a savings account at 5% simple interest. After 1 year, there is $924 in the account. How much was originally invested?

33. Board Cutting. An 8-m board is cut into two pieces. One piece is 2 m longer than the other. How long are the pieces?

34. Lengths of a Rectangle. The width of a rectangle is 96 yd. Find all possible lengths such that the perimeter of the rectangle will be at least 540 yd.

x + 2

8 m

x

35. Budgeting. Jason has budgeted an average of $95 per month for entertainment. For the first five months of the year, he has spent $98, $89, $110, $85, and $83. How much can Jason spend in the sixth month without exceeding his average budget?

36. Copy Machine Rental. A catalog publisher needs to lease a copy machine for use during a special project that they anticipate will take 3 months. It costs $225 per month plus 3.2¢ per copy to rent the machine. The company must stay within a budget of $4500 for copies. Determine (in terms of an inequality) the number of copies they can make and still remain within budget.

37. Solve $A = 2\pi rh$ for r.

38. Solve $y = 8x + b$ for x.

39. Senior Population. The number of Americans ages 65 and older is projected to grow from 40.4 million to 70.3 million between 2011 and 2030. Find the percent increase.

Source: U.S. Census Bureau

A. 42.5%   **B.** 47%
C. 57.5%   **D.** 74%

Synthesis

40. Solve $c = \dfrac{1}{a - d}$ for d.

41. Solve: $3|w| - 8 = 37$.

42. A movie theater had a certain number of tickets to give away. Five people got the tickets. The first got one-third of the tickets, the second got one-fourth of the tickets, and the third got one-fifth of the tickets. The fourth person got eight tickets, and there were five tickets left for the fifth person. Find the total number of tickets given away.

Evaluate.

1. $\dfrac{y - x}{4}$, when $y = 12$ and $x = 6$

2. $\dfrac{3x}{y}$, when $x = 5$ and $y = 4$

3. $x - 3$, when $x = 3$

4. Translate to an algebraic expression: Four less than twice w.

Use $<$ or $>$ for $\square$ to write a true sentence.

5. $-4 \;\square\; -6$

6. $0 \;\square\; -5$

7. $-8 \;\square\; 7$

8. Find the opposite and the reciprocal of $\dfrac{2}{5}$.

Find the absolute value.

9. $|3|$

10. $\left| -\dfrac{3}{4} \right|$

11. $|0|$

Compute and simplify.

12. $-6.7 + 2.3$

13. $-\dfrac{1}{6} - \dfrac{7}{3}$

14. $-\dfrac{5}{8}\left(-\dfrac{4}{3} \right)$

15. $(-7)(5)(-6)(-0.5)$

16. $81 \div (-9)$

17. $-10.8 \div 3.6$

18. $-\dfrac{4}{5} \div -\dfrac{25}{8}$

Multiply.

19. $5(3x + 5y + 2z)$

20. $4(-3x - 2)$

21. $-6(2y - 4x)$

Factor.

22. $64 + 18x + 24y$

23. $16y - 56$

24. $5a - 15b + 25$

Collect like terms.

25. $9b + 18y + 6b + 4y$

26. $3y + 4 + 6z + 6y$

27. $-4d - 6a + 3a - 5d + 1$

28. $3.2x + 2.9y - 5.8x - 8.1y$

Simplify.

29. $7 - 2x - (-5x) - 8$

30. $-3x - (-x + y)$

31. $-3(x - 2) - 4x$

32. $10 - 2(5 - 4x)$

33. $[3(x + 6) - 10] - [5 - 2(x - 8)]$

Solve.

34. $x + 1.75 = 6.25$

35. $\dfrac{5}{2}y = \dfrac{2}{5}$

36. $-2.6 + x = 8.3$

37. $4\dfrac{1}{2} + y = 8\dfrac{1}{3}$

38. $-\dfrac{3}{4}x = 36$

39. $\dfrac{2}{5}x = -\dfrac{3}{20}$

40. $5.8x = -35.96$

41. $-4x + 3 = 15$

42. $-3x + 5 = -8x - 7$

43. $4y - 4 + y = 6y + 20 - 4y$

44. $-3(x - 2) = -15$

45. $\dfrac{1}{3}x - \dfrac{5}{6} = \dfrac{1}{2} + 2x$

46. $-3.7x + 6.2 = -7.3x - 5.8$

47. $4(x + 2) = 4(x - 2) + 16$

48. $0(x + 3) + 4 = 0$

49. $3x - 1 < 2x + 1$

50. $3y + 7 > 5y + 13$

51. $5 - y \leq 2y - 7$

52. $H = 65 - m$, for m
(To determine the number of heating degree days H for a day with m degrees Fahrenheit as the average temperature)

53. $I = Prt$, for t
(Simple-interest formula, where I is interest, P is principal, r is interest rate, and t is time)

54. What number is 24% of 105?

55. 39.6 is what percent of 88?

56. $163.35 is 45% of what?

Solve.

57. *Price Reduction.* After a 25% reduction, a book is on sale for $18.45. What was the price before reduction?

58. *Rollerblade Costs.* Susan and Melinda purchased rollerblades for a total of $107. Susan paid $17 more for her rollerblades than Melinda did. What did Melinda pay?

59. *Savings Investment.* Money is invested in a savings account at 8% simple interest. After 1 year, there is $1134 in the account. How much was originally invested?

60. *Wire Cutting.* A 143-m wire is cut into three pieces. The second piece is 3 m longer than the first. The third is four-fifths as long as the first. How long is each piece?

61. *Grade Average.* Nadia is taking a mathematics course in which four tests are given. To get a B, a student must average at least 80 on the four tests. Nadia scored 82, 76, and 78 on the first three tests. What scores on the last test will earn her at least a B?

62. Simplify: $-125 \div 25 \cdot 625 \div 5$.
   A. $-390{,}625$ **B.** -125
   C. -625 **D.** 25

Synthesis

63. An engineer's salary at the end of a year is $48,418.24. This reflects a 4% salary increase and a later 3% cost-of-living adjustment during the year. What was the salary at the beginning of the year?

64. Ava needs to use a copier to reduce a drawing to fit on a page. The original drawing is 9 in. long and it must fit into a space that is 6.3 in. long. By what percent should she reduce the drawing?

Solve.

65. $4|x| - 13 = 3$

66. $\dfrac{2 + 5x}{4} = \dfrac{11}{28} + \dfrac{8x + 3}{7}$

67. $p = \dfrac{2}{m + Q}$, for Q

Graphs of Linear Equations

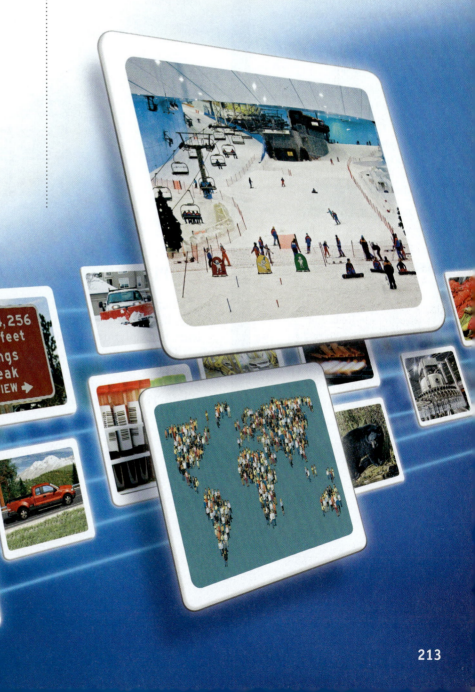

3.1 Graphs and Applications of Linear Equations

OBJECTIVES

a Plot points associated with ordered pairs of numbers; determine the quadrant in which a point lies.

b Find the coordinates of a point on a graph.

c Determine whether an ordered pair is a solution of an equation with two variables.

d Graph linear equations of the type $y = mx + b$ and $Ax + By = C$, identifying the y-intercept.

e Solve applied problems involving graphs of linear equations.

SKILL TO REVIEW

Objective 2.1a: Determine whether a given number is a solution of a given equation.

Determine whether −3 is a solution of each equation.

1. $8(w - 3) = 0$

2. $15 = -2y + 9$

You probably have seen bar graphs like the following in newspapers and magazines. Note that a straight line can be drawn along the tops of the bars. Such a line is a *graph of a linear equation*. In this chapter, we study how to graph linear equations and consider properties such as slope and intercepts. Many applications of these topics will also be considered.

Pieces of First-Class Mail

Number of pieces of first-class mail (in billions)

96.3, 90.7, 82.7, 77.6, 73.5, 68.7

2007 2008 2009 2010 2011 2012
Year

SOURCE: U. S. Postal Service

a PLOTTING ORDERED PAIRS

In Chapter 2, we graphed numbers and inequalities in one variable on a line. To enable us to graph an equation that contains two variables, we now learn to graph number pairs on a plane.

On the number line, each point is the graph of a number. On a plane, each point is the graph of a number pair. To form the plane, we use two perpendicular number lines called **axes**. They cross at a point called the **origin**. The arrows show the positive directions.

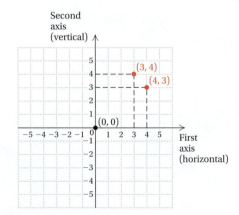

Second axis (vertical)

(3, 4)
(4, 3)
(0, 0)

First axis (horizontal)

Answers

Skill to Review:

1. −3 is not a solution.   **2.** −3 is a solution.

Consider the **ordered pair** $(3, 4)$. The numbers in an ordered pair are called **coordinates**. In $(3, 4)$, the **first coordinate** (the **abscissa**) is 3 and the **second coordinate** (the **ordinate**) is 4. To plot $(3, 4)$, we start at the origin and move *horizontally* to the 3. Then we move up *vertically* 4 units and make a "dot."

The point $(4, 3)$ is also plotted above. Note that $(3, 4)$ and $(4, 3)$ represent different points. The order of the numbers in the pair is important. We use the term *ordered* pairs because it makes a difference which number comes first. The coordinates of the origin are $(0, 0)$.

EXAMPLE 1 Plot the point $(-5, 2)$.

The first number, -5, is negative. Starting at the origin, we move -5 units in the horizontal direction (5 units to the left). The second number, 2, is positive. We move 2 units in the vertical direction (up).

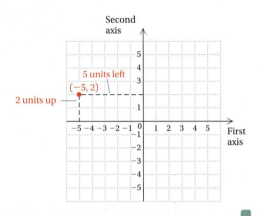

Caution!

The *first* coordinate of an ordered pair is always graphed in a *horizontal* direction, and the *second* coordinate is always graphed in a *vertical* direction.

Do Exercises 1–8.

The following figure shows some points and their coordinates. In region I (the *first quadrant*), both coordinates of any point are positive. In region II (the *second quadrant*), the first coordinate is negative and the second positive. In region III (the *third quadrant*), both coordinates are negative. In region IV (the *fourth quadrant*), the first coordinate is positive and the second is negative.

EXAMPLE 2 In which quadrant, if any, are the points $(-4, 5)$, $(5, -5)$, $(2, 4)$, $(-2, -5)$, and $(-5, 0)$ located?

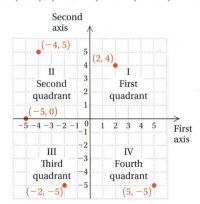

The point $(-4, 5)$ is in the second quadrant. The point $(5, -5)$ is in the fourth quadrant. The point $(2, 4)$ is in the first quadrant. The point $(-2, -5)$ is in the third quadrant. The point $(-5, 0)$ is on an axis and is *not in any quadrant*.

Do Exercises 9–16.

Plot these points on the grid below.

1. $(4, 5)$
2. $(5, 4)$
3. $(-2, 5)$
4. $(-3, -4)$
5. $(5, -3)$
6. $(-2, -1)$
7. $(0, -3)$
8. $(2, 0)$

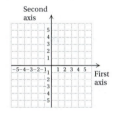

9. What can you say about the coordinates of a point in the third quadrant?

10. What can you say about the coordinates of a point in the fourth quadrant?

In which quadrant, if any, is each point located?

11. $(5, 3)$
12. $(-6, -4)$
13. $(10, -14)$
14. $(-13, 9)$
15. $(0, -3)$
16. $\left(-\dfrac{1}{2}, \dfrac{1}{4}\right)$

Answers

1.–8.
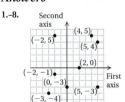

9. First, negative; second, negative
10. First, positive; second, negative **11.** I
12. III **13.** IV **14.** II **15.** On an axis, not in any quadrant **16.** II

b FINDING COORDINATES

To find the coordinates of a point, we see how far to the right or to the left of the origin it is located and how far up or down from the origin.

EXAMPLE 3 Find the coordinates of points A, B, C, D, E, F, and G.

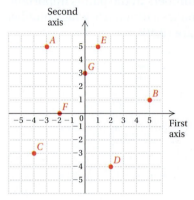

Point A is 3 units to the left (horizontal direction) and 5 units up (vertical direction). Its coordinates are $(-3, 5)$. Point D is 2 units to the right and 4 units down. Its coordinates are $(2, -4)$. The coordinates of the other points are as follows:

B: $(5, 1)$; C: $(-4, -3)$;

E: $(1, 5)$; F: $(-2, 0)$;

G: $(0, 3)$.

17. Find the coordinates of points A, B, C, D, E, F, and G on the graph below.

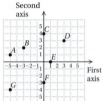

◀ **Do Exercise 17.**

c SOLUTIONS OF EQUATIONS

Now we begin to learn how graphs can be used to represent solutions of equations. When an equation contains two variables, the solutions of the equation are *ordered pairs* in which each number in the pair corresponds to a letter in the equation. Unless stated otherwise, to determine whether a pair is a solution, we use the first number in each pair to replace the variable that occurs first *alphabetically*.

EXAMPLE 4 Determine whether each of the following pairs is a solution of $4q - 3p = 22$: $(2, 7)$ and $(-1, 6)$.

For $(2, 7)$, we substitute 2 for p and 7 for q (using alphabetical order of variables):

$$\frac{4q - 3p = 22}{4 \cdot 7 - 3 \cdot 2 \;?\; 22}$$
$$28 - 6 \qquad$$
$$22 \;\bigm|\; \text{TRUE}$$

Thus, $(2, 7)$ *is* a solution of the equation.

For $(-1, 6)$, we substitute -1 for p and 6 for q:

$$\frac{4q - 3p = 22}{4 \cdot 6 - 3 \cdot (-1) \;?\; 22}$$
$$24 + 3 \qquad$$
$$27 \;\bigm|\; \text{FALSE}$$

Thus, $(-1, 6)$ *is not* a solution of the equation.

◀ **Do Exercises 18 and 19.**

18. Determine whether $(2, -4)$ is a solution of $4q - 3p = 22$. **GS**

$$\frac{4q - 3p = 22}{4 \cdot (\boxed{}) - 3 \cdot \boxed{} \;?\; 22}$$
$$-16 - \boxed{} \qquad$$
$$\boxed{} \;\bigm|\; \text{FALSE}$$

Thus, $(2, -4)$ _____ a solution.
$\underset{\text{is/is not}}{}$

19. Determine whether $(2, -4)$ is a solution of $7a + 5b = -6$.

EXAMPLE 5 Show that the pairs $(3, 7)$, $(0, 1)$, and $(-3, -5)$ are solutions of $y = 2x + 1$. Then graph the three points and use the graph to determine another pair that is a solution.

To show that a pair is a solution, we substitute, replacing x with the first coordinate and y with the second coordinate of each pair:

$$
\begin{array}{l}
y = 2x + 1 \\
\hline
7 \ ? \ 2 \cdot 3 + 1 \\
\quad\ \ 6 + 1 \\
\quad\ \ 7 \qquad\qquad \text{\textcolor{red}{TRUE}}
\end{array}
\qquad
\begin{array}{l}
y = 2x + 1 \\
\hline
1 \ ? \ 2 \cdot 0 + 1 \\
\quad\ \ 0 + 1 \\
\quad\ \ 1 \qquad\qquad \text{\textcolor{red}{TRUE}}
\end{array}
\qquad
\begin{array}{l}
y = 2x + 1 \\
\hline
-5 \ ? \ 2(-3) + 1 \\
\quad\ \ -6 + 1 \\
\quad\ \ -5 \qquad\quad \text{\textcolor{red}{TRUE}}
\end{array}
$$

In each of the three cases, the substitution results in a true equation. Thus the pairs are all solutions.

We plot the points as shown below. The order of the points follows the alphabetical order of the variables. That is, x is before y, so x-values are first coordinates and y-values are second coordinates. Similarly, we also label the horizontal axis as the x-axis and the vertical axis as the y-axis.

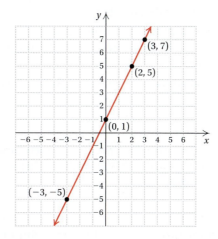

Note that the three points appear to "line up." That is, they appear to be on a straight line. Will other points that line up with these points also represent solutions of $y = 2x + 1$? To find out, we use a straightedge and sketch a line passing through $(3, 7)$, $(0, 1)$, and $(-3, -5)$.

The line appears to pass through $(2, 5)$ as well. Let's see if this pair is a solution of $y = 2x + 1$:

$$
\begin{array}{l}
y = 2x + 1 \\
\hline
5 \ ? \ 2 \cdot 2 + 1 \\
\quad\ \ 4 + 1 \\
\quad\ \ 5 \qquad\qquad \text{\textcolor{red}{TRUE}}
\end{array}
$$

Thus, $(2, 5)$ is a solution.

Do Exercise 20. ▶

Example 5 leads us to suspect that any point on the line that passes through $(3, 7)$, $(0, 1)$, and $(-3, -5)$ represents a solution of $y = 2x + 1$. In fact, every solution of $y = 2x + 1$ is represented by a point on that line and every point on that line represents a solution. The line is the *graph* of the equation.

20. Use the graph in Example 5 to find at least two more points that are solutions of $y = 2x + 1$.

Answer

20. $(-2, -3)$, $(1, 3)$; answers may vary

d GRAPHS OF LINEAR EQUATIONS

Equations like $y = 2x + 1$ and $4q - 3p = 22$ are said to be **linear** because the graph of each equation is a straight line. In general, any equation equivalent to one of the form $y = mx + b$ or $Ax + By = C$, where m, b, A, B, and C are constants (not variables) and A and B are not both 0, is linear.

To graph a linear equation:

1. Select a value for one variable and calculate the corresponding value of the other variable. Form an ordered pair using alphabetical order as indicated by the variables.

2. Repeat step (1) to obtain at least two other ordered pairs. Two points are essential to determine a straight line. A third point serves as a check.

3. Plot the ordered pairs and draw a straight line passing through the points.

In general, calculating three (or more) ordered pairs is not difficult for equations of the form $y = mx + b$. We simply substitute values for x and calculate the corresponding values for y.

EXAMPLE 6   Graph: $y = 2x$.

First, we find some ordered pairs that are solutions. We choose *any* number for x and then determine y by substitution. Since $y = 2x$, we find y by doubling x. Suppose that we choose 3 for x. Then

$$y = 2x = 2 \cdot 3 = 6.$$

We get a solution: the ordered pair $(3, 6)$.

Suppose that we choose 0 for x. Then

$$y = 2x = 2 \cdot 0 = 0.$$

We get another solution: the ordered pair $(0, 0)$.

For a third point, we make a negative choice for x. If x is -3, we have

$$y = 2x = 2 \cdot (-3) = -6.$$

This gives us the ordered pair $(-3, -6)$.

We now have enough points to plot the line, but if we wish, we can compute more. If a number takes us off the graph paper, we either do not use it or use larger paper or rescale the axes. Continuing in this manner, we create a table like the one shown on the following page.

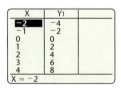

CALCULATOR CORNER

Finding Solutions of Equations   A table of values representing ordered pairs that are solutions of an equation can be displayed on a graphing calculator. To do this for the equation in Example 6, $y = 2x$, we first access the equation-editor screen. Then we clear any equations that are present. Next, we enter the equation, display the table set-up screen, and set both INDPNT and DEPEND to AUTO.

We will display a table of values that starts with $x = -2$ (TBLSTART) and adds 1 (ΔTBL) to the preceding x-value.

X	Y₁
−2	−4
−1	−2
0	0
1	2
2	4
3	6
4	8

X = −2

EXERCISE:

1. Create a table of ordered pairs that are solutions of the equations in Examples 7 and 8.

Now we plot these points. Then we draw the line, or graph, with a straightedge and label it $y = 2x$.

	y	
x	$y = 2x$	(x, y)
3	6	$(3, 6)$
1	2	$(1, 2)$
0	0	$(0, 0)$
-2	-4	$(-2, -4)$
-3	-6	$(-3, -6)$

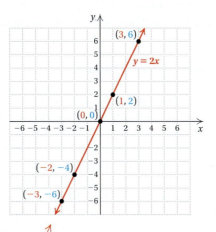

(1) Choose x.
(2) Compute y.
(3) Form the pair (x, y).
(4) Plot the points.

···················· **Caution!** ····················

Keep in mind that you can choose *any* number for x and then compute y. Our choice of certain numbers in the examples does not dictate those that you must choose.

Do Exercises 21 and 22. ▶

EXAMPLE 7 Graph: $y = -3x + 1$.

We select a value for x, compute y, and form an ordered pair. Then we repeat the process for other choices of x.

If $x = 2$, then $y = -3 \cdot 2 + 1 = -5$, and $(2, -5)$ is a solution.
If $x = 0$, then $y = -3 \cdot 0 + 1 = 1$, and $(0, 1)$ is a solution.
If $x = -1$, then $y = -3 \cdot (-1) + 1 = 4$, and $(-1, 4)$ is a solution.

Results are listed in the following table. The points corresponding to each pair are then plotted.

	y	
x	$y = -3x + 1$	(x, y)
2	-5	$(2, -5)$
0	1	$(0, 1)$
-1	4	$(-1, 4)$

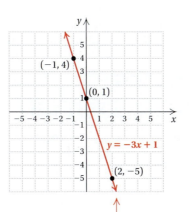

(1) Choose x.
(2) Compute y.
(3) Form the pair (x, y).
(4) Plot the points.

Complete each table and graph.

 21. $y = -2x$

x	y	(x, y)
-3	6	$(-3, 6)$
-1	☐	$(-1, ☐)$
0	0	$(0, 0)$
1	☐	$(☐, -2)$
3	☐	$(3, ☐)$

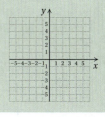

22. $y = \dfrac{1}{2}x$

x	y	(x, y)
4		
2		
0		
-2		
-4		
-1		

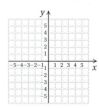

Answers

21.
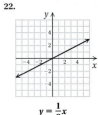
$y = -2x$

22.
$y = \dfrac{1}{2}x$

Guided Solution:
21. $2, 2, -2, 1, -6, -6$

Complete each table and graph.

23. $y = 2x + 3$

x	y	(x, y)

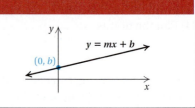

24. $y = -\dfrac{1}{2}x - 3$

x	y	(x, y)

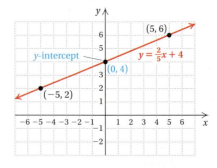

Note that all three points line up. If they did not, we would know that we had made a mistake. When only two points are plotted, a mistake is harder to detect. We use a ruler or another straightedge to draw a line through the points. Every point on the line represents a solution of $y = -3x + 1$.

◀ **Do Exercises 23 and 24.**

In Example 6, we saw that $(0, 0)$ is a solution of $y = 2x$. It is also the point at which the graph crosses the y-axis. Similarly, in Example 7, we saw that $(0, 1)$ is a solution of $y = -3x + 1$. It is also the point at which the graph crosses the y-axis. A generalization can be made: If x is replaced with 0 in the equation $y = mx + b$, then the corresponding y-value is $m \cdot 0 + b$, or b. Thus any equation of the form $y = mx + b$ has a graph that passes through the point $(0, b)$. Since $(0, b)$ is the point at which the graph crosses the y-axis, it is called the **y-intercept**. Sometimes, for convenience, we simply refer to b as the y-intercept.

---

y-INTERCEPT

The graph of the equation
$y = mx + b$ passes through the
y-intercept $(0, b)$.

EXAMPLE 8 Graph $y = \frac{2}{5}x + 4$ and identify the y-intercept.

We select a value for x, compute y, and form an ordered pair. Then we repeat the process for other choices of x. In this case, using multiples of 5 avoids fractions. We try to avoid graphing ordered pairs with fractions because they are difficult to graph accurately.

If $x = 0$,   then $y = \dfrac{2}{5} \cdot 0 + 4 = 4$,   and $(0, 4)$ is a solution.

If $x = 5$,   then $y = \dfrac{2}{5} \cdot 5 + 4 = 6$,   and $(5, 6)$ is a solution.

If $x = -5$,   then $y = \dfrac{2}{5} \cdot (-5) + 4 = 2$,   and $(-5, 2)$ is a solution.

The following table lists these solutions. Next, we plot the points and see that they form a line. Finally, we draw and label the line.

x	y = $\frac{2}{5}x$ + 4	(x, y)
0	4	(0, 4)
5	6	(5, 6)
-5	2	(-5, 2)

Answers

23.
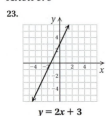
$y = 2x + 3$

24.
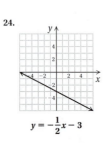
$y = -\dfrac{1}{2}x - 3$

We see that $(0, 4)$ is a solution of $y = \frac{2}{5}x + 4$. It is the y-intercept. Because the equation is in the form $y = mx + b$, we can read the y-intercept directly from the equation as follows:

$$y = \frac{2}{5}x + \underset{\uparrow}{4} \qquad (0, 4) \text{ is the } y\text{-intercept.}$$

Do Exercises 25 and 26. ▶

Calculating ordered pairs is generally easiest when y is isolated on one side of the equation, as in $y = mx + b$. To graph an equation in which y is not isolated, we can use the addition and multiplication principles to solve for y. (See Sections 2.3 and 2.4.)

EXAMPLE 9 Graph $3y + 5x = 0$ and identify the y-intercept.

To find an equivalent equation in the form $y = mx + b$, we solve for y:

$$3y + 5x = 0$$
$$3y + 5x - 5x = 0 - 5x \qquad \text{Subtracting } 5x$$
$$3y = -5x \qquad \text{Collecting like terms}$$
$$\frac{3y}{3} = \frac{-5x}{3} \qquad \text{Dividing by 3}$$
$$y = -\frac{5}{3}x.$$

Because all the equations above are equivalent, we can use $y = -\frac{5}{3}x$ to draw the graph of $3y + 5x = 0$. To graph $y = -\frac{5}{3}x$, we select x-values and compute y-values. In this case, if we select multiples of 3, we can avoid fractions.

If $x = 0$, then $y = -\frac{5}{3} \cdot 0 = 0.$

If $x = 3$, then $y = -\frac{5}{3} \cdot 3 = -5.$

If $x = -3$, then $y = -\frac{5}{3} \cdot (-3) = 5.$

We list these solutions in a table. Next, we plot the points and see that they form a line. Finally, we draw and label the line. The y-intercept is $(0, 0)$.

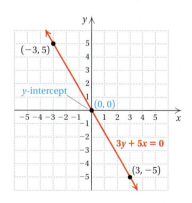

x	y
0	0	← y-intercept
3	−5
−3	5

Do Exercises 27 and 28. ▶

Graph each equation and identify the y-intercept.

25. $y = \frac{3}{5}x + 2$

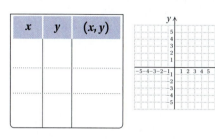

x	y	(x, y)

26. $y = -\frac{3}{5}x - 1$

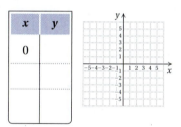

x	y	(x, y)

Graph each equation and identify the y-intercept.

27. $5y + 4x = 0$

x	y
0	

28. $4y = 3x$

x	y
0	

Answers

Answers to Margin Exercises 25–28 are on p. 222.

Graph each equation and identify the y-intercept.

29. $5y - 3x = -10$

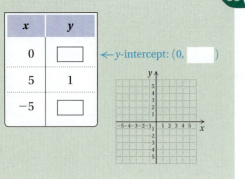

$\leftarrow$ y-intercept: (0, ▢)

x	y
0	▢
5	1
−5	▢

30. $5y + 3x = 20$

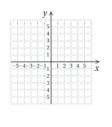

x	y

EXAMPLE 10 Graph $4y + 3x = -8$ and identify the y-intercept.

To find an equivalent equation in the form $y = mx + b$, we solve for y:

$$4y + 3x = -8$$
$$4y + 3x - 3x = -8 - 3x \qquad \text{Subtracting } 3x$$
$$4y = -3x - 8 \qquad \text{Simplifying}$$
$$\frac{1}{4} \cdot 4y = \frac{1}{4} \cdot (-3x - 8) \qquad \text{Multiplying by } \tfrac{1}{4} \text{ or dividing by } 4$$
$$y = \frac{1}{4} \cdot (-3x) - \frac{1}{4} \cdot 8 \qquad \text{Using the distributive law}$$
$$y = -\frac{3}{4}x - 2. \qquad \text{Simplifying}$$

Thus, $4y + 3x = -8$ is equivalent to $y = -\frac{3}{4}x - 2$. The y-intercept is $(0, -2)$. We find two other pairs using multiples of 4 for x to avoid fractions. We then complete and label the graph as shown.

x	y
0	−2	$\leftarrow$ y-intercept
4	−5
4	1

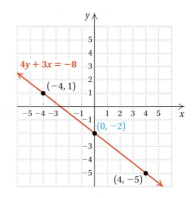

◀ **Do Exercises 29 and 30.**

Answers

25.

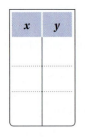

$y = \frac{3}{5}x + 2$

26.

$y = -\frac{3}{5}x - 1$

27.

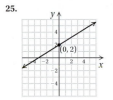

$5y + 4x = 0$

28.

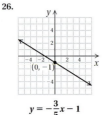

$4y = 3x$

29.

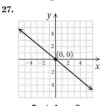

$5y - 3x = -10$

30.

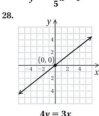

$5y + 3x = 20$

Guided Solution:
29. $-2, -5, -2$

e APPLICATIONS OF LINEAR EQUATIONS

Mathematical concepts become more understandable through visualization. Throughout this text, you will occasionally see the heading Algebraic–Graphical Connection, as in Example 11, which follows. In this feature, the algebraic approach is enhanced and expanded with a graphical connection. Relating a solution of an equation to a graph can often give added meaning to the algebraic solution.

EXAMPLE 11 *World Population.* The world population, in billions, is estimated and projected by the equation

$$y = 0.072x + 4.593,$$

where x is the number of years since 1980. That is, $x = 0$ corresponds to 1980, $x = 12$ corresponds to 1992, and so on.

Source: U.S. Census Bureau

a) Estimate the world population in 1980 and in 2005. Then project the population in 2030.

b) Graph the equation and then use the graph to estimate the world population in 2015.

c) In what year could we project the world population to be 7.761 billion?

a) The years 1980, 2005, and 2030 correspond to $x = 0$, $x = 25$, and $x = 50$, respectively. We substitute 0, 25, and 50 for x and then calculate y:

$$y = 0.072(0) + 4.593 = 0 + 4.593 = 4.593;$$
$$y = 0.072(25) + 4.593 = 1.8 + 4.593 = 6.393;$$
$$y = 0.072(50) + 4.593 = 3.6 + 4.593 = 8.193.$$

The world population in 1980, in 2005, and in 2030 is estimated and projected to be 4.593 billion, 6.393 billion, and 8.193 billion, respectively.

ALGEBRAIC ▶ ◀ **GRAPHICAL CONNECTION**

b) We have three ordered pairs from part (a). We plot these points and see that they line up. Thus our calculations are probably correct. Since we are considering only the year 2015 and the number of years since 1980 ($x \geq 0$) and since the population, in billions, for those years will be positive ($y > 0$), we need only the first quadrant for the graph. We use the three points we have plotted to draw a straight line. (See Figure 1.)

To use the graph to estimate world population in 2015, we first note in Figure 2 that this year corresponds to $x = 35$. We need to determine which y-value is paired with $x = 35$. We locate the point on the graph by moving up vertically from $x = 35$, and then find the value on the y-axis that corresponds to that point. It appears that the world population in 2015 will be about 7.1 billion.

To find a more accurate value, we can simply substitute into the equation:

$$y = 0.072(35) + 4.593 = 7.113.$$

The world population in 2015 is estimated to be 7.113 billion.

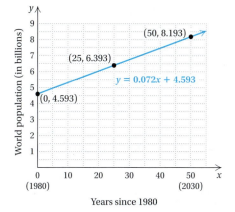

FIGURE 1

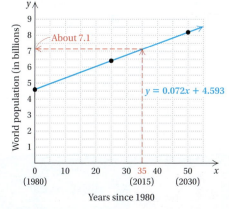

FIGURE 2

---

c) We substitute 7.761 for y and solve for x:

$$y = 0.072x + 4.593$$
$$7.761 = 0.072x + 4.593$$
$$3.168 = 0.072x$$
$$44 = x.$$

In 44 years after 1980, or in 2024, the world population is projected to be approximately 7.761 billion.

Do Exercise 31 on the following page. ▶

31. Milk Consumption. Milk consumption per capita (per person) in the United States can be estimated by

$$M = -0.183t + 27.776,$$

where M is the consumption, in gallons, t years since 1985.

Source: U.S. Department of Agriculture

a) Find the per capita consumption of milk in 1985, in 1995, and in 2015.

b) Graph the equation and use the graph to estimate milk consumption in 2010.

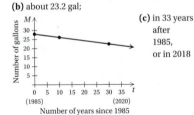

Years since 1985

c) In which year would the per capita consumption of milk be 21.737 gal?

Answers

31. (a) 27.776 gal; 25.946 gal; 22.286 gal;
(b) about 23.2 gal;

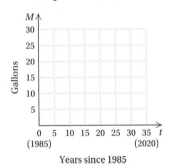

Number of years since 1985

(c) in 33 years after 1985, or in 2018

Many equations in two variables have graphs that are not straight lines. Three such nonlinear graphs are shown below. We will cover some such graphs in the optional Calculator Corners throughout the text and in Chapter 9.

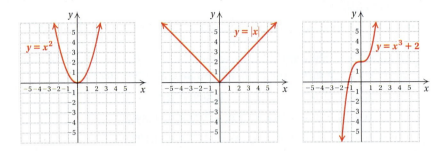

CALCULATOR CORNER

Graphing Equations Equations must be solved for y before they can be graphed on the TI-84 Plus. Consider the equation $3x + 2y = 6$. Solving for y, we get $y = \dfrac{6 - 3x}{2}$. We enter this equation as $y_1 = (6 - 3x)/2$ on the equation-editor screen. Then we select the standard viewing window and display the graph.

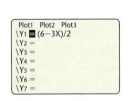

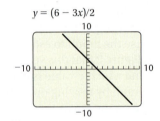

EXERCISES: Graph each equation in the standard viewing window $[-10, 10, -10, 10]$, with Xscl = 1 and Yscl = 1.

1. $y = -5x + 3$ **2.** $y = 4x - 5$

3. $4x - 5y = -10$ **4.** $5y + 5 = -3x$

3.1 Exercise Set

For Extra Help

MathXL®   MyMathLab®   PRACTICE   WATCH   READ   REVIEW

✓ Reading Check

Determine whether each statement is true or false.

RC1. The graph of a linear equation is always a straight line.

RC2. The point $(1, 0)$ is in quadrant I and in quadrant IV.

RC3. The ordered pairs $(4, -7)$ and $(-7, 4)$ name the same point.

RC4. To plot the point $(-3, 5)$, start at the origin and move horizontally to -3. Then move up vertically 5 units and make a "dot."

1. Plot these points.

$(2, 5)$ $(-1, 3)$ $(3, -2)$ $(-2, -4)$

$(0, 4)$ $(0, -5)$ $(5, 0)$ $(-5, 0)$

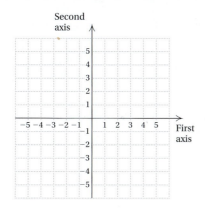

2. Plot these points.

$(4, 4)$ $(-2, 4)$ $(5, -3)$ $(-5, -5)$

$(0, 2)$ $(0, -4)$ $(3, 0)$ $(-4, 0)$

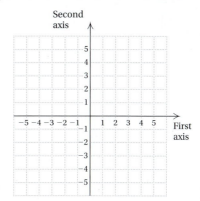

In which quadrant, if any, is each point located?

3. $(-5, 3)$

4. $(1, -12)$

5. $(100, -1)$

6. $(-2.5, 35.6)$

7. $(-6, -29)$

8. $(3.6, 105.9)$

9. $(3.8, 0)$

10. $(0, -492)$

11. $\left(-\dfrac{1}{3}, \dfrac{15}{7}\right)$

12. $\left(-\dfrac{2}{3}, -\dfrac{9}{8}\right)$

13. $\left(12\dfrac{7}{8}, -1\dfrac{1}{2}\right)$

14. $\left(23\dfrac{5}{8}, 81.74\right)$

In which quadrant(s) can the point described be located?

15. The first coordinate is negative and the second coordinate is positive.

16. The first and second coordinates are positive.

17. The first coordinate is positive.

18. The second coordinate is negative.

19. The first and second coordinates are equal.

20. The first coordinate is the additive inverse of the second coordinate.

b Find the coordinates of points *A, B, C, D,* and *E.*

21.

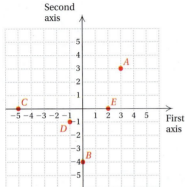

22.

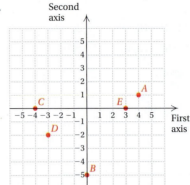

c Determine whether the given ordered pair is a solution of the equation.

23. $(2, 9)$; $y = 3x - 1$

24. $(1, 7)$; $y = 2x + 5$

25. $(4, 2)$; $2x + 3y = 12$

26. $(0, 5)$; $5x - 3y = 15$

27. $(3, -1)$; $3a - 4b = 13$

28. $(-5, 1)$; $2p - 3q = -13$

In each of Exercises 29–34, an equation and two ordered pairs are given. Show that each pair is a solution of the equation. Then use the graph of the equation to determine another solution. Answers may vary.

29. $y = x - 5$; $(4, -1)$ and $(1, -4)$

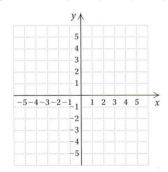

30. $y = x + 3$; $(-1, 2)$ and $(3, 6)$

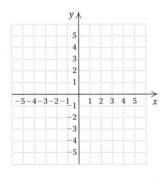

31. $y = \frac{1}{2}x + 3$; $(4, 5)$ and $(-2, 2)$

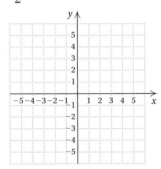

32. $3x + y = 7$; $(2, 1)$ and $(4, -5)$

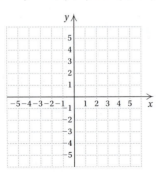

33. $4x - 2y = 10$; $(0, -5)$ and $(4, 3)$

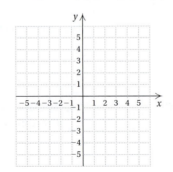

34. $6x - 3y = 3$; $(1, 1)$ and $(-1, -3)$

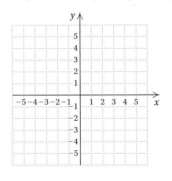

d Graph each equation and identify the *y*-intercept.

35. $y = x + 1$

x	y
−2	
−1	
0	
1	
2	
3	

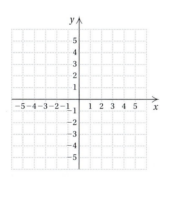

36. $y = x - 1$

x	y
−2	
−1	
0	
1	
2	
3	

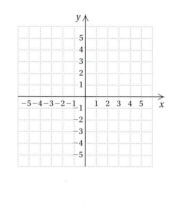

37. $y = x$

x	y
−2	
−1	
0	
1	
2	
3	

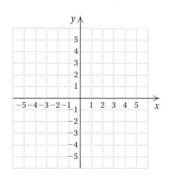

38. $y = -x$

x	y
−2	
−1	
0	
1	
2	
3	

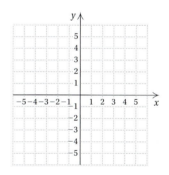

39. $y = \dfrac{1}{2}x$

x	y
−2	
0	
4	

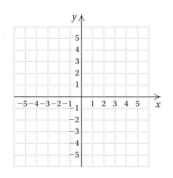

40. $y = \dfrac{1}{3}x$

x	y
−6	
0	
3	

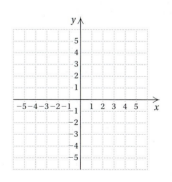

41. $y = x - 3$

x	y

42. $y = x + 3$

x	y

43. $y = 3x - 2$

x	y

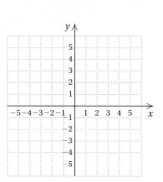

44. $y = 2x + 2$

x	y

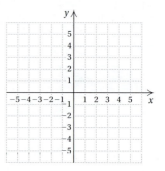

45. $y = \dfrac{1}{2}x + 1$

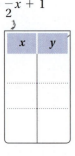

x	y

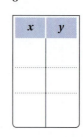

46. $y = \dfrac{1}{3}x - 4$

x	y

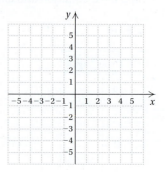

47. $x + y = -5$

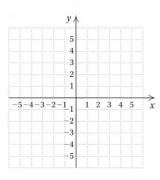

48. $x + y = 4$

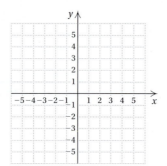

49. $y = \dfrac{5}{3}x - 2$

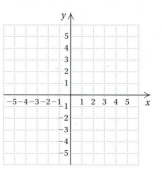

50. $y = \dfrac{5}{2}x + 3$

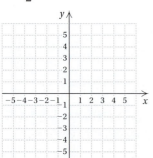

51. $x + 2y = 8$

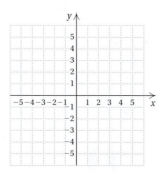

52. $x + 2y = -6$

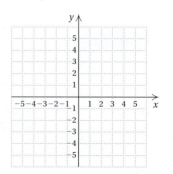

53. $y = \dfrac{3}{2}x + 1$

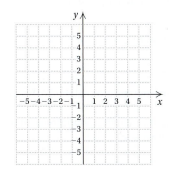

54. $y = -\dfrac{1}{2}x - 3$

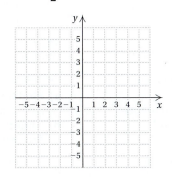

55. $8x - 2y = -10$

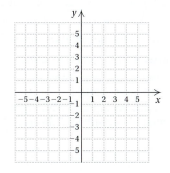

56. $6x - 3y = 9$

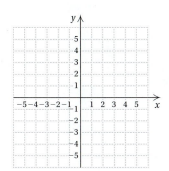

57. $8y + 2x = -4$

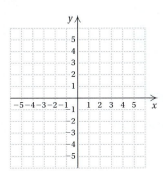

58. $6y + 2x = 8$

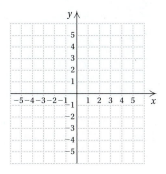

e Solve.

59. *Online Advertising.* Spending for online advertising is increasing. The amount A, in billions of dollars, spent worldwide for online advertising can be approximated and projected by

$$A = 12.83t + 68.38,$$

where t is the number of years since 2010.

a) Find the amount spent for online advertising in 2010, in 2014, and in 2015.

b) Graph the equation and use the graph to estimate the amount spent for online advertising in 2012.

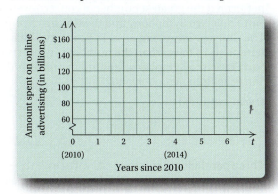

c) In what year will online advertising spending be about $158.19 billion?

60. *Price of a New Car.* The average price P of a new car, in dollars, can be approximated and projected by

$$P = 426t + 25{,}710,$$

where t is the number of years since 2002.

Source: Edmunds.com

a) Find the average price of a new car in 2002, in 2006, and in 2011.

b) Graph the equation and use the graph to estimate the price of a new car in 2010.

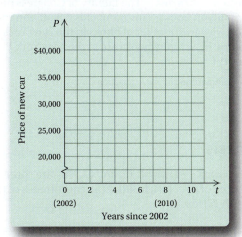

c) In what year will the average price of a new car be approximately $35,000?

61. Sheep and Lambs. The number of sheep and lambs S, in millions, on farms in the United States has declined in recent years and can be approximated and projected by

$$S = -0.125t + 6.898,$$

where t is the number of years since 2000.

Source: U.S. Department of Agriculture

a) Find the number of sheep and lambs in 2000, in 2007, and in 2012.

b) Graph the equation and then use the graph to estimate the number of sheep and lambs in 2010.

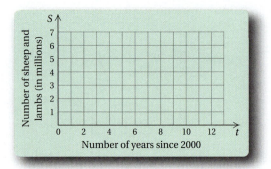

c) At this rate of decline, in what year will the number of sheep and lambs be 4.898 million?

62. Record Temperature Drop. On 22 January 1943, the temperature T, in degrees Fahrenheit, in Spearfish, South Dakota, could be approximated by

$$T = -2.15m + 54,$$

where m is the number of minutes since 9:00 that morning.

Source: Information Please Almanac

a) Find the temperature at 9:01 A.M., at 9:08 A.M., and at 9:20 A.M.

b) Graph the equation and use the graph to estimate the temperature at 9:15 A.M.

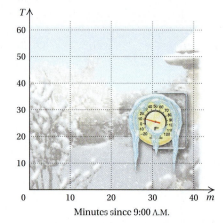

Minutes since 9:00 A.M.

c) The temperature stopped dropping when it reached $-4°F$. At what time did this occur?

Skill Maintenance

Find the absolute value. [1.2e]

63. $|-12|$

64. $|4.89|$

65. $|0|$

66. $\left|-\dfrac{4}{5}\right|$

Solve. [2.3a]

67. $2x - 14 = 29$

68. $\frac{1}{3}t + 6 = -12$

69. $-10 = 1.2y + 2$

70. $4 - 5w = -16$

Solve. [2.6a]

71. Books in Libraries. The Library of Congress houses 33.5 million books. This number is 0.3 million more than twice the number of books in the New York Public Library. How many books are in the New York Public Library?

Source: American Library Association

72. Busiest Orchestras. In 2011, the orchestras who performed the greatest number of concerts were the San Francisco Symphony and the Chicago Symphony. The Chicago Symphony held 133 concerts. This number is 24 fewer than the number of concerts by the San Francisco Symphony. How many concerts did the San Francisco Symphony hold?

Source: www.bachtrack.com

Synthesis

73. The points $(-1, 1)$, $(4, 1)$, and $(4, -5)$ are three vertices of a rectangle. Find the coordinates of the fourth vertex.

74. Three parallelograms share the vertices $(-2, -3)$, $(-1, 2)$, and $(4, -3)$. Find the fourth vertex of each parallelogram.

75. Graph eight points such that the sum of the coordinates in each pair is 6.

76. Graph eight points such that the first coordinate minus the second coordinate is 1.

77. Find the perimeter of a rectangle whose vertices have coordinates $(5, 3)$, $(5, -2)$, $(-3, -2)$, and $(-3, 3)$.

78. Find the area of a triangle whose vertices have coordinates $(0, 9)$, $(0, -4)$, and $(5, -4)$.

More with Graphing and Intercepts

a GRAPHING USING INTERCEPTS

In Section 3.1, we graphed linear equations of the form $Ax + By = C$ by first solving for y to find an equivalent equation in the form $y = mx + b$. We did so because it is then easier to calculate the y-value that corresponds to a given x-value. Another convenient way to graph $Ax + By = C$ is to use **intercepts**. Look at the graph of $-2x + y = 4$ shown below.

The y-intercept is $(0, 4)$. It occurs where the line crosses the y-axis and thus will always have 0 as the first coordinate. The x-intercept is $(-2, 0)$. It occurs where the line crosses the x-axis and thus will always have 0 as the second coordinate.

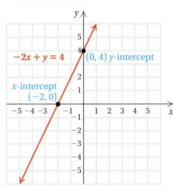

Do Margin Exercise 1. ▶

We find intercepts as follows.

OBJECTIVES

a Find the intercepts of a linear equation, and graph using intercepts.

b Graph equations equivalent to those of the type $x = a$ and $y = b$.

SKILL TO REVIEW

Objective 2.3a: Solve equations using both the addition principle and the multiplication principle.

Solve.

1. $5x - 7 = -10$

2. $-20 = \dfrac{7}{4}x + 8$

INTERCEPTS

The **y-intercept** is $(0, b)$. To find b, let $x = 0$ and solve the equation for y.

The **x-intercept** is $(a, 0)$. To find a, let $y = 0$ and solve the equation for x.

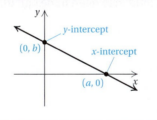

Now let's draw a graph using intercepts.

EXAMPLE 1 Consider $4x + 3y = 12$. Find the intercepts. Then graph the equation using the intercepts.

To find the y-intercept, we let $x = 0$. Then we solve for y:

$$4 \cdot 0 + 3y = 12$$
$$3y = 12$$
$$y = 4.$$

Thus, $(0, 4)$ is the y-intercept. Note that finding this intercept involves covering up the x-term and solving the rest of the equation for y.

To find the x-intercept, we let $y = 0$. Then we solve for x:

$$4x + 3 \cdot 0 = 12$$
$$4x = 12$$
$$x = 3.$$

1. Look at the graph shown below.

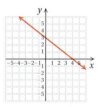

a) Find the coordinates of the y-intercept.

b) Find the coordinates of the x-intercept.

Answers

Skill to Review:

1. $-\dfrac{3}{5}$ **2.** -16

Margin Exercises:

1. **(a)** $(0, 3)$; **(b)** $(4, 0)$

For each equation, find the intercepts. Then graph the equation using the intercepts.

2. $2x + 3y = 6$ 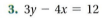 **GS**

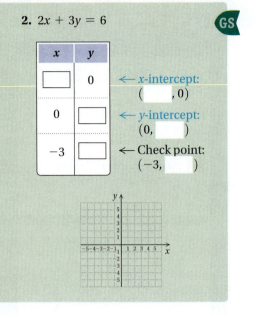

x	y
☐	0
0	☐
−3	☐

← x-intercept:
(☐ , 0)

← y-intercept:
(0, ☐)

← Check point:
(−3, ☐)

3. $3y - 4x = 12$

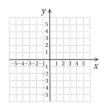

x	y

← x-intercept

← y-intercept

← Check point

Thus, $(3, 0)$ is the x-intercept. Note that finding this intercept involves covering up the y-term and solving the rest of the equation for x.

We plot these points and draw the line, or graph.

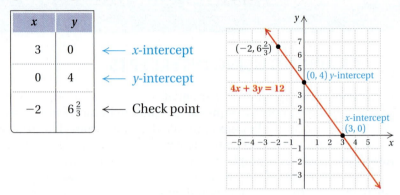

x	y
3	0	← x-intercept
0	4	← y-intercept
−2	$6\frac{2}{3}$	← Check point

A third point should be used as a check. We substitute any convenient value for x and solve for y. In this case, we choose $x = -2$. Then

$$4(-2) + 3y = 12 \qquad \text{Substituting } -2 \text{ for } x$$
$$-8 + 3y = 12$$
$$3y = 20 \qquad \text{Adding 8 on both sides}$$
$$y = \tfrac{20}{3}, \text{ or } 6\tfrac{2}{3}. \qquad \text{Solving for } y$$

It appears that the point $(-2, 6\frac{2}{3})$ is on the graph, though graphing fraction values can be inexact. The graph is probably correct.

◀ **Do Exercises 2 and 3.**

Graphs of equations of the type $y = mx$ pass through the origin. Thus the x-intercept and the y-intercept are the same, $(0, 0)$. In such cases, we must calculate another point in order to complete the graph. A third point would also need to be calculated if a check is desired.

EXAMPLE 2 Graph: $y = 3x$.

We know that $(0, 0)$ is both the x-intercept and the y-intercept. We calculate values at two other points and complete the graph, knowing that it passes through the origin $(0, 0)$.

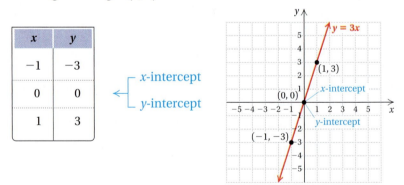

x	y
−1	−3
0	0
1	3

x-intercept
y-intercept

◀ **Do Exercises 4 and 5 on the following page.**

Answers

2. 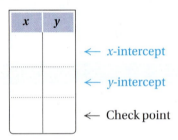   **3.**

$2x + 3y = 6$ $3y - 4x = 12$

Guided Solution:
2. 3, 2, 4, 3, 2, 4

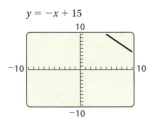
Viewing the Intercepts Knowing the intercepts of a linear equation helps us to determine a good viewing window for the graph of the equation. For example, when we graph the equation $y = -x + 15$ in the standard window, we see only a small portion of the graph in the upper right-hand corner of the screen, as shown on the left below.

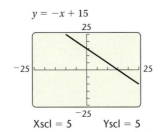

Using algebra, as we did in Example 1, we find that the intercepts of the graph of this equation are $(0, 15)$ and $(15, 0)$. This tells us that, if we are to see more of the graph than is shown on the left above, both Xmax and Ymax should be greater than 15. We can try different window settings until we find one that suits us. One good choice is $[-25, 25, -25, 25]$, with $Xscl = 5$ and $Yscl = 5$, shown on the right above.

EXERCISES: Find the intercepts of each equation algebraically. Then graph the equation on a graphing calculator, choosing window settings that allow the intercepts to be seen clearly. (Settings may vary.)

1. $y = -7.5x - 15$ **2.** $y - 2.15x = 43$

3. $6x - 5y = 150$ **4.** $y = 0.2x - 4$

5. $y = 1.5x - 15$ **6.** $5x - 4y = 2$

b EQUATIONS WHOSE GRAPHS ARE HORIZONTAL LINES OR VERTICAL LINES

EXAMPLE 3 Graph: $y = 3$.

The equation $y = 3$ tells us that y must be 3, but it doesn't give us any information about x. We can also think of this equation as $0 \cdot x + y = 3$. No matter what number we choose for x, we find that y is 3. We make up a table with all 3's in the y-column.

x	y
	3
	3
	3

Choose any number for x. $\rightarrow$

y must be 3.

x	y	
−2	3	
0	3	← y-intercept
4	3	

Graph.

4. $y = 2x$

x	y
−1	
0	
1	

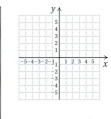

5. $y = -\dfrac{2}{3}x$

x	y

Answers

4.

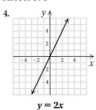

$y = 2x$

5.

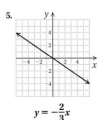

$y = -\dfrac{2}{3}x$

Graph.

6. $x = 5$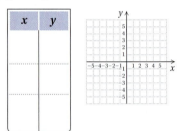

7. $y = -2$

8. $x = -3$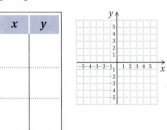

9. $x = 0$

When we plot the ordered pairs $(-2, 3)$, $(0, 3)$, and $(4, 3)$ and connect the points, we obtain a horizontal line. Any ordered pair $(x, 3)$ is a solution. So the line is parallel to the x-axis with y-intercept $(0, 3)$.

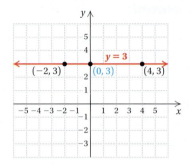

EXAMPLE 4   Graph: $x = -4$.

Consider $x = -4$. We can also think of this equation as $x + 0 \cdot y = -4$. We make up a table with all -4's in the x-column.

x	y
-4	
-4	
-4	
-4	

x must be -4.

x-intercept $\rightarrow$

x	y
-4	-5
-4	1
-4	3
-4	0

$\leftarrow$ Choose any number for y.

When we plot the ordered pairs $(-4, -5)$, $(-4, 1)$, $(-4, 3)$, and $(-4, 0)$ and connect the points, we obtain a vertical line. Any ordered pair $(-4, y)$ is a solution. So the line is parallel to the y-axis with x-intercept $(-4, 0)$.

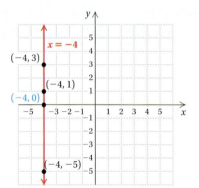

HORIZONTAL LINES AND VERTICAL LINES

The graph of $y = b$ is a **horizontal line**. The y-intercept is $(0, b)$.

The graph of $x = a$ is a **vertical line**. The x-intercept is $(a, 0)$.

◀ **Do Exercises 6–9.**

The following is a general procedure for graphing linear equations.

GRAPHING LINEAR EQUATIONS

1. If the equation is of the type $x = a$ or $y = b$, the graph will be a line parallel to an axis; $x = a$ is vertical and $y = b$ is horizontal.

Examples.

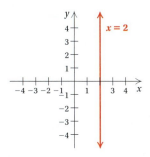

2. If the equation is of the type $y = mx$, both intercepts are the origin, $(0, 0)$. Plot $(0, 0)$ and two other points.

Example.

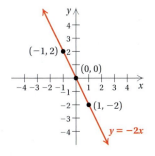

3. If the equation is of the type $y = mx + b$, plot the y-intercept $(0, b)$ and two other points.

Example.

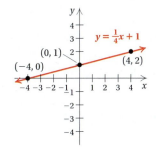

4. If the equation is of the type $Ax + By = C$, but not of the type $x = a$ or $y = b$, then either solve for y and proceed as with the equation $y = mx + b$, or graph using intercepts. If the intercepts are too close together, choose another point or points farther from the origin.

Examples.

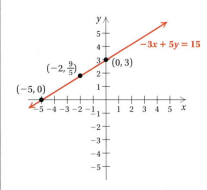

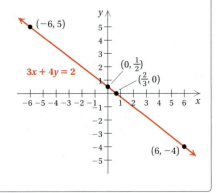

Answers

6.

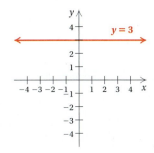

$x = 5$

7.

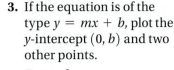

$y = -2$

8.

$x = -3$

9.

$x = 0$

Guided Solutions:
6. 5, 5   **7.** $-2, -2$

Visualizing for Success

A

B

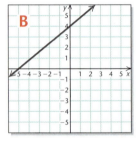

C

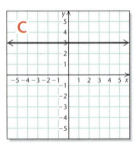

D

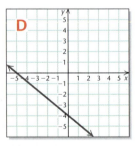

E

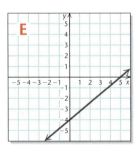

Match each equation with its graph.

1. $5y + 20 = 4x$

2. $y = 3$

3. $3x + 5y = 15$

4. $5y + 4x = 20$

5. $5y = 10 - 2x$

6. $4x + 5y + 20 = 0$

7. $5x - 4y = 20$

8. $4y + 5x + 20 = 0$

9. $5y - 4x = 20$

10. $x = -3$

Answers on page A-8

F

G

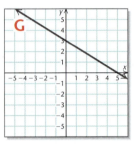

H

I

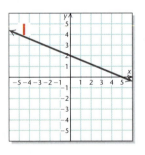

J

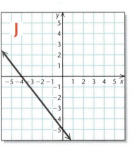

☑ Reading Check

Choose from the column on the right the word or the expression that best completes each statement. Not every choice will be used.

RC1. The graph of $y = -3$ is a(n) _____ line with $(0, -3)$ as its _____.

RC2. The x-intercept occurs when a line crosses the _____.

RC3. To find the x-intercept, let _____.

RC4. In the graph of $y = 2x$, the point _____ is both the x-intercept and the y-intercept.

RC5. To find the y-intercept, let _____.

RC6. The graph of $x = 4$ is a(n) _____ line with $(4, 0)$ as its _____.

$(0, 2)$
$(0, 0)$
horizontal
vertical
$x = 0$
$y = 0$
x-intercept
y-intercept
x-axis
y-axis

a For each of Exercises 1–4, find **(a)** the coordinates of the y-intercept and **(b)** the coordinates of the x-intercept.

1.
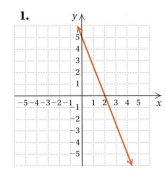

2.
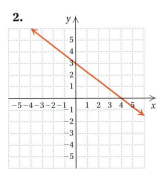

3.
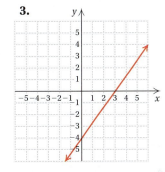

4.
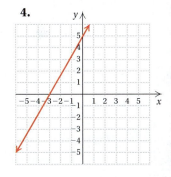

For each of Exercises 5–12, find **(a)** the coordinates of the y-intercept and **(b)** the coordinates of the x-intercept. Do not graph.

5. $3x + 5y = 15$

6. $5x + 2y = 20$

7. $7x - 2y = 28$

8. $3x - 4y = 24$

9. $-4x + 3y = 10$

10. $-2x + 3y = 7$

11. $6x - 3 = 9y$

12. $4y - 2 = 6x$

For each equation, find the intercepts. Then use the intercepts to graph the equation.

13. $x + 3y = 6$

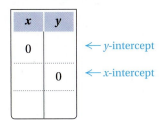
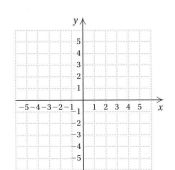

14. $x + 2y = 2$

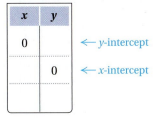
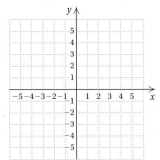

15. $-x + 2y = 4$

x	y
0	
	0

← *y*-intercept

← *x*-intercept

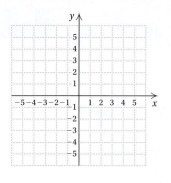

16. $-x + y = 5$

x	y
0	
	0

← *y*-intercept

← *x*-intercept

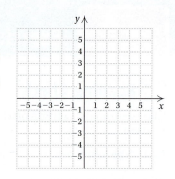

17. $3x + y = 6$

x	y
0	
	0

← *y*-intercept

← *x*-intercept

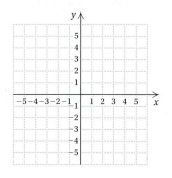

18. $2x + y = 6$

x	y
0	
	0

← *y*-intercept

← *x*-intercept

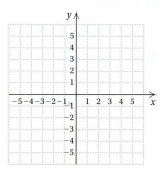

19. $2y - 2 = 6x$

x	y

← *y*-intercept

← *x*-intercept

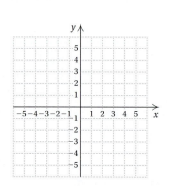

20. $3y - 6 = 9x$

x	y

← *y*-intercept

← *x*-intercept

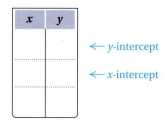
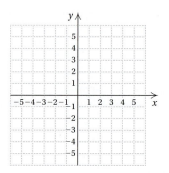

21. $3x - 9 = 3y$

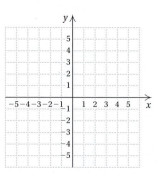

22. $5x - 10 = 5y$

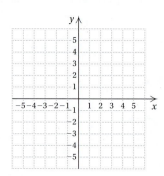

23. $2x - 3y = 6$

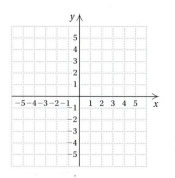

24. $2x - 5y = 10$

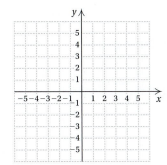

25. $4x + 5y = 20$

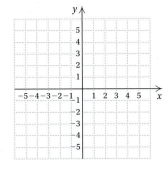

26. $2x + 6y = 12$

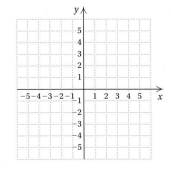

27. $2x + 3y = 8$

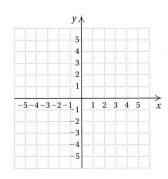

28. $x - 1 = y$

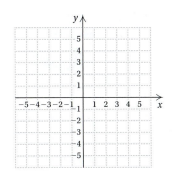

29. $3x + 4y = 5$

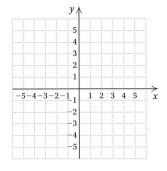

30. $2x - 1 = y$

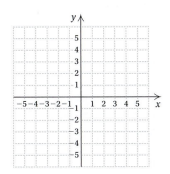

31. $3x - 2 = y$

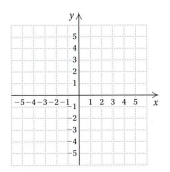

32. $4x - 3y = 12$

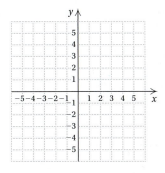

33. $6x - 2y = 12$

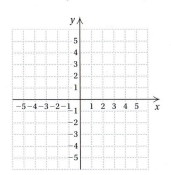

34. $7x + 2y = 6$

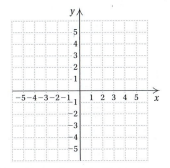

35. $y = -3 - 3x$

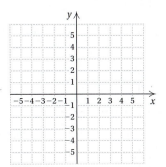

36. $-3x = 6y - 2$

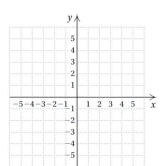

37. $y - 3x = 0$

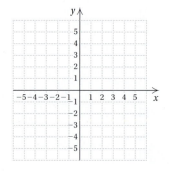

38. $x + 2y = 0$

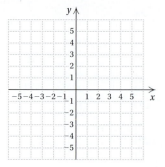

b   Graph.

39. $x = -2$

x	y
-2	
-2	
-2	

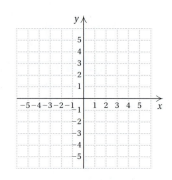

40. $x = 1$

x	y
1	
1	
1	

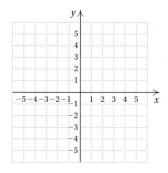

41. $y = 2$

x	y
	2
	2
	2

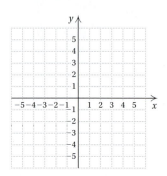

42. $y = -4$

x	y
	-4
	-4
	-4

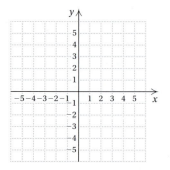

43. $x = 2$

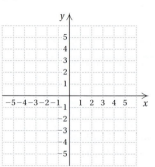

44. $x = 3$

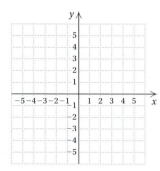

45. $y = 0$

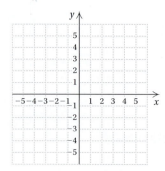

46. $y = -1$

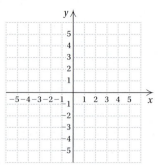

47. $x = \dfrac{3}{2}$

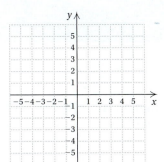

48. $x = -\dfrac{5}{2}$

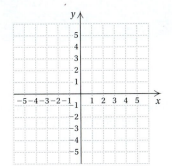

49. $3y = -5$

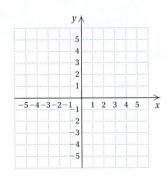

50. $12y = 45$

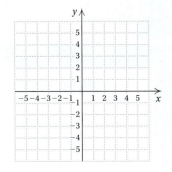

51. $4x + 3 = 0$

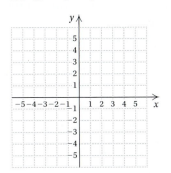

52. $-3x + 12 = 0$

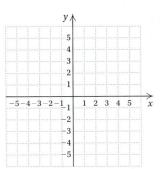

53. $48 - 3y = 0$

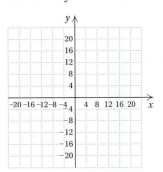

54. $63 + 7y = 0$

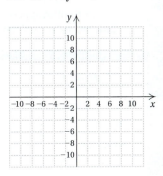

Write an equation for the graph shown.

55.

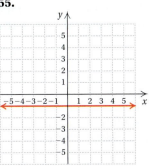

56.

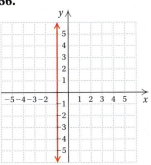

57.

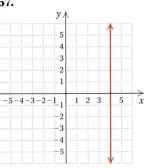

58.

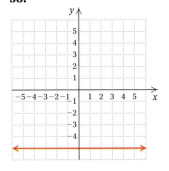

Skill Maintenance

Solve. [2.7e]

59. $x + (x - 1) < (x + 2) - (x + 1)$

60. $6 - 18x \le 4 - 12x - 5x$

61. $\dfrac{2x}{7} - 4 \le -2$

62. $\dfrac{1}{4} + \dfrac{x}{3} > \dfrac{7}{12}$

Synthesis

63. Write an equation of a line parallel to the x-axis and passing through $(-3, -4)$.

64. Find the value of m such that the graph of $y = mx + 6$ has an x-intercept of $(2, 0)$.

65. Find the value of k such that the graph of $3x + k = 5y$ has an x-intercept of $(-4, 0)$.

66. Find the value of k such that the graph of $4x = k - 3y$ has a y-intercept of $(0, -8)$.

3.3 Slope and Applications

OBJECTIVES

a Given the coordinates of two points on a line, find the slope of the line, if it exists.

b Find the slope of a line from an equation.

c Find the slope, or rate of change, in an applied problem involving slope.

SKILL TO REVIEW

Objective 1.4a: Subtract real numbers.

Subtract.

1. $-4 - 20$

2. $-21 - (-5)$

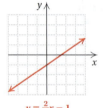

$y = \frac{2}{3}x - 1$

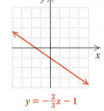

$y = -\frac{2}{3}x - 1$

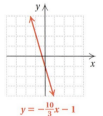

$y = -\frac{10}{3}x - 1$

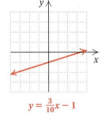

$y = \frac{3}{10}x - 1$

a SLOPE

We have considered two forms of a linear equation, $Ax + By = C$ and $y = mx + b$. We found that from the form of the equation $y = mx + b$, we know that the y-intercept of the line is $(0, b)$.

$$y = mx + b.$$

? The y-intercept is $(0, b)$.

What about the constant m? Does it give us information about the line? Look at the graphs in the margin and see if you can make any connection between the constant m and the "slant" of the line.

The graphs of some linear equations slant upward from left to right. Others slant downward. Some are vertical and some are horizontal. Some slant more steeply than others. We now look for a way to describe such possibilities with numbers.

Consider a line with two points marked P and Q. As we move from P to Q, the y-coordinate changes from 1 to 3 and the x-coordinate changes from 2 to 6. The change in y is $3 - 1$, or 2. The change in x is $6 - 2$, or 4.

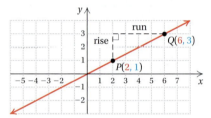

We call the change in y the **rise** and the change in x the **run**. The ratio rise/run is the same for any two points on a line. We call this ratio the **slope** of the line. Slope describes the slant of a line. The slope of the line in the graph above is given by

$$\frac{\text{rise}}{\text{run}} = \frac{\text{the change in } y}{\text{the change in } x}, \text{ or } \frac{2}{4}, \text{ or } \frac{1}{2}.$$

SLOPE

The **slope** of a line containing points (x_1, y_1) and (x_2, y_2) is given by

$$m = \frac{\text{rise}}{\text{run}} = \frac{\text{the change in } y}{\text{the change in } x} = \frac{y_2 - y_1}{x_2 - x_1}.$$

In the preceding definition, (x_1, y_1) and (x_2, y_2)—read "x sub-one, y sub-one and x sub-two, y sub-two"—represent two different points on a line. It does not matter which point is considered (x_1, y_1) and which is considered (x_2, y_2) so long as coordinates are subtracted in the same order in both the numerator and the denominator:

$$\frac{y_2 - y_1}{x_2 - x_1} = \frac{y_1 - y_2}{x_1 - x_2}.$$

EXAMPLE 1 Graph the line containing the points $(-4, 3)$ and $(2, -6)$ and find the slope.

The graph is shown below. We consider (x_1, y_1) to be $(-4, 3)$ and (x_2, y_2) to be $(2, -6)$. From $(-4, 3)$ and $(2, -6)$, we see that the change in y, or the rise, is $-6 - 3$, or -9. The change in x, or the run, is $2 - (-4)$, or 6.

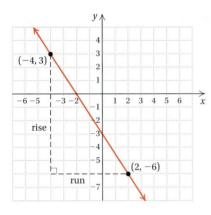

$$\text{Slope} = \frac{\text{rise}}{\text{run}} = \frac{\text{change in } y}{\text{change in } x}$$

$$= \frac{y_2 - y_1}{x_2 - x_1}$$

$$= \frac{-6 - 3}{2 - (-4)}$$

$$= \frac{-9}{6} = -\frac{9}{6}, \text{ or } -\frac{3}{2}$$

When we use the formula

$$m = \frac{y_2 - y_1}{x_2 - x_1},$$

we must remember to subtract the x-coordinates in the same order that we subtract the y-coordinates. Let's redo Example 1, where we consider (x_1, y_1) to be $(2, -6)$ and (x_2, y_2) to be $(-4, 3)$:

$$\text{Slope} = \frac{\text{change in } y}{\text{change in } x} = \frac{3 - (-6)}{-4 - 2} = \frac{9}{-6} = -\frac{9}{6} = -\frac{3}{2}.$$

Do Exercises 1 and 2. ▶

The slope of a line tells how it slants. A line with positive slope slants up from left to right. The larger the slope, the steeper the slant. A line with negative slope slants downward from left to right.

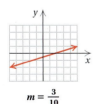  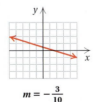

$m = \frac{3}{10}$ $\quad m = \frac{10}{3}$ $\quad m = -\frac{10}{3}$ $\quad m = -\frac{3}{10}$ $\quad m = 0$ $\quad m$ is not defined.

Later in this section, in Examples 7 and 8, we will discuss the slope of a horizontal line and of a vertical line.

Graph the line containing the points and find the slope in two different ways.

GS **1.** $(-2, 3)$ and $(3, 5)$

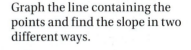

$$\frac{5 - \boxed{}}{\boxed{} - (-2)} = \frac{\boxed{}}{5}, \text{ or}$$

$$\frac{3 - \boxed{}}{\boxed{} - 3} = \frac{-2}{\boxed{}} = \frac{2}{\boxed{}}$$

2. $(0, -3)$ and $(-3, 2)$

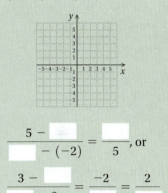

Answers

Answers to Margin Exercises 1 and 2 are on p. 244.

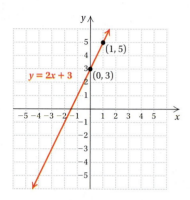

Find the slope of each line.

3. $y = 4x + 11$

4. $y = -17x + 8$

5. $y = -x + \dfrac{1}{2}$

6. $y = \dfrac{2}{3}x - 1$

Find the slope of each line.

7. $4x + 4y = 7$

8. $5x - 4y = 8$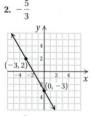

$$5x = \boxed{} + 8$$

$$5x - \boxed{} = 4y$$

$$\dfrac{5x - 8}{\boxed{}} = \dfrac{4y}{4}$$

$$\boxed{} \cdot x - 2 = y, \text{ or}$$

$$y = \boxed{} \cdot x - 2$$

$$\downarrow$$

Slope is $\boxed{}$.

Answers

1. $\dfrac{2}{5}$ **2.** $-\dfrac{5}{3}$

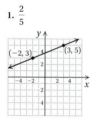

3. 4   **4.** −17   **5.** −1   **6.** $\dfrac{2}{3}$

7. −1   **8.** $\dfrac{5}{4}$

Guided Solutions

1. 3, 3, 2; 5, −2, −5, 5   **8.** 4y, 8, 4, $\dfrac{5}{4}, \dfrac{5}{4}, \dfrac{5}{4}$

b FINDING THE SLOPE FROM AN EQUATION

It is possible to find the slope of a line from its equation. Let's consider the equation $y = 2x + 3$, which is in the form $y = mx + b$. The graph of this equation is shown at right. We can find two points by choosing convenient values for x—say, 0 and 1—and substituting to find the corresponding y-values. We find the two points on the line to be $(0, 3)$ and $(1, 5)$. The slope of the line is found using the definition of slope:

$$m = \dfrac{\text{change in } y}{\text{change in } x} = \dfrac{5 - 3}{1 - 0} = \dfrac{2}{1} = 2.$$

The slope is 2. Note that this is also the coefficient of the x-term in the equation $y = 2x + 3$.

DETERMINING SLOPE FROM THE EQUATION $y = mx + b$

The slope of the line $y = mx + b$ is m. To find the slope of a non-vertical line, solve the linear equation in x and y for y and get the resulting equation in the form $y = mx + b$. The coefficient of the x-term, m, is the slope of the line.

EXAMPLES Find the slope of each line.

2. $y = -3x + \dfrac{2}{9}$

$\quad\quad m = -3 = \text{Slope}$

3. $y = \dfrac{4}{5}x$

$\quad\quad m = \dfrac{4}{5} = \text{Slope}$

4. $y = x + 6$

$\quad\quad m = 1 = \text{Slope}$

5. $y = -0.6x - 3.5$

$\quad\quad m = -0.6 = \text{Slope}$

◀ **Do Exercises 3–6.**

To find slope from an equation, we may need to first find an equivalent form of the equation.

EXAMPLE 6 Find the slope of the line $2x + 3y = 7$.

We solve for y to get the equation in the form $y = mx + b$:

$$2x + 3y = 7$$
$$3y = -2x + 7$$
$$y = \dfrac{1}{3}(-2x + 7)$$
$$y = -\dfrac{2}{3}x + \dfrac{7}{3}. \quad \text{This is } y = mx + b.$$

The slope is $-\dfrac{2}{3}$.

◀ **Do Exercises 7 and 8.**

What about the slope of a horizontal line or a vertical line?

EXAMPLE 7 Find the slope of the line $y = 5$.

We can think of $y = 5$ as $y = 0x + 5$. Then from this equation, we see that $m = 0$. Consider the points $(-3, 5)$ and $(4, 5)$, which are on the line. The change in $y = 5 - 5$, or 0. The change in $x = -3 - 4$, or -7. We have

$$m = \frac{5 - 5}{-3 - 4}$$

$$= \frac{0}{-7}$$

$$= 0.$$

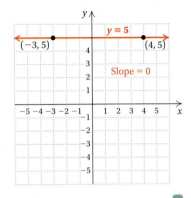

Any two points on a horizontal line have the same y-coordinate. The change in y is 0. Thus the slope of a horizontal line is 0.

EXAMPLE 8 Find the slope of the line $x = -4$.

Consider the points $(-4, 3)$ and $(-4, -2)$, which are on the line. The change in $y = 3 - (-2)$, or 5. The change in $x = -4 - (-4)$, or 0. We have

$$m = \frac{3 - (-2)}{-4 - (-4)}$$

$$= \frac{5}{0}. \qquad \text{Not defined}$$

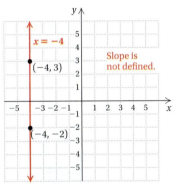

Since division by 0 is not defined, the slope of this line is not defined. The answer in this example is "The slope of this line is not defined."

SLOPE 0; SLOPE NOT DEFINED

The slope of a horizontal line is 0.
The slope of a vertical line is not defined.

Do Exercises 9 and 10. ▶

CALCULATOR CORNER

Visualizing Slope

EXERCISES: Graph each of the following sets of equations using the window settings $[-6, 6, -4, 4]$, with Xscl $= 1$ and Yscl $= 1$.

1. $y = x$, $y = 2x$,
   $y = 5x$, $y = 10x$
   What do you think the graph of $y = 123x$ will look like?

2. $y = x$, $y = \frac{3}{4}x$,
   $y = 0.38x$, $y = \frac{5}{32}x$
   What do you think the graph of $y = 0.000043x$ will look like?

3. $y = -x$, $y = -2x$,
   $y = -5x$, $y = -10x$
   What do you think the graph of $y = -123x$ will look like?

4. $y = -x$, $y = -\frac{3}{4}x$,
   $y = -0.38x$, $y = -\frac{5}{32}x$
   What do you think the graph of $y = -0.000043x$ will look like?

Find the slope, if it exists, of each line.

9. $x = 7$

10. $y = -5$

Answers

9. Not defined   10. 0

Slope has many real-world applications. For example, numbers like 2%, 3%, and 6% are often used to represent the *grade* of a road, a measure of how steep a road on a hill or mountain is. For example, a 3% grade $\left(3\% = \frac{3}{100}\right)$ means that for every horizontal distance of 100 ft, the road rises 3 ft, and a −3% grade means that for every horizontal distance of 100 ft, the road drops 3 ft. (Road signs do not include negative signs.)

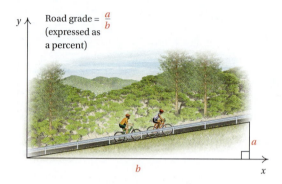

Road grade = $\frac{a}{b}$
(expressed as a percent)

The concept of grade also occurs in skiing or snowboarding, where a 7% grade is considered very tame, but a 70% grade is considered extremely steep.

EXAMPLE 9 *Dubai Ski Run.* Dubai Ski Resort has the fifth longest indoor ski run in the world. It drops 197 ft over a horizontal distance of 1297 ft. Find the grade of the ski run.

197 ft

1297 ft

11. *Grade of a Treadmill.* During a stress test, a physician may change the grade, or slope, of a treadmill to measure its effect on heart rate (number of beats per minute). Find the grade, or slope, of the treadmill shown below.

0.4 ft

5 ft

The grade of the ski run is its slope, expressed as a percent:

$$m = \frac{197}{1297} \begin{array}{l}\leftarrow \text{Vertical distance} \\ \leftarrow \text{Horizontal distance}\end{array}$$

$$\approx 0.15$$

$$\approx 15\%$$

◀ **Do Exercise 11.**

Answer

11. 8%

Slope can also be considered as a **rate of change**.

EXAMPLE 10 *Car Assembly Line.* Cameron, a supervisor in a car assembly plant, prepared the following graph to display data from a recent day's work. Use the graph to determine the slope, or the rate of change of the number of cars that came off an assembly line with respect to time.

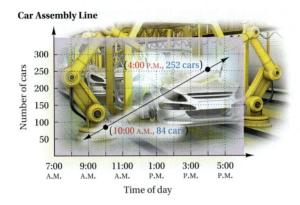

Car Assembly Line

(4:00 P.M., 252 cars)

(10:00 A.M., 84 cars)

Number of cars

Time of day

The vertical axis of the graph shows the number of cars, and the horizontal axis shows the time, in units of one hour. We can describe the rate of change of the number of cars with respect to time as

$$\frac{\text{Cars}}{\text{Hours}}, \quad \text{or} \quad \text{number of cars per hour.}$$

This value is the slope of the line. We determine two ordered pairs on the graph—in this case,

$$(10:00 \text{ A.M.}, 84 \text{ cars}) \quad \text{and} \quad (4:00 \text{ P.M.}, 252 \text{ cars}).$$

This tells us that in the 6 hr between 10:00 A.M. and 4:00 P.M., $252 - 84$, or 168, cars came off the assembly line. Thus,

$$\text{Rate of change} = \frac{252 \text{ cars} - 84 \text{ cars}}{4:00 \text{ P.M.} - 10:00 \text{ A.M.}}$$

$$= \frac{168 \text{ cars}}{6 \text{ hours}}$$

$$= 28 \text{ cars per hour.}$$

Do Exercise 12. ▶

EXAMPLE 11 *Advertising Revenue.* Print-newspaper advertising revenue has been decreasing since 2005. Use the following graph to determine the slope, or rate of change in the advertising revenue with respect to time.

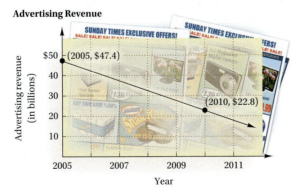

Advertising Revenue

(2005, $47.4)

(2010, $22.8)

Advertising revenue (in billions)

Year

SOURCE: Research Department, Newspaper Association of America

12. *Masonry.* Daryl, a mason, graphed data from a recent day's work. Use the following graph to determine the slope, or the rate of change of the number of bricks he can lay with respect to time.

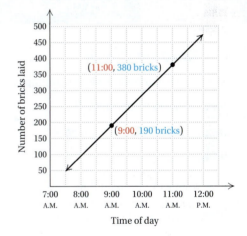

(11:00, 380 bricks)

(9:00, 190 bricks)

Number of bricks laid

Time of day

Answer

12. 95 bricks per hour

13. Sunday Newspapers. Use the following graph to determine the rate of change in the circulation of Sunday newspapers since 2005.

Sunday Newspaper

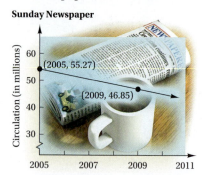

SOURCE: *Editor & Publisher International Yearbook,* 2010

Answer

13. About −2.11 million Sunday newspapers per year

The vertical axis of the graph shows the advertising revenue, in billions of dollars, and the horizontal axis shows the years. We can describe the rate of change in the advertising revenue with respect to time as

$$\frac{\text{Change in advertising revenue}}{\text{Years}}, \quad \text{or} \quad \begin{array}{l}\text{change in advertising}\\\text{revenue per year.}\end{array}$$

This value is the slope of the line. We determine two ordered pairs on the graph—in this case,

$$(2005, \$47.4) \quad \text{and} \quad (2010, \$22.8).$$

This tells us that in the 5 years from 2005 to 2010, newspaper advertising revenue dropped from $47.4 billion to $22.8 billion. Thus,

$$\text{Rate of change} = \frac{\$22.8 - \$47.4}{2010 - 2005}$$

$$= \frac{-\$24.6}{5} \approx -\$4.9 \text{ billion per year.}$$

◀ **Do Exercise 13.**

3.3 Exercise Set

☑ Reading Check

Match each expression with an appropriate description or value from the column on the right.

RC1. Slope of a horizontal line

RC2. *y*-intercept of $y = mx + b$

RC3. Change in x

RC4. Slope of a vertical line

RC5. Slope

RC6. Change in y

a) Rise
b) Run
c) Rise/run
d) 0
e) Not defined
f) $(0, b)$

a   Find the slope, if it exists, of each line.

1.

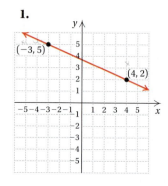

2.

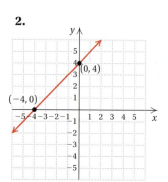

3.

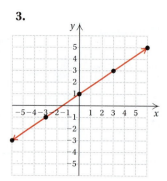

4.

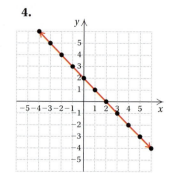

5.

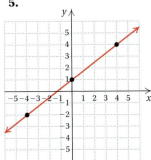

6.

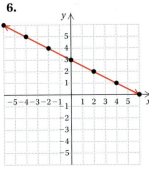

7.

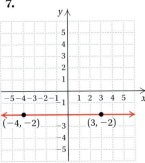

8.

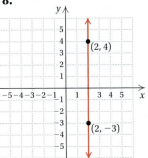

Graph the line containing the given pair of points and find the slope.

9. $(-2, 4), (3, 0)$

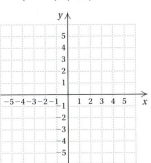

10. $(2, -4), (-3, 2)$

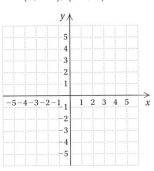

11. $(-4, 0), (-5, -3)$

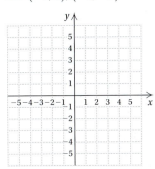

12. $(-3, 0), (-5, -2)$

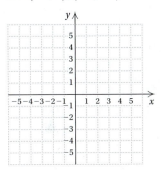

13. $(-4, 1), (2, -3)$

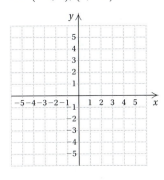

14. $(-3, 5), (4, -3)$

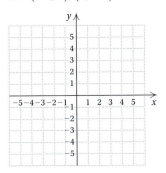

15. $(5, 3), (-3, -4)$

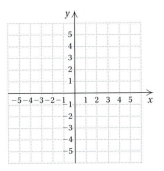

16. $(-4, -3), (2, 5)$

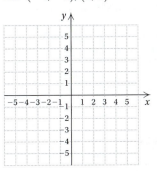

Find the slope, if it exists, of the line containing the given pair of points.

17. $\left(2, -\frac{1}{2}\right), \left(5, \frac{3}{2}\right)$

18. $\left(\frac{2}{3}, -1\right), \left(\frac{5}{3}, 2\right)$

19. $(4, -2), (4, 3)$

20. $(4, -3), (-2, -3)$

21. $(-11, 7), (15, -3)$

22. $(-13, 22), (8, -17)$

23. $\left(-\frac{1}{2}, \frac{3}{11}\right), \left(\frac{5}{4}, \frac{3}{11}\right)$

24. $(0.2, 4), (0.2, -0.04)$

Find the slope, if it exists, of each line.

25. $y = -10x$

26. $y = \dfrac{10}{3}x$

27. $y = 3.78x - 4$

28. $y = -\dfrac{3}{5}x + 28$

29. $3x - y = 4$

30. $-2x + y = 8$

31. $x + 5y = 10$

32. $x - 4y = 8$

33. $3x + 2y = 6$

34. $2x - 4y = 8$

35. $x = \dfrac{2}{15}$

36. $y = -\dfrac{1}{3}$

37. $y = 2 - x$

38. $y = \dfrac{3}{4} + x$

39. $9x = 3y + 5$

40. $4y = 9x - 7$

41. $5x - 4y + 12 = 0$

42. $16 + 2x - 8y = 0$

43. $y = 4$

44. $x = -3$

45. $x = \dfrac{3}{4}y - 2$

46. $3x - \dfrac{1}{5}y = -4$

47. $\dfrac{2}{3}y = -\dfrac{7}{4}x$

48. $-x = \dfrac{2}{11}y$

In each of Exercises 49–52, find the slope (or rate of change).

49. Find the slope (or pitch) of the roof.

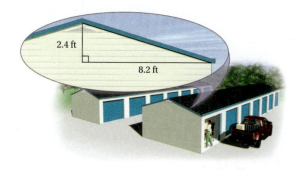

2.4 ft

8.2 ft

50. Find the slope (or grade) of the road.

148.8 m

2400 m

51. _Slope of a River._ When a river flows, the strength or force of the river depends on how far the river falls vertically compared to how far it flows horizontally. Find the slope of the river shown below.

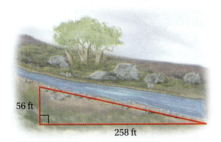

56 ft

258 ft

52. _Constructing Stairs._ Carpenters use slope when designing and building stairs. Public buildings normally include steps with 7-in. risers and 11-in. treads. Find the grade of such a stairway.

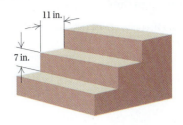

11 in.

7 in.

53. _Grade of a Transit System._ The maximum grade allowed between two stations in a rapid-transit rail system is 3.5%. Between station A and station B, which are 280 ft apart, the tracks rise $8\frac{1}{2}$ ft. What is the grade of the tracks between these two stations? Round the answer to the nearest tenth of a percent. Does this grade meet the rapid-transit rail standards?

Source: Brian Burell, _Merriam Webster's Guide to Everyday Math_, Merriam-Webster, Inc., Springfield MA

54. _Slope of Long's Peak._ From a base elevation of 9600 ft, Long's Peak in Colorado rises to a summit elevation of 14,256 ft over a horizontal distance of 15,840 ft. Find the grade of Long's Peak.

In each of Exercises 55–58, use the graph to calculate a rate of change in which the units of the horizontal axis are used in the denominator.

55. _Kindergarten in China._ The number of children enrolled in kindergarten in China is projected to increase from 26.58 million in 2009 to 34 million in 2015. Use the following graph to find the rate of change, rounded to the nearest hundredth of a million, in the number of children enrolled in kindergarten in China with respect to time.

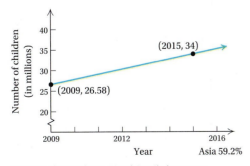

SOURCE: _China Daily_, Lin Qi and Guo Shuhan report, June 23, 2011, p.18

56. _Injuries on Farms._ In 1998, there were 37,775 injuries on farms to people under age 20. Since then, this number has steadily decreased. Using the following graph, find the rate of change, rounded to the nearest ten, in the number of farm injuries to people under age 20 with respect to time.

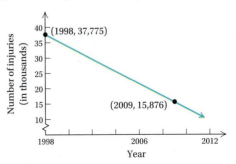

SOURCE: National Institute for Occupational Health and Safety

57. *Population Decrease of New Orleans.* The change in the population of New Orleans, Louisiana, is illustrated in the following graph. Find the rate of change, to the nearest hundred, in the population with respect to time.

Population Decrease of New Orleans

(2000, 484,949)
New Orleans
Louisiana
(2010, 343,829)

SOURCE: Decennial Census, U. S. Census Bureau

58. *Population Increase of Charlotte.* The change in the population of Charlotte, North Carolina, is illustrated in the following graph. Find the rate of change, to the nearest hundred, in the population with respect to time.

Population Increase of Charlotte

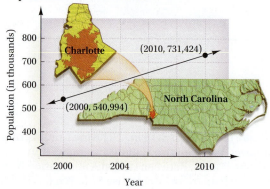

Charlotte
(2010, 731,424)
North Carolina
(2000, 540,994)

SOURCE: Decennial Census, U. S. Census Bureau

59. *Production of Blueberries.* U.S. production of blueberries is continually increasing. In 2006, 358,000,000 lb of blueberries were produced. By 2011, this amount had increased to 511,000,000 lb. Find the rate of change in the production of blueberries with respect to time.

Source: U.S. Department of Agriculture

60. *Bottled Water.* Bottled water consumption per person per year in the United States has increased from 16.7 gal in 2000 to 29.2 gal in 2011. Find the rate of change, rounded to the nearest tenth, in the number of gallons of bottled water consumed annually per person per year.

Sources: Beverage Marketing Corporation; International Bottled Water Association

Skill Maintenance

Solve. [2.3a]

61. $2x - 11 = 4$

62. $5 - \dfrac{1}{2}x = 11$

Collect like terms. [1.7e]

63. $\dfrac{1}{3}p - p$

64. $t - 6 + 4t + 5$

Synthesis

In each of Exercises 65–68, find an equation for the graph shown.

65.

66.

67.

68.

Equations of Lines

We have learned that the slope of a line and the y-intercept of the graph of the line can be read directly from the equation if it is in the form $y = mx + b$. Here we use slope and y-intercept in order to examine linear equations in more detail.

a FINDING AN EQUATION OF A LINE WHEN THE SLOPE AND THE y-INTERCEPT ARE GIVEN

We know from Sections 3.1 and 3.3 that in the equation $y = mx + b$, the **slope** is m and the **y-intercept** is $(0, b)$. Therefore, we call the equation $y = mx + b$ the **slope–intercept equation**.*

> ### THE SLOPE–INTERCEPT EQUATION: $y = mx + b$
>
> The equation $y = mx + b$ is called the **slope–intercept equation**. The slope is m and the y-intercept is $(0, b)$.

EXAMPLE 1 Find the slope and the y-intercept of $2x - 3y = 8$.

We first solve for y:

$$2x - 3y = 8$$
$$-3y = -2x + 8 \qquad \text{Subtracting } 2x$$
$$\frac{-3y}{-3} = \frac{-2x + 8}{-3} \qquad \text{Dividing by } -3$$
$$y = \frac{-2x}{-3} + \frac{8}{-3}$$

$$y = \frac{2}{3}x - \frac{8}{3}. \qquad \text{The slope is } \frac{2}{3}.$$
The y-intercept is $\left(0, -\frac{8}{3}\right)$.

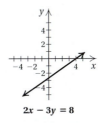

$2x - 3y = 8$

Do Margin Exercises 1–5. ▶

EXAMPLE 2 A line has slope -2.4 and y-intercept $(0, 11)$. Find an equation of the line.

We use the slope–intercept equation and substitute -2.4 for m and 11 for b:

$$y = mx + b$$
$$y = -2.4x + 11. \qquad \text{Substituting}$$

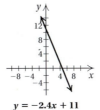

$y = -2.4x + 11$

---

*Equations of lines can also be found using the *point–slope equation*, $y - y_1 = m(x - x_1)$. See Appendix B.

OBJECTIVES

a Given an equation in the form $y = mx + b$, find the slope and the y-intercept; find an equation of a line when the slope and the y-intercept are given.

b Find an equation of a line when the slope and a point on the line are given.

c Find an equation of a line when two points on the line are given.

SKILL TO REVIEW

Objective 3.3a: Given the coordinates of two points on a line, find the slope of the line, if it exists.

Find the slope, if it exists, of the line containing the given pair of points.

1. $(3, 0), (0, 3)$
2. $(-8, 5), (-8, -5)$

Find the slope and the y-intercept.

1. $y = 5x$

2. $y = -\frac{3}{2}x - 6$

3. $3x + 4y = 15$

4. $y = 10 + x$

5. $-7x - 5y = 22$

Answers

Skill to Review:
1. -1 **2.** Not defined

Margin Exercises:
Answers to Margin Exercises 1–5 are on p. 254.

EXAMPLE 3 A line has slope 0 and y-intercept $(0, -6)$. Find an equation of the line.

We use the slope–intercept equation and substitute 0 for m and -6 for b:

$$y = mx + b$$
$$y = 0x + (-6) \qquad \text{Substituting}$$
$$y = -6.$$

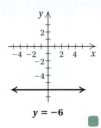

$$y = -6$$

EXAMPLE 4 A line has slope $-\frac{5}{3}$ and y-intercept $(0, 0)$. Find an equation of the line.

We use the slope–intercept equation and substitute $-\frac{5}{3}$ for m and 0 for b:

$$y = mx + b$$
$$y = -\frac{5}{3}x + 0 \qquad \text{Substituting}$$
$$y = -\frac{5}{3}x.$$

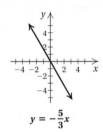

$$y = -\frac{5}{3}x$$

6. A line has slope 3.5 and y-intercept $(0, -23)$. Find an equation of the line.

7. A line has slope 0 and y-intercept $(0, 13)$. Find an equation of the line.

8. A line has slope -7.29 and y intercept $(0, 0)$. Find an equation of the line.

◀ Do Exercises 6–8.

b FINDING AN EQUATION OF A LINE WHEN THE SLOPE AND A POINT ARE GIVEN

Suppose we know the slope of a line and a certain point on that line. We can use the slope–intercept equation $y = mx + b$ to find an equation of the line. To write an equation in this form, we need to know the slope m and the y-intercept $(0, b)$.

EXAMPLE 5 Find an equation of the line with slope 3 that contains the point $(4, 1)$.

We know that the slope is 3, so the equation is $y = 3x + b$. This equation is true for $(4, 1)$. Using the point $(4, 1)$, we substitute 4 for x and 1 for y in $y = 3x + b$. Then we solve for b:

$$y = 3x + b \qquad \text{Substituting 3 for } m \text{ in } y = mx + b$$
$$1 = 3(4) + b \qquad \text{Substituting 4 for } x \text{ and 1 for } y$$
$$1 = 12 + b$$
$$-11 = b. \qquad \text{Solving for } b, \text{ we find that the } y\text{-intercept is } (0, -11).$$

We use the equation $y = mx + b$ and substitute 3 for m and -11 for b:

$$y = 3x - 11.$$

Answers

Margin Exercises:

1. Slope: 5; y-intercept: $(0, 0)$

2. Slope: $-\dfrac{3}{2}$; y-intercept: $(0, -6)$

3. Slope: $-\dfrac{3}{4}$; y-intercept: $\left(0, \dfrac{15}{4}\right)$

4. Slope: 1; y-intercept: $(0, 10)$

5. Slope: $-\dfrac{7}{5}$; y-intercept: $\left(0, -\dfrac{22}{5}\right)$

6. $y = 3.5x - 23$ **7.** $y = 13$

8. $y = -7.29x$

This is the equation of the line with slope 3 and y-intercept $(0, -11)$. Note that $(4, 1)$ is on the line.

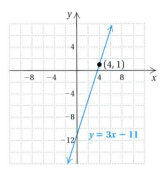

EXAMPLE 6 Find an equation of the line with slope -5 that contains the point $(-2, 3)$.

We know that the slope is -5, so the equation is $y = -5x + b$. This equation is true for all points on the line, including the point $(-2, 3)$. Using the point $(-2, 3)$, we substitute -2 for x and 3 for y in $y = -5x + b$. Then we solve for b:

$$y = -5x + b \qquad \text{Substituting } -5 \text{ for } m \text{ in } y = mx + b$$
$$3 = -5(-2) + b \qquad \text{Substituting } -2 \text{ for } x \text{ and } 3 \text{ for } y$$
$$3 = 10 + b$$
$$-7 = b. \qquad \text{Solving for } b$$

We use the equation $y = mx + b$ and substitute -5 for m and -7 for b:

$$y = -5x - 7.$$

This is the equation of the line with slope -5 and y-intercept $(0, -7)$.

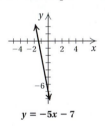

$y = -5x - 7$

Do Exercises 9–12. ▶

c FINDING AN EQUATION OF A LINE WHEN TWO POINTS ARE GIVEN

We can also use the slope–intercept equation to find an equation of a line when two points are given.

EXAMPLE 7 Find an equation of the line containing the points $(2, 3)$ and $(-2, 2)$.

First, we find the slope:

$$m = \frac{3 - 2}{2 - (-2)} = \frac{1}{4}.$$

Thus, $y = \frac{1}{4}x + b$. We then proceed as we did in Example 6, using either point to find b, since both points are on the line.

Find an equation of the line that contains the given point and has the given slope.

9. $(4, 2)$, $m = 5$

10. $(-2, 1)$, $m = -3$

GS **11.** $(3, 5)$, $m = 6$

$$y = mx + b$$
$$y = \boxed{}\, x + b$$
$$5 = 6 \cdot \boxed{} + b$$
$$5 = \boxed{} + b$$
$$\boxed{} = b$$

Thus, $y = \boxed{}\, x - \boxed{}$.

12. $(1, 4)$, $m = -\dfrac{2}{3}$

Find an equation of the line containing the given points.

13. $(2, 4)$ and $(3, 5)$

We choose $(2, 3)$ and substitute 2 for x and 3 for y:

$$y = \frac{1}{4}x + b \qquad \text{Substituting } \frac{1}{4} \text{ for } m \text{ in } y = mx + b$$

$$3 = \frac{1}{4} \cdot 2 + b \qquad \text{Substituting 2 for } x \text{ and 3 for } y$$

$$3 = \frac{1}{2} + b$$

$$\frac{5}{2} = b. \qquad \text{Solving for } b$$

We use the equation $y = mx + b$ and substitute $\frac{1}{4}$ for m and $\frac{5}{2}$ for b:

$$y = \frac{1}{4}x + \frac{5}{2}.$$

This is the equation of the line with slope $\frac{1}{4}$ and y-intercept $\left(0, \frac{5}{2}\right)$. Note that the line contains the points $(2, 3)$ and $(-2, 2)$.

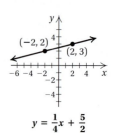

$$y = \frac{1}{4}x + \frac{5}{2}$$

Answers

13. $y = x + 2$ **14.** $y = 2x + 4$

Guided Solution:
14. $-1, -4, 2, -2, -3, 4, 4$

◀ **Do Exercises 13 and 14.**

3.4   **Exercise Set**

For Extra Help

MyMathLab®   MathXL®
PRACTICE   WATCH   READ   REVIEW

✓ **Reading Check**

Match each graph with the appropriate equation from the column on the right.

RC1.

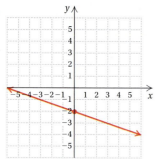

RC2.

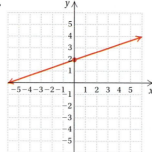

RC3.

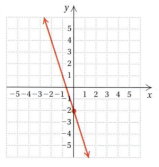

RC4.

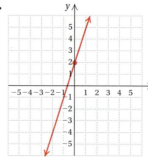

a) $y = -3x - 2$

b) $y = -3x + 2$

c) $y = \frac{1}{3}x - 2$

d) $y = -\frac{1}{3}x + 2$

e) $y = -\frac{1}{3}x - 2$

f) $y = \frac{1}{3}x + 2$

g) $y = 3x + 2$

h) $y = 3x - 2$

a Find the slope and the *y*-intercept.

1. $y = -4x - 9$

2. $y = 2x + 3$

3. $y = 1.8x$

4. $y = -27.4x$

5. $-8x - 7y = 21$

6. $-2x - 8y = 16$

7. $4x = 9y + 7$

8. $5x + 4y = 12$

9. $-6x = 4y + 2$

10. $4.8x - 1.2y = 36$

11. $y = -17$

12. $y = 28$

Find an equation of the line with the given slope and *y*-intercept.

13. Slope $= -7$,
  y-intercept $= (0, -13)$

14. Slope $= 73$,
  y-intercept $= (0, 54)$

15. Slope $= 1.01$,
  y-intercept $= (0, -2.6)$

16. Slope $= -\dfrac{3}{8}$,
  y-intercept $= \left(0, \dfrac{7}{11}\right)$

17. Slope $= 0$,
  y-intercept $= (0, -5)$

18. Slope $= \dfrac{6}{5}$,
  y-intercept $= (0, 0)$

b Find an equation of the line containing the given point and having the given slope.

19. $(-3, 0)$, $m = -2$

20. $(2, 5)$, $m = 5$

21. $(2, 4)$, $m = \dfrac{3}{4}$

22. $\left(\dfrac{1}{2}, 2\right)$, $m = -1$

23. $(2, -6)$, $m = 1$

24. $(4, -2)$, $m = 6$

25. $(0, 3)$, $m = -3$

26. $(-2, -4)$, $m = 0$

c Find an equation of the line that contains the given pair of points.

27. $(12, 16)$ and $(1, 5)$

28. $(-6, 1)$ and $(2, 3)$

29. $(0, 4)$ and $(4, 2)$

30. $(0, 0)$ and $(4, 2)$

31. $(3, 2)$ and $(1, 5)$ **32.** $(-4, 1)$ and $(-1, 4)$ **33.** $\left(4, -\dfrac{2}{5}\right)$ and $\left(4, \dfrac{2}{5}\right)$ **34.** $\left(\dfrac{3}{4}, -3\right)$ and $\left(\dfrac{1}{2}, -3\right)$

35. $(-4, 5)$ and $(-2, -3)$ **36.** $(-2, -4)$ and $(2, -1)$ **37.** $\left(-2, \dfrac{1}{4}\right)$ and $\left(3, \dfrac{1}{4}\right)$ **38.** $\left(\dfrac{3}{7}, -6\right)$ and $\left(\dfrac{3}{7}, -9\right)$

39. *Lottery Sales.* The following line graph shows increasing lottery sales in years t since 2000.

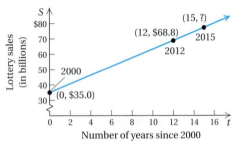

SOURCES: U.S. Census Bureau; National Association of State and Provincial Lotteries

a) Find an equation of the line.
b) What is the rate of change of lottery sales, in billions of dollars, with respect to time?
c) Use the equation to predict lottery sales in 2015.

40. *Aerobic Exercise.* The following line graph describes the *target heart rate T*, in number of beats per minute, of a person of age a, who is exercising. The goal is to get the number of beats per minute to this target level.

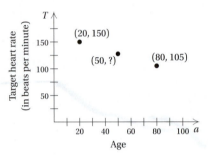

a) Find an equation of the line.
b) What is the rate of change of target heart rate with respect to time?
c) Use the equation to calculate the target heart rate of a person of age 50.

Skill Maintenance

Solve. [2.3c]

41. $3x - 4(9 - x) = 17$

42. $2(5 + 2y) + 4y = 13$

43. $4(a - 3) + 6 = 21 - \dfrac{1}{2}a$

44. $\dfrac{2}{3}(t - 3) = 6(9 - t)$

45. $40(2x - 7) = 50(4 - 6x)$

46. $\dfrac{2}{3}(x - 5) = \dfrac{3}{8}(x + 5)$

Solve. [2.5a]

47. What is 30% of 12?

48. 15 is 3% of what number?

49. What percent of 50 is 2.5?

50. 4240 is 106% of what number?

Synthesis

51. Find an equation of the line that contains the point $(2, -3)$ and has the same slope as the line $3x - y + 4 = 0$.

52. Find an equation of the line that has the same y-intercept as the line $x - 3y = 6$ and contains the point $(5, -1)$.

53. Find an equation of the line with the same slope as the line $3x - 2y = 8$ and the same y-intercept as the line $2y + 3x = -4$.

Mid-Chapter Review

Concept Reinforcement

Determine whether each statement is true or false.

_____ **1.** A slope of $-\frac{3}{4}$ is steeper than a slope of $-\frac{5}{2}$. [3.3a]

_____ **2.** The slope of the line that passes through (a, b) and (c, d) is $\dfrac{d - b}{c - a}$. [3.3a]

_____ **3.** The y-intercept of $Ax + By = C, B \neq 0$, is $\left(0, \dfrac{C}{B}\right)$. [3.2a]

_____ **4.** Both coordinates of points in quadrant IV are negative. [3.1a]

Guided Solutions

GS **5.** Given the graph of the line below, fill in the numbers that create correct statements. [3.2a], [3.3a], [3.4a]

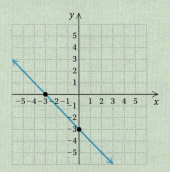

a) The ⬚-intercept is (⬚ , −3).

b) The ⬚-intercept is (⬚ , 0).

c) The slope is $\dfrac{-3 - \boxed{}}{\boxed{} - \boxed{}} = \dfrac{\boxed{}}{\boxed{}} = \boxed{}$.

d) The equation of the line in $y = mx + b$ form is
$$y = \boxed{}\, x + \boxed{}, \text{ or } y = -x - \boxed{}.$$

6. Given the graph of the line below, fill in the letters and the number 0 that create correct statements. [3.2a], [3.3a], [3.4a]

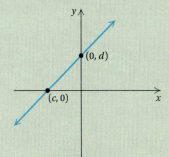

a) The x-intercept is (⬚ , ⬚).

b) The y-intercept is (⬚ , ⬚).

c) The slope is $\dfrac{\boxed{} - \boxed{}}{\boxed{} - c} = \dfrac{\boxed{}}{\boxed{}} = -\boxed{}$.

d) The equation of the line in $y = mx + b$ form is
$$y = \boxed{}\, x + \boxed{}.$$

Mixed Review

Determine whether the given ordered pair is a solution of the equation. [3.1c]

7. $(8, -5)$; $-2q - 7p = 19$

8. $\left(-1, \dfrac{2}{3}\right)$; $6y = -3x + 1$

Find the coordinates of the x-intercept and the y-intercept. [3.2a]

9. $-3x + 2y = 18$

10. $x - \dfrac{1}{2} = 10y$

Graph. [3.1d], [3.2a, b]

11. $-2x + y = -3$

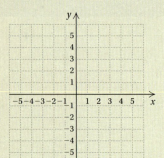

12. $y = -\dfrac{3}{2}$

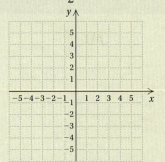

13. $y = -x + 4$

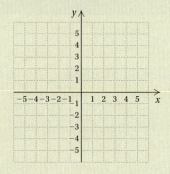

14. $x = 0$

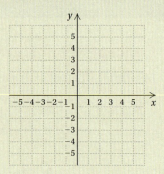

Find the slope, if it exists, of the line containing the given pair of points. [3.3a]

15. $\left(\dfrac{1}{4}, -6\right), (-2, 4)$

16. $(6, -3), (-6, 3)$

Find the slope, if it exists, of the line. [3.3b]

17. $y = 0.728$

18. $13x - y = -5$

19. $12x + 7 = 0$

20. The population of Texas in 2000 was 20,851,820. The population in 2010 was 25,145,561. Find the rate of change, to the nearest hundred, in the population with respect to time. [3.3c]

Match each equation with the appropriate characteristics from the column on the right. [3.2b], [3.4a, b]

21. $y = -1$

22. $x = 1$

23. $y = -x - 1$

24. $y = x - 1$

25. $y = x + 1$

A. The slope is 1 and the x-intercept is $(-1, 0)$.

B. The slope is -1 and the y-intercept is $(0, -1)$.

C. The slope is not defined and the x-intercept is $(1, 0)$.

D. The slope is 0 and the y-intercept is $(0, -1)$.

E. The slope is 1 and the x-intercept is $(1, 0)$.

26. Find an equation of the line with slope -3 that contains the point $\left(-\dfrac{1}{3}, 3\right)$. [3.4b]

Find an equation of the line that contains the given pair of points. [3.4c]

27. $\left(\dfrac{1}{2}, 6\right), \left(\dfrac{1}{2}, -6\right)$

28. $(3, -4), (-7, -2)$

29. $(3, -4), (2, -4)$

Understanding Through Discussion and Writing

30. Do all graphs of linear equations have y-intercepts? Why or why not? [3.2b]

31. The equations $3x + 4y = 8$ and $y = -\frac{3}{4}x + 2$ are equivalent. Which equation is easier to graph and why? [3.1d]

32. If the graph of the equation $Ax + By = C$ is a horizontal line, what can you conclude about A? Why? [3.2b]

33. Explain in your own words why the graph of $x = 7$ is a vertical line. [3.2b]

Graphing Using the Slope and the *y*-Intercept

3.5

a   GRAPHS USING THE SLOPE AND THE *y*-INTERCEPT

OBJECTIVE

a   Use the slope and the *y*-intercept to graph a line.

We can graph a line if we know the coordinates of two points on that line. We can also graph a line if we know the slope and the *y*-intercept.

EXAMPLE 1   Draw a line that has slope $\frac{1}{4}$ and *y*-intercept $(0, 2)$.

We plot $(0, 2)$ and from there move 1 unit *up* (since the numerator of $\frac{1}{4}$ is *positive* and corresponds to the change in *y*) and 4 units *to the right* (since the denominator is *positive* and corresponds to the change in *x*). This locates the point $(4, 3)$. We plot $(4, 3)$ and draw a line passing through $(0, 2)$ and $(4, 3)$. We are actually graphing the equation $y = \frac{1}{4}x + 2$.

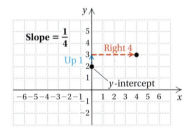 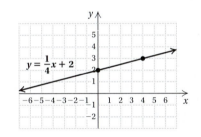

EXAMPLE 2   Draw a line that has slope $-\frac{2}{3}$ and *y*-intercept $(0, 4)$.

We can think of $-\frac{2}{3}$ as $\frac{-2}{3}$. We plot $(0, 4)$ and from there move 2 units *down* (since the numerator is *negative*) and 3 units *to the right* (since the denominator is *positive*). We plot the point $(3, 2)$ and draw a line passing through $(0, 4)$ and $(3, 2)$. We are actually graphing the equation $y = -\frac{2}{3}x + 4$.

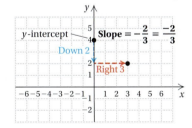 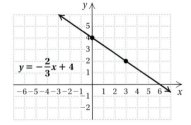

Do Exercises 1–3.   ▶

1. Draw a line that has slope $\frac{2}{5}$ and *y*-intercept $(0, -3)$. What equation is graphed?

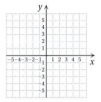

2. Draw a line that has slope $-\frac{2}{5}$ and *y*-intercept $(0, -3)$. What equation is graphed?

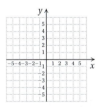

3. Draw a line that has slope 6 and *y*-intercept $(0, -3)$. Think of 6 as $\frac{6}{1}$. What equation is graphed?

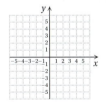

Answers

Answers to Margin Exercises 1–3 are on p. 262.

We now use our knowledge of the slope–intercept equation to graph linear equations.

EXAMPLE 3 Graph $y = \frac{3}{4}x + 5$ using the slope and the y-intercept.

From the equation $y = \frac{3}{4}x + 5$, we see that the slope of the graph is $\frac{3}{4}$ and the y-intercept is $(0, 5)$. We plot $(0, 5)$ and then consider the slope, $\frac{3}{4}$. Starting at $(0, 5)$, we plot a second point by moving 3 units *up* (since the numerator is *positive*) and 4 units *to the right* (since the denominator is *positive*). We reach a new point, $(4, 8)$.

We can also rewrite the slope as $\frac{-3}{-4}$. We again start at the y-intercept, $(0, 5)$, but move 3 units *down* (since the numerator is *negative* and corresponds to the change in y) and 4 units *to the left* (since the denominator is *negative* and corresponds to the change in x). We reach another point, $(-4, 2)$. Once two or three points have been plotted, the line representing all solutions of $y = \frac{3}{4}x + 5$ can be drawn.

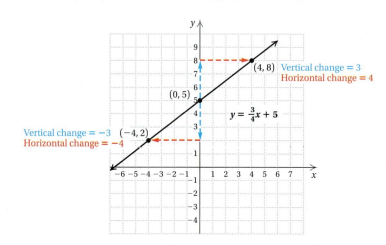

4. Graph $y = \frac{3}{5}x - 4$ using the slope and the y-intercept.

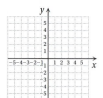

◀ **Do Exercise 4.**

EXAMPLE 4 Graph $2x + 3y = 3$ using the slope and the y-intercept.

To graph $2x + 3y = 3$, we first rewrite the equation in slope–intercept form:

$$2x + 3y = 3$$
$$3y = -2x + 3 \qquad \text{Adding } -2x$$
$$\tfrac{1}{3} \cdot 3y = \tfrac{1}{3}(-2x + 3) \qquad \text{Multiplying by } \tfrac{1}{3}$$
$$y = -\tfrac{2}{3}x + 1. \qquad \text{Simplifying}$$

To graph $y = -\frac{2}{3}x + 1$, we first plot the y-intercept, $(0, 1)$. We can think of the slope as $\frac{-2}{3}$. Starting at $(0, 1)$ and using the slope, we find a second point by moving 2 units *down* (since the numerator is *negative*) and 3 units *to the right* (since the denominator is *positive*). We plot the new point, $(3, -1)$. In a similar manner, we can move from the point $(3, -1)$ to locate a third point, $(6, -3)$. The line can then be drawn.

Answers

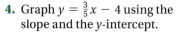

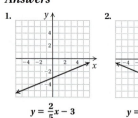

1.
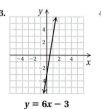
$y = \frac{2}{5}x - 3$

2.
$y = -\frac{2}{5}x - 3$

3.
$y = 6x - 3$

4.
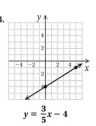
$y = \frac{3}{5}x - 4$

Since $-\frac{2}{3} = \frac{2}{-3}$, an alternative approach is to again plot $(0, 1)$, but this time we move 2 units *up* (since the numerator is *positive*) and 3 units *to the left* (since the denominator is *negative*). This leads to another point on the graph, $(-3, 3)$.

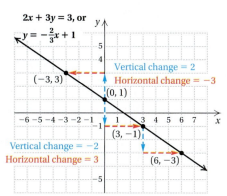

2x + 3y = 3, or
$y = -\frac{2}{3}x + 1$

Vertical change = 2
Horizontal change = −3
$(-3, 3)$
$(0, 1)$
$(3, -1)$
Vertical change = −2
Horizontal change = 3
$(6, -3)$

Do Exercise 5. ▶

5. Graph: $3x + 4y = 12$.

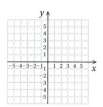

Answer

5.

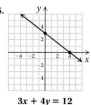

$3x + 4y = 12$

3.5 Exercise Set

✓ Reading Check

In each exercise, a line has been graphed using its *y*-intercept and its slope. Choose from the column on the right the appropriate slope of the line.

RC1.

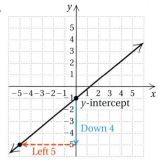

y-intercept
Down 4
Left 5

RC2.

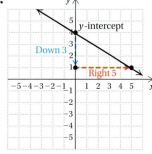

y-intercept
Down 3
Right 5

a) $\dfrac{3}{5}$

b) $-\dfrac{4}{5}$

c) $\dfrac{4}{5}$

d) $-\dfrac{3}{5}$

e) $-\dfrac{5}{4}$

f) $\dfrac{5}{3}$

RC3.

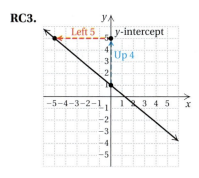

Left 5   y-intercept
Up 4

RC4.

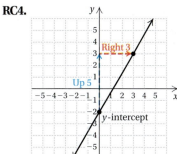

Right 3
Up 5
y-intercept

Draw a line that has the given slope and *y*-intercept.

1. Slope $\frac{2}{5}$; *y*-intercept $(0, 1)$

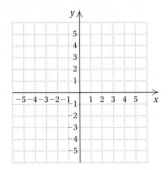

2. Slope $\frac{3}{5}$; *y*-intercept $(0, -1)$

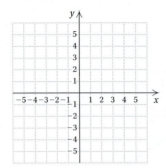

3. Slope $\frac{5}{3}$; *y*-intercept $(0, -2)$

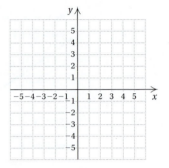

4. Slope $\frac{5}{2}$; *y*-intercept $(0, 1)$

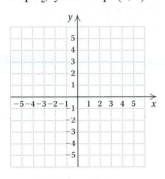

5. Slope $-\frac{3}{4}$; *y*-intercept $(0, 5)$

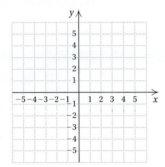

6. Slope $-\frac{4}{5}$; *y*-intercept $(0, 6)$

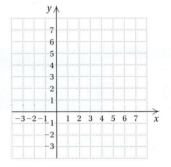

7. Slope $-\frac{1}{2}$; *y*-intercept $(0, 3)$

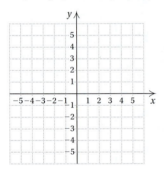

8. Slope $\frac{1}{3}$; *y*-intercept $(0, -4)$

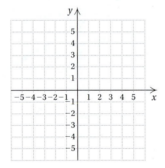

9. Slope 2; *y*-intercept $(0, -4)$

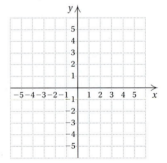

10. Slope -2; *y*-intercept $(0, -3)$

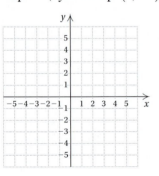

11. Slope -3; *y*-intercept $(0, 2)$

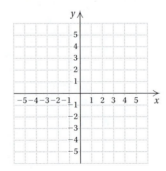

12. Slope 3; *y*-intercept $(0, 4)$

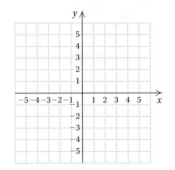

264   **CHAPTER 3**   Graphs of Linear Equations

Graph using the slope and the *y*-intercept.

13. $y = \frac{3}{5}x + 2$

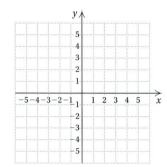

14. $y = -\frac{3}{5}x - 1$

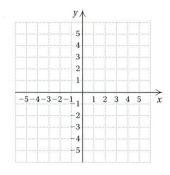

15. $y = -\frac{3}{5}x + 1$

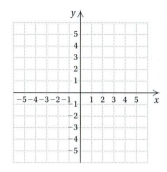

16. $y = \frac{3}{5}x - 2$

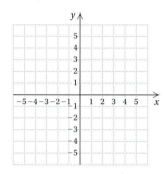

17. $y = \frac{5}{3}x + 3$

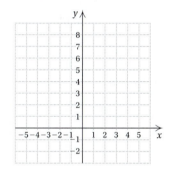

18. $y = \frac{5}{3}x - 2$

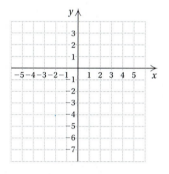

19. $y = -\frac{3}{2}x - 2$

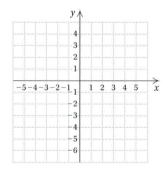

20. $y = -\frac{4}{3}x + 3$

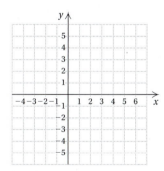

21. $2x + y = 1$

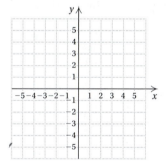

22. $3x + y = 2$

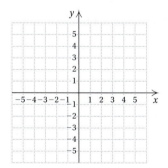

23. $3x - y = 4$

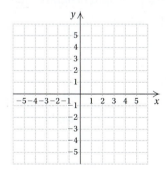

24. $2x - y = 5$

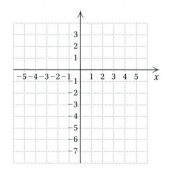

25. $2x + 3y = 9$

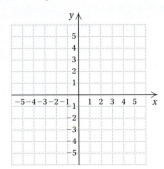

26. $4x + 5y = 15$

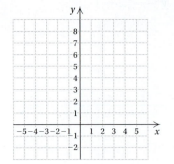

27. $x - 4y = 12$

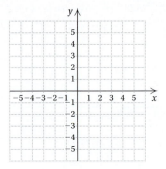

28. $x + 5y = 20$

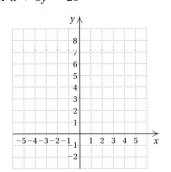

29. $x + 2y = 6$

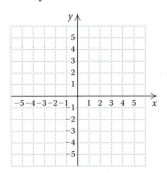

30. $x - 3y = 9$

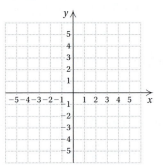

Skill Maintenance

Find the slope, if it exists, of the line containing the given pair of points.   [3.3a]

31. $(-2, -6), (8, 7)$

32. $(2, -6), (8, -7)$

33. $(4.5, -2.3), (14.5, 4.6)$

34. $(-0.8, -2.3), (-4.8, 0.1)$

35. $(-2, -6), (8, -6)$

36. $(-2, -6), (-2, 7)$

Simplify.   [1.4a]

37. $8 - (-11) + 23$

38. $-200 - 25 + 40$

39. $-10 - (-30) + 5 - 2$

40. $-40 - (-32) + 50 - 1$

Synthesis

41. Graph the line with slope 2 that passes through the point $(-3, 1)$.

42. Graph the line with slope -3 that passes through the point $(3, 0)$.

Parallel Lines and Perpendicular Lines

When we graph a pair of linear equations, there are three possibilities:

1. The graphs are the same.
2. The graphs intersect at exactly one point.
3. The graphs are parallel. (They do not intersect.)

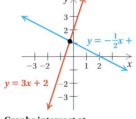

Equations have
the same graph.

Graphs intersect at
exactly one point.

Graphs are parallel.

OBJECTIVES

a   Determine whether the graphs of two linear equations are parallel.

b   Determine whether the graphs of two linear equations are perpendicular.

SKILL TO REVIEW

Objective 3.3b: Find the slope of a line from an equation.

Find the slope, if it exists, of each line.

1. $7y - 2x = 10$
2. $x - 13y = -1$

a   PARALLEL LINES

The graphs shown below are of the linear equations

$$y = 2x + 5 \quad \text{and} \quad y = 2x - 3.$$

The slope of each line is 2. The y-intercepts, $(0, 5)$ and $(0, -3)$, are different. The lines do not have the same graph, do not intersect, and are parallel.

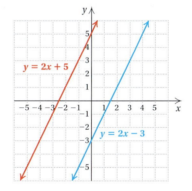

PARALLEL LINES

- Parallel nonvertical lines have the *same* slope, $m_1 = m_2$, and *different* y-intercepts, $b_1 \neq b_2$.
- Parallel horizontal lines have equations $y = p$ and $y = q$, where $p \neq q$.
- Parallel vertical lines have equations $x = p$ and $x = q$, where $p \neq q$.

By simply graphing, we may find it difficult to determine whether lines are parallel. Sometimes they may intersect very far from the origin. We can use the preceding statements about slopes, y-intercepts, and parallel lines to determine for certain whether lines are parallel.

Answers

Skill to Review:

1. $\frac{2}{7}$ 2. $\frac{1}{13}$

Determine whether the graphs of each pair of equations are parallel.

1. $3x - y = -5$,
$y - 3x = -2$

Solve each equation for y and then find the slope.

$3x - y = -5$

$-y = -3x - 5$

$y = \boxed{}\, x + 5$

The slope is $\boxed{}$.

$y - 3x = -2$

$y = \boxed{}\, x - 2$

The slope is 3.
The slope of each line is $\boxed{}$.
The y-intercepts, $(0, 5)$ and $(0, \boxed{})$, are different. Thus the lines are parallel.

2. $y - 3x = 1$,
$-2y = 3x + 2$

EXAMPLE 1 Determine whether the graphs of the lines $y = -3x + 4$ and $6x + 2y = -10$ are parallel.

The graphs of these equations are shown below. They appear to be parallel, but it is most accurate to determine this algebraically.

We first solve each equation for y. In this case, the first equation is already solved for y.

a) $y = -3x + 4$

b) $6x + 2y = -10$

$2y = -6x - 10$

$y = \frac{1}{2}(-6x - 10)$

$y = -3x - 5$

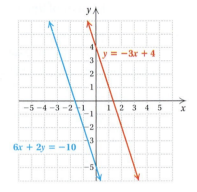

The slope of each line is -3. The y-intercepts are $(0, 4)$ and $(0, -5)$, which are different. The lines are parallel.

◀ **Do Exercises 1 and 2.**

b PERPENDICULAR LINES

Perpendicular lines in a plane are lines that intersect at a right, or 90°, angle. The lines whose graphs are shown below are perpendicular. You can check this approximately by using a protractor or placing the corner of a rectangular piece of paper at the intersection.

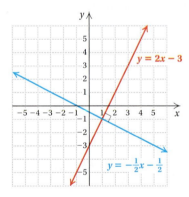

The slopes of the lines are 2 and $-\frac{1}{2}$. Note that $2\left(-\frac{1}{2}\right) = -1$. That is, the product of the slopes is -1.

> ### PERPENDICULAR LINES
>
> - Two nonvertical lines are perpendicular if the product of their slopes is -1, $m_1 \cdot m_2 = -1$. (If one line has slope m, the slope of the line perpendicular to it is $-1/m$.)
> - If one equation in a pair of perpendicular lines is vertical, then the other is horizontal. These equations are of the form $x = a$ and $y = b$.

Answers

1. Yes   **2.** No

Guided Solution:
1. 3, 3, 3, 3, −2

EXAMPLE 2   Determine whether the graphs of the lines $3y = 9x + 3$ and $6y + 2x = 6$ are perpendicular.

The graphs are shown below. They appear to be perpendicular, but it is most accurate to determine this algebraically.

We first solve each equation for y in order to determine the slopes:

a) $3y = 9x + 3$

$y = \frac{1}{3}(9x + 3)$

$y = 3x + 1;$

b) $6y + 2x = 6$

$6y = -2x + 6$

$y = \frac{1}{6}(-2x + 6)$

$y = -\frac{1}{3}x + 1.$

The slopes are 3 and $-\frac{1}{3}$. The product of the slopes is $3\left(-\frac{1}{3}\right) = -1$. The lines are perpendicular.

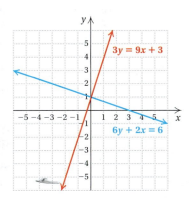

Do Exercises 3 and 4. ▶

Determine whether the graphs of each pair of equations are perpendicular.

 3. $y = -\frac{3}{4}x + 7,$

$y = \frac{4}{3}x - 9$

The slopes of the lines are $-\frac{3}{4}$ and $\boxed{}$.

The product of the slopes is $-\frac{3}{4} \cdot \frac{4}{3} = \boxed{}$.

The lines are perpendicular.

4. $4x - 5y = 8,$

$6x + 9y = -12$

┌───┐

CALCULATOR CORNER

Parallel Lines   Graph each pair of equations in Margin Exercises 1 and 2 in the standard viewing window, $[-10, 10, -10, 10]$. (Note that each equation must be solved for y so that it can be entered in Y = form on the graphing calculator.) Determine whether the lines appear to be parallel.

Perpendicular Lines   Graph each pair of equations in Margin Exercises 3 and 4 in the window $[-9, 9, -6, 6]$. (Note that the equations in Margin Exercise 4 must be solved for y so that they can be entered in Y = form on the graphing calculator.) Determine whether the lines appear to be perpendicular. Note (in the viewing window) that more of the x-axis is shown than the y-axis. The dimensions were chosen to more accurately reflect the slopes of the lines.

└───┘

Answers

3. Yes   **4.** No

Guided Solution:

3. $\frac{4}{3}, -1$

For Extra Help

MyMathLab®   MathXL®   PRACTICE   WATCH   READ   REVIEW

✓ Reading Check

The graphs of two of the equations listed below are parallel, and the graphs of two of the equations listed below are perpendicular.

RC1. Determine which two lines are parallel.

RC2. Determine which two lines are perpendicular.

a) $y = -4x - 1$

b) $y = -\frac{3}{4}x - 3$

c) $y = -3x + 1$

d) $y = \frac{1}{3}x + 1$

e) $y = -\frac{3}{4}x + 3$

f) $y = -\frac{1}{4}x - 3$

a   Determine whether the graphs of the equations are parallel lines.

1. $x + 4 = y,$
$y - x = -3$

2. $3x - 4 = y,$
$y - 3x = 8$

3. $y + 3 = 6x,$
$-6x - y = 2$

4. $y = -4x + 2,$
$-5 = -2y + 8x$

5. $10y + 32x = 16.4,$
$y + 3.5 = 0.3125x$

6. $y = 6.4x + 8.9,$
$5y - 32x = 5$

7. $y = 2x + 7,$
$5y + 10x = 20$

8. $y + 5x = -6,$
$3y + 5x = -15$

9. $3x - y = -9,$
$2y - 6x = -2$

10. $y - 6 = -6x,$
$-2x + y = 5$

11. $x = 3,$
$x = 4$

12. $y = 1,$
$y = -2$

13. $x = -6, y = 6$

14. $x = \frac{1}{3}, y = -\frac{1}{3}$

b   Determine whether the graphs of the equations are perpendicular lines.

15. $y = -4x + 3,$
$4y + x = -1$

16. $y = -\frac{2}{3}x + 4,$
$3x + 2y = 1$

17. $x + y = 6,$
$4y - 4x = 12$

18. $2x - 5y = -3,$
$5x + 2y = 6$

19. $y = -0.3125x + 11,$
$y - 3.2x = -14$

20. $y = -6.4x - 7,$
$64y - 5x = 32$

21. $y = -x + 8,$
$x - y = -1$

22. $2x + 6y = -3,$
$12y = 4x + 20$

23. $\dfrac{3}{8}x - \dfrac{y}{2} = 1,$

$\dfrac{4}{3}x - y + 1 = 0$

24. $\dfrac{1}{2}x + \dfrac{3}{4}y = 6,$

$-\dfrac{3}{2}x + y = 4$

25. $x = 0,$

$y = -2$

26. $x = -3,$

$y = 5$

27. $y = 4,$

$y = -\dfrac{1}{4}$

28. $x = -\dfrac{7}{8},$

$x = -\dfrac{8}{7}$

a , **b**   Determine whether the graphs of the equations are parallel, perpendicular, or neither.

29. $3y + 21 = 2x,$
$3y = 2x + 24$

30. $3y + 21 = 2x,$
$2y = 16 - 3x$

31. $3y = 2x - 21,$
$2y - 16 = 3x$

32. $3y + 2x + 7 = 0,$
$3y = 2x + 24$

Skill Maintenance

Determine whether the given ordered pair is a solution of the equation.   [3.1c]

33. $(1, -1);\ 2x - 15y = -17$

34. $(-14, -6);\ 16 - x = -5y$

Find the intercepts of each equation.   [3.2a]

35. $-40x + 5y = 80$

36. $y - 3 = 6x$

Synthesis

37. Find an equation of a line that contains the point $(0, 6)$ and is parallel to $y - 3x = 4$.

38. Find an equation of the line that contains the point $(-2, 4)$ and is parallel to $y = 2x - 3$.

39. Find an equation of the line that contains the point $(0, 2)$ and is perpendicular to $3y - x = 0$.

40. Find an equation of the line that contains the point $(1, 0)$ and is perpendicular to $2x + y = -4$.

41. Find an equation of the line that has x-intercept $(-2, 0)$ and is parallel to $4x - 8y = 12$.

42. Find the value of k such that $4y = kx - 6$ and $5x + 20y = 12$ are parallel.

43. Find the value of k such that $4y = kx - 6$ and $5x + 20y = 12$ are perpendicular.

The lines in the graphs in Exercises 44 and 45 are perpendicular, and the lines in the graph in Exercise 46 are parallel. Find an equation of each line.

44.

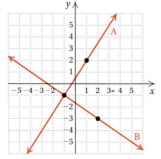

45.

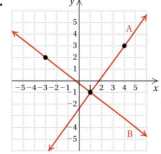

46.

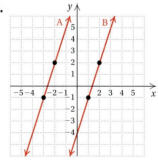

A graph of an inequality is a drawing that represents its solutions. An inequality in one variable can be graphed on the number line. An inequality in two variables can be graphed on a coordinate plane.

a SOLUTIONS OF INEQUALITIES IN TWO VARIABLES

The solutions of inequalities in two variables are ordered pairs.

EXAMPLE 1 Determine whether $(-3, 2)$ is a solution of $5x + 4y < 13$.

We use alphabetical order to replace x with -3 and y with 2.

$$\begin{array}{c|c} 5x + 4y < 13 \\ \hline 5(-3) + 4 \cdot 2 \;?\; 13 \\ -15 + 8 \\ -7 & \text{TRUE} \end{array}$$

Since $-7 < 13$ is true, $(-3, 2)$ is a solution.

EXAMPLE 2 Determine whether $(6, 8)$ is a solution of $5x + 4y < 13$.

We use alphabetical order to replace x with 6 and y with 8.

$$\begin{array}{c|c} 5x + 4y < 13 \\ \hline 5(6) + 4(8) \;?\; 13 \\ 30 + 32 \\ 62 & \text{FALSE} \end{array}$$

Since $62 < 13$ is false, $(6, 8)$ is not a solution.

◀ **Do Margin Exercises 1 and 2.**

1. Determine whether $(4, 3)$ is a solution of $3x - 2y < 1$.

2. Determine whether $(2, -5)$ is a solution of $4x + 7y \geq 12$.

b GRAPHING INEQUALITIES IN TWO VARIABLES

EXAMPLE 3 Graph: $y > x$.

We first graph the line $y = x$. Every solution of $y = x$ is an ordered pair like $(3, 3)$ in which the first coordinate and the second coordinate are the same. We draw the line $y = x$ dashed (as shown on the left below) because its points are *not* solutions of $y > x$.

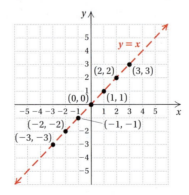 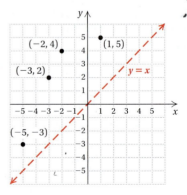

Now look at the graph on the right on the preceding page. Several ordered pairs are plotted in the **half-plane** above the line $y = x$. Each is a solution of $y > x$.

We can check a pair such as $(-2, 4)$ as follows:

$$\frac{y > x}{4 \;?\; -2} \;\text{TRUE}$$

It turns out that any point on the same side of $y = x$ as $(-2, 4)$ is also a solution. *If we know that one point in a half-plane is a solution, then all points in that half-plane are solutions.* We could have chosen other points to check. The graph of $y > x$ is shown below. (Solutions are indicated by color shading throughout.) We shade the half-plane above $y = x$.

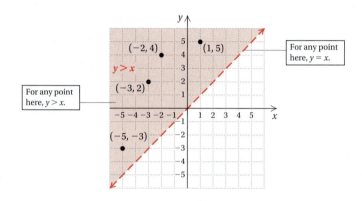

Do Exercise 3. ▶

A **linear inequality** is one that we can get from a linear equation by changing the equals symbol to an inequality symbol. Every linear equation has a graph that is a straight line. The graph of a linear inequality is a half-plane, sometimes including the line along the edge.

To graph an inequality in two variables:

1. Replace the inequality symbol with an equals sign and graph this related linear equation.

2. If the inequality symbol is $<$ or $>$, draw the line dashed. If the inequality symbol is $\leq$ or $\geq$, draw the line solid.

3. The graph consists of a half-plane, either above or below or left or right of the line, and, if the line is solid, the line as well. To determine which half-plane to shade, choose a point *not on the line* as a test point. Substitute to find whether that point is a solution of the *inequality*. If it is, shade the half-plane containing that point. If it is not, shade the half-plane on the opposite side of the line.

---

3. Graph: $y < x$.

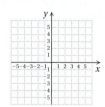

Answers

3.

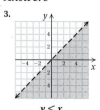

$y < x$

4. Graph: $2x + 4y < 8$.

Related equation:
$$2x + 4y \boxed{} 8$$

x-intercept: $(\boxed{}, 0)$

y-intercept: $(0, \boxed{})$

Draw the line dashed.

Test a point—try $(3, -1)$:

$$2 \cdot 3 + 4 \cdot (\boxed{}) < 8$$

$$\boxed{} < 8. \quad \text{TRUE}$$

The point $(3, -1)$ is a solution. We shade the half-plane that contains $(3, -1)$.

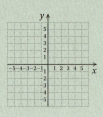

Graph.

5. $3x - 5y < 15$

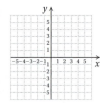

6. $2x + 3y \geq 12$

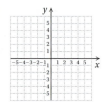

EXAMPLE 4 Graph: $5x - 2y < 10$.

1. We first graph the line $5x - 2y = 10$. The intercepts are $(0, -5)$ and $(2, 0)$. This line forms the boundary of the solutions of the inequality.

2. Since the inequality contains the $<$ symbol, points on the line are not solutions of the inequality, so we draw a dashed line.

3. To determine which half-plane to shade, we consider a test point *not* on the line. We try $(3, -2)$ and substitute:

$$\begin{array}{c|c} 5x - 2y < 10 & \\ \hline 5(3) - 2(-2) \; ? \; 10 & \\ 15 + 4 & \\ 19 & \text{FALSE} \end{array}$$

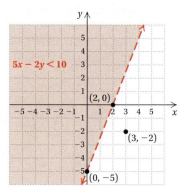

Since this inequality is false, the point $(3, -2)$ is *not* a solution; no point in the half-plane containing $(3, -2)$ is a solution. Thus the points in the opposite half-plane are solutions. The graph is shown above.

◀ **Do Exercise 4.**

EXAMPLE 5 Graph: $2x + 3y \leq 6$.

1. First, we graph the line $2x + 3y = 6$. The intercepts are $(0, 2)$ and $(3, 0)$.

2. Since the inequality contains the $\leq$ symbol, we draw the line solid to indicate that any pair on the line is a solution.

3. Next, we choose a test point that is not on the line. We substitute to determine whether this point is a solution. The origin $(0, 0)$ is generally an easy point to use:

$$\begin{array}{c|c} 2x + 3y \leq 6 & \\ \hline 2 \cdot 0 + 3 \cdot 0 \; ? \; 6 & \\ 0 & \text{TRUE} \end{array}$$

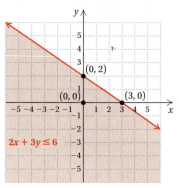

We see that $(0, 0)$ is a solution, so we shade the lower half-plane. Had the substitution given us a false inequality, we would have shaded the other half-plane.

◀ **Do Exercises 5 and 6.**

Answers

4.

$2x + 4y < 8$

5.

$3x - 5y < 15$

6.

$2x + 3y \geq 12$

Guided Solution:
4. $=, 4, 2, -1, 2$

EXAMPLE 6 Graph $x < 3$ on a plane.

There is no y-term in this inequality, but we can rewrite this inequality as $x + 0y < 3$. We use the same technique that we have used with the other examples.

1. We graph the related equation $x = 3$ on the plane.

2. Since the inequality symbol is $<$, we use a dashed line.

3. The graph is a half-plane either to the left or to the right of the line $x = 3$. To determine which, we consider a test point, $(-4, 5)$:

$$\frac{x + 0y < 3}{-4 + 0(5) \; ? \; 3}$$
$$-4 \;\big|\; \quad \text{TRUE}$$

We see that $(-4, 5)$ is a solution, so all the pairs in the half-plane containing $(-4, 5)$ are solutions. We shade that half-plane.

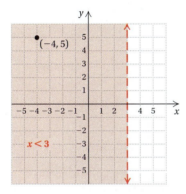

We see from the graph that the solutions of $x < 3$ are all those ordered pairs whose first coordinates are less than 3.

EXAMPLE 7 Graph: $y \geq -4$.

1. We first graph $y = -4$.

2. We use a solid line to indicate that all points on the line are solutions.

3. We then use $(2, 3)$ as a test point and substitute:

$$\frac{0x + y \geq -4}{0(2) + 3 \; ? \; -4}$$
$$3 \;\big|\; \quad \text{TRUE}$$

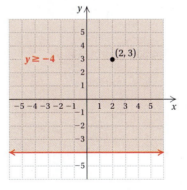

Since $(2, 3)$ is a solution, all points in the half-plane containing $(2, 3)$ are solutions. Note that this half-plane consists of all ordered pairs whose second coordinate is greater than or equal to -4.

Do Exercises 7 and 8. ▶

CALCULATOR CORNER

Graphs of Inequalities
We can graph inequalities on a graphing calculator, shading the region of the solution set. To graph the inequality in Example 5, $2x + 3y \leq 6$, we first graph the line $2x + 3y = 6$, or $y = (6 - 2x)/3$. After determining algebraically that the solution set consists of all points below the line, we use the graphing calculator's "shade below" graph style to shade this region. On the equation-editor screen, we position the cursor over the graph style icon to the left of the equation and press **ENTER** repeatedly until the ◣ icon appears. Then we press **GRAPH** to display the graph of the inequality. Some calculators also have an **APP** that graphs inequalities.

$$y = \frac{6 - 2x}{3}$$

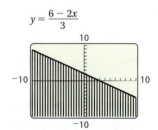

EXERCISE:

1. Use a graphing calculator to graph the inequalities in Margin Exercises 6 and 8 and in Example 7.

Graph.

7. $x > -3$ **8.** $y \leq 4$

Answers

7.

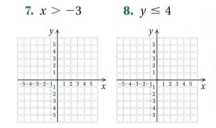

$x > -3$

8.

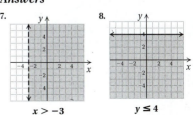

$y \leq 4$

Visualizing for Success

A

B

C

D

E

F

G

H

I

J

Match each equation or inequality with its graph.

1. $3x - 5y \leq 15$

2. $3x + 5y = 15$

3. $3x + 5y \leq 15$

4. $3x - 5y \geq 15$

5. $3x - 5y = 15$

6. $3x - 5y < 15$

7. $3x + 5y \geq 15$

8. $3x + 5y > 15$

9. $3x - 5y > 15$

10. $3x + 5y < 15$

Answers on page A-9

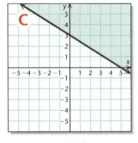

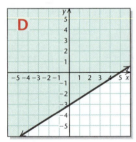

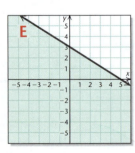

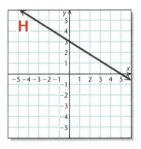

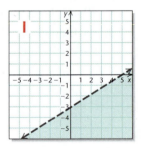

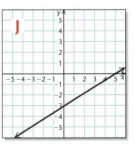

✓ Reading Check

The process for graphing the inequality $2x - 4y < -12$ is described in the following paragraph. Choose from the columns on the right the word or the symbol that completes each of Exercises RC1–RC9.

To graph $2x - 4y < -12$, we first replace $<$ with
RC1. _____ and graph the **RC2.** _____
equation. The **RC3.** _____ is $(-6, 0)$.
The **RC4.** _____ is $(0, 3)$. The inequality
symbol is **RC5.** _____, so we draw the line
RC6. _____. The graph consists of a
RC7. _____, either above or below the line.
To determine which half-plane to **RC8.** _____,
we test a point not on the line. Let's check
$(0, 0)$: $2 \cdot 0 - 4 \cdot 0 < -12$, or $0 < -12$ is
RC9. _____. We shade the other half-plane.

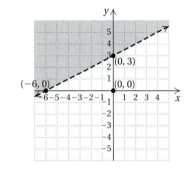

true	$<$
false	$\leq$
x-intercept	solid
y-intercept	dashed
half-plane	related
=	shade

a

1. Determine whether $(-3, -5)$ is a solution of
$-x - 3y < 18$.

2. Determine whether $(2, -3)$ is a solution of
$5x - 4y \geq 1$.

3. Determine whether $(1, -10)$ is a solution of
$7y - 9x \leq -3$.

4. Determine whether $(-8, 5)$ is a solution of
$x + 0 \cdot y > 4$.

b Graph on a plane.

5. $x > 2y$

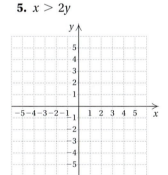

6. $x \geq 3y$

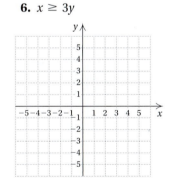

7. $y \leq x - 3$

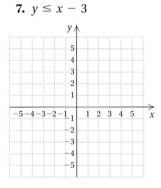

8. $y < x + 4$

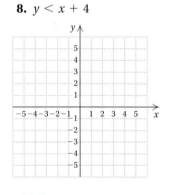

9. $x + y \leq 3$

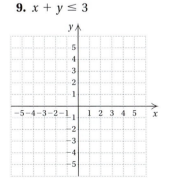

10. $x + y < 4$

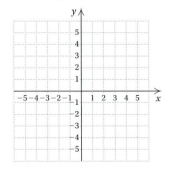

11. $y > x - 2$

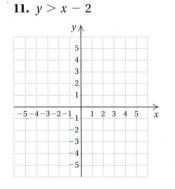

12. $y \geq x - 1$

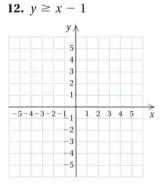

13. $x - y > 7$

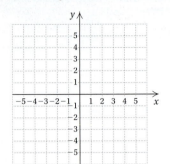

14. $x - y > -2$

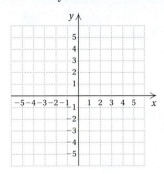

15. $y \geq 4x - 1$

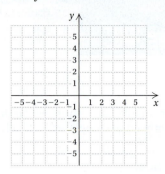

16. $y \geq 3x + 2$

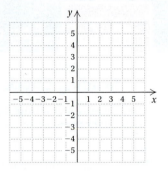

17. $y \geq 1 - 2x$

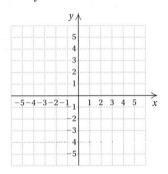

18. $y - 2x \leq -1$

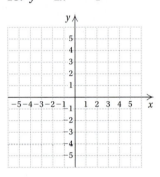

19. $2x + 3y \leq 12$

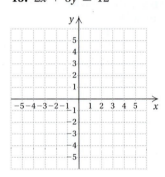

20. $5y - 2x > 10$

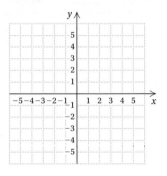

21. $y \leq 3$

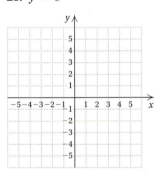

22. $y > -1$

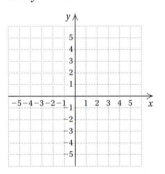

23. $x \geq -1$

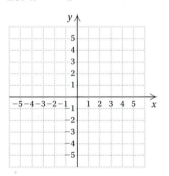

24. $x < 0$

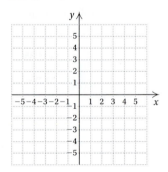

Skill Maintenance

Determine whether the graphs of the equations are parallel, perpendicular, or neither.   [3.6a, b]

25. $5y + 50 = 4x,$
$\quad 5y = 4x + 15$

26. $5x + 4y = 12,$
$\quad 5y + 50 = 4x$

27. $5y + 50 = 4x,$
$\quad 4y = 5x + 12$

28. $4x + 5y + 35 = 0,$
$\quad 5y = 4x + 40$

Synthesis

29. *Elevators.*   Many elevators have a capacity of 1 metric ton (1000 kg). Suppose c children, each weighing 35 kg, and a adults, each weighing 75 kg, are on an elevator. Find and graph an inequality that asserts that the elevator is overloaded.

30. *Hockey Wins and Losses.*   A hockey team determines that it needs at least 60 points for the season in order to make the playoffs. A win w is worth 2 points and a tie t is worth 1 point. Find and graph an inequality that describes the situation.

Vocabulary Reinforcement

Complete each statement with the correct term from the column on the right. Some of the choices may not be used.

1. The equation $y = mx + b$ is called the _____ equation. [3.4a]

2. _____ lines are graphs of equations of the type $y = b$. [3.2b]

3. _____ lines are graphs of equations of the type $x = a$. [3.2b]

4. The _____ of a line is a number that indicates how the line slants. [3.3a]

5. The _____ of a line, if it exists, indicates where the line crosses the x-axis, and thus will always have 0 as the _____ coordinate. [3.2a]

6. The _____ of a line, if it exists, indicates where the line crosses the y-axis, and thus will always have 0 as the _____ coordinate. [3.2a]

x-intercept

y-intercept

parallel

perpendicular

vertical

horizontal

slope–intercept

first

second

slope

Concept Reinforcement

Determine whether each statement is true or false.

_____ 1. The x- and y-intercepts of $y = mx$ are both $(0, 0)$. [3.2a]

_____ 2. Parallel lines have the same y-intercept. [3.6a]

_____ 3. The ordered pair $(0, 0)$ is a solution of $y > x$. [3.7a]

_____ 4. The second coordinate of all points in quadrant III is negative. [3.1a]

_____ 5. The x-intercept of $Ax + By = C$, $C \neq 0$, is $\left(\dfrac{A}{C}, 0\right)$. [3.2a]

Study Guide

Objective 3.1b Find the coordinates of a point on a graph.

Example Find the coordinates of points Q, R, and S.

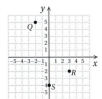

Point Q is 2 units left of the origin and 5 units up. Its coordinates are $(-2, 5)$.

Point R is 3 units right of the origin and 2 units down. Its coordinates are $(3, -2)$.

Point S is 0 units left or right of the origin and 4 units down. Its coordinates are $(0, -4)$.

Practice Exercise

1. Find the coordinates of points F, G, and H.

Objective 3.1d Graph linear equations of the type $y = mx + b$ and $Ax + By = C$, identifying the y-intercept.

Example Graph $2y + 2 = -3x$ and identify the y-intercept.

To find an equivalent equation in the form $y = mx + b$, we solve for y: $y = -\frac{3}{2}x - 1$. The y-intercept is $(0, -1)$.

We then find two other points using multiples of 2 for x to avoid fractions.

x	y
0	-1	← y-intercept
-2	2
2	-4

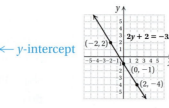

Practice Exercise

2. Graph $x + 2y = 8$ and identify the y-intercept.

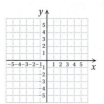

Objective 3.2a Find the intercepts of a linear equation, and graph using intercepts.

Example For $2x - y = -6$, find the intercepts. Then use the intercepts to graph the equation.

To find the y-intercept, we let $x = 0$ and solve for y:

$2 \cdot 0 - y = -6$ and $y = 6$.

The y-intercept is $(0, 6)$.

To find the x-intercept, we let $y = 0$ and solve for x:

$2x - 0 = -6$ and $x = -3$.

The x-intercept is $(-3, 0)$.

We find a third point as a check.

x	y
0	6	← y-intercept
-3	0	← x-intercept
-1	4

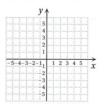

Practice Exercise

3. For $y - 2x = -4$, find the intercepts. Then use the intercepts to graph the equation.

Objective 3.2b Graph equations equivalent to those of the type $x = a$ and $y = b$.

Example Graph: $y = 1$ and $x = -\dfrac{3}{2}$.

For $y = 1$, no matter what number we choose for x, $y = 1$. The graph is a horizontal line. For $x = -\frac{3}{2}$, no matter what number we choose for y, $x = -\frac{3}{2}$. The graph is a vertical line.

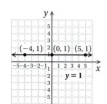

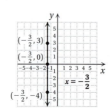

Practice Exercises

Graph.

4. $y = -\dfrac{5}{2}$

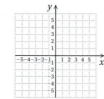

5. $x = 2$

Objective 3.3a Given the coordinates of two points on a line, find the slope of the line, if it exists.

Example Find the slope, if it exists, of the line containing the given points.

$(-9, 3)$ and $(5, -6)$: $m = \dfrac{-6 - 3}{5 - (-9)} = \dfrac{-9}{14} = -\dfrac{9}{14}$;

$\left(7, \dfrac{1}{2}\right)$ and $\left(-13, \dfrac{1}{2}\right)$: $m = \dfrac{\frac{1}{2} - \frac{1}{2}}{-13 - 7} = \dfrac{0}{-20} = 0$;

$(0.6, 1.5)$ and $(0.6, -1.5)$: $m = \dfrac{-1.5 - 1.5}{0.6 - 0.6} = \dfrac{-3}{0}$,

m is not defined.

Practice Exercises

Find the slope, if it exists, of the line containing the given points.

6. $(-8, 20), (-8, 14)$

7. $(2, -1), (16, 20)$

8. $(0.5, 2.8), (1.5, 2.8)$

Objective 3.3b Find the slope of a line from an equation.

Example Find the slope, if it exists, of each line.

a) $5x - 20y = -10$

We first solve for y: $y = \frac{1}{4}x + \frac{1}{2}$. The slope is $\frac{1}{4}$.

b) $y = -\frac{4}{5}$

Think: $y = 0 \cdot x - \frac{4}{5}$. This line is horizontal. The slope is 0.

c) $x = 6$

This line is vertical. The slope is not defined.

Practice Exercises

Find the slope, if it exists, of the line.

9. $x = 0.25$

10. $7y + 14x = -28$

11. $y = -5$

Objective 3.4b Find an equation of a line when the slope and a point on the line are given.

Example Find the equation of the line with slope -2 that contains the point $(3, -1)$.

$y = -2x + b$ Substituting -2 for m in $y = mx + b$

$-1 = -2 \cdot 3 + b$ Substituting 3 for x and -1 for y

$-1 = -6 + b$

$5 = b$ Solving for b

The equation is $y = -2x + 5$.

Practice Exercise

12. Find the equation of the line with slope 6 that contains the point $(-1, 1)$.

Objective 3.4c Find an equation of a line when two points on the line are given.

Example Find an equation of the line that contains $(-10, 5)$ and $(2, -5)$.

Slope $= m = \dfrac{-5 - 5}{2 - (-10)} = \dfrac{-10}{12} = -\dfrac{5}{6}$.

$-5 = -\dfrac{5}{6}(2) + b$ Substituting $-\frac{5}{6}$ for m, 2 for x, and -5 for y in $y = mx + b$

$-5 = -\dfrac{5}{3} + b$

$-\dfrac{10}{3} = b$ Solving for b

The equation is $y = -\dfrac{5}{6}x - \dfrac{10}{3}$.

Practice Exercise

13. Find an equation of the line that contains $(7, -3)$ and $(1, -2)$.

Objectives 3.6a, b Determine whether the graphs of two linear equations are parallel, perpendicular, or neither.

Example Determine whether the graphs of the equations are parallel, perpendicular, or neither:

$$2x - y = 8 \quad \text{and} \quad y + \frac{1}{2}x = -2.$$

We solve each equation for y and determine the slope of each:

$$y = 2x - 8 \quad \text{and} \quad y = -\frac{1}{2}x - 2.$$

The slopes are 2 and $-\frac{1}{2}$.

The slopes are not the same. The lines are not parallel. The product of the slopes, $2 \cdot \left(-\frac{1}{2}\right)$, is -1. The lines are perpendicular.

Practice Exercises

Determine whether the graphs of the equations are parallel, perpendicular, or neither.

14. $4y = -x - 12,$
$$y - 4x = \frac{1}{2}$$

15. $2y - x = -4,$
$$x - 2y = -12$$

Objective 3.7b Graph linear inequalities.

Example Graph: $3x - y < 3$.

We first graph the line $3x - y = 3$. The intercepts are $(0, -3)$ and $(1, 0)$. Since the inequality contains the $<$ symbol, points on the line are not solutions of the inequality, so we draw a dashed line.

To determine which half-plane to shade, we consider a test point not on the line. We try $(0, 0)$:

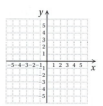

$$3 \cdot 0 - 0 < 3$$
$$0 < 3. \quad \text{TRUE}$$

We see that $(0, 0)$ is a solution, so we shade the upper half-plane.

Practice Exercise

16. Graph: $y - 3x \le -3$.

Review Exercises

Plot each point.   [3.1a]

1. $(2, 5)$ **2.** $(0, -3)$ **3.** $(-4, -2)$

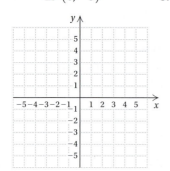

Find the coordinates of each point.   [3.1b]

4. A **5.** B **6.** C

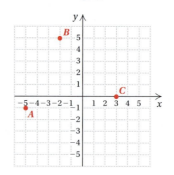

In which quadrant is each point located? [3.1a]

7. $(3, -8)$ **8.** $(-20, -14)$ **9.** $(4.9, 1.3)$

Determine whether each ordered pair is a solution of $2y - x = 10$. [3.1c]

10. $(2, -6)$ **11.** $(0, 5)$

12. Show that the ordered pairs $(0, -3)$ and $(2, 1)$ are solutions of the equation $2x - y = 3$. Then use the graph of the equation to determine another solution. Answers may vary. [3.1c]

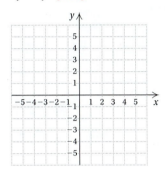

Graph each equation, identifying the *y*-intercept. [3.1d]

13. $y = 2x - 5$ **14.** $y = -\dfrac{3}{4}x$

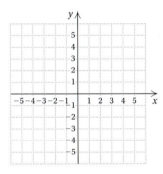 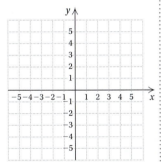

15. $y = -x + 4$ **16.** $y = 3 - 4x$

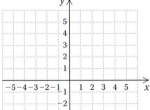 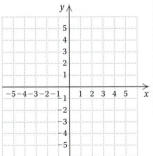

Solve. [3.1e]

17. *Kitchen Design.* Kitchen designers recommend that a refrigerator be selected on the basis of the number of people *n* in the household. The appropriate size *S*, in cubic feet, is given by

$$S = \dfrac{3}{2}n + 13.$$

a) Determine the recommended size of a refrigerator if the number of people is 1, 2, 5, and 10.

b) Graph the equation and use the graph to estimate the recommended size of a refrigerator for 4 people sharing an apartment.

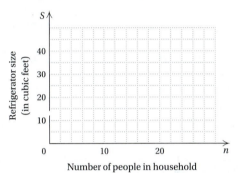

c) A refrigerator is 22 ft³. For how many residents is it the recommended size?

Find the intercepts of each equation. Then graph the equation. [3.2a]

18. $x - 2y = 6$ **19.** $5x - 2y = 10$

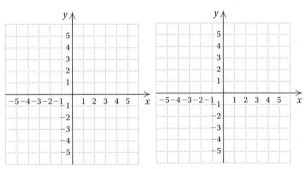

Graph each equation. [3.2b]

20. $y = 3$ **21.** $5x - 4 = 0$

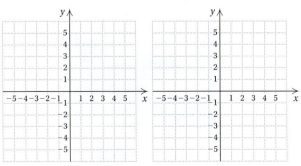

Find the slope. [3.3a]

22.

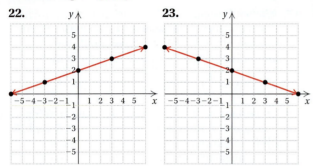

23.

Graph the line containing the given pair of points and find the slope. [3.3a]

24. $(-5, -2), (5, 4)$ **25.** $(-5, 5), (4, -4)$

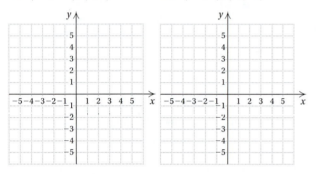

Find the slope, if it exists. [3.3b]

26. $y = -\dfrac{5}{8}x - 3$ **27.** $2x - 4y = 8$

28. $x = -2$ **29.** $y = 9$

30. *Snow Removal.* By 3:00 P.M., Erin had plowed 7 driveways and by 5:30 P.M., she had completed 13. [3.3c]

  a) Find Erin's plowing rate, in number of driveways per hour.

  b) Find Erin's plowing rate, in number of minutes per driveway.

31. *Road Grade.* At one point, Beartooth Highway in Yellowstone National Park rises 315 ft over a horizontal distance of 4500 ft. Find the slope, or grade, of the road. [3.3c]

32. *Organic Food.* Each year in the United States, the amount of sales in the organic food industry increases. Use the following graph to determine the slope, or rate of change, in the amount of sales, in billions of dollars, of organic food products with respect to time. [3.3a]

Organic Food

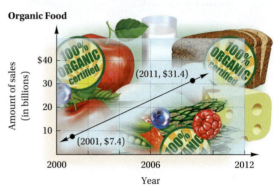

SOURCE: Organic Trade Association

Find the slope and the *y*-intercept. [3.4a]

33. $y = -9x + 46$

34. $x + y = 9$

35. $3x - 5y = 4$

Find an equation of the line with the given slope and *y*-intercept. [3.4a]

36. Slope: -2.8; *y*-intercept: $(0, 19)$

37. Slope: $\dfrac{5}{8}$; *y*-intercept: $\left(0, -\dfrac{7}{8}\right)$

Find an equation of the line containing the given point and with the given slope. [3.4b]

38. $(1, 2)$, $m = 3$

39. $(-2, -5)$, $m = \dfrac{2}{3}$

40. $(0, -4)$, $m = -2$

Find an equation of the line containing the given pair of points. [3.4c]

41. $(5, 7)$ and $(-1, 1)$

42. $(2, 0)$ and $(-4, -3)$

43. *Taking Cruises.* The following line graph illustrates the number of Americans, in millions, taking cruises for years since 2000. [3.4c]

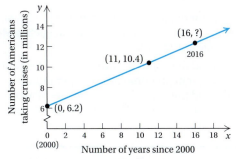

SOURCE: Cruise Lines International Association

a) Find an equation of the line. Let $x =$ the number of years since 2000.
b) What is the rate of change in the number of Americans taking cruises with respect to time?
c) Use the equation to estimate the number of Americans taking cruises in 2016.

44. Draw a line that has slope -1 and y-intercept $(0, 4)$. [3.5a]

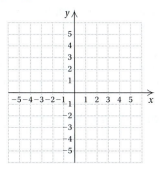

45. Draw a line that has slope $\frac{5}{3}$ and y-intercept $(0, -3)$. [3.5a]

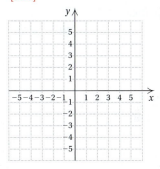

46. Graph $y = -\frac{3}{5}x + 2$ using the slope and the y-intercept. [3.5a]

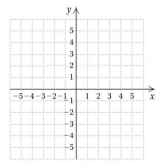

47. Graph $2y - 3x = 6$ using the slope and the y-intercept. [3.5a]

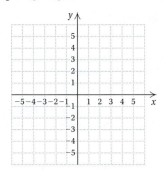

Determine whether the graphs of the equations are parallel, perpendicular, or neither. [3.6a, b]

48. $4x + y = 6$,
$4x + y = 8$

49. $2x + y = 10$,
$y = \frac{1}{2}x - 4$

50. $x + 4y = 8$,
$x = -4y - 10$

51. $3x - y = 6$,
$3x + y = 8$

Determine whether the given point is a solution of the inequality $x - 2y > 1$. [3.7a]

52. $(0, 0)$

53. $(1, 3)$

54. $(4, -1)$

Graph on a plane. [3.7b]

55. $x < y$

56. $x + 2y \geq 4$

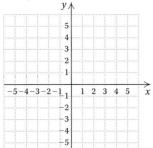

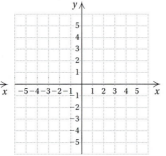

57. $x > -2$

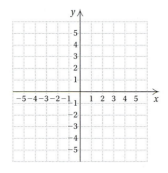

58. Select the statement that describes the graphs of the lines $-x + \frac{1}{2}y = -2$ and $2y + x - 8 = 0$. [3.6a, b]

   A. The lines are parallel.
   B. The lines are the same.
   C. The lines intersect and are not perpendicular.
   D. The lines are perpendicular.

59. Find the equation of the line with slope $-\frac{8}{3}$ and containing the point $(-3, 8)$. [3.4b]

   A. $y = -\frac{8}{3}x + \frac{55}{3}$
   B. $y = -\frac{3}{8}$
   C. $y = -\frac{8}{3}x$
   D. $8y + 3x = -3$

Synthesis

60. Find the area and the perimeter of a rectangle for which $(-2, 2), (7, 2)$, and $(7, -3)$ are three of the vertices. [3.1a]

61. *Gondola Aerial Lift.* In Telluride, Colorado, there is a free gondola ride that provides a spectacular view of the town and the surrounding mountains. The gondolas that begin in the town at an elevation of 8725 ft travel 5750 ft to Station St. Sophia, whose altitude is 10,550 ft. They then continue 3913 ft to Mountain Village, whose elevation is 9500 ft. [3.3c]

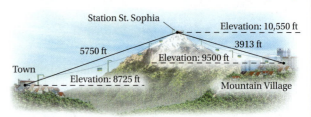

A visitor departs from the town at 11:55 A.M. and with no stop at Station St. Sophia reaches Mountain Village at 12:07 P.M.

   a) Find the gondola's average rate of ascent and descent, in number of feet per minute.
   b) Find the gondola's average rate of ascent and descent, in number of minutes per foot.

Understanding Through Discussion and Writing

1. Consider two equations of the type $Ax + By = C$. Explain how you would go about showing that their graphs are perpendicular. [3.6b]

2. Is the graph of any inequality in the form $y > mx + b$ shaded *above* the line $y = mx + b$? Why or why not? [3.7b]

3. Explain why the first coordinate of the y-intercept is always 0. [3.1d]

4. Graph $x < 1$ on both the number line and a plane, and explain the difference between the graphs. [3.7b]

5. Describe how you would graph $y = 0.37x + 2458$ using the slope and the y-intercept. You need not actually draw the graph. [3.5a]

6. Consider two equations of the type $Ax + By = C$. Explain how you would go about showing that their graphs are parallel. [3.6a]

CHAPTER

3 **Test**

For Extra Help For step-by-step test solutions, access the Chapter Test Prep Videos in MyMathLab® or on YouTube (search "BittingerIntro" and click on "Channels").

In which quadrant is each point located?

1. $\left(-\frac{1}{2}, 7\right)$

2. $(-5, -6)$

Find the coordinates of each point.

3. A

4. B

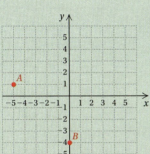

5. Show that the ordered pairs $(-4, -3)$ and $(-1, 3)$ are solutions of the equation $y - 2x = 5$. Then use the graph of the straight line containing the two points to determine another solution. Answers may vary.

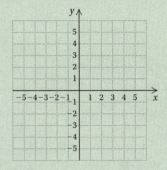

Graph each equation. Identify the y-intercept.

6. $y = 2x - 1$

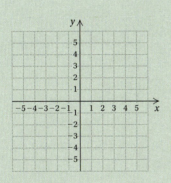

x	y

7. $y = -\frac{3}{2}x$

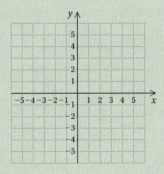

x	y

Find the intercepts of each equation. Then graph the equation.

8. $2x - 4y = -8$

x	y
	← x-intercept
	← y-intercept

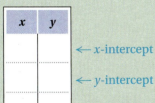

9. $2x - y = 3$

x	y
	← x-intercept
	← y-intercept

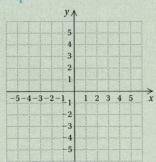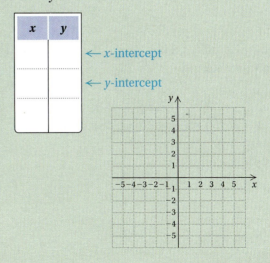

Graph each equation.

10. $2x + 8 = 0$

11. $y = 5$

12. *Health Insurance Cost.* The total annual cost, employer plus employee, of health insurance can be approximated by

$$C = 606t + 8593,$$

where t is the number of years since 2007. That is, $t = 0$ corresponds to 2007, $t = 3$ corresponds to 2010, and so on.

Source: Towers Watson

a) Find the total cost of health insurance in 2007, in 2009, and in 2012.

b) Graph the equation and then use the graph to estimate the cost of health insurance in 2016.

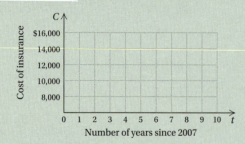

c) Predict the year in which the cost of health insurance will be $15,259.

13. *Train Travel.* The following graph shows data concerning a recent train ride from Denver to Kansas City. At what rate did the train travel?

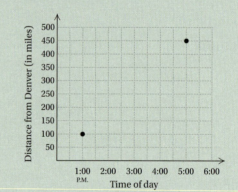

14. Find the slope.

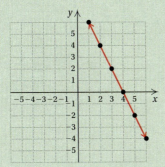

15. Graph the line containing $(-3, 1)$ and $(5, 4)$ and find the slope.

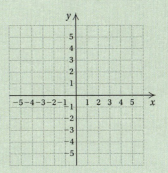

16. Find the slope, if it exists.

    a) $2x - 5y = 10$
    b) $x = -2$

17. *Navigation.*   Capital Rapids drops 54 ft vertically over a horizontal distance of 1080 ft. What is the slope of the rapids?

Find the slope and the *y*-intercept.

18. $y = 2x - \dfrac{1}{4}$

19. $-4x + 3y = -6$

Find an equation of the line with the given slope and *y*-intercept.

20. Slope: 1.8;   *y*-intercept: $(0, -7)$

21. Slope: $-\dfrac{3}{8}$;   *y*-intercept: $\left(0, -\dfrac{1}{8}\right)$

Find an equation of the line containing the given point and with the given slope.

22. $(3, 5)$, $m = 1$

23. $(-2, 0)$, $m = -3$

Find an equation of the line containing the given pair of points.

24. $(1, 1)$ and $(2, -2)$

25. $(4, -1)$ and $(-4, -3)$

26. Draw a graph of the line with slope $-\frac{3}{2}$ and y-intercept $(0, 1)$.

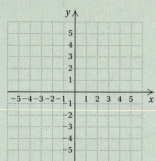

27. Graph $y = 2x - 3$ using the slope and the y-intercept.

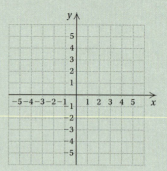

Determine whether the graphs of the equations are parallel, perpendicular, or neither.

28. $2x + y = 8,$
$2x + y = 4$

29. $2x + 5y = 2,$
$y = 2x + 4$

30. $x + 2y = 8,$
$-2x + y = 8$

Determine whether the given point is a solution of the inequality $3y - 2x < -2$.

31. $(0, 0)$

32. $(-4, -10)$

Graph on a plane.

33. $y > x - 1$

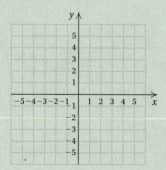

34. $2x - y \geq 4$

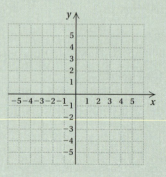

35. Select the statement that best describes the graphs of the lines $15x + 21y = 7$ and $35y + 14 = -25x$.

   A. The lines are parallel.
   C. The lines intersect and are not perpendicular.

   B. The lines are the same.
   D. The lines are perpendicular.

Synthesis

36. A diagonal of a square connects the points $(-3, -1)$ and $(2, 4)$. Find the area and the perimeter of the square.

37. Find the value of k such that $3x + 7y = 14$ and $ky - 7x = -3$ are perpendicular.

1. Evaluate $\dfrac{2m - n}{5}$ when $m = -1$ and $n = 2$.

2. Multiply: $-\dfrac{2}{3}(x - 6y + 3)$.

3. Factor: $18w - 24 + 9y$.

4. Find decimal notation: $-\dfrac{7}{9}$.

5. Find the absolute value: $\left| -2\dfrac{1}{5} \right|$.

6. Find the opposite of 8.17.

7. Find the reciprocal of $-\dfrac{8}{7}$.

8. Collect like terms: $2x - 5y + (-3x) + 4y$.

Simplify.

9. $-2.6 + (-0.4)$

10. $3 - [81 \div (1 + 2^3)]$

11. $\dfrac{5}{18} \div \left(-\dfrac{5}{12} \right)$

12. $6(x + 4) - 5[x - (2x - 3)]$

13. $\left(-\dfrac{1}{2} \right)(-1.1)(4.8)$

14. $-20 + 30 \div 10 \cdot 6$

Solve.

15. $\dfrac{4}{9}y = -36$

16. $-8 + w = w + 7$

17. $7.5 - 2x = 0.5x$

18. $4(x + 2) = 4(x - 2) + 16$

19. $2(x + 2) \geq 5(2x + 3)$

20. $x - \dfrac{5}{6} = \dfrac{1}{2}$

21. Find the slope, if it exists, of $9x - 12y = -3$.

22. Find the slope, if it exists, of $x = -\dfrac{15}{16}$.

23. Find an equation of a line with slope -20 and containing the point $(-8, -2)$.

24. Determine whether the graphs of $x = -5$ and $y = \frac{2}{7}$ are parallel, perpendicular, or neither.

25. Solve $A = \dfrac{1}{2}h(b + c)$ for h.

26. In which quadrant is the point $(3, -1)$ located?

27. Find the intercepts of $2x - 7y = 21$. Do not graph.

28. Graph on the number line: $-1 < x \leq 2$.

Graph.

29. $2x + 5y = 10$

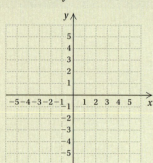

30. $y = -2$

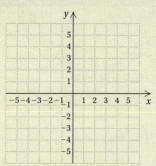

31. $y = -2x + 1$

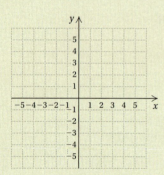

32. $3y + 6 = 2x$

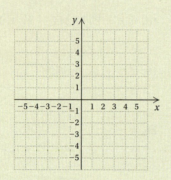

33. $y = -\dfrac{3}{2}x$

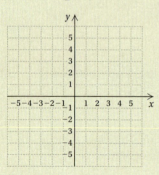

34. $x = 4.5$

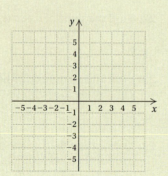

35. Find the slope of the line graphed below.

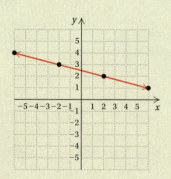

36. *Blood Types.* There are 134.6 million Americans with either O-positive or O-negative blood. Those with O-positive blood outnumber those with O-negative blood by 94.2 million. How many Americans have O-negative blood?

Source: Stanford University School of Medicine

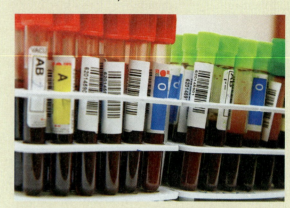

37. *Work Time.* Cory's contract stipulates that he cannot work more than 40 hr per week. For the first four days of one week, he worked 7, 10, 9, and 6 hr. Determine as an inequality the number of hours he can work on the fifth day without violating his contract.

38. *Wire Cutting.* A 143-m wire is cut into three pieces. The second piece is 3 m longer than the first. The third is four-fifths as long as the first. How long is each piece?

39. Compute and simplify: $1000 \div 100 \cdot 10 - 10$.

    A. 90   **B.** 0
    C. -9 **D.** -90

40. The slope of the line containing the points $(2, -7)$ and $(-4, 3)$ is which of the following?

    A. $-\dfrac{5}{2}$ **B.** $-\dfrac{3}{5}$
    C. $-\dfrac{2}{5}$ **D.** $-\dfrac{5}{3}$

Synthesis

Solve.

41. $4|x| - 13 = 3$

42. $\dfrac{2 + 5x}{4} = \dfrac{11}{28} + \dfrac{8x + 3}{7}$

43. $p = \dfrac{2}{m + Q}$, for Q

CHAPTER
4

Polynomials: Operations

4.1   Integers as Exponents

OBJECTIVES

a Tell the meaning of exponential notation.

b Evaluate exponential expressions with exponents of 0 and 1.

c Evaluate algebraic expressions containing exponents.

d Use the product rule to multiply exponential expressions with like bases.

e Use the quotient rule to divide exponential expressions with like bases.

f Express an exponential expression involving negative exponents with positive exponents.

SKILL TO REVIEW

Objective 1.1a: Evaluate algebraic expressions by substitution.

1. Evaluate $6y$ when $y = 4$.

2. Evaluate $\dfrac{m}{n}$ when $m = 48$ and $n = 8$.

a EXPONENTIAL NOTATION

An exponent of 2 or greater tells how many times the base is used as a factor. For example, $a \cdot a \cdot a \cdot a = a^4$. In this case, the **exponent** is 4 and the **base** is a. An expression for a power is called **exponential notation**.

$$\text{This is the base.} \longrightarrow a^n \longleftarrow \text{This is the exponent.}$$

EXAMPLE 1   What is the meaning of 3^5? of n^4? of $(2n)^3$? of $50x^2$? of $(-n)^3$? of $-n^3$?

3^5 means $3 \cdot 3 \cdot 3 \cdot 3 \cdot 3$; n^4 means $n \cdot n \cdot n \cdot n$;

$(2n)^3$ means $2n \cdot 2n \cdot 2n$; $50x^2$ means $50 \cdot x \cdot x$;

$(-n)^3$ means $(-n) \cdot (-n) \cdot (-n)$; $-n^3$ means $-1 \cdot n \cdot n \cdot n$

Do Margin Exercises 1–6 on the following page.

We read a^n as the **nth power of a**, or simply **a to the nth**, or **a to the n**. We often read x^2 as "**x-squared**" because the area of a square of side x is $x \cdot x$, or x^2. We often read x^3 as "**x-cubed**" because the volume of a cube with length, width, and height x is $x \cdot x \cdot x$, or x^3.

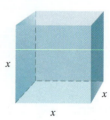

b ONE AND ZERO AS EXPONENTS

Look for a pattern in the following:

On each side, we **divide** by 8 at each step.	$8 \cdot 8 \cdot 8 \cdot 8 = 8^4$	On this side, the exponents **decrease** by 1 at each step.
	$8 \cdot 8 \cdot 8 = 8^3$	
	$8 \cdot 8 = 8^2$	
	$8 = 8^?$	
	$1 = 8^?$	

To continue the pattern, we would say that $8 = 8^1$ and $1 = 8^0$.

We make the following definition.

EXPONENTS OF 0 AND 1

$a^1 = a$, for any number a;

$a^0 = 1$, for any nonzero number a

We consider 0^0 to be not defined. We will explain why later in this section.

EXAMPLE 2 Evaluate 5^1, $(-8)^1$, 3^0, and $(-749.21)^0$.

$5^1 = 5;$ $(-8)^1 = -8;$

$3^0 = 1;$ $(-749.21)^0 = 1$

Do Exercises 7–12. ▶

c EVALUATING ALGEBRAIC EXPRESSIONS

Algebraic expressions can involve exponential notation. For example, the following are algebraic expressions:

$$x^4, \quad (3x)^3 - 2, \quad a^2 + 2ab + b^2.$$

We evaluate algebraic expressions by replacing variables with numbers and following the rules for order of operations.

EXAMPLE 3 Evaluate $1000 - x^4$ when $x = 5$.

$$\begin{aligned} 1000 - x^4 &= 1000 - 5^4 &&\text{Substituting} \\ &= 1000 - 625 &&\text{Evaluating } 5^4 \\ &= 375 \end{aligned}$$

EXAMPLE 4 *Area of a Circular Region.* The Richat Structure is a circular eroded geologic dome with a radius of 20 km. Find the area of the structure.

$$\begin{aligned} A &= \pi r^2 &&\text{Using the formula for the area of a circle} \\ &= \pi (20 \text{ km})^2 &&\text{Substituting} \\ &= \pi \cdot 20 \text{ km} \cdot 20 \text{ km} \\ &\approx 3.14 \times 400 \text{ km}^2 &&\text{Using 3.14 as an approximation for } \pi \\ &= 1256 \text{ km}^2 \end{aligned}$$

In Example 4, "km^2" means "square kilometers" and " $\approx$ " means "is approximately equal to."

EXAMPLE 5 Evaluate $(5x)^3$ when $x = -2$.

When we evaluate with a negative number, we often use extra parentheses to show the substitution.

$$\begin{aligned} (5x)^3 &= [5 \cdot (-2)]^3 &&\text{Substituting} \\ &= [-10]^3 &&\text{Multiplying within brackets first} \\ &= [-10] \cdot [-10] \cdot [-10] \\ &= -1000 &&\text{Evaluating the power} \end{aligned}$$

What is the meaning of each of the following?

1. 5^4 **2.** x^5

3. $(3t)^2$ **4.** $3t^2$

5. $(-x)^4$ **6.** $-y^3$

Evaluate.

7. 6^1 **8.** 7^0

9. $(8.4)^1$ **10.** 8654^0

11. $(-1.4)^1$ **12.** 0^1

Answers

1. $5 \cdot 5 \cdot 5 \cdot 5$ **2.** $x \cdot x \cdot x \cdot x \cdot x$
3. $3t \cdot 3t$ **4.** $3 \cdot t \cdot t$
5. $(-x) \cdot (-x) \cdot (-x) \cdot (-x)$ **6.** $-1 \cdot y \cdot y \cdot y$
7. 6 **8.** 1 **9.** 8.4 **10.** 1
11. -1.4 **12.** 0

13. Evaluate t^3 when $t = 5$.

14. Evaluate $-5x^5$ when $x = -2$.

15. Find the area of a circle when $r = 32$ cm. Use 3.14 for π.

16. Evaluate $200 - a^4$ when $a = 3$.

17. Evaluate $t^1 - 4$ and $t^0 - 4$ when $t = 7$.

18. a) Evaluate $(4t)^2$ when $t = -3$.

   b) Evaluate $4t^2$ when $t = -3$.

   c) Determine whether $(4t)^2$ and $4t^2$ are equivalent.

   a) $(4t)^2 = \left[4 \cdot (\boxed{})\right]^2$

           $= \left[\boxed{}\right]^2$

           $= \boxed{}$

   b) $4t^2 = 4 \cdot (\boxed{})^2$

        $= 4 \cdot (\boxed{})$

        $= \boxed{}$

   c) Since $144 \neq 36$, the expressions _____ are/are not equivalent.

Multiply and simplify.

19. $3^5 \cdot 3^5$

20. $x^4 \cdot x^6$

21. $p^4 p^{12} p^8$

22. $x \cdot x^4$

23. $(a^2 b^3)(a^7 b^5)$

EXAMPLE 6 Evaluate $5x^3$ when $x = -2$.

$$\begin{aligned}
5x^3 &= 5 \cdot (-2)^3 && \text{Substituting} \\
&= 5 \cdot (-2) \cdot (-2) \cdot (-2) && \text{Evaluating the power first} \\
&= 5(-8) && (-2)(-2)(-2) = -8 \\
&= -40
\end{aligned}$$

Recall that two expressions are equivalent if they have the same value for all meaningful replacements. Note that Examples 5 and 6 show that $(5x)^3$ and $5x^3$ are *not* equivalent—that is, $(5x)^3 \neq 5x^3$.

◀ Do Exercises 13–18.

d **MULTIPLYING POWERS WITH LIKE BASES**

We can multiply powers with like bases by adding exponents. For example,

$$a^3 \cdot a^2 = \underbrace{(a \cdot a \cdot a)}_{3 \text{ factors}}\underbrace{(a \cdot a)}_{2 \text{ factors}} = \underbrace{a \cdot a \cdot a \cdot a \cdot a}_{5 \text{ factors}} = a^5.$$

Note that the exponent in a^5 is the sum of those in $a^3 \cdot a^2$. That is, $3 + 2 = 5$. Likewise,

$$b^4 \cdot b^3 = (b \cdot b \cdot b \cdot b)(b \cdot b \cdot b) = b^7, \quad \text{where} \quad 4 + 3 = 7.$$

Adding the exponents gives the correct result.

THE PRODUCT RULE

For any number a and any positive integers m and n,

$$a^m \cdot a^n = a^{m+n}.$$

(When multiplying with exponential notation, if the bases are the same, keep the base and add the exponents.)

EXAMPLES Multiply and simplify.

7. $5^6 \cdot 5^2 = 5^{6+2}$ Adding exponents: $a^m \cdot a^n = a^{m+n}$
          $= 5^8$

8. $m^5 m^{10} m^3 = m^{5+10+3} = m^{18}$

9. $x \cdot x^8 = x^1 \cdot x^8$ Writing x as x^1
        $= x^{1+8}$
        $= x^9$

10. $(a^3 b^2)(a^3 b^5) = (a^3 a^3)(b^2 b^5)$
               $= a^6 b^7$

11. $(4y)^6 (4y)^3 = (4y)^{6+3} = (4y)^9$

◀ Do Exercises 19–23.

Answers

13. 125   **14.** 160   **15.** 3215.36 cm²   **16.** 119
17. 3; −3   **18. (a)** 144; **(b)** 36; **(c)** no
19. 3^{10} **20.** x^{10} **21.** p^{24} **22.** x^5 **23.** $a^9 b^8$

Guided Solution:
18. (a) −3, −12, 144; **(b)** −3, 9, 36; **(c)** are not

e DIVIDING POWERS WITH LIKE BASES

The following suggests a rule for dividing powers with like bases, such as a^5/a^2:

$$\frac{a^5}{a^2} = \frac{a \cdot a \cdot a \cdot a \cdot a}{a \cdot a} = \frac{a \cdot a \cdot a \cdot a \cdot a}{1 \cdot a \cdot a} = \frac{a \cdot a \cdot a}{1} \cdot \frac{a \cdot a}{a \cdot a}$$

$$= \frac{a \cdot a \cdot a}{1} \cdot 1 = a \cdot a \cdot a = a^3.$$

Note that the exponent in a^3 is the difference of those in $a^5 \div a^2$. That is, $5 - 2 = 3$. In a similar way, we have

$$\frac{t^9}{t^4} = \frac{t \cdot t \cdot t \cdot t \cdot t \cdot t \cdot t \cdot t \cdot t}{t \cdot t \cdot t \cdot t} = t^5, \quad \text{where} \quad 9 - 4 = 5.$$

Subtracting exponents gives the correct answer.

THE QUOTIENT RULE

For any nonzero number a and any positive integers m and n,

$$\frac{a^m}{a^n} = a^{m-n}.$$

(When dividing with exponential notation, if the bases are the same, keep the base and subtract the exponent of the denominator from the exponent of the numerator.)

EXAMPLES Divide and simplify.

12. $\dfrac{6^5}{6^3} = 6^{5-3}$ Subtracting exponents

$= 6^2$

13. $\dfrac{x^8}{x} = \dfrac{x^8}{x^1} = x^{8-1}$

$= x^7$

14. $\dfrac{(3t)^{12}}{(3t)^2} = (3t)^{12-2}$

$= (3t)^{10}$

15. $\dfrac{p^5 q^7}{p^2 q^5} = \dfrac{p^5}{p^2} \cdot \dfrac{q^7}{q^5} = p^{5-2} q^{7-5}$

$= p^3 q^2$

The quotient rule can also be used to explain the definition of 0 as an exponent. Consider the expression a^4/a^4, where a is nonzero:

$$\frac{a^4}{a^4} = \frac{a \cdot a \cdot a \cdot a}{a \cdot a \cdot a \cdot a} = 1.$$

This is true because the numerator and the denominator are the same. Now suppose we apply the rule for dividing powers with the same base:

$$\frac{a^4}{a^4} = a^{4-4} = a^0.$$

Since $a^4/a^4 = 1$ and $a^4/a^4 = a^0$, it follows that $a^0 = 1$, when $a \neq 0$.

We can explain why we do not define 0^0 using the quotient rule. We know that 0^0 is 0^{1-1}. But 0^{1-1} is also equal to $0^1/0^1$, or $0/0$. We have already seen that division by 0 is not defined, so 0^0 is also not defined.

Do Exercises 24–27. ▶

Divide and simplify.

24. $\dfrac{4^5}{4^2}$

25. $\dfrac{y^6}{y^2}$

26. $\dfrac{p^{10}}{p}$

27. $\dfrac{a^7 b^6}{a^3 b^4}$

Answers

24. 4^3 **25.** y^4 **26.** p^9 **27.** $a^4 b^2$

f NEGATIVE INTEGERS AS EXPONENTS

We can use the rule for dividing powers with like bases to lead us to a definition of exponential notation when the exponent is a negative integer. Consider $5^3/5^7$ and first simplify it using procedures we have learned for working with fractions:

$$\frac{5^3}{5^7} = \frac{5 \cdot 5 \cdot 5}{5 \cdot 5 \cdot 5 \cdot 5 \cdot 5 \cdot 5 \cdot 5} = \frac{5 \cdot 5 \cdot 5 \cdot 1}{5 \cdot 5 \cdot 5 \cdot 5 \cdot 5 \cdot 5 \cdot 5}$$

$$= \frac{5 \cdot 5 \cdot 5}{5 \cdot 5 \cdot 5} \cdot \frac{1}{5 \cdot 5 \cdot 5 \cdot 5} = \frac{1}{5^4}.$$

Now we apply the rule for dividing exponential expressions with the same bases. Then

$$\frac{5^3}{5^7} = 5^{3-7} = 5^{-4}.$$

From these two expressions for $5^3/5^7$, it follows that

$$5^{-4} = \frac{1}{5^4}.$$

This leads to our definition of negative exponents.

NEGATIVE EXPONENT

For any real number a that is nonzero and any integer n,

$$a^{-n} = \frac{1}{a^n}.$$

In fact, the numbers a^n and a^{-n} are reciprocals because

$$a^n \cdot a^{-n} = a^n \cdot \frac{1}{a^n} = \frac{a^n}{a^n} = 1.$$

The following is another way to arrive at the definition of negative exponents.

On each side, we **divide** by 5 at each step.

$$5 \cdot 5 \cdot 5 \cdot 5 = 5^4$$
$$5 \cdot 5 \cdot 5 = 5^3$$
$$5 \cdot 5 = 5^2$$
$$5 = 5^1$$
$$1 = 5^0$$
$$\frac{1}{5} = 5^?$$
$$\frac{1}{25} = 5^?$$

On this side, the exponents **decrease** by 1 at each step.

To continue the pattern, it should follow that

$$\frac{1}{5} = \frac{1}{5^1} = 5^{-1} \quad \text{and} \quad \frac{1}{25} = \frac{1}{5^2} = 5^{-2}.$$

EXAMPLES Express using positive exponents. Then simplify.

16. $4^{-2} = \dfrac{1}{4^2} = \dfrac{1}{16}$

17. $(-3)^{-2} = \dfrac{1}{(-3)^2} = \dfrac{1}{(-3)(-3)} = \dfrac{1}{9}$

18. $m^{-3} = \dfrac{1}{m^3}$

19. $ab^{-1} = a\left(\dfrac{1}{b^1}\right) = a\left(\dfrac{1}{b}\right) = \dfrac{a}{b}$

20. $\dfrac{1}{x^{-3}} = x^{-(-3)} = x^3$

21. $3c^{-5} = 3\left(\dfrac{1}{c^5}\right) = \dfrac{3}{c^5}$

Example 20 might also be done as follows:

$$\dfrac{1}{x^{-3}} = \dfrac{1}{\dfrac{1}{x^3}} = 1 \cdot \dfrac{x^3}{1} = x^3.$$

.... **Caution!** ....

As shown in Examples 16 and 17, a negative exponent does not necessarily mean that an expression is negative.
....

Do Exercises 28–33. ▶

The rules for multiplying and dividing powers with like bases hold when exponents are 0 or negative.

EXAMPLES Simplify. Write the result using positive exponents.

22. $7^{-3} \cdot 7^6 = 7^{-3+6}$ *Adding exponents*

    $= 7^3$

23. $x^4 \cdot x^{-3} = x^{4+(-3)} = x^1 = x$

24. $\dfrac{5^4}{5^{-2}} = 5^{4-(-2)}$ *Subtracting exponents*

    $= 5^{4+2} = 5^6$

25. $\dfrac{x}{x^7} = x^{1-7} = x^{-6} = \dfrac{1}{x^6}$

26. $\dfrac{b^{-4}}{b^{-5}} = b^{-4-(-5)}$

    $= b^{-4+5} = b^1 = b$

27. $y^{-4} \cdot y^{-8} = y^{-4+(-8)}$

    $= y^{-12} = \dfrac{1}{y^{12}}$

Do Exercises 34–38. ▶

The following is a summary of the definitions and rules for exponents that we have considered in this section.

┌───┐
DEFINITIONS AND RULES FOR EXPONENTS
....

1 as an exponent:	$a^1 = a$
0 as an exponent:	$a^0 = 1, a \neq 0$
Negative integers as exponents:	$a^{-n} = \dfrac{1}{a^n}, \dfrac{1}{a^{-n}} = a^n; a \neq 0$
Product Rule:	$a^m \cdot a^n = a^{m+n}$
Quotient Rule:	$\dfrac{a^m}{a^n} = a^{m-n}, a \neq 0$

└───┘

Express with positive exponents. Then simplify.

28. 4^{-3}

29. 5^{-2}

30. 2^{-4}

31. $(-2)^{-3}$

32. $\dfrac{1}{x^{-2}}$

GS **33.** $4p^{-3}$

$$= 4\left(\dfrac{1}{\boxed{}}\right) = \dfrac{4}{\boxed{}}$$

Simplify.

34. $5^{-2} \cdot 5^4$

35. $x^{-3} \cdot x^{-4}$

36. $\dfrac{7^{-2}}{7^3}$

37. $\dfrac{b^{-2}}{b^{-3}}$

38. $\dfrac{t}{t^{-5}}$

✓ Reading Check

Match each expression with the appropriate value from the column on the right. Choices may be used more than once or not at all.

RC1. _____ y^1 **RC2.** _____ $y^0, y \neq 0$

a) 1

b) 0

RC3. _____ $y^1 \cdot y^1$ **RC4.** _____ $\dfrac{y^9}{y^8}$

c) y

d) $\dfrac{1}{y}$

RC5. _____ $\dfrac{y^8}{y^9}$ **RC6.** _____ $\dfrac{1}{y^{-1}}$

e) y^2

a   What is the meaning of each of the following?

1. 3^4 **2.** 4^3 **3.** $(-1.1)^5$ **4.** $(87.2)^6$ **5.** $\left(\dfrac{2}{3}\right)^4$ **6.** $\left(-\dfrac{5}{8}\right)^3$

7. $(7p)^2$ **8.** $(11c)^3$ **9.** $8k^3$ **10.** $17x^2$ **11.** $-6y^4$ **12.** $-q^5$

b   Evaluate.

13. $a^0, a \neq 0$ **14.** $t^0, t \neq 0$ **15.** b^1 **16.** c^1 **17.** $\left(\dfrac{2}{3}\right)^0$

18. $\left(-\dfrac{5}{8}\right)^0$ **19.** $(-7.03)^1$ **20.** $\left(\dfrac{4}{5}\right)^1$ **21.** 8.38^0 **22.** 8.38^1

23. $(ab)^1$ **24.** $(ab)^0, a \neq 0, b \neq 0$ **25.** $ab^0, b \neq 0$ **26.** ab^1

c   Evaluate.

27. m^3, when $m = 3$ **28.** x^6, when $x = 2$ **29.** p^1, when $p = 19$ **30.** x^{19}, when $x = 0$

31. $-x^4$, when $x = -3$ **32.** $-2y^7$, when $y = 2$ **33.** x^4, when $x = 4$ **34.** y^{15}, when $y = 1$

35. $y^2 - 7$, when $y = -10$ **36.** $z^5 + 5$, when $z = -2$ **37.** $161 - b^2$, when $b = 5$ **38.** $325 - v^3$, when $v = -3$

39. $x^1 + 3$ and $x^0 + 3$, when $x = 7$ **40.** $y^0 - 8$ and $y^1 - 8$, when $y = -3$

41. Find the area of a circle when $r = 34$ ft. Use 3.14 for π.

42. The area A of a square with sides of length s is given by $A = s^2$. Find the area of a square with sides of length 24 m.

f Express using positive exponents. Then simplify.

43. 3^{-2}

44. 2^{-3}

45. 10^{-3}

46. 5^{-4}

47. a^{-3}

48. x^{-2}

49. $\dfrac{1}{8^{-2}}$

50. $\dfrac{1}{2^{-5}}$

51. $\dfrac{1}{y^{-4}}$

52. $\dfrac{1}{t^{-7}}$

53. $5z^{-4}$

54. $6n^{-5}$

55. xy^{-2}

56. ab^{-3}

Express using negative exponents.

57. $\dfrac{1}{4^3}$

58. $\dfrac{1}{5^2}$

59. $\dfrac{1}{x^3}$

60. $\dfrac{1}{y^2}$

61. $\dfrac{1}{a^5}$

62. $\dfrac{1}{b^7}$

d , **f** Multiply and simplify.

63. $2^4 \cdot 2^3$

64. $3^5 \cdot 3^2$

65. $9^{17} \cdot 9^{21}$

66. $7^{22} \cdot 7^{15}$

67. $x^4 \cdot x$

68. $y \cdot y^9$

69. $x^{14} \cdot x^3$

70. $x^9 \cdot x^4$

71. $(3y)^4(3y)^8$

72. $(2t)^8(2t)^{17}$

73. $(7y)^1(7y)^{16}$

74. $(8x)^0(8x)^1$

75. $3^{-5} \cdot 3^8$

76. $5^{-8} \cdot 5^9$

77. $x^{-2} \cdot x^2$

78. $x \cdot x^{-1}$

79. $x^{-7} \cdot x^{-6}$

80. $y^{-5} \cdot y^{-8}$

81. $a^{11} \cdot a^{-3} \cdot a^{-18}$

82. $a^{-11} \cdot a^{-3} \cdot a^{-7}$

83. $(x^4 y^7)(x^2 y^8)$

84. $(a^5 c^2)(a^3 c^9)$

85. $(s^2 t^3)(s t^4)$

86. $(m^4 n)(m^2 n^7)$

e , **f**   Divide and simplify.

87. $\dfrac{7^5}{7^2}$

88. $\dfrac{5^8}{5^6}$

89. $\dfrac{y^9}{y}$

90. $\dfrac{x^{11}}{x}$

91. $\dfrac{16^2}{16^8}$

92. $\dfrac{7^2}{7^9}$

93. $\dfrac{m^6}{m^{12}}$

94. $\dfrac{a^3}{a^4}$

95. $\dfrac{(8x)^6}{(8x)^{10}}$

96. $\dfrac{(8t)^4}{(8t)^{11}}$

97. $\dfrac{x}{x^{-1}}$

98. $\dfrac{t^8}{t^{-3}}$

99. $\dfrac{z^{-6}}{z^{-2}}$

100. $\dfrac{x^{-9}}{x^{-3}}$

101. $\dfrac{x^{-5}}{x^{-8}}$

102. $\dfrac{y^{-2}}{y^{-9}}$

103. $\dfrac{m^{-9}}{m^{-9}}$

104. $\dfrac{x^{-7}}{x^{-7}}$

105. $\dfrac{a^5 b^3}{a^2 b}$

106. $\dfrac{s^8 t^4}{s t^3}$

Matching.   In Exercises 107 and 108, match each item in the first column with the appropriate item in the second column by drawing connecting lines. Items in the second column may be used more than once.

107.

5^2	$-\dfrac{1}{10}$
5^{-2}	$\dfrac{1}{10}$
$\left(\dfrac{1}{5}\right)^2$	$-\dfrac{1}{25}$
$\left(\dfrac{1}{5}\right)^{-2}$	10
-5^2	25
$(-5)^2$	-25
$-\left(-\dfrac{1}{5}\right)^2$	$\dfrac{1}{25}$
$\left(-\dfrac{1}{5}\right)^{-2}$	-10

108.

$-\left(\dfrac{1}{8}\right)^2$	16
$\left(\dfrac{1}{8}\right)^{-2}$	-16
8^{-2}	64
8^2	-64
-8^2	$\dfrac{1}{64}$
$(-8)^2$	$-\dfrac{1}{64}$
$\left(-\dfrac{1}{8}\right)^{-2}$	$-\dfrac{1}{16}$
$\left(-\dfrac{1}{8}\right)^2$	$\dfrac{1}{16}$

Skill Maintenance

Solve.

109. A 12-in. submarine sandwich is cut into two pieces. One piece is twice as long as the other. How long are the pieces? [2.6a]

110. The first angle of a triangle is 24° more than the second. The third angle is twice the first. Find the measures of the angles of the triangle. [2.6a]

111. A warehouse stores 1800 lb of peanuts, 1500 lb of cashews, and 700 lb of almonds. What percent of the total is peanuts? cashews? almonds? [2.5a]

112. The width of a rectangle is fixed at 10 ft. For what lengths will the area be less than 25 ft²? [2.8b]

Solve.

113. $2x - 4 - 5x + 8 = x - 3$ [2.3b]

114. $8x + 7 - 9x = 12 - 6x + 5$ [2.3b]

115. $-6(2 - x) + 10(5x - 7) = 10$ [2.3c]

116. $-10(x - 4) = 5(2x + 5) - 7$ [2.3c]

Synthesis

Determine whether each of the following equations is true.

117. $(x + 1)^2 = x^2 + 1$

118. $(x - 1)^2 = x^2 - 2x + 1$

119. $(5x)^0 = 5x^0$

120. $\dfrac{x^3}{x^5} = x^2$

Simplify.

121. $(y^{2x})(y^{3x})$

122. $a^{5k} \div a^{3k}$

123. $\dfrac{a^{6t}(a^{7t})}{a^{9t}}$

124. $\dfrac{\left(\frac{1}{2}\right)^4}{\left(\frac{1}{2}\right)^5}$

125. $\dfrac{(0.8)^5}{(0.8)^3(0.8)^2}$

126. $\dfrac{(x - 3)^5}{x - 3}$

Use >, <, or = for ☐ to write a true sentence.

127. 3^5 ☐ 3^4

128. 4^2 ☐ 4^3

129. 4^3 ☐ 5^3

130. 4^3 ☐ 3^4

Evaluate.

131. $\dfrac{1}{-z^4}$, when $z = -10$

132. $\dfrac{1}{-z^5}$, when $z = -0.1$

133. Determine whether $(a + b)^2$ and $a^2 + b^2$ are equivalent. (*Hint*: Choose values for a and b and evaluate.)

4.2

OBJECTIVES

a Use the power rule to raise powers to powers.

b Raise a product to a power and a quotient to a power.

c Convert between scientific notation and decimal notation.

d Multiply and divide using scientific notation.

e Solve applied problems using scientific notation.

SKILL TO REVIEW

Objective 1.5a: Multiply real numbers.

Multiply.

1. $-5 \cdot 8$ **2.** $(-3)(-5)$

Simplify. Express the answers using positive exponents.

1. $(3^4)^5$ **2.** $(x^{-3})^4$

3. $(y^{-5})^{-3}$ **4.** $(x^4)^{-8}$

We now consider three rules used to simplify exponential expressions. We then apply our knowledge of exponents to *scientific notation*.

a RAISING POWERS TO POWERS

Consider an expression like $(3^2)^4$. We are raising 3^2 to the fourth power:

$$(3^2)^4 = (3^2)(3^2)(3^2)(3^2)$$
$$= (3 \cdot 3)(3 \cdot 3)(3 \cdot 3)(3 \cdot 3)$$
$$= 3 \cdot 3 \cdot 3 \cdot 3 \cdot 3 \cdot 3 \cdot 3 \cdot 3$$
$$= 3^8.$$

Note that in this case we could have multiplied the exponents:

$$(3^2)^4 = 3^{2 \cdot 4} = 3^8.$$

THE POWER RULE

For any real number a and any integers m and n,

$$(a^m)^n = a^{mn}.$$

(To raise a power to a power, multiply the exponents.)

EXAMPLES Simplify. Express the answers using positive exponents.

1. $(3^5)^4 = 3^{5 \cdot 4}$ Multiplying
$ = 3^{20}$ exponents

2. $(2^2)^5 = 2^{2 \cdot 5} = 2^{10}$

3. $(y^{-5})^7 = y^{-5 \cdot 7} = y^{-35} = \dfrac{1}{y^{35}}$

4. $(x^4)^{-2} = x^{4(-2)} = x^{-8} = \dfrac{1}{x^8}$

5. $(a^{-4})^{-6} = a^{(-4)(-6)} = a^{24}$

◀ **Do Margin Exercises 1–4.**

b RAISING A PRODUCT OR A QUOTIENT TO A POWER

When an expression inside parentheses is raised to a power, the inside expression is the base. Let's compare $2a^3$ and $(2a)^3$:

$$2a^3 = 2 \cdot a \cdot a \cdot a; \qquad \text{The base is } a.$$
$$(2a)^3 = (2a)(2a)(2a) \qquad \text{The base is } 2a.$$
$$ = (2 \cdot 2 \cdot 2)(a \cdot a \cdot a) \qquad \text{Using the associative and commutative}$$
$$ = 2^3 a^3 \qquad\qquad\qquad \text{laws of multiplication to regroup}$$
$$ = 8a^3. \qquad\qquad\qquad \text{the factors}$$

We see that $2a^3$ and $(2a)^3$ are *not* equivalent. We also see that we can evaluate the power $(2a)^3$ by raising each factor to the power 3. This leads us to a rule for raising a product to a power.

Answers

Skill to Review:
1. -40 **2.** 15

Margin Exercises:
1. 3^{20} **2.** $\dfrac{1}{x^{12}}$ **3.** y^{15} **4.** $\dfrac{1}{x^{32}}$

RAISING A PRODUCT TO A POWER

For any real numbers a and b and any integer n,

$$(ab)^n = a^n b^n.$$

(To raise a product to the nth power, raise each factor to the nth power.)

EXAMPLES Simplify.

6. $(4x^2)^3 = (4^1 x^2)^3$ $4 = 4^1$

$ = (4^1)^3 \cdot (x^2)^3$ Raising *each* factor to the third power

$ = 4^3 \cdot x^6 = 64x^6$ Using the power rule and simplifying

7. $(5x^3 y^5 z^2)^4 = 5^4 (x^3)^4 (y^5)^4 (z^2)^4$ Raising *each* factor to the fourth power

$ = 625 x^{12} y^{20} z^8$

8. $(-5x^4 y^3)^3 = (-5)^3 (x^4)^3 (y^3)^3$

$ = -125 x^{12} y^9$

9. $[(-x)^{25}]^2 = (-x)^{50}$ Using the power rule

$\phantom{[(-x)^{25}]^2} = (-1 \cdot x)^{50}$ Using the property of -1: $-x = -1 \cdot x$

$\phantom{[(-x)^{25}]^2} = (-1)^{50} x^{50}$ Raising each factor to the fiftieth power

$\phantom{[(-x)^{25}]^2} = 1 \cdot x^{50}$ The product of an even number of negative factors is positive.

$\phantom{[(-x)^{25}]^2} = x^{50}$

10. $(3x^3 y^{-5} z^2)^4 = 3^4 (x^3)^4 (y^{-5})^4 (z^2)^4 = 81 x^{12} y^{-20} z^8 = \dfrac{81 x^{12} z^8}{y^{20}}$

11. $(-x^4)^{-3} = (-1 \cdot x^4)^{-3} = (-1)^{-3} \cdot (x^4)^{-3} = (-1)^{-3} \cdot x^{-12}$

$\phantom{(-x^4)^{-3}} = \dfrac{1}{(-1)^3} \cdot \dfrac{1}{x^{12}} = \dfrac{1}{-1} \cdot \dfrac{1}{x^{12}} = -\dfrac{1}{x^{12}}$

12. $(-2x^{-5} y^4)^{-4} = (-2)^{-4} (x^{-5})^{-4} (y^4)^{-4} = \dfrac{1}{(-2)^4} \cdot x^{20} \cdot y^{-16}$

$\phantom{(-2x^{-5} y^4)^{-4}} = \dfrac{1}{16} \cdot x^{20} \cdot \dfrac{1}{y^{16}} = \dfrac{x^{20}}{16 y^{16}}$

Do Exercises 5–11. ▶

There is a similar rule for raising a quotient to a power.

RAISING A QUOTIENT TO A POWER

For any real numbers a and b, $b \neq 0$, and any integer n,

$$\left(\frac{a}{b} \right)^n = \frac{a^n}{b^n}.$$

(To raise a quotient to the nth power, raise both the numerator and the denominator to the nth power.)

Simplify.

5. $(2x^5 y^{-3})^4$

6. $(5x^5 y^{-6} z^{-3})^2$

7. $[(-x)^{37}]^2$

8. $(3y^{-2} x^{-5} z^8)^3$

9. $(-y^8)^{-3}$

GS **10.** $(-2x^4)^{-2}$

$= (-2)^{-2} ()^{-2}$

$= \dfrac{1}{(-2)^{}} \cdot x^{}$

$= \dfrac{1}{} \cdot \dfrac{1}{x^{}}$

$= \dfrac{1}{}$

11. $(-3x^2 y^{-5})^{-3}$

Answers

5. $\dfrac{16x^{20}}{y^{12}}$ 6. $\dfrac{25x^{10}}{y^{12} z^6}$ 7. x^{74} 8. $\dfrac{27 z^{24}}{y^6 x^{15}}$

9. $-\dfrac{1}{y^{24}}$ 10. $\dfrac{1}{4x^8}$ 11. $-\dfrac{y^{15}}{27 x^6}$

Guided Solution:
10. $x^4, 2, -8, 4, 8, 4x^8$

Simplify.

12. $\left(\dfrac{x^6}{5}\right)^2$

13. $\left(\dfrac{2t^5}{w^4}\right)^3$

14. $\left(\dfrac{a^4}{3b^{-2}}\right)^3$

15. $\left(\dfrac{x^4}{3}\right)^{-2}$ GS

Do this two ways.

$\left(\dfrac{x^4}{3}\right)^{-2} = \dfrac{(x^4)^{\square}}{3^{-2}} = \dfrac{x^{\square}}{3^{-2}}$

$\phantom{\left(\dfrac{x^4}{3}\right)^{-2}} = \dfrac{\dfrac{1}{x^{\square}}}{\dfrac{1}{3^2}} = \dfrac{1}{x^8} \div \dfrac{1}{3^2}$

$\phantom{\left(\dfrac{x^4}{3}\right)^{-2}} = \dfrac{1}{x^8} \cdot \dfrac{3^2}{\square} = \dfrac{9}{\square}$

This can be done a second way.

$\left(\dfrac{x^4}{3}\right)^{-2} = \left(\dfrac{3}{x^4}\right)^{\square}$

$\phantom{\left(\dfrac{x^4}{3}\right)^{-2}} = \dfrac{3^2}{(x^4)^{\square}} = \dfrac{9}{\square}$

EXAMPLES Simplify.

13. $\left(\dfrac{x^2}{4}\right)^3 = \dfrac{(x^2)^3}{4^3} = \dfrac{x^6}{64}$ Raising *both* the numerator and the denominator to the third power

14. $\left(\dfrac{3a^4}{b^3}\right)^2 = \dfrac{(3a^4)^2}{(b^3)^2} = \dfrac{3^2(a^4)^2}{b^{3\cdot2}} = \dfrac{9a^8}{b^6}$

15. $\left(\dfrac{y^2}{2z^{-5}}\right)^4 = \dfrac{(y^2)^4}{(2z^{-5})^4} = \dfrac{(y^2)^4}{2^4(z^{-5})^4} = \dfrac{y^8}{16z^{-20}} = \dfrac{y^8z^{20}}{16}$

16. $\left(\dfrac{y^3}{5}\right)^{-2} = \dfrac{(y^3)^{-2}}{5^{-2}} = \dfrac{y^{-6}}{5^{-2}} = \dfrac{\dfrac{1}{y^6}}{\dfrac{1}{5^2}} = \dfrac{1}{y^6} \div \dfrac{1}{5^2} = \dfrac{1}{y^6} \cdot \dfrac{5^2}{1} = \dfrac{25}{y^6}$

The following can often be used to simplify a quotient that is raised to a negative power.

For $a \neq 0$ and $b \neq 0$,

$$\left(\dfrac{a}{b}\right)^{-n} = \left(\dfrac{b}{a}\right)^{n}.$$

Example 16 might also be completed as follows:

$$\left(\dfrac{y^3}{5}\right)^{-2} = \left(\dfrac{5}{y^3}\right)^2 = \dfrac{5^2}{(y^3)^2} = \dfrac{25}{y^6}.$$

◀ **Do Exercises 12–15.**

C SCIENTIFIC NOTATION

We can write numbers using different types of notation, such as fraction notation, decimal notation, and percent notation. Another type, **scientific notation**, makes use of exponential notation. Scientific notation is especially useful when calculations involve very large or very small numbers. The following are examples of scientific notation.

The number of flamingos in Africa's Great Rift Valley:
$4 \times 10^6 = 4,000,000$

The length of an *E.coli* bacterium:
$2 \times 10^{-6}\,\text{m} = 0.000002\,\text{m}$

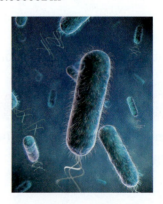

SCIENTIFIC NOTATION

Scientific notation for a number is an expression of the type

$$M \times 10^n,$$

where n is an integer, M is greater than or equal to 1 and less than 10 ($1 \leq M < 10$), and M is expressed in decimal notation. 10^n is also considered to be scientific notation when $M = 1$.

You should try to make conversions to scientific notation mentally as often as possible. Here is a handy mental device.

A positive exponent in scientific notation indicates a large number (greater than or equal to 10) and a negative exponent indicates a small number (between 0 and 1).

EXAMPLES Convert to scientific notation.

17. $78{,}000 = 7.8 \times 10^4$

7.8,000.   Large number, so the exponent is positive
    4 places

18. $0.0000057 = 5.7 \times 10^{-6}$

0.000005.7   Small number, so the exponent is negative
      6 places

Do Exercises 16 and 17. ▶

EXAMPLES Convert mentally to decimal notation.

19. $7.893 \times 10^5 = 789{,}300$

7.89300.   Positive exponent, so the answer is a large number
      5 places

20. $4.7 \times 10^{-8} = 0.000000047$

.00000004.7   Negative exponent, so the answer is a small number
      8 places

Do Exercises 18 and 19. ▶

Convert to scientific notation.

16. 0.000517

17. 523,000,000

Convert to decimal notation.

18. 6.893×10^{11}

19. 5.67×10^{-5}

d MULTIPLYING AND DIVIDING USING SCIENTIFIC NOTATION

Multiplying

Consider the product

$$400 \cdot 2000 = 800{,}000.$$

In scientific notation, this is

$$(4 \times 10^2) \cdot (2 \times 10^3) = (4 \cdot 2)(10^2 \cdot 10^3) = 8 \times 10^5.$$

Answers

16. 5.17×10^{-4} **17.** 5.23×10^8
18. 689,300,000,000   **19.** 0.0000567

Multiply and write scientific notation for the result.

20. $(1.12 \times 10^{-8})(5 \times 10^{-7})$

21. $(9.1 \times 10^{-17})(8.2 \times 10^3)$

Divide and write scientific notation for the result.

22. $\dfrac{4.2 \times 10^5}{2.1 \times 10^2}$

23. $\dfrac{1.1 \times 10^{-4}}{2.0 \times 10^{-7}}$

EXAMPLE 21 Multiply: $(1.8 \times 10^6) \cdot (2.3 \times 10^{-4})$.

We apply the commutative and associative laws to get

$$(1.8 \times 10^6) \cdot (2.3 \times 10^{-4}) = (1.8 \cdot 2.3) \times (10^6 \cdot 10^{-4})$$
$$= 4.14 \times 10^{6+(-4)}$$
$$= 4.14 \times 10^2.$$

We get 4.14 by multiplying 1.8 and 2.3. We get 10^2 by adding the exponents 6 and -4.

EXAMPLE 22 Multiply: $(3.1 \times 10^5) \cdot (4.5 \times 10^{-3})$.

$$(3.1 \times 10^5) \cdot (4.5 \times 10^{-3}) = (3.1 \times 4.5)(10^5 \cdot 10^{-3})$$

$$= 13.95 \times 10^2 \quad \text{Not scientific notation; 13.95 is greater than 10.}$$

$$= (1.395 \times 10^1) \times 10^2 \quad \text{Substituting } 1.395 \times 10^1 \text{ for } 13.95$$

$$= 1.395 \times (10^1 \times 10^2) \quad \text{Associative law}$$

$$= 1.395 \times 10^3 \quad \text{The answer is now in scientific notation.}$$

◀ **Do Exercises 20 and 21.**

Dividing

Consider the quotient $800,000 \div 400 = 2000$. In scientific notation, this is

$$(8 \times 10^5) \div (4 \times 10^2) = \frac{8 \times 10^5}{4 \times 10^2} = \frac{8}{4} \times \frac{10^5}{10^2} = 2 \times 10^3.$$

EXAMPLE 23 Divide: $(3.41 \times 10^5) \div (1.1 \times 10^{-3})$.

$$(3.41 \times 10^5) \div (1.1 \times 10^{-3}) = \frac{3.41 \times 10^5}{1.1 \times 10^{-3}} = \frac{3.41}{1.1} \times \frac{10^5}{10^{-3}}$$
$$= 3.1 \times 10^{5-(-3)}$$
$$= 3.1 \times 10^8$$

EXAMPLE 24 Divide: $(6.4 \times 10^{-7}) \div (8.0 \times 10^6)$.

$$(6.4 \times 10^{-7}) \div (8.0 \times 10^6) = \frac{6.4 \times 10^{-7}}{8.0 \times 10^6}$$

$$= \frac{6.4}{8.0} \times \frac{10^{-7}}{10^6}$$

$$= 0.8 \times 10^{-7-6}$$

$$= 0.8 \times 10^{-13} \quad \text{Not scientific notation; 0.8 is less than 1.}$$

$$= (8.0 \times 10^{-1}) \times 10^{-13} \quad \text{Substituting } 8.0 \times 10^{-1} \text{ for } 0.8$$

$$= 8.0 \times (10^{-1} \times 10^{-13}) \quad \text{Associative law}$$

$$= 8.0 \times 10^{-14} \quad \text{Adding exponents}$$

◀ **Do Exercises 22 and 23.**

Answers

20. 5.6×10^{-15} **21.** 7.462×10^{-13}
22. 2.0×10^3 **23.** 5.5×10^2

e APPLICATIONS WITH SCIENTIFIC NOTATION

EXAMPLE 25 *Distance from the Sun to Earth.* Light from the sun traveling at a rate of 300,000 kilometers per second (km/s) reaches Earth in 499 sec. Find the distance, expressed in scientific notation, from the sun to Earth.

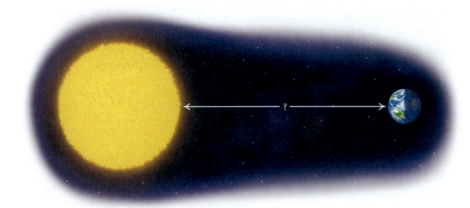

24. *Niagara Falls Water Flow.* On the Canadian side, the amount of water that spills over Niagara Falls in 1 min during the summer is about

$$1.3088 \times 10^8 \text{ L.}$$

How much water spills over the falls in one day? Express the answer in scientific notation.

The time t that it takes for light to reach Earth from the sun is 4.99×10^2 sec (s). The speed is 3.0×10^5 km/s. Recall that distance can be expressed in terms of speed and time as

$$\text{Distance} = \text{Speed} \cdot \text{Time}$$
$$d = rt.$$

We substitute 3.0×10^5 for r and 4.99×10^2 for t:

$$
\begin{aligned}
d &= rt \\
&= (3.0 \times 10^5)(4.99 \times 10^2) \qquad \text{Substituting} \\
&= 14.97 \times 10^7 \\
&= (1.497 \times 10^1) \times 10^7 \\
&= 1.497 \times (10^1 \times 10^7) \qquad \text{Converting to scientific notation} \\
&= 1.497 \times 10^8 \text{ km.}
\end{aligned}
$$

Thus the distance from the sun to Earth is 1.497×10^8 km.

Do Exercise 24. ▶

EXAMPLE 26 *Media Usage.* In January 2013, the 800 million active YouTube users viewed 120 billion videos on the site. On average, how many videos did each user view?

Source: YouTube

In order to find the average number of YouTube videos that each user viewed, we divide the total number viewed by the number of users. We first write each number using scientific notation:

$$800 \text{ million} = 800{,}000{,}000 = 8 \times 10^8,$$

$$120 \text{ billion} = 120{,}000{,}000{,}000 = 1.2 \times 10^{11}.$$

Answer

24. 1.884672×10^{11} L

25. DNA. The width of a DNA (deoxyribonucleic acid) double helix is about 2×10^{-9} m. If its length, fully stretched, is 5×10^{-2} m, how many times longer is the helix than it is wide?

Answer

25. The length of the helix is 2.5×10^7 times its width.

We then divide 1.2×10^{11} by 8×10^8:

$$\frac{1.2 \times 10^{11}}{8 \times 10^8} = \frac{1.2}{8} \times \frac{10^{11}}{10^8}$$
$$= 0.15 \times 10^3 = (1.5 \times 10^{-1}) \times 10^3 = 1.5 \times 10^2.$$

On average, each user viewed 1.5×10^2 videos.

◀ **Do Exercise 25.**

The following is a summary of the definitions and rules for exponents that we have considered in this section and the preceding one.

DEFINITIONS AND RULES FOR EXPONENTS

Exponent of 1:	$a^1 = a$
Exponent of 0:	$a^0 = 1, a \neq 0$
Negative exponents:	$a^{-n} = \dfrac{1}{a^n}, \dfrac{1}{a^{-n}} = a^n, a \neq 0$
Product Rule:	$a^m \cdot a^n = a^{m+n}$
Quotient Rule:	$\dfrac{a^m}{a^n} = a^{m-n}, a \neq 0$
Power Rule:	$(a^m)^n = a^{mn}$
Raising a product to a power:	$(ab)^n = a^n b^n$
Raising a quotient to a power:	$\left(\dfrac{a}{b}\right)^n = \dfrac{a^n}{b^n}, b \neq 0;$
	$\left(\dfrac{a}{b}\right)^{-n} = \left(\dfrac{b}{a}\right)^n, b \neq 0, a \neq 0$
Scientific notation:	$M \times 10^n$, where $1 \leq M < 10$

✓ Reading Check

Choose from the column on the right the appropriate word to complete each statement. Not every word will be used.

RC1. To raise a power to a power, _____ the exponents.

RC2. To raise a product to the nth power, raise each factor to the _____ power.

RC3. To convert a number less than 1 to scientific notation, move the decimal point to the _____.

RC4. A _____ exponent in scientific notation indicates a number greater than or equal to 10.

add
left
multiply
negative
nth
positive
right

a, **b**   Simplify.

1. $(2^3)^2$

2. $(5^2)^4$

3. $(5^2)^{-3}$

4. $(7^{-3})^5$

5. $(x^{-3})^{-4}$

6. $(a^{-5})^{-6}$

7. $(a^{-2})^9$

8. $(x^{-5})^6$

9. $(t^{-3})^{-6}$

10. $(a^{-4})^{-7}$

11. $(t^4)^{-3}$

12. $(t^5)^{-2}$

13. $(x^{-2})^{-4}$

14. $(t^{-6})^{-5}$

15. $(ab)^3$

16. $(xy)^2$

17. $(ab)^{-3}$

18. $(xy)^{-6}$

19. $(mn^2)^{-3}$

20. $(x^3y)^{-2}$

21. $(4x^3)^2$

22. $4(x^3)^2$

23. $(3x^{-4})^2$

24. $(2a^{-5})^3$

25. $(x^4y^5)^{-3}$

26. $(t^5x^3)^{-4}$

27. $(x^{-6}y^{-2})^{-4}$

28. $(x^{-2}y^{-7})^{-5}$

29. $(a^{-2}b^7)^{-5}$

30. $(q^5r^{-1})^{-3}$

31. $(5r^{-4}t^3)^2$

32. $(4x^5y^{-6})^3$

33. $(a^{-5}b^7c^{-2})^3$

34. $(x^{-4}y^{-2}z^9)^2$

35. $(3x^3y^{-8}z^{-3})^2$

36. $(2a^2y^{-4}z^{-5})^3$

37. $(-4x^3y^{-2})^2$

38. $(-8x^3y^{-2})^3$

39. $(-a^{-3}b^{-2})^{-4}$

40. $(-p^{-4}q^{-3})^{-2}$

41. $\left(\dfrac{y^3}{2}\right)^2$

42. $\left(\dfrac{a^5}{3}\right)^3$

43. $\left(\dfrac{a^2}{b^3}\right)^4$

44. $\left(\dfrac{x^3}{y^4}\right)^5$

45. $\left(\dfrac{y^2}{2}\right)^{-3}$

46. $\left(\dfrac{a^4}{3}\right)^{-2}$

47. $\left(\dfrac{7}{x^{-3}}\right)^2$

48. $\left(\dfrac{3}{a^{-2}}\right)^3$

49. $\left(\dfrac{x^2y}{z}\right)^3$

50. $\left(\dfrac{m}{n^4p}\right)^3$

51. $\left(\dfrac{a^2b}{cd^3}\right)^{-2}$

52. $\left(\dfrac{2a^2}{3b^4}\right)^{-3}$

c   Convert to scientific notation.

53. 28,000,000,000   **54.** 4,900,000,000,000   **55.** 907,000,000,000,000,000   **56.** 168,000,000,000,000

57. 0.00000304   **58.** 0.000000000865   **59.** 0.000000018   **60.** 0.00000000002

61. 100,000,000,000   **62.** 0.0000001

63. *Population of the United States.*   It is estimated that the population of the United States will be 419,854,000 in 2050. Convert 419,854,000 to scientific notation.

Source: U.S. Census Bureau

64. *Microprocessors.*   The minimum feature size of a microprocessor is the transistor gate length. In 2011, the transistor gate length for a new microprocessor was about 0.000000028 m. Convert 0.000000028 to scientific notation.

65. *Wavelength of Light.*   The wavelength of red light is 0.00000068 m. Convert 0.00000068 to scientific notation.

66. *Olympics.*   Great Britain spent about $15,000,000,000 to stage the 2012 Summer Olympics. Convert 15,000,000,000 to scientific notation.

Source: CNN.com

Convert to decimal notation.

67. 8.74×10^7 **68.** 1.85×10^8 **69.** 5.704×10^{-8} **70.** 8.043×10^{-4}

71. 10^7 **72.** 10^6 **73.** 10^{-5} **74.** 10^{-8}

d   Multiply or divide and write scientific notation for the result.

75. $(3 \times 10^4)(2 \times 10^5)$ **76.** $(3.9 \times 10^8)(8.4 \times 10^{-3})$ **77.** $(5.2 \times 10^5)(6.5 \times 10^{-2})$

78. $(7.1 \times 10^{-7})(8.6 \times 10^{-5})$ **79.** $(9.9 \times 10^{-6})(8.23 \times 10^{-8})$ **80.** $(1.123 \times 10^4) \times 10^{-9}$

81. $\dfrac{8.5 \times 10^{8}}{3.4 \times 10^{-5}}$

82. $\dfrac{5.6 \times 10^{-2}}{2.5 \times 10^{5}}$

83. $(3.0 \times 10^{6}) \div (6.0 \times 10^{9})$

84. $(1.5 \times 10^{-3}) \div (1.6 \times 10^{-6})$

85. $\dfrac{7.5 \times 10^{-9}}{2.5 \times 10^{12}}$

86. $\dfrac{4.0 \times 10^{-3}}{8.0 \times 10^{20}}$

e   Solve.

87. *River Discharge.*   The average discharge at the mouths of the Amazon River is 4,200,000 cubic feet per second. How much water is discharged from the Amazon River in 1 year? Express the answer in scientific notation.

88. *Coral Reefs.*   There are 10 million bacteria per square centimeter of coral in a coral reef. The coral reefs near the Hawaiian Islands cover 14,000 km². How many bacteria are there in Hawaii's coral reefs?

Sources: livescience.com; U.S. Geological Survey

89. *Stars.*   It is estimated that there are 10 billion trillion stars in the known universe. Express the number of stars in scientific notation. ($1 \text{ billion} = 10^{9}$; $1 \text{ trillion} = 10^{12}$)

90. *Water Contamination.*   Americans who change their own motor oil generate about 150 million gallons of used oil annually. If this oil is not disposed of properly, it can contaminate drinking water and soil. One gallon of used oil can contaminate one million gallons of drinking water. How many gallons of drinking water can 150 million gallons of oil contaminate? Express the answer in scientific notation. ($1 \text{ million} = 10^{6}$).

Source: *New Car Buying Guide*

91. *Earth vs. Jupiter.*   The mass of Earth is about 6×10^{21} metric tons. The mass of Jupiter is about 1.908×10^{24} metric tons. About how many times the mass of Earth is the mass of Jupiter? Express the answer in scientific notation.

92. *Office Supplies.*   A ream of copier paper weighs 2.25 kg. How much does a sheet of copier paper weigh?

93. *Information Technology.* In 2012, about 2.5×10^{18} bytes of information were generated each day by the worldwide online population of 2×10^9 people. Find the average amount of information generated per person per day in 2012.

Sources: IBM; internetworldstats.com

94. *Computer Technology.* Intel Corporation has developed silicon-based connections that use lasers to move data at a rate of 50 gigabytes per second. The printed collection of the U.S. Library of Congress contains 10 terabytes of information. How long would it take to copy the Library of Congress using these connections? *Note:* 1 gigabyte $= 10^9$ bytes and 1 terabyte $= 10^{12}$ bytes.

Sources: spie.org; newworldencyclopedia.org

95. *Gold Leaf.* Gold can be milled into a very thin film called gold leaf. This film is so thin that it took only 43 oz of gold to cover the dome of Georgia's state capitol building. The gold leaf used was 5×10^{-6} m thick. In contrast, a U.S. penny is 1.55×10^{-3} m thick. How many sheets of gold leaf are in a stack that is the height of a penny?

Source: georgiaencyclopedia.org

96. *Relative Size.* An influenza virus is about 1.2×10^{-7} m in diameter. A staphylococcus bacterium is about 1.5×10^{-6} m in diameter. How many influenza viruses would it take, laid side by side, to equal the diameter of the bacterium?

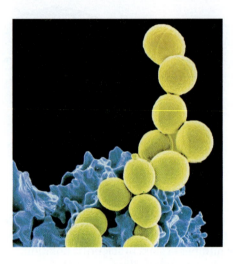

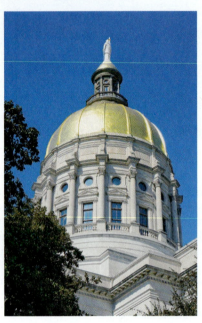

Space Travel. Use the following information for Exercises 97 and 98.

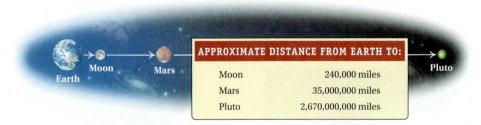

APPROXIMATE DISTANCE FROM EARTH TO:	
Moon	240,000 miles
Mars	35,000,000 miles
Pluto	2,670,000,000 miles

97. *Time to Reach Mars.* Suppose that it takes about 3 days for a space vehicle to travel from Earth to the moon. About how long would it take the same space vehicle traveling at the same speed to reach Mars? Express the answer in scientific notation.

98. *Time to Reach Pluto.* Suppose that it takes about 3 days for a space vehicle to travel from Earth to the moon. About how long would it take the same space vehicle traveling at the same speed to reach the dwarf planet Pluto? Express the answer in scientific notation.

Skill Maintenance

Graph.

99. $y = x - 5$ [3.2a]

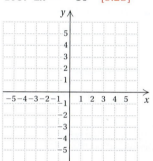

100. $2x + y = 4$ [3.2a]

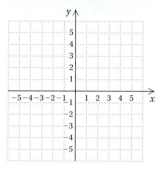

101. $3x - y = 3$ [3.2a]

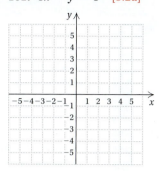

102. $y = -x$ [3.2a]

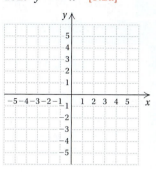

103. $2x = -10$ [3.2b]

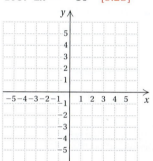

104. $y = -4$ [3.2b]

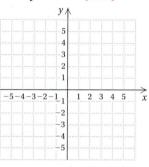

105. $8y - 16 = 0$ [3.2b]

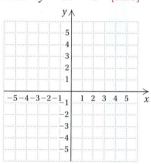

106. $x = 4$ [3.2b]

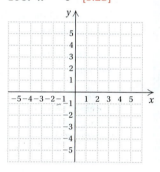

Synthesis

107. 🖩 Carry out the indicated operations. Express the result in scientific notation.

$$\frac{(5.2 \times 10^6)(6.1 \times 10^{-11})}{1.28 \times 10^{-3}}$$

108. Find the reciprocal and express it in scientific notation.

$$6.25 \times 10^{-3}$$

Simplify.

109. $\dfrac{(5^{12})^2}{5^{25}}$

110. $\dfrac{a^{22}}{(a^2)^{11}}$

111. $\dfrac{(3^5)^4}{3^5 \cdot 3^4}$

112. $\left(\dfrac{5x^{-2}}{3y^{-2}z}\right)^0$

113. $\dfrac{49^{18}}{7^{35}}$

114. $\left(\dfrac{1}{a}\right)^{-n}$

115. $\dfrac{(0.4)^5}{\left[(0.4)^3\right]^2}$

116. $\left(\dfrac{4a^3b^{-2}}{5c^{-3}}\right)^1$

Determine whether each equation is true or false for all pairs of integers m and n and all positive numbers x and y.

117. $x^m \cdot y^n = (xy)^{mn}$

118. $x^m \cdot y^m = (xy)^{2m}$

119. $(x - y)^m = x^m - y^m$

120. $-x^m = (-x)^m$

121. $(-x)^{2m} = x^{2m}$

122. $x^{-m} = \dfrac{-1}{x^m}$

OBJECTIVES

a Evaluate a polynomial for a given value of the variable.

b Identify the terms of a polynomial and classify a polynomial by its number of terms.

c Identify the coefficient and the degree of each term of a polynomial and the degree of the polynomial.

d Collect the like terms of a polynomial.

e Arrange a polynomial in descending order, or collect the like terms and then arrange in descending order.

f Identify the missing terms of a polynomial.

SKILL TO REVIEW

Objective 1.7e: Collect like terms.

Collect like terms.
1. $3x - 4y + 5x + y$
2. $2a - 7b + 6 - 3a + 4b - 1$

We have already learned to evaluate and to manipulate certain kinds of algebraic expressions. We will now consider algebraic expressions called *polynomials*.

The following are examples of *monomials in one variable*:

$$3x^2, \quad 2x, \quad -5, \quad 37p^4, \quad 0.$$

Each expression is a constant or a constant times some variable to a non-negative integer power.

> ### MONOMIAL
>
> A **monomial** is an expression of the type ax^n, where a is a real-number constant and n is a nonnegative integer.

Algebraic expressions like the following are **polynomials**:

$$\tfrac{3}{4}y^5, \quad -2, \quad 5y + 3, \quad 3x^2 + 2x - 5, \quad -7a^3 + \tfrac{1}{2}a, \quad 6x, \quad 37p^4, \quad x, \quad 0.$$

> ### POLYNOMIAL
>
> A **polynomial** is a monomial or a combination of sums and/or differences of monomials.

The following algebraic expressions are *not* polynomials:

$$\textbf{(1)} \ \frac{x+3}{x-4}, \qquad \textbf{(2)} \ 5x^3 - 2x^2 + \frac{1}{x}, \qquad \textbf{(3)} \ \frac{1}{x^3 - 2}.$$

Expressions (1) and (3) are not polynomials because they represent quotients, not sums or differences. Expression (2) is not a polynomial because

$$\frac{1}{x} = x^{-1},$$

and this is not a monomial because the exponent is negative.

◀ **Do Margin Exercise 1.**

a EVALUATING POLYNOMIALS AND APPLICATIONS

When we replace the variable in a polynomial with a number, the polynomial then represents a number called a **value** of the polynomial. Finding that number, or value, is called **evaluating the polynomial**. We evaluate a polynomial using the rules for order of operations.

1. Write three polynomials.

EXAMPLE 1 Evaluate the polynomial when $x = 2$.

a) $3x + 5 = 3 \cdot 2 + 5$
$= 6 + 5$
$= 11$

b) $2x^2 - 7x + 3 = 2 \cdot 2^2 - 7 \cdot 2 + 3$
$= 2 \cdot 4 - 7 \cdot 2 + 3$
$= 8 - 14 + 3$
$= -3$

Answers

Skill to Review:
1. $8x - 3y$ 2. $-a - 3b + 5$

Margin Exercise:
1. $4x^2 - 3x + \dfrac{5}{4}; 15y^3; -7x^3 + 1.1;$

answers may vary

EXAMPLE 2 Evaluate the polynomial when $x = -4$.

a) $2 - x^3 = 2 - (-4)^3 = 2 - (-64)$
$$= 2 + 64 = 66$$

b) $-x^2 - 3x + 1 = -(-4)^2 - 3(-4) + 1$
$$= -16 + 12 + 1 = -3$$

Do Exercises 2–5. ▶

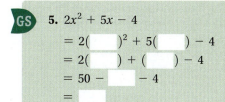

Evaluate each polynomial when $x = 3$.

  2. $-4x - 7$

  3. $-5x^3 + 7x + 10$

Evaluate each polynomial when $x = -5$.

  4. $5x + 7$

ALGEBRAIC ▶ ◀ GRAPHICAL CONNECTION

An equation like $y = 2x - 2$, which has a polynomial on one side and only y on the other, is called a **polynomial equation**. For such an equation, determining y is the same as evaluating the polynomial. Once the graph of such an equation has been drawn, we can evaluate the polynomial for a given x-value by finding the y-value that is paired with it on the graph.

GS   **5.** $2x^2 + 5x - 4$
$$= 2(\quad)^2 + 5(\quad) - 4$$
$$= 2(\quad) + (\quad) - 4$$
$$= 50 - \quad - 4$$
$$= \quad$$

EXAMPLE 3 Use *only* the given graph of $y = 2x - 2$ to evaluate the polynomial $2x - 2$ when $x = 3$.

First, we locate 3 on the x-axis. From there we move vertically to the graph of the equation and then horizontally to the y-axis. There we locate the y-value that is paired with 3. It appears that the y-value 4 is paired with 3. Thus the value of $2x - 2$ is 4 when $x = 3$. We can check this by evaluating $2x - 2$ when $x = 3$.

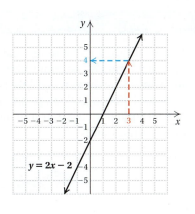

Do Exercise 6. ▶

6. Use *only* the graph shown in Example 3 to evaluate the polynomial $2x - 2$ when $x = 4$ and when $x = -1$.

Polynomial equations can be used to model many real-world situations.

EXAMPLE 4 *Games in a Sports League.* In a sports league of x teams in which each team plays every other team twice, the total number of games N to be played is given by the polynomial equation

$$N = x^2 - x.$$

A women's slow-pitch softball league has 10 teams and each team plays every other team twice. What is the total number of games to be played?

We evaluate the polynomial when $x = 10$:

$$N = x^2 - x = 10^2 - 10 = 100 - 10 = 90.$$

The league plays 90 games.

Do Exercise 7. ▶

7. Refer to Example 4. Determine the total number of games to be played in a league of 12 teams in which each team plays every other team twice.

Answers

2. -19 **3.** -104 **4.** -18 **5.** 21
6. $6; -4$ **7.** 132 games

Guided Solution:
5. $-5, -5, 25, -25, 25, 21$

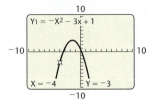

CALCULATOR CORNER

To use a graphing calculator to evaluate the polynomial in Example 2(b), $-x^2 - 3x + 1$, when $x = -4$, we can first graph $y_1 = -x^2 - 3x + 1$ in a window that includes the x-value -4. Then we can use the Value feature from the **CALC** menu, supplying the desired x-value and pressing **ENTER** to see $X = -4$, $Y = -3$ at the bottom of the screen. Thus, when $x = -4$, the value of $-x^2 - 3x + 1$ is -3.

```
          10
  Y₁ = -X² - 3X + 1
-10 |_____|_____| 10

  X = -4      Y = -3
         -10
```

EXERCISES: Use the Value feature to evaluate each polynomial for the given values of x.

1. $-x^2 - 3x + 1$, when $x = -2$, when $x = -0.5$, and when $x = 4$

2. $3x^2 - 5x + 2$, when $x = -3$, when $x = 1$, and when $x = 2.6$

8. **Medical Dosage.** Refer to Example 5.

a) Determine the concentration after 3 hr by evaluating the polynomial when $t = 3$.

b) Use *only* the graph showing medical dosage to check the value found in part (a).

9. **Medical Dosage.** Refer to Example 5. Use *only* the graph showing medical dosage to estimate the value of the polynomial when $t = 26$.

Answers

8. **(a)** 7.55 parts per million; **(b)** When $t = 3$, $C \approx 7.5$ so the value found in part (a) appears to be correct. **9.** 20 parts per million

EXAMPLE 5 *Medical Dosage.* The concentration C, in parts per million, of a certain antibiotic in the bloodstream after t hours is given by the polynomial equation

$$C = -0.05t^2 + 2t + 2.$$

Find the concentration after 2 hr.

To find the concentration after 2 hr, we evaluate the polynomial when $t = 2$:

$$\begin{aligned} C &= -0.05t^2 + 2t + 2 \\ &= -0.05(2)^2 + 2(2) + 2 \quad &&\text{Substituting 2 for } t \\ &= -0.05(4) + 2(2) + 2 \quad &&\text{Carrying out the calculation using} \\ & && \text{the rules for order of operations} \\ &= -0.2 + 4 + 2 \\ &= 3.8 + 2 \\ &= 5.8. \end{aligned}$$

The concentration after 2 hr is 5.8 parts per million.

ALGEBRAIC ▶◀ GRAPHICAL CONNECTION

The polynomial equation in Example 5 can be graphed if we evaluate the polynomial for several values of t. We list the values in a table and show the graph below. Note that the concentration peaks at the 20-hr mark and after slightly more than 40 hr, the concentration is 0. Since neither time nor concentration can be negative, our graph uses only the first quadrant.

t	C $C = -0.05t^2 + 2t + 2$
0	2
2	5.8 ← Example 5
10	17
20	22
30	17

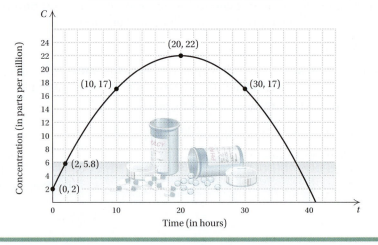

◀ **Do Exercises 8 and 9.**

b IDENTIFYING TERMS AND CLASSIFYING POLYNOMIALS

For any polynomial that has some subtractions, we can find an equivalent polynomial using only additions.

EXAMPLES Find an equivalent polynomial using only additions.

6. $-5x^2 - x = -5x^2 + (-x)$

7. $4x^5 - 2x^6 + 4x - 7 = 4x^5 + (-2x^6) + 4x + (-7)$

Do Exercises 10 and 11. ▶

When a polynomial is written using only additions, the monomials being added are called **terms**. In Example 6, the terms are $-5x^2$ and $-x$. In Example 7, the terms are $4x^5$, $-2x^6$, $4x$, and -7.

EXAMPLE 8 Identify the terms of the polynomial

$$4x^7 + 3x + 12 + 8x^3 + 5x.$$

Terms: $4x^7, 3x, 12, 8x^3$, and $5x$.

If there are subtractions, you can *think* of them as additions without rewriting.

EXAMPLE 9 Identify the terms of the polynomial

$$3t^4 - 5t^6 - 4t + 2.$$

Terms: $3t^4, -5t^6, -4t$, and 2.

Do Exercises 12 and 13. ▶

Polynomials with just one term are called **monomials**. Polynomials with just two terms are called **binomials**. Those with just three terms are called **trinomials**. Those with more than three terms are generally not specified with a name.

EXAMPLE 10

MONOMIALS	BINOMIALS	TRINOMIALS	NONE OF THESE
$-4x^2$	$2x + 4$	$3x^5 + 4x^4 + 7x$	$4x^3 - 5x^2 + x - 8$
9	$-9x^7 - 6x$	$4x^2 - 6x - \frac{1}{2}$	$z^5 + 2z^4 - z^3 + 7z + 3$

Do Exercises 14–17. ▶

c COEFFICIENTS AND DEGREES

The coefficient of the term $5x^3$ is 5. In the following polynomial, the red numbers are the **coefficients**, 3, -2, 5, and 4:

$$3x^5 - 2x^3 + 5x + 4.$$

Find an equivalent polynomial using only additions.

10. $-9x^3 - 4x^5$

11. $-2y^3 + 3y^7 - 7y - 9$

Identify the terms of each polynomial.

12. $3x^2 + 6x + \dfrac{1}{2}$

13. $-4y^5 + 7y^2 - 3y - 2$

Classify each polynomial as a monomial, a binomial, a trinomial, or none of these.

14. $3x^2 + x$

15. $5x^4$

16. $4x^3 - 3x^2 + 4x + 2$

17. $3x^2 + 2x - 4$

Answers

10. $-9x^3 + (-4x^5)$
11. $-2y^3 + 3y^7 + (-7y) + (-9)$
12. $3x^2, 6x, \dfrac{1}{2}$ **13.** $-4y^5, 7y^2, -3y, -2$
14. Binomial   **15.** Monomial
16. None of these   **17.** Trinomial

EXAMPLE 11 Identify the coefficient of each term in the polynomial

$$3x^4 - 4x^3 + \frac{1}{2}x^2 + x - 8.$$

The coefficient of $3x^4$ is 3.
The coefficient of $-4x^3$ is -4.
The coefficient of $\frac{1}{2}x^2$ is $\frac{1}{2}$.
The coefficient of x (or $1x$) is 1.
The coefficient of -8 is -8.

18. Identify the coefficient of each term in the polynomial
$2x^4 - 7x^3 - 8.5x^2 - x - 4.$

◀ **Do Exercise 18.**

The **degree** of a term is the exponent of the variable. The degree of the term $-5x^3$ is 3.

EXAMPLE 12 Identify the degree of each term of $8x^4 - 3x + 7$.

The degree of $8x^4$ is 4.
The degree of $-3x$ (or $-3x^1$) is 1. $x = x^1$
The degree of 7 (or $7x^0$) is 0. $7 = 7 \cdot 1 = 7 \cdot x^0$, since $x^0 = 1$

Because we can write 1 as x^0, the degree of any constant term (except 0) is 0. The term 0 is a special case. We agree that it has *no* degree because we can express 0 as $0 = 0x^5 = 0x^7$, and so on, using any exponent we wish.

> The degree of any nonzero constant term is 0.

The **degree of a polynomial** is the largest of the degrees of the terms, unless it is the polynomial 0.

EXAMPLE 13 Identify the degree of the polynomial $5x^3 - 6x^4 + 7$.

$$5x^3 - 6x^4 + 7.$$ The largest degree is 4.

The degree of the polynomial is 4.

Identify the degree of each term and the degree of the polynomial.

19. $-6x^4 + 8x^2 - 2x + 9$

20. $4 - x^3 + \frac{1}{2}x^6 - x^5$

◀ **Do Exercises 19 and 20.**

Let's summarize the terminology that we have learned, using the polynomial $3x^4 - 8x^3 + x^2 + 7x - 6$.

TERM	COEFFICIENT	DEGREE OF THE TERM	DEGREE OF THE POLYNOMIAL
$3x^4$	3	4	
$-8x^3$	-8	3	
x^2	1	2	4
$7x$	7	1	
-6	-6	0	

d COLLECTING LIKE TERMS

When terms have the same variable and the same exponent, we say that they are **like terms**.

EXAMPLES Identify the like terms in each polynomial.

14. $4x^3 + 5x - 4x^2 + 2x^3 + x^2$

    Like terms: $4x^3$ and $2x^3$ Same variable and exponent

    Like terms: $-4x^2$ and x^2 Same variable and exponent

15. $6 - 3a^2 - 8 - a - 5a$

    Like terms: 6 and -8 Constant terms are like terms; note that $6 = 6x^0$ and $-8 = -8x^0$.

    Like terms: $-a$ and $-5a$

> **Do Exercises 21–23.** ▶

We can often simplify polynomials by **collecting like terms**, or **combining like terms**. To do this, we use the distributive laws.

EXAMPLES Collect like terms.

16. $2x^3 - 6x^3 = (2 - 6)x^3 = -4x^3$

17. $5x^2 + 7 + 4x^4 + 2x^2 - 11 - 2x^4 = (5 + 2)x^2 + (4 - 2)x^4 + (7 - 11)$

                           $= 7x^2 + 2x^4 - 4$

Note that using the distributive laws in this manner allows us to collect like terms by adding or subtracting the coefficients. Often the middle step is omitted and we add or subtract mentally, writing just the answer. In collecting like terms, we may get 0.

EXAMPLE 18 Collect like terms: $3x^5 + 2x^2 - 3x^5 + 8$.

$$3x^5 + 2x^2 - 3x^5 + 8 = (3 - 3)x^5 + 2x^2 + 8$$
$$= 0x^5 + 2x^2 + 8$$
$$= 2x^2 + 8$$

> **Do Exercises 24–29.** ▶

Expressing a term like x^2 by showing 1 as a factor, $1 \cdot x^2$, may make it easier to understand how to factor or collect like terms.

EXAMPLES Collect like terms.

19. $5x^8 - 6x^5 - x^8 = 5x^8 - 6x^5 - 1x^8$ Replacing x^8 with $1x^8$

                  $= (5 - 1)x^8 - 6x^5$ Using a distributive law

                  $= 4x^8 - 6x^5$

20. $\frac{2}{3}x^4 - x^3 - \frac{1}{6}x^4 + \frac{2}{5}x^3 - \frac{3}{10}x^3$

    $= \left(\frac{2}{3} - \frac{1}{6}\right)x^4 + \left(-1 + \frac{2}{5} - \frac{3}{10}\right)x^3$ $-x^3 = -1 \cdot x^3$

    $= \left(\frac{4}{6} - \frac{1}{6}\right)x^4 + \left(-\frac{10}{10} + \frac{4}{10} - \frac{3}{10}\right)x^3$

    $= \frac{3}{6}x^4 - \frac{9}{10}x^3 = \frac{1}{2}x^4 - \frac{9}{10}x^3$

> **Do Exercises 30–32.** ▶

Identify the like terms in each polynomial.

21. $4x^3 - x^3 + 2$

22. $4t^4 - 9t^3 - 7t^4 + 10t^3$

23. $5x^2 + 3x - 10 + 7x^2 - 8x + 11$

Collect like terms.

24. $3x^2 + 5x^2$

25. $4x^3 - 2x^3 + 2 + 5$

26. $\frac{1}{2}x^5 - \frac{3}{4}x^5 + 4x^2 - 2x^2$

27. $24 - 4x^3 - 24$

28. $5x^3 - 8x^5 + 8x^5$

GS 29. $-2x^4 + 16 + 2x^4 + 9 - 3x^5$

    $= -3x^5 + (-2 + \boxed{})x^4 + (16 + \boxed{})$

    $= -3x^5 + 0x^4 + \boxed{}$

    $= -3x^5 + 25$

Collect like terms.

30. $5x^3 - x^3 + 4$

31. $\frac{3}{4}x^3 + 4x^2 - x^3 + 7$

32. $\frac{4}{5}x^4 - x^4 + x^5 - \frac{1}{5} - \frac{1}{4}x^4 + 10$

Answers

21. $4x^3$ and $-x^3$ **22.** $4t^4$ and $-7t^4$; $-9t^3$ and $10t^3$
23. $5x^2$ and $7x^2$; $3x$ and $-8x$; -10 and 11
24. $8x^2$ **25.** $2x^3 + 7$ **26.** $-\frac{1}{4}x^5 + 2x^2$
27. $-4x^3$ **28.** $5x^3$ **29.** $-3x^5 + 25$
30. $4x^3 + 4$ **31.** $-\frac{1}{4}x^3 + 4x^2 + 7$
32. $x^5 - \frac{9}{20}x^4 + \frac{49}{5}$

Guided Solution:
29. 2, 9, 25

e DESCENDING ORDER

A polynomial is written in **descending order** when the term with the largest degree is written first, the term with the next largest degree is written next, and so on, in order from left to right.

EXAMPLES Arrange each polynomial in descending order.

21. $6x^5 + 4x^7 + x^2 + 2x^3 = 4x^7 + 6x^5 + 2x^3 + x^2$

22. $\frac{2}{3} + 4x^5 - 8x^2 + 5x - 3x^3 = 4x^5 - 3x^3 - 8x^2 + 5x + \frac{2}{3}$

◀ **Do Exercises 33 and 34.**

EXAMPLE 23 Collect like terms and then arrange in descending order:

$$2x^2 - 4x^3 + 3 - x^2 - 2x^3.$$

$$2x^2 - 4x^3 + 3 - x^2 - 2x^3 = x^2 - 6x^3 + 3 \qquad \text{Collecting like terms}$$
$$= -6x^3 + x^2 + 3 \qquad \text{Arranging in descending order}$$

◀ **Do Exercises 35 and 36.**

The opposite of descending order is called **ascending order**. Generally, if an exercise is written in a certain order, we give the answer in that same order.

f MISSING TERMS

If a coefficient is 0, we generally do not write the term. We say that we have a **missing term**.

EXAMPLE 24 Identify the missing terms in the polynomial

$$8x^5 - 2x^3 + 5x^2 + 7x + 8.$$

There is no term with x^4. We say that the x^4-term is missing.

◀ **Do Exercises 37–39.**

We can either write missing terms with zero coefficients or leave space.

EXAMPLE 25 Write the polynomial $x^4 - 6x^3 + 2x - 1$ in two ways: with its missing term and by leaving space for it.

a) $x^4 - 6x^3 + 2x - 1 = x^4 - 6x^3 + 0x^2 + 2x - 1$ Writing with the missing x^2-term

b) $x^4 - 6x^3 + 2x - 1 = x^4 - 6x^3 \qquad + 2x - 1$ Leaving space for the missing x^2-term

EXAMPLE 26 Write the polynomial $y^5 - 1$ in two ways: with its missing terms and by leaving space for them.

a) $y^5 - 1 = y^5 + 0y^4 + 0y^3 + 0y^2 + 0y - 1$

b) $y^5 - 1 = y^5 \qquad\qquad\qquad - 1$

◀ **Do Exercises 40 and 41.**

Arrange each polynomial in descending order.

33. $4x^2 - 3 + 7x^5 + 2x^3 - 5x^4$

34. $-14 + 7t^2 - 10t^5 + 14t^7$

Collect like terms and then arrange in descending order.

35. $3x^2 - 2x + 3 - 5x^2 - 1 - x$

36. $-x + \dfrac{1}{2} + 14x^4 - 7x - 1 - 4x^4$

Identify the missing term(s) in each polynomial.

37. $2x^3 + 4x^2 - 2$

38. $-3x^4$

39. $x^3 + 1$

Write each polynomial in two ways: with its missing term(s) and by leaving space for them.

40. $2x^3 + 4x^2 - 2$

41. $a^4 + 10$

Answers

33. $7x^5 - 5x^4 + 2x^3 + 4x^2 - 3$
34. $14t^7 - 10t^5 + 7t^2 - 14$ **35.** $-2x^2 - 3x + 2$
36. $10x^4 - 8x - \dfrac{1}{2}$ **37.** x **38.** x^3, x^2, x, x^0
39. x^2, x **40.** $2x^3 + 4x^2 + 0x - 2;$
$\qquad\qquad 2x^3 + 4x^2 \qquad - 2$
41. $a^4 + 0a^3 + 0a^2 + 0a + 10;$
$\quad a^4 \qquad\qquad\qquad + 10$

☑ Reading Check

Choose from the column on the right the expression that best fits each description.

RC1. _____ The value of $x^2 - x$ when $x = -1$

RC2. _____ A polynomial written in ascending order

RC3. _____ A coefficient of $5x^4 - 3x + 7$

RC4. _____ A term of $5x^4 - 3x + 7$

RC5. _____ The degree of one of the terms of $5x^4 - 3x + 7$

RC6. _____ An example of a binomial

a) 0
b) 2
c) 5
d) $-3x$
e) $8x - 9$
f) $y + 6y^2 - 2y^8$

a Evaluate each polynomial when $x = 4$ and when $x = -1$.

1. $-5x + 2$

2. $-8x + 1$

3. $2x^2 - 5x + 7$

4. $3x^2 + x - 7$

5. $x^3 - 5x^2 + x$

6. $7 - x + 3x^2$

Evaluate each polynomial when $x = -2$ and when $x = 0$.

7. $\frac{1}{3}x + 5$

8. $8 - \frac{1}{4}x$

9. $x^2 - 2x + 1$

10. $5x + 6 - x^2$

11. $-3x^3 + 7x^2 - 3x - 2$

12. $-2x^3 + 5x^2 - 4x + 3$

13. _Skydiving._ During the first 13 sec of a jump, the distance S, in feet, that a skydiver falls in t seconds can be approximated by the polynomial equation
$$S = 11.12t^2.$$
In 2009, 108 U.S. skydivers fell headfirst in formation from a height of 18,000 ft. How far had they fallen 10 sec after having jumped from the plane?

Source: www.telegraph.co.uk

14. _Skydiving._ For jumps that exceed 13 sec, the polynomial equation
$$S = 173t - 369$$
can be used to approximate the distance S, in feet, that a skydiver has fallen in t seconds. Approximately how far has a skydiver fallen 20 sec after having jumped from a plane?

15. Stacking Spheres. In 2004, the journal *Annals of Mathematics* accepted a proof of the so-called Kepler Conjecture: that the most efficient way to pack spheres is in the shape of a square pyramid. The number N of balls in the stack is given by the polynomial equation

$$N = \frac{1}{3}x^3 + \frac{1}{2}x^2 + \frac{1}{6}x,$$

where x is the number of layers. A square pyramid with 3 layers is illustrated below. Find the number of oranges in a pyramid with 5 layers.

Source: *The New York Times* 4/6/04

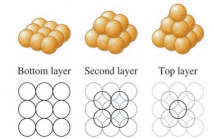

Bottom layer   Second layer   Top layer

16. SCAD Diving. The distance s, in feet, traveled by a body falling freely from rest in t seconds is approximated by the polynomial equation

$$s = 16t^2.$$

The SCAD thrill ride is a 2.5-sec free fall into a net. How far does the diver fall?

Source: www.scadfreefall.co.uk

17. The graph of the polynomial equation $y = 5 - x^2$ is shown below. Use *only* the graph to estimate the value of the polynomial when $x = -3$, $x = -1$, $x = 0$, $x = 1.5$, and $x = 2$.

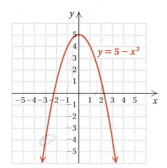

18. The graph of the polynomial equation $y = 6x^3 - 6x$ is shown below. Use *only* the graph to estimate the value of the polynomial when $x = -1$, $x = -0.5$, $x = 0.5$, $x = 1$, and $x = 1.1$.

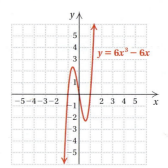

19. Solar Capacity. The average capacity C, in kilowatts (kW), of U.S. residential installations generating energy from the sun can be estimated by the polynomial equation

$$C = 0.27t + 2.97,$$

where t is the number of years since 2000.

Source: Based on data from IREC 2011 Updates & Trends

a) Use the equation to estimate the average capacity of a solar-energy residential installation in 2010.

b) Check the result of part (a) using the graph below.

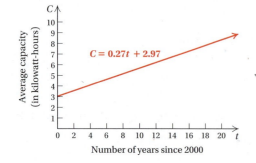

Number of years since 2000

20. Senior Population. The number N, in millions, of people in the United States ages 65 and older can be estimated by the polynomial equation

$$N = 1.24t + 28.4,$$

where t is the number of years since 2000.

Source: Based on data from U.S. Census Bureau

a) Use the equation to estimate the number of people in the United States ages 65 and older in 2030.

b) Check the result of part (a) using the graph below.

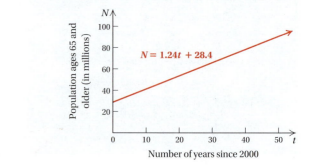

Number of years since 2000

Memorizing Words. Participants in a psychology experiment were able to memorize an average of M words in t minutes, where $M = -0.001t^3 + 0.1t^2$. Use the graph below for Exercises 21–26.

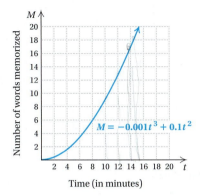

21. Estimate the number of words memorized after 10 min.

22. Estimate the number of words memorized after 14 min.

23. Find the approximate value of M for $t = 8$.

24. Find the approximate value of M for $t = 12$.

25. Estimate the value of M when t is 13.

26. Estimate the value of M when t is 7.

b Identify the terms of each polynomial.

27. $2 - 3x + x^2$

28. $2x^2 + 3x - 4$

29. $-2x^4 + \frac{1}{3}x^3 - x + 3$

30. $-\frac{2}{5}x^5 - x^3 + 6$

Classify each polynomial as a monomial, a binomial, a trinomial, or none of these.

31. $x^2 - 10x + 25$

32. $-6x^4$

33. $x^3 - 7x^2 + 2x - 4$

34. $x^2 - 9$

35. $4x^2 - 25$

36. $2x^4 - 7x^3 + x^2 + x - 6$

37. $40x$

38. $4x^2 + 12x + 9$

c Identify the coefficient of each term of the polynomial.

39. $-3x + 6$

40. $2x - 4$

41. $5x^2 + \frac{3}{4}x + 3$

42. $\frac{2}{3}x^2 - 5x + 2$

43. $-5x^4 + 6x^3 - 2.7x^2 + x - 2$

44. $7x^3 - x^2 - 4.2x + 5$

Identify the degree of each term of the polynomial and the degree of the polynomial.

45. $2x - 4$

46. $6 - 3x$

47. $3x^2 - 5x + 2$

48. $5x^3 - 2x^2 + 3$

49. $-7x^3 + 6x^2 + \frac{3}{5}x + 7$

50. $5x^4 + \frac{1}{4}x^2 - x + 2$

51. $x^2 - 3x + x^6 - 9x^4$

52. $8x - 3x^2 + 9 - 8x^3$

53. Complete the following table for the polynomial $-7x^4 + 6x^3 - x^2 + 8x - 2$.

TERM	COEFFICIENT	DEGREE OF THE TERM	DEGREE OF THE POLYNOMIAL
$-7x^4$			
$6x^3$	6		
		2	
$8x$		1	
	-2		

54. Complete the following table for the polynomial $3x^2 + x^5 - 46x^3 + 6x - 2.4 - \frac{1}{2}x^4$.

TERM	COEFFICIENT	DEGREE OF THE TERM	DEGREE OF THE POLYNOMIAL
		5	
$-\frac{1}{2}x^4$		4	
	-46		
$3x^2$		2	
	6		
-2.4			

d Identify the like terms in each polynomial.

55. $5x^3 + 6x^2 - 3x^2$

56. $3x^2 + 4x^3 - 2x^2$

57. $2x^4 + 5x - 7x - 3x^4$

58. $-3t + t^3 - 2t - 5t^3$

59. $3x^5 - 7x + 8 + 14x^5 - 2x - 9$

60. $8x^3 + 7x^2 - 11 - 4x^3 - 8x^2 - 29$

Collect like terms.

61. $2x - 5x$

62. $2x^2 + 8x^2$

63. $x - 9x$

64. $x - 5x$

65. $5x^3 + 6x^3 + 4$

66. $6x^4 - 2x^4 + 5$

67. $5x^3 + 6x - 4x^3 - 7x$

68. $3a^4 - 2a + 2a + a^4$

69. $6b^5 + 3b^2 - 2b^5 - 3b^2$

70. $2x^2 - 6x + 3x + 4x^2$

71. $\frac{1}{4}x^5 - 5 + \frac{1}{2}x^5 - 2x - 37$

72. $\frac{1}{3}x^3 + 2x - \frac{1}{6}x^3 + 4 - 16$

73. $6x^2 + 2x^4 - 2x^2 - x^4 - 4x^2$

74. $8x^2 + 2x^3 - 3x^3 - 4x^2 - 4x^2$

75. $\dfrac{1}{4}x^3 - x^2 - \dfrac{1}{6}x^2 + \dfrac{3}{8}x^3 + \dfrac{5}{16}x^3$

76. $\dfrac{1}{5}x^4 + \dfrac{1}{5} - 2x^2 + \dfrac{1}{10} - \dfrac{3}{15}x^4 + 2x^2 - \dfrac{3}{10}$

e   Arrange each polynomial in descending order.

77. $x^5 + x + 6x^3 + 1 + 2x^2$

78. $3 + 2x^2 - 5x^6 - 2x^3 + 3x$

79. $5y^3 + 15y^9 + y - y^2 + 7y^8$

80. $9p - 5 + 6p^3 - 5p^4 + p^5$

Collect like terms and then arrange in descending order.

81. $3x^4 - 5x^6 - 2x^4 + 6x^6$

82. $-1 + 5x^3 - 3 - 7x^3 + x^4 + 5$

83. $-2x + 4x^3 - 7x + 9x^3 + 8$

84. $-6x^2 + x - 5x + 7x^2 + 1$

85. $3x + 3x + 3x - x^2 - 4x^2$

86. $-2x - 2x - 2x + x^3 - 5x^3$

87. $-x + \dfrac{3}{4} + 15x^4 - x - \dfrac{1}{2} - 3x^4$

88. $2x - \dfrac{5}{6} + 4x^3 + x + \dfrac{1}{3} - 2x$

f   Identify the missing terms in each polynomial.

89. $x^3 - 27$

90. $x^5 + x$

91. $x^4 - x$

92. $5x^4 - 7x + 2$

93. $2x^3 - 5x^2 + x - 3$

94. $-6x^3$

Write each polynomial in two ways: with its missing terms and by leaving space for them.

95. $x^3 - 27$

96. $x^5 + x$

97. $x^4 - x$

98. $5x^4 - 7x + 2$

99. $2x^3 - 5x^2 + x - 3$

100. $-6x^3$

Skill Maintenance

Add. [1.3a]

101. $1 + (-20)$

102. $-\dfrac{2}{3} + \left(-\dfrac{1}{3}\right)$

103. $-4.2 + 1.95$

104. $-\dfrac{5}{8} + \dfrac{1}{4}$

Subtract. [1.4a]

105. $1 - 20$

106. $\dfrac{1}{8} - \dfrac{5}{6}$

107. $\dfrac{3}{8} - \left(-\dfrac{1}{4}\right)$

108. $5.6 - 8.2$

Multiply. [1.5a]

109. $(-6)(-3)$

110. $\left(-\dfrac{1}{2}\right)\left(\dfrac{2}{3}\right)$

111. $0.5\,(-1.2)$

112. $(-2)(-3)(-4)$

Divide, if possible. [1.6a, c]

113. $-600 \div (-30)$

114. $\dfrac{4}{5} \div \left(-\dfrac{1}{2}\right)$

115. $0 \div (-4)$

116. $\dfrac{-6.3}{-5 + 5}$

Synthesis

Collect like terms.

117. $6x^3 \cdot 7x^2 - (4x^3)^2 + (-3x^3)^2 - (-4x^2)(5x^3) - 10x^5 + 17x^6$

118. $(3x^2)^3 + 4x^2 \cdot 4x^4 - x^4(2x)^2 + ((2x)^2)^3 - 100x^2(x^2)^2$

119. Construct a polynomial in x (meaning that x is the variable) of degree 5 with four terms and coefficients that are integers.

120. What is the degree of $(5m^5)^2$?

121. A polynomial in x has degree 3. The coefficient of x^2 is 3 less than the coefficient of x^3. The coefficient of x is three times the coefficient of x^2. The remaining coefficient is 2 more than the coefficient of x^3. The sum of the coefficients is -4. Find the polynomial.

⚏ Use the CALC feature and choose VALUE on your graphing calculator to find the values in each of the following. (Refer to the Calculator Corner on p. 318.)

122. Exercise 18

123. Exercise 17

124. Exercise 22

125. Exercise 21

Addition and Subtraction of Polynomials

a ADDITION OF POLYNOMIALS

OBJECTIVES

a Add polynomials.

b Simplify the opposite of a polynomial.

c Subtract polynomials.

d Use polynomials to represent perimeter and area.

To add two polynomials, we can write a plus sign between them and then collect like terms. Depending on the situation, you may see polynomials written in descending order, ascending order, or neither. Generally, if an exercise is written in a particular order, we write the answer in that same order.

EXAMPLE 1 Add $(-3x^3 + 2x - 4)$ and $(4x^3 + 3x^2 + 2)$.

$(-3x^3 + 2x - 4) + (4x^3 + 3x^2 + 2)$
$= (-3 + 4)x^3 + 3x^2 + 2x + (-4 + 2)$ Collecting like terms
$= x^3 + 3x^2 + 2x - 2$

EXAMPLE 2 Add:

$$\left(\tfrac{2}{3}x^4 + 3x^2 - 2x + \tfrac{1}{2}\right) + \left(-\tfrac{1}{3}x^4 + 5x^3 - 3x^2 + 3x - \tfrac{1}{2}\right).$$

We have

$\left(\tfrac{2}{3}x^4 + 3x^2 - 2x + \tfrac{1}{2}\right) + \left(-\tfrac{1}{3}x^4 + 5x^3 - 3x^2 + 3x - \tfrac{1}{2}\right)$
$= \left(\tfrac{2}{3} - \tfrac{1}{3}\right)x^4 + 5x^3 + (3 - 3)x^2 + (-2 + 3)x + \left(\tfrac{1}{2} - \tfrac{1}{2}\right)$ Collecting like terms

$= \tfrac{1}{3}x^4 + 5x^3 + x.$

We can add polynomials as we do because they represent numbers. After some practice, you will be able to add mentally.

Do Margin Exercises 1–4. ▶

SKILL TO REVIEW

Objective 1.4a: Subtract real numbers and simplify combinations of additions and subtractions.

Simplify.
1. $-4 - (-8)$
2. $-5 - 6 + 4$

EXAMPLE 3 Add $(3x^2 - 2x + 2)$ and $(5x^3 - 2x^2 + 3x - 4)$.

$(3x^2 - 2x + 2) + (5x^3 - 2x^2 + 3x - 4)$
$= 5x^3 + (3 - 2)x^2 + (-2 + 3)x + (2 - 4)$ You might do this step mentally.
$= 5x^3 + x^2 + x - 2$ Then you would write only this.

Do Exercises 5 and 6 on the following page. ▶

We can also add polynomials by writing like terms in columns.

EXAMPLE 4 Add $9x^5 - 2x^3 + 6x^2 + 3$ and $5x^4 - 7x^2 + 6$ and $3x^6 - 5x^5 + x^2 + 5$.

We arrange the polynomials with the like terms in columns.

$$
\begin{array}{l}
\quad 9x^5 \qquad\quad - 2x^3 + 6x^2 + \;\;3 \\
\qquad\qquad 5x^4 \qquad\quad - 7x^2 + \;\;6 \\
\underline{3x^6 - 5x^5 \qquad\qquad\qquad + \;\;x^2 + \;\;5} \\
3x^6 + 4x^5 + 5x^4 - 2x^3 \qquad\quad + 14
\end{array}
$$

We leave spaces for missing terms.

Adding

We write the answer as $3x^6 + 4x^5 + 5x^4 - 2x^3 + 14$ without the space.

Add.

1. $(3x^2 + 2x - 2) + (-2x^2 + 5x + 5)$

2. $(-4x^5 + x^3 + 4) + (7x^4 + 2x^2)$

3. $(31x^4 + x^2 + 2x - 1) + (-7x^4 + 5x^3 - 2x + 2)$

4. $(17x^3 - x^2 + 3x + 4) + \left(-15x^3 + x^2 - 3x - \dfrac{2}{3}\right)$

Answers

Skill to Review:
1. 4   2. -7

Margin Exercises:
1. $x^2 + 7x + 3$
2. $-4x^5 + 7x^4 + x^3 + 2x^2 + 4$
3. $24x^4 + 5x^3 + x^2 + 1$
4. $2x^3 + \dfrac{10}{3}$

Add mentally. Try to write just the answer.

5. $(4x^2 - 5x + 3) + (-2x^2 + 2x - 4)$

6. $(3x^3 - 4x^2 - 5x + 3) + \left(5x^3 + 2x^2 - 3x - \dfrac{1}{2}\right)$

Add.

7.
$$\begin{array}{r} -2x^3 + 5x^2 - 2x + 4 \\ x^4 \qquad\quad + 6x^2 + 7x - 10 \\ -9x^4 + 6x^3 + x^2 \qquad\quad - 2 \\ \hline \end{array}$$

8. $-3x^3 + 5x + 2$ and $x^3 + x^2 + 5$ and $x^3 - 2x - 4$

Simplify.

9. $-(4x^3 - 6x + 3)$

10. $-(5x^4 + 3x^2 + 7x - 5)$

11. $-(14x^{10} - \frac{1}{2}x^5 + 5x^3 - x^2 + 3x)$

◀ **Do Exercises 7 and 8.**

b ## OPPOSITES OF POLYNOMIALS

We can use the property of -1 to write an equivalent expression for an opposite. For example, the opposite of $x - 2y + 5$ can be written as

$$-(x - 2y + 5).$$

We find an equivalent expression by changing the sign of every term:

$$-(x - 2y + 5) = -x + 2y - 5.$$

We use this concept when we subtract polynomials.

> ### OPPOSITES OF POLYNOMIALS
>
> To find an equivalent polynomial for the **opposite**, or **additive inverse**, of a polynomial, change the sign of every term. This is the same as multiplying by -1.

EXAMPLE 5 Simplify: $-(x^2 - 3x + 4)$.

$$-(x^2 - 3x + 4) = -x^2 + 3x - 4$$

EXAMPLE 6 Simplify: $-(-t^3 - 6t^2 - t + 4)$.

$$-(-t^3 - 6t^2 - t + 4) = t^3 + 6t^2 + t - 4$$

EXAMPLE 7 Simplify: $-(-7x^4 - \frac{5}{9}x^3 + 8x^2 - x + 67)$.

$$-(-7x^4 - \tfrac{5}{9}x^3 + 8x^2 - x + 67) = 7x^4 + \tfrac{5}{9}x^3 - 8x^2 + x - 67$$

◀ **Do Exercises 9–11.**

c ## SUBTRACTION OF POLYNOMIALS

Recall that we can subtract a real number by adding its opposite, or additive inverse: $a - b = a + (-b)$. This allows us to subtract polynomials.

EXAMPLE 8 Subtract:

$$(9x^5 + x^3 - 2x^2 + 4) - (2x^5 + x^4 - 4x^3 - 3x^2).$$

We have

$$(9x^5 + x^3 - 2x^2 + 4) - (2x^5 + x^4 - 4x^3 - 3x^2)$$
$$= 9x^5 + x^3 - 2x^2 + 4 + [-(2x^5 + x^4 - 4x^3 - 3x^2)] \qquad \text{Adding the opposite}$$
$$= 9x^5 + x^3 - 2x^2 + 4 - 2x^5 - x^4 + 4x^3 + 3x^2 \qquad \text{Finding the opposite by changing the sign of } each \text{ term}$$
$$= 7x^5 - x^4 + 5x^3 + x^2 + 4. \qquad \text{Adding (collecting like terms)}$$

Subtract.

12. $(7x^3 + 2x + 4) - (5x^3 - 4)$

13. $(-3x^2 + 5x - 4) - (-4x^2 + 11x - 2)$

◀ **Do Exercises 12 and 13.**

Answers

5. $2x^2 - 3x - 1$ **6.** $8x^3 - 2x^2 - 8x + \dfrac{5}{2}$
7. $-8x^4 + 4x^3 + 12x^2 + 5x - 8$
8. $-x^3 + x^2 + 3x + 3$ **9.** $-4x^3 + 6x - 3$
10. $-5x^4 - 3x^2 - 7x + 5$
11. $-14x^{10} + \dfrac{1}{2}x^5 - 5x^3 + x^2 - 3x$
12. $2x^3 + 2x + 8$ **13.** $x^2 - 6x - 2$

We combine steps by changing the sign of each term of the polynomial being subtracted and collecting like terms. Try to do this mentally as much as possible.

EXAMPLE 9 Subtract: $(9x^5 + x^3 - 2x) - (-2x^5 + 5x^3 + 6)$.

$$(9x^5 + x^3 - 2x) - (-2x^5 + 5x^3 + 6)$$
$$= 9x^5 + x^3 - 2x + 2x^5 - 5x^3 - 6 \qquad \text{Finding the opposite by changing the sign of each term}$$

$$= 11x^5 - 4x^3 - 2x - 6 \qquad \text{Collecting like terms}$$

Do Exercises 14 and 15. ▶

We can use columns to subtract. We replace coefficients with their opposites, as shown in Example 9.

EXAMPLE 10 Write in columns and subtract:

$$(5x^2 - 3x + 6) - (9x^2 - 5x - 3).$$

a)
$$5x^2 - 3x + 6 \qquad \text{Writing like terms in columns}$$
$$\underline{-(9x^2 - 5x - 3)}$$

b)
$$5x^2 - 3x + 6$$
$$\underline{-9x^2 + 5x + 3} \qquad \text{Changing signs}$$

c)
$$5x^2 - 3x + 6$$
$$\underline{-9x^2 + 5x + 3}$$
$$-4x^2 + 2x + 9 \qquad \text{Adding}$$

If you can do so without error, you can arrange the polynomials in columns and write just the answer, remembering to change the signs and add.

EXAMPLE 11 Write in columns and subtract:

$$(x^3 + x^2 + 2x - 12) - (-2x^3 + x^2 - 3x).$$

$$x^3 + x^2 + 2x - 12$$
$$\underline{-(-2x^3 + x^2 - 3x \qquad)} \qquad \text{Leaving space for the missing term}$$
$$3x^3 \qquad\quad + 5x - 12 \qquad \text{Changing the signs and adding}$$

Do Exercises 16 and 17. ▶

d POLYNOMIALS AND GEOMETRY

EXAMPLE 12 Find a polynomial for the sum of the areas of these four rectangles.

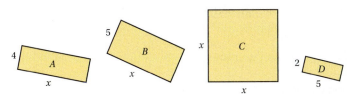

Recall that the area of a rectangle is the product of the length and the width. The sum of the areas is a sum of products. We find these products and then collect like terms.

Subtract.

14. GS $(-6x^4 + 3x^2 + 6) - (2x^4 + 5x^3 - 5x^2 + 7)$

$$= -6x^4 + 3x^2 + 6 - $$
$$2x^4 \;\boxed{}\; 5x^3 \;\boxed{}\; 5x^2 \;\boxed{}\; 7$$
$$= \boxed{}\; x^4 - 5x^3 + \boxed{}\; x^2$$
$$- \boxed{}$$

15. $\left(\dfrac{3}{2}x^3 - \dfrac{1}{2}x^2 + 0.3\right) -$
$\left(\dfrac{1}{2}x^3 + \dfrac{1}{2}x^2 + \dfrac{4}{3}x + 1.2\right)$

Write in columns and subtract.

16. $(4x^3 + 2x^2 - 2x - 3) - (2x^3 - 3x^2 + 2)$

17. $(2x^3 + x^2 - 6x + 2) - (x^5 + 4x^3 - 2x^2 - 4x)$

Answers

14. $-8x^4 - 5x^3 + 8x^2 - 1$

15. $x^3 - x^2 - \dfrac{4}{3}x - 0.9$

16. $2x^3 + 5x^2 - 2x - 5$

17. $-x^5 - 2x^3 + 3x^2 - 2x + 2$

Guided Solution:

14. $-, +, -, -8, 8, 1$

18. Find a polynomial for the sums of the perimeters and of the areas of the rectangles.

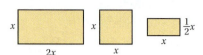

x	$2x$
x	x
	$\frac{1}{2}x$
	x

19. *Lawn Area.* An 8-ft by 8-ft shed is placed on a lawn x ft on a side. Find a polynomial for the remaining area.

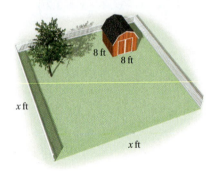

8 ft
8 ft
x ft
x ft

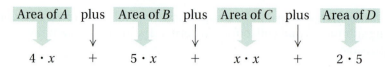

Area of A	plus	Area of B	plus	Area of C	plus	Area of D
$4 \cdot x$	$+$	$5 \cdot x$	$+$	$x \cdot x$	$+$	$2 \cdot 5$

We collect like terms:

$$4x + 5x + x^2 + 10 = x^2 + 9x + 10.$$

◀ **Do Exercise 18.**

EXAMPLE 13 *Lawn Area.* A new city park is to contain a square grassy area that is x ft on a side. Within that grassy area will be a circular playground, with a radius of 15 ft, that will be mulched. To determine the amount of sod needed, find a polynomial for the grassy area.

We make a drawing, reword the problem, and write the polynomial.

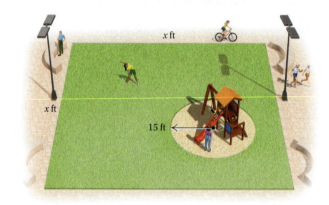

x ft

x ft

15 ft

Area of square	$-$	Area of playground	$=$	Area of grass
$x \cdot x \text{ ft}^2$	$-$	$\pi \cdot 15^2 \text{ ft}^2$	$=$	Area of grass

Then $(x^2 - 225\pi)\,\text{ft}^2 = $ Area of grass.

◀ **Do Exercise 19.**

Answers

18. Sum of perimeters: $13x$; sum of areas: $\frac{7}{2}x^2$

19. $(x^2 - 64)\,\text{ft}^2$

4.4 **Exercise Set**

For Extra Help

MyMathLab

MathXL® PRACTICE

WATCH

READ

REVIEW

✓ Reading Check

Determine whether each statement is true or false.

RC1. To find the opposite of a polynomial, we need change only the sign of the first term.

RC2. We can subtract a polynomial by adding its opposite.

RC3. The sum of two binomials is always a binomial.

RC4. The area of a rectangle is the sum of its length and its width.

1. $(3x + 2) + (-4x + 3)$

2. $(6x + 1) + (-7x + 2)$

3. $(-6x + 2) + \left(x^2 + \frac{1}{2}x - 3\right)$

4. $\left(x^2 - \frac{5}{3}x + 4\right) + (8x - 9)$

5. $(x^2 - 9) + (x^2 + 9)$

6. $(x^3 + x^2) + (2x^3 - 5x^2)$

7. $(3x^2 - 5x + 10) + (2x^2 + 8x - 40)$

8. $(6x^4 + 3x^3 - 1) + (4x^2 - 3x + 3)$

9. $(1.2x^3 + 4.5x^2 - 3.8x) + (-3.4x^3 - 4.7x^2 + 23)$

10. $(0.5x^4 - 0.6x^2 + 0.7) + (2.3x^4 + 1.8x - 3.9)$

11. $(1 + 4x + 6x^2 + 7x^3) + (5 - 4x + 6x^2 - 7x^3)$

12. $(3x^4 - 6x - 5x^2 + 5) + (6x^2 - 4x^3 - 1 + 7x)$

13. $\left(\frac{1}{4}x^4 + \frac{2}{3}x^3 + \frac{5}{8}x^2 + 7\right) + \left(-\frac{3}{4}x^4 + \frac{3}{8}x^2 - 7\right)$

14. $\left(\frac{1}{3}x^9 + \frac{1}{5}x^5 - \frac{1}{2}x^2 + 7\right) + \left(-\frac{1}{5}x^9 + \frac{1}{4}x^4 - \frac{3}{5}x^5 + \frac{3}{4}x^2 + \frac{1}{2}\right)$

15. $(0.02x^5 - 0.2x^3 + x + 0.08) + (-0.01x^5 + x^4 - 0.8x - 0.02)$

16. $(0.03x^6 + 0.05x^3 + 0.22x + 0.05) + \left(\frac{7}{100}x^6 - \frac{3}{100}x^3 + 0.5\right)$

17. $(9x^8 - 7x^4 + 2x^2 + 5) + (8x^7 + 4x^4 - 2x) + (-3x^4 + 6x^2 + 2x - 1)$

18. $(4x^5 - 6x^3 - 9x + 1) + (6x^3 + 9x^2 + 9x) + (-4x^3 + 8x^2 + 3x - 2)$

19.

$$
\begin{array}{rrrr}
0.15x^4 & + 0.10x^3 & - 0.9x^2 & \\
& - 0.01x^3 & + 0.01x^2 & + x \\
1.25x^4 & & + 0.11x^2 & + 0.01 \\
& 0.27x^3 & & + 0.99 \\
-0.35x^4 & & + 15x^2 & - 0.03 \\
\hline
\end{array}
$$

20.

$$
\begin{array}{rrrr}
0.05x^4 & + 0.12x^3 & - 0.5x^2 & \\
& - 0.02x^3 & + 0.02x^2 & + 2x \\
1.5x^4 & & + 0.01x^2 & + 0.15 \\
& 0.25x^3 & & + 0.85 \\
-0.25x^4 & & + 10x^2 & - 0.04 \\
\hline
\end{array}
$$

21. $-(-5x)$

22. $-(x^2 - 3x)$

23. $-\left(-x^2 + \frac{3}{2}x - 2\right)$

24. $-\left(-4x^3 - x^2 - \frac{1}{4}x\right)$

25. $-(12x^4 - 3x^3 + 3)$

26. $-(4x^3 - 6x^2 - 8x + 1)$

27. $-(3x - 7)$

28. $-(-2x + 4)$

29. $-(4x^2 - 3x + 2)$

30. $-(-6a^3 + 2a^2 - 9a + 1)$

31. $-\left(-4x^4 + 6x^2 + \frac{3}{4}x - 8\right)$

32. $-(-5x^4 + 4x^3 - x^2 + 0.9)$

c Subtract.

33. $(3x + 2) - (-4x + 3)$

34. $(6x + 1) - (-7x + 2)$

35. $(-6x + 2) - (x^2 + x - 3)$

36. $(x^2 - 5x + 4) - (8x - 9)$

37. $(x^2 - 9) - (x^2 + 9)$

38. $(x^3 + x^2) - (2x^3 - 5x^2)$

39. $(6x^4 + 3x^3 - 1) - (4x^2 - 3x + 3)$

40. $(-4x^2 + 2x) - (3x^3 - 5x^2 + 3)$

41. $(1.2x^3 + 4.5x^2 - 3.8x) - (-3.4x^3 - 4.7x^2 + 23)$

42. $(0.5x^4 - 0.6x^2 + 0.7) - (2.3x^4 + 1.8x - 3.9)$

43. $\left(\frac{5}{8}x^3 - \frac{1}{4}x - \frac{1}{3}\right) - \left(-\frac{1}{8}x^3 + \frac{1}{4}x - \frac{1}{3}\right)$

44. $\left(\frac{1}{5}x^3 + 2x^2 - 0.1\right) - \left(-\frac{2}{5}x^3 + 2x^2 + 0.01\right)$

45. $(0.08x^3 - 0.02x^2 + 0.01x) - (0.02x^3 + 0.03x^2 - 1)$

46. $(0.8x^4 + 0.2x - 1) - \left(\frac{7}{10}x^4 + \frac{1}{5}x - 0.1\right)$

Subtract.

47.
$$\begin{array}{r} x^2 + 5x + 6 \\ -(x^2 + 2x) \\ \hline \end{array}$$

48.
$$\begin{array}{r} x^3 + 1 \\ -(x^3 + x^2) \\ \hline \end{array}$$

49.
$$\begin{array}{r} 5x^4 + 6x^3 - 9x^2 \\ -(-6x^4 - 6x^3 + 8x + 9) \\ \hline \end{array}$$

50.
$$\begin{array}{r} 5x^4 + 6x^2 - 3x + 6 \\ -(6x^3 + 7x^2 - 8x - 9) \\ \hline \end{array}$$

51.
$$\begin{array}{r} x^5 - 1 \\ -(x^5 - x^4 + x^3 - x^2 + x - 1) \\ \hline \end{array}$$

52.
$$\begin{array}{r} x^5 + x^4 - x^3 + x^2 - x + 2 \\ -(x^5 - x^4 + x^3 - x^2 - x + 2) \\ \hline \end{array}$$

d Solve.

Find a polynomial for the perimeter of each figure.

53.

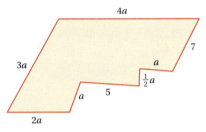

54.

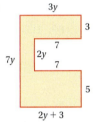

55. Find a polynomial for the sum of the areas of these rectangles.

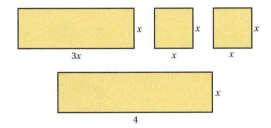

56. Find a polynomial for the sum of the areas of these circles.

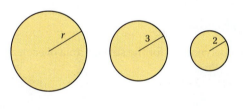

Find two algebraic expressions for the area of each figure. First, regard the figure as one large rectangle, and then regard the figure as a sum of four smaller rectangles.

57.

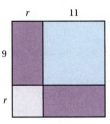

58.

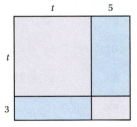

59.

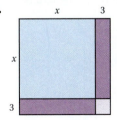

60.

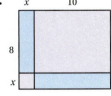

Find a polynomial for the shaded area of each figure.

61.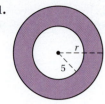

62.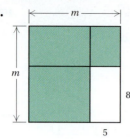

63.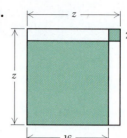

64.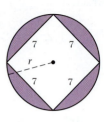

Skill Maintenance

Solve.   [2.3b]

65. $8x + 3x = 66$

66. $5x - 7x = 38$

67. $\frac{3}{8}x + \frac{1}{4} - \frac{3}{4}x = \frac{11}{16} + x$

68. $5x - 4 = 26 - x$

69. $1.5x - 2.7x = 22 - 5.6x$

70. $3x - 3 = -4x + 4$

Solve.   [2.3c]

71. $6(y - 3) - 8 = 4(y + 2) + 5$

72. $8(5x + 2) = 7(6x - 3)$

Solve.   [2.7e]

73. $3x - 7 \leq 5x + 13$

74. $2(x - 4) > 5(x - 3) + 7$

Synthesis

Find a polynomial for the surface area of each right rectangular solid.

75.

76.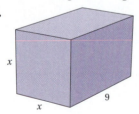

77.

78.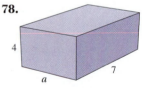

79. Find $(y - 2)^2$ using the four parts of this square.

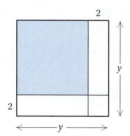

Simplify.

80. $(3x^2 - 4x + 6) - (-2x^2 + 4) + (-5x - 3)$

81. $(7y^2 - 5y + 6) - (3y^2 + 8y - 12) + (8y^2 - 10y + 3)$

82. $(-4 + x^2 + 2x^3) - (-6 - x + 3x^3) - (-x^2 - 5x^3)$

83. $(-y^4 - 7y^3 + y^2) + (-2y^4 + 5y - 2) - (-6y^3 + y^2)$

Mid-Chapter Review

Concept Reinforcement

Determine whether each statement is true or false.

_____ **1.** a^n and a^{-n} are reciprocals. [4.1f]

_____ **2.** $x^2 \cdot x^3 = x^6$ [4.1d]

_____ **3.** $-5y^4$ and $-5y^2$ are like terms. [4.3d]

_____ **4.** $4920^0 = 1$ [4.1b]

Guided Solutions

 Fill in each blank with the expression or operation sign that creates a correct statement or solution.

5. Collect like terms: $4w^3 + 6w - 8w^3 - 3w$. [4.3d]

$$4w^3 + 6w - 8w^3 - 3w = (4 - 8)\ \boxed{} + (6 - 3)\ \boxed{}$$
$$= \boxed{}\ w^3 + \boxed{}\ w$$

6. Subtract: $(3y^4 - y^2 + 11) - (y^4 - 4y^2 + 5)$. [4.4c]

$$(3y^4 - y^2 + 11) - (y^4 - 4y^2 + 5) = 3y^4 - y^2 + 11\ \boxed{}\ y^4\ \boxed{}\ 4y^2\ \boxed{}\ 5$$
$$= \boxed{}\ y^4 + \boxed{}\ y^2 + \boxed{}$$

Mixed Review

Evaluate. [4.1b, c]

7. z^1

8. 4.56^0

9. a^5, when $a = -2$

10. $-x^3$, when $x = -1$

Multiply and simplify. [4.1d, f]

11. $5^3 \cdot 5^4$

12. $(3a)^2(3a)^7$

13. $x^{-8} \cdot x^5$

14. $t^4 \cdot t^{-4}$

Divide and simplify. [4.1e, f]

15. $\dfrac{7^8}{7^4}$

16. $\dfrac{x}{x^3}$

17. $\dfrac{w^5}{w^{-3}}$

18. $\dfrac{y^{-6}}{y^{-2}}$

Simplify. [4.2a, b]

19. $(3^5)^3$

20. $(x^{-3}y^2)^{-6}$

21. $\left(\dfrac{a^4}{5}\right)^6$

22. $\left(\dfrac{2y^3}{xz^2}\right)^{-2}$

Convert to scientific notation. [4.2c]

23. $25{,}430{,}000$

24. 0.00012

Convert to decimal notation. [4.2c]

25. 3.6×10^{-5}

26. 1.44×10^8

Multiply or divide and write scientific notation for the result.   [4.2d]

27. $(3 \times 10^6)(2 \times 10^{-3})$ **28.** $\dfrac{1.2 \times 10^{-4}}{2.4 \times 10^2}$

Evaluate the polynomial when $x = -3$ and when $x = 2$.   [4.3a]

29. $-3x + 7$ **30.** $x^3 - 2x + 5$

Collect like terms and then arrange in descending order.   [4.3e]

31. $3x - 2x^5 + x - 5x^2 + 2$ **32.** $4x^3 - 9x^2 - 2x^3 + x^2 + 8x^6$

Identify the degree of each term of the polynomial and the degree of the polynomial.   [4.3c]

33. $5x^3 - x + 4$ **34.** $2x - x^4 + 3x^6$

Classify the polynomial as a monomial, a binomial, a trinomial, or none of these.   [4.3b]

35. $x - 9$ **36.** $x^5 - 2x^3 + 6x^2$

Add or subtract.   [4.4a, c]

37. $(3x^2 - 1) + (5x^2 + 6)$ **38.** $(x^3 + 2x - 5) + (4x^3 - 2x^2 - 6)$

39. $(5x - 8) - (9x + 2)$ **40.** $(0.1x^2 - 2.4x + 3.6) - (0.5x^2 + x - 5.4)$

41. Find a polynomial for the sum of the areas of these rectangles.   [4.4d]

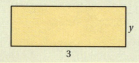

Understanding Through Discussion and Writing

42. Suppose that the length of a side of a square is three times the length of a side of a second square. How do the areas of the squares compare? Why?   [4.1d]

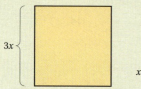

43. Suppose that the length of a side of a cube is twice the length of a side of a second cube. How do the volumes of the cubes compare? Why?   [4.1d]

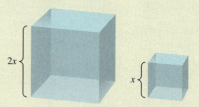

44. Explain in your own words when exponents should be added and when they should be multiplied. [4.1d], [4.2a]

45. Without performing actual computations, explain why 3^{-29} is smaller than 2^{-29}.   [4.1f]

46. Is it better to evaluate a polynomial before or after like terms have been collected? Why?   [4.3a, d]

47. Is the sum of two binomials ever a trinomial? Why or why not?   [4.3b], [4.4a]

Multiplication of Polynomials

4.5

We now multiply polynomials using techniques based, for the most part, on the distributive laws, but also on the associative and commutative laws.

a   MULTIPLYING MONOMIALS

Consider $(3x)(4x)$. We multiply as follows:

$$
\begin{aligned}
(3x)(4x) &= 3 \cdot x \cdot 4 \cdot x & \text{By the associative law of multiplication} \\
&= 3 \cdot 4 \cdot x \cdot x & \text{By the commutative law of multiplication} \\
&= (3 \cdot 4)(x \cdot x) & \text{By the associative law} \\
&= 12x^2. & \text{Using the product rule for exponents}
\end{aligned}
$$

> ### MULTIPLYING MONOMIALS
>
> To find an equivalent expression for the product of two monomials, multiply the coefficients and then multiply the variables using the product rule for exponents.

EXAMPLES   Multiply.

1. $5x \cdot 6x = (5 \cdot 6)(x \cdot x)$ By the associative and commutative laws
$= 30x^2$ Multiplying the coefficients and multiplying the variables

2. $(3x)(-x) = (3x)(-1x) = (3)(-1)(x \cdot x) = -3x^2$

3. $(-7y^5)(4y^3) = (-7 \cdot 4)(y^5 \cdot y^3)$
$= -28y^{5+3} = -28y^8$ Adding exponents

After some practice, you will be able to multiply mentally.

Do Margin Exercises 1–8. ▶

b   MULTIPLYING A MONOMIAL AND ANY POLYNOMIAL

To multiply a monomial, such as $2x$, and a binomial, such as $5x + 3$, we use a distributive law and multiply each term of $5x + 3$ by $2x$:

$$
\begin{aligned}
2x(5x + 3) &= (2x)(5x) + (2x)(3) & \text{Using a distributive law} \\
&= 10x^2 + 6x. & \text{Multiplying the monomials}
\end{aligned}
$$

OBJECTIVES

a	Multiply monomials.
b	Multiply a monomial and any polynomial.
c	Multiply two binomials.
d	Multiply any two polynomials.

SKILL TO REVIEW

Objective 1.7c: Use the distributive laws to multiply expressions like 8 and $x - y$.

Multiply.
1. $3(x - 5)$
2. $2(3y + 4z - 1)$

Multiply.

1. $(3x)(-5)$ **2.** $(-x) \cdot x$

3. $(-x)(-x)$ **4.** $(-x^2)(x^3)$

5. $3x^5 \cdot 4x^2$

6. $(4y^5)(-2y^6)$

7. $(-7y^4)(-y)$ **8.** $7x^5 \cdot 0$

EXAMPLE 4 Multiply: $5x(2x^2 - 3x + 4)$.

$$5x(2x^2 - 3x + 4) = (5x)(2x^2) - (5x)(3x) + (5x)(4)$$
$$= 10x^3 - 15x^2 + 20x$$

Multiply.

9. $4x(2x + 4)$

10. $3t^2(-5t + 2)$

11. $-5x^3(x^3 + 5x^2 - 6x + 8)$

> ### MULTIPLYING A MONOMIAL AND A POLYNOMIAL
>
> To multiply a monomial and a polynomial, multiply each term of the polynomial by the monomial.

EXAMPLE 5 Multiply: $-2x^2(x^3 - 7x^2 + 10x - 4)$.

$$-2x^2(x^3 - 7x^2 + 10x - 4)$$
$$= (-2x^2)(x^3) - (-2x^2)(7x^2) + (-2x^2)(10x) - (-2x^2)(4)$$
$$= -2x^5 + 14x^4 - 20x^3 + 8x^2$$

◀ **Do Exercises 9–11.**

12. a) Multiply: $(y + 2)(y + 7)$.

$(y + 2)(y + 7)$
$= y \cdot (y + 7) + 2 \cdot (\boxed{})$
$= y \cdot y + y \cdot \boxed{} + 2 \cdot y + 2 \cdot \boxed{}$
$= \boxed{} + 7y + 2y + \boxed{}$
$= y^2 + \boxed{} + 14$

b) Write an algebraic expression that represents the total area of the four smaller rectangles in the figure shown here.

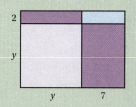

The area is $(y + 7)(y + \boxed{})$, or, from part (a), $y^2 + \boxed{} + 14$.

c MULTIPLYING TWO BINOMIALS

To find an equivalent expression for the product of two binomials, we use the distributive laws more than once. In Example 6, we use a distributive law three times.

EXAMPLE 6 Multiply: $(x + 5)(x + 4)$.

$$(x + 5)(x + 4) = x(x + 4) + 5(x + 4) \quad \text{Using a distributive law}$$
$$= x \cdot x + x \cdot 4 + 5 \cdot x + 5 \cdot 4 \quad \text{Using a distributive law twice}$$
$$= x^2 + 4x + 5x + 20 \quad \text{Multiplying the monomials}$$
$$= x^2 + 9x + 20 \quad \text{Collecting like terms}$$

To visualize the product in Example 6, consider a rectangle of length $x + 5$ and width $x + 4$.

Multiply.

13. $(x + 8)(x + 5)$

14. $(x + 5)(x - 4)$

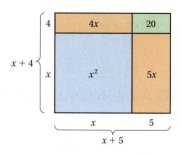

The total area can be expressed as $(x + 5)(x + 4)$ or, by adding the four smaller areas, $x^2 + 4x + 5x + 20$, or $x^2 + 9x + 20$.

◀ **Do Exercises 12–14.**

Answers

9. $8x^2 + 16x$ **10.** $-15t^3 + 6t^2$
11. $-5x^6 - 25x^5 + 30x^4 - 40x^3$
12. (a) $y^2 + 9y + 14$; (b) $(y + 2)(y + 7)$, or $y^2 + 2y + 7y + 14$, or $y^2 + 9y + 14$
13. $x^2 + 13x + 40$ **14.** $x^2 + x - 20$

Guided Solution:
12. (a) $y + 7, 7, 7, y^2, 14, 9y$; (b) $2, 9y$

EXAMPLE 7 Multiply: $(4x + 3)(x - 2)$.

$(4x + 3)(x - 2) = 4x(x - 2) + 3(x - 2)$ Using a distributive law

$= 4x \cdot x - 4x \cdot 2 + 3 \cdot x - 3 \cdot 2$ Using a distributive law twice

$= 4x^2 - 8x + 3x - 6$ Multiplying the monomials

$= 4x^2 - 5x - 6$ Collecting like terms

Do Exercises 15 and 16. ▶

Multiply.

15. $(5x + 3)(x - 4)$

16. $(2x - 3)(3x - 5)$

d   MULTIPLYING ANY TWO POLYNOMIALS

Let's consider the product of a binomial and a trinomial. We use a distributive law four times. You may see ways to skip some steps and do the work mentally.

EXAMPLE 8 Multiply: $(x^2 + 2x - 3)(x^2 + 4)$.

$(x^2 + 2x - 3)(x^2 + 4) = x^2(x^2 + 4) + 2x(x^2 + 4) - 3(x^2 + 4)$

$= x^2 \cdot x^2 + x^2 \cdot 4 + 2x \cdot x^2 + 2x \cdot 4 - 3 \cdot x^2 - 3 \cdot 4$

$= x^4 + 4x^2 + 2x^3 + 8x - 3x^2 - 12$

$= x^4 + 2x^3 + x^2 + 8x - 12$

Do Exercises 17 and 18. ▶

Multiply.

17. $(x^2 + 3x - 4)(x^2 + 5)$

GS **18.** $(3y^2 - 7)(2y^3 - 2y + 5)$

$= 3y^2(2y^3 - 2y + 5) - \boxed{}(2y^3 - 2y + 5)$

$= 6y^5 - \boxed{} + 15y^2 - 14y^3 + \boxed{} - 35$

$= 6y^5 - \boxed{} + 15y^2 + 14y - 35$

PRODUCT OF TWO POLYNOMIALS

To multiply two polynomials P and Q, select one of the polynomials—say, P. Then multiply each term of P by every term of Q and collect like terms.

To use columns for long multiplication, multiply each term in the top row by every term in the bottom row. We write like terms in columns, and then add the results. Such multiplication is like multiplying with whole numbers.

$$
\begin{array}{r}
3\ 2\ 1 \\
\times\ \ 1\ 2 \\
\hline
6\ 4\ 2 \\
3\ 2\ 1 \\
\hline
3\ 8\ 5\ 2
\end{array}
\qquad
\begin{array}{r}
300 + 20 + 1 \\
\times\ \ \ \ \ \ \ \ \ \ 10 + 2 \\
\hline
600 + 40 + 2 \\
3000 + 200 + 10 \\
\hline
3000 + 800 + 50 + 2
\end{array}
$$

Multiplying the top row by 2
Multiplying the top row by 10
Adding

EXAMPLE 9 Multiply: $(4x^3 - 2x^2 + 3x)(x^2 + 2x)$.

$$
\begin{array}{r}
4x^3 - 2x^2 + 3x \\
x^2 + 2x \\
\hline
8x^4 - 4x^3 + 6x^2 \\
4x^5 - 2x^4 + 3x^3 \\
\hline
4x^5 + 6x^4 -\ \ x^3 + 6x^2
\end{array}
$$

Multiplying the top row by $2x$
Multiplying the top row by x^2
Collecting like terms
Line up like terms in columns.

EXAMPLE 10 Multiply: $(2x^2 + 3x - 4)(2x^2 - x + 3)$.

$$
\begin{array}{r}
2x^2 + 3x - 4 \\
2x^2 - x + 3 \\
\hline
6x^2 + 9x - 12 \\
-2x^3 - 3x^2 + 4x \\
4x^4 + 6x^3 - 8x^2 \\
\hline
4x^4 + 4x^3 - 5x^2 + 13x - 12
\end{array}
$$

Multiplying by 3
Multiplying by $-x$
Multiplying by $2x^2$
Collecting like terms

19. Multiply.

$$
\begin{array}{r}
3x^2 - 2x - 5 \\
2x^2 + x - 2 \\
\hline
\end{array}
$$

◀ **Do Exercise 19.**

Multiply.

20. $3x^2 - 2x + 4$
$ x + 5$

21. $-5x^2 + 4x + 2$
$ -4x^2 - 8$

EXAMPLE 11 Multiply: $(5x^3 - 3x + 4)(-2x^2 - 3)$.

If terms are missing, it helps to leave spaces for them and align like terms in columns as we multiply.

$$
\begin{array}{r}
5x^3 - 3x + 4 \\
-2x^2 - 3 \\
\hline
-15x^3 + 9x - 12 \\
-10x^5 + 6x^3 - 8x^2 \\
\hline
-10x^5 - 9x^3 - 8x^2 + 9x - 12
\end{array}
$$

Multiplying by -3
Multiplying by $-2x^2$
Collecting like terms

◀ **Do Exercises 20 and 21.**

Answers

19. $6x^4 - x^3 - 18x^2 - x + 10$
20. $3x^3 + 13x^2 - 6x + 20$
21. $20x^4 - 16x^3 + 32x^2 - 32x - 16$

CALCULATOR CORNER

Checking Multiplication of Polynomials A partial check of multiplication of polynomials can be performed graphically. Consider the product $(x + 3)(x - 2) = x^2 + x - 6$. We will use two graph styles to determine whether this product is correct. First, we press **MODE** and select the **SEQUENTIAL** mode.

Next, on the Y= screen, we enter $y_1 = (x + 3)(x - 2)$ and $y_2 = x^2 + x - 6$. We will select the line-graph style for y_1 and the path style for y_2. To select these graph styles, we use ◁ to position the cursor over the icon to the left of the equation and press **ENTER** repeatedly until the desired style of icon appears, as shown below. Then we graph the equations.

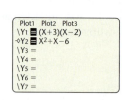
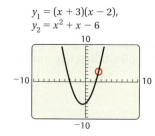

$y_1 = (x + 3)(x - 2),$
$y_2 = x^2 + x - 6$

The graphing calculator will graph y_1 first as a solid line. Then it will graph y_2 as the circular cursor traces the leading edge of the graph, allowing us to determine visually whether the graphs coincide. In this case, the graphs appear to coincide, so the factorization is probably correct.

A table of values can also be used as a check.

EXERCISES: Determine graphically whether each product is correct.

1. $(x + 5)(x + 4) = x^2 + 9x + 20$

2. $(4x + 3)(x - 2) = 4x^2 - 5x - 6$

3. $(5x + 3)(x - 4) = 5x^2 + 17x - 12$

4. $(2x - 3)(3x - 5) = 6x^2 - 19x - 15$

✓ Reading Check

Match each expression with an equivalent expression from the column on the right. Choices may be used more than once or not at all.

RC1. $8x \cdot 2x$

RC2. $(-16x)(-x)$

RC3. $2x(8x - 1)$

RC4. $(2x - 1)(8x + 1)$

a) $16x^2$
b) $-16x^2$
c) $16x^2 - 1$
d) $16x^2 - 2x$
e) $16x^2 - 6x - 1$

a Multiply.

1. $(8x^2)(5)$

2. $(4x^2)(-2)$

3. $(-x^2)(-x)$

4. $(-x^3)(x^2)$

5. $(8x^5)(4x^3)$

6. $(10a^2)(2a^2)$

7. $(0.1x^6)(0.3x^5)$

8. $(0.3x^4)(-0.8x^6)$

9. $\left(-\frac{1}{5}x^3\right)\left(-\frac{1}{3}x\right)$

10. $\left(-\frac{1}{4}x^4\right)\left(\frac{1}{5}x^8\right)$

11. $(-4x^2)(0)$

12. $(-4m^5)(-1)$

13. $(3x^2)(-4x^3)(2x^6)$

14. $(-2y^5)(10y^4)(-3y^3)$

b Multiply.

15. $2x(-x + 5)$

16. $3x(4x - 6)$

17. $-5x(x - 1)$

18. $-3x(-x - 1)$

19. $x^2(x^3 + 1)$

20. $-2x^3(x^2 - 1)$

21. $3x(2x^2 - 6x + 1)$

22. $-4x(2x^3 - 6x^2 - 5x + 1)$

23. $(-6x^2)(x^2 + x)$

24. $(-4x^2)(x^2 - x)$

25. $(3y^2)(6y^4 + 8y^3)$

26. $(4y^4)(y^3 - 6y^2)$

c Multiply.

27. $(x + 6)(x + 3)$

28. $(x + 5)(x + 2)$

29. $(x + 5)(x - 2)$

30. $(x + 6)(x - 2)$

31. $(x - 1)(x + 4)$

32. $(x - 8)(x + 7)$

33. $(x - 4)(x - 3)$

34. $(x - 7)(x - 3)$

35. $(x + 3)(x - 3)$

36. $(x + 6)(x - 6)$

37. $(x - 4)(x + 4)$

38. $(x - 9)(x + 9)$

39. $(3x + 5)(x + 2)$

40. $(2x + 6)(x + 3)$

41. $(5 - x)(5 - 2x)$

42. $(3 - 4x)(2 - x)$

43. $(2x + 5)(2x + 5)$ **44.** $(3x + 4)(3x + 4)$ **45.** $(x - 3)(x - 3)$ **46.** $(x - 6)(x - 6)$

47. $\left(x - \frac{5}{2}\right)\left(x + \frac{2}{5}\right)$ **48.** $\left(x + \frac{4}{3}\right)\left(x + \frac{3}{2}\right)$ **49.** $(x - 2.3)(x + 4.7)$ **50.** $(2x + 0.13)(2x - 0.13)$

Write an algebraic expression that represents the total area of the four smaller rectangles in each figure.

51. 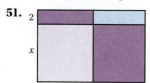 **52.** 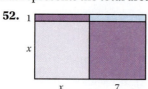 **53.** 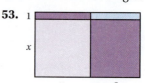 **54.**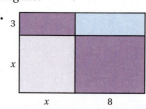

Draw and label rectangles similar to the one following Example 6 to illustrate each product.

55. $x(x + 5)$ **56.** $x(x + 2)$ **57.** $(x + 1)(x + 2)$

58. $(x + 3)(x + 1)$ **59.** $(x + 5)(x + 3)$ **60.** $(x + 4)(x + 6)$

d   Multiply.

61. $(x^2 + x + 1)(x - 1)$ **62.** $(x^2 + x - 2)(x + 2)$ **63.** $(2x + 1)(2x^2 + 6x + 1)$

64. $(3x - 1)(4x^2 - 2x - 1)$ **65.** $(y^2 - 3)(3y^2 - 6y + 2)$ **66.** $(3y^2 - 3)(y^2 + 6y + 1)$

67. $(x^3 + x^2)(x^3 + x^2 - x)$ **68.** $(x^3 - x^2)(x^3 - x^2 + x)$ **69.** $(-5x^3 - 7x^2 + 1)(2x^2 - x)$

70. $(-4x^3 + 5x^2 - 2)(5x^2 + 1)$ **71.** $(1 + x + x^2)(-1 - x + x^2)$ **72.** $(1 - x + x^2)(1 - x + x^2)$

73. $(2t^2 - t - 4)(3t^2 + 2t - 1)$ **74.** $(3a^2 - 5a + 2)(2a^2 - 3a + 4)$ **75.** $(x - x^3 + x^5)(x^2 - 1 + x^4)$

76. $(x - x^3 + x^5)(3x^2 + 3x^6 + 3x^4)$ **77.** $(x + 1)(x^3 + 7x^2 + 5x + 4)$ **78.** $(x + 2)(x^3 + 5x^2 + 9x + 3)$

79. $\left(x - \frac{1}{2}\right)\left(2x^3 - 4x^2 + 3x - \frac{2}{5}\right)$ **80.** $\left(x + \frac{1}{3}\right)\left(6x^3 - 12x^2 - 5x + \frac{1}{2}\right)$

Skill Maintenance

Simplify.

81. $x - (2x - 3)$ [1.8b]

82. $5 - 2[3 - 4(8 - 2)]$
    [1.8c]

83. $(10 - 2)(10 + 2)$
    [1.8d]

84. $10 - 2 + (-6)^2 \div 3 \cdot 2$
    [1.8d]

Factor.   [1.7d]

85. $15x - 18y + 12$

86. $16x - 24y + 36$

87. $-9x - 45y + 15$

88. $100x - 100y + 1000a$

Synthesis

Find a polynomial for the shaded area of each figure.

89.

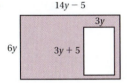

90.

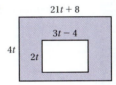

For each figure, determine what the missing number must be in order for the figure to have the given area.

91. Area $= x^2 + 8x + 15$

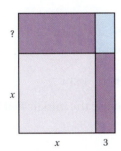

92. Area $= x^2 + 7x + 10$

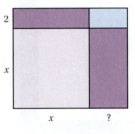

93. Find a polynomial for the volume of the solid shown below.

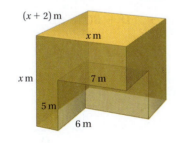

94. An open wooden box is a cube with side x cm. The box, including its bottom, is made of wood that is 1 cm thick. Find a polynomial for the interior volume of the cube.

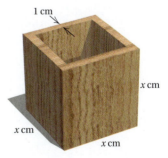

Compute and simplify.

95. $(x - 2)(x - 7) - (x - 7)(x - 2)$

96. $(x + 5)^2 - (x - 3)^2$

97. Extend the pattern and simplify:
$$(x - a)(x - b)(x - c)(x - d) \cdots (x - z).$$

98. 📈 Use a graphing calculator to check your answers to Exercises 15, 29, and 61. Use graphs, tables, or both, as directed by your instructor.

4.6

Special Products

OBJECTIVES

a   Multiply two binomials mentally using the FOIL method.

b   Multiply the sum and the difference of the same two terms mentally.

c   Square a binomial mentally.

d   Find special products when polynomial products are mixed together.

SKILL TO REVIEW

Objective 1.7e: Collect like terms.

Collect like terms.

1. $\dfrac{2}{3}x - \dfrac{2}{3}x$

2. $-12n + 12n$

We encounter certain products so often that it is helpful to have efficient methods of computing them. Such techniques are called *special products*.

a   PRODUCTS OF TWO BINOMIALS USING FOIL

To multiply two binomials, we can select one binomial and multiply each term of that binomial by every term of the other. Then we collect like terms. Consider the product $(x + 3)(x + 7)$:

$$(x + 3)(x + 7) = x(x + 7) + 3(x + 7)$$
$$= x \cdot x + x \cdot 7 + 3 \cdot x + 3 \cdot 7$$
$$= x^2 + 7x + 3x + 21$$
$$= x^2 + 10x + 21.$$

This example illustrates a special technique for finding the product of two binomials:

First terms   Outside terms   Inside terms   Last terms

$$(x + 3)(x + 7) = x \cdot x + 7 \cdot x + 3 \cdot x + 3 \cdot 7.$$

To remember this method of multiplying, we use the initials **FOIL**.

THE FOIL METHOD

To multiply two binomials, $A + B$ and $C + D$, multiply the First terms AC, the Outside terms AD, the Inside terms BC, and then the Last terms BD. Then collect like terms, if possible.

$$(A + B)(C + D) = AC + AD + BC + BD$$

1. Multiply First terms: AC.

2. Multiply Outside terms: AD.

3. Multiply Inside terms: BC.

4. Multiply Last terms: BD.

    ↓

    FOIL

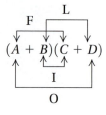

EXAMPLE 1   Multiply: $(x + 8)(x^2 - 5)$.

We have

$$\overset{F}{}\quad\overset{O}{}\quad\overset{I}{}\quad\overset{L}{}$$
$$(x + 8)(x^2 - 5) = x \cdot x^2 + x \cdot (-5) + 8 \cdot x^2 + 8(-5)$$
$$= x^3 - 5x + 8x^2 - 40$$
$$= x^3 + 8x^2 - 5x - 40.$$

Answers

Skill to Review:
1. 0   **2.** 0

Since each of the original binomials is in descending order, we write the product in descending order, as is customary, but this is not a "must." ◼

Often we can collect like terms after we have multiplied.

EXAMPLES Multiply.

2. $(x + 6)(x - 6) = x^2 - 6x + 6x - 36$ Using FOIL
$= x^2 - 36$ Collecting like terms

3. $(x + 7)(x + 4) = x^2 + 4x + 7x + 28$
$= x^2 + 11x + 28$

4. $(y - 3)(y - 2) = y^2 - 2y - 3y + 6$
$= y^2 - 5y + 6$

5. $(x^3 - 1)(x^3 + 5) = x^6 + 5x^3 - x^3 - 5$
$= x^6 + 4x^3 - 5$

6. $(4t^3 + 5)(3t^2 - 2) = 12t^5 - 8t^3 + 15t^2 - 10$

Do Exercises 1–8. ▶

EXAMPLES Multiply.

7. $\left(x - \frac{2}{3}\right)\left(x + \frac{2}{3}\right) = x^2 + \frac{2}{3}x - \frac{2}{3}x - \frac{4}{9}$
$= x^2 - \frac{4}{9}$

8. $(x^2 - 0.3)(x^2 - 0.3) = x^4 - 0.3x^2 - 0.3x^2 + 0.09$
$= x^4 - 0.6x^2 + 0.09$

9. $(3 - 4x)(7 - 5x^3) = 21 - 15x^3 - 28x + 20x^4$
$= 21 - 28x - 15x^3 + 20x^4$

(*Note*: If the original polynomials are in ascending order, it is natural to write the product in ascending order, but this is not a "must.")

10. $(5x^4 + 2x^3)(3x^2 - 7x) = 15x^6 - 35x^5 + 6x^5 - 14x^4$
$= 15x^6 - 29x^5 - 14x^4$

Do Exercises 9–12. ▶

We can show the FOIL method geometrically as follows. One way to write the area of the large rectangle below is $(A + B)(C + D)$. To find another expression for the area of the large rectangle, we add the areas of the smaller rectangles.

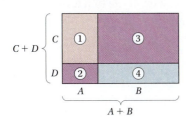

The area of rectangle ① is AC.
The area of rectangle ② is AD.
The area of rectangle ③ is BC.
The area of rectangle ④ is BD.

The area of the large rectangle is the sum of the areas of the smaller rectangles. Thus,

$$(A + B)(C + D) = AC + AD + BC + BD.$$

Multiply mentally, if possible. If you need extra steps, be sure to use them.

1. $(x + 3)(x + 4)$

2. $(x + 3)(x - 5)$

3. $(2x - 1)(x - 4)$

4. $(2x^2 - 3)(x^2 - 2)$

5. $(6x^2 + 5)(2x^3 + 1)$

6. $(y^3 + 7)(y^3 - 7)$

7. $(t + 2)(t + 3)$

8. $(2x^4 + x^2)(-x^3 + x)$

Multiply.

9. $\left(x + \frac{4}{5}\right)\left(x - \frac{4}{5}\right)$

10. $(x^3 - 0.5)(x^2 + 0.5)$

11. $(2 + 3x^2)(4 - 5x^2)$

12. $(6x^3 - 3x^2)(5x^2 - 2x)$

Answers

1. $x^2 + 7x + 12$ 2. $x^2 - 2x - 15$
3. $2x^2 - 9x + 4$ 4. $2x^4 - 7x^2 + 6$
5. $12x^5 + 10x^3 + 6x^2 + 5$ 6. $y^6 - 49$
7. $t^2 + 5t + 6$ 8. $-2x^7 + x^5 + x^3$
9. $x^2 - \dfrac{16}{25}$ 10. $x^5 + 0.5x^3 - 0.5x^2 - 0.25$
11. $8 + 2x^2 - 15x^4$ 12. $30x^5 - 27x^4 + 6x^3$

b MULTIPLYING SUMS AND DIFFERENCES OF TWO TERMS

Consider the product of the sum and the difference of the same two terms, such as

$$(x + 2)(x - 2).$$

Since this is the product of two binomials, we can use FOIL. This type of product occurs so often, however, that it would be valuable if we could use an even faster method. To find a faster way to compute such a product, look for a pattern in the following:

a) $(x + 2)(x - 2) = x^2 - 2x + 2x - 4$ Using FOIL
$$= x^2 - 4;$$

b) $(3x - 5)(3x + 5) = 9x^2 + 15x - 15x - 25$
$$= 9x^2 - 25.$$

◀ **Do Exercises 13 and 14.**

Perhaps you discovered in each case that when you multiply the two binomials, two terms are opposites, or additive inverses, which add to 0 and "drop out."

PRODUCT OF THE SUM AND THE DIFFERENCE OF TWO TERMS

The product of the sum and the difference of the same two terms is the square of the first term minus the square of the second term:

$$(A + B)(A - B) = A^2 - B^2.$$

It is helpful to memorize this rule in both words and symbols. (If you do forget it, you can, of course, use FOIL.)

EXAMPLES Multiply. (Carry out the rule and say the words as you go.)

$$(A + B)(A - B) = A^2 - B^2$$

11. $(x + 4)(x - 4) = x^2 - 4^2$ "The square of the first term, x^2, minus the square of the second, 4^2"
$$= x^2 - 16 \quad \text{Simplifying}$$

12. $(5 + 2w)(5 - 2w) = 5^2 - (2w)^2$
$$= 25 - 4w^2$$

13. $(3x^2 - 7)(3x^2 + 7) = (3x^2)^2 - 7^2$
$$= 9x^4 - 49$$

14. $(-4x - 10)(-4x + 10) = (-4x)^2 - 10^2$
$$= 16x^2 - 100$$

15. $\left(x + \dfrac{3}{8}\right)\left(x - \dfrac{3}{8}\right) = x^2 - \left(\dfrac{3}{8}\right)^2 = x^2 - \dfrac{9}{64}$

◀ **Do Exercises 15–19.**

Multiply.

13. $(x + 5)(x - 5)$

14. $(2x - 3)(2x + 3)$

Multiply.

15. $(x + 8)(x - 8)$

16. $(x - 7)(x + 7)$

17. $(6 - 4y)(6 + 4y)$ GS

$(\quad)^2 - (\quad)^2 = 36 - \boxed{}$

18. $(2x^3 - 1)(2x^3 + 1)$

19. $\left(x - \dfrac{2}{5}\right)\left(x + \dfrac{2}{5}\right)$

Answers

13. $x^2 - 25$ **14.** $4x^2 - 9$ **15.** $x^2 - 64$
16. $x^2 - 49$ **17.** $36 - 16y^2$ **18.** $4x^6 - 1$
19. $x^2 - \dfrac{4}{25}$

Guided Solution:
17. $6, 4y, 16y^2$

C SQUARING BINOMIALS

Consider the square of a binomial, such as $(x + 3)^2$. This can be expressed as $(x + 3)(x + 3)$. Since this is the product of two binomials, we can use FOIL. But again, this type of product occurs so often that we would like to use an even faster method. Look for a pattern in the following.

a)
$$(x + 3)^2 = (x + 3)(x + 3)$$
$$= x^2 + 3x + 3x + 9$$
$$= x^2 + 6x + 9;$$

b)
$$(x - 3)^2 = (x - 3)(x - 3)$$
$$= x^2 - 3x - 3x + 9$$
$$= x^2 - 6x + 9$$

Do Exercises 20 and 21. ▶

Multiply.

20. $(x + 8)(x + 8)$

21. $(x - 5)(x - 5)$

When squaring a binomial, we multiply a binomial by itself. Perhaps you noticed that two terms are the same and when added give twice the product of the terms in the binomial. The other two terms are squares.

SQUARE OF A BINOMIAL

The square of a sum or a difference of two terms is the square of the first term, plus twice the product of the two terms, plus the square of the last term:

$$(A + B)^2 = A^2 + 2AB + B^2; \qquad (A - B)^2 = A^2 - 2AB + B^2.$$

It is helpful to memorize this rule in both words and symbols.

EXAMPLES Multiply. (Carry out the rule and say the words as you go.)

$$(A + B)^2 = A^2 + 2 \cdot A \cdot B + B^2$$

16. $(x + 3)^2 = x^2 + 2 \cdot x \cdot 3 + 3^2$ "x^2 plus 2 times x times 3 plus 3^2"
$$= x^2 + 6x + 9$$

$$(A - B)^2 = A^2 - 2 \cdot A \cdot B + B^2$$

17. $(t - 5)^2 = t^2 - 2 \cdot t \cdot 5 + 5^2$
$$= t^2 - 10t + 25$$

18. $(2x + 7)^2 = (2x)^2 + 2 \cdot 2x \cdot 7 + 7^2 = 4x^2 + 28x + 49$

19. $(5x - 3x^2)^2 = (5x)^2 - 2 \cdot 5x \cdot 3x^2 + (3x^2)^2 = 25x^2 - 30x^3 + 9x^4$

20. $(2.3 - 5.4m)^2 = 2.3^2 - 2(2.3)(5.4m) + (5.4m)^2$
$$= 5.29 - 24.84m + 29.16m^2$$

Do Exercises 22–27. ▶

Multiply.

22. $(x + 2)^2$

23. $(a - 4)^2$

24. $(2x + 5)^2$

25. $(4x^2 - 3x)^2$

26. $(7.8 + 1.2y)(7.8 + 1.2y)$

GS 27. $(3x^2 - 5)(3x^2 - 5)$
$$(3x^2)^2 - 2(3x^2)(\boxed{}) + 5^2$$
$$= \boxed{} x^4 - \boxed{} x^2 + 25$$

.... **Caution!** ....

Although the square of a product is the product of the squares, the square of a sum is *not* the sum of the squares. That is, $(AB)^2 = A^2B^2$, but

The term $2AB$ is missing.
↓
$$(A + B)^2 \neq A^2 + B^2.$$

To illustrate this inequality, note, using the rules for order of operations, that $(7 + 5)^2 = 12^2 = 144$, whereas $7^2 + 5^2 = 49 + 25 = 74$, and $74 \neq 144$.

Answers

20. $x^2 + 16x + 64$ **21.** $x^2 - 10x + 25$
22. $x^2 + 4x + 4$ **23.** $a^2 - 8a + 16$
24. $4x^2 + 20x + 25$ **25.** $16x^4 - 24x^3 + 9x^2$
26. $60.84 + 18.72y + 1.44y^2$
27. $9x^4 - 30x^2 + 25$

We can look at the rule for finding $(A + B)^2$ geometrically as follows. The area of the large square is

$$(A + B)(A + B) = (A + B)^2.$$

This is equal to the sum of the areas of the smaller rectangles:

$$A^2 + AB + AB + B^2 = A^2 + 2AB + B^2.$$

Thus, $(A + B)^2 = A^2 + 2AB + B^2$.

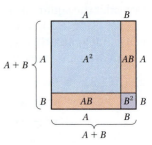

◀ **Do Exercise 28.**

28. In the figure at right, describe in terms of area the sum $A^2 + B^2$. How can the figure be used to verify that $(A + B)^2 \neq A^2 + B^2$?

d   MULTIPLICATION OF VARIOUS TYPES

Let's now try several types of multiplications mixed together so that we can learn to sort them out. When you multiply, first see what kind of multiplication you have. Then use the best method.

MULTIPLYING TWO POLYNOMIALS
....

1. Is it the product of a monomial and a polynomial? If so, multiply each term of the polynomial by the monomial.

   Example: $5x(x + 7) = 5x \cdot x + 5x \cdot 7 = 5x^2 + 35x$

2. Is it the product of the sum and the difference of the *same* two terms? If so, use the following:

   $$(A + B)(A - B) = A^2 - B^2.$$

   Example: $(x + 7)(x - 7) = x^2 - 7^2 = x^2 - 49$

3. Is the product the square of a binomial? If so, use the following:

   $$(A + B)(A + B) = (A + B)^2 = A^2 + 2AB + B^2,$$
   or $(A - B)(A - B) = (A - B)^2 = A^2 - 2AB + B^2.$

   Example: $(x + 7)(x + 7) = (x + 7)^2$
   $$= x^2 + 2 \cdot x \cdot 7 + 7^2$$
   $$= x^2 + 14x + 49$$

4. Is it the product of two binomials other than those above? If so, use FOIL.

   Example: $(x + 7)(x - 4) = x^2 - 4x + 7x - 28$
   $$= x^2 + 3x - 28$$

5. Is it the product of two polynomials other than those above? If so, multiply each term of one by every term of the other. Use columns if you wish.

   Example:

   $$(x^2 - 3x + 2)(x + 7) = x^2(x + 7) - 3x(x + 7) + 2(x + 7)$$
   $$= x^2 \cdot x + x^2 \cdot 7 - 3x \cdot x - 3x \cdot 7$$
   $$+ 2 \cdot x + 2 \cdot 7$$
   $$= x^3 + 7x^2 - 3x^2 - 21x + 2x + 14$$
   $$= x^3 + 4x^2 - 19x + 14$$

Answers

28. $(A + B)^2$ represents the area of the large square. This includes all four sections. $A^2 + B^2$ represents the area of only two of the sections.

Guided Solution:
27. 5, 9, 30

Remember that FOIL will *always* work for two binomials. You can use it instead of either of rules (2) and (3), but those rules will make your work go faster.

EXAMPLE 21   Multiply: $(x + 3)(x - 3)$.

$$
\begin{aligned}
(x + 3)(x - 3) &= x^2 - 3^2 \\
&= x^2 - 9
\end{aligned}
$$

This is the product of the sum and the difference of the same two terms. We use $(A + B)(A - B) = A^2 - B^2$. ◼

EXAMPLE 22   Multiply: $(t + 7)(t - 5)$.

$$
(t + 7)(t - 5) = t^2 + 2t - 35
$$

This is the product of two binomials, but neither the square of a binomial nor the product of the sum and the difference of two terms. We use FOIL. ◼

EXAMPLE 23   Multiply: $(x + 6)(x + 6)$.

$$
\begin{aligned}
(x + 6)(x + 6) &= x^2 + 2(6)x + 6^2 \\
&= x^2 + 12x + 36
\end{aligned}
$$

This the square of a binomial. We use $(A + B)(A + B) = A^2 + 2AB + B^2$. ◼

EXAMPLE 24   Multiply: $2x^3(9x^2 + x - 7)$.

$$
2x^3(9x^2 + x - 7) = 18x^5 + 2x^4 - 14x^3
$$

This is the product of a monomial and a trinomial. We multiply each term of the trinomial by the monomial. ◼

EXAMPLE 25   Multiply: $(5x^3 - 7x)^2$.

$$
\begin{aligned}
(5x^3 - 7x)^2 &= (5x^3)^2 - 2(5x^3)(7x) + (7x)^2 \\
&= 25x^6 - 70x^4 + 49x^2
\end{aligned}
$$

$(A - B)^2 = A^2 - 2AB + B^2$ ◼

EXAMPLE 26   Multiply: $\left(3x + \frac{1}{4}\right)^2$.

$$
\begin{aligned}
\left(3x + \tfrac{1}{4}\right)^2 &= (3x)^2 + 2(3x)\left(\tfrac{1}{4}\right) + \left(\tfrac{1}{4}\right)^2 \\
&= 9x^2 + \tfrac{3}{2}x + \tfrac{1}{16}
\end{aligned}
$$

$(A + B)^2 = A^2 + 2AB + B^2$ ◼

EXAMPLE 27   Multiply: $\left(4x - \frac{3}{4}\right)\left(4x + \frac{3}{4}\right)$.

$$
\begin{aligned}
\left(4x - \tfrac{3}{4}\right)\left(4x + \tfrac{3}{4}\right) &= (4x)^2 - \left(\tfrac{3}{4}\right)^2 \\
&= 16x^2 - \tfrac{9}{16}
\end{aligned}
$$

$(A + B)(A - B) = A^2 - B^2$ ◼

EXAMPLE 28   Multiply: $(p + 3)(p^2 + 2p - 1)$.

$$
\begin{array}{r}
p^2 + 2p - 1 \\
p + 3 \\
\hline
3p^2 + 6p - 3 \\
p^3 + 2p^2 - p \\
\hline
p^3 + 5p^2 + 5p - 3
\end{array}
$$

Finding the product of two polynomials

Multiplying by 3

Multiplying by p

Do Exercises 29–36. ▶

Multiply.

29. $(x + 5)(x + 6)$

30. $(t - 4)(t + 4)$

31. $4x^2(-2x^3 + 5x^2 + 10)$

32. $(9x^2 + 1)^2$

33. $(2a - 5)(2a + 8)$

34. $\left(5x + \dfrac{1}{2}\right)^2$

35. $\left(2x - \dfrac{1}{2}\right)^2$

36. $(x^2 - x + 4)(x - 2)$

Answers

29. $x^2 + 11x + 30$ **30.** $t^2 - 16$
31. $-8x^5 + 20x^4 + 40x^2$ **32.** $81x^4 + 18x^2 + 1$
33. $4a^2 + 6a - 40$ **34.** $25x^2 + 5x + \dfrac{1}{4}$
35. $4x^2 - 2x + \dfrac{1}{4}$ **36.** $x^3 - 3x^2 + 6x - 8$

Visualizing for Success

1

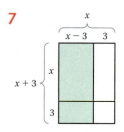

2

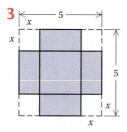

3

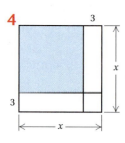

4

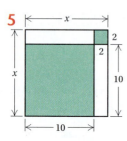

In each of Exercises 1–10, find two algebraic expressions for the shaded area of the figure from the list below.

A. $9 - 4x^2$

B. $x^2 - (x - 6)^2$

C. $(x + 3)(x - 3)$

D. $10^2 + 2^2$

E. $x^2 + 8x + 15$

F. $(x + 5)(x + 3)$

G. $x^2 - 6x + 9$

H. $(3 - 2x)^2 + 4x(3 - 2x)$

I. $(x + 3)^2$

J. $(5x + 3)^2$

K. $(5 - 2x)^2 + 4x(5 - 2x)$

L. $x^2 - 9$

M. 104

N. $x^2 - 15$

O. $12x - 36$

P. $25x^2 + 30x + 9$

Q. $(x - 5)(x - 3)$
$+ 3(x - 5) + 5(x - 3)$

R. $(x - 3)^2$

S. $25 - 4x^2$

T. $x^2 + 6x + 9$

Answers on page A-14

6

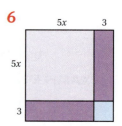

7

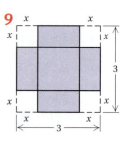

8

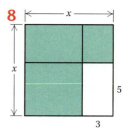

9

10

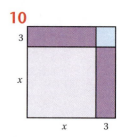

✓ Reading Check

Complete each statement with the appropriate word from the column on the right. A word may be used more than once or not at all.

RC1. For the FOIL multiplication method, the initials F O I L represent the words first, _____ , inside, and _____ .

RC2. If the polynomials being multiplied are written in descending order, we generally write the product in _____ order.

RC3. The expression $(A + B)(A - B)$ is the product of the sum and the _____ of the same two terms.

RC4. The expression $(A + B)^2$ is the _____ of a _____ .

RC5. We can find the product of any two _____ using the FOIL method.

RC6. The product of the sum and the difference of the same two terms is the _____ of their squares.

ascending
binomial(s)
descending
difference
last
outside
product
square

a Multiply. Try to write only the answer. If you need more steps, be sure to use them.

1. $(x + 1)(x^2 + 3)$

2. $(x^2 - 3)(x - 1)$

3. $(x^3 + 2)(x + 1)$

4. $(x^4 + 2)(x + 10)$

5. $(y + 2)(y - 3)$

6. $(a + 2)(a + 3)$

7. $(3x + 2)(3x + 2)$

8. $(4x + 1)(4x + 1)$

9. $(5x - 6)(x + 2)$

10. $(x - 8)(x + 8)$

11. $(3t - 1)(3t + 1)$

12. $(2m + 3)(2m + 3)$

13. $(4x - 2)(x - 1)$

14. $(2x - 1)(3x + 1)$

15. $\left(p - \frac{1}{4}\right)\left(p + \frac{1}{4}\right)$

16. $\left(q + \frac{3}{4}\right)\left(q + \frac{3}{4}\right)$

17. $(x - 0.1)(x + 0.1)$

18. $(x + 0.3)(x - 0.4)$

19. $(2x^2 + 6)(x + 1)$

20. $(2x^2 + 3)(2x - 1)$

21. $(-2x + 1)(x + 6)$

22. $(3x + 4)(2x - 4)$

23. $(a + 7)(a + 7)$

24. $(2y + 5)(2y + 5)$

25. $(1 + 2x)(1 - 3x)$

26. $(-3x - 2)(x + 1)$

27. $\left(\frac{3}{8}y - \frac{5}{6}\right)\left(\frac{3}{8}y - \frac{5}{6}\right)$

28. $\left(\frac{1}{5}x - \frac{2}{7}\right)\left(\frac{1}{5}x + \frac{2}{7}\right)$

29. $(x^2 + 3)(x^3 - 1)$

30. $(x^4 - 3)(2x + 1)$

31. $(3x^2 - 2)(x^4 - 2)$

32. $(x^{10} + 3)(x^{10} - 3)$

33. $(2.8x - 1.5)(4.7x + 9.3)$

34. $\left(x - \frac{3}{8}\right)\left(x + \frac{4}{7}\right)$

35. $(3x^5 + 2)(2x^2 + 6)$

36. $(1 - 2x)(1 + 3x^2)$

37. $(4x^2 + 3)(x - 3)$

38. $(7x - 2)(2x - 7)$

39. $(4y^4 + y^2)(y^2 + y)$

40. $(5y^6 + 3y^3)(2y^6 + 2y^3)$

b Multiply mentally, if possible. If you need extra steps, be sure to use them.

41. $(x + 4)(x - 4)$

42. $(x + 1)(x - 1)$

43. $(2x + 1)(2x - 1)$

44. $(x^2 + 1)(x^2 - 1)$

45. $(5m - 2)(5m + 2)$

46. $(3x^4 + 2)(3x^4 - 2)$

47. $(2x^2 + 3)(2x^2 - 3)$

48. $(6x^5 - 5)(6x^5 + 5)$

49. $(3x^4 - 4)(3x^4 + 4)$

50. $(t^2 - 0.2)(t^2 + 0.2)$

51. $(x^6 - x^2)(x^6 + x^2)$

52. $(2x^3 - 0.3)(2x^3 + 0.3)$

53. $(x^4 + 3x)(x^4 - 3x)$

54. $\left(\frac{3}{4} + 2x^3\right)\left(\frac{3}{4} - 2x^3\right)$

55. $(x^{12} - 3)(x^{12} + 3)$

56. $(12 - 3x^2)(12 + 3x^2)$

57. $(2y^8 + 3)(2y^8 - 3)$

58. $\left(m - \frac{2}{3}\right)\left(m + \frac{2}{3}\right)$

59. $\left(\frac{5}{8}x - 4.3\right)\left(\frac{5}{8}x + 4.3\right)$

60. $(10.7 - x^3)(10.7 + x^3)$

c Multiply mentally, if possible. If you need extra steps, be sure to use them.

61. $(x + 2)^2$

62. $(2x - 1)^2$

63. $(3x^2 + 1)^2$

64. $\left(3x + \frac{3}{4}\right)^2$

65. $\left(a - \frac{1}{2}\right)^2$

66. $\left(2a - \frac{1}{5}\right)^2$

67. $(3 + x)^2$

68. $(x^3 - 1)^2$

69. $(x^2 + 1)^2$

70. $(8x - x^2)^2$

71. $(2 - 3x^4)^2$

72. $(6x^3 - 2)^2$

73. $(5 + 6t^2)^2$

74. $(3p^2 - p)^2$

75. $\left(x - \frac{5}{8}\right)^2$

76. $(0.3y + 2.4)^2$

d   Multiply mentally, if possible.

77. $(3 - 2x^3)^2$

78. $(x - 4x^3)^2$

79. $4x(x^2 + 6x - 3)$

80. $8x(-x^5 + 6x^2 + 9)$

81. $\left(2x^2 - \frac{1}{2}\right)\left(2x^2 - \frac{1}{2}\right)$

82. $(-x^2 + 1)^2$

83. $(-1 + 3p)(1 + 3p)$

84. $(-3q + 2)(3q + 2)$

85. $3t^2(5t^3 - t^2 + t)$

86. $-6x^2(x^3 + 8x - 9)$

87. $(6x^4 + 4)^2$

88. $(8a + 5)^2$

89. $(3x + 2)(4x^2 + 5)$

90. $(2x^2 - 7)(3x^2 + 9)$

91. $(8 - 6x^4)^2$

92. $\left(\frac{1}{5}x^2 + 9\right)\left(\frac{3}{5}x^2 - 7\right)$

93. $(t - 1)(t^2 + t + 1)$

94. $(y + 5)(y^2 - 5y + 25)$

Compute each of the following and compare.

95. $3^2 + 4^2;\ (3 + 4)^2$

96. $6^2 + 7^2;\ (6 + 7)^2$

97. $9^2 - 5^2;\ (9 - 5)^2$

98. $11^2 - 4^2;\ (11 - 4)^2$

Find the total area of all the shaded rectangles.

99.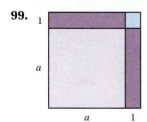

100.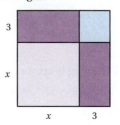

101.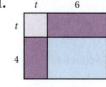

102.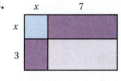

Skill Maintenance

Solve. [2.3c]

103. $3x - 8x = 4(7 - 8x)$

104. $3(x - 2) = 5(2x + 7)$

105. $5(2x - 3) - 2(3x - 4) = 20$

Solve. [2.4b]

106. $3x - 2y = 12$, for y

107. $C = ab - r$, for b

108. $3a - 5d = 4$, for a

Synthesis

Multiply.

109. $5x(3x - 1)(2x + 3)$

110. $[(2x - 3)(2x + 3)](4x^2 + 9)$

111. $[(a - 5)(a + 5)]^2$

112. $(a - 3)^2(a + 3)^2$
(*Hint*: Examine Exercise 111.)

113. $(3t^4 - 2)^2(3t^4 + 2)^2$
(*Hint*: Examine Exercise 111.)

114. $[3a - (2a - 3)][3a + (2a - 3)]$

Solve.

115. $(x + 2)(x - 5) = (x + 1)(x - 3)$

116. $(2x + 5)(x - 4) = (x + 5)(2x - 4)$

117. *Factors and Sums.* To *factor* a number is to express it as a product. Since $12 = 4 \cdot 3$, we say that 12 is *factored* and that 4 and 3 are *factors* of 12. In the following table, the top number has been factored in such a way that the sum of the factors is the bottom number. For example, in the first column, 40 has been factored as $5 \cdot 8$, and $5 + 8 = 13$, the bottom number. Such thinking is important in algebra when we factor trinomials of the type $x^2 + bx + c$. Find the missing numbers in the table.

PRODUCT	40	63	36	72	−140	−96	48	168	110			
FACTOR	5									−9	−24	−3
FACTOR	8									−10	18	
SUM	13	16	−20	−38	−4	4	−14	−29	−21			18

118. Consider the rectangle below.

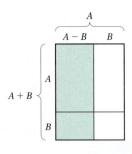

a) Find a polynomial for the area of the entire rectangle.

b) Find a polynomial for the sum of the areas of the two small unshaded rectangles.

c) Find a polynomial for the area in part (a) minus the area in part (b).

d) Find a polynomial for the area of the shaded region and compare this with the polynomial found in part (c).

Use the TABLE or GRAPH feature to check whether each of the following is correct.

119. $(x - 1)^2 = x^2 - 2x + 1$

120. $(x - 2)^2 = x^2 - 4x - 4$

121. $(x - 3)(x + 3) = x^2 - 6$

122. $(x - 3)(x + 2) = x^2 - x - 6$

Operations with Polynomials in Several Variables

The polynomials that we have been studying have only one variable. A **polynomial in several variables** is an expression like those you have already seen, but with more than one variable. Here are two examples:

$$3x + xy^2 + 5y + 4, \qquad 8xy^2z - 2x^3z - 13x^4y^2 + 15.$$

a EVALUATING POLYNOMIALS

EXAMPLE 1 Evaluate the polynomial

$$4 + 3x + xy^2 + 8x^3y^3$$

when $x = -2$ and $y = 5$.

We replace x with -2 and y with 5:

$$
\begin{aligned}
4 + 3x + xy^2 + 8x^3y^3 &= 4 + 3(-2) + (-2) \cdot 5^2 + 8(-2)^3 \cdot 5^3 \\
&= 4 + 3(-2) + (-2) \cdot 25 + 8(-8)(125) \\
&= 4 - 6 - 50 - 8000 \\
&= -8052.
\end{aligned}
$$

EXAMPLE 2 *Zoology.* The weight, in kilograms, of an elephant with a girth of g centimeters at the heart, a length of l centimeters, and a footpad circumference of f centimeters can be estimated by the polynomial

$$11.5g + 7.55l + 12.5f - 4016.$$

A field zoologist finds that the girth of a 3-year-old female elephant is 231 cm, the length is 135 cm, and the footpad circumference is 86 cm. Approximately how much does the elephant weigh?

Source: "How Much Does That Elephant Weigh?" by Mark MacAllister on fieldtripearth.org

We evaluate the polynomial for $g = 231$, $l = 135$, and $f = 86$:

$$
\begin{aligned}
11.5g + 7.55l + 12.5f - 4016 &= 11.5(231) + 7.55(135) + 12.5(86) - 4016 \\
&= 734.75.
\end{aligned}
$$

The elephant weighs about 735 kg.

Do Exercises 1–3. ▶

OBJECTIVES

a Evaluate a polynomial in several variables for given values of the variables.

b Identify the coefficients and the degrees of the terms of a polynomial and the degree of a polynomial.

c Collect like terms of a polynomial.

d Add polynomials.

e Subtract polynomials.

f Multiply polynomials.

1. Evaluate the polynomial
$$4 + 3x + xy^2 + 8x^3y^3$$
when $x = 2$ and $y = -5$.

2. Evaluate the polynomial
$$8xy^2 - 2x^3z - 13x^4y^2 + 5$$
when $x = -1$, $y = 3$, and $z = 4$.

3. *Zoology.* Refer to Example 2. A 25-year-old female elephant has a girth of 366 cm, a length of 226 cm, and a footpad circumference of 117 cm. How much does the elephant weigh?

Answers

1. -7940 **2.** -176 **3.** About 3362 kg

b COEFFICIENTS AND DEGREES

The **degree** of a term is the sum of the exponents of the variables. For example, the degree of $3x^5y^2$ is $5 + 2$, or 7. The **degree of a polynomial** is the degree of the term of highest degree.

EXAMPLE 3 Identify the coefficient and the degree of each term and the degree of the polynomial

$$9x^2y^3 - 14xy^2z^3 + xy + 4y + 5x^2 + 7.$$

TERM	COEFFICIENT	DEGREE	DEGREE OF THE POLYNOMIAL
$9x^2y^3$	9	5	
$-14xy^2z^3$	-14	6	6
xy	1	2	
$4y$	4	1	
$5x^2$	5	2	
7	7	0	

Think: $4y = 4y^1$.

Think: $7 = 7x^0$, or $7x^0y^0z^0$.

◀ **Do Exercises 4 and 5.**

4. Identify the coefficient of each term:

$-3xy^2 + 3x^2y - 2y^3 + xy + 2$.

5. Identify the degree of each term and the degree of the polynomial

$4xy^2 + 7x^2y^3z^2 - 5x + 2y + 4$.

c COLLECTING LIKE TERMS

Like terms have exactly the same variables with exactly the same exponents. For example,

$3x^2y^3$ and $-7x^2y^3$ are like terms;

$9x^4z^7$ and $12x^4z^7$ are like terms.

But

$13xy^5$ and $-2x^2y^5$ are *not* like terms, because the x-factors have different exponents;

and

$3xyz^2$ and $4xy$ are *not* like terms, because there is no factor of z^2 in the second expression.

Collecting like terms is based on the distributive laws.

EXAMPLES Collect like terms.

4. $5x^2y + 3xy^2 - 5x^2y - xy^2 = (5 - 5)x^2y + (3 - 1)xy^2 = 2xy^2$

5. $8a^2 - 2ab + 7b^2 + 4a^2 - 9ab - 17b^2 = 12a^2 - 11ab - 10b^2$

6. $7xy - 5xy^2 + 3xy^2 - 7 + 6x^3 + 9xy - 11x^3 + y - 1$
$= 16xy - 2xy^2 - 5x^3 + y - 8$

◀ **Do Exercises 6 and 7.**

Collect like terms.

6. $4x^2y + 3xy - 2x^2y$

7. $-3pt - 5ptr^3 - 12 + 8pt + 5ptr^3 + 4$ **GS**

The like terms are $-3pt$ and

☐ , $-5ptr^3$ and ☐ ,

and -12 and ☐ .

Collecting like terms, we have

$(-3 + $ ☐ $)pt +$

$(-5 + $ ☐ $) ptr^3 +$

$(-12 + $ ☐ $)$

$= $ ☐ $- 8$.

d ADDITION

We can find the sum of two polynomials in several variables by writing a plus sign between them and then collecting like terms.

EXAMPLE 7 Add: $(-5x^3 + 3y - 5y^2) + (8x^3 + 4x^2 + 7y^2)$.

$$(-5x^3 + 3y - 5y^2) + (8x^3 + 4x^2 + 7y^2)$$
$$= (-5 + 8)x^3 + 4x^2 + 3y + (-5 + 7)y^2$$
$$= 3x^3 + 4x^2 + 3y + 2y^2$$

EXAMPLE 8 Add:

$$(5xy^2 - 4x^2y + 5x^3 + 2) + (3xy^2 - 2x^2y + 3x^3y - 5).$$

We have

$$(5xy^2 - 4x^2y + 5x^3 + 2) + (3xy^2 - 2x^2y + 3x^3y - 5)$$
$$= (5 + 3)xy^2 + (-4 - 2)x^2y + 5x^3 + 3x^3y + (2 - 5)$$
$$= 8xy^2 - 6x^2y + 5x^3 + 3x^3y - 3.$$

Do Exercises 8–10. ▶

e SUBTRACTION

We subtract a polynomial by adding its opposite, or additive inverse. The opposite of the polynomial $4x^2y - 6x^3y^2 + x^2y^2 - 5y$ is

$$-(4x^2y - 6x^3y^2 + x^2y^2 - 5y) = -4x^2y + 6x^3y^2 - x^2y^2 + 5y.$$

EXAMPLE 9 Subtract:

$$(4x^2y + x^3y^2 + 3x^2y^3 + 6y + 10) - (4x^2y - 6x^3y^2 + x^2y^2 - 5y - 8).$$

We have

$$(4x^2y + x^3y^2 + 3x^2y^3 + 6y + 10) - (4x^2y - 6x^3y^2 + x^2y^2 - 5y - 8)$$
$$= 4x^2y + x^3y^2 + 3x^2y^3 + 6y + 10 - 4x^2y + 6x^3y^2 - x^2y^2 + 5y + 8$$

Finding the opposite by changing the sign of each term

$$= 7x^3y^2 + 3x^2y^3 - x^2y^2 + 11y + 18.$$

Collecting like terms. (Try to write just the answer!)

.... **Caution!** ....

Do *not* add exponents when collecting like terms—that is,

$$7x^3 + 8x^3 \neq 15x^6; \leftarrow \text{Adding exponents is incorrect.}$$
$$7x^3 + 8x^3 = 15x^3. \leftarrow \text{Correct}$$

....

Do Exercises 11 and 12. ▶

Add.

8. $(4x^3 + 4x^2 - 8y - 3) + (-8x^3 - 2x^2 + 4y + 5)$

9. $(13x^3y + 3x^2y - 5y) + (x^3y + 4x^2y - 3xy + 3y)$

10. $(-5p^2q^4 + 2p^2q^2 + 3q) + (6pq^2 + 3p^2q + 5)$

Subtract.

11. $(-4s^4t + s^3t^2 + 2s^2t^3) - (4s^4t - 5s^3t^2 + s^2t^2)$

12. $(-5p^4q + 5p^3q^2 - 3p^2q^3 - 7q^4 - 2) - (4p^4q - 4p^3q^2 + p^2q^3 + 2q^4 - 7)$

Answers

8. $-4x^3 + 2x^2 - 4y + 2$
9. $14x^3y + 7x^2y - 3xy - 2y$
10. $-5p^2q^4 + 2p^2q^2 + 3p^2q + 6pq^2 + 3q + 5$
11. $-8s^4t + 6s^3t^2 + 2s^2t^3 - s^2t^2$
12. $-9p^4q + 9p^3q^2 - 4p^2q^3 - 9q^4 + 5$

f MULTIPLICATION

To multiply polynomials in several variables, we can multiply each term of one by every term of the other. We can use columns for long multiplications as with polynomials in one variable. We multiply each term at the top by every term at the bottom. We write like terms in columns, and then we add.

EXAMPLE 10 Multiply: $(3x^2y - 2xy + 3y)(xy + 2y)$.

$$
\begin{array}{r}
3x^2y - 2xy + 3y \\
xy + 2y \\
\hline
6x^2y^2 - 4xy^2 + 6y^2 \\
3x^3y^2 - 2x^2y^2 + 3xy^2 \\
\hline
3x^3y^2 + 4x^2y^2 - xy^2 + 6y^2
\end{array}
$$

Multiplying by $2y$

Multiplying by xy

Adding

◀ **Do Exercises 13 and 14.**

Where appropriate, we use the special products that we have learned.

EXAMPLES Multiply.

11. $(x^2y + 2x)(xy^2 + y^2) = x^3y^3 + x^2y^3 + 2x^2y^2 + 2xy^2$ Using FOIL

12. $(p + 5q)(2p - 3q) = 2p^2 - 3pq + 10pq - 15q^2$ Using FOIL
$$= 2p^2 + 7pq - 15q^2$$

$(A + B)^2 = A^2 + 2 \cdot A \cdot B + B^2$

13. $(3x + 2y)^2 = (3x)^2 + 2(3x)(2y) + (2y)^2 = 9x^2 + 12xy + 4y^2$

$(A - B)^2 = A^2 - 2 \cdot A \cdot B + B^2$

14. $(2y^2 - 5x^2y)^2 = (2y^2)^2 - 2(2y^2)(5x^2y) + (5x^2y)^2$
$$= 4y^4 - 20x^2y^3 + 25x^4y^2$$

$(A + B)(A - B) = A^2 - B^2$

15. $(3x^2y + 2y)(3x^2y - 2y) = (3x^2y)^2 - (2y)^2 = 9x^4y^2 - 4y^2$

16. $(-2x^3y^2 + 5t)(2x^3y^2 + 5t) = (5t - 2x^3y^2)(5t + 2x^3y^2)$

The sum and the difference of the same two terms

$$= (5t)^2 - (2x^3y^2)^2 = 25t^2 - 4x^6y^4$$

$(A - B)(A + B) = A^2 - B^2$

17. $(2x + 3 - 2y)(2x + 3 + 2y) = (2x + 3)^2 - (2y)^2$
$$= 4x^2 + 12x + 9 - 4y^2$$

Remember that FOIL will always work when you are multiplying binomials. You can use it instead of the rules for special products, but those rules will make your work go faster.

◀ **Do Exercises 15–22.**

Multiply.

13. $(x^2y^3 + 2x)(x^3y^2 + 3x)$

14. $(p^4q - 2p^3q^2 + 3q^3)(p + 2q)$

Multiply.

15. $(3xy + 2x)(x^2 + 2xy^2)$

16. $(x - 3y)(2x - 5y)$

17. $(4x + 5y)^2$

18. $(3x^2 - 2xy^2)^2$

19. $(2xy^2 + 3x)(2xy^2 - 3x)$

20. $(3xy^2 + 4y)(-3xy^2 + 4y)$

21. $(3y + 4 - 3x)(3y + 4 + 3x)$

22. $(2a + 5b + c)(2a - 5b - c)$ **GS**

$= [2a + (5b + c)][2a - (\boxed{})]$

$= (2a)^2 - (\boxed{})^2$

$= \boxed{} - (25b^2 + 10bc + \boxed{})$

$= 4a^2 - 25b^2 - 10bc - \boxed{}$

Answers

13. $x^5y^5 + 2x^4y^2 + 3x^3y^3 + 6x^2$
14. $p^5q - 4p^3q^3 + 3pq^3 + 6q^4$
15. $3x^3y + 6x^2y^3 + 2x^3 + 4x^2y^2$
16. $2x^2 - 11xy + 15y^2$
17. $16x^2 + 40xy + 25y^2$
18. $9x^4 - 12x^3y^2 + 4x^2y^4$
19. $4x^2y^4 - 9x^2$
20. $16y^2 - 9x^2y^4$
21. $9y^2 + 24y + 16 - 9x^2$
22. $4a^2 - 25b^2 - 10bc - c^2$

Guided Solution:
22. $5b + c, 5b + c, 4a^2, c^2, c^2$

✓ Reading Check

Determine whether each sentence is true or false.

RC1. The variables in the polynomial $8x - xy + t^2 - xy^2$ are t, x, and y.

RC2. The degree of the term $4xy$ is 4.

RC3. The terms $3x^2y$ and $3xy^2$ are like terms.

RC4. When we collect like terms, we add the exponents of the variables.

a Evaluate the polynomial when $x = 3$, $y = -2$, and $z = -5$.

1. $x^2 - y^2 + xy$

2. $x^2 + y^2 - xy$

3. $x^2 - 3y^2 + 2xy$

4. $x^2 - 4xy + 5y^2$

5. $8xyz$

6. $-3xyz^2$

7. $xyz^2 - z$

8. $xy - xz + yz$

9. *Lung Capacity.* The polynomial equation
$$C = 0.041h - 0.018A - 2.69$$
can be used to estimate the lung capacity C, in liters, of a person of height h, in centimeters, and age A, in years. Find the lung capacity of a 20-year-old person who is 165 cm tall.

10. *Altitude of a Launched Object.* The altitude h, in meters, of a launched object is given by the polynomial equation
$$h = h_0 + vt - 4.9t^2,$$
where h_0 is the height, in meters, from which the launch occurs, v is the initial upward speed (or velocity), in meters per second (m/s), and t is the number of seconds for which the object is airborne. A rock is thrown upward from the top of the Lands End Arch, near San Lucas, Baja, Mexico, 32 m above the ground. The upward speed is 10 m/s. How high will the rock be 3 sec after it has been thrown?

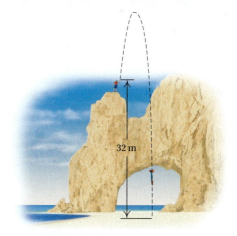

32 m

11. *Male Caloric Needs.* The number of calories needed each day by a moderately active man who weighs w kilograms, is h centimeters tall, and is a years old can be estimated by the polynomial
$$19.18w + 7h - 9.52a + 92.4.$$
Steve is moderately active, weighs 82 kg, is 185 cm tall, and is 67 years old. What is his daily caloric need?

Source: Parker, M., *She Does Math.* Mathematical Association of America

12. *Female Caloric Needs.* The number of calories needed each day by a moderately active woman who weighs w pounds, is h inches tall, and is a years old can be estimated by the polynomial
$$917 + 6w + 6h - 6a.$$
Christine is moderately active, weighs 125 lb, is 64 in. tall, and is 27 years old. What is her daily caloric need?

Source: Parker, M., *She Does Math.* Mathematical Association of America

Surface Area of a Right Circular Cylinder. The surface area S of a right circular cylinder is given by the polynomial equation $S = 2\pi rh + 2\pi r^2$, where h is the height and r is the radius of the base. Use this formula for Exercises 13 and 14.

13. A 12-oz beverage can has a height of 4.7 in. and a radius of 1.2 in. Find the surface area of the can. Use 3.14 for π.

14. A 26-oz coffee can has a height of 6.5 in. and a radius of 2.5 in. Find the surface area of the can. Use 3.14 for π.

Surface Area of a Silo. A silo is a structure that is shaped like a right circular cylinder with a half sphere on top. The surface area S of a silo of height h and radius r (including the area of the base) is given by the polynomial equation $S = 2\pi rh + \pi r^2$. Note that h is the height of the entire silo.

15. A coffee grinder is shaped like a silo, with a height of 7 in. and a radius of $1\frac{1}{2}$ in. Find the surface area " of the coffee grinder. Use 3.14 for π.

16. A $1\frac{1}{2}$-oz bottle of roll-on deodorant has a height of 4 in. and a radius of $\frac{3}{4}$ in. Find the surface area of the bottle if the bottle is shaped like a silo. Use 3.14 for π.

b Identify the coefficient and the degree of each term of the polynomial. Then find the degree of the polynomial.

17. $x^3y - 2xy + 3x^2 - 5$

18. $5x^2y^2 - y^2 + 15xy + 1$

19. $17x^2y^3 - 3x^3yz - 7$

20. $6 - xy + 8x^2y^2 - y^5$

c Collect like terms.

21. $a + b - 2a - 3b$

22. $xy^2 - 1 + y - 6 - xy^2$

23. $3x^2y - 2xy^2 + x^2$

24. $m^3 + 2m^2n - 3m^2 + 3mn^2$

25. $6au + 3av + 14au + 7av$

26. $3x^2y - 2z^2y + 3xy^2 + 5z^2y$

27. $2u^2v - 3uv^2 + 6u^2v - 2uv^2$

28. $3x^2 + 6xy + 3y^2 - 5x^2 - 10xy - 5y^2$

d Add.

29. $(2x^2 - xy + y^2) + (-x^2 - 3xy + 2y^2)$

30. $(2zt - z^2 + 5t^2) + (z^2 - 3zt + t^2)$

31. $(r - 2s + 3) + (2r + s) + (s + 4)$

32. $(ab - 2a + 3b) + (5a - 4b) + (3a + 7ab - 8b)$

33. $(b^3a^2 - 2b^2a^3 + 3ba + 4)$
$+ (b^2a^3 - 4b^3a^2 + 2ba - 1)$

34. $(2x^2 - 3xy + y^2) + (-4x^2 - 6xy - y^2)$
$+ (x^2 + xy - y^2)$

e Subtract.

35. $(a^3 + b^3) - (a^2b - ab^2 + b^3 + a^3)$

36. $(x^3 - y^3) - (-2x^3 + x^2y - xy^2 + 2y^3)$

37. $(xy - ab - 8) - (xy - 3ab - 6)$

38. $(3y^4x^2 + 2y^3x - 3y - 7)$
$- (2y^4x^2 + 2y^3x - 4y - 2x + 5)$

39. $(-2a + 7b - c) - (-3b + 4c - 8d)$

40. Subtract $5a + 2b$ from the sum of $2a + b$ and $3a - b$.

f Multiply.

41. $(3z - u)(2z + 3u)$

42. $(a - b)(a^2 + b^2 + 2ab)$

43. $(a^2b - 2)(a^2b - 5)$

44. $(xy + 7)(xy - 4)$

45. $(a^3 + bc)(a^3 - bc)$

46. $(m^2 + n^2 - mn)(m^2 + mn + n^2)$

47. $(y^4x + y^2 + 1)(y^2 + 1)$

48. $(a - b)(a^2 + ab + b^2)$

49. $(3xy - 1)(4xy + 2)$

50. $(m^3n + 8)(m^3n - 6)$

51. $(3 - c^2d^2)(4 + c^2d^2)$

52. $(6x - 2y)(5x - 3y)$

53. $(m^2 - n^2)(m + n)$

54. $(pq + 0.2) \times$
$(0.4pq - 0.1)$

55. $(xy + x^5 y^5) \times$
$(x^4 y^4 - xy)$

56. $(x - y^3)(2y^3 + x)$

57. $(x + h)^2$

58. $(y - a)^2$

59. $(3a + 2b)^2$

60. $(2ab - cd)^2$

61. $(r^3 t^2 - 4)^2$

62. $(3a^2 b - b^2)^2$

63. $(p^4 + m^2 n^2)^2$

64. $\left(2a^3 - \frac{1}{2}b^3\right)^2$

65. $3a(a - 2b)^2$

66. $-3x(x + 8y)^2$

67. $(m + n - 3)^2$

68. $(a^2 + b + 2)^2$

69. $(a + b)(a - b)$

70. $(x - y)(x + y)$

71. $(2a - b)(2a + b)$

72. $(w + 3z)(w - 3z)$

73. $(c^2 - d)(c^2 + d)$

74. $(p^3 - 5q)(p^3 + 5q)$

75. $(ab + cd^2) \times$
$(ab - cd^2)$

76. $(xy + pq) \times$
$(xy - pq)$

77. $(x + y - 3)(x + y + 3)$

78. $(p + q + 4)(p + q - 4)$

79. $[x + y + z][x - (y + z)]$

80. $[a + b + c][a - (b + c)]$

81. $(a + b + c)(a + b - c)$

82. $(3x + 2 - 5y)(3x + 2 + 5y)$

83. $(x^2 - 4y + 2)(3x^2 + 5y - 3)$

84. $(2x^2 - 7y + 4)(x^2 + y - 3)$

Skill Maintenance

In which quadrant is each point located? [3.1a]

85. $(2, -5)$ **86.** $(-8, -9)$ **87.** $(16, 23)$ **88.** $(-3, 2)$

89. Find the absolute value: $|-39|$. [1.2e]

90. Convert $\dfrac{9}{8}$ to decimal notation. [1.2c]

91. Use either $<$ or $>$ for $\square$ to write a true sentence:
$-17 \,\square\, -5$. [1.2d]

92. Evaluate $-(-x)$ when $x = -3$. [1.3b]

Synthesis

Find a polynomial for each shaded area. (Leave results in terms of π where appropriate.)

93.
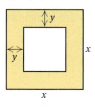

94.
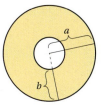

95.
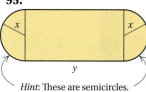

Hint: These are semicircles.

96.
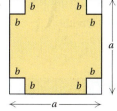

Find a formula for the surface area of each solid object. Leave results in terms of π.

97.

98.
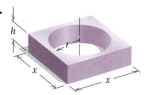

99. *Observatory Paint Costs.* The observatory at Danville University is shaped like a silo that is 40 ft high and 30 ft wide (see Exercise 15). The Heavenly Bodies Astronomy Club is to paint the exterior of the observatory using paint that covers 250 ft² per gallon. How many gallons should they purchase?

100. *Interest Compounded Annually.* An amount of money P that is invested at the yearly interest rate r grows to the amount

$$P(1 + r)^t$$

after t years. Find a polynomial that can be used to determine the amount to which P will grow after 2 years.

101. Suppose that $10,400 is invested at 3.5%, compounded annually. How much is in the account at the end of 5 years? (See Exercise 100.)

102. Multiply: $(x + a)(x - b)(x - a)(x + b)$.

Division of Polynomials

SKILL TO REVIEW

Objective 1.6a: Divide integers.

Divide.

1. $\dfrac{20}{4}$ **2.** $\dfrac{-30}{5}$

Divide.

1. $\dfrac{20x^3}{5x}$ **2.** $\dfrac{-28x^{14}}{4x^3}$

3. $\dfrac{-56p^5q^7}{2p^2q^6}$ **4.** $\dfrac{x^5}{4x}$

In this section, we consider division of polynomials. You will see that such division is similar to what is done in arithmetic.

a   DIVIDING BY A MONOMIAL

We first consider division by a monomial. When dividing a monomial by a monomial, we use the quotient rule to subtract exponents when the bases are the same. We also divide the coefficients.

EXAMPLES   Divide.

Caution!

The coefficients are divided, but the exponents are subtracted.

1. $\dfrac{10x^2}{2x} = \dfrac{10}{2} \cdot \dfrac{x^2}{x} = 5x^{2-1} = 5x$

2. $\dfrac{x^9}{3x^2} = \dfrac{1x^9}{3x^2} = \dfrac{1}{3} \cdot \dfrac{x^9}{x^2} = \dfrac{1}{3}x^{9-2} = \dfrac{1}{3}x^7$

3. $\dfrac{-18x^{10}}{3x^3} = \dfrac{-18}{3} \cdot \dfrac{x^{10}}{x^3} = -6x^{10-3} = -6x^7$

4. $\dfrac{42a^2b^5}{-3ab^2} = \dfrac{42}{-3} \cdot \dfrac{a^2}{a} \cdot \dfrac{b^5}{b^2} = -14a^{2-1}b^{5-2} = -14ab^3$

◀ **Do Margin Exercises 1–4.**

To divide a polynomial by a monomial, we note that since

$$\frac{A}{C} + \frac{B}{C} = \frac{A+B}{C},$$

it follows that

$$\frac{A+B}{C} = \frac{A}{C} + \frac{B}{C}.$$ Switching the left and right sides of the equation

This is actually the procedure we use when performing divisions like $86 \div 2$. Although we might write

$$\frac{86}{2} = 43,$$

we could also calculate as follows:

$$\frac{86}{2} = \frac{80+6}{2} = \frac{80}{2} + \frac{6}{2} = 40 + 3 = 43.$$

Similarly, to divide a polynomial by a monomial, we divide each term by the monomial.

EXAMPLE 5   Divide: $(9x^8 + 12x^6) \div (3x^2)$.

We have

$$(9x^8 + 12x^6) \div (3x^2) = \frac{9x^8 + 12x^6}{3x^2}$$

$$= \frac{9x^8}{3x^2} + \frac{12x^6}{3x^2}.$$ To see this, add and get the original expression.

Answers

Skill to Review:
1. 5   **2.** −6

Margin Exercises:
1. $4x^2$ **2.** $-7x^{11}$ **3.** $-28p^3q$ **4.** $\frac{1}{4}x^4$

We now perform the separate divisions:

$$\frac{9x^8}{3x^2} + \frac{12x^6}{3x^2} = \frac{9}{3} \cdot \frac{x^8}{x^2} + \frac{12}{3} \cdot \frac{x^6}{x^2}$$

$$= 3x^{8-2} + 4x^{6-2}$$

$$= 3x^6 + 4x^4.$$

·············· **Caution!** ··············

The coefficients are *divided*, but the exponents are *subtracted*.

···

To check, we multiply the quotient, $3x^6 + 4x^4$, by the divisor, $3x^2$:

$$3x^2(3x^6 + 4x^4) = (3x^2)(3x^6) + (3x^2)(4x^4) = 9x^8 + 12x^6.$$

This is the polynomial that was being divided, so our answer is $3x^6 + 4x^4$.

Do Exercises 5–7. ▶

EXAMPLE 6 Divide and check: $(10a^5b^4 - 2a^3b^2 + 6a^2b) \div (-2a^2b)$.

$$\frac{10a^5b^4 - 2a^3b^2 + 6a^2b}{-2a^2b} = \frac{10a^5b^4}{-2a^2b} - \frac{2a^3b^2}{-2a^2b} + \frac{6a^2b}{-2a^2b}$$

$$= \frac{10}{-2} \cdot a^{5-2}b^{4-1} - \frac{2}{-2} \cdot a^{3-2}b^{2-1} + \frac{6}{-2}$$

$$= -5a^3b^3 + ab - 3$$

Check: $-2a^2b(-5a^3b^3 + ab - 3) = (-2a^2b)(-5a^3b^3) + (-2a^2b)(ab) - (-2a^2b)(3)$

$$= 10a^5b^4 - 2a^3b^2 + 6a^2b$$

Our answer, $-5a^3b^3 + ab - 3$, checks.

> To divide a polynomial by a monomial, divide each term of the polynomial by the monomial.

Do Exercises 8 and 9. ▶

b DIVIDING BY A BINOMIAL

Let's first consider long division as it is performed in arithmetic. When we divide, we repeat the procedure at right.

We review this by considering the division $3711 \div 8$.

$$\begin{array}{r} 4 \\ 8{\overline{\smash{\big)}\,3711}} \\ \underline{32} \\ 51 \end{array}$$

① Divide: $37 \div 8 \approx 4$.

② Multiply: $4 \times 8 = 32$.

③ Subtract: $37 - 32 = 5$.

④ Bring down the 1.

$$\begin{array}{r} 463 \\ 8{\overline{\smash{\big)}\,3711}} \\ \underline{32} \\ 51 \\ \underline{48} \\ 31 \\ \underline{24} \\ 7 \end{array}$$

Next, we repeat the process two more times. We obtain the complete division as shown on the right above. The quotient is 463. The remainder is 7, expressed as R $= 7$. We write the answer as

$$463 \,\text{R}\, 7 \quad \text{or} \quad 463 + \frac{7}{8} = 463\frac{7}{8}.$$

Divide. Check the result.

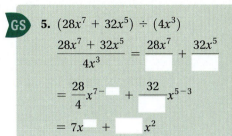

GS **5.** $(28x^7 + 32x^5) \div (4x^3)$

$$\frac{28x^7 + 32x^5}{4x^3} = \frac{28x^7}{\boxed{}} + \frac{32x^5}{\boxed{}}$$

$$= \frac{28}{4}x^{7-\boxed{}} + \frac{32}{\boxed{}}x^{5-3}$$

$$= 7x^{\boxed{}} + \boxed{}x^2$$

6. $(2x^3 + 6x^2 + 4x) \div (2x)$

7. $(6x^2 + 3x - 2) \div 3$

Divide and check.

8. $(8x^2 - 3x + 1) \div (-2)$

9. $\dfrac{2x^4y^6 - 3x^3y^4 + 5x^2y^3}{x^2y^2}$

To carry out long division:

1. Divide,
2. Multiply,
3. Subtract, and
4. Bring down the next number or term.

Answers

5. $7x^4 + 8x^2$ **6.** $x^2 + 3x + 2$

7. $2x^2 + x - \dfrac{2}{3}$ **8.** $-4x^2 + \dfrac{3}{2}x - \dfrac{1}{2}$

9. $2x^2y^4 - 3xy^2 + 5y$

Guided Solution:

5. $4x^3, 4x^3, 3, 4, 4, 8$

We check the answer, 463 R 7, by multiplying the quotient, 463, by the divisor, 8, and adding the remainder, 7:

$$8 \cdot 463 + 7 = 3704 + 7 = 3711.$$

Now let's look at long division with polynomials. We use this procedure when the divisor is not a monomial. We write polynomials in descending order and then write in missing terms, if necessary.

EXAMPLE 7 Divide $x^2 + 5x + 6$ by $x + 2$.

$$
\begin{array}{r}
x \\
x + 2 \overline{\smash{)}x^2 + 5x + 6} \\
x^2 + 2x \\
\hline
3x
\end{array}
$$

Divide the first term by the first term: $x^2/x = x$. Ignore the term 2 for this step.

Multiply x above by the divisor, $x + 2$.

Subtract: $(x^2 + 5x) - (x^2 + 2x) = x^2 + 5x - x^2 - 2x = 3x.$

We now "bring down" the next term of the dividend—in this case, 6.

$$
\begin{array}{r}
x + 3 \\
x + 2 \overline{\smash{)}x^2 + 5x + 6} \\
x^2 + 2x \\
\hline
3x + 6 \\
3x + 6 \\
\hline
0
\end{array}
$$

Divide the first term of $3x + 6$ by the first term of the divisor: $3x/x = 3$.

The 6 has been "brought down."

Multiply 3 above by the divisor, $x + 2$.

Subtract: $(3x + 6) - (3x + 6) = 3x + 6 - 3x - 6 = 0.$

The quotient is $x + 3$. The remainder is 0. A remainder of 0 is generally not included in an answer.

To check, we multiply the quotient by the divisor and add the remainder, if any, to see if we get the dividend:

$$
\underbrace{(x + 2)}_{\text{Divisor}} \cdot \underbrace{(x + 3)}_{\text{Quotient}} + \underbrace{0}_{\text{Remainder}} = \underbrace{x^2 + 5x + 6.}_{\text{Dividend}} \quad \text{The division checks.}
$$

◀ **Do Exercise 10.**

EXAMPLE 8 Divide and check: $(x^2 + 2x - 12) \div (x - 3)$.

$$
\begin{array}{r}
x \\
x - 3 \overline{\smash{)}x^2 + 2x - 12} \\
x^2 - 3x \\
\hline
5x
\end{array}
$$

Divide the first term by the first term: $x^2/x = x$.

Multiply x above by the divisor, $x - 3$.

Subtract: $(x^2 + 2x) - (x^2 - 3x) = x^2 + 2x - x^2 + 3x = 5x.$

We now "bring down" the next term of the dividend—in this case, -12.

$$
\begin{array}{r}
x + 5 \\
x - 3 \overline{\smash{)}x^2 + 2x - 12} \\
x^2 - 3x \\
\hline
5x - 12 \\
5x - 15 \\
\hline
3
\end{array}
$$

Divide the first term of $5x - 12$ by the first term of the divisor: $5x/x = 5$.

Bring down the -12.

Multiply 5 above by the divisor, $x - 3$.

Subtract: $(5x - 12) - (5x - 15) = 5x - 12 - 5x + 15 = 3.$

---

10. Divide and check: $(x^2 + x - 6) \div (x + 3)$.

$$
\begin{array}{r}
\boxed{} - \boxed{} \\
x + 3 \overline{\smash{)}x^2 + x - 6} \\
x^2 + \boxed{} \\
\hline
\boxed{} - 6 \\
-2x - \boxed{} \\
\hline
\boxed{}
\end{array}
$$

---

10. $x - 2$

Guided Solution:

10.
$$
\begin{array}{r}
x - 2 \\
x + 3 \overline{\smash{)}x^2 + x - 6} \\
x^2 + 3x \\
\hline
-2x - 6 \\
-2x - 6 \\
\hline
0
\end{array}
$$

The answer is $x + 5$ with R $= 3$, or

Quotient $\underbrace{x + 5}$ $+ \dfrac{3\; \longrightarrow \text{Remainder}}{\underbrace{x - 3}}$.

$\longrightarrow$ Divisor

(This is the way answers will be given at the back of the book.)

Check: We can check by multiplying the divisor by the quotient and adding the remainder, as follows:

$$(x - 3)(x + 5) + 3 = x^2 + 2x - 15 + 3$$
$$= x^2 + 2x - 12.$$

When dividing, an answer may "come out even" (that is, have a remainder of 0, as in Example 7), or it may not (as in Example 8). **If a remainder is not 0, we continue dividing until the degree of the remainder is less than the degree of the divisor.**

Do Exercises 11 and 12. ▶

Divide and check.

11. $x - 2\overline{\smash{)}x^2 + 2x - 8}$

12. $x + 3\overline{\smash{)}x^2 + 7x + 10}$

EXAMPLE 9 Divide and check: $(x^3 + 1) \div (x + 1)$.

$$
\begin{array}{r}
x^2 - x + 1 \\
x + 1\overline{\smash{)}x^3 + 0x^2 + 0x + 1} \\
\end{array}
$$
← Fill in the missing terms. (See Section 4.3.)

$\underline{x^3 + x^2}$ ← Subtract: $x^3 - (x^3 + x^2) = -x^2.$

$-x^2 + 0x$

$\underline{-x^2 - x}$ ← Subtract: $-x^2 - (-x^2 - x) = x.$

$x + 1$

$\underline{x + 1}$ ← Subtract: $(x + 1) - (x + 1) = 0.$

0

The answer is $x^2 - x + 1$. The check is left to the student.

EXAMPLE 10 Divide and check: $(9x^4 - 7x^2 - 4x + 13) \div (3x - 1)$.

$$
\begin{array}{r}
3x^3 + x^2 - 2x - 2 \\
3x - 1\overline{\smash{)}9x^4 + 0x^3 - 7x^2 - 4x + 13} \\
\end{array}
$$
← Fill in the missing term.

$\underline{9x^4 - 3x^3}$ ← Subtract: $9x^4 - (9x^4 - 3x^3) = 3x^3.$

$3x^3 - 7x^2$

$\underline{3x^3 - x^2}$ ← Subtract:

$-6x^2 - 4x$ $(3x^3 - 7x^2) - (3x^3 - x^2) = -6x^2.$

$\underline{-6x^2 + 2x}$ ← Subtract:

$-6x + 13$ $(-6x^2 - 4x) - (-6x^2 + 2x) = -6x.$

$\underline{-6x + 2}$ ← Subtract:

11 $(-6x + 13) - (-6x + 2) = 11.$

The answer is $3x^3 + x^2 - 2x - 2$ with R $= 11$, or

$$3x^3 + x^2 - 2x - 2 + \dfrac{11}{3x - 1}.$$

Check: $(3x - 1)(3x^3 + x^2 - 2x - 2) + 11$
$$= 9x^4 + 3x^3 - 6x^2 - 6x - 3x^3 - x^2 + 2x + 2 + 11$$
$$= 9x^4 - 7x^2 - 4x + 13$$

Divide and check.

13. $(x^3 - 1) \div (x - 1)$

14. $(8x^4 + 10x^2 + 2x + 9) \div (4x + 2)$

Answers

11. $x + 4$ **12.** $x + 4$ with R $= -2$, or $x + 4 + \dfrac{-2}{x + 3}$ **13.** $x^2 + x + 1$

14. $2x^3 - x^2 + 3x - 1$ with R $= 11$, or $2x^3 - x^2 + 3x - 1 + \dfrac{11}{4x + 2}$

Do Exercises 13 and 14. ▶

4.8 Exercise Set

For Extra Help

MyMathLab®

MathXL®

PRACTICE

WATCH

READ

REVIEW

✓ Reading Check

Complete each statement with the appropriate word(s) from the column on the right. A word may be used more than once.

RC1. When dividing a monomial by a monomial, we _____ exponents and _____ coefficients.

RC2. To divide a polynomial by a monomial, we _____ each term by the monomial.

RC3. To carry out long division, we repeat the following process: divide, _____, _____, and bring down the next term.

RC4. To check division, we _____ the divisor and the quotient, and then _____ the remainder.

add

subtract

multiply

divide

a Divide and check.

1. $\dfrac{24x^4}{8}$

2. $\dfrac{-2u^2}{u}$

3. $\dfrac{25x^3}{5x^2}$

4. $\dfrac{16x^7}{-2x^2}$

5. $\dfrac{-54x^{11}}{-3x^8}$

6. $\dfrac{-75a^{10}}{3a^2}$

7. $\dfrac{64a^5b^4}{16a^2b^3}$

8. $\dfrac{-34p^{10}q^{11}}{-17pq^9}$

9. $\dfrac{24x^4 - 4x^3 + x^2 - 16}{8}$

10. $\dfrac{12a^4 - 3a^2 + a - 6}{6}$

11. $\dfrac{u - 2u^2 - u^5}{u}$

12. $\dfrac{50x^5 - 7x^4 + x^2}{x}$

13. $(15t^3 + 24t^2 - 6t) \div (3t)$

14. $(25t^3 + 15t^2 - 30t) \div (5t)$

15. $(20x^6 - 20x^4 - 5x^2) \div (-5x^2)$

16. $(24x^6 + 32x^5 - 8x^2) \div (-8x^2)$

17. $(24x^5 - 40x^4 + 6x^3) \div (4x^3)$

18. $(18x^6 - 27x^5 - 3x^3) \div (9x^3)$

19. $\dfrac{18x^2 - 5x + 2}{2}$

20. $\dfrac{15x^2 - 30x + 6}{3}$

21. $\dfrac{12x^3 + 26x^2 + 8x}{2x}$

22. $\dfrac{2x^4 - 3x^3 + 5x^2}{x^2}$

23. $\dfrac{9r^2s^2 + 3r^2s - 6rs^2}{3rs}$

24. $\dfrac{4x^4y - 8x^6y^2 + 12x^8y^6}{4x^4y}$

b   Divide.

25. $(x^2 + 4x + 4) \div (x + 2)$

26. $(x^2 - 6x + 9) \div (x - 3)$

27. $(x^2 - 10x - 25) \div (x - 5)$

28. $(x^2 + 8x - 16) \div (x + 4)$

29. $(x^2 + 4x - 14) \div (x + 6)$

30. $(x^2 + 5x - 9) \div (x - 2)$

31. $\dfrac{x^2 - 9}{x + 3}$

32. $\dfrac{x^2 - 25}{x - 5}$

33. $\dfrac{x^5 + 1}{x + 1}$

34. $\dfrac{x^4 - 81}{x - 3}$

35. $\dfrac{8x^3 - 22x^2 - 5x + 12}{4x + 3}$

36. $\dfrac{2x^3 - 9x^2 + 11x - 3}{2x - 3}$

37. $(x^6 - 13x^3 + 42) \div (x^3 - 7)$

38. $(x^6 + 5x^3 - 24) \div (x^3 - 3)$

39. $(t^3 - t^2 + t - 1) \div (t - 1)$

40. $(y^3 + 3y^2 - 5y - 15) \div (y + 3)$

41. $(y^3 - y^2 - 5y - 3) \div (y + 2)$

42. $(t^3 - t^2 + t - 1) \div (t + 1)$

43. $(15x^3 + 8x^2 + 11x + 12) \div (5x + 1)$

44. $(20x^4 - 2x^3 + 5x + 3) \div (2x - 3)$

45. $(12y^3 + 42y^2 - 10y - 41) \div (2y + 7)$

46. $(15y^3 - 27y^2 - 35y + 60) \div (5y - 9)$

Skill Maintenance

Solve.

47. $-13 = 8d - 5$ [2.3a]

48. $x + \dfrac{1}{2}x = 5$ [2.3b]

49. $4(x - 3) = 5(2 - 3x) + 1$ [2.3c]

50. $3(r + 1) - 5(r + 2) \geq 15 - (r + 7)$ [2.7e]

51. The number of patients with the flu who were treated at Riverview Clinic increased from 25 one week to 60 the next week. What was the percent increase?   [2.5a]

52. Todd's quiz grades are 82, 88, 93, and 92. Determine (in terms of an inequality) what scores on the last quiz will allow him to get an average quiz grade of at least 90.   [2.8b]

53. The perimeter of a rectangle is 640 ft. The length is 15 ft more than the width. Find the area of the rectangle.   [2.6a]

54. *Book Pages.*   The sum of the page numbers on the facing pages of a book is 457. Find the page numbers.   [2.6a]

Synthesis

Divide.

55. $(x^4 + 9x^2 + 20) \div (x^2 + 4)$

56. $(y^4 + a^2) \div (y + a)$

57. $(5a^3 + 8a^2 - 23a - 1) \div (5a^2 - 7a - 2)$

58. $(15y^3 - 30y + 7 - 19y^2) \div (3y^2 - 2 - 5y)$

59. $(6x^5 - 13x^3 + 5x + 3 - 4x^2 + 3x^4) \div (3x^3 - 2x - 1)$

60. $(5x^7 - 3x^4 + 2x^2 - 10x + 2) \div (x^2 - x + 1)$

61. $(a^6 - b^6) \div (a - b)$

62. $(x^5 + y^5) \div (x + y)$

If the remainder is 0 when one polynomial is divided by another, the divisor is a *factor* of the dividend. Find the value(s) of c for which $x - 1$ is a factor of the polynomial.

63. $x^2 + 4x + c$

64. $2x^2 + 3cx - 8$

65. $c^2x^2 - 2cx + 1$

Vocabulary Reinforcement

Complete each statement with the correct word from the column on the right. Some of the choices may not be used.

ascending
descending
degree
fraction
scientific
base
exponent
product
quotient
monomial
binomial
trinomial

1. In the expression 7^5, the number 5 is the _____. [4.1a]

2. The _____ rule asserts that when multiplying with exponential notation, if the bases are the same, we keep the base and add the exponent. [4.1d]

3. An expression of the type ax^n, where a is a real-number constant and n is a nonnegative integer, is a(n) _____. [4.3a, b]

4. A(n) _____ is a polynomial with three terms, such as $5x^4 - 7x^2 + 4$. [4.3b]

5. The _____ rule asserts that when dividing with exponential notation, if the bases are the same, we keep the base and subtract the exponent of the denominator from the exponent of the numerator. [4.1e]

6. If the exponents in a polynomial decrease from left to right, the polynomial is arranged in _____ order. [4.3e]

7. The _____ of a term is the sum of the exponents of the variables. [4.7b]

8. The number 2.3×10^{-5} is written in _____ notation. [4.2c]

Concept Reinforcement

Determine whether each statement is true or false.

_____   **1.** All trinomials are polynomials. [4.3b]

_____   **2.** $(x + y)^2 = x^2 + y^2$ [4.6c]

_____   **3.** The square of the difference of two expressions is the difference of the squares of the two expressions. [4.6c]

_____   **4.** The product of the sum and the difference of the same two expressions is the difference of the squares of the expressions. [4.6b]

Study Guide

Objective 4.1d   Use the product rule to multiply exponential expressions with like bases.

Example Multiply and simplify: $x^3 \cdot x^4$. $$x^3 \cdot x^4 = x^{3+4} = x^7$$	**Practice Exercise** **1.** Multiply and simplify: $z^5 \cdot z^3$.

Objective 4.1e   Use the quotient rule to divide exponential expressions with like bases.

Example   Divide and simplify: $\dfrac{x^6y^5}{xy^3}$.

$$\dfrac{x^6y^5}{xy^3} = \dfrac{x^6}{x} \cdot \dfrac{y^5}{y^3}$$
$$= x^{6-1}y^{5-3} = x^5y^2$$

Practice Exercise

2. Divide and simplify: $\dfrac{a^4b^7}{a^2b}$.

Objective 4.1f   Express an exponential expression involving negative exponents with positive exponents.

Objective 4.2a   Use the power rule to raise powers to powers.

Objective 4.2b   Raise a product to a power and a quotient to a power.

Example   Simplify: $\left(\dfrac{2a^3b^{-2}}{c^4}\right)^5$.

$$\left(\dfrac{2a^3b^{-2}}{c^4}\right)^5 = \dfrac{(2a^3b^{-2})^5}{(c^4)^5}$$
$$= \dfrac{2^5(a^3)^5(b^{-2})^5}{(c^4)^5} = \dfrac{32a^{3\cdot5}b^{-2\cdot5}}{c^{4\cdot5}}$$
$$= \dfrac{32a^{15}b^{-10}}{c^{20}} = \dfrac{32a^{15}}{b^{10}c^{20}}$$

Practice Exercise

3. Simplify: $\left(\dfrac{x^{-4}y^2}{3z^3}\right)^3$.

Objective 4.2c   Convert between scientific notation and decimal notation.

Example   Convert 0.00095 to scientific notation.

0.0009.5
$\underset{\text{4 places}}{\llcorner\!\!\uparrow}$

The number is small, so the exponent is negative.

$$0.00095 = 9.5 \times 10^{-4}$$

Practice Exercises

4. Convert to scientific notation: 763,000.

Example   Convert 3.409×10^6 to decimal notation.

3.409000.
$\underset{\text{6 places}}{\llcorner\!\!\uparrow}$

The exponent is positive, so the number is large.

$$3.409 \times 10^6 = 3,409,000$$

5. Convert to decimal notation: 3×10^{-4}.

Objective 4.2d   Multiply and divide using scientific notation.

Example   Multiply and express the result in scientific notation: $(5.3 \times 10^9) \cdot (2.4 \times 10^{-5})$.

$$(5.3 \times 10^9) \cdot (2.4 \times 10^{-5}) = (5.3 \cdot 2.4) \times (10^9 \cdot 10^{-5})$$
$$= 12.72 \times 10^4$$

We convert 12.72 to scientific notation and simplify:

$$12.72 \times 10^4 = (1.272 \times 10) \times 10^4$$
$$= 1.272 \times (10 \times 10^4)$$
$$= 1.272 \times 10^5.$$

Practice Exercise

6. Divide and express the result in scientific notation:

$$\dfrac{3.6 \times 10^3}{6.0 \times 10^{-2}}.$$

Objective 4.3d Collect the like terms of a polynomial.

Example Collect like terms:
$$4x^3 - 2x^2 + 5 + 3x^2 - 12.$$
$$4x^3 - 2x^2 + 5 + 3x^2 - 12$$
$$= 4x^3 + (-2 + 3)x^2 + (5 - 12)$$
$$= 4x^3 + x^2 - 7$$

Practice Exercise

7. Collect like terms: $5x^4 - 6x^2 - 3x^4 + 2x^2 - 3$.

Objective 4.4a Add polynomials.

Example Add: $(4x^3 + x^2 - 8) + (2x^3 - 5x + 1)$.
$$(4x^3 + x^2 - 8) + (2x^3 - 5x + 1)$$
$$= (4 + 2)x^3 + x^2 - 5x + (-8 + 1)$$
$$= 6x^3 + x^2 - 5x - 7$$

Practice Exercise

8. Add: $(3x^4 - 5x^2 - 4) + (x^3 + 3x^2 + 6)$.

Objective 4.5d Multiply any two polynomials.

Example Multiply: $(z^2 - 2z + 3)(z - 1)$.

We use columns. First, we multiply the top row by -1 and then by z, placing like terms of the product in the same column. Finally, we collect like terms.

$$
\begin{array}{r}
z^2 - 2z + 3 \\
z - 1 \\
\hline
-z^2 + 2z - 3 \\
z^3 - 2z^2 + 3z \\
\hline
z^3 - 3z^2 + 5z - 3
\end{array}
$$

Practice Exercise

9. Multiply: $(x^4 - 3x^2 + 2)(x^2 - 3)$.

Objective 4.6a Multiply two binomials mentally using the FOIL method.

Example Multiply: $(3x + 5)(x - 1)$.
$$
\begin{array}{cccc}
\text{F} & \text{O} & \text{I} & \text{L}
\end{array}
$$
$$(3x + 5)(x - 1) = 3x \cdot x + 3x \cdot (-1) + 5 \cdot x + 5 \cdot (-1)$$
$$= 3x^2 - 3x + 5x - 5$$
$$= 3x^2 + 2x - 5$$

Practice Exercise

10. Multiply: $(y + 4)(2y + 3)$.

Objective 4.6b Multiply the sum and the difference of the same two terms mentally.

Example Multiply: $(3y + 2)(3y - 2)$.
$$(3y + 2)(3y - 2) = (3y)^2 - 2^2$$
$$= 9y^2 - 4$$

Practice Exercise

11. Multiply: $(x + 5)(x - 5)$.

Objective 4.6c Square a binomial mentally.

Example Multiply: $(2x - 3)^2$.
$$(2x - 3)^2 = (2x)^2 - 2 \cdot 2x \cdot 3 + 3^2$$
$$= 4x^2 - 12x + 9$$

Practice Exercise

12. Multiply: $(3w + 4)^2$.

Objective 4.7e Subtract polynomials.

Example Subtract:
$(m^4n + 2m^3n^2 - m^2n^3) - (3m^4n + 2m^3n^2 - 4m^2n^2)$.
$(m^4n + 2m^3n^2 - m^2n^3) - (3m^4n + 2m^3n^2 - 4m^2n^2)$
$= m^4n + 2m^3n^2 - m^2n^3 - 3m^4n - 2m^3n^2 + 4m^2n^2$
$= -2m^4n - m^2n^3 + 4m^2n^2$

Practice Exercise

13. Subtract:
$(a^3b^2 - 5a^2b + 2ab) - (3a^3b^2 - ab^2 + 4ab)$.

Objective 4.8a Divide a polynomial by a monomial.

Example Divide: $(6x^3 - 8x^2 + 15x) \div (3x)$.
$$\frac{6x^3 - 8x^2 + 15x}{3x} = \frac{6x^3}{3x} - \frac{8x^2}{3x} + \frac{15x}{3x}$$
$$= \frac{6}{3}x^{3-1} - \frac{8}{3}x^{2-1} + \frac{15}{3}x^{1-1}$$
$$= 2x^2 - \frac{8}{3}x + 5$$

Practice Exercise

14. Divide: $(5y^2 - 20y + 8) \div 5$.

Objective 4.8b Divide a polynomial by a divisor that is a binomial.

Example Divide $x^2 - 3x + 7$ by $x + 1$.

$$
\begin{array}{r}
x - 4 \\
x + 1 \overline{) x^2 - 3x + 7} \\
\underline{x^2 + x} \\
-4x + 7 \\
\underline{-4x - 4} \\
11
\end{array}
$$

The answer is $x - 4 + \dfrac{11}{x + 1}$.

Practice Exercise

15. Divide: $(x^2 - 4x + 3) \div (x + 5)$.

Review Exercises

Multiply and simplify. [4.1d, f]

1. $7^2 \cdot 7^{-4}$

2. $y^7 \cdot y^3 \cdot y$

3. $(3x)^5 \cdot (3x)^9$

4. $t^8 \cdot t^0$

Divide and simplify. [4.1e, f]

5. $\dfrac{4^5}{4^2}$

6. $\dfrac{a^5}{a^8}$

7. $\dfrac{(7x)^4}{(7x)^4}$

Simplify.

8. $(3t^4)^2$ [4.2a, b]

9. $(2x^3)^2(-3x)^2$
[4.1d], [4.2a, b]

10. $\left(\dfrac{2x}{y}\right)^{-3}$ [4.2b]

11. Express using a negative exponent: $\dfrac{1}{t^5}$. [4.1f]

12. Express using a positive exponent: y^{-4}. [4.1f]

13. Convert to scientific notation: 0.0000328. [4.2c]

14. Convert to decimal notation: 8.3×10^6. [4.2c]

Multiply or divide and write scientific notation for the result. [4.2d]

15. $(3.8 \times 10^4)(5.5 \times 10^{-1})$

16. $\dfrac{1.28 \times 10^{-8}}{2.5 \times 10^{-4}}$

17. *Pizza Consumption.* Each man, woman, and child in the United States eats an average of 46 slices of pizza per year. The U.S. population is projected to be about 340 million in 2020. At this rate, how many slices of pizza would be consumed in 2020? Express the answer in scientific notation. [4.2e]

Sources: Packaged Facts; U.S. Census Bureau

18. Evaluate the polynomial $x^2 - 3x + 6$ when $x = -1$. [4.3a]

19. Identify the terms of the polynomial $-4y^5 + 7y^2 - 3y - 2$. [4.3b]

20. Identify the missing terms in $x^3 + x$. [4.3f]

21. Identify the degree of each term and the degree of the polynomial $4x^3 + 6x^2 - 5x + \frac{5}{3}$. [4.3c]

Classify the polynomial as a monomial, a binomial, a trinomial, or none of these. [4.3b]

22. $4x^3 - 1$

23. $4 - 9t^3 - 7t^4 + 10t^2$

24. $7y^2$

Collect like terms and then arrange in descending order. [4.3e]

25. $3x^2 - 2x + 3 - 5x^2 - 1 - x$

26. $-x + \frac{1}{2} + 14x^4 - 7x^2 - 1 - 4x^4$

Add. [4.4a]

27. $(3x^4 - x^3 + x - 4) + (x^5 + 7x^3 - 3x^2 - 5) + (-5x^4 + 6x^2 - x)$

28. $(3x^5 - 4x^4 + x^3 - 3) + (3x^4 - 5x^3 + 3x^2) + (-5x^5 - 5x^2) + (-5x^4 + 2x^3 + 5)$

Subtract. [4.4c]

29. $(5x^2 - 4x + 1) - (3x^2 + 1)$

30. $(3x^5 - 4x^4 + 3x^2 + 3) - (2x^5 - 4x^4 + 3x^3 + 4x^2 - 5)$

31. Find a polynomial for the perimeter and for the area. [4.4d], [4.5b]

32. Find two algebraic expressions for the area of this figure. First, regard the figure as one large rectangle, and then regard the figure as a sum of four smaller rectangles. [4.4d]

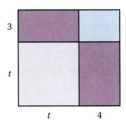

Multiply.

33. $\left(x + \frac{2}{3}\right)\left(x + \frac{1}{2}\right)$ [4.6a] **34.** $(7x + 1)^2$ [4.6c]

35. $(4x^2 - 5x + 1)(3x - 2)$ [4.5d] **36.** $(3x^2 + 4)(3x^2 - 4)$ [4.6b]

37. $5x^4(3x^3 - 8x^2 + 10x + 2)$ [4.5b]

38. $(x + 4)(x - 7)$ [4.6a]

39. $(3y^2 - 2y)^2$ [4.6c] **40.** $(2t^2 + 3)(t^2 - 7)$ [4.6a]

41. Evaluate the polynomial
$$2 - 5xy + y^2 - 4xy^3 + x^6$$
when $x = -1$ and $y = 2$. [4.7a]

42. Identify the coefficient and the degree of each term of the polynomial
$$x^5y - 7xy + 9x^2 - 8.$$
Then find the degree of the polynomial. [4.7b]

Collect like terms. [4.7c]

43. $y + w - 2y + 8w - 5$

44. $m^6 - 2m^2n + m^2n^2 + n^2m - 6m^3 + m^2n^2 + 7n^2m$

45. Add: [4.7d]
$(5x^2 - 7xy + y^2) + (-6x^2 - 3xy - y^2) + (x^2 + xy - 2y^2)$.

46. Subtract: [4.7e]
$(6x^3y^2 - 4x^2y - 6x) - (-5x^3y^2 + 4x^2y + 6x^2 - 6)$.

Multiply. [4.7f]

47. $(p - q)(p^2 + pq + q^2)$ **48.** $(3a^4 - \frac{1}{3}b^3)^2$

Divide.

49. $(10x^3 - x^2 + 6x) \div (2x)$ [4.8a]

50. $(6x^3 - 5x^2 - 13x + 13) \div (2x + 3)$ [4.8b]

51. The graph of the polynomial equation $y = 10x^3 - 10x$ is shown below. Use *only* the graph to estimate the value of the polynomial when $x = -1$, $x = -0.5$, $x = 0.5$, and $x = 1$. [4.3a]

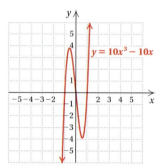

52. Subtract: $(2x^2 - 3x + 4) - (x^2 + 2x)$. [4.4c]
  A. $x^2 - 3x - 2$ **B.** $x^2 - 5x + 4$
  C. $x^2 - x + 4$ **D.** $3x^2 - x + 4$

53. Multiply: $(x - 1)^2$. [4.6c]
  A. $x^2 - 1$ **B.** $x^2 + 1$
  C. $x^2 - 2x - 1$ **D.** $x^2 - 2x + 1$

Synthesis

Find a polynomial for each shaded area. [4.4d], [4.6b]

54. **55.**

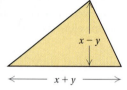

56. Collect like terms: [4.1d], [4.2a], [4.3d]
$-3x^5 \cdot 3x^3 - x^6(2x)^2 + (3x^4)^2 + (2x^2)^4 - 40x^2(x^3)^2$.

57. Solve: [2.3b], [4.6a]
$(x - 7)(x + 10) = (x - 4)(x - 6)$.

58. The product of two polynomials is $x^5 - 1$. One of the polynomials is $x - 1$. Find the other. [4.8b]

59. A rectangular garden is twice as long as it is wide and is surrounded by a sidewalk that is 4 ft wide. The area of the sidewalk is 1024 ft^2. Find the dimensions of the garden. [2.3b], [4.4d], [4.5a], [4.6a]

Understanding Through Discussion and Writing

1. Explain why the expression 578.6×10^{-7} is not written in scientific notation. [4.2c]

2. Explain why an understanding of the rules for order of operations is essential when evaluating polynomials. [4.3a]

3. How can the following figure be used to show that $(x + 3)^2 \neq x^2 + 9$? [4.5c]

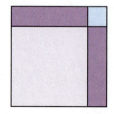

4. On an assignment, Emma *incorrectly* writes
$$\frac{12x^3 - 6x}{3x} = 4x^2 - 6x.$$
What mistake do you think she is making and how might you convince her that a mistake has been made? [4.8a]

5. Can the sum of two trinomials in several variables be a trinomial in one variable? Why or why not? [4.7d]

6. Is it possible for a polynomial in four variables to have a degree less than 4? Why or why not? [4.7b]

CHAPTER

4   Test

For Extra Help   For step-by-step test solutions, access the Chapter Test Prep Videos in MyMathLab® or on YouTube (search "BittingerIntro" and click on "Channels").

Multiply and simplify.

1. $6^{-2} \cdot 6^{-3}$

2. $x^6 \cdot x^2 \cdot x$

3. $(4a)^3 \cdot (4a)^8$

Divide and simplify.

4. $\dfrac{3^5}{3^2}$

5. $\dfrac{x^3}{x^8}$

6. $\dfrac{(2x)^5}{(2x)^5}$

Simplify.

7. $(x^3)^2$

8. $(-3y^2)^3$

9. $(2a^3b)^4$

10. $\left(\dfrac{ab}{c}\right)^3$

11. $(3x^2)^3(-2x^5)^3$

12. $3(x^2)^3(-2x^5)^3$

13. $2x^2(-3x^2)^4$

14. $(2x)^2(-3x^2)^4$

15. Express using a positive exponent: 5^{-3}.

16. Express using a negative exponent: $\dfrac{1}{y^8}$.

17. Convert to scientific notation: 3,900,000,000.

18. Convert to decimal notation: 5×10^{-8}.

Multiply or divide and write scientific notation for the answer.

19. $\dfrac{5.6 \times 10^6}{3.2 \times 10^{-11}}$

20. $(2.4 \times 10^5)(5.4 \times 10^{16})$

21. *Earth vs. Saturn.* The mass of Earth is about 6×10^{21} metric tons. The mass of Saturn is about 5.7×10^{23} metric tons. About how many times the mass of Earth is the mass of Saturn? Express the answer in scientific notation.

22. Evaluate the polynomial $x^5 + 5x - 1$ when $x = -2$.

23. Identify the coefficient of each term of the polynomial $\frac{1}{3}x^5 - x + 7$.

24. Identify the degree of each term and the degree of the polynomial $2x^3 - 4 + 5x + 3x^6$.

25. Classify the polynomial $7 - x$ as a monomial, a binomial, a trinomial, or none of these.

Collect like terms.

26. $4a^2 - 6 + a^2$

27. $y^2 - 3y - y + \dfrac{3}{4}y^2$

28. Collect like terms and then arrange in descending order:
$$3 - x^2 + 2x^3 + 5x^2 - 6x - 2x + x^5.$$

Add.

29. $(3x^5 + 5x^3 - 5x^2 - 3) +$
$(x^5 + x^4 - 3x^3 - 3x^2 + 2x - 4)$

30. $\left(x^4 + \dfrac{2}{3}x + 5\right) + \left(4x^4 + 5x^2 + \dfrac{1}{3}x\right)$

Subtract.

31. $(2x^4 + x^3 - 8x^2 - 6x - 3) - (6x^4 - 8x^2 + 2x)$

32. $(x^3 - 0.4x^2 - 12) - (x^5 + 0.3x^3 + 0.4x^2 + 9)$

Multiply.

33. $-3x^2(4x^2 - 3x - 5)$

34. $\left(x - \dfrac{1}{3}\right)^2$

35. $(3x + 10)(3x - 10)$

36. $(3b + 5)(b - 3)$

37. $(x^6 - 4)(x^8 + 4)$

38. $(8 - y)(6 + 5y)$

39. $(2x + 1)(3x^2 - 5x - 3)$

40. $(5t + 2)^2$

41. Collect like terms:

$x^3y - y^3 + xy^3 + 8 - 6x^3y - x^2y^2 + 11.$

42. Subtract:

$(8a^2b^2 - ab + b^3) - (-6ab^2 - 7ab - ab^3 + 5b^3).$

43. Multiply: $(3x^5 - 4y^5)(3x^5 + 4y^5).$

Divide.

44. $(12x^4 + 9x^3 - 15x^2) \div (3x^2)$

45. $(6x^3 - 8x^2 - 14x + 13) \div (3x + 2)$

46. The graph of the polynomial equation

$$y = x^3 - 5x - 1$$

is shown at right. Use *only* the graph to estimate the value of the polynomial when $x = -1$, $x = -0.5$, $x = 0.5$, $x = 1$, and $x = 1.1$.

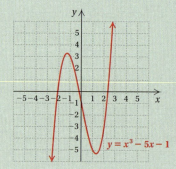

$y = x^3 - 5x - 1$

47. Find two algebraic expressions for the area of this figure. First, regard the figure as one large rectangle, and then regard the figure as a sum of four smaller rectangles.

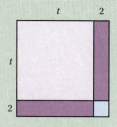

48. Which of the following is a polynomial for the surface area of this right rectangular solid?

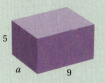

A. $28a$

B. $28a + 90$

C. $14a + 45$

D. $45a$

Synthesis

49. The height of a box is 1 less than its length, and the length is 2 more than its width. Find the volume in terms of the length.

50. Solve: $(x - 5)(x + 5) = (x + 6)^2.$

1. Evaluate $\dfrac{x}{2y}$ when $x = 10$ and $y = 2$.

2. Evaluate $2x^3 + x^2 - 3$ when $x = -1$.

3. Evaluate $x^3y^2 + xy + 2xy^2$ when $x = -1$ and $y = 2$.

4. Find the absolute value: $|-4|$.

5. Find the reciprocal of 5.

Compute and simplify.

6. $-\dfrac{3}{5} + \dfrac{5}{12}$

7. $3.4 - (-0.8)$

8. $(-2)(-1.4)(2.6)$

9. $\dfrac{3}{8} \div \left(-\dfrac{9}{10}\right)$

10. $(1.1 \times 10^{10})(2 \times 10^{12})$

11. $(3.2 \times 10^{-10}) \div (8 \times 10^{-6})$

Simplify.

12. $\dfrac{-9x}{3x}$

13. $y - (3y + 7)$

14. $3(x - 1) - 2[x - (2x + 7)]$

15. $2 - [32 \div (4 + 2^2)]$

Add.

16. $(x^4 + 3x^3 - x + 7) + (2x^5 - 3x^4 + x - 5)$

17. $(x^2 + 2xy) + (y^2 - xy) + (2x^2 - 3y^2)$

Subtract.

18. $(x^3 + 3x^2 - 4) - (-2x^2 + x + 3)$

19. $\left(\dfrac{1}{3}x^2 - \dfrac{1}{4}x - \dfrac{1}{5}\right) - \left(\dfrac{2}{3}x^2 + \dfrac{1}{2}x - \dfrac{1}{5}\right)$

Multiply.

20. $3(4x - 5y + 7)$

21. $(-2x^3)(-3x^5)$

22. $2x^2(x^3 - 2x^2 + 4x - 5)$

23. $(y^2 - 2)(3y^2 + 5y + 6)$

24. $(2p^3 + p^2q + pq^2)(p - pq + q)$

25. $(2x + 3)(3x + 2)$

26. $(3x^2 + 1)^2$

27. $\left(t + \dfrac{1}{2}\right)\left(t - \dfrac{1}{2}\right)$

28. $(2y^2 + 5)(2y^2 - 5)$

29. $(2x^4 - 3)(2x^2 + 3)$

30. $(t - 2t^2)^2$

31. $(3p + q)(5p - 2q)$

Divide.

32. $(18x^3 + 6x^2 - 9x) \div (3x)$

33. $(3x^3 + 7x^2 - 13x - 21) \div (x + 3)$

Solve.

34. $1.5 = 2.7 + x$

35. $\dfrac{2}{7}x = -6$

36. $5x - 9 = 36$

37. $\dfrac{2}{3} = \dfrac{-m}{10}$

38. $5.4 - 1.9x = 0.8x$

39. $x - \dfrac{7}{8} = \dfrac{3}{4}$

40. $2(2 - 3x) = 3(5x + 7)$

41. $\dfrac{1}{4}x - \dfrac{2}{3} = \dfrac{3}{4} + \dfrac{1}{3}x$

42. $y + 5 - 3y = 5y - 9$

43. $\dfrac{1}{4}x - 7 < 5 - \dfrac{1}{2}x$

44. $2(x + 2) \geq 5(2x + 3)$

45. $A = Qx + P$, for x

Solve.

46. **Markup.** A bookstore sells books at a price that is 80% higher than the price the store pays for the books. A book is priced for sale at $6.30. How much did the store pay for the book?

47. A 6-ft by 3-ft raft is floating in a swimming pool of radius r. Find a polynomial for the area of the surface of the pool not covered by the raft.

48. *Consecutive Page Numbers.* The sum of the page numbers on the facing pages of a book is 37. What are the page numbers?

49. *Room Perimeter.* The perimeter of a room is 88 ft. The width is 4 ft less than the length. Find the width and the length.

50. The second angle of a triangle is five times as large as the first. The third angle is twice the sum of the other two angles. Find the measure of the first angle.

Simplify.

51. $y^2 \cdot y^{-6} \cdot y^8$

52. $\dfrac{x^6}{x^7}$

53. $(-3x^3y^{-2})^3$

54. $\dfrac{x^3x^{-4}}{x^{-5}x}$

55. Find the intercepts of $4x - 5y = 20$ and then graph the equation using the intercepts.

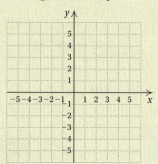

56. *Matching.* Match each item in the first column with the appropriate item in the second column by drawing connecting lines.

3^2	$\dfrac{1}{6}$
3^{-2}	$-\dfrac{1}{9}$
$\left(\dfrac{1}{3}\right)^2$	6
$\left(\dfrac{1}{3}\right)^{-2}$	9
-3^2	-9
$(-3)^2$	$\dfrac{1}{9}$
$\left(-\dfrac{1}{3}\right)^2$	-6
$\left(-\dfrac{1}{3}\right)^{-2}$	12

Synthesis

57. A picture frame is x in. square. The picture that it frames is 2 in. shorter than the frame in both length and width. Find a polynomial for the area of the frame.

Add.

58. $[(2x)^2 - (3x)^3 + 2x^2x^3 + (x^2)^2] +$
$[5x^2(2x^3) - ((2x)^2)^2]$

59. $(x - 3)^2 + (2x + 1)^2$

Solve.

60. $(x + 3)(2x - 5) + (x - 1)^2 = (3x + 1)(x - 3)$

61. $(2x^2 + x - 6) \div (2x - 3) = (2x^2 - 9x - 5) \div (x - 5)$

62. $20 - 3|x| = 5$

63. $(x - 3)(x + 4) = (x^3 - 4x^2 - 17x + 60) \div (x - 5)$

CHAPTER
5

Polynomials: Factoring

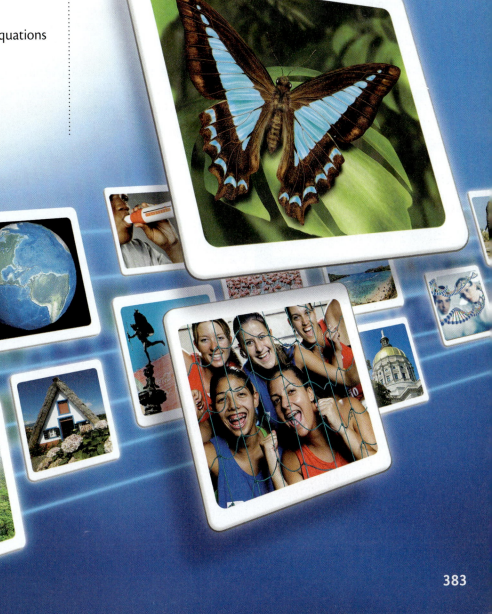

5.1   Introduction to Factoring

OBJECTIVES

a Find the greatest common factor, the GCF, of monomials.

b Factor polynomials when the terms have a common factor, factoring out the greatest common factor.

c Factor certain expressions with four terms using factoring by grouping.

SKILL TO REVIEW

Objective 1.7d: Use the distributive laws to factor expressions like $4x - 12 + 24y$.

Factor.

1. $15m - 10c + 5$

2. $ax - ay + az$

We introduce factoring with a review of factoring natural numbers. Consider the product $3 \cdot 5 = 15$. We say that 3 and 5 are **factors** of 15 and that $3 \cdot 5$ is a **factorization** of 15. Since $15 = 15 \cdot 1$, we also know that 15 and 1 are factors of 15 and that $15 \cdot 1$ is a factorization of 15.

a   FINDING THE GREATEST COMMON FACTOR

The numbers 20 and 30 have several factors in common, among them 2 and 5. The greatest of the common factors is called the **greatest common factor**, **GCF**. We can find the GCF of a set of numbers using prime factorizations.

EXAMPLE 1   Find the GCF of 20 and 30.

We find the prime factorization of each number. Then we draw lines between the common factors.

$$20 = 2 \cdot 2 \cdot 5$$
$$30 = 2 \cdot 3 \cdot 5$$

The GCF $= 2 \cdot 5 = 10$.

EXAMPLE 2   Find the GCF of 180 and 420.

We find the prime factorization of each number. Then we draw lines between the common factors.

$$180 = 2 \cdot 2 \cdot 3 \cdot 3 \cdot 5 = 2^2 \cdot 3^2 \cdot 5^1$$
$$420 = 2 \cdot 2 \cdot 3 \cdot 5 \cdot 7 = 2^2 \cdot 3^1 \cdot 5^1 \cdot 7^1$$

The GCF $= 2 \cdot 2 \cdot 3 \cdot 5 = 2^2 \cdot 3^1 \cdot 5^1 = 60$. Note how we can use the exponents to determine the GCF. There are 2 lines for the 2's, 1 line for the 3, 1 line for the 5, and no line for the 7.

EXAMPLE 3   Find the GCF of 30 and 77.

We find the prime factorization of each number. Then we draw lines between the common factors, if any exist.

$$30 = 2 \cdot 3 \cdot 5 = 2^1 \cdot 3^1 \cdot 5^1$$
$$77 = 7 \cdot 11 = 7^1 \cdot 11^1$$

Since there is no common prime factor, the GCF is 1.

Answers

Skill to Review:
1. $5(3m - 2c + 1)$ **2.** $a(x - y + z)$

EXAMPLE 4 Find the GCF of 54, 90, and 252.

We find the prime factorization of each number. Then we draw lines between the common factors.

$$54 = 2 \cdot 3 \cdot 3 \cdot 3 = 2^1 \cdot 3^3,$$

$$90 = 2 \cdot 3 \cdot 3 \cdot 5 = 2^1 \cdot 3^2 \cdot 5^1,$$

$$252 = 2 \cdot 2 \cdot 3 \cdot 3 \cdot 7 = 2^2 \cdot 3^2 \cdot 7^1$$

The GCF $= 2^1 \cdot 3^2 = 18$.

Do Exercises 1–4. ▶

Find the GCF.

1. 40, 100

2. 7, 21

3. 72, 360, 432

4. 3, 5, 22

Consider the product

$$12x^3(x^2 - 6x + 2) = 12x^5 - 72x^4 + 24x^3.$$

To factor the polynomial on the right-hand side, we reverse the process of multiplication:

$$12x^5 - 72x^4 + 24x^3 = \underline{12x^3(x^2 - 6x + 2)}.$$

This is a *factorization*. The *factors* are $(12x^3)$ and $(x^2 - 6x + 2)$.

FACTOR; FACTORIZATION

To **factor** a polynomial is to express it as a product.

A **factor** of a polynomial P is a polynomial that can be used to express P as a product.

A **factorization** of a polynomial is an expression that names that polynomial as a product.

In the factorization

$$12x^5 - 72x^4 + 24x^3 = 12x^3(x^2 - 6x + 2),$$

the monomial $12x^3$ is called the GCF of the terms, $12x^5$, $-72x^4$, and $24x^3$. The first step in factoring polynomials is to find the GCF of the terms. To do this, we find the greatest positive common factor of the coefficients and the greatest common factors of the powers of any variables.

EXAMPLE 5 Find the GCF of $15x^5$, $-12x^4$, $27x^3$, and $-3x^2$.

First, we find a prime factorization of the coefficients, including a factor of -1 for the negative coefficients.

$$15x^5 = \qquad 3 \cdot 5 \cdot x^5,$$

$$-12x^4 = -1 \cdot 2 \cdot 2 \cdot 3 \cdot x^4,$$

$$27x^3 = \qquad 3 \cdot 3 \cdot 3 \cdot x^3,$$

$$-3x^2 = \qquad -1 \cdot 3 \cdot x^2$$

The greatest *positive* common factor of the coefficients is 3.

Next, we find the GCF of the powers of x. That GCF is x^2, because 2 is the smallest exponent of x. Thus the GCF of the set of monomials is $3x^2$. ■

Answers

1. 20   2. 7   3. 72   4. 1

EXAMPLE 6 Find the GCF of $14p^2y^3$, $-8py^2$, $2py$, and $4p^3$.

We have

$$14p^2y^3 = 2 \cdot 7 \cdot p^2 \cdot y^3,$$

$$-8py^2 = -1 \cdot 2 \cdot 2 \cdot 2 \cdot p \cdot y^2,$$

$$2py = 2 \cdot p \cdot y,$$

$$4p^3 = 2 \cdot 2 \cdot p^3.$$

The greatest positive common factor of the coefficients is 2, the GCF of the powers of p is p, and the GCF of the powers of y is 1 since there is no y-factor in the last monomial. Thus the GCF is $2p$.

Find the GCF.

5. $12x^2$, $-16x^3$

6. $3y^6$, $-5y^3$, $2y^2$

7. $-24m^5n^6$, $12mn^3$, $-16m^2n^2$, $8m^4n^4$ **GS**

The coefficients are $-24, 12, -16,$ and ☐.

The greatest positive common factor of the coefficients is ☐.

The smallest exponent of the variable m is ☐.

The smallest exponent of the variable n is ☐.

The GCF $= 4mn$ ☐.

8. $-35x^7$, $-49x^6$, $-14x^5$, $-63x^3$

TO FIND THE GCF OF TWO OR MORE MONOMIALS

1. Find the prime factorization of the coefficients, including -1 as a factor if any coefficient is negative.

2. Determine the greatest positive common factor of the coefficients. (If the coefficients have no prime factors in common, the GCF of the coefficients is 1.)

3. Determine the greatest common factor of the powers of any variables. If any variable appears as a factor of all the monomials, include it as a factor, using the smallest exponent of the variable. (If none occurs in all the monomials, the GCF of the variables is 1.)

4. The GCF of the monomials is the product of the results of steps (2) and (3).

◀ Do Exercises 5–8.

b FACTORING WHEN TERMS HAVE A COMMON FACTOR

To multiply a monomial and a polynomial with more than one term, we multiply each term of the polynomial by the monomial using the distributive laws:

$$a(b + c) = ab + ac \quad \text{and} \quad a(b - c) = ab - ac.$$

To factor, we express a polynomial as a product using the distributive laws in reverse:

$$ab + ac = a(b + c) \quad \text{and} \quad ab - ac = a(b - c).$$

Compare.

9. a) Multiply: $3(x + 2)$.

b) Factor: $3x + 6$.

10. a) Multiply: $2x(x^2 + 5x + 4)$.

b) Factor: $2x^3 + 10x^2 + 8x$.

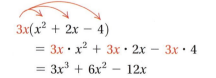

Multiply

$$3x(x^2 + 2x - 4)$$
$$= 3x \cdot x^2 + 3x \cdot 2x - 3x \cdot 4$$
$$= 3x^3 + 6x^2 - 12x$$

Factor

$$3x^3 + 6x^2 - 12x$$
$$= 3x \cdot x^2 + 3x \cdot 2x - 3x \cdot 4$$
$$= 3x(x^2 + 2x - 4)$$

◀ Do Exercises 9 and 10.

Answers

5. $4x^2$ **6.** y^2 **7.** $4mn^2$ **8.** $7x^3$
9. (a) $3x + 6$; **(b)** $3(x + 2)$
10. (a) $2x^3 + 10x^2 + 8x$; **(b)** $2x(x^2 + 5x + 4)$

Guided Solution:
7. $8, 4, 1, 2, 2$

EXAMPLE 7 Factor: $7x^2 + 14$.

We have

$$7x^2 + 14 = 7 \cdot x^2 + 7 \cdot 2 \qquad \text{Factoring each term}$$
$$= 7(x^2 + 2). \qquad \text{Factoring out the GCF, 7}$$

Check: We multiply to check:

$$7(x^2 + 2) = 7 \cdot x^2 + 7 \cdot 2 = 7x^2 + 14.$$

EXAMPLE 8 Factor: $16x^3 + 20x^2$.

$$16x^3 + 20x^2 = (4x^2)(4x) + (4x^2)(5) \qquad \text{Factoring each term}$$
$$= 4x^2(4x + 5) \qquad \text{Factoring out the GCF, } 4x^2$$

Although it is always more efficient to begin by finding the GCF, suppose in Example 8 that you had not recognized the GCF and removed only part of it, as follows:

$$16x^3 + 20x^2 = (2x^2)(8x) + (2x^2)(10)$$
$$= 2x^2(8x + 10).$$

Note that $8x + 10$ still has a common factor of 2. You need not begin again. Just continue factoring out common factors, as follows, until finished:

$$= 2x^2(2 \cdot 4x + 2 \cdot 5)$$
$$= 2x^2[2(4x + 5)]$$
$$= (2x^2 \cdot 2)(4x + 5)$$
$$= 4x^2(4x + 5).$$

EXAMPLE 9 Factor: $15x^5 - 12x^4 + 27x^3 - 3x^2$.

$$15x^5 - 12x^4 + 27x^3 - 3x^2 = (3x^2)(5x^3) - (3x^2)(4x^2) + (3x^2)(9x) - (3x^2)(1)$$
$$= 3x^2(5x^3 - 4x^2 + 9x - 1) \qquad \begin{array}{l}\text{Factoring out}\\ \text{the GCF, } 3x^2\end{array}$$

················· **Caution!** ·················

Don't forget the term -1.

As you become more familiar with factoring, you will be able to spot the GCF without factoring each term. Then you can write just the answer.

EXAMPLES Factor.

10. $24x^2 + 12x - 36 = 12(2x^2 + x - 3)$

11. $8m^3 - 16m = 8m(m^2 - 2)$

12. $14p^2y^3 - 8py^2 + 2py = 2py(7py^2 - 4y + 1)$

13. $\dfrac{4}{5}x^2 + \dfrac{1}{5}x + \dfrac{2}{5} = \dfrac{1}{5}(4x^2 + x + 2)$

Do Exercises 11–16.

················· **Caution!** ·················

Consider the following:

$$7x^2 + 14 = 7 \cdot x \cdot x + 7 \cdot 2.$$

The terms of the polynomial have been factored, but the polynomial itself has not been factored. This is not what we mean by the factorization of the polynomial. The *factorization* is

$$7(x^2 + 2). \qquad \leftarrow \text{A product}$$

The expressions 7 and $x^2 + 2$ are *factors* of $7x^2 + 14$.

Factor. Check by multiplying.

11. $x^2 + 3x$

12. $3y^6 - 5y^3 + 2y^2$

13. $9x^4y^2 - 15x^3y + 3x^2y$

14. $\dfrac{3}{4}t^3 + \dfrac{5}{4}t^2 + \dfrac{7}{4}t + \dfrac{1}{4}$

15. $35x^7 - 49x^6 + 14x^5 - 63x^3$

16. $84x^2 - 56x + 28$

Answers

11. $x(x + 3)$ **12.** $y^2(3y^4 - 5y + 2)$
13. $3x^2y(3x^2y - 5x + 1)$
14. $\dfrac{1}{4}(3t^3 + 5t^2 + 7t + 1)$
15. $7x^3(5x^4 - 7x^3 + 2x^2 - 9)$
16. $28(3x^2 - 2x + 1)$

C FACTORING BY GROUPING: FOUR TERMS

Certain polynomials with four terms can be factored using a method called *factoring by grouping.*

EXAMPLE 14 Factor: $x^2(x + 1) + 2(x + 1)$.

The binomial $x + 1$ is a common factor. We factor it out:

$$x^2(x + 1) + 2(x + 1) = (x + 1)(x^2 + 2).$$

The factorization is $(x + 1)(x^2 + 2)$.

◀ **Do Exercises 17 and 18.**

Consider the four-term polynomial

$$x^3 + x^2 + 2x + 2.$$

There is no factor other than 1 that is common to all the terms. We can, however, factor $x^3 + x^2$ and $2x + 2$ separately:

$$x^3 + x^2 = x^2(x + 1);\qquad \text{Factoring } x^3 + x^2$$
$$2x + 2 = 2(x + 1).\qquad \text{Factoring } 2x + 2$$

When we group the terms as shown above and factor each polynomial separately, we see that $(x + 1)$ appears in *both* factorizations. Thus we can factor out the common binomial factor as in Example 14:

$$\begin{aligned}x^3 + x^2 + 2x + 2 &= (x^3 + x^2) + (2x + 2)\\ &= x^2(x + 1) + 2(x + 1)\\ &= (x + 1)(x^2 + 2).\end{aligned}$$

This method of factoring is called **factoring by grouping**.

Not all polynomials with four terms can be factored by grouping, but it does give us a method to try.

EXAMPLES Factor by grouping.

15. $6x^3 - 9x^2 + 4x - 6$

$$\begin{aligned}&= (6x^3 - 9x^2) + (4x - 6)\qquad &&\text{Grouping the terms}\\ &= 3x^2(2x - 3) + 2(2x - 3)\qquad &&\text{Factoring each binomial}\\ &= (2x - 3)(3x^2 + 2)\qquad &&\text{Factoring out the common factor } 2x - 3\end{aligned}$$

We think through this process as follows:

$$6x^3 - 9x^2 + 4x - 6 = \underline{3x^2(2x - 3)}\ \square\ (2x - 3)$$

(1) Factor the first two terms.

(2) The factor $2x - 3$ gives us a hint for factoring of the last two terms.

(3) Now we ask ourselves, "What times $2x - 3$ is $4x - 6$?" The answer is $+ 2$.

Factor.

17. $x^2(x + 7) + 3(x + 7)$

18. $x^3(a + b) - 5(a + b)$

16. $x^3 + x^2 + x + 1 = (x^3 + x^2) + (x + 1)$ ········ **Caution!** ·······

Don't forget the 1.

$\quad\quad\quad = x^2(x + 1) + 1(x + 1)$ Factoring each binomial

$\quad\quad\quad = (x + 1)(x^2 + 1)$ Factoring out the common factor $x + 1$

17. $2x^3 - 6x^2 - x + 3$

$\quad = (2x^3 - 6x^2) + (-x + 3)$ Grouping as two binomials

$\quad = 2x^2(x - 3) - 1(x - 3)$ *Check:* $-1(x - 3) = -x + 3$.

$\quad = (x - 3)(2x^2 - 1)$ Factoring out the common factor $x - 3$

We can think through this process as follows.

(1) Factor the first two terms: $2x^3 - 6x^2 = 2x^2(x - 3)$.

(2) The factor $x - 3$ gives us a hint for factoring the last two terms:

$\downarrow$

$$2x^3 - 6x^2 - x + 3 = 2x^2(x - 3) \,\square\, (x - 3).$$

$\uparrow$

(3) We ask, "What times $x - 3$ is $-x + 3$?" The answer is -1.

18. $12x^5 + 20x^2 - 21x^3 - 35 = 4x^2(3x^3 + 5) - 7(3x^3 + 5)$

$\quad\quad\quad\quad\quad\quad\quad\quad = (3x^3 + 5)(4x^2 - 7)$

19. $x^3 + x^2 + 2x - 2 = x^2(x + 1) + 2(x - 1)$

This polynomial is not factorable using factoring by grouping. It may be factorable, but not by methods that we will consider in this text.

Do Exercises 19–24. ▶

There are two important points to keep in mind when factoring.

TIPS FOR FACTORING
···
• Before doing any other kind of factoring, first try to factor out the GCF.
• Always check the result of factoring by multiplying.

Factor by grouping.

GS **19.** $x^3 + 7x^2 + 3x + 21$

$\quad = x^2(\;\;\;\;\;) + 3(\;\;\;\;\;)$

$\quad = (\;\;\;\;\;)(x^2 + 3)$

20. $8t^3 + 2t^2 + 12t + 3$

21. $3m^5 - 15m^3 + 2m^2 - 10$

22. $3x^3 - 6x^2 - x + 2$

23. $4x^3 - 6x^2 - 6x + 9$

24. $y^4 - 2y^3 - 2y - 10$

Answers

19. $(x + 7)(x^2 + 3)$ **20.** $(4t + 1)(2t^2 + 3)$
21. $(m^2 - 5)(3m^3 + 2)$
22. $(x - 2)(3x^2 - 1)$ **23.** $(2x - 3)(2x^2 - 3)$
24. Not factorable using factoring by grouping

Guided Solution:
19. $x + 7, x + 7, x + 7$

5.1 **Exercise Set**

For Extra Help

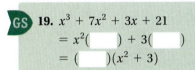

☑ Reading Check

Choose from the column on the right the expression that fits each description.

RC1. ___ A factorization of $36x^2$

RC2. ___ A factorization of $27x$

RC3. ___ The greatest common factor of $36x^2 - 27x$

RC4. ___ A factorization of $36x^2 - 27x$

a) $9x(4x - 3)$
b) $(9x)(4x)$
c) $(9x)(3)$
d) $9x$

Find the GCF.

1. 36, 42

2. 60, 75

3. 48, 72, 120

4. 90, 135, 225

5. 8, 15, 40

6. 12, 20, 75

7. x^2, $-6x$

8. x^2, $5x$

9. $3x^4$, x^2

10. $8x^4$, $-24x^2$

11. $2x^2$, $2x$, -8

12. $8x^2$, $-4x$, -20

13. $-17x^5y^3$, $34x^3y^2$, $51xy$

14. $16p^6q^4$, $32p^3q^3$, $-48pq^2$

15. $-x^2$, $-5x$, $-20x^3$

16. $-x^2$, $-6x$, $-24x^5$

17. x^5y^5, x^4y^3, x^3y^3, $-x^2y^2$

18. $-x^9y^6$, $-x^7y^5$, x^4y^4, x^3y^3

Factor. Check by multiplying.

19. $x^2 - 6x$

20. $x^2 + 5x$

21. $2x^2 + 6x$

22. $8y^2 - 8y$

23. $x^3 + 6x^2$

24. $3x^4 - x^2$

25. $8x^4 - 24x^2$

26. $5x^5 + 10x^3$

27. $2x^2 + 2x - 8$

28. $8x^2 - 4x - 20$

29. $17x^5y^3 + 34x^3y^2 + 51xy$

30. $16p^6q^4 + 32p^5q^3 - 48pq^2$

31. $6x^4 - 10x^3 + 3x^2$

32. $5x^5 + 10x^2 - 8x$

33. $x^5y^5 + x^4y^3 + x^3y^3 - x^2y^2$

34. $x^9y^6 - x^7y^5 + x^4y^4 + x^3y^3$

35. $2x^7 - 2x^6 - 64x^5 + 4x^3$

36. $8y^3 - 20y^2 + 12y - 16$

37. $1.6x^4 - 2.4x^3 + 3.2x^2 + 6.4x$

38. $2.5x^6 - 0.5x^4 + 5x^3 + 10x^2$

39. $\dfrac{5}{3}x^6 + \dfrac{4}{3}x^5 + \dfrac{1}{3}x^4 + \dfrac{1}{3}x^3$

40. $\dfrac{5}{9}x^7 + \dfrac{2}{9}x^5 - \dfrac{4}{9}x^3 - \dfrac{1}{9}x$

Factor.

41. $x^2(x + 3) + 2(x + 3)$

42. $y^2(y + 4) + 6(y + 4)$

43. $4z^2(3z - 1) + 7(3z - 1)$

44. $2x^2(4x - 3) + 5(4x - 3)$

45. $2x^2(3x + 2) + (3x + 2)$

46. $3z^2(2z + 7) + (2z + 7)$

47. $5a^3(2a - 7) - (2a - 7)$

48. $m^4(8 - 3m) - 3(8 - 3m)$

Factor by grouping.

49. $x^3 + 3x^2 + 2x + 6$

50. $6z^3 + 3z^2 + 2z + 1$

51. $2x^3 + 6x^2 + x + 3$

52. $3x^3 + 2x^2 + 3x + 2$

53. $8x^3 - 12x^2 + 6x - 9$

54. $10x^3 - 25x^2 + 4x - 10$

55. $12p^3 - 16p^2 + 3p - 4$

56. $18x^3 - 21x^2 + 30x - 35$

57. $5x^3 - 5x^2 - x + 1$

58. $7x^3 - 14x^2 - x + 2$

59. $x^3 + 8x^2 - 3x - 24$

60. $2x^3 + 12x^2 - 5x - 30$

61. $2x^3 - 8x^2 - 9x + 36$

62. $20g^3 - 4g^2 - 25g + 5$

Skill Maintenance

Multiply.   [4.5b], [4.6d], [4.7f]

63. $(y + 5)(y + 7)$

64. $(y + 7)^2$

65. $(y + 7)(y - 7)$

66. $(y - 7)^2$

67. $8x(2x^2 - 6x + 1)$

68. $(7w + 6)(4w - 11)$

69. $(7w + 6)^2$

70. $(4w - 11)^2$

71. $(4w - 11)(4w + 11)$

72. $-y(-y^2 + 3y - 5)$

73. $(3x - 5y)(2x + 7y)$

74. $(5x - t)^2$

Synthesis

Factor.

75. $4x^5 + 6x^3 + 6x^2 + 9$

76. $x^6 + x^4 + x^2 + 1$

77. $x^{12} + x^7 + x^5 + 1$

78. $x^3 - x^2 - 2x + 5$

79. $p^3 + p^2 - 3p + 10$

80. $4y^6 + 2y^4 - 12y^3 - 6y$

a FACTORING $x^2 + bx + c$

We now begin a study of the factoring of trinomials. We first factor trinomials like

$$x^2 + 5x + 6 \quad \text{and} \quad x^2 + 3x - 10$$

by a refined *trial-and-error process*. In this section, we restrict our attention to trinomials of the type $ax^2 + bx + c$, where the **leading coefficient** a is 1.

Compare the following multiplications:

$$
\begin{array}{cccc}
\text{F} & \text{O} & \text{I} & \text{L} \\
\downarrow & \downarrow & \downarrow & \downarrow
\end{array}
$$

$$(x + 2)(x + 5) = x^2 + 5x + 2x + 2 \cdot 5$$
$$= x^2 + 7x + 10;$$

$$(x - 2)(x - 5) = x^2 - 5x - 2x + (-2)(-5)$$
$$= x^2 - 7x + 10;$$

$$(x + 3)(x - 7) = x^2 - 7x + 3x + 3(-7)$$
$$= x^2 - 4x - 21;$$

$$(x - 3)(x + 7) = x^2 + 7x - 3x + (-3)7$$
$$= x^2 + 4x - 21.$$

Note that for all four products:

- The product of the two binomials is a trinomial.
- The coefficient of x in the trinomial is the sum of the constant terms in the binomials.
- The constant term in the trinomial is the product of the constant terms in the binomials.

These observations lead to a method for factoring certain trinomials. The first type we consider has a positive constant term, just as in the first two multiplications above.

Constant Term Positive

To factor $x^2 + 7x + 10$, we think of FOIL in reverse. Since $x \cdot x - x^2$, the first term of each binomial is x.

Next, we look for numbers p and q such that

$$x^2 + 7x + 10 = (x + p)(x + q).$$

To get the middle term and the last term of the trinomial, we look for two numbers p and q whose product is 10 and whose sum is 7. Those numbers are 2 and 5. Thus the factorization is

$$(x + 2)(x + 5).$$

Check: $(x + 2)(x + 5) = x^2 + 5x + 2x + 10$
$$= x^2 + 7x + 10.$$

EXAMPLE 1 Factor: $x^2 + 5x + 6$.

Think of FOIL in reverse. The first term of each factor is x:

$$(x + \square)(x + \square).$$

Next, we look for two numbers whose product is 6 and whose sum is 5. All the pairs of factors of 6 are shown in the table on the left below. Since both the product, 6, and the sum, 5, of the pair of numbers must be positive, we need consider only the positive factors, listed in the table on the right.

PAIRS OF FACTORS	SUMS OF FACTORS
1, 6	7
−1, −6	−7
2, 3	5
−2, −3	−5

PAIRS OF FACTORS	SUMS OF FACTORS
1, 6	7
2, 3	**5**

The numbers we need are 2 and 3.

The factorization is $(x + 2)(x + 3)$. We can check by multiplying to see whether we get the original trinomial.

Check: $(x + 2)(x + 3) = x^2 + 3x + 2x + 6 = x^2 + 5x + 6.$

Do Exercises 1 and 2. ▶

Compare these multiplications:

$$(x - 2)(x - 5) = x^2 - 5x - 2x + 10 = x^2 - 7x + 10;$$
$$(x + 2)(x + 5) = x^2 + 5x + 2x + 10 = x^2 + 7x + 10.$$

TO FACTOR $x^2 + bx + c$ WHEN c IS POSITIVE

When the constant term of a trinomial is positive, look for two numbers with the same sign. The sign is that of the middle term:

$$x^2 - 7x + 10 = (x - 2)(x - 5);$$

$$x^2 + 7x + 10 = (x + 2)(x + 5).$$

EXAMPLE 2 Factor: $y^2 - 8y + 12$.

Since the constant term, 12, is positive and the coefficient of the middle term, −8, is negative, we look for a factorization of 12 in which both factors are negative. Their sum must be −8.

PAIRS OF FACTORS	SUMS OF FACTORS	
−1, −12	−13	
−2, −6	**−8**	← The numbers we need
−3, −4	−7	are −2 and −6.

The factorization is $(y - 2)(y - 6)$. The student should check by multiplying.

Do Exercises 3–5. ▶

Factor.

 1. $x^2 + 7x + 12$

Complete the following table.

PAIRS OF FACTORS	SUMS OF FACTORS
1, 12	13
−1, −12	$\square$
2, 6	$\square$
−2, −6	$\square$
3, 4	$\square$
−3, −4	$\square$

Because both 7 and 12 are positive, we need consider only the ____ factors in the table above.

$x^2 + 7x + 12$
$= (x + 3)(\quad)$

2. $x^2 + 13x + 36$

3. Explain why you would *not* consider the pairs of factors listed below in factoring $y^2 - 8y + 12$.

PAIRS OF FACTORS	SUMS OF FACTORS
1, 12	
2, 6	
3, 4	

Factor.

4. $x^2 - 8x + 15$

5. $t^2 - 9t + 20$

Answers

1. $(x + 3)(x + 4)$ **2.** $(x + 4)(x + 9)$
3. The coefficient of the middle term, −8, is negative. **4.** $(x - 5)(x - 3)$
5. $(t - 5)(t - 4)$

Guided Solution:
1. −13, 8, −8, 7, −7; positive; $x + 4$

Constant Term Negative

As we saw in two of the multiplications earlier in this section, the product of two binomials can have a negative constant term:

$$(x + 3)(x - 7) = x^2 - 4x - 21$$

and

$$(x - 3)(x + 7) = x^2 + 4x - 21.$$

Note that when the signs of the constants in the binomials are reversed, only the sign of the middle term in the product changes.

EXAMPLE 3 Factor: $x^2 - 8x - 20$.

The constant term, -20, must be expressed as the product of a negative number and a positive number. Since the sum of these two numbers must be negative (specifically, -8), the negative number must have the greater absolute value.

PAIRS OF FACTORS	SUMS OF FACTORS
1, -20	-19
2, -10	**-8**
4, -5	-1
5, -4	1
10, -2	8
20, -1	19

The numbers we need are 2 and -10.

Because these sums are all positive, for this problem all the corresponding pairs can be disregarded. Note that in all three pairs, the positive number has the greater absolute value.

The numbers that we are looking for are 2 and -10. The factorization is $(x + 2)(x - 10)$.

Check: $(x + 2)(x - 10) = x^2 - 10x + 2x - 20$
$= x^2 - 8x - 20.$

TO FACTOR $x^2 + bx + c$ WHEN c IS NEGATIVE

When the constant term of a trinomial is negative, look for two numbers whose product is negative. One must be positive and the other negative:

$$x^2 - 4x - 21 = (x + 3)(x - 7);$$

$$x^2 + 4x - 21 = (x - 3)(x + 7).$$

Consider pairs of numbers for which the number with the larger absolute value has the same sign as b, the coefficient of the middle term.

---

6. Consider $x^2 - 5x - 24$.

a) Explain why you would *not* consider the pairs of factors listed below in factoring $x^2 - 5x - 24$.

PAIRS OF FACTORS	SUMS OF FACTORS
$-1, 24$	
$-2, 12$	
$-3, 8$	
$-4, 6$	

b) Explain why you *would* consider the pairs of factors listed below in factoring $x^2 - 5x - 24$.

PAIRS OF FACTORS	SUMS OF FACTORS
$1, -24$	
$2, -12$	
$3, -8$	
$4, -6$	

c) Factor: $x^2 - 5x - 24$.

◀ **Do Exercises 6 and 7. (Exercise 7 is on the following page.)**

Answers

6. (a) The positive factor has the larger absolute value. **(b)** The negative factor has the larger absolute value. **(c)** $(x + 3)(x - 8)$

EXAMPLE 4 Factor: $t^2 - 24 + 5t$.

We first write the trinomial in descending order: $t^2 + 5t - 24$. Since the constant term, -24, is negative, factorizations of -24 will have one positive factor and one negative factor. The sum of the factors must be 5, so we consider only pairs of factors in which the positive factor has the larger absolute value.

PAIRS OF FACTORS	SUMS OF FACTORS	
$-1, 24$	23	
$-2, 12$	10	
$-3, \ 8$	5	← The numbers we need are -3 and 8.
$-4, \ 6$	2	

The factorization is $(t - 3)(t + 8)$. The check is left to the student.

Do Exercises 8 and 9. ▶

EXAMPLE 5 Factor: $x^4 - x^2 - 110$.

Consider this trinomial as $(x^2)^2 - x^2 - 110$. We look for numbers p and q such that

$$x^4 - x^2 - 110 = (x^2 + p)(x^2 + q).$$

We look for two numbers whose product is -110 and whose sum is -1. The middle-term coefficient, -1, is small compared to -110. This tells us that the desired factors are close to each other in absolute value. The numbers we want are 10 and -11. The factorization is $(x^2 + 10)(x^2 - 11)$.

EXAMPLE 6 Factor: $a^2 + 4ab - 21b^2$.

We consider the trinomial in the equivalent form

$$a^2 + 4ba - 21b^2,$$

and think of $-21b^2$ as the "constant" term and $4b$ as the "coefficient" of the middle term. Then we try to express $-21b^2$ as a product of two factors whose sum is $4b$. Those factors are $-3b$ and $7b$. The factorization is $(a - 3b)(a + 7b)$.

Check: $(a - 3b)(a + 7b) = a^2 + 7ab - 3ba - 21b^2$
$$= a^2 + 4ab - 21b^2.$$

There are polynomials that are not factorable.

EXAMPLE 7 Factor: $x^2 - x + 5$.

Since 5 has very few factors, we can easily check all possibilities.

PAIRS OF FACTORS	SUMS OF FACTORS
$5, \ \ 1$	6
$-5, -1$	-6

There are no factors of 5 whose sum is -1. Thus the polynomial is *not* factorable into factors that are polynomials with rational-number coefficients.

---

7. Consider $x^2 + 5x - 6$.

a) Explain why you would *not* consider the pairs of factors listed below in factoring $x^2 + 5x - 6$.

PAIRS OF FACTORS	SUMS OF FACTORS
$1, -6$	
$2, -3$	

b) Explain why you *would* consider the pairs of factors listed below in factoring $x^2 + 5x - 6$.

PAIRS OF FACTORS	SUMS OF FACTORS
$-1, 6$	
$-2, 3$	

c) Factor: $x^2 + 5x - 6$.

Factor.

GS **8.** $a^2 - 40 + 3a$

First, rewrite in descending order: $a^2 + 3a - \boxed{}$.

PAIRS OF FACTORS	SUMS OF FACTORS
$-1, 40$	$\boxed{}$
$-2, 30$	$\boxed{}$
$-4, 10$	$\boxed{}$
$-5, \ \ 8$	$\boxed{}$

The factorization is $(a - 5) (\boxed{})$.

9. $-18 - 3t + t^2$

Answers

7. (a) The negative factor has the larger absolute value. **(b)** The positive factor has the larger absolute value. **(c)** $(x - 1)(x + 6)$
8. $(a - 5)(a + 8)$ **9.** $(t - 6)(t + 3)$

Guided Solution:
8. 40, 39, 28, 6, 3, $a + 8$

Factor.

10. $y^2 - 12 - 4y$

11. $t^4 + 5t^2 - 14$

12. $x^2 + 2x + 7$

In this text, a polynomial like $x^2 - x + 5$ that cannot be factored further is said to be **prime**. In more advanced courses, polynomials like $x^2 - x + 5$ can be factored and are not considered prime.

◀ **Do Exercises 10–12.**

Often factoring requires two or more steps. In general, when told to factor, we should be sure to *factor completely*. This means that the final factorization should not contain any factors that can be factored further.

EXAMPLE 8 Factor: $2x^3 - 20x^2 + 50x$.

Always look first for a common factor. This time there is one, $2x$, which we factor out first:

$$2x^3 - 20x^2 + 50x = 2x(x^2 - 10x + 25).$$

Now consider $x^2 - 10x + 25$. Since the constant term is positive and the coefficient of the middle term is negative, we look for a factorization of 25 in which both factors are negative. Their sum must be -10.

PAIRS OF FACTORS	SUMS OF FACTORS	
$-25, -1$	-26	
$-5, -5$	**-10** ←	The numbers we need are -5 and -5.

The factorization of $x^2 - 10x + 25$ is $(x - 5)(x - 5)$, or $(x - 5)^2$. The final factorization is $2x(x - 5)^2$. We check by multiplying:

$$\begin{aligned} 2x(x - 5)^2 &= 2x(x^2 - 10x + 25) \\ &= (2x)(x^2) - (2x)(10x) + (2x)(25) \\ &= 2x^3 - 20x^2 + 50x. \end{aligned}$$

Factor.

13. $x^3 + 4x^2 - 12x$

14. $p^2 - pq - 3pq^2$

15. $3x^3 + 24x^2 + 48x$

◀ **Do Exercises 13–15.**

Once any common factors have been factored out, the following summary can be used to factor $x^2 + bx + c$.

TO FACTOR $x^2 + bx + c$

1. First arrange the polynomial in descending order.
2. Use a trial-and-error process that looks for factors of c whose sum is b.
3. If c is positive, the signs of the factors are the same as the sign of b.
4. If c is negative, one factor is positive and the other is negative. If the sum of two factors is the opposite of b, changing the sign of each factor will give the desired factors whose sum is b.
5. Check by multiplying.

Leading Coefficient -1

EXAMPLE 9 Factor: $10 - 3x - x^2$.

Note that the polynomial is written in ascending order. When we write it in descending order, we get

$$-x^2 - 3x + 10,$$

which has a leading coefficient of -1. Before factoring in such a case, we can factor out a -1, as follows:

$$-x^2 - 3x + 10 = -1 \cdot x^2 + (-1)(3x) + (-1)(-10)$$
$$= -1(x^2 + 3x - 10).$$

Then we proceed to factor $x^2 + 3x - 10$. We get

$$-x^2 - 3x + 10 = -1(x^2 + 3x - 10) = -1(x + 5)(x - 2).$$

We can also express this answer in two other ways by multiplying either binomial by -1. Thus each of the following is a correct answer:

$$10 - 3x - x^2 = -1(x + 5)(x - 2)$$
$$= (-x - 5)(x - 2) \quad \text{Multiplying } x + 5 \text{ by } -1$$
$$= (x + 5)(-x + 2). \quad \text{Multiplying } x - 2 \text{ by } -1$$

Do Exercises 16 and 17. ▶

Factor.

16. $14 + 5x - x^2$

17. $-x^2 + 3x + 18$

Answers

16. $-1(x + 2)(x - 7)$, or $(-x - 2)(x - 7)$, or $(x + 2)(-x + 7)$
17. $-1(x + 3)(x - 6)$, or $(-x - 3)(x - 6)$, or $(x + 3)(-x + 6)$

5.2 **Exercise Set**

For Extra Help
MyMathLab®
MathXL®
PRACTICE   WATCH   READ   REVIEW

☑ Reading Check

Determine whether each statement is true or false.

RC1. The leading coefficient of $x^2 - 3x - 10$ is 1.

RC2. To factor $x^2 - 3x - 10$, we look for two numbers whose product is -10.

RC3. To factor $x^2 - 3x - 10$, we look for two numbers whose sum is -3.

RC4. The factorization of $x^2 - 3x - 10$ is $(x + 5)(x - 2)$.

a Factor. Remember that you can check by multiplying.

1. $x^2 + 8x + 15$

PAIRS OF FACTORS	SUMS OF FACTORS

2. $x^2 + 5x + 6$

PAIRS OF FACTORS	SUMS OF FACTORS

3. $x^2 + 7x + 12$

PAIRS OF FACTORS	SUMS OF FACTORS

4. $x^2 + 9x + 8$

PAIRS OF FACTORS	SUMS OF FACTORS

5. $x^2 - 6x + 9$

PAIRS OF FACTORS	SUMS OF FACTORS

6. $y^2 - 11y + 28$

PAIRS OF FACTORS	SUMS OF FACTORS

7. $x^2 - 5x - 14$

PAIRS OF FACTORS	SUMS OF FACTORS

8. $a^2 + 7a - 30$

PAIRS OF FACTORS	SUMS OF FACTORS

9. $b^2 + 5b + 4$

PAIRS OF FACTORS	SUMS OF FACTORS

10. $z^2 - 8z + 7$

PAIRS OF FACTORS	SUMS OF FACTORS

11. $t^2 + 3t - 18$

PAIRS OF FACTORS	SUMS OF FACTORS

12. $t^2 + 8t + 16$

PAIRS OF FACTORS	SUMS OF FACTORS

13. $d^2 - 7d + 10$

14. $t^2 - 12t + 35$

15. $y^2 - 11y + 10$

16. $x^2 - 4x - 21$

17. $x^2 + x + 1$

18. $x^2 + 5x + 3$

19. $x^2 - 7x - 18$

20. $y^2 - 3y - 28$

21. $x^3 - 6x^2 - 16x$

22. $x^3 - x^2 - 42x$

23. $y^3 - 4y^2 - 45y$

24. $x^3 - 7x^2 - 60x$

25. $-2x - 99 + x^2$

26. $x^2 - 72 + 6x$

27. $c^4 + c^2 - 56$

28. $b^4 + 5b^2 - 24$

29. $a^4 + 2a^2 - 35$

30. $x^4 - x^2 - 6$

31. $x^2 + x - 42$

32. $x^2 + 2x - 15$

33. $7 - 2p + p^2$

34. $11 - 3w + w^2$

35. $x^2 + 20x + 100$

36. $a^2 + 19a + 88$

37. $2z^3 - 2z^2 - 24z$

38. $5w^4 - 20w^3 - 25w^2$

39. $3t^4 + 3t^3 + 3t^2$

40. $4y^5 - 4y^4 - 4y^3$

41. $x^4 - 21x^3 - 100x^2$ **42.** $x^4 - 20x^3 + 96x^2$ **43.** $x^2 - 21x - 72$ **44.** $4x^2 + 40x + 100$

45. $x^2 - 25x + 144$ **46.** $y^2 - 21y + 108$ **47.** $a^2 + a - 132$ **48.** $a^2 + 9a - 90$

49. $3t^2 + 6t + 3$ **50.** $2y^2 + 24y + 72$ **51.** $w^4 - 8w^3 + 16w^2$ **52.** $z^5 - 6z^4 + 9z^3$

53. $30 + 7x - x^2$ **54.** $45 + 4x - x^2$ **55.** $24 - a^2 - 10a$ **56.** $-z^2 + 36 - 9z$

57. $120 - 23x + x^2$ **58.** $96 + 22d + d^2$ **59.** $108 - 3x - x^2$ **60.** $112 + 9y - y^2$

61. $y^2 - 0.2y - 0.08$ **62.** $t^2 - 0.3t - 0.10$ **63.** $p^2 + 3pq - 10q^2$ **64.** $a^2 + 2ab - 3b^2$

65. $84 - 8t - t^2$ **66.** $72 - 6m - m^2$ **67.** $m^2 + 5mn + 4n^2$ **68.** $x^2 + 11xy + 24y^2$

69. $s^2 - 2st - 15t^2$ **70.** $p^2 + 5pq - 24q^2$ **71.** $6a^{10} - 30a^9 - 84a^8$ **72.** $7x^9 - 28x^8 - 35x^7$

Skill Maintenance

Solve.

73. $2x + 7 = 0$ [2.3a]

74. $-2y + 11y = 108$ [2.3b]

75. $3x - 2 - x = 5x - 9x$ [2.3b]

76. $\dfrac{1}{2}x - \dfrac{1}{3}x = \dfrac{2}{3} + \dfrac{5}{6}x$ [2.3b]

77. $5(t - 1) - 3 = 4t - (7t - 2)$ [2.3c]

78. $10 - (x - 7) = 4x - (1 + 5x)$ [2.3c]

79. $-2x < 48$ [2.7d]

80. $4x - 8x + 16 \geq 6(x - 2)$ [2.7e]

81. $-6(x - 4) + 8(4 - x) \leq 3(x - 7)$ [2.7e]

Solve. [2.4b]

82. $A = \dfrac{p + w}{2}$, for p

83. $y = mx + b$, for x

84. $a - c + r = 2$, for c

85. *Major League Baseball Attendance.* Total attendance at Major League baseball games was about 74.97 million in 2012. This was a 2% increase over the attendance in 2011. What was the attendance in 2011? [2.5a]

Source: Based on information from baseball-reference.com

86. The first angle of a triangle is four times as large as the second. The measure of the third angle is 30° greater than that of the second. Find the angle measures. [2.6a]

Synthesis

87. Find all integers m for which $y^2 + my + 50$ can be factored.

88. Find all integers b for which $a^2 + ba - 50$ can be factored.

Factor completely.

89. $x^2 - \dfrac{1}{2}x - \dfrac{3}{16}$

90. $x^2 - \dfrac{1}{4}x - \dfrac{1}{8}$

91. $x^2 + \dfrac{2}{3}x + \dfrac{1}{9}$

92. $x^2 - \dfrac{2}{5}x + \dfrac{1}{25}$

93. $x^2 + \dfrac{30}{7}x - \dfrac{25}{7}$

94. $\dfrac{1}{3}x^3 + \dfrac{1}{3}x^2 - 2x$

95. $b^{2n} + 7b^n + 10$

96. $a^{2m} - 11a^m + 28$

Find a polynomial in factored form for the shaded area in each figure. (Leave answers in terms of π.)

97.

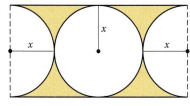

98.

OBJECTIVE

a Factor trinomials of the type $ax^2 + bx + c, a \neq 1$, using the FOIL method.

SKILL TO REVIEW

Objective 4.6a: Multiply two binomials mentally using the FOIL method.

Multiply.

1. $(2x + 3)(x + 1)$

2. $(3x - 4)(2x - 1)$

The ac-method in Section 5.4

To the student: In Section 5.4, we will consider an alternative method for the same kind of factoring. It involves factoring by grouping and is called the *ac*-method.

To the instructor: We present two ways to factor general trinomials in Sections 5.3 and 5.4: the FOIL method in Section 5.3 and the *ac*-method in Section 5.4. You can teach both methods and let the student use the one that he or she prefers or you can select just one.

Answers

Skill to Review:
1. $2x^2 + 5x + 3$ **2.** $6x^2 - 11x + 4$

In this section, we factor trinomials in which the coefficient of the leading term x^2 is not 1. The procedure we use is a refined trial-and-error method.

a THE FOIL METHOD

We want to factor trinomials of the type $ax^2 + bx + c$. Consider the following multiplication:

$$
\begin{array}{ccccc}
 & \text{F} & \text{O} & \text{I} & \text{L} \\
(2x + 5)(3x + 4) = & 6x^2 + & 8x + & 15x + & 20 \\
 = & 6x^2 + & & 23x & + 20
\end{array}
$$

F	**O + I**	**L**
$2 \cdot 3$	$2 \cdot 4 \quad 5 \cdot 3$	$5 \cdot 4$

To factor $6x^2 + 23x + 20$, we reverse the above multiplication, using what we might call an "unFOIL" process. We look for two binomials $rx + p$ and $sx + q$ whose product is $(rx + p)(sx + q) = 6x^2 + 23x + 20$. The product of the First terms must be $6x^2$. The product of the Outside terms plus the product of the Inside terms must be $23x$. The product of the Last terms must be 20. We know from the preceding discussion that the answer is $(2x + 5)(3x + 4)$. Generally, however, finding such an answer is a refined trial-and-error process. It turns out that $(-2x - 5)(-3x - 4)$ is also a correct answer, but we generally choose an answer in which the first coefficients are positive.

We will use the following trial-and-error method.

THE FOIL METHOD

To factor $ax^2 + bx + c, a \neq 1$, using the FOIL method:

1. Factor out the largest common factor, if one exists.

2. Find two First terms whose product is ax^2.

$$(\square x + \;\;)(\square x + \;\;) = ax^2 + bx + c.$$

$\qquad\qquad\qquad$ FOIL

3. Find two Last terms whose product is c:

$$(\;\;x + \square)(\;\;x + \square) = ax^2 + bx + c.$$

$\qquad\qquad\qquad$ FOIL

4. Look for Outer and Inner products resulting from steps (2) and (3) for which the sum is bx:

$$(\square x + \square)(\square x + \square) = ax^2 + bx + c.$$

$\qquad\qquad$ I $\qquad\qquad\qquad$ FOIL

$\qquad\qquad$ O

5. Always check by multiplying.

EXAMPLE 1 Factor: $3x^2 - 10x - 8$.

1) First, we check for a common factor. Here there is none (other than 1 or −1).

2) Find two **First** terms whose product is $3x^2$.

   The only possibilities for the First terms are $3x$ and x, so any factorization must be of the form

   $(3x + \Box)(x + \Box)$.

3) Find two **Last** terms whose product is −8.

   Possible factorizations of −8 are

   $(-8) \cdot 1, \quad 8 \cdot (-1), \quad (-2) \cdot 4, \quad \text{and} \quad 2 \cdot (-4).$

   Since the First terms are not identical, we must also consider

   $1 \cdot (-8), \quad (-1) \cdot 8, \quad 4 \cdot (-2), \quad \text{and} \quad (-4) \cdot 2.$

4) Inspect the **Outside** and **Inside** products resulting from steps (2) and (3). Look for a combination in which the sum of the products is the middle term, $-10x$:

Trial	*Product*	
$(3x - 8)(x + 1)$	$3x^2 + 3x - 8x - 8$	
	$= 3x^2 - 5x - 8$	← Wrong middle term
$(3x + 8)(x - 1)$	$3x^2 - 3x + 8x - 8$	
	$= 3x^2 + 5x - 8$	← Wrong middle term
$(3x - 2)(x + 4)$	$3x^2 + 12x - 2x - 8$	
	$= 3x^2 + 10x - 8$	← Wrong middle term
$(3x + 2)(x - 4)$	$3x^2 - 12x + 2x - 8$	
	$= 3x^2 - 10x - 8$	← **Correct middle term!**
$(3x + 1)(x - 8)$	$3x^2 - 24x + x - 8$	
	$= 3x^2 - 23x - 8$	← Wrong middle term
$(3x - 1)(x + 8)$	$3x^2 + 24x - x - 8$	
	$= 3x^2 + 23x - 8$	← Wrong middle term
$(3x + 4)(x - 2)$	$3x^2 - 6x + 4x - 8$	
	$= 3x^2 - 2x - 8$	← Wrong middle term
$(3x - 4)(x + 2)$	$3x^2 + 6x - 4x - 8$	
	$= 3x^2 + 2x - 8$	← Wrong middle term

The correct factorization is $(3x + 2)(x - 4)$.

5) Check: $(3x + 2)(x - 4) = 3x^2 - 10x - 8$.

Two observations can be made from Example 1. First, we listed all possible trials even though we could have stopped after having found the correct factorization. We did this to show that each trial differs only in the middle term of the product. **Second, note that only the sign of the middle term changes when the signs in the binomials are reversed:**

Plus   Minus

$(3x + 4)(x - 2) = 3x^2 - 2x - 8$

Minus   Plus ————— Middle term changes sign

$(3x - 4)(x + 2) = 3x^2 + 2x - 8.$

CALCULATOR CORNER

A partial check of a factorization can be performed using a table or a graph. To check the factorization $6x^3 - 9x^2 + 4x - 6 = (2x - 3)(3x^2 + 2)$, for example, we enter $y_1 = 6x^3 - 9x^2 + 4x - 6$ and $y_2 = (2x - 3)(3x^2 + 2)$ on the equation-editor screen. Then we set up a table in AUTO mode. If the factorization is correct, the values of y_1 and y_2 will be the same regardless of the table settings used.

X	Y₁	Y₂
−3	−261	−261
−2	−98	−98
−1	−25	−25
0	−6	−6
1	−5	−5
2	14	14
3	87	87
X = −3		

We can also graph $y_1 = 6x^3 - 9x^2 + 4x - 6$ and $y_2 = (2x - 3)(3x^2 + 2)$. If the graphs appear to coincide, the factorization is probably correct.

$y_1 = 6x^3 - 9x^2 + 4x - 6,$
$y_2 = (2x - 3)(3x^2 + 2)$

Yscl = 2

Keep in mind that these procedures provide only a partial check since we cannot view all possible values of x in a table or see the entire graph.

EXERCISES: Use a table or a graph to determine whether each factorization is correct.

1. $24x^2 - 76x + 40 = 4(3x - 2)(2x - 5)$

2. $4x^2 - 5x - 6 = (4x + 3)(x - 2)$

3. $5x^2 + 17x - 12 = (5x + 3)(x - 4)$

4. $10x^2 + 37x + 7 = (5x - 1)(2x + 7)$

Factor.

1. $2x^2 - x - 15$

2. $12x^2 - 17x - 5$

◀ **Do Exercises 1 and 2.**

EXAMPLE 2 Factor: $24x^2 - 76x + 40$.

1) First, we factor out the largest common factor, 4:

$$4(6x^2 - 19x + 10).$$

Now we factor the trinomial $6x^2 - 19x + 10$.

2) Because $6x^2$ can be factored as $3x \cdot 2x$ or $6x \cdot x$, we have these possibilities for factorizations:

$$(3x + \boxed{})(2x + \boxed{}) \quad \text{or} \quad (6x + \boxed{})(x + \boxed{}).$$

3) There are four pairs of factors of 10 and each pair can be listed in two ways:

$$10, 1 \qquad -10, -1 \qquad 5, 2 \qquad -5, -2$$

and

$$1, 10 \qquad -1, -10 \qquad 2, 5 \qquad -2, -5.$$

4) The two possibilities from step (2) and the eight possibilities from step (3) give $2 \cdot 8$, or 16 possibilities for factorizations. We look for **O**utside and **I**nside products resulting from steps (2) and (3) for which the sum is the middle term, $-19x$. Since the sign of the middle term is negative, but the sign of the last term, 10, is positive, both factors of 10 must be negative. This means only four pairings from step (3) need be considered. We first try these factors with

$$(3x + \boxed{})(2x + \boxed{}).$$

If none gives the correct factorization, we will consider

$$(6x + \boxed{})(x + \boxed{}).$$

Trial	*Product*	
$(3x - 10)(2x - 1)$	$6x^2 - 3x - 20x + 10$	
	$= 6x^2 - 23x + 10$	← Wrong middle term
$(3x - 1)(2x - 10)$	$6x^2 - 30x - 2x + 10$	
	$= 6x^2 - 32x + 10$	← Wrong middle term
$(3x - 5)(2x - 2)$	$6x^2 - 6x - 10x + 10$	
	$= 6x^2 - 16x + 10$	← Wrong middle term
$(3x - 2)(2x - 5)$	$6x^2 - 15x - 4x + 10$	
	$= 6x^2 - 19x + 10$	← **Correct middle term!**

Since we have a correct factorization, we need not consider

$$(6x + \boxed{})(x + \boxed{}).$$

The factorization of $6x^2 - 19x + 10$ is $(3x - 2)(2x - 5)$, but *do not forget the common factor*! We must include it in order to factor the original trinomial:

$$24x^2 - 76x + 40 = 4(6x^2 - 19x + 10)$$
$$= 4(3x - 2)(2x - 5).$$

5) Check: $4(3x - 2)(2x - 5) = 4(6x^2 - 19x + 10) = 24x^2 - 76x + 40.$

······································ **Caution!** ······································

When factoring any polynomial, always look for a common factor first. Failure to do so is such a common error that this caution bears repeating.

···

Answers

1. $(2x + 5)(x - 3)$ **2.** $(4x + 1)(3x - 5)$

In Example 2, look again at the possibility $(3x - 5)(2x - 2)$. Without multiplying, we can reject such a possibility. To see why, consider the following:

$$(3x - 5)(2x - 2) = (3x - 5)(2)(x - 1) = 2(3x - 5)(x - 1).$$

The expression $2x - 2$ has a common factor, 2. But we removed the *largest* common factor in the first step. If $2x - 2$ were one of the factors, then 2 would have to be a common factor in addition to the original 4. Thus, $(2x - 2)$ cannot be part of the factorization of the original trinomial.

> Given that the largest common factor is factored out at the outset, we need not consider factorizations that have a common factor.

Do Exercises 3 and 4. ▶

Factor.

3. $3x^2 - 19x + 20$

4. $20x^2 - 46x + 24$

EXAMPLE 3 Factor: $10x^2 + 37x + 7$.

1) There is no common factor (other than 1 or -1).

2) Because $10x^2$ factors as $10x \cdot x$ or $5x \cdot 2x$, we have these possibilities for factorizations:

$$(10x + \ \square \ )(x + \ \square \ ) \quad \text{or} \quad (5x + \ \square \ )(2x + \ \square \ ).$$

3) There are two pairs of factors of 7 and each pair can be listed in two ways:

$$1, 7 \quad -1, -7 \qquad \text{and} \qquad 7, 1 \quad -7, -1.$$

4) From steps (2) and (3), we see that there are 8 possibilities for factorizations. Look for **O**uter and **I**nner products for which the sum is the middle term. Because all coefficients in $10x^2 + 37x + 7$ are positive, we need consider only positive factors of 7. The possibilities are

$$(10x + 1)(x + 7) = 10x^2 + 71x + 7,$$
$$(10x + 7)(x + 1) = 10x^2 + 17x + 7,$$
$$(5x + 7)(2x + 1) = 10x^2 + 19x + 7,$$
$$(5x + 1)(2x + 7) = 10x^2 + 37x + 7. \quad \leftarrow \textbf{Correct middle term}$$

The factorization is $(5x + 1)(2x + 7)$.

5) Check: $(5x + 1)(2x + 7) = 10x^2 + 37x + 7$.

Do Exercise 5. ▶

5. Factor: $6x^2 + 7x + 2$.

TIPS FOR FACTORING $ax^2 + bx + c, a \neq 1$

- Always factor out the largest common factor first, if one exists.
- Once the common factor has been factored out of the original trinomial, no binomial factor can contain a common factor (other than 1 or -1).
- If c is positive, then the signs in both binomial factors must match the sign of b. (This assumes that $a > 0$.)
- Reversing the signs in the binomials reverses the sign of the middle term of their product.
- Organize your work so that you can keep track of which possibilities have or have not been checked.
- Always check by multiplying.

Answers

3. $(3x - 4)(x - 5)$ **4.** $2(5x - 4)(2x - 3)$
5. $(2x + 1)(3x + 2)$

EXAMPLE 4 Factor: $10x + 8 - 3x^2$.

An important problem-solving strategy is to find a way to make new problems look like problems we already know how to solve. The factoring tips on the preceding page apply only to trinomials of the form $ax^2 + bx + c$, with $a > 0$. This leads us to rewrite $10x + 8 - 3x^2$ in descending order:

$$10x + 8 - 3x^2 = -3x^2 + 10x + 8. \qquad \text{Writing in descending order}$$

Although $-3x^2 + 10x + 8$ looks similar to the trinomials we have factored, the factoring tips require a positive leading coefficient, so we factor out -1:

$$-3x^2 + 10x + 8 = -1(3x^2 - 10x - 8) \qquad \text{Factoring out } -1 \text{ changes the signs of the coefficients.}$$
$$= -1(3x + 2)(x - 4). \qquad \text{Using the result from Example 1}$$

The factorization of $10x + 8 - 3x^2$ is $-1(3x + 2)(x - 4)$. Other correct answers are

Factor.

6. $2 - x - 6x^2$

7. $2x + 8 - 6x^2$

$$10x + 8 - 3x^2 = (3x + 2)(-x + 4) \qquad \text{Multiplying } x - 4 \text{ by } -1$$
$$= (-3x - 2)(x - 4). \qquad \text{Multiplying } 3x + 2 \text{ by } -1$$

◀ **Do Exercises 6 and 7.**

EXAMPLE 5 Factor: $6p^2 - 13pv - 28v^2$.

1) Factor out a common factor, if any.

There is none (other than 1 or -1).

2) Factor the first term, $6p^2$.

Possibilities are $2p, 3p$ and $6p, p$. We have these as possibilities for factorizations:

$$(2p + \square)(3p + \square) \quad \text{or} \quad (6p + \square)(p + \square).$$

3) Factor the last term, $-28v^2$, which has a negative coefficient.

There are six pairs of factors and each can be listed in two ways:

$$-28v, v \qquad 28v, -v \qquad -14v, 2v \qquad 14v, -2v \qquad -7v, 4v \qquad 7v, -4v$$

and

$$v, -28v \qquad -v, 28v \qquad 2v, -14v \qquad -2v, 14v \qquad 4v, -7v \qquad -4v, 7v.$$

4) The coefficient of the middle term is negative, so we look for combinations of factors from steps (2) and (3) such that the sum of their products has a negative coefficient. We try some possibilities:

$$(2p + v)(3p - 28v) = 6p^2 - 53pv - 28v^2,$$
$$(2p - 7v)(3p + 4v) = 6p^2 - 13pv - 28v^2. \quad \leftarrow \textbf{Correct middle term}$$

Factor.

8. $6a^2 - 5ab + b^2$

9. $6x^2 + 15xy + 9y^2$

The factorization of $6p^2 - 13pv - 28v^2$ is $(2p - 7v)(3p + 4v)$.

5) The check is left to the student.

◀ **Do Exercises 8 and 9.**

Answers

6. $-1(2x - 1)(3x + 2)$, or $(2x - 1)(-3x - 2)$, or $(-2x + 1)(3x + 2)$ **7.** $-2(3x - 4)(x + 1)$, or $2(3x - 4)(-x - 1)$, or $2(-3x + 4)(x + 1)$ **8.** $(2a - b)(3a - b)$ **9.** $3(2x + 3y)(x + y)$

✓ Reading Check

Determine whether each statement is true or false.

RC1. When factoring a polynomial, we always look for a common factor first.

RC2. We can check any factorization by multiplying.

RC3. When we are factoring $10x^2 + 21x + 2$, the only choices for the First terms in the binomial factors are $2x$ and $5x$.

RC4. The factorization of $10x^2 + 21x + 2$ is $(2x + 1)(5x + 2)$.

a Factor.

1. $2x^2 - 7x - 4$

2. $3x^2 - x - 4$

3. $5x^2 - x - 18$

4. $4x^2 - 17x + 15$

5. $6x^2 + 23x + 7$

6. $6x^2 - 23x + 7$

7. $3x^2 + 4x + 1$

8. $7x^2 + 15x + 2$

9. $4x^2 + 4x - 15$

10. $9x^2 + 6x - 8$

11. $2x^2 - x - 1$

12. $15x^2 - 19x - 10$

13. $9x^2 + 18x - 16$

14. $2x^2 + 5x + 2$

15. $3x^2 - 5x - 2$

16. $18x^2 - 3x - 10$

17. $12x^2 + 31x + 20$

18. $15x^2 + 19x - 10$

19. $14x^2 + 19x - 3$

20. $35x^2 + 34x + 8$

21. $9x^2 + 18x + 8$

22. $6 - 13x + 6x^2$

23. $49 - 42x + 9x^2$

24. $16 + 36x^2 + 48x$

25. $24x^2 + 47x - 2$ **26.** $16p^2 - 78p + 27$ **27.** $35x^2 - 57x - 44$ **28.** $9a^2 + 12a - 5$

29. $20 + 6x - 2x^2$ **30.** $15 + x - 2x^2$ **31.** $12x^2 + 28x - 24$ **32.** $6x^2 + 33x + 15$

33. $30x^2 - 24x - 54$ **34.** $18t^2 - 24t + 6$ **35.** $4y + 6y^2 - 10$ **36.** $-9 + 18x^2 - 21x$

37. $3x^2 - 4x + 1$ **38.** $6t^2 + 13t + 6$ **39.** $12x^2 - 28x - 24$ **40.** $6x^2 - 33x + 15$

41. $-1 + 2x^2 - x$ **42.** $-19x + 15x^2 + 6$ **43.** $9x^2 - 18x - 16$ **44.** $14y^2 + 35y + 14$

45. $15x^2 - 25x - 10$ **46.** $18x^2 + 3x - 10$ **47.** $12p^3 + 31p^2 + 20p$ **48.** $15x^3 + 19x^2 - 10x$

49. $16 + 18x - 9x^2$ **50.** $33t - 15 - 6t^2$ **51.** $-15x^2 + 19x - 6$ **52.** $1 + p - 2p^2$

53. $14x^4 + 19x^3 - 3x^2$ **54.** $70x^4 + 68x^3 + 16x^2$ **55.** $168x^3 - 45x^2 + 3x$ **56.** $144x^5 + 168x^4 + 48x^3$

57. $15x^4 - 19x^2 + 6$ **58.** $9x^4 + 18x^2 + 8$ **59.** $25t^2 + 80t + 64$ **60.** $9x^2 - 42x + 49$

61. $6x^3 + 4x^2 - 10x$ **62.** $18x^3 - 21x^2 - 9x$ **63.** $25x^2 + 79x + 64$ **64.** $9y^2 + 42y + 47$

65. $6x^2 - 19x - 5$

66. $2x^2 + 11x - 9$

67. $12m^2 - mn - 20n^2$

68. $12a^2 - 17ab + 6b^2$

69. $6a^2 - ab - 15b^2$

70. $3p^2 - 16pw - 12w^2$

71. $9a^2 + 18ab + 8b^2$

72. $10s^2 + 4st - 6t^2$

73. $35p^2 + 34pt + 8t^2$

74. $30a^2 + 87ab + 30b^2$

75. $18x^2 - 6xy - 24y^2$

76. $15a^2 - 5ab - 20b^2$

Skill Maintenance

Graph.

77. $y = \dfrac{2}{5}x - 1$ [3.1d]

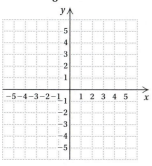

78. $2x = 6$ [3.2b]

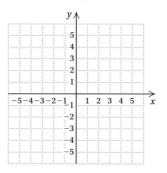

79. $x = 4 - 2y$ [3.1d]

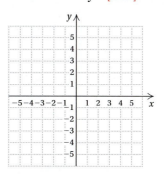

80. $y = -3$ [3.2b]

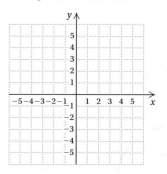

Find the intercepts of each equation. Then graph the equation.   [3.2a]

81. $x + y = 4$

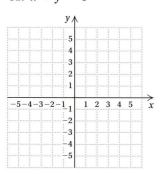

82. $x - y = 3$

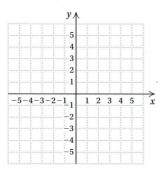

83. $5x - 3y = 15$

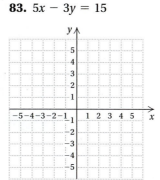

84. $y - 3x = 3$

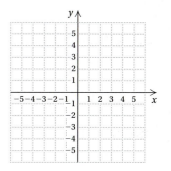

Synthesis

Factor.

85. $20x^{2n} + 16x^n + 3$

86. $-15x^{2m} + 26x^m - 8$

87. $3x^{6a} - 2x^{3a} - 1$

88. $x^{2n+1} - 2x^{n+1} + x$

89.–94. Use the TABLE feature to check the factoring in Exercises 15–20. (See the Calculator Corner on p. 403.)

OBJECTIVE

a Factor trinomials of the type $ax^2 + bx + c, a \neq 1$, using the ac-method.

SKILL TO REVIEW

Objective 5.1c: Factor certain expressions with four terms using factoring by grouping.

Factor by grouping.

1. $2x^3 - 3x^2 + 10x - 15$

2. $n^3 + 3n^2 + n + 3$

a THE ac-METHOD

Another method for factoring trinomials of the type $ax^2 + bx + c, a \neq 1$, involves the product, ac, of the leading coefficient a and the last term c. It is called the **ac-method**. Because it uses factoring by grouping, it is also referred to as the **grouping method**.

We know how to factor the trinomial $x^2 + 5x + 6$. We look for factors of the constant term, 6, whose sum is the coefficient of the middle term, 5. What happens when the leading coefficient is not 1? To factor a trinomial like $3x^2 - 10x - 8$, we can use a method similar to the one that we used for $x^2 + 5x + 6$. That method is outlined as follows.

THE ac-METHOD

To factor a trinomial using the ac-method:

1. Factor out a common factor, if any. We refer to the remaining trinomial as $ax^2 + bx + c$.

2. Multiply the leading coefficient a and the constant c.

3. Try to factor the product ac so that the sum of the factors is b. That is, find integers p and q such that $pq = ac$ and $p + q = b$.

4. Split the middle term, writing it as a sum using the factors found in step (3).

5. Factor by grouping.

6. Check by multiplying.

EXAMPLE 1 Factor: $3x^2 - 10x - 8$.

1) First, we factor out a common factor, if any. There is none (other than 1 or −1).

2) We multiply the leading coefficient, 3, and the constant, −8:

$$3(-8) = -24.$$

3) Then we look for a factorization of −24 in which the sum of the factors is the coefficient of the middle term, −10.

PAIRS OF FACTORS	SUMS OF FACTORS	
−1, 24	23	
1, −24	−23	
−2, 12	10	
2, −12	**−10**	← $2 + (-12) = -10$
−3, 8	5	We show these sums for completeness. In practice, we stop when we find the correct sum.
3, −8	−5	
−4, 6	2	
4, −6	−2	

4) Next, we split the middle term as a sum or a difference using the factors found in step (3): $-10x = 2x - 12x$.

5) Finally, we factor by grouping, as follows:

$$3x^2 - 10x - 8 = 3x^2 + 2x - 12x - 8 \qquad \text{Substituting } 2x - 12x \text{ for } -10x$$

$$\begin{aligned} &= (3x^2 + 2x) + (-12x - 8) \\ &= x(3x + 2) - 4(3x + 2) \qquad \text{Factoring by grouping} \\ &= (3x + 2)(x - 4). \end{aligned}$$

We can also split the middle term as $-12x + 2x$. We still get the same factorization, although the factors may be in a different order. Note the following:

$$3x^2 - 10x - 8 = 3x^2 - 12x + 2x - 8 \qquad \text{Substituting } -12x + 2x \text{ for } -10x$$

$$\begin{aligned} &= (3x^2 - 12x) + (2x - 8) \\ &= 3x(x - 4) + 2(x - 4) \qquad \text{Factoring by grouping} \\ &= (x - 4)(3x + 2). \end{aligned}$$

6) Check: $(3x + 2)(x - 4) = 3x^2 - 10x - 8.$

Do Exercises 1 and 2. ▶

EXAMPLE 2 Factor: $8x^2 + 8x - 6.$

1) First, we factor out a common factor, if any. The number 2 is common to all three terms, so we factor it out: $2(4x^2 + 4x - 3).$

2) Next, we factor the trinomial $4x^2 + 4x - 3$. We multiply the leading coefficient and the constant, 4 and -3: $4(-3) = -12.$

3) We try to factor -12 so that the sum of the factors is 4.

PAIRS OF FACTORS	SUMS OF FACTORS
$-1,\ \ 12$	11
$1,\ -12$	-11
$-2,\ \ \ 6$	4 ←
$2,\ -6$	·
$-3,\ \ \ 4$	·
$3,\ -4$	·

$-2 + 6 = 4$

We have found the correct sum, so there is no need to complete the table.

4) Then we split the middle term, $4x$, as follows: $4x = -2x + 6x.$

5) Finally, we factor by grouping:

$$4x^2 + 4x - 3 = 4x^2 - 2x + 6x - 3 \qquad \text{Substituting } -2x + 6x \text{ for } 4x$$

$$\begin{aligned} &= (4x^2 - 2x) + (6x - 3) \\ &= 2x(2x - 1) + 3(2x - 1) \qquad \text{Factoring by grouping} \\ &= (2x - 1)(2x + 3). \end{aligned}$$

The factorization of $4x^2 + 4x - 3$ is $(2x - 1)(2x + 3)$. *But don't forget the common factor!* We must include it to get a factorization of the original trinomial: $8x^2 + 8x - 6 = 2(2x - 1)(2x + 3).$

6) Check: $2(2x - 1)(2x + 3) = 2(4x^2 + 4x - 3) = 8x^2 + 8x - 6.$

Do Exercises 3 and 4. ▶

Factor.

1. $6x^2 + 7x + 2$

 2. $12x^2 - 17x - 5$
  1) There is no common factor.
  2) Multiplying the leading coefficient and the constant:
    $12(-5) = \boxed{}.$
  3) Look for a pair of factors of -60 whose sum is -17. Those factors are 3 and $\boxed{}$.
  4) Split the middle term:
    $-17x = 3x - \boxed{}.$
  5) Factor by grouping:
    $12x^2 + 3x - 20x - 5$
    $= 3x(4x + 1) - 5(\boxed{})$
    $= (\boxed{})(3x - 5).$
  6) Check:
    $(4x + 1)(3x - 5)$
    $= 12x^2 - 17x - 5.$

Factor.

3. $6x^2 + 15x + 9$

4. $20x^2 - 46x + 24$

Answers

1. $(2x + 1)(3x + 2)$ **2.** $(4x + 1)(3x - 5)$
3. $3(2x + 3)(x + 1)$ **4.** $2(5x - 4)(2x - 3)$

Guided Solution:
2. $-60, -20, 20x, 4x + 1, 4x + 1$

✓ Reading Check

Complete each step in the process to factor $10x^2 + 21x + 2$.

RC1. Using the *ac*-method, multiply the _____ 10 and the constant 2. The product is 20.

RC2. Find two integers whose _____ is 20 and whose _____ is 21. The integers are 20 and 1.

RC3. Split the middle term, $21x$, writing it as the _____ of $20x$ and x.

RC4. Factor by _____: $10x^2 + 20x + x + 2 = 10x(x + 2) + 1(x + 2) = (x + 2)(10x + 1)$.

a   Factor. Note that the middle term has already been split.

1. $x^2 + 2x + 7x + 14$

2. $x^2 + 3x + x + 3$

3. $x^2 - 4x - x + 4$

4. $a^2 + 5a - 2a - 10$

5. $6x^2 + 4x + 9x + 6$

6. $3x^2 - 2x + 3x - 2$

7. $3x^2 - 4x - 12x + 16$

8. $24 - 18y - 20y + 15y^2$

9. $35x^2 - 40x + 21x - 24$

10. $8x^2 - 6x - 28x + 21$

11. $4x^2 + 6x - 6x - 9$

12. $2x^4 - 6x^2 - 5x^2 + 15$

13. $2x^4 + 6x^2 + 5x^2 + 15$

14. $9x^4 - 6x^2 - 6x^2 + 4$

Factor using the *ac*-method.

15. $2x^2 + 7x + 6$ **16.** $5x^2 + 17x + 6$ **17.** $3x^2 - 4x - 15$ **18.** $3x^2 + x - 4$

19. $5x^2 + 11x + 2$ **20.** $3x^2 + 16x + 5$ **21.** $3x^2 - 4x + 1$ **22.** $7x^2 - 15x + 2$

23. $6x^2 + 23x + 7$ **24.** $6x^2 + 13x + 6$ **25.** $4x^2 - 4x - 15$ **26.** $9x^2 - 6x - 8$

27. $15x^2 + 19x - 10$ **28.** $6 - 13x + 6x^2$ **29.** $9x^2 - 18x - 16$ **30.** $18x^2 + 3x - 10$

31. $3x^2 + 5x - 2$ **32.** $2x^2 - 5x + 2$ **33.** $12x^2 - 31x + 20$ **34.** $35x^2 - 34x + 8$

35. $14x^2 - 19x - 3$ **36.** $15x^2 - 19x - 10$ **37.** $49 - 42x + 9x^2$ **38.** $25x^2 + 40x + 16$

39. $9x^2 + 18x + 8$ **40.** $24x^2 - 47x - 2$ **41.** $5 - 9a^2 - 12a$ **42.** $17x - 4x^2 + 15$

43. $20 + 6x - 2x^2$ **44.** $15 + x - 2x^2$ **45.** $12x^2 + 28x - 24$ **46.** $6x^2 + 33x + 15$

47. $30x^2 - 24x - 54$ **48.** $18t^2 - 24t + 6$ **49.** $4y + 6y^2 - 10$ **50.** $-9 + 18x^2 - 21x$

51. $3x^2 - 4x + 1$

52. $6t^2 + t - 15$

53. $12x^2 - 28x - 24$

54. $6x^2 - 33x + 15$

55. $-1 + 2x^2 - x$

56. $-19x + 15x^2 + 6$

57. $9x^2 + 18x - 16$

58. $14y^2 + 35y + 14$

59. $15x^2 - 25x - 10$

60. $18x^2 + 3x - 10$

61. $12p^3 + 31p^2 + 20p$

62. $15x^3 + 19x^2 - 10x$

63. $4 - x - 5x^2$

64. $1 - p - 2p^2$

65. $33t - 15 - 6t^2$

66. $-15x^2 - 19x - 6$

67. $14x^4 + 19x^3 - 3x^2$

68. $70x^4 + 68x^3 + 16x^2$

69. $168x^3 - 45x^2 + 3x$

70. $144x^5 + 168x^4 + 48x^3$

71. $15x^4 - 19x^2 + 6$

72. $9x^4 + 18x^2 + 8$

73. $25t^2 + 80t + 64$

74. $9x^2 - 42x + 49$

75. $6x^3 + 4x^2 - 10x$

76. $18x^3 - 21x^2 - 9x$

77. $3x^2 + 9x + 5$

78. $4x^2 + 6x + 3$

79. $6x^2 - 19x - 5$

80. $2x^2 + 11x - 9$

81. $12m^2 - mn - 20n^2$

82. $12a^2 - 17ab + 6b^2$

83. $6a^2 - ab - 15b^2$

84. $3p^2 - 16pq - 12q^2$

85. $9a^2 - 18ab + 8b^2$

86. $10s^2 + 4st - 6t^2$

87. $35p^2 + 34pq + 8q^2$

88. $30a^2 + 87ab + 30b^2$

89. $18x^2 - 6xy - 24y^2$

90. $15a^2 - 5ab - 20b^2$

91. $60x + 18x^2 - 6x^3$

92. $60x + 4x^2 - 8x^3$

93. $35x^5 - 57x^4 - 44x^3$

94. $15x^3 + 33x^4 + 6x^5$

Skill Maintenance

Simplify. Express the result using positive exponents.

95. $(3x^4)^3$ [4.2a, b]

96. $5^{-6} \cdot 5^{-8}$ [4.1d, f]

97. $(x^2y)(x^3y^5)$ [4.1d]

98. $\dfrac{a^{-7}}{a^{-8}}$ [4.1e, f]

99. Convert to scientific notation: 30,080,000,000.
[4.2c]

100. Convert to decimal notation: 1.5×10^{-5}.
[4.2c]

Solve.   [2.6a]

101. The earth is a sphere (or ball) that is about 40,000 km
in circumference. Find the radius of the earth,
in kilometers and in miles. Use 3.14 for π.
(*Hint*: 1 km $\approx$ 0.62 mi.)

102. The second angle of a triangle is 10° less than
twice the first. The third angle is 15° more than
four times the first. Find the measure of the second
angle.

Synthesis

Factor.

103. $9x^{10} - 12x^5 + 4$

104. $24x^{2n} + 22x^n + 3$

105. $16x^{10} + 8x^5 + 1$

106. $(a + 4)^2 - 2(a + 4) + 1$

107.–112. 〰 Use graphs to check the factoring in Exercises 15–20. (See the Calculator Corner on p. 403.)

Mid-Chapter Review

Concept Reinforcement

Determine whether each statement is true or false.

_____ **1.** The greatest common factor (GCF) of a set of natural numbers is at least 1 and always less than or equal to the smallest number in the set.   [5.1a]

_____ **2.** To factor $x^2 + bx + c$, we use a trial-and-error process that looks for factors of b whose sum is c.   [5.2a]

_____ **3.** A prime polynomial has no common factor other than 1 and -1.   [5.2a]

_____ **4.** When factoring $x^2 - 14x + 45$, we need consider only positive pairs of factors of 45.   [5.2a]

Guided Solutions

 Fill in each blank with the number, variable, or expression that creates a correct statement or solution.

5. Factor: $10y^3 - 18y^2 + 12y$.   [5.1b]
$$10y^3 - 18y^2 + 12y = \boxed{} \cdot 5y^2 - \boxed{} \cdot 9y + \boxed{} \cdot 6$$
$$= 2y(\boxed{})$$

6. Factor $2x^2 - x - 6$ using the ac-method.   [5.4a]

$a \cdot c = \boxed{} \cdot \boxed{} = -12;$ Multiplying the leading coefficient and the constant

$-x = \boxed{} + 3x;$ Splitting the middle term

$2x^2 - x - 6 = 2x^2 - 4x + \boxed{} - 6$
$ = \boxed{}(x - 2) + \boxed{}(x - 2)$
$ = (x - 2)(\boxed{})$

Mixed Review

Find the GCF.   [5.1a]

7. $x^3, 3x$

8. $5x^4, x^2$

9. $6x^5, -12x^3$

10. $-8x, -12, 16x^2$

11. $15x^3y^2, 5x^2y, 40x^4y^3$

12. $x^2y^4, -x^3y^3, x^3y^2, x^5y^4$

Factor completely.   [5.1b, c], [5.2a], [5.3a], [5.4a]

13. $x^3 - 8x$

14. $3x^2 + 12x$

15. $2y^2 + 8y - 4$

16. $3t^6 - 5t^4 - 2t^3$

17. $x^2 + 4x + 3$

18. $z^2 - 4z + 4$

19. $x^3 + 4x^2 + 3x + 12$

20. $8y^5 - 48y^3$

21. $6x^3y + 24x^2y^2 - 42xy^3$

22. $6 - 11t + 4t^2$

23. $z^2 + 4z - 5$

24. $2z^3 + 8z^2 + 5z + 20$

25. $3p^3 - 2p^2 - 9p + 6$

26. $10x^8 - 25x^6 - 15x^5 + 35x^3$

27. $2w^3 + 3w^2 - 6w - 9$

28. $4x^4 - 5x^3 + 3x^2$

29. $6y^2 + 7y - 10$

30. $3x^2 - 3x - 18$

31. $6x^3 + 4x^2 + 3x + 2$

32. $15 - 8w + w^2$

33. $8x^3 + 20x^2 + 2x + 5$

34. $10z^2 - 21z - 10$

35. $6x^2 + 7x + 2$

36. $x^2 - 10xy + 24y^2$

37. $6z^3 + 3z^2 + 2z + 1$

38. $a^3b^7 + a^4b^5 - a^2b^3 + a^5b^6$

39. $4y^2 - 7yz - 15z^2$

40. $3x^3 + 21x^2 + 30x$

41. $x^3 - 3x^2 - 2x + 6$

42. $9y^2 + 6y + 1$

43. $y^2 + 6y + 8$

44. $6y^2 + 33y + 45$

45. $x^3 - 7x^2 + 4x - 28$

46. $4 + 3y - y^2$

47. $16x^2 - 16x - 60$

48. $10a^2 - 11ab + 3b^2$

49. $6w^3 - 15w^2 - 10w + 25$

50. $y^3 + 9y^2 + 18y$

51. $4x^2 + 11xy + 6y^2$

52. $6 - 5z - 6z^2$

53. $12t^3 + 8t^2 - 9t - 6$

54. $y^2 + yz - 20z^2$

55. $9x^2 - 6xy - 8y^2$

56. $-3 + 8z + 3z^2$

57. $m^2 - 6mn - 16n^2$

58. $2w^2 - 12w + 18$

59. $18t^3 - 18t^2 + 4t$

60. $5z^3 + 15z^2 + z + 3$

61. $-14 + 5t + t^2$

62. $4t^2 - 20t + 25$

63. $t^2 + 4t - 12$

64. $12 + 5z - 2z^2$

65. $12 + 4y - y^2$

Understanding Through Discussion and Writing

66. Explain how one could construct a polynomial with four terms that can be factored by grouping. [5.1c], [5.4a]

67. When searching for a factorization, why do we list pairs of numbers with the correct *product* instead of pairs of numbers with the correct *sum*? [5.2a]

68. Without multiplying $(x - 17)(x - 18)$, explain why it cannot possibly be a factorization of $x^2 + 35x + 306$. [5.2a]

69. A student presents the following work:
$$4x^2 + 28x + 48 = (2x + 6)(2x + 8)$$
$$= 2(x + 3)(x + 4).$$
Is it correct? Explain. [5.3a], [5.4a]

☐   If you find applied problems challenging, don't give up! Your skill will improve with each problem that you solve.

☐   Make applications real! Look for applications of the math that you are studying in newspapers and magazines.

5.5

Factoring Trinomial Squares and Differences of Squares

OBJECTIVES

a Recognize trinomial squares.

b Factor trinomial squares.

c Recognize differences of squares.

d Factor differences of squares, being careful to factor completely.

It would be helpful to memorize this table of perfect squares.

NUMBER, N	PERFECT SQUARE, N^2
1	1
2	4
3	9
4	16
5	25
6	36
7	49
8	64
9	81
10	100
11	121
12	144
13	169
14	196
15	225
16	256
20	400
25	625

In this section, we first learn to factor trinomials that are squares of binomials. Then we factor binomials that are differences of squares.

a   RECOGNIZING TRINOMIAL SQUARES

Some trinomials are squares of binomials. For example, the trinomial $x^2 + 10x + 25$ is the square of the binomial $x + 5$. To see this, we can calculate $(x + 5)^2$. It is $x^2 + 2 \cdot x \cdot 5 + 5^2$, or $x^2 + 10x + 25$. A trinomial that is the square of a binomial is called a **trinomial square**, or a **perfect-square trinomial**.

We can use the following special-product rules in reverse to factor trinomial squares:

$$(A + B)^2 = A^2 + 2AB + B^2;$$
$$(A - B)^2 = A^2 - 2AB + B^2.$$

How can we recognize when an expression to be factored is a trinomial square? Look at $A^2 + 2AB + B^2$ and $A^2 - 2AB + B^2$. In order for an expression to be a trinomial square:

a) The two expressions A^2 and B^2 must be squares, such as

$$4, \quad x^2, \quad 25x^4, \quad 16t^2.$$

When the coefficient is a perfect square and the power(s) of the variable(s) is (are) even, then the expression is a perfect square.

b) There must be no minus sign before A^2 or B^2.

c) The remaining term is either twice the product of A and B, $2AB$, or its opposite, $-2AB$.

If a number c can be multiplied by itself to get a number n, then c is a **square root** of n. Thus, 3 is a square root of 9 because $3 \cdot 3$, or 3^2, is 9. Similarly, A is a square root of A^2 and B is a square root of B^2.

EXAMPLE 1   Determine whether $x^2 + 6x + 9$ is a trinomial square.

a) We know that x^2 and 9 are squares.

b) There is no minus sign before x^2 or 9.

c) If we multiply the square roots, x and 3, and double the product, we get the remaining term: $2 \cdot x \cdot 3 = 6x$.

Thus, $x^2 + 6x + 9$ is the square of a binomial. In fact, $x^2 + 6x + 9 = (x + 3)^2$.

EXAMPLE 2 Determine whether $x^2 + 6x + 11$ is a trinomial square.

The answer is no, because only one term, x^2, is a square.

EXAMPLE 3 Determine whether $16x^2 + 49 - 56x$ is a trinomial square.

It helps to first write the trinomial in descending order:

$$16x^2 - 56x + 49.$$

a) We know that $16x^2$ and 49 are squares.

b) There is no minus sign before $16x^2$ or 49.

c) We multiply the square roots, $4x$ and 7, and double the product to get $2 \cdot 4x \cdot 7 = 56x$. The remaining term, $-56x$, is the opposite of this product.

Thus, $16x^2 + 49 - 56x$ is a trinomial square.

Do Exercises 1–8. ▶

b FACTORING TRINOMIAL SQUARES

We can use the factoring methods from Sections 5.2–5.4 to factor trinomial squares, but there is a faster method using the following equations.

FACTORING TRINOMIAL SQUARES

$$A^2 + 2AB + B^2 = (A + B)^2;$$
$$A^2 - 2AB + B^2 = (A - B)^2$$

We use square roots of the squared terms and the sign of the remaining term to factor a trinomial square.

EXAMPLE 4 Factor: $x^2 + 6x + 9$.

$$x^2 + 6x + 9 = x^2 + 2 \cdot x \cdot 3 + 3^2 = (x + 3)^2$$

The sign of the middle term is positive.

$$A^2 + 2 \; A \; B + B^2 = (A + B)^2$$

EXAMPLE 5 Factor: $x^2 + 49 - 14x$.

$$x^2 + 49 - 14x = x^2 - 14x + 49$$ Changing to descending order

$$= x^2 - 2 \cdot x \cdot 7 + 7^2$$ The sign of the middle term is negative.

$$= (x - 7)^2$$

EXAMPLE 6 Factor: $16x^2 - 40x + 25$.

$$16x^2 - 40x + 25 = (4x)^2 - 2 \cdot 4x \cdot 5 + 5^2 = (4x - 5)^2$$

$$A^2 \; - 2 \; A \; B + B^2 = (A - B)^2$$

Do Exercises 9–13. ▶

Determine whether each is a trinomial square. Write "yes" or "no."

1. $x^2 + 8x + 16$

2. $25 - x^2 + 10x$

3. $t^2 - 12t + 4$

4. $25 + 20y + 4y^2$

5. $5x^2 + 16 - 14x$

6. $16x^2 + 40x + 25$

7. $p^2 + 6p - 9$

8. $25a^2 + 9 - 30a$

Factor.

9. $x^2 + 2x + 1$

10. $1 - 2x + x^2$

11. $4 + t^2 + 4t$

12. $25x^2 - 70x + 49$

GS **13.** $49 - 56y + 16y^2$

Write in descending order:
$$16y^2 - 56y + 49.$$

Factor as a trinomial square:
$$(4y)^2 - 2 \cdot 4y \cdot \boxed{} + (\boxed{})^2$$
$$= (4y - \boxed{})^2.$$

Answers

1. Yes **2.** No **3.** No **4.** Yes **5.** No
6. Yes **7.** No **8.** Yes **9.** $(x + 1)^2$
10. $(x - 1)^2$, or $(1 - x)^2$ **11.** $(t + 2)^2$
12. $(5x - 7)^2$ **13.** $(4y - 7)^2$, or $(7 - 4y)^2$

Guided Solution:
13. 7, 7, 7

EXAMPLE 7   Factor: $t^4 + 20t^2 + 100$.

$$t^4 + 20t^2 + 100 = (t^2)^2 + 2(t^2)(10) + 10^2$$
$$= (t^2 + 10)^2$$

EXAMPLE 8   Factor: $75m^3 + 210m^2 + 147m$.

Always look first for a common factor. This time there is one, $3m$:

$$75m^3 + 210m^2 + 147m = \textcolor{red}{3m}(25m^2 + 70m + 49)$$
$$= 3m[(5m)^2 + 2(5m)(7) + 7^2]$$
$$= 3m(5m + 7)^2.$$

EXAMPLE 9   Factor: $4p^2 - 12pq + 9q^2$.

$$4p^2 - 12pq + 9q^2 = (2p)^2 - 2(2p)(3q) + (3q)^2$$
$$= (2p - 3q)^2$$

◀ **Do Exercises 14–17.**

Factor.

14. $48m^2 + 75 + 120m$

15. $p^4 + 18p^2 + 81$

16. $4z^5 - 20z^4 + 25z^3$

17. $9a^2 + 30ab + 25b^2$

C   RECOGNIZING DIFFERENCES OF SQUARES

The following polynomials are *differences of squares:*

$$x^2 - 9, \quad 4t^2 - 49, \quad a^2 - 25b^2.$$

To factor a difference of squares such as $x^2 - 9$, we will use the following special-product rule in reverse:

$$(A + B)(A - B) = A^2 - B^2.$$

A **difference of squares** is an expression like the following:

$$A^2 - B^2.$$

How can we recognize such expressions? Look at $A^2 - B^2$. In order for a binomial to be a difference of squares:

a) There must be two expressions, both squares, such as

$$4x^2, \quad 9, \quad 25t^4, \quad 1, \quad x^6, \quad 49y^8.$$

b) The terms must have different signs.

EXAMPLE 10   Is $9x^2 - 64$ a difference of squares?

a) The first expression is a square: $9x^2 = (3x)^2$.
The second expression is a square: $64 = 8^2$.

b) The terms have different signs, $+9x^2$ and -64.

Thus we have a difference of squares, $(3x)^2 - 8^2$.

EXAMPLE 11   Is $25 - t^3$ a difference of squares?

a) The expression t^3 is not a square.

The expression is not a difference of squares.

Answers

14. $3(4m + 5)^2$ **15.** $(p^2 + 9)^2$
16. $z^3(2z - 5)^2$ **17.** $(3a + 5b)^2$

EXAMPLE 12 Is $-4x^2 + 16$ a difference of squares?

a) The expressions $4x^2$ and 16 are squares: $4x^2 = (2x)^2$ and $16 = 4^2$.

b) The terms have different signs, $-4x^2$ and $+16$.

Thus we have a difference of squares. We can also see this by rewriting in the equivalent form: $16 - 4x^2$.

Do Exercises 18–24. ▶

Do Exercises 18–24.

d   FACTORING DIFFERENCES OF SQUARES

To factor a difference of squares, we use the following equation.

FACTORING A DIFFERENCE OF SQUARES

$$A^2 - B^2 = (A + B)(A - B)$$

To factor a difference of squares $A^2 - B^2$, we find A and B, which are square roots of the expressions A^2 and B^2. We then use A and B to form two factors. One is the sum $A + B$, and the other is the difference $A - B$.

EXAMPLE 13 Factor: $x^2 - 4$.

$$x^2 - 4 = x^2 - 2^2 = (x + 2)(x - 2)$$
$$A^2 - B^2 = (A + B)(A - B)$$

EXAMPLE 14 Factor: $9 - 16t^4$.

$$9 - 16t^4 = 3^2 - (4t^2)^2 = (3 + 4t^2)(3 - 4t^2)$$
$$A^2 - B^2 = (A + B)\ (A - B)$$

EXAMPLE 15 Factor: $m^2 - 4p^2$.

$$m^2 - 4p^2 = m^2 - (2p)^2 = (m + 2p)(m - 2p)$$

EXAMPLE 16 Factor: $x^2 - \dfrac{1}{9}$.

$$x^2 - \frac{1}{9} = x^2 - \left(\frac{1}{3}\right)^2 = \left(x + \frac{1}{3}\right)\left(x - \frac{1}{3}\right)$$

EXAMPLE 17 Factor: $18x^2 - 50x^6$.

Always look first for a factor common to all terms. This time there is one, $2x^2$.

$$18x^2 - 50x^6 = 2x^2(9 - 25x^4)$$
$$= 2x^2[3^2 - (5x^2)^2]$$
$$= 2x^2(3 + 5x^2)(3 - 5x^2)$$

Determine whether each is a difference of squares. Write "yes" or "no."

18. $x^2 - 25$

19. $t^2 - 24$

20. $y^2 + 36$

21. $4x^2 - 15$

22. $16x^4 - 49$

23. $9w^6 - 1$

24. $-49 + 25t^2$

Answers

18. Yes   **19.** No   **20.** No   **21.** No
22. Yes   **23.** Yes   **24.** Yes

Factor.

25. $x^2 - 9$

26. $4t^2 - 64$

27. $a^2 - 25b^2$ **GS**
$= a^2 - (\quad)^2$
$= (a + \quad)(a - \quad)$

28. $64x^4 - 25x^6$

29. $5 - 20t^6$
[*Hint:* $t^6 = (t^3)^2$.]

EXAMPLE 18 Factor: $36x^{10} - 4x^2$.

Although this expression is a difference of squares, the terms have a common factor. We always begin factoring by factoring out the greatest common factor.

$$36x^{10} - 4x^2 = 4x^2(9x^8 - 1)$$
$$= 4x^2[(3x^4)^2 - 1^2] \qquad \text{Note that } x^8 = (x^4)^2 \text{ and } 1 = 1^2.$$
$$= 4x^2(3x^4 + 1)(3x^4 - 1)$$

◀ **Do Exercises 25–29.**

·· **Caution!** ································

Note carefully in these examples that a difference of squares is *not* the square of the difference; that is,

$$A^2 - B^2 \neq (A - B)^2.$$

For example,

$$(45 - 5)^2 = 40^2 = 1600,$$

but

$$45^2 - 5^2 = 2025 - 25 = 2000.$$

··

Factoring Completely

If a factor with more than one term can still be factored, you should do so. When no factor can be factored further, you have **factored completely**. Always factor completely whenever told to factor.

EXAMPLE 19 Factor: $p^4 - 16$.

$$p^4 - 16 = (p^2)^2 - 4^2$$
$$= (p^2 + 4)(p^2 - 4) \qquad \text{Factoring a difference of squares}$$
$$= (p^2 + 4)(p + 2)(p - 2) \qquad \text{Factoring further; } p^2 - 4 \text{ is a difference of squares.}$$

The polynomial $p^2 + 4$ cannot be factored further into polynomials with real coefficients.

·· **Caution!** ································

Apart from possibly removing a common factor, we cannot, in general, factor a sum of squares. In particular,

$$A^2 + B^2 \neq (A + B)^2.$$

Consider $25x^2 + 100$. In this case, a sum of squares has a common factor, 25. Factoring, we get $25(x^2 + 4)$, where $x^2 + 4$ is prime.

··

Answers

25. $(x + 3)(x - 3)$
26. $4(t + 4)(t - 4)$
27. $(a + 5b)(a - 5b)$
28. $x^4(8 + 5x)(8 - 5x)$
29. $5(1 + 2t^3)(1 - 2t^3)$

Guided Solution:
27. 5b, 5b, 5b

EXAMPLE 20 Factor: $y^4 - 16x^{12}$.

$$y^4 - 16x^{12} = (y^2 + 4x^6)(y^2 - 4x^6)$$ Factoring a difference of squares

$$= (y^2 + 4x^6)(y + 2x^3)(y - 2x^3)$$ Factoring further. The factor $y^2 - 4x^6$ is a difference of squares.

The polynomial $y^2 + 4x^6$ cannot be factored further into polynomials with real coefficients.

EXAMPLE 21 Factor: $\frac{1}{16}x^8 - 81$.

$$\frac{1}{16}x^8 - 81 = \left(\frac{1}{4}x^4 + 9\right)\left(\frac{1}{4}x^4 - 9\right)$$ Factoring a difference of squares

$$= \left(\frac{1}{4}x^4 + 9\right)\left(\frac{1}{2}x^2 + 3\right)\left(\frac{1}{2}x^2 - 3\right)$$ Factoring further. The factor $\frac{1}{4}x^4 - 9$ is a difference of squares.

Factor completely.

30. $81x^4 - 1$

31. $16 - \frac{1}{81}y^8$

32. $49p^4 - 25q^6$

TIPS FOR FACTORING
....

- Always look first for a common factor. If there is one, factor it out.
- Be alert for trinomial squares and differences of squares. Once recognized, they can be factored without trial and error.
- Always factor completely.
- Check by multiplying.

Answers

30. $(9x^2 + 1)(3x + 1)(3x - 1)$

31. $\left(4 + \frac{1}{9}y^4\right)\left(2 + \frac{1}{3}y^2\right)\left(2 - \frac{1}{3}y^2\right)$

32. $(7p^2 + 5q^3)(7p^2 - 5q^3)$

Do Exercises 30–32. ▶

 5.5 **Exercise Set**

For Extra Help

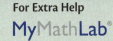 MathXL®
PRACTICE WATCH READ REVIEW

 Reading Check

Determine whether each statement is true or false.

RC1. A trinomial can be considered a trinomial square if only one term is a perfect square.

RC2. A trinomial square is the square of a binomial.

RC3. In order for a binomial to be a difference of squares, the terms in the binomial must have the same sign.

RC4. A binomial cannot have a common factor.

a Determine whether each of the following is a trinomial square. Answer "yes" or "no."

1. $x^2 - 14x + 49$

2. $x^2 - 16x + 64$

3. $x^2 + 16x - 64$

4. $x^2 - 14x - 49$

5. $x^2 - 2x + 4$

6. $x^2 + 3x + 9$

7. $9x^2 - 24x + 16$

8. $25x^2 + 30x + 9$

b Factor completely. Remember to look first for a common factor and to check by multiplying.

9. $x^2 - 14x + 49$

10. $x^2 - 20x + 100$

11. $x^2 + 16x + 64$

12. $x^2 + 20x + 100$

13. $x^2 - 2x + 1$

14. $x^2 + 2x + 1$

15. $4 + 4x + x^2$

16. $4 + x^2 - 4x$

17. $y^2 + 12y + 36$

18. $y^2 + 18y + 81$

19. $16 + t^2 - 8t$

20. $9 + t^2 - 6t$

21. $q^4 - 6q^2 + 9$

22. $64 + 16a^2 + a^4$

23. $49 + 56y + 16y^2$

24. $75 + 48a^2 - 120a$

25. $2x^2 - 4x + 2$

26. $2x^2 - 40x + 200$

27. $x^3 - 18x^2 + 81x$

28. $x^3 + 24x^2 + 144x$

29. $12q^2 - 36q + 27$

30. $20p^2 + 100p + 125$

31. $49 - 42x + 9x^2$

32. $64 - 112x + 49x^2$

33. $5y^4 + 10y^2 + 5$

34. $a^4 + 14a^2 + 49$

35. $1 + 4x^4 + 4x^2$

36. $1 - 2a^5 + a^{10}$

37. $4p^2 + 12pt + 9t^2$

38. $25m^2 + 20mn + 4n^2$

39. $a^2 - 6ab + 9b^2$

40. $x^2 - 14xy + 49y^2$

41. $81a^2 - 18ab + b^2$

42. $64p^2 + 16pt + t^2$

43. $36a^2 + 96ab + 64b^2$

44. $16m^2 - 40mn + 25n^2$

c   Determine whether each of the following is a difference of squares. Answer "yes" or "no."

45. $x^2 - 4$

46. $x^2 - 36$

47. $x^2 + 25$

48. $x^2 + 9$

49. $x^2 - 45$

50. $x^2 - 80y^2$

51. $-25y^2 + 16x^2$

52. $-1 + 36x^2$

d   Factor completely. Remember to look first for a common factor.

53. $y^2 - 4$

54. $q^2 - 1$

55. $p^2 - 1$

56. $x^2 - 36$

57. $-49 + t^2$

58. $-64 + m^2$

59. $a^2 - b^2$

60. $p^2 - v^2$

61. $25t^2 - m^2$

62. $w^2 - 49z^2$

63. $100 - k^2$

64. $81 - w^2$

65. $16a^2 - 9$

66. $25x^2 - 4$

67. $4x^2 - 25y^2$

68. $9a^2 - 16b^2$

69. $8x^2 - 98$

70. $24x^2 - 54$

71. $36x - 49x^3$

72. $16x - 81x^3$

73. $\dfrac{1}{16} - 49x^8$

74. $\dfrac{1}{625}x^8 - 49$

75. $0.09y^2 - 0.0004$

76. $0.16p^2 - 0.0025$

77. $49a^4 - 81$

78. $25a^4 - 9$

79. $a^4 - 16$

80. $y^4 - 1$

81. $5x^4 - 405$

82. $4x^4 - 64$

83. $1 - y^8$

84. $x^8 - 1$

85. $x^{12} - 16$

86. $x^8 - 81$

87. $y^2 - \dfrac{1}{16}$

88. $x^2 - \dfrac{1}{25}$

89. $25 - \dfrac{1}{49}x^2$

90. $\dfrac{1}{4} - 9q^2$

91. $16m^4 - t^4$

92. $p^4t^4 - 1$

Skill Maintenance

Find the intercepts of each equation.   [3.2a]

93. $4x + 16y = 64$

94. $x - 1.3y = 6.5$

95. $y = 2x - 5$

Find the intercepts. Then graph each equation.   [3.2a]

96. $y - 5x = 5$

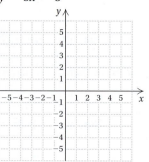

97. $2x + 5y = 10$

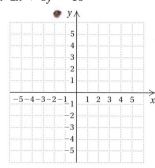

98. $3x - 5y = 15$

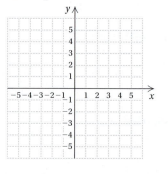

Find a polynomial for the shaded area in each figure. (Leave results in terms of π where appropriate.) [4.4d]

99.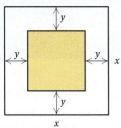

100.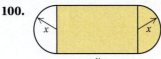

Synthesis

Factor completely, if possible.

101. $49x^2 - 216$

102. $27x^3 - 13x$

103. $x^2 + 22x + 121$

104. $x^2 - 5x + 25$

105. $18x^3 + 12x^2 + 2x$

106. $162x^2 - 82$

107. $x^8 - 2^8$

108. $4x^4 - 4x^2$

109. $3x^5 - 12x^3$

110. $3x^2 - \dfrac{1}{3}$

111. $18x^3 - \dfrac{8}{25}x$

112. $x^2 - 2.25$

113. $0.49p - p^3$

114. $3.24x^2 - 0.81$

115. $0.64x^2 - 1.21$

116. $1.28x^2 - 2$

117. $(x + 3)^2 - 9$

118. $(y - 5)^2 - 36q^2$

119. $x^2 - \left(\dfrac{1}{x}\right)^2$

120. $a^{2n} - 49b^{2n}$

121. $81 - b^{4k}$

122. $9x^{18} + 48x^9 + 64$

123. $9b^{2n} + 12b^n + 4$

124. $(x + 7)^2 - 4x - 24$

125. $(y + 3)^2 + 2(y + 3) + 1$

126. $49(x + 1)^2 - 42(x + 1) + 9$

Find c such that the polynomial is the square of a binomial.

127. $cy^2 + 6y + 1$

128. $cy^2 - 24y + 9$

Use the TABLE feature or graphs to determine whether each factorization is correct. (See the Calculator Corner on p. 403.)

129. $x^2 + 9 = (x + 3)(x + 3)$

130. $x^2 - 49 = (x - 7)(x + 7)$

131. $x^2 + 9 = (x + 3)^2$

132. $x^2 - 49 = (x - 7)^2$

SKILL TO REVIEW

Objective 4.7f: Multiply polynomials.

Multiply.

1. $(aw + 5y)(aw - 3y)$
2. $(c + d)(p + t)$

a We now combine all of our factoring techniques and consider a general strategy for factoring polynomials. Here we will encounter polynomials of all the types we have considered, in random order, so you will have the opportunity to determine which method to use.

> **FACTORING STRATEGY**
>
> To factor a polynomial:
>
> **a)** Always look first for a common factor. If there is one, factor out the largest common factor.
>
> **b)** Then look at the number of terms.
>
> *Two terms*: Determine whether you have a difference of squares, $A^2 - B^2$. Do not try to factor a sum of squares: $A^2 + B^2$.
>
> *Three terms*: Determine whether the trinomial is a square. If it is, you know how to factor. If not, try trial and error, using FOIL or the *ac*-method.
>
> *Four terms*: Try factoring by grouping.
>
> **c)** *Always factor completely.* If a factor with more than one term can still be factored, you should factor it. When no factor can be factored further, you have finished.
>
> **d)** Check by multiplying.

EXAMPLE 1 Factor: $5t^4 - 80$.

a) We look for a common factor. There is one, 5.

$$5t^4 - 80 = 5(t^4 - 16)$$

b) The factor $t^4 - 16$ has only two terms. It is a difference of squares: $(t^2)^2 - 4^2$. We factor $t^4 - 16$ and then include the common factor:

$$5(t^2 + 4)(t^2 - 4).$$

c) We see that one of the factors, $t^2 - 4$, is again a difference of squares. We factor it:

$$5(t^2 + 4)(t + 2)(t - 2).$$

This is a sum of squares. It cannot be factored.

We have factored completely because no factor with more than one term can be factored further.

d) Check: $5(t^2 + 4)(t + 2)(t - 2) = 5(t^2 + 4)(t^2 - 4)$
$$= 5(t^4 - 16)$$
$$= 5t^4 - 80.$$

Answers:

Skill to Review:
1. $a^2w^2 + 2awy - 15y^2$
2. $cp + ct + dp + dt$

EXAMPLE 2 Factor: $2x^3 + 10x^2 + x + 5$.

a) We look for a common factor. There isn't one.

b) There are four terms. We try factoring by grouping:

$$2x^3 + 10x^2 + x + 5$$
$$= (2x^3 + 10x^2) + (x + 5) \qquad \text{Separating into two binomials}$$
$$= 2x^2(x + 5) + 1(x + 5) \qquad \text{Factoring each binomial}$$
$$= (x + 5)(2x^2 + 1). \qquad \text{Factoring out the common factor } x + 5$$

c) None of these factors can be factored further, so we have factored completely.

d) Check: $(x + 5)(2x^2 + 1) = x \cdot 2x^2 + x \cdot 1 + 5 \cdot 2x^2 + 5 \cdot 1$
$$= 2x^3 + x + 10x^2 + 5, \text{ or}$$
$$2x^3 + 10x^2 + x + 5.$$

EXAMPLE 3 Factor: $x^5 - 2x^4 - 35x^3$.

a) We look first for a common factor. This time there is one, x^3:

$$x^5 - 2x^4 - 35x^3 = x^3(x^2 - 2x - 35).$$

b) The factor $x^2 - 2x - 35$ has three terms, but it is not a trinomial square. We factor it using trial and error:

$$x^5 - 2x^4 - 35x^3 = x^3(x^2 - 2x - 35) = x^3(x - 7)(x + 5).$$

> Don't forget to include the common factor in the final answer!

c) No factor with more than one term can be factored further, so we have factored completely.

d) Check: $x^3(x - 7)(x + 5) = x^3(x^2 - 2x - 35)$
$$= x^5 - 2x^4 - 35x^3.$$

EXAMPLE 4 Factor: $x^4 - 10x^2 + 25$.

a) We look first for a common factor. There isn't one.

b) There are three terms. We see that this polynomial is a trinomial square. We factor it:

$$x^4 - 10x^2 + 25 = (x^2)^2 - 2 \cdot x^2 \cdot 5 + 5^2 = (x^2 - 5)^2.$$

We could use trial and error if we have not recognized that we have a trinomial square.

c) Since $x^2 - 5$ cannot be factored further, we have factored completely.

d) Check: $(x^2 - 5)^2 = (x^2)^2 - 2(x^2)(5) + 5^2 = x^4 - 10x^2 + 25.$

Do Exercises 1–5. ▶

Factor.

1. $3m^4 - 3$

2. $x^6 + 8x^3 + 16$

3. $2x^4 + 8x^3 + 6x^2$

4. $3x^3 + 12x^2 - 2x - 8$

GS **5.** $8x^3 - 200x$

a) Factor out the largest common factor:

$8x^3 - 200x$

$= \boxed{}(x^2 - 25).$

b) There are two terms inside the parentheses. Factor the difference of squares:

$8x(x^2 - 25)$

$= 8x(x + \boxed{})(x - \boxed{}).$

c) We have factored completely.

d) Check:

$8x(x + 5)(x - 5)$

$= 8x(x^2 - 25)$

$= 8x^3 - 200x.$

Answers

1. $3(m^2 + 1)(m + 1)(m - 1)$ **2.** $(x^3 + 4)^2$
3. $2x^2(x + 1)(x + 3)$ **4.** $(x + 4)(3x^2 - 2)$
5. $8x(x + 5)(x - 5)$

Guided Solution:
5. $8x$, 5, 5

EXAMPLE 5 Factor: $6x^2y^4 - 21x^3y^5 + 3x^2y^6$.

a) We look first for a common factor:

$$6x^2y^4 - 21x^3y^5 + 3x^2y^6 = 3x^2y^4(2 - 7xy + y^2).$$

b) There are three terms in $2 - 7xy + y^2$. Since only y^2 is a square, we do not have a trinomial square. Can the trinomial be factored by trial and error? A key to the answer is that x is in only the term $-7xy$. The polynomial might be in a form like $(1 - y)(2 + y)$, but there would be no x in the middle term. Thus, $2 - 7xy + y^2$ cannot be factored.

c) Have we factored completely? Yes, because no factor with more than one term can be factored further.

d) The check is left to the student.

EXAMPLE 6 Factor: $(p + q)(x + 2) + (p + q)(x + y)$.

a) We look for a common factor:

$$(p + q)(x + 2) + (p + q)(x + y) = (p + q)[(x + 2) + (x + y)]$$
$$= (p + q)(2x + y + 2).$$

b) The trinomial $2x + y + 2$ cannot be factored further.

c) Neither factor can be factored further, so we have factored completely.

d) The check is left to the student.

EXAMPLE 7 Factor: $px + py + qx + qy$.

a) We look first for a common factor. There isn't one.

b) There are four terms. We try factoring by grouping:

$$px + py + qx + qy = p(x + y) + q(x + y)$$
$$= (x + y)(p + q).$$

c) Since neither factor can be factored further, we have factored completely.

d) Check: $(x + y)(p + q) = px + qx + py + qy$, or
$$px + py + qx + qy.$$

EXAMPLE 8 Factor: $25x^2 + 20xy + 4y^2$.

a) We look first for a common factor. There isn't one.

b) There are three terms. We determine whether the trinomial is a square. The first term and the last term are squares:

$$25x^2 = (5x)^2 \quad \text{and} \quad 4y^2 = (2y)^2.$$

Since twice the product of $5x$ and $2y$ is the other term,

$$2 \cdot 5x \cdot 2y = 20xy,$$

the trinomial is a perfect square.

We factor by writing the square roots of the square terms and the sign of the middle term:

$$25x^2 + 20xy + 4y^2 = (5x + 2y)^2.$$

c) Since $5x + 2y$ cannot be factored further, we have factored completely.

d) Check: $(5x + 2y)^2 = (5x)^2 + 2(5x)(2y) + (2y)^2$
$$= 25x^2 + 20xy + 4y^2.$$

EXAMPLE 9   Factor: $p^2q^2 + 7pq + 12$.

a) We look first for a common factor. There isn't one.

b) There are three terms. We determine whether the trinomial is a square. The first term is a square, but neither of the other terms is a square, so we do not have a trinomial square. We factor, thinking of the product pq as a single variable. We consider this possibility for factorization:

$$(pq + \boxed{})(pq + \boxed{}).$$

We factor the last term, 12. All the signs are positive, so we consider only positive factors. Possibilities are 1, 12 and 2, 6 and 3, 4. The pair 3, 4 gives a sum of 7 for the coefficient of the middle term. Thus,

$$p^2q^2 + 7pq + 12 = (pq + 3)(pq + 4).$$

c) No factor with more than one term can be factored further, so we have factored completely.

d) Check: $(pq + 3)(pq + 4) = (pq)(pq) + 4 \cdot pq + 3 \cdot pq + 3 \cdot 4$
$$= p^2q^2 + 7pq + 12.$$

EXAMPLE 10   Factor: $8x^4 - 20x^2y - 12y^2$.

a) We look first for a common factor:

$$8x^4 - 20x^2y - 12y^2 = 4(2x^4 - 5x^2y - 3y^2).$$

b) There are three terms in $2x^4 - 5x^2y - 3y^2$. Since none of the terms is a square, we do not have a trinomial square. The x^2 in the middle term, $-5x^2y$, leads us to factor $2x^4$ as $2x^2 \cdot x^2$. We also factor the last term, $-3y^2$. Possibilities are $3y$, $-y$ and $-3y$, y and others. We look for factors such that the sum of their products is the middle term. We try some possibilities:

$$(2x^2 - y)(x^2 + 3y) = 2x^4 + 5x^2y - 3y^2,$$
$$(2x^2 + y)(x^2 - 3y) = 2x^4 - 5x^2y - 3y^2.$$

c) No factor with more than one term can be factored further, so we have factored completely. The factorization, including the common factor, is

$$4(2x^2 + y)(x^2 - 3y).$$

d) Check: $4(2x^2 + y)(x^2 - 3y) = 4[(2x^2)(x^2) + 2x^2(-3y) + yx^2 + y(-3y)]$
$$= 4[2x^4 - 6x^2y + x^2y - 3y^2]$$
$$= 4(2x^4 - 5x^2y - 3y^2)$$
$$= 8x^4 - 20x^2y - 12y^2.$$

EXAMPLE 11   Factor: $a^4 - 16b^4$.

a) We look first for a common factor. There isn't one.

b) There are two terms. Since $a^4 = (a^2)^2$ and $16b^4 = (4b^2)^2$, we see that we have a difference of squares. Thus,

$$a^4 - 16b^4 = (a^2 + 4b^2)(a^2 - 4b^2).$$

c) The last factor can be factored further. It is also a difference of squares.

$$a^4 - 16b^4 = (a^2 + 4b^2)(a + 2b)(a - 2b)$$

d) Check: $(a^2 + 4b^2)(a + 2b)(a - 2b) = (a^2 + 4b^2)(a^2 - 4b^2)$
$$= a^4 - 16b^4.$$

Do Exercises 6–12. ▶

Factor.

6. $15x^4 + 5x^2y - 10y^2$

7. $10p^6q^2 + 4p^5q^3 + 2p^4q^4$

8. $(a - b)(x + 5) + (a - b)(x + y^2)$

9. $ax^2 + ay + bx^2 + by$

GS **10.** $x^4 + 2x^2y^2 + y^4$

    a) There is no common factor.

    b) There are three terms. Factor the trinomial square:
$$x^4 + 2x^2y^2 + y^4 = (x^2 + \boxed{})^2.$$

    c) We have factored completely.

    d) Check:
$$(x^2 + y^2)^2$$
$$= (x^2)^2 + 2(x^2)(y^2) + (y^2)^2$$
$$= x^4 + 2x^2y^2 + y^4.$$

Factor.

11. $x^2y^2 + 5xy + 4$

12. $p^4 - 81q^4$

Answers

6. $5(3x^2 - 2y)(x^2 + y)$
7. $2p^4q^2(5p^2 + 2pq + q^2)$
8. $(a - b)(2x + 5 + y^2)$
9. $(x^2 + y)(a + b)$ 10. $(x^2 + y^2)^2$
11. $(xy + 1)(xy + 4)$
12. $(p^2 + 9q^2)(p + 3q)(p - 3q)$

Guided Solution:
10. y^2

For Extra Help
MyMathLab®   MathXL®   PRACTICE   WATCH   READ   REVIEW

✓ Reading Check

Complete each step in the following factoring strategy with the appropriate word(s) from the column on the right.

RC1. Always look first for a _____ factor.

RC2. If there are two terms, determine whether the binomial is a _____ of squares.

RC3. If there are three terms, determine whether the trinomial is a _____.

RC4. If there are four terms, try factoring by _____.

RC5. Always factor _____.

RC6. Always _____ by multiplying.

check
completely
grouping
sum
common
product
square
difference

a   Factor completely.

1. $3x^2 - 192$

2. $2t^2 - 18$

3. $a^2 + 25 - 10a$

4. $y^2 + 49 + 14y$

5. $2x^2 - 11x + 12$

6. $8y^2 - 18y - 5$

7. $x^3 + 24x^2 + 144x$

8. $x^3 - 18x^2 + 81x$

9. $x^3 + 3x^2 - 4x - 12$

10. $x^3 - 5x^2 - 25x + 125$

11. $48x^2 - 3$

12. $50x^2 - 32$

13. $9x^3 + 12x^2 - 45x$

14. $20x^3 - 4x^2 - 72x$

15. $x^2 + 4$

16. $t^2 + 25$

17. $x^4 + 7x^2 - 3x^3 - 21x$

18. $m^4 + 8m^3 + 8m^2 + 64m$

19. $x^5 - 14x^4 + 49x^3$

20. $2x^6 + 8x^5 + 8x^4$

21. $20 - 6x - 2x^2$

22. $45 - 3x - 6x^2$

23. $x^2 - 6x + 1$

24. $x^2 + 8x + 5$

25. $4x^4 - 64$

26. $5x^5 - 80x$

27. $1 - y^8$

28. $t^8 - 1$

29. $x^5 - 4x^4 + 3x^3$

30. $x^6 - 2x^5 + 7x^4$

31. $\dfrac{1}{81}x^6 - \dfrac{8}{27}x^3 + \dfrac{16}{9}$

32. $36a^2 - 15a + \dfrac{25}{16}$

33. $mx^2 + my^2$

34. $12p^2 + 24q^3$

35. $9x^2y^2 - 36xy$

36. $x^2y - xy^2$

37. $2\pi rh + 2\pi r^2$

38. $10p^4q^4 + 35p^3q^3 + 10p^2q^2$

39. $(a + b)(x - 3) + (a + b)(x + 4)$

40. $5c(a^3 + b) - (a^3 + b)$

41. $(x - 1)(x + 1) - y(x + 1)$

42. $3(p - q) - q^2(p - q)$

43. $n^2 + 2n + np + 2p$

44. $a^2 - 3a + ay - 3y$

45. $6q^2 - 3q + 2pq - p$

46. $2x^2 - 4x + xy - 2y$

47. $4b^2 + a^2 - 4ab$

48. $x^2 + y^2 - 2xy$

49. $16x^2 + 24xy + 9y^2$

50. $9c^2 + 6cd + d^2$

51. $49m^4 - 112m^2n + 64n^2$

52. $4x^2y^2 + 12xyz + 9z^2$

53. $y^4 + 10y^2z^2 + 25z^4$

54. $0.01x^4 - 0.1x^2y^2 + 0.25y^4$

55. $\dfrac{1}{4}a^2 + \dfrac{1}{3}ab + \dfrac{1}{9}b^2$

56. $4p^2q + pq^2 + 4p^3$

57. $a^2 - ab - 2b^2$

58. $3b^2 - 17ab - 6a^2$

59. $2mn - 360n^2 + m^2$

60. $15 + x^2y^2 + 8xy$

61. $m^2n^2 - 4mn - 32$

62. $p^2q^2 + 7pq + 6$

63. $r^5s^2 - 10r^4s + 16r^3$

64. $p^5q^2 + 3p^4q - 10p^3$

65. $a^5 + 4a^4b - 5a^3b^2$

66. $2s^6t^2 + 10s^3t^3 + 12t^4$

67. $a^2 - \dfrac{1}{25}b^2$

68. $p^2 - \dfrac{1}{49}b^2$

69. $x^2 - y^2$

70. $p^2q^2 - r^2$

71. $16 - p^4q^4$

72. $15a^4 - 15b^4$

73. $1 - 16x^{12}y^{12}$

74. $81a^4 - b^4$

75. $q^3 + 8q^2 - q - 8$

76. $m^3 - 7m^2 - 4m + 28$

77. $6a^3b^3 - a^2b^2 - 2ab$

78. $4ab^5 - 32b^4 + a^2b^6$

79. $m^4 - 5m^2 + 4$

80. $8x^3y^3 - 6x^2y^2 - 5xy$

Skill Maintenance

Compute and simplify.

81. $1.2 - 9.87$ [1.4a]

82. $-3 + (-5) + 12 + (-7)$ [1.3a]

83. $\left(-\dfrac{1}{3}\right)\left(-\dfrac{3}{5}\right)$ [1.5a]

84. $-3.86 \div 0.5$ [1.6c]

85. $-50 \div (-5)(-2) - 18 \div (-3)^2$ [1.8d]

86. $3(-2) - 2 + |-4 - (-1)|$ [1.8d]

87. Evaluate $-x$ when $x = -7$.   [1.3b]

88. Use either $<$ or $>$ for $\square$ to write a true sentence:
$$-\frac{1}{3} \,\square\, -\frac{1}{2}. \quad [1.2d]$$

Synthesis

Factor completely.

89. $t^4 - 2t^2 + 1$

90. $x^4 + 9$

91. $x^3 + 20 - (5x^2 + 4x)$

92. $\dfrac{1}{5}x^2 - x + \dfrac{4}{5}$

93. $12.25x^2 - 7x + 1$

94. $x^3 + x^2 - (4x + 4)$

95. $18 + y^3 - 9y - 2y^2$

96. $3x^4 - 15x^2 + 12$

97. $y^2(y - 1) - 2y(y - 1) + (y - 1)$

98. $y^2(y + 1) - 4y(y + 1) - 21(y + 1)$

99. $(y + 4)^2 + 2x(y + 4) + x^2$

OBJECTIVES

a Solve equations (already factored) using the principle of zero products.

b Solve quadratic equations by factoring and then using the principle of zero products.

SKILL TO REVIEW

Objective 2.3a: Solve equations using both the addition principle and the multiplication principle.

Solve.

1. $3x - 7 = 2$
2. $4y + 5 = 1$

Second-degree equations like $x^2 + x - 156 = 0$ and $9 - x^2 = 0$ are examples of *quadratic equations*.

QUADRATIC EQUATION

A **quadratic equation** is an equation equivalent to an equation of the type

$$ax^2 + bx + c = 0, \ a \neq 0.$$

In order to solve quadratic equations, we need a new equation-solving principle.

a   THE PRINCIPLE OF ZERO PRODUCTS

The product of two numbers is 0 if one or both of the numbers is 0. Furthermore, *if any product is 0, then a factor must be* 0. For example:

If $7x = 0$, then we know that $x = 0$.

If $x(2x - 9) = 0$, then we know that $x = 0$ *or* $2x - 9 = 0$.

If $(x + 3)(x - 2) = 0$, then we know that $x + 3 = 0$ *or* $x - 2 = 0$.

.... **Caution!** ....

In a product such as $ab = 24$, we cannot conclude with certainty that a is 24 or that b is 24, but if $ab = 0$, we can conclude that $a = 0$ or $b = 0$.

....

EXAMPLE 1   Solve: $(x + 3)(x - 2) = 0$.

We have a product of 0. This equation will be true when either factor is 0. Thus it is true when

$$x + 3 = 0 \quad or \quad x - 2 = 0.$$

Here we have two simple equations that we know how to solve:

$$x = -3 \quad or \quad x = 2.$$

Each of the numbers -3 and 2 is a solution of the original equation, as we can see in the following checks.

Check: For -3:

$$\frac{(x + 3)(x - 2) = 0}{(-3 + 3)(-3 - 2) \ ? \ 0}$$
$$0(-5)$$
$$0 \quad \text{TRUE}$$

For 2:

$$\frac{(x + 3)(x - 2) = 0}{(2 + 3)(2 - 2) \ ? \ 0}$$
$$5(0)$$
$$0 \quad \text{TRUE}$$

Answers

Skill to Review:
1. 3   2. -1

We now have a principle to help in solving quadratic equations.

THE PRINCIPLE OF ZERO PRODUCTS

An equation $ab = 0$ is true if and only if $a = 0$ is true or $b = 0$ is true, or both are true. (A product is 0 if and only if one or both of the factors is 0.)

EXAMPLE 2 Solve: $(5x + 1)(x - 7) = 0$.

We have

$(5x + 1)(x - 7) = 0$

$5x + 1 = 0 \quad or \quad x - 7 = 0 \qquad$ Using the principle of zero products

$\quad\quad 5x = -1 \quad or \quad\quad x = 7 \qquad$ Solving the two equations separately

$\quad\quad\quad x = -\frac{1}{5} \quad or \quad\quad x = 7.$

Check: For $-\frac{1}{5}$:

$$(5x + 1)(x - 7) = 0$$
$$\overline{\left(5\left(-\frac{1}{5}\right) + 1\right)\left(-\frac{1}{5} - 7\right)} \;?\; 0$$
$$(-1 + 1)\left(-7\frac{1}{5}\right)$$
$$0\left(-7\frac{1}{5}\right)$$
$$0 \;\bigg|\; \text{TRUE}$$

For 7:

$$(5x + 1)(x - 7) = 0$$
$$\overline{(5(7) + 1)(7 - 7)} \;?\; 0$$
$$(35 + 1) \cdot 0$$
$$36 \cdot 0$$
$$0 \;\bigg|\; \text{TRUE}$$

The solutions are $-\frac{1}{5}$ and 7.

When some factors have only one term, you can still use the principle of zero products.

EXAMPLE 3 Solve: $x(2x - 9) = 0$.

We have

$x(2x - 9) = 0$

$x = 0 \quad or \quad 2x - 9 = 0 \qquad$ Using the principle of zero products

$x = 0 \quad or \quad\quad\quad 2x = 9$

$x = 0 \quad or \quad\quad\quad\; x = \dfrac{9}{2}.$

Check: For 0:

$$x(2x - 9) = 0$$
$$\overline{0 \cdot (2 \cdot 0 - 9)} \;?\; 0$$
$$0 \cdot (-9)$$
$$0 \;\bigg|\; \text{TRUE}$$

For $\frac{9}{2}$:

$$x(2x - 9) = 0$$
$$\overline{\frac{9}{2} \cdot \left(2 \cdot \frac{9}{2} - 9\right)} \;?\; 0$$
$$\frac{9}{2} \cdot (9 - 9)$$
$$\frac{9}{2} \cdot 0$$
$$0 \;\bigg|\; \text{TRUE}$$

When you solve an equation using the principle of zero products, a check by substitution will detect errors in solving.

Do Exercises 1–4. ▶

Solving Quadratic Equations   We can solve quadratic equations graphically. Consider the equation $x^2 + 2x = 8$. First, we write the equation with 0 on one side: $x^2 + 2x - 8 = 0$. Next, we graph $y = x^2 + 2x - 8$ in a window that shows the x-intercepts. The standard window works well in this case.

The solutions of the equation are the values of x for which $x^2 + 2x - 8 = 0$. These are also the first coordinates of the x-intercepts of the graph. We use the **ZERO** feature from the **CALC** menu to find these numbers. For each x-intercept, we choose an x-value to the left of the intercept as a Left Bound, an x-value to the right of the intercept as a Right Bound, and an x-value near the intercept as a Guess. Beginning with the intercept on the left, we can read its coordinates, $(-4, 0)$, from the resulting screen.

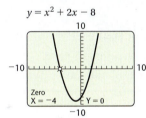

$$y = x^2 + 2x - 8$$

We can repeat this procedure to find the coordinates of the other x-intercept, $(2, 0)$. The solutions of $x^2 + 2x - 8 = 0$ are -4 and 2.

EXERCISE:

1. Solve each of the equations in Examples 4–6 graphically.

Solve using the principle of zero products.

1. $(x - 3)(x + 4) = 0$

2. $(x - 7)(x - 3) = 0$

3. $(4t + 1)(3t - 2) = 0$

4. $y(3y - 17) = 0$

Answers

1. $3, -4$ **2.** $7, 3$ **3.** $-\dfrac{1}{4}, \dfrac{2}{3}$ **4.** $0, \dfrac{17}{3}$

b USING FACTORING TO SOLVE EQUATIONS

Using factoring and the principle of zero products, we can solve some new kinds of equations. Thus we have extended our equation-solving abilities.

EXAMPLE 4 Solve: $x^2 + 5x + 6 = 0$.

There are no like terms to collect, and we have a squared term. We first factor the polynomial. Then we use the principle of zero products.

$$x^2 + 5x + 6 = 0$$
$$(x + 2)(x + 3) = 0 \qquad \text{Factoring}$$
$$x + 2 = 0 \quad or \quad x + 3 = 0 \qquad \text{Using the principle of zero products}$$
$$x = -2 \quad or \qquad x = -3$$

Check: For -2:

$$x^2 + 5x + 6 = 0$$
$$(-2)^2 + 5(-2) + 6 \ ? \ 0$$
$$4 - 10 + 6$$
$$-6 + 6$$
$$0 \quad | \quad \text{TRUE}$$

For -3:

$$x^2 + 5x + 6 = 0$$
$$(-3)^2 + 5(-3) + 6 \ ? \ 0$$
$$9 - 15 + 6$$
$$-6 + 6$$
$$0 \quad | \quad \text{TRUE}$$

The solutions are -2 and -3.

·· **Caution!** ··

Keep in mind that you *must* have 0 on one side of the equation before you can use the principle of zero products. Get all nonzero terms on one side and 0 on the other.

··

◀ **Do Exercise 5.**

EXAMPLE 5 Solve: $x^2 - 8x = -16$.

We first add 16 to get 0 on one side:

$$x^2 - 8x = -16$$
$$x^2 - 8x + 16 = 0 \qquad \text{Adding 16}$$
$$(x - 4)(x - 4) = 0 \qquad \text{Factoring}$$
$$x - 4 = 0 \quad or \quad x - 4 = 0 \qquad \text{Using the principle of zero products}$$
$$x = 4 \quad or \qquad x = 4. \qquad \text{Solving each equation}$$

There is only one solution, 4. The check is left to the student.

◀ **Do Exercises 6 and 7.**

EXAMPLE 6 Solve: $x^2 + 5x = 0$.

$$x^2 + 5x = 0$$
$$x(x + 5) = 0 \qquad \text{Factoring out a common factor}$$
$$x = 0 \quad or \quad x + 5 = 0 \qquad \text{Using the principle of zero products}$$
$$x = 0 \quad or \qquad x = -5$$

The solutions are 0 and -5. The check is left to the student.

---

5. Solve: $x^2 - x - 6 = 0$. **GS**

$$x^2 - x - 6 = 0$$
$$(x + 2)(\boxed{}) = 0$$
$$x + 2 = 0 \quad or \quad \boxed{} = 0$$
$$x = -2 \quad or \qquad x = \boxed{}$$

Both numbers check.

The solutions are -2 and $\boxed{}$.

Solve.

6. $x^2 - 3x = 28$

7. $x^2 = 6x - 9$

Answers

5. $-2, 3$ **6.** $-4, 7$ **7.** 3

Guided Solution:

5. $x - 3, x - 3, 3, 3$

438 | **CHAPTER 5** Polynomials: Factoring

EXAMPLE 7 Solve: $4x^2 = 25$.

$$4x^2 = 25$$
$$4x^2 - 25 = 0 \qquad \text{Subtracting 25 on both sides to get 0 on one side}$$
$$(2x - 5)(2x + 5) = 0 \qquad \text{Factoring a difference of squares}$$
$$2x - 5 = 0 \quad \text{or} \quad 2x + 5 = 0 \qquad \text{Using the principle of zero products}$$
$$2x = 5 \quad \text{or} \qquad 2x = -5 \qquad \text{Solving each equation}$$
$$x = \frac{5}{2} \quad \text{or} \qquad x = -\frac{5}{2}$$

The solutions are $\frac{5}{2}$ and $-\frac{5}{2}$. The check is left to the student.

Do Exercises 8 and 9. ▶

EXAMPLE 8 Solve: $-5x^2 + 2x + 3 = 0$.

In this case, the leading coefficient of the trinomial is negative. Thus we first multiply by -1 and then proceed as we have in Examples 4–7.

$$-5x^2 + 2x + 3 = 0$$
$$-1(-5x^2 + 2x + 3) = -1 \cdot 0 \qquad \text{Multiplying by } -1$$
$$5x^2 - 2x - 3 = 0 \qquad \text{Simplifying}$$
$$(5x + 3)(x - 1) = 0 \qquad \text{Factoring}$$
$$5x + 3 = 0 \quad \text{or} \quad x - 1 = 0 \qquad \text{Using the principle of zero products}$$
$$5x = -3 \quad \text{or} \qquad x = 1$$
$$x = -\frac{3}{5} \quad \text{or} \qquad x = 1$$

The solutions are $-\frac{3}{5}$ and 1. The check is left to the student.

Do Exercises 10 and 11. ▶

EXAMPLE 9 Solve: $(x + 2)(x - 2) = 5$.

Be careful with an equation like this one! It might be tempting to set each factor equal to 5. **Remember: We must have 0 on one side.** We first carry out the multiplication on the left. Next, we subtract 5 on both sides to get 0 on one side. Then we proceed using the principle of zero products.

$$(x + 2)(x - 2) = 5$$
$$x^2 - 4 = 5 \qquad \text{Multiplying on the left}$$
$$x^2 - 4 - 5 = 5 - 5 \qquad \text{Subtracting 5}$$
$$x^2 - 9 = 0 \qquad \text{Simplifying}$$
$$(x + 3)(x - 3) = 0 \qquad \text{Factoring}$$
$$x + 3 = 0 \quad \text{or} \quad x - 3 = 0 \qquad \text{Using the principle of zero products}$$
$$x = -3 \quad \text{or} \qquad x = 3$$

The solutions are -3 and 3. The check is left to the student.

Do Exercise 12. ▶

Solve.

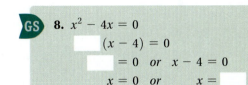

GS **8.** $x^2 - 4x = 0$

$$\boxed{}(x - 4) = 0$$
$$\boxed{} = 0 \quad \text{or} \quad x - 4 = 0$$
$$x = 0 \quad \text{or} \qquad x = \boxed{}$$

Both numbers check.

The solutions are 0 and $\boxed{}$.

9. $9x^2 = 16$

Solve.

10. $-2x^2 + 13x - 21 = 0$

11. $10 - 3x - x^2 = 0$

12. Solve: $(x + 1)(x - 1) = 8$.

Answers

8. $0, 4$ **9.** $-\dfrac{4}{3}, \dfrac{4}{3}$ **10.** $3, \dfrac{7}{2}$ **11.** $-5, 2$

12. $-3, 3$

Guided Solution:

8. $x, x, 4, 4$

To find the x-intercept of a linear equation, we replace y with 0 and solve for x. This procedure can also be used to find the x-intercepts of a quadratic equation.

The graph of $y = ax^2 + bx + c$, $a \neq 0$, is shaped like one of the following curves. Note that each x-intercept represents a solution of $ax^2 + bx + c = 0$.

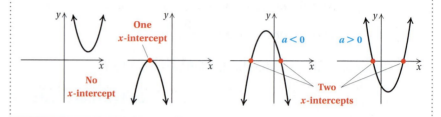

EXAMPLE 10 Find the x-intercepts of the graph of $y = x^2 - 4x - 5$ shown at right. (The grid is intentionally not included.)

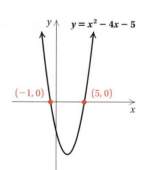

To find the x-intercepts, we let $y = 0$ and solve for x:

$$y = x^2 - 4x - 5$$
$$0 = x^2 - 4x - 5 \qquad \text{Substituting 0 for } y$$
$$0 = (x - 5)(x + 1) \qquad \text{Factoring}$$
$$x - 5 = 0 \quad or \quad x + 1 = 0 \qquad \text{Using the principle of zero products}$$
$$x = 5 \quad or \qquad x = -1.$$

The solutions of the equation $0 = x^2 - 4x - 5$ are 5 and -1. Thus the x-intercepts of the graph of $y = x^2 - 4x - 5$ are $(5, 0)$ and $(-1, 0)$. We can now label them on the graph.

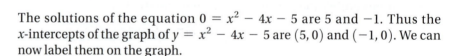

13. Find the x-intercepts of the graph shown below.

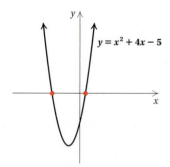

14. Use *only* the graph shown below to solve $3x - x^2 = 0$.

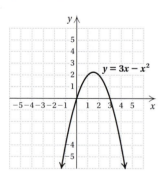

◀ **Do Exercises 13 and 14.**

Answers

13. $(-5, 0), (1, 0)$ **14.** $0, 3$

✓ Reading Check

Determine whether each statement is true or false.

RC1. If $(x + 2)(x + 3) = 10$, then $x + 2 = 10$ *or* $x + 3 = 10$.

RC2. A quadratic equation always has two different solutions.

RC3. The number 0 is never a solution of a quadratic equation.

RC4. If $ax^2 + bx + c = 0$ has no real-number solution, then the graph of $y = ax^2 + bx + c$ has no x-intercept.

a   Solve using the principle of zero products.

1. $(x + 4)(x + 9) = 0$

2. $(x + 2)(x - 7) = 0$

3. $(x + 3)(x - 8) = 0$

4. $(x + 6)(x - 8) = 0$

5. $(x + 12)(x - 11) = 0$

6. $(x - 13)(x + 53) = 0$

7. $x(x + 3) = 0$

8. $y(y + 5) = 0$

9. $0 = y(y + 18)$

10. $0 = x(x - 19)$

11. $(2x + 5)(x + 4) = 0$

12. $(2x + 9)(x + 8) = 0$

13. $(5x + 1)(4x - 12) = 0$

14. $(4x + 9)(14x - 7) = 0$

15. $(7x - 28)(28x - 7) = 0$

16. $(13x + 14)(6x - 5) = 0$

17. $2x(3x - 2) = 0$

18. $55x(8x - 9) = 0$

19. $\left(\frac{1}{5} + 2x\right)\left(\frac{1}{9} - 3x\right) = 0$

20. $\left(\frac{7}{4}x - \frac{1}{16}\right)\left(\frac{2}{3}x - \frac{16}{15}\right) = 0$

21. $(0.3x - 0.1)(0.05x + 1) = 0$

22. $(0.1x + 0.3)(0.4x - 20) = 0$

23. $9x(3x - 2)(2x - 1) = 0$

24. $(x + 5)(x - 75)(5x - 1) = 0$

b   Solve by factoring and using the principle of zero products. Remember to check.

25. $x^2 + 6x + 5 = 0$

26. $x^2 + 7x + 6 = 0$

27. $x^2 + 7x - 18 = 0$

28. $x^2 + 4x - 21 = 0$

29. $x^2 - 8x + 15 = 0$

30. $x^2 - 9x + 14 = 0$

31. $x^2 - 8x = 0$

32. $x^2 - 3x = 0$

33. $x^2 + 18x = 0$

34. $x^2 + 16x = 0$

35. $x^2 = 16$

36. $100 = x^2$

37. $9x^2 - 4 = 0$

38. $4x^2 - 9 = 0$

39. $0 = 6x + x^2 + 9$

40. $0 = 25 + x^2 + 10x$

41. $x^2 + 16 = 8x$

42. $1 + x^2 = 2x$

43. $5x^2 = 6x$

44. $7x^2 = 8x$

45. $6x^2 - 4x = 10$

46. $3x^2 - 7x = 20$

47. $12y^2 - 5y = 2$

48. $2y^2 + 12y = -10$

49. $t(3t + 1) = 2$

50. $x(x - 5) = 14$

51. $100y^2 = 49$

52. $64a^2 = 81$

53. $x^2 - 5x = 18 + 2x$

54. $3x^2 + 8x = 9 + 2x$

55. $10x^2 - 23x + 12 = 0$

56. $12x^2 + 17x - 5 = 0$

Find the *x*-intercepts of the graph of each equation. (The grids are intentionally not included.)

57.

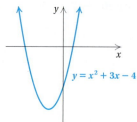

$y = x^2 + 3x - 4$

58.

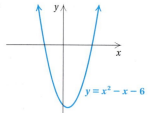

$y = x^2 - x - 6$

59.

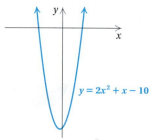

$y = 2x^2 + x - 10$

60.

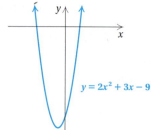

$y = 2x^2 + 3x - 9$

61.

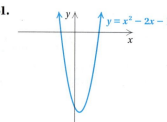

$y = x^2 - 2x - 15$

62.

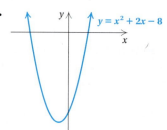

$y = x^2 + 2x - 8$

63. Use the following graph to solve $x^2 - 3x - 4 = 0$.

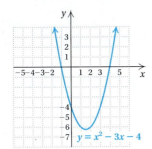

64. Use the following graph to solve $x^2 + x - 6 = 0$.

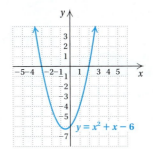

65. Use the following graph to solve $-x^2 + 2x + 3 = 0$.

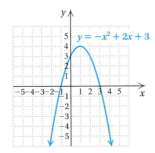

66. Use the following graph to solve $-x^2 - x + 6 = 0$.

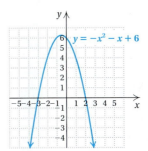

Skill Maintenance

Translate to an algebraic expression.   [1.1b]

67. The square of the sum of a and b

68. The sum of the squares of a and b

Solve.   [2.7d, e]

69. $-10x > 1000$

70. $6 - 3x \geq -18$

71. $3 - 2x - 4x > -9$

72. $\frac{1}{2}x - 6x + 10 \leq x - 5x$

Synthesis

Solve.

73. $b(b + 9) = 4(5 + 2b)$

74. $y(y + 8) = 16(y - 1)$

75. $(t - 3)^2 = 36$

76. $(t - 5)^2 = 2(5 - t)$

77. $x^2 - \frac{1}{64} = 0$

78. $x^2 - \frac{25}{36} = 0$

79. $\frac{5}{16}x^2 = 5$

80. $\frac{27}{25}x^2 = \frac{1}{3}$

Use a graphing calculator to find the solutions of each equation. Round solutions to the nearest hundredth.

81. $x^2 - 9.10x + 15.77 = 0$

82. $-x^2 + 0.63x + 0.22 = 0$

83. Find an equation that has the given numbers as solutions. For example, 3 and -2 are solutions of $x^2 - x - 6 = 0$.

a) $-3, 4$ **b)** $-3, -4$ **c)** $\frac{1}{2}, \frac{1}{2}$

d) $5, -5$ **e)** $0, 0.1, \frac{1}{4}$

OBJECTIVE

a Solve applied problems involving quadratic equations that can be solved by factoring.

SKILL TO REVIEW

Objective 2.6a: Solve applied problems by translating to equations.

Solve.

1. The length of a rectangular garden is twice as long as the width. The perimeter of the garden is 48 ft. Find the length and the width of the garden.

2. The second angle of a triangle is 25° more than the first angle. The third angle is twice as large as the second angle. Find the measures of the angles of the triangle.

1. **Dimensions of a Picture.** A rectangular picture is twice as long as it is wide. If the area of the picture is 288 in², what are its dimensions?

$2w$

Answers

Skill to Review:
1. Length: 16 ft; width: 8 ft
2. 26.25°, 51.25°, 102.5°

Margin Exercise:
1. Length: 24 in.; width: 12 in.

a APPLIED PROBLEMS, QUADRATIC EQUATIONS, AND FACTORING

We can solve problems that translate to quadratic equations using the five steps for solving problems.

EXAMPLE 1   *Kitchen Island.* Lisa buys a kitchen island with a butcher-block top as part of a remodeling project. The top of the island is a rectangle that is twice as long as it is wide and that has an area of 800 in². What are the dimensions of the top of the island?

1. **Familiarize.** We first make a drawing. Recall that the area of a rectangle is Length · Width. We let x = the width of the top, in inches. The length is then $2x$.

2. **Translate.** We reword and translate as follows:

   Rewording: The area of the rectangle is 800 in²

   Translating: $2x \cdot x$ = 800

3. **Solve.** We solve the equation as follows:

$$2x \cdot x = 800$$
$$2x^2 = 800$$

$2x^2 - 800 = 0$ Subtracting 800 to get 0 on one side

$2(x^2 - 400) = 0$ Removing a common factor of 2

$2(x - 20)(x + 20) = 0$ Factoring a difference of squares

$(x - 20)(x + 20) = 0$ Dividing by 2

$x - 20 = 0$ *or* $x + 20 = 0$ Using the principle of zero products

$x = 20$ *or* $x = -20.$ Solving each equation

4. **Check.** The solutions of the equation are 20 and −20. Since the width must be positive, −20 cannot be a solution. To check 20 in., we note that if the width is 20 in., then the length is 2 · 20 in., or 40 in., and the area is 20 in. · 40 in., or 800 in². Thus the solution 20 checks.

5. **State.** The top of the island is 20 in. wide and 40 in. long.

◀ **Do Margin Exercise 1.**

EXAMPLE 2 *Butterfly Wings.* The *Graphium sarpedon* butterfly has areas of light blue on each wing. When the wings are joined, the blue areas form a triangle, giving rise to the butterfly's common name, Blue Triangle Butterfly. On one butterfly, the base of the blue triangle is 6 cm longer than the height of the triangle. The area of the triangle is 8 cm². Find the base and the height of the triangle.

Source: Based on information from australianmuseum.net.au

1. **Familiarize.** We first make a drawing, letting h = the height of the triangle, in centimeters. Then $h + 6$ = the base. We also recall or look up the formula for the area of a triangle: Area $= \frac{1}{2}$ (base)(height).

2. **Translate.** We reword the problem and translate:

Rewording: $\frac{1}{2}$ times   Base   times   Height   is   8
              ↓   ↓   ↓   ↓   ↓   ↓   ↓
Translating: $\frac{1}{2}$ · $(h + 6)$ · h =   8.

3. **Solve.** We solve the quadratic equation using the principle of zero products:

$$\frac{1}{2} \cdot (h + 6) \cdot h = 8$$

$$\frac{1}{2}(h^2 + 6h) = 8 \qquad \text{Multiplying } h + 6 \text{ and } h$$

$$2 \cdot \frac{1}{2}(h^2 + 6h) = 2 \cdot 8 \qquad \text{Multiplying by 2}$$

$$h^2 + 6h = 16 \qquad \text{Simplifying}$$

$$h^2 + 6h - 16 = 16 - 16 \qquad \text{Subtracting 16 to get 0 on one side}$$

$$h^2 + 6h - 16 = 0$$

$$(h - 2)(h + 8) = 0 \qquad \text{Factoring}$$

$$h - 2 = 0 \quad or \quad h + 8 = 0 \qquad \text{Using the principle of zero products}$$

$$h = 2 \quad or \qquad h = -8.$$

4. **Check.** The height of a triangle cannot have a negative length, so -8 cannot be a solution. Suppose the height is 2 cm. The base is 6 cm more than the height, so the base is 2 cm + 6 cm, or 8 cm, and the area is $\frac{1}{2}(8)(2)$, or 8 cm². The numbers check in the original problem.

5. **State.** The base of the blue triangle is 8 cm and the height is 2 cm.

Do Exercise 2. ▶

---

2. *Dimensions of a Sail.* The triangular mainsail on Stacey's lightning-styled sailboat has an area of 125 ft². The height of the sail is 15 ft more than the base. Find the height and the base of the sail.

b + 15

b

Answer

2. Height: 25 ft; base: 10 ft

EXAMPLE 3 *Games in a Sports League.* In a sports league of x teams in which each team plays every other team twice, the total number N of games to be played is given by

$$x^2 - x = N.$$

Maggie's volleyball league plays a total of 240 games. How many teams are in the league?

1., 2. Familiarize and **Translate.** We are given that x is the number of teams in a league and N is the number of games. To find the number of teams x in a league in which 240 games are played, we substitute 240 for N in the equation:

$$x^2 - x = 240. \quad \text{Substituting 240 for } N$$

3. Solve. We solve the equation as follows:

$$x^2 - x = 240$$
$$x^2 - x - 240 = 240 - 240 \qquad \text{Subtracting 240 to get 0 on one side}$$
$$x^2 - x - 240 = 0$$
$$(x - 16)(x + 15) = 0 \qquad \text{Factoring}$$
$$x - 16 = 0 \quad or \quad x + 15 = 0 \qquad \text{Using the principle of zero products}$$
$$x = 16 \quad or \qquad x = -15.$$

4. Check. The solutions of the equation are 16 and -15. Since the number of teams cannot be negative, -15 cannot be a solution. But 16 checks, since $16^2 - 16 = 256 - 16 = 240$.

5. State. There are 16 teams in the league.

3. Use $N = x^2 - x$ for each of the following.

a) *Basketball League.* Amy's basketball league has 19 teams. What is the total number of games to be played if each team plays every other team twice?

b) *Softball League.* Barry's slow-pitch softball league plays a total of 72 games. How many teams are in the league if each team plays every other team twice?

◀ **Do Exercise 3.**

EXAMPLE 4 *Race Numbers.* Terry and Jody registered their boats in the Lakeport Race at the same time. The racing numbers assigned to their boats were consecutive integers, the product of which was 156. Find the integers.

1. Familiarize. Consecutive integers are one unit apart, like 49 and 50. Let $x =$ the first boat number; then $x + 1 =$ the next boat number.

2. Translate. We reword the problem before translating:

Rewording: First integer times Second integer is 156

Translating: $\qquad x \qquad \cdot \qquad (x + 1) \qquad = \qquad 156.$

3. Solve. We solve the equation as follows:

$$x(x + 1) = 156$$
$$x^2 + x = 156 \qquad \text{Multiplying}$$
$$x^2 + x - 156 = 156 - 156 \qquad \text{Subtracting 156 to get 0 on one side}$$
$$x^2 + x - 156 = 0 \qquad \text{Simplifying}$$
$$(x - 12)(x + 13) = 0 \qquad \text{Factoring}$$
$$x - 12 = 0 \quad or \quad x + 13 = 0 \qquad \text{Using the principle of zero products}$$
$$x = 12 \quad or \qquad x = -13.$$

4. Check. The solutions of the equation are 12 and -13. Since race numbers are not negative, -13 must be rejected. On the other hand, if x is 12, then $x + 1$ is 13 and $12 \cdot 13 = 156$. Thus the solution 12 checks.

5. State. The boat numbers for Terry and Jody were 12 and 13.

Do Exercise 4. ▶

4. *Page Numbers.* The product of the page numbers on two facing pages of a book is 506. Find the page numbers.

The Pythagorean Theorem

The Pythagorean theorem states a relationship involving the lengths of the sides of a *right* triangle. A triangle is a **right triangle** if it has a 90°, or *right*, angle. The side opposite the 90° angle is called the **hypotenuse**. The other sides are called **legs**.

THE PYTHAGOREAN THEOREM

In any right triangle, if a and b are the lengths of the legs and c is the length of the hypotenuse, then

$$a^2 + b^2 = c^2.$$

The symbol $\llcorner$ denotes a 90° angle.

EXAMPLE 5 *Wood Scaffold.* Jonah is building a wood scaffold to use for a home improvement project. The scaffold has diagonal braces that are 5 ft long and that span a distance of 3 ft. How high does each brace reach vertically?

1. Familiarize. We make a drawing as shown above and let $h = $ the height, in feet, to which each brace rises vertically.

2. Translate. A right triangle is formed, so we can use the Pythagorean theorem:

$$a^2 + b^2 = c^2$$
$$3^2 + h^2 = 5^2. \qquad \text{Substituting}$$

Answer

4. 22 and 23

3. Solve. We solve the equation as follows:

$$3^2 + h^2 = 5^2$$
$$9 + h^2 = 25 \qquad \text{Squaring 3 and 5}$$
$$9 + h^2 - 25 = 25 - 25 \qquad \text{Subtracting 25 to get 0 on one side}$$
$$h^2 - 16 = 0 \qquad \text{Simplifying}$$
$$(h - 4)(h + 4) = 0 \qquad \text{Factoring}$$
$$h - 4 = 0 \quad or \quad h + 4 = 0 \qquad \text{Using the principle of zero products}$$
$$h = 4 \quad or \qquad h = -4.$$

4. Check. Since height cannot be negative, -4 cannot be a solution. If the height is 4 ft, we have $3^2 + 4^2 = 9 + 16 = 25$, which is 5^2. Thus, 4 checks and is the solution.

5. State. Each brace reaches a height of 4 ft.

◀ **Do Exercise 5.**

5. *Reach of a Ladder.* Twila has a 26-ft ladder leaning against her house. If the bottom of the ladder is 10 ft from the base of the house, how high does the ladder reach?

EXAMPLE 6 *Ladder Settings.* A ladder of length 13 ft is placed against a building in such a way that the distance from the top of the ladder to the ground is 7 ft more than the distance from the bottom of the ladder to the building. Find both distances.

1. Familiarize. We first make a drawing. The ladder and the missing distances form the hypotenuse and the legs of a right triangle. We let $x =$ the length of the side (leg) across the bottom, in feet. Then $x + 7 =$ the length of the other side (leg). The hypotenuse has length 13 ft.

2. Translate. Since a right triangle is formed, we can use the Pythagorean theorem:

$$a^2 + b^2 = c^2$$
$$x^2 + (x + 7)^2 = 13^2. \qquad \text{Substituting}$$

$x + 7$ 13 ft

x

3. Solve. We solve the equation as follows:

$$x^2 + (x^2 + 14x + 49) = 169 \qquad \text{Squaring the binomial and 13}$$
$$2x^2 + 14x + 49 = 169 \qquad \text{Collecting like terms}$$
$$2x^2 + 14x + 49 - 169 = 169 - 169 \qquad \text{Subtracting 169 to get 0 on one side}$$
$$2x^2 + 14x - 120 = 0 \qquad \text{Simplifying}$$
$$2(x^2 + 7x - 60) = 0 \qquad \text{Factoring out a common factor}$$
$$x^2 + 7x - 60 = 0 \qquad \text{Dividing by 2}$$
$$(x + 12)(x - 5) = 0 \qquad \text{Factoring}$$
$$x + 12 = 0 \quad or \quad x - 5 = 0 \qquad \text{Using the principle of zero products}$$
$$x = -12 \quad or \qquad x = 5.$$

4. Check. The negative integer -12 cannot be the length of a side. When $x = 5$, $x + 7 = 12$, and $5^2 + 12^2 = 13^2$. Thus, 5 and 12 check.

5. State. The distance from the top of the ladder to the ground is 12 ft. The distance from the bottom of the ladder to the building is 5 ft.

◀ **Do Exercise 6.**

6. *Right-Triangle Geometry.* The length of one leg of a right triangle is 1 m longer than the other. The length of the hypotenuse is 5 m. Find the lengths of the legs.

Answers

5. 24 ft   **6.** 3 m, 4 m

Translating for Success

1. **Angle Measures.** The measures of the angles of a triangle are three consecutive integers. Find the measures of the angles.

2. **Rectangle Dimensions.** The area of a rectangle is 3599 ft^2. The length is 2 ft longer than the width. Find the dimensions of the rectangle.

3. **Sales Tax.** Claire paid $40,704 for a new hybrid car. This included 6% for sales tax. How much did the vehicle cost before tax?

4. **Wire Cutting.** A 180-m wire is cut into three pieces. The third piece is 2 m longer than the first. The second is two-thirds as long as the first. How long is each piece?

5. **Perimeter.** The perimeter of a rectangle is 240 ft. The length is 2 ft greater than the width. Find the length and the width.

The goal of these matching questions is to practice step (2), Translate, of the five-step problem-solving process. Translate each word problem to an equation and select a correct translation from equations A–O.

A. $2x \cdot x = 288$

B. $x(x + 60) = 7021$

C. $59 = x \cdot 60$

D. $x^2 + (x + 2)^2 = 3599$

E. $x^2 + (x + 70)^2 = 130^2$

F. $6\% \cdot x = 40{,}704$

G. $2(x + 2) + 2x = 240$

H. $\dfrac{1}{2}x(x - 1) = 1770$

I. $x + \dfrac{2}{3}x + (x + 2) = 180$

J. $59\% \cdot x = 60$

K. $x + 6\% \cdot x = 40{,}704$

L. $2x^2 + x = 288$

M. $x(x + 2) = 3599$

N. $x^2 + 60 = 7021$

O. $x + (x + 1) + (x + 2) = 180$

Answers on page A-18

6. **Cell-Phone Tower.** A guy wire on a cell-phone tower is 130 ft long and is attached to the top of the tower. The height of the tower is 70 ft longer than the distance from the point on the ground where the wire is attached to the bottom of the tower. Find the height of the tower.

7. **Sales Meeting Attendance.** PTQ Corporation holds a sales meeting in Tucson. Of the 60 employees, 59 of them attend the meeting. What percent attend the meeting?

8. **Dimensions of a Pool.** A rectangular swimming pool is twice as long as it is wide. The area of the surface is 288 ft^2. Find the dimensions of the pool.

9. **Dimensions of a Triangle.** The height of a triangle is 1 cm less than the length of the base. The area of the triangle is 1770 cm^2. Find the height and the length of the base.

10. **Width of a Rectangle.** The length of a rectangle is 60 ft longer than the width. Find the width if the area of the rectangle is 7021 ft^2.

☑ Reading Check

Match each statement with an appropriate translation from the column on the right. Choices may be used more than once or not at all.

RC1. The product of two consecutive integers is 20.

RC2. The length of a rectangle is 1 cm longer than the width. The area of the rectangle is 20 cm².

RC3. One leg of a right triangle is 1 cm longer than the other leg. The length of the hypotenuse is 20 cm.

RC4. One leg of a right triangle is 1 cm longer than the other leg. The area of the triangle is 20 cm².

a) $x + (x + 1) = 20$

b) $x(x + 1) = 20$

c) $\frac{1}{2}x(x + 1) = 20$

d) $x^2 + (x + 1)^2 = 20^2$

 a   Solve.

1. *Dimensions of a Painting.* A rectangular painting is three times as long as it is wide. The area of the picture is 588 in². Find the dimensions of the painting.

w

$3w$

2. *Area of a Garden.* The length of a rectangular garden is 4 m greater than the width. The area of the garden is 96 m². Find the length and the width.

$w + 4$ w

3. *Design.* The screen of the TI-84 Plus graphing calculator is nearly rectangular. The length of the rectangle is 2 cm more than the width. If the area of the rectangle is 24 cm², find the length and the width.

$w + 2$

w

4. *Construction.* The front porch on Trent's new home is five times as long as it is wide. If the area of the porch is 320 ft², find the dimensions.

w $5w$

5. *Dimensions of a Triangle.* A triangle is 10 cm wider than it is tall. The area is 28 cm². Find the height and the base.

6. *Dimensions of a Triangle.* The height of a triangle is 3 cm less than the length of the base. The area of the triangle is 35 cm². Find the height and the length of the base.

7. *Road Design.* A triangular traffic island has a base half as long as its height. The island has an area of 64 m². Find the base and the height.

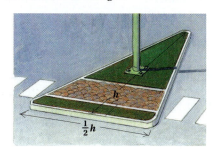

8. *Dimensions of a Sail.* The height of the jib sail on a Lightning sailboat is 5 ft greater than the length of its "foot." The area of the sail is 42 ft². Find the length of the foot and the height of the sail.

Games in a League. Use $x^2 - x = N$ for Exercises 9–12.

9. A Scrabble league has 14 teams. What is the total number of games to be played if each team plays every other team twice?

10. A chess league has 23 teams. What is the total number of games to be played if each team plays every other team twice?

11. A slow-pitch softball league plays a total of 132 games. How many teams are in the league if each team plays every other team twice?

12. A basketball league plays a total of 90 games. How many teams are in the league if each team plays every other team twice?

Handshakes. Dr. Benton wants to investigate the potential spread of germs by contact. She knows that the number of possible handshakes within a group of x people, assuming each person shakes every other person's hand only once, is given by

$$N = \tfrac{1}{2}(x^2 - x).$$

Use this formula for Exercises 13–16.

13. There are 100 people at a party. How many handshakes are possible?

14. There are 40 people at a meeting. How many handshakes are possible?

15. Everyone at a meeting shook hands with each other. There were 300 handshakes in all. How many people were at the meeting?

16. Everyone at a party shook hands with each other. There were 153 handshakes in all. How many people were at the party?

17. *Consecutive Page Numbers.* The product of the page numbers on two facing pages of a book is 210. Find the page numbers.

18. *Consecutive Page Numbers.* The product of the page numbers on two facing pages of a book is 420. Find the page numbers.

19. The product of two consecutive even integers is 168. Find the integers. (Consecutive even integers are two units apart.)

20. The product of two consecutive even integers is 224. Find the integers. (Consecutive even integers are two units apart.)

21. The product of two consecutive odd integers is 255. Find the integers. (Consecutive odd integers are two units apart.)

22. The product of two consecutive odd integers is 143. Find the integers. (Consecutive odd integers are two units apart.)

23. *Roadway Design.* Elliott Street is 24 ft wide when it ends at Main Street in Brattleboro, Vermont. A 40-ft long diagonal crosswalk allows pedestrians to cross Main Street to or from either corner of Elliott Street (see the figure). Determine the width of Main Street.

24. *Lookout Tower.* The diagonal braces in a lookout tower are 15 ft long and span a distance of 12 ft. How high does each brace reach vertically?

25. *Right-Triangle Geometry.* The length of one leg of a right triangle is 8 ft. The length of the hypotenuse is 2 ft longer than the other leg. Find the lengths of the hypotenuse and the other leg.

26. *Right-Triangle Geometry.* The length of one leg of a right triangle is 24 ft. The length of the other leg is 16 ft shorter than the hypotenuse. Find the lengths of the hypotenuse and the other leg.

27. Archaeology. Archaeologists have discovered that the 18th-century garden of the Charles Carroll House in Annapolis, Maryland, was a right triangle. One leg of the triangle was formed by a 400-ft long sea wall. The hypotenuse of the triangle was 200 ft longer than the other leg. What were the dimensions of the garden?

Source: www.bsos.umd.edu

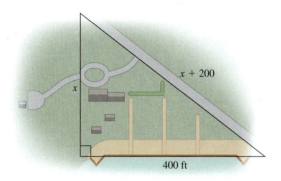

28. Guy Wire. The height of a wind power assessment tower is 5 m shorter than the guy wire that supports it. If the guy wire is anchored 15 m from the foot of the antenna, how tall is the antenna?

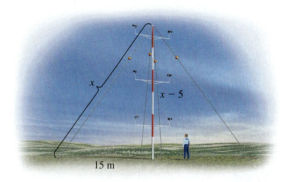

29. Right Triangle. The shortest side of a right triangle measures 7 m. The lengths of the other two sides are consecutive integers. Find the lengths of the other two sides.

30. Right Triangle. The shortest side of a right triangle measures 8 cm. The lengths of the other two sides are consecutive odd integers. Find the lengths of the other two sides.

31. Architecture. An architect has allocated a rectangular space of 264 ft^2 for a square dining room and a 10-ft wide kitchen, as shown in the figure. Find the dimensions of each room.

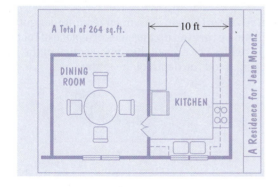

32. Design. A window panel for a sun porch consists of a 7-ft tall rectangular window stacked above a square window. The windows have the same width. If the total area of the window panel is 18 ft^2, find the dimensions of each window.

Height of a Rocket. For Exercises 33 and 34, assume that a water rocket is launched upward with an initial velocity of 48 ft/sec. Its height h, in feet, after t seconds, is given by $h = 48t - 16t^2$.

33. When will the rocket be exactly 32 ft above the ground?

34. When will the rocket crash into the ground?

35. The sum of the squares of two consecutive odd positive integers is 74. Find the integers.

36. The sum of the squares of two consecutive odd positive integers is 130. Find the integers.

Skill Maintenance

Compute and simplify.

37. $-3.57 + 8.1$ [1.3a]

38. $-\dfrac{2}{3} - \dfrac{1}{6}$ [1.4a]

39. $(-2)(-4)(-5)$ [1.5a]

40. $2 \cdot 6^2 \div (-2) \cdot 3 - 8$ [1.8d]

41. $\dfrac{2 - |3 - 8|}{(-1 - 4)^2}$ [1.8d]

42. $1.2 + (-2)^3 + 3.4$ [1.8d]

Remove parentheses and simplify.

43. $2(y - 7) - (6y - 1)$ [1.8b]

44. $2\{x - 3[4 - (x - 1)] + x\}$ [1.8c]

Synthesis

45. *Pool Sidewalk.* A cement walk of constant width is built around a 20-ft by 40-ft rectangular pool. The total area of the pool and the walk is 1500 ft². Find the width of the walk.

46. *Roofing.* A *square* of shingles covers 100 ft² of surface area. How many squares will be needed to reshingle the roof of the house shown?

25 ft 16 ft 24 ft 32 ft

47. *Dimensions of an Open Box.* A rectangular piece of cardboard is twice as long as it is wide. A 4-cm square is cut out of each corner, and the sides are turned up to make a box with an open top. The volume of the box is 616 cm³. Find the original dimensions of the cardboard.

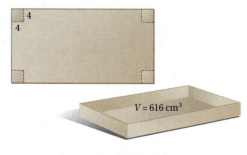

4
4
$V = 616\ \text{cm}^3$

48. *Rain-Gutter Design.* An open rectangular gutter is made by turning up the sides of a piece of metal 20 in. wide. The area of the cross-section of the gutter is 50 in². Find the depth of the gutter.

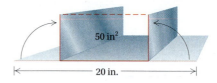

50 in²
20 in.

49. *Right Triangle.* The longest side of a right triangle is 5 yd shorter than six times the length of the shortest side. The other side of the triangle is 5 yd longer than five times the length of the shortest side. Find the lengths of the sides of the triangle.

50. Solve for x.

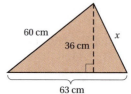

60 cm 36 cm x
63 cm

Vocabulary Reinforcement

Complete each statement with the correct term from the column on the right. Some of the choices may be used more than once or not at all.

1. To _____ a polynomial is to express it as a product.   [5.1a]

2. A(n) _____ of a polynomial P is a polynomial that can be used to express P as a product.   [5.1a]

3. A(n) _____ of a polynomial is an expression that names that polynomial as a product.   [5.1a]

4. When factoring, always look first for a(n) _____ factor.   [5.1b]

5. When factoring a polynomial with four terms, try factoring by _____.   [5.6a]

6. A trinomial square is the square of a(n) _____.   [5.5a]

7. The principle of _____ products states that if $ab = 0$, then $a = 0$ or $b = 0$.   [5.7a]

8. The factorization of a _____ of squares is the product of the sum and the difference of two terms.   [5.5d]

common
similar
product
difference
factor
factorization
grouping
monomial
binomial
trinomial
zero

Concept Reinforcement

Determine whether each statement is true or false.

_____ 1. Every polynomial with four terms can be factored by grouping.   [5.1c]

_____ 2. When factoring $x^2 + 5x + 6$, we need consider only positive pairs of factors of 6.   [5.2a]

_____ 3. A product is 0 if and only if all the factors are 0.   [5.7a]

_____ 4. If the principle of zero products is to be used, one side of the equation must be 0.   [5.7b]

Study Guide

Objective 5.1a   Find the greatest common factor, the GCF, of monomials.

Example   Find the GCF of $15x^4y^2$, $-18x$, and $12x^3y$.

$15x^4y^2 = 3 \cdot 5 \cdot x^4 \cdot y^2;$

$-18x = -1 \cdot 2 \cdot 3 \cdot 3 \cdot x;$

$12x^3y = 2 \cdot 2 \cdot 3 \cdot x^3 \cdot y$

The GCF of the coefficients is 3. The GCF of the powers of x is x because 1 is the smallest exponent of x. The GCF of the powers of y is 1 because $-18x$ has no y-factor. Thus the GCF is $3 \cdot x \cdot 1$, or $3x$.

Practice Exercise

1. Find the GCF of $8x^3y^2$, $-20xy^3$, and $32x^2y$.

Objective 5.1b   Factor polynomials when the terms have a common factor, factoring out the greatest common factor.

Example   Factor: $16y^4 + 8y^3 - 24y^2$.

The *largest* common factor is $8y^2$.

$16y^4 + 8y^3 - 24y^2 = (8y^2)(2y^2) + (8y^2)(y) - (8y^2)(3)$
$= 8y^2(2y^2 + y - 3)$

Practice Exercise

2. Factor $27x^5 - 9x^3 + 18x^2$, factoring out the largest common factor.

Objective 5.1c   Factor certain expressions with four terms using factoring by grouping.

Example   Factor $6x^3 + 4x^2 - 15x - 10$ by grouping.

$6x^3 + 4x^2 - 15x - 10 = (6x^3 + 4x^2) + (-15x - 10)$
$= 2x^2(3x + 2) - 5(3x + 2)$
$= (3x + 2)(2x^2 - 5)$

Practice Exercise

3. Factor $z^3 - 3z^2 + 4z - 12$ by grouping.

Objective 5.2a   Factor trinomials of the type $x^2 + bx + c$ by examining the constant term c.

Example   Factor: $x^2 - x - 12$.

Since the constant term, -12, is negative, we look for a factorization of -12 in which one factor is positive and one factor is negative. The sum of the factors must be the coefficient of the middle term, -1, so the negative factor must have the larger absolute value. The possible pairs of factors that meet these criteria are $1, -12$ and $2, -6$ and $3, -4$. The numbers we need are 3 and -4:

$x^2 - x - 12 = (x + 3)(x - 4)$.

Practice Exercise

4. Factor: $x^2 + 6x + 8$.

Objective 5.3a   Factor trinomials of the type $ax^2 + bx + c, a \neq 1$, using the FOIL method.

Example   Factor $2y^3 + 5y^2 - 3y$.

1) Factor out the largest common factor, y:

$y(2y^2 + 5y - 3)$.

Now we factor $2y^2 + 5y - 3$.

2) Because $2y^2$ factors as $2y \cdot y$, we have this possibility for a factorization:

$(2y + \quad)(y + \quad)$.

3) There are two pairs of factors of -3 and each can be written in two ways:

$3, -1 \qquad -3, 1$

and $\quad -1, 3 \qquad 1, -3$.

4) From steps (2) and (3), we see that there are 4 possibilities for factorizations. We look for **O**utside and **I**nside products for which the sum is the middle term, $5y$. We try some possibilities and find that the factorization of $2y^2 + 5y - 3$ is $(2y - 1)(y + 3)$.

We must include the common factor to get a factorization of the original trinomial:

$2y^3 + 5y^2 - 3y = y(2y - 1)(y + 3)$.

Practice Exercise

5. Factor: $6z^2 - 21z - 12$.

Objective 5.4a Factor trinomials of the type $ax^2 + bx + c, a \neq 1$, using the ac-method.

Example Factor $5x^2 + 7x - 6$ using the ac-method.

1) There is no common factor (other than 1 or -1).
2) Multiply the leading coefficient 5 and the constant, -6:
$$5(-6) = -30.$$
3) Look for a factorization of -30 in which the sum of the factors is the coefficient of the middle term, 7. One number will be positive and the other will be negative. Since their sum, 7, is positive, the positive number will have the larger absolute value. The numbers we need are 10 and -3.
4) Split the middle term, writing it as a sum or a difference using the factors found in step (3):
$$7x = 10x - 3x.$$
5) Factor by grouping:
$$5x^2 + 7x - 6 = 5x^2 + 10x - 3x - 6$$
$$= 5x(x + 2) - 3(x + 2)$$
$$= (x + 2)(5x - 3).$$
6) Check: $(x + 2)(5x - 3) = 5x^2 + 7x - 6.$

Practice Exercise

6. Factor $6y^2 + 7y - 3$ using the ac-method.

Objective 5.5b Factor trinomial squares.

Example Factor: $9x^2 - 12x + 4$.
$$9x^2 - 12x + 4 = (3x)^2 - 2 \cdot 3x \cdot 2 + 2^2 = (3x - 2)^2$$

Practice Exercise

7. Factor: $4x^2 + 4x + 1$.

Objective 5.5d Factor differences of squares, being careful to factor completely.

Example Factor: $b^6 - b^2$.
$$b^6 - b^2 = b^2(b^4 - 1) = b^2(b^2 + 1)(b^2 - 1)$$
$$= b^2(b^2 + 1)(b + 1)(b - 1)$$

Practice Exercise

8. Factor $18x^2 - 8$ completely.

Objective 5.7b Solve quadratic equations by factoring and then using the principle of zero products.

Example Solve: $x^2 - 3x = 28$.
$$x^2 - 3x = 28$$
$$x^2 - 3x - 28 = 28 - 28$$
$$x^2 - 3x - 28 = 0$$
$$(x + 4)(x - 7) = 0$$
$$x + 4 = 0 \quad or \quad x - 7 = 0$$
$$x = -4 \quad or \quad x = 7$$
The solutions are -4 and 7.

Practice Exercise

9. Solve: $x^2 + 4x = 5$.

Review Exercises

Find the GCF. [5.1a]

1. $-15y^2$, $25y^6$

2. $12x^3$, $-60x^2y$, $36xy$

Factor completely. [5.6a]

3. $5 - 20x^6$

4. $x^2 - 3x$

$x(x-3)$

5. $9x^2 - 4$

6. $x^2 + 4x - 12$

7. $x^2 + 14x + 49$

8. $6x^3 + 12x^2 + 3x$

9. $x^3 + x^2 + 3x + 3$

10. $6x^2 - 5x + 1$

11. $x^4 - 81$

12. $9x^3 + 12x^2 - 45x$

13. $2x^2 - 50$

14. $x^4 + 4x^3 - 2x - 8$

15. $16x^4 - 1$

16. $8x^6 - 32x^5 + 4x^4$

17. $75 + 12x^2 + 60x$

18. $x^2 + 9$

19. $x^3 - x^2 - 30x$

20. $4x^2 - 25$

21. $9x^2 + 25 - 30x$

22. $6x^2 - 28x - 48$

23. $x^2 - 6x + 9$

24. $2x^2 - 7x - 4$

25. $18x^2 - 12x + 2$

26. $3x^2 - 27$

27. $15 - 8x + x^2$

28. $25x^2 - 20x + 4$

29. $49b^{10} + 4a^8 - 28a^4b^5$

30. $x^2y^2 + xy - 12$

31. $12a^2 + 84ab + 147b^2$

32. $m^2 + 5m + mt + 5t$

33. $32x^4 - 128y^4z^4$

Solve. [5.7a, b]

34. $(x - 1)(x + 3) = 0$

35. $x^2 + 2x - 35 = 0$

36. $x^2 + 4x = 0$

37. $3x^2 + 2 = 5x$

38. $x^2 = 64$

39. $16 = x(x - 6)$

Find the x-intercepts of the graph of each equation.
[5.7b]

40.

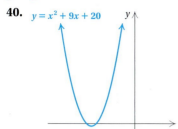

$y = x^2 + 9x + 20$

41.

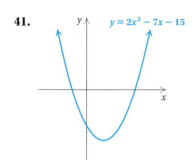

$y = 2x^2 - 7x - 15$

Solve. [5.8a]

42. *Sharks' Teeth.* Sharks' teeth are shaped like triangles. The height of a tooth of a great white shark is 1 cm longer than the base. The area is 15 cm². Find the height and the base.

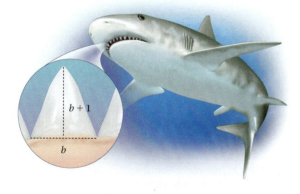

$b + 1$

b

43. The product of two consecutive even integers is 288. Find the integers.

44. *Zipline.* On one zipline in a canopy tour in Costa Rica, riders drop 58 ft while covering a distance of 840 ft along the ground. How long is the zipline?

45. *Tree Supports.* A duckbill-anchor system is used to support a newly planted Bradford pear tree. Each cable is 5 ft long. The distance from the base of the tree to the point on the ground where each cable is anchored is 1 ft more than the distance from the base of the tree to the point where the cable is attached to the tree. Find both distances.

46. If the sides of a square are lengthened by 3 km, the area increases to 81 km². Find the length of a side of the original square.

47. Factor: $x^2 - 9x + 8$. Which of the following is one factor? [5.2a], [5.6a]

A. $(x + 1)$ **B.** $(x - 1)$
C. $(x + 8)$ **D.** $(x - 4)$

48. Factor $15x^2 + 5x - 20$ completely. Which of the following is one factor? [5.3a], [5.4a], [5.6a]

A. $(3x + 4)$ **B.** $(3x - 4)$
C. $(5x - 5)$ **D.** $(15x + 20)$

Solve. [5.8a]

49. The pages of a book measure 15 cm by 20 cm. Margins of equal width surround the printing on each page and constitute one-half of the area of the page. Find the width of the margins.

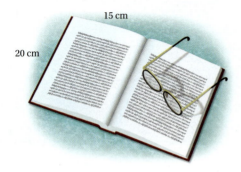

15 cm

20 cm

50. The cube of a number is the same as twice the square of the number. Find all such numbers.

51. The length of a rectangle is two times its width. When the length is increased by 20 in. and the width is decreased by 1 in., the area is 160 in². Find the original length and width.

52. Use the information in the figure below to determine the height of the telephone pole.

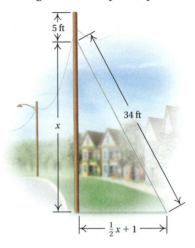

5 ft

x

34 ft

$\frac{1}{2}x + 1$

Solve. [5.7b]

53. $x^2 + 25 = 0$

54. $(x - 2)(x + 3)(2x - 5) = 0$

55. $(x - 3)4x^2 + 3x(x - 3) - (x - 3)10 = 0$

Understanding Through Discussion and Writing

1. Gwen factors $x^3 - 8x^2 + 15x$ as $(x^2 - 5x)(x - 3)$. Is she wrong? Why or why not? What advice would you offer? [5.2a]

2. After a test, Josh told a classmate that he was sure he had not written any incorrect factorizations. How could he be certain? [5.6a]

3. Kelly factored $16 - 8x + x^2$ as $(x - 4)^2$, while Tony factored it as $(4 - x)^2$. Evaluate each expression for several values of x. Then explain why both answers are correct. [5.5b]

4. What is wrong with the following? Explain the correct method of solution. [5.7b]

$$(x - 3)(x + 4) = 8$$
$$x - 3 = 8 \quad or \quad x + 4 = 8$$
$$x = 11 \quad or \quad x = 4$$

5. What is incorrect about solving $x^2 = 3x$ by dividing by x on both sides? [5.7b]

6. An archaeologist has measuring sticks of 3 ft, 4 ft, and 5 ft. Explain how she could draw a 7-ft by 9-ft rectangle on a piece of land being excavated. [5.8a]

CHAPTER

5 Test

For Extra Help For step-by-step test solutions, access the Chapter Test Prep Videos in MyMathLab® or on YouTube (search "BittingerIntro" and click on "Channels").

1. Find the GCF: $28x^3, 48x^7$.

Factor completely.

2. $x^2 - 7x + 10$

3. $x^2 + 25 - 10x$

4. $6y^2 - 8y^3 + 4y^4$

5. $x^3 + x^2 + 2x + 2$

6. $x^2 - 5x$

7. $x^3 + 2x^2 - 3x$

8. $28x - 48 + 10x^2$

9. $4x^2 - 9$

10. $x^2 - x - 12$

11. $6m^3 + 9m^2 + 3m$

12. $3w^2 - 75$

13. $60x + 45x^2 + 20$

14. $3x^4 - 48$

15. $49x^2 - 84x + 36$

16. $5x^2 - 26x + 5$

17. $x^4 + 2x^3 - 3x - 6$

18. $80 - 5x^4$

19. $6t^3 + 9t^2 - 15t$

Solve.

20. $x^2 - 3x = 0$

21. $2x^2 = 32$

22. $x^2 - x - 20 = 0$

23. $2x^2 + 7x = 15$

24. $x(x - 3) = 28$

Find the *x*-intercepts of the graph of each equation.

25.

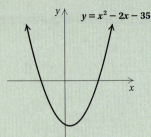

$$y = x^2 - 2x - 35$$

26.

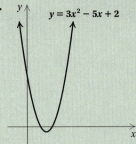

$$y = 3x^2 - 5x + 2$$

Solve.

27. The length of a rectangle is 2 m more than the width. The area of the rectangle is 48 m². Find the length and the width.

28. The base of a triangle is 6 cm greater than twice the height. The area is 28 cm². Find the height and the base.

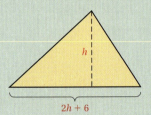

29. *Masonry Corner.* A mason wants to be sure that he has a right-angle corner of a building's foundation. He marks a point 3 ft from the corner along one wall and another point 4 ft from the corner along the other wall. If the corner is a right angle, what should the distance be between the two marked points?

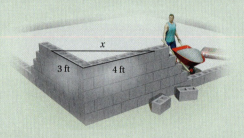

30. Factor $2y^4 - 32$ completely. Which of the following is one factor?

A. $(y + 2)$ **B.** $(y + 4)$
C. $(y^2 - 4)$ **D.** $(2y^2 + 8)$

Synthesis

31. The length of a rectangle is five times its width. When the length is decreased by 3 m and the width is increased by 2 m, the area of the new rectangle is 60 m². Find the original length and width.

32. Factor: $(a + 3)^2 - 2(a + 3) - 35$.

33. Solve: $20x(x + 2)(x - 1) = 5x^3 - 24x - 14x^2$.

34. If $x + y = 4$ and $x - y = 6$, then $x^2 - y^2$ equals which of the following?

A. 2 **B.** 10
C. 34 **D.** 24

Use either $<$ or $>$ for $\square$ to write a true sentence.

1. $\dfrac{2}{3} \ \square \ \dfrac{5}{7}$ **2.** $-\dfrac{4}{7} \ \square \ -\dfrac{8}{11}$

Compute and simplify.

3. $2.06 + (-4.79) - (-3.08)$

4. $5.652 \div (-3.6)$

5. $\left(\dfrac{2}{9}\right)\left(-\dfrac{3}{8}\right)\left(\dfrac{6}{7}\right)$ **6.** $\dfrac{21}{5} \div \left(-\dfrac{7}{2}\right)$

Simplify.

7. $[3x + 2(x - 1)] - [2x - (x + 3)]$

8. $1 - [14 + 28 \div 7 - (6 + 9 \div 3)]$

9. $(2x^2 y^{-1})^3$ **10.** $\dfrac{3x^5}{4x^3} \cdot \dfrac{-2x^{-3}}{9x^2}$

11. Add: $(2x^2 - 3x^3 + x - 4) + (x^4 - x - 5x^2)$.

12. Subtract: $(2x^2 y^2 + xy - 2xy^2) - (2xy - 2xy^2 + x^2 y)$.

13. Divide: $(x^3 + 2x^2 - x + 1) \div (x - 1)$.

Multiply.

14. $(2t - 3)^2$ **15.** $(x^2 - 3)(x^2 + 3)$

16. $(2x + 4)(3x - 4)$ **17.** $2x(x^3 + 3x^2 + 4x)$

18. $(2y - 1)(2y^2 + 3y + 4)$

19. $\left(x + \dfrac{2}{3}\right)\left(x - \dfrac{2}{3}\right)$

Factor.

20. $x^2 + 2x - 8$ **21.** $4x^2 - 25$

22. $3x^3 - 4x^2 + 3x - 4$ **23.** $x^2 - 26x + 169$

24. $75x^2 - 108y^2$ **25.** $6x^2 - 13x - 63$

26. $x^4 - 2x^2 - 3$ **27.** $4y^3 - 6y^2 - 4y + 6$

28. $6p^2 + pq - q^2$ **29.** $10x^3 + 52x^2 + 10x$

30. $49x^3 - 42x^2 + 9x$ **31.** $3x^2 + 5x - 4$

32. $75x^3 + 27x$ **33.** $3x^8 - 48y^8$

34. $14x^2 + 28 + 42x$ **35.** $2x^5 - 2x^3 + x^2 - 1$

Solve.

36. $3x - 5 = 2x + 10$ **37.** $3y + 4 > 5y - 8$

38. $(x - 15)\left(x + \dfrac{1}{4}\right) = 0$ **39.** $-98x(x + 37) = 0$

40. $x^3 + x^2 = 25x + 25$ **41.** $2x^2 = 72$

42. $9x^2 + 1 = 6x$ **43.** $x^2 + 17x + 70 = 0$

44. $14y^2 = 21y$ **45.** $1.6 - 3.5x = 0.9$

46. $(x + 3)(x - 4) = 8$ **47.** $1.5x - 3.6 \leq 1.3x + 0.4$

48. $2x - [3x - (2x + 3)] = 3x + [4 - (2x + 1)]$

49. $y = mx + b$, for m

Solve.

50. The sum of two consecutive even integers is 102. Find the integers.

51. The product of two consecutive even integers is 360. Find the integers.

52. The length of a rectangular window is 3 ft longer than the height. The area of the window is 18 ft². Find the length and the height.

53. The length of a rectangular lot is 200 m longer than the width. The perimeter of the lot is 1000 m. Find the dimensions of the lot.

54. Money is borrowed at 12% simple interest. After 1 year, $7280 pays off the loan. How much was originally borrowed?

55. The length of one leg of a right triangle is 15 m. The length of the other leg is 9 m shorter than the length of the hypotenuse. Find the length of the hypotenuse.

56. A 100-m wire is cut into three pieces. The second piece is twice as long as the first piece. The third piece is one-third as long as the first piece. How long is each piece?

57. After a 25% price reduction, a pair of shoes is on sale for $21.75. What was the price before reduction?

58. The front of a house is a triangle that is as wide as it is tall. Its area is 98 ft². Find the height and the base.

59. Find the intercepts. Then graph the equation.
$$3x + 4y = -12$$

Synthesis

Solve.

60. $(x + 3)(x - 5) \leq (x + 2)(x - 1)$

61. $\dfrac{x - 3}{2} - \dfrac{2x + 5}{26} = \dfrac{4x + 11}{13}$

62. $(x + 1)^2 = 25$

Factor.

63. $x^2(x - 3) - x(x - 3) - 2(x - 3)$

64. $4a^2 - 4a + 1 - 9b^2 - 24b - 16$

Solve.

65. Find c such that the polynomial will be the square of a binomial: $cx^2 - 40x + 16$.

66. The length of the radius of a circle is increased by 2 cm to form a new circle. The area of the new circle is four times the area of the original circle. Find the length of the radius of the original circle.

Rational Expressions and Equations

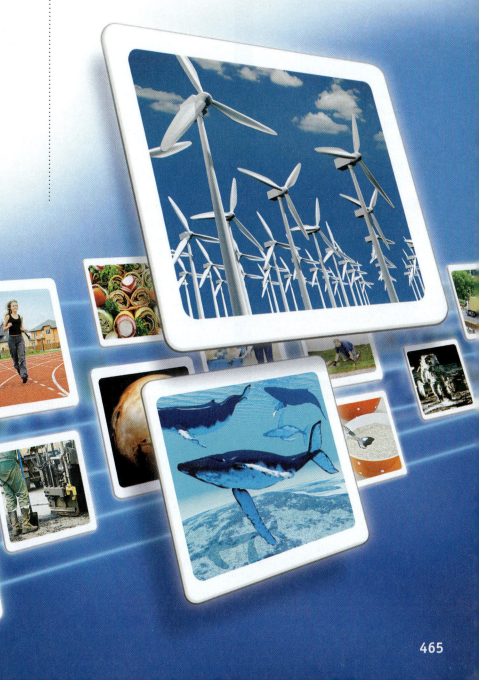

STUDYING FOR SUCCESS   *Quiz and Test Follow-Up*

☐ Immediately after completing a chapter quiz or test, write out a step-by-step solution of each question that you missed.

☐ Visit your instructor or tutor for help with problems that are still giving you trouble.

☐ Keep your tests and quizzes, along with their corrections, to use as a study guide for the final examination.

6.1 Multiplying and Simplifying Rational Expressions

OBJECTIVES

a Find all numbers for which a rational expression is not defined.

b Multiply a rational expression by 1, using an expression such as A/A.

c Simplify rational expressions by factoring the numerator and the denominator and removing factors of 1.

d Multiply rational expressions and simplify.

SKILL TO REVIEW

Objective 5.3a: Factor trinomials of the type $ax^2 + bx + c, a \neq 1$, using the FOIL method.

Factor.

1. $2x^2 - x - 21$

2. $40x^2 - 43x - 6$

a RATIONAL EXPRESSIONS AND REPLACEMENTS

Rational numbers are quotients of integers. Some examples are

$$\frac{2}{3}, \quad \frac{4}{-5}, \quad \frac{-8}{17}, \quad \frac{563}{1}.$$

The following are called **rational expressions** or **fraction expressions**. They are quotients, or ratios, of polynomials:

$$\frac{3}{4}, \quad \frac{z}{6}, \quad \frac{5}{x+2}, \quad \frac{t^2 + 3t - 10}{7t^2 - 4}.$$

A rational expression is also a division. For example,

$$\frac{3}{4} \quad \text{means} \quad 3 \div 4 \quad \text{and} \quad \frac{x-8}{x+2} \quad \text{means} \quad (x-8) \div (x+2).$$

Because rational expressions indicate division, we must be careful to avoid denominators of zero. When a variable is replaced with a number that produces a denominator equal to zero, the rational expression is not defined. For example, in the expression

$$\frac{x-8}{x+2},$$

when x is replaced with -2, the denominator is 0, and the expression is *not* defined:

$$\frac{x-8}{x+2} = \frac{-2-8}{-2+2} = \frac{-10}{0}. \leftarrow \text{Division by 0 is not defined.}$$

When x is replaced with a number other than -2, such as 3, the expression *is* defined because the denominator is nonzero:

$$\frac{x-8}{x+2} = \frac{3-8}{3+2} = \frac{-5}{5} = -1.$$

EXAMPLE 1 Find all numbers for which the rational expression

$$\frac{x+4}{x^2 - 3x - 10}$$

is not defined.

Answers

Skill to Review:

1. $(2x-7)(x+3)$ **2.** $(5x-6)(8x+1)$

The value of the numerator has no bearing on whether or not a rational expression is defined. To determine which numbers make the rational expression not defined, we set the *denominator* equal to 0 and solve:

$$x^2 - 3x - 10 = 0$$

$$(x - 5)(x + 2) = 0 \qquad \text{Factoring}$$

$$x - 5 = 0 \quad or \quad x + 2 = 0 \qquad \begin{array}{l}\text{Using the principle}\\ \text{of zero products (See}\\ \text{Section 5.8.)}\end{array}$$

$$x = 5 \quad or \qquad x = -2.$$

The rational expression is not defined for the replacement numbers 5 and −2.

Do Exercises 1–3. ▶

b MULTIPLYING BY 1

We multiply rational expressions in the same way that we multiply fraction notation in arithmetic. For review, we see that

$$\frac{3}{7} \cdot \frac{2}{5} = \frac{3 \cdot 2}{7 \cdot 5} = \frac{6}{35}.$$

MULTIPLYING RATIONAL EXPRESSIONS

To multiply rational expressions, multiply numerators and multiply denominators:

$$\frac{A}{B} \cdot \frac{C}{D} = \frac{AC}{BD}.$$

For example,

$$\frac{x - 2}{3} \cdot \frac{x + 2}{x + 7} = \frac{(x - 2)(x + 2)}{3(x + 7)}. \qquad \begin{array}{l}\text{Multiplying the numerators}\\ \text{and the denominators}\end{array}$$

Note that we leave the numerator, $(x - 2)(x + 2)$, and the denominator, $3(x + 7)$, in factored form because it is easier to simplify if we do not multiply. In order to learn to simplify, we first need to consider multiplying the rational expression by 1.

Any rational expression with the same numerator and denominator (except $0/0$) is a symbol for 1:

$$\frac{19}{19} = 1, \qquad \frac{x + 8}{x + 8} = 1, \qquad \frac{3x^2 - 4}{3x^2 - 4} = 1, \qquad \frac{-1}{-1} = 1.$$

EQUIVALENT EXPRESSIONS

Expressions that have the same value for all allowable (or meaningful) replacements are called **equivalent expressions**.

Find all numbers for which the rational expression is not defined.

1. $\dfrac{16}{x - 3}$

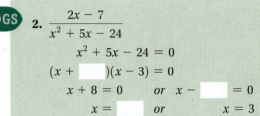

GS 2. $\dfrac{2x - 7}{x^2 + 5x - 24}$

$$x^2 + 5x - 24 = 0$$

$$(x + \boxed{})(x - 3) = 0$$

$$x + 8 = 0 \qquad or \quad x - \boxed{} = 0$$

$$x = \boxed{} \qquad or \qquad x = 3$$

The rational expression is not defined for replacements −8 and $\boxed{}$.

3. $\dfrac{x + 5}{8}$

Answers

1. 3 **2.** −8, 3 **3.** None

Guided Solution:

2. 8, 3, −8, 3

We can multiply by 1 to obtain an *equivalent expression*. At this point, we select expressions for 1 arbitrarily. Later, we will have a system for our choices when we add and subtract.

EXAMPLES Multiply.

2. $\dfrac{3x + 2}{x + 1} \cdot 1 = \dfrac{3x + 2}{x + 1} \cdot \dfrac{2x}{2x}$ Using the identity property of 1. We arbitrarily choose $(2x)/(2x)$ as a symbol for 1.

$= \dfrac{(3x + 2)2x}{(x + 1)2x}$

3. $\dfrac{x + 2}{x - 7} \cdot \dfrac{x + 3}{x + 3} = \dfrac{(x + 2)(x + 3)}{(x - 7)(x + 3)}$ We arbitrarily choose $(x + 3)/(x + 3)$ as a symbol for 1.

4. $\dfrac{2 + x}{2 - x} \cdot \dfrac{-1}{-1} = \dfrac{(2 + x)(-1)}{(2 - x)(-1)}$ Using $(-1)/(-1)$ as a symbol for 1

◀ **Do Exercises 4–6.**

Multiply.

4. $\dfrac{2x + 1}{3x - 2} \cdot \dfrac{x}{x}$

5. $\dfrac{x + 1}{x - 2} \cdot \dfrac{x + 2}{x + 2}$

6. $\dfrac{x - 8}{x - y} \cdot \dfrac{-1}{-1}$

c SIMPLIFYING RATIONAL EXPRESSIONS

Simplifying rational expressions is similar to simplifying fraction expressions in arithmetic. For example, an expression like $\frac{15}{40}$ can be simplified as follows:

$\dfrac{15}{40} = \dfrac{3 \cdot 5}{8 \cdot 5}$ Factoring the numerator and the denominator. Note the common factor, 5.

$= \dfrac{3}{8} \cdot \dfrac{5}{5}$ Factoring the fraction expression

$= \dfrac{3}{8} \cdot 1 \qquad \dfrac{5}{5} = 1$

$= \dfrac{3}{8}.$ Using the identity property of 1, or "removing" a factor of 1

Similar steps are followed when simplifying rational expressions: We factor and remove a factor of 1, using the fact that

$$\dfrac{ab}{cb} = \dfrac{a}{c} \cdot \dfrac{b}{b} = \dfrac{a}{c} \cdot 1 = \dfrac{a}{c}.$$

In algebra, instead of simplifying

$$\dfrac{15}{40},$$

we may need to simplify an expression like

$$\dfrac{x^2 - 16}{x + 4}.$$

Just as factoring is important in simplifying in arithmetic, so too is it important in simplifying rational expressions. The factoring we use most is the factoring of polynomials, which we studied in Chapter 5.

Answers

4. $\dfrac{(2x + 1)x}{(3x - 2)x}$ 5. $\dfrac{(x + 1)(x + 2)}{(x - 2)(x + 2)}$

6. $\dfrac{(x - 8)(-1)}{(x - y)(-1)}$

To simplify, we can do the reverse of multiplying. We factor the numerator and the denominator and "remove" a factor of 1.

EXAMPLE 5   Simplify: $\dfrac{8x^2}{24x}$.

$\dfrac{8x^2}{24x} = \dfrac{8 \cdot x \cdot x}{3 \cdot 8 \cdot x}$ Factoring the numerator and the denominator.
Note the common factor, 8x.

$= \dfrac{8x}{8x} \cdot \dfrac{x}{3}$ Factoring the rational expression

$= 1 \cdot \dfrac{x}{3}$ $\dfrac{8x}{8x} = 1$

$= \dfrac{x}{3}$ We removed a factor of 1.

Do Exercises 7 and 8. ▶

Simplify.

7. $\dfrac{5y}{y}$ 8. $\dfrac{9x^2}{36x}$

EXAMPLES   Simplify.

6. $\dfrac{5a + 15}{10} = \dfrac{5(a + 3)}{5 \cdot 2}$ Factoring the numerator and the denominator

$= \dfrac{5}{5} \cdot \dfrac{a + 3}{2}$ Factoring the rational expression

$= 1 \cdot \dfrac{a + 3}{2}$ $\dfrac{5}{5} = 1$

$= \dfrac{a + 3}{2}$ Removing a factor of 1

7. $\dfrac{6a + 12}{7a + 14} = \dfrac{6(a + 2)}{7(a + 2)}$ Factoring the numerator and the denominator

$= \dfrac{6}{7} \cdot \dfrac{a + 2}{a + 2}$ Factoring the rational expression

$= \dfrac{6}{7} \cdot 1$ $\dfrac{a + 2}{a + 2} = 1$

$= \dfrac{6}{7}$ Removing a factor of 1

8. $\dfrac{6x^2 + 4x}{2x^2 + 2x} = \dfrac{2x(3x + 2)}{2x(x + 1)}$ Factoring the numerator and the denominator

$= \dfrac{2x}{2x} \cdot \dfrac{3x + 2}{x + 1}$ Factoring the rational expression

$= 1 \cdot \dfrac{3x + 2}{x + 1}$ $\dfrac{2x}{2x} = 1$

$= \dfrac{3x + 2}{x + 1}$ Removing a factor of 1

.... **Caution!** ....

Note that you *cannot* simplify further by removing the x's because x is not a *factor* of the entire numerator, $3x + 2$, and the entire denominator, $x + 1$.
....

Answers

7. 5   8. $\dfrac{x}{4}$

$$9. \quad \frac{x^2 + 3x + 2}{x^2 - 1} = \frac{(x + 2)(x + 1)}{(x + 1)(x - 1)} \quad \text{Factoring the numerator and the denominator}$$

$$= \frac{x + 1}{x + 1} \cdot \frac{x + 2}{x - 1} \quad \text{Factoring the rational expression}$$

$$= 1 \cdot \frac{x + 2}{x - 1} \qquad \frac{x + 1}{x + 1} = 1$$

$$= \frac{x + 2}{x - 1} \qquad \text{Removing a factor of 1}$$

Canceling

You may have encountered canceling when working with rational expressions. With great concern, we mention it as a possible way to speed up your work. Our concern is that canceling be done with care and understanding. Example 9 might have been done faster as follows:

Simplify.

9. $\dfrac{2x^2 + x}{3x^2 + 2x}$

10. $\dfrac{x^2 - 1}{2x^2 - x - 1}$

11. $\dfrac{7x + 14}{7}$

12. $\dfrac{12y + 24}{48}$

$$\frac{x^2 + 3x + 2}{x^2 - 1} = \frac{(x + 2)(x + 1)}{(x + 1)(x - 1)} \quad \text{Factoring the numerator and the denominator}$$

$$= \frac{(x + 2)\cancel{(x + 1)}}{\cancel{(x + 1)}(x - 1)} \quad \text{When a factor of 1 is noted, it is canceled, as shown: } \frac{x + 1}{x + 1} = 1.$$

$$= \frac{x + 2}{x - 1}. \qquad \text{Simplifying}$$

◀ **Do Exercises 9–12.**

Opposites in Rational Expressions

Expressions of the form $a - b$ and $b - a$ are **opposites** of each other. When either of these binomials is multiplied by -1, the result is the other binomial:

$$\left.\begin{array}{l} -1(a - b) = -a + b = b + (-a) = b - a; \\ -1(b - a) = -b + a = a + (-b) = a - b. \end{array}\right\} \begin{array}{l} \text{Multiplication by } -1 \\ \text{reverses the order in} \\ \text{which subtraction occurs.} \end{array}$$

Consider, for example,

$$\frac{x - 4}{4 - x}.$$

At first glance, it appears as though the numerator and the denominator do not have any common factors other than 1. But $x - 4$ and $4 - x$ are opposites, or additive inverses, of each other. Thus we can rewrite one as the opposite of the other by factoring out a -1.

Simplify.

13. $\dfrac{x - 8}{8 - x}$ **GS**

$$= \frac{x - 8}{-(x - \boxed{})}$$

$$= \frac{1(x - 8)}{-1(x - 8)} = \frac{1}{\boxed{}} \cdot \frac{x - 8}{x - 8}$$

$$= -1 \cdot \boxed{} = \boxed{}$$

14. $\dfrac{c - d}{d - c}$

15. $\dfrac{-x - 7}{x + 7}$

EXAMPLE 10 Simplify: $\dfrac{x - 4}{4 - x}$.

$$\frac{x - 4}{4 - x} = \frac{x - 4}{-(x - 4)} = \frac{1(x - 4)}{-1(x - 4)} \quad \begin{array}{l} 4 - x = -(x - 4); 4 - x \text{ and} \\ x - 4 \text{ are opposites.} \end{array}$$

$$= \frac{1}{-1} \cdot \frac{x - 4}{x - 4}$$

$$= -1 \cdot 1 \qquad 1/(-1) = -1$$

$$= -1$$

◀ **Do Exercises 13–15.**

Answers

9. $\dfrac{2x + 1}{3x + 2}$ **10.** $\dfrac{x + 1}{2x + 1}$ **11.** $x + 2$

12. $\dfrac{y + 2}{4}$ **13.** -1 **14.** -1 **15.** -1

Guided Solution:
13. 8, −1, 1, −1

d MULTIPLYING AND SIMPLIFYING

We try to simplify after we multiply. That is why we leave the numerator and the denominator in factored form.

EXAMPLE 11 Multiply and simplify: $\dfrac{5a^3}{4} \cdot \dfrac{2}{5a}$.

$\dfrac{5a^3}{4} \cdot \dfrac{2}{5a} = \dfrac{5a^3(2)}{4(5a)}$ Multiplying the numerators and the denominators

$= \dfrac{5 \cdot a \cdot a \cdot a \cdot 2}{2 \cdot 2 \cdot 5 \cdot a}$ Factoring the numerator and the denominator

$= \dfrac{5 \cdot a \cdot a \cdot a \cdot 2}{2 \cdot 2 \cdot 5 \cdot a}$ Removing a factor of 1: $\dfrac{2 \cdot 5 \cdot a}{2 \cdot 5 \cdot a} = 1$

$= \dfrac{a^2}{2}$ Simplifying

EXAMPLE 12 Multiply and simplify: $\dfrac{x^2 + 6x + 9}{x^2 - 4} \cdot \dfrac{x - 2}{x + 3}$.

$\dfrac{x^2 + 6x + 9}{x^2 - 4} \cdot \dfrac{x - 2}{x + 3} = \dfrac{(x^2 + 6x + 9)(x - 2)}{(x^2 - 4)(x + 3)}$ Multiplying the numerators and the denominators

$= \dfrac{(x + 3)(x + 3)(x - 2)}{(x + 2)(x - 2)(x + 3)}$ Factoring the numerator and the denominator

$= \dfrac{(x + 3)(x + 3)(x - 2)}{(x + 2)(x - 2)(x + 3)}$ Removing a factor of 1: $\dfrac{(x + 3)(x - 2)}{(x + 3)(x - 2)} = 1$

$= \dfrac{x + 3}{x + 2}$ Simplifying

Do Exercise 16. ▶

EXAMPLE 13 Multiply and simplify: $\dfrac{x^2 + x - 2}{15} \cdot \dfrac{5}{2x^2 - 3x + 1}$.

$\dfrac{x^2 + x - 2}{15} \cdot \dfrac{5}{2x^2 - 3x + 1} = \dfrac{(x^2 + x - 2)5}{15(2x^2 - 3x + 1)}$ Multiplying the numerators and the denominators

$= \dfrac{(x + 2)(x - 1)5}{5(3)(x - 1)(2x - 1)}$ Factoring the numerator and the denominator

$= \dfrac{(x + 2)(x - 1)5}{5(3)(x - 1)(2x - 1)}$ Removing a factor of 1: $\dfrac{(x - 1)5}{(x - 1)5} = 1$

$= \dfrac{x + 2}{3(2x - 1)}$ Simplifying

| You need not carry out this multiplication. |

Do Exercise 17. ▶

GS **16.** Multiply and simplify:

$\dfrac{a^2 - 4a + 4}{a^2 - 9} \cdot \dfrac{a + 3}{a - 2}$.

$\dfrac{a^2 - 4a + 4}{a^2 - 9} \cdot \dfrac{a + 3}{a - 2}$

$= \dfrac{(a^2 - 4a + 4)(a + \boxed{})}{(a^2 - \boxed{})(a - 2)}$

$= \dfrac{(a - \boxed{})(a - 2)(a + 3)}{(a + 3)(a - \boxed{})(a - 2)}$

$= \dfrac{(a - 2)(a - 2)(a + 3)}{(a + 3)(a - 3)(a - 2)}$

$= \dfrac{a - \boxed{}}{a - \boxed{}}$

17. Multiply and simplify:

$\dfrac{x^2 - 25}{6} \cdot \dfrac{3}{x + 5}$.

Answers

16. $\dfrac{a - 2}{a - 3}$ **17.** $\dfrac{x - 5}{2}$

Guided Solution:
16. 3, 9, 2, 3, 2, 3

6.1

Exercise Set

For Extra Help

MyMathLab®

MathXL®
PRACTICE   WATCH   READ   REVIEW

✓ Reading Check

Choose the word below each blank that best completes the statement.

RC1. Expressions that have the same value for all allowable replacements are called _____ expressions.
rational/equivalent

RC2. A rational expression is undefined when the _____ is zero.
denominator/numerator

RC3. A rational expression can be written as a _____ of two polynomials.
product/quotient

RC4. A rational expression is simplified when the numerator and the denominator have no _____ (other than 1) in common.
factors/terms

a   Find all numbers for which each rational expression is not defined.

1. $\dfrac{-3}{2x}$

2. $\dfrac{24}{-8y}$

3. $\dfrac{5}{x-8}$

4. $\dfrac{y-4}{y+6}$

5. $\dfrac{3}{2y+5}$

6. $\dfrac{x^2-9}{4x-15}$

7. $\dfrac{x^2+11}{x^2-3x-28}$

8. $\dfrac{p^2-9}{p^2-7p+10}$

9. $\dfrac{m^3-2m}{m^2-25}$

10. $\dfrac{7-3x+x^2}{49-x^2}$

11. $\dfrac{x-4}{3}$

12. $\dfrac{x^2-25}{14}$

b   Multiply. Do not simplify. Note that in each case you are multiplying by 1.

13. $\dfrac{4x}{4x}\cdot\dfrac{3x^2}{5y}$

14. $\dfrac{5x^2}{5x^2}\cdot\dfrac{6y^3}{3z^4}$

15. $\dfrac{2x}{2x}\cdot\dfrac{x-1}{x+4}$

16. $\dfrac{2a-3}{5a+2}\cdot\dfrac{a}{a}$

17. $\dfrac{3-x}{4-x}\cdot\dfrac{-1}{-1}$

18. $\dfrac{x-5}{5-x}\cdot\dfrac{-1}{-1}$

19. $\dfrac{y+6}{y+6}\cdot\dfrac{y-7}{y+2}$

20. $\dfrac{x^2+1}{x^3-2}\cdot\dfrac{x-4}{x-4}$

c Simplify.

21. $\dfrac{8x^3}{32x}$

22. $\dfrac{4x^2}{20x}$

23. $\dfrac{48p^7q^5}{18p^5q^4}$

24. $\dfrac{-76x^8y^3}{-24x^4y^3}$

25. $\dfrac{4x-12}{4x}$

26. $\dfrac{5a-40}{5}$

27. $\dfrac{3m^2+3m}{6m^2+9m}$

28. $\dfrac{4y^2-2y}{5y^2-5y}$

29. $\dfrac{a^2-9}{a^2+5a+6}$

30. $\dfrac{t^2-25}{t^2+t-20}$

31. $\dfrac{a^2-10a+21}{a^2-11a+28}$

32. $\dfrac{x^2-2x-8}{x^2-x-6}$

33. $\dfrac{x^2-25}{x^2-10x+25}$

34. $\dfrac{x^2+8x+16}{x^2-16}$

35. $\dfrac{a^2-1}{a-1}$

36. $\dfrac{t^2-1}{t+1}$

37. $\dfrac{x^2+1}{x+1}$

38. $\dfrac{m^2+9}{m+3}$

39. $\dfrac{6x^2-54}{4x^2-36}$

40. $\dfrac{8x^2-32}{4x^2-16}$

41. $\dfrac{6t+12}{t^2-t-6}$

42. $\dfrac{4x+32}{x^2+9x+8}$

43. $\dfrac{2t^2+6t+4}{4t^2-12t-16}$

44. $\dfrac{3a^2-9a-12}{6a^2+30a+24}$

45. $\dfrac{t^2 - 4}{(t + 2)^2}$

46. $\dfrac{m^2 - 36}{(m - 6)^2}$

47. $\dfrac{6 - x}{x - 6}$

48. $\dfrac{t - 3}{3 - t}$

49. $\dfrac{a - b}{b - a}$

50. $\dfrac{y - x}{-x + y}$

51. $\dfrac{6t - 12}{2 - t}$

52. $\dfrac{5a - 15}{3 - a}$

53. $\dfrac{x^2 - 1}{1 - x}$

54. $\dfrac{a^2 - b^2}{b^2 - a^2}$

55. $\dfrac{6qt - 3t^4}{t^3 - 2q}$

56. $\dfrac{2z - w^5}{5w^{10} - 10zw^5}$

d   Multiply and simplify.

57. $\dfrac{4x^3}{3x} \cdot \dfrac{14}{x}$

58. $\dfrac{18}{x^3} \cdot \dfrac{5x^2}{6}$

59. $\dfrac{3c}{d^2} \cdot \dfrac{4d}{6c^3}$

60. $\dfrac{3x^2y}{2} \cdot \dfrac{4}{xy^3}$

61. $\dfrac{x + 4}{x} \cdot \dfrac{x^2 - 3x}{x^2 + x - 12}$

62. $\dfrac{t^2}{t^2 - 4} \cdot \dfrac{t^2 - 5t + 6}{t^2 - 3t}$

63. $\dfrac{a^2 - 9}{a^2} \cdot \dfrac{a^2 - 3a}{a^2 + a - 12}$

64. $\dfrac{x^2 + 10x - 11}{x^2 - 1} \cdot \dfrac{x + 1}{x + 11}$

65. $\dfrac{4a^2}{3a^2 - 12a + 12} \cdot \dfrac{3a - 6}{2a}$

66. $\dfrac{5v + 5}{v - 2} \cdot \dfrac{v^2 - 4v + 4}{v^2 - 1}$

67. $\dfrac{t^4 - 16}{t^4 - 1} \cdot \dfrac{t^2 + 1}{t^2 + 4}$

68. $\dfrac{x^4 - 1}{x^4 - 81} \cdot \dfrac{x^2 + 9}{x^2 + 1}$

69. $\dfrac{(x + 4)^3}{(x + 2)^3} \cdot \dfrac{x^2 + 4x + 4}{x^2 + 8x + 16}$

70. $\dfrac{(t - 2)^3}{(t - 1)^3} \cdot \dfrac{t^2 - 2t + 1}{t^2 - 4t + 4}$

71. $\dfrac{5a^2 - 180}{10a^2 - 10} \cdot \dfrac{20a + 20}{2a - 12}$

72. $\dfrac{2t^2 - 98}{4t^2 - 4} \cdot \dfrac{8t + 8}{16t - 112}$

Skill Maintenance

Graph.

73. $x + y = -1$ [3.2a]

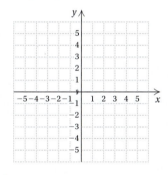

74. $y = -\dfrac{7}{2}$ [3.2b]

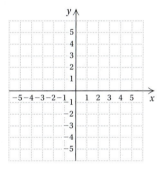

Factor. [5.6a]

75. $x^5 - 2x^4 - 35x^3$

76. $2y^3 - 10y^2 + y - 5$

77. $16 - t^4$

78. $10x^2 + 80x + 70$

Synthesis

Simplify.

79. $\dfrac{x^2 - y^2}{(x - y)^2} \cdot \dfrac{x^2 - 2xy + y^2}{x^2 - 4xy - 5y^2}$

80. $\dfrac{(t + 2)^3}{(t + 1)^3} \cdot \dfrac{t^2 + 2t + 1}{t^2 + 4t + 4} \cdot \dfrac{t + 1}{t + 2}$

81. $\dfrac{x - 1}{x^2 + 1} \cdot \dfrac{x^4 - 1}{(x - 1)^2} \cdot \dfrac{x^2 - 1}{x^4 - 2x^2 + 1}$

82. Select any number x, multiply by 2, add 5, multiply by 5, subtract 25, and divide by 10. What do you get? Explain how this procedure can be used for a number trick.

SKILL TO REVIEW

Objective 5.6a: Factor polynomials.

Factor.

1. $x^2 - 2x$

2. $5y^2 - 11y - 12$

Find the reciprocal.

1. $\dfrac{7}{2}$

2. $\dfrac{x^2 + 5}{2x^3 - 1}$

3. $x - 5$

4. $\dfrac{1}{x^2 - 3}$

There is a similarity between what we do with rational expressions and what we do with rational numbers. In fact, after variables have been replaced with rational numbers, a rational expression represents a rational number.

a FINDING RECIPROCALS

Two expressions are **reciprocals** of each other if their product is 1. The reciprocal of a rational expression is found by interchanging the numerator and the denominator.

EXAMPLES

1. The reciprocal of $\dfrac{2}{5}$ is $\dfrac{5}{2}$. $\left(\text{This is because } \dfrac{2}{5} \cdot \dfrac{5}{2} = \dfrac{10}{10} = 1.\right)$

2. The reciprocal of $\dfrac{2x^2 - 3}{x + 4}$ is $\dfrac{x + 4}{2x^2 - 3}$.

3. The reciprocal of $x + 2$ is $\dfrac{1}{x + 2}$. $\left(\text{Think of } x + 2 \text{ as } \dfrac{x + 2}{1}.\right)$

◀ **Do Margin Exercises 1–4.**

b DIVISION

We divide rational expressions in the same way that we divide fraction notation in arithmetic.

DIVIDING RATIONAL EXPRESSIONS

To divide by a rational expression, multiply by its reciprocal:

$$\frac{A}{B} \div \frac{C}{D} = \frac{A}{B} \cdot \frac{D}{C} = \frac{AD}{BC}.$$

Then factor and, if possible, simplify.

EXAMPLE 4 Divide and simplify: $\dfrac{3}{4} \div \dfrac{9}{5}$.

5. Divide and simplify: $\dfrac{3}{5} \div \dfrac{7}{10}$.

$$\frac{3}{4} \div \frac{9}{5} = \frac{3}{4} \cdot \frac{5}{9} \qquad \text{Multiplying by the reciprocal of the divisor}$$

$$= \frac{3 \cdot 5}{4 \cdot 9} = \frac{3 \cdot 5}{2 \cdot 2 \cdot 3 \cdot 3} \qquad \text{Factoring}$$

$$= \frac{\cancel{3} \cdot 5}{2 \cdot 2 \cdot \cancel{3} \cdot 3} \qquad \text{Removing a factor of 1: } \frac{3}{3} = 1$$

$$= \frac{5}{12} \qquad \text{Simplifying}$$

◀ **Do Margin Exercise 5.**

Answers

Skill to Review:
1. $x(x - 2)$ **2.** $(5y + 4)(y - 3)$

Margin Exercises:
1. $\dfrac{2}{7}$ **2.** $\dfrac{2x^3 - 1}{x^2 + 5}$ **3.** $\dfrac{1}{x - 5}$

4. $x^2 - 3$ **5.** $\dfrac{6}{7}$

EXAMPLE 5 Divide and simplify: $\dfrac{2}{x} \div \dfrac{3}{x}$.

$$\dfrac{2}{x} \div \dfrac{3}{x} = \dfrac{2}{x} \cdot \dfrac{x}{3}$$ Multiplying by the reciprocal of the divisor

$$= \dfrac{2 \cdot x}{x \cdot 3} = \dfrac{2 \cdot \cancel{x}}{\cancel{x} \cdot 3}$$ Removing a factor of 1: $\dfrac{x}{x} = 1$

$$= \dfrac{2}{3}$$

Do Exercise 6. ▶

EXAMPLE 6 Divide and simplify: $\dfrac{x+1}{x+2} \div \dfrac{x-1}{x+3}$.

$$\dfrac{x+1}{x+2} \div \dfrac{x-1}{x+3} = \dfrac{x+1}{x+2} \cdot \dfrac{x+3}{x-1}$$ Multiplying by the reciprocal of the divisor

$$= \left.\dfrac{(x+1)(x+3)}{(x+2)(x-1)}\right\} \longleftarrow$$

> We usually do not carry out the multiplication in the numerator or the denominator. It is not wrong to do so, but the factored form is often more useful.

Do Exercise 7. ▶

EXAMPLE 7 Divide and simplify: $\dfrac{4}{x^2 - 7x} \div \dfrac{28x}{x^2 - 49}$.

$$\dfrac{4}{x^2 - 7x} \div \dfrac{28x}{x^2 - 49} = \dfrac{4}{x^2 - 7x} \cdot \dfrac{x^2 - 49}{28x}$$ Multiplying by the reciprocal

$$= \dfrac{4(x^2 - 49)}{(x^2 - 7x)(28x)}$$

$$= \dfrac{2 \cdot 2 \cdot (x-7)(x+7)}{x(x-7) \cdot 2 \cdot 2 \cdot 7 \cdot x}$$ Factoring the numerator and the denominator

$$= \dfrac{2 \cdot 2 \cdot (\cancel{x-7})(x+7)}{x(\cancel{x-7}) \cdot 2 \cdot 2 \cdot 7 \cdot x}$$ Removing a factor of 1: $\dfrac{2 \cdot 2 \cdot (x-7)}{2 \cdot 2 \cdot (x-7)} = 1$

$$= \dfrac{x+7}{7x^2}$$

Do Exercise 8. ▶

EXAMPLE 8 Divide and simplify: $\dfrac{x+1}{x^2 - 1} \div \dfrac{x+1}{x^2 - 2x + 1}$.

$$\dfrac{x+1}{x^2 - 1} \div \dfrac{x+1}{x^2 - 2x + 1}$$

$$= \dfrac{x+1}{x^2 - 1} \cdot \dfrac{x^2 - 2x + 1}{x+1}$$ Multiplying by the reciprocal

---

GS 6. Divide and simplify: $\dfrac{x}{8} \div \dfrac{x}{5}$.

$$\dfrac{x}{8} \div \dfrac{x}{5} = \dfrac{x}{8} \cdot \dfrac{5}{\boxed{}}$$

$$= \dfrac{x \cdot \boxed{}}{8 \cdot x}$$

$$= \dfrac{\cancel{x} \cdot 5}{8 \cdot \cancel{x}}$$

$$= \dfrac{\boxed{}}{8}$$

7. Divide and simplify:
$$\dfrac{x-3}{x+5} \div \dfrac{x+5}{x-2}.$$

8. Divide and simplify:
$$\dfrac{a^2 + 5a}{6} \div \dfrac{a^2 - 25}{18a}.$$

Answers

6. $\dfrac{5}{8}$ **7.** $\dfrac{(x-3)(x-2)}{(x+5)(x+5)}$ **8.** $\dfrac{3a^2}{a-5}$

Guided Solution:
6. $x, 5, 5$

Then we multiply numerators and multiply denominators. We have

$$= \frac{(x + 1)(x^2 - 2x + 1)}{(x^2 - 1)(x + 1)}$$

$$= \frac{(x + 1)(x - 1)(x - 1)}{(x - 1)(x + 1)(x + 1)} \quad \text{Factoring the numerator and the denominator}$$

$$= \frac{\cancel{(x + 1)}\cancel{(x - 1)}(x - 1)}{\cancel{(x - 1)}\cancel{(x + 1)}(x + 1)} \quad \begin{array}{l}\text{Removing a factor of 1:}\\ \dfrac{(x + 1)(x - 1)}{(x + 1)(x - 1)} = 1\end{array}$$

$$= \frac{x - 1}{x + 1}.$$

Divide and simplify.

9. $\dfrac{x - 3}{x + 5} \div \dfrac{x + 2}{x + 5}$

10. $\dfrac{x^2 - 5x + 6}{x + 5} \div \dfrac{x + 2}{x + 5}$

11. **GS** $\dfrac{y^2 - 1}{y + 1} \div \dfrac{y^2 - 2y + 1}{y + 1}$

$$= \frac{y^2 - 1}{y + 1} \cdot \frac{\boxed{} + 1}{y^2 - \boxed{} + 1}$$

$$= \frac{(y^2 - 1)(y + 1)}{(y + 1)(y^2 - 2y + 1)}$$

$$= \frac{(y + \boxed{})(y - \boxed{})(y + 1)}{(y + 1)(y - \boxed{})(y - \boxed{})}$$

$$= \frac{(y + 1)\cancel{(y - 1)}(y + 1)}{(y + 1)\cancel{(y - 1)}(y - 1)}$$

$$= \frac{y + \boxed{}}{\boxed{} - 1}$$

EXAMPLE 9 Divide and simplify: $\dfrac{x^2 - 2x - 3}{x^2 - 4} \div \dfrac{x + 1}{x + 5}.$

$$\frac{x^2 - 2x - 3}{x^2 - 4} \div \frac{x + 1}{x + 5}$$

$$= \frac{x^2 - 2x - 3}{x^2 - 4} \cdot \frac{x + 5}{x + 1} \quad \text{Multiplying by the reciprocal}$$

$$= \frac{(x^2 - 2x - 3)(x + 5)}{(x^2 - 4)(x + 1)}$$

$$= \frac{(x - 3)(x + 1)(x + 5)}{(x - 2)(x + 2)(x + 1)} \quad \begin{array}{l}\text{Factoring the numerator and the}\\ \text{denominator}\end{array}$$

$$= \frac{(x - 3)\cancel{(x + 1)}(x + 5)}{(x - 2)(x + 2)\cancel{(x + 1)}} \quad \text{Removing a factor of 1: } \dfrac{x + 1}{x + 1} = 1$$

$$= \frac{(x - 3)(x + 5)}{(x - 2)(x + 2)} \left] \leftarrow \boxed{\begin{array}{l}\text{You need not carry out the}\\ \text{multiplications in the numerator}\\ \text{and the denominator.}\end{array}}\right.$$

Answers

9. $\dfrac{x - 3}{x + 2}$ 10. $\dfrac{(x - 3)(x - 2)}{x + 2}$ 11. $\dfrac{y + 1}{y - 1}$

Guided Solution:
11. $y, 2y, 1, 1, 1, 1, 1, y$

◀ **Do Exercises 9–11.**

6.2 Exercise Set

For Extra Help   MathXL®   MyMathLab®   PRACTICE   WATCH   READ   REVIEW

✓ Reading Check

Choose from the choices on the right an equivalent expression.

RC1. The reciprocal of $\dfrac{2}{x - 2}$

RC2. The reciprocal of $x - 2$

RC3. $\dfrac{2}{x} \cdot \dfrac{x}{2}$

RC4. $\dfrac{x}{2} \cdot \dfrac{2}{x + 1}$

RC5. $\dfrac{1}{x} \div \dfrac{1}{2}$

RC6. $\dfrac{1}{x} \div \dfrac{2}{x}$

a) 1

b) $\dfrac{2}{x}$

c) $\dfrac{1}{x} \cdot \dfrac{x}{2}$

d) $\dfrac{x - 2}{2}$

e) $\dfrac{1}{x - 2}$

f) $\dfrac{x}{2} \div \dfrac{x + 1}{2}$

Find the reciprocal.

1. $\dfrac{4}{x}$

2. $\dfrac{a+3}{a-1}$

3. $x^2 - y^2$

4. $x^2 - 5x + 7$

5. $\dfrac{1}{a+b}$

6. $\dfrac{x^2}{x^2-3}$

7. $\dfrac{x^2+2x-5}{x^2-4x+7}$

8. $\dfrac{(a-b)(a+b)}{(a+4)(a-5)}$

b Divide and simplify.

9. $\dfrac{2}{5} \div \dfrac{4}{3}$

10. $\dfrac{3}{10} \div \dfrac{3}{2}$

11. $\dfrac{2}{x} \div \dfrac{8}{x}$

12. $\dfrac{t}{3} \div \dfrac{t}{15}$

13. $\dfrac{a}{b^2} \div \dfrac{a^2}{b^3}$

14. $\dfrac{x^2}{y} \div \dfrac{x^3}{y^3}$

15. $\dfrac{a+2}{a-3} \div \dfrac{a-1}{a+3}$

16. $\dfrac{x-8}{x+9} \div \dfrac{x+2}{x-1}$

17. $\dfrac{x^2-1}{x} \div \dfrac{x+1}{x-1}$

18. $\dfrac{4y-8}{y+2} \div \dfrac{y-2}{y^2-4}$

19. $\dfrac{x+1}{6} \div \dfrac{x+1}{3}$

20. $\dfrac{a}{a-b} \div \dfrac{b}{a-b}$

21. $\dfrac{5x-5}{16} \div \dfrac{x-1}{6}$

22. $\dfrac{4y-12}{12} \div \dfrac{y-3}{3}$

23. $\dfrac{-6+3x}{5} \div \dfrac{4x-8}{25}$

24. $\dfrac{-12+4x}{4} \div \dfrac{-6+2x}{6}$

25. $\dfrac{a+2}{a-1} \div \dfrac{3a+6}{a-5}$

26. $\dfrac{t-3}{t+2} \div \dfrac{4t-12}{t+1}$

27. $\dfrac{x^2-4}{x} \div \dfrac{x-2}{x+2}$

28. $\dfrac{x+y}{x-y} \div \dfrac{x^2+y}{x^2-y^2}$

29. $\dfrac{x^2-9}{4x+12} \div \dfrac{x-3}{6}$

30. $\dfrac{a-b}{2a} \div \dfrac{a^2-b^2}{8a^3}$

31. $\dfrac{c^2+3c}{c^2+2c-3} \div \dfrac{c}{c+1}$

32. $\dfrac{y+5}{2y} \div \dfrac{y^2-25}{4y^2}$

33. $\dfrac{2y^2 - 7y + 3}{2y^2 + 3y - 2} \div \dfrac{6y^2 - 5y + 1}{3y^2 + 5y - 2}$

34. $\dfrac{x^2 + x - 20}{x^2 - 7x + 12} \div \dfrac{x^2 + 10x + 25}{x^2 - 6x + 9}$

35. $\dfrac{x^2 - 1}{4x + 4} \div \dfrac{2x^2 - 4x + 2}{8x + 8}$

36. $\dfrac{5t^2 + 5t - 30}{10t + 30} \div \dfrac{2t^2 - 8}{6t^2 + 36t + 54}$

Skill Maintenance

Solve.

37. Camila is taking an astronomy course. In order to receive an A, she must average at least 90 after four exams. Camila scored 96, 98, and 89 on the first three tests. Determine (in terms of an inequality) what scores on the last test will earn her an A. [2.8b]

38. *Triangle Dimensions.* The base of a triangle is 4 in. less than twice the height. The area is 35 in². Find the height and the base. [5.8a]

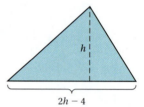

$2h - 4$

Subtract. [4.4c]

39. $(8x^3 - 3x^2 + 7) - (8x^2 + 3x - 5)$

40. $(3p^2 - 6pq + 7q^2) - (5p^2 - 10pq + 11q^2)$

Simplify. [4.2a, b]

41. $(2x^{-3}y^4)^2$

42. $(5x^6y^{-4})^3$

43. $\left(\dfrac{2x^3}{y^5}\right)^2$

44. $\left(\dfrac{a^{-3}}{b^4}\right)^5$

Synthesis

Simplify.

45. $\dfrac{3a^2 - 5ab - 12b^2}{3ab + 4b^2} \div (3b^2 - ab)$

46. $\dfrac{3x + 3y + 3}{9x} \div \dfrac{x^2 + 2xy + y^2 - 1}{x^4 + x^2}$

47. $\dfrac{a^2b^2 + 3ab^2 + 2b^2}{a^2b^4 + 4b^4} \div (5a^2 + 10a)$

48. The volume of this rectangular solid is $x - 3$. What is its height?

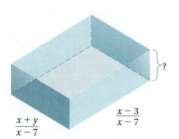

$\dfrac{x + y}{x - 7}$ $\dfrac{x - 3}{x - 7}$

Least Common Multiples and Denominators

a LEAST COMMON MULTIPLES

To add when denominators are different, we first find a common denominator. For example, to add $\frac{5}{12}$ and $\frac{7}{30}$, we first look for the **least common multiple, LCM**, of 12 and 30. That number becomes the **least common denominator, LCD**. To find the LCM of 12 and 30, we factor:

$$12 = 2 \cdot 2 \cdot 3;$$
$$30 = 2 \cdot 3 \cdot 5.$$

The LCM is the number that has 2 as a factor twice, 3 as a factor once, and 5 as a factor once:

12 is a factor of the LCM.

$$\text{LCM} = 2 \cdot 2 \cdot 3 \cdot 5 = 60.$$

30 is a factor of the LCM.

FINDING LCMS

To find the LCM, use each factor the greatest number of times that it appears in any one factorization.

EXAMPLE 1 Find the LCM of 24 and 36.

$$\left. \begin{array}{l} 24 = 2 \cdot 2 \cdot 2 \cdot 3 \\ 36 = 2 \cdot 2 \cdot 3 \cdot 3 \end{array} \right\} \quad \text{LCM} = 2 \cdot 2 \cdot 2 \cdot 3 \cdot 3, \text{ or } 72$$

Do Margin Exercises 1–4. ▶

b ADDING USING THE LCD

Let's finish adding $\frac{5}{12}$ and $\frac{7}{30}$:

$$\frac{5}{12} + \frac{7}{30} = \frac{5}{2 \cdot 2 \cdot 3} + \frac{7}{2 \cdot 3 \cdot 5}.$$

The least common denominator, LCD, is $2 \cdot 2 \cdot 3 \cdot 5$. To get the LCD in the first denominator, we need a 5. To get the LCD in the second denominator, we need another 2. We get these numbers by multiplying by forms of 1:

$$\frac{5}{12} + \frac{7}{30} = \frac{5}{2 \cdot 2 \cdot 3} \cdot \frac{5}{5} + \frac{7}{2 \cdot 3 \cdot 5} \cdot \frac{2}{2} \quad \textcolor{red}{\text{Multiplying by 1}}$$

$$= \frac{25}{2 \cdot 2 \cdot 3 \cdot 5} + \frac{14}{2 \cdot 3 \cdot 5 \cdot 2} \quad \textcolor{red}{\text{Each denominator is now the LCD.}}$$

$$= \frac{39}{2 \cdot 2 \cdot 3 \cdot 5} \quad \textcolor{red}{\text{Adding the numerators and keeping the LCD}}$$

$$= \frac{3 \cdot 13}{2 \cdot 2 \cdot 3 \cdot 5} \quad \textcolor{red}{\text{Factoring the numerator and removing a factor of 1: } \frac{3}{3} = 1}$$

$$= \frac{13}{20}. \quad \textcolor{red}{\text{Simplifying}}$$

OBJECTIVES

a Find the LCM of several numbers by factoring.

b Add fractions, first finding the LCD.

c Find the LCM of algebraic expressions by factoring.

SKILL TO REVIEW

Objective 5.5b: Factor trinomial squares.

Factor.

1. $x^2 - 16x + 64$

2. $4x^2 + 12x + 9$

Find the LCM by factoring.

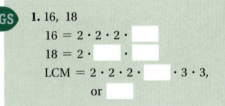

GS **1.** 16, 18

$$16 = 2 \cdot 2 \cdot 2 \cdot \boxed{}$$
$$18 = 2 \cdot \boxed{} \cdot \boxed{}$$
$$\text{LCM} = 2 \cdot 2 \cdot 2 \cdot \boxed{} \cdot 3 \cdot 3,$$
$$\text{or } \boxed{}$$

2. 6, 12

3. 2, 5

4. 24, 30, 20

Add, first finding the LCD. Simplify if possible.

5. $\dfrac{3}{16} + \dfrac{1}{18}$

$$= \frac{3}{2 \cdot 2 \cdot 2 \cdot \boxed{}} + \frac{1}{2 \cdot \boxed{} \cdot 3}$$

$$= \frac{3}{2 \cdot 2 \cdot 2 \cdot 2} \cdot \frac{3 \cdot 3}{3 \cdot \boxed{}}$$

$$+ \frac{1}{2 \cdot 3 \cdot 3} \cdot \frac{2 \cdot \boxed{} \cdot 2}{2 \cdot 2 \cdot 2}$$

$$= \frac{27 + \boxed{}}{2 \cdot 2 \cdot 2 \cdot 2 \cdot 3 \cdot 3}$$

$$= \frac{35}{\boxed{}}$$

6. $\dfrac{1}{6} + \dfrac{1}{12}$

7. $\dfrac{1}{2} + \dfrac{3}{5}$

8. $\dfrac{1}{24} + \dfrac{1}{30} + \dfrac{3}{20}$

Find the LCM.

9. $12xy^2$, $15x^3y$

10. $y^2 + 5y + 4$, $y^2 + 2y + 1$

11. $t^2 + 16$, $t - 2$, 7

12. $x^2 + 2x + 1$, $3x^2 - 3x$, $x^2 - 1$

EXAMPLE 2 Add: $\dfrac{5}{12} + \dfrac{11}{18}$.

$$\left. \begin{array}{l} 12 = 2 \cdot 2 \cdot 3 \\ 18 = 2 \cdot 3 \cdot 3 \end{array} \right\} \quad \text{LCD} = 2 \cdot 2 \cdot 3 \cdot 3, \text{ or } 36$$

$$\frac{5}{12} + \frac{11}{18} = \frac{5}{2 \cdot 2 \cdot 3} \cdot \frac{3}{3} + \frac{11}{2 \cdot 3 \cdot 3} \cdot \frac{2}{2} = \frac{15 + 22}{2 \cdot 2 \cdot 3 \cdot 3} = \frac{37}{36}$$

◀ Do Exercises 5–8.

C LCMs OF ALGEBRAIC EXPRESSIONS

To find the LCM of two or more algebraic expressions, we factor them. Then we use each factor the greatest number of times that it occurs in any one expression. In Section 6.4, each LCM will become an LCD used to add rational expressions.

EXAMPLE 3 Find the LCM of $12x$, $16y$, and $8xyz$.

$$\left. \begin{array}{l} 12x = 2 \cdot 2 \cdot 3 \cdot x \\ 16y = 2 \cdot 2 \cdot 2 \cdot 2 \cdot y \\ 8xyz = 2 \cdot 2 \cdot 2 \cdot x \cdot y \cdot z \end{array} \right\} \quad \begin{array}{l} \text{LCM} = 2 \cdot 2 \cdot 2 \cdot 2 \cdot 3 \cdot x \cdot y \cdot z \\ = 48xyz \end{array}$$

EXAMPLE 4 Find the LCM of $x^2 + 5x - 6$ and $x^2 - 1$.

$$\left. \begin{array}{l} x^2 + 5x - 6 = (x + 6)(x - 1) \\ x^2 - 1 = (x + 1)(x - 1) \end{array} \right\} \quad \text{LCM} = (x + 6)(x - 1)(x + 1)$$

EXAMPLE 5 Find the LCM of $x^2 + 4$, $x + 1$, and 5.

These expressions do not share a common factor other than 1, so the LCM is their product:

$$5(x^2 + 4)(x + 1).$$

EXAMPLE 6 Find the LCM of $x^2 - 25$ and $2x - 10$.

$$\left. \begin{array}{l} x^2 - 25 = (x + 5)(x - 5) \\ 2x - 10 = 2(x - 5) \end{array} \right\} \quad \text{LCM} = 2(x + 5)(x - 5)$$

EXAMPLE 7 Find the LCM of $x^2 - 4y^2$, $x^2 - 4xy + 4y^2$, and $x - 2y$.

$$\left. \begin{array}{l} x^2 - 4y^2 = (x - 2y)(x + 2y) \\ x^2 - 4xy + 4y^2 = (x - 2y)(x - 2y) \\ x - 2y = x - 2y \end{array} \right\} \quad \begin{array}{l} \text{LCM} = (x + 2y)(x - 2y)(x - 2y) \\ = (x + 2y)(x - 2y)^2 \end{array}$$

◀ Do Exercises 9–12.

✓ Reading Check

Complete each statement with the best choice from the column on the right. Some choices will not be used.

$2 \cdot 2 \cdot 2 \cdot 3$
$2 \cdot 2 \cdot 2 \cdot 2 \cdot 3$
common
multiple
numerator
denominator
greatest
least

To add $\dfrac{5}{16} + \dfrac{7}{24}$, we begin by finding a **RC1.** _____ denominator.

We first look for the least common **RC2.** _____ of 16 and 24. That number becomes the least common **RC3.** _____ of the two fractions. We factor 16 and 24: $16 = 2 \cdot 2 \cdot 2 \cdot 2$ and $24 = 2 \cdot 2 \cdot 2 \cdot 3$. Then to find the LCM of 16 and 24, we use each factor the **RC4.** _____ number of times that it appears in any one factorization. The LCM is **RC5.** _____.

a Find the LCM.

1. 12, 27

2. 10, 15

3. 8, 9

4. 12, 18

5. 6, 9, 21

6. 8, 36, 40

7. 24, 36, 40

8. 4, 5, 20

9. 10, 100, 500

10. 28, 42, 60

b Add, first finding the LCD. Simplify, if possible.

11. $\dfrac{7}{24} + \dfrac{11}{18}$

12. $\dfrac{7}{60} + \dfrac{2}{25}$

13. $\dfrac{1}{6} + \dfrac{3}{40}$

14. $\dfrac{5}{24} + \dfrac{3}{20}$

15. $\dfrac{1}{20} + \dfrac{1}{30} + \dfrac{2}{45}$

16. $\dfrac{2}{15} + \dfrac{5}{9} + \dfrac{3}{20}$

c Find the LCM.

17. $6x^2, 12x^3$

18. $2a^2b, 8ab^3$

19. $2x^2, 6xy, 18y^2$

20. p^3q, p^2q, pq^2

21. $2(y - 3), 6(y - 3)$

22. $5(m + 2), 15(m + 2)$

23. $t, t + 2, t - 2$

24. $y, y - 5, y + 5$

25. $x^2 - 4, x^2 + 5x + 6$

26. $x^2 - 4, x^2 - x - 2$

27. $t^3 + 4t^2 + 4t, t^2 - 4t$

28. $m^4 - m^2, m^3 - m^2$

29. $a + 1$, $(a - 1)^2$, $a^2 - 1$

30. $a^2 - 2ab + b^2$, $a^2 - b^2$, $3a + 3b$

31. $m^2 - 5m + 6$, $m^2 - 4m + 4$

32. $2x^2 + 5x + 2$, $2x^2 - x - 1$

33. $2 + 3x$, $4 - 9x^2$, $2 - 3x$

34. $9 - 4x^2$, $3 + 2x$, $3 - 2x$

35. $10v^2 + 30v$, $5v^2 + 35v + 60$

36. $12a^2 + 24a$, $4a^2 + 20a + 24$

37. $9x^3 - 9x^2 - 18x$, $6x^5 - 24x^4 + 24x^3$

38. $x^5 - 4x^3$, $x^3 + 4x^2 + 4x$

39. $x^5 + 4x^4 + 4x^3$, $3x^2 - 12$, $2x + 4$

40. $x^5 + 2x^4 + x^3$, $2x^3 - 2x$, $5x - 5$

41. $24w^4$, w^2, $10w^3$, w^6

42. t, $6t^4$, t^2, $15t^{15}$, $2t^3$

Skill Maintenance

Complete the tables below, finding the LCM, the GCF, and the product of each pair of expressions.   [4.5a], [5.1a], [6.3a]

	Expressions	LCM	GCF	Product
Example	$12x^3$, $8x^2$	$24x^3$	$4x^2$	$96x^5$
43.	$40x^3$, $24x^4$			
45.	$16x^5$, $48x^6$			
47.	$20x^2$, $10x$			

	Expressions	LCM	GCF	Product
44.	$12ab$, $16ab^3$			
46.	$10x^2$, $24x^3$			
48.	a^5, a^{15}			

Synthesis

49. *Running.*   Gabriela and Madison leave the starting point of a fitness loop at the same time. Gabriela jogs a lap in 6 min and Madison jogs one in 8 min. Assuming they continue to run at the same pace, when will they next meet at the starting place?

Adding Rational Expressions

a ADDING RATIONAL EXPRESSIONS

We add rational expressions as we do rational numbers.

> **ADDING RATIONAL EXPRESSIONS WITH LIKE DENOMINATORS**
>
> To add when the denominators are the same, add the numerators and keep the same denominator. Then simplify, if possible.

EXAMPLES Add.

1. $\dfrac{x}{x+1} + \dfrac{2}{x+1} = \dfrac{x+2}{x+1}$

2. $\dfrac{2x^2+3x-7}{2x+1} + \dfrac{x^2+x-8}{2x+1} = \dfrac{(2x^2+3x-7)+(x^2+x-8)}{2x+1}$

$$= \dfrac{3x^2+4x-15}{2x+1}$$ Factoring the numerator to determine if we can simplify

$$= \dfrac{(x+3)(3x-5)}{2x+1}$$

3. $\dfrac{x-5}{x^2-9} + \dfrac{2}{x^2-9} = \dfrac{(x-5)+2}{x^2-9} = \dfrac{x-3}{x^2-9}$

$$= \dfrac{x-3}{(x-3)(x+3)}$$ Factoring

$$= \dfrac{1(x-3)}{(x-3)(x+3)}$$ Removing a factor of 1: $\dfrac{x-3}{x-3} = 1$

$$= \dfrac{1}{x+3}$$ Simplifying

Do Margin Exercises 1–3. ▶

When denominators are different, we find the least common denominator, LCD. The procedure we use follows.

> **ADDING RATIONAL EXPRESSIONS WITH DIFFERENT DENOMINATORS**
>
> To add rational expressions with different denominators:
>
> **1.** Find the LCM of the denominators. This is the least common denominator (LCD).
> **2.** For each rational expression, find an equivalent expression with the LCD. Multiply by 1 using an expression for 1 made up of factors of the LCD that are missing from the original denominator.
> **3.** Add the numerators. Write the sum over the LCD.
> **4.** Simplify, if possible.

OBJECTIVE

a Add rational expressions.

SKILL TO REVIEW

Objective 6.1c: Simplify rational expressions by factoring the numerator and the denominator and removing factors of 1.

Simplify.

1. $\dfrac{a^2-b^2}{a+b}$

2. $\dfrac{x^2-x-6}{x^2+2x-15}$

Add.

1. $\dfrac{5}{9} + \dfrac{2}{9}$

2. $\dfrac{3}{x-2} + \dfrac{x}{x-2}$

3. $\dfrac{4x+5}{x-1} + \dfrac{2x-1}{x-1}$

Answers

Skill to Review:

1. $a - b$ 2. $\dfrac{x+2}{x+5}$

Margin Exercises:

1. $\dfrac{7}{9}$ 2. $\dfrac{3+x}{x-2}$ 3. $\dfrac{6x+4}{x-1}$

EXAMPLE 4 Add: $\dfrac{5x^2}{8} + \dfrac{7x}{12}$.

First, we find the LCD:

$$\left. \begin{array}{l} 8 = 2 \cdot 2 \cdot 2 \\ 12 = 2 \cdot 2 \cdot 3 \end{array} \right\} \quad \text{LCD} = 2 \cdot 2 \cdot 2 \cdot 3, \text{ or } 24.$$

Compare the factorization $8 = 2 \cdot 2 \cdot 2$ with the factorization of the LCD, $24 = 2 \cdot 2 \cdot 2 \cdot 3$. The factor of 24 that is missing from 8 is 3. Compare $12 = 2 \cdot 2 \cdot 3$ and $24 = 2 \cdot 2 \cdot 2 \cdot 3$. The factor of 24 that is missing from 12 is 2.

We multiply each term by a symbol for 1 to get the LCD in each expression, and then add and, if possible, simplify:

$$\begin{aligned} \frac{5x^2}{8} + \frac{7x}{12} &= \frac{5x^2}{2 \cdot 2 \cdot 2} + \frac{7x}{2 \cdot 2 \cdot 3} \\ &= \frac{5x^2}{2 \cdot 2 \cdot 2} \cdot \frac{3}{3} + \frac{7x}{2 \cdot 2 \cdot 3} \cdot \frac{2}{2} \qquad \text{Multiplying by 1 to get the same denominators} \\ &= \frac{15x^2}{24} + \frac{14x}{24} = \frac{15x^2 + 14x}{24} = \frac{x(15x + 14)}{24}. \end{aligned}$$

Add.

4. $\dfrac{3x}{16} + \dfrac{5x^2}{24}$

EXAMPLE 5 Add: $\dfrac{3}{8x} + \dfrac{5}{12x^2}$.

First, we find the LCD:

$$\left. \begin{array}{l} 8x = 2 \cdot 2 \cdot 2 \cdot x \\ 12x^2 = 2 \cdot 2 \cdot 3 \cdot x \cdot x \end{array} \right\} \quad \text{LCD} = 2 \cdot 2 \cdot 2 \cdot 3 \cdot x \cdot x, \text{ or } 24x^2.$$

The factors of the LCD missing from $8x$ are 3 and x. The factor of the LCD missing from $12x^2$ is 2. We multiply each term by 1 to get the LCD in each expression, and then add and, if possible, simplify:

$$\begin{aligned} \frac{3}{8x} + \frac{5}{12x^2} &= \frac{3}{8x} \cdot \frac{3 \cdot x}{3 \cdot x} + \frac{5}{12x^2} \cdot \frac{2}{2} \\ &= \frac{9x}{24x^2} + \frac{10}{24x^2} = \frac{9x + 10}{24x^2}. \end{aligned}$$

GS

5. $\dfrac{3}{16x} + \dfrac{5}{24x^2}$

$$16x = 2 \cdot 2 \cdot 2 \cdot \boxed{} \cdot x$$
$$24x^2 = 2 \cdot 2 \cdot 2 \cdot 3 \cdot \boxed{} \cdot x$$
$$\text{LCD} = 2 \cdot 2 \cdot 2 \cdot \boxed{} \cdot 3 \cdot x \cdot x,$$
$$\text{or } 48x^2$$

$$\frac{3}{16x} \cdot \frac{3x}{\boxed{}} + \frac{5}{24x^2} \cdot \frac{\boxed{}}{2}$$

$$= \frac{\boxed{}}{48x^2} + \frac{10}{\boxed{}}$$

$$= \frac{9x + \boxed{}}{48x^2}$$

◀ **Do Exercises 4 and 5.**

EXAMPLE 6 Add: $\dfrac{2a}{a^2 - 1} + \dfrac{1}{a^2 + a}$.

First, we find the LCD:

$$\left. \begin{array}{l} a^2 - 1 = (a - 1)(a + 1) \\ a^2 + a = a(a + 1) \end{array} \right\} \quad \text{LCD} = a(a - 1)(a + 1).$$

We multiply each term by 1 to get the LCD in each expression, and then add and, if possible, simplify:

$$\begin{aligned} \frac{2a}{(a - 1)(a + 1)} \cdot \frac{a}{a} &+ \frac{1}{a(a + 1)} \cdot \frac{a - 1}{a - 1} \\ &= \frac{2a^2}{a(a - 1)(a + 1)} + \frac{a - 1}{a(a - 1)(a + 1)} \\ &= \frac{2a^2 + a - 1}{a(a - 1)(a + 1)} \\ &= \frac{(a + 1)(2a - 1)}{a(a - 1)(a + 1)}. \qquad \text{Factoring the numerator in order to simplify} \end{aligned}$$

Answers

4. $\dfrac{x(10x + 9)}{48}$ **5.** $\dfrac{9x + 10}{48x^2}$

Guided Solution:
5. 2, x, 2, $3x$, 2, $9x$, $48x^2$, 10

Then

$$= \frac{\cancel{(a+1)}(2a-1)}{a(a-1)\cancel{(a+1)}}$$ Removing a factor of 1: $\dfrac{a+1}{a+1} = 1$

$$= \frac{2a-1}{a(a-1)}.$$

Do Exercise 6. ▶

EXAMPLE 7 Add: $\dfrac{x+4}{x-2} + \dfrac{x-7}{x+5}.$

First, we find the LCD. It is just the product of the denominators:

LCD $= (x-2)(x+5).$

We multiply by 1 to get the LCD in each expression, and then add and simplify:

$$\frac{x+4}{x-2} \cdot \frac{x+5}{x+5} + \frac{x-7}{x+5} \cdot \frac{x-2}{x-2}$$

$$= \frac{(x+4)(x+5)}{(x-2)(x+5)} + \frac{(x-7)(x-2)}{(x-2)(x+5)}$$

$$= \frac{x^2 + 9x + 20}{(x-2)(x+5)} + \frac{x^2 - 9x + 14}{(x-2)(x+5)}$$

$$= \frac{x^2 + 9x + 20 + x^2 - 9x + 14}{(x-2)(x+5)}$$

$$= \frac{2x^2 + 34}{(x-2)(x+5)} = \frac{2(x^2+17)}{(x-2)(x+5)}.$$

Do Exercise 7. ▶

EXAMPLE 8 Add: $\dfrac{x}{x^2+11x+30} + \dfrac{-5}{x^2+9x+20}.$

$$\frac{x}{x^2+11x+30} + \frac{-5}{x^2+9x+20}$$

$$= \frac{x}{(x+5)(x+6)} + \frac{-5}{(x+5)(x+4)}$$ Factoring the denominators in order to find the LCD. The LCD is $(x+4)(x+5)(x+6)$.

$$= \frac{x}{(x+5)(x+6)} \cdot \frac{x+4}{x+4} + \frac{-5}{(x+5)(x+4)} \cdot \frac{x+6}{x+6}$$ Multiplying by 1

$$= \frac{x(x+4) + (-5)(x+6)}{(x+4)(x+5)(x+6)} = \frac{x^2 + 4x - 5x - 30}{(x+4)(x+5)(x+6)}$$

$$= \frac{x^2 - x - 30}{(x+4)(x+5)(x+6)}$$

$$= \frac{(x-6)\cancel{(x+5)}}{(x+4)\cancel{(x+5)}(x+6)}$$ Always simplify at the end if possible: $\dfrac{x+5}{x+5} = 1.$

$$= \frac{x-6}{(x+4)(x+6)}$$

Do Exercise 8. ▶

Denominators That Are Opposites

When one denominator is the opposite of the other, we can first multiply either expression by 1 using $-1/-1.$

6. Add:

$$\frac{3}{x^3-x} + \frac{4}{x^2+2x+1}.$$

7. Add:

$$\frac{x-2}{x+3} + \frac{x+7}{x+8}.$$

8. Add:

$$\frac{5}{x^2+17x+16} + \frac{3}{x^2+9x+8}.$$

Answers

6. $\dfrac{4x^2-x+3}{x(x-1)(x+1)^2}$ **7.** $\dfrac{2x^2+16x+5}{(x+3)(x+8)}$

8. $\dfrac{8(x+11)}{(x+16)(x+1)(x+8)}$

Add.

9. $\dfrac{x}{4} + \dfrac{5}{-4}$

10. $\dfrac{2x + 1}{x - 3} + \dfrac{x + 2}{3 - x}$

$= \dfrac{2x + 1}{x - 3} + \dfrac{x + 2}{3 - x} \cdot \dfrac{-1}{\boxed{}}$

$= \dfrac{2x + 1}{x - 3} + \dfrac{\boxed{} - 2}{x - 3}$

$= \dfrac{(2x + 1) + (-x - 2)}{x - \boxed{}}$

$= \dfrac{\boxed{} - 1}{x - 3}$

EXAMPLES

9. $\dfrac{x}{2} + \dfrac{3}{-2} = \dfrac{x}{2} + \dfrac{3}{-2} \cdot \dfrac{-1}{-1}$ Multiplying by 1 using $\dfrac{-1}{-1}$

$\qquad\quad = \dfrac{x}{2} + \dfrac{-3}{2}$ The denominators are now the same.

$\qquad\quad = \dfrac{x + (-3)}{2} = \dfrac{x - 3}{2}$

10. $\dfrac{3x + 4}{x - 2} + \dfrac{x - 7}{2 - x} = \dfrac{3x + 4}{x - 2} + \dfrac{x - 7}{2 - x} \cdot \dfrac{-1}{-1}$

> We could have chosen to multiply this expression by $-1/-1$. We multiply only one expression, *not* both.

$\qquad\quad = \dfrac{3x + 4}{x - 2} + \dfrac{-x + 7}{x - 2}$ *Note:* $(2 - x)(-1) = -2 + x$
                                                                  $= x - 2.$

$\qquad\quad = \dfrac{(3x + 4) + (-x + 7)}{x - 2} = \dfrac{2x + 11}{x - 2}$

◀ **Do Exercises 9 and 10.**

Factors That Are Opposites

Suppose that when we factor to find the LCD, we find factors that are opposites. The easiest way to handle this is to first go back and multiply by $-1/-1$ appropriately to change factors so that they are not opposites.

EXAMPLE 11   Add: $\dfrac{x}{x^2 - 25} + \dfrac{3}{10 - 2x}$.

First, we factor to find the LCD:

$$x^2 - 25 = (x - 5)(x + 5);$$
$$10 - 2x = 2(5 - x).$$

We note that $x - 5$ is one factor of $x^2 - 25$ and $5 - x$ is one factor of $10 - 2x$. If the denominator of the second expression were $2x - 10$, then $x - 5$ would be a factor of both denominators. To rewrite the second expression with a denominator of $2x - 10$, we multiply by 1 using $-1/-1$, and then continue as before:

$\dfrac{x}{x^2 - 25} + \dfrac{3}{10 - 2x} = \dfrac{x}{(x - 5)(x + 5)} + \dfrac{3}{10 - 2x} \cdot \dfrac{-1}{-1}$

$\qquad\qquad\qquad\qquad = \dfrac{x}{(x - 5)(x + 5)} + \dfrac{-3}{2x - 10}$

$\qquad\qquad\qquad\qquad = \dfrac{x}{(x - 5)(x + 5)} + \dfrac{-3}{2(x - 5)}$ LCD $= 2(x - 5)(x + 5)$

$\qquad\qquad\qquad\qquad = \dfrac{x}{(x - 5)(x + 5)} \cdot \dfrac{2}{2} + \dfrac{-3}{2(x - 5)} \cdot \dfrac{x + 5}{x + 5}$

$\qquad\qquad\qquad\qquad = \dfrac{2x}{2(x - 5)(x + 5)} + \dfrac{-3(x + 5)}{2(x - 5)(x + 5)}$

$\qquad\qquad\qquad\qquad = \dfrac{2x - 3(x + 5)}{2(x - 5)(x + 5)} = \dfrac{2x - 3x - 15}{2(x - 5)(x + 5)}$

$\qquad\qquad\qquad\qquad = \dfrac{-x - 15}{2(x - 5)(x + 5)}.$ Collecting like terms

◀ **Do Exercise 11.**

11. Add:

$\dfrac{x + 3}{x^2 - 16} + \dfrac{5}{12 - 3x}$.

Answers

9. $\dfrac{x - 5}{4}$ **10.** $\dfrac{x - 1}{x - 3}$ **11.** $\dfrac{-2x - 11}{3(x + 4)(x - 4)}$

Guided Solution:
10. $-1, -x, 3, x$

✓ Reading Check

From the choices on the right, select the names for 1 that result in the same denominator in each addition. Some choices may be used more than once. Some choices may not be used. Do not complete the addition.

RC1. $\dfrac{2x^2}{15} + \dfrac{3x}{10} = \dfrac{2x^2}{3\cdot 5}\cdot\left(\right) + \dfrac{3x}{2\cdot 5}\cdot\left(\right)$

RC2. $\dfrac{3}{2} + \dfrac{2x}{x+3} = \dfrac{3}{2}\cdot\left(\right) + \dfrac{2x}{x+3}\cdot\left(\right)$

RC3. $\dfrac{3x}{x+2} + \dfrac{2}{x-3} = \dfrac{3x}{x+2}\cdot\left(\right) + \dfrac{2}{x-3}\cdot\left(\right)$

RC4. $\dfrac{2}{15x} + \dfrac{3}{10x^2} = \dfrac{2}{3\cdot 5\cdot x}\cdot\left(\right) + \dfrac{3}{2\cdot 5\cdot x\cdot x}\cdot\left(\right)$

$\dfrac{x+3}{x+3}$ $\quad$ $\dfrac{2}{2}$

$\dfrac{3x}{3x}$ $\quad$ $\dfrac{x-2}{x-2}$

$\dfrac{x-3}{x-3}$ $\quad$ $\dfrac{3}{3}$

$\dfrac{2x}{2x}$ $\quad$ $\dfrac{x+2}{x+2}$

a Add. Simplify, if possible.

1. $\dfrac{5}{8} + \dfrac{3}{8}$

2. $\dfrac{3}{16} + \dfrac{5}{16}$

3. $\dfrac{1}{3+x} + \dfrac{5}{3+x}$

4. $\dfrac{x^2+7x}{x^2-5x} + \dfrac{x^2-4x}{x^2-5x}$

5. $\dfrac{4x+6}{2x-1} + \dfrac{5-8x}{-1+2x}$

6. $\dfrac{4}{x+y} + \dfrac{9}{y+x}$

7. $\dfrac{2}{x} + \dfrac{5}{x^2}$

8. $\dfrac{3}{y^2} + \dfrac{6}{y}$

9. $\dfrac{5}{6r} + \dfrac{7}{8r}$

10. $\dfrac{13}{18x} + \dfrac{7}{24x}$

11. $\dfrac{4}{xy^2} + \dfrac{6}{x^2y}$

12. $\dfrac{8}{ab^3} + \dfrac{3}{a^2b}$

13. $\dfrac{2}{9t^3} + \dfrac{1}{6t^2}$

14. $\dfrac{5}{c^2d^3} + \dfrac{-4}{7cd^2}$

15. $\dfrac{x+y}{xy^2} + \dfrac{3x+y}{x^2y}$

16. $\dfrac{2c - d}{c^2 d} + \dfrac{c + d}{cd^2}$

17. $\dfrac{3}{x - 2} + \dfrac{3}{x + 2}$

18. $\dfrac{2}{y + 1} + \dfrac{2}{y - 1}$

19. $\dfrac{3}{x + 1} + \dfrac{2}{3x}$

20. $\dfrac{4}{5y} + \dfrac{7}{y - 2}$

21. $\dfrac{2x}{x^2 - 16} + \dfrac{x}{x - 4}$

22. $\dfrac{4x}{x^2 - 25} + \dfrac{x}{x + 5}$

23. $\dfrac{5}{z + 4} + \dfrac{3}{3z + 12}$

24. $\dfrac{t}{t - 3} + \dfrac{5}{4t - 12}$

25. $\dfrac{3}{x - 1} + \dfrac{2}{(x - 1)^2}$

26. $\dfrac{8}{(y + 3)^2} + \dfrac{5}{y + 3}$

27. $\dfrac{4a}{5a - 10} + \dfrac{3a}{10a - 20}$

28. $\dfrac{9x}{6x - 30} + \dfrac{3x}{4x - 20}$

29. $\dfrac{x + 4}{x} + \dfrac{x}{x + 4}$

30. $\dfrac{a}{a - 3} + \dfrac{a - 3}{a}$

31. $\dfrac{4}{a^2 - a - 2} + \dfrac{3}{a^2 + 4a + 3}$

32. $\dfrac{a}{a^2 - 2a + 1} + \dfrac{1}{a^2 - 5a + 4}$

33. $\dfrac{x + 3}{x - 5} + \dfrac{x - 5}{x + 3}$

34. $\dfrac{3x}{2y - 3} + \dfrac{2x}{3y - 2}$

35. $\dfrac{a}{a^2 - 1} + \dfrac{2a}{a^2 - a}$

36. $\dfrac{3x + 2}{3x + 6} + \dfrac{x - 2}{x^2 - 4}$

37. $\dfrac{7}{8} + \dfrac{5}{-8}$

38. $\dfrac{5}{-3} + \dfrac{11}{3}$

39. $\dfrac{3}{t} + \dfrac{4}{-t}$

40. $\dfrac{5}{-a} + \dfrac{8}{a}$

41. $\dfrac{2x + 7}{x - 6} + \dfrac{3x}{6 - x}$

42. $\dfrac{2x - 7}{5x - 8} + \dfrac{6 + 10x}{8 - 5x}$

43. $\dfrac{y^2}{y - 3} + \dfrac{9}{3 - y}$

44. $\dfrac{t^2}{t - 2} + \dfrac{4}{2 - t}$

45. $\dfrac{b - 7}{b^2 - 16} + \dfrac{7 - b}{16 - b^2}$

46. $\dfrac{a - 3}{a^2 - 25} + \dfrac{a - 3}{25 - a^2}$

47. $\dfrac{a^2}{a - b} + \dfrac{b^2}{b - a}$

48. $\dfrac{x^2}{x - 7} + \dfrac{49}{7 - x}$

49. $\dfrac{x + 3}{x - 5} + \dfrac{2x - 1}{5 - x} + \dfrac{2(3x - 1)}{x - 5}$

50. $\dfrac{3(x - 2)}{2x - 3} + \dfrac{5(2x + 1)}{2x - 3} + \dfrac{3(x + 1)}{3 - 2x}$

51. $\dfrac{2(4x + 1)}{5x - 7} + \dfrac{3(x - 2)}{7 - 5x} + \dfrac{-10x - 1}{5x - 7}$

52. $\dfrac{5(x - 2)}{3x - 4} + \dfrac{2(x - 3)}{4 - 3x} + \dfrac{3(5x + 1)}{4 - 3x}$

53. $\dfrac{x + 1}{(x + 3)(x - 3)} + \dfrac{4(x - 3)}{(x - 3)(x + 3)} + \dfrac{(x - 1)(x - 3)}{(3 - x)(x + 3)}$

54. $\dfrac{2(x + 5)}{(2x - 3)(x - 1)} + \dfrac{3x + 4}{(2x - 3)(1 - x)} + \dfrac{x - 5}{(3 - 2x)(x - 1)}$

55. $\dfrac{6}{x-y} + \dfrac{4x}{y^2 - x^2}$

56. $\dfrac{a-2}{3-a} + \dfrac{4-a^2}{a^2-9}$

57. $\dfrac{4-a}{25-a^2} + \dfrac{a+1}{a-5}$

58. $\dfrac{x+2}{x-7} + \dfrac{3-x}{49-x^2}$

59. $\dfrac{2}{t^2+t-6} + \dfrac{3}{t^2-9}$

60. $\dfrac{10}{a^2-a-6} + \dfrac{3a}{a^2+4a+4}$

Skill Maintenance

Subtract. [4.4c]

61. $(x^2 + x) - (x + 1)$

62. $(4y^3 - 5y^2 + 7y - 24) - (-9y^3 + 9y^2 - 5y + 49)$

Simplify. [4.2a, b]

63. $\left(\dfrac{x^{-4}}{y^7}\right)^3$

64. $(5x^{-2}y^{-3})^2$

Solve.

65. $3x - 7 = 5x + 9$ [2.3b]

66. $x^2 - 7x = 18$ [5.7b]

Graph.

67. $y = \dfrac{1}{2}x - 5$ [3.2a]

68. $2y + x + 10 = 0$ [3.2a]

69. $y = 3$ [3.2b]

70. $x = -5$ [3.2b]

Synthesis

Find the perimeter and the area of each figure.

71.
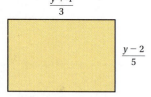

72.
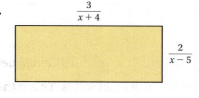

Add. Simplify, if possible.

73. $\dfrac{5}{z+2} + \dfrac{4z}{z^2-4} + 2$

74. $\dfrac{-2}{y^2-9} + \dfrac{4y}{(y-3)^2} + \dfrac{6}{3-y}$

75. $\dfrac{3z^2}{z^4-4} + \dfrac{5z^2-3}{2z^4+z^2-6}$

Subtracting Rational Expressions

6.5

a SUBTRACTING RATIONAL EXPRESSIONS

We subtract rational expressions as we do rational numbers.

> ### SUBTRACTING RATIONAL EXPRESSIONS WITH LIKE DENOMINATORS
>
> To subtract when the denominators are the same, subtract the numerators and keep the same denominator. Then simplify, if possible.

EXAMPLE 1 Subtract: $\dfrac{8}{x} - \dfrac{3}{x}$.

$$\dfrac{8}{x} - \dfrac{3}{x} = \dfrac{8 - 3}{x} = \dfrac{5}{x}$$

EXAMPLE 2 Subtract: $\dfrac{3x}{x + 2} - \dfrac{x - 2}{x + 2}$.

$$\dfrac{3x}{x + 2} - \dfrac{x - 2}{x + 2} = \dfrac{3x - (x - 2)}{x + 2}$$

Caution!

The parentheses are important to make sure that you subtract the entire numerator.

$$= \dfrac{3x - x + 2}{x + 2}$$

Removing parentheses

$$= \dfrac{2x + 2}{x + 2} = \dfrac{2(x + 1)}{x + 2}$$

Do Margin Exercises 1–3. ▶

To subtract rational expressions with different denominators, we use a procedure similar to what we used for addition, except that we subtract numerators and write the difference over the LCD.

> ### SUBTRACTING RATIONAL EXPRESSIONS WITH DIFFERENT DENOMINATORS
>
> To subtract rational expressions with different denominators:
>
> 1. Find the LCM of the denominators. This is the least common denominator (LCD).
> 2. For each rational expression, find an equivalent expression with the LCD. To do so, multiply by 1 using a symbol for 1 made up of factors of the LCD that are missing from the original denominator.
> 3. Subtract the numerators. Write the difference over the LCD.
> 4. Simplify, if possible.

OBJECTIVES

a Subtract rational expressions.

b Simplify combined additions and subtractions of rational expressions.

Subtract.

1. $\dfrac{7}{11} - \dfrac{3}{11}$

2. $\dfrac{7}{y} - \dfrac{2}{y}$

3. $\dfrac{2x^2 + 3x - 7}{2x + 1} - \dfrac{x^2 + x - 8}{2x + 1}$

Answers

Skill to Review:
1. $-3x + 11$ 2. $x - 8$

Margin Exercises:
1. $\dfrac{4}{11}$ 2. $\dfrac{5}{y}$ 3. $\dfrac{(x + 1)^2}{2x + 1}$

4. Subtract:
$$\frac{x-2}{3x} - \frac{2x-1}{5x}.$$

$$\frac{x-2}{3x} - \frac{2x-1}{5x}$$

$$\text{LCD} = 3 \cdot x \cdot 5 = 15x$$

$$= \frac{x-2}{3x} \cdot \frac{5}{\boxed{}} - \frac{2x-1}{5x} \cdot \frac{\boxed{}}{3}$$

$$= \frac{5x - \boxed{}}{15x} - \frac{\boxed{} - 3}{15x}$$

$$= \frac{5x - 10 - (6x - \boxed{})}{15x}$$

$$= \frac{5x - 10 - \boxed{} + 3}{15x}$$

$$= \frac{\boxed{} - 7}{15x}$$

EXAMPLE 3 Subtract: $\dfrac{x+2}{x-4} - \dfrac{x+1}{x+4}.$

The LCD $= (x-4)(x+4)$.

$$\frac{x+2}{x-4} \cdot \frac{x+4}{x+4} - \frac{x+1}{x+4} \cdot \frac{x-4}{x-4} \qquad \text{Multiplying by 1}$$

$$= \frac{(x+2)(x+4)}{(x-4)(x+4)} - \frac{(x+1)(x-4)}{(x-4)(x+4)}$$

$$= \frac{x^2 + 6x + 8}{(x-4)(x+4)} - \frac{x^2 - 3x - 4}{(x-4)(x+4)}$$

Subtracting this numerator. Don't forget the parentheses.

$$= \frac{x^2 + 6x + 8 - (x^2 - 3x - 4)}{(x-4)(x+4)}$$

$$= \frac{x^2 + 6x + 8 - x^2 + 3x + 4}{(x-4)(x+4)} \qquad \text{Removing parentheses}$$

$$= \frac{9x + 12}{(x-4)(x+4)} = \frac{3(3x+4)}{(x-4)(x+4)}$$

◀ **Do Exercise 4.**

EXAMPLE 4 Subtract: $\dfrac{x}{x^2 + 5x + 6} - \dfrac{2}{x^2 + 3x + 2}.$

$$\frac{x}{x^2 + 5x + 6} - \frac{2}{x^2 + 3x + 2}$$

$$= \frac{x}{(x+2)(x+3)} - \frac{2}{(x+2)(x+1)} \qquad \text{LCD} = (x+1)(x+2)(x+3)$$

$$= \frac{x}{(x+2)(x+3)} \cdot \frac{x+1}{x+1} - \frac{2}{(x+2)(x+1)} \cdot \frac{x+3}{x+3}$$

$$= \frac{x^2 + x}{(x+1)(x+2)(x+3)} - \frac{2x + 6}{(x+1)(x+2)(x+3)}$$

Subtracting this numerator. Don't forget the parentheses.

$$= \frac{x^2 + x - (2x + 6)}{(x+1)(x+2)(x+3)}$$

$$= \frac{x^2 + x - 2x - 6}{(x+1)(x+2)(x+3)} = \frac{x^2 - x - 6}{(x+1)(x+2)(x+3)}$$

$$= \frac{(x+2)(x-3)}{(x+1)(x+2)(x+3)}$$

$$= \frac{(x+2)(x-3)}{(x+1)(x+2)(x+3)} \qquad \text{Simplifying by removing a factor}$$

of 1: $\dfrac{x+2}{x+2} = 1$

$$= \frac{x-3}{(x+1)(x+3)}$$

5. Subtract:

$$\frac{x}{x^2 + 15x + 56} - \frac{6}{x^2 + 13x + 42}.$$

◀ **Do Exercise 5.**

Denominators That Are Opposites

When one denominator is the opposite of the other, we can first multiply one expression by $-1/-1$ to obtain a common denominator.

EXAMPLE 5 Subtract: $\dfrac{x}{5} - \dfrac{3x-4}{-5}$.

$$\dfrac{x}{5} - \dfrac{3x-4}{-5} = \dfrac{x}{5} - \dfrac{3x-4}{-5} \cdot \dfrac{-1}{-1} \quad \text{Multiplying by 1 using } \dfrac{-1}{-1}$$

This is equal to 1 (not −1).

$$= \dfrac{x}{5} - \dfrac{(3x-4)(-1)}{(-5)(-1)}$$

$$= \dfrac{x}{5} - \dfrac{4-3x}{5}$$

$$= \dfrac{x-(4-3x)}{5} \quad \text{Remember the parentheses!}$$

$$= \dfrac{x-4+3x}{5} = \dfrac{4x-4}{5} = \dfrac{4(x-1)}{5}$$

EXAMPLE 6 Subtract: $\dfrac{5y}{y-5} - \dfrac{2y-3}{5-y}$.

$$\dfrac{5y}{y-5} - \dfrac{2y-3}{5-y} = \dfrac{5y}{y-5} - \dfrac{2y-3}{5-y} \cdot \dfrac{-1}{-1}$$

$$= \dfrac{5y}{y-5} - \dfrac{(2y-3)(-1)}{(5-y)(-1)}$$

$$= \dfrac{5y}{y-5} - \dfrac{3-2y}{y-5}$$

$$= \dfrac{5y-(3-2y)}{y-5} \quad \text{Remember the parentheses!}$$

$$= \dfrac{5y-3+2y}{y-5} = \dfrac{7y-3}{y-5}$$

Do Exercises 6 and 7. ▶

Subtract.

6. $\dfrac{x}{3} - \dfrac{2x-1}{-3}$

7. $\dfrac{3x}{x-2} - \dfrac{x-3}{2-x}$

Factors That Are Opposites

Suppose that when we factor to find the LCD, we find factors that are opposites. Then we multiply by $-1/-1$ appropriately to change factors so that they are not opposites.

EXAMPLE 7 Subtract: $\dfrac{p}{64-p^2} - \dfrac{5}{p-8}$.

Factoring $64 - p^2$, we get $(8-p)(8+p)$. Note that the factors $8 - p$ in the first denominator and $p - 8$ in the second denominator are opposites. We multiply the first expression by $-1/-1$ to avoid this situation. Then we proceed as before.

$$\dfrac{p}{64-p^2} - \dfrac{5}{p-8} = \dfrac{p}{64-p^2} \cdot \dfrac{-1}{-1} - \dfrac{5}{p-8}$$

$$= \dfrac{-p}{p^2-64} - \dfrac{5}{p-8}$$

$$= \dfrac{-p}{(p-8)(p+8)} - \dfrac{5}{p-8} \quad \text{LCD} = (p-8)(p+8)$$

$$= \dfrac{-p}{(p-8)(p+8)} - \dfrac{5}{p-8} \cdot \dfrac{p+8}{p+8}$$

8. Subtract:

$$\frac{y}{16 - y^2} - \frac{7}{y - 4}.$$

$$\frac{y}{16 - y^2} - \frac{7}{y - 4}$$

$$= \frac{y}{16 - y^2} \cdot \frac{-1}{\boxed{}} - \frac{7}{y - 4}$$

$$= \frac{-y}{\boxed{} - 16} - \frac{7}{y - 4}$$

$$= \frac{-y}{(y + 4)(y - \boxed{})} - \frac{7}{y - 4} \cdot \frac{\boxed{} + 4}{y + 4}$$

$$= \frac{-y}{(y + 4)(y - 4)} - \frac{7y + \boxed{}}{(y + 4)(y - 4)}$$

$$= \frac{-y - (7y + 28)}{(y + 4)(y - 4)} = \frac{-y - 7y - \boxed{}}{(y + 4)(y - 4)}$$

$$= \frac{-\boxed{} - 28}{(y + 4)(y - 4)} = \frac{\boxed{}(2y + 7)}{(y + 4)(y - 4)}$$

9. Perform the indicated operations and simplify:

$$\frac{x + 2}{x^2 - 9} - \frac{x - 7}{9 - x^2} + \frac{-8 - x}{x^2 - 9}.$$

10. Perform the indicated operations and simplify:

$$\frac{1}{x} - \frac{5}{3x} + \frac{2x}{x + 1}.$$

Answers

8. $\dfrac{-4(2y + 7)}{(y + 4)(y - 4)}$ 9. $\dfrac{x - 13}{(x + 3)(x - 3)}$

10. $\dfrac{2(3x^2 - x - 1)}{3x(x + 1)}$

Guided Solution:

8. $-1, y^2, 4, y, 28, 28, 8y, -4$

Multiplying, we have

$$\frac{-p}{(p - 8)(p + 8)} - \frac{5p + 40}{(p - 8)(p + 8)}$$

Subtracting this numerator. Don't forget the parentheses.

$$= \frac{-p - (5p + 40)}{(p - 8)(p + 8)}$$

$$= \frac{-p - 5p - 40}{(p - 8)(p + 8)} = \frac{-6p - 40}{(p - 8)(p + 8)} = \frac{-2(3p + 20)}{(p - 8)(p + 8)}.$$

◀ **Do Exercise 8.**

b | COMBINED ADDITIONS AND SUBTRACTIONS

Now let's look at some combined additions and subtractions.

EXAMPLE 8 Perform the indicated operations and simplify:

$$\frac{x + 9}{x^2 - 4} + \frac{5 - x}{4 - x^2} - \frac{2 + x}{x^2 - 4}.$$

$$\frac{x + 9}{x^2 - 4} + \frac{5 - x}{4 - x^2} - \frac{2 + x}{x^2 - 4}$$

$$= \frac{x + 9}{x^2 - 4} + \frac{5 - x}{4 - x^2} \cdot \frac{-1}{-1} - \frac{2 + x}{x^2 - 4}$$

$$= \frac{x + 9}{x^2 - 4} + \frac{x - 5}{x^2 - 4} - \frac{2 + x}{x^2 - 4} = \frac{(x + 9) + (x - 5) - (2 + x)}{x^2 - 4}$$

$$= \frac{x + 9 + x - 5 - 2 - x}{x^2 - 4} = \frac{x + 2}{x^2 - 4} = \frac{(x + 2) \cdot 1}{(x + 2)(x - 2)} = \frac{1}{x - 2}$$

◀ **Do Exercise 9.**

EXAMPLE 9 Perform the indicated operations and simplify:

$$\frac{1}{x} - \frac{1}{x^2} + \frac{2}{x + 1}.$$

The LCD $= x \cdot x(x + 1)$, or $x^2(x + 1)$.

$$\frac{1}{x} \cdot \frac{x(x + 1)}{x(x + 1)} - \frac{1}{x^2} \cdot \frac{(x + 1)}{(x + 1)} + \frac{2}{x + 1} \cdot \frac{x^2}{x^2}$$

$$= \frac{x(x + 1)}{x^2(x + 1)} - \frac{x + 1}{x^2(x + 1)} + \frac{2x^2}{x^2(x + 1)}$$

Subtracting this numerator. Don't forget the parentheses.

$$= \frac{x(x + 1) - (x + 1) + 2x^2}{x^2(x + 1)}$$

$$= \frac{x^2 + x - x - 1 + 2x^2}{x^2(x + 1)}$$ Removing parentheses

$$= \frac{3x^2 - 1}{x^2(x + 1)}$$

◀ **Do Exercise 10.**

✓ Reading Check

When subtracting rational expressions, parentheses are important to make sure that you subtract the entire numerator. In Exercises RC1–RC3, complete each numerator by **(a)** filling in the parentheses, **(b)** removing the parentheses, and **(c)** collecting like terms.

$$
\textbf{RC1.}\quad \frac{10x}{x-7} - \frac{3x+5}{x-7} = \overset{\textbf{(a)}}{\frac{10x-(\quad\quad)}{x-7}} = \overset{\textbf{(b)}}{\frac{\rule{1.5cm}{0.4pt}}{x-7}} = \overset{\textbf{(c)}}{\frac{\rule{1.5cm}{0.4pt}}{x-7}}
$$

$$
\textbf{RC2.}\quad \frac{7}{4+a} - \frac{4-9a}{4+a} = \overset{\textbf{(a)}}{\frac{7-(\quad\quad)}{4+a}} = \overset{\textbf{(b)}}{\frac{\rule{1.5cm}{0.4pt}}{4+a}} = \overset{\textbf{(c)}}{\frac{\rule{1.5cm}{0.4pt}}{4+a}}
$$

$$
\textbf{RC3.}\quad \frac{9y-2}{y^2-10} - \frac{y+1}{y^2-10} = \overset{\textbf{(a)}}{\frac{9y-2-(\quad\quad)}{y^2-10}} = \overset{\textbf{(b)}}{\frac{\rule{1.5cm}{0.4pt}}{y^2-10}} = \overset{\textbf{(c)}}{\frac{\rule{1.5cm}{0.4pt}}{y^2-10}}
$$

a   Subtract. Simplify, if possible.

1. $\dfrac{7}{x} - \dfrac{3}{x}$

2. $\dfrac{5}{a} - \dfrac{8}{a}$

3. $\dfrac{y}{y-4} - \dfrac{4}{y-4}$

4. $\dfrac{t^2}{t+5} - \dfrac{25}{t+5}$

5. $\dfrac{2x-3}{x^2+3x-4} - \dfrac{x-7}{x^2+3x-4}$

6. $\dfrac{x+1}{x^2-2x+1} - \dfrac{5-3x}{x^2-2x+1}$

7. $\dfrac{a-2}{10} - \dfrac{a+1}{5}$

8. $\dfrac{y+3}{2} - \dfrac{y-4}{4}$

9. $\dfrac{4z-9}{3z} - \dfrac{3z-8}{4z}$

10. $\dfrac{a-1}{4a} - \dfrac{2a+3}{a}$

11. $\dfrac{4x+2t}{3xt^2} - \dfrac{5x-3t}{x^2t}$

12. $\dfrac{5x+3y}{2x^2y} - \dfrac{3x+4y}{xy^2}$

13. $\dfrac{5}{x+5} - \dfrac{3}{x-5}$

14. $\dfrac{3t}{t-1} - \dfrac{8t}{t+1}$

15. $\dfrac{3}{2t^2-2t} - \dfrac{5}{2t-2}$

16. $\dfrac{11}{x^2 - 4} - \dfrac{8}{x + 2}$

17. $\dfrac{2s}{t^2 - s^2} - \dfrac{s}{t - s}$

18. $\dfrac{3}{12 + x - x^2} - \dfrac{2}{x^2 - 9}$

19. $\dfrac{y - 5}{y} - \dfrac{3y - 1}{4y}$

20. $\dfrac{3x - 2}{4x} - \dfrac{3x + 1}{6x}$

21. $\dfrac{a}{x + a} - \dfrac{a}{x - a}$

22. $\dfrac{a}{a - b} - \dfrac{a}{a + b}$

23. $\dfrac{11}{6} - \dfrac{5}{-6}$

24. $\dfrac{5}{9} - \dfrac{7}{-9}$

25. $\dfrac{5}{a} - \dfrac{8}{-a}$

26. $\dfrac{8}{x} - \dfrac{3}{-x}$

27. $\dfrac{4}{y - 1} - \dfrac{4}{1 - y}$

28. $\dfrac{5}{a - 2} - \dfrac{3}{2 - a}$

29. $\dfrac{3 - x}{x - 7} - \dfrac{2x - 5}{7 - x}$

30. $\dfrac{t^2}{t - 2} - \dfrac{4}{2 - t}$

31. $\dfrac{a - 2}{a^2 - 25} - \dfrac{6 - a}{25 - a^2}$

32. $\dfrac{x - 8}{x^2 - 16} - \dfrac{x - 8}{16 - x^2}$

33. $\dfrac{4 - x}{x - 9} - \dfrac{3x - 8}{9 - x}$

34. $\dfrac{4x - 6}{x - 5} - \dfrac{7 - 2x}{5 - x}$

35. $\dfrac{5x}{x^2 - 9} - \dfrac{4}{3 - x}$

36. $\dfrac{8x}{16 - x^2} - \dfrac{5}{x - 4}$

37. $\dfrac{t^2}{2t^2 - 2t} - \dfrac{1}{2t - 2}$

38. $\dfrac{4}{5a^2 - 5a} - \dfrac{2}{5a - 5}$

39. $\dfrac{x}{x^2 + 5x + 6} - \dfrac{2}{x^2 + 3x + 2}$

40. $\dfrac{a}{a^2 + 11a + 30} - \dfrac{5}{a^2 + 9a + 20}$

b Perform the indicated operations and simplify.

41. $\dfrac{3(2x + 5)}{x - 1} - \dfrac{3(2x - 3)}{1 - x} + \dfrac{6x - 1}{x - 1}$

42. $\dfrac{a - 2b}{b - a} - \dfrac{3a - 3b}{a - b} + \dfrac{2a - b}{a - b}$

43. $\dfrac{x - y}{x^2 - y^2} + \dfrac{x + y}{x^2 - y^2} - \dfrac{2x}{x^2 - y^2}$

44. $\dfrac{x - 3y}{2(y - x)} + \dfrac{x + y}{2(x - y)} - \dfrac{2x - 2y}{2(x - y)}$

45. $\dfrac{2(x - 1)}{2x - 3} - \dfrac{3(x + 2)}{2x - 3} - \dfrac{x - 1}{3 - 2x}$

46. $\dfrac{5(2y + 1)}{2y - 3} - \dfrac{3(y - 1)}{3 - 2y} - \dfrac{3(y - 2)}{2y - 3}$

47. $\dfrac{10}{2y - 1} - \dfrac{6}{1 - 2y} + \dfrac{y}{2y - 1} + \dfrac{y - 4}{1 - 2y}$

48. $\dfrac{(x + 1)(2x - 1)}{(2x - 3)(x - 3)} - \dfrac{(x - 3)(x + 1)}{(3 - x)(3 - 2x)} + \dfrac{(2x + 1)(x + 3)}{(3 - 2x)(x - 3)}$

49. $\dfrac{a + 6}{4 - a^2} - \dfrac{a + 3}{a + 2} + \dfrac{a - 3}{2 - a}$

50. $\dfrac{4t}{t^2 - 1} - \dfrac{2}{t} - \dfrac{2}{t + 1}$

51. $\dfrac{2z}{1 - 2z} + \dfrac{3z}{2z + 1} - \dfrac{3}{4z^2 - 1}$

52. $\dfrac{1}{x - y} - \dfrac{2x}{x^2 - y^2} + \dfrac{1}{x + y}$

53. $\dfrac{1}{x + y} - \dfrac{1}{x - y} + \dfrac{2x}{x^2 - y^2}$

54. $\dfrac{2b}{a^2 - b^2} - \dfrac{1}{a + b} + \dfrac{1}{a - b}$

Skill Maintenance

Simplify.

55. $(a^2 b^{-5})^{-4}$ [4.2a, b]

56. $\dfrac{54x^{10}}{3x^7}$ [4.1e]

57. $3x^4 \cdot 10x^8$ [4.1d]

Solve.   [2.3b]

58. $2.5x + 15.5 = 0.5 + 4x$

59. $\dfrac{4}{7} + 3x = \dfrac{1}{2}x - \dfrac{3}{14}$

60. $6x - 0.5 = 6 - 0.5x$

Find a polynomial for the shaded area of each figure.   [4.4d]

61.

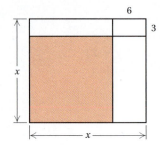

62.

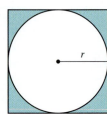

Synthesis

63. The perimeter of the following right triangle is $2a + 5$. Find the missing length of the third side and the area.

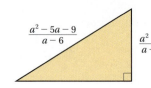

$$\dfrac{a^2 - 5a - 9}{a - 6} \qquad \dfrac{a^2 - 6}{a - 6}$$

Concept Reinforcement

Determine whether each statement is true or false.

_____ **1.** The reciprocal of $\dfrac{3 - w}{w + 2}$ is $\dfrac{w - 3}{w + 2}$. [6.2a]

_____ **2.** The value of the numerator has no bearing on whether or not a rational expression is defined. [6.1a]

_____ **3.** To add or subtract rational expressions when the denominators are the same, add or subtract the numerators and keep the same denominator. [6.4a], [6.5a]

_____ **4.** For the rational expression $\dfrac{x(x - 2)}{x + 3}$, x is a factor of the numerator and a factor of the denominator. [6.1c]

_____ **5.** To find the LCM, use each factor the greatest number of times that it appears in any one factorization. [6.3a, c]

Guided Solutions

GS Fill in each blank with the number or expression that creates a correct solution.

6. Subtract: $\dfrac{x - 1}{x - 2} - \dfrac{x + 1}{x + 2} - \dfrac{x - 6}{4 - x^2}$. [6.5b]

$$\dfrac{x - 1}{x - 2} - \dfrac{x + 1}{x + 2} - \dfrac{x - 6}{4 - x^2} = \dfrac{x - 1}{x - 2} - \dfrac{x + 1}{x + 2} - \dfrac{x - 6}{4 - x^2} \cdot \dfrac{\boxed{}}{\boxed{}} = \dfrac{x - 1}{x - 2} - \dfrac{x + 1}{x + 2} - \dfrac{6 - \boxed{}}{\boxed{} - 4}$$

$$= \dfrac{x - 1}{x - 2} - \dfrac{x + 1}{x + 2} - \dfrac{6 - x}{(x - \boxed{})(\boxed{} + 2)}$$

$$= \dfrac{x - 1}{x - 2} \cdot \dfrac{\boxed{}}{\boxed{}} - \dfrac{x + 1}{x + 2} \cdot \dfrac{\boxed{}}{\boxed{}} - \dfrac{6 - x}{(x - 2)(x + 2)}$$

$$= \dfrac{x^2 + \boxed{} - 2}{(x - 2)(x + 2)} - \dfrac{\boxed{} - x - 2}{(x - 2)(x + 2)} - \dfrac{6 - x}{(x - 2)(x + 2)}$$

$$= \dfrac{x^2 + x - \boxed{} - x^2 + \boxed{} + 2 - \boxed{} + x}{(x - 2)(x + 2)}$$

$$= \dfrac{\boxed{} - \boxed{}}{(x - 2)(x + 2)} = \dfrac{3(\boxed{} - \boxed{})}{(x - 2)(x + 2)} = \dfrac{\boxed{}}{\boxed{}} \cdot \dfrac{\boxed{}}{x + 2} = \dfrac{3}{\boxed{}}$$

Mixed Review

Find all numbers for which the rational expression is not defined. [6.1a]

7. $\dfrac{t^2 - 16}{3}$

8. $\dfrac{x - 8}{x^2 - 11x + 24}$

9. $\dfrac{7}{2w - 7}$

Simplify. [6.1c]

10. $\dfrac{x^2 + 2x - 3}{x^2 - 9}$

11. $\dfrac{6y^2 + 12y - 48}{3y^2 - 9y + 6}$

12. $\dfrac{r - s}{s - r}$

13. Find the reciprocal of $-x + 3$.   [6.2a]

14. Find the LCM of
$$x^2 - 100, 10x^3, \text{ and } x^2 - 20x + 100.   \text{[6.3c]}$$

Add, subtract, multiply, or divide and simplify, if possible.

15. $\dfrac{a^2 - a - 2}{a^2 - a - 6} \div \dfrac{a^2 - 2a}{2a + a^2}$ [6.2b]

16. $\dfrac{3y}{y^2 - 7y + 10} - \dfrac{2y}{y^2 - 8y + 15}$ [6.5a]

17. $\dfrac{x^2}{x - 11} + \dfrac{121}{11 - x}$ [6.4a]

18. $\dfrac{x^2 - y^2}{(x - y)^2} \cdot \dfrac{1}{x + y}$ [6.1d]

19. $\dfrac{3a - b}{a^2 b} + \dfrac{a + 2b}{ab^2}$ [6.4a]

20. $\dfrac{5x}{x^2 - 4} - \dfrac{3}{x} + \dfrac{4}{x + 2}$ [6.5b]

Matching.   Perform the indicated operation and simplify. Then select the correct answer from selections A–G listed in the second column.   [6.1d], [6.2b], [6.4a], [6.5a]

21. $\dfrac{2}{x - 2} \div \dfrac{1}{x + 3}$

A. $\dfrac{-x - 8}{(x - 2)(x + 3)}$

22. $\dfrac{1}{x + 3} - \dfrac{2}{x - 2}$

B. $\dfrac{x - 2}{2(x + 3)}$

23. $\dfrac{2}{x - 2} - \dfrac{1}{x + 3}$

C. $\dfrac{2}{(x - 2)(x + 3)}$

24. $\dfrac{1}{x + 3} \div \dfrac{2}{x - 2}$

D. $\dfrac{x + 8}{(x - 2)(x + 3)}$

25. $\dfrac{2}{x - 2} + \dfrac{1}{x + 3}$

E. $\dfrac{2(x + 3)}{x - 2}$

26. $\dfrac{2}{x - 2} \cdot \dfrac{1}{x + 3}$

F. $\dfrac{3x + 4}{(x - 2)(x + 3)}$

G. $\dfrac{x + 3}{x - 2}$

Understanding Through Discussion and Writing

27. Explain why the product of two numbers is not always their least common multiple.   [6.3a]

28. Is the reciprocal of a product the product of the reciprocals? Why or why not?   [6.2a]

29. A student insists on finding a common denominator by always multiplying the denominators of the expressions being added. How could this approach be improved?   [6.4a]

30. Explain why the expressions
$$\dfrac{1}{3 - x} \text{ and } \dfrac{1}{x - 3}$$
are opposites.   [6.4a]

31. Explain why 5, -1, and 7 are *not* allowable replacements in the division
$$\dfrac{x + 3}{x - 5} \div \dfrac{x - 7}{x + 1}.   \text{[6.1a], [6.2a, b]}$$

32. If the LCM of a binomial and a trinomial is the trinomial, what relationship exists between the two expressions?   [6.3c]

Complex Rational Expressions

6.6

a   SIMPLIFYING COMPLEX RATIONAL EXPRESSIONS

A **complex rational expression**, or **complex fraction expression**, is a rational expression that has one or more rational expressions within its numerator or denominator. Here are some examples:

$$\frac{1 + \dfrac{2}{x}}{3}, \quad \frac{\dfrac{x + y}{2}}{\dfrac{2x}{x + 1}}, \quad \frac{\dfrac{1}{3} + \dfrac{1}{5}}{\dfrac{2}{x} - \dfrac{x}{y}}.$$

These are rational expressions within the complex rational expression.

There are two methods used to simplify complex rational expressions.

OBJECTIVE

a   Simplify complex rational expressions.

SKILL TO REVIEW

Objective 6.3c: Find the LCM of algebraic expressions by factoring.

1. Find the LCM of 2, 4, 6, and 8.

2. Find the LCM of x, x^2, and $5x$.

METHOD 1: MULTIPLYING BY THE LCM OF ALL THE DENOMINATORS

To simplify a complex rational expression:

1. First, find the LCM of all the denominators of all the rational expressions occurring *within* both the numerator and the denominator of the complex rational expression.

2. Then multiply by 1 using LCM/LCM.

3. If possible, simplify by removing a factor of 1.

To the instructor and the student: Students can be instructed either to try both methods and then choose the one that works best for them or to use the method chosen by the instructor.

EXAMPLE 1   Simplify: $\dfrac{\dfrac{1}{2} + \dfrac{3}{4}}{\dfrac{5}{6} - \dfrac{3}{8}}$.

$\dfrac{\dfrac{1}{2} + \dfrac{3}{4}}{\dfrac{5}{6} - \dfrac{3}{8}}$ { The denominators *within* the complex rational expression are 2, 4, 6, and 8. The LCM of these denominators is 24. We multiply by 1 using $\frac{24}{24}$. This amounts to multiplying both the numerator *and* the denominator by 24.

$= \dfrac{\dfrac{1}{2} + \dfrac{3}{4}}{\dfrac{5}{6} - \dfrac{3}{8}} \cdot \dfrac{24}{24}$ Multiplying by 1

Answers

Skill to Review:
1. 24   2. $5x^2$

Using the distributive laws, we carry out the multiplications:

$$= \frac{\left(\dfrac{1}{2} + \dfrac{3}{4}\right)24}{\left(\dfrac{5}{6} - \dfrac{3}{8}\right)24} = \frac{\dfrac{1}{2}(24) + \dfrac{3}{4}(24)}{\dfrac{5}{6}(24) - \dfrac{3}{8}(24)}$$ ← Multiplying the numerator by 24
← Multiplying the denominator by 24

$$= \frac{12 + 18}{20 - 9}$$ Simplifying

$$= \frac{30}{11}.$$

Multiplying in this manner has the effect of clearing fractions in both the numerator and the denominator of the complex rational expression.

◀ **Do Exercise 1.**

1. Simplify: $\dfrac{\dfrac{1}{3} + \dfrac{4}{5}}{\dfrac{7}{8} - \dfrac{5}{6}}$.

EXAMPLE 2 Simplify: $\dfrac{\dfrac{3}{x} + \dfrac{1}{2x}}{\dfrac{1}{3x} - \dfrac{3}{4x}}$.

The denominators within the complex expression are x, $2x$, $3x$, and $4x$. The LCM of these denominators is $12x$. We multiply by 1 using $12x/12x$.

$$\frac{\dfrac{3}{x} + \dfrac{1}{2x}}{\dfrac{1}{3x} - \dfrac{3}{4x}} \cdot \frac{12x}{12x} = \frac{\left(\dfrac{3}{x} + \dfrac{1}{2x}\right)12x}{\left(\dfrac{1}{3x} - \dfrac{3}{4x}\right)12x} = \frac{\dfrac{3}{x}(12x) + \dfrac{1}{2x}(12x)}{\dfrac{1}{3x}(12x) - \dfrac{3}{4x}(12x)}$$

$$= \frac{36 + 6}{4 - 9} = \frac{42}{-5} = -\frac{42}{5}$$

◀ **Do Exercise 2.**

2. Simplify: $\dfrac{\dfrac{x}{2} + \dfrac{2x}{3}}{\dfrac{1}{x} - \dfrac{x}{2}}$.

$\dfrac{\dfrac{x}{2} + \dfrac{2x}{3}}{\dfrac{1}{x} - \dfrac{x}{2}}$

LCM of denominators $= 6x$

$= \dfrac{\dfrac{x}{2} + \dfrac{2x}{3}}{\dfrac{1}{x} - \dfrac{x}{2}} \cdot \dfrac{6x}{\boxed{}}$

$= \dfrac{\left(\dfrac{x}{2} + \dfrac{2x}{3}\right) \cdot \boxed{}}{\left(\dfrac{1}{x} - \dfrac{x}{2}\right) \cdot 6x}$

$= \dfrac{3x^2 + \boxed{}}{\boxed{} - 3x^2} = \dfrac{7x^2}{3(\boxed{} - x^2)}$

EXAMPLE 3 Simplify: $\dfrac{1 - \dfrac{1}{x}}{1 - \dfrac{1}{x^2}}$.

The denominators within the complex expression are x and x^2. The LCM of these denominators is x^2. We multiply by 1 using x^2/x^2. Then, after obtaining a single rational expression, we simplify:

$$\frac{1 - \dfrac{1}{x}}{1 - \dfrac{1}{x^2}} \cdot \frac{x^2}{x^2} = \frac{\left(1 - \dfrac{1}{x}\right)x^2}{\left(1 - \dfrac{1}{x^2}\right)x^2} = \frac{1(x^2) - \dfrac{1}{x}(x^2)}{1(x^2) - \dfrac{1}{x^2}(x^2)} = \frac{x^2 - x}{x^2 - 1}$$

$$= \frac{x(x - 1)}{(x + 1)(x - 1)} = \frac{x}{x + 1}.$$

3. Simplify: $\dfrac{1 + \dfrac{1}{x}}{1 - \dfrac{1}{x^2}}$.

◀ **Do Exercise 3.**

Answers

1. $\dfrac{136}{5}$ 2. $\dfrac{7x^2}{3(2 - x^2)}$ 3. $\dfrac{x}{x - 1}$

Guided Solution:
2. $6x, 6x, 4x^2, 6, 2$

METHOD 2: ADDING IN THE NUMERATOR AND THE DENOMINATOR

To simplify a complex rational expression:

1. Add or subtract, as necessary, to get a single rational expression in the numerator.

2. Add or subtract, as necessary, to get a single rational expression in the denominator.

3. Divide the numerator by the denominator.

4. If possible, simplify by removing a factor of 1.

We will redo Examples 1–3 using this method.

EXAMPLE 4 Simplify: $\dfrac{\dfrac{1}{2} + \dfrac{3}{4}}{\dfrac{5}{6} - \dfrac{3}{8}}$.

The LCM of 2 and 4 in the numerator is 4. The LCM of 6 and 8 in the denominator is 24. We have

$$\frac{\dfrac{1}{2} + \dfrac{3}{4}}{\dfrac{5}{6} - \dfrac{3}{8}} = \frac{\dfrac{1}{2} \cdot \dfrac{2}{2} + \dfrac{3}{4}}{\dfrac{5}{6} \cdot \dfrac{4}{4} - \dfrac{3}{8} \cdot \dfrac{3}{3}}$$

$\left.\begin{array}{l}\end{array}\right\}$ ← Multiplying $\frac{1}{2}$ by 1 to get the common denominator, 4

$\left.\begin{array}{l}\end{array}\right\}$ ← Multiplying $\frac{5}{6}$ and $\frac{3}{8}$ by 1 to get the common denominator, 24

$$= \frac{\dfrac{2}{4} + \dfrac{3}{4}}{\dfrac{20}{24} - \dfrac{9}{24}}$$

$$= \frac{\dfrac{5}{4}}{\dfrac{11}{24}}$$ Adding in the numerator; subtracting in the denominator

$$= \frac{5}{4} \div \frac{11}{24}$$

$$= \frac{5}{4} \cdot \frac{24}{11}$$ Multiplying by the reciprocal of the divisor

$$= \frac{5 \cdot 2 \cdot 2 \cdot 2 \cdot 3}{2 \cdot 2 \cdot 11}$$ Factoring

$$= \frac{5 \cdot 2 \cdot 2 \cdot 2 \cdot 3}{2 \cdot 2 \cdot 11}$$ Removing a factor of 1: $\dfrac{2 \cdot 2}{2 \cdot 2} = 1$

$$= \frac{5 \cdot 2 \cdot 3}{11}$$

$$= \frac{30}{11}.$$

Do Exercise 4. ▶

4. Simplify. Use method 2.

$$\frac{\dfrac{1}{3} + \dfrac{4}{5}}{\dfrac{7}{8} - \dfrac{5}{6}}$$

Answer

4. $\dfrac{136}{5}$

5. Simplify. Use method 2.

$$\dfrac{\dfrac{x}{2} + \dfrac{2x}{3}}{\dfrac{1}{x} - \dfrac{x}{2}}$$

$$\dfrac{\dfrac{x}{2} + \dfrac{2x}{3}}{\dfrac{1}{x} - \dfrac{x}{2}} \quad \begin{array}{l} \leftarrow \text{LCD} = 6 \\[2em] \leftarrow \text{LCD} = 2x \end{array}$$

$$= \dfrac{\dfrac{x}{2} \cdot \dfrac{3}{\boxed{}} + \dfrac{2x}{3} \cdot \dfrac{\boxed{}}{2}}{\dfrac{1}{x} \cdot \dfrac{\boxed{}}{2} - \dfrac{x}{2} \cdot \dfrac{x}{\boxed{}}}$$

$$= \dfrac{\dfrac{\boxed{}}{6} + \dfrac{4x}{6}}{\dfrac{2}{2x} - \dfrac{\boxed{}}{2x}} = \dfrac{\dfrac{3x + 4x}{\boxed{}}}{\dfrac{2 - x^2}{\boxed{}}}$$

$$= \dfrac{7x}{6} \cdot \dfrac{\boxed{}}{2 - x^2} = \dfrac{7 \cdot 2 \cdot x \cdot x}{2 \cdot 3(2 - x^2)}$$

$$= \dfrac{\boxed{}}{3(2 - x^2)}$$

6. Simplify. Use method 2.

$$\dfrac{1 + \dfrac{1}{x}}{1 - \dfrac{1}{x^2}}$$

Answers

5. $\dfrac{7x^2}{3(2 - x^2)}$ 6. $\dfrac{x}{x - 1}$

Guided Solution:
5. 3, 2, 2, x, 3x, x², 6, 2x, 2x, 7x²

GS   **EXAMPLE 5**   Simplify: $\dfrac{\dfrac{3}{x} + \dfrac{1}{2x}}{\dfrac{1}{3x} - \dfrac{3}{4x}}$.

$$\dfrac{\dfrac{3}{x} + \dfrac{1}{2x}}{\dfrac{1}{3x} - \dfrac{3}{4x}} = \dfrac{\dfrac{3}{x} \cdot \dfrac{2}{2} + \dfrac{1}{2x}}{\dfrac{1}{3x} \cdot \dfrac{4}{4} - \dfrac{3}{4x} \cdot \dfrac{3}{3}} \left.\begin{array}{l} \\[1em] \end{array}\right\} \begin{array}{l} \leftarrow \text{Finding the LCD, } 2x \text{, and multiplying} \\ \text{by 1 in the numerator} \\[0.5em] \leftarrow \text{Finding the LCD, } 12x \text{, and multiplying} \\ \text{by 1 in the denominator} \end{array}$$

$$= \dfrac{\dfrac{6}{2x} + \dfrac{1}{2x}}{\dfrac{4}{12x} - \dfrac{9}{12x}} = \dfrac{\dfrac{7}{2x}}{\dfrac{-5}{12x}} \quad \begin{array}{l} \leftarrow \text{Adding in the numerator and} \\ \leftarrow \text{subtracting in the denominator} \end{array}$$

$$= \dfrac{7}{2x} \div \dfrac{-5}{12x}$$

$$= \dfrac{7}{2x} \cdot \dfrac{12x}{-5} \qquad \begin{array}{l} \text{Multiplying by the reciprocal} \\ \text{of the divisor} \end{array}$$

$$= \dfrac{7 \cdot 6 \cdot \cancel{(2x)}}{\cancel{(2x)}(-5)} \qquad \begin{array}{l} \text{Multiplying, factoring, and} \\ \text{removing a factor of 1: } \dfrac{2x}{2x} = 1 \end{array}$$

$$= \dfrac{42}{-5} = -\dfrac{42}{5}$$

◀ **Do Exercise 5.**

EXAMPLE 6   Simplify: $\dfrac{1 - \dfrac{1}{x}}{1 - \dfrac{1}{x^2}}$.

$$\dfrac{1 - \dfrac{1}{x}}{1 - \dfrac{1}{x^2}} = \dfrac{1 \cdot \dfrac{x}{x} - \dfrac{1}{x}}{1 \cdot \dfrac{x^2}{x^2} - \dfrac{1}{x^2}} \left.\begin{array}{l} \\[1em] \end{array}\right\} \begin{array}{l} \leftarrow \text{Finding the LCD, } x \text{, and} \\ \text{multiplying by 1 in the numerator} \\[0.5em] \leftarrow \text{Finding the LCD, } x^2 \text{, and} \\ \text{multiplying by 1 in the denominator} \end{array}$$

$$= \dfrac{\dfrac{x - 1}{x}}{\dfrac{x^2 - 1}{x^2}} \quad \begin{array}{l} \leftarrow \text{Subtracting in the numerator and} \\ \leftarrow \text{subtracting in the denominator} \end{array}$$

$$= \dfrac{x - 1}{x} \div \dfrac{x^2 - 1}{x^2}$$

$$= \dfrac{x - 1}{x} \cdot \dfrac{x^2}{x^2 - 1} \qquad \begin{array}{l} \text{Multiplying by the reciprocal} \\ \text{of the divisor} \end{array}$$

$$= \dfrac{\cancel{(x - 1)}x \cdot x}{\cancel{x}\cancel{(x - 1)}(x + 1)} \qquad \begin{array}{l} \text{Multiplying, factoring, and removing} \\ \text{a factor of 1: } \dfrac{x(x - 1)}{x(x - 1)} = 1 \end{array}$$

$$= \dfrac{x}{x + 1}$$

◀ **Do Exercise 6.**

506   **CHAPTER 6**   Rational Expressions and Equations

✓ Reading Check

Consider the expression $\dfrac{\dfrac{8}{x} - \dfrac{5}{9}}{\dfrac{2}{x}}$. Complete each statement with the correct word(s) from the column on the right.

RC1. The expression given above is a(n) _____ rational expression.

RC2. The expression $\dfrac{8}{x} - \dfrac{5}{9}$ is the _____ of the expression.

RC3. The _____ of the rational expressions $\dfrac{8}{x}, \dfrac{5}{9},$ and $\dfrac{2}{x}$ is $9x$.

RC4. To simplify the expression, we can multiply the numerator $\dfrac{8}{x} - \dfrac{5}{9}$ by the _____ of the denominator, $\dfrac{2}{x}$.

numerator

denominator

opposite

reciprocal

complex

least common denominator

a Simplify.

1. $\dfrac{1 + \dfrac{9}{16}}{1 - \dfrac{3}{4}}$

2. $\dfrac{6 - \dfrac{3}{8}}{4 + \dfrac{5}{6}}$

3. $\dfrac{1 - \dfrac{3}{5}}{1 + \dfrac{1}{5}}$

4. $\dfrac{2 + \dfrac{2}{3}}{2 - \dfrac{2}{3}}$

5. $\dfrac{\dfrac{1}{2} + \dfrac{3}{4}}{\dfrac{5}{8} - \dfrac{5}{6}}$

6. $\dfrac{\dfrac{3}{4} + \dfrac{7}{8}}{\dfrac{2}{3} - \dfrac{5}{6}}$

7. $\dfrac{\dfrac{1}{x} + 3}{\dfrac{1}{x} - 5}$

8. $\dfrac{2 - \dfrac{1}{a}}{4 + \dfrac{1}{a}}$

9. $\dfrac{4 - \dfrac{1}{x^2}}{2 - \dfrac{1}{x}}$

10. $\dfrac{\dfrac{2}{y} + \dfrac{1}{2y}}{y + \dfrac{y}{2}}$

11. $\dfrac{8 + \dfrac{8}{d}}{1 + \dfrac{1}{d}}$

12. $\dfrac{3 + \dfrac{2}{t}}{3 - \dfrac{2}{t}}$

13. $\dfrac{\dfrac{x}{8} - \dfrac{8}{x}}{\dfrac{1}{8} + \dfrac{1}{x}}$

14. $\dfrac{\dfrac{2}{m} + \dfrac{m}{2}}{\dfrac{m}{3} - \dfrac{3}{m}}$

15. $\dfrac{1 + \dfrac{1}{y}}{1 - \dfrac{1}{y^2}}$

16. $\dfrac{\dfrac{1}{q^2} - 1}{\dfrac{1}{q} + 1}$

17. $\dfrac{\dfrac{1}{5} - \dfrac{1}{a}}{\dfrac{5 - a}{5}}$

18. $\dfrac{\dfrac{4}{t}}{4 + \dfrac{1}{t}}$

19. $\dfrac{\dfrac{1}{a} + \dfrac{1}{b}}{\dfrac{1}{a^2} - \dfrac{1}{b^2}}$

20. $\dfrac{\dfrac{1}{x^2} - \dfrac{1}{y^2}}{\dfrac{2}{x} - \dfrac{2}{y}}$

21. $\dfrac{\dfrac{p}{q} + \dfrac{q}{p}}{\dfrac{1}{p} + \dfrac{1}{q}}$

22. $\dfrac{x - 3 + \dfrac{2}{x}}{x - 4 + \dfrac{3}{x}}$

23. $\dfrac{\dfrac{2}{a} + \dfrac{4}{a^2}}{\dfrac{5}{a^3} - \dfrac{3}{a}}$

24. $\dfrac{\dfrac{5}{x^3} - \dfrac{1}{x^2}}{\dfrac{2}{x} + \dfrac{3}{x^2}}$

25. $\dfrac{\dfrac{2}{7a^4} - \dfrac{1}{14a}}{\dfrac{3}{5a^2} + \dfrac{2}{15a}}$

26. $\dfrac{\dfrac{5}{4x^3} - \dfrac{3}{8x}}{\dfrac{3}{2x} + \dfrac{3}{4x^3}}$

27. $\dfrac{\dfrac{a}{b} + \dfrac{c}{d}}{\dfrac{b}{a} + \dfrac{d}{c}}$

28. $\dfrac{\dfrac{a}{b} - \dfrac{c}{d}}{\dfrac{b}{a} - \dfrac{d}{c}}$

29. $\dfrac{\dfrac{x}{5y^3} + \dfrac{3}{10y}}{\dfrac{3}{10y} + \dfrac{x}{5y^3}}$

30. $\dfrac{\dfrac{a}{6b^3} + \dfrac{4}{9b^2}}{\dfrac{5}{6b} - \dfrac{1}{9b^3}}$

31. $\dfrac{\dfrac{3}{x + 1} + \dfrac{1}{x}}{\dfrac{2}{x + 1} + \dfrac{3}{x}}$

32. $\dfrac{x - 7 + \dfrac{5}{x - 1}}{x - 3 + \dfrac{1}{x - 1}}$

Skill Maintenance

Solve. [2.7e]

33. $4 - \dfrac{1}{6}x \geq -12$

34. $3(b - 8) > -2(3b + 1)$

35. $1.5x + 19.2 < 4.2 - 3.5x$

Solve. [5.8a]

36. *Ladder Distances.*   A ladder of length 13 ft is placed against a building in such a way that the distance from the top of the ladder to the ground is 7 ft more than the distance from the bottom of the ladder to the building. Find these distances.

37. *Perimeter of a Rectangle.*   The length of a rectangle is 3 yd greater than the width. The area of the rectangle is 10 yd². Find the perimeter.

Synthesis

Simplify.

38. $\left[\dfrac{\dfrac{x + 1}{x - 1} + 1}{\dfrac{x + 1}{x - 1} - 1} \right]^5$

39. $1 + \dfrac{1}{1 + \dfrac{1}{1 + \dfrac{1}{1 + \dfrac{1}{x}}}}$

40. $\dfrac{\dfrac{z}{1 - \dfrac{z}{2 + 2z}} - 2z}{\dfrac{2z}{5z - 2} - 3}$

Solving Rational Equations

a RATIONAL EQUATIONS

In Sections 6.1–6.6, we studied operations with *rational expressions*. These expressions have no equals signs. We can add, subtract, multiply, or divide and simplify expressions, but we cannot solve if there are no equals signs—as, for example, in

$$\frac{x^2 + 6x + 9}{x^2 - 4} \cdot \frac{x - 2}{x + 3}, \quad \frac{x + y}{x - y} \div \frac{x^2 + y}{x^2 - y^2}, \quad \text{and} \quad \frac{a + 3}{a^2 - 16} + \frac{5}{12 - 3a}.$$

Operation signs occur. There are no equals signs!

Most often, the result of our calculation is another rational expression that has not been cleared of fractions.

Equations *do have* equals signs, and we can clear them of fractions as we did in Section 2.3. A **rational**, or **fraction**, **equation**, is an equation containing one or more rational expressions. Here are some examples:

$$\frac{2}{3} + \frac{5}{6} = \frac{x}{9}, \quad x + \frac{6}{x} = -5, \quad \text{and} \quad \frac{x^2}{x - 1} = \frac{1}{x - 1}.$$

There are equals signs as well as operation signs.

SOLVING RATIONAL EQUATIONS

To solve a rational equation, the first step is to clear the equation of fractions. To do this, multiply all terms on both sides of the equation by the LCM of all the denominators. Then carry out the equation-solving process as we learned it in Chapters 2 and 5.

When clearing an equation of fractions, we use the terminology LCM instead of LCD because we are *not* adding or subtracting rational expressions.

EXAMPLE 1 Solve: $\frac{2}{3} + \frac{5}{6} = \frac{x}{9}$.

The LCM of all denominators is $2 \cdot 3 \cdot 3$, or 18. We multiply all terms on both sides by 18:

$$18\left(\frac{2}{3} + \frac{5}{6}\right) = 18 \cdot \frac{x}{9} \quad \text{Multiplying by the LCM on both sides}$$

$$18 \cdot \frac{2}{3} + 18 \cdot \frac{5}{6} = 18 \cdot \frac{x}{9} \quad \text{Multiplying each term by the LCM to remove parentheses}$$

$$12 + 15 = 2x \quad \text{Simplifying. Note that we have now cleared fractions.}$$

$$27 = 2x$$

$$\frac{27}{2} = x.$$

The check is left to the student. The solution is $\frac{27}{2}$.

Do Margin Exercise 1. ▶

OBJECTIVE

a Solve rational equations.

SKILL TO REVIEW

Objective 2.3b: Solve equations in which like terms may need to be collected.

Solve. Clear fractions first.

1. $4 - \frac{5}{6}y = y + \frac{7}{12}$

2. $\frac{2}{5}x + \frac{1}{3} = \frac{7}{10}x - 2$

Caution!

We are introducing a new use of the LCM in this section. We previously used the LCM in adding or subtracting rational expressions. *Now* we have equations with equals signs. We clear fractions by multiplying by the LCM on both sides of the equation. This eliminates the denominators. Do *not* make the mistake of trying to clear fractions when you do not have an equation.

1. Solve: $\frac{3}{4} + \frac{5}{8} = \frac{x}{12}$.

Answers

Skill to Review:
1. $\frac{41}{22}$ 2. $\frac{70}{9}$

Margin Exercise:
1. $\frac{33}{2}$

We can obtain a visual check of the solutions of a rational equation by graphing. For example, consider the equation

$$\frac{x}{4} + \frac{x}{2} = 6.$$

We can examine the solution by graphing the equations

$$y = \frac{x}{4} + \frac{x}{2} \quad \text{and} \quad y = 6$$

using the same set of axes.

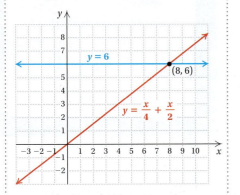

The first coordinate of the point of intersection of the graphs is the value of x for which $\frac{x}{4} + \frac{x}{2} = 6$, so it is the solution of the equation. It appears from the graph that when $x = 8$, the value of $x/4 + x/2$ is 6. We can check by substitution:

$$\frac{x}{4} + \frac{x}{2} = \frac{8}{4} + \frac{8}{2} = 2 + 4 = 6.$$

Thus the solution is 8.

Solve.

2. $\dfrac{x}{4} - \dfrac{x}{6} = \dfrac{1}{8}$

3. $\dfrac{1}{x} = \dfrac{1}{6 - x}$

Answers

2. $\dfrac{3}{2}$ 3. 3

EXAMPLE 2   Solve: $\dfrac{x}{6} - \dfrac{x}{8} = \dfrac{1}{12}.$

The LCM is 24. We multiply all terms on both sides by 24:

$$\frac{x}{6} - \frac{x}{8} = \frac{1}{12}$$

$$24\left(\frac{x}{6} - \frac{x}{8}\right) = 24 \cdot \frac{1}{12} \qquad \text{Multiplying by the LCM on both sides}$$

$$24 \cdot \frac{x}{6} - 24 \cdot \frac{x}{8} = 24 \cdot \frac{1}{12} \qquad \text{Multiplying to remove parentheses}$$

> Be sure to multiply each term by the LCM.

$$4x - 3x = 2 \qquad \text{Simplifying}$$

$$x = 2.$$

Check: $\dfrac{x}{6} - \dfrac{x}{8} = \dfrac{1}{12}$

$$\begin{array}{c|c} \dfrac{2}{6} - \dfrac{2}{8} & \dfrac{1}{12} \\[2mm] \dfrac{1}{3} - \dfrac{1}{4} & \\[2mm] \dfrac{4}{12} - \dfrac{3}{12} & \\[2mm] \dfrac{1}{12} & \text{TRUE} \end{array}$$

This checks, so the solution is 2.

EXAMPLE 3   Solve: $\dfrac{1}{x} = \dfrac{1}{4 - x}.$

The LCM is $x(4 - x)$. We multiply all terms on both sides by $x(4 - x)$:

$$\frac{1}{x} = \frac{1}{4 - x}$$

$$x(4 - x) \cdot \frac{1}{x} = x(4 - x) \cdot \frac{1}{4 - x} \qquad \text{Multiplying by the LCM on both sides}$$

$$4 - x = x \qquad \text{Simplifying}$$

$$4 = 2x$$

$$x = 2.$$

Check: $\dfrac{1}{x} = \dfrac{1}{4 - x}$

$$\begin{array}{c|c} \dfrac{1}{2} & \dfrac{1}{4 - 2} \\[2mm] & \dfrac{1}{2} \quad \text{TRUE} \end{array}$$

This checks, so the solution is 2.

◀ **Do Exercises 2 and 3.**

EXAMPLE 4 Solve: $\dfrac{2}{3x} + \dfrac{1}{x} = 10$.

The LCM is $3x$. We multiply all terms on both sides by $3x$:

$$\frac{2}{3x} + \frac{1}{x} = 10$$

$$3x\left(\frac{2}{3x} + \frac{1}{x}\right) = 3x \cdot 10 \qquad \text{Multiplying by the LCM on both sides}$$

$$3x \cdot \frac{2}{3x} + 3x \cdot \frac{1}{x} = 3x \cdot 10 \qquad \text{Multiplying to remove parentheses}$$

$$2 + 3 = 30x \qquad \text{Simplifying}$$

$$5 = 30x$$

$$\frac{5}{30} = x$$

$$\frac{1}{6} = x.$$

The check is left to the student. The solution is $\frac{1}{6}$.

Do Exercise 4. ▶

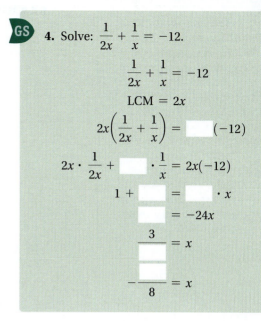

GS **4.** Solve: $\dfrac{1}{2x} + \dfrac{1}{x} = -12$.

$$\frac{1}{2x} + \frac{1}{x} = -12$$

$$\text{LCM} = 2x$$

$$2x\left(\frac{1}{2x} + \frac{1}{x}\right) = \boxed{}(-12)$$

$$2x \cdot \frac{1}{2x} + \boxed{} \cdot \frac{1}{x} = 2x(-12)$$

$$1 + \boxed{} = \boxed{} \cdot x$$

$$\boxed{} = -24x$$

$$\frac{3}{\boxed{}} = x$$

$$-\frac{\boxed{}}{8} = x$$

EXAMPLE 5 Solve: $x + \dfrac{6}{x} = -5$.

The LCM is x. We multiply all terms on both sides by x:

$$x + \frac{6}{x} = -5$$

$$x\left(x + \frac{6}{x}\right) = x \cdot (-5) \qquad \text{Multiplying by } x \text{ on both sides}$$

$$x \cdot x + x \cdot \frac{6}{x} = -5x \qquad \text{Note that each rational expression on the left is now multiplied by } x.$$

$$x^2 + 6 = -5x \qquad \text{Simplifying}$$

$$x^2 + 5x + 6 = 0 \qquad \text{Adding } 5x \text{ to get 0 on one side}$$

$$(x + 3)(x + 2) = 0 \qquad \text{Factoring}$$

$$x + 3 = 0 \quad \text{or} \quad x + 2 = 0 \qquad \text{Using the principle of zero products}$$

$$x = -3 \quad \text{or} \qquad x = -2.$$

> **CHECKING POSSIBLE SOLUTIONS**
>
> When we multiply by the LCM on both sides of an equation, the resulting equation might have solutions that are *not* solutions of the original equation. Thus we must *always* check possible solutions in the original equation.

Check:

For -3:

$$\begin{array}{c|c} x + \dfrac{6}{x} = -5 & \\ \hline -3 + \dfrac{6}{-3} \; ? \; -5 & \\ -3 - 2 & \\ -5 & \text{TRUE} \end{array}$$

For -2:

$$\begin{array}{c|c} x + \dfrac{6}{x} = -5 & \\ \hline -2 + \dfrac{6}{-2} \; ? \; -5 & \\ -2 - 3 & \\ -5 & \text{TRUE} \end{array}$$

Both of these check, so there are two solutions, -3 and -2.

Do Exercise 5. ▶

5. Solve: $x + \dfrac{1}{x} = 2$.

Answers

4. $-\dfrac{1}{8}$ **5.** 1

Guided Solution:
4. $2x$, $2x$, 2, -24, 3, -24, 1

Example 6 illustrates the importance of checking all possible solutions.

EXAMPLE 6 Solve: $\dfrac{x^2}{x-1} = \dfrac{1}{x-1}$.

The LCM is $x - 1$. We multiply all terms on both sides by $x - 1$:

$$\frac{x^2}{x-1} = \frac{1}{x-1}$$

$$(x-1) \cdot \frac{x^2}{x-1} = (x-1) \cdot \frac{1}{x-1} \quad \text{Multiplying by } x-1 \text{ on both sides}$$

$$x^2 = 1 \quad \text{Simplifying}$$

$$x^2 - 1 = 0 \quad \text{Subtracting 1 to get 0 on one side}$$

$$(x-1)(x+1) = 0 \quad \text{Factoring}$$

$$x - 1 = 0 \quad \text{or} \quad x + 1 = 0 \quad \text{Using the principle of zero products}$$

$$x = 1 \quad \text{or} \qquad x = -1.$$

The numbers 1 and -1 are possible solutions.

Check: For 1:

$$\frac{x^2}{x-1} = \frac{1}{x-1}$$

$$\frac{1^2}{1-1} \;\overset{?}{\;}\; \frac{1}{1-1}$$

$$\frac{1}{0} \;\Big|\; \frac{1}{0} \qquad \text{NOT DEFINED}$$

For -1:

$$\frac{x^2}{x-1} = \frac{1}{x-1}$$

$$\frac{(-1)^2}{(-1)-1} \;\overset{?}{\;}\; \frac{1}{(-1)-1}$$

$$-\frac{1}{2} \;\Big|\; -\frac{1}{2} \qquad \text{TRUE}$$

We look at the original equation and see that 1 makes a denominator 0 and is thus not a solution. The number -1 checks and is a solution.

Solve.

6. $\dfrac{x^2}{x+2} = \dfrac{4}{x+2}$

7. $\dfrac{4}{x-2} + \dfrac{1}{x+2} = \dfrac{26}{x^2-4}$

$\text{LCM} = (x-2)(x+2)$

$$(x-2)\left(x + \boxed{}\right)\left(\frac{4}{x-2} + \frac{1}{x+2}\right)$$

$$= (x-2)(x+2) \cdot \frac{26}{x^2-4}$$

$$\boxed{}(x+2) + \boxed{}(x-2) = 26$$

$$\boxed{} + 8 + x - \boxed{} = 26$$

$$\boxed{} + 6 = 26$$

$$5x = \boxed{}$$

$$x = \boxed{}$$

EXAMPLE 7 Solve: $\dfrac{3}{x-5} + \dfrac{1}{x+5} = \dfrac{2}{x^2-25}$.

The LCM is $(x-5)(x+5)$. We multiply all terms on both sides by $(x-5)(x+5)$:

$$(x-5)(x+5)\left(\frac{3}{x-5} + \frac{1}{x+5}\right) = (x-5)(x+5)\left(\frac{2}{x^2-25}\right)$$

Multiplying by the LCM on both sides

$$(x-5)(x+5) \cdot \frac{3}{x-5} + (x-5)(x+5) \cdot \frac{1}{x+5}$$

$$= (x-5)(x+5) \cdot \frac{2}{x^2-25}$$

$$3(x+5) + (x-5) = 2 \quad \text{Simplifying}$$

$$3x + 15 + x - 5 = 2 \quad \text{Removing parentheses}$$

$$4x + 10 = 2$$

$$4x = -8$$

$$x = -2.$$

The check is left to the student. The number -2 checks and is the solution.

◀ **Do Exercises 6 and 7.**

Answers

6. 2   7. 4

Guided Solution:
7. 2, 4, 1, 4x, 2, 5x, 20, 4

✓ Reading Check

One of the common difficulties with this chapter is being sure about the task at hand. Are you combining expressions using operations to get another *rational expression*, or are you solving equations for which the results are numbers that are *solutions* of an equation? To learn to make these decisions, determine for each of the following exercises the type of answer you should get: "Rational expression" or "Solutions." You need not complete the mathematical operations.

RC1. Add: $\dfrac{5a}{a^2 - 1} + \dfrac{a}{a^2 - a}$.

RC2. Solve: $\dfrac{5}{y - 3} - \dfrac{30}{y^2 - 9} = 1$.

RC3. Subtract: $\dfrac{4}{x - 2} - \dfrac{1}{x + 2}$.

RC4. Divide: $\dfrac{x + 4}{x - 2} \div \dfrac{6x}{x^2 - 4}$.

RC5. Solve: $\dfrac{x^2}{x - 1} = \dfrac{1}{x - 1}$.

RC6. Solve: $\dfrac{10}{x} + x = -2$.

RC7. Multiply: $\dfrac{2t^2}{t^2 - 25} \cdot \dfrac{t^2 + 10t + 25}{t^8}$.

RC8. Solve: $\dfrac{7}{x - 4} - \dfrac{2}{x + 4} = \dfrac{1}{x^2 - 16}$.

a   Solve. Don't forget to check!

1. $\dfrac{4}{5} - \dfrac{2}{3} = \dfrac{x}{9}$

2. $\dfrac{x}{20} = \dfrac{3}{8} - \dfrac{4}{5}$

3. $\dfrac{3}{5} + \dfrac{1}{8} = \dfrac{1}{x}$

4. $\dfrac{2}{3} + \dfrac{5}{6} = \dfrac{1}{x}$

5. $\dfrac{3}{8} + \dfrac{4}{5} = \dfrac{x}{20}$

6. $\dfrac{3}{5} + \dfrac{2}{3} = \dfrac{x}{9}$

7. $\dfrac{1}{x} = \dfrac{2}{3} - \dfrac{5}{6}$

8. $\dfrac{1}{x} = \dfrac{1}{8} - \dfrac{3}{5}$

9. $\dfrac{1}{6} + \dfrac{1}{8} = \dfrac{1}{t}$

10. $\dfrac{1}{8} + \dfrac{1}{12} = \dfrac{1}{t}$

11. $x + \dfrac{4}{x} = -5$

12. $\dfrac{10}{x} - x = 3$

13. $\dfrac{x}{4} - \dfrac{4}{x} = 0$

14. $\dfrac{x}{5} - \dfrac{5}{x} = 0$

15. $\dfrac{5}{x} = \dfrac{6}{x} - \dfrac{1}{3}$

16. $\dfrac{4}{x} = \dfrac{5}{x} - \dfrac{1}{2}$

17. $\dfrac{5}{3x} + \dfrac{3}{x} = 1$

18. $\dfrac{5}{2y} + \dfrac{8}{y} = 1$

19. $\dfrac{t-2}{t+3} = \dfrac{3}{8}$

20. $\dfrac{x-7}{x+2} = \dfrac{1}{4}$

21. $\dfrac{2}{x+1} = \dfrac{1}{x-2}$

22. $\dfrac{8}{y-3} = \dfrac{6}{y+4}$

23. $\dfrac{x}{6} - \dfrac{x}{10} = \dfrac{1}{6}$

24. $\dfrac{x}{8} - \dfrac{x}{12} = \dfrac{1}{8}$

25. $\dfrac{t+2}{5} - \dfrac{t-2}{4} = 1$

26. $\dfrac{x+1}{3} - \dfrac{x-1}{2} = 1$

27. $\dfrac{5}{x-1} = \dfrac{3}{x+2}$

28. $\dfrac{x-7}{x-9} = \dfrac{2}{x-9}$

29. $\dfrac{a-3}{3a+2} = \dfrac{1}{5}$

30. $\dfrac{x+7}{8x-5} = \dfrac{2}{3}$

31. $\dfrac{x-1}{x-5} = \dfrac{4}{x-5}$

32. $\dfrac{y+11}{y+8} = \dfrac{3}{y+8}$

33. $\dfrac{2}{x+3} = \dfrac{5}{x}$

34. $\dfrac{6}{y} = \dfrac{5}{y-8}$

35. $\dfrac{x-2}{x-3} = \dfrac{x-1}{x+1}$

36. $\dfrac{t+5}{t-2} = \dfrac{t-2}{t+4}$

37. $\dfrac{1}{x+3} + \dfrac{1}{x-3} = \dfrac{1}{x^2-9}$

38. $\dfrac{4}{x-3} + \dfrac{2x}{x^2-9} = \dfrac{1}{x+3}$

39. $\dfrac{x}{x+4} - \dfrac{4}{x-4} = \dfrac{x^2+16}{x^2-16}$

40. $\dfrac{5}{y-3} - \dfrac{30}{y^2-9} = 1$

41. $\dfrac{4 - a}{8 - a} = \dfrac{4}{a - 8}$

42. $\dfrac{3}{x - 7} = \dfrac{x + 10}{x - 7}$

43. $2 - \dfrac{a - 2}{a + 3} = \dfrac{a^2 - 4}{a + 3}$

44. $\dfrac{5}{x - 1} + x + 1 = \dfrac{5x + 4}{x - 1}$

45. $\dfrac{x + 1}{x + 2} = \dfrac{x + 3}{x + 4}$

46. $\dfrac{x^2}{x^2 - 4} = \dfrac{x}{x + 2} - \dfrac{2x}{2 - x}$

47. $4a - 3 = \dfrac{a + 13}{a + 1}$

48. $\dfrac{3x - 9}{x - 3} = \dfrac{5x - 4}{2}$

49. $\dfrac{4}{y - 2} - \dfrac{2y - 3}{y^2 - 4} = \dfrac{5}{y + 2}$

50. $\dfrac{y^2 - 4}{y + 3} = 2 - \dfrac{y - 2}{y + 3}$

Skill Maintenance

Add. [4.4a]

51. $(2x^3 - 4x^2 + x - 7) + (4x^4 + x^3 + 4x^2 + x)$

52. $(2x^3 - 4x^2 + x - 7) + (-2x^3 + 4x^2 - x + 7)$

Factor. [5.6a]

53. $50p^2 - 100$

54. $5p^2 - 40p - 100$

Solve.

55. *Consecutive Even Integers.* The product of two consecutive even integers is 360. Find the integers. [5.8a]

56. *Chemistry.* About 5 L of oxygen can be dissolved in 100 L of water at 0°C. This is 1.6 times the amount that can be dissolved in the same volume of water at 20°C. How much oxygen can be dissolved in 100 L at 20°C? [2.6a]

Synthesis

57. Solve: $\dfrac{x}{x^2 + 3x - 4} + \dfrac{x + 1}{x^2 + 6x + 8} = \dfrac{2x}{x^2 + x - 2}$.

58. Use a graphing calculator to check the solutions to Exercises 13, 15, and 25.

In many areas of study, applications involving rates, proportions, or reciprocals translate to rational equations. By using the five steps for problem solving and the skills of Sections 6.1–6.7, we can now solve such problems.

a SOLVING APPLIED PROBLEMS

Problems Involving Motion

Problems that deal with distance, speed (or rate), and time are called **motion problems**. Translation of these problems involves the distance formula, $d = r \cdot t$, and/or the equivalent formulas $r = d/t$ and $t = d/r$.

OBJECTIVES

a Solve applied problems using rational equations.

b Solve proportion problems.

SKILL TO REVIEW

Objective 2.4b: Solve a formula for a specified letter.

Solve for the indicated letter.
1. $x = w \cdot y$, for w
2. $A = c - bt$, for t

MOTION FORMULAS

The following are the formulas for motion problems:

$d = rt$;   Distance = Rate · Time (basic formula)

$r = \dfrac{d}{t}$;   Rate = Distance/Time

$t = \dfrac{d}{r}$.   Time = Distance/Rate

EXAMPLE 1 *Speed of Sea Animals.* The shortfin Mako shark is known to have the fastest speed of all sharks. The sailfish has the fastest speed of all fish. The top speed recorded for a sailfish is approximately 25 mph faster than the fastest speed of a Mako shark. A sailfish can swim 14 mi in the same time that a Mako shark can swim 9 mi. Find the speed of each sea animal.

Sources: International Union for the Conservation of Nature; theshark.dk/en/records.php; thetravelalmanac.com/lists/fish-speed.htm

1. **Familiarize.** We first make a drawing. We let $r =$ the speed of the shark. Then $r + 25 =$ the speed of the sailfish.

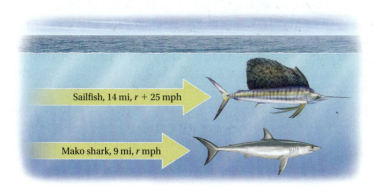

Sailfish, 14 mi, $r + 25$ mph

Mako shark, 9 mi, r mph

Answers

Skill to Review:

1. $w = \dfrac{x}{y}$ 2. $t = \dfrac{c - A}{b}$, or $\dfrac{A - c}{-b}$

Recall that sometimes we need to use a formula in order to solve an application. As we see above, a formula that relates the notions of distance, speed, and time is $d = rt$, or *Distance* = *Speed* · *Time*.

Since each sea animal travels for the same length of time, we can use just t for time. We organize the information in a chart, as follows.

	DISTANCE	SPEED	TIME	
MAKO SHARK	9	r	t	→ $9 = rt$
SAILFISH	14	$r + 25$	t	→ $14 = (r + 25)t$

d = r · t

2. **Translate.** We can apply the formula $d = rt$ along the rows of the table to obtain two equations:

$$9 = rt \quad \text{and} \quad 14 = (r + 25)t.$$

We know that the sea animals travel for the same length of time. Thus if we solve each equation for t and set the results equal to each other, we get an equation in terms of r.

Solving $9 = rt$ for t: $\qquad\qquad t = \dfrac{9}{r}$

Solving $14 = (r + 25)t$ for t: $\quad t = \dfrac{14}{r + 25}$

Since the times are the same, we have the following equation:

$$\frac{9}{r} = \frac{14}{r + 25}.$$

3. **Solve.** To solve the equation, we first multiply on both sides by the LCM, which is $r(r + 25)$:

$$r(r + 25) \cdot \frac{9}{r} = r(r + 25) \cdot \frac{14}{r + 25} \qquad \text{Multiplying on both sides by the LCM, which is } r(r + 25)$$

$$9(r + 25) = 14r \qquad \text{Simplifying}$$

$$9r + 225 = 14r \qquad \text{Removing parentheses}$$

$$225 = 5r$$

$$45 = r.$$

We now have a possible solution. The speed of the shark is 45 mph, and the speed of the sailfish is $r + 25 = 45 + 25$, or 70 mph.

4. **Check.** We check the speeds of 45 mph for the shark and 70 mph for the sailfish. The sailfish does swim 25 mph faster than the shark. If the sailfish swims 14 mi at 70 mph, the time that it has traveled is $\frac{14}{70}$, or $\frac{1}{5}$ hr. If the shark swims 9 mi at 45 mph, the time that it has traveled is $\frac{9}{45}$, or $\frac{1}{5}$ hr. Since the times are the same, the speeds check.

5. **State.** The speed of the Mako shark is 45 mph, and the speed of the sailfish is 70 mph.

◄ **Do Exercise 1.**

1. *Driving Speed.* Catherine drives 20 mph faster than her father, Gary. In the same time that Catherine travels 180 mi, her father travels 120 mi. Find their speeds.

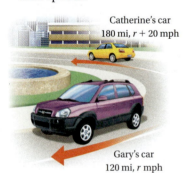

Catherine's car
180 mi, $r + 20$ mph

Gary's car
120 mi, r mph

Answer

1. Gary: 40 mph; Catherine: 60 mph

Problems Involving Work

EXAMPLE 2 *Sodding a Yard.* Charlie's Lawn Care has two three-person crews who lay sod. Crew A can lay 7 skids of sod in 4 hr, while crew B requires 6 hr to do the same job. How long would it take the two crews working together to lay 7 skids of sod?

1. **Familiarize.** A common *incorrect* way to translate the problem is to add the two times: 4 hr + 6 hr = 10 hr. Let's think about this. Crew A can do the job in 4 hr. If crew A and crew B work together, the time that it takes them should be *less* than 4 hr. Thus we reject 10 hr as a solution, but we do have a partial check on any answer we get. The answer should be less than 4 hr.

   We proceed to a translation by considering how much of the job is finished in 1 hr, 2 hr, 3 hr, and so on. It takes crew A 4 hr to do the sodding job alone. Then, in 1 hr, crew A can do $\frac{1}{4}$ of the job. It takes crew B 6 hr to do the job alone. Then, in 1 hr, crew B can do $\frac{1}{6}$ of the job. Working together for 1 hr (see Fig. 1), the crews can do

$$\frac{1}{4} + \frac{1}{6}, \text{ or } \frac{3}{12} + \frac{2}{12}, \text{ or } \frac{5}{12} \text{ of the job in 1 hr.}$$

In 2 hr, crew A can do $2\left(\frac{1}{4}\right)$ of the job and crew B can do $2\left(\frac{1}{6}\right)$ of the job. Working together for 2 hr (see Fig. 2), they can do

$$2\left(\frac{1}{4}\right) + 2\left(\frac{1}{6}\right), \text{ or } \frac{6}{12} + \frac{4}{12}, \text{ or } \frac{10}{12}, \text{ or } \frac{5}{6} \text{ of the job in 2 hr.}$$

In one hour:
Crew A   Crew B

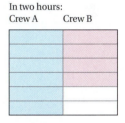

FIGURE 1

In two hours:
Crew A   Crew B

FIGURE 2

	FRACTION OF THE JOB COMPLETED		
TIME	CREW A	CREW B	TOGETHER
1 hr	$\frac{1}{4}$	$\frac{1}{6}$	$\frac{1}{4} + \frac{1}{6}$, or $\frac{5}{12}$
2 hr	$2\left(\frac{1}{4}\right)$	$2\left(\frac{1}{6}\right)$	$2\left(\frac{1}{4}\right) + 2\left(\frac{1}{6}\right)$, or $\frac{5}{6}$
3 hr	$3\left(\frac{1}{4}\right)$	$3\left(\frac{1}{6}\right)$	$3\left(\frac{1}{4}\right) + 3\left(\frac{1}{6}\right)$, or $1\frac{1}{4}$
t hr	$t\left(\frac{1}{4}\right)$	$t\left(\frac{1}{6}\right)$	$t\left(\frac{1}{4}\right) + t\left(\frac{1}{6}\right)$

We see that the answer is somewhere between 2 hr and 3 hr. What we want is a number t such that the fraction of the job that gets completed is 1; that is, the job is just completed.

2. **Translate.** From the table, we see that the time we want is some number t for which

$$t\left(\frac{1}{4}\right) + t\left(\frac{1}{6}\right) = 1, \quad \text{or} \quad \frac{t}{4} + \frac{t}{6} = 1,$$

where 1 represents the idea that the entire job is completed in time t.

3. Solve. We solve the equation:

$$12\left(\frac{t}{4} + \frac{t}{6}\right) = 12 \cdot 1 \qquad \textcolor{red}{\text{Multiplying by the LCM, which is } 2 \cdot 2 \cdot 3, \text{ or } 12}$$

$$12 \cdot \frac{t}{4} + 12 \cdot \frac{t}{6} = 12$$

$$3t + 2t = 12$$

$$5t = 12$$

$$t = \frac{12}{5}, \text{ or } 2\frac{2}{5} \text{ hr.}$$

4. Check. In $\frac{12}{5}$ hr, crew A does $\frac{12}{5} \cdot \frac{1}{4}$, or $\frac{3}{5}$, of the job and crew B does $\frac{12}{5} \cdot \frac{1}{6}$, or $\frac{2}{5}$, of the job. Together, they do $\frac{3}{5} + \frac{2}{5}$, or 1 entire job. The answer, $2\frac{2}{5}$ hr, is between 2 hr and 3 hr (see the table), and it is less than 4 hr, the time it takes crew A working alone. The answer checks.

5. State. It takes $2\frac{2}{5}$ hr for crew A and crew B working together to lay 7 skids of sod.

THE WORK PRINCIPLE

Suppose $a = $ the time that it takes A to do a job, $b = $ the time that it takes B to do the same job, and $t = $ the time that it takes them to do the job working together. Then

$$\frac{t}{a} + \frac{t}{b} = 1.$$

◀ **Do Exercise 2.**

b APPLICATIONS INVOLVING PROPORTIONS

We now consider applications with proportions. A **proportion** involves ratios. A **ratio** of two quantities is their quotient. For example, 73% is the ratio of 73 to 100, $\frac{73}{100}$. The ratio of two different kinds of measure is called a **rate**. Suppose an animal travels 2720 ft in 2.5 hr. Its **rate**, or **speed**, is then

$$\frac{2720 \text{ ft}}{2.5 \text{ hr}} = 1088 \frac{\text{ft}}{\text{hr}}.$$

◀ **Do Exercises 3–6.**

PROPORTION

An equality of ratios,

$$\frac{A}{B} = \frac{C}{D},$$

is called a **proportion**. The numbers within a proportion are said to be **proportional** to each other.

---

2. *Work Recycling.* Emma and Evan work as volunteers at a community recycling center. Emma can sort a morning's accumulation of recyclable objects in 3 hr, while Evan requires 5 hr to do the same job. How long would it take them, working together, to sort the recyclable material?

3. Find the ratio of 145 km to 2.5 liters (L).

4. *Batting Average.* Recently, a baseball player got 7 hits in 25 times at bat. What was the rate, or batting average, in number of hits per times at bat?

5. Impulses in nerve fibers travel 310 km in 2.5 hr. What is the rate, or speed, in kilometers per hour?

6. A lake of area 550 yd^2 contains 1320 fish. What is the population density of the lake, in number of fish per square yard?

Answers

2. $1\frac{7}{8}$ hr   **3.** 58 km/L
4. 0.28 hit per times at bat   **5.** 124 km/h
6. 2.4 fish/yd^2

EXAMPLE 3 *Mileage.* A 2013 Fiat 500 Turbo can travel 306 mi in highway driving on 9 gal of gas. Find the amount of gas required for 425 mi of highway driving.

Source: *Car and Driver,* January 2013

1. **Familiarize.** We know that the Fiat can travel 306 mi on 9 gal of gas. Thus we can set up a proportion, letting $x =$ the number of gallons of gas required to drive 425 mi.

2. **Translate.** We assume that the car uses gas at the same rate in all highway driving. Thus the ratios are the same and we can write a proportion. Note that the units of *mileage* are in the numerators and the units of *gasoline* are in the denominators.

$$\text{Miles} \rightarrow \frac{306}{9} = \frac{425}{x} \leftarrow \text{Miles}$$
$$\text{Gas} \rightarrow \qquad\qquad \leftarrow \text{Gas}$$

3. **Solve.** To solve for x, we multiply on both sides by the LCM, which is $9x$:

$$9x \cdot \frac{306}{9} = 9x \cdot \frac{425}{x} \qquad \text{Multiplying by } 9x$$

$$306x = 3825 \qquad \text{Simplifying}$$

$$\frac{306x}{306} = \frac{3825}{306} \qquad \text{Dividing by 306}$$

$$x = 12.5. \qquad \text{Simplifying}$$

We can also use cross products to solve the proportion:

$$\frac{306}{9} = \frac{425}{x} \qquad 306 \cdot x \text{ and } 9 \cdot 425 \text{ are cross products.}$$

$$306 \cdot x = 9 \cdot 425 \qquad \text{Equating cross products}$$

$$\frac{306x}{306} = \frac{3825}{306} \qquad \text{Dividing by 306}$$

$$x = 12.5.$$

4. **Check.** The check is left to the student.

5. **State.** The Fiat will require 12.5 gal of gas for 425 mi of highway driving.

Do Exercise 7. ▶

EXAMPLE 4 *Fruit Quality.* A company that prepares and sells gift boxes and baskets of fruit must order quantities of fruit larger than what they need to allow for selecting fruit that meets their quality standards. The packing-room supervisor keeps records and notes that approximately 87 pears from a shipment of 1000 do not meet the company standards. Over the holidays, a shipment of 3200 pears is ordered. How many pears can the company expect will not meet the quality required?

7. *Mileage.* In city driving, a 2013 Ford Mustang GT can travel 105 mi on 7 gal of gas. How much gas will be required for 243 mi of city driving?

Source: *Motor Trend,* March 2013

Answer

7. 16.2 gal

1. **Familiarize.** The ratio of the number of pears P that do not meet the standards to the total order of 3200 is $P/3200$. The ratio of the average number of pears that do not meet the standard in an order of 1000 pears is $\frac{87}{1000}$.

2. **Translate.** Assuming that the two ratios are the same, we can translate to a proportion:

$$\frac{P}{3200} = \frac{87}{1000}.$$

3. **Solve.** We solve the proportion. We multiply by the LCM, which is 16,000.

$$16{,}000 \cdot \frac{P}{3200} = 16{,}000 \cdot \frac{87}{1000}$$

$$5 \cdot P = 16 \cdot 87$$

$$P = \frac{16 \cdot 87}{5}$$

$$P \approx 278.4$$

4. **Check.** The check is left to the student.

5. **State.** We estimate that in an order of 3200 pears, there are about 278 pears that do not meet the quality standards.

◀ **Do Exercise 8.**

Similar Triangles

Proportions arise in geometry when we are studying *similar triangles*. If two triangles are **similar**, then their corresponding angles have the same measure and their corresponding sides are proportional. To illustrate, if triangle *ABC* is similar to triangle *RST*, then angles *A* and *R* have the same measure, angles *B* and *S* have the same measure, angles *C* and *T* have the same measure, and

$$\frac{a}{r} = \frac{b}{s} = \frac{c}{t}.$$

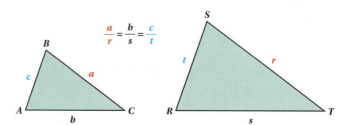

8. *Chlorine for a Pool.* XYZ Pools and Spas, Inc., adds 2 gal of chlorine per 8000 gal of water in a newly constructed pool. How much chlorine is needed for a pool requiring 20,500 gal of water? Round the answer to the nearest tenth of a gallon.

Answer

8. 5.1 gal

SIMILAR TRIANGLES

In **similar triangles**, corresponding angles have the same measure and the lengths of corresponding sides are proportional.

EXAMPLE 5 *Similar Triangles.* Triangles *ABC* and *XYZ* below are similar triangles. Solve for *z* if $a = 8$, $c = 5$, and $x = 10$.

We make a drawing, write a proportion, and then solve. Note that side *a* is always opposite angle *A*, side *x* is always opposite angle *X*, and so on.

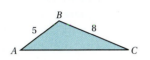 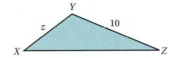

We have

$$\frac{z}{5} = \frac{10}{8}$$ The proportion $\frac{5}{z} = \frac{8}{10}$ could also be used.

$$40 \cdot \frac{z}{5} = 40 \cdot \frac{10}{8}$$ Multiplying by 40

$$8z = 50$$

$$z = \frac{50}{8}$$ Dividing by 8

$$z = \frac{25}{4}, \text{ or } 6.25.$$

Do Exercise 9. ▶

EXAMPLE 6 *Rafters of a House.* Carpenters use similar triangles to determine the lengths of rafters for a house. They first choose the pitch of the roof, or the ratio of the rise over the run. Then using a triangle with that ratio, they calculate the length of the rafter needed for the house. Loren is constructing rafters for a roof with a 6/12 pitch on a house that is 30 ft wide. Using a rafter guide (see the figure at right), Loren knows that the rafter length corresponding to a 6-unit rise and a 12-unit run is 13.4. Find the length *x* of the rafter of the house.

We have the proportion

Length of rafter in 6/12 triangle → $\frac{13.4}{x} = \frac{12}{15}$ ← Run in 6/12 triangle

Length of rafter on the house → ← Run in similar triangle on the house

Solve: $13.4 \cdot 15 = x \cdot 12$ Equating cross products

$$\frac{13.4 \cdot 15}{12} = \frac{x \cdot 12}{12}$$ Dividing by 12 on both sides

$$\frac{13.4 \cdot 15}{12} = x$$

$$16.75 = x$$

The length of the rafter *x* of the house is about 16.75 ft, or 16 ft 9 in.

Do Exercise 10. ▶

9. *Height of a Flagpole.* How high is a flagpole that casts a 45-ft shadow at the same time that a 5.5-ft woman casts a 10-ft shadow?

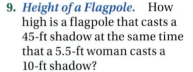

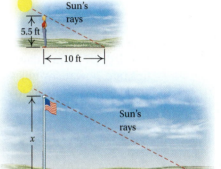

10. *Rafters of a House.* Refer to Example 6. Find the length *y* in the rafter of the house.

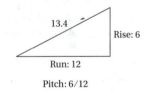

Answers

9. 24.75 ft   **10.** 7.5 ft

SECTION 6.8 Applications Using Rational Equations and Proportions   **523**

Translating for Success

1. **Pharmaceutical Research.** In 2013, a pharmaceutical firm spent $3.6 million on marketing a new drug. This was a 25% increase over the amount spent for marketing in 2012. How much was spent in 2012?

2. **Cycling Distance.** A bicyclist traveled 197 mi in 7 days. At this rate, how many miles could the cyclist travel in 30 days?

3. **Bicycling.** The speed of one bicyclist is 2 km/h faster than the speed of another bicyclist. The first bicyclist travels 60 km in the same amount of time that it takes the second to travel 50 km. Find the speed of each bicyclist.

4. **Filling Time.** A swimming pool can be filled in 5 hr by hose A alone and in 6 hr by hose B alone. How long would it take to fill the tank if both hoses were working?

5. **Office Budget.** Emma has $36 budgeted for office stationery. Engraved stationery costs $20 for the first 25 sheets and $0.08 for each additional sheet. How many engraved sheets of stationery can Emma order and still stay within her budget?

The goal of these matching questions is to practice step (2), Translate, of the five-step problem-solving process. Translate each word problem to an equation and select a correct translation from equations A–O.

A. $2x + 2(x + 1) = 613$

B. $x^2 + (x + 1)^2 = 613$

C. $\dfrac{60}{x + 2} = \dfrac{50}{x}$

D. $20 + 0.08(x - 25) = 36$

E. $\dfrac{197}{7} = \dfrac{x}{30}$

F. $x + (x + 1) = 613$

G. $\dfrac{7}{197} = \dfrac{x}{30}$

H. $x^2 + (x + 2)^2 = 612$

I. $x^2 + (x + 1)^2 = 612$

J. $\dfrac{50}{x + 2} = \dfrac{60}{x}$

K. $x + 25\% \cdot x = 3.6$

L. $t + 5 = 7$

M. $x^2 + (x + 1)^2 = 452$

N. $\dfrac{1}{5} + \dfrac{1}{6} = \dfrac{1}{t}$

O. $x^2 + (x + 2)^2 = 452$

Answers on page A-21

6. **Sides of a Square.** If each side of a square is increased by 2 ft, the area of the original square plus the area of the enlarged square is 452 ft^2. Find the length of a side of the original square.

7. **Consecutive Integers.** The sum of two consecutive integers is 613. Find the integers.

8. **Sums of Squares.** The sum of the squares of two consecutive odd integers is 612. Find the integers.

9. **Sums of Squares.** The sum of the squares of two consecutive integers is 613. Find the integers.

10. **Rectangle Dimensions.** The length of a rectangle is 1 ft longer than its width. Find the dimensions of the rectangle such that the perimeter of the rectangle is 613 ft.

✓ Reading Check

Complete each statement with the appropriate word(s) from the column on the right.

RC1. If two triangles are similar, then their _____ angles have the _____ measures and their corresponding sides are _____ .

RC2. A ratio of two quantities is their _____ .

RC3. An equality of ratios, $\dfrac{A}{B} = \dfrac{C}{D}$, is called a(n) _____ .

RC4. Distance equals _____ times time.

RC5. Rate equals _____ divided by time.

same
product
distance
proportion
similar
different
quotient
rate
proportional
corresponding

a   Solve.

1. Car Speed. Rick drives his four-wheel-drive truck 40 km/h faster than Sarah drives her Kia. While Sarah travels 150 km, Rick travels 350 km. Find their speeds.

Complete this table as part of the *Familiarize* step.

d	$=$	r	$\cdot$	t
	DISTANCE	SPEED	TIME
Car	150	r	
Truck	350		t

2. Train Speed. The speed of a CSW freight train is 14 mph slower than the speed of an Amtrak passenger train. The freight train travels 330 mi in the same time that it takes the passenger train to travel 400 mi. Find the speed of each train.

Complete this table as part of the *Familiarize* step.

d	$=$	r	$\cdot$	t
	DISTANCE	SPEED	TIME
CSW	330		t
Amtrak	400	r	

3. Animal Speeds. An ostrich can run 8 mph faster than a giraffe. An ostrich can run 5 mi in the same time that a giraffe can run 4 mi. Find the speed of each animal.

Source: www.infoplease.com

4. Animal Speeds. A cheetah can run 28 mph faster than a gray fox. A cheetah can run 10 mi in the same time that a gray fox can run 6 mi. Find the speed of each animal.

Source: www.infoplease.com

5. Bicycle Speed. Hank bicycles 5 km/h slower than Kelly. In the time that it takes Hank to bicycle 42 km, Kelly can bicycle 57 km. How fast does each bicyclist travel?

6. Driving Speed. Kaylee's Lexus travels 30 mph faster than Gavin's Harley. In the same time that Gavin travels 75 mi, Kaylee travels 120 mi. Find their speeds.

7. *Trucking Speed.* A long-distance trucker traveled 120 mi in one direction during a snowstorm. The return trip in rainy weather was accomplished at double the speed and took 3 hr less time. Find the speed going.

120 mi, r, t

120 mi, $2r$, $t - 3$

8. *Car Speed.* After driving 126 mi, Syd found that the drive would have taken 1 hr less time by increasing the speed by 8 mph. What was the actual speed?

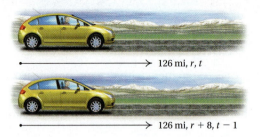

126 mi, r, t

126 mi, $r + 8$, $t - 1$

9. *Walking Speed.* Bonnie power walks 3 km/h faster than Ralph. In the time that it takes Ralph to walk 7.5 km, Bonnie walks 12 km. Find their speeds.

10. *Cross-Country Skiing.* Gerard skis cross-country 4 km/h faster than Sally. In the time that it takes Sally to ski 18 km, Gerard skis 24 km. Find their speeds.

11. *Boat Speed.* Tory and Emilio's motorboats travel at the same speed. Tory pilots her boat 40 km before docking. Emilio continues for another 2 hr, traveling a total of 100 km before docking. How long did it take Tory to navigate the 40 km?

12. *Tractor Speed.* Hobart's tractor is just as fast as Evan's. It takes Hobart 1 hr more than it takes Evan to drive to town. If Hobart is 20 mi from town and Evan is 15 mi from town, how long does it take Evan to drive to town?

13. *Gardening.* Nicole can weed her vegetable garden in 50 min. Glen can weed the same garden in 40 min. How long would it take if they worked together?

14. *Harvesting.* Bobbi can pick a quart of raspberries in 20 min. Blanche can pick a quart in 25 min. How long would it take if Bobbi and Blanche worked together?

15. *Shoveling.* Vern can shovel the snow from his driveway in 45 min. Nina can do the same job in 60 min. How long would it take Nina and Vern to shovel the driveway if they worked together?

16. *Raking.* Zoë can rake her yard in 4 hr. Steffi does the same job in 3 hr. How long would it take them, working together, to rake the yard?

17. Deli Trays. A grocery needs to prepare a large order of deli trays for Super Bowl weekend. It would take Henry 8.5 hr to prepare the trays. Carly can complete the job in 10.4 hr. How long would it take them, working together, to prepare the trays? Round the time to the nearest tenth of an hour.

18. School Photos. Rebecca can take photos for an elementary school with 325 students in 11.5 days. Jack can complete the same job in 9.2 days. How long would it take them working together? Round the time to the nearest tenth of a day.

19. Wiring. By checking work records, a contractor finds that Peggyann can wire a home theater in 9 hr. It takes Matthew 7 hr to wire the same room. How long would it take if they worked together?

20. Plumbing. By checking work records, a plumber finds that Raul can plumb a house in 48 hr. Mira can do the same job in 36 hr. How long would it take if they worked together?

21. Office Printers. The HP Officejet 4215 All-In-One printer, fax, scanner, and copier can print one black-and-white copy of a company's year-end report in 10 min. The HP Officejet 7410 All-In-One can print the same report in 6 min. How long would it take the two printers, working together, to print one copy of the report?

22. Office Copiers. The HP Officejet 7410 All-In-One printer, fax, scanner, and copier can make a color copy of a staff training manual in 9 min. The HP Officejet 4215 All-In-One can copy the same report in 15 min. How long would it take the two copiers, working together, to make one copy of the manual?

b Find the ratio of each of the following. Simplify, if possible.

23. 60 students, 18 teachers

24. 800 mi, 50 gal

25. Speed of Black Racer. A black racer snake travels 4.6 km in 2 hr. What is the speed, in kilometers per hour?

26. Speed of Light. Light travels 558,000 mi in 3 sec. What is the speed, in miles per second?

Solve.

27. *Protein Needs.* A 120-lb person should eat a minimum of 44 g of protein each day. How much protein should a 180-lb person eat each day?

28. *Coffee Beans.* The coffee beans from 14 trees are required to produce 7.7 kg of coffee. (This is the amount that the average person in the United States drinks each year.) How many trees are required to produce 320 kg of coffee?

29. *Hemoglobin.* A normal 10-cc specimen of human blood contains 1.2 g of hemoglobin. How much hemoglobin would 16 cc of the same blood contain?

30. *Walking Speed.* Wanda walked 234 km in 14 days. At this rate, how far would she walk in 42 days?

31. *Mileage.* A 2014 Subaru Forester can travel 242 mi of city driving on 11 gal of gas. Find the amount of gas required for 176 mi of city driving.

Source: *Car and Driver,* January 2013

32. *Mileage.* A 2013 RAM 1500 SLT crew cab can travel 575 mi of highway driving on 23 gal of gas. Find the amount of gas required for 850 mi of highway driving.

Source: *Car and Driver,* February 2013

33. *Estimating Trout Population.* To determine the number of trout in a lake, a conservationist catches 112 trout, tags them, and throws them back into the lake. Later, 82 trout are caught; 32 of them are tagged. Estimate the number of trout in the lake.

34. *Grass Seed.* It takes 60 oz of grass seed to seed 3000 ft^2 of lawn. At this rate, how much would be needed to seed 5000 ft^2 of lawn?

35. *Quality Control.* A sample of 144 firecrackers contained 9 "duds." How many duds would you expect in a sample of 3200 firecrackers?

36. *Frog Population.* To estimate how many frogs there are in a rain forest, a research team tags 600 frogs and then releases them. Later, the team catches 300 frogs and notes that 25 of them have been tagged. Estimate the total frog population in the rain forest.

37. *Mishandled Baggage.* The rate of the number of bags mishandled by the airlines fell to 3.39 bags per 1000 passengers in 2011. At this rate, estimate how many bags would be mishandled for 1450 passengers. Round the answer to the nearest hundredth.

Source: U.S. Department of Transportation, Bureau of Transportation Statistics

38. *Left-Handed People.* Research shows that 3 in 20 people are left-handed. At a recent NFL game, 63,240 people were in attendance. How many are left-handed?

Source: *Scientific American*

39. *Honey Bees.* Making 1 lb of honey requires 20,000 trips by bees to flowers to gather nectar. How many pounds of honey would 35,000 trips produce?

Source: Tom Turpin, Professor of Entomology, Purdue University

40. *Money.* The ratio of the weight of copper to the weight of zinc in a U.S. penny is $\frac{1}{39}$. If 50 kg of zinc is being turned into pennies, how much copper is needed?

Geometry. For each pair of similar triangles, find the length of the indicated side.

41. *b*:

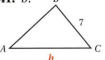

42. *a*:

43. *f*:

44. *r*:

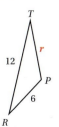

45. *h*:

46. *n*:

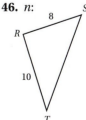

47. Environmental Science. The Fish and Wildlife Division of the Indiana Department of Natural Resources recently completed a study that determined the number of largemouth bass in Lake Monroe, near Bloomington, Indiana. For this project, anglers caught 300 largemouth bass, tagged them, and threw them back into the lake. Later, they caught 85 largemouth bass and found that 15 of them were tagged. Estimate how many largemouth bass are in the lake.

Source: Department of Natural Resources, Fish and Wildlife Division, Kevin Hoffman

Lake Monroe

48. Environmental Science. To determine the number of humpback whales in a pod, a marine biologist, using tail markings, identifies 27 members of the pod. Several weeks later, 40 whales from the pod are randomly sighted. Of the 40 sighted, 12 are from the 27 originally identified. Estimate the number of whales in the pod.

Skill Maintenance

Find the slope, if it exists, of the line containing the given pair of points. [3.4a]

49. $(7, -6), (0, -6)$

50. $(3, -11), (-4, 3)$

Simplify. [4.1d, f]

51. $x^5 \cdot x^6$

52. $x^{-5} \cdot x^6$

53. $x^{-5} \cdot x^{-6}$

54. $x^5 \cdot x^{-6}$

Graph.

55. $y = -\dfrac{3}{4}x + 2$ [3.2a]

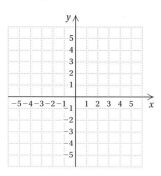

56. $y = \dfrac{2}{5}x - 4$ [3.2a]

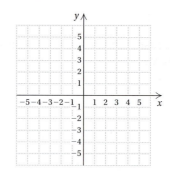

57. $x = -3$ [3.2b]

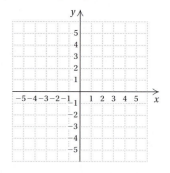

Synthesis

58. Rachel allows herself 1 hr to reach a sales appointment 50 mi away. After she has driven 30 mi, she realizes that she must increase her speed by 15 mph in order to arrive on time. What was her speed for the first 30 mi?

59. How soon, in minutes, after 5 o'clock will the hands on a clock first be together?

Direct Variation and Inverse Variation

a EQUATIONS OF DIRECT VARIATION

OBJECTIVES

a Find an equation of direct variation given a pair of values of the variables.

b Solve applied problems involving direct variation.

c Find an equation of inverse variation given a pair of values of the variables.

d Solve applied problems involving inverse variation.

A bicycle is traveling at a speed of 15 km/h. In 1 hr, it goes 15 km; in 2 hr, it goes 30 km; in 3 hr, it goes 45 km; and so on. We can form a set of ordered pairs using the number of hours as the first coordinate and the number of kilometers traveled as the second coordinate. These determine a set of ordered pairs:

$$(1, 15), \quad (2, 30), \quad (3, 45), \quad (4, 60), \quad \text{and so on.}$$

Note that the second coordinate is always 15 times the first.

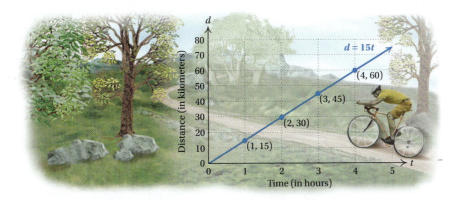

In this example, distance is a constant multiple of time, so we say that there is *direct variation* and that distance *varies directly* as time. The *equation of variation* is $d = 15t$.

DIRECT VARIATION

When a situation translates to an equation described by $y = kx$, with k a positive constant, we say that **y varies directly as x**. The equation $y = kx$ is called an **equation of direct variation**.

In direct variation, as one variable increases, the other variable increases as well. This is shown in the graph above.

The terminologies

"y varies as x,"

"y is directly proportional to x," and

"y is proportional to x"

also imply direct variation and are used in many situations. The constant k is called the **constant of proportionality**, or the **variation constant**. It can be found if one pair of values of x and y is known. Once k is known, other pairs can be determined.

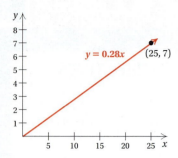

$y = 0.28x$

$(25, 7)$

EXAMPLE 1 Find an equation of variation in which y varies directly as x, and $y = 7$ when $x = 25$.

We first substitute to find k:

$$y = kx$$

$$7 = k \cdot 25 \qquad \text{Substituting 25 for } x \text{ and 7 for } y$$

$$\frac{7}{25} = k, \quad \text{or} \quad k = 0.28. \qquad \text{Solving for } k, \text{ the variation constant}$$

Then the equation of variation is

$$y = 0.28x.$$

The answer is the equation $y = 0.28x$, *not* simply $k = 0.28$. We can visualize the example by looking at the graph at left. ▪

We see that when y varies directly as x, the constant of proportionality is also the slope of the associated graph—the rate at which y changes with respect to x.

EXAMPLE 2 Find an equation in which s varies directly as t, and $s = 10$ when $t = 15$. Then find the value of s when $t = 32$.

We have

$$s = kt \qquad \text{We know that } s \text{ varies directly as } t.$$

$$10 = k \cdot 15 \qquad \text{Substituting 10 for } s \text{ and 15 for } t$$

$$\tfrac{10}{15} = k, \quad \text{or} \quad k = \tfrac{2}{3}. \qquad \text{Solving for } k$$

Thus the equation of variation is $s = \frac{2}{3}t$.

$$s = \tfrac{2}{3}t$$

$$= \tfrac{2}{3} \cdot 32 \qquad \text{Substituting 32 for } t$$

$$= \tfrac{64}{3}, \text{ or } 21\tfrac{1}{3}$$

The value of s is $21\frac{1}{3}$ when $t = 32$.

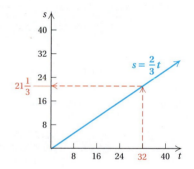

1. Find an equation of variation in which y varies directly as x, and $y = 84$ when $x = 12$.

$$y = kx$$

$$84 = k \cdot \boxed{}$$

$$k = \frac{\boxed{}}{12} = \boxed{}$$

$$y = 7 \cdot \boxed{}$$

$$y = 7x = 7 \cdot \boxed{}$$

$$= \boxed{}$$

2. Find an equation of variation in which y varies directly as x, and $y = 50$ when $x = 80$. Then find the value of y when $x = 20$.

◀ **Do Exercises 1 and 2.**

b APPLICATIONS OF DIRECT VARIATION

EXAMPLE 3 *Oatmeal Servings.* The number of servings S of oatmeal varies directly as the net weight W of the container purchased. A 42-oz box of oatmeal contains 30 servings. How many servings does a 63-oz box of oatmeal contain?

1., 2. Familiarize and **Translate.** The problem states that we have direct variation between the variables, S and W. Thus an equation $S = kW$, $k > 0$, applies. As the weight of the container increases, the number of servings increases.

Answers

1. $y = 7x$; 287   **2.** $y = \dfrac{5}{8}x$; $\dfrac{25}{2}$

Guided Solution:
1. 12, 84, 7, x; 41, 287

3. Solve. The mathematical manipulation has two parts. First, we determine the equation of variation by substituting known values for S and W to find the variation constant k. Second, we compute the number of servings in a 63-oz box of oatmeal.

a) First, we find an equation of variation:

$$S = kW$$

$$\color{red}{30} = k(\color{red}{42}) \qquad \text{Substituting 30 for } S \text{ and 42 for } W$$

$$\frac{30}{42} = k$$

$$\frac{5}{7} = k. \qquad \text{Simplifying}$$

The equation of variation is $S = \dfrac{5}{7}W$.

b) We then use the equation to find the number of servings in a 63-oz box of oatmeal:

$$S = \frac{5}{7}W$$

$$= \frac{5}{7} \cdot 63 \qquad \text{Substituting 63 for } W$$

$$= 45.$$

4. Check. The check might be done by repeating the computations. You might also do some reasoning about the answer. The number of servings increased from 30 to 45. Similarly, the weight increased from 42 oz to 63 oz. The answer seems reasonable.

5. State. A 63-oz box of oatmeal contains 45 servings.

Do Exercises 3 and 4. ▶

Let's consider direct variation from the standpoint of a graph. The graph of $y = kx$, $k > 0$, always goes through the origin and rises from left to right. Note that as x increases, y increases; and as x decreases, y decreases. This is why the terminology "direct" is used. What one variable does, the other does as well.

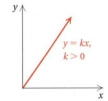

$y = kx,$
$k > 0$

c EQUATIONS OF INVERSE VARIATION

A car is traveling a distance of 20 mi. At a speed of 5 mph, it will take 4 hr; at 20 mph, it will take 1 hr; at 40 mph, it will take $\frac{1}{2}$ hr; and so on. We use the speed as the first coordinate and the time as the second coordinate. These determine a set of ordered pairs:

$$(5, 4), \quad (20, 1), \quad \left(40, \tfrac{1}{2}\right), \quad \left(60, \tfrac{1}{3}\right), \quad \text{and so on.}$$

Note that the product of speed and time for each of these pairs is 20. Note too that as the speed *increases,* the time *decreases.*

In this case, the product of speed and time is constant so we say that there is *inverse variation* and that time *varies inversely* as speed. The equation of variation is

$$rt = 20 \, \text{(a constant)}, \quad \text{or} \quad t = \frac{20}{r}.$$

3. Gold. The karat rating of a gold object varies directly as the percentage of gold in the object. A 14-karat gold chain is 58.25% gold. What is the percentage of gold in a 10-karat gold chain?

4. Weight on Venus. The weight V of an object on Venus varies directly as its weight E on Earth. A person weighing 165 lb on Earth would weigh 145.2 lb on Venus.

a) Find an equation of variation.

b) How much would a person weighing 198 lb on Earth weigh on Venus?

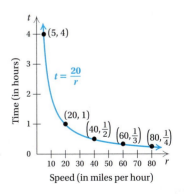

Answers

3. About 41.6%
4. **(a)** $V = 0.88E$; **(b)** 174.24 lb

INVERSE VARIATION

When a situation translates to an equation described by $y = k/x$, with k a positive constant, we say that **y varies inversely as x**. The equation $y = k/x$ is called an **equation of inverse variation**.

In inverse variation, as one variable increases, the other variable decreases.

The terminology

"y is inversely proportional to x"

also implies inverse variation and is used in some situations. The constant k is again called the **constant of proportionality**, or the **variation constant**.

EXAMPLE 4 Find an equation of variation in which y varies inversely as x, and $y = 145$ when $x = 0.8$. Then find the value of y when $x = 25$.

We first substitute to find k:

$$y = \frac{k}{x}$$

$$145 = \frac{k}{0.8}$$

$$(0.8)145 = k$$

$$116 = k.$$

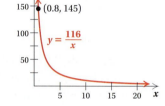

The equation of variation is $y = 116/x$. The answer is the equation $y = 116/x$, *not* simply $k = 116$.

When $x = 25$, we have

$$y = \frac{116}{x}$$

$$= \frac{116}{25} \qquad \text{Substituting 25 for } x$$

$$= 4.64.$$

The value of y is 4.64 when $x = 25$.

◀ **Do Exercises 5 and 6.**

The graph of $y = k/x$, $k > 0$, is shaped like the figure at right for positive values of x. (You need not know how to graph such equations at this time.) Note that as x increases, y decreases; and as x decreases, y increases. This is why the terminology "inverse" is used. One variable does the opposite of what the other does.

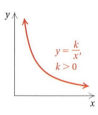

---

5. Find an equation of variation in which y varies inversely as x, and $y = 105$ when $x = 0.6$.

$$y = \frac{k}{x}$$

$$105 = \frac{k}{\boxed{}}$$

$$k = \boxed{} \cdot 105$$

$$k = \boxed{}$$

$$y = \frac{63}{\boxed{}}$$

$$y = \frac{63}{x} = \frac{63}{\boxed{}}$$

$$= \boxed{}$$

6. Find an equation of variation in which y varies inversely as x, and $y = 45$ when $x = 20$. Then find the value of y when $x = 1.6$.

Answers

5. $y = \dfrac{63}{x}$; 3.15   **6.** $y = \dfrac{900}{x}$; 562.5

Guided Solution:
5. 0.6, 0.6, 63, x; 20, 3.15

d APPLICATIONS OF INVERSE VARIATION

Often in an applied situation we must decide which kind of variation, if any, might apply to the problem.

EXAMPLE 5 *Trash Removal.* The day after the Indianapolis 500 race, local organizations are assigned the task of cleaning up the grandstands. It takes approximately 8 hr for 30 people to remove the trash from one grandstand. How long would it take 42 people to do the job?

Source: Indianapolis Motor Speedway

1. **Familiarize.** Think about the problem situation. What kind of variation would be used? It seems reasonable that the more people there are working on the job, the less time it will take to finish. Thus inverse variation might apply. We let $T = $ the time to do the job, in hours, and $N = $ the number of people. Assuming inverse variation, we know that an equation $T = k/N$, $k > 0$, applies. As the number of people increases, the time it takes to do the job decreases.

2. **Translate.** We write an equation of variation:

$$T = \frac{k}{N}.$$

Time varies inversely as the number of people involved.

3. **Solve.** The mathematical manipulation has two parts. First, we find the equation of variation by substituting known values for T and N to find k. Second, we compute the amount of time it would take 42 people to do the job.

   a) First, we find an equation of variation:

$$T = \frac{k}{N}$$

$$8 = \frac{k}{30} \qquad \text{Substituting 8 for } T \text{ and } 30 \text{ for } N$$

$$30 \cdot 8 = k$$

$$240 = k.$$

   The equation of variation is $T = \dfrac{240}{N}$.

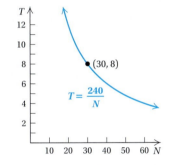

   b) We then use the equation to find the amount of time that it takes 42 people to do the job:

$$T = \frac{240}{N}$$

$$= \frac{240}{42} \qquad \text{Substituting 42 for } N$$

$$\approx 5.7.$$

4. **Check.** The check might be done by repeating the computations. We might also analyze the results. The number of people increased from 30 to 42. Did the time decrease? It did, and this confirms what we expect with inverse variation.

5. **State.** It should take 5.7 hr for 42 people to complete the job.

Do Exercises 7 and 8. ▶

7. Refer to Example 5. Determine how long it would take 25 people to do the job.

8. *Time of Travel.* The time t required to drive a fixed distance varies inversely as the speed r. It takes 5 hr at 60 km/h to drive a fixed distance.

   a) Find an equation of variation.

   b) How long would it take at 40 km/h?

Answers

7. 9.6 hr

8. **(a)** $t = \dfrac{300}{r}$; **(b)** 7.5 hr

✓ Reading Check

Match each description of variation with the appropriate equation of variation from the column on the right.

RC1. r varies directly as s.

RC2. x is inversely proportional to z.

RC3. m varies inversely as n.

RC4. n is directly proportional to m.

RC5. z varies directly as w.

RC6. r varies inversely as s.

a) $r = \dfrac{k}{s}$

b) $r = ks$

c) $z = kw$

d) $x = \dfrac{k}{z}$

e) $n = km$

f) $m = \dfrac{k}{n}$

Find an equation of variation in which y varies directly as x and the following are true. Then find the value of y when $x = 20$.

1. $y = 36$ when $x = 9$ **2.** $y = 60$ when $x = 16$ **3.** $y = 0.8$ when $x = 0.5$ **4.** $y = 0.7$ when $x = 0.4$

5. $y = 630$ when $x = 175$ **6.** $y = 400$ when $x = 125$ **7.** $y = 500$ when $x = 60$ **8.** $y = 200$ when $x = 300$

Solve.

9. *Wages and Work Time.* A person's paycheck P varies directly as the number H of hours worked. For working 15 hr, the pay is $180.

  a) Find an equation of variation.
  b) Find the pay for 35 hr of work.

10. *Interest and Interest Rate.* The interest I earned in 1 year on a fixed principal varies directly as the interest rate r. An investment earns $53.55 at an interest rate of 4.25%.

  a) Find an equation of variation.
  b) How much will the investment earn at a rate of 5.75%?

11. *Cost of Sand.* The cost C, in dollars, to fill a sandbox varies directly as the depth S, in inches, of the sand. The director of Creekside Daycare checks at her local hardware store and finds that it would cost $67.50 to fill the daycare's box with 6 in. of sand. She decides to fill the sandbox to a depth of 9 in.

  a) Find an equation of variation.
  b) How much will the sand cost?

12. *Cost of Cement.* The cost C, in dollars, of cement needed to pave a driveway varies directly as the depth D, in inches, of the driveway. John checks at his local building materials store and finds that it costs $1000 to install his driveway with a depth of 8 in. He decides to build a stronger driveway at a depth of 12 in.

  a) Find an equation of variation.
  b) How much will it cost for the cement?

13. *Lunar Weight.* The weight M of an object on the moon varies directly as its weight E on Earth. Jared weighs 192 lb, but would weigh only 32 lb on the moon.

   a) Find an equation of variation.

   b) Jared's wife, Elizabeth, weighs 110 lb on Earth. How much would she weigh on the moon?

   c) Jared's granddaughter, Jasmine, would weigh only 5 lb on the moon. How much does Jasmine weigh on Earth?

14. *Mars Weight.* The weight M of an object on Mars varies directly as its weight E on Earth. In 1999, Chen Yanqing, who weighs 128 lb, set a world record for her weight class with a lift (snatch) of 231 lb. On Mars, this lift would be only 88 lb.

Source: *The Guinness Book of Records*, 2001

   a) Find an equation of variation.

   b) How much would Yanqing weigh on Mars?

15. *Computer Megahertz.* The number of instructions N performed per second by a computer varies directly as the speed S of the computer's internal processor. A processor with a speed of 25 megahertz can perform 2,000,000 instructions per second.

   a) Find an equation of variation.

   b) How many instructions per second will the same processor perform if it is running at a speed of 200 megahertz?

16. *Water in Human Body.* The number of kilograms W of water in a human body varies directly as the total body weight B. A person who weighs 75 kg contains 54 kg of water.

   a) Find an equation of variation.

   b) How many kilograms of water are in a person who weighs 95 kg?

17. *Steak Servings.* The number of servings S of meat that can be obtained from round steak varies directly as the weight W. From 9 kg of round steak, one can get 70 servings of meat. How many servings can one get from 12 kg of round steak?

18. *Turkey Servings.* A chef is planning meals in a refreshment tent at a golf tournament. The number of servings S of meat that can be obtained from a turkey varies directly as its weight W. From a turkey weighing 30.8 lb, one can get 40 servings of meat. How many servings can be obtained from a 19.8-lb turkey?

C Find an equation of variation in which y varies inversely as x and the following are true. Then find the value of y when $x = 10$.

19. $y = 3$ when $x = 25$ **20.** $y = 2$ when $x = 45$ **21.** $y = 10$ when $x = 8$ **22.** $y = 10$ when $x = 7$

23. $y = 6.25$ when $x = 0.16$ **24.** $y = 0.125$ when $x = 8$ **25.** $y = 50$ when $x = 42$ **26.** $y = 25$ when $x = 42$

27. $y = 0.2$ when $x = 0.3$ **28.** $y = 0.4$ when $x = 0.6$

b , d Solve.

29. *Production and Time.* A production line produces 15 CD players every 8 hr. How many players can it produce in 37 hr?

a) What kind of variation might apply to this situation?
b) Solve the problem.

30. *Wages and Work Time.* A person works for 15 hr and makes $251.25. How much will the person make by working 35 hr?

a) What kind of variation might apply to this situation?
b) Solve the problem.

31. *Cooking Time.* It takes 4 hr for 9 cooks to prepare the food for a wedding rehearsal dinner. How long will it take 8 cooks to prepare the dinner?

a) What kind of variation might apply to this situation?
b) Solve the problem.

32. *Work Time.* It takes 16 hr for 2 people to resurface a tennis court. How long will it take 6 people to do the job?

a) What kind of variation might apply to this situation?
b) Solve the problem.

33. *Miles per Gallon.* To travel a fixed distance, the number of gallons N of gasoline needed is inversely proportional to the miles-per-gallon rating P of the car. A car that gets 20 miles per gallon (mpg) needs 14 gal to travel the distance.

a) Find an equation of variation.
b) How much gas will be needed for a car that gets 28 mpg?

34. *Miles per Gallon.* To travel a fixed distance, the number of gallons N of gasoline needed is inversely proportional to the miles-per-gallon rating P of the car. A car that gets 25 miles per gallon (mpg) needs 12 gal to travel the distance.

a) Find an equation of variation.
b) How much gas will be needed for a car that gets 20 mpg?

35. *Electrical Current.* The current I in an electrical conductor varies inversely as the resistance R of the conductor. The current is 96 amperes when the resistance is 20 ohms.

a) Find an equation of variation.
b) What is the current when the resistance is 60 ohms?

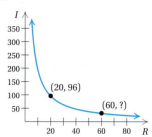

36. *Gas Volume.* The volume V of a gas varies inversely as the pressure P on it. The volume of a gas is 200 cm³ under a pressure of 32 kg/cm².

a) Find an equation of variation.
b) What will be its volume under a pressure of 20 kg/cm²?

37. *Answering Questions.* For a fixed time limit for a quiz, the number of minutes m that a student should allow for each question on a quiz (assuming they are of equal difficulty) is inversely proportional to the number of questions n on the quiz. For a given time limit on a 16-question quiz, students have 2.5 min per question.

   a) Find an equation of variation.

   b) How many questions would appear on a quiz in which students have the same time limit and have 4 min per question?

38. *Pumping Time.* The time t required to empty a tank varies inversely as the rate r of pumping. A pump can empty a tank in 90 min at a rate of 1200 L/min.

   a) Find an equation of variation.

   b) How long will it take the pump to empty the tank at a rate of 2000 L/min?

39. *Apparent Size.* The apparent size A of an object varies inversely as the distance d of the object from the eye. A flagpole 30 ft from an observer appears to be 27.5 ft tall. How tall will the same flagpole appear to be if it is 100 ft from the eye?

40. *Driving Time.* The time t required to drive a fixed distance varies inversely as the speed r. It takes 5 hr at 55 mph to drive a fixed distance. How long would it take at 40 mph?

Skill Maintenance

Solve.   [6.7a]

41. $\dfrac{x + 2}{x + 5} = \dfrac{x - 4}{x - 6}$

42. $\dfrac{x - 3}{x - 5} = \dfrac{x + 5}{x + 1}$

Calculate.   [1.8d]

43. $3^7 \div 3^4 \div 3^3 \div 3$

44. $\dfrac{37 - 5(4 - 6)}{2 \cdot 6 + 8}$

45. $-5^2 + 4 \cdot 6$

46. $(-5)^2 + 4 \cdot 6$

Synthesis

Write an equation of variation for each situation.

47. The square of the pitch P of a vibrating string varies directly as the tension t on the string.

48. In a stream, the amount S of salt carried varies directly as the sixth power of the speed V of the stream.

49. The power P in a windmill varies directly as the cube of the wind speed V.

50. The volume V of a sphere varies directly as the cube of the radius r.

Vocabulary Reinforcement

Complete each statement with the correct term from the column on the right.

reciprocals
proportion
rational
equivalent
directly
complex
direct variation
similar
inversely
opposites
inverse variation

1. A _____ rational expression is a rational expression that has one or more rational expressions within its numerator or its denominator. [6.6a]

2. An equality of ratios, $\dfrac{A}{B} = \dfrac{C}{D}$, is called a(n) _____. [6.8b]

3. Two expressions are _____ of each other if their product is 1. [6.2a]

4. Expressions that have the same value for all allowable replacements are called _____ expressions. [6.1b]

5. Expressions of the form $a - b$ and $b - a$ are _____ of each other. [6.1c]

6. In _____ triangles, corresponding angles have the same measure and the lengths of corresponding sides are proportional. [6.8b]

7. When a situation translates to an equation described by $y = \dfrac{k}{x}$, with k a positive constant, we say that y varies _____ as x. The equation $y = \dfrac{k}{x}$ is called an equation of _____. [6.9c]

8. When a situation translates to an equation described by $y = kx$, with k a positive constant, we say that y varies _____ as x. The equation $y = kx$ is called an equation of _____. [6.9a]

Concept Reinforcement

Determine whether each statement is true or false.

_____ 1. To determine the numbers for which a rational expression is not defined, we set the denominator equal to 0 and solve. [6.1a]

_____ 2. The expressions $y + 5$ and $y - 5$ are opposites of each other. [6.1c]

_____ 3. The opposite of $2 - x$ is $x - 2$. [6.1c]

Study Guide

Objective 6.1a Find all numbers for which a rational expression is not defined.

Example Find all numbers for which the rational expression $\dfrac{2 - y}{y^2 + 3y - 28}$ is not defined.

$$y^2 + 3y - 28 = 0$$
$$(y + 7)(y - 4) = 0$$
$$y + 7 = 0 \quad or \quad y - 4 = 0$$
$$y = -7 \quad or \quad y = 4$$

The rational expression is not defined for -7 and 4.

Practice Exercise

1. Find all numbers for which the rational expression $\dfrac{c + 8}{c^2 - 11c + 30}$ is not defined.

Objective 6.1c Simplify rational expressions by factoring the numerator and the denominator and removing factors of 1.

Example Simplify: $\dfrac{6y - 12}{2y^2 + y - 10}$.

$$\dfrac{6y - 12}{2y^2 + y - 10} = \dfrac{6(y - 2)}{(2y + 5)(y - 2)}$$

$$= \dfrac{y - 2}{y - 2} \cdot \dfrac{6}{2y + 5}$$

$$= 1 \cdot \dfrac{6}{2y + 5} = \dfrac{6}{2y + 5}$$

Practice Exercise

2. Simplify:
$$\dfrac{2x^2 - 2}{4x^2 + 24x + 20}.$$

Objective 6.1d Multiply rational expressions and simplify.

Example Multiply and simplify:
$$\dfrac{x^2 + 14x + 49}{x^2 - 25} \cdot \dfrac{x + 5}{x + 7}.$$

$$\dfrac{x^2 + 14x + 49}{x^2 - 25} \cdot \dfrac{x + 5}{x + 7} = \dfrac{(x^2 + 14x + 49)(x + 5)}{(x^2 - 25)(x + 7)}$$

$$= \dfrac{(x + 7)(x + 7)(x + 5)}{(x + 5)(x - 5)(x + 7)}$$

$$= \dfrac{x + 7}{x - 5}$$

Practice Exercise

3. Multiply and simplify:
$$\dfrac{2y^2 + 7y - 15}{5y^2 - 45} \cdot \dfrac{y - 3}{2y - 3}.$$

Objective 6.2b Divide rational expressions and simplify.

Example Divide and simplify: $\dfrac{a^2 - 9a}{a^2 - a - 6} \div \dfrac{a}{a + 2}$.

$$\dfrac{a^2 - 9a}{a^2 - a - 6} \div \dfrac{a}{a + 2} = \dfrac{a^2 - 9a}{a^2 - a - 6} \cdot \dfrac{a + 2}{a}$$

$$= \dfrac{(a^2 - 9a)(a + 2)}{(a^2 - a - 6)a}$$

$$= \dfrac{a(a - 9)(a + 2)}{(a + 2)(a - 3)a}$$

$$= \dfrac{a - 9}{a - 3}$$

Practice Exercise

4. Divide and simplify:
$$\dfrac{b^2 + 3b - 28}{b^2 + 5b - 24} \div \dfrac{b - 4}{b - 3}.$$

Objective 6.3b Add fractions, first finding the LCD.

Example Add: $\dfrac{13}{30} + \dfrac{11}{24}$.

$$\left.\begin{array}{l} 30 = 2 \cdot 3 \cdot 5 \\ 24 = 2 \cdot 2 \cdot 2 \cdot 3 \end{array}\right\} \quad \text{LCD} = 2 \cdot 2 \cdot 2 \cdot 3 \cdot 5, \text{ or } 120$$

$$\dfrac{13}{30} + \dfrac{11}{24} = \dfrac{13}{2 \cdot 3 \cdot 5} \cdot \dfrac{2 \cdot 2}{2 \cdot 2} + \dfrac{11}{2 \cdot 2 \cdot 2 \cdot 3} \cdot \dfrac{5}{5}$$

$$= \dfrac{52 + 55}{2 \cdot 2 \cdot 2 \cdot 3 \cdot 5} = \dfrac{107}{120}$$

Practice Exercise

5. Add: $\dfrac{5}{18} + \dfrac{7}{60}$.

Objective 6.3c Find the LCM of algebraic expressions by factoring.

Example Find the LCM of
$$x^2 - 36 \quad \text{and} \quad x^2 - 5x - 6.$$
$$x^2 - 36 = (x + 6)(x - 6)$$
$$x^2 - 5x - 6 = (x - 6)(x + 1)$$
$$\text{LCM} = (x + 6)(x - 6)(x + 1)$$

Practice Exercise

6. Find the LCM of
$$x^2 - 7x - 18 \quad \text{and} \quad x^2 - 81.$$

Objective 6.4a Add rational expressions.

Example Add and simplify: $\dfrac{6x - 5}{x - 1} + \dfrac{x}{1 - x}$.

$$\frac{6x - 5}{x - 1} + \frac{x}{1 - x} = \frac{6x - 5}{x - 1} + \frac{x}{1 - x} \cdot \frac{-1}{-1}$$
$$= \frac{6x - 5}{x - 1} + \frac{-x}{x - 1}$$
$$= \frac{6x - 5 - x}{x - 1}$$
$$= \frac{5x - 5}{x - 1}$$
$$= \frac{5(x - 1)}{x - 1} = 5$$

Practice Exercise

7. Add and simplify:
$$\frac{x}{x - 4} + \frac{2x - 4}{4 - x}.$$

Objective 6.5a Subtract rational expressions.

Example Subtract: $\dfrac{3}{x^2 - 1} - \dfrac{2x - 1}{x^2 + x - 2}$.

$$\frac{3}{x^2 - 1} - \frac{2x - 1}{x^2 + x - 2}$$
$$= \frac{3}{(x + 1)(x - 1)} - \frac{2x - 1}{(x + 2)(x - 1)}$$

The LCM is $(x + 1)(x - 1)(x + 2)$.

$$= \frac{3}{(x + 1)(x - 1)} \cdot \frac{x + 2}{x + 2} - \frac{2x - 1}{(x + 2)(x - 1)} \cdot \frac{x + 1}{x + 1}$$
$$= \frac{3(x + 2)}{(x + 1)(x - 1)(x + 2)} - \frac{(2x - 1)(x + 1)}{(x + 2)(x - 1)(x + 1)}$$
$$= \frac{3x + 6 - (2x^2 + x - 1)}{(x + 1)(x - 1)(x + 2)}$$
$$= \frac{3x + 6 - 2x^2 - x + 1}{(x + 1)(x - 1)(x + 2)}$$
$$= \frac{-2x^2 + 2x + 7}{(x + 1)(x - 1)(x + 2)}$$

Practice Exercise

8. Subtract:
$$\frac{x}{x^2 + x - 2} - \frac{5}{x^2 - 1}.$$

Objective 6.6a Simplify complex rational expressions.

Example Simplify $\dfrac{\dfrac{1}{3} - \dfrac{1}{x}}{\dfrac{1}{x} - \dfrac{1}{2}}$ using method 1.

The LCM of 3, x, and 2 is $6x$.

$$\frac{\dfrac{1}{3} - \dfrac{1}{x}}{\dfrac{1}{x} - \dfrac{1}{2}} = \frac{\dfrac{1}{3} - \dfrac{1}{x}}{\dfrac{1}{x} - \dfrac{1}{2}} \cdot \frac{6x}{6x} = \frac{\dfrac{1}{3} \cdot 6x - \dfrac{1}{x} \cdot 6x}{\dfrac{1}{x} \cdot 6x - \dfrac{1}{2} \cdot 6x}$$

$$= \frac{2x - 6}{6 - 3x} = \frac{2(x - 3)}{3(2 - x)}$$

Practice Exercise

9. Simplify: $\dfrac{\dfrac{2}{5} - \dfrac{1}{y}}{\dfrac{3}{y} - \dfrac{1}{3}}$.

Objective 6.7a Solve rational equations.

Example Solve: $12 = \dfrac{1}{5x} + \dfrac{4}{x}$.

The LCM of the denominators is $5x$. We multiply by $5x$ on both sides.

$$12 = \frac{1}{5x} + \frac{4}{x}$$

$$5x \cdot 12 = 5x\left(\frac{1}{5x} + \frac{4}{x}\right)$$

$$5x \cdot 12 = 5x \cdot \frac{1}{5x} + 5x \cdot \frac{4}{x}$$

$$60x = 1 + 20$$

$$60x = 21$$

$$x = \frac{21}{60} = \frac{7}{20}$$

This checks, so the solution is $\dfrac{7}{20}$.

Practice Exercise

10. Solve: $\dfrac{1}{x} = \dfrac{2}{3 - x}$.

Objective 6.9a Find an equation of direct variation given a pair of values of the variables.

Example Find an equation of variation in which y varies directly as x, and $y = 30$ when $x = 200$. Then find the value of y when $x = \frac{1}{2}$.

$$y = kx \qquad \text{Direct variation}$$

$$30 = k \cdot 200 \qquad \text{Substituting 30 for } y \text{ and 200 for } x$$

$$\frac{30}{200} = k, \text{ or } k = \frac{3}{20}$$

The equation of variation is $y = \frac{3}{20}x$.

Next, we substitute $\frac{1}{2}$ for x in $y = \frac{3}{20}x$ and solve for y:

$$y = \frac{3}{20}x = \frac{3}{20} \cdot \frac{1}{2} = \frac{3}{40}.$$

When $x = \dfrac{1}{2}$, $y = \dfrac{3}{40}$.

Practice Exercise

11. Find an equation of variation in which y varies directly as x, and $y = 60$ when $x = 0.4$. Then find the value of y when $x = 2$.

Objective 6.9c   Find an equation of inverse variation given a pair of values of the variables.

Example   Find an equation of variation in which y varies inversely as x, and $y = 0.5$ when $x = 20$. Then find the value of y when $x = 6$.

$$y = \frac{k}{x} \qquad \text{Inverse variation}$$

$$0.5 = \frac{k}{20} \qquad \text{Substituting 0.5 for } y \text{ and 20 for } x$$

$$10 = k$$

The equation of variation is $y = \dfrac{10}{x}$.

Next, we substitute 6 for x in $y = 10/x$ and solve for y:

$$y = \frac{10}{x} = \frac{10}{6} = \frac{5}{3}.$$

When $x = 6$, $y = \dfrac{5}{3}$.

Practice Exercise

12. Find an equation of variation in which y varies inversely as x, and $y = 150$ when $x = 1.5$. Then find the value of y when $x = 10$.

Review Exercises

Find all numbers for which the rational expression is not defined.   [6.1a]

1. $\dfrac{3}{x}$

2. $\dfrac{4}{x - 6}$

3. $\dfrac{x + 5}{x^2 - 36}$

4. $\dfrac{x^2 - 3x + 2}{x^2 + x - 30}$

5. $\dfrac{-4}{(x + 2)^2}$

6. $\dfrac{x - 5}{5}$

Simplify.   [6.1c]

7. $\dfrac{4x^2 - 8x}{4x^2 + 4x}$

8. $\dfrac{14x^2 - x - 3}{2x^2 - 7x + 3}$

9. $\dfrac{(y - 5)^2}{y^2 - 25}$

Multiply and simplify.   [6.1d]

10. $\dfrac{a^2 - 36}{10a} \cdot \dfrac{2a}{a + 6}$

11. $\dfrac{6t - 6}{2t^2 + t - 1} \cdot \dfrac{t^2 - 1}{t^2 - 2t + 1}$

Divide and simplify.   [6.2b]

12. $\dfrac{10 - 5t}{3} \div \dfrac{t - 2}{12t}$

13. $\dfrac{4x^4}{x^2 - 1} \div \dfrac{2x^3}{x^2 - 2x + 1}$

Find the LCM.   [6.3c]

14. $3x^2,\ 10xy,\ 15y^2$

15. $a - 2,\ 4a - 8$

16. $y^2 - y - 2,\ y^2 - 4$

Add and simplify.   [6.4a]

17. $\dfrac{x + 8}{x + 7} + \dfrac{10 - 4x}{x + 7}$

18. $\dfrac{3}{3x - 9} + \dfrac{x - 2}{3 - x}$

19. $\dfrac{2a}{a + 1} + \dfrac{4a}{a^2 - 1}$

20. $\dfrac{d^2}{d - c} + \dfrac{c^2}{c - d}$

Subtract and simplify. [6.5a]

21. $\dfrac{6x - 3}{x^2 - x - 12} - \dfrac{2x - 15}{x^2 - x - 12}$

22. $\dfrac{3x - 1}{2x} - \dfrac{x - 3}{x}$

23. $\dfrac{x + 3}{x - 2} - \dfrac{x}{2 - x}$

24. $\dfrac{1}{x^2 - 25} - \dfrac{x - 5}{x^2 - 4x - 5}$

25. Perform the indicated operations and simplify: [6.5b]

$$\frac{3x}{x + 2} - \frac{x}{x - 2} + \frac{8}{x^2 - 4}.$$

Simplify. [6.6a]

26. $\dfrac{\dfrac{1}{z} + 1}{\dfrac{1}{z^2} - 1}$

27. $\dfrac{\dfrac{c}{d} - \dfrac{d}{c}}{\dfrac{1}{c} + \dfrac{1}{d}}$

Solve. [6.7a]

28. $\dfrac{3}{y} - \dfrac{1}{4} = \dfrac{1}{y}$

29. $\dfrac{15}{x} - \dfrac{15}{x + 2} = 2$

Solve. [6.8a]

30. *Highway Work.* In checking records, a contractor finds that crew A can pave a certain length of highway in 9 hr, while crew B can do the same job in 12 hr. How long would it take if they worked together?

31. *Airplane Speed.* One plane travels 80 mph faster than another. While one travels 1750 mi, the other travels 950 mi. Find the speed of each plane.

32. *Train Speed.* A manufacturer is testing two high-speed trains. One train travels 40 km/h faster than the other. While one train travels 70 km, the other travels 60 km. Find the speed of each train.

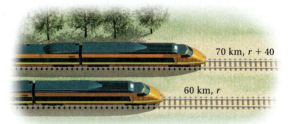

70 km, $r + 40$

60 km, r

Solve. [6.8b]

33. *Quality Control.* A sample of 250 calculators contained 8 defective calculators. How many defective calculators would you expect to find in a sample of 5000?

34. *Pizza Proportions.* At Finnelli's Pizzeria, the following ratios are used: 5 parts sausage to 7 parts cheese, 6 parts onion to 13 parts green pepper, and 9 parts pepperoni to 14 parts cheese.

a) Finnelli's makes several pizzas with green pepper and onion. They use 2 cups of green pepper. How much onion would they use?

b) Finnelli's makes several pizzas with sausage and cheese. They use 3 cups of sausage. How much cheese would they use?

c) Finnelli's makes several pizzas with pepperoni and cheese. They use 6 cups of pepperoni. How much cheese would they use?

35. *Estimating Whale Population.* To determine the number of blue whales in the world's oceans, marine biologists tag 500 blue whales in various parts of the world. Later, 400 blue whales are checked, and it is found that 20 of them are tagged. Estimate the blue whale population.

36. Triangles ABC and XYZ below are similar. Find the value of x.

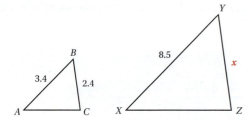

Find an equation of variation in which y varies directly as x and the following are true. Then find the value of y when $x = 20$. [6.9a]

37. $y = 12$ when $x = 4$

38. $y = 0.4$ when $x = 0.5$

Find an equation of variation in which y varies inversely as x and the following are true. Then find the value of y when $x = 5$. [6.9c]

39. $y = 5$ when $x = 6$

40. $y = 0.5$ when $x = 2$

41. $y = 1.3$ when $x = 0.5$

Solve.

42. *Wages.* A person's paycheck P varies directly as the number H of hours worked. The pay is \$165.00 for working 20 hr. Find the pay for 35 hr of work. [6.9b]

43. *WashingTime.* It takes 5 hr for 2 washing machines to wash a fixed amount of laundry. How long would it take 10 washing machines to do the same job? (The number of hours varies inversely as the number of washing machines.) [6.9d]

44. Find all numbers for which

$$\frac{3x^2 - 2x - 1}{3x^2 + x}$$

is not defined. [6.1a]

A. $1, -\dfrac{1}{3}$ **B.** $-\dfrac{1}{3}$

C. $0, -\dfrac{1}{3}$ **D.** $0, \dfrac{1}{3}$

45. Subtract: $\dfrac{1}{x - 5} - \dfrac{1}{x + 5}$. [6.5a]

A. $\dfrac{10}{(x - 5)(x + 5)}$ **B.** 0

C. $\dfrac{5}{x - 5}$ **D.** $\dfrac{10}{x + 5}$

Synthesis

46. Simplify: [6.1d], [6.2b]

$$\frac{2a^2 + 5a - 3}{a^2} \cdot \frac{5a^3 + 30a^2}{2a^2 + 7a - 4} \div \frac{a^2 + 6a}{a^2 + 7a + 12}.$$

47. Compare

$$\frac{A + B}{B} = \frac{C + D}{D}$$

with the proportion

$$\frac{A}{B} = \frac{C}{D}. \quad [6.8b]$$

Understanding Through Discussion and Writing

1. Are parentheses as important when adding rational expressions as they are when subtracting? Why or why not? [6.4a], [6.5a]

2. How can a graph be used to determine how many solutions an equation has? [6.7a]

3. How is the process of canceling related to the identity property of 1? [6.1c]

4. Determine whether the following situation represents direct variation, inverse variation, or neither. Give a reason for your answer. [6.9a, c]

The number of plays that it takes to go 80 yd for a touchdown and the average gain per play

5. Explain how a rational expression can be formed for which -3 and 4 are not allowable replacements. [6.1a]

6. Why is it especially important to check the possible solutions to a rational equation? [6.7a]

CHAPTER

6 Test

For Extra Help For step-by-step test solutions, access the Chapter Test Prep Videos in MyMathLab® or on YouTube (search "BittingerIntro" and click on "Channels").

Find all numbers for which the rational expression is not defined.

1. $\dfrac{8}{2x}$

2. $\dfrac{5}{x + 8}$

3. $\dfrac{x - 7}{x^2 - 49}$

4. $\dfrac{x^2 + x - 30}{x^2 - 3x + 2}$

5. $\dfrac{11}{(x - 1)^2}$

6. $\dfrac{x + 2}{2}$

7. Simplify:

$$\dfrac{6x^2 + 17x + 7}{2x^2 + 7x + 3}.$$

8. Multiply and simplify:

$$\dfrac{a^2 - 25}{6a} \cdot \dfrac{3a}{a - 5}.$$

9. Divide and simplify:

$$\dfrac{25x^2 - 1}{9x^2 - 6x} \div \dfrac{5x^2 + 9x - 2}{3x^2 + x - 2}.$$

10. Find the LCM:

$$y^2 - 9, \ y^2 + 10y + 21, \ y^2 + 4y - 21.$$

Add or subtract. Simplify, if possible.

11. $\dfrac{16 + x}{x^3} + \dfrac{7 - 4x}{x^3}$

12. $\dfrac{5 - t}{t^2 + 1} - \dfrac{t - 3}{t^2 + 1}$

13. $\dfrac{x - 4}{x - 3} + \dfrac{x - 1}{3 - x}$

14. $\dfrac{x - 4}{x - 3} - \dfrac{x - 1}{3 - x}$

15. $\dfrac{5}{t - 1} + \dfrac{3}{t}$

16. $\dfrac{1}{x^2 - 16} - \dfrac{x + 4}{x^2 - 3x - 4}$

17. $\dfrac{1}{x - 1} + \dfrac{4}{x^2 - 1} - \dfrac{2}{x^2 - 2x + 1}$

18. Simplify: $\dfrac{9 - \dfrac{1}{y^2}}{3 - \dfrac{1}{y}}.$

Solve.

19. $\dfrac{7}{y} - \dfrac{1}{3} = \dfrac{1}{4}$

20. $\dfrac{15}{x} - \dfrac{15}{x - 2} = -2$

Find an equation of variation in which y varies directly as x and the following are true. Then find the value of y when $x = 25$.

21. $y = 6$ when $x = 3$

22. $y = 1.5$ when $x = 3$

Find an equation of variation in which y varies inversely as x and the following are true. Then find the value of y when $x = 100$.

23. $y = 6$ when $x = 3$

24. $y = 11$ when $x = 2$

Solve.

25. *Train Travel.* The distance d traveled by a train varies directly as the time t that it travels. The train travels 60 km in $\frac{1}{2}$ hr. How far will it travel in 2 hr?

26. *Concrete Work.* It takes 3 hr for 2 concrete mixers to mix a fixed amount of concrete. The number of hours varies inversely as the number of concrete mixers used. How long would it take 5 concrete mixers to do the same job?

27. *Quality Control.* A sample of 125 spark plugs contained 4 defective spark plugs. How many defective spark plugs would you expect to find in a sample of 500?

28. *Zebra Population.* A game warden catches, tags, and then releases 15 zebras. A month later, a sample of 20 zebras is collected and 6 of them have tags. Use this information to estimate the size of the zebra population in that area.

29. *Copying Time.* Kopy Kwik has 2 copiers. One can copy a year-end report in 20 min. The other can copy the same document in 30 min. How long would it take both machines, working together, to copy the report?

30. *Driving Speed.* Craig drives 20 km/h faster than Marilyn. In the same time that Marilyn drives 225 km, Craig drives 325 km. Find the speed of each car.

31. This pair of triangles is similar. Find the missing length x.

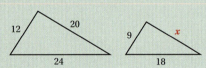

32. Solve: $\dfrac{2}{x-4} + \dfrac{2x}{x^2-16} = \dfrac{1}{x+4}$.

A. -4 **B.** 4

C. $4, -4$ **D.** No solution

Synthesis

33. Reggie and Rema work together to mulch the flower beds around an office complex in $2\frac{6}{7}$ hr. Working alone, it would take Reggie 6 hr more than it would take Rema. How long would it take each of them to complete the landscaping working alone?

34. Simplify: $1 + \dfrac{1}{1 + \dfrac{1}{1 + \dfrac{1}{a}}}$.

1. Find the absolute value: $|3.5|$.

2. Identify the degree of each term and the degree of the polynomial:
$$x^3 - 2x^2 + x - 1.$$

3. *Free-Range Eggs.* In an egg-testing project, it was found that eggs produced by hens raised on pasture contain approximately 35% less cholesterol than factory farm eggs. If a standard factory egg contains 423 mg of cholesterol, how much cholesterol does a free-range egg contain?

   Source: "Meet Real Free-Range Eggs," by Cheryl Long and Tabitha Alterman, *Mother Earth News*, October/November 2007

4. *Square Footage.* In the third quarter of 2008, the size of new single-family homes averaged 2438 ft^2, down from 2629 ft^2 in the second quarter. What was the percent of decrease?

   Source: Gopal Ahluwalia, Director of Research, National Association of Home Builders

5. *Principal Borrowed.* Money is borrowed at 6% simple interest. After 1 year, $2650 pays off the loan. How much was originally borrowed?

6. *Car Travel.* One car travels 105 mi in the same time that a car traveling 10 mph slower travels 75 mi. Find the speed of each car.

7. *Areas.* If each side of a square is increased by 2 ft, the sum of the areas of the two squares is 452 ft^2. Find the length of a side of the original square.

8. *Muscle Weight.* The number of pounds of muscle M in the human body varies directly as body weight B. A person who weighs 175 lb has a muscle weight of 70 lb.
   a) Write an equation of variation that describes this situation.
   b) Mike weighs 192 lb. What is his muscle weight?

9. Collect like terms: $x^2 - 3x^3 - 4x^2 + 5x^3 - 2$.

Simplify.

10. $\dfrac{1}{2}x - \left[\dfrac{3}{8}x - \left(\dfrac{2}{3} + \dfrac{1}{4}x\right) - \dfrac{1}{3}\right]$

11. $\left(\dfrac{2x^3}{3x^{-1}}\right)^{-2}$

12. $\dfrac{\dfrac{4}{x} - \dfrac{6}{x^2}}{\dfrac{5}{x} + \dfrac{7}{2x}}$

Perform the indicated operations. Simplify, if possible.

13. $(5xy^2 - 6x^2y^2 - 3xy^3) - (-4xy^3 + 7xy^2 - 2x^2y^2)$

14. $(4x^4 + 6x^3 - 6x^2 - 4) +$
    $\qquad (2x^5 + 2x^4 - 4x^3 - 4x^2 + 3x - 5)$

15. $\dfrac{2y + 4}{21} \cdot \dfrac{7}{y^2 + 4y + 4}$

16. $\dfrac{x^2 - 9}{x^2 + 8x + 15} \div \dfrac{x - 3}{2x + 10}$

17. $\dfrac{x^2}{x - 4} + \dfrac{16}{4 - x}$

18. $\dfrac{5x}{x^2 - 4} - \dfrac{-3}{2 - x}$

Multiply.

19. $(2.5a + 7.5)(0.4a - 1.2)$

20. $(6x - 5)^2$

21. $(2x^3 + 1)(2x^3 - 1)$

Factor.

22. $9a^2 + 52a - 12$

23. $9x^2 - 30xy + 25y^2$

24. $49x^2 - 1$

Solve.

25. $x - [x - (x - 1)] = 2$

26. $2x^2 + 7x = 4$

27. $x^2 = 10x$

28. $3(x - 2) \le 4(x + 5)$

29. $\dfrac{5x - 2}{4} - \dfrac{4x - 5}{3} = 1$

30. $t = ax + ay$, for a

Find the slope, if it exists, of the line containing the given pair of points.

31. $(-2, 6)$ and $(-2, -1)$

32. $(-4, 1)$ and $(3, -2)$

33. $\left(-\dfrac{1}{2}, 4\right)$ and $\left(3\dfrac{1}{2}, -5\right)$

34. $\left(-7, \dfrac{3}{4}\right)$ and $\left(-4, \dfrac{3}{4}\right)$

For each equation, find the coordinates of the *y*-intercept and the *x*-intercept. Do not graph.

35. $-8x - 24y = 48$

36. $15 - 40x = -120y$

37. $y = 25$

38. $x = -\dfrac{1}{4}$

Graph on a plane.

39. $x = -3$

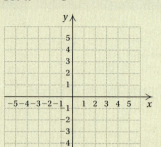

40. $y = -3$

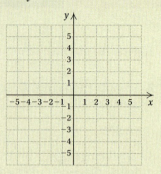

41. $3x - 5y = 15$

42. $2x - 6y = 12$

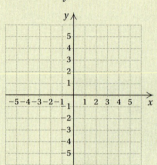

43. $y = -\dfrac{1}{3}x - 2$

44. $x - y = -5$

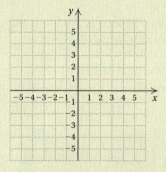

Synthesis

45. Find all numbers for which the following complex rational expression is not defined:

$$\dfrac{\dfrac{1}{x} + x}{2 + \dfrac{1}{x - 3}}.$$

550   CHAPTER 6   Rational Expressions and Equations

Systems of Equations

7.1 Systems of Equations in Two Variables

OBJECTIVES

a Determine whether an ordered pair is a solution of a system of equations.

b Solve systems of two linear equations in two variables by graphing.

SKILL TO REVIEW

Objective 3.1d: Graph linear equations of the type $y = mx + b$ and $Ax + By = C$.

Graph.

1. $y = 2x + 1$

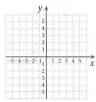

2. $3x - 4y = 12$

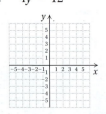

a SYSTEMS OF EQUATIONS AND SOLUTIONS

Many problems can be solved more easily by translating to two equations in two variables. The following is such a **system of equations**:

$$x + y = 8,$$
$$2x - y = 1.$$

SOLUTION OF A SYSTEM OF EQUATIONS

A **solution** of a system of two equations is an ordered pair that makes both equations true.

Look at the graphs shown below. Recall that a graph of an equation is a drawing that represents its solution set. Each point on the graph corresponds to a solution of that equation. Which points (ordered pairs) are solutions of *both* equations?

The graph shows that there is only one. It is the point P where the graphs cross, or intersect. This point looks as if its coordinates are $(3, 5)$. We check to see whether $(3, 5)$ is a solution of *both* equations, substituting 3 for x and 5 for y.

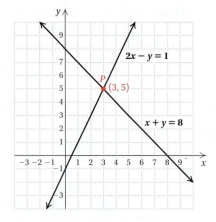

Check:

$$\frac{x + y = 8}{3 + 5 \ ? \ 8}$$
$$8 \ \Big| \ \text{TRUE}$$

$$\frac{2x - y = 1}{2 \cdot 3 - 5 \ ? \ 1}$$
$$6 - 5 \ \Big|$$
$$1 \ \Big| \ \text{TRUE}$$

There is just one solution of the system of equations. It is $(3, 5)$. In other words, $x = 3$ and $y = 5$.

Answers

Answers to Skill to Review Exercises are on p. 553.

EXAMPLE 1 Determine whether $(1, 2)$ is a solution of the system

$$y = x + 1,$$
$$2x + y = 4.$$

We check by substituting alphabetically 1 for x and 2 for y.

Check:

$$\begin{array}{c|c}
y = x + 1 & 2x + y = 4 \\
\hline
2 \;?\; 1 + 1 & 2 \cdot 1 + 2 \;?\; 4 \\
2 \quad \text{TRUE} & 2 + 2 \\
& 4 \quad \text{TRUE}
\end{array}$$

This checks, so $(1, 2)$ is a solution of the system of equations.

EXAMPLE 2 Determine whether $(-3, 2)$ is a solution of the system

$$p + q = -1,$$
$$q + 3p = 4.$$

We check by substituting alphabetically -3 for p and 2 for q.

Check:

$$\begin{array}{c|c}
p + q = -1 & q + 3p = 4 \\
\hline
-3 + 2 \;?\; -1 & 2 + 3(-3) \;?\; 4 \\
-1 \quad \text{TRUE} & 2 - 9 \\
& -7 \quad \text{FALSE}
\end{array}$$

The point $(-3, 2)$ is not a solution of $q + 3p = 4$. Thus it is not a solution of the system of equations.

▸ **Do Exercises 1 and 2.** ▶

b GRAPHING SYSTEMS OF EQUATIONS

When we solve a system of two equations by graphing, we graph both equations and find the coordinates of the points of intersection, if any exist.

EXAMPLE 3 Solve this system of equations by graphing:

$$x + y = 6,$$
$$x = y + 2.$$

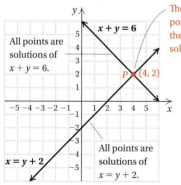

All points are solutions of $x + y = 6$.

$x + y = 6$

The common point gives the common solution.

P (4, 2)

All points are solutions of $x = y + 2$.

$x = y + 2$

Point P with coordinates $(4, 2)$ looks as if it is the solution. We check the pair as follows.

Check:

$$\begin{array}{c}
x + y = 6 \\
\hline
4 + 2 \;?\; 6 \\
6 \quad \text{TRUE}
\end{array}$$

$$\begin{array}{c}
x = y + 2 \\
\hline
4 \;?\; 2 + 2 \\
4 \quad \text{TRUE}
\end{array}$$

The solution is $(4, 2)$.

Do Exercise 3 on the following page. ▶

Determine whether the given ordered pair is a solution of the system of equations.

1. $(2, -3)$; $x = 2y + 8,$
$\qquad\qquad\;\; 2x + y = 1$

Check:

$$\begin{array}{c|c}
x = 2y + 8 & 2x + y = 1 \\
\hline
? & ? \\
| & |
\end{array}$$

 GS **2.** $(20, 40)$; $a = \dfrac{1}{2}b,$
$\qquad\qquad\qquad b - a = 60$

Check:

$$\begin{array}{c}
a = \dfrac{1}{2}b \\
\hline
\boxed{} \;?\; \dfrac{1}{2}(\boxed{}) \\
\boxed{}
\end{array}$$

$$\begin{array}{c}
b - a = 60 \\
\hline
\boxed{} - \boxed{} \;?\; 60 \\
\boxed{}
\end{array}$$

$(20, 40)$ $\boxed{}$ a solution of $a = \dfrac{1}{2}b$.

$(20, 40)$ $\boxed{}$ a solution of $b - a = 60$.

Therefore, $(20, 40)$ $\boxed{}$ a solution of the system.

---

3. Solve this system by graphing:
$$2x + y = 1,$$
$$x = 2y + 8.$$

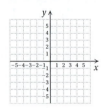

4. Solve this system by graphing:
$$x = -4,$$
$$y = 3.$$

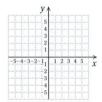

5. Solve this system by graphing:
$$y + 4 = x,$$
$$x - y = -2.$$

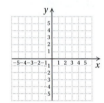

EXAMPLE 4 Solve this system of equations by graphing:
$$x = 2,$$
$$y = -3.$$

The graph of $x = 2$ is a vertical line, and the graph of $y = -3$ is a horizontal line. They intersect at the point $(2, -3)$.
The solution is $(2, -3)$.

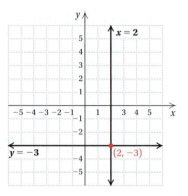

◀ **Do Exercise 4.**

Sometimes the equations in a system have graphs that are parallel lines.

EXAMPLE 5 Solve this system of equations by graphing:
$$y = 3x + 4,$$
$$y = 3x - 3.$$

The lines have the same slope, 3, and different y-intercepts, $(0, 4)$ and $(0, -3)$, so they are parallel.
There is no point at which the lines intersect, so the system has no solution. The solution set is the empty set, denoted $\varnothing$, or $\{\ \}$.

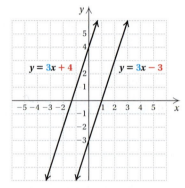

◀ **Do Exercise 5.**

Sometimes the equations in a system have the same graph.

EXAMPLE 6 Solve this system of equations by graphing:
$$2x + 3y = 6,$$
$$-8x - 12y = -24.$$

We graph the equations and see that the graphs are the same. Thus any solution of one of the equations is a solution of the other. Each equation has an infinite number of solutions, some of which are indicated on the graph.
On the following page, we check one such solution, $(0, 2)$, the y-intercept of each equation.

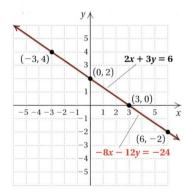

Check:
$$2x + 3y = 6$$
$$\frac{\quad\quad\quad\quad\quad}{2(0) + 3(2) \; ? \; 6}$$
$$0 + 6$$
$$6 \quad | \quad \text{TRUE}$$

$$-8x - 12y = -24$$
$$\frac{\quad\quad\quad\quad\quad\quad\quad}{-8(0) - 12(2) \; ? \; -24}$$
$$0 - 24$$
$$-24 \quad | \quad \text{TRUE}$$

We leave it to the student to check that $(-3, 4)$ is also a solution of the system. If $(0, 2)$ and $(-3, 4)$ are solutions, then all points on the line containing them are solutions. The system has an infinite number of solutions.

Do Exercise 6. ▶

When we graph a system of two equations in two variables, we obtain one of the following three results.

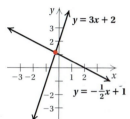

One solution
Graphs intersect.

No solution
Graphs are parallel.

Infinitely many solutions
Equations have the same graph.

ALGEBRAIC ▶◀ GRAPHICAL CONNECTION

Let's take an algebraic–graphical look at equation solving. Such interpretation is useful when using a graphing calculator.

Consider the equation $6 - x = x - 2$. Let's solve it algebraically:

$$6 - x = x - 2$$
$$6 = 2x - 2 \qquad \text{Adding } x$$
$$8 = 2x \qquad \text{Adding } 2$$
$$4 = x. \qquad \text{Dividing by } 2$$

Can we also solve the equation graphically? We can, as we see in the following two methods.

Method 1. Solve $6 - x = x - 2$ graphically.

We let $y = 6 - x$ and $y = x - 2$. Graphing the system of equations gives us the graph at right. The point of intersection is $(4, 2)$. Note that the x-coordinate of the point of intersection is 4. This value for x is also the *solution* of the equation $6 - x = x - 2$.

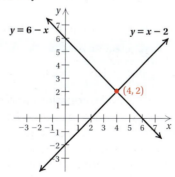

Do Exercise 7. ▶

6. Solve this system by graphing:
$$2x + y = 4,$$
$$-6x - 3y = -12.$$

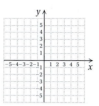

7. a) Solve $2x - 1 = 8 - x$ algebraically.

b) Solve $2x - 1 = 8 - x$ graphically using method 1.

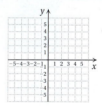

c) Compare your answers to parts (a) and (b).

Answers

6. Infinite number of solutions
7. (a) 3; **(b)** 3; **(c)** They are the same.

8. a) Solve $2x - 1 = 8 - x$ graphically using method 2.

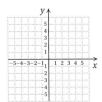

b) Compare your answers to Margin Exercises 7(a), 7(b), and 8(a).

Method 2. Solve $6 - x = x - 2$ graphically.

Adding x and -6 on both sides, we obtain the form $0 = 2x - 8$. In this case, we let $y = 0$ and $y = 2x - 8$. Since $y = 0$ is the x-axis, we need graph only $y = 2x - 8$ and see where it crosses the x-axis. Note that the x-intercept of $y = 2x - 8$ is $(4, 0)$. The x-coordinate of this ordered pair is also the *solution* of the equation $6 - x = x - 2$.

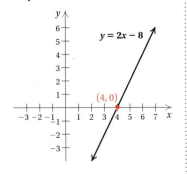

Answers

8. (a) 3; **(b)** They are the same.

◀ **Do Exercise 8.**

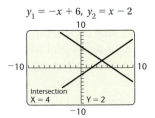

CALCULATOR CORNER

Solving Systems of Equations We can solve a system of two equations in two variables on a graphing calculator. Consider the system of equations in Example 3,

$$x + y = 6,$$
$$x = y + 2.$$

First, we solve the equations for y, obtaining $y = -x + 6$ and $y = x - 2$. Then we enter $y_1 = -x + 6$ and $y_2 = x - 2$ on the equation-editor screen and graph the equations. We can use the standard viewing window, $[-10, 10, -10, 10]$.

We will use the **INTERSECT** feature to find the coordinates of the point of intersection of the lines. To use this feature, we select the two graphs, called First curve and Second curve, and then choose a Guess close to the point of intersection. The coordinates of the point of intersection of the graphs, $x = 4, y = 2$, appear at the bottom of the screen. Thus the solution of the system of equations is $(4, 2)$.

EXERCISES: Use a graphing calculator to solve each system of equations.

1. $x + y = 2,$
$\quad y = x + 4$

2. $x + 3y = -1,$
$\quad x - y = -5$

3. $3x + 5y = 19,$
$\quad 4x = 10 + y$

| 7.1 | **Exercise Set** |

For Extra Help

MyMathLab® · MathXL® PRACTICE · WATCH · READ · REVIEW

✓ Reading Check

Determine whether each statement is true or false.

RC1. A solution of a system of two equations in two variables is an ordered pair.

RC2. To check whether $(1, 3)$ is a solution of $y - 3x = 0$, we substitute 1 for x and 3 for y.

RC3. Graphs of two lines may have one point, no points, or an infinite number of points in common.

RC4. Every system of equations has at least one solution.

a Determine whether the given ordered pair is a solution of the system of equations. Use alphabetical order of the variables.

1. $(1, 5);$ $5x - 2y = -5,$
$\qquad\qquad 3x - 7y = -32$

2. $(3, 2);$ $2x + 3y = 12,$
$\qquad\qquad x - 4y = -5$

3. $(4, 2);$ $3b - 2a = -2,$
$\qquad\qquad b + 2a = 8$

4. $(6, -6);$ $t + 2s = 6,$
$\qquad\qquad\;\; t - s = -12$

5. $(15, 20);$ $3x - 2y = 5,$
$\qquad\qquad\;\; 6x - 5y = -10$

6. $(-1, -5);$ $4r + s = -9,$
$\qquad\qquad\;\;\; 3r = 2 + s$

7. $(-1, 1);$ $x = -1,$
$\qquad\qquad x - y = -2$

8. $(-3, 4);$ $2x = -y - 2,$
$\qquad\qquad\;\; y = -4$

9. $(18, 3);$ $y = \dfrac{1}{6}x,$
$\qquad\qquad 2x - y = 33$

10. $(-3, 1);$ $y = -\dfrac{1}{3}x,$
$\qquad\qquad\;\; 3y = -5x - 12$

b Solve each system of equations by graphing.

11. $x - y = 2,$
$\quad\;\; x + y = 6$

12. $x + y = 3,$
$\quad\;\; x - y = 1$

13. $8x - y = 29,$
$\quad\;\; 2x + y = 11$

14. $4x - y = 10,$
$\quad\;\; 3x + 5y = 19$

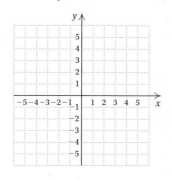 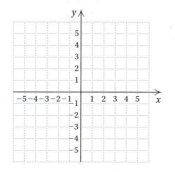 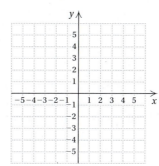 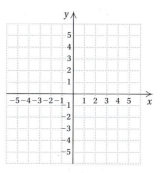

15. $t = v,$
$\quad\;\; 4t = 2v - 6$

16. $x = 3y,$
$\quad\;\; 3y - 6 = 2x$

17. $x = -y,$
$\quad\;\; x + y = 4$

18. $-3x = 5 - y,$
$\quad\;\; 2y = 6x + 10$

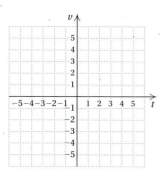 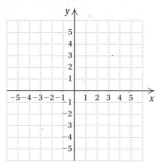 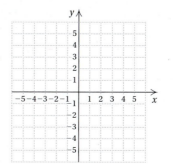 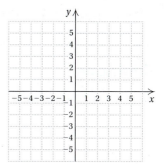

19. $a = \dfrac{1}{2}b + 1,$
$a - 2b = -2$

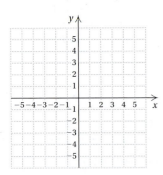

20. $x = \dfrac{1}{3}y + 2,$
$-2x - y = 1$

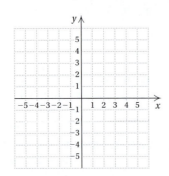

21. $y - 2x = 0,$
$y = 6x - 2$

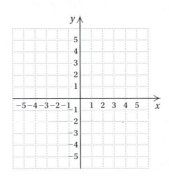

22. $y = 3x,$
$y = -3x + 2$

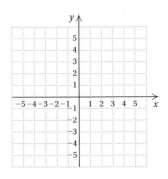

23. $x + y = 9,$
$3x + 3y = 27$

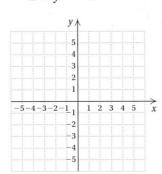

24. $x + y = 4,$
$x + y = -4$

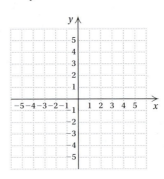

25. $x = 5,$
$y = -3$

26. $y = 2,$
$y = -4$

Skill Maintenance

Simplify. [6.1c]

27. $\dfrac{x^2 - 25}{x^2 - 10x + 25}$

28. $\dfrac{8d^2 + 16d}{40d^3 - 8d}$

29. $\dfrac{3y - 12}{4 - y}$

30. $\dfrac{2x^2 - x - 15}{x^2 - 9}$

Classify each polynomial as a monomial, a binomial, a trinomial, or none of these. [4.3b]

31. $5x^2 - 3x + 7$

32. $4x^3 - 2x^2$

33. $1.8x^5$

34. $x^3 + 2x^2 - 3x + 1$

Synthesis

35. The solution of the following system is $(2, -3)$. Find A and B.

$Ax - 3y = 13,$
$x - By = 8$

36. Find an equation to pair with $5x + 2y = 11$ such that the solution of the system is $(3, -2)$. Answers may vary.

37. Find a system of equations with $(6, -2)$ as a solution. Answers may vary.

38.–41. 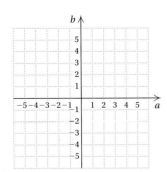 Use a graphing calculator to do Exercises 15–18.

The Substitution Method

7.2

Consider the following system of equations:

$$3x + 7y = 5,$$
$$6x - 7y = 1.$$

Suppose we try to solve this system graphically. We obtain the graph shown at right.

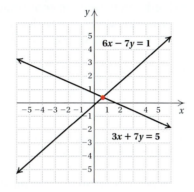

What is the solution? It is rather difficult to tell exactly. It would appear that the coordinates of the point are not integers. It turns out that the solution is $\left(\frac{2}{3}, \frac{3}{7}\right)$. Graphing helps us picture the solution of a system of equations, but solving by graphing, though practical in many applications, is not always fast or accurate. We now learn **algebraic** methods that can be used to determine solutions exactly.

a SOLVING BY THE SUBSTITUTION METHOD

One nongraphical method for solving systems is known as the **substitution method**.

EXAMPLE 1 Solve the system

$$x + y = 6, \quad \textbf{(1)}$$
$$x = y + 2. \quad \textbf{(2)}$$

Equation (2) says that x and $y + 2$ name the same number. Thus in equation (1), we can substitute $y + 2$ for x:

$$x + y = 6 \qquad \text{Equation (1)}$$
$$(y + 2) + y = 6. \qquad \text{Substituting } y + 2 \text{ for } x$$

This last equation has only one variable. We solve it:

$$y + 2 + y = 6 \qquad \text{Removing parentheses}$$
$$2y + 2 = 6 \qquad \text{Collecting like terms}$$
$$2y + 2 - 2 = 6 - 2 \qquad \text{Subtracting 2 on both sides}$$
$$2y = 4 \qquad \text{Simplifying}$$
$$\frac{2y}{2} = \frac{4}{2} \qquad \text{Dividing by 2}$$
$$y = 2. \qquad \text{Simplifying}$$

We have found the y-value of the solution. To find the x-value, we return to the original pair of equations. Substituting into either equation will give us the x-value.

Answers

Skill to Review:
1. $x = -3y + 5$ **2.** $y = 2x - 9$

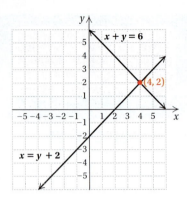

We choose equation (2) because it has x alone on one side:

$$x = y + 2 \qquad \text{Equation (2)}$$
$$= 2 + 2 \qquad \text{Substituting 2 for } y$$
$$= 4.$$

The ordered pair $(4, 2)$ may be a solution. Note that we are using alphabetical order in listing the coordinates in an ordered pair. That is, since x precedes y alphabetically, we list 4 before 2 in the pair $(4, 2)$.

We check as follows.

Check:

$$
\begin{array}{c|c}
x + y = 6 & x = y + 2 \\
\hline
4 + 2 \ ? \ 6 & 4 \ ? \ 2 + 2 \\
6 \quad \text{TRUE} & 4 \quad \text{TRUE}
\end{array}
$$

Since $(4, 2)$ checks, we have the solution. The graphical solution shown at left provides another check. ■

Note in Example 1 that substituting 2 for y in equation (1) will also give us the x-value of the solution:

$$x + y = 6$$
$$x + 2 = 6$$
$$x = 4.$$

◀ **Do Exercise 1.**

EXAMPLE 2 Solve the system

$$t = 1 - 3s, \qquad \text{(1)}$$
$$s - t = 11. \qquad \text{(2)}$$

We substitute $1 - 3s$ for t in equation (2):

$$s - t = 11 \qquad \text{Equation (2)}$$
$$s - (1 - 3s) = 11. \qquad \text{Substituting } 1 - 3s \text{ for } t$$

> Remember to use parentheses when you substitute.

Now we solve for s:

$$s - 1 + 3s = 11 \qquad \text{Removing parentheses}$$
$$4s - 1 = 11 \qquad \text{Collecting like terms}$$
$$4s = 12 \qquad \text{Adding 1}$$
$$s = 3. \qquad \text{Dividing by 4}$$

Next, we substitute 3 for s in equation (1) of the original system:

$$t = 1 - 3s \qquad \text{Equation (1)}$$
$$= 1 - 3 \cdot 3 \qquad \text{Substituting 3 for } s$$
$$= -8.$$

The pair $(3, -8)$ checks and is the solution. Remember: We list the answer in alphabetical order, (s, t). That is, since s comes before t in the alphabet, 3 is listed first and -8 second.

◀ **Do Exercise 2.**

---

1. Solve by the substitution method. Do not graph. **GS**

$$x + y = 5, \qquad \textbf{(1)}$$
$$x = y + 1 \qquad \textbf{(2)}$$

Substitute $y + 1$ for x in equation (1) and solve for y.

$$x + y = 5$$
$$(\boxed{}) + y = 5$$
$$\boxed{} + 1 = 5$$
$$2y = \boxed{}$$
$$y = \boxed{}$$

Substitute $\boxed{}$ for y in equation (2) and solve for x.

$$x = y + 1$$
$$= \boxed{} + 1$$
$$= \boxed{}$$

The numbers check. The solution is $(\boxed{}, \boxed{})$.

2. Solve by the substitution method:

$$a - b = 4,$$
$$b = 2 - a.$$

Answers

1. $(3, 2)$ **2.** $(3, -1)$

Guided Solution:
1. $y + 1, 2y, 4, 2, 2, 2, 3, 3, 2$

b SOLVING FOR THE VARIABLE FIRST

Sometimes neither equation of a pair has a variable alone on one side. Then we solve one equation for one of the variables and proceed as before, substituting into the *other* equation. If possible, we solve in either equation for a variable that has a coefficient of 1.

EXAMPLE 3 Solve the system

$$x - 2y = 6, \quad \textbf{(1)}$$
$$3x + 2y = 4. \quad \textbf{(2)}$$

We solve one equation for one variable. Since the coefficient of x is 1 in equation (1), it is easier to solve that equation for x:

$$x - 2y = 6 \qquad \text{Equation (1)}$$
$$x = 6 + 2y. \qquad \text{Adding } 2y \qquad \textbf{(3)}$$

We substitute $6 + 2y$ for x in equation (2) of the original pair and solve for y:

$$3x + 2y = 4 \qquad \text{Equation (2)}$$
$$3(6 + 2y) + 2y = 4 \qquad \text{Substituting } 6 + 2y \text{ for } x$$
$$18 + 6y + 2y = 4 \qquad \text{Removing parentheses}$$
$$18 + 8y = 4 \qquad \text{Collecting like terms}$$
$$8y = -14 \qquad \text{Subtracting 18}$$
$$y = \frac{-14}{8}, \text{ or } -\frac{7}{4}. \qquad \text{Dividing by 8}$$

To find x, we go back to either of the original equations, (1) or (2), or to equation (3), which is solved for x. It is generally easier to use an equation like equation (3) where we have solved for a specific variable. We substitute $-\frac{7}{4}$ for y in equation (3) and compute x:

$$x = 6 + 2y \qquad \text{Equation (3)}$$
$$= 6 + 2\left(-\frac{7}{4}\right) \qquad \text{Substituting } -\frac{7}{4} \text{ for } y$$
$$= 6 - \frac{7}{2} = \frac{5}{2}.$$

We check the ordered pair $\left(\frac{5}{2}, -\frac{7}{4}\right)$.

Check:

$$\frac{x - 2y = 6}{\frac{5}{2} - 2\left(-\frac{7}{4}\right) \; ? \; 6}$$
$$\frac{5}{2} + \frac{7}{2}$$
$$\frac{12}{2}$$
$$6 \quad | \quad \text{TRUE}$$

$$\frac{3x + 2y = 4}{3 \cdot \frac{5}{2} + 2\left(-\frac{7}{4}\right) \; ? \; 4}$$
$$\frac{15}{2} - \frac{7}{2}$$
$$\frac{8}{2}$$
$$4 \quad | \quad \text{TRUE}$$

Since $\left(\frac{5}{2}, -\frac{7}{4}\right)$ checks, it is the solution. This solution would have been difficult to find graphically because it involves fractions.

Do Exercise 3. ▶

c SOLVING APPLIED PROBLEMS

Now let's solve an applied problem using systems of equations and the substitution method.

Caution!

A solution of a system of equations in two variables is an ordered *pair* of numbers. Once you have solved for one variable, don't forget the other. A common mistake is to solve for only one variable.

GS **3.** Solve:

$$x - 2y = 8, \quad \textbf{(1)}$$
$$2x + y = 8. \quad \textbf{(2)}$$

Solve for y in equation (2).

$$2x + y = 8$$
$$y = 8 - \boxed{} \qquad \textbf{(3)}$$

Substitute $8 - 2x$ for y in equation (1) and solve for x.

$$x - 2y = 8$$
$$x - 2\left(\boxed{}\right) = 8$$
$$x - 16 + \boxed{} = 8$$
$$\boxed{} - 16 = 8$$
$$5x = \boxed{}$$
$$x = \boxed{}$$

Substitute $\boxed{}$ for x in equation (3) and solve for y.

$$y = 8 - 2x$$
$$= 8 - 2\left(\boxed{}\right)$$
$$= \frac{40}{5} - \frac{\boxed{}}{5}$$
$$= \boxed{}$$

The numbers check. The solution is $\left(\boxed{}, \boxed{}\right)$.

Answer

3. $\left(\frac{24}{5}, -\frac{8}{5}\right)$

Guided Solution:

3. $2x, 8 - 2x, 4x, 5x, 24, \frac{24}{5}, \frac{24}{5}, \frac{24}{5}, 48, -\frac{8}{5}, \frac{24}{5}, -\frac{8}{5}$

EXAMPLE 4   *Standard Billboard.*   A standard rectangular highway billboard has a perimeter of 124 ft. The length is 34 ft more than the width. Find the length and the width.

Source: Eller Sign Company

1. **Familiarize.**   We make a drawing and label it. We let l = the length and w = the width.

2. **Translate.**   The perimeter of a rectangle is given by the formula $2l + 2w$. We translate each statement, as follows.

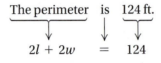

The perimeter   is   124 ft.

$$2l + 2w = 124$$

The length   is   34 ft longer than the width.

$$l = 34 + w$$

We now have a system of equations:

$$2l + 2w = 124, \quad \textbf{(1)}$$
$$l = 34 + w. \quad \textbf{(2)}$$

3. **Solve.**   We solve the system. To begin, we substitute $34 + w$ for l in the first equation and solve:

$$2(34 + w) + 2w = 124 \qquad \text{Substituting } 34 + w \text{ for } l \text{ in equation (1)}$$
$$2 \cdot 34 + 2 \cdot w + 2w = 124 \qquad \text{Removing parentheses}$$
$$4w + 68 = 124 \qquad \text{Collecting like terms}$$
$$4w = 56 \qquad \text{Subtracting 68}$$
$$w = 14. \qquad \text{Dividing by 4}$$

We go back to one of the original equations and substitute 14 for w:

$$l = 34 + w = 34 + 14 = 48. \qquad \text{Substituting in equation (2)}$$

4. **Check.**   If the length is 48 ft and the width is 14 ft, then the length is 34 ft more than the width ($48 - 14 = 34$), and the perimeter is $2(48 \text{ ft}) + 2(14 \text{ ft})$, or 124 ft. Thus these dimensions check.

5. **State.**   The width is 14 ft and the length is 48 ft.

Example 4 illustrates that many problems that can be solved by translating to *one* equation in *one* variable may actually be easier to solve by translating to *two* equations in *two* variables.

◀ **Do Exercise 4.**

---

4. *Community Garden.*   A rectangular community garden is to be enclosed with 92 m of fencing. In order to allow for compost storage, the garden must be 4 m longer than it is wide. Determine the dimensions of the garden.

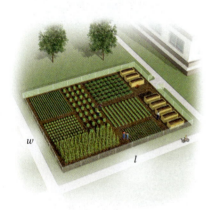

Answer

4. Length: 25 m; width: 21 m

✓ Reading Check

Determine whether each statement is true or false.

RC1. The substitution method is an algebraic method for solving systems of equations.

RC2. We can find solutions of systems of equations involving fractions using the substitution method.

RC3. When writing the solution of a system, we write the value that we found first as the first number in the ordered pair.

RC4. When solving using substitution, we may have to solve for a variable before substituting.

a Solve using the substitution method.

1. $x = -2y,$
$x + 4y = 2$

2. $r = -3s,$
$r + 4s = 10$

3. $y = x - 6,$
$x + y = -2$

4. $y = x + 1,$
$2x + y = 4$

5. $y = 2x - 5,$
$3y - x = 5$

6. $y = 2x + 1,$
$x + y = -2$

7. $x = y + 5,$
$2x + y = 1$

8. $x = y - 3,$
$x + 2y = 9$

9. $x + y = 10,$
$y = x + 8$

10. $x + y = 4,$
$y = 2x + 1$

11. $2x + y = 5,$
$x = y + 7$

12. $3x + y = -1,$
$x = 2y - 5$

b Solve using the substitution method. First, solve one equation for one variable.

13. $x - y = 6,$
$x + y = -2$

14. $s + t = -4,$
$s - t = 2$

15. $y - 2x = -6,$
$2y - x = 5$

16. $x - y = 5,$
$x + 2y = 7$

17. $r - 2s = 0,$
$4r - 3s = 15$

18. $y - 2x = 0,$
$3x + 7y = 17$

19. $2x + 3y = -2,$
$2x - y = 9$

20. $3x - 6y = 4,$
$5x + y = 3$

21. $x + 3y = 5,$
$3x + 5y = 3$

22. $x + 2y = 10,$
$3x + 4y = 8$

23. $x - y = -3,$
$2x + 3y = -6$

24. $x - 2y = 8,$
$2x + 3y = 2$

Solve.

25. *Two-by-Four.*   The perimeter of a cross section of a "two-by-four" piece of lumber is 10 in. The length is 2 in. more than the width. Find the actual dimensions of a cross section of a two-by-four.

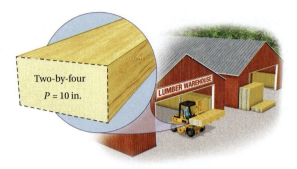

26. *Billboards.*   As an advertisement for chocolate candy sales in anticipation of Valentine's Day 2011, the Meiji Seika Kaisha confectionary factory in Takatsuki, Osaka Prefecture, Japan, built a giant billboard in the shape of a chocolate bar. The perimeter of the billboard was 388 m, and the length was 138 m more than the width. Find the length and the width.

Source: www.worldrecordsacademy.org

27. *Dimensions of Wyoming.*   The state of Wyoming is roughly a rectangle with a perimeter of 1280 mi. The width is 90 mi less than the length. Find the length and the width.

28. *Dimensions of Colorado.*   The state of Colorado is roughly a rectangle whose perimeter is 1300 mi. The width is 110 mi less than the length. Find the length and the width.

29. *Lacrosse.*   The perimeter of a lacrosse field is 340 yd. The length is 10 yd less than twice the width. Find the length and the width.

30. *Soccer.*   The perimeter of a soccer field is 280 yd. The width is 5 more than half the length. Find the length and the width.

31. The sum of two numbers is 37. One number is 5 more than the other. Find the numbers.

32. The sum of two numbers is 26. One number is 12 more than the other. Find the numbers.

33. Find two numbers whose sum is 52 and whose difference is 28.

34. Find two numbers whose sum is 63 and whose difference is 5.

35. The difference of two numbers is 12. Two times the larger is five times the smaller. What are the numbers?

36. The difference of two numbers is 18. Twice the smaller number plus three times the larger is 74. What are the numbers?

37. On average, Americans spend 12.5 hr per month watching videos on both Netflix and Hulu. They spend 7.5 more hours watching videos on Netflix than they do on Hulu. How much time do they spend on each?

Source: *2011 State of Media Report*, Nielsen

38. On average, Americans spend 33.25 hr per week watching both traditional TV and video online. They spend 32.25 more hours watching traditional TV than they do watching video online. How much time do they spend on each?

Source: *2011 State of Media Report*, Nielsen

Skill Maintenance

Graph.   [3.2a, b]

39. $2x - 3y = 6$

40. $2x + 3y = 6$

41. $y = 2x - 5$

42. $y = 4$

Factor completely.   [5.6a]

43. $6x^2 - 13x + 6$

44. $4p^2 - p - 3$

45. $4x^2 + 3x + 2$

46. $9a^2 - 25$

Simplify.   [4.1d, e, f]

47. $\dfrac{x^{-2}}{x^{-5}}$

48. $x^2 \cdot x^5$

49. $x^{-2} \cdot x^{-5}$

50. $\dfrac{a^2 b^{-3}}{a^5 b^{-6}}$

Synthesis

Solve using the INTERSECT feature on a graphing calculator. Then solve algebraically and decide which method you prefer to use.

51. $x - y = 5$,
$x + 2y = 7$

52. $y - 2x = -6$,
$2y - x = 5$

53. $y - 2.35x = -5.97$,
$2.14y - x = 4.88$

54. $y = 1.2x - 32.7$,
$y = -0.7x + 46.15$

55. *Softball.*   The perimeter of a softball diamond is two-thirds of the perimeter of a baseball diamond. Together, the two perimeters measure 200 yd. Find the distance between the bases in each sport.

56. Write a system of two linear equations that can be solved more quickly—but still precisely—by a graphing calculator than by substitution. Time yourself using both methods to solve the system.

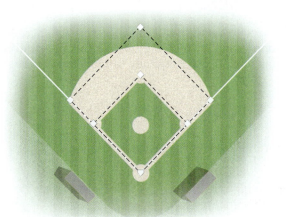

SKILL TO REVIEW

Objective 2.3c: Solve equations with an infinite number of solutions and equations with no solutions.

Solve.

1. $6(1 - x) = 6 - 6x$

2. $2(x - 1) = 2x + 3$

a SOLVING BY THE ELIMINATION METHOD

The **elimination method** for solving systems of equations makes use of the *addition principle*. Some systems are much easier to solve using this method rather than the substitution method. For example, to solve the system

$$2x + 3y = 13, \quad \textbf{(1)}$$
$$4x - 3y = 17 \quad \textbf{(2)}$$

by substitution, we would need to first solve for a variable in one of the equations. Were we to solve equation (1) for y, we would find (after several steps) that $y = \frac{13}{3} - \frac{2}{3}x$. We could then use the expression $\frac{13}{3} - \frac{2}{3}x$ in equation (2) as a replacement for y:

$$4x - 3\left(\tfrac{13}{3} - \tfrac{2}{3}x\right) = 17.$$

As you can see, although substitution could be used to solve this system, doing so involves working with fractions. Fortunately, another method, elimination, can be used to solve systems and, for problems like this, is simpler to use.

EXAMPLE 1 Solve the system

$$2x + 3y = 13, \quad \textbf{(1)}$$
$$4x - 3y = 17. \quad \textbf{(2)}$$

The key to the advantage of the elimination method for solving this system involves the $3y$ in one equation and the $-3y$ in the other. These terms are opposites. If we add the terms on the left sides of the equations, the y-terms will add to 0, and in effect, the variable y will be eliminated.

We will use the addition principle for equations. According to equation (2), $4x - 3y$ and 17 are the same number. Thus we can use a vertical form and add $4x - 3y$ on the left side of equation (1) and 17 on the right side—in effect, adding the same number on both sides of equation (1):

$$2x + 3y = 13 \qquad \textbf{(1)}$$
$$\underline{4x - 3y = 17} \qquad \textbf{(2)}$$
$$6x + 0y = 30, \text{ or } \qquad \text{Adding}$$
$$6x \qquad\quad = 30.$$

We have "eliminated" one variable. This is why we call this the **elimination method**. We now have an equation with just one variable that can be solved for x:

$$6x = 30$$
$$x = 5.$$

Next, we substitute 5 for x in either of the original equations:

$$2x + 3y = 13 \qquad \text{Equation (1)}$$
$$2(5) + 3y = 13 \qquad \text{Substituting 5 for } x$$
$$10 + 3y = 13$$
$$3y = 3$$
$$y = 1. \qquad \text{Solving for } y$$

Answers

Skill to Review:

1. All real numbers **2.** No solution

We check the ordered pair $(5, 1)$.

Check:

$$\begin{array}{c} 2x + 3y = 13 \\ \hline 2(5) + 3(1) \overset{?}{\ } 13 \\ 10 + 3 \\ 13 \qquad \text{TRUE} \end{array}$$

$$\begin{array}{c} 4x - 3y = 17 \\ \hline 4(5) - 3(1) \overset{?}{\ } 17 \\ 20 - 3 \\ 17 \qquad \text{TRUE} \end{array}$$

Since $(5, 1)$ checks, it is the solution. We can see the solution in the graph shown at right.

Do Exercises 1 and 2. ▶

b USING THE MULTIPLICATION PRINCIPLE FIRST

The elimination method allows us to eliminate a variable. We may need to multiply by certain numbers first, however, so that terms become opposites.

EXAMPLE 2 Solve the system

$$2x + 3y = 8, \qquad \textbf{(1)}$$
$$x + 3y = 7. \qquad \textbf{(2)}$$

If we add, we will not eliminate a variable. However, if the $3y$ were $-3y$ in one equation, we could eliminate y. Thus we multiply by -1 on both sides of equation (2) and then add, using a vertical form:

$$\begin{array}{ll} 2x + 3y = 8 & \text{Equation (1)} \\ \underline{-x - 3y = -7} & \text{Multiplying equation (2) by } -1 \\ x = 1. & \text{Adding} \end{array}$$

Next, we substitute 1 for x in one of the original equations:

$$\begin{array}{ll} x + 3y = 7 & \text{Equation (2)} \\ 1 + 3y = 7 & \text{Substituting 1 for } x \\ 3y = 6 & \\ y = 2. & \text{Solving for } y \end{array}$$

We check the ordered pair $(1, 2)$.

Check:

$$\begin{array}{c} 2x + 3y = 8 \\ \hline 2 \cdot 1 + 3 \cdot 2 \overset{?}{\ } 8 \\ 2 + 6 \\ 8 \qquad \text{TRUE} \end{array}$$

$$\begin{array}{c} x + 3y = 7 \\ \hline 1 + 3 \cdot 2 \overset{?}{\ } 7 \\ 1 + 6 \\ 7 \qquad \text{TRUE} \end{array}$$

Since $(1, 2)$ checks, it is the solution. We can see the solution in the graph shown at right.

Do Exercises 3 and 4. ▶

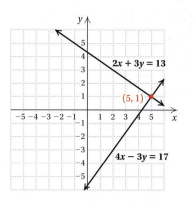

Solve using the elimination method.

1. $x + y = 5,$
   $2x - y = 4$

2. $-2x + y = -4,$
   $2x - 5y = 12$

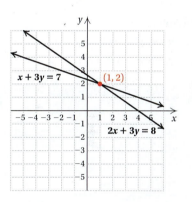

3. Solve. Multiply one equation by -1 first.
   $$5x + 3y = 17,$$
   $$5x - 2y = -3$$

4. Solve the system
   $$3x - 2y = -30,$$
   $$5x - 2y = -46.$$

Answers

1. $(3, 2)$ **2.** $(1, -2)$ **3.** $(1, 4)$ **4.** $(-8, 3)$

In Example 2, we used the multiplication principle, multiplying by -1. However, we often need to multiply by something other than -1.

EXAMPLE 3 Solve the system

$$3x + 6y = -6, \quad \textbf{(1)}$$
$$5x - 2y = 14. \quad \textbf{(2)}$$

Looking at the terms with variables, we see that if $-2y$ were $-6y$, we would have terms that are opposites. We can achieve this by multiplying by 3 on both sides of equation (2). Then we add and solve for x:

$$
\begin{array}{ll}
3x + 6y = -6 & \text{Equation (1)} \\
\underline{15x - 6y = 42} & \text{Multiplying by 3 on both sides of equation (2)} \\
18x = 36 & \text{Adding} \\
x = 2. & \text{Solving for } x
\end{array}
$$

Next, we substitute 2 for x in either of the original equations. We choose the first:

$$
\begin{array}{ll}
3x + 6y = -6 & \text{Equation (1)} \\
3 \cdot 2 + 6y = -6 & \text{Substituting 2 for } x \\
6 + 6y = -6 & \\
6y = -12 & \\
y = -2. & \text{Solving for } y
\end{array}
$$

We check the ordered pair $(2, -2)$.

Check:
$$
\begin{array}{c|c}
3x + 6y = -6 & \\
\hline
3 \cdot 2 + 6 \cdot (-2) \;?\; -6 & \\
6 + (-12) & \\
-6 & \text{TRUE}
\end{array}
\qquad
\begin{array}{c|c}
5x - 2y = 14 & \\
\hline
5 \cdot 2 - 2 \cdot (-2) \;?\; 14 & \\
10 - (-4) & \\
14 & \text{TRUE}
\end{array}
$$

Since $(2, -2)$ checks, it is the solution. (See the graph at left.)

◀ **Do Exercises 5 and 6.**

Part of the strategy in using the elimination method is making a decision about which variable to eliminate. So long as the algebra has been carried out correctly, the solution can be found by eliminating *either* variable. We multiply so that terms involving the variable to be eliminated are opposites. It is helpful to first get each equation in a form equivalent to $Ax + By = C$.

EXAMPLE 4 Solve the system

$$3y + 1 + 2x = 0, \quad \textbf{(1)}$$
$$5x = 7 - 4y. \quad \textbf{(2)}$$

We first rewrite each equation in a form equivalent to $Ax + By = C$:

$$
\begin{array}{lll}
2x + 3y = -1, & \textbf{(1)} & \text{Subtracting 1 on both sides and rearranging terms} \\
5x + 4y = 7. & \textbf{(2)} & \text{Adding } 4y \text{ on both sides}
\end{array}
$$

We decide to eliminate the x-term. We do so by multiplying by 5 on both sides of equation (1) and by -2 on both sides of equation (2). Then we add and solve for y.

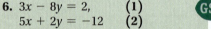

Solve each system.

5. $4a + 7b = 11,$
$\quad 2a + 3b = 5$

6. $3x - 8y = 2, \quad \textbf{(1)}$ GS
$\quad 5x + 2y = -12 \quad \textbf{(2)}$

Multiply equation (2) by 4, add, and solve for x.

$$
\begin{array}{l}
3x - 8y = 2 \\
\underline{20x + \boxed{} = \boxed{}} \\
23x = \boxed{} \\
x = \boxed{}
\end{array}
$$

Substitute $\boxed{}$ for x in equation (1) and solve for y.

$$
\begin{array}{l}
3x - 8y = 2 \\
3\left(\boxed{}\right) - 8y = 2 \\
\boxed{} - 8y = 2 \\
-8y = \boxed{} \\
y = \boxed{}
\end{array}
$$

The numbers check. The solution is $\left(\boxed{}, \boxed{}\right)$.

Answers

5. $(1, 1)$ **6.** $(-2, -1)$

Guided Solution:
6. $8y, -48, -46, -2, -2, -2, -6, 8, -1, -2, -1$

$$10x + 15y = -5 \qquad \text{Multiplying by 5 on both sides of equation (1)}$$
$$\underline{-10x - 8y = -14} \qquad \text{Multiplying by } -2 \text{ on both sides of equation (2)}$$
$$7y = -19 \qquad \text{Adding}$$
$$y = \frac{-19}{7}, \text{ or } -\frac{19}{7} \qquad \text{Solving for } y$$

Next, we substitute $-\frac{19}{7}$ for y in one of the original equations:

$$2x + 3y = -1 \qquad \text{Equation (1)}$$
$$2x + 3\left(-\frac{19}{7}\right) = -1 \qquad \text{Substituting } -\frac{19}{7} \text{ for } y$$
$$2x - \frac{57}{7} = -1$$
$$2x = -1 + \frac{57}{7}$$
$$2x = -\frac{7}{7} + \frac{57}{7}$$
$$2x = \frac{50}{7}$$
$$\frac{1}{2} \cdot 2x = \frac{1}{2} \cdot \frac{50}{7} \qquad \text{Multiplying by } \frac{1}{2} \text{ on both sides of the equation}$$
$$x = \frac{50}{14}$$
$$x = \frac{25}{7}. \qquad \text{Simplifying}$$

We check the ordered pair $\left(\frac{25}{7}, -\frac{19}{7}\right)$.

Check:

$$\begin{array}{c|c}
\multicolumn{1}{c}{3y + 1 + 2x = 0} \\
\hline
3\left(-\frac{19}{7}\right) + 1 + 2\left(\frac{25}{7}\right) \;?\; 0 \\
-\frac{57}{7} + \frac{7}{7} + \frac{50}{7} \quad \Big| \\
0 \quad \Big| \quad \text{TRUE}
\end{array}$$

$$\begin{array}{c|c}
\multicolumn{1}{c}{5x = 7 - 4y} \\
\hline
5\left(\frac{25}{7}\right) \;?\; 7 - 4\left(-\frac{19}{7}\right) \\
\frac{125}{7} \quad \Big| \quad \frac{49}{7} + \frac{76}{7} \\
\quad \Big| \quad \frac{125}{7} \qquad \text{TRUE}
\end{array}$$

The solution is $\left(\frac{25}{7}, -\frac{19}{7}\right)$.

Do Exercise 7. ▶

Let's consider a system with no solution and see what happens when we apply the elimination method.

EXAMPLE 5 Solve the system

$$y - 3x = 2, \quad \textbf{(1)}$$
$$y - 3x = 1. \quad \textbf{(2)}$$

We multiply by -1 on both sides of equation (2) and then add:

$$y - 3x = 2 \qquad \text{Equation (1)}$$
$$\underline{-y + 3x = -1} \qquad \text{Multiplying by } -1 \text{ on both sides of equation (2)}$$
$$0 = 1. \qquad \text{Adding}$$

We obtain a false equation, $0 = 1$, so there is no solution. The slope–intercept forms of these equations are

$$y = 3x + 2,$$
$$y = 3x + 1.$$

The slopes, 3, are the same and the y-intercepts, $(0, 2)$ and $(0, 1)$, are different. Thus the lines are parallel. They do not intersect. (See the graph at right.)

Do Exercise 8. ▶

7. Solve the system

$$3x = 5 + 2y,$$
$$2x + 3y - 1 = 0.$$

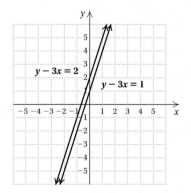

8. Solve the system

$$2x + y = 15,$$
$$4x + 2y = 23.$$

Answers

7. $\left(\frac{17}{13}, -\frac{7}{13}\right)$ **8.** No solution

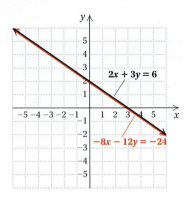

Sometimes there is an infinite number of solutions.

EXAMPLE 6   Solve the system

$$2x + 3y = 6, \qquad \textbf{(1)}$$
$$-8x - 12y = -24. \qquad \textbf{(2)}$$

We multiply by 4 on both sides of equation (1) and then add the two equations:

$$8x + 12y = 24 \qquad \text{Multiplying by 4 on both sides of equation (1)}$$
$$\underline{-8x - 12y = -24}$$
$$0 = 0. \qquad \text{Adding}$$

We have eliminated both variables, and what remains, $0 = 0$, is an equation easily seen to be true. If this happens when we use the elimination method, we have an infinite number of solutions. The equations in the system have the same graph. (See the graph at left.) Any point on the line gives a solution of the system.

◀ **Do Exercise 9.**

When decimals or fractions appear, we can first multiply to clear them. Then we proceed as before.

EXAMPLE 7   Solve the system

$$\frac{1}{3}x + \frac{1}{2}y = -\frac{1}{6}, \qquad \textbf{(1)}$$
$$\frac{1}{2}x + \frac{2}{5}y = \frac{7}{10}. \qquad \textbf{(2)}$$

The number 6 is the least common multiple of all the denominators of equation (1). The number 10 is the least common multiple of all the denominators of equation (2). We multiply by 6 on both sides of equation (1) and by 10 on both sides of equation (2):

$$6\left(\frac{1}{3}x + \frac{1}{2}y\right) = 6\left(-\frac{1}{6}\right) \qquad 10\left(\frac{1}{2}x + \frac{2}{5}y\right) = 10\left(\frac{7}{10}\right)$$
$$6 \cdot \frac{1}{3}x + 6 \cdot \frac{1}{2}y = -1 \qquad 10 \cdot \frac{1}{2}x + 10 \cdot \frac{2}{5}y = 7$$
$$2x + 3y = -1; \qquad 5x + 4y = 7.$$

The resulting system is

$$2x + 3y = -1,$$
$$5x + 4y = 7.$$

As we saw in Example 4, the solution of this system is $\left(\frac{25}{7}, -\frac{19}{7}\right)$.

◀ **Do Exercises 10 and 11.**

---

9. Solve the system

$$5x - 2y = 3, \qquad \textbf{(1)}$$
$$-15x + 6y = -9. \qquad \textbf{(2)}$$

Multiply equation (1) by 3 and add.

$$15x - \boxed{} = \boxed{}$$
$$\underline{-15x + 6y = -9}$$
$$0 = \boxed{}$$

The system has a(n) $\boxed{}$ number of solutions.

Solve each system.

10. $\frac{1}{2}x + \frac{3}{10}y = \frac{1}{5}$,
$\frac{3}{5}x + y = -\frac{2}{5}$

11. $3.3x + 6.6y = -6.6$,
$0.1x - 0.04y = 0.28$

Answers

9. Infinite number of solutions
10. $(1, -1)$ **11.** $(2, -2)$

Guided Solution:
9. $6y$, 9, 0, infinite

The following is a summary that compares the graphical, substitution, and elimination methods for solving systems of equations.

METHOD	STRENGTHS	WEAKNESSES
Graphical	Can "see" solution.	Inexact when solution involves numbers that are not integers or are very large and off the graph.
Substitution	Works well when solutions are not integers. Easy to use when a variable is alone on one side.	Introduces extensive computations with fractions for more complicated systems where coefficients are not 1 or -1. Cannot "see" solution.
Elimination	Works well when solutions are not integers, when coefficients are not 1 or -1, and when coefficients involve decimals or fractions.	Cannot "see" solution.

 7.3 **Exercise Set**

 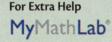
 Reading Check

Determine whether each statement is true or false.

RC1. When we are solving a system of equations in x and y using the elimination method, only x can be eliminated.

RC2. Before we add to eliminate a variable, the coefficients of the terms containing that variable should be opposites.

RC3. When solving a system of equations using the elimination method, we may need to multiply before adding in order to eliminate a variable.

RC4. When solving a system of equations using the elimination method, we never need to use substitution.

RC5. Solutions of systems of equations containing fractions cannot be found using the elimination method.

RC6. When we are solving a system of equations algebraically, if we obtain a false equation, the system has infinitely many solutions.

a   Solve using the elimination method.

1. $x - y = 7,$
$x + y = 5$

2. $x + y = 11,$
$x - y = 7$

3. $x + y = 8,$
$-x + 2y = 7$

4. $x + y = 6,$
$-x + 3y = -2$

5. $5x - y = 5,$
$3x + y = 11$

6. $2x - y = 8,$
$3x + y = 12$

7. $4a + 3b = 7,$
$-4a + b = 5$

8. $7c + 5d = 18,$
$c - 5d = -2$

9. $8x - 5y = -9,$
$3x + 5y = -2$

10. $3a - 3b = -15,$
$-3a - 3b = -3$

11. $4x - 5y = 7,$
$-4x + 5y = 7$

12. $2x + 3y = 4,$
$-2x - 3y = -4$

b   Solve using the multiplication principle first. Then add.

13. $x + y = -7,$
$3x + y = -9$

14. $-x - y = 8,$
$2x - y = -1$

15. $3x - y = 8,$
$x + 2y = 5$

16. $x + 3y = 19,$
$x - y = -1$

17. $x - y = 5,$
$4x - 5y = 17$

18. $x + y = 4,$
$5x - 3y = 12$

19. $2w - 3z = -1,$
$3w + 4z = 24$

20. $7p + 5q = 2,$
$8p - 9q = 17$

21. $2a + 3b = -1,$
$3a + 5b = -2$

22. $3x - 4y = 16,$
$5x + 6y = 14$

23. $x = 3y,$
$5x + 14 = y$

24. $5a = 2b,$
$2a + 11 = 3b$

25. $2x + 5y = 16,$
$3x - 2y = 5$

26. $3p - 2q = 8,$
$5p + 3q = 7$

27. $p = 32 + q,$
$3p = 8q + 6$

28. $3x = 8y + 11,$
$x + 6y - 8 = 0$

29. $3x - 2y = 10,$
$-6x + 4y = -20$

30. $2x + y = 13,$
$4x + 2y = 23$

31. $0.06x + 0.05y = 0.07,$
$0.4x - 0.3y = 1.1$

32. $1.8x - 2y = 0.9,$
$0.04x + 0.18y = 0.15$

33. $\frac{1}{3}x + \frac{3}{2}y = \frac{5}{4}$,

$\frac{3}{4}x - \frac{5}{6}y = \frac{3}{8}$

34. $x - \frac{3}{2}y = 13$,

$\frac{3}{2}x - y = 17$

35. $-4.5x + 7.5y = 6$,

$-x + 1.5y = 5$

36. $0.75x + 0.6y = -0.3$,

$3.9x + 5.2y = 96.2$

Skill Maintenance

Solve.

37. $2t - 13 - t = 5t + 7$ [2.3b]

38. $\frac{2}{3}x - \frac{1}{4} = \frac{x}{2}$ [2.3b]

39. $m - (5 - m) = 2(m + 1)$
   [2.3c]

40. $-20y \le 10$ [2.7d]

41. $2x^2 = x$ [5.7b]

42. $x^2 - x - 20 = 0$ [5.7b]

43. $\frac{a - 3}{a - 1} = \frac{2}{5}$ [6.7a]

44. $\frac{1}{x + 1} - \frac{1}{x} = \frac{2}{x - 2}$ [6.7a]

Synthesis

45.–48. Use the INTERSECT feature on a graphing calculator to solve the systems of equations in Exercises 5–8.

49.–54. Use the INTERSECT feature on a graphing calculator to solve the systems of equations in Exercises 25–30.

Solve using the substitution method, the elimination method, or the graphing method.

55. $3(x - y) = 9$,

$x + y = 7$

56. $2(x - y) = 3 + x$,

$x = 3y + 4$

57. $2(5a - 5b) = 10$,

$-5(6a + 2b) = 10$

58. $\frac{x}{3} + \frac{y}{2} = 1\frac{1}{3}$,

$x + 0.05y = 4$

59. $y = -\frac{2}{7}x + 3$,

$y = \frac{4}{5}x + 3$

60. $y = \frac{2}{5}x - 7$,

$y = \frac{2}{5}x + 4$

Solve for x and y.

61. $y = ax + b$,

$y = x + c$

62. $ax + by + c = 0$,

$ax + cy + b = 0$

Mid-Chapter Review

Concept Reinforcement

Determine whether each statement is true or false.

_____ **1.** A solution of a system of two equations is an ordered pair that makes at least one equation true.   [7.1a]

_____ **2.** Every system of two equations has one and only one ordered pair as a solution.   [7.1b]

_____ **3.** The system of equations $y = ax + b$ and $y = ax - b, b \neq 0$, has no solution.   [7.1b]

_____ **4.** The solution of the system of equations $x = a$ and $y = b$ is (a, b).   [7.1b]

Guided Solutions

 Fill in each blank with the number or expression that creates a correct solution.

Solve.

5. $x + y = -1,$ **(1)**
$\quad\,\, y = x - 3$ **(2)**   [7.2a]

$x + \boxed{} = -1$ Substituting for y in equation (1)

$\boxed{} = -1$ Simplifying

$2x = -1 + \boxed{}$

$2x = \boxed{}$ Simplifying

$x = \boxed{}$

$y = \boxed{} - 3$ Substituting for x in equation (2)

$y = \boxed{}$ Simplifying

The solution is $\left(\boxed{}, \boxed{} \right)$.

6. $2x - 3y = 7,$ **(1)**
$\quad\, x + 3y = -10$ **(2)**   [7.3a]

$2x - \quad 3y = \quad 7$
$\underline{\quad\, x + \quad 3y = -10}$

$\boxed{} x + \boxed{} y = \boxed{}$, or   Adding

$\boxed{} x \qquad\quad = \boxed{}$

$x = \boxed{}$

$\boxed{} + 3y = -10$ Substituting for x in equation (2)

$3y = \boxed{}$

$y = \boxed{}$

The solution is $\left(\boxed{}, \boxed{} \right)$.

Mixed Review

Determine whether the given ordered pair is a solution of the system of equations. Use alphabetical order of the variables.   [7.1a]

7. $(-4, 5);$ $x + y = 1,$
$\qquad\qquad\,\, x = y - 9$

8. $(6, -4);$ $x = y + 10,$
$\qquad\qquad\,\, x - y = 2$

9. $(-1, 1);$ $3x + 5y = 2,$
$\qquad\qquad\,\, 2x - y = -1$

10. $(2, -3);$ $2x + y = 1,$
$\qquad\qquad\,\, 3x - 2y = 12$

Solve each system of equations by graphing. [7.1b]

11. $x + y = 1,$
$x - y = 5$

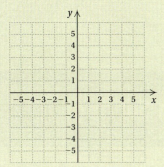

12. $2x + y = -1,$
$x + 2y = 4$

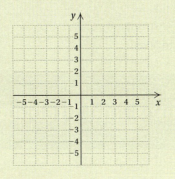

13. $2y = x - 1,$
$3x = 3 + 6y$

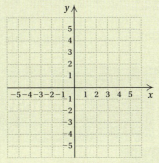

14. $x = y + 3,$
$y = x + 2$

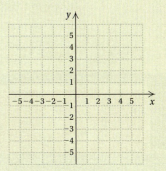

Solve using the substitution method. [7.2a, b]

15. $x + y = 2,$
$y = x - 8$

16. $x = y - 1,$
$2x - 5y = 1$

17. $x + y = 1,$
$3x + 6y = 1$

18. $2x + y = 2,$
$2x - y = -1$

Solve using the elimination method. [7.3a, b]

19. $x + y = 3,$
$-x - y = 5$

20. $3x - 2y = 2,$
$5x + 2y = -2$

21. $2x + 3y = 1,$
$3x + 2y = -6$

22. $2x - 3y = 6,$
$-4x + 6y = -12$

Solve. [7.2c]

23. *Dimensions of an Area Rug.* Lily buys an area rug with a perimeter of 18 ft. The width is 1 ft shorter than the length. Find the dimensions of the rug.

24. Find two numbers whose sum is 18 and whose difference is 86.

25. The difference of two numbers is 4. Two times the larger number is three times the smaller. What are the numbers?

Understanding Through Discussion and Writing

26. Suppose you have shown that the solution of the equation $3x - 1 = 9 - 2x$ is 2. How can this result be used to determine where the graphs of $y = 3x - 1$ and $y = 9 - 2x$ intersect? [7.1b]

27. Graph this system of equations. What happens when you try to determine a solution from the graph? [7.1b]
$$x - 2y = 6,$$
$$3x + 2y = 4$$

28. Janine can tell by inspection that the system
$$x = 2y - 1,$$
$$x = 2y + 3$$
has no solution. How does she know this? [7.1b]

29. Joel solves every system of two equations (in x and y) by first solving for y in the first equation and then substituting into the second equation. Is he using the best approach? Why or why not? [7.2b]

- ☐ Choose to improve your attitude and raise your goals.
- ☐ Choose to make a strong commitment to learning.
- ☐ Choose to place the primary responsibility for learning on yourself.
- ☐ Choose to allocate the proper amount of time to learn.

7.4 Applications and Problem Solving

OBJECTIVE

a Solve applied problems by translating to a system of two equations in two variables.

SKILL TO REVIEW

Objective 1.1b: Translate phrases to algebraic expressions.

Translate each phrase to an algebraic expression.

1. 52% of x liters
2. 26 less than a number

a

We now use systems of equations to solve applied problems that involve two equations in two variables.

EXAMPLE 1 *Produce Prices.*   Shelby and Donna buy much of their produce at the City Farmer's Market. One Saturday, Shelby bought 3 ears of corn and 2 sweet peppers for \$3.20. Donna bought 8 ears of corn and 1 sweet pepper for \$4.85. Determine the price of one ear of corn and one sweet pepper.

1. **Familiarize.**   We let c = the price of one ear of corn and p = the price of one sweet pepper.

2. **Translate.**   Each purchase gives us one equation.

Shelby's purchase: 3 ears of corn and 2 sweet peppers cost \$3.20

$$3c + 2p = 3.20$$

Donna's purchase: 8 ears of corn and 1 sweet pepper cost \$4.85

$$8c + p = 4.85$$

3. **Solve.**   We solve the system of equations

$$3c + 2p = 3.20 \quad \textbf{(1)}$$
$$8c + p = 4.85. \quad \textbf{(2)}$$

Although we could solve this system using graphing, substitution, or elimination, we decide to use elimination. If we multiply each side of equation (2) by -2 and add, the p-terms can be eliminated and we can solve for c:

$$
\begin{array}{ll}
3c + 2p = 3.20 & \text{Equation (1)} \\
\underline{-16c - 2p = -9.70} & \text{Multiplying equation (2) by } -2 \\
-13c = -6.50 & \text{Adding} \\
c = 0.50. &
\end{array}
$$

Next, we substitute 0.50 for c in equation (2) and solve for p:

$$
\begin{aligned}
8c + p &= 4.85 \\
8(0.50) + p &= 4.85 \\
4 + p &= 4.85 \\
p &= 0.85.
\end{aligned}
$$

Answers

Skill to Review:
1. 52% x, or $0.52x$
2. $n - 26$

4. **Check.** If one ear of corn cost $0.50 and one sweet pepper cost $0.85, then Shelby would have spent $3(\$0.50) + 2(\$0.85) = \$3.20$. Donna would have spent $8(\$0.50) + \$0.85 = \$4.85$. The prices check.

5. **State.** The price of one ear of corn was $0.50, and the price of one sweet pepper was $0.85.

Do Exercise 1. ▶

EXAMPLE 2 *IMAX Movie Prices.* There were 322 people at a showing of the IMAX 3D movie *To the Arctic*. Admission was $16.25 each for adults and $12.00 each for children, and receipts totaled $4790.50. How many adults and how many children attended?

1. **Familiarize.** To familiarize ourselves with the problem situation, let's make and check a guess.

The total number of people at the movie was 322, so we choose numbers that total 322. Let's try 242 adults and 80 children. How much money was taken in?

Money from adults: $242(\$16.25)$, or $3932.50
Money from children: $80(\$12.00)$, or $960
Total receipts: $3932.50 + \$960$, or $4892.50

Our guess is not the answer to the problem because the total taken in, according to the problem, was $4790.50. The steps we have used to see if our guess is correct, however, help us to understand the actual steps involved in solving the problem.

Let's list the information in a table. We let $a =$ the number of adults and $c =$ the number of children.

	ADULTS	CHILDREN	TOTAL	
ADMISSION	$16.25	$12.00		
NUMBER ATTENDING	a	c	322	→ $a + c = 322$
MONEY TAKEN IN	$16.25a$	$12.00c$	$4790.50	→ $16.25a + 12.00c$ $= 4790.50$

2. **Translate.** The total number of people attending was 322, so

$a + c = 322$.

The amount taken in from the adults was $16.25a$, and the amount taken in from the children was $12.00c$. These amounts are in dollars. The total was $4790.50, so we have

$16.25a + 12.00c = 4790.50$.

We can multiply by 100 on both sides to clear decimals. Thus we have a translation to a system of equations:

$$a + c = 322, \quad \textbf{(1)}$$
$$1625a + 1200c = 479{,}050. \quad \textbf{(2)} \quad \text{Multiplying by 100}$$

1. *Chicken and Hamburger Prices.* Fast Good Food offers a special two-and-one promotion. The price of one hamburger and two pieces of chicken is $5.39, and the price of two hamburgers and one piece of chicken is $5.68. Find the price of one hamburger and the price of one piece of chicken.

Answer

1. Hamburger: $1.99; chicken: $1.70

2. Game Admissions. There were 166 paid admissions to a high school basketball game. The price was $3.10 each for adults and $1.75 each for children. The amount collected was $459.25. How many adults and how many children attended? **GS**

Complete the following table to aid with the familiarization.

	ADULTS	CHILDREN	TOTAL
PAID ADMISSION	☐	$1.75	
NUMBER ATTENDING	a	c	☐
MONEY TAKEN IN	$3.10a$	☐	$459.25

$a + c = $ ☐

$3.10a + $ ☐ $= 459.25$

3. Solve. We solve the system using the elimination method since the equations are both in the form $Ax + By = C$. (A case can certainly be made for using the substitution method since we can solve for one of the variables quite easily in the first equation. Very often a decision is just a matter of preference.) We multiply by -1200 on both sides of equation (1) and then add and solve for a:

$$
\begin{array}{rl}
-1200a - 1200c = -386{,}400 & \text{Multiplying by } -1200 \\
\underline{1625a + 1200c = 479{,}050} & \\
425a = 92{,}650 & \text{Adding} \\
a = \dfrac{92{,}650}{425} & \text{Dividing by 425} \\
a = 218. &
\end{array}
$$

Next, we go back to equation (1), substituting 218 for a, and solve for c:

$$
\begin{aligned}
a + c &= 322 \\
218 + c &= 322 \\
c &= 104.
\end{aligned}
$$

4. Check. The check is left to the student. It is similar to what we did in the *Familiarize* step.

5. State. Attending the showing were 218 adults and 104 children.

◀ **Do Exercise 2.**

EXAMPLE 3 *Mixture of Solutions.* A chemist has one solution that is 80% acid (that is, 8 parts are acid and 2 parts are water) and another solution that is 30% acid. What is needed is 200 L of a solution that is 62% acid. The chemist will prepare it by mixing the two solutions. How much of each should be used?

1. **Familiarize.** We can make a drawing of the situation. The chemist uses x liters of the first solution and y liters of the second solution. We can also arrange the information in a table.

x liters y liters

80% solution   30% solution

$x + y$ liters

62% mixture

	FIRST SOLUTION	SECOND SOLUTION	MIXTURE	
AMOUNT OF SOLUTION	x	y	200 L	→ $x + y = 200$
PERCENT OF ACID	80%	30%	62%	
AMOUNT OF ACID IN SOLUTION	80%x	30%y	62% × 200, or 124 L	→ 80%x + 30%y = 124

2. Translate. The chemist uses x liters of the first solution and y liters of the second. Since the total is to be 200 L, we have

Total amount of solution: $x + y = 200$.

The amount of acid in the new mixture is to be 62% of 200 L, or 0.62(200 L), or 124 L. The amounts of acid from the two solutions are 80%x and 30%y. Thus,

Total amount of acid: $\quad 80\%x + 30\%y = 124$

or $\qquad\qquad\qquad\qquad 0.8x + 0.3y = 124$.

We clear decimals by multiplying by 10 on both sides of the second equation:

$$10(0.8x + 0.3y) = 10 \cdot 124$$
$$8x + 3y = 1240.$$

Thus we have a translation to a system of equations:

$$x + y = 200, \qquad \textbf{(1)}$$
$$8x + 3y = 1240. \qquad \textbf{(2)}$$

3. Solve. We solve the system. We use the elimination method, again because equations are in the form $Ax + By = C$ and a multiplication in one equation will allow us to eliminate a variable, but substitution would also work. We multiply by -3 on both sides of equation (1) and then add and solve for x:

$$
\begin{array}{ll}
-3x - 3y = -600 & \text{Multiplying by } -3 \\
\underline{8x + 3y = 1240} & \\
5x = 640 & \text{Adding} \\
x = \dfrac{640}{5} & \text{Dividing by 5} \\
x = 128. &
\end{array}
$$

Next, we go back to equation (1) and substitute 128 for x:

$$
\begin{array}{l}
x + y = 200 \\
128 + y = 200 \\
\phantom{128 + {}}y = 72.
\end{array}
$$

The solution is $x = 128$ and $y = 72$.

4. Check. The sum of 128 and 72 is 200. Also, 80% of 128 is 102.4 and 30% of 72 is 21.6. These add up to 124. The numbers check.

5. State. The chemist should use 128 L of the 80%-acid solution and 72 L of the 30%-acid solution.

Do Exercise 3. ▶

EXAMPLE 4 *Candy Mixtures.* A bulk wholesaler wishes to mix some candy worth 45 cents per pound and some worth 80 cents per pound in order to make 350 lb of a mixture worth 65 cents per pound. How much of each type of candy should be used?

1. Familiarize. Arranging the information in a table will help. We let $x =$ the amount of 45-cent candy and $y =$ the amount of 80-cent candy.

 3. *Mixture of Solutions.* One solution is 50% alcohol and a second is 70% alcohol. How much of each should be mixed in order to make 30 L of a solution that is 55% alcohol?

Complete the following table to aid in the familiarization.

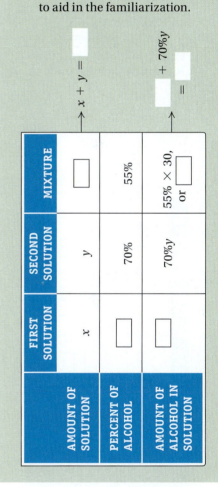

Answer

	INEXPENSIVE CANDY	EXPENSIVE CANDY	MIXTURE	
COST OF CANDY	45 cents	80 cents	65 cents	
AMOUNT (in pounds)	x	y	350	→ $x + y = 350$
TOTAL COST	$45x$	$80y$	65 cents · (350), or 22,750 cents	→ $45x + 80y = 22,750$

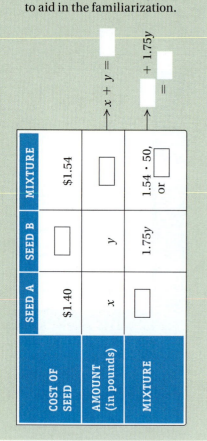

4. *Mixture of Grass Seeds.* **GS**

Grass seed A is worth $1.40 per pound and seed B is worth $1.75 per pound. How much of each should be mixed in order to make 50 lb of a mixture worth $1.54 per pound?

Complete the following table to aid in the familiarization.

Note the similarity of this problem to Example 2. Here we consider types of candy instead of groups of people.

2. Translate. We translate as follows. From the second row of the table, we have

Total amount of candy: $x + y = 350.$

Our second equation will come from the costs. The value of the inexpensive candy, in cents, is $45x$ (x pounds at 45 cents per pound). The value of the expensive candy is $80y$, and the value of the mixture is 65×350, or 22,750 cents. Thus we have

Total cost of mixture: $45x + 80y = 22,750.$

Remember the problem-solving tip about dimension symbols. In this last equation, all expressions are given in cents. We could have expressed them all in dollars, but we do not want some in cents and some in dollars. Thus we have a translation to a system of equations:

$$x + y = 350, \quad \textbf{(1)}$$
$$45x + 80y = 22,750. \quad \textbf{(2)}$$

3. Solve. We solve the system using the elimination method again. We multiply by -45 on both sides of equation (1) and then add and solve for y:

$$
\begin{array}{ll}
-45x - 45y = -15,750 & \text{Multiplying by } -45 \\
\underline{45x + 80y = 22,750} & \\
35y = 7,000 & \text{Adding} \\
y = \dfrac{7000}{35} & \\
y = 200. &
\end{array}
$$

Next, we go back to equation (1), substituting 200 for y, and solve for x:

$$
\begin{array}{l}
x + y = 350 \\
x + 200 = 350 \\
x = 150.
\end{array}
$$

4. Check. We consider $x = 150$ lb and $y = 200$ lb. The sum is 350 lb. The value of the candy is $45(150) + 80(200)$, or 22,750 cents and each pound of the mixture is worth $22,750 \div 350$, or 65 cents. These values check.

5. State. The grocer should mix 150 lb of the 45-cent candy with 200 lb of the 80-cent candy.

◄ **Do Exercise 4.**

Answer

4. Seed A: 30 lb; seed B: 20 lb

Guided Solution:

4.

$1.40	$1.75	$1.54
x	y	50
$1.40x$	$1.75y$	1.54 · 50, or 77

$x + y = 50,$
$1.40x + 1.75y = 77$

EXAMPLE 5 *Coin Value.* A student assistant at the university copy center has some nickels and dimes to use for change when students make copies. The value of the coins is $7.40. There are 26 more dimes than nickels. How many of each kind of coin are there?

1. **Familiarize.** We let d = the number of dimes and n = the number of nickels.

2. **Translate.** There are 26 more dimes than nickels, so we have

$$d = n + 26.$$

The value of the nickels, in cents, is $5n$, since each coin is worth 5 cents. The value of the dimes, in cents, is $10d$, since each coin is worth 10 cents. The total value is given as $7.40. Since we have the values of the nickels and dimes *in cents,* we must use cents for the total value. We express $7.40 as 740 cents. This gives us another equation:

$$10d + 5n = 740.$$

We now have a system of equations:

$$d = n + 26, \quad \textbf{(1)}$$
$$10d + 5n = 740. \quad \textbf{(2)}$$

3. **Solve.** Since we have d alone on one side of one equation, we use the substitution method. We substitute $n + 26$ for d in equation (2):

$10d + 5n = 740$	Equation (2)
$10(n + 26) + 5n = 740$	Substituting $n + 26$ for d
$10n + 260 + 5n = 740$	Removing parentheses
$15n + 260 = 740$	Collecting like terms
$15n = 480$	Subtracting 260
$n = \dfrac{480}{15}$, or 32.	Dividing by 15

Next, we substitute 32 for n in either of the original equations to find d. We use equation (1):

$$d = n + 26 = 32 + 26 = 58.$$

4. **Check.** We have 58 dimes and 32 nickels. There are 26 more dimes than nickels. The value of the coins is $58(\$0.10) + 32(\$0.05)$, which is $7.40. This checks.

5. **State.** The student assistant has 58 dimes and 32 nickels.

Do Exercise 5. ▶

Look back over Examples 2–5. The problems are quite similar in their structure. Compare them and try to see the similarities. The problems in Examples 2–5 are often called *mixture problems.* These problems provide a pattern, or model, for many related problems.

PROBLEM-SOLVING TIP

When solving a problem, first see if it is patterned or modeled after other problems that you have studied.

---

5. *Coin Value.* On a table are 20 coins, quarters and dimes. Their total value is $3.05. How many of each kind of coin are there?

Answer

5. Quarters: 7; dimes: 13

✓ Reading Check

Consider the following mixture problem.

Cherry Breeze is 30% fruit juice and Berry Choice is 15% fruit juice. How much of each should be used in order to make 10 gal of a drink that is 20% fruit juice?

The following table can be used to translate the problem. Choose the expression from the column to the right of the table that best fits each numbered space.

	CHERRY BREEZE	BERRY CHOICE	MIXTURE
NUMBER OF GALLONS	x	y	RC1. ___
PERCENT OF FRUIT JUICE	30%	RC2. ___%	20%
AMOUNT OF FRUIT JUICE	$0.3x$	RC3. ___y	RC4. ___

2
10
15
0.15

a   Solve.

1. *Assignments.*   The professor teaching Introduction to Sociology gives points for each discussion-board post and points for each reply to a post. Ana wrote 3 posts and 10 replies and received 95 points. Jae wrote 8 posts and 1 reply and received 125 points. Determine how many points a discussion post is worth and how many points a reply is worth.

2. *Video Games.*   Ethan and Ruth are competing in an online hidden objects game. Ethan found 1 magic ring and 8 four-leaf clovers for a total of 5950 points. Ruth found 4 magic rings and 5 four-leaf clovers for a total of 6250 points. Determine how many points a magic ring is worth and how many points a four-leaf clover is worth.

3. *Butterflies.*   Admission prices to the butterflies exhibit at the Brookfield Zoo in Illinois are $4 each for adults and $2 each for children. One day, 320 people visited the exhibit, and receipts totaled $820. How many adults and how many children visited that day?

4. *Zoo Admissions.*   The Bronx Zoo charges $17 admission for each adult and $13 for each child. One day, a total of $12,700 was collected from 860 admissions. How many adults and how many children were admitted that day?

5. *Photo Prints.*   Lucy paid $16.50 for 36 prints from Photo World. Some prints were 4 × 6 and the rest were 5 × 7, and they were priced as shown in the following table. How many prints of each size did she order?

PhotoWorld

PHOTO PRICES	
4 × 6	$0.19
5 × 7	$1.80
8 × 10	$3.90
11 × 14	$6.50

6. *Baseball Admissions.*   Members of the Benton Youth Club attended a baseball game, buying a total of 29 bleacher and lower reserved seats. Ticket prices are shown in the following table. The total cost of the tickets was $913. How many of each kind of ticket was bought?

TICKET INFORMATION	
Lower Box	$36
Upper Box	$21
Lower Reserved	$32
Upper Reserved	$17
Bleacher	$31

7. Basketball Scoring. In a game against the Orlando Magic, the Portland Trail Blazers scored 85 of their points on a combination of 40 two- and three-point baskets. How many of each type of shot was made?

Source: National Basketball Association

8. Basketball Scoring. Tony Parker of the San Antonio Spurs once scored 29 points on 17 shots in an NBA game, shooting only two-point shots and free throws (one point each). How many of each type of shot did he make?

Source: National Basketball Association

9. Investments. Cassandra has a number of $50 and $100 savings bonds to use for part of her college expenses. The total value of the bonds is $1250. There are 7 more $50 bonds than $100 bonds. How many of each type of bond does she have?

10. Commercial Lengths. During a football game, a television network aired both 30-sec commercials and 60-sec commercials. The total commercial time was 25 min, and there were 11 more 30-sec commercials shown than 60-sec commercials. How many of each length commercial were shown?

11. Mixture of Solutions. Solution A is 50% acid and solution B is 80% acid. How many liters of each should be used in order to make 100 L of a solution that is 68% acid? Complete the following table to aid in the familiarization.

	SOLUTION A	SOLUTION B	MIXTURE
AMOUNT OF SOLUTION	x	y	
PERCENT OF ACID	50%		68%
AMOUNT OF ACID IN SOLUTION		80%y	68% × 100, or

→ $x + y = (\quad)$

→ $50\%x + (\quad) = (\quad)$

12. Production. Clear Shine window cleaner is 12% alcohol and Sunstream window cleaner is 30% alcohol. How much of each should be used to make 90 oz of a cleaner that is 20% alcohol?

13. Grain Mixtures for Horses. Brianna needs to calculate the correct mix of grain and hay to feed her horse. On the basis of her horse's age, weight, and workload, she determines that he needs to eat 15 lb of feed per day, with an average protein content of 8%. Hay contains 6% protein, whereas grain has a 12% protein content. How many pounds of hay and grain should she feed her horse each day?

Source: Michael Plumb's Horse Journal, February 1996, pp. 26–29

14. Paint Mixtures. At a local "paint swap," Kari found large supplies of Skylite Pink (12.5% red pigment) and MacIntosh Red (20% red pigment). How many gallons of each color should Kari pick up in order to mix a gallon of Summer Rose (17% red pigment)?

15. **Coin Value.** A parking meter contains dimes and quarters worth $15.25. There are 103 coins in all. How many of each type of coin are there?

16. **Coin Value.** A vending machine contains nickels and dimes worth $14.50. There are 95 more nickels than dimes. How many of each type of coin are there?

17. **Coffee Blends.** Carolla's Coffee Shop mixes Brazilian coffee worth $19 per pound with Turkish coffee worth $22 per pound. The mixture is to sell for $20 per pound. How much of each type of coffee should be used in order to make a 300-lb mixture? Complete the following table to aid in the familiarization.

	BRAZILIAN COFFEE	TURKISH COFFEE	MIXTURE	
COST OF COFFEE	$19		$20	
AMOUNT (in pounds)	x	y	300	$\rightarrow x + y$ $= (\ \ )$
MIXTURE		$22y$	20(300), or $6000	$\rightarrow 19x + (\ \ )$ $= 6000$

18. **Coffee Blends.** The Java Joint wishes to mix organic Kenyan coffee beans that sell for $7.25 per pound with organic Venezuelan beans that sell for $8.50 per pound in order to form a 50-lb batch of Morning Blend that sells for $8.00 per pound. How many pounds of each type of bean should be used to make the blend?

19. **Mixed Nuts.** A customer has asked a caterer to provide 60 lb of nuts, 60% of which are to be cashews. The caterer has available mixtures of 70% cashews and 45% cashews. How many pounds of each mixture should be used?

20. **Mixture of Grass Seeds.** Grass seed A is worth $2.50 per pound and seed B is worth $1.75 per pound. How much of each would you use in order to make 75 lb of a mixture worth $2.14 per pound?

21. **Cough Syrup.** Dr. Zeke's cough syrup is 2% alcohol. Vitabrite cough syrup is 5% alcohol. How much of each type should be used in order to prepare an 80-oz batch of cough syrup that is 3% alcohol?

22. **Mixture of Solutions.** Solution A is 30% alcohol and solution B is 75% alcohol. How much of each should be used in order to make 100 L of a solution that is 50% alcohol?

23. **Test Scores.** Anna is taking a test in which items of type A are worth 10 points and items of type B are worth 15 points. It takes 3 min to complete each item of type A and 6 min to complete each item of type B. The total time allowed is 60 min and Anna answers exactly 16 questions. How many questions of each type did she complete? Assuming that all her answers were correct, what was her score?

24. **Gold Alloys.** A goldsmith has two alloys that are different purities of gold. The first is three-fourths pure gold and the second is five-twelfths pure gold. How many ounces of each should be melted and mixed in order to obtain a 6-oz mixture that is two-thirds pure gold?

25. **Ages.** The Kuyatts' house is twice as old as the Marconis' house. Eight years ago, the Kuyatts' house was three times as old as the Marconis' house. How old is each house?

26. **Ages.** David is twice as old as his daughter. In 4 years, David's age will be three times what his daughter's age was 6 years ago. How old are they now?

27. _Ages._ Randy is four times as old as Marie. In 12 years, Marie's age will be half of Randy's. How old are they now?

28. _Ages._ Jennifer is twice as old as Ramon. The sum of their ages 7 years ago was 13. How old are they now?

29. _Supplementary Angles._ **Supplementary angles** are angles whose sum is 180°. Two supplementary angles are such that one is 30° more than two times the other. Find the angles.

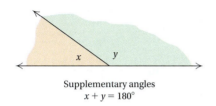

Supplementary angles
$x + y = 180°$

30. _Supplementary Angles._ Two supplementary angles are such that one is 8° less than three times the other. Find the angles.

31. _Complementary Angles._ **Complementary angles** are angles whose sum is 90°. Two complementary angles are such that their difference is 34°. Find the angles.

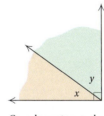

Complementary angles
$x + y = 90°$

32. _Complementary Angles._ Two angles are complementary. One angle is 42° more than one-half the other. Find the angles.

33. _Octane Ratings._ In most areas of the United States, gas stations offer three grades of gasoline, indicated by octane ratings on the pumps, such as 87, 89, and 93. When a tanker delivers gas, it brings only two grades of gasoline, the highest and the lowest, filling two large underground tanks. If you purchase the middle grade, the pump's computer mixes the other two grades appropriately. How much 87-octane gas and 93-octane gas should be blended in order to make 18 gal of 89-octane gas?

Source: Exxon

34. _Octane Ratings._ Refer to Exercise 33. Suppose the pump grades offered are 85, 87, and 91. How much 85-octane gas and 91-octane gas should be blended in order to make 12 gal of 87-octane gas?

Source: Exxon

35. _Printing._ A printer knows that a page of print contains 830 words if large type is used and 1050 words if small type is used. A document containing 11,720 words fills exactly 12 pages. How many pages are in the large type? in the small type?

36. _Paint Mixture._ A merchant has two kinds of paint. If 9 gal of the inexpensive paint is mixed with 7 gal of the expensive paint, the mixture will be worth $19.70 per gallon. If 3 gal of the inexpensive paint is mixed with 5 gal of the expensive paint, the mixture will be worth $19.825 per gallon. What is the price per gallon of each type of paint?

Skill Maintenance

Perform the indicated operations and simplify.

37. $(2x^2 - 3) - (x^2 - x - 3)$
[4.4c]

38. $\left(a + \dfrac{1}{2}\right)\left(a - \dfrac{1}{2}\right)$ [4.6b]

39. $(t^2 + 1.2)^2$ [4.6c]

40. $(3x + 5)(2x - 7)$ [4.6a]

41. $(x - 1)(x^2 + x + 1)$ [4.5d]

42. $(3mn - m^2n - n^2) + (mn^2 + n^2)$
[4.7d]

43. $\dfrac{x + 7}{x^2 - 1} - \dfrac{3}{x + 1}$ [6.5a]

44. $\dfrac{3 - x}{x - 2} - \dfrac{x - 7}{2 - x}$ [6.5a]

45. $\dfrac{1}{x} - \dfrac{1}{x^2} + \dfrac{1}{x + 1}$ [6.5b]

46. $\dfrac{a^2 + a - 20}{a^2 - 4} \cdot \dfrac{2a^2 + 3a - 2}{a^2 + 10a + 25}$
[6.1d]

47. $\dfrac{10c^2d}{3x^2} \div \dfrac{30cd}{x^3}$ [6.2b]

48. $\dfrac{t^4 - 16}{t^4 - 1} \div \dfrac{t^2 + 4}{t^2 + 1}$ [6.2b]

Find the intercepts. Then graph the equation.   [3.2a]

49. $y = -2x - 3$

50. $y = -0.1x + 0.4$

51. $5x - 2y = -10$

52. $2.5x + 4y = 10$

Synthesis

53. *Milk Mixture.*   A farmer has 100 L of milk that is 4.6% butterfat. How much skim milk (no butterfat) should be mixed with it in order to make milk that is 3.2% butterfat?

54. One year, Shannon made $85 from two investments: $1100 was invested at one yearly rate and $1800 at a rate that was 1.5% higher. Find the two rates of interest.

55. *Automobile Maintenance.*   An automobile radiator contains 16 L of antifreeze and water. This mixture is 30% antifreeze. How much of this mixture should be drained and replaced with pure antifreeze so that the mixture will be 50% antifreeze?

56. *Employer Payroll.*   An employer has a daily payroll of $1225 when employing some workers at $80 per day and others at $85 per day. When the number of $80 workers is increased by 50% and the number of $85 workers is decreased by $\frac{1}{5}$, the new daily payroll is $1540. How many were originally employed at each rate?

57. A two-digit number is six times the sum of its digits. The tens digit is 1 more than the ones digit. Find the number.

Applications with Motion

OBJECTIVE

a Solve motion problems using the formula $d = rt$.

a We first studied problems involving motion in Chapter 6. Here we extend our problem-solving skills by solving certain motion problems whose solutions can be found using systems of equations. Recall the motion formula.

THE MOTION FORMULA

Distance = Rate (or speed) · Time
$$d = rt$$

SKILL TO REVIEW

Objective 2.3c: Solve equations by first removing parentheses and collecting like terms.

Solve.

1. $(x + 3)2 = (x - 1)5$
2. $18(y + 1) = 20y$

We use five steps for problem solving. The tips in the margin at right are also helpful when solving motion problems.

As we saw in Chapter 6, there are motion problems that can be solved with just one equation. Let's start with another such problem.

EXAMPLE 1 *Car Travel.* Two cars leave Ashland at the same time traveling in opposite directions. One travels at 60 mph and the other at 30 mph. In how many hours will they be 150 mi apart?

1. Familiarize. We first make a drawing.

30 mph   60 mph
Distance of slower car   Distance of faster car
150 miles

TIPS FOR SOLVING MOTION PROBLEMS

1. Draw a diagram using an arrow or arrows to represent distance and the direction of each object in motion.
2. Organize the information in a chart.
3. Look for as many things as you can that are the same so that you can write equations.

From the wording of the problem and the drawing, we see that the distances may *not* be the same. But the times that the cars travel are the same, so we can use just t for time. We can organize the information in a chart.

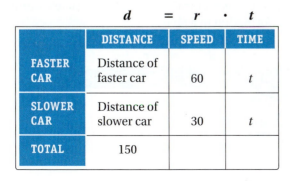

$$d = r \cdot t$$

	DISTANCE	SPEED	TIME
FASTER CAR	Distance of faster car	60	t
SLOWER CAR	Distance of slower car	30	t
TOTAL	150		

Answers

Skill to Review:

1. $\frac{11}{3}$ 2. 9

2. **Translate.** From the drawing, we see that

(Distance of faster car) + (Distance of slower car) = 150.

Then using $d = rt$ in each row of the table, we get

$60t + 30t = 150.$

3. **Solve.** We solve the equation:

$$60t + 30t = 150$$
$$90t = 150 \qquad \text{Collecting like terms}$$
$$t = \frac{150}{90}, \text{ or } \frac{5}{3}, \text{ or } 1\frac{2}{3}\text{ hr.} \qquad \text{Dividing by 90}$$

4. **Check.** When $t = \frac{5}{3}$ hr,

(Distance of faster car) + (Distance of slower car) $= 60\left(\dfrac{5}{3}\right) + 30\left(\dfrac{5}{3}\right)$

$= 100 + 50, \text{ or } 150 \text{ mi.}$

Thus the time of $\frac{5}{3}$ hr, or $1\frac{2}{3}$ hr, checks.

5. **State.** In $1\frac{2}{3}$ hr, the cars will be 150 mi apart.

◀ **Do Exercise 1 and 2.**

Now let's solve some motion problems using systems of equations.

EXAMPLE 2 *Train Travel.* A train leaves Stanton traveling east at 35 miles per hour (mph). An hour later, another train leaves Stanton on a parallel track at 40 mph. How far from Stanton will the second (or faster) train catch up with the first (or slower) train?

1. **Familiarize.** We first make a drawing.

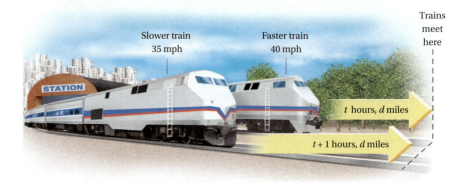

From the drawing, we see that the distances are the same. Let's call the distance d. We don't know the times. We let $t =$ the time for the faster train. Then the time for the slower train $= t + 1$, since it left 1 hr earlier. We can organize the information in a chart.

$$d \; = \; r \; \cdot \; t$$

	DISTANCE	SPEED	TIME	
SLOWER TRAIN	d	35	$t + 1$	→ $d = 35(t + 1)$
FASTER TRAIN	d	40	t	→ $d = 40t$

---

1. *Car Travel.* Two cars leave town at the same time traveling in opposite directions. One travels at 48 mph and the other at 60 mph. How far apart will they be 3 hr later? (*Hint*: The times are the same. Be *sure* to make a drawing.)

2. *Car Travel.* Two cars leave town at the same time traveling in the same direction. One travels at 35 mph and the other at 40 mph. In how many hours will they be 15 mi apart? (*Hint*: The times are the same. Be *sure* to make a drawing.)

Answers

1. 324 mi   **2.** 3 hr

2. Translate. In motion problems, we look for quantities that are the same so that we can write equations. From each row of the chart, we get an equation, $d = rt$. Thus we have two equations:

$$d = 35(t + 1), \quad \textbf{(1)}$$
$$d = 40t. \quad \textbf{(2)}$$

3. Solve. Since we have a variable alone on one side, we solve the system using the substitution method:

$$35(t + 1) = 40t \quad \text{Using the substitution method (substituting } 35(t+1) \text{ for } d \text{ in equation 2)}$$
$$35t + 35 = 40t \quad \text{Removing parentheses}$$
$$35 = 5t \quad \text{Subtracting } 35t$$
$$\frac{35}{5} = t \quad \text{Dividing by 5}$$
$$7 = t.$$

The problem asks us to find how far from Stanton the faster train catches up with the other. Thus we need to find d. We can do this by substituting 7 for t in the equation $d = 40t$:

$$d = 40(7)$$
$$= 280.$$

4. Check. If the time is 7 hr, then the distance that the slower train travels is $35(7 + 1)$, or 280 mi. The faster train travels $40(7)$, or 280 mi. Since the distances are the same, we know how far from Stanton the trains will be when the faster train catches up with the other.

5. State. The faster train will catch up with the slower train 280 mi from Stanton.

Do Exercise 3. ▶

EXAMPLE 3 *Boat Travel.* A motorboat took 3 hr to make a downstream trip with a 6-km/h current. The return trip against the same current took 5 hr. Find the speed of the boat in still water.

Upstream, $r - 6$
6-km/h current, 5 hours,
d kilometers

Downstream, $r + 6$
6-km/h current, 3 hours,
d kilometers

1. Familiarize. We first make a drawing. From the drawing, we see that the distances are the same. Let's call the distance d. We let $r =$ the speed of the boat in still water. Then, when the boat is traveling downstream, its speed is $r + 6$. (The current helps the boat along.) When it is traveling upstream, its speed is $r - 6$. (The current holds the boat back.)

---

3. *Car Travel.* A car leaves Spokane traveling north at 56 km/h. Another car leaves Spokane 1 hr later traveling north at 84 km/h. How far from Spokane will the second car catch up with the first? (*Hint*: The cars travel the same distance.)

1. Familiarize. Let $t =$ the time for the first car. Then ☐ $=$ the time for the second car.

2. Translate.
$$d = 56t, \quad \text{First car}$$
$$d = 84(\ \ ) \quad \text{Second car}$$

3. Solve.
$$84(t - 1) = 56t$$
$$84t - \boxed{} = 56t$$
$$-84 = \boxed{}$$
$$\boxed{} = t$$

If $t = 3$, then
$$d = 56(\boxed{}) = \boxed{}.$$

4. Check. The first car travels 168 km in ☐ hr, and the second car travels 168 km in ☐ hr.

5. State. The second car will catch up with the first car in ☐ km.

Answer

3. 168 km

Guided Solution:
3. $t - 1, t - 1, 84, -28t, 3, 3, 168, 3, 2, 168$

We can organize the information in a chart. In this case, the distances are the same, so we use the formula $d = rt$.

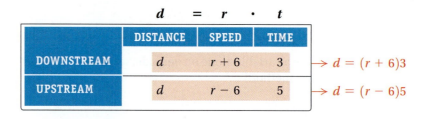

d	$=$	r	$\cdot$	t
	DISTANCE	SPEED	TIME	
DOWNSTREAM	d	$r + 6$	3	$\rightarrow d = (r + 6)3$
UPSTREAM	d	$r - 6$	5	$\rightarrow d = (r - 6)5$

2. Translate.   From each row of the chart, we get an equation, $d = rt$:

$$d = (r + 6)3, \quad \textbf{(1)}$$
$$d = (r - 6)5. \quad \textbf{(2)}$$

3. Solve.   Since there is a variable alone on one side of an equation, we solve the system using substitution:

$(r + 6)3 = (r - 6)5$	Substituting $(r + 6)3$ for d in equation (2)
$3r + 18 = 5r - 30$	Removing parentheses
$-2r + 18 = -30$	Subtracting $5r$
$-2r = -48$	Subtracting 18
$r = \dfrac{-48}{-2}$, or 24.	Dividing by -2

4. Check.   When $r = 24$, $r + 6 = 24 + 6$, or 30, and $30 \cdot 3 = 90$, the distance downstream. When $r = 24$, $r - 6 = 24 - 6$, or 18, and $18 \cdot 5 = 90$, the distance upstream. In both cases, we get the same distance so the answer checks.

5. State.   The speed in still water is 24 km/h.

◀ **Do Exercise 4.**

4. Air Travel.   An airplane flew for 5 hr with a 25-km/h tail wind. The return flight against the same wind took 6 hr. Find the speed of the airplane in still air. (*Hint*: The distance is the same both ways. The speeds are $r + 25$ and $r - 25$, where r is the speed in still air.)

Wind
$r + 25$
5 hr

Wind
$r - 25$
6 hr

MORE TIPS FOR SOLVING MOTION PROBLEMS

1. Translating to a system of equations eases the solution of many motion problems.

2. At the end of the problem, always ask yourself, "Have I found what the problem asked for?" You might have solved for a certain variable but still not have answered the question of the original problem. For instance, in Example 2 we solve for t but the question of the original problem asks for d. Thus we need to continue the *Solve* step.

Answer

4. 275 km/h

Translating for Success

1. **Car Travel.** Two cars leave town at the same time traveling in opposite directions. One travels 50 mph and the other travels 55 mph. In how many hours will they be 500 mi apart?

2. **Mixture of Solutions.** Solution A is 20% alcohol and solution B is 60% alcohol. How much of each should be used in order to make 10 L of a solution that is 50% alcohol?

3. **Triangle Dimensions.** The height of a triangle is 3 cm less than the base. The area is 27 cm². Find the height and the base.

4. **Fish Population.** To determine the number of fish in a lake, a conservationist catches 85 fish, tags them, and throws them back into the lake. Later, 60 fish are caught, 25 of which are tagged. How many fish are in the lake?

5. **Supplementary Angles.** Two angles are supplementary. One angle measures 36° more than three times the measure of the other. Find the measure of each angle.

The goal of these matching questions is to practice step (2), Translate, of the five-step problem-solving process. Translate each word problem to an equation or a system of equations and select a correct translation from A–O.

A. $20\%x + 60\%y = 50\% \cdot 10,$
   $x + y = 10$

B. $18 + 0.35x = 100$

C. $55x + 50x = 500$

D. $11x + 9x = 1$

E. $\dfrac{85}{x} = \dfrac{25}{60}$

F. $\dfrac{x}{11} + \dfrac{x}{9} = 1$

G. $\dfrac{1}{2}x(x - 3) = 27$

H. $x^2 + (x + 4)^2 = 8^2$

I. $8^2 + x^2 = (x + 4)^2$

J. $x + (3x + 36) = 180$

K. $20x + 60y = 5,$
   $x + y = 10$

L. $x + (3x + 36) + (x - 7)$
   $= 180$

M. $18 + 35x = 100$

N. $\dfrac{x}{85} = \dfrac{25}{60}$

O. $x + (3x + 36) = 90$

Answers on page A–24

6. **Triangle Dimensions.** The length of one leg of a right triangle is 8 m. The length of the hypotenuse is 4 m longer than the length of the other leg. Find the lengths of the hypotenuse and the other leg.

7. **Costs of Promotional Buttons.** The vice-president of the Spanish club has $100 to spend on promotional buttons for membership week. There is a setup fee of $18 and a cost of 35¢ per button. How many buttons can he purchase?

8. **Triangle Measures.** The second angle of a triangle measures 36° more than three times the measure of the first. The measure of the third angle is 7° less than the measure of the first. Find the measure of each angle of the triangle.

9. **Complementary Angles.** Two angles are complementary. One angle measures 36° more than three times the measure of the other. Find the measure of each angle.

10. **Work Time.** It takes Maggie 11 hr to paint a room. It takes Claire 9 hr to paint the same room. How long would it take to paint the room if they worked together?

 Reading Check

Complete each statement with the appropriate expression from the choices under each blank.

RC1. If Troy drove t hr at 60 mph, he traveled _____ mi.
$$60t, 60/t, t/60$$

RC2. Sophia paddles in still water at a rate of r mph. If she is paddling downstream in a river with a current of 2 mph, she is moving at a rate of _____ mph.
$$r + 2, r - 2, 2 - r$$

RC3. Rosa's motorboat travels r mph in still water. If she is motoring upstream in a river with a current of 4 mph, she is moving at a rate of _____ mph.
$$r + 4, r - 4, 4 - r$$

RC4. Jay's plane travels 125 mph in still air. If he is flying against a head wind of r mph, he is moving at a rate of _____ mph.
$$r + 125, r - 125, 125 - r$$

a   Solve. In Exercises 1–6, complete the chart to aid the translation.

1. *Car Travel.* Two cars leave town at the same time going in the same direction. One travels at 30 mph and the other travels at 46 mph. In how many hours will they be 72 mi apart?

	d	$=$ r	$\cdot$ t
	DISTANCE	**SPEED**	**TIME**
SLOWER CAR	Distance of slow car		t
FASTER CAR	Distance of fast car	46	

2. *Car and Truck Travel.* A truck and a car leave a service station at the same time and travel in the same direction. The truck travels at 55 mph and the car at 40 mph. They can maintain CB radio contact within a range of 10 mi. When will they lose contact?

	d	$=$ r	$\cdot$ t
	DISTANCE	**SPEED**	**TIME**
TRUCK	Distance of truck	55	
CAR	Distance of car		t

3. *Train Travel.* A train leaves a station and travels east at 72 mph. Three hours later, a second train leaves on a parallel track and travels east at 120 mph. When will it overtake the first train?

	d	$=$ r	$\cdot$ t
	DISTANCE	**SPEED**	**TIME**
SLOWER TRAIN	d		$t + 3$
FASTER TRAIN	d	120	

→ $d = 72(\quad)$

→ $d = (\quad)t$

4. *Airplane Travel.* A private airplane leaves an airport and flies due south at 192 mph. Two hours later, a jet leaves the same airport and flies due south at 960 mph. When will the jet overtake the plane?

	d	$=$ r	$\cdot$ t
	DISTANCE	**SPEED**	**TIME**
PRIVATE PLANE	d	192	
JET	d		t

→ $d = 192(\quad)$

→ $d = (\quad)t$

5. Canoeing. A canoeist paddled for 4 hr with a 6-km/h current to reach a campsite. The return trip against the same current took 10 hr. Find the speed of the canoe in still water.

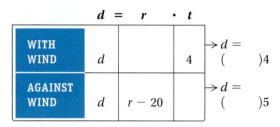

d	$=$	r	$\cdot$	t
DOWN-STREAM	d	$r + 6$		$\rightarrow d = (\quad)4$
UPSTREAM	d		10	$\rightarrow \quad = (r - 6)10$

6. Airplane Travel. An airplane flew for 4 hr with a 20-km/h tail wind. The return flight against the same wind took 5 hr. Find the speed of the plane in still air.

d	$=$	r	$\cdot$	t
WITH WIND	d		4	$\rightarrow d = (\quad)4$
AGAINST WIND	d	$r - 20$		$\rightarrow d = (\quad)5$

7. Train Travel. It takes a passenger train 2 hr less time than it takes a freight train to make the trip from Central City to Clear Creek. The passenger train averages 96 km/h, while the freight train averages 64 km/h. How far is it from Central City to Clear Creek?

8. Airplane Travel. It takes a small jet 4 hr less time than it takes a propeller-driven plane to travel from Glen Rock to Oakville. The jet averages 637 km/h, while the propeller plane averages 273 km/h. How far is it from Glen Rock to Oakville?

9. Motorboat Travel. On a weekend outing, Antoine rents a motorboat for 8 hr to travel down the river and back. The rental operator tells him to go downstream for 3 hr, leaving him 5 hr to return upstream.
  a) If the river current flows at a speed of 6 mph, how fast must Antoine travel in order to return in 8 hr?
  b) How far downstream did Antoine travel before he turned back?

10. Airplane Travel. For spring break, a group of students flew to Cancun. From Mexico City, the airplane took 2 hr to fly 600 mi against a head wind. The return trip with the wind took $1\frac{2}{3}$ hr. Find the speed of the plane in still air.

11. Running. A toddler starts running down a sidewalk at 230 ft/min. One minute later, a worried mother runs after the child at 660 ft/min. When will the mother overtake the toddler?

12. Airplane Travel. Two airplanes start at the same time and fly toward each other from points 1000 km apart at rates of 420 km/h and 330 km/h. When will they meet?

13. Motorcycle Travel. A motorcycle breaks down and the rider must jog the rest of the way to work. The motorcycle was being driven at 45 mph, and the rider jogs at a speed of 6 mph. The distance from home to work is 25 mi, and the total time for the trip was 2 hr. How far did the motorcycle go before it broke down?

14. Walking and Jogging. A student walks and jogs to college each day. She averages 5 km/h walking and 9 km/h jogging. The distance from home to college is 8 km, and she makes the trip in 1 hr. How far does the student jog?

Skill Maintenance

Factor completely. [5.6a]

15. $25x^2 - 81$

16. $12a^2 + 16a - 3$

17. $9y^3 - 12y^2 + 4y$

18. $7x^3 + 7x^2 + 14x + 14$

Synthesis

19. Lindbergh's Flight. Charles Lindbergh flew the Spirit of St. Louis in 1927 from New York to Paris at an average speed of 107.4 mph. Eleven years later, Howard Hughes flew the same route, averaged 217.1 mph, and took 16 hr and 57 min less time. Find the length of their route.

20. River Cruising. An afternoon sightseeing cruise up river and back down river is scheduled to last 1 hr. The speed of the current is 4 mph, and the speed of the riverboat in still water is 12 mph. How far upstream should the pilot travel before turning around?

Vocabulary Reinforcement

Complete each statement with the correct term from the column on the right. Some of the choices may be used more than once and some may not be used at all.

algebraic
graphical
intersection
union
infinitely many solutions
no solution
pair
variable

1. A solution of a system of two equations in two variables is an ordered _____ that makes both equations true. [7.1a]

2. To solve a system of equations graphically, we graph both equations and find the coordinates of any points of _____. [7.1b]

3. The substitution method is a(n) _____ method for solving systems of equations. [7.2a]

4. If, when solving a system algebraically, we obtain a false equation, then the system has _____. [7.3b]

5. If the graphs of the equations in a system of two equations are parallel, then the system has _____. [7.1b]

6. If the graphs of the equations in a system of two equations are the same line, then the system has _____. [7.1b]

Concept Reinforcement

Determine whether each statement is true or false.

_____ 1. A system of two linear equations can have exactly two solutions. [7.1b]

_____ 2. The solution(s) of a system of two equations can be found by determining where the graphs of the equations intersect. [7.1b]

_____ 3. When we obtain a false equation when solving a system of equations, the system has no solution. [7.3b]

_____ 4. If a system of equations has infinitely many solutions, then *any* ordered pair is a solution. [7.1b]

Study Guide

Objective 7.1a Determine whether an ordered pair is a solution of a system of equations.

Example Determine whether $(2, -3)$ is a solution of the system of equations

$$y = x - 5,$$
$$2x + y = 3.$$

Using alphabetical order of the variables, we substitute 2 for x and -3 for y in both equations.

$$\begin{array}{c|c} y = x - 5 \\ \hline -3 \ ? \ 2 - 5 \\ \quad\ -3 \quad \text{TRUE} \end{array}$$

$$\begin{array}{c|c} 2x + y = 3 \\ \hline 2 \cdot 2 + (-3) \ ? \ 3 \\ 4 - 3 \\ 1 \quad \text{FALSE} \end{array}$$

The pair $(2, -3)$ is not a solution of $2x + y = 3$, so it is not a solution of the system of equations.

Practice Exercise

1. Determine whether $(-2, 1)$ is a solution of the system of equations

$$x + 3y = 1,$$
$$y = x + 3.$$

Objective 7.1b   Solve systems of two linear equations in two variables by graphing.

Example   Solve this system of equations by graphing:

$$x - y = 1,$$
$$y = 2x - 4.$$

We graph the equations.

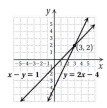

The point of intersection appears to be $(3, 2)$. We check this pair.

$$
\begin{array}{c|c}
x - y = 1 & \\
\hline
3 - 2 \; ? \; 1 & \\
1 \;\big| & \text{TRUE}
\end{array}
\qquad
\begin{array}{c|c}
y = 2x - 4 & \\
\hline
2 \; ? \; 2 \cdot 3 - 4 & \\
6 - 4 & \\
2 & \text{TRUE}
\end{array}
$$

The pair $(3, 2)$ checks in both equations. It is the solution.

Practice Exercise

2. Solve this system of equations by graphing:

$$2x + 3y = 2,$$
$$x + \; y = 2.$$

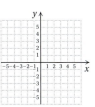

Objective 7.2b   Solve a system of two equations in two variables by the substitution method when neither equation has a variable alone on one side.

Example   Solve the system

$$x - 2y = 1, \quad \textbf{(1)}$$
$$2x - 3y = 3. \quad \textbf{(2)}$$

We solve equation (1) for x, because the coefficient of x is 1 in that equation:

$$x - 2y = 1$$
$$x = 2y + 1. \quad \textbf{(3)}$$

Now we substitute $2y + 1$ for x in equation (2) and solve for y:

$$2x - 3y = 3$$
$$2(2y + 1) - 3y = 3$$
$$4y + 2 - 3y = 3$$
$$y + 2 = 3$$
$$y = 1.$$

Next, we substitute 1 for y in either equation (1), (2), or (3) and find x. We choose equation (3) since it is already solved for x:

$$x = 2y + 1 = 2 \cdot 1 + 1 = 2 + 1 = 3.$$

We check the ordered pair $(3, 1)$ in both equations.

$$
\begin{array}{c|c}
x - 2y = 1 & \\
\hline
3 - 2 \cdot 1 \; ? \; 1 & \\
3 - 2 & \\
1 \;\big| & \text{TRUE}
\end{array}
\qquad
\begin{array}{c|c}
2x - 3y = 3 & \\
\hline
2 \cdot 3 - 3 \cdot 1 \; ? \; 3 & \\
6 - 3 & \\
3 \;\big| & \text{TRUE}
\end{array}
$$

The pair $(3, 1)$ checks in both equations. It is the solution.

Practice Exercise

3. Solve the system of equations

$$x + \; y = -1,$$
$$2x + 5y = 1.$$

Objective 7.3b Solve a system of two equations in two variables using the elimination method when multiplication is necessary.

Example Solve the system

$$2a - 3b = 7, \quad \textbf{(1)}$$
$$3a - 2b = 8. \quad \textbf{(2)}$$

We could eliminate either a or b. Here we decide to eliminate the a-terms.

$$6a - 9b = 21 \qquad \text{Multiplying equation (1) by 3}$$
$$\underline{-6a + 4b = -16} \qquad \text{Multiplying equation (2) by } -2$$
$$-5b = 5 \qquad \text{Adding}$$
$$b = -1 \qquad \text{Solving for } b$$

Next, we substitute -1 for b in either of the original equations:

$$2a - 3b = 7 \qquad \text{Equation (1)}$$
$$2a - 3(-1) = 7$$
$$2a + 3 = 7$$
$$2a = 4$$
$$a = 2.$$

The ordered pair $(2, -1)$ checks in both equations, so it is the solution of the system of equations.

Practice Exercise

4. Solve the system of equations

$$3x + 2y = 6,$$
$$x - y = 7.$$

Review Exercises

Determine whether the given ordered pair is a solution of the system of equations. [7.1a]

1. $(6, -1)$; $x - y = 3,$
   $2x + 5y = 6$

2. $(2, -3)$; $2x + y = 1,$
   $x - y = 5$

3. $(-2, 1)$; $x + 3y = 1,$
   $2x - y = -5$

4. $(-4, -1)$; $x - y = 3,$
   $x + y = -5$

Solve each system by graphing. [7.1b]

5. $x + y = 3,$
   $x - y = 7$

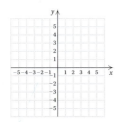

6. $x - 3y = 3,$
   $2x - 6y = 6$

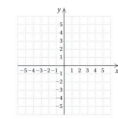

7. $3x - 2y = -4,$
   $2y - 3x = -2$

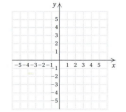

Solve each system using the substitution method. [7.2a, b]

8. $y = 5 - x,$
   $3x - 4y = -20$

9. $x + y = 6,$
   $y = 3 - 2x$

10. $x - y = 4,$
    $y = 2 - x$

11. $s + t = 5,$
    $s = 13 - 3t$

12. $x + 2y = 6,$
    $2x + 3y = 8$

13. $3x + y = 1,$
    $x - 2y = 5$

Solve each system using the elimination method. [7.3a, b]

14. $x + y = 4,$
$2x - y = 5$

15. $x + 2y = 9,$
$3x - 2y = -5$

16. $x - y = 8,$
$2x - 2y = 7$

17. $2x + 3y = 8,$
$5x + 2y = -2$

18. $5x - 2y = 2,$
$3x - 7y = 36$

19. $-x - y = -5,$
$2x - y = 4$

20. $6x + 2y = 4,$
$10x + 7y = -8$

21. $-6x - 2y = 5,$
$12x + 4y = -10$

22. $\frac{2}{3}x + y = -\frac{5}{3},$
$x - \frac{1}{3}y = -\frac{13}{3}$

Solve. [7.2c], [7.4a]

23. *Rectangle Dimensions.* The perimeter of a rectangle is 96 cm. The length is 27 cm more than the width. Find the length and the width.

24. *Paid Admissions.* There were 508 people at a choral concert. Orchestra seats cost $25 each and balcony seats cost $18 each. The total receipts were $11,223. Find the number of orchestra seats and the number of balcony seats sold for the concert.

25. *Window Cleaner.* Clear Shine window cleaner is 30% alcohol, whereas Sunstream window cleaner is 60% alcohol. How much of each is needed to make 80 L of a cleaner that is 45% alcohol?

26. *Weights of Elephants.* A zoo has both an Asian elephant and an African elephant. The African elephant weighs 2400 kg more than the Asian elephant. Together, they weigh 12,000 kg. How much does each elephant weigh?

Asian elephant   African elephant

27. *Mixed Nuts.* Sandy's Catering needs to provide 13 lb of mixed nuts for a wedding reception. The wedding couple has allocated $71 for nuts. Peanuts cost $4.50 per pound and fancy nuts cost $7.00 per pound. How many pounds of each type should be mixed?

28. *Octane Ratings.* The octane rating of a gasoline is a measure of the amount of isooctane in the gas. How much 87-octane gas and 95-octane gas should be blended in order to make a 10-gal batch of 93-octane gas?

Source: Champlain Electric and Petroleum Equipment

29. *Age.* Jeff is three times as old as his son. In 13 years, Jeff will be twice as old as his son. How old is each now?

30. *Complementary Angles.* Two angles are complementary. Their difference is 26°. Find the measure of each angle.

31. *Supplementary Angles.* Two angles are supplementary. Their difference is 26°. Find the measure of each angle.

Solve. [7.5a]

32. *Air Travel.* An airplane flew for 4 hr with a 15-km/h tail wind. The return flight against the wind took 5 hr. Find the speed of the airplane in still air.

d =	r	·	t
	DISTANCE	SPEED	TIME
WITH WIND			
AGAINST WIND			

33. *Car Travel.* One car leaves Phoenix, Arizona, on Interstate highway I-10 traveling at a speed of 55 mph. Two hours later, another car leaves Phoenix traveling in the same direction on I-10 at a speed of 75 mph. How far from Phoenix will the second car catch up to the first?

d =	r	·	t
	DISTANCE	SPEED	TIME
SLOWER CAR			
FASTER CAR			

Solve each system of equations. [7.1b], [7.2a, b], [7.3a, b]

34. $y = x - 2,$
$x - 2y = 6$

   A. The y-value is 0.
   B. The y-value is -12.
   C. The y-value is -2.
   D. The y-value is -4.

35. $3x + 2y = 5,$
$x - y = 5$

   A. The x-value is 3.
   B. The x-value is 2.
   C. The x-value is -2.
   D. The x-value is -3.

Synthesis

36. The solution of the following system is $(6, 2)$. Find C and D. [7.1a]

$$2x - Dy = 6,$$
$$Cx + 4y = 14$$

37. Solve: [7.2a]

$$3(x - y) = 4 + x,$$
$$x = 5y + 2.$$

38. *Value of a Horse.* Stephanie agreed to work as a stablehand for 1 year. At the end of that time, she was to receive \$2400 and a horse. After 7 months, she quit the job, but still received the horse and \$1000. What was the value of the horse? [7.4a]

Each of the following shows the graph of a system of equations. Find the equations. [3.4c], [7.1b]

39.

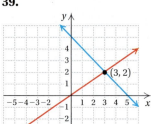

40.

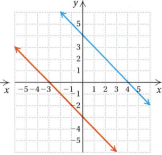

41. *Ancient Chinese Math Problem.* Several ancient Chinese books included problems that can be solved by translating to systems of equations. *Arithmetical Rules in Nine Sections* is a book of 246 problems compiled by a Chinese mathematician, Chang Tsang, who died in 152 B.C. One of the problems is: Suppose there are a number of rabbits and pheasants confined in a cage. In all, there are 35 heads and 94 feet. How many rabbits and how many pheasants are there? Solve the problem. [7.4a]

Understanding Through Discussion and Writing

1. James can tell by inspection that the system
$$2x - y = 3,$$
$$-4x + 2y = -6$$
has an infinite number of solutions. How did he determine this? [7.1b]

2. Explain how the addition and multiplication principles are used to solve systems of equations using the elimination method. [7.3a, b]

3. Which of the five problem-solving steps have you found the most challenging? Why? [7.4a], [7.5a]

4. Discuss the advantages of using a chart to organize information when solving a motion problem. [7.5a]

CHAPTER

7   **Test**

For Extra Help   For step-by-step test solutions, access the Chapter Test Prep Videos in
MyMathLab® or on YouTube (search "BittingerIntro" and click on "Channels").

1. Determine whether the given ordered pair is a solution of the system of equations.

$$(-2, -1);\ 2x - 3y = 4,$$
$$x = 4 + 2y$$

2. Solve this system by graphing. Show your work.

$$x - y = 3,$$
$$x - 2y = 4$$

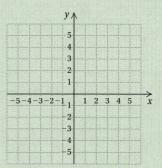

Solve each system using the substitution method.

3. $y = 6 - x,$
  $2x - 3y = 22$

4. $x + 2y = 5,$
  $x + y = 2$

5. $y = 5x - 2,$
  $y - 2 = x$

Solve each system using the elimination method.

6. $x - y = 6,$
  $3x + y = -2$

7. $\dfrac{1}{2}x - \dfrac{1}{3}y = 8,$
  $\dfrac{1}{3}x - \dfrac{2}{9}y = 1$

8. $-4x - 9y = 4,$
  $6x + 3y = 1$

9. $2x + 3y = 13,$
  $3x - 5y = 10$

Solve.

10. *Rectangle Dimensions.* The perimeter of a rectangular field is 8266 yd. The length is 84 yd more than the width. Find the length and the width.

11. *Mixture of Solutions.* Solution A is 25% acid, and solution B is 40% acid. How much of each is needed to make 60 L of a solution that is 30% acid?

12. *Motorboat Travel.* A motorboat traveled for 2 hr with an 8-km/h current. The return trip against the same current took 3 hr. Find the speed of the motorboat in still water.

13. *Carnival Income.* A traveling carnival has receipts of $4275 one day. Twice as much was made on concessions as on the rides. How much did the concessions bring in? How much did the rides bring in?

14. *Farm Acreage.* The Rolling Velvet Horse Farm allots 650 acres to plant hay and oats. The owners know that their needs are best met if they plant 180 acres more of hay than of oats. How many acres of each should they plant?

15. *Supplementary Angles.* Two angles are supplementary. One angle measures 45° more than twice the measure of the other. Find the measure of each angle.

16. *Octane Ratings.* The octane rating of a gasoline is a measure of the amount of isooctane in the gas. How much 87-octane gas and 93-octane gas should be blended in order to make 12 gal of 91-octane gas?

Source: Champlain Electric and Petroleum Equipment

17. *Phone Rates.* A telephone company offers a domestic calling plan for $2.95 per month plus 10¢ per minute. Another plan charges $1.95 per month plus 15¢ per minute. For what number of minutes will the two plans cost the same?

18. *Ski Trip.* A group of students drove both a car and an SUV on a ski trip. The car left first and traveled at 55 mph. The SUV left 2 hr later and traveled at 65 mph. How long did it take the SUV to catch up to the car?

19. Solve: $x - 2y = 4$,
$2x - 3y = 3$.

A. Both x and y are positive.
B. x is positive; y is negative.
C. x is negative; y is positive.
D. Both x and y are negative.

Synthesis

20. Find the numbers C and D such that $(-2, 3)$ is a solution of the system

$$Cx - 4y = 7,$$
$$3x + Dy = 8.$$

21. *Ticket Line.* Lily is in line at a ticket window. There are two more people ahead of her than there are behind her. In the entire line, there are three times as many people as there are behind her. How many are ahead of Lily in line?

Each of the following shows the graph of a system of equations. Find the equations.

22.

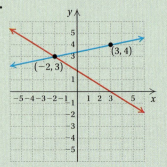

23.

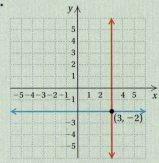

Compute and simplify.

1. $-2[1.4 - (-0.8 - 1.2)]$

2. $(1.3 \times 10^8)(2.4 \times 10^{-10})$

3. $\left(-\dfrac{1}{6}\right) \div \left(\dfrac{2}{9}\right)$

4. $\dfrac{2^{12}\,2^{-7}}{2^8}$

Simplify.

5. $\dfrac{x^2 - 9}{2x^2 - 7x + 3}$

6. $\dfrac{t^2 - 16}{(t + 4)^2}$

7. $\dfrac{x - \dfrac{x}{x + 2}}{\dfrac{2}{x} - \dfrac{1}{x + 2}}$

Perform the indicated operations and simplify.

8. $(1 - 3x^2)(2 - 4x^2)$

9. $(2a^2b - 5ab^2)^2$

10. $(3x^2 + 4y)(3x^2 - 4y)$

11. $-2x^2(x - 2x^2 + 3x^3)$

12. $(1 + 2x)(4x^2 - 2x + 1)$

13. $\left(8 - \dfrac{1}{3}x\right)\left(8 + \dfrac{1}{3}x\right)$

14. $(-8y^2 - y + 2) - (y^3 - 6y^2 + y - 5)$

15. $(2x^3 - 3x^2 - x - 1) \div (2x - 1)$

16. $\dfrac{7}{5x - 25} + \dfrac{x + 7}{5 - x}$

17. $\dfrac{2x - 1}{x - 2} - \dfrac{2x}{2 - x}$

18. $\dfrac{y^2 + y}{y^2 + y - 2} \cdot \dfrac{y + 2}{y^2 - 1}$

19. $\dfrac{7x + 7}{x^2 - 2x} \div \dfrac{14}{3x - 6}$

Factor completely.

20. $6x^5 - 36x^3 + 9x^2$

21. $16y^4 - 81$

22. $3x^2 + 10x - 8$

23. $4x^4 - 12x^2y + 9y^2$

24. $3m^3 + 6m^2 - 45m$

25. $x^3 + x^2 - x - 1$

Solve.

26. $3x - 4(x + 1) = 5$

27. $x(2x - 5) = 0$

28. $5x + 3 \geq 6(x - 4) + 7$

29. $1.5x - 2.3x = 0.4(x - 0.9)$

30. $2x^2 = 338$

31. $3x^2 + 15 = 14x$

32. $\dfrac{2}{x} - \dfrac{3}{x - 2} = \dfrac{1}{x}$

33. $1 + \dfrac{3}{x} + \dfrac{x}{x + 1} = \dfrac{1}{x^2 + x}$

34. $y = 2x - 9,$
$2x + 3y = -3$

35. $6x + 3y = -6,$
$-2x + 5y = 14$

36. $2x = y - 2,$
$3y - 6x = 6$

37. $N = rx - t$, for x

Solve.

38. *Digital Photo Frame.* Joel paid $37.10, including 6% sales tax, for a digital photo frame. What was the price of the frame itself?

39. Roofing Time. It takes David 15 hr to put a roof on a house. It takes Loren 12 hr to put a roof on the same type of house. How long would it take to complete the job if they worked together?

40. Triangle Dimensions. The length of one leg of a right triangle is 12 in. The length of the hypotenuse is 8 in. longer than the length of the other leg. Find the lengths of the hypotenuse and the other leg.

41. Quality Control. A sample of 120 computer chips contained 5 defective chips. How many defective chips would you expect to find in a batch of 1800 chips?

42. Triangle Dimensions. The height of a triangle is 5 ft more than the base. The area is 18 ft². Find the height and the base.

43. Height of a Parallelogram. The height h of a parallelogram of fixed area varies inversely as the base b. Suppose that the height is 24 ft when the base is 15 ft. Find the height when the base is 5 ft. What is the variation constant?

44. Travel Time. Two trains leave Brookston at the same time going in opposite directions. One travels 55 mph and the other travels 65 mph. In how many hours will they be 180 mi apart?

45. Mixing Solutions. Solution A is 10% salt, and solution B is 40% salt. How much of each should be used in order to make 100 mL of a solution that is 22% salt?

46. Find an equation of variation in which y varies directly as x, and $y = 2.4$ when $x = 12$.

47. Find the slope of the line containing the points $(2, 3)$ and $(-1, 3)$.

48. Find the slope and the y-intercept of the line $2x + 3y = 6$.

49. Find an equation of the line that contains the points $(-5, 6)$ and $(2, -4)$.

50. Find an equation of the line containing the point $(0, -3)$ and having the slope $m = 6$.

Graph on a plane.

51. $y = -2$

52. $2x + 5y = 10$

53. $y \leq 5x$

54. $5x - 1 < 24$

55. Solve by graphing:
$$3x - y = 4,$$
$$x + 3y = -2.$$

Synthesis

56. The solution of the following system of equations is $(-5, 2)$. Find A and B.
$$3x - Ay = -7,$$
$$Bx + 4y = 15$$

57. Solve: $x^2 + 2 < 0$.

58. Simplify:
$$\frac{x - 5}{x + 3} - \frac{x^2 - 6x + 5}{x^2 + x - 2} \div \frac{x^2 + 4x + 3}{x^2 + 3x + 2}.$$

59. Find the value of k such that $y - kx = 4$ and $10x - 3y = -12$ are perpendicular.

Radical Expressions and Equations

STUDYING FOR SUCCESS   *Working with Others*

☐ Try being a tutor for a fellow student. You may find that you understand concepts better after you explain them to someone else.

☐ Consider forming a study group.

☐ Verbalize the math. Often simply talking about a concept with a classmate can help clarify the material.

8.1

Introduction to Radical Expressions

OBJECTIVES

a Find the principal square roots and their opposites of the whole numbers from 0^2 to 25^2.

b Approximate square roots of real numbers using a calculator.

c Solve applied problems involving square roots.

d Identify radicands of radical expressions.

e Determine whether a radical expression represents a real number.

f Simplify a radical expression with a perfect-square radicand.

SKILL TO REVIEW

Objective R.5b: Evaluate exponential expressions.

Evaluate.

1. 7^2 **2.** $\left(\dfrac{1}{2}\right)^2$

a SQUARE ROOTS

When we raise a number to the second power, we have squared the number. Sometimes we may need to find the number that was squared. We call this process finding a square root of a number.

SQUARE ROOT

The number c is a **square root** of a if $c^2 = a$.

Every positive number has two square roots. For example, the square roots of 25 are 5 and -5 because $5^2 = 25$ and $(-5)^2 = 25$. The positive square root is also called the **principal square root**. The symbol $\sqrt{\ }$ is called a **radical*** (or **square root**) symbol. The radical symbol represents only the principal square root. Thus, $\sqrt{25} = 5$. To name the negative square root of a number, we use $-\sqrt{\ }$. The number 0 has only one square root, 0.

EXAMPLE 1   Find the square roots of 81.

The square roots are 9 and -9.

EXAMPLE 2   Find $\sqrt{225}$.

There are two square roots of 225, 15 and -15. We want the principal, or positive, square root since this is what $\sqrt{\ }$ represents. Thus,

$$\sqrt{225} = 15.$$

EXAMPLE 3   Find $-\sqrt{64}$.

The symbol $\sqrt{64}$ represents the positive square root. Then $-\sqrt{64}$ represents the negative square root. That is, $\sqrt{64} = 8$, so

$$-\sqrt{64} = -8.$$

◀ **Do Margin Exercises 1–10 on the following page.**

Answers

Skill to Review:

1. 49 **2.** $\dfrac{1}{4}$

*Radicals can be other than square roots, but we will consider only square-root radicals in Chapter 8. See Appendix C for other types of radicals.

We can think of the processes of "squaring" and "finding square roots" as inverses of each other. We square a number and get one answer. When we find the square roots of the answer, we get the original number *and* its opposite.

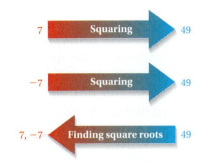

Find the square roots.

1. 36   **2.** 64

3. 121   **4.** 144

Find the following.

5. $\sqrt{16}$ **6.** $\sqrt{49}$

7. $\sqrt{100}$ **8.** $\sqrt{441}$

9. $-\sqrt{49}$ **10.** $-\sqrt{169}$

b APPROXIMATING SQUARE ROOTS

We often need to use rational numbers to *approximate* square roots that are irrational. Such approximations can be found using a calculator with a square-root key $\sqrt{}$.

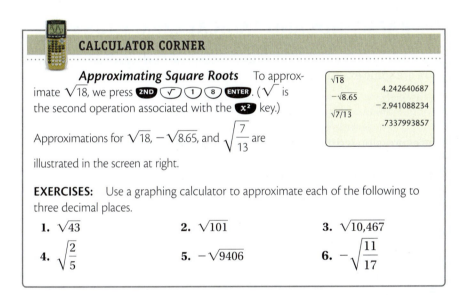

CALCULATOR CORNER

Approximating Square Roots   To approximate $\sqrt{18}$, we press **2ND** $\boxed{\sqrt{}}$ **1** **8** **ENTER**. ($\sqrt{}$ is the second operation associated with the $\boxed{x^2}$ key.)

Approximations for $\sqrt{18}$, $-\sqrt{8.65}$, and $\sqrt{\dfrac{7}{13}}$ are illustrated in the screen at right.

$\sqrt{18}$
 4.242640687
$-\sqrt{8.65}$
 −2.941088234
$\sqrt{7/13}$
 .7337993857

EXERCISES: Use a graphing calculator to approximate each of the following to three decimal places.

1. $\sqrt{43}$ **2.** $\sqrt{101}$ **3.** $\sqrt{10{,}467}$

4. $\sqrt{\dfrac{2}{5}}$ **5.** $-\sqrt{9406}$ **6.** $-\sqrt{\dfrac{11}{17}}$

EXAMPLES   Use a calculator to approximate each of the following.

Number	Using a calculator with a 10-digit readout	Rounded to three decimal places
4. $\sqrt{10}$	3.162277660	3.162
5. $-\sqrt{583.8}$	−24.16195356	−24.162
6. $\sqrt{\dfrac{48}{55}}$	0.934198733	0.934

Do Exercises 11–16. ▶

Use a calculator to approximate each of the following square roots to three decimal places.

11. $\sqrt{15}$ **12.** $\sqrt{30}$

13. $\sqrt{980}$ **14.** $-\sqrt{667.8}$

15. $\sqrt{\dfrac{2}{3}}$ **16.** $-\sqrt{\dfrac{203.4}{67.82}}$

Answers

1. 6, −6   **2.** 8, −8   **3.** 11, −11
4. 12, −12   **5.** 4   **6.** 7   **7.** 10
8. 21   **9.** −7   **10.** −13
11. 3.873   **12.** 5.477   **13.** 31.305
14. −25.842   **15.** 0.816   **16.** −1.732

c APPLICATIONS OF SQUARE ROOTS

We now consider an application that involves a formula with a radical expression.

EXAMPLE 7   *Speed of a Skidding Car.*   After an accident, how do police determine the speed at which the car had been traveling? The formula $r = 2\sqrt{5L}$ can be used to approximate the speed r, in miles per hour, of a car that has left a skid mark of length L, in feet. What was the speed of a car that left skid marks of length **(a)** 30 ft? **(b)** 150 ft?

a) We substitute 30 for L and find an approximation:

$$r = 2\sqrt{5L} = 2\sqrt{5 \cdot 30} = 2\sqrt{150} \approx 24.495.$$

The speed of the car was about 24.5 mph.

b) We substitute 150 for L and find an approximation:

$$r = 2\sqrt{5L} = 2\sqrt{5 \cdot 150} \approx 54.772.$$

The speed of the car was about 54.8 mph.

◀ **Do Exercise 17.**

17. *Speed of a Skidding Car.* Refer to Example 7. Determine the speed of a car that left skid marks of length **(a)** 40 ft; **(b)** 123 ft.

d RADICANDS AND RADICAL EXPRESSIONS

When an expression is written under a radical, we have a **radical expression**. Here are some examples:

$$\sqrt{14}, \qquad \sqrt{x}, \qquad 8\sqrt{x^2 + 4}, \qquad \sqrt{\dfrac{x^2 - 5}{2}}.$$

The expression written under the radical is called the **radicand**.

EXAMPLES   Identify the radicand in each expression.

8. $-\sqrt{105}$ — The radicand is 105.

9. $\sqrt{x} + 2$ — The radicand is x.

10. $\sqrt{x + 2}$ — The radicand is $x + 2$.

11. $6\sqrt{y^2 - 5}$ — The radicand is $y^2 - 5$.

12. $\sqrt{\dfrac{a - b}{a + b}}$ — The radicand is $\dfrac{a - b}{a + b}$.

◀ **Do Exercises 18–21.**

Identify the radicand.

18. $\sqrt{227}$

19. $-\sqrt{45 + x}$

20. $\sqrt{\dfrac{x}{x + 2}}$

21. $8\sqrt{x^2 + 4}$

e EXPRESSIONS THAT ARE MEANINGFUL AS REAL NUMBERS

The square of any nonzero number is always positive. For example, $8^2 = 64$ and $(-11)^2 = 121$. There are no real numbers that when squared yield negative numbers. Thus, $\sqrt{-100}$ does not represent a real number because there is no real number that when squared yields -100. We can try to square 10 and -10, but we know that $10^2 = 100$ and $(-10)^2 = 100$. Neither square is -100. Thus the following expressions do not represent real numbers (they are meaningless as real numbers):

$$\sqrt{-100}, \qquad \sqrt{-49}, \qquad -\sqrt{-3}.$$

Answers

17. (a) About 28.3 mph; **(b)** about 49.6 mph
18. 227   **19.** $45 + x$ **20.** $\dfrac{x}{x + 2}$
21. $x^2 + 4$

> **EXCLUDING NEGATIVE RADICANDS**
> ....
> Radical expressions with negative radicands do not represent real numbers.

Later in your study of mathematics, you may encounter a number system called the **complex numbers** in which negative numbers have defined square roots.

Do Exercises 22–25. ▶

Determine whether each expression represents a real number. Write "yes" or "no."

22. $-\sqrt{25}$ **23.** $\sqrt{-25}$

24. $-\sqrt{-36}$ **25.** $-\sqrt{36}$

f ⬛ PERFECT-SQUARE RADICANDS

The expression $\sqrt{x^2}$, with a perfect-square radicand, x^2, can be troublesome to simplify. Recall that $\sqrt{\ }$ denotes the principal square root. That is, the answer is nonnegative (either positive or zero). If x represents a nonnegative number, $\sqrt{x^2}$ simplifies to x. If x represents a negative number, $\sqrt{x^2}$ simplifies to $-x$ (the opposite of x), which is positive.

Suppose that $x = 3$. Then

$$\sqrt{x^2} = \sqrt{3^2} = \sqrt{9} = 3.$$

Suppose that $x = -3$. Then

$$\sqrt{x^2} = \sqrt{(-3)^2} = \sqrt{9} = 3, \quad \text{the } opposite \text{ of } -3.$$

Note that 3 is the *absolute value* of both 3 and -3. In general, when replacements for x are considered to be *any* real numbers, it follows that

$$\sqrt{x^2} = |x|,$$

and when $x = 3$ or $x = -3$,

$$\sqrt{x^2} = \sqrt{3^2} = |3| = 3 \quad \text{and} \quad \sqrt{x^2} = \sqrt{(-3)^2} = |-3| = 3.$$

> **PRINCIPAL SQUARE ROOT OF A^2**
>
> For any real number A,
> $$\sqrt{A^2} = |A|.$$
> (That is, for any real number A, the principal square root of A^2 is the absolute value of A.)

EXAMPLES Simplify. Assume that expressions under radicals represent any real number.

13. $\sqrt{10^2} = |10| = 10$

14. $\sqrt{(-7)^2} = |-7| = 7$

15. $\sqrt{(3x)^2} = |3x| = 3|x|$ Absolute-value notation is necessary.

16. $\sqrt{a^2b^2} = \sqrt{(ab)^2} = |ab|$

17. $\sqrt{x^2 + 2x + 1} = \sqrt{(x+1)^2} = |x+1|$

Do Exercises 26–31. ▶

Simplify. Assume that expressions under radicals represent any real number.

26. $\sqrt{(-13)^2}$ **27.** $\sqrt{(7w)^2}$

28. $\sqrt{(xy)^2}$ **29.** $\sqrt{x^2y^2}$

30. $\sqrt{(x-11)^2}$

GS **31.** $\sqrt{x^2 + 8x + 16}$
$= \sqrt{(x + \boxed{})^2}$
$= |\boxed{}|$

Answers

22. Yes   **23.** No   **24.** No   **25.** Yes
26. 13   **27.** $7|w|$ **28.** $|xy|$
29. $|xy|$ **30.** $|x-11|$ **31.** $|x+4|$

Guided Solution:
31. $4, x+4$

Fortunately, in many cases, it can be assumed that radicands that are variable expressions do not represent the square of a negative number. When this assumption is made, the need for absolute-value symbols disappears. Then for $x \geq 0$, $\sqrt{x^2} = x$, since x is nonnegative.

PRINCIPAL SQUARE ROOT OF A^2
....

For any *nonnegative* real number A,

$$\sqrt{A^2} = A.$$

(That is, for any *nonnegative* real number A, the principal square root of A^2 is A.)

Simplify. Assume that radicands do not represent the square of a negative number.

32. $\sqrt{(xy)^2}$ **33.** $\sqrt{x^2 y^2}$

34. $\sqrt{25y^2}$ **35.** $\sqrt{\frac{1}{4}t^2}$

36. $\sqrt{(x-11)^2}$

37. $\sqrt{x^2 + 8x + 16}$ **GS**
$= \sqrt{(x + \boxed{})^2}$
$= \boxed{} + 4$

EXAMPLES Simplify. Assume that radicands do not represent the square of a negative number.

18. $\sqrt{(3x)^2} = 3x$ Since $3x$ is assumed to be nonnegative, $|3x| = 3x$.

19. $\sqrt{a^2 b^2} = \sqrt{(ab)^2} = ab$ Since ab is assumed to be nonnegative, $|ab| = ab$.

20. $\sqrt{x^2 + 2x + 1} = \sqrt{(x+1)^2} = x + 1$ Since $x + 1$ is assumed to be nonnegative

◀ **Do Exercises 32–37.**

RADICALS AND ABSOLUTE VALUE
....

Henceforth, in this text we will assume that no radicands are formed by raising negative quantities to even powers.

We make this assumption in order to eliminate some confusion and because it is valid in many applications. As you study further in mathematics, however, you will frequently need to make a determination about expressions under radicals being nonnegative or positive. This will often be necessary in calculus.

Answers

32. xy **33.** xy **34.** $5y$
35. $\frac{1}{2}t$ **36.** $x - 11$ **37.** $x + 4$

Guided Solution:
37. $4, x$

8.1 **Exercise Set**

For Extra Help

MyMathLab®   MathXL®

PRACTICE   WATCH   READ   REVIEW

 Reading Check

Determine whether each statement is true or false.

RC1. The radical symbol $\sqrt{}$ represents only the principal square root.

RC2. For any real number A, $\sqrt{A^2} = A$.

RC3. The radicand in the expression $\sqrt{y} + 3$ is $y + 3$.

RC4. There are no real numbers that when squared yield negative numbers.

RC5. For any nonnegative real number A, $\sqrt{A^2} = A$.

RC6. The number c is a square root of a if $c^2 = \sqrt{a}$.

a Find the square roots.

1. 4

2. 1

3. 9

4. 16

5. 100

6. 121

7. 169

8. 144

9. 256

10. 625

Simplify.

11. $\sqrt{4}$

12. $\sqrt{1}$

13. $-\sqrt{9}$

14. $-\sqrt{25}$

15. $-\sqrt{36}$

16. $-\sqrt{81}$

17. $-\sqrt{225}$

18. $\sqrt{400}$

19. $\sqrt{361}$

20. $-\sqrt{441}$

b Use a calculator to approximate each square root. Round to three decimal places.

21. $\sqrt{5}$

22. $\sqrt{8}$

23. $\sqrt{432}$

24. $-\sqrt{8196}$

25. $-\sqrt{347.7}$

26. $-\sqrt{204.788}$

27. $\sqrt{\dfrac{278}{36}}$

28. $-\sqrt{\dfrac{567}{788}}$

29. $-5\sqrt{189 \cdot 6}$

30. $2\sqrt{18 \cdot 3}$

c Solve.

31. *Water Flow of Fire Hose.* The number of gallons per minute discharged from a fire hose depends on the diameter of the hose and the nozzle pressure. For a 2-in. diameter solid bore nozzle, the water flow W, in gallons per minute (GPM), is given by $W = 118.8\sqrt{P}$, where P is the nozzle pressure, in pounds per square inch (psi). Find the water flow, in GPM, when the pressure is **(a)** 650 psi; **(b)** 1500 psi.

Source: www.firetactics.com

32. *Parking-Lot Arrival Spaces.* The attendants at a parking lot park cars in temporary spaces before the cars are taken to long-term parking spaces. The number N of such spaces needed is approximated by the formula $N = 2.5\sqrt{A}$, where A is the average number of arrivals during peak hours. Find the number of spaces needed when the average number of arrivals is **(a)** 25; **(b)** 62.

Hang Time. An athlete's *hang time* T (the time airborne for a jump), in seconds, is given by $T = 0.144\sqrt{V}$, where V is the athlete's vertical leap, in inches.

Source: Peter Brancazio

33. Vince Carter of the Dallas Mavericks can jump 43 in. vertically. Find his hang time.

34. Jason Richardson of the Philadelphia 76ers can jump 46 in. vertically. Find his hang time.

35. Paul Pierce of the Brooklyn Nets can jump 38 in. vertically. Find his hang time.

36. Shawn Merion of the Dallas Mavericks can jump 42 in. vertically. Find his hang time.

d Identify the radicand.

37. $\sqrt{200}$

38. $\sqrt{16z}$

39. $\sqrt{x} - 4$

40. $\sqrt{3t + 10} + 8$

41. $5\sqrt{t^2 + 1}$

42. $-9\sqrt{x^2 + 16}$

43. $x^2y\sqrt{\dfrac{3}{x + 2}}$

44. $ab^2\sqrt{\dfrac{a}{a + b}}$

e Determine whether each expression represents a real number. Write "yes" or "no."

45. $\sqrt{-16}$

46. $\sqrt{-81}$

47. $-\sqrt{81}$

48. $-\sqrt{64}$

49. $-\sqrt{-25}$

50. $\sqrt{-(-49)}$

f Simplify. Remember that we have assumed that radicands do not represent the square of a negative number.

51. $\sqrt{c^2}$

52. $\sqrt{x^2}$

53. $\sqrt{9x^2}$

54. $\sqrt{16y^2}$

55. $\sqrt{(8p)^2}$

56. $\sqrt{(7pq)^2}$

57. $\sqrt{(ab)^2}$

58. $\sqrt{(6y)^2}$

59. $\sqrt{(34d)^2}$ **60.** $\sqrt{(53b)^2}$ **61.** $\sqrt{(x+3)^2}$ **62.** $\sqrt{(d-3)^2}$

63. $\sqrt{a^2 - 10a + 25}$ **64.** $\sqrt{x^2 + 2x + 1}$ **65.** $\sqrt{4a^2 - 20a + 25}$ **66.** $\sqrt{9p^2 + 12p + 4}$

67. $\sqrt{121y^2 - 198y + 81}$ **68.** $\sqrt{49b^2 + 140b + 100}$

Skill Maintenance

Solve. [7.4a]

69. *Supplementary Angles.* Two angles are supplementary. One angle is 3° less than twice the other. Find the measures of the angles.

70. *Complementary Angles.* Two angles are complementary. The sum of the measure of the first angle and half the measure of the second is 64°. Find the measures of the angles.

71. *Food Expenses.* The amount F that a family spends on food varies directly as its income I. A family making $39,200 per year will spend $10,192 on food. At this rate, how much would a family making $41,000 per year spend on food? [6.9b]

Divide and simplify. [6.2b]

72. $\dfrac{x-3}{x+4} \div \dfrac{x^2-9}{x+4}$

73. $\dfrac{x^2+10x-11}{x^2-1} \div \dfrac{x+11}{x+1}$

74. $\dfrac{x^4-16}{x^4-1} \div \dfrac{x^2+4}{x^2+1}$

Synthesis

75. Use only the graph of $y = \sqrt{x}$, shown below, to approximate $\sqrt{3}$, $\sqrt{5}$, and $\sqrt{7}$. Answers may vary.

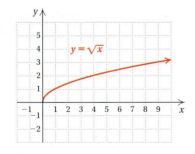

76. *Wind Chill Temperature.* When the temperature is T degrees Celsius and the wind speed is V meters per second, the *wind chill temperature*, T_W, is the temperature that it feels like. Here is a formula for finding wind chill temperature:

$$T_W = 13.112 + 0.6215T - 11.37V^{0.16} + 0.3965TV^{0.16}.$$

Estimate the wind chill temperature (to the nearest tenth of a degree) for the given actual temperatures and wind speeds.

a) $T = 7°C,\ V = 8\,\text{m/sec}$
b) $T = -5°C,\ V = 14\,\text{m/sec}$

Solve.

77. $\sqrt{x^2} = 16$ **78.** $\sqrt{y^2} = -7$ **79.** $t^2 = 49$

80. Suppose that the area of a square is 3. Find the length of a side.

8.2

Multiplying and Simplifying with Radical Expressions

1. Simplify.

  a) $\sqrt{25} \cdot \sqrt{16}$

  b) $\sqrt{25 \cdot 16}$

Multiply.

2. $\sqrt{3}\sqrt{11}$ **3.** $\sqrt{5}\sqrt{5}$

4. $\sqrt{\dfrac{5}{11}}\sqrt{\dfrac{6}{7}}$ **5.** $\sqrt{x}\sqrt{x+1}$

6. $\sqrt{x+2}\sqrt{x-2}$

a SIMPLIFYING BY FACTORING

To see how to multiply with radical notation, consider the following.

a) $\sqrt{9} \cdot \sqrt{4} = 3 \cdot 2 = 6$ This is a product of square roots.

b) $\sqrt{9 \cdot 4} = \sqrt{36} = 6$ This is the square root of a product.

Note that

$$\sqrt{9} \cdot \sqrt{4} = \sqrt{9 \cdot 4}.$$

◀ **Do Margin Exercise 1.**

We can multiply radical expressions by multiplying the radicands.

THE PRODUCT RULE FOR RADICALS

For any nonnegative radicands A and B,

$$\sqrt{A} \cdot \sqrt{B} = \sqrt{A \cdot B}.$$

(The product of square roots is the square root of the product of the radicands.)

EXAMPLES Multiply.

1. $\sqrt{5}\sqrt{7} = \sqrt{5 \cdot 7} = \sqrt{35}$

2. $\sqrt{8}\sqrt{8} = \sqrt{8 \cdot 8} = \sqrt{64} = 8$

3. $\sqrt{\dfrac{2}{3}}\sqrt{\dfrac{4}{5}} = \sqrt{\dfrac{2}{3} \cdot \dfrac{4}{5}} = \sqrt{\dfrac{8}{15}}$

4. $\sqrt{2x}\sqrt{3x-1} = \sqrt{2x(3x-1)} = \sqrt{6x^2 - 2x}$

◀ **Do Margin Exercises 2–6.**

To factor radical expressions, we can use the product rule for radicals in reverse.

FACTORING RADICAL EXPRESSIONS

$$\sqrt{AB} = \sqrt{A}\sqrt{B}$$

In some cases, we can simplify after factoring.

A square-root radical expression is simplified when its radicand has no factors that are perfect squares.

When simplifying a square-root radical expression, we first determine whether the radicand is a perfect square. Then we determine whether it has perfect-square factors. If so, the radicand is then factored and the radical expression simplified using the preceding rule.

Compare the following:

$$\sqrt{50} = \sqrt{10 \cdot 5} = \sqrt{10}\,\sqrt{5};$$
$$\sqrt{50} = \sqrt{25 \cdot 2} = \sqrt{25}\,\sqrt{2} = 5\sqrt{2}.$$

In the second case, the radicand is written using the perfect-square factor 25. If you do not recognize perfect-square factors, try factoring the radicand into its prime factors. For example,

$$\sqrt{50} = \sqrt{2 \cdot \underbrace{5 \cdot 5}} = 5\sqrt{2}.$$

Perfect square (a pair of the same factors)

Square-root radical expressions in which the radicand has no perfect-square factors, such as $5\sqrt{2}$, are considered to be in simplest form.

EXAMPLES   Simplify by factoring.

5. $\sqrt{18} = \sqrt{9 \cdot 2}$ Identifying a perfect-square factor and factoring the radicand. The factor 9 is a perfect square.

$\quad\quad = \sqrt{9} \cdot \sqrt{2}$ Factoring into a product of radicals

$\quad\quad = 3\sqrt{2}$ Simplifying $\sqrt{9}$

The radicand has no factors that are perfect squares.

6. $\sqrt{48t} = \sqrt{16 \cdot 3 \cdot t}$ Identifying a perfect-square factor and factoring the radicand. The factor 16 is a perfect square.

$\quad\quad = \sqrt{16}\,\sqrt{3t}$ Factoring into a product of radicals

$\quad\quad = 4\sqrt{3t}$ Taking a square root

7. $\sqrt{20t^2} = \sqrt{4 \cdot 5 \cdot t^2}$ Identifying perfect-square factors and factoring the radicand. The factors 4 and t^2 are perfect squares.

$\quad\quad = \sqrt{4}\,\sqrt{t^2}\,\sqrt{5}$ Factoring into a product of several radicals

$\quad\quad = 2t\sqrt{5}$ Taking square roots. No absolute-value signs are necessary since we have assumed that expressions under radicals do not represent the square of a negative number.

8. $\sqrt{x^2 - 6x + 9} = \sqrt{(x-3)^2} = x - 3$ No absolute-value signs are necessary since we have assumed that expressions under radicals do not represent the square of a negative number.

9. $\sqrt{36x^2} = \sqrt{36}\,\sqrt{x^2} = 6x$, or $\sqrt{36x^2} = \sqrt{(6x)^2} = 6x$

10. $\sqrt{3x^2 + 6x + 3} = \sqrt{3(x^2 + 2x + 1)}$ Factoring the radicand

$\quad\quad = \sqrt{3(x + 1)^2}$ Factoring further

$\quad\quad = \sqrt{3}\,\sqrt{(x + 1)^2}$ Factoring into a product of radicals

$\quad\quad = \sqrt{3}(x + 1)$, or $(x + 1)\sqrt{3}$ Taking the square root

Do Exercises 7–14. ▶

Simplify by factoring.

7. $\sqrt{32}$ **8.** $\sqrt{92}$

GS **9.** $\sqrt{363q}$

$\quad = \sqrt{\boxed{} \cdot 3 \cdot q}$

$\quad = \sqrt{121}\,\sqrt{\boxed{}}$

$\quad = \boxed{}\,\sqrt{3q}$

10. $\sqrt{128t}$ **11.** $\sqrt{63x^2}$

12. $\sqrt{81m^2}$

13. $\sqrt{x^2 + 14x + 49}$

14. $\sqrt{3x^2 - 60x + 300}$

Answers

7. $4\sqrt{2}$ 8. $2\sqrt{23}$ 9. $11\sqrt{3q}$ 10. $8\sqrt{2t}$
11. $3x\sqrt{7}$ 12. $9m$ 13. $x + 7$
14. $\sqrt{3}(x - 10)$, or $(x - 10)\sqrt{3}$

Guided Solution:
9. 121, 3q, 11

b SIMPLIFYING SQUARE ROOTS OF POWERS

To take the square root of an even power such as x^{10}, we note that $x^{10} = (x^5)^2$. Then

$$\sqrt{x^{10}} = \sqrt{(x^5)^2} = x^5.$$

We can find the answer by taking half the exponent. That is,

$$\sqrt{x^{10}} = x^5. \quad \frac{1}{2}(10) = 5$$

EXAMPLES Simplify.

11. $\sqrt{x^6} = \sqrt{(x^3)^2} = x^3 \quad \frac{1}{2}(6) = 3$

12. $\sqrt{x^8} = x^4$

13. $\sqrt{t^{22}} = t^{11}$

◀ **Do Exercises 15–18.**

If an odd power occurs, we express the power in terms of the largest even power. Then we simplify the even power as in Examples 11–13.

EXAMPLE 14 Simplify by factoring: $\sqrt{x^9}$.

$$\sqrt{x^9} = \sqrt{x^8 \cdot x}$$
$$= \sqrt{x^8}\sqrt{x}$$
$$= x^4\sqrt{x} \longleftarrow$$

········ **Caution!** ········

Note that $\sqrt{x^9} \neq x^3$.

EXAMPLE 15 Simplify by factoring: $\sqrt{32x^{15}}$.

$$\sqrt{32x^{15}} = \sqrt{16 \cdot 2 \cdot x^{14} \cdot x}$$
We factor the radicand, looking for perfect-square factors. The largest even power of x is 14.

$$= \sqrt{16}\sqrt{x^{14}}\sqrt{2x}$$
Factoring into a product of radicals. Perfect-square factors are usually listed first.

$$= 4x^7\sqrt{2x}$$
Simplifying

◀ **Do Exercises 19 and 20.**

c MULTIPLYING AND SIMPLIFYING

Sometimes we can simplify after multiplying. We leave the radicand in factored form and factor further to determine perfect-square factors. Then we simplify the perfect-square factors.

EXAMPLE 16 Multiply and then simplify by factoring: $\sqrt{2}\sqrt{14}$.

$$\sqrt{2}\sqrt{14} = \sqrt{2 \cdot 14} \quad \text{Multiplying}$$
$$= \sqrt{2 \cdot 2 \cdot 7} \quad \text{Factoring}$$
$$= \sqrt{2 \cdot 2}\sqrt{7} \quad \text{Looking for perfect-square factors, pairs of factors}$$
$$= 2\sqrt{7}$$

◀ **Do Exercises 21 and 22.**

Simplify.

15. $\sqrt{t^4}$

16. $\sqrt{t^{20}}$

17. $\sqrt{h^{46}}$

18. $\sqrt{x^{100}}$

Simplify by factoring.

19. $\sqrt{x^7}$

20. $\sqrt{24x^{11}}$
$$= \sqrt{4 \cdot 6 \cdot \boxed{} \cdot x}$$
$$= \sqrt{4}\sqrt{x^{10}}\sqrt{\boxed{}}$$
$$= 2\boxed{}\sqrt{6x}$$
GS

Multiply and simplify.

21. $\sqrt{3}\sqrt{6}$

22. $\sqrt{2}\sqrt{50}$

EXAMPLE 17 Multiply and then simplify by factoring: $\sqrt{3x^2}\sqrt{9x^3}$.

$$\sqrt{3x^2}\sqrt{9x^3} = \sqrt{3x^2 \cdot 9x^3} \qquad \text{Multiplying}$$
$$= \sqrt{3 \cdot x^2 \cdot 9 \cdot x^2 \cdot x} \qquad \begin{array}{l}\text{Looking for perfect-square}\\\text{factors or largest even powers}\end{array}$$
$$= \sqrt{9 \cdot x^2 \cdot x^2 \cdot 3x}$$

$\qquad\qquad\qquad\qquad$ Perfect-square factors are usually listed first.

$$= \sqrt{9}\sqrt{x^2}\sqrt{x^2}\sqrt{3x}$$
$$= 3 \cdot x \cdot x \cdot \sqrt{3x}$$
$$= 3x^2\sqrt{3x}$$

In doing an example like the preceding one, it might be helpful to do more factoring, as follows:

$$\sqrt{3x^2} \cdot \sqrt{9x^3} = \sqrt{3 \cdot \underline{x \cdot x} \cdot \underline{3 \cdot 3} \cdot \underline{x \cdot x} \cdot x}.$$

Then we look for pairs of factors, as shown, and simplify perfect-square factors:

$$= 3 \cdot x \cdot x\sqrt{3x}$$
$$= 3x^2\sqrt{3x}.$$

EXAMPLE 18 Simplify: $\sqrt{20cd^2}\sqrt{35cd^5}$.

$$\sqrt{20cd^2}\sqrt{35cd^5}$$
$$= \sqrt{20cd^2 \cdot 35cd^5} \qquad \text{Multiplying}$$
$$= \sqrt{2 \cdot 2 \cdot 5 \cdot c \cdot d \cdot d \cdot 5 \cdot 7 \cdot c \cdot d \cdot d \cdot d \cdot d \cdot d} \qquad \begin{array}{l}\text{Looking}\\\text{for pairs of}\\\text{factors}\end{array}$$

$$= \sqrt{2 \cdot 2 \cdot 5 \cdot 5 \cdot c \cdot c \cdot d \cdot d \cdot d \cdot d \cdot d \cdot d \cdot 7d}$$
$$= 2 \cdot 5 \cdot c \cdot d \cdot d \cdot d\sqrt{7d}$$
$$= 10cd^3\sqrt{7d}$$

Do Exercises 23–25. ▶

Multiply and simplify.

23. $\sqrt{2x^3}\sqrt{8x^3y^4}$

24. $\sqrt{10xy^2}\sqrt{5x^2y^3}$

25. $\sqrt{28q^2r} \cdot \sqrt{21q^3r^7}$

We know that $\sqrt{AB} = \sqrt{A}\sqrt{B}$. That is, the square root of a product is the product of the square roots. What about the square root of a sum? That is, is the square root of a sum equal to the sum of the square roots? To check, consider $\sqrt{A+B}$ and $\sqrt{A} + \sqrt{B}$ when $A = 16$ and $B = 9$:

$$\sqrt{A+B} = \sqrt{16+9} = \sqrt{25} = 5;$$

and

$$\sqrt{A} + \sqrt{B} = \sqrt{16} + \sqrt{9} = 4 + 3 = 7.$$

Thus we see the following.

·················· **Caution!** ··················

The square root of a sum is not the sum of the square roots.

$$\sqrt{A+B} \neq \sqrt{A} + \sqrt{B}$$

CALCULATOR CORNER

Simplifying Radical Expressions

EXERCISES: Use a table or a graph to determine whether each of the following is true.

1. $\sqrt{x+4} = \sqrt{x} + 2$
2. $\sqrt{3+x} = \sqrt{3} + x$
3. $\sqrt{x-2} = \sqrt{x} - \sqrt{2}$
4. $\sqrt{9x} = 3\sqrt{x}$

Answers

23. $4x^3y^2$ 24. $5xy^2\sqrt{2xy}$ 25. $14q^2r^4\sqrt{3q}$

✓ Reading Check

Square-root radical expressions in which the radicand has no perfect-square factors are considered to be in simplest form. Determine whether each radical expression is in simplest form. Answer "yes" or "no."

RC1. $\sqrt{49w^2}$ **RC2.** $5\sqrt{15}$ **RC3.** $\sqrt{121q}$

RC4. $4\sqrt{25}$ **RC5.** $\sqrt{t^2 + 1}$ **RC6.** $\sqrt{900}$

RC7. $\dfrac{\sqrt{30}}{4}$ **RC8.** $\sqrt{221}$ **RC9.** $\sqrt{x^3 + x^2}$

 Simplify by factoring.

1. $\sqrt{12}$ **2.** $\sqrt{8}$ **3.** $\sqrt{75}$ **4.** $\sqrt{50}$ **5.** $\sqrt{20}$

6. $\sqrt{45}$ **7.** $\sqrt{600}$ **8.** $\sqrt{300}$ **9.** $\sqrt{486}$ **10.** $\sqrt{567}$

11. $\sqrt{9x}$ **12.** $\sqrt{4y}$ **13.** $\sqrt{48x}$ **14.** $\sqrt{40m}$

15. $\sqrt{16a}$ **16.** $\sqrt{49b}$ **17.** $\sqrt{64y^2}$ **18.** $\sqrt{9x^2}$

19. $\sqrt{13x^2}$ **20.** $\sqrt{23s^2}$ **21.** $\sqrt{8t^2}$ **22.** $\sqrt{125a^2}$

23. $\sqrt{180}$

24. $\sqrt{320}$

25. $\sqrt{288y}$

26. $\sqrt{363p}$

27. $\sqrt{28x^2}$

28. $\sqrt{20x^2}$

29. $\sqrt{x^2 - 6x + 9}$

30. $\sqrt{t^2 + 22t + 121}$

31. $\sqrt{8x^2 + 8x + 2}$

32. $\sqrt{20x^2 - 20x + 5}$

33. $\sqrt{36y + 12y^2 + y^3}$

34. $\sqrt{x - 2x^2 + x^3}$

b Simplify by factoring.

35. $\sqrt{t^6}$

36. $\sqrt{x^{18}}$

37. $\sqrt{x^{12}}$

38. $\sqrt{x^{16}}$

39. $\sqrt{x^5}$

40. $\sqrt{x^3}$

41. $\sqrt{t^{19}}$

42. $\sqrt{p^{17}}$

43. $\sqrt{(y - 2)^8}$

44. $\sqrt{(x + 3)^6}$

45. $\sqrt{4(x + 5)^{10}}$

46. $\sqrt{16(a - 7)^4}$

47. $\sqrt{36m^3}$

48. $\sqrt{250y^3}$

49. $\sqrt{8a^5}$

50. $\sqrt{12b^7}$

51. $\sqrt{104p^{17}}$

52. $\sqrt{284m^{23}}$

53. $\sqrt{448x^6y^3}$

54. $\sqrt{243x^5y^4}$

C   Multiply and then, if possible, simplify by factoring.

55. $\sqrt{3}\,\sqrt{18}$ **56.** $\sqrt{5}\,\sqrt{10}$ **57.** $\sqrt{15}\,\sqrt{6}$ **58.** $\sqrt{3}\,\sqrt{27}$

59. $\sqrt{18}\,\sqrt{14x}$ **60.** $\sqrt{12}\,\sqrt{18x}$ **61.** $\sqrt{3x}\,\sqrt{12y}$ **62.** $\sqrt{7x}\,\sqrt{21y}$

63. $\sqrt{13}\,\sqrt{13}$ **64.** $\sqrt{11}\,\sqrt{11x}$ **65.** $\sqrt{5b}\,\sqrt{15b}$ **66.** $\sqrt{6a}\,\sqrt{18a}$

67. $\sqrt{2t}\,\sqrt{2t}$ **68.** $\sqrt{7a}\,\sqrt{7a}$ **69.** $\sqrt{ab}\,\sqrt{ac}$ **70.** $\sqrt{xy}\,\sqrt{xz}$

71. $\sqrt{2x^2y}\,\sqrt{4xy^2}$ **72.** $\sqrt{15mn^2}\,\sqrt{5m^2n}$ **73.** $\sqrt{18}\,\sqrt{18}$ **74.** $\sqrt{16}\,\sqrt{16}$

75. $\sqrt{5}\,\sqrt{2x-1}$ **76.** $\sqrt{3}\,\sqrt{4x+2}$ **77.** $\sqrt{x+2}\,\sqrt{x+2}$

78. $\sqrt{x-9}\,\sqrt{x-9}$ **79.** $\sqrt{18x^2y^3}\,\sqrt{6xy^4}$ **80.** $\sqrt{12x^3y^2}\,\sqrt{8xy}$

81. $\sqrt{50x^4y^6}\,\sqrt{10xy}$ **82.** $\sqrt{10xy^2}\,\sqrt{5x^2y^3}$ **83.** $\sqrt{99p^4q^3}\,\sqrt{22p^5q^2}$

84. $\sqrt{75m^8n^9}\,\sqrt{50m^5n^7}$ **85.** $\sqrt{24a^2b^3c^4}\,\sqrt{32a^5b^4c^7}$ **86.** $\sqrt{18p^5q^2r^{11}}\,\sqrt{108p^3q^6r^9}$

Solve. [7.3a, b]

87. $x - y = -6,$
$x + y = 2$

88. $3x + 5y = 6,$
$5x + 3y = 4$

89. $3x - 2y = 4,$
$2x + 5y = 9$

90. $4a - 5b = 25,$
$a - b = 7$

Solve.

91. *Insecticide Mixtures.* A solution containing 30% insecticide is to be mixed with a solution containing 50% insecticide in order to make 200 L of a solution containing 42% insecticide. How much of each solution should be used? [7.4a]

92. *Storage Area Dimensions.* The perimeter of a rectangular storage area is 84 ft. The length is 18 ft greater than the width. Find the area of the rectangle. [7.4a]

93. *Canoe Travel.* Greg and Beth paddled to a picnic spot downriver in 2 hr. It took them 3 hr to return against the current. If the speed of the current was 2 mph, at what speed were they paddling the canoe? [7.5a]

94. *Fund-Raiser Attendance.* As part of a fund-raiser, 382 people attended a dinner and tour of a space museum. Tickets were $24 each for adults and $9 each for children, and receipts totaled $6603. How many adults and how many children attended? [7.4a]

Synthesis

Factor.

95. $\sqrt{5x - 5}$

96. $\sqrt{x^2 - x - 2}$

97. $\sqrt{x^2 - 36}$

98. $\sqrt{2x^2 - 5x - 12}$

99. $\sqrt{x^3 - 2x^2}$

100. $\sqrt{a^2 - b^2}$

Simplify.

101. $\sqrt{0.25}$

102. $\sqrt{0.01}$

103. $\sqrt{\sqrt{\sqrt{256}}}$

Multiply and then simplify by factoring.

104. $(\sqrt{2y})(\sqrt{3})(\sqrt{8y})$

105. $\sqrt{18(x - 2)} \, \sqrt{20(x - 2)^3}$

106. $\sqrt{27(x + 1)} \, \sqrt{12y(x + 1)^2}$

107. $\sqrt{2^{109}} \, \sqrt{x^{306}} \, \sqrt{x^{11}}$

108. $\sqrt{x} \, \sqrt{2x} \, \sqrt{10x^5}$

109. $\sqrt{a}(\sqrt{a^3} - 5)$

OBJECTIVES

a Divide radical expressions.

b Simplify square roots of quotients.

c Rationalize the denominator of a radical expression.

SKILL TO REVIEW

Objective 6.1c: Simplify rational expressions by factoring the numerator and the denominator and removing factors of 1.

Simplify.

1. $\dfrac{10x^8}{15x^3}$ **2.** $\dfrac{64a^5b}{24a^2b^6}$

a DIVIDING RADICAL EXPRESSIONS

Consider the expressions

$$\frac{\sqrt{25}}{\sqrt{16}} \quad \text{and} \quad \sqrt{\frac{25}{16}}.$$

Let's evaluate them separately:

a) $\dfrac{\sqrt{25}}{\sqrt{16}} = \dfrac{5}{4}$ because $\sqrt{25} = 5$ and $\sqrt{16} = 4$;

b) $\sqrt{\dfrac{25}{16}} = \dfrac{5}{4}$ because $\dfrac{5}{4} \cdot \dfrac{5}{4} = \dfrac{25}{16}$.

Both expressions represent the same number. This suggests that the quotient of two square roots is the square root of the quotient of the radicands.

THE QUOTIENT RULE FOR RADICALS

For any nonnegative number A and any positive number B,

$$\frac{\sqrt{A}}{\sqrt{B}} = \sqrt{\frac{A}{B}}.$$

(The quotient of two square roots is the square root of the quotient of the radicands.)

EXAMPLES   Divide and simplify.

1. $\dfrac{\sqrt{27}}{\sqrt{3}} = \sqrt{\dfrac{27}{3}} = \sqrt{9} = 3$

2. $\dfrac{\sqrt{30a^5}}{\sqrt{6a^2}} = \sqrt{\dfrac{30a^5}{6a^2}} = \sqrt{5a^3} = \sqrt{5 \cdot a^2 \cdot a} = \sqrt{a^2} \cdot \sqrt{5a} = a\sqrt{5a}$

◀ **Do Margin Exercises 1–4.**

Divide and simplify.

1. $\dfrac{\sqrt{96}}{\sqrt{6}}$ **2.** $\dfrac{\sqrt{75}}{\sqrt{3}}$

3. $\dfrac{\sqrt{x^{14}}}{\sqrt{x^3}}$ **4.** $\dfrac{\sqrt{42x^5}}{\sqrt{7x^2}}$

b SQUARE ROOTS OF QUOTIENTS

To find the square root of certain quotients, we can reverse the quotient rule for radicals. We can take the square root of a quotient by taking the square roots of the numerator and the denominator separately.

SQUARE ROOTS OF QUOTIENTS

For any nonnegative number A and any positive number B,

$$\sqrt{\frac{A}{B}} = \frac{\sqrt{A}}{\sqrt{B}}.$$

(We can take the square roots of the numerator and the denominator separately.)

Answers

Skill to Review

1. $\dfrac{2x^5}{3}$ 2. $\dfrac{8a^3}{3b^5}$

Margin Exercises:

1. 4   2. 5   3. $x^5\sqrt{x}$ 4. $x\sqrt{6x}$

EXAMPLES Simplify by taking the square roots of the numerator and the denominator separately.

3. $\sqrt{\dfrac{25}{9}} = \dfrac{\sqrt{25}}{\sqrt{9}} = \dfrac{5}{3}$ Taking the square root of the numerator and the square root of the denominator

4. $\sqrt{\dfrac{1}{16}} = \dfrac{\sqrt{1}}{\sqrt{16}} = \dfrac{1}{4}$ Taking the square root of the numerator and the square root of the denominator

5. $\sqrt{\dfrac{49}{t^2}} = \dfrac{\sqrt{49}}{\sqrt{t^2}} = \dfrac{7}{t}$

Do Exercises 5–8. ▶

We are assuming that expressions for numerators are nonnegative and expressions for denominators are positive. Thus we need not be concerned about absolute-value signs or zero denominators.

Sometimes a rational expression can be simplified to one that has a perfect-square numerator and a perfect-square denominator.

EXAMPLES Simplify.

6. $\sqrt{\dfrac{18}{50}} = \sqrt{\dfrac{9 \cdot 2}{25 \cdot 2}} = \sqrt{\dfrac{9}{25} \cdot \dfrac{2}{2}} = \sqrt{\dfrac{9}{25} \cdot 1}$

$\quad = \sqrt{\dfrac{9}{25}} = \dfrac{\sqrt{9}}{\sqrt{25}} = \dfrac{3}{5}$

7. $\sqrt{\dfrac{2560}{2890}} = \sqrt{\dfrac{256 \cdot 10}{289 \cdot 10}} = \sqrt{\dfrac{256}{289} \cdot \dfrac{10}{10}} = \sqrt{\dfrac{256}{289} \cdot 1}$

$\quad = \sqrt{\dfrac{256}{289}} = \dfrac{\sqrt{256}}{\sqrt{289}} = \dfrac{16}{17}$

8. $\dfrac{\sqrt{48x^3}}{\sqrt{3x^7}} = \sqrt{\dfrac{48x^3}{3x^7}} = \sqrt{\dfrac{16}{x^4}}$ Simplifying the radicand

$\quad = \dfrac{\sqrt{16}}{\sqrt{x^4}} = \dfrac{4}{x^2}$

Do Exercises 9–12. ▶

C RATIONALIZING DENOMINATORS

Sometimes in mathematics it is useful to find an equivalent expression without a radical in the denominator. This provides a standard notation for expressing results. The procedure for finding such an expression is called **rationalizing the denominator**. We carry this out by multiplying by 1 in either of two ways.

> To rationalize a denominator:
>
> **Method 1.** Multiply by 1 under the radical to make the denominator of the radicand a perfect square.
>
> **Method 2.** Multiply by 1 outside the radical to make the radicand in the denominator a perfect square.

Simplify.

5. $\sqrt{\dfrac{16}{9}}$ **6.** $\sqrt{\dfrac{1}{25}}$

7. $\sqrt{\dfrac{36}{x^2}}$ **8.** $\sqrt{\dfrac{b^2}{121}}$

Simplify.

9. $\sqrt{\dfrac{18}{32}}$ **10.** $\sqrt{\dfrac{2250}{2560}}$

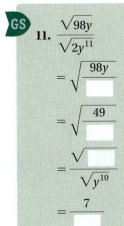

GS 11. $\dfrac{\sqrt{98y}}{\sqrt{2y^{11}}}$

$= \sqrt{\dfrac{98y}{\boxed{}}}$

$= \sqrt{\dfrac{49}{\boxed{}}}$

$= \dfrac{\sqrt{\boxed{}}}{\sqrt{y^{10}}}$

$= \dfrac{7}{\boxed{}}$

12. $\sqrt{\dfrac{108a^{11}}{3a^{37}}}$

Answers

5. $\dfrac{4}{3}$ **6.** $\dfrac{1}{5}$ **7.** $\dfrac{6}{x}$ **8.** $\dfrac{b}{11}$ **9.** $\dfrac{3}{4}$

10. $\dfrac{15}{16}$ **11.** $\dfrac{7}{y^5}$ **12.** $\dfrac{6}{a^{13}}$

Guided Solution:
11. $2y^{11}, y^{10}, 49, y^5$

EXAMPLE 9 Rationalize the denominator: $\sqrt{\dfrac{2}{3}}$.

Method 1: We multiply by 1, choosing $\frac{3}{3}$ for 1. This makes the denominator of the radicand a perfect square:

$$\sqrt{\dfrac{2}{3}} = \sqrt{\dfrac{2}{3} \cdot \dfrac{3}{3}} \qquad \text{Multiplying by 1}$$

$$= \sqrt{\dfrac{6}{9}} = \dfrac{\sqrt{6}}{\sqrt{9}} \qquad \begin{array}{l}\text{The radicand in the denominator, 9,}\\ \text{is a perfect square.}\end{array}$$

$$= \dfrac{\sqrt{6}}{3}.$$

Method 2: We can also rationalize by first taking the square roots of the numerator and the denominator. Then we multiply by 1, using $\sqrt{3}/\sqrt{3}$:

$$\sqrt{\dfrac{2}{3}} = \dfrac{\sqrt{2}}{\sqrt{3}}$$

$$= \dfrac{\sqrt{2}}{\sqrt{3}} \cdot \dfrac{\sqrt{3}}{\sqrt{3}} \qquad \text{Multiplying by 1}$$

$$= \dfrac{\sqrt{2} \cdot \sqrt{3}}{\sqrt{3} \cdot \sqrt{3}} = \dfrac{\sqrt{6}}{\sqrt{9}} \qquad \begin{array}{l}\text{The radicand, 9, in the denominator}\\ \text{is a perfect square.}\end{array}$$

$$= \dfrac{\sqrt{6}}{3}.$$

13. Rationalize the denominator:

$$\sqrt{\dfrac{3}{5}}.$$

a) Use method 1.

b) Use method 2.

◀ **Do Exercise 13.**

We can always multiply by 1 to make a denominator a perfect square. Then we can take the square root of the denominator.

EXAMPLE 10 Rationalize the denominator: $\sqrt{\dfrac{5}{18}}$.

The denominator, 18, is not a perfect square. Factoring, we get $18 = 3 \cdot 3 \cdot 2$. If we had another factor of 2, however, we would have a perfect square, 36. Thus we multiply by 1, choosing $\frac{2}{2}$. This makes the denominator a perfect square.

$$\sqrt{\dfrac{5}{18}} = \sqrt{\dfrac{5}{3 \cdot 3 \cdot 2}} = \sqrt{\dfrac{5}{3 \cdot 3 \cdot 2} \cdot \dfrac{2}{2}} = \sqrt{\dfrac{10}{36}} = \dfrac{\sqrt{10}}{\sqrt{36}} = \dfrac{\sqrt{10}}{6} \quad \blacksquare$$

EXAMPLE 11 Rationalize the denominator: $\dfrac{8}{\sqrt{7}}$.

This time we obtain an expression without a radical in the denominator by multiplying by 1, choosing $\sqrt{7}/\sqrt{7}$:

$$\dfrac{8}{\sqrt{7}} = \dfrac{8}{\sqrt{7}} \cdot \dfrac{\sqrt{7}}{\sqrt{7}} = \dfrac{8\sqrt{7}}{\sqrt{49}} = \dfrac{8\sqrt{7}}{7}.$$

◀ **Do Exercises 14 and 15.**

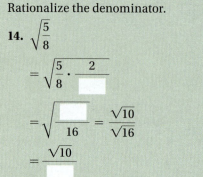

Rationalize the denominator. **GS**

14. $\sqrt{\dfrac{5}{8}}$

$$= \sqrt{\dfrac{5}{8} \cdot \dfrac{2}{\boxed{}}}$$

$$= \sqrt{\dfrac{\boxed{}}{16}} = \dfrac{\sqrt{10}}{\sqrt{16}}$$

$$= \dfrac{\sqrt{10}}{\boxed{}}$$

15. $\dfrac{10}{\sqrt{3}}$

$$= \dfrac{10}{\sqrt{3}} \cdot \dfrac{\boxed{}}{\sqrt{3}}$$

$$= \dfrac{\boxed{}}{\sqrt{9}} \cdot \sqrt{3}$$

$$= \dfrac{10\sqrt{3}}{\boxed{}}$$

Answers

13. (a) $\dfrac{\sqrt{15}}{5}$; (b) $\dfrac{\sqrt{15}}{5}$ 14. $\dfrac{\sqrt{10}}{4}$ 15. $\dfrac{10\sqrt{3}}{3}$

Guided Solutions:
14. 2, 10, 4   **15.** $\sqrt{3}$, 10, 3

EXAMPLE 12 Rationalize the denominator: $\dfrac{\sqrt{3}}{\sqrt{2}}$.

We look at the denominator. It is $\sqrt{2}$. We multiply by 1, choosing $\sqrt{2}/\sqrt{2}$:

$$\frac{\sqrt{3}}{\sqrt{2}} = \frac{\sqrt{3}}{\sqrt{2}} \cdot \frac{\sqrt{2}}{\sqrt{2}} = \frac{\sqrt{3} \cdot \sqrt{2}}{\sqrt{2} \cdot \sqrt{2}} = \frac{\sqrt{6}}{\sqrt{4}} = \frac{\sqrt{6}}{2}, \text{ or } \frac{1}{2}\sqrt{6}. \quad \blacksquare$$

EXAMPLES Rationalize the denominator.

13. $\dfrac{\sqrt{5}}{\sqrt{x}} = \dfrac{\sqrt{5}}{\sqrt{x}} \cdot \dfrac{\sqrt{x}}{\sqrt{x}}$ Multiplying by 1

$= \dfrac{\sqrt{5}\sqrt{x}}{\sqrt{x}\sqrt{x}}$

$= \dfrac{\sqrt{5x}}{x}$ $\sqrt{x} \cdot \sqrt{x} = x$ by the definition of square root

14. $\dfrac{\sqrt{49a^5}}{\sqrt{12}} = \dfrac{\sqrt{49a^5}}{\sqrt{12}} \cdot \dfrac{\sqrt{3}}{\sqrt{3}}$ Factoring 12, we get $2 \cdot 2 \cdot 3$, so we need another factor of 3 in order for the radicand in the denominator to be a perfect square. We multiply by $\sqrt{3}/\sqrt{3}$.

$= \dfrac{\sqrt{49a^5}\sqrt{3}}{\sqrt{12}\sqrt{3}}$

$= \dfrac{\sqrt{49 \cdot a^4 \cdot a \cdot 3}}{\sqrt{36}} = \dfrac{\sqrt{49}\sqrt{a^4}\sqrt{3a}}{\sqrt{36}}$

$= \dfrac{7a^2\sqrt{3a}}{6}$

Rationalize the denominator.

16. $\dfrac{\sqrt{3}}{\sqrt{7}}$ **17.** $\dfrac{\sqrt{5}}{\sqrt{r}}$

18. $\dfrac{\sqrt{64y^2}}{\sqrt{7}}$ **19.** $\dfrac{\sqrt{64x^9}}{\sqrt{15}}$

Answers

16. $\dfrac{\sqrt{21}}{7}$ **17.** $\dfrac{\sqrt{5r}}{r}$ **18.** $\dfrac{8y\sqrt{7}}{7}$

19. $\dfrac{8x^4\sqrt{15x}}{15}$

Do Exercises 16–19. ▶

8.3 Exercise Set

☑ Reading Check

Choose from the columns on the right the symbol for 1 that you would use to rationalize the denominator. Some choices may not be used and others may be used more than once.

RC1. $\dfrac{\sqrt{x^3}}{\sqrt{10}}$ **RC2.** $\dfrac{\sqrt{5}}{\sqrt{3}}$

RC3. $\dfrac{\sqrt{6}}{\sqrt{x^5}}$ **RC4.** $\dfrac{\sqrt{x}}{\sqrt{5}}$

RC5. $\dfrac{\sqrt{3}}{\sqrt{2}}$ **RC6.** $\dfrac{\sqrt{3}}{\sqrt{x}}$

a) $\dfrac{\sqrt{2}}{\sqrt{2}}$ b) $\dfrac{\sqrt{x^2}}{\sqrt{x^2}}$

c) $\dfrac{\sqrt{3}}{\sqrt{3}}$ d) $\dfrac{\sqrt{5}}{\sqrt{5}}$

e) $\dfrac{\sqrt{10}}{\sqrt{10}}$ f) $\dfrac{\sqrt{x}}{\sqrt{x}}$

g) $\dfrac{\sqrt{3x}}{\sqrt{3x}}$ h) $\dfrac{\sqrt{6}}{\sqrt{6}}$

a   Divide and simplify.

1. $\dfrac{\sqrt{18}}{\sqrt{2}}$

2. $\dfrac{\sqrt{20}}{\sqrt{5}}$

3. $\dfrac{\sqrt{108}}{\sqrt{3}}$

4. $\dfrac{\sqrt{60}}{\sqrt{15}}$

5. $\dfrac{\sqrt{65}}{\sqrt{13}}$

6. $\dfrac{\sqrt{45}}{\sqrt{15}}$

7. $\dfrac{\sqrt{3}}{\sqrt{75}}$

8. $\dfrac{\sqrt{3}}{\sqrt{48}}$

9. $\dfrac{\sqrt{12}}{\sqrt{75}}$

10. $\dfrac{\sqrt{18}}{\sqrt{32}}$

11. $\dfrac{\sqrt{8x}}{\sqrt{2x}}$

12. $\dfrac{\sqrt{18b}}{\sqrt{2b}}$

13. $\dfrac{\sqrt{63y^3}}{\sqrt{7y}}$

14. $\dfrac{\sqrt{48x^3}}{\sqrt{3x}}$

b   Simplify.

15. $\sqrt{\dfrac{16}{49}}$

16. $\sqrt{\dfrac{9}{49}}$

17. $\sqrt{\dfrac{1}{36}}$

18. $\sqrt{\dfrac{1}{4}}$

19. $-\sqrt{\dfrac{16}{81}}$

20. $-\sqrt{\dfrac{25}{49}}$

21. $\sqrt{\dfrac{64}{289}}$

22. $\sqrt{\dfrac{81}{361}}$

23. $\sqrt{\dfrac{1690}{1960}}$

24. $\sqrt{\dfrac{1210}{6250}}$

25. $\sqrt{\dfrac{25}{x^2}}$

26. $\sqrt{\dfrac{36}{a^2}}$

27. $\sqrt{\dfrac{9a^2}{625}}$

28. $\sqrt{\dfrac{x^2y^2}{256}}$

29. $\dfrac{\sqrt{50y^{15}}}{\sqrt{2y^{25}}}$

30. $\dfrac{\sqrt{3t^{15}}}{\sqrt{12t}}$

31. $\dfrac{\sqrt{7x^{23}}}{\sqrt{343x^5}}$

32. $\dfrac{\sqrt{125q^3}}{\sqrt{5q^{19}}}$

33. $\sqrt{\dfrac{2}{5}}$

34. $\sqrt{\dfrac{2}{7}}$

35. $\sqrt{\dfrac{7}{8}}$

36. $\sqrt{\dfrac{3}{8}}$

37. $\sqrt{\dfrac{1}{12}}$

38. $\sqrt{\dfrac{7}{12}}$

39. $\sqrt{\dfrac{5}{18}}$

40. $\sqrt{\dfrac{1}{18}}$

41. $\dfrac{3}{\sqrt{5}}$

42. $\dfrac{4}{\sqrt{3}}$

43. $\sqrt{\dfrac{8}{3}}$

44. $\sqrt{\dfrac{12}{5}}$

45. $\sqrt{\dfrac{3}{x}}$

46. $\sqrt{\dfrac{2}{x}}$

47. $\sqrt{\dfrac{x}{y}}$

48. $\sqrt{\dfrac{a}{b}}$

49. $\sqrt{\dfrac{x^2}{20}}$

50. $\sqrt{\dfrac{x^2}{18}}$

51. $\dfrac{1}{\sqrt{3}}$

52. $\dfrac{1}{\sqrt{2}}$

53. $\dfrac{\sqrt{9}}{\sqrt{8}}$

54. $\dfrac{\sqrt{4}}{\sqrt{27}}$

55. $\dfrac{\sqrt{11}}{\sqrt{5}}$

56. $\dfrac{\sqrt{2}}{\sqrt{5}}$

57. $\dfrac{2}{\sqrt{2}}$

58. $\dfrac{3}{\sqrt{3}}$

59. $\dfrac{\sqrt{5}}{\sqrt{11}}$

60. $\dfrac{\sqrt{7}}{\sqrt{27}}$

61. $\dfrac{\sqrt{7}}{\sqrt{12}}$

62. $\dfrac{\sqrt{5}}{\sqrt{18}}$

63. $\dfrac{\sqrt{48}}{\sqrt{32}}$

64. $\dfrac{\sqrt{56}}{\sqrt{40}}$

65. $\dfrac{\sqrt{450}}{\sqrt{18}}$

66. $\dfrac{\sqrt{224}}{\sqrt{14}}$

67. $\dfrac{\sqrt{3}}{\sqrt{x}}$

68. $\dfrac{\sqrt{2}}{\sqrt{y}}$

69. $\dfrac{4y}{\sqrt{5}}$

70. $\dfrac{8x}{\sqrt{3}}$

71. $\dfrac{\sqrt{a^3}}{\sqrt{8}}$

72. $\dfrac{\sqrt{x^3}}{\sqrt{27}}$

73. $\dfrac{\sqrt{56}}{\sqrt{12x}}$

74. $\dfrac{\sqrt{45}}{\sqrt{8a}}$

75. $\dfrac{\sqrt{27c}}{\sqrt{32c^3}}$

76. $\dfrac{\sqrt{7x^3}}{\sqrt{12x}}$

77. $\dfrac{\sqrt{y^5}}{\sqrt{xy^2}}$

78. $\dfrac{\sqrt{x^3}}{\sqrt{xy}}$

79. $\dfrac{\sqrt{45mn^2}}{\sqrt{32m}}$

80. $\dfrac{\sqrt{16a^4b^6}}{\sqrt{128a^6b^6}}$

Skill Maintenance

Solve. [7.3a, b]

81. $x + y = -7,$
    $x - y = 2$

82. $2x - 3y = 7,$
    $-4x + 6y = -14$

83. $2x - 3y = 7,$
    $2x - 3y = 9$

Divide and simplify. [6.2b]

84. $\dfrac{x - 2}{x + 3} \div \dfrac{x^2 - 4x + 4}{x^2 - 9}$

85. $\dfrac{a^2 - 25}{6} \div \dfrac{a + 5}{3}$

86. $\dfrac{x - 2}{x - 3} \div \dfrac{x - 4}{x - 5}$

Synthesis

Periods of Pendulums.   The period T of a pendulum is the time it takes the pendulum to move from one side to the other and back. A formula for the period is

$$T = 2\pi\sqrt{\dfrac{L}{32}},$$

where T is in seconds and L is the length of the pendulum, in feet. Use 3.14 for π.

A Foucault pendulum, shown here at the California Academy of Sciences, demonstrates the rotation of the earth. The pendulum knocks down pins at different positions as the earth rotates.

87. Find the periods of pendulums of lengths 2 ft, 8 ft, and 10 in.

88. The pendulum of a grandfather clock is $(32/\pi^2)$ ft long. How long does it take to swing from one side to the other?

89. Rationalize the denominator: $\sqrt{\dfrac{3x^2y}{a^2x^5}}$.

90. Simplify: $\sqrt{\dfrac{1}{x^2} - \dfrac{2}{xy} + \dfrac{1}{y^2}}$.

Mid-Chapter Review

Concept Reinforcement

Determine whether each statement is true or false.

_____ **1.** The radical symbol $\sqrt{}$ represents only the principal square root. [8.1a]

_____ **2.** For any nonnegative real number A, the principal square root of A^2 is $-A$. [8.1f]

_____ **3.** Every nonnegative number has two square roots. [8.1a]

_____ **4.** There are no real numbers that when squared yield negative numbers. [8.1e]

Guided Solutions

 Fill in each blank with the number or the expression that creates a correct solution.

5. Simplify by factoring:

$\sqrt{3x^2 - 48x + 192}$. [8.2a]

$\sqrt{3x^2 - 48x + 192} = \sqrt{\boxed{}\,(x^2 - 16x + 64)}$

$\phantom{\sqrt{3x^2 - 48x + 192}} = \sqrt{3\,(\boxed{})^2}$

$\phantom{\sqrt{3x^2 - 48x + 192}} = \sqrt{\boxed{}}\,\sqrt{(x-8)^2}$

$\phantom{\sqrt{3x^2 - 48x + 192}} = \sqrt{3}(x - \boxed{})$

6. Multiply and simplify by factoring:

$\sqrt{30}\,\sqrt{40y}$. [8.2c]

$\sqrt{30}\,\sqrt{40y} = \sqrt{30 \cdot \boxed{}\,y}$

$\phantom{\sqrt{30}\,\sqrt{40y}} = \sqrt{\boxed{}\,y}$

$\phantom{\sqrt{30}\,\sqrt{40y}} = \sqrt{100 \cdot \boxed{} \cdot y}$

$\phantom{\sqrt{30}\,\sqrt{40y}} = \sqrt{100 \cdot \boxed{} \cdot 3 \cdot y}$

$\phantom{\sqrt{30}\,\sqrt{40y}} = \sqrt{100}\,\sqrt{4}\,\sqrt{\boxed{}}$

$\phantom{\sqrt{30}\,\sqrt{40y}} = 10 \cdot \boxed{}\,\sqrt{3y}$

$\phantom{\sqrt{30}\,\sqrt{40y}} = \boxed{}\,\sqrt{3y}$

7. Multiply and simplify by factoring:

$\sqrt{18ab^2}\,\sqrt{14a^2 b^4}$. [8.2c]

$\sqrt{18ab^2}\,\sqrt{14a^2 b^4} = \sqrt{18ab^2 \cdot 14\,\boxed{}\,b^4}$

$\phantom{\sqrt{18ab^2}\,\sqrt{14a^2 b^4}} = \sqrt{2 \cdot 3 \cdot 3 \cdot 2 \cdot 7 \cdot \boxed{} \cdot b^6}$

$\phantom{\sqrt{18ab^2}\,\sqrt{14a^2 b^4}} = \sqrt{2^2 \cdot 3^2 \cdot 7 \cdot a^2 \cdot \boxed{} \cdot b^6}$

$\phantom{\sqrt{18ab^2}\,\sqrt{14a^2 b^4}} = \sqrt{2^2}\,\sqrt{3^2}\,\sqrt{a^2}\,\sqrt{b^6}\,\sqrt{\boxed{}}$

$\phantom{\sqrt{18ab^2}\,\sqrt{14a^2 b^4}} = 2 \cdot 3 \cdot a \cdot \boxed{}\,\sqrt{7a}$

$\phantom{\sqrt{18ab^2}\,\sqrt{14a^2 b^4}} = 6\,\boxed{}\,b^3\,\sqrt{7a}$

8. Rationalize the denominator: $\sqrt{\dfrac{3y^2}{44}}$. [8.3c]

$\sqrt{\dfrac{3y^2}{44}} = \sqrt{\dfrac{3y^2}{2 \cdot \boxed{} \cdot 11}}$

$\phantom{\sqrt{\dfrac{3y^2}{44}}} = \sqrt{\dfrac{3y^2}{2 \cdot 2 \cdot 11} \cdot \dfrac{\boxed{}}{\boxed{}}}$

$\phantom{\sqrt{\dfrac{3y^2}{44}}} = \sqrt{\dfrac{33y^2}{\boxed{}^2 \cdot 11^2}}$

$\phantom{\sqrt{\dfrac{3y^2}{44}}} = \dfrac{\boxed{}\,\sqrt{33}}{2 \cdot 11} = \dfrac{y\sqrt{33}}{\boxed{}}$

Mixed Review

9. Find the square roots of 121. [8.1a]

10. Identify the radicand: $2x\sqrt{\dfrac{x-3}{7}}$. [8.1d]

11. Determine whether each expression represents a real number. Write "yes" or "no." [8.1e]

a) $\sqrt{-100}$ b) $-\sqrt{9}$

Simplify.

12. $\sqrt{128r^7s^6}$ [8.2b]

13. $\sqrt{25(x-3)^2}$ [8.2b]

14. $\sqrt{\dfrac{1}{100}}$ [8.3b]

15. $-\sqrt{36}$ [8.1a]

16. $-\sqrt{\dfrac{6250}{490}}$ [8.3b]

17. $\sqrt{225}$ [8.1a]

18. $\sqrt{(10y)^2}$ [8.1f]

19. $\sqrt{4x^2 - 4x + 1}$ [8.2a]

20. $\sqrt{800x}$ [8.2a]

21. $\dfrac{\sqrt{6}}{\sqrt{96}}$ [8.3a]

22. $\sqrt{32q^{11}}$ [8.2b]

23. $\sqrt{\dfrac{81}{z^2}}$ [8.3b]

Multiply or divide and, if possible, simplify.

24. $\sqrt{25}\,\sqrt{25}$ [8.2c]

25. $\dfrac{\sqrt{18}}{\sqrt{98}}$ [8.3a]

26. $\dfrac{\sqrt{192x}}{\sqrt{3x}}$ [8.3a]

27. $\sqrt{40c^2d^7}\,\sqrt{15c^3d^3}$ [8.2c]

28. $\sqrt{24x^5y^8z^2}\,\sqrt{60xy^3z}$ [8.2c]

29. $\sqrt{2x}\,\sqrt{30y}$ [8.2c]

30. $\sqrt{21a}\,\sqrt{35a}$ [8.2c]

31. $\dfrac{\sqrt{3y^{29}}}{\sqrt{75y^5}}$ [8.3a]

32. Rationalize the denominator and simplify. Match each expression in the first column with an equivalent expression in the second column by drawing connecting lines.   [8.3c]

$\dfrac{x}{\sqrt{3}}$ $\dfrac{3\sqrt{x}}{x}$

$\sqrt{\dfrac{3}{x}}$ $\dfrac{\sqrt{3x}}{3}$

$\dfrac{3}{\sqrt{x}}$ $\dfrac{x\sqrt{3}}{3}$

$\dfrac{3x}{\sqrt{3}}$ $\sqrt{3}$

$\dfrac{3}{\sqrt{3}}$ $\dfrac{\sqrt{3x}}{x}$

$\sqrt{\dfrac{x}{3}}$ $x\sqrt{3}$

Understanding Through Discussion and Writing

33. What is the difference between "**the** square root of 100" and "**a** square root of 100"?   [8.1a]

34. Explain why the following is incorrect:

$$\sqrt{\dfrac{9 + 100}{25}} = \dfrac{3 + 10}{5}. \quad [8.3b]$$

35. Explain the error(s) in the following:

$$\sqrt{x^2 - 25} = \sqrt{x^2} - \sqrt{25} = x - 5. \quad [8.2a]$$

36. Describe a method that could be used to rationalize the *numerator* of a radical expression.   [8.3c]

Addition, Subtraction, and More Multiplication

8.4

a   ADDITION AND SUBTRACTION

We can add any two real numbers. The sum of 5 and $\sqrt{2}$ can be expressed as $5 + \sqrt{2}$. We cannot simplify this unless we use rational approximations such as $5 + \sqrt{2} \approx 5 + 1.414 = 6.414$. However, when we have *like radicals*, a sum can be simplified using the distributive laws and collecting like terms. **Like radicals** have the same radicands.

EXAMPLE 1   Add: $3\sqrt{5} + 4\sqrt{5}$.

Suppose we were considering $3x + 4x$. Recall that to add, we use a distributive law as follows:

$$3x + 4x = (3 + 4)x = 7x.$$

The situation is similar in this example, but we let $x = \sqrt{5}$:

$$3\sqrt{5} + 4\sqrt{5} = (3 + 4)\sqrt{5} \qquad \text{Using a distributive law to factor out } \sqrt{5}$$
$$= 7\sqrt{5}.$$

If we wish to add or subtract as we did in Example 1, the radicands must be the same. Sometimes after simplifying the radical terms, we discover that we have like radicals.

EXAMPLES   Add or subtract. Simplify, if possible, by collecting like radical terms.

2. $5\sqrt{2} - \sqrt{18} = 5\sqrt{2} - \sqrt{9 \cdot 2}$ Factoring 18
$$= 5\sqrt{2} - \sqrt{9}\sqrt{2}$$
$$= 5\sqrt{2} - 3\sqrt{2}$$
$$= (5 - 3)\sqrt{2} \qquad \text{Using a distributive law to factor out the common factor, } \sqrt{2}$$
$$= 2\sqrt{2}$$

3. $\sqrt{4x^3} + 7\sqrt{x} = \sqrt{4 \cdot x^2 \cdot x} + 7\sqrt{x}$
$$= 2x\sqrt{x} + 7\sqrt{x}$$
$$= (2x + 7)\sqrt{x} \qquad \text{Using a distributive law to factor out } \sqrt{x}$$

Don't forget the parentheses!

OBJECTIVES

a   Add or subtract with radical notation, using the distributive laws to simplify.

b   Multiply expressions involving radicals, where some of the expressions contain more than one term.

c   Rationalize denominators having two terms.

SKILL TO REVIEW

Objective 4.6d: Find special products when polynomial products are mixed together.

Multiply.

1. $(3x - 7)(3x + 7)$

2. $\left(4x - \dfrac{1}{2}\right)^2$

Answers

Skill to Review:

1. $9x^2 - 49$ **2.** $16x^2 - 4x + \dfrac{1}{4}$

Add or subtract and, if possible, simplify by collecting like radical terms.

1. $3\sqrt{2} + 9\sqrt{2}$

2. $8\sqrt{5} - 3\sqrt{5}$

3. $2\sqrt{10} - 7\sqrt{40}$

4. $\sqrt{24} + \sqrt{54}$

$= \sqrt{\boxed{} \cdot 6} + \sqrt{\boxed{} \cdot 6}$

$= \boxed{}\sqrt{6} + 3\sqrt{6}$

$= (2 + \boxed{})\sqrt{6}$

$= 5\sqrt{\boxed{}}$

5. $\sqrt{9x + 9} - \sqrt{4x + 4}$

Add or subtract.

6. $\sqrt{2} + \sqrt{\dfrac{1}{2}}$

7. $\sqrt{\dfrac{5}{3}} + \sqrt{\dfrac{3}{5}}$

4. $\sqrt{x^3 - x^2} + \sqrt{4x - 4} = \sqrt{x^2(x - 1)} + \sqrt{4(x - 1)}$ Factoring radicands

$= \sqrt{x^2}\sqrt{x - 1} + \sqrt{4}\sqrt{x - 1}$

$= x\sqrt{x - 1} + 2\sqrt{x - 1}$

$= (x + 2)\sqrt{x - 1}$ Using a distributive law to factor out the common factor, $\sqrt{x - 1}$. Don't forget the parentheses!

◀ **Do Exercises 1–5.**

Sometimes rationalizing denominators enables us to combine like radicals.

EXAMPLE 5   Add: $\sqrt{3} + \sqrt{\dfrac{1}{3}}$.

$\sqrt{3} + \sqrt{\dfrac{1}{3}} = \sqrt{3} + \sqrt{\dfrac{1}{3} \cdot \dfrac{3}{3}}$ Multiplying by 1 in order to rationalize the denominator

$= \sqrt{3} + \sqrt{\dfrac{3}{9}} = \sqrt{3} + \dfrac{\sqrt{3}}{\sqrt{9}}$

$= \sqrt{3} + \dfrac{\sqrt{3}}{3} = 1 \cdot \sqrt{3} + \dfrac{1}{3}\sqrt{3}$

$= \left(1 + \dfrac{1}{3}\right)\sqrt{3}$ Factoring out the common factor, $\sqrt{3}$

$= \dfrac{4}{3}\sqrt{3}$, or $\dfrac{4\sqrt{3}}{3}$

◀ **Do Exercises 6 and 7.**

b   MULTIPLICATION

Now let's multiply where some of the expressions may contain more than one term. To do this, we use procedures already studied in this chapter as well as the distributive laws and special products for multiplying with polynomials.

EXAMPLE 6   Multiply: $\sqrt{2}(\sqrt{3} + \sqrt{7})$.

$\sqrt{2}(\sqrt{3} + \sqrt{7}) = \sqrt{2}\sqrt{3} + \sqrt{2}\sqrt{7}$ Multiplying using a distributive law

$= \sqrt{6} + \sqrt{14}$ Using the rule for multiplying with radicals

EXAMPLE 7   Multiply: $(2 + \sqrt{3})(5 - 4\sqrt{3})$.

$(2 + \sqrt{3})(5 - 4\sqrt{3}) = 2 \cdot 5 - 2 \cdot 4\sqrt{3} + \sqrt{3} \cdot 5 - \sqrt{3} \cdot 4\sqrt{3}$

                                     Using FOIL

$= 10 - 8\sqrt{3} + 5\sqrt{3} - 4 \cdot 3$

$= 10 - 8\sqrt{3} + 5\sqrt{3} - 12$

$= -2 - 3\sqrt{3}$

Answers

1. $12\sqrt{2}$ **2.** $5\sqrt{5}$ **3.** $-12\sqrt{10}$

4. $5\sqrt{6}$ **5.** $\sqrt{x + 1}$ **6.** $\dfrac{3}{2}\sqrt{2}$, or $\dfrac{3\sqrt{2}}{2}$

7. $\dfrac{8}{15}\sqrt{15}$, or $\dfrac{8\sqrt{15}}{15}$

Guided Solution:
4. 4, 9, 2, 3, 6

EXAMPLE 8 Multiply: $(\sqrt{3} - \sqrt{x})(\sqrt{3} + \sqrt{x})$.

$(\sqrt{3} - \sqrt{x})(\sqrt{3} + \sqrt{x}) = (\sqrt{3})^2 - (\sqrt{x})^2$ Using
$(A - B)(A + B) = A^2 - B^2$

$= 3 - x$

EXAMPLE 9 Multiply: $(3 - \sqrt{p})^2$.

$(3 - \sqrt{p})^2 = 3^2 - 2 \cdot 3 \cdot \sqrt{p} + (\sqrt{p})^2$ Using
$(A - B)^2 = A^2 - 2AB + B^2$

$= 9 - 6\sqrt{p} + p$

EXAMPLE 10 Multiply: $(2 + \sqrt{5})^2$.

$(2 + \sqrt{5})^2 = 2^2 + 2 \cdot 2\sqrt{5} + (\sqrt{5})^2$ Using
$(A + B)^2 = A^2 + 2AB + B^2$

$= 4 + 4\sqrt{5} + 5$

$= 9 + 4\sqrt{5}$

Do Exercises 8–12. ▶

Multiply.

8. $\sqrt{3}(\sqrt{5} + \sqrt{2})$

9. $(1 - \sqrt{2})(4 + 3\sqrt{5})$

10. $(\sqrt{2} + \sqrt{a})(\sqrt{2} - \sqrt{a})$

11. $(5 + \sqrt{x})^2$

12. $(3 - \sqrt{7})(3 + \sqrt{7})$

c MORE ON RATIONALIZING DENOMINATORS

Note in Example 8 that the result has no radicals. This will happen whenever we multiply expressions such as $\sqrt{a} - \sqrt{b}$ and $\sqrt{a} + \sqrt{b}$. We see this in the following:

$$(\sqrt{a} + \sqrt{b})(\sqrt{a} - \sqrt{b}) = (\sqrt{a})^2 - (\sqrt{b})^2 = a - b.$$

Expressions such as $\sqrt{3} - \sqrt{x}$ and $\sqrt{3} + \sqrt{x}$ are known as **conjugates**; so too are $2 + \sqrt{5}$ and $2 - \sqrt{5}$. We can use conjugates to rationalize a denominator that involves a sum or a difference of two terms, where one or both are radicals. To do so, we multiply by 1 using the conjugate to form the expression for 1.

Do Exercises 13–15. ▶

Find the conjugate of each expression.

13. $7 + \sqrt{5}$

14. $\sqrt{5} - \sqrt{2}$

15. $1 - \sqrt{x}$

EXAMPLE 11 Rationalize the denominator: $\dfrac{3}{2 + \sqrt{5}}$.

We multiply by 1 using the conjugate of $2 + \sqrt{5}$, which is $2 - \sqrt{5}$, as the numerator and the denominator of the expression for 1:

$\dfrac{3}{2 + \sqrt{5}} = \dfrac{3}{2 + \sqrt{5}} \cdot \dfrac{2 - \sqrt{5}}{2 - \sqrt{5}}$ Multiplying by 1

$= \dfrac{3(2 - \sqrt{5})}{(2 + \sqrt{5})(2 - \sqrt{5})}$ Multiplying

$= \dfrac{6 - 3\sqrt{5}}{2^2 - (\sqrt{5})^2}$ Using $(A + B)(A - B) = A^2 - B^2$

$= \dfrac{6 - 3\sqrt{5}}{4 - 5}$

$= \dfrac{6 - 3\sqrt{5}}{-1}$

$= -6 + 3\sqrt{5}$, or $3\sqrt{5} - 6$.

EXAMPLE 12 Rationalize the denominator: $\dfrac{\sqrt{3} + \sqrt{5}}{\sqrt{3} - \sqrt{5}}$.

We multiply by 1 using the conjugate of $\sqrt{3} - \sqrt{5}$, which is $\sqrt{3} + \sqrt{5}$, as the numerator and the denominator of the expression for 1:

$$\dfrac{\sqrt{3} + \sqrt{5}}{\sqrt{3} - \sqrt{5}} = \dfrac{\sqrt{3} + \sqrt{5}}{\sqrt{3} - \sqrt{5}} \cdot \dfrac{\sqrt{3} + \sqrt{5}}{\sqrt{3} + \sqrt{5}} \qquad \text{Multiplying by 1}$$

$$= \dfrac{(\sqrt{3} + \sqrt{5})^2}{(\sqrt{3} - \sqrt{5})(\sqrt{3} + \sqrt{5})}$$

$$= \dfrac{(\sqrt{3})^2 + 2\sqrt{3}\sqrt{5} + (\sqrt{5})^2}{(\sqrt{3})^2 - (\sqrt{5})^2} \qquad \begin{array}{l}\text{Using } (A + B)^2 = A^2 + 2AB + B^2 \\ \text{and } (A + B)(A - B) = A^2 - B^2\end{array}$$

$$= \dfrac{3 + 2\sqrt{15} + 5}{3 - 5}$$

$$= \dfrac{8 + 2\sqrt{15}}{-2}$$

$$= \dfrac{2(4 + \sqrt{15})}{2(-1)} \qquad \text{Factoring in order to simplify}$$

$$= \dfrac{2}{2} \cdot \dfrac{4 + \sqrt{15}}{-1}$$

$$= \dfrac{4 + \sqrt{15}}{-1}$$

$$= -4 - \sqrt{15}.$$

◀ **Do Exercises 16 and 17.**

EXAMPLE 13 Rationalize the denominator: $\dfrac{5}{2 + \sqrt{x}}$.

We multiply by 1 using the conjugate of $2 + \sqrt{x}$, which is $2 - \sqrt{x}$, as the numerator and the denominator of the expression for 1:

$$\dfrac{5}{2 + \sqrt{x}} = \dfrac{5}{2 + \sqrt{x}} \cdot \dfrac{2 - \sqrt{x}}{2 - \sqrt{x}} \qquad \text{Multiplying by 1}$$

$$= \dfrac{5(2 - \sqrt{x})}{(2 + \sqrt{x})(2 - \sqrt{x})}$$

$$= \dfrac{5 \cdot 2 - 5 \cdot \sqrt{x}}{2^2 - (\sqrt{x})^2} \qquad \text{Using } (A + B)(A - B) = A^2 - B^2$$

$$= \dfrac{10 - 5\sqrt{x}}{4 - x}.$$

◀ **Do Exercise 18.**

Rationalize the denominator.

16. $\dfrac{3}{7 + \sqrt{5}}$ **GS**

$$= \dfrac{3}{7 + \sqrt{5}} \cdot \dfrac{7 - \sqrt{5}}{\boxed{} - \sqrt{5}}$$

$$= \dfrac{\boxed{}(7 - \sqrt{5})}{7^2 - (\boxed{})^2}$$

$$= \dfrac{\boxed{} - 3\sqrt{5}}{49 - \boxed{}}$$

$$= \dfrac{21 - 3\sqrt{5}}{\boxed{}}$$

17. $\dfrac{\sqrt{5} + \sqrt{7}}{\sqrt{5} - \sqrt{7}}$

18. Rationalize the denominator:

$$\dfrac{7}{1 - \sqrt{x}}.$$

Answers

16. $\dfrac{21 - 3\sqrt{5}}{44}$ **17.** $-6 - \sqrt{35}$

18. $\dfrac{7 + 7\sqrt{x}}{1 - x}$

Guided Solution:
16. $7, 3, \sqrt{5}, 21, 5, 44$

✓ Reading Check

Determine whether the two given expressions are conjugates. Answer "yes" or "no."

RC1. $5 - \sqrt{3}, 5 + \sqrt{3}$

RC2. $\sqrt{3} \cdot \sqrt{5}, \sqrt{5} \cdot \sqrt{3}$

RC3. $\sqrt{3} - \sqrt{5}, \sqrt{5} - \sqrt{3}$

RC4. $\sqrt{3} - \sqrt{5}, \sqrt{3} + \sqrt{5}$

RC5. $-\sqrt{5} + 3, \sqrt{5} + 3$

RC6. $\sqrt{5} - 3, 3 - \sqrt{5}$

RC7. $\dfrac{\sqrt{3}}{\sqrt{5}}, \dfrac{\sqrt{5}}{\sqrt{3}}$

RC8. $3 - \sqrt{5}, 5 - \sqrt{3}$

RC9. $5\sqrt{3} + 3\sqrt{5}, 3\sqrt{5} - 5\sqrt{3}$

a   Add or subtract. Simplify by collecting like radical terms, if possible.

1. $7\sqrt{3} + 9\sqrt{3}$

2. $6\sqrt{2} + 8\sqrt{2}$

3. $7\sqrt{5} - 3\sqrt{5}$

4. $8\sqrt{2} - 5\sqrt{2}$

5. $6\sqrt{x} + 7\sqrt{x}$

6. $9\sqrt{y} + 3\sqrt{y}$

7. $4\sqrt{d} - 13\sqrt{d}$

8. $2\sqrt{a} - 17\sqrt{a}$

9. $5\sqrt{8} + 15\sqrt{2}$

10. $3\sqrt{12} + 2\sqrt{3}$

11. $\sqrt{27} - 2\sqrt{3}$

12. $7\sqrt{50} - 3\sqrt{2}$

13. $\sqrt{45} - \sqrt{20}$

14. $\sqrt{27} - \sqrt{12}$

15. $\sqrt{72} + \sqrt{98}$

16. $\sqrt{45} + \sqrt{80}$

17. $2\sqrt{12} + \sqrt{27} - \sqrt{48}$

18. $9\sqrt{8} - \sqrt{72} + \sqrt{98}$

19. $\sqrt{18} - 3\sqrt{8} + \sqrt{50}$

20. $3\sqrt{18} - 2\sqrt{32} - 5\sqrt{50}$

21. $2\sqrt{27} - 3\sqrt{48} + 3\sqrt{12}$

22. $3\sqrt{48} - 2\sqrt{27} - 3\sqrt{12}$

23. $\sqrt{4x} + \sqrt{81x^3}$

24. $\sqrt{12x^2} + \sqrt{27}$

25. $\sqrt{27} - \sqrt{12x^2}$

26. $\sqrt{81x^3} - \sqrt{4x}$

27. $\sqrt{8x + 8} + \sqrt{2x + 2}$

28. $\sqrt{12x + 12} + \sqrt{3x + 3}$

29. $\sqrt{x^5 - x^2} + \sqrt{9x^3 - 9}$

30. $\sqrt{16x - 16} + \sqrt{25x^3 - 25x^2}$

31. $4a\sqrt{a^2b} + a\sqrt{a^2b^3} - 5\sqrt{b^3}$

32. $3x\sqrt{y^3x} - x\sqrt{yx^3} + y\sqrt{y^3x}$

33. $\sqrt{3} - \sqrt{\dfrac{1}{3}}$

34. $\sqrt{2} - \sqrt{\dfrac{1}{2}}$

35. $5\sqrt{2} + 3\sqrt{\dfrac{1}{2}}$

36. $4\sqrt{3} + 2\sqrt{\dfrac{1}{3}}$

37. $\sqrt{\dfrac{2}{3}} - \sqrt{\dfrac{1}{6}}$

38. $\sqrt{\dfrac{1}{2}} - \sqrt{\dfrac{1}{8}}$

b Multiply.

39. $\sqrt{3}(\sqrt{5} - 1)$

40. $\sqrt{2}(\sqrt{2} + \sqrt{3})$

41. $(2 + \sqrt{3})(5 - \sqrt{7})$

42. $(\sqrt{5} + \sqrt{7})(2\sqrt{5} - 3\sqrt{7})$

43. $(2 - \sqrt{5})^2$

44. $(\sqrt{3} + \sqrt{10})^2$

45. $(\sqrt{2} + 8)(\sqrt{2} - 8)$

46. $(1 + \sqrt{7})(1 - \sqrt{7})$

47. $(\sqrt{6} - \sqrt{5})(\sqrt{6} + \sqrt{5})$

48. $(\sqrt{3} + \sqrt{10})(\sqrt{3} - \sqrt{10})$

49. $(3\sqrt{5} - 2)(\sqrt{5} + 1)$

50. $(\sqrt{5} - 2\sqrt{2})(\sqrt{10} - 1)$

51. $(\sqrt{x} - \sqrt{y})^2$

52. $(\sqrt{w} + 11)^2$

c Rationalize the denominator.

53. $\dfrac{2}{\sqrt{3} - \sqrt{5}}$

54. $\dfrac{5}{3 + \sqrt{7}}$

55. $\dfrac{\sqrt{3} - \sqrt{2}}{\sqrt{3} + \sqrt{2}}$

56. $\dfrac{2 - \sqrt{7}}{\sqrt{3} - \sqrt{2}}$

57. $\dfrac{4}{\sqrt{10} + 1}$

58. $\dfrac{6}{\sqrt{11} - 3}$

59. $\dfrac{1 - \sqrt{7}}{3 + \sqrt{7}}$

60. $\dfrac{2 + \sqrt{8}}{1 - \sqrt{5}}$

61. $\dfrac{3}{4 + \sqrt{x}}$

62. $\dfrac{8}{2 - \sqrt{x}}$

63. $\dfrac{3 + \sqrt{2}}{8 - \sqrt{x}}$

64. $\dfrac{4 - \sqrt{3}}{6 + \sqrt{y}}$

65. $\dfrac{\sqrt{a} - 1}{1 + \sqrt{a}}$

66. $\dfrac{12 + \sqrt{w}}{\sqrt{w} - 12}$

67. $\dfrac{4 + \sqrt{3}}{\sqrt{a} - \sqrt{t}}$

68. $\dfrac{\sqrt{2} - 1}{\sqrt{w} + \sqrt{b}}$

Skill Maintenance

Solve.

69. $3x + 5 + 2(x - 3) = 4 - 6x$ [2.3c]

70. $3(x - 4) - 2 = 8(2x + 3)$ [2.3c]

71. $x^2 - 5x = 6$ [5.7b]

72. $x^2 + 10 = 7x$ [5.7b]

73. Multiply and simplify:

$$\frac{3}{x^2 - 9} \cdot \frac{x^2 - 6x + 9}{12}.$$ [6.1d]

74. The graph of the polynomial equation

$$y = x^3 - 5x^2 + x - 2$$

is shown below. Use either the graph or the equation to estimate or find the value of the polynomial when $x = -1$, $x = 0$, $x = 1$, $x = 3$, and $x = 4.85$.   [4.3a]

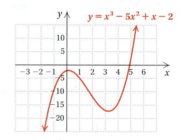

75. *Continental Divide.*   The Continental Divide in the Americas divides the flow of water between the Pacific Ocean and the Atlantic Ocean. The Continental Divide National Scenic Trail in the United States runs through five states: Montana, Idaho, Wyoming, Colorado, and New Mexico. The Trail's highest altitude is 9990 ft higher than its lowest altitude of 4280 ft. What is the highest altitude of the Trail?   [2.6a]

Source: www.continental-divide.net

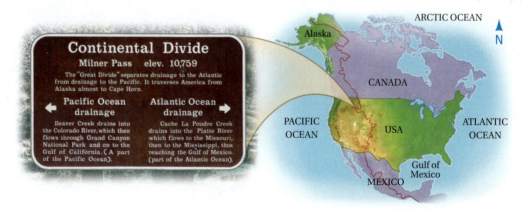

Synthesis

76. Evaluate $\sqrt{a^2 + b^2}$ and $\sqrt{a^2} + \sqrt{b^2}$ when $a = 2$ and $b = 3$. Then determine whether $\sqrt{a^2 + b^2}$ and $\sqrt{a^2} + \sqrt{b^2}$ are equivalent.

Use the TABLE feature to determine whether each of the following is correct.

77. $\sqrt{9x^3} + \sqrt{x} = \sqrt{9x^3 + x}$

78. $\sqrt{x^2 + 4} = x + 2$

Add or subtract as indicated.

79. $\frac{3}{5}\sqrt{24} + \frac{2}{5}\sqrt{150} - \sqrt{96}$

80. $\frac{1}{3}\sqrt{27} + \sqrt{8} + \sqrt{300} - \sqrt{18} - \sqrt{162}$

Radical Equations

a SOLVING RADICAL EQUATIONS

The following are examples of *radical equations*:

$$\sqrt{2x} - 4 = 7, \qquad \sqrt{x + 1} = \sqrt{2x - 5}.$$

A **radical equation** has variables in one or more radicands. To solve radical equations, we first convert them to equations without radicals. We do this for square-root radical equations by squaring both sides of the equation, using the following principle.

THE PRINCIPLE OF SQUARING

If an equation $a = b$ is true, then the equation $a^2 = b^2$ is true.

To solve square-root radical equations, we first try to get a radical by itself. That is, we try to isolate the radical. Then we use the principle of squaring. This allows us to eliminate one radical.

EXAMPLE 1 Solve: $\sqrt{2x} - 4 = 7$.

$$\sqrt{2x} - 4 = 7$$

$\sqrt{2x} = 11$	Adding 4 to isolate the radical
$(\sqrt{2x})^2 = 11^2$	Squaring both sides
$2x = 121$	$\sqrt{2x} \cdot \sqrt{2x} = 2x$, by the definition of square root
$x = \dfrac{121}{2}$	Dividing by 2

Check:
$$\begin{array}{c|c}
\sqrt{2x} - 4 = 7 & \\
\hline
\sqrt{2 \cdot \dfrac{121}{2} - 4} \;\; ? \;\; 7 & \\
\sqrt{121} - 4 & \\
11 - 4 & \\
7 & \text{TRUE}
\end{array}$$

The solution is $\frac{121}{2}$.

Do Margin Exercise 1. ▶

EXAMPLE 2 Solve: $2\sqrt{x + 2} = \sqrt{x + 10}$.

Each radical is isolated. We proceed with the principle of squaring.

$(2\sqrt{x + 2})^2 = (\sqrt{x + 10})^2$	Squaring both sides
$2^2(\sqrt{x + 2})^2 = (\sqrt{x + 10})^2$	Raising each factor of the product on the left to the second power
$4(x + 2) = x + 10$	Simplifying
$4x + 8 = x + 10$	Removing parentheses
$3x = 2$	Subtracting x and 8
$x = \dfrac{2}{3}$	Dividing by 3

OBJECTIVES

a Solve radical equations with one or two radical terms isolated, using the principle of squaring once.

b Solve radical equations with two radical terms, using the principle of squaring twice.

c Solve applied problems using radical equations.

SKILL TO REVIEW

Objective 5.7b: Solve quadratic equations by factoring and then using the principle of zero products.

Solve.
1. $x^2 + 4x - 45 = 0$
2. $1 + x^2 = -2x$

---

1. Solve: $\sqrt{3x} - 5 = 3$.

Answers

Skill to Review:
1. $-9, 5$ 2. -1

Margin Exercise:
1. $\dfrac{64}{3}$

Solve.

2. $\sqrt{3x+1} = \sqrt{2x+3}$ **GS**
$(\sqrt{3x+1})^2 = (\boxed{})^2$
$\boxed{} + 1 = 2x + \boxed{}$
$\boxed{} = 2$

3. $3\sqrt{x+1} = \sqrt{x+12}$

ALGEBRAIC ▶◀ **GRAPHICAL CONNECTION**

Consider the equation of Example 3:

$$x - 5 = \sqrt{x+7}.$$

We can visualize the solutions by graphing the equations

$$y = x - 5 \quad \text{and} \quad y = \sqrt{x+7}$$

using the same set of axes.

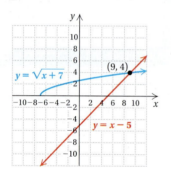

It appears that when $x = 9$, the values of $y = x - 5$ and $y = \sqrt{x+7}$ are the same, 4. We can check this as we did in Example 3. Note also that the graphs *do not* intersect at $x = 2$.

4. Solve: $x - 1 = \sqrt{x+5}$.

Answers

2. 2 **3.** $\dfrac{3}{8}$ **4.** 4

Guided Solution:
2. $\sqrt{2x+3}, 3x, 3, x$

Check:
$$2\sqrt{x+2} = \sqrt{x+10}$$

$$2\sqrt{\dfrac{2}{3}+2} \overset{?}{=} \sqrt{\dfrac{2}{3}+10}$$

$$2\sqrt{\dfrac{8}{3}} \quad \sqrt{\dfrac{32}{3}}$$

$$2\sqrt{\dfrac{4\cdot 2}{3}} \quad \sqrt{\dfrac{16\cdot 2}{3}}$$

$$4\sqrt{\dfrac{2}{3}} \quad \Big| \quad 4\sqrt{\dfrac{2}{3}} \quad \text{TRUE}$$

The number $\frac{2}{3}$ checks. The solution is $\frac{2}{3}$.

◀ **Do Exercises 2 and 3.**

It is necessary to check when using the principle of squaring. This principle may not produce equivalent equations. When we square both sides of an equation, the new equation may have solutions that the first one does not. For example, the equation

$$x = 1 \qquad \textbf{(1)}$$

has just **one** solution, the number 1. When we square both sides, we get

$$x^2 = 1, \qquad \textbf{(2)}$$

which has **two** solutions, 1 and -1. The equations $x = 1$ and $x^2 = 1$ do not have the same solutions and thus are not equivalent. Whereas it is true that any solution of equation (1) is a solution of equation (2), it is *not* true that any solution of equation (2) is a solution of equation (1).

·············· **Caution!** ··············

When the principle of squaring is used to solve an equation, all possible solutions *must* be checked in the original equation!

Sometimes we may need to apply the principle of zero products after squaring. (See Section 5.7.)

EXAMPLE 3 Solve: $x - 5 = \sqrt{x+7}$.

$$x - 5 = \sqrt{x+7}$$
$$(x-5)^2 = (\sqrt{x+7})^2 \qquad \text{Using the principle of squaring}$$
$$x^2 - 10x + 25 = x + 7$$
$$x^2 - 11x + 18 = 0 \qquad \text{Subtracting } x \text{ and } 7$$
$$(x-9)(x-2) = 0 \qquad \text{Factoring}$$
$$x - 9 = 0 \quad or \quad x - 2 = 0 \qquad \text{Using the principle of zero products}$$
$$x = 9 \quad or \qquad x = 2$$

Check: For 9:

$$x - 5 = \sqrt{x+7}$$
$$9 - 5 \overset{?}{=} \sqrt{9+7}$$
$$4 \quad \Big| \quad 4 \qquad \text{TRUE}$$

For 2:

$$x - 5 = \sqrt{x+7}$$
$$2 - 5 \overset{?}{=} \sqrt{2+7}$$
$$-3 \quad \Big| \quad 3 \qquad \text{FALSE}$$

The number 9 checks, but 2 does not. Thus the solution is 9.

◀ **Do Exercise 4.**

EXAMPLE 4 Solve: $3 + \sqrt{27 - 3x} = x$.

In this case, we must first isolate the radical.

$$3 + \sqrt{27 - 3x} = x$$

$\qquad \sqrt{27 - 3x} = x - 3$ Subtracting 3 to isolate the radical

$\qquad (\sqrt{27 - 3x})^2 = (x - 3)^2$ Using the principle of squaring

$\qquad 27 - 3x = x^2 - 6x + 9$ Squaring on each side

$\qquad 0 = x^2 - 3x - 18$ Adding $3x$ and subtracting 27 to obtain 0 on the left

$\qquad 0 = (x - 6)(x + 3)$ Factoring

$\qquad x - 6 = 0 \ \ or \ \ x + 3 = 0$ Using the principle of zero products

$\qquad x = 6 \ \ or \qquad x = -3$

Check: For 6:

$$3 + \sqrt{27 - 3x} = x$$
$$\overline{3 + \sqrt{27 - 3 \cdot 6}} \ ? \ 6$$
$$3 + \sqrt{27 - 18}$$
$$3 + \sqrt{9}$$
$$3 + 3$$
$$6 \ \bigg| \ \ \text{TRUE}$$

For -3:

$$3 + \sqrt{27 - 3x} = x$$
$$\overline{3 + \sqrt{27 - 3 \cdot (-3)}} \ ? \ -3$$
$$3 + \sqrt{27 + 9}$$
$$3 + \sqrt{36}$$
$$3 + 6$$
$$9 \ \bigg| \ \ \text{FALSE}$$

The number 6 checks, but -3 does not. The solution is 6.

Do Exercise 5. ▶

b USING THE PRINCIPLE OF SQUARING MORE THAN ONCE

Sometimes when we have two radical terms, we may need to apply the principle of squaring a second time.

EXAMPLE 5 Solve: $\sqrt{x} - 1 = \sqrt{x - 5}$.

We have

$$\sqrt{x} - 1 = \sqrt{x - 5}$$

$\qquad (\sqrt{x} - 1)^2 = (\sqrt{x - 5})^2$ Using the principle of squaring

$\qquad (\sqrt{x})^2 - 2 \cdot \sqrt{x} \cdot 1 + 1^2 = x - 5$ Using $(A - B)^2 = A^2 - 2AB + B^2$ on the left side

$\qquad x - 2\sqrt{x} + 1 = x - 5$ Simplifying. Only one radical term remains.

$\qquad -2\sqrt{x} = -6$ Isolating the radical by subtracting x and 1

$\qquad \sqrt{x} = 3$ Dividing by -2

$\qquad (\sqrt{x})^2 = 3^2$ Using the principle of squaring

$\qquad x = 9.$

The check is left to the student. The number 9 checks and is the solution.

5. Solve: $1 + \sqrt{1 - x} = x$.

CALCULATOR CORNER

Solving Radical Equations We can solve radical equations on a graphing calculator. Consider the equation in Example 3: $x - 5 = \sqrt{x + 7}$. We first graph each side of the equation. We enter $y_1 = x - 5$ and $y_2 = \sqrt{x + 7}$ on the equation-editor screen and graph the equations, using the window $[-2, 12, -6, 6]$. Note that there is one point of intersection. Use the INTERSECT feature to find its coordinates. (See the Calculator Corner on p. 558 for the procedure.)

The first coordinate, 9, is the value of x for which $y_1 = y_2$, or $x - 5 = \sqrt{x + 7}$. It is the solution of the equation. Note that the graph shows a single solution whereas the algebraic solution in Example 3 yields two possible solutions, 9 and 2, that must be checked. The check shows that 9 is the only solution.

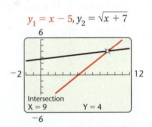

$y_1 = x - 5, y_2 = \sqrt{x + 7}$

EXERCISES:

1. Solve the equations in Examples 4 and 5 graphically.

2. Solve the equations in Margin Exercises 1–6 graphically.

Answer

5. 1

The following is a procedure for solving square-root radical equations.

> ## SOLVING SQUARE-ROOT RADICAL EQUATIONS
> ....
> To solve square-root radical equations:
>
> 1. Isolate one of the radical terms.
> 2. Use the principle of squaring.
> 3. If a radical term remains, perform steps (1) and (2) again.
> 4. Solve the equation and check possible solutions.

6. Solve: $\sqrt{x} - 1 = \sqrt{x - 3}$.

$$\sqrt{x} - 1 = \sqrt{x - 3}$$
$$(\boxed{})^2 = (\sqrt{x - 3})^2$$
$$(\boxed{})^2 - \boxed{}\sqrt{x} + 1 = x - 3$$
$$\boxed{} - 2\sqrt{x} + 1 = x - 3$$
$$-2\sqrt{x} + 1 = -3$$
$$-2\sqrt{x} = \boxed{}$$
$$\sqrt{x} = 2$$
$$(\sqrt{x})^2 = \boxed{}^2$$
$$\boxed{} = 4$$

GS   ◀ Do Exercise 6.

C APPLICATIONS

Sighting to the Horizon.   How far can you see from a given height? The equation

$$D = \sqrt{2h}$$

can be used to approximate the distance D, in miles, that a person can see to the horizon from a height h, in feet.

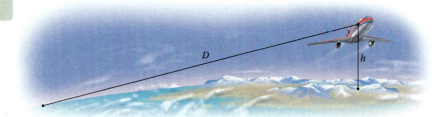

7. How far to the horizon can you see through an airplane window at a height, or altitude, of 38,000 ft?

8. A sailor climbs 40 ft up the mast of a ship to the crow's nest. How far can he see to the horizon?

EXAMPLE 6   How far to the horizon can you see through an airplane window at a height, or altitude, of 30,000 ft?

We substitute 30,000 for h in $D = \sqrt{2h}$ and find an approximation using a calculator:

$$D = \sqrt{2 \cdot 30{,}000} \approx 245 \text{ mi.}$$

You can see for about 245 mi to the horizon.

◀ Do Exercises 7 and 8.

Answers

6. 4   **7.** About 276 mi   **8.** About 9 mi

Guided Solution:
6. $\sqrt{x} - 1$, $\sqrt{x}$, 2, x, -4, 2, x

EXAMPLE 7 *Height of a Ranger Station.* How high is a ranger station if the ranger is able to see out to a fire on the horizon 15.4 mi away?

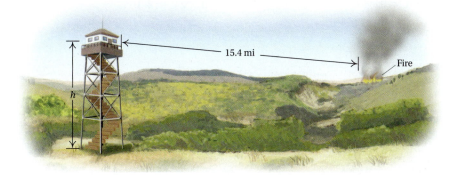

15.4 mi

Fire

h

We substitute 15.4 for D in $D = \sqrt{2h}$ and solve:

$$15.4 = \sqrt{2h}$$
$$(15.4)^2 = (\sqrt{2h})^2 \qquad \text{Using the principle of squaring}$$
$$237.16 = 2h$$
$$\frac{237.16}{2} = h$$
$$118.58 = h.$$

The height of the ranger tower must be about 119 ft in order for the ranger to see out to a fire 15.4 mi away.

Do Exercise 9. ▶

9. How far above sea level must a sailor climb on the mast of a ship in order to see 10.2 mi out to an iceberg?

Answer

9. About 52 ft

  Exercise Set

For Extra Help

MyMathLab®   MathXL®   PRACTICE   WATCH   READ   REVIEW

✓ Reading Check

Determine the number of times that the principle of squaring must be used in order to solve the equation. Do not solve the equation.

RC1. $x + 4 = 4\sqrt{x + 1}$

RC2. $\sqrt{x} - 1 = \sqrt{x - 31}$

RC3. $\sqrt{4x - 5} = \sqrt{x + 9}$

RC4. $\sqrt{x + 7} = x - 5$

RC5. $\sqrt{x + 9} = 1 + \sqrt{x}$

RC6. $1 + \sqrt{x + 7} = \sqrt{3x - 2}$

Solve.

1. $\sqrt{x} = 6$

2. $\sqrt{x} = 1$

3. $\sqrt{x} = 4.3$

4. $\sqrt{x} = 6.2$

5. $\sqrt{y + 4} = 13$

6. $\sqrt{y - 5} = 21$

7. $\sqrt{2x + 4} = 25$

8. $\sqrt{2x + 1} = 13$

9. $3 + \sqrt{x - 1} = 5$

10. $4 + \sqrt{y - 3} = 11$

11. $6 - 2\sqrt{3n} = 0$

12. $8 - 4\sqrt{5n} = 0$

13. $\sqrt{5x - 7} = \sqrt{x + 10}$

14. $\sqrt{4x - 5} = \sqrt{x + 9}$

15. $\sqrt{x} = -7$

16. $\sqrt{x} = -5$

17. $\sqrt{2y + 6} = \sqrt{2y - 5}$

18. $2\sqrt{3x - 2} = \sqrt{2x - 3}$

19. $x - 7 = \sqrt{x - 5}$

20. $\sqrt{x + 7} = x - 5$

21. $x - 9 = \sqrt{x - 3}$

22. $\sqrt{x + 18} = x - 2$

23. $2\sqrt{x - 1} = x - 1$

24. $x + 4 = 4\sqrt{x + 1}$

25. $\sqrt{5x + 21} = x + 3$

26. $\sqrt{27 - 3x} = x - 3$

27. $\sqrt{2x - 1} + 2 = x$

28. $x = 1 + 6\sqrt{x - 9}$

29. $\sqrt{x^2 + 6} - x + 3 = 0$

30. $\sqrt{x^2 + 5} - x + 2 = 0$

31. $\sqrt{x^2 - 4} - x = 6$

32. $\sqrt{x^2 - 5x + 7} = x - 3$

33. $\sqrt{(p + 6)(p + 1)} - 2 = p + 1$

34. $\sqrt{(4x + 5)(x + 4)} = 2x + 5$

35. $\sqrt{4x - 10} = \sqrt{2 - x}$

36. $\sqrt{2 - x} = \sqrt{3x - 7}$

b   Solve. Use the principle of squaring twice.

37. $\sqrt{x - 5} = 5 - \sqrt{x}$

38. $\sqrt{x + 9} = 1 + \sqrt{x}$

39. $\sqrt{y + 8} - \sqrt{y} = 2$

40. $\sqrt{3x + 1} = 1 - \sqrt{x + 4}$

41. $\sqrt{x - 4} + \sqrt{x + 1} = 5$

42. $1 + \sqrt{x + 7} = \sqrt{3x - 2}$

43. $\sqrt{x - 1} = \sqrt{x - 31}$

44. $\sqrt{2x - 5} - 1 = \sqrt{x - 3}$

C Solve.

The speed v, in meters per second, of a wave on the surface of the ocean can be approximated by the formula $v = 3.1\sqrt{d}$, where d is the depth of the water, in meters. Use this formula for Exercises 45–48.

Source: myweb.dal.ca

45. What is the speed, in meters per second, of a wave on the surface of the ocean where the depth of the water is 400 m?

46. What is the speed, in meters per second, of a wave on the surface of the ocean where the depth of the water is 225 m?

47. A wave is traveling at a speed of 34.1 m/sec. What is the water depth?

48. A wave is traveling at a speed of 38.75 m/sec. What is the water depth?

Speed of a Skidding Car. How do police determine how fast a car had been traveling after an accident has occurred? The formula
$$r = 2\sqrt{5L}$$
can be used to approximate the speed r, in miles per hour, of a car that has left a skid mark of length L, in feet. (See Example 7 in Section 8.1.) Use this formula for Exercises 49 and 50.

49. How far will a car skid at 65 mph? at 75 mph?

50. How far will a car skid at 55 mph? at 90 mph?

Skill Maintenance

51. Solve $R = \dfrac{s + t}{2}$ for s.   [2.4b]

52. Solve: $4 - \dfrac{1}{4}x < \dfrac{1}{2}x + 10$.   [2.7e]

53. Evaluate $(3x)^3$ when $x = -4$.   [4.1c]

54. Divide and simplify: $\dfrac{y^8 w^3}{y^3 w}$.   [4.1e]

Factor.

55. $2x^2 + 11x + 5$ [5.4a]   **56.** $y^2 - 36$ [5.5d]

57. $9t^2 + 24t + 16$ [5.5b]   **58.** $1 - x^8$ [5.6a]

Synthesis

Solve.

59. $\sqrt{5x^2 + 5} = 5$

60. $\sqrt{x} = -x$

61. $4 + \sqrt{19 - x} = 6 + \sqrt{4 - x}$

62. $x = (x - 2)\sqrt{x}$

63. $\sqrt{x + 3} = \dfrac{8}{\sqrt{x - 9}}$

64. $\dfrac{12}{\sqrt{5x + 6}} = \sqrt{2x + 5}$

8.6

a   RIGHT TRIANGLES

A **right triangle** is a triangle with a 90° angle, as shown in the figure below. The small square in the corner indicates the 90° angle.

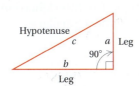

In a right triangle, the longest side is called the **hypotenuse**. It is also the side opposite the right angle. The other two sides are called **legs**. We generally use the letters a and b for the lengths of the legs and c for the length of the hypotenuse. They are related as follows.

THE PYTHAGOREAN THEOREM

In any right triangle, if a and b are the lengths of the legs and c is the length of the hypotenuse, then

$$a^2 + b^2 = c^2.$$

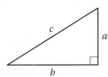

The equation $a^2 + b^2 = c^2$ is called the **Pythagorean equation**.

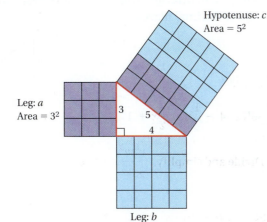

Hypotenuse: c
Area = 5^2

Leg: a
Area = 3^2

$$a^2 + b^2 = c^2$$
$$3^2 + 4^2 = 5^2$$
$$9 + 16 = 25$$

Leg: b
Area = 4^2

The Pythagorean theorem is named after the ancient Greek mathematician Pythagoras (569?–500? B.C.). It is uncertain who actually proved this result the first time. A proof can be found in most geometry books.

If we know the lengths of any two sides of a right triangle, we can find the length of the third side.

OBJECTIVES

a   Given the lengths of any two sides of a right triangle, find the length of the third side.

b   Solve applied problems involving right triangles.

SKILL TO REVIEW

Objective 8.1b: Approximate square roots of real numbers using a calculator.

Use a calculator to approximate each square root. Round to three decimal places.

1. $\sqrt{217}$ **2.** $\sqrt{29}$

Answers

Skill to Review:
1. 14.731   2. 5.385

1. Find the length of the hypotenuse of this right triangle. Give an exact answer and an approximation to three decimal places.

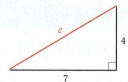

2. Find the length a in this right triangle. Give an exact answer and an approximation to three decimal places.

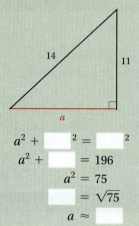

$$a^2 + \boxed{}^2 = \boxed{}^2$$
$$a^2 + \boxed{} = 196$$
$$a^2 = 75$$
$$\boxed{} = \sqrt{75}$$
$$a \approx \boxed{}$$

Find the missing length of a leg in the right triangle. Give an exact answer and an approximation to three decimal places.

3.

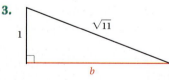

4.

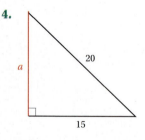

EXAMPLE 1 Find the length of the hypotenuse of this right triangle. Give an exact answer and an approximation to three decimal places.

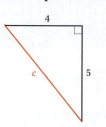

$$4^2 + 5^2 = c^2 \qquad \text{Substituting in the Pythagorean equation}$$
$$16 + 25 = c^2$$
$$41 = c^2$$
$$c = \sqrt{41} \qquad \text{Exact answer}$$
$$c \approx 6.403 \qquad \text{Using a calculator}$$

EXAMPLE 2 Find the length b in this right triangle. Give an exact answer and an approximation to three decimal places.

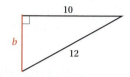

$$10^2 + b^2 = 12^2 \qquad \text{Substituting in the Pythagorean equation}$$
$$100 + b^2 = 144$$
$$b^2 = 144 - 100$$
$$b^2 = 44$$
$$b = \sqrt{44} \qquad \text{Exact answer}$$
$$b \approx 6.633 \qquad \text{Using a calculator}$$

◀ **Do Exercises 1 and 2.**

EXAMPLE 3 Find the length b in this right triangle. Give an exact answer and an approximation to three decimal places.

$$1^2 + b^2 = (\sqrt{7})^2 \qquad \text{Substituting in the Pythagorean equation}$$
$$1 + b^2 = 7$$
$$b^2 = 7 - 1 = 6$$
$$b = \sqrt{6} \qquad \text{Exact answer}$$
$$b \approx 2.449 \qquad \text{Using a calculator}$$

EXAMPLE 4 Find the length a in this right triangle. Give an exact answer and an approximation to three decimal places.

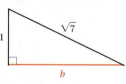

$$a^2 + 10^2 = 15^2$$
$$a^2 + 100 = 225$$
$$a^2 = 225 - 100$$
$$a^2 = 125$$
$$a = \sqrt{125} \qquad \text{Exact answer}$$
$$a \approx 11.180 \qquad \text{Using a calculator}$$

◀ **Do Exercises 3 and 4.**

Answers

1. $\sqrt{65} \approx 8.062$ 2. $\sqrt{75} \approx 8.660$
3. $\sqrt{10} \approx 3.162$ 4. $\sqrt{175} \approx 13.229$

Guided Solution:
2. 11, 14, 121, a, 8.660

b APPLICATIONS

EXAMPLE 5 *Skateboard Ramp.* Ramps.com of America sells Landwave ramps and decks that can be combined to create a skateboard ramp as high or as wide as one wants. The dimensions of the basic ramp unit are 28 in. wide, 38.5 in. long, and 12 in. high.
Source: www.ramps.com

a) What is the length of the skating surface of one ramp unit?

b) How many ramp units are needed for a 10-ft long skating surface?

a)

1. **Familiarize.** We first make a drawing and label it with the given dimensions. The base and the end of the ramp unit form a right angle. We label the length of the skating surface *r*.

2. **Translate.** We use the Pythagorean equation:

$$a^2 + b^2 = c^2 \qquad \text{Pythagorean equation}$$
$$(38.5)^2 + 12^2 = r^2. \qquad \text{Substituting 38.5 for } a, \text{ 12 for } b, \text{ and } c \text{ for } r$$

3. **Solve.** We solve as follows:

$$(38.5)^2 + 12^2 = r^2$$
$$1482.25 + 144 = r^2 \qquad \text{Squaring}$$
$$1626.25 = r^2$$
$$\sqrt{1626.25} = r \qquad \text{Exact answer}$$
$$40.327 \approx r. \qquad \text{Approximate answer}$$

4. **Check.** We check the calculations using the Pythagorean equation: $38.5^2 + 12^2 = 1626.25$ and $(40.327)^2 \approx 1626$. The length checks. (Remember that we estimated the value of *r*.)

5. **State.** The length of the skating surface of a single ramp unit is about 40.327 in.

b) In inches, the length of a 10-ft skating surface is 10 × 12 in., or 120 in. Each ramp unit is about 40 in. long. Thus it will take 120 ÷ 40, or 3, ramp units for a ramp that has a 10-ft long surface.

Do Exercise 5. ▶

5. *Christmas Tree.* Each Christmas season since 1962, the Soldiers' and Sailors' Monument in the center of Indianapolis is decorated as a giant Christmas tree. The tree is composed of 52 strands of lights that are attached at the top of the monument, which is 242 ft tall. How long is each garland of lights if each strand is attached to the ground 39 ft from the center of the monument? Give an exact answer and an approximation to three decimal places.
Source: Indianapolis Downtown Inc.

Answer
5. $\sqrt{60{,}085}$ ft ≈ 245.122 ft

Translating for Success

1. **Coin Mixture.** A collection of nickels and quarters is worth $9.35. There are 59 coins in all. How many of each coin are there?

2. **Diagonal of a Square.** Find the length of a diagonal of a square whose sides are 8 ft long.

3. **Shoveling Time.** It takes Mark 55 min to shovel 4 in. of snow from his driveway. It takes Eric 75 min to do the same job. How long would it take if they worked together?

4. **Angles of a Triangle.** The second angle of a triangle is three times as large as the first. The third is 17° less than the sum of the other angles. Find the measures of the angles.

5. **Perimeter.** The perimeter of a rectangle is 568 ft. The length is 26 ft greater than the width. Find the length and the width.

The goal of these matching questions is to practice step (2), Translate, of the five-step problem-solving process. Translate each word problem to an equation or a system of equations and select a correct translation from equations A–O.

A. $5x + 25y = 9.35,$
$x + y = 59$

B. $4^2 + x^2 = 8^2$

C. $x(x + 26) = 568$

D. $8 = x \cdot 24$

E. $\dfrac{75}{x} = \dfrac{105}{x + 5}$

F. $\dfrac{75}{x} = \dfrac{55}{x + 5}$

G. $2x + 2(x + 26) = 568$

H. $x + 3x + (x + 3x - 17) = 180$

I. $x + 3x + (3x - 17) = 180$

J. $0.05x + 0.25y = 9.35,$
$x + y = 59$

K. $8^2 + 8^2 = x^2$

L. $x^2 + (x + 26)^2 = 568$

M. $x - 5\% \cdot x = 8568$

N. $\dfrac{1}{55} + \dfrac{1}{75} = \dfrac{1}{x}$

O. $x + 5\% \cdot x = 8568$

Answers on page A-26

6. **Car Travel.** One horse travels 75 km in the same time that a horse traveling 5 km/h faster travels 105 km. Find the speed of each horse.

7. **Money Borrowed.** Emma borrows some money at 5% simple interest. After 1 year, $8568 pays off her loan. How much did she originally borrow?

8. **TV Time.** The average amount of time per day that TV sets in the United States are turned on is 8 hr. What percent of the time are our TV sets on?

Source: Nielsen Media Research

9. **Ladder Height.** An 8-ft plank is leaning against a shed. The bottom of the plank is 4 ft from the building. How high is the top of the plank?

10. **Lengths of a Rectangle.** The area of a rectangle is 568 ft². The length is 26 ft greater than the width. Find the length and the width.

✓ Reading Check

Choose from the column on the right the equation whose solution is the length of the missing side of the right triangle.

RC1.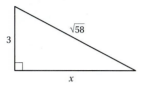

RC2.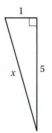

a) $3^2 = x^2 + (\sqrt{58})^2$
b) $1^2 = x^2 + 5^2$
c) $1^2 + 5^2 = x^2$
d) $x^2 + 3^2 = (\sqrt{58})^2$

a Find the length of the third side of each right triangle. Where appropriate, give both an exact answer and an approximation to three decimal places.

1.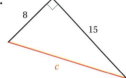

2.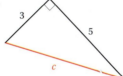

3.

4.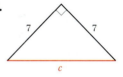

5.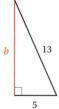

6.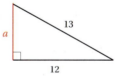

7.

8.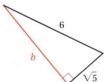

In a right triangle, find the length of the side not given. Where appropriate, give both an exact answer and an approximation to three decimal places. Standard lettering has been used.

9. $a = 10$, $b = 24$

10. $a = 5$, $b = 12$

11. $a = 9$, $c = 15$

12. $a = 18$, $c = 30$

13. $b = 1$, $c = \sqrt{5}$

14. $b = 1$, $c = \sqrt{2}$

15. $a = 1$, $c = \sqrt{3}$

16. $a = \sqrt{3}$, $b = \sqrt{5}$

17. $c = 10$, $b = 5\sqrt{3}$

18. $a = 5$, $b = 5$

19. $a = \sqrt{2}$, $b = \sqrt{7}$

20. $c = \sqrt{7}$, $a = \sqrt{2}$

b Solve. Don't forget to use a drawing. Give an exact answer and an approximation to three decimal places.

21. Panama Canal. A photographer is assigned a feature story on a new lock at the Panama Canal and needs to determine the distance between points *A* and *B* on opposite sides of the canal. What is the distance between *A* and *B*?

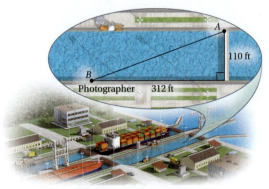

22. Rope Course. An outdoor rope course consists of a cable that slopes downward from a height of 37 ft to a resting place 30 ft above the ground. The trees that the cable connects are 24 ft apart. How long is the cable?

23. Ladder Height. A 10-m ladder is leaning against a building. The bottom of the ladder is 5 m from the building. How high is the top of the ladder?

24. Diagonal of a Square. Find the length of a diagonal of a square whose sides are 3 cm long.

25. Diagonal of a Soccer Field. The largest regulation soccer field is 100 yd wide and 130 yd long. Find the length of a diagonal of such a field.

26. Guy Wire. How long is a guy wire reaching from the top of a 12-ft pole to a point on the ground 8 ft from the base of the pole?

Skill Maintenance

Solve. [7.3a, b]

27. $5x + 7 = 8y,$
$3x = 8y - 4$

28. $5x + y = 17,$
$-5x + 2y = 10$

29. $3x - 4y = -11,$
$5x + 6y = 12$

30. $x + y = -9,$
$x - y = -11$

31. Find the slope of the line $4 - x = 3y$. [3.4a]

32. Find the slope of the line containing the points $(8, -3)$ and $(0, -8)$. [3.3a]

Synthesis

33. Find x.

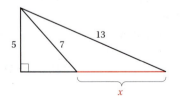

34. Skateboard Ramp. Ramps.com of America sells Landwave ramps and decks that can be combined to create a skateboard tower, as shown in Example 5. The dimensions of the Landwave ramp are 28 in. × 38.5 in. × 12 in. The Landwave deck that is a rectangular prism measures 28 in. × 38.5 in. × 12 in. How many ramps and how many decks are needed to build a tower that is 7 ft wide and 7 ft high? (*Hint*: For safety, the tallest column of the tower is built of only decks.)

Source: www.ramps.com

Vocabulary Reinforcement

Complete each statement with the correct term from the column on the right. Some of the choices may not be used.

principal
Pythagorean
radical
legs
square
radicands
hypotenuse
conjugates
root
rationalizing

1. The positive square root is called the _____ square root. [8.1a]

2. The symbol $\sqrt{}$ is called a(n) _____ symbol. [8.1a]

3. The number c is a(n) _____ root of a if $c^2 = a$. [8.1a]

4. The procedure to find an equivalent expression without a radical in the denominator is called _____ the denominator. [8.3c]

5. Like radicals have the same _____. [8.4a]

6. Expressions such as $\sqrt{6} - \sqrt{a}$ and $\sqrt{6} + \sqrt{a}$ are known as _____. [8.4c]

7. In any right triangle, if a and b are the lengths of the legs and c is the length of the _____, then $a^2 + b^2 = c^2$. The equation $a^2 + b^2 = c^2$ is called the _____ equation. [8.6a]

Concept Reinforcement

Determine whether each statement is true or false.

_____ 1. When both sides of an equation are squared, the new equation may have solutions that the first equation does not. [8.5a]

_____ 2. The square root of a sum is not the sum of the square roots. [8.2c]

_____ 3. If an equation $a = b$ is true, then the equation $a^2 = b^2$ is true. [8.5a]

_____ 4. If an equation $a^2 = b^2$ is true, then the equation $a = b$ is true. [8.5a]

Study Guide

Objective 8.1d   Identify radicands of radical expressions.

Example   Identify the radicand in the radical expression.

$$\sqrt{a + 2} + \frac{1}{4}.$$

The radicand is $a + 2$.

Practice Exercise

1. Identify the radicand in the radical expression $10y + \sqrt{y^2 - 3}$.

Objective 8.1e   Determine whether a radical expression represents a real number.

Example   Determine whether each expression represents a real number.

a) $\sqrt{-11}$ **b)** $-\sqrt{134}$

a) The radicand, -11, is negative; $\sqrt{-11}$ *is not* a real number.

b) The radicand, 134, is positive; $-\sqrt{134}$ *is* a real number.

Practice Exercise

2. Determine whether each expression represents a real number. Write "yes" or "no."

a) $-\sqrt{-(-3)}$ **b)** $\sqrt{-200}$

Objective 8.2a   Simplify radical expressions.

Example   Simplify by factoring: $\sqrt{162x^2}$.

$$\sqrt{162x^2} = \sqrt{81 \cdot 2 \cdot x^2}$$
$$= \sqrt{81}\sqrt{x^2}\sqrt{2} = 9x\sqrt{2}$$

Practice Exercise

3. Simplify by factoring: $\sqrt{1200y^2}$.

Objective 8.2b   Simplify radical expressions where radicands are powers.

Example   Simplify by factoring: $\sqrt{98x^7y^8}$.

$$\sqrt{98x^7y^8} = \sqrt{49 \cdot 2 \cdot x^6 \cdot x \cdot y^8}$$
$$= \sqrt{49}\sqrt{x^6}\sqrt{y^8}\sqrt{2x} = 7x^3y^4\sqrt{2x}$$

Practice Exercise

4. Simplify by factoring: $\sqrt{175a^{12}b^9}$.

Objective 8.2c   Multiply radical expressions and, if possible, simplify.

Example   Multiply and then, if possible, simplify: $\sqrt{6cd^3}\sqrt{30c^3d^2}$.

$$\sqrt{6cd^3}\sqrt{30c^3d^2} = \sqrt{6cd^3 \cdot 30c^3d^2}$$
$$= \sqrt{2 \cdot 3 \cdot 2 \cdot 3 \cdot 5 \cdot c^4 \cdot d^4 \cdot d}$$
$$= \sqrt{4}\sqrt{9}\sqrt{c^4}\sqrt{d^4}\sqrt{5d}$$
$$= 2 \cdot 3 \cdot c^2 \cdot d^2 \cdot \sqrt{5d}$$
$$= 6c^2d^2\sqrt{5d}$$

Practice Exercise

5. Multiply and then, if possible, simplify: $\sqrt{8x^3y}\sqrt{12x^4y^3}$.

Objective 8.3a   Divide radical expressions.

Example   Divide and simplify: $\dfrac{\sqrt{108y^5}}{\sqrt{3y^2}}$.

$$\frac{\sqrt{108y^5}}{\sqrt{3y^2}} = \sqrt{\frac{108y^5}{3y^2}} = \sqrt{36y^3}$$
$$= \sqrt{36 \cdot y^2 \cdot y} = \sqrt{36}\sqrt{y^2}\sqrt{y} = 6y\sqrt{y}$$

Practice Exercise

6. Divide and simplify: $\dfrac{\sqrt{15b^7}}{\sqrt{5b^4}}$.

Objective 8.3b   Simplify square roots of quotients.

Example   Simplify: $\sqrt{\dfrac{320}{500}}$.

$$\sqrt{\frac{320}{500}} = \sqrt{\frac{16 \cdot 20}{25 \cdot 20}} = \sqrt{\frac{16}{25} \cdot \frac{20}{20}}$$
$$= \sqrt{\frac{16}{25} \cdot 1} = \sqrt{\frac{16}{25}} = \frac{\sqrt{16}}{\sqrt{25}} = \frac{4}{5}$$

Practice Exercise

7. Simplify: $\sqrt{\dfrac{50}{162}}$.

Objective 8.3c   Rationalize the denominator of a radical expression.

Example   Rationalize the denominator: $\dfrac{7x}{\sqrt{18}}$.

$$\frac{7x}{\sqrt{18}} = \frac{7x}{\sqrt{2 \cdot 3 \cdot 3}} \cdot \frac{\sqrt{2}}{\sqrt{2}}$$
$$= \frac{7x \cdot \sqrt{2}}{\sqrt{2 \cdot 3 \cdot 3} \cdot \sqrt{2}} = \frac{7x\sqrt{2}}{\sqrt{2 \cdot 3 \cdot 3 \cdot 2}}$$
$$= \frac{7x\sqrt{2}}{\sqrt{36}} = \frac{7x\sqrt{2}}{6}$$

Practice Exercise

8. Rationalize the denominator: $\dfrac{2a}{\sqrt{50}}$.

Objective 8.4a   Add or subtract with radical notation, using the distributive laws to simplify.

Example   Add and, if possible, simplify.
$$\sqrt{9x - 18} + \sqrt{16x^3 - 32x^2}.$$

$\sqrt{9x - 18} + \sqrt{16x^3 - 32x^2}$

$\quad = \sqrt{9(x - 2)} + \sqrt{16x^2(x - 2)}$

$\quad = \sqrt{9}\sqrt{x - 2} + \sqrt{16x^2}\sqrt{x - 2}$

$\quad = 3\sqrt{x - 2} + 4x\sqrt{x - 2}$

$\quad = (3 + 4x)\sqrt{x - 2}$

Practice Exercise

9. Add and, if possible, simplify:
$$\sqrt{x^3 - x^2} + \sqrt{36x - 36}.$$

Objective 8.4b   Multiply expressions involving radicals, where some of the expressions contain more than one term.

Example   Multiply: $(\sqrt{3} + 4\sqrt{5})(\sqrt{3} - \sqrt{5})$.

$(\sqrt{3} + 4\sqrt{5})(\sqrt{3} - \sqrt{5})$

$\quad = \sqrt{3} \cdot \sqrt{3} - \sqrt{3} \cdot \sqrt{5} + 4\sqrt{5} \cdot \sqrt{3} - 4\sqrt{5} \cdot \sqrt{5}$

$\quad = 3 - \sqrt{15} + 4\sqrt{15} - 4 \cdot 5$

$\quad = 3 - \sqrt{15} + 4\sqrt{15} - 20 = 3\sqrt{15} - 17$

Practice Exercise

10. Multiply: $(\sqrt{13} - \sqrt{2})(\sqrt{13} + 2\sqrt{2})$.

Objective 8.4c   Rationalize denominators having two terms.

Example   Rationalize the denominator: $\dfrac{1 + \sqrt{3}}{5 - \sqrt{3}}$.

$\dfrac{1 + \sqrt{3}}{5 - \sqrt{3}} = \dfrac{1 + \sqrt{3}}{5 - \sqrt{3}} \cdot \dfrac{5 + \sqrt{3}}{5 + \sqrt{3}}$

$\quad = \dfrac{(1 + \sqrt{3})(5 + \sqrt{3})}{(5 - \sqrt{3})(5 + \sqrt{3})}$

$\quad = \dfrac{1 \cdot 5 + 1 \cdot \sqrt{3} + 5 \cdot \sqrt{3} + (\sqrt{3})^2}{5^2 - (\sqrt{3})^2}$

$\quad = \dfrac{5 + \sqrt{3} + 5\sqrt{3} + 3}{25 - 3} = \dfrac{8 + 6\sqrt{3}}{22}$

$\quad = \dfrac{2(4 + 3\sqrt{3})}{2 \cdot 11} = \dfrac{4 + 3\sqrt{3}}{11}$

Practice Exercise

11. Rationalize the denominator: $\dfrac{5 - \sqrt{2}}{9 + \sqrt{2}}$.

Objective 8.5a   Solve radical equations with one or two radical terms isolated, using the principle of squaring once.

Example   Solve: $x - 3 = \sqrt{x - 1}$.

$x - 3 = \sqrt{x - 1}$

$(x - 3)^2 = (\sqrt{x - 1})^2$ Squaring both sides

$x^2 - 6x + 9 = x - 1$

$x^2 - 7x + 10 = 0$

$(x - 5)(x - 2) = 0$

$x - 5 = 0 \quad or \quad x - 2 = 0$

$x = 5 \quad or \quad x = 2$

The number 5 checks, but 2 does not. Thus the solution is 5.

Practice Exercise

12. Solve: $x - 4 = \sqrt{x - 2}$.

Objective 8.5b Solve radical equations with two radical terms, using the principle of squaring twice.

Example Solve: $\sqrt{x + 20} = 10 - \sqrt{x}$.

$\sqrt{x + 20} = 10 - \sqrt{x}$

$(\sqrt{x + 20})^2 = (10 - \sqrt{x})^2$ Squaring both sides

$x + 20 = 100 - 20\sqrt{x} + x$

$20\sqrt{x} = 80$

$\sqrt{x} = 4$

$(\sqrt{x})^2 = 4^2$ Squaring both sides

$x = 16$

The number 16 checks and is the solution.

Practice Exercise

13. Solve: $12 - \sqrt{x} = \sqrt{90 - x}$.

Objective 8.6a Given the lengths of any two sides of a right triangle, find the length of the third side.

Example Find the length of the third side of this right triangle.

$a^2 + b^2 = c^2$ Pythagorean equation

$7^2 + b^2 = 17^2$ Substituting 7 for a and 17 for c

$49 + b^2 = 289$

$b^2 = 240$

$b = \sqrt{240} \approx 15.492$

Practice Exercise

14. Find the length of the third side of this triangle.

Review Exercises

Find the square roots.   [8.1a]

1. 64

2. 400

Simplify.   [8.1a]

3. $\sqrt{36}$

4. $-\sqrt{169}$

Use a calculator to approximate each of the following square roots to three decimal places.   [8.1b]

5. $\sqrt{3}$

6. $\sqrt{99}$

7. $-\sqrt{320.12}$

8. $\sqrt{\dfrac{11}{20}}$

9. $-\sqrt{\dfrac{47.3}{11.2}}$

10. $18\sqrt{11} \cdot 43.7$

Identify the radicand.   [8.1d]

11. $\sqrt{x^2 + 4}$

12. $\sqrt{x} + 2$

13. $3\sqrt{4 - x}$

14. $\sqrt{\dfrac{2}{y - 7}}$

Determine whether the expression represents a real number. Write " yes" or "no."   [8.1e]

15. $-\sqrt{49}$

16. $-\sqrt{-4}$

17. $\sqrt{-36}$

18. $\sqrt{(-3)(-27)}$

Simplify.   [8.1f]

19. $\sqrt{m^2}$

20. $\sqrt{(x - 4)^2}$

21. $\sqrt{16x^2}$

22. $\sqrt{4p^2 - 12p + 9}$

Simplify by factoring.   [8.2a]

23. $\sqrt{48}$

24. $\sqrt{32t^2}$

25. $\sqrt{t^2 - 14t + 49}$

26. $\sqrt{x^2 + 16x + 64}$

Simplify by factoring. [8.2b]

27. $\sqrt{x^8}$

28. $\sqrt{75a^7}$

Multiply. [8.2c]

29. $\sqrt{3}\sqrt{7}$

30. $\sqrt{x-3}\sqrt{x+3}$

Multiply and simplify. [8.2c]

31. $\sqrt{6}\sqrt{10}$

32. $\sqrt{5x}\sqrt{8x}$

33. $\sqrt{5x}\sqrt{10xy^2}$

34. $\sqrt{20a^3b}\sqrt{5a^2b^2}$

Simplify. [8.3a, b]

35. $\sqrt{\dfrac{25}{64}}$

36. $\sqrt{\dfrac{49}{t^2}}$

37. $\dfrac{\sqrt{2c^9}}{\sqrt{32c}}$

Rationalize the denominator. [8.3c]

38. $\sqrt{\dfrac{1}{2}}$

39. $\dfrac{\sqrt{x^3}}{\sqrt{15}}$

40. $\sqrt{\dfrac{5}{y}}$

41. $\dfrac{\sqrt{b^9}}{\sqrt{ab^2}}$

42. $\dfrac{\sqrt{27}}{\sqrt{45}}$

43. $\dfrac{\sqrt{45x^2y}}{\sqrt{54y}}$

Simplify. [8.4a]

44. $10\sqrt{5} + 3\sqrt{5}$

45. $\sqrt{80} - \sqrt{45}$

46. $3\sqrt{2} - 5\sqrt{\dfrac{1}{2}}$

Simplify. [8.4b]

47. $(2 + \sqrt{3})^2$

48. $(2 + \sqrt{3})(2 - \sqrt{3})$

49. Rationalize the denominator:

$$\dfrac{4}{2 + \sqrt{3}}.\quad [8.4c]$$

Solve. [8.5a]

50. $\sqrt{x-3} = 7$

51. $\sqrt{5x+3} = \sqrt{2x-1}$

52. $1 + x = \sqrt{1 + 5x}$

53. Solve: $\sqrt{x} = \sqrt{x-5} + 1$. [8.5b]

Solve. [8.1c], [8.5c]

54. *Speed of a Skidding Car.* The formula $r = 2\sqrt{5L}$ can be used to approximate the speed r, in miles per hour, of a car that has left a skid mark of length L, in feet.

   a) What was the speed of a car that left skid marks of length 200 ft?
   b) How far will a car skid at 90 mph?

In a right triangle, find the length of the side not given. Give an exact answer and an approximation to three decimal places where appropriate. Standard lettering has been used. [8.6a]

55. $a = 15,\ c = 25$

56. $a = 1,\ b = \sqrt{2}$

Solve. [8.6b]

57. *Airplane Descent.* A pilot is instructed to descend from 30,000 ft to 20,000 ft over a horizontal distance of 50,000 ft. What distance will the plane travel during this descent?

30,000 ft

?

20,000 ft

50,000 ft

58. *Lookout Tower.* The diagonal braces in a lookout tower are 15 ft long and span a distance of 12 ft. How high does each brace reach vertically?

12 ft

15 ft

59. Solve: $x - 2 = \sqrt{4 - 9x}$. [8.5a]

  A. -5 **B.** No solution
  C. 0 **D.** $0, -5$

60. Simplify: $(2\sqrt{7} + \sqrt{2})(\sqrt{7} - \sqrt{2})$. [8.4b]

  A. $12 - \sqrt{7}$ **B.** 12
  C. $12 - \sqrt{14}$ **D.** $3\sqrt{7} - 2$

Synthesis

61. *Distance Driven.* Two cars leave a service station at the same time. One car travels east at a speed of 50 mph, and the other travels south at a speed of 60 mph. After one-half hour, how far apart are they? [8.6b]

50 mph

60 mph

62. Solve $A = \sqrt{a^2 + b^2}$ for b. [8.5a]

63. Find x. [8.6a]

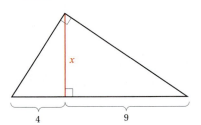

x

4   9

Understanding Through Discussion and Writing

1. Explain why it is necessary for the signs within a pair of conjugates to differ. [8.4c]

2. Determine whether the statement below is true or false and explain your answer. [8.1a], [8.5a]

    The solution of $\sqrt{11 - 2x} = -3$ is 1.

3. Why are the rules for manipulating expressions with exponents important when simplifying radical expressions? [8.2b]

4. Can a carpenter use a 28-ft ladder to repair clapboard that is 28 ft above ground level? Why or why not? [8.6b]

5. Explain why possible solutions of radical equations must be checked. [8.5a]

6. Determine whether each of the following is true for all real numbers. Explain why or why not. [8.1f], [8.2a]

  a) $\sqrt{5x^2} = |x|\sqrt{5}$
  b) $\sqrt{b^2 - 4} = b - 2$
  c) $\sqrt{x^2 + 16} = x + 4$

CHAPTER

8 **Test**

For Extra Help For step-by-step test solutions, access the Chapter Test Prep Videos in MyMathLab® or on YouTube (search "BittingerIntro" and click on "Channels").

1. Find the square roots of 81.

Simplify.

2. $\sqrt{64}$

3. $-\sqrt{25}$

Approximate the expression involving square roots to three decimal places.

4. $\sqrt{116}$

5. $-\sqrt{87.4}$

6. $4\sqrt{5} \cdot 6$

7. Identify the radicand in $8\sqrt{4 - y^3}$.

Determine whether each expression represents a real number. Write "yes" or "no."

8. $\sqrt{24}$

9. $\sqrt{-23}$

Simplify.

10. $\sqrt{a^2}$

11. $\sqrt{36y^2}$

Multiply.

12. $\sqrt{5}\sqrt{6}$

13. $\sqrt{x-8}\sqrt{x+8}$

Simplify by factoring.

14. $\sqrt{27}$

15. $\sqrt{25x - 25}$

16. $\sqrt{t^5}$

Multiply and simplify.

17. $\sqrt{5}\sqrt{10}$

18. $\sqrt{3ab}\sqrt{6ab^3}$

Simplify.

19. $\sqrt{\dfrac{27}{12}}$

20. $\sqrt{\dfrac{144}{a^2}}$

Rationalize the denominator.

21. $\sqrt{\dfrac{2}{5}}$

22. $\sqrt{\dfrac{2x}{y}}$

Divide and simplify.

23. $\dfrac{\sqrt{27}}{\sqrt{32}}$

24. $\dfrac{\sqrt{35x}}{\sqrt{80xy^2}}$

Add or subtract.

25. $3\sqrt{18} - 5\sqrt{18}$

26. $\sqrt{5} + \sqrt{\dfrac{1}{5}}$

Simplify.

27. $(4 - \sqrt{5})^2$

28. $(4 - \sqrt{5})(4 + \sqrt{5})$

29. Rationalize the denominator: $\dfrac{10}{4 - \sqrt{5}}$.

30. In a right triangle, $a = 8$ and $b = 4$. Find c. Give an exact answer and an approximation to three decimal places.

Solve.

31. $\sqrt{3x} + 2 = 14$

32. $\sqrt{6x + 13} = x + 3$

33. $\sqrt{1 - x} + 1 = \sqrt{6 - x}$

34. *Sighting to the Horizon.* The equation $D = \sqrt{2h}$ can be used to approximate the distance D, in miles, that a person can see to the horizon from a height h, in feet.

   a) How far to the horizon can you see through an airplane window at a height of 28,000 ft?
   b) Christina can see about 261 mi to the horizon through an airplane window. How high is the airplane?

35. *Lacrosse.* A regulation lacrosse field is 60 yd wide and 110 yd long. Find the length of a diagonal of such a field.

36. Rationalize the denominator: $\sqrt{\dfrac{2a}{5b}}$.

   A. $\dfrac{\sqrt{10ab}}{5b}$

   B. $\dfrac{a}{b}\sqrt{\dfrac{2b}{5a}}$

   C. $\dfrac{\sqrt{10}}{5}$

   D. $\dfrac{\sqrt{6a^3b}}{15ab}$

Synthesis

Simplify.

37. $\sqrt{\sqrt{\sqrt{625}}}$

38. $\sqrt{y^{16n}}$

1. Find all numbers for which the expression is not defined:

$$\frac{x - 6}{2x + 1}.$$

2. Determine whether the expression represents a real number. Write "yes" or "no."

$$\sqrt{-24}$$

Perform the indicated operations and simplify.

3. $(2 + \sqrt{3})(2 - \sqrt{3})$

4. $-\sqrt{196}$

5. $\sqrt{3}\sqrt{75}$

6. $(1 - \sqrt{2})^2$

7. $\dfrac{\sqrt{162}}{\sqrt{125}}$

8. $2\sqrt{45} + 3\sqrt{20}$

9. $(3x^4 - 2y^5)(3x^4 + 2y^5)$

10. $(x^2 + 4)^2$

11. $\left(2x + \dfrac{1}{4}\right)\left(4x - \dfrac{1}{2}\right)$

12. $\dfrac{x}{2x - 1} - \dfrac{3x + 2}{1 - 2x}$

13. $(3x^2 - 2x^3) - (x^3 - 2x^2 + 5) + (3x^2 - 5x + 5)$

14. $\dfrac{2x + 2}{3x - 9} \cdot \dfrac{x^2 - 8x + 15}{x^2 - 1}$

15. $\dfrac{2x^2 - 2}{2x^2 + 7x + 3} \div \dfrac{4x - 4}{2x^2 - 5x - 3}$

16. $(3x^3 - 2x^2 + x - 5) \div (x - 2)$

Simplify.

17. $\sqrt{2x^2 - 4x + 2}$

18. $x^{-9} \cdot x^{-3}$

19. $\sqrt{\dfrac{50}{2x^8}}$

20. $\dfrac{x - \dfrac{1}{x}}{1 - \dfrac{x - 1}{2x}}$

Factor completely.

21. $3 - 12x^8$

22. $12t - 4t^2 - 48t^4$

23. $6x^2 - 28x + 16$

24. $4x^3 + 4x^2 - x - 1$

25. $16x^4 - 56x^2 + 49$

26. $x^2 + 3x - 180$

Solve.

27. $x^2 = -17x$

28. $-4(x + 5) \geq 2(x + 5) - 3$

29. $\dfrac{1}{x} + \dfrac{2}{3} = \dfrac{1}{4}$

30. $x^2 - 30 = x$

31. $\sqrt{2x - 1} + 5 = 14$

32. $\dfrac{1}{4}x + \dfrac{2}{3}x = \dfrac{2}{3} - \dfrac{3}{4}x$

33. $\dfrac{x}{x - 1} - \dfrac{x}{x + 1} = \dfrac{1}{2x - 2}$

34. $x = y + 3,$
    $3y - 4x = -13$

35. $2x - 3y = 30,$
    $5y - 2x = -46$

36. Solve $4A = pr + pq$ for p.

Graph on a plane.

37. $3y - 3x > -6$

38. $x = 5$

39. $2x - 6y = 12$

40. Find an equation of the line containing the points $(1, -2)$ and $(5, 9)$.

41. Find the slope and the y-intercept of the line $5x - 3y = 9$.

42. The graph of the polynomial equation $y = x^3 - 4x - 2$ is shown below. Use either the graph or the equation to estimate the value of the polynomial when $x = -2$, $x = -1$, $x = 0$, $x = 1$, and $x = 2$.

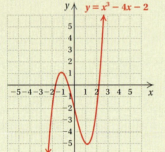

Solve.

43. *Cost Breakdown.* The cost of 6 hamburgers and 4 milkshakes is $27.70. Three hamburgers and 1 milkshake cost $11.35. Find the cost of a hamburger and the cost of a milkshake.

44. *Apparent Size.* The apparent size A of an object varies inversely as the distance d of the object from the eye. You are sitting at a concert 100 ft from the stage. The musicians appear to be 4 ft tall. How tall would they appear to be if you were sitting 1000 ft away in the lawn seats?

45. *Angles of a Triangle.* The second angle of a triangle is twice as large as the first. The third angle is $48°$ less than the sum of the other two angles. Find the measures of the angles.

46. *Quality Control.* A sample of 150 resistors contained 12 defective resistors. How many defective resistors would you expect to find in a sample of 250 resistors?

47. *Rectangle Dimensions.* The length of a rectangle is 3 m greater than the width. The area of the rectangle is 180 m². Find the length and the width.

48. *Coin Mixture.* A collection of dimes and quarters is worth $19.00. There are 115 coins in all. How many of each are there?

49. *Amount Invested.* Money is invested in an account at 4.5% simple interest. At the end of 1 year, there is $2717 in the account. How much was originally invested?

50. *Car Travel.* Andrew traveled by car 600 mi in one direction. The return trip took 2 hr longer at a speed that was 10 mph less. Find the speed going.

Synthesis

51. *Salt Solutions.* A tank contains 200 L of a 30%-salt solution. How much pure water should be added in order to make a solution that is 12% salt?

52. Solve: $\sqrt{x} + 1 = y$,
$\sqrt{x} + \sqrt{y} = 5$.

Quadratic Equations

9.1   Introduction to Quadratic Equations

OBJECTIVES

a Write a quadratic equation in standard form $ax^2 + bx + c = 0$, $a > 0$, and determine the coefficients a, b, and c.

b Solve quadratic equations of the type $ax^2 + bx = 0$, where $b \neq 0$, by factoring.

c Solve quadratic equations of the type $ax^2 + bx + c = 0$, where $b \neq 0$ and $c \neq 0$, by factoring.

d Solve applied problems involving quadratic equations.

SKILL TO REVIEW

Objective 5.7b: Solve quadratic equations by factoring and then using the principle of zero products.

Solve by factoring and using the principle of zero products.

1. $x^2 + 15x = 0$
2. $x^2 + 8x = 33$

ALGEBRAIC ▶◀ GRAPHICAL CONNECTION

Before we begin this chapter, let's look back at some algebraic–graphical equation-solving connections. In Chapter 3, we considered the graph of a *linear equation* $y = mx + b$. For example, the graph of the equation $y = \frac{5}{2}x - 4$ and its x-intercept are shown below.

If $y = 0$, then $x = \frac{8}{5}$. Thus the x-intercept is $\left(\frac{8}{5}, 0\right)$. This point is also the intersection of the graphs of $y = \frac{5}{2}x - 4$ and $y = 0$.

We can solve the linear equation $\frac{5}{2}x - 4 = 0$ algebraically:

$$\frac{5}{2}x - 4 = 0$$
$$\frac{5}{2}x = 4 \qquad \text{Adding 4}$$
$$x = \frac{8}{5}. \qquad \text{Multiplying by } \frac{2}{5}$$

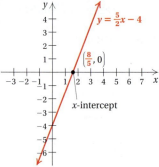

We see that $\frac{8}{5}$, the solution of $\frac{5}{2}x - 4 = 0$, is the first coordinate of the x-intercept of the graph of $y = \frac{5}{2}x - 4$.

◀ **Do Margin Exercise 1 on the following page.**

In this chapter, we build on these ideas by applying them to quadratic equations. In Section 5.7, we briefly considered the graph of a *quadratic equation* $y = ax^2 + bx + c$, $a \neq 0$. For example, the graph of the equation $y = x^2 + 6x + 8$ and its x-intercepts are shown below.

The x-intercepts are $(-4, 0)$ and $(-2, 0)$. We will develop in detail the creation of such graphs in Section 9.6. The points $(-4, 0)$ and $(-2, 0)$ are the intersections of the graphs of $y = x^2 + 6x + 8$ and $y = 0$.

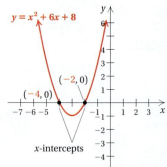

Answers

Skill to Review:
1. $-15, 0$ 2. $-11, 3$

We can solve the quadratic equation $x^2 + 6x + 8 = 0$ by factoring and using the principle of zero products:

$$x^2 + 6x + 8 = 0$$
$$(x + 4)(x + 2) = 0 \qquad \text{Factoring}$$
$$x + 4 = 0 \quad or \quad x + 2 = 0 \qquad \text{Using the principle of zero products}$$
$$x = -4 \quad or \qquad x = -2.$$

We see that the solutions of $x^2 + 6x + 8 = 0$, -4 and -2, are the first coordinates of the x-intercepts, $(-4, 0)$ and $(-2, 0)$, of the graph of $y = x^2 + 6x + 8$.

Do Exercise 2. ▶

a STANDARD FORM

The following are **quadratic equations**. They contain polynomials of second degree.

$$4x^2 + 7x - 5 = 0,$$
$$3t^2 - \tfrac{1}{2}t = 9,$$
$$5y^2 = -6y,$$
$$5m^2 = 15$$

The quadratic equation $4x^2 + 7x - 5 = 0$ is said to be in **standard form**. Although the quadratic equation $4x^2 = 5 - 7x$ is equivalent to the preceding equation, it is *not* in standard form.

QUADRATIC EQUATION

A **quadratic equation** is an equation equivalent to an equation of the type

$$ax^2 + bx + c = 0, \ a > 0,$$

where a, b, and c are real-number constants. We say that the preceding is the **standard form of a quadratic equation**.

We define $a > 0$ to ease the proof of the quadratic formula, which we consider later, and to ease solving by factoring. Suppose we are studying an equation like $-3x^2 + 8x - 2 = 0$. It is not in standard form because $-3 < 0$. We can find an equivalent equation that is in standard form by multiplying by -1 on both sides:

$$-1(-3x^2 + 8x - 2) = -1(0)$$
$$3x^2 - 8x + 2 = 0. \qquad \text{Standard form}$$

When a quadratic equation is written in standard form, we can readily determine a, b, and c.

$$ax^2 + bx + c = 0$$
$$3x^2 - 8x + 2 = 0$$

For this equation, $a = 3$, $b = -8$, and $c = 2$.

---

1. **a)** Consider the linear equation $y = -\tfrac{2}{3}x - 3$. Find the intercepts and graph the equation.

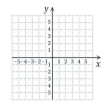

b) Solve the equation

$$0 = -\frac{2}{3}x - 3.$$

c) Complete: The solution of the equation $0 = -\tfrac{2}{3}x - 3$ is ____. This value is the ____ ____ of the x-intercept, (____, ____), of the graph of $y = -\tfrac{2}{3}x - 3$.

2. Consider the quadratic equation $y = x^2 - 2x - 3$ and its graph shown below.

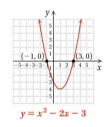

$$y = x^2 - 2x - 3$$

a) Solve the equation

$$x^2 - 2x - 3 = 0.$$

(*Hint*: Use the principle of zero products.)

b) Complete: The solutions of the equation $x^2 - 2x - 3 = 0$ are ____ and ____. These values are the ____ ____ of the x-intercepts, (____, ____) and (____, ____), of the graph of $y = x^2 - 2x - 3$.

Answers

1. (a) y-intercept: $(0, -3)$; x-intercept: $\left(-\tfrac{9}{2}, 0\right)$;
(b) $-\tfrac{9}{2}$; (c) $-\tfrac{9}{2}$; first coordinate; $\left(-\tfrac{9}{2}, 0\right)$
2. (a) $-1, 3$; (b) $-1, 3$; first coordinates; $(-1, 0)$, $(3, 0)$

$$y = -\frac{2}{3}x - 3$$

EXAMPLES Write in standard form and determine a, b, and c.

1. $4x^2 + 7x - 5 = 0$ The equation is already in standard form.

$a = 4; \quad b = 7; \quad c = -5$

2. $3x^2 - 0.5x = 9$

$3x^2 - 0.5x - 9 = 0$ Subtracting 9. This is standard form.

$a = 3; \quad b = -0.5; \quad c = -9$

3. $-4y^2 = 5y$

$-4y^2 - 5y = 0$ Subtracting $5y$

Not positive

$4y^2 + 5y = 0$ Multiplying by -1. This is standard form.

$a = 4; \quad b = 5; \quad c = 0$

◀ **Do Exercises 3–7.**

Write in standard form and determine a, b, and c.

3. $6x^2 = 3 - 7x$

4. $y^2 = 8y$

5. $3 - x^2 = 9x$

6. $3x + 5x^2 = x^2 - 4 + 2x$

7. $5x^2 = 21$

b SOLVING QUADRATIC EQUATIONS OF THE TYPE $ax^2 + bx = 0$

Sometimes we can use factoring and the principle of zero products to solve quadratic equations. In particular, when $c = 0$ and $b \neq 0$, we can always factor and use the principle of zero products.

EXAMPLE 4 Solve: $7x^2 + 2x = 0$.

$7x^2 + 2x = 0$

$x(7x + 2) = 0$ Factoring

$x = 0 \quad or \quad 7x + 2 = 0$ Using the principle of zero products

$x = 0 \quad or \quad 7x = -2$

$x = 0 \quad or \quad x = -\frac{2}{7}$

Solve.

8. $2x^2 + 8x = 0$ **GS**

$2x(\boxed{}) = 0$

$2x = 0 \quad or \quad \boxed{} = 0$

$x = 0 \quad or \quad x = \boxed{}$

Both numbers check.
The solutions are 0 and $\boxed{}$.

9. $10x^2 - 6x = 0$

Check: For 0:

$$\frac{7x^2 + 2x = 0}{7 \cdot 0^2 + 2 \cdot 0 \;?\; 0}$$

$0 \;|\;$ **TRUE**

For $-\frac{2}{7}$:

$$\frac{7x^2 + 2x = 0}{7\left(-\frac{2}{7}\right)^2 + 2\left(-\frac{2}{7}\right) \;?\; 0}$$

$7\left(\frac{4}{49}\right) - \frac{4}{7}$

$\frac{4}{7} - \frac{4}{7}$

$0 \;|\;$ **TRUE**

The solutions are 0 and $-\frac{2}{7}$.

·········· **Caution!** ··········

You may be tempted to divide each term in an equation like the one in Example 4 by x. This method would yield the equation

$7x + 2 = 0$,

whose only solution is $-\frac{2}{7}$. In effect, since 0 is also a solution of the original equation, we have divided by 0. The error of such division results in the loss of one of the solutions.

Answers

3. $6x^2 + 7x - 3 = 0; a = 6, b = 7, c = -3$
4. $y^2 - 8y = 0; a = 1, b = -8, c = 0$
5. $x^2 + 9x - 3 = 0; a = 1, b = 9, c = -3$
6. $4x^2 + x + 4 = 0; a = 4, b = 1, c = 4$
7. $5x^2 - 21 = 0; a = 5, b = 0, c = -21$

8. $0, -4$ **9.** $0, \dfrac{3}{5}$

Guided Solution:
8. $x + 4, x + 4, -4, -4$

EXAMPLE 5 Solve: $4x^2 - 8x = 0$.

We have

$$4x^2 - 8x = 0$$
$$4x(x - 2) = 0 \qquad \text{Factoring}$$
$$4x = 0 \quad or \quad x - 2 = 0 \qquad \text{Using the principle of zero products}$$
$$x = 0 \quad or \qquad x = 2.$$

The solutions are 0 and 2.

A quadratic equation of the type $ax^2 + bx = 0$, where $c = 0$ and $b \neq 0$, will always have 0 as one solution and a nonzero number as the other solution.

Do Exercises 8 and 9 on the preceding page. ▶

c SOLVING QUADRATIC EQUATIONS OF THE TYPE $ax^2 + bx + c = 0$

When neither b nor c is 0, we can sometimes solve by factoring.

EXAMPLE 6 Solve: $2x^2 - x - 21 = 0$.

We have

$$2x^2 - x - 21 = 0$$
$$(2x - 7)(x + 3) = 0 \qquad \text{Factoring}$$
$$2x - 7 = 0 \quad or \quad x + 3 = 0 \qquad \text{Using the principle of zero products}$$
$$2x = 7 \quad or \qquad x = -3$$
$$x = \tfrac{7}{2} \quad or \qquad x = -3.$$

The solutions are $\frac{7}{2}$ and -3.

EXAMPLE 7 Solve: $(y - 3)(y - 2) = 6(y - 3)$.

We write the equation in standard form and then factor:

$$y^2 - 5y + 6 = 6y - 18 \qquad \text{Multiplying}$$
$$y^2 - 11y + 24 = 0 \qquad \text{Standard form}$$
$$(y - 8)(y - 3) = 0 \qquad \text{Factoring}$$
$$y - 8 = 0 \quad or \quad y - 3 = 0 \qquad \text{Using the principle of zero products}$$
$$y = 8 \quad or \qquad y = 3.$$

The solutions are 8 and 3.

Do Exercises 10 and 11. ▶

Recall that to solve a rational equation, we multiply both sides by the LCM of all the denominators. We may obtain a quadratic equation after a few steps. When that happens, we solve the quadratic equation, but we must remember to check possible solutions because a replacement may result in division by 0.

ALGEBRAIC ▶◀ GRAPHICAL CONNECTION

Let's visualize the solutions in Example 5.

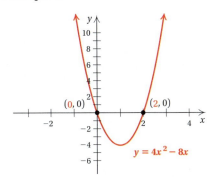

The solutions of $4x^2 - 8x = 0$, 0 and 2, are the first coordinates of the x-intercepts, $(0, 0)$ and $(2, 0)$, of the graph of $y = 4x^2 - 8x$.

ALGEBRAIC ▶◀ GRAPHICAL CONNECTION

Let's visualize the solutions in Example 6.

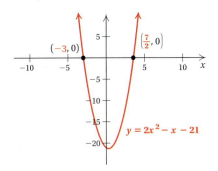

The solutions of $2x^2 - x - 21 = 0$, -3 and $\frac{7}{2}$, are the first coordinates of the x-intercepts, $(-3, 0)$ and $\left(\frac{7}{2}, 0\right)$, of the graph of $y = 2x^2 - x - 21$.

Solve.

10. $4x^2 + 5x - 6 = 0$

11. $(x - 1)(x + 1) = 5(x - 1)$

Answers

10. $-2, \frac{3}{4}$ **11.** $1, 4$

EXAMPLE 8 Solve: $\dfrac{3}{x-1} + \dfrac{5}{x+1} = 2$.

We multiply by the LCM, which is $(x-1)(x+1)$:

$$(x-1)(x+1) \cdot \left(\frac{3}{x-1} + \frac{5}{x+1} \right) = 2 \cdot (x-1)(x+1).$$

We use the distributive law on the left:

$$(x-1)(x+1) \cdot \frac{3}{x-1} + (x-1)(x+1) \cdot \frac{5}{x+1} = 2(x-1)(x+1)$$

$$3(x+1) + 5(x-1) = 2(x-1)(x+1)$$

$$3x + 3 + 5x - 5 = 2(x^2 - 1)$$

$$8x - 2 = 2x^2 - 2$$

$$0 = 2x^2 - 8x$$

$$0 = 2x(x-4) \qquad \text{Factoring}$$

$$2x = 0 \quad or \quad x - 4 = 0$$

$$x = 0 \quad or \qquad x = 4.$$

Check: For 0:

$$\dfrac{3}{x-1} + \dfrac{5}{x+1} = 2$$

$$\dfrac{3}{0-1} + \dfrac{5}{0+1} \;?\; 2$$

$$\dfrac{3}{-1} + \dfrac{5}{1}$$

$$-3 + 5$$

$$2 \qquad \text{TRUE}$$

For 4:

$$\dfrac{3}{x-1} + \dfrac{5}{x+1} = 2$$

$$\dfrac{3}{4-1} + \dfrac{5}{4+1} \;?\; 2$$

$$\dfrac{3}{3} + \dfrac{5}{5}$$

$$1 + 1$$

$$2 \qquad \text{TRUE}$$

The solutions are 0 and 4.

◀ **Do Exercise 12.**

12. Solve: **GS**

$$\dfrac{20}{x+5} - \dfrac{1}{x-4} = 1.$$

$$(x+5)(x-4) \cdot \left(\dfrac{20}{x+5} - \dfrac{1}{x-4} \right)$$

$$= (x+5)(x-4) \cdot 1$$

$$(x+5)(x-4) \cdot \dfrac{20}{x+5}$$

$$- (x+5)(x-4) \cdot \dfrac{1}{x-4}$$

$$= (x+5)(x-4)$$

$$20(x-4) - 1(\boxed{})$$

$$= (x+5)(x-4)$$

$$20x - 80 - x - \boxed{}$$

$$= x^2 + x - \boxed{}$$

$$19x - \boxed{} = x^2 + x - 20$$

$$0 = x^2 - 18x + \boxed{}$$

$$0 = (x-5)(\boxed{})$$

$$x - 5 = 0 \quad or \quad \boxed{} = 0$$

$$x = 5 \quad or \qquad x = \boxed{}$$

Both numbers check in the original equation. The solutions are 5 and $\boxed{}$.

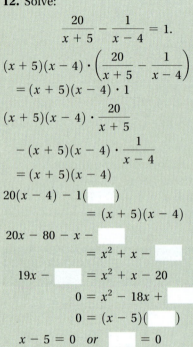

d   SOLVING APPLIED PROBLEMS

EXAMPLE 9 *Diagonals of a Polygon.* The number of diagonals d of a polygon of n sides is given by the formula

$$d = \frac{n^2 - 3n}{2}.$$

If a polygon has 27 diagonals, how many sides does it have?

1. **Familiarize.** To familiarize ourselves with the problem, we draw an octagon (8 sides) with its diagonals, as shown at left, and see that there are 20 diagonals. To check this, we evaluate the formula for $n = 8$:

$$d = \frac{8^2 - 3(8)}{2} = \frac{64 - 24}{2} = \frac{40}{2} = 20.$$

2. **Translate.** We substitute 27 for d:

$$27 = \frac{n^2 - 3n}{2}. \qquad \text{The number of diagonals is 27.}$$

Answer

12. 5, 13

Guided Solution:

12. $x + 5$, 5, 20, 85, 65, $x - 13$, $x - 13$, 13, 13

3. Solve. We solve the equation for n:

$$\frac{n^2 - 3n}{2} = 27$$

$$2 \cdot \frac{n^2 - 3n}{2} = 2 \cdot 27 \qquad \text{Multiplying by 2 to clear fractions}$$

$$n^2 - 3n = 54$$

$$n^2 - 3n - 54 = 0 \qquad \text{Subtracting 54}$$

$$(n - 9)(n + 6) = 0 \qquad \text{Factoring}$$

$$n - 9 = 0 \quad or \quad n + 6 = 0$$

$$n = 9 \quad or \qquad n = -6.$$

4. Check. Since the number of sides cannot be negative, -6 cannot be a solution. The answer 9 for the number of sides checks in the formula.

5. State. The polygon has 9 sides. (It is a nonagon.)

Do Exercise 13. ▶

13. A polygon has 44 diagonals. How many sides does it have?

Answer

13. 11 sides

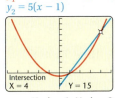 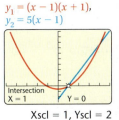

CALCULATOR CORNER

Solving Quadratic Equations We can use the **INTERSECT** feature to solve a quadratic equation. Consider the equation in Margin Exercise 11, $(x - 1)(x + 1) = 5(x - 1)$. First, we enter $y_1 = (x - 1)(x + 1)$ and $y_2 = 5(x - 1)$ on the equation-editor screen and graph the equations, using the window $[-5, 5, -5, 20]$, Yscl $= 2$. We see that there are two points of intersection, so the equation has two solutions.

Next, we use the **INTERSECT** feature to find the coordinates of the left-hand point of intersection. The first coordinate of this point, 1, is one solution of the equation. We use the **INTERSECT** feature again to find the other solution, 4.

Note that we could use the **ZERO** feature to solve this equation if we first write it with 0 on one side, that is, $(x - 1)(x + 1) - 5(x - 1) = 0$.

EXERCISES: Solve.

1. $5x^2 - 8x + 3 = 0$

2. $2x^2 - 7x - 15 = 0$

3. $6(x - 3) = (x - 3)(x - 2)$

4. $(x + 1)(x - 4) = 3(x - 4)$

9.1 Exercise Set

For Extra Help

MyMathLab MathXL® PRACTICE   WATCH READ   REVIEW

☑ Reading Check

Determine whether each statement is true or false.

RC1. The solutions of $x^2 - x - 12 = 0$ are the second coordinates of the x-intercepts of the graph of $y = x^2 - x - 12$.

RC2. The quadratic equation $14x^2 - 3x - 58 = 0$ is written in standard form.

RC3. One of the solutions of $85x^2 - 96x = 0$ is 0.

RC4. Every quadratic equation can be written in standard form.

1. $x^2 - 3x + 2 = 0$

2. $x^2 - 8x - 5 = 0$

3. $7x^2 = 4x - 3$

4. $9x^2 = x + 5$

5. $5 = -2x^2 + 3x$

6. $3x - 1 = 5x^2 + 9$

b   Solve.

7. $x^2 + 5x = 0$

8. $x^2 + 7x = 0$

9. $3x^2 + 6x = 0$

10. $4x^2 + 8x = 0$

11. $5x^2 = 2x$

12. $11x = 3x^2$

13. $4x^2 + 4x = 0$

14. $8x^2 - 8x = 0$

15. $0 = 10x^2 - 30x$

16. $0 = 10x^2 - 50x$

17. $11x = 55x^2$

18. $33x^2 = -11x$

19. $14t^2 = 3t$

20. $6m = 19m^2$

21. $5y^2 - 3y^2 = 72y + 9y$

22. $63p - 16p^2 = 17p + 58p^2$

c   Solve.

23. $x^2 + 8x - 48 = 0$

24. $x^2 - 16x + 48 = 0$

25. $5 + 6x + x^2 = 0$

26. $x^2 + 10 + 11x = 0$

27. $18 = 7p + p^2$

28. $t^2 + 14t = -24$

29. $-15 = -8y + y^2$

30. $q^2 + 14 = 9q$

31. $x^2 + 10x + 25 = 0$

32. $x^2 + 6x + 9 = 0$

33. $r^2 = 8r - 16$

34. $x^2 + 1 = 2x$

35. $6x^2 + x - 2 = 0$

36. $2x^2 - 11x + 15 = 0$

37. $3a^2 = 10a + 8$

38. $15b - 9b^2 = 4$

39. $6x^2 - 4x = 10$

40. $3x^2 - 7x = 20$

41. $2t^2 + 12t = -10$

42. $12w^2 - 5w = 2$

43. $t(t - 5) = 14$

44. $6z^2 + z - 1 = 0$

45. $t(9 + t) = 4(2t + 5)$

46. $3y^2 + 8y = 12y + 15$

47. $16(p - 1) = p(p + 8)$

48. $(2x - 3)(x + 1) = 4(2x - 3)$

49. $(t - 1)(t + 3) = t - 1$

50. $(x - 2)(x + 2) = x + 2$

Solve.

51. $\dfrac{24}{x-2} + \dfrac{24}{x+2} = 5$

52. $\dfrac{8}{x+2} + \dfrac{8}{x-2} = 3$

53. $\dfrac{1}{x} + \dfrac{1}{x+6} = \dfrac{1}{4}$

54. $\dfrac{1}{x} + \dfrac{1}{x+9} = \dfrac{1}{20}$

55. $1 + \dfrac{12}{x^2-4} = \dfrac{3}{x-2}$

56. $\dfrac{5}{t-3} - \dfrac{30}{t^2-9} = 1$

57. $\dfrac{r}{r-1} + \dfrac{2}{r^2-1} = \dfrac{8}{r+1}$

58. $\dfrac{x+2}{x^2-2} = \dfrac{2}{3-x}$

59. $\dfrac{x-1}{1-x} = -\dfrac{x+8}{x-8}$

60. $\dfrac{4-x}{x-4} + \dfrac{x+3}{x-3} = 0$

61. $\dfrac{5}{y+4} - \dfrac{3}{y-2} = 4$

62. $\dfrac{2z+11}{2z+8} = \dfrac{3z-1}{z-1}$

d Solve.

63. *Diagonals.* A decagon is a figure with 10 sides. How many diagonals does a decagon have?

64. *Diagonals.* A hexagon is a figure with 6 sides. How many diagonals does a hexagon have?

65. *Diagonals.* A polygon has 14 diagonals. How many sides does it have?

66. *Diagonals.* A polygon has 9 diagonals. How many sides does it have?

Skill Maintenance

Factor completely. [5.6a]

67. $18t - 9p + 9$

68. $2x^3 - 18x^2 + 16x$

69. $x^2 + 14x + 49$

70. $36x^2 - 60xy + 25y^2$

71. $t^4 - 81$

72. $6a^2 - a - 15$

73. $a^2 + ac + a + c$

74. $20a^2b + ab - b$

75. $x^2y^2 - 1$

76. $c^2 - \dfrac{1}{4}$

77. $3x^4 - 6x^2 + 3$

78. $\dfrac{1}{4}x^2 - \dfrac{1}{4}xy + \dfrac{1}{16}y^2$

Synthesis

Solve.

79. $4m^2 - (m+1)^2 = 0$

80. $x^2 + \sqrt{22}x = 0$

81. $\sqrt{5}x^2 - x = 0$

82. $\sqrt{7}x^2 + \sqrt{3}x = 0$

Use a graphing calculator to solve each equation.

83. $3x^2 - 7x = 20$

84. $x(x-5) = 14$

85. $3x^2 + 8x = 12x + 15$

86. $(x-2)(x+2) = x + 2$

87. $(x-2)^2 + 3(x-2) = 4$

88. $(x+3)^2 = 4$

89. $16(x-1) = x(x+8)$

90. $x^2 + 2.5x + 1.5625 = 9.61$

9.2

Solving Quadratic Equations by Completing the Square

OBJECTIVES

a Solve quadratic equations of the type $ax^2 = p$.

b Solve quadratic equations of the type $(x + c)^2 = d$.

c Solve quadratic equations by completing the square.

d Solve certain applied problems involving quadratic equations of the type $ax^2 = p$.

SKILL TO REVIEW

Objective 5.5b: Factor trinomial squares.

Factor.

1. $x^2 - 18x + 81$

2. $x^2 + 2x + 1$

---

1. Solve: $x^2 = 10$.

2. Solve: $6x^2 = 0$.

---

a SOLVING QUADRATIC EQUATIONS OF THE TYPE $ax^2 = p$

For equations of the type $ax^2 = p$, we first solve for x^2 and then apply the *principle of square roots*, which states that a positive number has two square roots. The number 0 has one square root, 0.

> **THE PRINCIPLE OF SQUARE ROOTS**
>
> - The equation $x^2 = d$ has two real solutions when $d > 0$. The solutions are $\sqrt{d}$ and $-\sqrt{d}$.
> - The equation $x^2 = d$ has no real-number solution when $d < 0$.
> - The equation $x^2 = 0$ has 0 as its only solution.

EXAMPLE 1 Solve: $x^2 = 3$.

$$x^2 = 3$$
$$x = \sqrt{3} \quad or \quad x = -\sqrt{3} \qquad \text{Using the principle of square roots}$$

Check: For $\sqrt{3}$:   For $-\sqrt{3}$:

$$\begin{array}{c|c} x^2 = 3 & x^2 = 3 \\ \hline (\sqrt{3})^2 \ ? \ 3 & (-\sqrt{3})^2 \ ? \ 3 \\ 3 \ \big| \quad \text{TRUE} & 3 \ \big| \quad \text{TRUE} \end{array}$$

The solutions are $\sqrt{3}$ and $-\sqrt{3}$.

◀ **Do Margin Exercise 1.**

EXAMPLE 2 Solve: $\frac{1}{8}x^2 = 0$.

$$\frac{1}{8}x^2 = 0$$
$$x^2 = 0 \qquad \text{Multiplying by 8}$$
$$x = 0 \qquad \text{Using the principle of square roots}$$

The solution is 0.

◀ **Do Margin Exercise 2.**

EXAMPLE 3 Solve: $-3x^2 + 7 = 0$.

$$-3x^2 + 7 = 0$$
$$-3x^2 = -7 \qquad\qquad \text{Subtracting 7}$$
$$x^2 = \frac{-7}{-3} = \frac{7}{3} \qquad \text{Dividing by } -3$$
$$x = \sqrt{\frac{7}{3}} \quad or \quad x = -\sqrt{\frac{7}{3}} \qquad \text{Using the principle of square roots}$$
$$x = \sqrt{\frac{7}{3} \cdot \frac{3}{3}} \quad or \quad x = -\sqrt{\frac{7}{3} \cdot \frac{3}{3}} \qquad \text{Rationalizing the denominators}$$
$$x = \frac{\sqrt{21}}{3} \quad or \quad x = -\frac{\sqrt{21}}{3}$$

Answers

Skill to Review:
1. $(x - 9)^2$ **2.** $(x + 1)^2$

Margin Exercises:
1. $\sqrt{10}, -\sqrt{10}$ **2.** 0

Check: For $\dfrac{\sqrt{21}}{3}$:

$$-3x^2 + 7 = 0$$

$$-3\left(\dfrac{\sqrt{21}}{3}\right)^2 + 7 \;?\; 0$$

$$-3 \cdot \dfrac{21}{9} + 7$$

$$-7 + 7$$

$$0 \quad | \quad \text{TRUE}$$

For $-\dfrac{\sqrt{21}}{3}$:

$$-3x^2 + 7 = 0$$

$$-3\left(-\dfrac{\sqrt{21}}{3}\right)^2 + 7 \;?\; 0$$

$$-3 \cdot \dfrac{21}{9} + 7$$

$$-7 + 7$$

$$0 \quad | \quad \text{TRUE}$$

The solutions are $\dfrac{\sqrt{21}}{3}$ and $-\dfrac{\sqrt{21}}{3}$.

Do Exercise 3. ▶

b SOLVING QUADRATIC EQUATIONS OF THE TYPE $(x + c)^2 = d$

In an equation of the type $(x + c)^2 = d$, we have the square of a binomial equal to a constant. We can use the principle of square roots to solve such an equation.

EXAMPLE 4 Solve: $(x - 5)^2 = 9$.

$$(x - 5)^2 = 9$$

$$x - 5 = 3 \quad or \quad x - 5 = -3 \qquad \text{Using the principle of square roots}$$

$$x = 8 \quad or \qquad x = 2$$

The solutions are 8 and 2.

EXAMPLE 5 Solve: $(x + 2)^2 = 7$.

$$(x + 2)^2 = 7$$

$$x + 2 = \sqrt{7} \quad or \quad x + 2 = -\sqrt{7} \qquad \text{Using the principle of square roots}$$

$$x = -2 + \sqrt{7} \quad or \qquad x = -2 - \sqrt{7}$$

The solutions are $-2 + \sqrt{7}$ and $-2 - \sqrt{7}$, or simply $-2 \pm \sqrt{7}$ (read "-2 plus or minus $\sqrt{7}$").

Do Exercises 4 and 5. ▶

In Examples 4 and 5, the left sides of the equations are squares of binomials. If we can express an equation in such a form, we can proceed as we did in those examples.

EXAMPLE 6 Solve: $x^2 + 8x + 16 = 49$.

$$x^2 + 8x + 16 = 49 \qquad \text{The left side is the square of a binomial; } A^2 + 2AB + B^2 = (A + B)^2.$$

$$(x + 4)^2 = 49$$

$$x + 4 = 7 \quad or \quad x + 4 = -7 \qquad \text{Using the principle of square roots}$$

$$x = 3 \quad or \qquad x = -11$$

The solutions are 3 and -11.

Do Exercises 6 and 7. ▶

GS **3.** Solve: $2x^2 - 3 = 0$.

$$2x^2 = \boxed{}$$

$$x^2 = \dfrac{3}{2}$$

$$x = \sqrt{\dfrac{3}{2}} \quad or \quad x = -\sqrt{\boxed{}}$$

$$x = \sqrt{\dfrac{3}{2} \cdot \dfrac{2}{2}} \quad or \quad x = -\sqrt{\dfrac{3}{2} \cdot \boxed{}}$$

$$x = \dfrac{\sqrt{6}}{2} \quad or \quad x = \boxed{}$$

Both numbers check.

The solutions are $\dfrac{\sqrt{6}}{2}$ and $\boxed{}$.

Solve.

4. $(x - 3)^2 = 16$

5. $(x + 4)^2 = 11$

Solve.

6. $x^2 - 6x + 9 = 64$

7. $x^2 - 2x + 1 = 5$

Answers

3. $\dfrac{\sqrt{6}}{2}, -\dfrac{\sqrt{6}}{2}$ **4.** $7, -1$ **5.** $-4 \pm \sqrt{11}$
6. $-5, 11$ **7.** $1 \pm \sqrt{5}$

Guided Solution:
3. $3, \dfrac{3}{2}, \dfrac{2}{2}, \dfrac{\sqrt{6}}{2}, -\dfrac{\sqrt{6}}{2}$

C COMPLETING THE SQUARE

We have seen that a quadratic equation like $(x - 5)^2 = 9$ can be solved by using the principle of square roots. We also noted that an equation like $x^2 + 8x + 16 = 49$ can be solved in the same manner because the expression on the left side is the square of a binomial, $(x + 4)^2$. This second procedure is the basis for a method of solving quadratic equations called **completing the square**. *It can be used to solve any quadratic equation.*

Suppose we have the following quadratic equation:

$$x^2 + 10x = 4.$$

If we could add to both sides of the equation a constant that would make the expression on the left the square of a binomial, we could then solve the equation using the principle of square roots.

How can we determine what to add to $x^2 + 10x$ in order to construct the square of a binomial? We want to find a number a such that the following equation is satisfied:

$$x^2 + 10x + a^2 = (x + a)(x + a) = x^2 + 2ax + a^2.$$

Thus, a is such that $2a = 10$. Solving for a, we get $a = 5$; that is, a is half of the coefficient of x in $x^2 + 10x$. Since $a^2 = \left(\frac{10}{2}\right)^2 = 5^2 = 25$, we add 25 to our original expression:

$$x^2 + 10x + 25 \quad \text{is the square of} \quad x + 5;$$

that is,

$$x^2 + 10x + 25 = (x + 5)^2.$$

COMPLETING THE SQUARE

To **complete the square** for an expression like $x^2 + bx$, we take half of the coefficient of x and square it. Then we add that number, which is $(b/2)^2$.

Returning to solve our original equation, we first add 25 on *both* sides to complete the square on the left and find an equation equivalent to our original equation. Then we solve as follows:

$$x^2 + 10x \qquad\quad = 4 \qquad\qquad \text{Original equation}$$
$$x^2 + 10x + 25 = 4 + 25 \qquad \text{Adding 25:}$$
$$\left(\tfrac{10}{2}\right)^2 = 5^2 = 25$$

$$(x + 5)^2 = 29$$
$$x + 5 = \sqrt{29} \qquad or \quad x + 5 = -\sqrt{29} \qquad \text{Using the principle of square roots}$$

$$x = -5 + \sqrt{29} \quad or \qquad x = -5 - \sqrt{29}.$$

The solutions are $-5 \pm \sqrt{29}$.

We have seen that a quadratic equation $(x + c)^2 = d$ can be solved by using the principle of square roots. Any quadratic equation can be put in this form by completing the square. Then we can solve as before.

EXAMPLE 7 Solve: $x^2 + 6x + 8 = 0$.

We have

$x^2 + 6x + 8 = 0$

$x^2 + 6x \quad\ = -8.$ Subtracting 8

We take half of 6, $\frac{6}{2} = 3$, and square it, to get 3^2, or 9. Then we add 9 on *both* sides of the equation. This makes the left side the square of a binomial. We have now completed the square.

$x^2 + 6x + 9 = -8 + 9$ Adding 9. The left side is the square of a binomial.

$(x + 3)^2 = 1$

$x + 3 = 1 \quad or \quad x + 3 = -1$ Using the principle of square roots

$x = -2 \quad or \qquad x = -4$

The solutions are -2 and -4.

Do Exercises 8 and 9. ▶

Solve.

8. $x^2 - 6x + 8 = 0$

9. $x^2 + 8x - 20 = 0$

EXAMPLE 8 Solve $x^2 - 4x - 7 = 0$ by completing the square.

We have

$x^2 - 4x - 7 = 0$

$x^2 - 4x \qquad = 7$ Adding 7

$x^2 - 4x + 4 = 7 + 4$ Adding 4: $\left(\frac{-4}{2}\right)^2 = (-2)^2 = 4$

$(x - 2)^2 = 11$

$x - 2 = \sqrt{11} \qquad or \quad x - 2 = -\sqrt{11}$ Using the principle of square roots

$x = 2 + \sqrt{11} \quad or \qquad x = 2 - \sqrt{11}.$

The solutions are $2 \pm \sqrt{11}$.

Do Exercise 10. ▶

Example 7, as well as the following example, can be solved more easily by factoring. We solve them by completing the square only to illustrate that completing the square can be used to solve *any* quadratic equation.

EXAMPLE 9 Solve $x^2 + 3x - 10 = 0$ by completing the square.

We have

$x^2 + 3x - 10 = 0$

$x^2 + 3x \qquad = 10$

$x^2 + 3x + \frac{9}{4} = 10 + \frac{9}{4}$ Adding $\frac{9}{4}$: $\left(\frac{3}{2}\right)^2 = \frac{9}{4}$

$\left(x + \frac{3}{2}\right)^2 = \frac{40}{4} + \frac{9}{4} = \frac{49}{4}$

$x + \frac{3}{2} = \frac{7}{2} \quad or \quad x + \frac{3}{2} = -\frac{7}{2}$ Using the principle of square roots

$x = \frac{4}{2} \quad or \qquad x = -\frac{10}{2}$

$x = 2 \quad or \qquad x = -5.$

The solutions are 2 and -5.

Do Exercise 11. ▶

GS **10.** Solve: $x^2 - 12x + 23 = 0$.

$x^2 - 12x \qquad\ = -23$

$x^2 - 12x + 36 = -23 + \boxed{}$

$\left(\boxed{}\right)^2 = 13$

$x - 6 = \sqrt{13} \qquad or \quad x - 6 = -\boxed{}$

$x = 6 + \sqrt{13} \quad or \qquad x = \boxed{}$

The solutions are $6 \pm \boxed{}$.

11. Solve: $x^2 - 3x - 10 = 0$.

Answers

8. 2, 4   **9.** $-10, 2$ **10.** $6 \pm \sqrt{13}$
11. $-2, 5$

Guided Solution:
10. 36, $x - 6$, $\sqrt{13}$, $6 - \sqrt{13}$, $\sqrt{13}$

When the coefficient of x^2 is not 1, we can make it 1, as shown in the following example.

EXAMPLE 10   Solve $2x^2 = 3x + 1$ by completing the square.

We first obtain standard form. Then we multiply by $\frac{1}{2}$ on both sides to make the x^2-coefficient 1.

$$2x^2 = 3x + 1$$

$$2x^2 - 3x - 1 = 0 \qquad \text{Finding standard form}$$

$$\frac{1}{2}(2x^2 - 3x - 1) = \frac{1}{2} \cdot 0 \qquad \text{Multiplying by } \frac{1}{2} \text{ to make the}$$
$$x^2\text{-coefficient 1}$$

$$x^2 - \frac{3}{2}x - \frac{1}{2} = 0$$

$$x^2 - \frac{3}{2}x = \frac{1}{2} \qquad \text{Adding } \frac{1}{2}$$

$$x^2 - \frac{3}{2}x + \frac{9}{16} = \frac{1}{2} + \frac{9}{16} \qquad \text{Adding } \frac{9}{16}\text{:}\ \left[\frac{1}{2}\left(-\frac{3}{2}\right)\right]^2 = \left[-\frac{3}{4}\right]^2 = \frac{9}{16}$$

$$\left(x - \frac{3}{4}\right)^2 = \frac{8}{16} + \frac{9}{16}$$

$$\left(x - \frac{3}{4}\right)^2 = \frac{17}{16}$$

$$x - \frac{3}{4} = \frac{\sqrt{17}}{4} \quad \textit{or} \quad x - \frac{3}{4} = -\frac{\sqrt{17}}{4} \qquad \text{Using the principle of square roots}$$

$$x = \frac{3}{4} + \frac{\sqrt{17}}{4} \quad \textit{or} \quad x = \frac{3}{4} - \frac{\sqrt{17}}{4}$$

The solutions are $\dfrac{3 \pm \sqrt{17}}{4}$.

SOLVING BY COMPLETING THE SQUARE

To solve a quadratic equation $ax^2 + bx + c = 0$ by completing the square:

1. If $a \neq 1$, multiply by $1/a$ so that the x^2-coefficient is 1.
2. If the x^2-coefficient is 1, add so that the equation is in the form

$$x^2 + bx = -c, \quad \text{or} \quad x^2 + \frac{b}{a}x = -\frac{c}{a} \text{ if step (1) has been applied.}$$

3. Take half of the x-coefficient and square it. Add the result on both sides of the equation.
4. Express the side with the variables as the square of a binomial.
5. Use the principle of square roots and complete the solution.

Completing the square provides a basis for the quadratic formula, which we will discuss in Section 9.3. It also has other uses in later mathematics courses.

◀ **Do Exercise 12.**

12. Solve: $2x^2 + 3x - 3 = 0$.

Answer

12. $\dfrac{-3 \pm \sqrt{33}}{4}$

d APPLICATIONS

EXAMPLE 11 *Falling Object.* The Grand Canyon Skywalk, a horseshoe-shaped glass observation deck, extends 70 ft off the South Rim of the Grand Canyon. This structure, completed in 2007, can support a few hundred people, but the number of visitors allowed on the skywalk at any one time is 120. The Skywalk is approximately 4000 ft above the ground. If a tourist accidentally drops a camera from the observation deck, how long will it take the camera to fall to the ground?

Source: The Grand Canyon Skywalk

1. **Familiarize.** If we did not know anything about this problem, we might consider looking up a formula in a mathematics or physics book. A formula that fits this situation is

$$s = 16t^2,$$

where s is the distance, in feet, traveled by a body falling freely from rest in t seconds. This formula is actually an approximation in that it does not account for air resistance. In this problem, we know the distance s to be 4000 ft. We want to determine the time t that it takes the object to reach the ground.

2. **Translate.** We know that the distance is 4000 ft and that we need to solve for t. We substitute 4000 for s: $4000 = 16t^2$. This gives us a translation.

3. **Solve.** We solve the equation:

$$4000 = 16t^2$$

$$\frac{4000}{16} = t^2 \qquad \text{Solving for } t^2$$

$$\sqrt{\frac{4000}{16}} = t \quad or \quad -\sqrt{\frac{4000}{16}} = t \qquad \text{Using the principle of square roots}$$

$$15.8 \approx t \quad or \quad -15.8 \approx t. \qquad \text{Using a calculator to find the square root and rounding to the nearest tenth}$$

4. **Check.** The number -15.8 cannot be a solution because time cannot be negative in this situation. We substitute 15.8 in the original equation:

$$s = 16(15.8)^2 = 16(249.64) = 3994.24.$$

This answer is close: $3994.24 \approx 4000$. Remember that we rounded to approximate our solution, $t \approx 15.8$. Thus we have a check.

5. **State.** It would take about 15.8 sec for the camera to fall to the ground from the Grand Canyon Skywalk.

Do Exercise 13. ▶

13. *Falling Object.* The CN Tower in Toronto is 1815 ft tall. How long would it take an object to fall to the ground from the top?

Source: *World Almanac*, 2008

Answer

13. About 10.7 sec

✓ **Reading Check**

Determine whether each statement is true or false.

RC1. Every quadratic equation has two different solutions.

RC2. The solutions of $(x + 3)^2 = 5$ are $\sqrt{5}$ and $-\sqrt{5}$.

RC3. To complete the square for $x^2 + 6x$, we need to add 36.

RC4. Every quadratic equation can be solved by completing the square.

a Solve.

1. $x^2 = 121$

2. $x^2 = 100$

3. $5x^2 = 35$

4. $5x^2 = 45$

5. $5x^2 = 3$

6. $2x^2 = 9$

7. $4x^2 - 25 = 0$

8. $9x^2 - 4 = 0$

9. $3x^2 - 49 = 0$

10. $5x^2 - 16 = 0$

11. $4y^2 - 3 = 9$

12. $36y^2 - 25 = 0$

13. $49y^2 - 64 = 0$

14. $8x^2 - 400 = 0$

b Solve.

15. $(x + 3)^2 = 16$

16. $(x - 4)^2 = 25$

17. $(x + 3)^2 = 21$

18. $(x - 3)^2 = 6$

19. $(x + 13)^2 = 8$

20. $(x - 13)^2 = 64$

21. $(x - 7)^2 = 12$

22. $(x + 1)^2 = 14$

23. $(x + 9)^2 = 34$

24. $(t + 5)^2 = 49$

25. $\left(x + \frac{3}{2}\right)^2 = \frac{7}{2}$

26. $\left(y - \frac{3}{4}\right)^2 = \frac{17}{16}$

27. $x^2 - 6x + 9 = 64$

28. $p^2 - 10p + 25 = 100$

29. $x^2 + 14x + 49 = 64$

30. $t^2 + 8t + 16 = 36$

Solve by completing the square. Show your work.

31. $x^2 - 6x - 16 = 0$

32. $x^2 + 8x + 15 = 0$

33. $x^2 + 22x + 21 = 0$

34. $x^2 + 14x - 15 = 0$

35. $x^2 - 2x - 5 = 0$

36. $x^2 - 4x - 11 = 0$

37. $x^2 - 22x + 102 = 0$

38. $x^2 - 18x + 74 = 0$

39. $x^2 + 10x - 4 = 0$

40. $x^2 - 10x - 4 = 0$

41. $x^2 - 7x - 2 = 0$

42. $x^2 + 7x - 2 = 0$

43. $x^2 + 3x - 28 = 0$

44. $x^2 - 3x - 28 = 0$

45. $x^2 + \frac{3}{2}x - \frac{1}{2} = 0$

46. $x^2 - \frac{3}{2}x - 2 = 0$

47. $2x^2 + 3x - 17 = 0$

48. $2x^2 - 3x - 1 = 0$

49. $3x^2 + 4x - 1 = 0$

50. $3x^2 - 4x - 3 = 0$

51. $2x^2 = 9x + 5$

52. $2x^2 = 5x + 12$

53. $6x^2 + 11x = 10$

54. $4x^2 + 12x = 7$

55. *Burj Khalifa.* The Burj Khalifa in Dubai, The United Arab Emirates, is the tallest structure in the world. It stands at 2723 ft. How long would it take an object to fall from the top?

Source: http://www.skyscrapercenter.com

2723 ft

56. *Petronas Towers.* At a height of 1483 ft, the Petronas Towers in Kuala Lumpur, Malaysia, is one of the tallest buildings in the world. How long would it take an object to fall from the top?

Source: *The New York Times Almanac*

1483 ft

57. *Willis Tower.* The Willis Tower in Chicago, formerly called the Sears Tower, is 1451 ft tall. The Willis Tower Skydeck, an observation deck, is 1353 ft above ground. How long would it take an object to fall from the observation deck?

Source: The Willis Tower

58. *Taipei 101.* The Taipei 101 building in Taipei, Taiwan, is 1670 ft tall. How long would it take an object to fall from the top?

Source: *World Almanac*

Skill Maintenance

59. Add: $(3x^4 - x^2 + 1) + (4x^3 - x^2)$. [4.4a]

60. Subtract: $(3x^4 - x^2 + 1) - (4x^3 - x^2)$. [4.4c]

Multiply.

61. $(3x^2 + x)^2$ [4.6c]

62. $(x^2 - x + 1)(2x + 1)$ [4.5d]

Divide.

63. $(12x^4 - 15x^3 - x^2) \div (3x)$ [4.8a]

64. $(5x^2 - x - 3) \div (x - 1)$ [4.8b]

Synthesis

Find b such that the trinomial is a square.

65. $x^2 + bx + 36$

66. $x^2 + bx + 55$

67. $x^2 + bx + 128$

68. $4x^2 + bx + 16$

69. $x^2 + bx + c$

70. $ax^2 + bx + c$

Solve.

71. 📈 $4.82x^2 = 12,000$

72. $\dfrac{x}{2} = \dfrac{32}{x}$

73. $\dfrac{x}{9} = \dfrac{36}{4x}$

74. $\dfrac{4}{m^2 - 7} = 1$

The Quadratic Formula

We learn to complete the square to prove a general formula that can be used to solve quadratic equations even when they cannot be solved by factoring.

a SOLVING USING THE QUADRATIC FORMULA

Each time you solve by completing the square, you perform nearly the same steps. When we repeat the same kind of computation many times, we look for a formula so we can speed up our work. Consider

$$ax^2 + bx + c = 0, \quad a > 0.$$

Let's solve by completing the square. As we carry out the steps, compare them with Example 10 in the preceding section.

$$x^2 + \frac{b}{a}x + \frac{c}{a} = 0 \qquad \text{Multiplying by } \frac{1}{a}$$

$$x^2 + \frac{b}{a}x \qquad = -\frac{c}{a} \qquad \text{Adding } -\frac{c}{a}$$

Half of $\frac{b}{a}$ is $\frac{b}{2a}$. The square is $\frac{b^2}{4a^2}$. Thus we add $\frac{b^2}{4a^2}$ on both sides.

$$x^2 + \frac{b}{a}x + \frac{b^2}{4a^2} = -\frac{c}{a} + \frac{b^2}{4a^2} \qquad \text{Adding } \frac{b^2}{4a^2}$$

$$\left(x + \frac{b}{2a}\right)^2 = -\frac{4ac}{4a^2} + \frac{b^2}{4a^2} \qquad \begin{array}{l}\text{Factoring the left side and finding}\\ \text{a common denominator on the}\\ \text{right}\end{array}$$

$$\left(x + \frac{b}{2a}\right)^2 = \frac{b^2 - 4ac}{4a^2}$$

$$x + \frac{b}{2a} = \sqrt{\frac{b^2 - 4ac}{4a^2}} \quad or \quad x + \frac{b}{2a} = -\sqrt{\frac{b^2 - 4ac}{4a^2}} \qquad \begin{array}{l}\text{Using the principle}\\ \text{of square roots}\end{array}$$

Since $a > 0$, $\sqrt{4a^2} = 2a$, so we can simplify as follows:

$$x + \frac{b}{2a} = \frac{\sqrt{b^2 - 4ac}}{2a} \quad or \quad x + \frac{b}{2a} = -\frac{\sqrt{b^2 - 4ac}}{2a}.$$

Thus,

$$x = -\frac{b}{2a} + \frac{\sqrt{b^2 - 4ac}}{2a} \quad or \quad x = -\frac{b}{2a} - \frac{\sqrt{b^2 - 4ac}}{2a},$$

so $\quad x = -\frac{b}{2a} \pm \frac{\sqrt{b^2 - 4ac}}{2a}, \quad or \quad x = \frac{-b \pm \sqrt{b^2 - 4ac}}{2a}.$

We now have the following.

THE QUADRATIC FORMULA

The solutions of $ax^2 + bx + c = 0$ are given by

$$x = \frac{-b \pm \sqrt{b^2 - 4ac}}{2a}.$$

OBJECTIVES

a Solve quadratic equations using the quadratic formula.

b Find approximate solutions of quadratic equations using a calculator.

SKILL TO REVIEW

Objective 1.1a: Evaluate algebraic expressions by substitution.

Evaluate.

1. $-\frac{b}{2a}$, when $a = -2$ and $b = 4$

2. $-x + y$, when $x = -1$ and $y = -10$

Answers

Skill to Review:
1. 1   2. −9

The formula also holds when $a < 0$. A similar proof would show this, but we will not consider it here.

EXAMPLE 1 Solve $5x^2 - 8x = -3$ using the quadratic formula.

We first find standard form and determine a, b, and c:

$$5x^2 - 8x + 3 = 0;$$
$$a = 5, \quad b = -8, \quad c = 3.$$

We then use the quadratic formula:

$$x = \frac{-b \pm \sqrt{b^2 - 4ac}}{2a}$$

$$x = \frac{-(-8) \pm \sqrt{(-8)^2 - 4 \cdot 5 \cdot 3}}{2 \cdot 5} \qquad \text{Substituting}$$

$$x = \frac{8 \pm \sqrt{64 - 60}}{10} \qquad \text{Caution!}$$

Be sure to write the fraction bar all the way across.

$$x = \frac{8 \pm \sqrt{4}}{10}$$

$$x = \frac{8 \pm 2}{10}$$

$$x = \frac{8 + 2}{10} \quad or \quad x = \frac{8 - 2}{10}$$

$$x = \frac{10}{10} \quad or \quad x = \frac{6}{10}$$

$$x = 1 \quad or \quad x = \frac{3}{5}.$$

The solutions are 1 and $\frac{3}{5}$.

◀ **Do Exercise 1.**

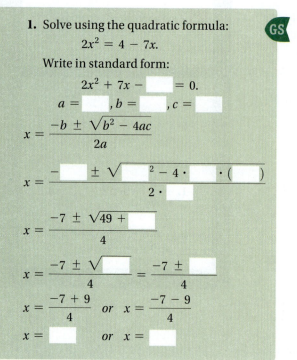

1. Solve using the quadratic formula:
$$2x^2 = 4 - 7x.$$
Write in standard form:
$$2x^2 + 7x - \boxed{} = 0.$$
$$a = \boxed{}, b = \boxed{}, c = \boxed{}$$
$$x = \frac{-b \pm \sqrt{b^2 - 4ac}}{2a}$$
$$x = \frac{-\boxed{} \pm \sqrt{\boxed{}^2 - 4 \cdot \boxed{} \cdot (\boxed{})}}{2 \cdot \boxed{}}$$
$$x = \frac{-7 \pm \sqrt{49 + \boxed{}}}{4}$$
$$x = \frac{-7 \pm \sqrt{\boxed{}}}{4} = \frac{-7 \pm \boxed{}}{4}$$
$$x = \frac{-7 + 9}{4} \quad or \quad x = \frac{-7 - 9}{4}$$
$$x = \boxed{} \quad or \quad x = \boxed{}$$

It would have been easier to solve the equation in Example 1 by factoring. We used the quadratic formula only to illustrate that it can be used to solve any quadratic equation. The following is a general procedure for solving a quadratic equation.

SOLVING QUADRATIC EQUATIONS

To solve a quadratic equation:

1. Check to see if it is in the form $ax^2 = p$ or $(x + c)^2 = d$. If it is, use the principle of square roots.

2. If it is not in the form of (1), write it in standard form, $ax^2 + bx + c = 0$ with a and b nonzero.

3. Then try factoring.

4. If it is not possible to factor or if factoring seems difficult, use the quadratic formula.

The solutions of a quadratic equation can always be found using the quadratic formula. They cannot always be found by factoring. (When the radicand $b^2 - 4ac \geq 0$, the equation has real-number solutions. When $b^2 - 4ac < 0$, the equation has no real-number solutions.)

Answer

1. $-4, \dfrac{1}{2}$

Guided Solution:

1. $4, 2, 7, -4$; $\dfrac{-7 \pm \sqrt{7^2 - 4 \cdot 2 \cdot (-4)}}{2 \cdot 2}$;

$32, 81, 9, \dfrac{1}{2}, -4$

EXAMPLE 2 Solve $x^2 + 3x - 10 = 0$ using the quadratic formula.

The equation is in standard form, so we determine a, b, and c:

$$x^2 + 3x - 10 = 0;$$
$$a = 1, \quad b = 3, \quad c = -10.$$

We then use the quadratic formula:

$$x = \frac{-b \pm \sqrt{b^2 - 4ac}}{2a}$$

$$= \frac{-3 \pm \sqrt{3^2 - 4 \cdot 1 \cdot (-10)}}{2 \cdot 1} \qquad \text{Substituting}$$

$$= \frac{-3 \pm \sqrt{9 + 40}}{2}$$

$$= \frac{-3 \pm \sqrt{49}}{2} = \frac{-3 \pm 7}{2}.$$

Thus,

$$x = \frac{-3 + 7}{2} = \frac{4}{2} = 2 \quad or \quad x = \frac{-3 - 7}{2} = \frac{-10}{2} = -5.$$

The solutions are 2 and −5.

Note that when the radicand is a perfect square, as in this example, we could have solved using factoring.

Do Exercise 2. ▶

EXAMPLE 3 Solve $x^2 = 4x + 7$ using the quadratic formula. Compare using the quadratic formula here with completing the square as we did in Example 8 of Section 9.2.

We first find standard form and determine a, b, and c:

$$x^2 - 4x - 7 = 0;$$
$$a = 1, \quad b = -4, \quad c = -7.$$

We then use the quadratic formula:

$$x = \frac{-b \pm \sqrt{b^2 - 4ac}}{2a}$$

$$= \frac{-(-4) \pm \sqrt{(-4)^2 - 4 \cdot 1 \cdot (-7)}}{2 \cdot 1} \qquad \text{Substituting}$$

$$= \frac{4 \pm \sqrt{16 + 28}}{2} = \frac{4 \pm \sqrt{44}}{2}$$

$$= \frac{4 \pm \sqrt{4 \cdot 11}}{2} = \frac{4 \pm \sqrt{4}\sqrt{11}}{2}$$

$$= \frac{4 \pm 2\sqrt{11}}{2} = \frac{2 \cdot 2 \pm 2\sqrt{11}}{2 \cdot 1}$$

$$= \frac{2(2 \pm \sqrt{11})}{2 \cdot 1} = \frac{2}{2} \cdot \frac{2 \pm \sqrt{11}}{1} \qquad \text{Factoring out 2 in the numerator and the denominator}$$

$$= 2 \pm \sqrt{11}.$$

The solutions are $2 + \sqrt{11}$ and $2 - \sqrt{11}$, or $2 \pm \sqrt{11}$.

Do Exercise 3. ▶

Let's visualize the solutions in Example 2.

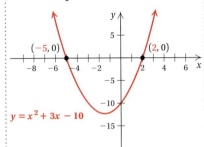

The solutions of $x^2 + 3x - 10 = 0$, −5 and 2, are the first coordinates of the x-intercepts, $(-5, 0)$ and $(2, 0)$, of the graph of $y = x^2 + 3x - 10$.

2. Solve using the quadratic formula:
$$x^2 - 3x - 10 = 0.$$

GS **3.** Solve using the quadratic formula:
$$x^2 + 4x = 7.$$

Write in standard form:
$$x^2 + 4x - \boxed{} = 0.$$
$$a = \boxed{}, b = \boxed{}, c = \boxed{}$$

$$x = \frac{-b \pm \sqrt{b^2 - 4ac}}{2a}$$

$$= \frac{-\boxed{} \pm \sqrt{\boxed{}^2 - 4 \cdot \boxed{} \cdot (\boxed{})}}{2 \cdot \boxed{}}$$

$$= \frac{-4 \pm \sqrt{16 + \boxed{}}}{2} = \frac{-4 \pm \sqrt{\boxed{}}}{2}$$

$$= \frac{-4 \pm 2\sqrt{\boxed{}}}{2} = \frac{2(-2 \pm \sqrt{11})}{2 \cdot 1}$$

$$= \boxed{} \pm \boxed{}$$

Answers

2. −2, 5   **3.** $-2 \pm \sqrt{11}$

Guided Solution:

3. 7, 1, 4, −7; $\dfrac{-4 \pm \sqrt{4^2 - 4 \cdot 1 \cdot (-7)}}{2 \cdot 1}$;

28, 44, 11; −2, $\sqrt{11}$

CALCULATOR CORNER

Visualizing Solutions of Quadratic Equations To see that there are no real-number solutions of the equation in Example 4,

$$x^2 + x = -1,$$

we graph $y_1 = x^2 + x$ and $y_2 = -1$.

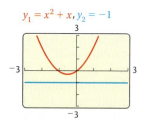

$y_1 = x^2 + x, \ y_2 = -1$

We see that the graphs do not intersect. Thus there is no real number for which $y_1 = y_2$, or $x^2 + x = -1$.

EXERCISE:

1. Explain how the graph of $y = x^2 + x + 1$ shows that the equation in Example 4, $x^2 + x = -1$, has no real-number solutions.

Solve using the quadratic formula.

4. $x^2 = x - 1$

5. $5x^2 - 8x = 3$

6. Approximate the solutions to the equation in Margin Exercise 5. Round to the nearest tenth.

Answers

4. No real-number solutions   **5.** $\dfrac{4 \pm \sqrt{31}}{5}$

6. $-0.3, 1.9$

EXAMPLE 4   Solve $x^2 + x = -1$ using the quadratic formula.

We first find standard form and determine a, b, and c:

$$x^2 + x + 1 = 0;$$
$$a = 1, \quad b = 1, \quad c = 1.$$

We then use the quadratic formula:

$$x = \frac{-b \pm \sqrt{b^2 - 4ac}}{2a} = \frac{-1 \pm \sqrt{1^2 - 4 \cdot 1 \cdot 1}}{2 \cdot 1} = \frac{-1 \pm \sqrt{-3}}{2}.$$

Note that the radicand ($b^2 - 4ac = -3$) in the quadratic formula is negative. Thus there are no real-number solutions because square roots of negative numbers do not exist as real numbers.

EXAMPLE 5   Solve $3x^2 = 7 - 2x$ using the quadratic formula.

We first find standard form and determine a, b, and c:

$$3x^2 + 2x - 7 = 0;$$
$$a = 3, \quad b = 2, \quad c = -7.$$

We then use the quadratic formula:

$$
\begin{aligned}
x &= \frac{-b \pm \sqrt{b^2 - 4ac}}{2a} = \frac{-2 \pm \sqrt{2^2 - 4 \cdot 3 \cdot (-7)}}{2 \cdot 3} \\
&= \frac{-2 \pm \sqrt{4 + 84}}{2 \cdot 3} \\
&= \frac{-2 \pm \sqrt{88}}{6} = \frac{-2 \pm \sqrt{4 \cdot 22}}{6} \\
&= \frac{-2 \pm \sqrt{4}\sqrt{22}}{6} = \frac{-2 \pm 2\sqrt{22}}{6} \\
&= \frac{2(-1 \pm \sqrt{22})}{2 \cdot 3} \\
&= \frac{2}{2} \cdot \frac{-1 \pm \sqrt{22}}{3} = \frac{-1 \pm \sqrt{22}}{3}.
\end{aligned}
$$

The solutions are $\dfrac{-1 + \sqrt{22}}{3}$ and $\dfrac{-1 - \sqrt{22}}{3}$, or $\dfrac{-1 \pm \sqrt{22}}{3}$.

◀ **Do Exercises 4 and 5.**

b APPROXIMATE SOLUTIONS

A calculator can be used to approximate solutions of quadratic equations.

EXAMPLE 6   Use a calculator to approximate to the nearest tenth the solutions to the equation in Example 5.

Using a calculator, we have

$$\frac{-1 + \sqrt{22}}{3} \approx 1.230138587 \approx 1.2 \text{ to the nearest tenth,} \quad \text{and}$$

$$\frac{-1 - \sqrt{22}}{3} \approx -1.896805253 \approx -1.9 \text{ to the nearest tenth.}$$

The approximate solutions are 1.2 and -1.9.

◀ **Do Exercise 6.**

✓ Reading Check

Choose from the column on the right the quadratic equation that has the given values of a, b, and c.

RC1. $a = 1$, $b = -7$, $c = 2$

RC2. $a = 1$, $b = 7$, $c = -2$

RC3. $a = 3$, $b = 7$, $c = -2$

RC4. $a = 3$, $b = -7$, $c = -2$

RC5. $a = 3$, $b = -7$, $c = 0$

RC6. $a = 3$, $b = 0$, $c = -7$

a) $3x^2 + 7x - 2 = 0$

b) $3x^2 - 2 = 7x$

c) $7x - 2 = x^2$

d) $2 - 7x - x^2 = 0$

e) $3x^2 = 7x$

f) $3x^2 = 7$

a Solve. Try factoring first. If factoring is not possible or is difficult, use the quadratic formula.

1. $x^2 - 4x = 21$

2. $x^2 + 8x = 9$

3. $x^2 = 6x - 9$

4. $x^2 = 24x - 144$

5. $3y^2 - 2y - 8 = 0$

6. $3y^2 - 7y + 4 = 0$

7. $4x^2 + 4x = 15$

8. $4x^2 + 12x = 7$

9. $x^2 - 9 = 0$

10. $x^2 - 16 = 0$

11. $x^2 - 2x - 2 = 0$

12. $x^2 - 2x - 11 = 0$

13. $y^2 - 10y + 22 = 0$

14. $y^2 + 6y - 1 = 0$

15. $x^2 + 4x + 4 = 7$

16. $x^2 - 2x + 1 = 5$

17. $3x^2 + 8x + 2 = 0$

18. $3x^2 - 4x - 2 = 0$

19. $2x^2 - 5x = 1$

20. $4x^2 + 4x = 5$

21. $2y^2 - 2y - 1 = 0$

22. $4y^2 + 4y - 1 = 0$

23. $2t^2 + 6t + 5 = 0$

24. $4y^2 + 3y + 2 = 0$

25. $3x^2 = 5x + 4$

26. $2x^2 + 3x = 1$

27. $2y^2 - 6y = 10$

28. $5m^2 = 3 + 11m$

29. $\dfrac{x^2}{x+3} - \dfrac{5}{x+3} = 0$

30. $\dfrac{x^2}{x-4} - \dfrac{7}{x-4} = 0$

31. $x + 2 = \dfrac{3}{x+2}$

32. $x - 3 = \dfrac{5}{x-3}$

33. $\dfrac{1}{x} + \dfrac{1}{x+1} = \dfrac{1}{3}$

34. $\dfrac{1}{x} + \dfrac{1}{x+6} = \dfrac{1}{5}$

b   Solve using the quadratic formula. Use a calculator to approximate the solutions to the nearest tenth.

35. $x^2 - 4x - 7 = 0$

36. $x^2 + 2x - 2 = 0$

37. $y^2 - 6y - 1 = 0$

38. $y^2 + 10y + 22 = 0$

3.9. $4x^2 + 4x = 1$

40. $4x^2 = 4x + 1$

41. $3x^2 - 8x + 2 = 0$

42. $3x^2 + 4x - 2 = 0$

Skill Maintenance

Solve.

43. $2(x + 3) - (x - 5) = 6x$ [2.3c]

44. $-9x \geq \dfrac{1}{2}$ [2.7d]

45. $\dfrac{1}{3}y + \dfrac{1}{2} = \dfrac{2}{3} - \dfrac{5}{6}y$ [2.3b]

46. $t^2 - 7t - 8 = 0$ [5.7b]

47. $3d^2 - 3 = 0$ [5.7b]

48. $\sqrt{3x} = 15$ [8.5a]

49. $\sqrt{2x - 1} = x - 2$ [8.5a]

50. $y^2 = -3y$ [5.7b]

51. $2n^2 - n = 3$ [5.7b]

52. $\dfrac{2}{x} = \dfrac{x}{8}$ [6.6a]

53. $\dfrac{x}{x-1} + \dfrac{1}{x} = 1$ [6.6a]

54. $\dfrac{5x}{x+2} - \dfrac{1}{x-1} = 3$ [6.6a]

Synthesis

Solve.

55. $5x + x(x - 7) = 0$

56. $x(3x + 7) - 3x = 0$

57. $3 - x(x - 3) = 4$

58. $x(5x - 7) = 1$

59. $(y + 4)(y + 3) = 15$

60. $(y + 5)(y - 1) = 27$

61. $x^2 + (x + 2)^2 = 7$

62. $x^2 + (x + 1)^2 = 5$

63. ◪ Use a graphing calculator to determine whether the equation $x^2 + x = 1$ has real-number solutions.

64. ◪ Use a graphing calculator to determine whether the equation $x^2 = 2x - 3$ has real-number solutions.

65.–72. ◪ Use a graphing calculator to approximate the solutions of the equations in Exercises 35–42. Compare your answers with those found using the quadratic formula.

Mid-Chapter Review

Concept Reinforcement

Determine whether each statement is true or false.

_____ **1.** The equation $x^2 = -4$ has no real-number solutions.   [9.2a]

_____ **2.** The solutions of $ax^2 + bx + c = 0$ are the first coordinates of the y-intercepts of the graph of $y = ax^2 + bx + c$.   [9.1c]

_____ **3.** A quadratic equation of the type $ax^2 + bx = 0$, where $c = 0$ and $b \neq 0$, will always have 0 as one solution and a nonzero number as the other solution.   [9.1b]

Guided Solutions

 Fill in each blank with the number or the expression that creates a correct solution.

4. Solve $x^2 - 6x - 2 = 0$ by completing the square.   [9.2c]

$$x^2 - 6x - 2 = 0$$
$$x^2 - 6x = \boxed{}$$
$$x^2 - 6x + \boxed{} = 2 + \boxed{}$$
$$\left(x - \boxed{}\right)^2 = \boxed{}$$
$$x - \boxed{} = \pm \sqrt{\boxed{}}$$
$$x = \boxed{} \pm \sqrt{11}$$

5. Solve $3x^2 = 8x - 2$ using the quadratic formula.   [9.3a]

$$3x^2 = 8x - 2$$
$$3x^2 - \boxed{} + \boxed{} = 0 \qquad \text{Standard form}$$
$$a = \boxed{}, \quad b = \boxed{}, \quad c = \boxed{}$$

We substitute for a, b, and c in the quadratic formula:

$$x = \frac{-b \pm \sqrt{b^2 - 4ac}}{2a} \qquad \text{Quadratic formula}$$

$$= \frac{-\boxed{} \pm \sqrt{(\boxed{})^2 - 4 \cdot \boxed{} \cdot \boxed{}}}{2 \cdot \boxed{}} \qquad \text{Substituting}$$

$$= \frac{\boxed{} \pm \sqrt{64 - \boxed{}}}{\boxed{}} = \frac{8 \pm \sqrt{\boxed{}}}{6} = \frac{8 \pm \sqrt{\boxed{} \cdot 10}}{6}$$

$$= \frac{8 \pm \boxed{} \sqrt{10}}{6} = \frac{2(\boxed{} \pm \sqrt{10})}{2 \cdot \boxed{}} = \frac{\boxed{} \pm \sqrt{10}}{\boxed{}}.$$

Mixed Review

Write in standard form and determine a, b, and c.   [9.1a]

6. $q^2 - 5q + 10 = 0$

7. $6 - x^2 = 14x + 2$

8. $17z = 3z^2$

Solve by factoring.

9. $16x = 48x^2$ [9.1b]

10. $x(x - 3) = 10$ [9.1c]

11. $20x^2 - 20x = 0$ [9.1b]

12. $x^2 = 14x - 49$ [9.1c]

13. $t^2 + 2t = 0$ [9.1b]

14. $18w^2 + 21w = 4$ [9.1c]

15. $9y^2 - 5y^2 = 82y + 6y$ [9.1b]

16. $2(s - 3) = s(s - 3)$ [9.1c]

17. $8y^2 - 40y = -7y + 35$ [9.1c]

Solve by completing the square. [9.2c]

18. $x^2 + 2x - 3 = 0$

19. $x^2 - 9x + 6 = 0$

20. $2x^2 = 7x + 8$

21. $y^2 + 80 = 18y$

22. $t^2 + \dfrac{3}{2}t - \dfrac{3}{2} = 0$

23. $x + 7 = -3x^2$

Solve.

24. $6x^2 = 384$ [9.2a]

25. $5y^2 + 2y + 3 = 0$ [9.3a]

26. $6(x - 3)^2 = 12$ [9.2b]

27. $4x^2 + 4x = 3$ [9.3a]

28. $8y^2 - 5 = 19$ [9.2a]

29. $a^2 = a + 1$ [9.3a]

30. $(w - 2)^2 = 100$ [9.2b]

31. $5m^2 + 2m = -3$ [9.3a]

32. $\left(y - \dfrac{1}{2}\right)^2 = \dfrac{5}{4}$ [9.2b]

33. $3x^2 - 75 = 0$ [9.2a]

34. $2x^2 - 2x - 5 = 0$ [9.3a]

35. $(x + 2)^2 = -5$ [9.2b]

Solve and use a calculator to approximate the solutions to the nearest tenth. [9.3b]

36. $y^2 - y - 8 = 0$

37. $2x^2 + 7x + 1 = 0$

For each equation in Exercises 38–42, select from the column on the right the correct description of the solutions of the equation.

38. $x^2 - x - 6 = 0$ [9.3a]

A. Two real-number solutions

B. No real-number solutions

39. $x^2 = -9$ [9.2a]

C. 0 is the only solution.

40. $x^2 = 31$ [9.2a]

41. $x^2 = 0$ [9.2a]

42. $x^2 - x + 6 = 0$ [9.3a]

43. Solve: $(x - 3)^2 = 36$. [9.2b]

  A. $-9, 3$ **B.** $-33, 39$
  C. $-3, 9$ **D.** $\sqrt{6}$

44. Simplify: $\dfrac{-24 \pm \sqrt{720}}{18}$. [9.3a]

  A. $\dfrac{-8 \pm 4\sqrt{5}}{6}$ **B.** $\dfrac{-4 \pm 2\sqrt{5}}{3}$

  C. $\dfrac{-4 \pm \sqrt{20}}{3}$ **D.** $-2 \pm 2\sqrt{5}$

Understanding Through Discussion and Writing

45. Mark asserts that the solution of a quadratic equation is $3 \pm \sqrt{14}$ and states that there is only one solution. What mistake is being made? [9.2b]

46. Find the errors in the following solution of the equation $x^2 + x = 6$. [9.1c]

$$x^2 + x = 6$$
$$x(x + 1) = 6$$
$$x = 6 \quad or \quad x + 1 = 6$$
$$x = 6 \quad or \qquad x = 5$$

47. Explain how the graph of $y = (x - 2)(x + 3)$ is related to the solutions of the equation $(x - 2)(x + 3) = 0$. [9.1a]

48. Under what condition(s) would using the quadratic formula *not* be the easiest way to solve a quadratic equation? [9.3a]

49. Write a quadratic equation in the form $y = ax^2 + bx + c$ that does not cross the x-axis. [9.3a]

50. Explain how you might go about constructing a quadratic equation whose solutions are -5 and 7. [9.1c]

Formulas

9.4

a SOLVING FORMULAS

Formulas arise frequently in the natural and social sciences, business, engineering, and health care. The same steps that are used to solve a linear, rational, radical, or quadratic equation can also be used to solve a formula that appears in one of these forms.

EXAMPLE 1 *Intelligence Quotient.* The formula $Q = \dfrac{100m}{c}$ is used to determine the intelligence quotient, Q, of a person of mental age m and chronological age c. Solve for c.

$$Q = \frac{100m}{c}$$

$$c \cdot Q = c \cdot \frac{100m}{c} \qquad \text{Multiplying by } c \text{ on both sides to clear the fraction}$$

$$cQ = 100m \qquad \text{Simplifying}$$

$$c = \frac{100m}{Q}. \qquad \text{Dividing by } Q \text{ on both sides}$$

This formula can be used to determine a person's chronological, or actual, age from his or her mental age and intelligence quotient.

Do Margin Exercise 1. ▶

EXAMPLE 2 Solve for x: $y = ax + bx - 4$.

$$y = ax + bx - 4 \qquad \text{We want this letter alone on one side.}$$

$$y + 4 = ax + bx \qquad \text{Adding 4. All terms containing } x \text{ are on the right side of the equation.}$$

$$y + 4 = (a + b)x \qquad \text{Factoring out the } x$$

$$\frac{y + 4}{(a + b)} = \frac{(a + b)x}{(a + b)} \qquad \text{Dividing by } a + b \text{ on both sides}$$

$$\frac{y + 4}{a + b} = x \qquad \text{Simplifying. The answer can also be written as } x = \frac{y + 4}{a + b}.$$

Do Margin Exercise 2. ▶

OBJECTIVE

a Solve a formula for a specified letter.

SKILL TO REVIEW

Objective 2.4b: Solve a formula for a specified letter.

Solve for the indicated letter.

1. $Q = mx + y$, for x

2. $R = \dfrac{x + y + z}{2}$, for z

1. **a)** Solve for I: $E = \dfrac{9R}{I}$.

   b) Solve for R: $E = \dfrac{9R}{I}$.

2. Solve for x: $y = ax - bx + 5$.

Answers

Skill to Review:

1. $x = \dfrac{Q - y}{m}$ 2. $z = 2R - x - y$

Margin Exercises:

1. **(a)** $I = \dfrac{9R}{E}$; **(b)** $R = \dfrac{EI}{9}$ 2. $x = \dfrac{y - 5}{a - b}$

.... **Caution!** ....

Had we performed the following steps in Example 2, we would *not* have solved for x:

$$y = ax + bx - 4$$

$$y - ax + 4 = bx \qquad \text{Subtracting } ax \text{ and adding 4}$$

x occurs on both sides of the equals sign.

$$\frac{y - ax + 4}{b} = x. \qquad \text{Dividing by } b$$

The mathematics of each step is correct, but since x occurs on both sides of the formula, *we have not solved the formula for x*. Remember that the letter being solved for should be **alone** on one side of the equation, with no occurrence of that letter on the other side!

....

EXAMPLE 3 Solve the following work formula for t:

$$\frac{t}{a} + \frac{t}{b} = 1.$$

We clear fractions by multiplying by the LCM, which is ab:

$$ab \cdot \left(\frac{t}{a} + \frac{t}{b} \right) = ab \cdot 1 \qquad \text{Multiplying by } ab$$

$$ab \cdot \frac{t}{a} + ab \cdot \frac{t}{b} = ab \qquad \begin{array}{l}\text{Using a distributive law to}\\\text{remove parentheses}\end{array}$$

$$bt + at = ab \qquad \text{Simplifying}$$

$$(b + a)t = ab \qquad \text{Factoring out } t$$

$$t = \frac{ab}{b + a}. \qquad \text{Dividing by } b + a$$

◀ **Do Exercise 3.**

EXAMPLE 4 *Distance to the Horizon.* Solve $D = \sqrt{2h}$ for h, where D is the approximate distance, in miles, that a person can see to the horizon from a height h, in feet.

This is a radical equation. Recall that we first isolate the radical. Then we use the principle of squaring.

$$D = \sqrt{2h}$$

$$D^2 = (\sqrt{2h})^2 \qquad \text{Using the principle of squaring (Section 8.5)}$$

$$D^2 = 2h \qquad \text{Simplifying}$$

$$\frac{D^2}{2} = h \qquad \text{Dividing by 2}$$

EXAMPLE 5 *Period of a Pendulum.* Solve $T = 2\pi\sqrt{\dfrac{L}{g}}$ for g, where T is

the period, in seconds, of a pendulum of length L, in feet, and g is a gravitational constant.

We have

$$\frac{T}{2\pi} = \sqrt{\frac{L}{g}} \qquad \text{Dividing by } 2\pi \text{ to isolate the radical}$$

$$\left(\frac{T}{2\pi}\right)^2 = \left(\sqrt{\frac{L}{g}}\right)^2. \qquad \text{Using the principle of squaring}$$

---

3. *Optics Formula.* Solve for f:

$$\frac{1}{p} + \frac{1}{q} = \frac{1}{f}.$$

The LCM is ▢.

$$pqf\left(\frac{1}{p} + \frac{1}{q}\right) = pqf\left(\frac{1}{f}\right)$$

$$pqf \cdot \frac{1}{p} + pqf\left(\frac{1}{q}\right) = pqf\left(\frac{1}{f}\right)$$

$$qf + ▢ = pq$$

$$f(▢) = pq$$

$$f = \frac{pq}{▢}$$

Answer

3. $f = \dfrac{pq}{q + p}$

Guided Solution:
3. $pqf, pf, q + p, q + p$

Then

$$\frac{T^2}{4\pi^2} = \frac{L}{g}$$

$$gT^2 = 4\pi^2 L \qquad \text{Multiplying by } 4\pi^2 g \text{ to clear fractions}$$

$$g = \frac{4\pi^2 L}{T^2}. \qquad \text{Dividing by } T^2 \text{ to get } g \text{ alone}$$

Do Exercises 4–6. ▶

In most formulas, the letters represent nonnegative numbers, so we need not use absolute values when taking square roots.

EXAMPLE 6 *Torricelli's Theorem.* The speed v of a liquid leaving a water cooler from an opening is related to gravity g and the height h of the top of the liquid above the opening by the formula

$$h = \frac{v^2}{2g}.$$

Solve for v.

Since v^2 appears by itself and there is no expression involving v, we first solve for v^2. Then we use the principle of square roots, taking only the nonnegative square root because v is nonnegative.

$$2gh = v^2 \qquad \text{Multiplying by } 2g \text{ to clear the fraction}$$

$$\sqrt{2gh} = v \qquad \text{Using the principle of square roots. Assume that } v \text{ is nonnegative.}$$

Do Exercise 7. ▶

EXAMPLE 7 Solve $d = \dfrac{n^2 - 3n}{2}$ for n, where d is the number of diagonals of an n-sided polygon.

In this case, there is a term involving n as well as an n^2-term. Thus we must use the quadratic formula.

$$d = \frac{n^2 - 3n}{2}$$

$$n^2 - 3n = 2d \qquad \text{Multiplying by 2 to clear fractions}$$

$$n^2 - 3n - 2d = 0 \qquad \text{Finding standard form}$$

$$a = 1, \quad b = -3, \quad c = -2d \qquad \text{The variable is } n; d \text{ represents a constant.}$$

$$n = \frac{-b \pm \sqrt{b^2 - 4ac}}{2a} \qquad \text{Quadratic formula}$$

$$= \frac{-(-3) \pm \sqrt{(-3)^2 - 4 \cdot 1 \cdot (-2d)}}{2 \cdot 1} \qquad \text{Substituting into the quadratic formula}$$

$$= \frac{3 + \sqrt{9 + 8d}}{2} \qquad \text{Using the positive root}$$

Do Exercise 8. ▶

---

4. Solve for L: $r = 2\sqrt{5L}$ (the speed of a skidding car).

5. Solve for L: $T = 2\pi\sqrt{\dfrac{L}{g}}$.

6. Solve for m: $c = \sqrt{\dfrac{E}{m}}$.

GS 7. Solve for r: $A = \pi r^2$ (the area of a circle).

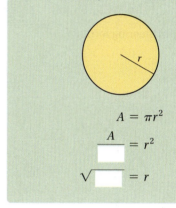

$$A = \pi r^2$$

$$\frac{A}{\boxed{}} = r^2$$

$$\sqrt{\boxed{}} = r$$

8. Solve for n: $N = n^2 - n$.

Answers

4. $L = \dfrac{r^2}{20}$ 5. $L = \dfrac{T^2 g}{4\pi^2}$ 6. $m = \dfrac{E}{c^2}$

7. $r = \sqrt{\dfrac{A}{\pi}}$ 8. $n = \dfrac{1 + \sqrt{1 + 4N}}{2}$

Guided Solution:

7. $\pi, \dfrac{A}{\pi}$

Reading Check

For each formula given, choose from the column on the right the best step to take first when solving the formula for x. Choices may be used more than once or not at all.

RC1. $Q = \dfrac{50x}{n}$

RC2. $t = ax + bx$

RC3. $\dfrac{x}{a} + 2 = \dfrac{x}{b}$

RC4. $N = \sqrt{3x}$

RC5. $x^2 = \dfrac{v}{50n}$

RC6. $p = \dfrac{5x^2 + x}{2}$

a) Factor out the x.

b) Divide both sides by x.

c) Use the principle of squaring.

d) Use the principle of square roots.

e) Clear fractions by multiplying both sides of the equation by the LCM of the denominators.

a Solve for the indicated letter.

1. $q = \dfrac{VQ}{I}$, for I
(An engineering formula)

2. $y = \dfrac{4A}{a}$, for a

3. $S = \dfrac{kmM}{d^2}$, for m

4. $S = \dfrac{kmM}{d^2}$, for M

5. $S = \dfrac{kmM}{d^2}$, for d^2

6. $T = \dfrac{10t}{W^2}$, for W^2

7. $T = \dfrac{10t}{W^2}$, for W

8. $S = \dfrac{kmM}{d^2}$, for d

9. $A = at + bt$, for t

10. $S = rx + sx$, for x

11. $y = ax + bx + c$, for x

12. $y = ax - bx - c$, for x

13. $\dfrac{t}{a} + \dfrac{t}{b} = 1$, for a
(A work formula)

14. $\dfrac{t}{a} + \dfrac{t}{b} = 1$, for b
(A work formula)

15. $\dfrac{1}{p} + \dfrac{1}{q} = \dfrac{1}{f}$, for p
(An optics formula)

16. $\dfrac{1}{p} + \dfrac{1}{q} = \dfrac{1}{f}$, for q
(An optics formula)

17. $A = \dfrac{1}{2}bh$, for b
(The area of a triangle)

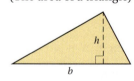

18. $s = \dfrac{1}{2}gt^2$, for g

19. $S = 2\pi r(r + h)$, for h
(The surface area of a right circular cylinder)

20. $S = 2\pi(r + h)$, for r

21. $\dfrac{1}{R} = \dfrac{1}{r_1} + \dfrac{1}{r_2}$, for R

(An electricity formula)

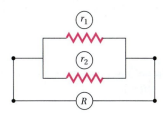

22. $\dfrac{1}{R} = \dfrac{1}{r_1} + \dfrac{1}{r_2}$, for r_1

23. $P = 17\sqrt{Q}$, for Q

24. $A = 1.4\sqrt{t}$, for t

25. $v = \sqrt{\dfrac{2gE}{m}}$, for E

26. $Q = \sqrt{\dfrac{aT}{c}}$, for T

27. $S = 4\pi r^2$, for r

28. $E = mc^2$, for c

29. $P = kA^2 + mA$, for A

30. $Q = ad^2 - cd$, for d

31. $c^2 = a^2 + b^2$, for a

32. $c = \sqrt{a^2 + b^2}$, for b

33. $s = 16t^2$, for t

34. $V = \pi r^2 h$, for r

35. $A = \pi r^2 + 2\pi rh$, for r

36. $A = 2\pi r^2 + 2\pi rh$, for r

37. $F = \dfrac{Av^2}{400}$, for v

38. $A = \dfrac{\pi r^2 S}{360}$, for r

39. $c = \sqrt{a^2 + b^2}$, for a

40. $c^2 = a^2 + b^2$, for b

41. $h = \dfrac{a}{2}\sqrt{3}$, for a

(The height of an equilateral triangle with sides of length a)

42. $d = s\sqrt{2}$, for s

(The hypotenuse of an isosceles right triangle with s the length of the legs)

43. $n = aT^2 - 4T + m$, for T

44. $y = ax^2 + bx + c$, for x

45. $v = 2\sqrt{\dfrac{2kT}{\pi m}}$, for T

46. $E = \dfrac{1}{2}mv^2 + mgy$, for v

47. $3x^2 = d^2$, for x

48. $c = \sqrt{\dfrac{E}{m}}$, for E

49. $N = \dfrac{n^2 - n}{2}$, for n

50. $M = \dfrac{m}{\sqrt{1 - \left(\dfrac{v}{c}\right)^2}}$, for c

51. $S = \dfrac{a + b}{3b}$, for b

52. $Q = \dfrac{a - b}{2b}$, for b

53. $\dfrac{A - B}{AB} = Q$, for B

54. $L = \dfrac{Mt + g}{t}$, for t

55. $S = 180(n - 2)$, for n

56. $S = \dfrac{n}{2}(a + 1)$, for a

57. $A = P(1 + rt)$, for t
(An interest formula)

58. $A = P(1 + rt)$, for r
(An interest formula)

59. $\dfrac{A}{B} = \dfrac{C}{D}$, for D

60. $\dfrac{A}{B} = \dfrac{C}{D}$, for B

Skill Maintenance

In a right triangle, where a and b represent the lengths of the legs and c represents the length of the hypotenuse, find the length of the side not given. Give an exact answer and an approximation to three decimal places. [8.6a]

61. $a = 4, b = 7$

62. $b = 11, c = 14$

63. $a = 4, b = 5$

64. $a = 10, c = 12$

65. $c = 8\sqrt{17}, a = 2$

66. $a = \sqrt{2}, b = \sqrt{3}$

Solve. [8.6b]

67. *Guy Wire.* How long is a guy wire reaching from the top of an 18-ft pole to a point on the ground 10 ft from the pole? Give an exact answer and an approximation to three decimal places.

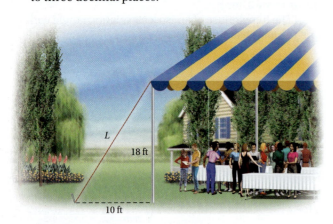

68. *Soccer Fields.* The smallest regulation soccer field is 50 yd wide and 100 yd long. Find the length of a diagonal of such a field.

Simplify.

69. $\sqrt{80}$ [8.2a]

70. $\sqrt{9000x^{10}}$ [8.2b]

Multiply and simplify. [8.2c]

71. $3\sqrt{t} \cdot \sqrt{t}$

72. $\sqrt{8x^2} \cdot \sqrt{24x^3}$

Add or subtract. [8.4a]

73. $\sqrt{40} - 2\sqrt{10} + \sqrt{90}$

74. $\sqrt{18} + \sqrt{50} - 3\sqrt{8}$

Synthesis

75. The circumference C of a circle is given by $C = 2\pi r$.

  a) Solve $C = 2\pi r$ for r.
  b) The area is given by $A = \pi r^2$. Express the area in terms of the circumference C.
  c) Express the circumference C in terms of the area A.

76. Solve $3ax^2 - x - 3ax + 1 = 0$ for x.

Applications and Problem Solving

a USING QUADRATIC EQUATIONS TO SOLVE APPLIED PROBLEMS

EXAMPLE 1 *Blueberry Farming.* Kevon is investing in blueberry farming. The area of his rectangular blueberry field is 4800 yd². The length is 10 yd longer than five times the width. Find the dimensions of the blueberry field.

1. **Familiarize.** We first make a drawing and label it with both known and unknown information. We let $w =$ the width of the rectangle. The length of the rectangle is 10 yd longer than five times the width. Thus the length is $5w + 10$.

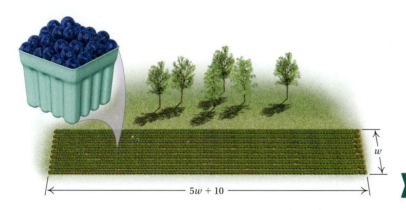

2. **Translate.** Recall that area is length × width. Thus we have two expressions for the area of the rectangle: $(5w + 10)(w)$ and 4800. This gives us a translation:

$$(5w + 10)(w) = 4800.$$

3. **Solve.** We solve the equation:

$$5w^2 + 10w = 4800$$
$$5w^2 + 10w - 4800 = 0$$
$$w^2 + 2w - 960 = 0 \qquad \text{Dividing by 5}$$
$$(w + 32)(w - 30) = 0 \qquad \text{Factoring (the quadratic formula could also be used)}$$
$$w + 32 = 0 \quad or \quad w - 30 = 0 \qquad \text{Using the principle of zero products}$$
$$w = -32 \quad or \qquad w = 30.$$

4. **Check.** We check in the original problem. We know that -32 is not a solution because width cannot be negative. When $w = 30$, $5w + 10 = 160$, and the area is 30×160, or 4800. This checks.

5. **State.** The width of the rectangular blueberry field is 30 yd, and the length is 160 yd.

Do Exercise 1. ▶

GS **1.** *Mural Dimensions.* The area of a rectangular mural is 52 ft². The length is 5 ft longer than twice the width. Find the dimensions of the mural.

1. **Familiarize.** Let $w =$ the width of the mural. Then the length is $2w + \boxed{}$.

2. **Translate.**
$$(2w + 5)(w) = \boxed{}$$

3. **Solve.**
$$2w^2 + 5w = 52$$
$$2w^2 + 5w - \boxed{} = 0$$
$$(2w + 13)(\boxed{}) = 0$$
$$2w + 13 = 0 \quad or \quad \boxed{} = 0$$
$$w = -\tfrac{13}{2} \quad or \qquad w = \boxed{}$$

4. **Check.** Only $\boxed{}$ checks. When the width is $\boxed{}$ ft, the length is $2(\boxed{}) + 5 = \boxed{}$ ft.

5. **State.** The width is 4 ft, and the length is $\boxed{}$ ft.

Answer

1. Length: 13 ft; width: 4 ft

Guided Solution:
1. 5, 52, 52, $w - 4$, $w - 4$, 4, 4, 4, 4, 13, 13

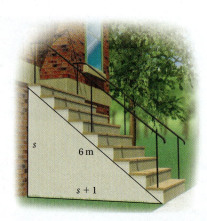

EXAMPLE 2 *Staircase.* A mason builds a staircase in such a way that the portion underneath the stairs forms a right triangle. The hypotenuse is 6 m long. The leg across the ground is 1 m longer than the leg next to the wall at the back. Find the lengths of the legs. Round to the nearest tenth.

1. **Familiarize.** We first make a drawing, letting $s =$ the length of the shorter leg. Then $s + 1 =$ the length of the other leg.

2. **Translate.** To translate, we use the Pythagorean equation:
$$s^2 + (s + 1)^2 = 6^2.$$

3. **Solve.** We solve the equation:
$$s^2 + (s + 1)^2 = 6^2$$
$$s^2 + s^2 + 2s + 1 = 36$$
$$2s^2 + 2s - 35 = 0.$$

Since we cannot factor, we use the quadratic formula:

$$a = 2, \quad b = 2, \quad c = -35;$$
$$s = \frac{-b \pm \sqrt{b^2 - 4ac}}{2a} = \frac{-2 \pm \sqrt{2^2 - 4 \cdot 2(-35)}}{2 \cdot 2}$$
$$= \frac{-2 \pm \sqrt{4 + 280}}{4} = \frac{-2 \pm \sqrt{284}}{4}$$
$$= \frac{-2 \pm \sqrt{4 \cdot 71}}{4} = \frac{-2 \pm 2 \cdot \sqrt{71}}{2 \cdot 2}$$
$$= \frac{2(-1 \pm \sqrt{71})}{2 \cdot 2} = \frac{2}{2} \cdot \frac{-1 \pm \sqrt{71}}{2} = \frac{-1 \pm \sqrt{71}}{2}.$$

Using a calculator, we get approximations:
$$\frac{-1 + \sqrt{71}}{2} \approx 3.7 \quad or \quad \frac{-1 - \sqrt{71}}{2} \approx -4.7.$$

4. **Check.** Since the length of a leg cannot be negative, -4.7 does not check. But 3.7 does check. If the smaller leg s is 3.7, the other leg is $s + 1$, or 4.7. Then
$$(3.7)^2 + (4.7)^2 = 13.69 + 22.09 = 35.78.$$

Using a calculator, we get $\sqrt{35.78} \approx 5.98 \approx 6$, the length of the hypotenuse. Note that our check is not exact because we are using an approximation for $\sqrt{71}$.

5. **State.** One leg is about 3.7 m long, and the other is about 4.7 m long.

◀ **Do Exercise 2.**

2. *Construction.* Gil and Hal dug a trench across the diagonal of their rectangular backyard in order to install a drainage pipe. If the pipe is 20 m long and the yard is 5 m longer than it is wide, find the dimensions of the yard. Round to the nearest tenth of a meter.

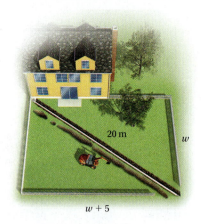

EXAMPLE 3 *Kayak Speed.* The current in a stream moves at a speed of 2 km/h. A kayak travels 24 km upstream and 24 km downstream in a total time of 5 hr. What is the speed of the kayak in still water?

1. **Familiarize.** We first make a drawing. The distances are the same. We let $r =$ the speed of the kayak in still water. Then when the kayak is traveling upstream, its speed is $r - 2$. When it is traveling downstream, its speed is $r + 2$. We let t_1 represent the time it takes the kayak to go upstream and t_2 the time it takes to go downstream. We summarize in a table.

Answer

2. 11.4 m by 16.4 m

Upstream, $r - 2$
t_1 hours, 24 km

Downstream, $r + 2$
t_2 hours, 24 km

	d	r	t
UPSTREAM	24	$r - 2$	t_1
DOWNSTREAM	24	$r + 2$	t_2
TOTAL TIME			5

$$\rightarrow t_1 = \frac{24}{r - 2}$$
$$\rightarrow t_2 = \frac{24}{r + 2}$$

2. Translate. Recall the basic formula for motion: $d = rt$. From it we can obtain an equation for time: $t = d/r$. Total time consists of the time to go upstream, t_1, plus the time to go downstream, t_2. Using $t = d/r$ and the rows of the table, we have

$$t_1 = \frac{24}{r - 2} \quad \text{and} \quad t_2 = \frac{24}{r + 2}.$$

Since the total time is 5 hr, $t_1 + t_2 = 5$, and we have

$$\frac{24}{r - 2} + \frac{24}{r + 2} = 5.$$ We have translated to an equation with one variable.

3. Solve. We solve the equation. We multiply on both sides by the LCM, which is $(r - 2)(r + 2)$:

$$(r - 2)(r + 2) \cdot \left(\frac{24}{r - 2} + \frac{24}{r + 2} \right) = (r - 2)(r + 2)5$$

$$(r - 2)(r + 2) \cdot \frac{24}{r - 2} + (r - 2)(r + 2) \cdot \frac{24}{r + 2} = (r^2 - 4)5$$

$$24(r + 2) + 24(r - 2) = 5r^2 - 20$$

$$24r + 48 + 24r - 48 = 5r^2 - 20$$

$$-5r^2 + 48r + 20 = 0$$

$$5r^2 - 48r - 20 = 0 \quad \text{Multiplying by −1}$$

$$(5r + 2)(r - 10) = 0 \quad \text{Factoring}$$

$$5r + 2 = 0 \quad or \quad r - 10 = 0$$
Using the principle of zero products

$$5r = -2 \quad or \quad r = 10$$

$$r = -\tfrac{2}{5} \quad or \quad r = 10.$$

4. Check. Since speed cannot be negative, $-\frac{2}{5}$ cannot be a solution. But suppose the speed of the kayak in still water is 10 km/h. The speed upstream is then $10 - 2$, or 8 km/h. The speed downstream is $10 + 2$, or 12 km/h. The time upstream, using $t = d/r$, is $24/8$, or 3 hr. The time downstream is $24/12$, or 2 hr. The total time is 5 hr. This checks.

5. State. The speed of the kayak in still water is 10 km/h.

Do Exercise 3. ▶

GS **3.** *Speed of a Stream.* The speed of a boat in still water is 12 km/h. The boat travels 45 km upstream and 45 km downstream in a total time of 8 hr. What is the speed of the stream? (*Hint*: Let $s =$ the speed of the stream. Then $12 - s$ is the speed upstream and $12 + s$ is the speed downstream.)

1. Familiarize.

d	r	t
45	$12 - s$	t_1
☐	$12 + s$	t_2
		☐

2. Translate.

$$t_1 = \frac{45}{12 - s}, \; t_2 = \frac{45}{\boxed{}};$$

$$\frac{45}{12 - s} + \frac{45}{12 + s} = \boxed{}$$

3. Solve.

$$(12 - s)(12 + s)\left(\frac{45}{12 - s} + \frac{45}{12 + s} \right)$$
$$= (12 - s)(12 + s)(8)$$

$$45(12 + s) + 45(12 - s)$$
$$= (144 - s^2)(8)$$

$$1080 = 1152 - \boxed{}$$

$$8s^2 - 72 = 0$$

$$s^2 - \boxed{} = 0$$

$$s + 3 = 0 \quad or \quad s - 3 = 0$$

$$s = -3 \quad or \quad s = \boxed{}$$

4. Check. The speed of the stream cannot be negative. A speed of 3 km/h checks.

5. State. The speed of the stream is ☐ km/h.

Answer

3. 3 km/h

Guided Solution:
1. 45, 8, $12 + s$, 8, $8s^2$, 9, 3, 3

Translating for Success

1. **Guy Wire.** How long is a guy wire that reaches from the top of a 75-ft cell-phone tower to a point on the ground 21 ft from the pole?

2. **Coin Mixture.** A collection of dimes and quarters is worth $16.95. There are 90 coins in all. How many of each coin are there?

3. **Wire Cutting.** A 486-in. wire is cut into three pieces. The second piece is 5 in. longer than the first. The third is one-half as long as the first. How long is each piece?

4. **Amount Invested.** Money is invested at 3.2% simple interest. At the end of 1 year, there is $27,864 in the account. How much was originally invested?

5. **Foreign Languages.** Last year, 3.2% of the 27,864 students at East End Community College took a foreign language course. How many students took a foreign language course?

The goal of these matching questions is to practice step (2), Translate, of the five-step problem-solving process. Translate each word problem to an equation or a system of equations and select a correct translation from A–O.

A. $x^2 + (x - 1)^2 = 7$

B. $\dfrac{600}{x} = \dfrac{600}{x + 2} + 10$

C. $13{,}932 = x \cdot 27{,}864$

D. $x = 3.2\% \cdot 27{,}864$

E. $2x + 2(x - 1) = 49$

F. $x + (x + 5) + \dfrac{1}{2}x = 486$

G. $0.10x + 0.25y = 16.95,$
$x + y = 90$

H. $x + 25y = 16.95,$
$x + y = 90$

I. $3.2x = 27{,}864 - x$

J. $x^2 + (x - 1)^2 = 49$

K. $x^2 + 21^2 = 75^2$

L. $x + 3.2\%x = 27{,}864$

M. $75^2 + 21^2 = x^2$

N. $x + (x + 1)$
$+ (x + 2) = 894$

O. $\dfrac{600}{x} + \dfrac{600}{x - 2} = 10$

Answers on page A-29

---

6. **Locker Numbers.** The numbers on three adjoining lockers are consecutive integers whose sum is 894. Find the integers.

7. **Triangle Dimensions.** The hypotenuse of a right triangle is 7 ft. The length of one leg is 1 ft shorter than the other. Find the lengths of the legs.

8. **Rectangle Dimensions.** The perimeter of a rectangle is 49 ft. The length is 1 ft shorter than the width. Find the length and the width.

9. **Car Travel.** Maggie drove her car 600 mi to see her friend. The return trip was 2 hr faster at a speed that was 10 mph greater. Find the time for the return trip.

10. **Literature.** Last year, 13,932 of the 27,864 students at East End Community College took a literature course. What percent of the students took a literature course?

✓ Reading Check

Determine whether each statement is true or false.

RC1. To find the area of a rectangle, multiply the length of the rectangle by the width.

RC2. The Pythagorean equation is true for all triangles.

RC3. Lengths of sides of rectangles and triangles are positive numbers.

RC4. The speed of a boat moving upstream is the speed of the boat in still water plus the speed of the current in the stream.

a   Solve.

1. *Pool Dimensions.*   The area of a rectangular swimming pool is 80 yd². The length is 1 yd longer than three times the width. Find the dimensions of the swimming pool.

2. The length of a rectangular area rug is 3 ft greater than the width. The area is 70 ft². Find the length and the width.

$w + 3$

w

Area = 70 ft²

$w + 3$

w

3. *Carpenter's Square.*   A *square* is a carpenter's tool in the shape of a right triangle. One side, or leg, of a square is 8 in. longer than the other. The length of the hypotenuse is $8\sqrt{13}$ in. Find the lengths of the legs of the square.

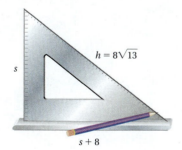

$h = 8\sqrt{13}$

s

$s + 8$

4. *HDTV Dimensions.*   When we say that a television is 42 in., we mean that the diagonal is 42 in. For a 42-in. television, the width is 15 in. more than the height. Find the dimensions of a 42-in. high-definition television. Round to the nearest inch.

$h + 15$

h

42 in.

5. Rectangle Dimensions. The length of a rectangular garden is three times the width. The area is 300 ft². Find the length and the width of the garden.

6. Rectangle Dimensions. The length of a rectangular lobby in a hotel is twice the width. The area is 50 m². Find the length and the width of the lobby.

7. Rectangle Dimensions. The width of a rectangle is 4 cm less than the length. The area is 320 cm². Find the length and the width.

8. Rectangle Dimensions. The width of a rectangle is 3 cm less than the length. The area is 340 cm². Find the length and the width.

Find the approximate answers for Exercises 9–14. Round to the nearest tenth.

9. Right-Triangle Dimensions. The hypotenuse of a right triangle is 8 m long. One leg is 2 m longer than the other. Find the lengths of the legs.

10. Right-Triangle Dimensions. The hypotenuse of a right triangle is 5 cm long. One leg is 2 cm longer than the other. Find the lengths of the legs.

11. Rectangle Dimensions. The length of a rectangle is 2 in. greater than the width. The area is 20 in². Find the length and the width.

12. Rectangle Dimensions. The length of a rectangle is 3 ft greater than the width. The area is 15 ft². Find the length and the width.

13. Rectangle Dimensions. The length of a rectangle is twice the width. The area is 20 cm². Find the length and the width.

14. Rectangle Dimensions. The length of a rectangle is twice the width. The area is 10 m². Find the length and the width.

15. Picture Frame. A picture frame measures 25 cm by 20 cm. There is 266 cm² of picture showing. The frame is of uniform width. Find the width of the frame.

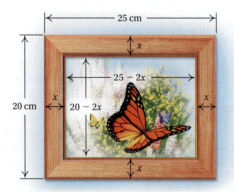

16. Tablecloth. A rectangular tablecloth measures 96 in. by 72 in. It is laid on a tabletop with an area of 5040 in², and hangs over the edge by the same amount on all sides. By how many inches does the cloth hang over the edge?

For Exercises 17–24, complete the table to help with the familiarization.

17. *Boat Speed.* The current in a stream moves at a speed of 3 km/h. A boat travels 40 km upstream and 40 km downstream in a total time of 14 hr. What is the speed of the boat in still water?

	d	r	t
UPSTREAM		$r - 3$	t_1
DOWNSTREAM	40		t_2
TOTAL TIME			

Upstream, $r - 3$
t_1 hours, 40 km

Downstream, $r + 3$
t_2 hours, 40 km

18. *Wind Speed.* An airplane flies 1449 mi against the wind and 1539 mi with the wind in a total time of 5 hr. The speed of the airplane in still air is 600 mph. What is the speed of the wind?

	d	r	t
WITH WIND	1539		
AGAINST WIND		$600 - r$	
TOTAL TIME			5

19. *Speed of a Stream.* The speed of a boat in still water is 8 km/h. The boat travels 60 km upstream and 60 km downstream in a total time of 16 hr. What is the speed of the stream?

	d	r	t
UPSTREAM			
DOWNSTREAM			
TOTAL TIME			

20. *Boat Speed.* The current in a stream moves at a speed of 4 mph. A boat travels 5 mi upstream and 13 mi downstream in a total time of 2 hr. What is the speed of the boat in still water?

	d	r	t
UPSTREAM		$r - 4$	t_1
DOWNSTREAM	13		t_2
TOTAL TIME			

21. *Wind Speed.* An airplane flies 520 km against the wind and 680 km with the wind in a total time of 4 hr. The speed of the airplane in still air is 300 km/h. What is the speed of the wind?

	d	r	t
WITH WIND		$300 + r$	
AGAINST WIND	520		
TOTAL TIME			4

22. *Speed of a Stream.* The speed of a boat in still water is 10 km/h. The boat travels 12 km upstream and 28 km downstream in a total time of 4 hr. What is the speed of the stream?

	d	r	t
UPSTREAM			
DOWNSTREAM			
TOTAL TIME			

23. Boat Speed. The current in a stream moves at a speed of 4 mph. A boat travels 4 mi upstream and 12 mi downstream in a total time of 2 hr. What is the speed of the boat in still water?

	d	r	t
UPSTREAM			
DOWNSTREAM			
TOTAL TIME			

24. Boat Speed. The current in a stream moves at a speed of 3 mph. A boat travels 45 mi upstream and 45 mi downstream in a total time of 8 hr. What is the speed of the boat in still water?

	d	r	t
UPSTREAM			
DOWNSTREAM			
TOTAL TIME			

25. Speed of a Stream. The speed of a boat in still water is 9 km/h. The boat travels 80 km upstream and 80 km downstream in a total time of 18 hr. What is the speed of the stream?

26. Speed of a Stream. The speed of a boat in still water is 10 km/h. The boat travels 48 km upstream and 48 km downstream in a total time of 10 hr. What is the speed of the stream?

Skill Maintenance

Find the coordinates of the *y*-intercept and of the *x*-intercept. Do not graph.   [3.2a]

27. $8x = 4 - y$

28. $5y - 3x = -45$

Graph.

29. $y = 3x - 5$ [3.5a]

30. $x = -2$ [3.2b]

31. $y = 1$ [3.2b]

32. $2x - y = 2$ [3.2a]

Find an equation of the line with the given slope and *y*-intercept.   [3.4a]

33. Slope: -2;
   y-intercept: $(0, -5)$

34. Slope: $\frac{1}{2}$;
   y-intercept: $(0, 1)$

Determine whether each pair of equations represents parallel lines.   [3.6a]

35. $y = \frac{3}{4}x - 7$,
   $3x + 4y = 7$

36. $y = \frac{3}{5}$,
   $y = -\frac{5}{3}$

Synthesis

37. Pizza. What should the diameter *d* of a pizza be so that it has the same area as two 12-in. pizzas? Which provides more servings: a 16-in. pizza or two 12-in. pizzas?

38. ▦ **Golden Rectangle.** The *golden rectangle* is said to be extremely pleasing visually and was used often by ancient Greek and Roman architects. The length of a golden rectangle is approximately 1.6 times the width. Find the dimensions of a golden rectangle if its area is 9000 m².

Graphs of Quadratic Equations

In this section, we will graph equations of the form

$$y = ax^2 + bx + c, \quad a \neq 0.$$

The polynomial on the right side of the equation is of second degree, or **quadratic**. Examples of the types of equations we are going to graph are

$$y = x^2, \quad y = x^2 + 2x - 3, \quad y = -2x^2 + 3.$$

a GRAPHING QUADRATIC EQUATIONS OF THE TYPE $y = ax^2 + bx + c$

Graphs of quadratic equations of the type $y = ax^2 + bx + c$ (where $a \neq 0$) are always cup-shaped. They have a **line of symmetry** like the dashed lines shown in the figures below. If we fold on this line, the two halves will match exactly. The curve goes on forever. The highest or lowest point on the curve is called the **vertex**. The second coordinate of the vertex is either the smallest value of y or the largest value of y. The vertex is also thought of as a turning point. Graphs of quadratic equations are called **parabolas**.

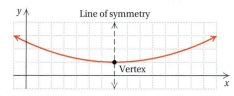

To graph a quadratic equation, we begin by choosing some numbers for x and computing the corresponding values of y.

EXAMPLE 1   Graph: $y = x^2$.

We choose numbers for x and find the corresponding values for y. Then we plot the ordered pairs (x, y) resulting from the computations and connect them with a smooth curve.

For $x = -3, y = x^2 = (-3)^2 = 9$.
For $x = -2, y = x^2 = (-2)^2 = 4$.
For $x = -1, y = x^2 = (-1)^2 = 1$.
For $x = 0, y = x^2 = (0)^2 = 0$.
For $x = 1, y = x^2 = (1)^2 = 1$.
For $x = 2, y = x^2 = (2)^2 = 4$.
For $x = 3, y = x^2 = (3)^2 = 9$.

x	y	(x, y)
-3	9	$(-3, 9)$
-2	4	$(-2, 4)$
-1	1	$(-1, 1)$
0	0	$(0, 0)$
1	1	$(1, 1)$
2	4	$(2, 4)$
3	9	$(3, 9)$

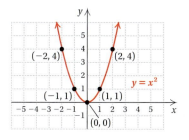

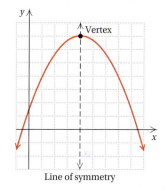

Answers

Skill to Review:

1.
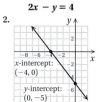
x-intercept:
$(2, 0)$
y-intercept:
$(0, -4)$

$2x - y = 4$

2.
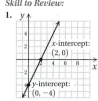
x-intercept:
$(-4, 0)$
y-intercept:
$(0, -5)$

$4y + 20 = -5x$

In Example 1, the vertex is the point $(0, 0)$. The second coordinate of the vertex, 0, is the smallest y-value. The y-axis $(x = 0)$ is the line of symmetry.

Parabolas whose equations are $y = ax^2$ always have the origin $(0, 0)$ as the vertex and the y-axis as the line of symmetry.

A key to graphing a parabola is knowing the vertex. By graphing it and then choosing x-values on both sides of the vertex, we can compute more points and complete the graph.

FINDING THE VERTEX

For a parabola given by the quadratic equation $y = ax^2 + bx + c$:

1. The x-coordinate of the vertex is $-\dfrac{b}{2a}$.

   The line of symmetry is $x = -\dfrac{b}{2a}$.

2. The second coordinate of the vertex is found by substituting the x-coordinate into the equation and computing y.

The proof that the vertex can be found in this way can be shown by completing the square in a manner similar to the proof of the quadratic formula, but it will not be considered here.

EXAMPLE 2 Graph: $y = -2x^2 + 3$.

We first find the vertex. The x-coordinate of the vertex is

$$-\frac{b}{2a} = -\frac{0}{2(-2)} = 0.$$

We next find the second coordinate of the vertex:

$$y = -2x^2 + 3 = -2(0)^2 + 3 = 3. \qquad \text{Substituting 0 for } x$$

The vertex is $(0, 3)$. The line of symmetry is the y-axis $(x = 0)$. We choose some x-values on both sides of the vertex and graph the parabola.

For $x = 1$, $y = -2x^2 + 3 = -2(1)^2 + 3 = -2 + 3 = 1$.
For $x = -1$, $y = -2x^2 + 3 = -2(-1)^2 + 3 = -2 + 3 = 1$.
For $x = 2$, $y = -2x^2 + 3 = -2(2)^2 + 3 = -8 + 3 = -5$.
For $x = -2$, $y = -2x^2 + 3 = -2(-2)^2 + 3 = -8 + 3 = -5$.

x	y
0	3
1	1
−1	1
2	−5
−2	−5

← This is the vertex.

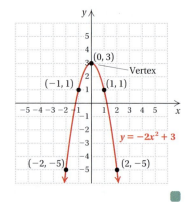

There are two other tips you might use when graphing quadratic equations. The first involves the coefficient of x^2. The a in $y = ax^2 + bx + c$ tells us whether the graph opens up or down. When a is positive, as in Example 1, the graph opens up; when a is negative, as in Example 2, the graph opens down. It is also helpful to plot the y-intercept. It occurs when $x = 0$.

TIPS FOR GRAPHING QUADRATIC EQUATIONS

1. Graphs of quadratic equations $y = ax^2 + bx + c$ are all parabolas. They are *smooth* cup-shaped symmetric curves, with no sharp points or kinks in them.

2. Find the vertex and the line of symmetry.

3. The graph of $y = ax^2 + bx + c$ opens up if $a > 0$. It opens down if $a < 0$.

4. Find the y-intercept. It occurs when $x = 0$, and it is easy to compute.

EXAMPLE 3 Graph: $y = x^2 + 2x - 3$.

We first find the vertex. The x-coordinate of the vertex is

$$-\frac{b}{2a} = -\frac{2}{2(1)} = -1.$$

We substitute -1 for x into the equation to find the second coordinate of the vertex:

$$
\begin{aligned}
y &= x^2 + 2x - 3 \\
&= (-1)^2 + 2(-1) - 3 \\
&= 1 - 2 - 3 \\
&= -4.
\end{aligned}
$$

The vertex is $(-1, -4)$. The line of symmetry is $x = -1$.

We choose some x-values on both sides of $x = -1$—say, $-2, -3, -4$ and $0, 1, 2$—and graph the parabola. Since the coefficient of x^2 is 1, which is positive, we know that the graph opens up. Be sure to find y when $x = 0$. This gives the y-intercept.

x	y	
-1	-4	← Vertex
0	-3	← y-intercept
-2	-3	
1	0	
-3	0	
2	5	
-4	5	

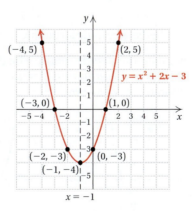

Do Exercises 1–3. ▶

Graph. Label the ordered pairs for the vertex and the y-intercept.

1. $y = x^2 - 3$

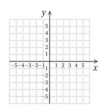

2. $y = -3x^2 + 6x$

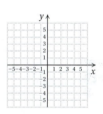

3. $y = x^2 - 4x + 4$

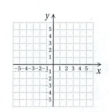

Answers

1.

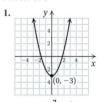

$y = x^2 - 3$

2.

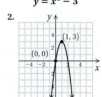

$y = -3x^2 + 6x$

3.

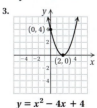

$y = x^2 - 4x + 4$

The *x*-intercepts of the graph of
$y = ax^2 + bx + c$ occur at those
values of *x* for which $y = 0$. Thus the
first coordinates of the *x*-intercepts
are solutions of the equation

$$0 = ax^2 + bx + c.$$

We have been studying how to find
such numbers in Sections 9.1–9.3.

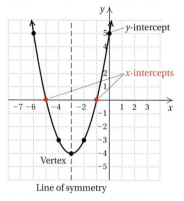

EXAMPLE 4 Find the *x*-intercepts of the graph of $y = x^2 - 4x + 1$.

We solve the equation $x^2 - 4x + 1 = 0$ using the quadratic formula.

$$a = 1, \quad b = -4, \quad c = 1$$

$$x = \frac{-b \pm \sqrt{b^2 - 4ac}}{2a}$$

$$= \frac{-(-4) \pm \sqrt{(-4)^2 - 4(1)(1)}}{2(1)}$$

$$= \frac{4 \pm \sqrt{16 - 4}}{2}$$

$$= \frac{4 \pm \sqrt{12}}{2} = \frac{4 \pm \sqrt{4 \cdot 3}}{2}$$

$$= \frac{4 \pm 2\sqrt{3}}{2} = \frac{2 \cdot 2 \pm 2\sqrt{3}}{2 \cdot 1}$$

$$= \frac{2}{2} \cdot \frac{2 \pm \sqrt{3}}{1} = 2 \pm \sqrt{3}$$

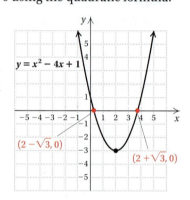

The *x*-intercepts are $(2 - \sqrt{3}, 0)$ and $(2 + \sqrt{3}, 0)$.

In the quadratic formula $x = \dfrac{-b \pm \sqrt{b^2 - 4ac}}{2a}$, the radicand $b^2 - 4ac$
is called the **discriminant**. The discriminant tells how many real-number
solutions the equation $0 = ax^2 + bx + c$ has, so it also tells how many
x-intercepts there are.

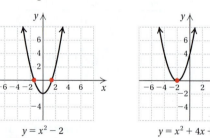

$y = x^2 - 2$
$b^2 - 4ac = 8 > 0$
Two real solutions
Two *x*-intercepts

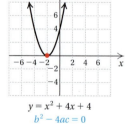

$y = x^2 + 4x + 4$
$b^2 - 4ac = 0$
One real solution
One *x*-intercept

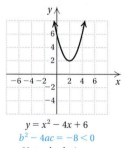

$y = x^2 - 4x + 6$
$b^2 - 4ac = -8 < 0$
No real solutions
No *x*-intercepts

◀ **Do Exercises 4–7.**

Find the *x*-intercepts.

4. $y = x^2 - 3$

5. $y = x^2 + 6x + 8$ **GS**
$$x^2 + 6x + 8 = 0$$
$$(x + 4)(\boxed{}) = 0$$
$$x + 4 = 0 \quad or \quad \boxed{} = 0$$
$$x = -4 \quad or \quad x = \boxed{}$$
The *x*-intercepts are $(-4, 0)$
and $(\boxed{}, 0)$.

6. $y = -2x^2 - 4x + 1$

7. $y = x^2 + 3$ **GS**
$$x^2 + 3 = 0$$
$$x^2 = \boxed{}$$
Since -3 is negative, the
equation has no real-number
solutions. There are no
x-intercepts.

Answers

4. $(-\sqrt{3}, 0); (\sqrt{3}, 0)$ **5.** $(-4, 0); (-2, 0)$
6. $\left(\dfrac{-2 - \sqrt{6}}{2}, 0\right); \left(\dfrac{-2 + \sqrt{6}}{2}, 0\right)$ **7.** None

Guided Solutions:
5. $x + 2, x + 2, -2, -2$ **7.** -3

Visualizing for Success

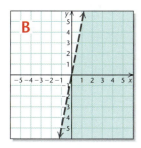

A

B

C

D

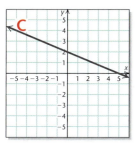

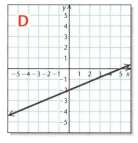

E

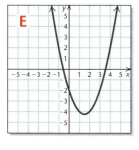

Match each equation or inequality with its graph.

1. $y = -4 + 4x - x^2$

2. $y = 5 - x^2$

3. $5x + 2y = -10$

4. $5x + 2y \leq 10$

5. $y < 5x$

6. $y = x^2 - 3x - 2$

7. $2x - 5y = 10$

8. $5x - 2y = 10$

9. $2x + 5y = 10$

10. $y = x^2 + 3x - 2$

Answers on page A-29

F

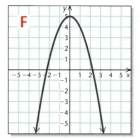

G

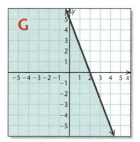

H

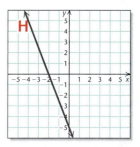

I

J

✓ **Reading Check**

Choose from the column on the right the word that best completes each statement.

RC1. The equation $x^2 - 9x + 8 = 0$ is an example of a(n) _____ equation.

RC2. The graph of $y = x^2 - 9x + 8$ is a(n) _____.

RC3. The turning point of the graph of $y = x^2 - 9x + 8$ is the _____.

RC4. The graph of $y = x^2 - 9x + 8$ could be folded in half along its _____ of symmetry.

line

parabola

quadratic

vertex

a Graph the quadratic equation. In Exercises 1–8, label the ordered pairs for the vertex and the y-intercept.

1. $y = x^2 + 1$

x	y
-2	
-1	
0	
1	
2	
3	

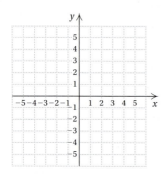

2. $y = 2x^2$

x	y
-2	
-1	
0	
1	
2	
3	

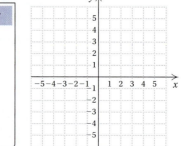

3. $y = -1 \cdot x^2$

x	y

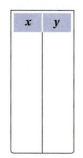

4. $y = x^2 - 1$

x	y

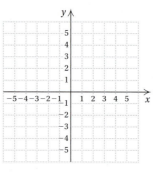

5. $y = -x^2 + 2x$

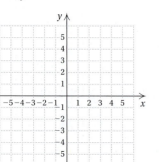

6. $y = x^2 + x - 2$

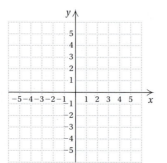

7. $y = 5 - x - x^2$

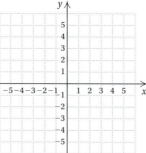

8. $y = x^2 + 2x + 1$

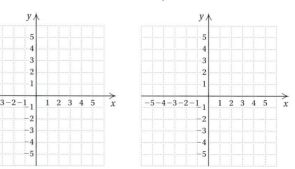

9. $y = x^2 - 2x + 1$

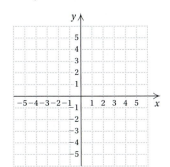

10. $y = -\frac{1}{2}x^2$

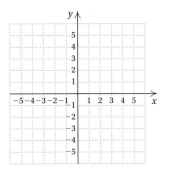

11. $y = -x^2 + 2x + 3$

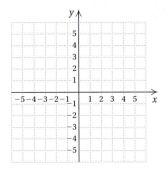

12. $y = -x^2 - 2x + 3$

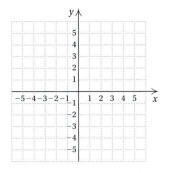

13. $y = -2x^2 - 4x + 1$

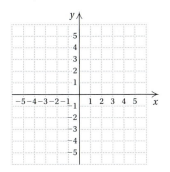

14. $y = 2x^2 + 4x - 1$

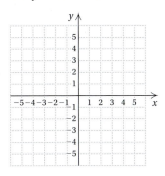

15. $y = 5 - x^2$

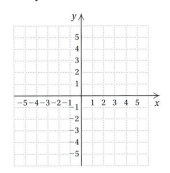

16. $y = 4 - x^2$

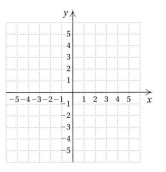

17. $y = \frac{1}{4}x^2$

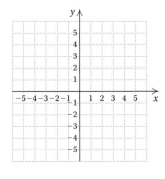

18. $y = -0.1x^2$

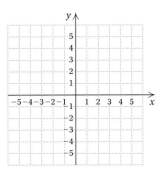

19. $y = -x^2 + x - 1$

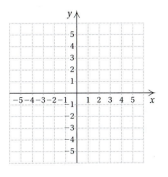

20. $y = x^2 + 2x$

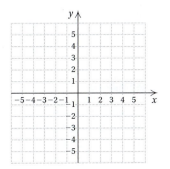

21. $y = -2x^2$

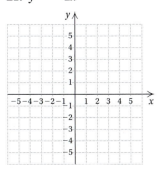

22. $y = -x^2 - 1$

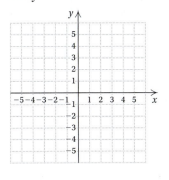

23. $y = x^2 - x - 6$

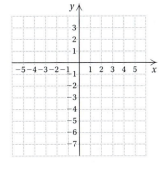

24. $y = 6 + x - x^2$

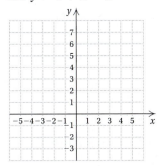

b Find the *x*-intercepts.

25. $y = x^2 - 2$

26. $y = x^2 - 7$

27. $y = x^2 + 5x$

28. $y = x^2 - 4x$

29. $y = 8 - x - x^2$

30. $y = 8 + x - x^2$

31. $y = x^2 - 6x + 9$

32. $y = x^2 + 10x + 25$

33. $y = -x^2 - 4x + 1$

34. $y = x^2 + 4x - 1$

35. $y = x^2 + 9$

36. $y = x^2 + 1$

Skill Maintenance

37. Add: $\sqrt{8} + \sqrt{50} + \sqrt{98} + \sqrt{128}$.   [8.4a]

38. Multiply and simplify: $\sqrt{5y^4}\sqrt{125y}$.   [8.2c]

39. Find an equation of variation in which *y* varies inversely as *x*, and $y = 12.4$ when $x = 2.4$.   [6.9c]

40. Evaluate $3x^4 + 3x - 7$ when $x = -2$.   [4.3a]

Simplify.   [1.4a]

41. $-\dfrac{1}{5} + \dfrac{7}{10} - \left(-\dfrac{4}{15}\right) + \dfrac{1}{60}$

42. $-0.63 - 3.4 + 11.08 - (-42.5)$

Synthesis

43. *Height of a Projectile.*   The height *H*, in feet, of a projectile with an initial velocity of 96 ft/sec is given by the equation

$$H = -16t^2 + 96t,$$

where *t* is the time, in seconds. Use the graph of this equation, shown here, or any equation-solving technique to answer the following questions.

a) How many seconds after launch is the projectile 128 ft above ground?

b) When does the projectile reach its maximum height?

c) How many seconds after launch does the projectile return to the ground?

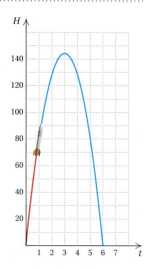

For each equation in Exercises 44–47, evaluate the discriminant $b^2 - 4ac$. Then use the answer to state how many real-number solutions exist for the equation.

44. $y = x^2 + 8x + 16$

45. $y = x^2 + 2x - 3$

46. $y = -2x^2 + 4x - 3$

47. $y = -0.02x^2 + 4.7x - 2300$

Functions

a IDENTIFYING FUNCTIONS

We now develop one of the most important concepts in mathematics: **functions**. We have actually been studying functions all through this text; we just haven't identified them as such. Ordered pairs form a correspondence between first and second coordinates. A function is a special correspondence from one set to another. For example:

To each student in a college, there corresponds his or her student ID number.

To each item in a store, there corresponds its price.

To each real number, there corresponds the cube of that number.

In each case, the first set is called the **domain** and the second set is called the **range**. Given a member of the domain, there is *just one* member of the range to which it corresponds. This kind of correspondence is called a **function**.

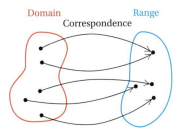

EXAMPLE 1 Determine whether the correspondence is a function.

Domain	Range
1	→ $107.40
f: 2	→ $ 34.10
3	→ $ 29.60
4	→ $ 19.60

Domain	Range
3	→ 5
g: 4	→ 9
5	
6	→ −7

Domain	Range
New York	→ Mets
h:	→ Yankees
St. Louis	→ Cardinals
San Diego	→ Padres

Domain	Range
Mets	→ New York
p: Yankees	
Cardinals	→ St. Louis
Padres	→ San Diego

The correspondence *f* is a function because each member of the domain is matched to only one member of the range.

The correspondence *g* is also a function because each member of the domain is matched to only one member of the range.

The correspondence *h* is not a function because one member of the domain, New York, is matched to more than one member of the range.

The correspondence *p* is a function because each member of the domain is paired with only one member of the range. Note that a function can pair a number of the range with more than one member of the domain.

OBJECTIVES

a Determine whether a correspondence is a function.

b Given a function described by an equation, find function values (outputs) for specified values (inputs).

c Draw a graph of a function.

d Determine whether a graph is that of a function.

e Solve applied problems involving functions and their graphs.

SKILL TO REVIEW

Objective 4.3a: Evaluate a polynomial for a given value of the variable.

Evaluate each polynomial for the indicated value.

1. $10 - \frac{1}{8}x$, when $x = 16$

2. $3x - 5 + x^2$, when $x = -3$

Answers

Skill to Review:
1. 8 **2.** −5

Determine whether each correspondence is a function.

1. *Domain*   *Range*

2. *Domain*   *Range*

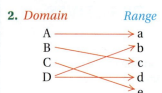

3. *Domain*   *Range*

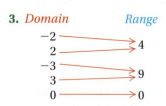

4. *Domain*   *Range*

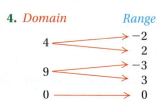

Determine whether each of the following is a function.

5. *Domain*
A set of numbers

Correspondence
10 less than the square of each number

Range
A set of numbers

6. *Domain*
A set of polygons

Correspondence
The perimeter of each polygon

Range
A set of numbers

◀ **Do Exercises 1–4.**

EXAMPLE 2 Determine whether each correspondence is a function.

Domain	*Correspondence*	*Range*
a) A family	Each person's weight	A set of positive numbers
b) The natural numbers	Each number's square	A set of natural numbers
c) The set of all states	Each state's members of the U.S. Senate	A set of U.S. Senators

a) The correspondence *is* a function because each person has *only one* weight.

b) The correspondence *is* a function because each natural number has *only one* square.

c) The correspondence *is not* a function because each state has two U.S. Senators.

◀ **Do Exercises 5 and 6.**

When a correspondence between two sets is not a function, it may still be an example of a *relation*.

Thus, although the correspondences of Examples 1 and 2 are not all functions, they *are* all relations. A function is a special type of relation—one in which each member of the domain is paired with *exactly one* member of the range.

b   FINDING FUNCTION VALUES

Most functions considered in mathematics are described by equations. A function described by a linear equation like $y = 2x + 3$ is called a **linear function**. A function described by a quadratic equation like $y = 4 - x^2$ is called a **quadratic function**.

Answers

1. Yes   **2.** No   **3.** Yes   **4.** No   **5.** Yes
6. Yes

Recall that when graphing $y = 2x + 3$, we chose x-values and then found corresponding y-values. For example, when $x = 4$,

$$y = 2x + 3 = 2 \cdot 4 + 3 = 11.$$

When thinking of functions, we call the number 4 an **input** and the number 11 an **output**.

The function $y = 2x + 3$ can be named f and described by the equation $f(x) = 2x + 3$. We call the input x and the output $f(x)$. This is read "f of x," or "f at x," or "the value of f at x."

Caution!

The notation $f(x)$ *does not mean "f times x"* and should not be read that way.

It helps to think of a function as a machine; that is, think of putting a member of the domain (an input) into the machine. The machine knows the correspondence and produces a member of the range (the output).

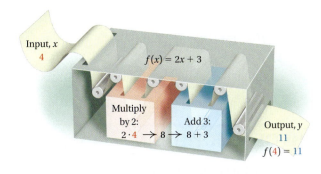

The equation $f(x) = 2x + 3$ describes the function that takes an input x, multiplies it by 2, and then adds 3.

Input

$$f(x) = 2x \quad + 3$$

Multiply by 2   Add 3

To find the output $f(4)$, we take the input 4, double it, and add 3 to get 11. That is, we substitute 4 into the formula for $f(x)$:

$$f(4) = 2 \cdot 4 + 3 = 11.$$

Outputs of functions are also called **function values**. For $f(x) = 2x + 3$, we know that $f(4) = 11$. We can say that "the function value at 4 is 11."

EXAMPLE 3   Find the indicated function value.

a) $f(5)$, for $f(x) = 3x + 2$ **b)** $g(3)$, for $g(z) = 5z^2 - 4$
c) $A(-2)$, for $A(r) = 3r^2 + 2r$ **d)** $f(-5)$, for $f(x) = x^2 + 3x - 4$

a) $f(5) = 3 \cdot 5 + 2 = 15 + 2 = 17$
b) $g(3) = 5(3)^2 - 4 = 5(9) - 4 = 45 - 4 = 41$
c) $A(-2) = 3(-2)^2 + 2(-2) = 3(4) - 4 = 12 - 4 = 8$
d) $f(-5) = (-5)^2 + 3(-5) - 4 = 25 - 15 - 4 = 6$

Do Exercises 7–13. ▶

Find the indicated function value.

7. $f(1)$, for $f(x) = 5x - 3$

8. $g(-4)$, for $g(x) = \dfrac{1}{2}x + 7$

GS **9.** $p(0)$, for $p(x) = x^4 - 5x^2 + 8$

$$p(0) = \boxed{}^4 - 5 \cdot \boxed{}^2 + 8$$
$$= 0 - 0 + \boxed{}$$
$$= \boxed{}$$

10. $h\left(\dfrac{1}{2}\right)$, for $h(x) = 10x$

11. $f(-3)$, for $f(t) = t^2 - t + 1$

12. $g(-94)$, for $g(z) = |z|$

GS **13.** $F(100)$, for $F(r) = \sqrt{r} + 9$

$$F(100) = \sqrt{\boxed{}} + 9$$
$$= \boxed{} + 9$$
$$= \boxed{}$$

Answers

7. 2   **8.** 5   **9.** 8   **10.** 5   **11.** 13
12. 94   **13.** 19

Guided Solutions:
9. 0, 0, 8, 8   **13.** 100, 10, 19

Finding Function Values We can find function values on a graphing calculator by substituting inputs directly into the formula. After we have entered a function on the equation-editor screen, there are several other methods that we can use to find function values.

Consider the function in Example 3(d), $f(x) = x^2 + 3x - 4$. We enter $y_1 = x^2 + 3x - 4$ and then use a table set in **ASK** mode and enter $x = -5$. We see that the function value, y_1, is 6. We can also use the **VALUE** feature to evaluate the function. To do this, we first graph the function in a window that includes $x = -5$ and then press **2ND** **CALC** to access the **VALUE** feature in order to find the value of y when $x = -5$. A third method uses function notation. We choose Y_1 from the **VARS Y-VARS** function menu, enclose the input in parentheses, and press **ENTER**. All three methods indicate that $f(-5) = 6$.

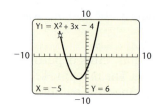
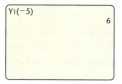

EXERCISES: Find each function value.

1. $f(-3.4)$, for $f(x) = 2x - 6$

2. $f(4)$, for $f(x) = -2.3x$

3. $f(-1)$, for $f(x) = x^2 - 3$

4. $f(3)$, for $f(x) = 2x^2 - x + 5$

c GRAPHS OF FUNCTIONS

To graph a function, we find ordered pairs (x, y) or $(x, f(x))$, plot them, and connect the points. Note that y and $f(x)$ are used interchangeably when we are working with functions and their graphs.

EXAMPLE 4 Graph: $f(x) = x + 2$.

A list of some function values is shown in this table. We plot the points and connect them. The graph is a straight line.

x	$f(x)$
-4	-2
-3	-1
-2	0
-1	1
0	2
1	3
2	4
3	5
4	6

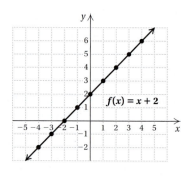

EXAMPLE 5 Graph: $g(x) = 4 - x^2$.

Recall from Section 9.6 that the graph is a parabola. We calculate some function values and draw the curve.

$$g(0) = 4 - 0^2 = 4 - 0 = 4,$$
$$g(-1) = 4 - (-1)^2 = 4 - 1 = 3,$$
$$g(2) = 4 - (2)^2 = 4 - 4 = 0,$$
$$g(-3) = 4 - (-3)^2 = 4 - 9 = -5$$

x	$g(x)$
-3	-5
-2	0
-1	3
0	4
1	3
2	0
3	-5

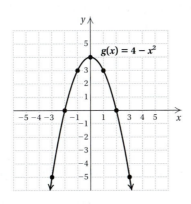

EXAMPLE 6 Graph: $h(x) = |x|$.

A list of some function values is shown in the following table. We plot the points and connect them. The graph is a V-shaped "curve" that rises on either side of the vertical axis.

x	$h(x)$
-3	3
-2	2
-1	1
0	0
1	1
2	2
3	3

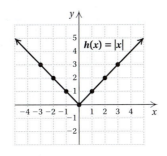

Do Exercises 14–16. ▶

d THE VERTICAL-LINE TEST

Consider the function f described by $f(x) = x^2 - 5$. Its graph is shown at right. It is also the graph of the equation $y = x^2 - 5$.

To find a function value, like $f(3)$, from a graph, we locate the input on the horizontal axis, move vertically to the graph of the function, and then move horizontally to find the output on the vertical axis, where members of the range can be found. As shown, $f(3) = 4$.

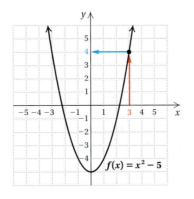

Graph.

14. $f(x) = x - 4$

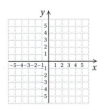

15. $g(x) = 5 - x^2$

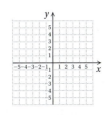

16. $t(x) = 3 - |x|$

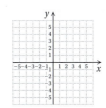

Answers

14.

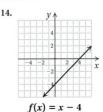

$f(x) = x - 4$

15.

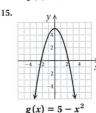

$g(x) = 5 - x^2$

16.

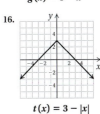

$t(x) = 3 - |x|$

Recall that when one member of the domain is paired with two or more different members of the range, the correspondence is *not* a function. Thus, when a graph contains two or more different points with the same first coordinate, the graph cannot represent a function. Points sharing a common first coordinate are vertically above or below each other. (See the following graph.)

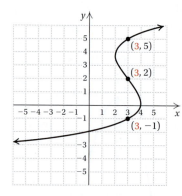

Since 3 is paired with more than one member of the range, the graph does not represent a function.

This observation leads to the *vertical-line test*.

THE VERTICAL-LINE TEST

A graph represents a function if it is impossible to draw a vertical line that intersects the graph more than once.

EXAMPLE 7   Determine whether each of the following is the graph of a function.

a)
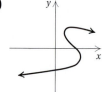

b)
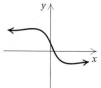

c)
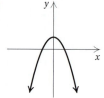

a) The graph *is not* that of a function because a vertical line crosses the graph at more than one point.

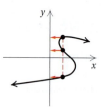

b) The graph *is* that of a function because no vertical line can cross the graph at more than one point. This can be confirmed with a ruler or a straightedge.

c) The graph *is* that of a function.

◀ **Do Exercises 17–20.**

Determine whether each of the following is the graph of a function.

17.

18.
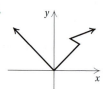

19.
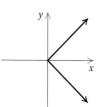

20.
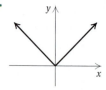

Answers

17. Yes   **18.** No   **19.** No   **20.** Yes

 **APPLICATIONS OF FUNCTIONS AND THEIR GRAPHS**

Functions are often described by graphs, whether or not an equation is given.

EXAMPLE 8 *Movie Revenue.* The graph shown approximates the weekly revenue, in millions of dollars, from a movie. The revenue is a function of the week, and no equation is given for the function. Use the graph to answer the following.

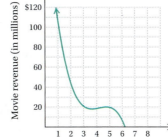

a) What was the movie revenue for week 1?

b) What was the movie revenue for week 5?

a) To estimate the revenue for week 1, we locate 1 on the horizontal axis and move directly up until we reach the graph. Then we move across to the vertical axis. We estimate that value to be about $105 million.

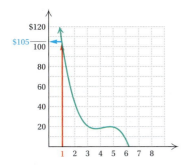 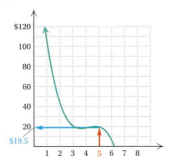

b) To estimate the revenue for week 5, we locate 5 on the horizontal axis and move directly up until we reach the graph. Then we move across to the vertical axis. We estimate that value to be about $19.5 million.

Do Exercises 21 and 22. ▶

Refer to the graph in Example 8.

21. What was the movie revenue for week 2?

22. What was the movie revenue for week 6?

Answers

21. About $43 million   **22.** About $6 million

9.7 Exercise Set

MyMathLab   MathXL®   PRACTICE   WATCH   READ   REVIEW

☑ Reading Check

Choose from the columns on the right the best word or words to complete each sentence. Not all words will be used.

RC1. A function is a special kind of correspondence between a first set, called the _____, and a second set, called the _____.

RC2. When we write $f(4) = 9$, we mean that the number 4 is a(n) _____ of the function.

RC3. We use the _____-line test to test whether a graph represents a function.

RC4. The function given by $f(x) = 5x + 7$ is an example of a(n) _____ function.

domain	linear
range	quadratic
horizontal	input
vertical	output

SECTION 9.7 Functions   **715**

Determine whether each correspondence is a function.

1. Domain   Range

2 ⟶ 9
5 ⟶ 8
19

2. Domain   Range

5 ⟶ 3
−3 ⟶ 7
7
−7

3. Domain   Range

−5 ⟶ 1
5
8

4. Domain   Range

6 ⟶ −6
7 ⟶ −7
3 ⟶ −3

5. Domain   Range

Texas ⟷ Austin
    Houston
    Dallas
    Cleveland
Ohio ⟷ Toledo
    Cincinnati

6. Domain   Range

Austin
Houston ⟶ Texas
Dallas
Cleveland
Toledo ⟶ Ohio
Cincinnati

7. Domain   Range

STATE	NUMBER OF COLLEGES IN STATE
Alaska	⟶ 7
California	⟶ 436
Kentucky	⟶ 77
Texas	⟶ 230
Vermont	⟶ 25

Source: http://www.publicagendaarchives.org

8. Domain   Range

CEREAL	CALORIES $\left(\frac{3}{4}\text{-cup serving}\right)$
Cinnamon Life	⟶ 160
Life (Regular)	
Lucky Charms	⟶ 120
Kellogg's Complete	
Wheaties	⟶ 110

Sources: Quaker Oats; General Mills; Kellogg's

Determine whether each of the following is a function. Identify any relations that are not functions.

Domain	Correspondence	Range
9. A math class	Each person's seat number	A set of numbers
10. A set of numbers	4 more than the square of each number	A set of numbers
11. A set of shapes	The area of each shape	A set of numbers
12. A family	Each person's eye color	A set of colors
13. The people in a town	Each person's neighbor	A set of people
14. Students in a college	Each student's classes	A set of classes

Find the function values.

15. $f(x) = x + 5$

a) $f(4)$ b) $f(7)$
c) $f(-3)$ d) $f(0)$
e) $f(2.4)$ f) $f\left(\frac{2}{3}\right)$

16. $g(t) = t - 6$

a) $g(0)$ b) $g(6)$
c) $g(13)$ d) $g(-1)$
e) $g(-1.08)$ f) $g\left(\frac{7}{8}\right)$

17. $h(p) = 3p$

a) $h(-7)$ b) $h(5)$
c) $h(14)$ d) $h(0)$
e) $h\left(\frac{2}{3}\right)$ f) $h(-54.2)$

18. $f(x) = -4x$
  a) $f(6)$ **b)** $f\left(-\frac{1}{2}\right)$
  c) $f(20)$ **d)** $f(11.8)$
  e) $f(0)$ **f)** $f(-1)$

19. $g(s) = 3s + 4$
  a) $g(1)$ **b)** $g(-7)$
  c) $g(6.7)$ **d)** $g(0)$
  e) $g(-10)$ **f)** $g\left(\frac{2}{3}\right)$

20. $h(x) = 19$, a constant function
  a) $h(4)$ **b)** $h(-6)$
  c) $h(12.5)$ **d)** $h(0)$
  e) $h\left(\frac{2}{3}\right)$ **f)** $h(1234)$

21. $f(x) = 2x^2 - 3x$
  a) $f(0)$ **b)** $f(-1)$
  c) $f(2)$ **d)** $f(10)$
  e) $f(-5)$ **f)** $f(-10)$

22. $f(x) = 3x^2 - 2x + 1$
  a) $f(0)$ **b)** $f(1)$
  c) $f(-1)$ **d)** $f(10)$
  e) $f(2)$ **f)** $f(-3)$

23. $f(x) = |x| + 1$
  a) $f(0)$ **b)** $f(-2)$
  c) $f(2)$ **d)** $f(-3)$
  e) $f(-10)$ **f)** $f(22)$

24. $g(t) = \sqrt{t}$
  a) $g(4)$ **b)** $g(25)$
  c) $g(16)$ **d)** $g(100)$
  e) $g(50)$ **f)** $g(84)$

25. $f(x) = x^3$
  a) $f(0)$ **b)** $f(-1)$
  c) $f(2)$ **d)** $f(10)$
  e) $f(-5)$ **f)** $f(-10)$

26. $f(x) = x^4 - 3$
  a) $f(1)$ **b)** $f(-1)$
  c) $f(0)$ **d)** $f(2)$
  e) $f(-2)$ **f)** $f(10)$

27. *Life Span.* The function given by $l(x) = \dfrac{1700}{x}$ can be used to approximate the life span, in years, of an animal with a pulse rate of x beats per minute.
  a) Find the approximate life span of a horse with a pulse rate of 50 beats per minute.
  b) Find the approximate life span of a seal with a pulse rate of 85 beats per minute.

28. *Temperature as a Function of Depth.* The function $T(d) = 10d + 20$ gives the temperature, in degrees Celsius, inside the earth as a function of the depth d, in kilometers. Find the temperature at 5 km, 20 km, and 1000 km.

29. *Growth in the Number of Facebook Users.* The function $F(t) = 22t^2 - 41t + 2.7$ can be used to approximate the number of monthly active Facebook users, in millions, t years after 2004. Estimate the number of monthly active Facebook users in 2004 $(t = 0)$, in 2010 $(t = 6)$, and in 2012 $(t = 8)$.

Source: Based on data from Facebook

30. *Female Physicians.* The function $p(t) = \frac{1}{8}t^2 + 3t + 35$ can be used to approximate the number of female physicians, in thousands, in the United States t years after 1975. Estimate the number of female physicians in the United States in 1975 $(t = 0)$, in 2000 $(t = 25)$, and in 2010 $(t = 35)$.

Source: Based on data from AMA Physician Characteristics and Distribution in the US, 2012 Edition

31. *Pressure at Sea Depth.* The function $P(d) = 1 + (d/33)$ gives the pressure, in *atmospheres* (atm), at a depth of d feet in the sea. Note that $P(0) = 1$ atm, $P(33) = 2$ atm, and so on. Find the pressure at 20 ft, 30 ft, and 100 ft.

32. *Temperature Conversions.* The function $C(F) = \frac{5}{9}(F - 32)$ determines the Celsius temperature that corresponds to F degrees Fahrenheit. Find the Celsius temperature that corresponds to 62°F, 77°F, and 23°F.

C   Graph each function.

33. $f(x) = 3x - 1$

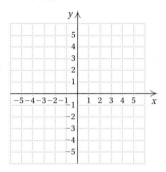

34. $g(x) = 2x + 5$

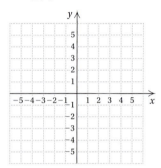

35. $g(x) = -2x + 3$

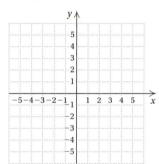

36. $f(x) = -\frac{1}{2}x + 2$

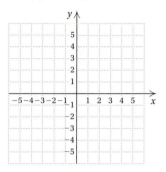

37. $f(x) = \frac{1}{2}x + 1$

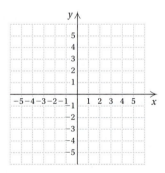

38. $f(x) = -\frac{3}{4}x - 2$

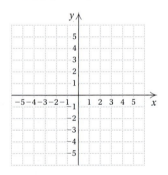

39. $f(x) = 2 - |x|$

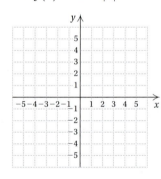

40. $f(x) = |x| - 4$

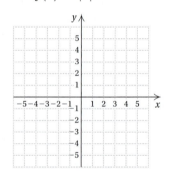

41. $f(x) = x^2$

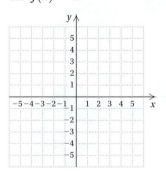

42. $f(x) = x^2 - 1$

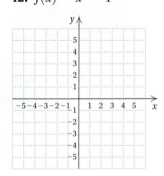

43. $f(x) = x^2 - x - 2$

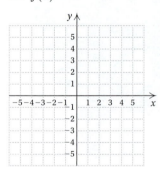

44. $f(x) = x^2 + 6x + 5$

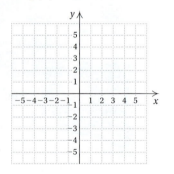

d Determine whether each of the following is the graph of a function.

45.
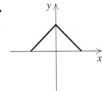

46.
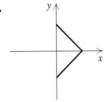

47.
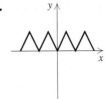

48.
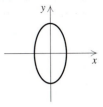

49.
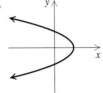

50.
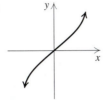

51.

52.

e *Las Vegas Conventions.* The number of conventions held in Las Vegas can be modeled as a function of the number of years after 2000. The graph of this function is shown below.

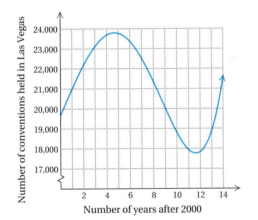

Data from Las Vegas Convention and Visitors Authority

53. Approximate the number of conventions held in Las Vegas in 2005.

54. Approximate the number of conventions held in Las Vegas in 2012.

Skill Maintenance

Solve each system using the substitution method. [7.2b]

55. $x = 2 - y,$
$3x + y = 5$

56. $3x - y = 9,$
$2x = 6 + y$

Solve each system using the elimination method. [7.3a], [7.3b]

57. $2x - 5y = 7,$
$x + 5y = 2$

58. $x - 3y = 2,$
$3x - 9y = 6$

Synthesis

Graph.

59. $g(x) = x^3$

60. $f(x) = 2 + \sqrt{x}$

61. $f(x) = |x| + x$

62. $g(x) = |x| - x$

Key Formulas and Principles

Standard Form of a Quadratic Equation: $ax^2 + bx + c = 0, a > 0$

Principle of Square Roots: The equation $x^2 = d$, where $d > 0$, has two solutions, $\sqrt{d}$ and $-\sqrt{d}$. The solution of $x^2 = 0$ is 0.

The x-coordinate of the vertex of a parabola: $-\dfrac{b}{2a}$

Quadratic Formula: $x = \dfrac{-b \pm \sqrt{b^2 - 4ac}}{2a}$

Discriminant: $b^2 - 4ac$

Vocabulary Reinforcement

Complete each statement with the correct term from the column on the right. Some of the choices may be used more than once and some may not be used at all.

domain
range
function
relation
quadratic
linear
vertical-line
horizontal-line
square
parabola
line of symmetry
vertex
discriminant

1. The equation $ax^2 + bx + c = 0$ is the standard form of a(n) _____ equation.   [9.1a]

2. When we add 25 to $x^2 + 10x$, we are completing the _____.
[9.2c]

3. The expression $b^2 - 4ac$ is called the _____.   [9.6b]

4. The turning point of the graph of a quadratic equation is the _____.   [9.6a]

5. The graph of a quadratic equation is called a(n) _____.
[9.6a]

6. The function given by $f(x) = 7x^2 - 3x$ is an example of a(n) _____ function.   [9.7b]

7. A graph is the graph of a function if it passes the _____ test.   [9.7d]

8. The set of all inputs of a function is called the _____.
[9.7a]

Concept Reinforcement

Determine whether each statement is true or false.

_____ 1. A graph represents a function if it is possible to draw a vertical line that intersects the graph more than once.   [9.7d]

_____ 2. All graphs of quadratic equations, $y = ax^2 + bx + c$, have a y-intercept.   [9.6a]

_____ 3. If (p, q) is the vertex of the graph of $y = ax^2 + bx + c, a < 0$, then q is the largest value of y.   [9.6a]

_____ 4. If a quadratic equation $ax^2 + bx + c = 0$ has no real-number solutions, then the graph of $y = ax^2 + bx + c$ does not have an x-intercept.   [9.6b]

Study Guide

Objective 9.1c Solve quadratic equations of the type $ax^2 + bx + c = 0$, where $b \neq 0$ and $c \neq 0$, by factoring.

Example Solve: $\dfrac{1}{x} + \dfrac{2}{x + 3} = \dfrac{3}{2}$.

We multiply by the LCM, which is $2x(x + 3)$.

$$2x(x + 3) \cdot \left(\dfrac{1}{x} + \dfrac{2}{x + 3} \right) = \dfrac{3}{2} \cdot 2x(x + 3)$$

$$2x(x + 3) \cdot \dfrac{1}{x} + 2x(x + 3) \cdot \dfrac{2}{x + 3} = 3x(x + 3)$$

$$2(x + 3) + 2x \cdot 2 = 3x^2 + 9x$$

$$2x + 6 + 4x = 3x^2 + 9x$$

$$0 = 3x^2 + 3x - 6$$

$$0 = 3(x^2 + x - 2)$$

$$0 = 3(x + 2)(x - 1)$$

$$x + 2 = 0 \quad or \quad x - 1 = 0$$

$$x = -2 \quad or \quad x = 1$$

Both numbers check. The solutions are -2 and 1.

Practice Exercise

1. Solve: $\dfrac{3}{x + 2} + \dfrac{1}{x} = \dfrac{5}{4}$.

Objective 9.2a Solve quadratic equations of the type $ax^2 = p$.

Example Solve: $5x^2 - 2 = 12$.

$$5x^2 - 2 = 12$$

$$5x^2 = 14 \qquad \text{Adding 2}$$

$$x^2 = \dfrac{14}{5} \qquad \text{Dividing by 5}$$

$$x = \sqrt{\dfrac{14}{5}} \quad or \quad x = -\sqrt{\dfrac{14}{5}} \qquad \text{Using the principle of square roots}$$

$$x = \sqrt{\dfrac{14}{5} \cdot \dfrac{5}{5}} \quad or \quad x = -\sqrt{\dfrac{14}{5} \cdot \dfrac{5}{5}} \qquad \text{Rationalizing the denominator}$$

$$x = \dfrac{\sqrt{70}}{5} \quad or \quad x = -\dfrac{\sqrt{70}}{5}$$

The solutions are $\dfrac{\sqrt{70}}{5}$ and $-\dfrac{\sqrt{70}}{5}$.

Practice Exercise

2. Solve: $7x^2 - 3 = 8$.

Objective 9.2c Solve quadratic equations by completing the square.

Example Solve $x^2 - 10x + 8 = 0$ by completing the square.

$$x^2 - 10x + 8 = 0$$

$$x^2 - 10x = -8 \qquad \text{Subtracting 8}$$

$$x^2 - 10x + 25 = -8 + 25 \qquad \left(\dfrac{-10}{2} \right)^2 = 25$$

$$(x - 5)^2 = 17$$

$$x - 5 = \sqrt{17} \quad or \quad x - 5 = -\sqrt{17}$$

$$x = 5 + \sqrt{17} \quad or \quad x = 5 - \sqrt{17}$$

The solutions are $5 \pm \sqrt{17}$.

Practice Exercise

3. Solve $x^2 - 4x + 1 = 0$ by completing the square.

Objective 9.3a Solve quadratic equations using the quadratic formula.

Example Solve $6x^2 = 4x + 5$ using the quadratic formula.

$$6x^2 - 4x - 5 = 0 \quad \text{Standard form}$$
$$a = 6, \quad b = -4, \quad c = -5$$
$$x = \frac{-b \pm \sqrt{b^2 - 4ac}}{2a} \quad \text{Quadratic formula}$$
$$= \frac{-(-4) \pm \sqrt{(-4)^2 - 4 \cdot 6 \cdot (-5)}}{2 \cdot 6} \quad \text{Substituting}$$
$$= \frac{4 \pm \sqrt{16 + 120}}{12} = \frac{4 \pm \sqrt{136}}{12}$$
$$= \frac{4 \pm \sqrt{4 \cdot 34}}{12} = \frac{4 \pm 2\sqrt{34}}{12}$$
$$= \frac{2(2 \pm \sqrt{34})}{2 \cdot 6} = \frac{2 \pm \sqrt{34}}{6}$$

The solutions are $\dfrac{2 \pm \sqrt{34}}{6}$.

Practice Exercise

4. Solve: $4y^2 = 6y + 3$.

Objective 9.6a Graph quadratic equations.

Example Graph: $y = 2x^2 + 4x - 1$.

We first find the vertex. The x-coordinate of the vertex is

$$-\frac{b}{2a} = -\frac{4}{2 \cdot 2} = -1.$$

We substitute -1 for x into the equation to find the second coordinate of the vertex:

$$y = 2(-1)^2 + 4(-1) - 1 = -3.$$

The vertex is $(-1, -3)$. The line of symmetry is $x = -1$. We choose x-values on both sides of $x = -1$ and graph the parabola.

x	y	
-1	-3	← Vertex
-2	-1	
0	-1	
-3	5	
1	5	

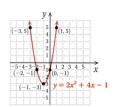

Practice Exercise

5. Graph: $y = x^2 - 4x + 2$.

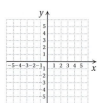

Objective 9.7b Given a function described by an equation, find function values (outputs) for specified values (inputs).

Example Find $f(-2)$, for $f(x) = -\dfrac{1}{2}x + 7$.

$$f(-2) = -\frac{1}{2}(-2) + 7 = 1 + 7 = 8$$

Practice Exercise

6. Find the indicated function value.

a) $h(5)$, for $h(x) = \dfrac{1}{5}x^2 + x - 1$

b) $f(0)$, for $f(x) = -3x - 4$

Review Exercises

Solve.

1. $8x^2 = 24$ [9.2a]

2. $40 = 5y^2$ [9.2a]

3. $5x^2 - 8x + 3 = 0$ [9.1c]

4. $3y^2 + 5y = 2$ [9.1c]

5. $(x + 8)^2 = 13$ [9.2b]

6. $9x^2 = 0$ [9.2a]

7. $5t^2 - 7t = 0$ [9.1b]

Solve. [9.3a]

8. $x^2 - 2x - 10 = 0$

9. $9x^2 - 6x - 9 = 0$

10. $x^2 + 6x = 9$

11. $1 + 4x^2 = 8x$

12. $6 + 3y = y^2$

13. $3m = 4 + 5m^2$

14. $3x^2 = 4x$

Solve. [9.1c]

15. $\dfrac{15}{x} - \dfrac{15}{x + 2} = 2$

16. $x + \dfrac{1}{x} = 2$

Solve by completing the square. Show your work. [9.2c]

17. $x^2 - 4x + 2 = 0$

18. $3x^2 - 2x - 5 = 0$

Approximate the solutions to the nearest tenth. [9.3b]

19. $x^2 - 5x + 2 = 0$

20. $4y^2 + 8y + 1 = 0$

21. Solve for T: $V = \dfrac{1}{2}\sqrt{1 + \dfrac{T}{L}}$. [9.4a]

Graph each quadratic equation. Label the ordered pairs for the vertex and the y-intercept. [9.6a]

22. $y = 2 - x^2$

23. $y = x^2 - 4x - 2$

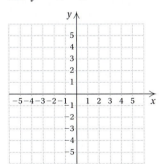

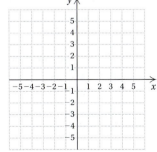

Find the x-intercepts. [9.6b]

24. $y = 2 - x^2$

25. $y = x^2 - 4x - 2$

Solve.

26. *Right-Triangle Dimensions.* The hypotenuse of a right triangle is 5 cm long. One leg is 3 cm longer than the other. Find the lengths of the legs. Round to the nearest tenth. [9.5a]

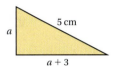

27. *Car-Loading Ramp.* The length of a loading ramp on a car hauler is 25 ft. This ramp and its height form the hypotenuse and one leg of a right triangle. The height of the ramp is 5 ft less than the length of the other leg. Find the height of the loading ramp. [9.5a]

28. *Falling Object.* The Royal Gorge Bridge above Colorado's Arkansas River is the highest suspension bridge in the United States. It hangs 1053 ft above the river. How long would it take an object to fall to the water from the bridge? [9.2d]

Find the function values. [9.7b]

29. If $f(x) = 2x - 5$, find $f(2)$, $f(-1)$, and $f(3.5)$.

30. If $g(x) = |x| - 1$, find $g(1)$, $g(-1)$, and $g(-20)$.

31. *Caloric Needs.* If you are moderately active, you need to consume about 15 calories per pound of body weight each day. The function $C(p) = 15p$ approximates the number of calories C that are needed to maintain body weight p, in pounds. How many calories are needed to maintain a body weight of 180 lb? [9.7e]

Graph each function. [9.7c]

32. $g(x) = 4 - x$

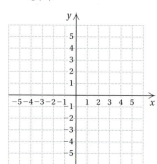

33. $f(x) = x^2 - 3$

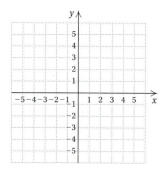

34. $h(x) = |x| - 5$

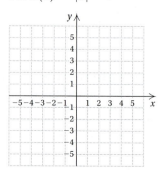

35. $f(x) = x^2 - 2x + 1$

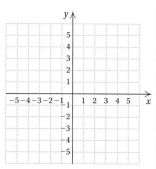

Determine whether each of the following is the graph of a function. [9.7d]

36.

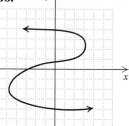

37.

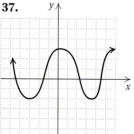

38. Solve: $40x - x^2 = 0$. [9.1b]

   A. 40   **B.** $2\sqrt{10}$
   C. $-2\sqrt{10}$ **D.** 0, 40

39. Solve: $\dfrac{1}{2}c^2 + c - \dfrac{1}{2} = 0$. [9.3a]

   A. $-1 \pm \sqrt{2}$ **B.** $-1 \pm \sqrt{5}$
   C. $1 \pm \sqrt{2}$ **D.** $-3, 1$

Synthesis

40. Two consecutive integers have squares that differ by 63. Find the integers. [9.5a]

41. A square with sides of length s has the same area as a circle with a radius of 5 in. Find s. [9.5a]

42. Solve: $x - 4\sqrt{x} - 5 = 0$. [9.1c]

Use the graph of
$$y = (x + 3)^2$$
to solve each equation.
[9.6b]

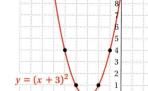

$y = (x + 3)^2$

43. $(x + 3)^2 = 1$

44. $(x + 3)^2 = 4$

45. $(x + 3)^2 = 9$

46. $(x + 3)^2 = 0$

Understanding Through Discussion and Writing

1. Find and explain the error(s) in the following solution of a quadratic equation. [9.2b]

$$(x + 6)^2 = 16$$
$$x + 6 = \sqrt{16}$$
$$x + 6 = 4$$
$$x = -2$$

2. Is it possible for a function to have more numbers as outputs than as inputs? Why or why not? [9.7b]

3. Suppose that the x-intercepts of a parabola are $(a_1, 0)$ and $(a_2, 0)$. What is the easiest way to find an equation for the line of symmetry? to find the coordinates of the vertex? [9.6b]

4. Discuss the effect of the sign of a on the graph of $y = ax^2 + bx + c$. [9.6a]

5. If a quadratic equation can be solved by factoring, what type of number(s) will generally be solutions? [9.1c]

CHAPTER

9 Test

For Extra Help For step-by-step test solutions, access the Chapter Test Prep Videos in
MyMathLab® or on YouTube (search "BittingerIntro" and click on "Channels").

Solve.

1. $7x^2 = 35$

2. $7x^2 + 8x = 0$

3. $48 = t^2 + 2t$

4. $3y^2 - 5y = 2$

5. $(x - 8)^2 = 13$

6. $x^2 = x + 3$

7. $m^2 - 3m = 7$

8. $10 = 4x + x^2$

9. $3x^2 - 7x + 1 = 0$

10. $x - \dfrac{2}{x} = 1$

11. $\dfrac{4}{x} - \dfrac{4}{x + 2} = 1$

12. Solve $x^2 - 4x - 10 = 0$ by completing the square. Show your work.

13. Approximate the solutions to $x^2 - 4x - 10 = 0$ to the nearest tenth.

14. Solve for n: $d = an^2 + bn$.

15. Find the x-intercepts: $y = -x^2 + x + 5$.

Graph. Label the ordered pairs for the vertex and the y-intercept.

16. $y = 4 - x^2$

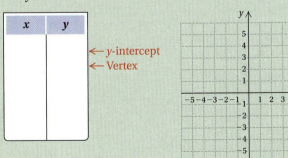

17. $y = -x^2 + x + 5$

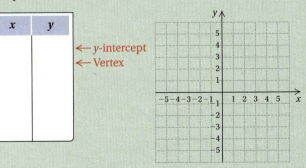

18. If $f(x) = \frac{1}{2}x + 1$, find $f(0)$, $f(1)$, and $f(2)$.

19. If $g(t) = -2|t| + 3$, find $g(-1)$, $g(0)$, and $g(3)$.

Solve.

20. *Rug Dimensions.* The width of a rectangular area rug is 4 m less than the length. The area is 16.25 m². Find the length and the width.

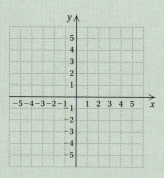

21. *Boat Speed.* The current in a stream moves at a speed of 2 km/h. A boat travels 44 km upstream and 52 km downstream in a total of 4 hr. What is the speed of the boat in still water?

22. *World Record for 10,000-m Run.* The world record for the 10,000-m run has been decreasing steadily since 1940. The function $R(t) = 30.18 - 0.06t$ estimates the record R, in minutes, as a function of t, the time in years since 1940. Estimate the record in 2012.

Graph.

23. $h(x) = x - 4$

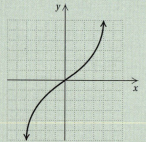

24. $g(x) = x^2 - 4$

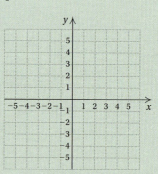

Determine whether each of the following is the graph of a function.

25.

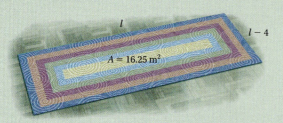

26.

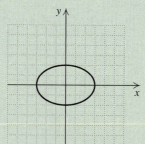

27. Given $g(x) = -2x - x^2$, find $g(-8)$.

  A. -80 **B.** 48

  C. -66 **D.** -48

Synthesis

28. Find the side of a square whose diagonal is 5 ft longer than a side.

29. Solve:
$$x - y = 2,$$
$$xy = 4.$$

1. What is the meaning of x^3?

2. Evaluate $(x - 3)^2 + 5$ when $x = 10$.

3. Find decimal notation: $-\dfrac{3}{11}$.

4. Find the LCM of 15 and 48.

5. Find the absolute value: $|-7|$.

Compute and simplify.

6. $-6 + 12 + (-4) + 7$　　　**7.** $2.8 - (-12.2)$

8. $-\dfrac{3}{8} \div \dfrac{5}{2}$　　　**9.** $13 \cdot 6 \div 3 \cdot 2 \div 13$

10. Remove parentheses and simplify:
$$4m + 9 - (6m + 13).$$

Solve.

11. $3x = -24$　　　**12.** $3x + 7 = 2x - 5$

13. $3(y - 1) - 2(y + 2) = 0$　　**14.** $x^2 - 8x + 15 = 0$

15. $y - x = 1,$
$\quad y = 3 - x$
16. $x + y = 17,$
$\quad x - y = 7$

17. $4x - 3y = 3,$
$\quad 3x - 2y = 4$
18. $x^2 - x - 6 = 0$

19. $x^2 + 3x = 5$　　　**20.** $3 - x = \sqrt{x^2 - 3}$

21. $5 - 9x \le 19 + 5x$　　**22.** $-\dfrac{7}{8}x + 7 = \dfrac{3}{8}x - 3$

23. $0.6x - 1.8 = 1.2x$　　**24.** $-3x > 24$

25. $23 - 19y - 3y \ge -12$　　**26.** $3y^2 = 30$

27. $(x - 3)^2 = 6$　　**28.** $\dfrac{6x - 2}{2x - 1} = \dfrac{9x}{3x + 1}$

29. $\dfrac{2x}{x + 1} = 2 - \dfrac{5}{2x}$

30. $\dfrac{2x}{x + 3} + \dfrac{6}{x} + 7 = \dfrac{18}{x^2 + 3x}$

31. $\sqrt{x + 9} = \sqrt{2x - 3}$

Solve the formula for the given letter.

32. $A = \dfrac{4b}{t}$, for b

33. $\dfrac{1}{t} = \dfrac{1}{m} - \dfrac{1}{n}$, for m

34. $r = \sqrt{\dfrac{A}{\pi}}$, for A

35. $y = ax^2 - bx$, for x

Simplify.

36. $x^{-6} \cdot x^2$

37. $\dfrac{y^3}{y^{-4}}$

38. $(2y^6)^2$

39. Collect like terms and arrange in descending order:
$$2x - 3 + 5x^3 - 2x^3 + 7x^3 + x.$$

Compute and simplify.

40. $(4x^3 + 3x^2 - 5) + (3x^3 - 5x^2 + 4x - 12)$

41. $(6x^2 - 4x + 1) - (-2x^2 + 7)$

42. $-2y^2(4y^2 - 3y + 1)$

43. $(2t - 3)(3t^2 - 4t + 2)$

44. $\left(t - \dfrac{1}{4}\right)\left(t + \dfrac{1}{4}\right)$

45. $(3m - 2)^2$

46. $(15x^2y^3 + 10xy^2 + 5) - (5xy^2 - x^2y^2 - 2)$

47. $(x^2 - 0.2y)(x^2 + 0.2y)$

48. $(3p + 4q^2)^2$

49. $\dfrac{4}{2x - 6} \cdot \dfrac{x - 3}{x + 3}$

50. $\dfrac{3a^4}{a^2 - 1} \div \dfrac{2a^3}{a^2 - 2a + 1}$

51. $\dfrac{3}{3x - 1} + \dfrac{4}{5x}$

52. $\dfrac{2}{x^2 - 16} - \dfrac{x - 3}{x^2 - 9x + 20}$

Factor.

53. $8x^2 - 4x$

54. $25x^2 - 4$

55. $6y^2 - 5y - 6$

56. $m^2 - 8m + 16$

57. $x^3 - 8x^2 - 5x + 40$

58. $3a^4 + 6a^2 - 72$

59. $16x^4 - 1$

60. $49a^2b^2 - 4$

61. $9x^2 + 30xy + 25y^2$

62. $2ac - 6ab - 3db + dc$

63. $15x^2 + 14xy - 8y^2$

Simplify.

64. $\dfrac{\dfrac{3}{x} + \dfrac{1}{2x}}{\dfrac{1}{3x} - \dfrac{3}{4x}}$

65. $\sqrt{49}$

66. $-\sqrt{625}$

67. $\sqrt{64x^2}$

68. Multiply: $\sqrt{a+b}\,\sqrt{a-b}$.

69. Multiply and simplify: $\sqrt{32ab}\,\sqrt{6a^4b^2}$.

Simplify.

70. $\sqrt{150}$

71. $\sqrt{243x^3y^2}$

72. $\sqrt{\dfrac{100}{81}}$

73. $\sqrt{\dfrac{64}{x^2}}$

74. $4\sqrt{12}+2\sqrt{12}$

75. Divide and simplify: $\dfrac{\sqrt{72}}{\sqrt{45}}$.

76. In a right triangle, where a and b represent the legs and c represents the hypotenuse, $a=9$ and $c=41$. Find b.

Graph.

77. $y=\dfrac{1}{3}x-2$

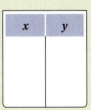

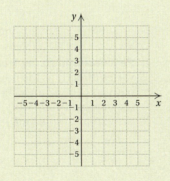

78. $2x+3y=-6$

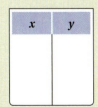

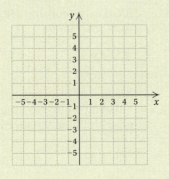

79. $y=-3$

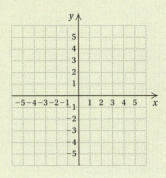

80. $x\geq-3$

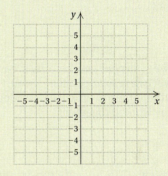

81. $4x-3y>12$

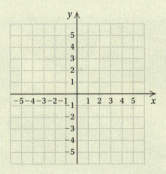

82. Graph $y=x^2+2x+1$. Label the vertex and the y-intercept.

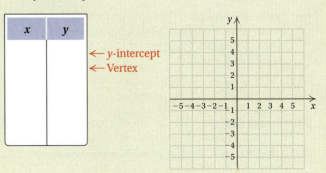

$\leftarrow$ y-intercept
$\leftarrow$ Vertex

83. Solve $9x^2 - 12x - 2 = 0$ by completing the square. Show your work.

84. Approximate the solutions of $4x^2 = 4x + 1$ to the nearest tenth.

Solve.

85. What percent of 52 is 13?

86. 12 is 20% of what?

87. *Work Time.* In checking records, a contractor finds that crew A can resurface a tennis court in 8 hr. Crew B can do the same job in 10 hr. How long would they take if they worked together?

88. *Movie Screen.* The area of a rectangular movie screen is 96 ft². The length is 4 ft longer than the width. Find the length and the width of the movie screen.

89. *Speed of a Stream.* The speed of a boat in still water is 8 km/h. It travels 60 km upstream and 60 km downstream in a total time of 16 hr. What is the speed of the stream?

90. *Garage Length.* The length of a rectangular garage floor is 7 m more than the width. The length of a diagonal is 13 m. Find the length of the garage floor.

91. *Consecutive Odd Integers.* The sum of the squares of two consecutive odd integers is 74. Find the integers.

92. *Alcohol Solutions.* Solution A is 75% alcohol and solution B is 50% alcohol. How much of each is needed in order to make 60 L of a solution that is $66\frac{2}{3}$% alcohol?

93. *Big Ben.* The Elizabeth Tower in London, Great Britain, is 315 ft high. How long would it take an object to fall to the ground from the top? Use $s = 16t^2$.

94. *Paycheck and Hours Worked.* A student's paycheck varies directly as the number of hours worked. The pay was $242.52 for 43 hr of work. What would be the pay for 80 hr of work? Explain the meaning of the variation constant.

95. *Parking Spaces.* Three-fifths of the automobiles entering the city each morning will be parked in city parking lots. There are 3654 such parking spaces filled each morning. How many cars enter the city each morning?

96. *Air Travel.* An airplane flew for 3 hr with a 20-mph tail wind. The return flight against the same wind took 4 hr. Find the speed of the plane in still air.

97. *Candy Mixture.* A candy shop wants to mix nuts worth $9.90 per pound with candy worth $7.20 per pound in order to make 42 lb of a mixture worth $8.10 per pound. How many pounds of each should be used?

98. Use *only* the graph below to solve $x^2 + x - 6 = 0$.

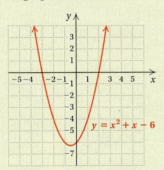

99. Find the x-intercepts of $y = x^2 + 4x + 1$.

100. Find the slope and the y-intercept:
$$-6x + 3y = -24.$$

101. Determine whether the graphs of the following equations are parallel, perpendicular, or neither.
$$y - x = 4,$$
$$3y + x = 8$$

102. Find the slope of the line containing the points $(-5, -6)$ and $(-4, 9)$.

103. Find an equation of variation in which y varies directly as x, and $y = 100$ when $x = 10$. Then find the value of y when $x = 64$.

104. Find an equation of variation in which y varies inversely as x, and $y = 100$ when $x = 10$. Then find the value of y when $x = 125$.

Determine whether each of the following is the graph of a function.

105.

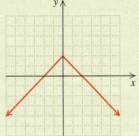

106.

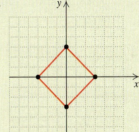

Graph the function.

107. $f(x) = x^2 + x - 2$

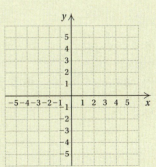

108. $g(x) = |x + 2|$

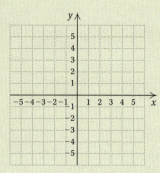

109. For the function f described by $f(x) = 2x^2 + 7x - 4$, find $f(0)$, $f(-4)$, and $f\left(\frac{1}{2}\right)$.

110. An airplane flies 408 mi against the wind and 492 mi with the wind in a total time of 3 hr. The speed of the airplane in still air is 300 mph. If we assume there is some wind, the speed of the wind is between which of the following?

A. 8 and 15 mph   **B.** 15 and 22 mph
C. 22 and 29 mph   **D.** 29 and 36 mph

111. Solve: $2x^2 + 6x + 5 = 4$.

A. $-3 \pm \sqrt{7}$ **B.** $-3 \pm 2\sqrt{7}$

C. No real solutions   **D.** $\dfrac{-3 \pm \sqrt{7}}{2}$

112. Solve for b: $S = \dfrac{a + b}{3b}$.

A. $b = 3bS - a$ **B.** $b = \dfrac{a + b}{3S}$

C. $a = b(3S - 1)$ **D.** $b = \dfrac{a}{3S - 1}$

113. Which of the following is the graph of $3x - 4y = 12$?

A.
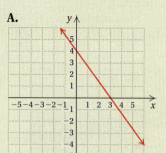

B.
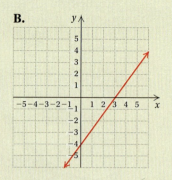

C.
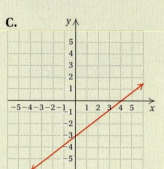

D.
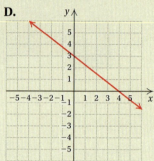

Synthesis

114. Solve: $|x| = 12$.

115. Simplify: $\sqrt{\sqrt{\sqrt{81}}}$.

116. Find b such that the trinomial $x^2 - bx + 225$ is a square.

117. Find x.

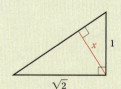

Determine whether each pair of expressions is equivalent.

118. $x^2 - 9$, $(x - 3)(x + 3)$

119. $\dfrac{x + 3}{3}$, x

120. $(x + 5)^2$, $x^2 + 25$

121. $\sqrt{x^2 + 16}$, $x + 4$

122. $\sqrt{x^2}$, $|x|$

Appendixes

Factoring Sums or Differences of Cubes

N	N^3
0.1	0.001
0.2	0.008
0	0
1	1
2	8
3	27
4	64
5	125
6	216
7	343
8	512
9	729
10	1000

a FACTORING SUMS OR DIFFERENCES OF CUBES

We can factor the sum or the difference of two expressions that are cubes. Consider the following products:

$$(A + B)(A^2 - AB + B^2) = A(A^2 - AB + B^2) + B(A^2 - AB + B^2)$$
$$= A^3 - A^2B + AB^2 + A^2B - AB^2 + B^3$$
$$= A^3 + B^3$$

and

$$(A - B)(A^2 + AB + B^2) = A(A^2 + AB + B^2) - B(A^2 + AB + B^2)$$
$$= A^3 + A^2B + AB^2 - A^2B - AB^2 - B^3$$
$$= A^3 - B^3.$$

The above equations (reversed) show how we can factor a sum or a difference of two cubes.

FACTORING SUMS OR DIFFERENCES OF CUBES

$A^3 + B^3 = (A + B)(A^2 - AB + B^2)$,
$A^3 - B^3 = (A - B)(A^2 + AB + B^2)$

Note that what we are considering here is a sum or a difference of cubes. We are not cubing a binomial. For example, $(A + B)^3$ is *not* the same as $A^3 + B^3$. The table of cubes in the margin is helpful.

EXAMPLE 1 Factor: $x^3 - 8$.

We have

$$x^3 - 8 = x^3 - 2^3 = (x - 2)(x^2 + x \cdot 2 + 2^2).$$

$$A^3 - B^3 = (A - B)(A^2 + A \ B + B^2)$$

This tells us that $x^3 - 8 = (x - 2)(x^2 + 2x + 4)$. Note that we cannot factor $x^2 + 2x + 4$. (It is not a trinomial square nor can it be factored by trial and error or the *ac*-method.) The check is left to the student.

◀ **Do Exercises 1 and 2.**

EXAMPLE 2 Factor: $x^3 + 125$.

We have

$$x^3 + 125 = x^3 + 5^3 = (x + 5)(x^2 - x \cdot 5 + 5^2).$$

$$A^3 + B^3 = (A + B)(A^2 - A \ B + B^2)$$

Thus, $x^3 + 125 = (x + 5)(x^2 - 5x + 25)$. The check is left to the student.

◀ **Do Exercises 3 and 4.**

Factor.

1. $x^3 - 27$ **2.** $64 - y^3$

Factor.

3. $y^3 + 8$ **4.** $125 + t^3$

Answers

1. $(x - 3)(x^2 + 3x + 9)$
2. $(4 - y)(16 + 4y + y^2)$
3. $(y + 2)(y^2 - 2y + 4)$
4. $(5 + t)(25 - 5t + t^2)$

EXAMPLE 3 Factor: $x^3 - 27t^3$.

We have

$$x^3 - 27t^3 = x^3 - (3t)^3 = (x - 3t)(x^2 + x \cdot 3t + (3t)^2)$$

$$A^3 - B^3 = (A - B)(A^2 + A \quad B + B^2)$$

$$= (x - 3t)(x^2 + 3xt + 9t^2)$$

Do Exercises 5 and 6. ▶

Factor.

5. $27x^3 - y^3$ **6.** $8y^3 + z^3$

EXAMPLE 4 Factor: $128y^7 - 250x^6y$.

We first look for a common factor:

$$128y^7 - 250x^6y = 2y(64y^6 - 125x^6) = 2y[(4y^2)^3 - (5x^2)^3]$$
$$= 2y(4y^2 - 5x^2)(16y^4 + 20x^2y^2 + 25x^4).$$

EXAMPLE 5 Factor: $a^6 - b^6$.

We can express this polynomial as a difference of squares:

$$(a^3)^2 - (b^3)^2.$$

We factor as follows:

$$a^6 - b^6 = (a^3 + b^3)(a^3 - b^3).$$

One factor is a sum of two cubes, and the other factor is a difference of two cubes. We factor them:

$$(a + b)(a^2 - ab + b^2)(a - b)(a^2 + ab + b^2).$$

We have now factored completely.

In Example 5, had we thought of factoring first as a difference of two cubes, we would have had

$$(a^2)^3 - (b^2)^3 = (a^2 - b^2)(a^4 + a^2b^2 + b^4)$$
$$= (a + b)(a - b)(a^4 + a^2b^2 + b^4).$$

In this case, we might have missed some factors; $a^4 + a^2b^2 + b^4$ can be factored as $(a^2 - ab + b^2)(a^2 + ab + b^2)$, but we probably would not have known to do such factoring.

EXAMPLE 6 Factor: $64a^6 - 729b^6$.

$$64a^6 - 729b^6 = (8a^3 - 27b^3)(8a^3 + 27b^3) \qquad \text{Factoring a difference of squares}$$

$$= [(2a)^3 - (3b)^3][(2a)^3 + (3b)^3].$$

Each factor is a sum or a difference of cubes. We factor each:

$$= (2a - 3b)(4a^2 + 6ab + 9b^2)(2a + 3b)(4a^2 - 6ab + 9b^2)$$

Sum of cubes:	$A^3 + B^3 = (A + B)(A^2 - AB + B^2)$;
Difference of cubes:	$A^3 - B^3 = (A - B)(A^2 + AB + B^2)$;
Difference of squares:	$A^2 - B^2 = (A + B)(A - B)$;
Sum of squares:	$A^2 + B^2$ cannot be factored using real numbers if the largest common factor has been removed.

Do Exercises 7–10. ▶

Factor.

7. $m^6 - n^6$

8. $16x^7y + 54xy^7$

9. $729x^6 - 64y^6$

10. $x^3 - 0.027$

Answers

5. $(3x - y)(9x^2 + 3xy + y^2)$
6. $(2y + z)(4y^2 - 2yz + z^2)$
7. $(m + n)(m^2 - mn + n^2) \times$
   $(m - n)(m^2 + mn + n^2)$
8. $2xy(2x^2 + 3y^2)(4x^4 - 6x^2y^2 + 9y^4)$
9. $(3x + 2y)(9x^2 - 6xy + 4y^2) \times$
   $(3x - 2y)(9x^2 + 6xy + 4y^2)$
10. $(x - 0.3)(x^2 + 0.3x + 0.09)$

a   Factor.

1. $z^3 + 27$

2. $a^3 + 8$

3. $x^3 - 1$

4. $c^3 - 64$

5. $y^3 + 125$

6. $x^3 + 1$

7. $8a^3 + 1$

8. $27x^3 + 1$

9. $y^3 - 8$

10. $p^3 - 27$

11. $8 - 27b^3$

12. $64 - 125x^3$

13. $64y^3 + 1$

14. $125x^3 + 1$

15. $8x^3 + 27$

16. $27y^3 + 64$

17. $a^3 - b^3$

18. $x^3 - y^3$

19. $a^3 + \dfrac{1}{8}$

20. $b^3 + \dfrac{1}{27}$

21. $2y^3 - 128$

22. $3z^3 - 3$

23. $24a^3 + 3$

24. $54x^3 + 2$

25. $rs^3 + 64r$

26. $ab^3 + 125a$

27. $5x^3 - 40z^3$

28. $2y^3 - 54z^3$

29. $x^3 + 0.001$

30. $y^3 + 0.125$

31. $64x^6 - 8t^6$

32. $125c^6 - 8d^6$

33. $2y^4 - 128y$

34. $3z^5 - 3z^2$

35. $z^6 - 1$

36. $t^6 + 1$

37. $t^6 + 64y^6$

38. $p^6 - q^6$

Synthesis

Consider these polynomials:

$(a + b)^3$; $a^3 + b^3$; $(a + b)(a^2 - ab + b^2)$;
$(a + b)(a^2 + ab + b^2)$; $(a + b)(a + b)(a + b)$.

39. Evaluate each polynomial when $a = -2$ and $b = 3$.

40. Evaluate each polynomial when $a = 4$ and $b = -1$.

Factor. Assume that variables in exponents represent natural numbers.

41. $x^{6a} + y^{3b}$

42. $a^3x^3 - b^3y^3$

43. $3x^{3a} + 24y^{3b}$

44. $\frac{8}{27}x^3 + \frac{1}{64}y^3$

45. $\frac{1}{24}x^3y^3 + \frac{1}{3}z^3$

46. $7x^3 - \frac{7}{8}$

47. $(x + y)^3 - x^3$

48. $(1 - x)^3 + (x - 1)^6$

49. $(a + 2)^3 - (a - 2)^3$

50. $y^4 - 8y^3 - y + 8$

OBJECTIVES

a   Find an equation of a line when the slope and a point are given.

b   Find an equation of a line when two points are given.

We can use the slope–intercept equation, $y = mx + b$, to find an equation of a line. Here we introduce another form, the *point–slope equation,* and find equations of lines using both forms.

a   FINDING AN EQUATION OF A LINE WHEN THE SLOPE AND A POINT ARE GIVEN

Suppose we know the slope of a line and the coordinates of one point on the line. We can use the slope–intercept equation to find an equation of the line. Or, we can use what is called a **point–slope equation**. We first develop a formula for such a line.

Suppose that a line of slope m passes through the point (x_1, y_1). For any other point (x, y) to lie on this line, we must have

$$\frac{y - y_1}{x - x_1} = m.$$

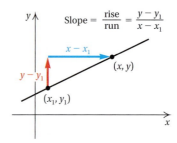

It is tempting to use this last equation as an equation of the line of slope m that passes through (x_1, y_1). The only problem with this form is that when x and y are replaced with x_1 and y_1, we have $\frac{0}{0} = m$, a false equation. To avoid this difficulty, we multiply by $x - x_1$ on both sides and simplify:

$$(x - x_1)\frac{y - y_1}{x - x_1} = m(x - x_1) \qquad \text{Multiplying by } x - x_1 \text{ on both sides}$$

$$y - y_1 = m(x - x_1). \qquad \text{Removing a factor of 1: } \frac{x - x_1}{x - x_1} = 1$$

This is the *point–slope* form of a linear equation.

POINT–SLOPE EQUATION

The **point–slope equation** of a line with slope m, passing through (x_1, y_1), is

$$y - y_1 = m(x - x_1).$$

If we know the slope of a line and a certain point on the line, we can find an equation of the line using either the point–slope equation,

$$y - y_1 = m(x - x_1),$$

or the slope–intercept equation,

$$y = mx + b.$$

EXAMPLE 1 Find an equation of the line with slope -2 and containing the point $(-1, 3)$.

Using the Point–Slope Equation: We consider $(-1, 3)$ to be (x_1, y_1) and -2 to be the slope m, and substitute:

$$y - y_1 = m(x - x_1)$$
$$y - 3 = -2[x - (-1)] \qquad \text{Substituting}$$
$$y - 3 = -2(x + 1)$$
$$y - 3 = -2x - 2$$
$$y = -2x - 2 + 3$$
$$y = -2x + 1.$$

Using the Slope–Intercept Equation: The point $(-1, 3)$ is on the line, so it is a solution. Thus we can substitute -1 for x and 3 for y in $y = mx + b$. We also substitute -2 for m, the slope. Then we solve for b:

$$y = mx + b$$
$$3 = -2 \cdot (-1) + b \qquad \text{Substituting}$$
$$3 = 2 + b$$
$$1 = b. \qquad \text{Solving for } b$$

We then use the equation $y = mx + b$ and substitute -2 for m and 1 for b:

$$y = -2x + 1.$$

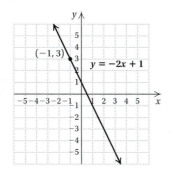

Do Exercises 1–4. ▶

Find an equation of the line with the given slope and containing the given point.

1. $m = -3$, $(-5, 4)$

2. $m = 5$, $(-2, 1)$

3. $m = 6$, $(3, -5)$

4. $m = -\dfrac{2}{3}$, $(1, 2)$

Answers

1. $y = -3x - 11$ **2.** $y = 5x + 11$
3. $y = 6x - 23$ **4.** $y = -\dfrac{2}{3}x + \dfrac{8}{3}$

b FINDING AN EQUATION OF A LINE WHEN TWO POINTS ARE GIVEN

We can also use the point-slope equation or the slope-intercept equation to find an equation of a line when two points are given.

EXAMPLE 2 Find an equation of the line containing the points $(3, 4)$ and $(-5, 2)$.

First, we find the slope:

$$m = \frac{4 - 2}{3 - (-5)} = \frac{2}{8}, \text{ or } \frac{1}{4}.$$

Now we have the slope and two points. We then proceed as we did in Example 1, using either point, and either the point-slope equation or the slope-intercept equation.

Using the Point–Slope Equation: We choose $(3, 4)$ and substitute 3 for x_1, 4 for y_1, and $\frac{1}{4}$ for m:

$$y - y_1 = m(x - x_1)$$
$$y - 4 = \tfrac{1}{4}(x - 3) \qquad \text{Substituting}$$
$$y - 4 = \tfrac{1}{4}x - \tfrac{3}{4}$$
$$y = \tfrac{1}{4}x - \tfrac{3}{4} + 4$$
$$y = \tfrac{1}{4}x - \tfrac{3}{4} + \tfrac{16}{4}$$
$$y = \tfrac{1}{4}x + \tfrac{13}{4}.$$

Using the Slope–Intercept Equation: We choose $(3, 4)$ and substitute 3 for x, 4 for y, and $\frac{1}{4}$ for m and solve for b:

$$y = mx + b$$
$$4 = \tfrac{1}{4} \cdot 3 + b \qquad \text{Substituting}$$
$$4 = \tfrac{3}{4} + b$$
$$4 - \tfrac{3}{4} = b$$
$$\tfrac{16}{4} - \tfrac{3}{4} = b$$
$$\tfrac{13}{4} = b. \qquad \text{Solving for } b$$

Finally, we use the equation $y = mx + b$ and substitute $\frac{1}{4}$ for m and $\frac{13}{4}$ for b:

$$y = \tfrac{1}{4}x + \tfrac{13}{4}.$$

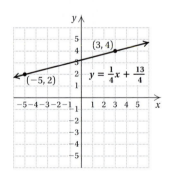

5. Find an equation of the line containing the points $(3, -5)$ and $(-1, 4)$.

6. Find an equation of the line containing the points $(-3, 11)$ and $(-4, 20)$.

◀ **Do Exercises 5 and 6.**

Answers

5. $y = -\dfrac{9}{4}x + \dfrac{7}{4}$ **6.** $y = -9x - 16$

a   Find an equation of the line having the given slope and containing the given point.

1. $m = 4,\ (5, 2)$

2. $m = 5,\ (4, 3)$

3. $m = -2,\ (2, 8)$

4. $m = -3,\ (9, 6)$

5. $m = 3,\ (-2, -2)$

6. $m = 1,\ (-1, -7)$

7. $m = -3,\ (-2, 0)$

8. $m = -2,\ (8, 0)$

9. $m = 0,\ (0, 4)$

10. $m = 0,\ (0, -7)$

11. $m = -\frac{4}{5},\ (2, 3)$

12. $m = \frac{2}{3},\ (1, -2)$

b   Find an equation of the line containing the given pair of points.

13. $(2, 5)$ and $(4, 7)$

14. $(1, 4)$ and $(5, 6)$

15. $(-1, -1)$ and $(9, 9)$

16. $(-3, -3)$ and $(2, 2)$

17. $(0, -5)$ and $(3, 0)$

18. $(-4, 0)$ and $(0, 7)$

19. $(-4, -7)$ and $(-2, -1)$

20. $(-2, -3)$ and $(-4, -6)$

21. $(0, 0)$ and $(-4, 7)$

22. $(0, 0)$ and $(6, 1)$

23. $\left(\frac{2}{3}, \frac{3}{2}\right)$ and $\left(-3, \frac{5}{6}\right)$

24. $\left(\frac{1}{4}, -\frac{1}{2}\right)$ and $\left(\frac{3}{4}, 6\right)$

Synthesis

25. Find an equation of the line that has the same y-intercept as the line $2x - y = -3$ and contains the point $(-1, -2)$.

26. Find an equation of the line with the same slope as the line $\frac{1}{2}x - \frac{1}{3}y = 10$ and the same y-intercept as the line $\frac{1}{4}x + 3y = -2$.

C

Higher Roots

OBJECTIVES

OBJECTIVES

a Find higher roots of real numbers.

b Simplify radical expressions using the product rule and the quotient rule.

In this appendix, we study *higher* roots, such as cube roots or fourth roots.

a HIGHER ROOTS

Recall that c is a square root of a if $c^2 = a$. A similar definition can be made for *cube roots*.

CUBE ROOT

The number c is the **cube root** of a if $c^3 = a$.

Every real number has exactly *one* real-number cube root. The symbolism $\sqrt[3]{a}$ is used to represent the cube root of a. In the radical $\sqrt[3]{a}$, the number 3 is called the **index** and a is called the **radicand**.

EXAMPLE 1 Find $\sqrt[3]{8}$.

The cube root of 8 is the number whose cube is 8. Since $2^3 = 2 \cdot 2 \cdot 2 = 8$, the cube root of 8 is 2, so $\sqrt[3]{8} = 2$.

EXAMPLE 2 Find $\sqrt[3]{-125}$.

The cube root of -125 is the number whose cube is -125. Since $(-5)^3 = (-5)(-5)(-5) = -125$, the cube root of -125 is -5, so $\sqrt[3]{-125} = -5$.

◀ **Do Exercises 1–3.**

Find each of the following.

1. $\sqrt[3]{27}$

2. $\sqrt[3]{-8}$

3. $\sqrt[3]{216}$

Positive real numbers always have *two* nth roots (one positive and one negative) when n is even, but we refer to the *positive nth root* of a positive number a as the *nth root* and denote it $\sqrt[n]{a}$. For example, although both -3 and 3 are fourth roots of 81, since $(-3)^4 = 81$ and $3^4 = 81$, 3 is considered to be *the* fourth root of 81. In symbols, $\sqrt[4]{81} = 3$.

nTH ROOT

The number c is the **nth root** of a if $c^n = a$.

If n is odd, then there is exactly one real-number nth root of a and $\sqrt[n]{a}$ represents that root.

If n is even and a is positive, then $\sqrt[n]{a}$ represents the nonnegative nth root.

Even roots of negative numbers are not real numbers.

EXAMPLES Find the root of each of the following.

3. $\sqrt[4]{16} = 2$ Since $2^4 = 2 \cdot 2 \cdot 2 \cdot 2 = 16$

4. $\sqrt[4]{-16}$ is not a real number, because it is an even root of a negative number.

Answers

1. 3   2. -2 3. 6

5. $\sqrt[5]{32} = 2$ Since $2^5 = 2 \cdot 2 \cdot 2 \cdot 2 \cdot 2 = 32$

6. $\sqrt[5]{-32} = -2$ Since $(-2)^5 = (-2)(-2)(-2)(-2)(-2) = -32$

7. $-\sqrt[3]{64} = -(\sqrt[3]{64})$ This is the opposite of $\sqrt[3]{64}$.

$\quad\quad\quad = -4$ Since $4^3 = 4 \cdot 4 \cdot 4 = 64$

Do Exercises 4–9. ▶

Find the root, if it exists, of each of the following.

4. $\sqrt[5]{1}$ **5.** $\sqrt[5]{-1}$

6. $\sqrt[4]{-81}$ **7.** $\sqrt[4]{81}$

8. $\sqrt[3]{-216}$ **9.** $-\sqrt[3]{216}$

Some roots occur so frequently that you may want to memorize them.

SQUARE ROOTS		CUBE ROOTS	FOURTH ROOTS	FIFTH ROOTS
$\sqrt{1} = 1$	$\sqrt{4} = 2$	$\sqrt[3]{1} = 1$	$\sqrt[4]{1} = 1$	$\sqrt[5]{1} = 1$
$\sqrt{9} = 3$	$\sqrt{16} = 4$	$\sqrt[3]{8} = 2$	$\sqrt[4]{16} = 2$	$\sqrt[5]{32} = 2$
$\sqrt{25} = 5$	$\sqrt{36} = 6$	$\sqrt[3]{27} = 3$	$\sqrt[4]{81} = 3$	$\sqrt[5]{243} = 3$
$\sqrt{49} = 7$	$\sqrt{64} = 8$	$\sqrt[3]{64} = 4$	$\sqrt[4]{256} = 4$	
$\sqrt{81} = 9$	$\sqrt{100} = 10$	$\sqrt[3]{125} = 5$	$\sqrt[4]{625} = 5$	
$\sqrt{121} = 11$	$\sqrt{144} = 12$	$\sqrt[3]{216} = 6$		

b PRODUCTS AND QUOTIENTS INVOLVING HIGHER ROOTS

The rules for working with products and quotients of square roots can be extended to products and quotients of nth roots.

THE PRODUCT AND QUOTIENT RULES FOR RADICALS

For any nonnegative real numbers A and B and any index n, $n \geq 2$,

$$\sqrt[n]{AB} = \sqrt[n]{A} \cdot \sqrt[n]{B} \quad \text{and} \quad \sqrt[n]{\frac{A}{B}} = \frac{\sqrt[n]{A}}{\sqrt[n]{B}}.$$

EXAMPLES Simplify.

8. $\sqrt[3]{40} = \sqrt[3]{8 \cdot 5}$ Factoring the radicand. 8 is a perfect cube.

$\quad\quad = \sqrt[3]{8} \cdot \sqrt[3]{5}$ Using the product rule

$\quad\quad = 2\sqrt[3]{5}$

9. $\sqrt[3]{\frac{125}{27}} = \frac{\sqrt[3]{125}}{\sqrt[3]{27}}$ Using the quotient rule

$\quad\quad\quad = \frac{5}{3}$ Simplifying. 125 and 27 are perfect cubes.

10. $\sqrt[4]{1250} = \sqrt[4]{2 \cdot 625}$ Factoring the radicand. 625 is a perfect fourth power.

$\quad\quad\quad = \sqrt[4]{2 \cdot 5 \cdot 5 \cdot 5 \cdot 5}$

$\quad\quad\quad = 5\sqrt[4]{2}$ Simplifying

11. $\sqrt[5]{\frac{2}{243}} = \frac{\sqrt[5]{2}}{\sqrt[5]{243}}$ Using the quotient rule

$\quad\quad\quad = \frac{\sqrt[5]{2}}{3}$ Simplifying. 243 is a perfect fifth power.

Do Exercises 10–13. ▶

Simplify.

10. $\sqrt[3]{24}$ **11.** $\sqrt[4]{\frac{81}{256}}$

12. $\sqrt[5]{96}$ **13.** $\sqrt[3]{\frac{4}{125}}$

Answers

4. 1   **5.** −1   **6.** Not a real number   **7.** 3

8. −6   **9.** −6   **10.** $2\sqrt[3]{3}$ **11.** $\frac{3}{4}$

12. $2\sqrt[5]{3}$ **13.** $\frac{\sqrt[3]{4}}{5}$

a Simplify. If an expression does not represent a real number, state this.

1. $\sqrt[3]{125}$

2. $\sqrt[3]{-27}$

3. $\sqrt[3]{-1000}$

4. $\sqrt[3]{8}$

5. $\sqrt[4]{1}$

6. $-\sqrt[5]{32}$

7. $\sqrt[4]{-256}$

8. $\sqrt[6]{-1}$

9. $-\sqrt[3]{-216}$

10. $\sqrt[3]{-125}$

11. $\sqrt[4]{256}$

12. $-\sqrt[3]{-8}$

13. $\sqrt[4]{10{,}000}$

14. $\sqrt[3]{-64}$

15. $-\sqrt[4]{81}$

16. $-\sqrt[3]{1}$

17. $-\sqrt[4]{-16}$

18. $\sqrt[6]{64}$

19. $-\sqrt[3]{125}$

20. $\sqrt[3]{1000}$

21. $\sqrt[5]{t^5}$ t

22. $\sqrt[7]{y^7}$ y

23. $-\sqrt[3]{x^3}$

24. $-\sqrt[9]{a^9}$

25. $\sqrt[3]{64}$

26. $-\sqrt[3]{216}$

27. $\sqrt[3]{-343}$

28. $\sqrt[5]{-243}$

29. $\sqrt[5]{-3125}$

30. $\sqrt[4]{625}$

31. $\sqrt[6]{1{,}000{,}000}$

32. $\sqrt[5]{243}$

33. $-\sqrt[5]{-100{,}000}$

34. $-\sqrt[4]{-10{,}000}$

35. $-\sqrt[3]{343}$

36. $\sqrt[3]{512}$

37. $\sqrt[8]{-1}$

38. $\sqrt[6]{-64}$

39. $\sqrt[5]{3125}$

40. $\sqrt[4]{-625}$

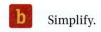 Simplify.

41. $\sqrt[3]{54}$

42. $\sqrt[5]{64}$

43. $\sqrt[4]{324}$

44. $\sqrt[3]{81}$

45. $\sqrt[3]{\dfrac{27}{64}}$

46. $\sqrt[3]{\dfrac{125}{64}}$

47. $\sqrt[4]{512}$

48. $\sqrt[3]{375}$

49. $\sqrt[5]{128}$

50. $\sqrt[4]{112}$

51. $\sqrt[4]{\dfrac{256}{625}}$

52. $\sqrt[5]{\dfrac{243}{32}}$

53. $\sqrt[3]{\dfrac{17}{8}}$

54. $\sqrt[5]{\dfrac{11}{32}}$

55. $\sqrt[3]{250}$

56. $\sqrt[5]{160}$

57. $\sqrt[5]{486}$

58. $\sqrt[3]{128}$

59. $\sqrt[4]{\dfrac{13}{81}}$

60. $\sqrt[3]{\dfrac{10}{27}}$

61. $\sqrt[4]{\dfrac{7}{16}}$

62. $\sqrt[4]{\dfrac{27}{256}}$

63. $\sqrt[4]{\dfrac{16}{625}}$

64. $\sqrt[3]{\dfrac{216}{27}}$

Synthesis

Simplify. If an expression does not represent a real number, state this.

65. $\sqrt[3]{\sqrt{64}}$

66. $\sqrt{\sqrt[3]{-64}}$

67. $\sqrt[3]{\sqrt[3]{1,000,000,000}}$

68. $\sqrt{-\sqrt[3]{-1}}$

D Sets

a NAMING SETS

To name the set of whole numbers less than 6, we can use the **roster method**, as follows: $\{0, 1, 2, 3, 4, 5\}$.

The set of real numbers x such that x is less than 6 cannot be named by listing all its members because there are infinitely many. We name such a set using **set-builder notation**, as follows: $\{x | x < 6\}$. This is read "The set of all x such that x is less than 6."

◀ **Do Exercises 1 and 2.**

b SET MEMBERSHIP AND SUBSETS

The symbol $\in$ means **is a member of** or **belongs to**, or **is an element of**. Thus, $x \in A$ means x is a member of A or x belongs to A or x is an element of A.

EXAMPLE 1 Classify each of the following as true or false.

a) $1 \in \{1, 2, 3\}$

b) $1 \in \{2, 3\}$

c) $4 \in \{x | x \text{ is an even whole number}\}$

d) $5 \in \{x | x \text{ is an even whole number}\}$

a) Since 1 *is* listed as a member of the set, $1 \in \{1, 2, 3\}$ is true.

b) Since 1 *is not* a member of $\{2, 3\}$, the statement $1 \in \{2, 3\}$ is false.

c) Since 4 *is* an even whole number, $4 \in \{x | x \text{ is an even whole number}\}$ is a true statement.

d) Since 5 *is not* even, $5 \in \{x | x \text{ is an even whole number}\}$ is false.

Set membership can be illustrated with a diagram, as shown here.

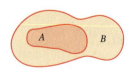

◀ **Do Exercises 3 and 4.**

If every element of A is an element of B, then A is a **subset** of B. This is denoted $A \subseteq B$. The set of whole numbers is a subset of the set of integers. The set of rational numbers is a subset of the set of real numbers.

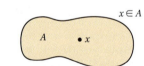

EXAMPLE 2 Classify each of the following as true or false.

a) $\{1, 2\} \subseteq \{1, 2, 3, 4\}$ **b)** $\{p, q, r, w\} \subseteq \{a, p, r, z\}$

c) $\{x | x < 6\} \subseteq \{x | x \leq 11\}$

a) Since every element of $\{1, 2\}$ is in the set $\{1, 2, 3, 4\}$, the statement $\{1, 2\} \subseteq \{1, 2, 3, 4\}$ is true.

Name each set using the roster method.

1. The set of whole numbers 0 through 7

2. $\{x | \text{ the square of } x \text{ is } 25\}$

Determine whether each of the following is true or false.

3. $8 \in \{x | x \text{ is an even whole number}\}$

4. $2 \in \{x | x \text{ is a prime number}\}$

Answers

1. $\{0, 1, 2, 3, 4, 5, 6, 7\}$ **2.** $\{-5, 5\}$
3. True   **4.** True

b) Since q ∈ {p, q, r, w}, but q ∉ {a, p, r, z}, the statement {p, q, r, w} ⊆ {a, p, r, z} is false.

c) Since every number that is less than 6 is also less than or equal to 11, the statement $\{x \,|\, x < 6\} \subseteq \{x \,|\, x \leq 11\}$ is true.

Do Exercises 5–7. ▶

Determine whether each of the following is true or false.

5. $\{-2, -3, 4\} \subseteq \{-5, -4, -2, 7, -3, 5, 4\}$

6. {a, e, i, o, u} ⊆ The set of all consonants

7. $\{x \,|\, x \leq -8\} \subseteq \{x \,|\, x \leq -7\}$

C INTERSECTIONS AND UNIONS

The **intersection** of sets A and B, denoted $A \cap B$, is the set of members that are common to both sets.

EXAMPLE 3 Find the intersection.

a) {0, 1, 3, 5, 25} ∩ {2, 3, 4, 5, 6, 7, 9}   **b)** {a, p, q, w} ∩ {p, q, t}

a) {0, 1, 3, 5, 25} ∩ {2, 3, 4, 5, 6, 7, 9} = {3, 5}

b) {a, p, q, w} ∩ {p, q, t} = {p, q}

Set intersection can be illustrated with a diagram, as shown here.

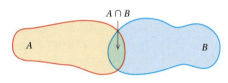

The set without members is known as the **empty set**, and is often named ∅, and sometimes { }. Each of the following is a description of the empty set:

{2, 3} ∩ {5, 6, 7};

{x | x is an even natural number} ∩ {x | x is an odd natural number}.

Do Exercises 8–10. ▶

Two sets A and B can be combined to form a set that contains the members of A as well as those of B. The new set is called the **union** of A and B, denoted $A \cup B$.

EXAMPLE 4 Find the union.

a) {0, 5, 7, 13, 27} ∪ {0, 2, 3, 4, 5}   **b)** {a, c, e, g} ∪ {b, d, f}

a) {0, 5, 7, 13, 27} ∪ {0, 2, 3, 4, 5} = {0, 2, 3, 4, 5, 7, 13, 27}
    Note that the 0 and the 5 are *not* listed twice in the solution.

b) {a, c, e, g} ∪ {b, d, f} = {a, b, c, d, e, f, g}

Set union can be illustrated with a diagram, as shown here.

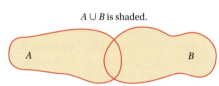

The solution set of the equation $(x - 3)(x + 2) = 0$ is {3, −2}. This set is the union of the solution sets of $x - 3 = 0$ and $x + 2 = 0$, which are {3} and {−2}.

Do Exercises 11–13. ▶

Find the intersection.

8. {−2, −3, 4, −4, 8} ∩ {−5, −4, −2, 7, −3, 5, 4}

9. {a, e, i, o, u} ∩ {m, a, r, v, i, n}

10. {a, e, i, o, u} ∩ The set of all consonants

Find the union.

11. {−2, −3, 4, −4, 8} ∪ {−5, −4, −2, 7, −3, 5, 4}

12. {a, e, i, o, u} ∪ {m, a, r, v, i, n}

13. {a, e, i, o, u} ∪ The set of all consonants

Answers

5. True   **6.** False   **7.** True
8. {−2, −3, 4, −4}   **9.** {a, i}   **10.** { }, or ∅
11. {−2, −3, 4, −4, 8, −5, 7, 5}
12. {a, e, i, o, u, m, r, v, n}
13. {a, b, c, d, e, f, g, h, i, j, k, l, m, n, o, p, q, r, s, t, u, v, w, x, y, z}

a   Name each set using the roster method.

1. The set of whole numbers 3 through 8

2. The set of whole numbers 101 through 107

3. The set of odd numbers between 40 and 50

4. The set of multiples of 5 between 11 and 39

5. $\{x \mid$ the square of x is 9$\}$

6. $\{x \mid x$ is the cube of 0.2$\}$

b   Classify each statement as true or false.

7. $2 \in \{x \mid x$ is an odd number$\}$

8. $7 \in \{x \mid x$ is an odd number$\}$

9. Kyle Busch $\in$ The set of all NASCAR drivers

10. Apple $\in$ The set of all fruit

11. $-3 \in \{-4, -3, 0, 1\}$

12. $0 \in \{-4, -3, 0, 1\}$

13. $\frac{2}{3} \in \{x \mid x$ is a rational number$\}$

14. Heads $\in$ The set of outcomes of flipping a penny

15. $\{4, 5, 8,\} \subseteq \{1, 3, 4, 5, 6, 7, 8, 9\}$

16. The set of vowels $\subseteq$ The set of consonants

17. $\{-1, -2, -3, -4, -5\} \subseteq \{-1, 2, 3, 4, 5\}$

18. The set of integers $\subseteq$ The set of rational numbers

c   Find the intersection.

19. $\{a, b, c, d, e\} \cap \{c, d, e, f, g\}$

20. $\{a, e, i, o, u\} \cap \{q, u, i, c, k\}$

21. $\{1, 2, 5, 10\} \cap \{0, 1, 7, 10\}$

22. $\{0, 1, 7, 10\} \cap \{0, 1, 2, 5\}$ **23.** $\{1, 2, 5, 10\} \cap \{3, 4, 7, 8\}$ **24.** $\{a, e, i, o, u\} \cap \{m, n, f, g, h\}$

Find the union.

25. $\{a, e, i, o, u\} \cup \{q, u, i, c, k\}$ **26.** $\{a, b, c, d, e\} \cup \{c, d, e, f, g\}$

27. $\{0, 1, 7, 10\} \cup \{0, 1, 2, 5\}$ **28.** $\{1, 2, 5, 10\} \cup \{0, 1, 7, 10\}$

29. $\{a, e, i, o, u\} \cup \{m, n, f, g, h\}$ **30.** $\{1, 2, 5, 10\} \cup \{a, b\}$

Synthesis

31. Find the union of the set of integers and the set of whole numbers.

32. Find the intersection of the set of odd integers and the set of even integers.

33. Find the union of the set of rational numbers and the set of irrational numbers.

34. Find the intersection of the set of even integers and the set of positive rational numbers.

35. Find the intersection of the set of rational numbers and the set of irrational numbers.

36. Find the union of the set of negative integers, the set of positive integers, and the set containing 0.

37. For a set A, find each of the following.

    a) $A \cup \varnothing$ **b)** $A \cup A$
    c) $A \cap A$ **d)** $A \cap \varnothing$

38. A set is *closed* under an operation if, when the operation is performed on its members, the result is in the set. For example, the set of real numbers is closed under the operation of addition since the sum of any two real numbers is a real number.

    a) Is the set of even numbers closed under addition?

    b) Is the set of odd numbers closed under addition?

    c) Is the set $\{0, 1\}$ closed under addition?

    d) Is the set $\{0, 1\}$ closed under multiplication?

    e) Is the set of real numbers closed under multiplication?

    f) Is the set of integers closed under division?

39. Experiment with sets of various types and determine whether the following distributive law for sets is true:

$$A \cap (B \cup C) = (A \cap B) \cup (A \cap C).$$

Mean, Median, and Mode

OBJECTIVE

a Find the mean (average), the median, and the mode of a set of data and solve related applied problems.

a MEAN, MEDIAN, AND MODE

One way to analyze data is to look for a single representative number, called a **center point** or **measure of central tendency**. Those most often used are the **mean** (or **average**), the **median**, and the **mode**.

Mean

> **MEAN, OR AVERAGE**
>
> The **mean**, or **average**, of a set of numbers is the sum of the numbers divided by the number of addends.

Find the mean. Round to the nearest tenth.

1. 28, 103, 39

2. 85, 46, 105.7, 22.1

3. A student scored the following on five tests:

    78, 95, 84, 100, 82.

What was the average score?

EXAMPLE 1 Consider the number of prescriptions, in millions, filled in supermarkets in 1997, 1999, 2001, 2003, 2005, and 2007:

    269, 357, 418, 462, 465, 478.

What is the mean, or average, of the numbers?
Source: IMS HEALTH and NACDS Economics Department

First, we add the numbers:

$$269 + 357 + 418 + 462 + 465 + 478 = 2449.$$

Then we divide by the number of addends, 6:

$$\frac{2449}{6} \approx 408. \quad \text{Rounding to the nearest one}$$

The mean, or average, number of prescriptions filled in supermarkets in those six years is about 408 million.

Note that if the number of prescriptions had been the average (same) for each of the six years, we would have

$$408 + 408 + 408 + 408 + 408 + 408 = 2448 \approx 2449.$$

The number 408 is called the mean, or average, of the set of numbers.

◀ **Do Exercises 1–3.**

Median

The *median* is useful when we wish to de-emphasize extreme values. For example, suppose five workers in a technology company manufactured the following number of computers during one month's work:

    Sarah: 88   Jen: 94   Matt: 92
    Mark: 91   Pat: 66

Let's first list the values in order from smallest to largest:

    66 88 91 92 94.
            ↑
        Middle number

The middle number—in this case, 91—is the **median**.

Answers

1. 56.7   **2.** 64.7   **3.** 87.8

MEDIAN

Once a set of data has been arranged from smallest to largest, the **median** of the set of data is the middle number if there is an odd number of data numbers. If there is an even number of data numbers, then there are two middle numbers and the median is the *average* of the two middle numbers.

EXAMPLE 2 What is the median of the following set of yearly salaries?

$76,000, $58,000, $87,000, $32,500, $64,800, $62,500

We first rearrange the numbers in order from smallest to largest.

$32,500, $58,000, $62,500, $64,800, $76,000, $87,000
↑
Median

There is an even number of numbers. We look for the middle two, which are $62,500 and $64,800. In this case, the median is the average of $62,500 and $64,800:

$$\frac{\$62{,}500 + \$64{,}800}{2} = \$63{,}650.$$

Do Exercises 4–6. ▶

Mode

The last center point we consider is called the *mode*. A number that occurs most often in a set of data can be considered a representative number, or center point.

MODE

The **mode** of a set of data is the number or numbers that occur most often. If each number occurs the same number of times, there is *no* mode.

EXAMPLE 3 Find the mode of the following data:

23, 24, 27, 18, 19, 27

The number that occurs most often is 27. Thus the mode is 27. ■

EXAMPLE 4 Find the mode of the following data:

83, 84, 84, 84, 85, 86, 87, 87, 87, 88, 89, 90.

There are two numbers that occur most often, 84 and 87. Thus the modes are 84 and 87. ■

EXAMPLE 5 Find the mode of the following data:

115, 117, 211, 213, 219.

Each number occurs the same number of times. The set of data has *no* mode.

Do Exercises 7–10. ▶

Find the median.

4. 17, 13, 18, 14, 19

5. 17, 18, 16, 19, 13, 14

6. 122, 102, 103, 91, 83, 81, 78, 119, 88

Find any modes that exist.

7. 33, 55, 55, 88, 55

8. 90, 54, 88, 87, 87, 54

9. 23.7, 27.5, 54.9, 17.2, 20.1

10. In conducting laboratory tests, Carole discovers bacteria in different lab dishes grew to the following areas, in square millimeters:

25, 19, 29, 24, 28.

a) What is the mean?
b) What is the median?
c) What is the mode?

Answers

4. 17 **5.** 16.5 **6.** 91 **7.** 55 **8.** 54, 87
9. No mode exists. **10. (a)** 25 mm^2;
(b) 25 mm^2; **(c)** no mode exists.

E Exercise Set

a For each set of numbers, find the mean (average), the median, and any modes that exist.

1. 17, 19, 29, 18, 14, 29

2. 13, 32, 25, 27, 13

3. 4.3, 7.4, 1.2, 5.7, 8.3

4. 13.4, 13.4, 12.6, 42.9

5. 234, 228, 234, 229, 234, 278

6. $29.95, $28.79, $30.95, $28.79

7. *Atlantic Storms and Hurricanes.* The following bar graph shows the number of Atlantic storms or hurricanes that formed in various months from 1980 to 2007. What is the average number for the 9 months given? the median? the mode?

Atlantic Storms and Hurricanes

Tropical storm and hurricane formation in 1980–2007, by month

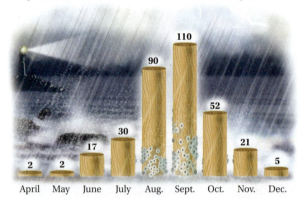

SOURCE: Colorado State University, Department of Atmospheric Science, Phil Klotzbach, Ph.D., Research Scientist

8. *Tornadoes.* The following bar graph shows the average number of tornado deaths by month since 1950. What is the average number of tornado deaths for the 12 months? the median? the mode?

Average Number of Deaths by Tornado by Month

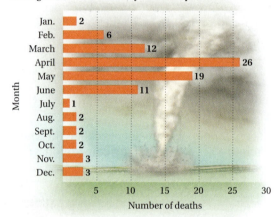

SOURCE: National Weather Service's Storm Prediction Center

9. *Brussels Sprouts.* The following prices per stalk of Brussels sprouts were found at five supermarkets:

$3.99, $4.49, $4.99, $3.99, $3.49.

What was the average price per stalk? the median price? the mode?

10. *Cheddar Cheese Prices.* The following prices per pound of sharp cheddar cheese were found at five supermarkets:

$5.99, $6.79, $5.99, $6.99, $6.79.

What was the average price per pound? the median price? the mode?

Inequalities and Interval Notation

a INEQUALITIES AND INTERVAL NOTATION

The **graph** of an inequality is a drawing that represents its solutions. An inequality in one variable can be graphed on the number line.

EXAMPLE 1 Graph $x < 4$ on the number line.

The solutions are all real numbers less than 4, so we shade all numbers less than 4 on the number line. To indicate that 4 is not a solution, we use a right parenthesis ")" at 4.

We can write the solution set for $x < 4$ using **set-builder notation**: $\{x \mid x < 4\}$. This is read "The set of all x such that x is less than 4."

Another way to write solutions of an inequality in one variable is to use **interval notation**. Interval notation uses parentheses ( ) and brackets [ ].

If a and b are real numbers such that $a < b$, we define the interval (a, b) as the set of all numbers between but not including a and b—that is, the set of all x for which $a < x < b$. Thus,

$$(a, b) = \{x \mid a < x < b\}.$$

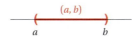

The points a and b are the **endpoints** of the interval. The parentheses indicate that the endpoints are *not* included in the graph.

The interval $[a, b]$ is defined as the set of all numbers x for which $a \le x \le b$. Thus,

$$[a, b] = \{x \mid a \le x \le b\}.$$

The brackets indicate that the endpoints *are* included in the graph.

The following intervals include one endpoint and exclude the other:

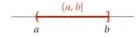

$(a, b] = \{x \mid a < x \le b\}.$ The graph excludes a and includes b.

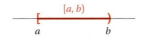

$[a, b) = \{x \mid a \le x < b\}.$ The graph includes a and excludes b.

Some intervals extend without bound in one or both directions. We use the symbols ∞, read "infinity," and $-\infty$, read "negative infinity," to name these intervals. The notation (a, ∞) represents the set of all numbers greater than a—that is,

$$(a, \infty) = \{x \mid x > a\}.$$

OBJECTIVES

a Write interval notation for the solution set or the graph of an inequality.

b Solve an inequality expressing the solution in interval notation and then graph the inequality.

Caution!

Do not confuse the *interval* (a, b) with the *ordered pair* (a, b), which denotes a point in the plane. The context in which the notation appears usually makes the meaning clear.

Similarly, the notation $(-\infty, a)$ represents the set of all numbers less than a—that is,

$$(-\infty, a) = \{x \mid x < a\}.$$

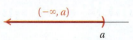

The notations $[a, \infty)$ and $(-\infty, a]$ are used when we want to include the endpoint a. The interval $(-\infty, \infty)$ names the set of all real numbers.

$$(-\infty, \infty) = \{x \mid x \text{ is a real number}\}$$

Interval notation is summarized in the following table.

Intervals: Notation and Graphs

INTERVAL NOTATION	SET NOTATION	GRAPH
(a, b)	$\{x \mid a < x < b\}$	
$[a, b]$	$\{x \mid a \leq x \leq b\}$	
$[a, b)$	$\{x \mid a \leq x < b\}$	
$(a, b]$	$\{x \mid a < x \leq b\}$	
(a, ∞)	$\{x \mid x > a\}$	
$[a, \infty)$	$\{x \mid x \geq a\}$	
$(-\infty, b)$	$\{x \mid x < b\}$	
$(-\infty, b]$	$\{x \mid x \leq b\}$	
$(-\infty, \infty)$	$\{x \mid x \text{ is a real number}\}$	

········· **Caution!** ·········

Whenever the symbol ∞ is included in interval notation, a right parenthesis ")" is used. Similarly, when $-\infty$ is included, a left parenthesis "(" is used.

··

Write interval notation for the given set or graph.

1. $\{x \mid -4 \leq x < 5\}$

2. $\{x \mid x \leq -2\}$

3. $\{x \mid 6 \geq x > 2\}$

4.
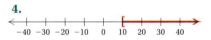

5.
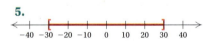

EXAMPLES Write interval notation for the given set or graph.

2. $\{x \mid -4 < x < 5\} = (-4, 5)$
3. $\{x \mid x \geq -2\} = [-2, \infty)$
4. $\{x \mid 7 > x \geq 1\} = \{x \mid 1 \leq x < 7\} = [1, 7)$
5.
$$\underset{-6\ -5\ -4\ -3\ -2\ -1\ \ 0\ \ 1\ \ 2\ \ 3\ \ 4\ \ 5\ \ 6}{\longleftrightarrow}$$
$$(-2, 4]$$
6.
$$\underset{-6\ -5\ -4\ -3\ -2\ -1\ \ 0\ \ 1\ \ 2\ \ 3\ \ 4\ \ 5\ \ 6}{\longleftrightarrow}$$
$$(-\infty, -1)$$

◀ **Do Exercises 1–5.**

Answers

1. $[-4, 5)$ 2. $(-\infty, -2]$ 3. $(2, 6]$
4. $[10, \infty)$ 5. $[-30, 30]$

b SOLVING INEQUALITIES

We now express the solution set in both set-builder notation and interval notation.

EXAMPLE 7 Solve and graph: $x + 5 > 1$.

We have

$$x + 5 > 1$$

$$x + 5 - 5 > 1 - 5 \qquad \text{Using the addition principle:}$$
$$\text{adding } -5 \text{ or subtracting } 5$$

$$x > -4.$$

The solution set is $\{x \mid x > -4\}$, or $(-4, \infty)$. The graph is as follows:

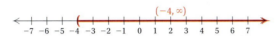

Do Exercises 6 and 7. ▶

EXAMPLE 8 Solve and graph: $4x - 1 \geq 5x - 2$.

We have

$$4x - 1 \geq 5x - 2$$

$$4x - 1 + 2 \geq 5x - 2 + 2 \qquad \text{Adding 2}$$

$$4x + 1 \geq 5x \qquad \text{Simplifying}$$

$$4x + 1 - 4x \geq 5x - 4x \qquad \text{Subtracting } 4x$$

$$1 \geq x, \text{ or } x \leq 1. \qquad \text{Simplifying}$$

The solution set is $\{x \mid x \geq 1\}$, or $(-\infty, 1]$.

EXAMPLE 9 Solve: $16 - 7y \geq 10y - 4$.

We have

$$16 - 7y \geq 10y - 4$$

$$-16 + 16 - 7y \geq -16 + 10y - 4 \qquad \text{Adding } -16$$

$$-7y \geq 10y - 20 \qquad \text{Collecting like terms}$$

$$-10y + (-7y) \geq -10y + 10y - 20 \qquad \text{Adding } -10y$$

$$-17y \geq -20 \qquad \text{Collecting like terms}$$

$$\frac{-17y}{-17} \leq \frac{-20}{-17} \qquad \begin{array}{l}\text{Dividing by } -17. \text{ The}\\ \text{symbol must be reversed.}\end{array}$$

$$y \leq \frac{20}{17}. \qquad \text{Simplifying}$$

The solution set is $\left\{y \mid y \leq \frac{20}{17}\right\}$, or $\left(-\infty, \frac{20}{17}\right]$.

Do Exercises 8 and 9. ▶

Solve and graph.

6. $x + 6 > 9$

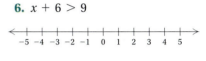

7. $x + 4 \leq 7$

8. Solve and graph:
$$2x - 3 \geq 3x - 1.$$

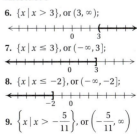

9. Solve: $3 - 9x < 2x + 8$.

Answers

6. $\{x \mid x > 3\}$, or $(3, \infty)$;

7. $\{x \mid x \leq 3\}$, or $(-\infty, 3]$;

8. $\{x \mid x \leq -2\}$, or $(-\infty, -2]$;

9. $\left\{x \mid x > -\dfrac{5}{11}\right\}$, or $\left(-\dfrac{5}{11}, \infty\right)$

a   Write interval notation for the given set or graph.

1. $\{x \mid x < 5\}$

2. $\{t \mid t \geq -5\}$

3. $\{x \mid -3 \leq x \leq 3\}$

4. $\{t \mid -10 < t \leq 10\}$

5. $\{x \mid -4 > x > -8\}$

6. $\{x \mid 13 > x \geq 5\}$

7.

8.

9. $-\sqrt{2}$

10.

b   Solve and graph.

11. $x + 2 > 1$

12. $y + 4 < 10$

13. $\frac{2}{3}x > 2$

14. $0.6x < 30$

15. $0.3x < -18$

16. $8x \geq 24$

17. $a - 9 \leq -31$

18. $y - 9 > -18$

Solve.

19. $-9x \geq -8.1$

20. $-5y \leq 3.5$

21. $-\frac{3}{4}x \geq -\frac{5}{8}$

22. $-\frac{1}{8}y \leq -\frac{9}{8}$

23. $2x + 7 < 19$

24. $5y + 13 > 28$

25. $5y + 2y \leq -21$

26. $-9x + 3x \geq -24$

27. $2y - 7 < 5y - 9$

28. $8x - 9 < 3x - 11$

Answers

CHAPTER R

Exercise Set R.1, p. 7

RC1. Smaller   **RC2.** Composite   **RC3.** Larger   **RC4.** Prime
1. 1, 2, 4, 5, 10, 20   **3.** 1, 2, 3, 4, 6, 8, 9, 12, 18, 24, 36, 72   **5.** $3 \cdot 5$
7. $7 \cdot 7$ **9.** $2 \cdot 3 \cdot 3$ **11.** $2 \cdot 2 \cdot 2 \cdot 5$ **13.** $2 \cdot 3 \cdot 3 \cdot 5$
15. $2 \cdot 3 \cdot 5 \cdot 7$ **17.** $7 \cdot 13$ **19.** 12   **21.** 13   **23.** 1
25. 18   **27.** 6   **29.** 5   **31.** 35   **33.** $2 \cdot 2 \cdot 2 \cdot 3$;
$2 \cdot 2 \cdot 3 \cdot 3$; 72   **35.** $3; 3 \cdot 5$; 15   **37.** $2 \cdot 3 \cdot 5; 2 \cdot 2 \cdot 2 \cdot 5$; 120
39. 13; 23; 299   **41.** $2 \cdot 3 \cdot 3; 2 \cdot 3 \cdot 5$; 90
43. $2 \cdot 3 \cdot 5; 2 \cdot 2 \cdot 3 \cdot 3$; 180   **45.** $2 \cdot 2 \cdot 3; 2 \cdot 2 \cdot 7$; 84
47. $2 \cdot 2 \cdot 2 \cdot 3; 2 \cdot 2 \cdot 3 \cdot 3; 2 \cdot 2 \cdot 3$; 72
49. $5; 2 \cdot 3 \cdot 3; 3 \cdot 5$; 60   **51.** $2 \cdot 3; 2 \cdot 2 \cdot 3; 2 \cdot 3 \cdot 3$; 36
53. Every 60 years   **55.** Every 420 years   **57.** 30 strands
59. (a) No; not a multiple of 8; **(b)** no; it is a multiple of both 8 and 12, but it is not the least common multiple; **(c)** no; not a multiple of 8 or 12; **(d)** yes; it is a multiple of both 8 and 12 and is the smallest such multiple.

Calculator Corner, p. 15

1. $\frac{41}{24}$ **2.** $\frac{27}{112}$ **3.** $\frac{35}{16}$ **4.** $\frac{2}{3}$

Exercise Set R.2, p. 16

RC1. (d)   **RC2.** (a)   **RC3.** (c)   **RC4.** (f)   **RC5.** (e)
RC6. (b)
1. $\frac{9}{12}$ **3.** $\frac{60}{100}$ **5.** $\frac{104}{160}$ **7.** $\frac{21}{24}$ **9.** $\frac{20}{16}$ **11.** $\frac{391}{437}$ **13.** $\frac{2}{3}$
15. 4   **17.** $\frac{1}{7}$ **19.** 8   **21.** $\frac{1}{4}$ **23.** 5   **25.** $\frac{17}{21}$ **27.** $\frac{13}{7}$
29. $\frac{4}{3}$ **31.** $\frac{1}{12}$ **33.** $\frac{45}{16}$ **35.** $\frac{2}{3}$ **37.** $\frac{7}{6}$ **39.** $\frac{5}{6}$ **41.** $\frac{13}{20}$
43. $\frac{1}{2}$ **45.** $\frac{13}{24}$ **47.** $\frac{31}{16}$ **49.** $\frac{10}{3}$ **51.** $\frac{10}{3}$ **53.** $\frac{1}{2}$ **55.** $\frac{5}{36}$
57. 500   **59.** $\frac{3}{40}$ **61.** $\frac{99,999}{100}$ **63.** 900   **65.** $2 \cdot 2 \cdot 7$
66. $2 \cdot 2 \cdot 2 \cdot 7$ **67.** $2 \cdot 2 \cdot 2 \cdot 5 \cdot 5 \cdot 5$
68. $2 \cdot 2 \cdot 2 \cdot 2 \cdot 2 \cdot 2 \cdot 3$ **69.** 14   **70.** 5   **71.** 8   **72.** 18
73. 48   **74.** 392   **75.** 192   **76.** 150   **77.** $\frac{3}{4}$ **79.** 4   **81.** 1

Calculator Corner, p. 22

1. 40.42   **2.** 3.33   **3.** 0.69324   **4.** 2.38

Exercise Set R.3, p. 24

RC1. Thousandths   **RC2.** Tenths   **RC3.** Tens
RC4. Ten-thousandths   **RC5.** Hundredths
RC6. Thousands
1. $\frac{53}{10}$ **3.** $\frac{67}{100}$ **5.** $\frac{20,007}{10,000}$ **7.** $\frac{78,898}{10}$ **9.** 0.1   **11.** 0.0001
13. 9.999   **15.** 0.4578   **17.** 444.94   **19.** 390.617
21. 155.724   **23.** 63.79   **25.** 32.234   **27.** 26.835   **29.** 47.91
31. 1.9193   **33.** 13.212   **35.** 0.7998   **37.** 179.5   **39.** 1.40756
41. 3.60558   **43.** 2.3   **45.** 5.2   **47.** 0.023   **49.** 18.75
51. 660   **53.** 0.68   **55.** 0.34375   **57.** $1.1\overline{8}$ **59.** $0.\overline{5}$
61. $2.\overline{1}$ **63.** 745.07; 745.1; 745; 750; 700   **65.** 6780.51; 6780.5;
6781; 6780; 6800   **67.** $17.99; $18   **69.** $346.08; $346
71. $17   **73.** $190   **75.** 0.2857; 0.286; 0.29; 0.3; 0

---

77. 0.5897; 0.590; 0.59; 0.6; 1   **79.** $\frac{33}{32}$ **80.** $\frac{1}{48}$ **81.** $\frac{55}{64}$ **82.** $\frac{5}{4}$
83. $\frac{139}{210}$ **84.** $\frac{449}{336}$ **85.** $\frac{1023}{1000}$ **86.** $\frac{259}{210}$, or $\frac{37}{30}$ **87.** $2 \cdot 2 \cdot 2 \cdot 2 \cdot 13$
88. $2 \cdot 2 \cdot 2 \cdot 2 \cdot 2 \cdot 2 \cdot 2 \cdot 2 \cdot 5$ **89.** 6   **90.** 100
91. $0.\overline{714285}$ **93.** $0.\overline{6428571}$

Exercise Set R.4, p. 30

RC1. (g)   **RC2.** (f)   **RC3.** (c)   **RC4.** (b)   **RC5.** (h)
RC6. (e)   **RC7.** (d)   **RC8.** (a)
1. 0.13   **3.** 0.354; 0.201   **5.** 0.63   **7.** 0.941   **9.** 0.01
11. 0.0061   **13.** 2.4   **15.** 0.0325   **17.** $\frac{93}{100}; \frac{6}{100}$, or $\frac{3}{50}; \frac{1}{100}$
19. $\frac{39}{100}$ **21.** $\frac{88}{100}$ **23.** $\frac{60}{100}$ **25.** $\frac{289}{100}$ **27.** $\frac{110}{100}$ **29.** $\frac{42}{100,000}$
31. $\frac{250}{100}$ **33.** $\frac{347}{10,000}$ **35.** 21.4%   **37.** 14%   **39.** 99%
41. 100%   **43.** 0.47%   **45.** 7.2%   **47.** 920%   **49.** 0.68%
51. $16.\overline{6}$%, or $16\frac{2}{3}$%   **53.** 65%   **55.** 29%   **57.** 80%   **59.** 60%
61. $66.\overline{6}$%, or $66\frac{2}{3}$%   **63.** 175%   **65.** 75%   **67.** 40%   **69.** 2.25
70. $1.5\overline{4}$ **71.** 164.90974   **72.** 56.43   **73.** $\frac{123}{10,000}$ **74.** $\frac{507}{100}$ **75.** $\frac{29}{8}$
76. $\frac{1}{12}$ **77.** 18.23   **78.** 0.10   **79.** $\frac{343}{1000}$ **80.** $\frac{4}{5}$ **81.** 32%
83. 70%   **85.** 105%   **87.** 345%   **89.** 2.5%

Calculator Corner, p. 34

1. 40,353,607   **2.** 10.4976   **3.** 12,812.904   **4.** $\frac{64}{729}$

Calculator Corner, p. 35

1. 81   **2.** 2   **3.** 5932   **4.** 743.027   **5.** 783   **6.** 228,112.96

Exercise Set R.5, p. 36

RC1. Addition   **RC2.** Division   **RC3.** Multiplication
RC4. Subtraction
1. 5^4 **3.** 10^3 **5.** 10^6 **7.** 49   **9.** 59,049   **11.** 1   **13.** 5.29
15. 0.008   **17.** 416.16   **19.** $\frac{9}{64}$ **21.** 125   **23.** $\frac{27}{8}$
25. 1061.208   **27.** 25   **29.** 114   **31.** 33   **33.** 5   **35.** 324
37. 100   **39.** 1000   **41.** 22   **43.** 4   **45.** 102   **47.** 96
49. 24   **51.** 90   **53.** 8   **55.** 1   **57.** 50,000   **59.** 5
61. 27   **63.** $\frac{22}{45}$ **65.** $\frac{19}{66}$ **67.** 9   **69.** 31.25%
70. $183.\overline{3}$%, or $183\frac{1}{3}$%   **71.** $\frac{3}{667}$ **72.** $\frac{401}{728}$ **73.** $2 \cdot 2 \cdot 2 \cdot 2 \cdot 3$
74. 168   **75.** 10^2 **77.** 5^6

Summary and Review: Chapter R, p. 38

Vocabulary Reinforcement

1. Base   **2.** LCM   **3.** Factor   **4.** Numerator
5. Identity property of 1   **6.** Composite   **7.** Equivalent
8. Identity property of 0   **9.** Prime   **10.** GCF
11. Reciprocal   **12.** Exponent

Concept Reinforcement

1. True   **2.** False   **3.** True   **4.** False   **5.** True   **6.** False

Review Exercises

1. $2 \cdot 2 \cdot 23$ **2.** $2 \cdot 2 \cdot 2 \cdot 5 \cdot 5 \cdot 7$ **3.** 10 **4.** 9 **5.** 416
6. 90 **7.** $\frac{12}{30}$ **8.** $\frac{96}{184}$ **9.** $\frac{40}{64}$ **10.** $\frac{91}{84}$ **11.** $\frac{5}{12}$ **12.** $\frac{51}{91}$
13. $\frac{31}{36}$ **14.** $\frac{1}{4}$ **15.** $\frac{3}{5}$ **16.** $\frac{72}{25}$ **17.** $\frac{205}{144}$ **18.** $\frac{139}{72}$ **19.** $\frac{101}{54}$
20. $\frac{109}{84}$ **21.** $\frac{13}{72}$ **22.** $\frac{29}{144}$ **23.** $\frac{1}{12}$ **24.** $\frac{23}{90}$ **25.** $\frac{1797}{100}$
26. 0.2337 **27.** 2442.905 **28.** 86.0298 **29.** 9.342
30. 133.264 **31.** 430.8 **32.** 110.483 **33.** 55.6 **34.** 0.45
35. $1.58\overline{3}$ **36.** 34.1 **37.** 0.105 **38.** $\frac{357}{1000}$ **39.** 5.02%
40. 39.6% **41.** 62.5%, or $62\frac{1}{2}$% **42.** 116% **43.** 6^3
44. 1.1236 **45.** 119 **46.** 4 **47.** 29 **48.** 1 **49.** 7
50. 64 **51.** $\frac{103}{17}$

Understanding Through Discussion and Writing

1. Canceling is possible only when the numerator and the denominator of a fraction have common factors. **2.** The parentheses are not necessary. If we use the rules for order of operations, the multiplication will be performed first regardless of whether it is in parentheses. **3.** One approach would be to express 0.1 as $\frac{1}{10}$ and divide as follows: $5.367 \div \frac{1}{10} = 5.367 \cdot 10 = 53.67$.

Test: Chapter R, p. 41

1. [R.1a] $2 \cdot 2 \cdot 3 \cdot 5 \cdot 5$ **2.** [R.1b] 14 **3.** [R.1c] 120
4. [R.2a] $\frac{21}{49}$ **5.** [R.2a] $\frac{33}{48}$ **6.** [R.2b] $\frac{2}{3}$ **7.** [R.2b] $\frac{37}{61}$
8. [R.2c] $\frac{5}{36}$ **9.** [R.2c] $\frac{11}{40}$ **10.** [R.2c] $\frac{67}{36}$ **11.** [R.2c] $\frac{5}{36}$
12. [R.3a] $\frac{678}{100}$ **13.** [R.3a] 1.895 **14.** [R.3b] 99.0187
15. [R.3b] 1796.58 **16.** [R.3b] 435.072 **17.** [R.3b] 1.6
18. [R.3b] $2.0\overline{9}$ **19.** [R.3c] 234.7 **20.** [R.3c] 234.728
21. [R.4a] 0.007 **22.** [R.4b] $\frac{91}{100}$ **23.** [R.4d] 44%
24. [R.5b] 625 **25.** [R.5b] 1.44 **26.** [R.5c] 242
27. [R.5c] 20,000 **28.** [R.4a] 0.054 **29.** [R.4d] 1.2%
30. [R.2b] $19y$

CHAPTER 1

Exercise Set 1.1, p. 48

RC1. Division **RC2.** Multiplication **RC3.** Multiplication
RC4. Division
1. 32 min; 69 min; 81 min **3.** 260 mi **5.** 576 in^2 **7.** 1935 m^2
9. 56 **11.** 8 **13.** 1 **15.** 6 **17.** 2 **19.** $b + 7$, or $7 + b$
21. $c - 12$ **23.** $a + b$, or $b + a$ **25.** $x \div y$, or $\frac{x}{y}$, or x/y
27. $x + w$, or $w + x$ **29.** $n - m$ **31.** $2z$ **33.** $3m$, or $m \cdot 3$
35. $4a + 6$, or $6 + 4a$ **37.** $xy - 8$ **39.** $2t - 5$ **41.** $3n + 11$, or $11 + 3n$ **43.** $4x + 3y$, or $3y + 4x$ **45.** $s + 0.05s$
47. $65t$ miles **49.** $\$50 - x$ **51.** $\$12.50n$ **53.** $2 \cdot 3 \cdot 3 \cdot 3$
54. $2 \cdot 2 \cdot 2 \cdot 2 \cdot 2 \cdot 3$ **55.** $\frac{41}{56}$ **56.** $\frac{31}{54}$ **57.** 0.0515
58. 43,500 **59.** 96 **60.** 396 **61.** $\frac{1}{4}$ **63.** 0

Calculator Corner, p. 55

1. 8.717797887 **2.** 17.80449381 **3.** 67.08203932
4. 35.4807407 **5.** 3.141592654 **6.** 91.10618695
7. 530.9291585 **8.** 138.8663978

Calculator Corner, p. 56

1. -0.75 **2.** -0.45 **3.** -0.125 **4.** -1.8 **5.** -0.675
6. -0.6875 **7.** -3.5 **8.** -0.76

Calculator Corner, p. 58

1. 5 **2.** 17 **3.** 0 **4.** 6.48 **5.** 12.7 **6.** 0.9 **7.** $\frac{5}{7}$ **8.** $\frac{4}{3}$

Exercise Set 1.2, p. 59

RC1. H **RC2.** E **RC3.** J **RC4.** D **RC5.** B **RC6.** G
RC7. True **RC8.** True **RC9.** False **RC10.** False
1. 24; -2 **3.** 7,200,000,000,000; -460 **5.** 1454; -55

7. (number line, $\frac{10}{3}$ marked between 3 and 4) **9.** (number line, -5.2 marked near -5)

11. (number line, $-4\frac{2}{5}$ marked near -4) **13.** -0.875 **15.** $0.8\overline{3}$

17. $-1.1\overline{6}$ **19.** $0.\overline{6}$ **21.** 0.1 **23.** -0.5 **25.** 0.16 **27.** $>$
29. $<$ **31.** $<$ **33.** $<$ **35.** $>$ **37.** $>$ **39.** $<$ **41.** $>$
43. $<$ **45.** $<$ **47.** $x < -6$ **49.** $y \geq -10$ **51.** False
53. True **55.** True **57.** False **59.** 3 **61.** 11 **63.** $\frac{2}{3}$
65. 0 **67.** 2.65 **69.** 1.1 **70.** 0.238 **71.** 52%
72. 59.375%, or $59\frac{3}{8}$% **73.** 81 **74.** 1 **75.** 45 **76.** 0
77. $-\frac{2}{3}, -\frac{2}{5}, -\frac{1}{3}, -\frac{2}{7}, -\frac{1}{7}, \frac{1}{3}, \frac{2}{5}, \frac{3}{8}$ **79.** $\frac{1}{1}$ **81.** $\frac{50}{9}$

Exercise Set 1.3, p. 67

RC1. Right; right **RC2.** Left; left **RC3.** Right; left
RC4. Left; right
1. -7 **3.** -6 **5.** 0 **7.** -8 **9.** -7 **11.** -27 **13.** 0
15. -42 **17.** 0 **19.** 0 **21.** 3 **23.** -9 **25.** 7 **27.** 0
29. 35 **31.** -3.8 **33.** -8.1 **35.** $-\frac{1}{5}$ **37.** $-\frac{7}{9}$ **39.** $-\frac{3}{8}$
41. $-\frac{19}{24}$ **43.** $\frac{1}{24}$ **45.** $\frac{8}{15}$ **47.** $\frac{16}{45}$ **49.** 37 **51.** 50 **53.** -24
55. 26.9 **57.** -8 **59.** $\frac{13}{8}$ **61.** -43 **63.** $\frac{4}{3}$ **65.** 24 **67.** $\frac{3}{8}$
69. 13,796 ft **71.** $-3°$F **73.** Profit of $4300 **75.** He owes $85.
77. 39 **78.** $y > -3$ **79.** $-0.08\overline{3}$ **80.** 0.625 **81.** 0 **82.** 21.4
83. All positive numbers **85.** B

Exercise Set 1.4, p. 73

RC1. (c) **RC2.** (b) **RC3.** (d) **RC4.** (a)
1. -7 **3.** -6 **5.** 0 **7.** -4 **9.** -7 **11.** -6 **13.** 0
15. 14 **17.** 11 **19.** -14 **21.** 5 **23.** -1 **25.** 18 **27.** -3
29. -21 **31.** 5 **33.** -8 **35.** 12 **37.** -23 **39.** -68
41. -73 **43.** 116 **45.** 0 **47.** -1 **49.** $\frac{1}{12}$ **51.** $-\frac{17}{12}$ **53.** $\frac{1}{8}$
55. 19.9 **57.** -8.6 **59.** -0.01 **61.** -193 **63.** 500
65. -2.8 **67.** -3.53 **69.** $-\frac{1}{2}$ **71.** $\frac{6}{7}$ **73.** $-\frac{41}{30}$ **75.** $-\frac{2}{15}$
77. $-\frac{1}{48}$ **79.** $-\frac{43}{60}$ **81.** 37 **83.** -62 **85.** -139 **87.** 6
89. 108.5 **91.** $\frac{1}{4}$ **93.** 30,383 ft **95.** $347.94 **97.** 3780 m
99. 381 ft **101.** 1130°F **103.** $y + 7$, or $7 + y$ **104.** $t - 41$
105. $a - h$ **106.** $6c$, or $c \cdot 6$ **107.** $r + s$, or $s + r$ **108.** $y - x$
109. False; $3 - 0 \neq 0 - 3$ **111.** True **113.** True

Mid-Chapter Review: Chapter 1, p. 77

1. True **2.** False **3.** True **4.** False **5.** $-x = -(-4) = 4$; $-(-x) = -(-(-4)) = -(4) = -4$
6. $5 - 13 = 5 + (-13) = -8$ **7.** $-6 - 7 = -6 + (-7) = -13$
8. 4 **9.** 11 **10.** $3y$ **11.** $n - 5$ **12.** 450; -79
13. (number line, -3.5 marked between -4 and -3) **14.** -0.8 **15.** $2.\overline{3}$ **16.** $<$
17. $>$ **18.** False **19.** True **20.** $5 > y$ **21.** $t \leq -3$
22. 15.6 **23.** 18 **24.** 0 **25.** $\frac{12}{5}$ **26.** 5.6 **27.** $-\frac{7}{4}$ **28.** 0
29. 49 **30.** 19 **31.** 2.3 **32.** -2 **33.** $-\frac{1}{8}$ **34.** 0 **35.** -17
36. $-\frac{11}{24}$ **37.** -8.1 **38.** -9 **39.** -2 **40.** -10.4 **41.** 16
42. $\frac{7}{20}$ **43.** -12 **44.** -4 **45.** $-\frac{4}{3}$ **46.** -1.8 **47.** 13
48. 9 **49.** -23 **50.** 75 **51.** 14 **52.** 33°C **53.** $54.80
54. Answers may vary. Three examples are $\frac{6}{13}$, -23.8, and $\frac{43}{5}$. These are rational numbers because they can be named in the form $\frac{a}{b}$, where a and b are integers and b is not 0. They are not integers, however, because they are neither whole numbers nor the opposites of whole numbers. **55.** Answers may vary. Three examples are π, $-\sqrt{7}$, and $0.31311311131111\ldots$. Irrational numbers cannot be

written as the quotient of two integers. Real numbers that are not rational are irrational. Decimal notation for rational numbers either terminates or repeats. Decimal notation for irrational numbers neither terminates nor repeats. **56.** Answers may vary. If we think of the addition on the number line, we start at 0, move to the left to a negative number, and then move to the left again. This always brings us to a point on the negative portion of the number line.

57. Yes; consider $m - (-n)$, where both m and n are positive. Then $m - (-n) = m + n$. Now $m + n$, the sum of two positive numbers, is positive.

Exercise Set 1.5, p. 82

RC1. Negative **RC2.** Positive **RC3.** Positive
RC4. Negative **RC5.** -9 **RC6.** 9 **RC7.** $-\frac{1}{4}$ **RC8.** $-\frac{1}{4}$
1. -8 **3.** -24 **5.** -72 **7.** 16 **9.** 42 **11.** -120 **13.** -238
15. 1200 **17.** 98 **19.** -72 **21.** -12.4 **23.** 30 **25.** 21.7
27. $-\frac{2}{5}$ **29.** $\frac{1}{12}$ **31.** -17.01 **33.** 420 **35.** $\frac{2}{7}$ **37.** -60
39. 150 **41.** 50.4 **43.** $\frac{10}{189}$ **45.** -960 **47.** 17.64 **49.** $-\frac{5}{784}$
51. 0 **53.** -720 **55.** $-30{,}240$ **57.** 1 **59.** 16, -16; 16, -16
61. $\frac{4}{25}$, $-\frac{4}{25}$; $\frac{4}{25}$, $-\frac{4}{25}$ **63.** $-9, -9$; $-9, -9$ **65.** 441, -147; 441, -147
67. 20; 20 **69.** -2; 2 **71.** $-24°C$ **73.** -20 lb **75.** \$12.71
77. -32 m **79.** $38°F$ **81.** 2 **82.** $\frac{4}{15}$ **83.** $-\frac{1}{3}$ **84.** -4.3
85. 44 **86.** $-\frac{1}{12}$ **87.** True **88.** False **89.** False
90. False **91.** A **93.** Largest quotient: $10 \div \frac{1}{5} = 50$; smallest quotient: $-5 \div \frac{1}{5} = -25$

Calculator Corner, p. 91

1. -4 **2.** -2 **3.** -32 **4.** 1.4 **5.** 2.7 **6.** -9.5
7. -0.8 **8.** 14.44

Exercise Set 1.6, p. 91

RC1. Opposites **RC2.** 1 **RC3.** 0 **RC4.** Reciprocals
1. -8 **3.** -14 **5.** -3 **7.** 3 **9.** -8 **11.** 2 **13.** -12
15. -8 **17.** Not defined **19.** 0 **21.** $\frac{7}{15}$ **23.** $-\frac{13}{47}$ **25.** $\frac{1}{13}$
27. $-\frac{1}{32}$ **29.** -7.1 **31.** 9 **33.** $4y$ **35.** $\frac{3b}{2a}$ **37.** $4 \cdot \left(\frac{1}{17}\right)$
39. $8 \cdot \left(-\frac{1}{13}\right)$ **41.** $13.9 \cdot \left(-\frac{1}{1.5}\right)$ **43.** $\frac{2}{3} \cdot \left(-\frac{5}{4}\right)$ **45.** $x \cdot y$
47. $(3x + 4)\left(\frac{1}{5}\right)$ **49.** $-\frac{9}{8}$ **51.** $\frac{5}{3}$ **53.** $\frac{9}{14}$ **55.** $\frac{9}{64}$ **57.** $-\frac{5}{4}$
59. $-\frac{27}{5}$ **61.** $\frac{11}{13}$ **63.** -2 **65.** -16.2 **67.** -2.5 **69.** -1.25
71. Not defined **73.** Percent increase is about 44%.
75. Percent decrease is about -21%. **77.** $-\frac{1}{4}$ **78.** 5
79. -42 **80.** -48 **81.** 8.5 **82.** $-\frac{1}{8}$ **83.** $-0.0\overline{9}$ **84.** $0.91\overline{6}$
85. 3.75 **86.** $-3.\overline{3}$ **87.** $\frac{1}{-10.5}$; -10.5, the reciprocal of the reciprocal is the original number. **89.** Negative **91.** Positive
93. Negative

Exercise Set 1.7, p. 103

RC1. (g) **RC2.** (h) **RC3.** (f) **RC4.** (e) **RC5.** (d)
RC6. (a) **RC7.** (b) **1.** $\frac{3y}{5y}$ **3.** $\frac{10x}{15x}$ **5.** $\frac{2x}{x^2}$ **7.** $-\frac{3}{2}$ **9.** $-\frac{7}{6}$
11. $\frac{4s}{3}$ **13.** $8 + y$ **15.** nm **17.** $xy + 9$, or $9 + yx$, or $yx + 9$
19. $c + ab$, or $ba + c$, or $c + ba$ **21.** $(a + b) + 2$ **23.** $8(xy)$
25. $a + (b + 3)$ **27.** $(3a)b$ **29.** $2 + (b + a), (2 + a) + b$,
$(b + 2) + a$; answers may vary **31.** $(5 + w) + v, (v + 5) + w$,
$(w + v) + 5$; answers may vary **33.** $(3x)y, y(x \cdot 3), 3(yx)$;
answers may vary **35.** $a(7b), b(7a), (7b)a$; answers may vary
37. $2b + 10$ **39.** $7 + 7t$ **41.** $30x + 12$ **43.** $7x + 28 + 42y$
45. $7x - 21$ **47.** $-3x + 21$ **49.** $\frac{2}{3}b - 4$ **51.** $7.3x - 14.6$
53. $-\frac{3}{5}x + \frac{3}{5}y - 6$ **55.** $45x + 54y - 72$ **57.** $-4x + 12y + 8z$
59. $-3.72x + 9.92y - 3.41$ **61.** $4x, 3z$ **63.** $7x, 8y, -9z$
65. $2(x + 2)$ **67.** $5(6 + y)$ **69.** $7(2x + 3y)$ **71.** $7(2t - 1)$
73. $8(x - 3)$ **75.** $6(3a - 4b)$ **77.** $-4(y - 8)$, or $4(-y + 8)$
79. $5(x + 2 + 3y)$ **81.** $8(2m - 4n + 1)$ **83.** $4(3a + b - 6)$
85. $2(4x + 5y - 11)$ **87.** $a(x - 1)$ **89.** $a(x - y - z)$

91. $-6(3x - 2y - 1)$, or $6(-3x + 2y + 1)$ **93.** $\frac{1}{3}(2x - 5y + 1)$
95. $6(6x - y + 3z)$ **97.** $19a$ **99.** $9a$ **101.** $8x + 9z$
103. $7x + 15y^2$ **105.** $-19a + 88$ **107.** $4t + 6y - 4$ **109.** b
111. $\frac{13}{4}y$ **113.** $8x$ **115.** $5n$ **117.** $-16y$ **119.** $17a - 12b - 1$
121. $4x + 2y$ **123.** $7x + y$ **125.** $0.8x + 0.5y$ **127.** $\frac{35}{6}a + \frac{3}{2}b - 42$
129. 38 **130.** -4.9 **131.** 4 **132.** 10 **133.** $-\frac{4}{13}$ **134.** -106
135. 1 **136.** -34 **137.** 180 **138.** $\frac{4}{13}$ **139.** True **140.** False
141. True **142.** True **143.** Not equivalent;
$3 \cdot 2 + 5 \neq 3 \cdot 5 + 2$ **145.** Equivalent; commutative law of addition **147.** $q(1 + r + rs + rst)$

Calculator Corner, p. 112

1. -16 **2.** 9 **3.** 117,649 **4.** $-1{,}419{,}857$ **5.** $-117{,}649$
6. $-1{,}419{,}857$ **7.** -4 **8.** -2

Exercise Set 1.8, p. 113

RC1. Multiplication **RC2.** Addition **RC3.** Subtraction
RC4. Division **RC5.** Division **RC6.** Multiplication
1. $-2x - 7$ **3.** $-8 + x$ **5.** $-4a + 3b - 7c$
7. $-6x + 8y - 5$ **9.** $-3x + 5y + 6$ **11.** $8x + 6y + 43$
13. $5x - 3$ **15.** $-3a + 9$ **17.** $5x - 6$ **19.** $-19x + 2y$
21. $9y - 25z$ **23.** $-7x + 10y$ **25.** $37a - 23b + 35c$
27. 7 **29.** -40 **31.** 19 **33.** $12x + 30$ **35.** $3x + 30$
37. $9x - 18$ **39.** $-4x - 64$ **41.** -7 **43.** -1 **45.** -16
47. -334 **49.** 14 **51.** 1880 **53.** 12 **55.** 8 **57.** -86
59. 37 **61.** -1 **63.** -10 **65.** -67 **67.** -7988 **69.** -3000
71. 60 **73.** 1 **75.** 10 **77.** $-\frac{13}{45}$ **79.** $-\frac{23}{18}$ **81.** -122
83. 18 **84.** 35 **85.** 0.4 **86.** $\frac{15}{2}$ **87.** $-\frac{1}{9}$ **88.** $\frac{3}{7}$
89. -25 **90.** -35 **91.** 25 **92.** 35 **93.** $-2x - f$
95. **(a)** 52; 52; 28.130169; **(b)** -24; -24; -108.307025 **97.** -6

Summary and Review: Chapter 1, p. 117

Vocabulary Reinforcement

1. Integers **2.** Additive inverses **3.** Commutative law
4. Identity property of 1 **5.** Associative law
6. Multiplicative inverses **7.** Identity property of 0

Concept Reinforcement

1. True **2.** True **3.** False **4.** False

Study Guide

1. 14 **2.** $<$ **3.** $\frac{5}{4}$ **4.** -8.5 **5.** -2 **6.** 56 **7.** -8
8. $\frac{9}{20}$ **9.** $\frac{5}{3}$ **10.** $5x + 15y - 20z$ **11.** $9(3x + y - 4z)$
12. $5a - 2b$ **13.** $4a - 4b$ **14.** -2

Review Exercises

1. 4 **2.** $19\%x$, or $0.19x$ **3.** 620, -125 **4.** 38 **5.** 126
6.

7.

8. $<$ **9.** $>$ **10.** $>$ **11.** $<$ **12.** $x > -3$ **13.** True
14. False **15.** -3.8 **16.** $\frac{3}{4}$ **17.** $\frac{8}{3}$ **18.** $-\frac{1}{7}$ **19.** 34
20. 5 **21.** -3 **22.** -4 **23.** -5 **24.** 1 **25.** $-\frac{7}{5}$
26. -7.9 **27.** 54 **28.** -9.18 **29.** $-\frac{2}{7}$ **30.** -210 **31.** -7
32. -3 **33.** $\frac{3}{4}$ **34.** 24.8 **35.** -2 **36.** 2 **37.** -2
38. 8-yd gain **39.** $-\$360$ **40.** \$4.64 **41.** \$18.95
42. $15x - 35$ **43.** $-8x + 10$ **44.** $4x + 15$ **45.** $-24 + 48x$
46. $2(x - 7)$ **47.** $-6(x - 1)$, or $6(-x + 1)$ **48.** $5(x + 2)$
49. $-3(x - 4y + 4)$, or $3(-x + 4y - 4)$ **50.** $7a - 3b$
51. $-2x + 5y$ **52.** $5x - y$ **53.** $-a + 8b$ **54.** $-3a + 9$
55. $-2b + 21$ **56.** 6 **57.** $12y - 34$ **58.** $5x + 24$
59. $-15x + 25$ **60.** D **61.** B **62.** $-\frac{5}{8}$ **63.** -2.1
64. 1000 **65.** $4a + 2b$

Understanding Through Discussion and Writing

1. The sum of each pair of opposites such as -50 and 50, -49 and 49, and so on, is 0. The sum of these sums and the remaining integer, 0, is 0.　**2.** The product of an even number of negative numbers is positive, and the product of an odd number of negative numbers is negative. Now $(-7)^8$ is the product of 8 factors of -7 so it is positive, and $(-7)^{11}$ is the product of 11 factors of -7 so it is negative.

3. Consider $\dfrac{a}{b} = q$, where a and b are both negative numbers. Then $q \cdot b = a$, so q must be a positive number in order for the product to be negative.　**4.** Consider $\dfrac{a}{b} = q$, where a is a negative number and b is a positive number. Then $q \cdot b = a$, so q must be a negative number in order for the product to be negative.　**5.** We use the distributive law when we collect like terms even though we might not always write this step.　**6.** Jake expects the calculator to multiply 2 and 3 first and then divide 18 by that product. This procedure does not follow the rules for order of operations.

Test: Chapter 1, p. 123

1. [1.1a] 6　**2.** [1.1b] $x - 9$　**3.** [1.2d] $>$　**4.** [1.2d] $<$
5. [1.2d] $>$　**6.** [1.2d] $-2 > x$　**7.** [1.2d] True　**8.** [1.2e] 7
9. [1.2e] $\frac{9}{4}$　**10.** [1.2e] 2.7　**11.** [1.3b] $-\frac{2}{3}$　**12.** [1.3b] 1.4
13. [1.6b] $-\frac{1}{2}$　**14.** [1.6b] $\frac{7}{4}$　**15.** [1.3b] 8　**16.** [1.4a] 7.8
17. [1.3a] -8　**18.** [1.3a] $\frac{7}{40}$　**19.** [1.4a] 10　**20.** [1.4a] -2.5
21. [1.4a] $\frac{7}{8}$　**22.** [1.5a] -48　**23.** [1.5a] $\frac{3}{16}$　**24.** [1.6a] -9
25. [1.6c] $\frac{3}{4}$　**26.** [1.6c] -9.728　**27.** [1.2e], [1.8d] -173
28. [1.8d] -5　**29.** [1.3c], [1.4b] Up 15 points　**30.** [1.4b] 2244 m
31. [1.5b] 16,080　**32.** [1.6d] $-0.75°$C each minute
33. [1.7c] $18 - 3x$　**34.** [1.7c] $-5y + 5$　**35.** [1.7d] $2(6 - 11x)$
36. [1.7d] $7(x + 3 + 2y)$　**37.** [1.4a] 12　**38.** [1.8b] $2x + 7$
39. [1.8b] $9a - 12b - 7$　**40.** [1.8c] $68y - 8$　**41.** [1.8d] -4
42. [1.8d] 448　**43.** [1.2d] B　**44.** [1.2e], [1.8d] 15
45. [1.8c] $4a$　**46.** [1.7e] $4x + 4y$

CHAPTER 2

Exercise Set 2.1, p. 130

RC1. (f)　**RC2.** (c)　**RC3.** (e)　**RC4.** (a)
1. Yes　**3.** No　**5.** No　**7.** Yes　**9.** Yes　**11.** No　**13.** 4
15. -20　**17.** -14　**19.** -18　**21.** 15　**23.** -14　**25.** 2
27. 20　**29.** -6　**31.** $6\frac{1}{2}$　**33.** 19.9　**35.** $\frac{7}{3}$　**37.** $-\frac{7}{4}$　**39.** $\frac{41}{24}$
41. $-\frac{1}{20}$　**43.** 5.1　**45.** 12.4　**47.** -5　**49.** $1\frac{5}{6}$　**51.** $-\frac{10}{21}$
53. $-\frac{3}{2}$　**54.** -5.2　**55.** $-\frac{1}{24}$　**56.** 172.72　**57.** $\$83 - x$
58. $65t$ miles　**59.** $-\frac{26}{15}$　**61.** -10　**63.** All real numbers

Exercise Set 2.2, p. 135

RC1. (f)　**RC2.** (d)　**RC3.** (a)　**RC4.** (b)
1. 6　**3.** 9　**5.** 12　**7.** -40　**9.** 1　**11.** -7　**13.** -6
15. 6　**17.** -63　**19.** -48　**21.** 36　**23.** -9　**25.** -21
27. $-\frac{3}{5}$　**29.** $-\frac{3}{2}$　**31.** $\frac{9}{2}$　**33.** 7　**35.** -7　**37.** 8　**39.** 15.9
41. -50　**43.** -14　**45.** $7x$　**46.** $-x + 5$　**47.** $8x + 11$
48. $-32y$　**49.** $x - 4$　**50.** $-5x - 23$　**51.** $-10y - 42$
52. $-22a + 4$　**53.** $8r$ miles　**54.** $\frac{1}{2}b \cdot 10$ m^2, or $5b$ m^2
55. -8655　**57.** No solution　**59.** No solution　**61.** $\dfrac{b}{3a}$
63. $\dfrac{4b}{a}$

Calculator Corner, p. 141

1. Left to the student

Exercise Set 2.3, p. 145

RC1. (d)　**RC2.** (a)　**RC3.** (c)　**RC4.** (e)　**RC5.** (b)
1. 5　**3.** 8　**5.** 10　**7.** 14　**9.** -8　**11.** -8　**13.** -7
15. 12　**17.** 6　**19.** 4　**21.** 6　**23.** -3　**25.** 1　**27.** 6
29. -20　**31.** 7　**33.** 2　**35.** 5　**37.** 2　**39.** 10　**41.** 4
43. 0　**45.** -1　**47.** $-\frac{4}{3}$　**49.** $\frac{2}{5}$　**51.** -2　**53.** -4　**55.** $\frac{4}{5}$
57. $-\frac{28}{27}$　**59.** 6　**61.** 2　**63.** No solution　**65.** All real numbers
67. 6　**69.** 8　**71.** 1　**73.** 17　**75.** $-\frac{5}{3}$　**77.** All real numbers
79. No solution　**81.** -3　**83.** 2　**85.** $\frac{4}{7}$　**87.** No solution
89. All real numbers　**91.** $-\frac{51}{31}$　**93.** -6.5　**94.** -75.14
95. $7(x - 3 - 2y)$　**96.** $8(y - 11x + 1)$　**97.** -160
98. $-17x + 18$　**99.** $91x - 242$　**100.** 0.25　**101.** $-\frac{5}{32}$　**103.** $\frac{52}{45}$

Exercise Set 2.4, p. 153

RC1. (b)　**RC2.** (c)　**RC3.** (d)　**RC4.** (a)
1. $14\frac{1}{3}$ meters per cycle　**3.** 10.5 calories per ounce

5. (a) 337.5 mi; **(b)** $t = \dfrac{d}{r}$　**7. (a)** 1423 students; **(b)** $n = 15f$

9. $x = \dfrac{y}{5}$　**11.** $c = \dfrac{a}{b}$　**13.** $m = n - 11$　**15.** $x = y + \dfrac{3}{5}$

17. $x = y - 13$　**19.** $x = y - b$　**21.** $x = 5 - y$

23. $x = a - y$　**25.** $y = \dfrac{5x}{8}$, or $\dfrac{5}{8}x$　**27.** $x = \dfrac{By}{A}$

29. $t = \dfrac{W - b}{m}$　**31.** $x = \dfrac{y - c}{b}$　**33.** $h = \dfrac{A}{b}$

35. $w = \dfrac{P - 2l}{2}$, or $\dfrac{1}{2}P - l$　**37.** $a = 2A - b$

39. $b = 3A - a - c$　**41.** $t = \dfrac{A - b}{a}$　**43.** $x = \dfrac{c - By}{A}$

45. $a = \dfrac{F}{m}$　**47.** $c^2 = \dfrac{E}{m}$　**49.** $t = \dfrac{3k}{v}$　**51.** 7

52. $-21a + 12b$　**53.** -13.2　**54.** $-\frac{3}{2}$　**55.** $-35\frac{1}{2}$　**56.** $-\frac{1}{6}$
57. -9.325　**58.** $3\frac{3}{4}$　**59.** $\frac{11}{8}$　**60.** -1　**61.** -3　**62.** $\frac{9}{7}$

63. $b = \dfrac{Ha - 2}{H}$, or $a - \dfrac{2}{H}$; $a = \dfrac{2 + Hb}{H}$, or $\dfrac{2}{H} + b$

65. A quadruples.　**67.** A increases by $2h$ units.

Mid-Chapter Review: Chapter 2, p. 157

1. False　**2.** True　**3.** True　**4.** False
5. $x + 5 = -3$ 　 **6.** $-6x = 42$
$x + 5 - 5 = -3 - 5$ 　 $\dfrac{-6x}{-6} = \dfrac{42}{-6}$
$x + 0 = -8$ 　 $1 \cdot x = -7$
$x = -8$ 　 $x = -7$

7. $5y + z = t$
$5y + z - z = t - z$
$5y = t - z$
$\dfrac{5y}{5} = \dfrac{t - z}{5}$
$y = \dfrac{t - z}{5}$

8. 6　**9.** -12　**10.** 7　**11.** -10　**12.** 20　**13.** 5　**14.** $\frac{3}{4}$
15. -1.4　**16.** 6　**17.** -17　**18.** -9　**19.** 17　**20.** 21
21. 18　**22.** -15　**23.** $-\frac{3}{2}$　**24.** 1　**25.** -3　**26.** $\frac{3}{2}$　**27.** -1
28. 3　**29.** -7　**30.** 4　**31.** 2　**32.** $\frac{9}{8}$　**33.** $-\frac{21}{5}$　**34.** 9
35. -2　**36.** 0　**37.** All real numbers　**38.** No solution
39. $-\frac{13}{2}$　**40.** All real numbers　**41.** $b = \dfrac{A}{4}$　**42.** $x = y + 1.5$

43. $m = s - n$　**44.** $t = \dfrac{9w}{4}$　**45.** $t = \dfrac{B + c}{a}$

46. $y = 2M - x - z$ **47.** Equivalent expressions have the same value for all possible replacements for the variable(s). Equivalent equations have the same solution(s). **48.** The equations are not equivalent because they do not have the same solutions. Although 5 is a solution of both equations, -5 is a solution of $x^2 = 25$ but not of $x = 5$. **49.** For an equation $x + a = b$, add the opposite of a (or subtract a) on both sides of the equation. **50.** The student probably added $\frac{1}{3}$ on both sides of the equation rather than adding $-\frac{1}{3}$ (or subtracting $\frac{1}{3}$) on both sides. The correct solution is -2. **51.** For an equation $ax = b$, multiply by $1/a$ (or divide by a) on both sides of the equation. **52.** Answers may vary. A walker who knows how far and how long she walks each day wants to know her average speed each day.

Exercise Set 2.5, p. 163

RC1. (d) **RC2.** (b) **RC3.** (e) **RC4.** (a) **RC5.** (f) **RC6.** (c)
1. 20% **3.** 150 **5.** 546 **7.** 24% **9.** 2.5 **11.** 5%
13. 25% **15.** 84 **17.** 24% **19.** 16% **21.** $46\frac{2}{3}$ **23.** 0.8
25. 5 **27.** 40 **29.** 811 million **31.** 5274 million
33. 1764 million **35.** $968 million **37.** $221 **39.** 21%
41. (a) 12.5%; **(b)** $13.50 **43. (a)** $31; **(b)** $35.65 **45.** About 85,821 acres **47.** About 82.4% **49.** About 53.1% decrease
51. About 53.4% **53.** About 19.2% decrease **55.** $12 + 3q$
56. $5x - 21$ **57.** $\dfrac{15w}{8}$ **58.** $-\frac{3}{2}$ **59.** 44 **60.** $x + 8$
61. 6 ft 7 in.

Translating for Success, p. 178

1. B **2.** H **3.** G **4.** N **5.** J **6.** C **7.** L **8.** E
9. F **10.** D

Exercise Set 2.6, p. 179

RC1. Familiarize **RC2.** Translate **RC3.** Solve
RC4. Check **RC5.** State
1. 1522 Medals of Honor **3.** 180 in.; 60 in. **5.** 21.8 million
7. 4.37 mi **9.** 1204 and 1205 **11.** 41, 42, 43 **13.** 61, 63, 65
15. 36 in. × 110 in. **17.** $63 **19.** $24.95 **21.** 11 visits
23. 28°, 84°, 68° **25.** 33°, 38°, 109° **27.** $350 **29.** $852.94
31. 18 mi **33.** $38.60 **35.** 89 and 96 **37.** -12 **39.** $-\frac{47}{40}$
40. $-\frac{17}{40}$ **41.** $-\frac{3}{10}$ **42.** $-\frac{32}{15}$ **43.** -10 **44.** 1.6 **45.** 409.6
46. -9.6 **47.** -41.6 **48.** 0.1 **49.** $yz + 12$, $zy + 12$, or $12 + zy$ **50.** $c + (4 + d)$ **51.** 120 apples **53.** About 0.65 in.

Exercise Set 2.7, p. 192

RC1. Not equivalent **RC2.** Not equivalent **RC3.** Equivalent
RC4. Equivalent
1. (a) Yes; **(b)** yes; **(c)** no; **(d)** yes; **(e)** yes **3. (a)** No; **(b)** no; **(c)** no; **(d)** yes; **(e)** no
5.
$x > 4$
7.
$t < -3$
9.
$m \geq -1$
11.
$-3 < x \leq 4$
13.
$0 < x < 3$
15. $\{x \mid x > -5\}$
17. $\{x \mid x \leq -18\}$;
19. $\{y \mid y > -5\}$
21. $\{x \mid x > 2\}$ **23.** $\{x \mid x \leq -3\}$ **25.** $\{x \mid x < 4\}$
27. $\{t \mid t > 14\}$ **29.** $\{y \mid y \leq \frac{1}{4}\}$ **31.** $\{x \mid x > \frac{7}{12}\}$
33. $\{x \mid x < 7\}$;

35. $\{x \mid x < 3\}$;
37. $\{y \mid y \geq -\frac{2}{5}\}$ **39.** $\{x \mid x \geq -6\}$ **41.** $\{y \mid y \leq 4\}$
43. $\{x \mid x > \frac{17}{3}\}$ **45.** $\{y \mid y < -\frac{1}{14}\}$ **47.** $\{x \mid x \leq \frac{3}{10}\}$
49. $\{x \mid x < 8\}$ **51.** $\{x \mid x \leq 6\}$ **53.** $\{x \mid x < -3\}$
55. $\{x \mid x > -3\}$ **57.** $\{x \mid x \leq 7\}$ **59.** $\{x \mid x > -10\}$
61. $\{y \mid y < 2\}$ **63.** $\{y \mid y \geq 3\}$ **65.** $\{y \mid y > -2\}$
67. $\{x \mid x > -4\}$ **69.** $\{x \mid x \leq 9\}$ **71.** $\{y \mid y \leq -3\}$
73. $\{y \mid y \leqslant 6\}$ **75.** $\{m \mid m \geq 6\}$ **77.** $\{t \mid t < -\frac{5}{3}\}$
79. $\{r \mid r > -3\}$ **81.** $\{x \mid x \geq -\frac{57}{34}\}$ **83.** $\{x \mid x > -2\}$
85. $-\frac{5}{8}$ **86.** -1.11 **87.** -9.4 **88.** $-\frac{7}{8}$ **89.** 140 **90.** 41
91. $-2x - 23$ **92.** $37x - 1$ **93. (a)** Yes; **(b)** yes; **(c)** no; **(d)** no; **(e)** no; **(f)** yes; **(g)** yes **95.** No solution

Exercise Set 2.8, p. 199

RC1. $r \leq q$ **RC2.** $q \leq r$ **RC3.** $r < q$ **RC4.** $q \leq r$
RC5. $r < q$ **RC6.** $r \leq q$
1. $n \geq 7$ **3.** $w > 2$ kg **5.** 90 mph $< s < 110$ mph
7. $w \leq 20$ hr **9.** $c \geq \$3.20$ **11.** $x > 8$ **13.** $y \leq -4$
15. $n \geq 1300$ **17.** $W \leq 500$ L **19.** $3x + 2 < 13$
21. $\{x \mid x \geq 84\}$ **23.** $\{C \mid C < 1063°\}$ **25.** $\{Y \mid Y \geq 1935\}$
27. 15 or fewer copies **29.** $\{L \mid L \geq 5 \text{ in.}\}$ **31.** 5 min or more
33. 2 courses **35.** 4 servings or more **37.** Lengths greater than or equal to 92 ft; lengths less than or equal to 92 ft
39. Lengths less than 21.5 cm **41.** The blue-book value is greater than or equal to $10,625. **43.** It has at least 16 g of fat.
45. Heights greater than or equal to 4 ft **47.** Dates at least 6 weeks after July 1 **49.** 21 calls or more **51.** 40 **52.** -22
53. 12 **54.** 6 **55.** All real numbers **56.** No solution **57.** 7.5%
58. 31 **59.** 1250 **60.** $83.\overline{3}\%$, or $83\frac{1}{3}\%$ **61.** Temperatures between $-15°$C and $-9\frac{4°}{9}$C **63.** They contain at least 7.5 g of fat per serving.

Summary and Review: Chapter 2, p. 204

Vocabulary Reinforcement

1. Solution **2.** Addition principle **3.** Multiplication principle
4. Inequality **5.** Equivalent

Concept Reinforcement

1. True **2.** True **3.** False **4.** True

Study Guide

1. -12 **2.** All real numbers **3.** No solution **4.** $b = \dfrac{2A}{h}$
5.
$x > 1$
6.
$x \leq -1$
7. $\{y \mid y > -4\}$

Review Exercises

1. -22 **2.** 1 **3.** 25 **4.** 9.99 **5.** $\frac{1}{4}$ **6.** 7 **7.** -192
8. $-\frac{7}{3}$ **9.** $-\frac{15}{64}$ **10.** -8 **11.** 4 **12.** -5 **13.** $-\frac{1}{3}$
14. 3 **15.** 4 **16.** 16 **17.** All real numbers **18.** 6
19. -3 **20.** 28 **21.** 4 **22.** No solution **23.** Yes **24.** No
25. Yes **26.** $\{y \mid y \geq -\frac{1}{2}\}$ **27.** $\{x \mid x \geq 7\}$ **28.** $\{y \mid y > 2\}$
29. $\{y \mid y \leq -4\}$ **30.** $\{x \mid x < -11\}$ **31.** $\{y \mid y > -7\}$
32. $\{x \mid x > -\frac{9}{11}\}$ **33.** $\{x \mid x \geq -\frac{1}{12}\}$
34.
$x < 3$
35.
$-2 < x \leq 5$
36.
$y > 0$
37. $d = \dfrac{C}{\pi}$ **38.** $B = \dfrac{3V}{h}$

39. $a = 2A - b$ **40.** $x = \dfrac{y - b}{m}$ **41.** Length: 365 mi; width: 275 mi   **42.** 345, 346   **43.** \$2117   **44.** 27 subscriptions   **45.** $35°, 85°, 60°$ **46.** 15   **47.** 18.75%   **48.** 600   **49.** About 87.1%   **50.** About 28.2% decrease   **51.** \$220   **52.** \$53,400   **53.** \$138.95   **54.** 86   **55.** $\{w\,|\,w > 17\text{ cm}\}$ **56.** C   **57.** A   **58.** 23, −23   **59.** 20, −20   **60.** $a = \dfrac{y - 3}{2 - b}$

Understanding Through Discussion and Writing

1. The end result is the same either way. If s is the original salary, the new salary after a 5% raise followed by an 8% raise is $1.08(1.05s)$. If the raises occur the other way around, the new salary is $1.05(1.08s)$. By the commutative and associative laws of multiplication, we see that these are equal. However, it would be better to receive the 8% raise first, because this increase yields a higher salary initially than a 5% raise.   **2.** No; Erin paid 75% of the original price and was offered credit for 125% of this amount, not to be used on sale items. Now, 125% of 75% is 93.75%, so Erin would have a credit of 93.75% of the original price. Since this credit can be applied only to nonsale items, she has less purchasing power than if the amount she paid were refunded and she could spend it on sale items.   **3.** The inequalities are equivalent by the multiplication principle for inequalities. If we multiply on both sides of one inequality by −1, the other inequality results.   **4.** For any pair of numbers, their relative position on the number line is reversed when both are multiplied by the same negative number. For example, −3 is to the left of 5 on the number line $(-3 < 5)$, but 12 is to the right of −20 $(-3(-4) > 5(-4))$.   **5.** Answers may vary. Fran is more than 3 years older than Todd.   **6.** Let n represent "a number." Then "five more than a number" translates to the *expression* $n + 5$, or $5 + n$, and "five is more than a number" translates to the *inequality* $5 > n$.

Test: Chapter 2, p. 209

1. [2.1b] 8   **2.** [2.1b] 26   **3.** [2.2a] −6   **4.** [2.2a] 49   **5.** [2.3b] −12   **6.** [2.3a] 2   **7.** [2.3a] −8   **8.** [2.1b] $-\dfrac{7}{20}$ **9.** [2.3c] 7   **10.** [2.3c] $\dfrac{5}{3}$ **11.** [2.3b] $\dfrac{5}{2}$ **12.** [2.3c] No solution   **13.** [2.3c] All real numbers   **14.** [2.7c] $\{x\,|\,x \le -4\}$ **15.** [2.7c] $\{x\,|\,x > -13\}$ **16.** [2.7d] $\{x\,|\,x \le 5\}$ **17.** [2.7d] $\{y\,|\,y \le -13\}$ **18.** [2.7d] $\{y\,|\,y \ge 8\}$ **19.** [2.7d] $\{x\,|\,x \le -\frac{1}{20}\}$ **20.** [2.7e] $\{x\,|\,x < -6\}$ **21.** [2.7e] $\{x\,|\,x \le -1\}$

22. [2.7b]

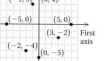

23. [2.7b, e]

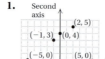

24. [2.7b]

25. [2.5a] 18

26. [2.5a] 16.5%   **27.** [2.5a] 40,000   **28.** [2.5a] About 60.4%   **29.** [2.6a] Width: 7 cm; length: 11 cm   **30.** [2.5a] About \$230,556   **31.** [2.6a] 2509, 2510, 2511   **32.** [2.6a] \$880   **33.** [2.6a] 3 m, 5 m   **34.** [2.8b] $\{l\,|\,l \ge 174\text{ yd}\}$ **35.** [2.8b] $\{b\,|\,b \le \$105\}$ **36.** [2.8b] $\{c\,|\,c \le 119{,}531\}$ **37.** [2.4b] $r = \dfrac{A}{2\pi h}$ **38.** [2.4b] $x = \dfrac{y - b}{8}$ **39.** [2.5a] D   **40.** [2.4b] $d = \dfrac{1 - ca}{-c}$, or $\dfrac{ca - 1}{c}$ **41.** [1.2e], [2.3a] 15, −15   **42.** [2.6a] 60 tickets

Cumulative Review: Chapters 1–2, p. 211

1. [1.1a] $\dfrac{3}{2}$ **2.** [1.1a] $\dfrac{15}{4}$ **3.** [1.1a] 0   **4.** [1.1b] $2w - 4$ **5.** [1.2d] $>$ **6.** [1.2d] $>$ **7.** [1.2d] $<$ **8.** [1.3b], [1.6b] $-\frac{2}{5}, \frac{5}{2}$ **9.** [1.2e] 3   **10.** [1.2e] $\frac{3}{4}$ **11.** [1.2e] 0   **12.** [1.3a] −4.4

13. [1.4a] $-\dfrac{5}{2}$ **14.** [1.5a] $\dfrac{5}{6}$ **15.** [1.5a] −105   **16.** [1.6a] −9   **17.** [1.6c] −3   **18.** [1.6c] $\dfrac{32}{125}$ **19.** [1.7c] $15x + 25y + 10z$ **20.** [1.7c] $-12x - 8$ **21.** [1.7c] $-12y + 24x$ **22.** [1.7d] $2(32 + 9x + 12y)$ **23.** [1.7d] $8(2y - 7)$ **24.** [1.7d] $5(a - 3b + 5)$ **25.** [1.7e] $15b + 22y$ **26.** [1.7e] $4 + 9y + 6z$ **27.** [1.7e] $1 - 3a - 9d$ **28.** [1.7e] $-2.6x - 5.2y$ **29.** [1.8b] $3x - 1$ **30.** [1.8b] $-2x - y$ **31.** [1.8b] $-7x + 6$ **32.** [1.8b] $8x$ **33.** [1.8c] $5x - 13$ **34.** [2.1b] 4.5   **35.** [2.2a] $\dfrac{4}{25}$ **36.** [2.1b] 10.9   **37.** [2.1b] $3\frac{5}{6}$ **38.** [2.2a] −48   **39.** [2.2a] $-\frac{3}{8}$ **40.** [2.2a] −6.2   **41.** [2.3a] −3   **42.** [2.3b] $-\dfrac{12}{5}$ **43.** [2.3b] 8   **44.** [2.3c] 7   **45.** [2.3b] $-\frac{4}{5}$ **46.** [2.3b] $-\dfrac{10}{3}$ **47.** [2.3c] All real numbers   **48.** [2.3c] No solution   **49.** [2.7c] $\{x\,|\,x < 2\}$ **50.** [2.7e] $\{y\,|\,y < -3\}$ **51.** [2.7e] $\{y\,|\,y \ge 4\}$ **52.** [2.4b] $m = 65 - H$ **53.** [2.4b] $t = \dfrac{I}{Pr}$ **54.** [2.5a] 25.2   **55.** [2.5a] 45%   **56.** [2.5a] \$363   **57.** [2.6a] \$24.60   **58.** [2.6a] \$45   **59.** [2.6a] \$1050   **60.** [2.6a] 50 m, 53 m, 40 m   **61.** [2.8b] $\{s\,|\,s \ge 84\}$ **62.** [1.8d] C   **63.** [2.6a] \$45,200   **64.** [2.6a] 30%   **65.** [1.2e], [2.3a] 4, −4   **66.** [2.3b] 3   **67.** [2.4b] $Q = \dfrac{2 - pm}{p}$

CHAPTER 3

Calculator Corner, p. 218

1. Left to the student

Calculator Corner, p. 224

1. $y = -5x + 3$

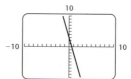

2. $y = 4x - 5$

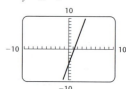

3. $y = \dfrac{4}{5}x + 2$

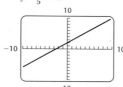

4. $y = -\dfrac{3}{5}x - 1$

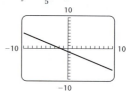

Exercise Set 3.1, p. 224

RC1. True   **RC2.** False   **RC3.** False   **RC4.** True

1.

3. II   **5.** IV   **7.** III   **9.** On an axis, not in a quadrant   **11.** II   **13.** IV   **15.** II   **17.** I, IV   **19.** I, III

21. A: $(3, 3)$; B: $(0, -4)$; C: $(-5, 0)$; D: $(-1, -1)$; E: $(2, 0)$ **23.** No   **25.** No   **27.** Yes

29. $y = x - 5$

$\dfrac{-1\ ?\ 4 - 5}{\quad|\ -1}$ TRUE

$y = x - 5$

$\dfrac{-4\ ?\ 1 - 5}{\quad|\ -4}$ TRUE

31. $y = \frac{1}{2}x + 3$

$$\frac{y = \frac{1}{2}x + 3}{5 \; ? \; \frac{1}{2} \cdot 4 + 3}$$
$$2 + 3$$
$$5 \qquad \text{TRUE}$$

$$\frac{y = \frac{1}{2}x + 3}{2 \; ? \; \frac{1}{2}(-2) + 3}$$
$$-1 + 3$$
$$2 \qquad \text{TRUE}$$

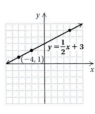

33. $4x - 2y = 10$

$$\frac{4x - 2y = 10}{4 \cdot 0 - 2(-5) \; ? \; 10}$$
$$0 + 10$$
$$10 \qquad \text{TRUE}$$

$$\frac{4x - 2y = 10}{4 \cdot 4 - 2 \cdot 3 \; ? \; 10}$$
$$16 - 6$$
$$10 \qquad \text{TRUE}$$

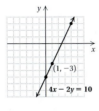

35.

x	y
-2	-1
-1	0
0	1
1	2
2	3
3	4

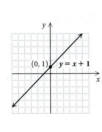

37.

x	y
-2	-2
-1	-1
0	0
1	1
2	2
3	3

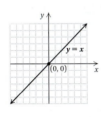

39.

x	y
-2	-1
0	0
4	2

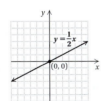

41.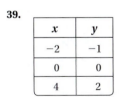

43.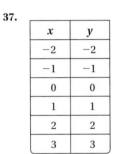

45.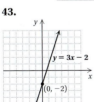

47.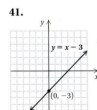

49.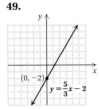

51.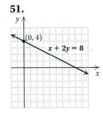

53.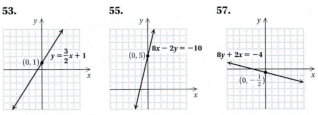

55.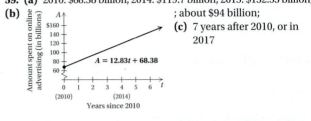

57.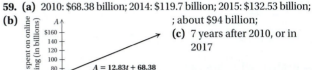

59. **(a)** 2010: \$68.38 billion; 2014: \$119.7 billion; 2015: \$132.53 billion;
(b) 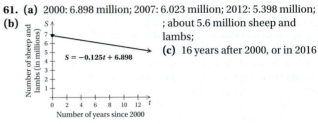 ; about \$94 billion;
(c) 7 years after 2010, or in 2017

61. **(a)** 2000: 6.898 million; 2007: 6.023 million; 2012: 5.398 million;
(b) 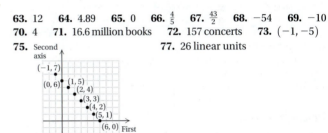 ; about 5.6 million sheep and lambs;
(c) 16 years after 2000, or in 2016

63. 12   **64.** 4.89   **65.** 0   **66.** $\frac{4}{5}$ **67.** $\frac{43}{2}$ **68.** -54 **69.** -10
70. 4   **71.** 16.6 million books   **72.** 157 concerts   **73.** $(-1, -5)$
75.  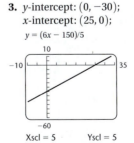   **77.** 26 linear units

Calculator Corner, p. 233

1. y-intercept: $(0, -15)$;
x-intercept: $(-2, 0)$;
$y = -7.5x - 15$

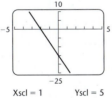

Xscl = 1   Yscl = 5

2. y-intercept: $(0, 43)$;
x-intercept: $(-20, 0)$;
$y = 2.15x + 43$

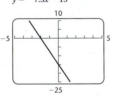

Xscl = 5   Yscl = 5

3. y-intercept: $(0, -30)$;
x-intercept: $(25, 0)$;
$y = (6x - 150)/5$

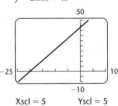 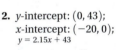

Xscl = 5   Yscl = 5

4. y-intercept: $(0, -4)$;
x-intercept: $(20, 0)$;
$y = 0.2x - 4$

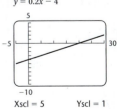 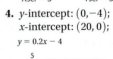

Xscl = 5   Yscl = 1

5. y-intercept: $(0,-15)$;
  x-intercept: $(10,0)$;
  $y = 1.5x - 15$

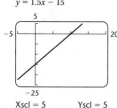

Xscl = 5   Yscl = 5

6. y-intercept: $(0, -\frac{1}{2})$;
  x-intercept: $(\frac{2}{5}, 0)$;
  $y = (5x - 2)/4$

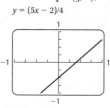

Xscl = 0.25   Yscl = 0.25

Visualizing for Success, p. 236

1. E   **2.** C   **3.** G   **4.** A   **5.** I   **6.** D   **7.** F   **8.** J
9. B   **10.** H

Exercise Set 3.2, p. 237

RC1. Horizontal; y-intercept   **RC2.** x-axis   **RC3.** $y = 0$
RC4. $(0,0)$ **RC5.** $x = 0$ **RC6.** Vertical; x-intercept
1. (a) $(0,5)$; (b) $(2,0)$ **3.** (a) $(0,-4)$; (b) $(3,0)$
5. (a) $(0,3)$; (b) $(5,0)$ **7.** (a) $(0,-14)$; (b) $(4,0)$
9. (a) $\left(0, \frac{10}{3}\right)$; (b) $\left(-\frac{5}{2}, 0\right)$ **11.** (a) $\left(0, -\frac{1}{3}\right)$; (b) $\left(\frac{1}{2}, 0\right)$

13.

15.

17.

19.

21.

23.

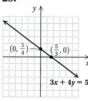

25.

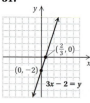

27.

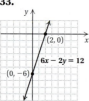

29.

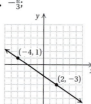

31.

33.

35.

37.

39.

41.

43.

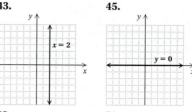

45.

47.

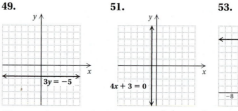

49.

51.

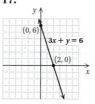

53.

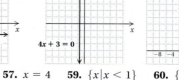

55. $y = -1$ **57.** $x = 4$ **59.** $\{x \mid x < 1\}$ **60.** $\{x \mid x \geq 2\}$
61. $\{x \mid x \leq 7\}$ **62.** $\{x \mid x > 1\}$ **63.** $y = -4$ **65.** $k = 12$

Calculator Corner, p. 245

1. This line will pass through the origin and slant up from left to right. This line will be steeper than $y = 10x$.   **2.** This line will pass through the origin and slant up from left to right. This line will be less steep than $y = \frac{5}{32}x$.   **3.** This line will pass through the origin and slant down from left to right. This line will be steeper than $y = -10x$.   **4.** This line will pass through the origin and slant down from left to right. This line will be less steep than $y = -\frac{5}{32}x$.

Exercise Set 3.3, p. 248

RC1. (d)   **RC2.** (f)   **RC3.** (b)   **RC4.** (e)
RC5. (c)   **RC6.** (a)
1. $-\frac{3}{7}$ **3.** $\frac{2}{3}$ **5.** $\frac{3}{4}$ **7.** 0

9. $-\frac{4}{5}$;

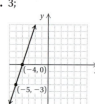

11. 3;

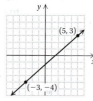

13. $-\frac{2}{3}$;

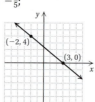

15. $\frac{7}{8}$;

17. $\frac{2}{3}$ **19.** Not defined   **21.** $-\frac{5}{13}$ **23.** 0   **25.** -10
27. 3.78   **29.** 3   **31.** $-\frac{1}{5}$ **33.** $-\frac{3}{2}$ **35.** Not defined
37. -1 **39.** 3   **41.** $\frac{5}{4}$ **43.** 0   **45.** $\frac{4}{3}$ **47.** $-\frac{21}{8}$ **49.** $\frac{12}{41}$
51. $\frac{28}{129}$ **53.** 3.0%; yes   **55.** About 1.24 million children per year
57. About $-14,100$ people per year   **59.** 30,600,000 lb per year
61. $\frac{15}{2}$ **62.** -12 **63.** $-\frac{2}{3}p$ **64.** $5t - 1$
65. $y = -x + 5$ **67.** $y = x + 2$

Exercise Set 3.4, p. 256

RC1. (e)   **RC2.** (f)   **RC3.** (a)   **RC4.** (g)
1. Slope: -4; y-intercept: $(0, -9)$ **3.** Slope: 1.8; y-intercept: $(0, 0)$
5. Slope: $-\frac{8}{7}$; y-intercept: $(0, -3)$ **7.** Slope: $\frac{4}{9}$; y-intercept: $\left(0, -\frac{7}{9}\right)$
9. Slope: $-\frac{3}{2}$; y-intercept: $\left(0, -\frac{1}{2}\right)$ **11.** Slope: 0; y-intercept: $(0, -17)$

13. $y = -7x - 13$ **15.** $y = 1.01x - 2.6$ **17.** $y = -5$
19. $y = -2x - 6$ **21.** $y = \frac{3}{4}x + \frac{5}{2}$ **23.** $y = x - 8$
25. $y = -3x + 3$ **27.** $y = x + 4$ **29.** $y = -\frac{1}{2}x + 4$
31. $y = -\frac{3}{2}x + \frac{13}{2}$ **33.** $x = 4$ **35.** $y = -4x - 11$ **37.** $y = \frac{1}{4}$
39. (a) $S = 2.82t + 35$; (b) an increase of \$2.82 billion per year;
(c) \$77.3 billion **41.** $\frac{53}{7}$ **42.** $\frac{3}{8}$ **43.** 6 **44.** $\frac{42}{5}$ **45.** $\frac{24}{19}$
46. $\frac{125}{7}$ **47.** 3.6 **48.** 500 **49.** 5% **50.** 4000
51. $y = 3x - 9$ **53.** $y = \frac{3}{2}x - 2$

Mid-Chapter Review: Chapter 3, p. 259

1. False **2.** True **3.** True **4.** False
5. (a) The y-intercept is $(0, -3)$. (b) The x-intercept is $(-3, 0)$.
(c) The slope is $\dfrac{-3 - 0}{0 - (-3)} = \dfrac{-3}{3} = -1$. (d) The equation of the line
in $y = mx + b$ form is $y = -1x + -3$, or $y = x - 3$.
6. (a) The x-intercept is $(c, 0)$. (b) The y-intercept is $(0, d)$.
(c) The slope is $\dfrac{d - 0}{0 - c} = \dfrac{d}{-c} = -\dfrac{d}{c}$. (d) The equation of the line in
$y = mx + b$ form is $y = -\dfrac{d}{c}x + d$.
7. No **8.** Yes **9.** x-intercept: $(-6, 0)$; y-intercept: $(0, 9)$
10. x-intercept: $\left(\frac{1}{2}, 0\right)$; y-intercept: $\left(0, -\frac{1}{20}\right)$
11.
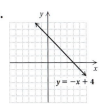
12.
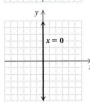
13.

14.
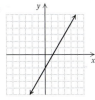
15. $-\frac{40}{9}$ **16.** $-\frac{1}{2}$ **17.** 0 **18.** 13 **19.** Not defined
20. 429,400 people per year **21.** D **22.** C **23.** B **24.** E
25. A **26.** $y = -3x + 2$ **27.** $x = \frac{1}{2}$ **28.** $y = -\frac{1}{5}x - \frac{17}{5}$
29. $y = -4$ **30.** No; an equation $x = a$, $a \neq 0$, does not have
a y-intercept. **31.** Most would probably say that the second
equation would be easier to graph because it has been solved for
y. This makes it more efficient to find the y-value that corresponds
to a given x-value. **32.** $A = 0$. If the line is horizontal, then the
equation is of the form $y =$ a constant. Thus, Ax must be 0 and,
hence, $A = 0$. **33.** Any ordered pair $(7, y)$ is a solution of $x = 7$.
Thus all points on the graph are 7 units to the right of the y-axis, so
they lie on a vertical line.

Exercise Set 3.5, p. 263

RC1. (c) **RC2.** (d) **RC3.** (b) **RC4.** (f)

1.

3.
5.
7.

9.
11.

13.
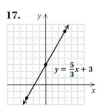
15.
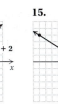
$y = \frac{3}{5}x + 2$
$y = -\frac{3}{5}x + 1$
17.

$y = \frac{5}{3}x + 3$

19.
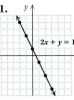
$y = -\frac{3}{2}x - 2$
21.

$2x + y = 1$
23.
$3x - y = 4$

25.

$2x + 3y = 9$
27.

$x - 4y = 12$
29.
$x + 2y = 6$

31. $\frac{13}{10}$ **32.** $-\frac{1}{6}$ **33.** $\frac{69}{100}$, or 0.69 **34.** $-\frac{3}{5}$, or -0.6 **35.** 0
36. Not defined **37.** 42 **38.** -185 **39.** 23 **40.** 41
41.

Exercise Set 3.6, p. 270

RC1. (b) and (e) **RC2.** (c) and (d)
1. Yes **3.** No **5.** No **7.** No **9.** Yes **11.** Yes **13.** No
15. No **17.** Yes **19.** Yes **21.** Yes **23.** No **25.** Yes
27. No **29.** Parallel **31.** Neither **33.** No **34.** Yes
35. x-intercept: $(-2, 0)$; y-intercept: $(0, 16)$ **36.** x-intercept:
$\left(-\frac{1}{2}, 0\right)$; y-intercept: $(0, 3)$ **37.** $y = 3x + 6$ **39.** $y = -3x + 2$
41. $y = \frac{1}{2}x + 1$ **43.** 16 **45.** A: $y = \frac{4}{3}x - \frac{7}{3}$; B: $y = -\frac{3}{4}x - \frac{1}{4}$

Calculator Corner, p. 275

1. Left to the student

Visualizing for Success, p. 276

1. D **2.** H **3.** E **4.** A **5.** J **6.** F **7.** C **8.** B
9. I **10.** G

Exercise Set 3.7, p. 277

RC1. $=$ **RC2.** Related **RC3.** x-intercept
RC4. y-intercept **RC5.** $<$ **RC6.** Dashed **RC7.** Half-plane
RC8. Shade **RC9.** False
1. No **3.** Yes
5.
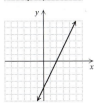
$x > 2y$
7.
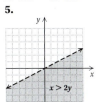
$y \leq x - 3$
9.

$x + y \leq 3$

11.

13. 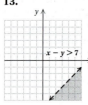 $x - y > 7$

15. $y \geq 4x - 1$

17. 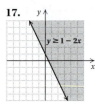 $y \geq 1 - 2x$

19.

21. $y \leq 3$

$2x + 3y \leq 12$

23. $x \geq -1$

25. Parallel   **26.** Perpendicular
27. Neither   **28.** Neither

29. $35c + 75a > 1000$ 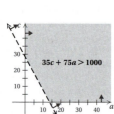 $35c + 75a > 1000$

Summary and Review: Chapter 3, p. 279

Vocabulary Reinforcement

1. Slope-intercept   **2.** Horizontal   **3.** Vertical   **4.** Slope
5. x-intercept; second   **6.** y-intercept; first

Concept Reinforcement

1. True   **2.** False   **3.** False   **4.** True   **5.** False

Study Guide

1. F: $(2, 4)$; G: $(-2, 0)$; H: $(-3, -5)$

2. 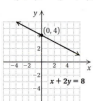 $x + 2y = 8$

3. 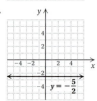 $y - 2x = -4$

4. $y = -\frac{5}{2}$

5. 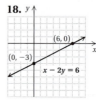 $x = 2$

6. m is not defined.   **7.** $\frac{3}{2}$ **8.** 0   **9.** m is not defined.
10. -2 **11.** 0   **12.** $y = 6x + 7$ **13.** $y = -\frac{1}{6}x - \frac{11}{6}$
14. Perpendicular   **15.** Parallel
16.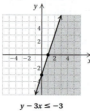

$y - 3x \leq -3$

Review Exercises

1.–3.   **4.** $(-5, -1)$ **5.** $(-2, 5)$ **6.** $(3, 0)$
7. IV   **8.** III   **9.** I   **10.** No
11. Yes

12.

$$2x - y = 3$$
$$2 \cdot 0 - (-3) \; ? \; 3$$
$$0 + 3 \;\big|$$
$$3 \;\big| \quad \text{TRUE}$$

$$2x - y = 3$$
$$2 \cdot 2 - 1 \; ? \; 3$$
$$4 - 1 \;\big|$$
$$3 \;\big| \quad \text{TRUE}$$

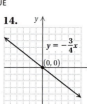 $2x - y = 3$

13. 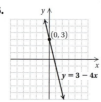 $y = 2x - 5$

14. 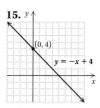 $y = -\frac{3}{4}x$

15. $y = -x + 4$

16. $y = 3 - 4x$

17. **(a)** $14\frac{1}{2}$ ft^3, 16 ft^3, $20\frac{1}{2}$ ft^3, 28 ft^3;
(b) $S = \frac{3}{2}n + 13$ 19 ft^3;
(c) 6 residents

18. 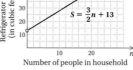 $x - 2y = 6$

19. $5x - 2y = 10$

20.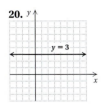

21.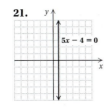

22. $\frac{1}{3}$ **23.** $-\frac{1}{3}$

24. $\frac{3}{5}$;

25. -1;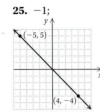

26. $-\frac{5}{8}$ **27.** $\frac{1}{2}$ **28.** Not defined   **29.** 0   **30.** (a) 2.4 driveways per hour; (b) 25 minutes per driveway   **31.** 7%   **32.** $2.4 billion per year   **33.** Slope: -9; y-intercept: $(0, 46)$ **34.** Slope: -1; y-intercept: $(0, 9)$ **35.** Slope: $\frac{3}{5}$; y-intercept: $\left(0, -\frac{4}{5}\right)$
36. $y = -2.8x + 19$ **37.** $y = \frac{5}{8}x - \frac{7}{8}$ **38.** $y = 3x - 1$
39. $y = \frac{2}{3}x - \frac{11}{3}$ **40.** $y = -2x - 4$ **41.** $y = x + 2$
42. $y = \frac{1}{2}x - 1$ **43.** (a) $y = 0.38x + 6.2$; (b) increase of 0.38 million, or 380,000, people per year; (c) 12.28 million, or 12,280,000, people

44.

45.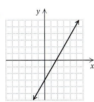

46.

47.

48. Parallel   **49.** Perpendicular   **50.** Parallel   **51.** Neither
52. No   **53.** No   **54.** Yes

55.

56.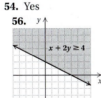

57.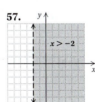

58. D   **59.** C   **60.** 45 square units; 28 linear units
61. (a) 239.58$\overline{3}$ ft per minute; (b) about 0.004 min per foot

Understanding Through Discussion and Writing

1. If one equation represents a vertical line (that is, it is of the form $x = a$) and the other represents a horizontal line (that is, it is of the form $y = b$), then the graphs are perpendicular. If neither line is of one of the forms above, then solve each equation for y in order to determine the slope of each. Then, if the product of the slopes is -1, the graphs are perpendicular.   **2.** If $b > 0$, then the y-intercept of $y = mx + b$ is on the positive y-axis and the graph of $y = mx + b$ lies "above" the origin. Using $(0, 0)$ as a test point, we have the false inequality $0 > b$, so the region above $y = mx + b$ is shaded.

If $b = 0$, the line $y = mx + b$ or $y = mx$ passes through the origin. Testing a point above the line, such as $(1, m + 1)$, we have the true inequality $m + 1 > m$, so the region above the line is shaded.

If $b < 0$, then the y-intercept of $y = mx + b$ is on the negative y-axis and the graph of $y = mx + b$ lies "below" the origin. Using $(0, 0)$ as a test point, we get the true inequality $0 > b$, so the region above $y = mx + b$ is shaded.

Thus we see that in any case the graph of any inequality of the form $y > mx + b$ is always shaded above the line $y = mx + b$.
3. The y-intercept is the point at which the graph crosses the y-axis. Since a point on the y-axis is neither left nor right of the origin, the first or x-coordinate of the point is 0.
4.

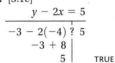

The graph of $x < 1$ on the number line consists of the points in the set $\{x \mid x < 1\}$. The graph of $x < 1$ on a plane consists of the points, or ordered pairs, in the set $\{(x, y) \mid x + 0 \cdot y < 1\}$. This is the set of ordered pairs with first coordinate less than 1.

5. First, plot the y-intercept, $(0, 2458)$. Then, thinking of the slope as $\frac{37}{100}$, plot a second point on the line by moving up 37 units and to the right 100 units from the y-intercept. Next, thinking of the slope as $\frac{-37}{-100}$, start at the y-intercept and plot a third point by moving down 37 units and to the left 100 units. Finally, draw a line through the three points.   **6.** If the equations are of the form $x = p$ and $x = q$, where $p \neq q$, then the graphs are parallel vertical lines. If neither equation is of the form $x = p$, then solve each for y in order to determine the slope and the y-intercept of each. If the slopes are the same and the y-intercepts are different, the lines are parallel.

Test: Chapter 3, p. 287

1. [3.1a] II   **2.** [3.1a] III   **3.** [3.1b] $(-5, 1)$
4. [3.1b] $(0, -4)$
5. [3.1c]

$$
\begin{array}{c|c}
y - 2x = 5 & \\
\hline
-3 - 2(-4) \ ? \ 5 & \\
-3 + 8 & \\
5 & \text{TRUE}
\end{array}
$$

$$
\begin{array}{c|c}
y - 2x = 5 & \\
\hline
3 - 2(-1) \ ? \ 5 & \\
3 + 2 & \\
5 & \text{TRUE}
\end{array}
$$

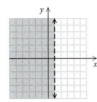

6. [3.1d]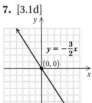

7. [3.1d]

8. [3.2a]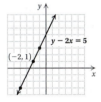

9. [3.2a]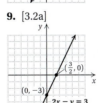

10. [3.2b]

11. [3.2b]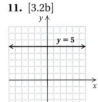

12. [3.1e] **(a)** 2007: $8593; 2009: $9805; 2012: $11,623;
(b)

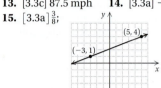

; approximately $14,000;
(c) 11 years after 2007, or in 2018

13. [3.3c] 87.5 mph   **14.** [3.3a] -2
15. [3.3a] $\frac{3}{8}$;

16. [3.3b] **(a)** $\frac{2}{5}$; **(b)** not defined   **17.** [3.3c] $-\frac{1}{20}$, or -0.05
18. [3.4a] Slope: 2; y-intercept: $(0, -\frac{1}{4})$ **19.** [3.4a] Slope: $\frac{4}{3}$; y-intercept: $(0, -2)$ **20.** [3.4a] $y = 1.8x - 7$
21. [3.4a] $y = -\frac{3}{8}x - \frac{1}{8}$ **22.** [3.4b] $y = x + 2$
23. [3.4b] $y = -3x - 6$ **24.** [3.4c] $y = -3x + 4$
25. [3.4c] $y = \frac{1}{4}x - 2$
26. [3.5a]

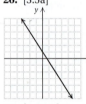

27. [3.5a]

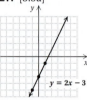

$y = 2x - 3$

28. [3.6a, b] Parallel   **29.** [3.6a, b] Neither
30. [3.6a, b] Perpendicular   **31.** [3.7a] No   **32.** [3.7a] Yes
33. [3.7b]

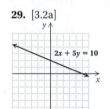

$y > x - 1$

34. [3.7b]

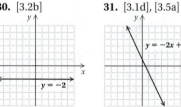

$2x - y \geq 4$

35. [3.6a, b] A   **36.** [3.1a] 25 square units; 20 linear units
37. [3.6b] 3

Cumulative Review: Chapters 1–3, p. 291

1. [1.1a] $-\frac{4}{5}$ **2.** [1.7c] $-\frac{2}{3}x + 4y - 2$ **3.** [1.7d] $3(6w - 8 + 3y)$
4. [1.2c] $-0.\overline{7}$ **5.** [1.2e] $2\frac{1}{5}$ **6.** [1.3b] -8.17 **7.** [1.6b] $-\frac{7}{8}$
8. [1.7e] $-x - y$ **9.** [1.3a] -3 **10.** [1.8d] -6 **11.** [1.6c] $-\frac{2}{3}$
12. [1.8c] $11x + 9$ **13.** [1.5a] 2.64   **14.** [1.8d] -2
15. [2.2a] -81 **16.** [2.3c] No solution   **17.** [2.3b] 3
18. [2.3c] All real numbers   **19.** [2.7e] $\{x \mid x \leq -\frac{11}{8}\}$
20. [2.1b] $\frac{4}{3}$ **21.** [3.3b] $\frac{3}{4}$ **22.** [3.3b] Not defined
23. [3.4b] $y = -20x - 162$ **24.** [3.6a, b] Perpendicular
25. [2.4b] $h = \dfrac{2A}{b + c}$ **26.** [3.1a] IV
27. [3.2a] y-intercept: $(0, -3)$; x-intercept: $\left(\frac{21}{2}, 0\right)$
28. [2.7b]

29. [3.2a]

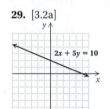

$2x + 5y = 10$

30. [3.2b]

$y = -2$

31. [3.1d], [3.5a]

$y = -2x + 1$

---

32. [3.1d]

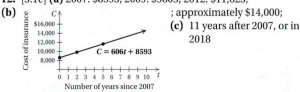

$3y + 6 = 2x$

33. [3.1d]

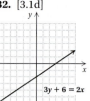

$y = -\frac{3}{2}x$

34. [3.2b]

$x = 4.5$

35. [3.3a] $-\frac{1}{4}$ **36.** [2.6a] 20.2 million people
37. [2.8b] $\{x \mid x \leq 8\}$ **38.** [2.6a] First: 50 m; second: 53 m; third: 40 m
39. [1.8d] A   **40.** [3.3a] D   **41.** [2.3a], [1.2e] $-4, 4$ **42.** [2.3b] 3
43. [2.4b] $Q = \dfrac{2 - pm}{p}$, or $Q = \dfrac{2}{p} - m$

CHAPTER 4

Exercise Set 4.1, p. 300

RC1. (c)   **RC2.** (a)   **RC3.** (e)   **RC4.** (c)   **RC5.** (d)
RC6. (c)
1. $3 \cdot 3 \cdot 3 \cdot 3$ **3.** $(-1.1)(-1.1)(-1.1)(-1.1)(-1.1)$
5. $\left(\frac{2}{3}\right)\left(\frac{2}{3}\right)\left(\frac{2}{3}\right)\left(\frac{2}{3}\right)$ **7.** $(7p)(7p)$ **9.** $8 \cdot k \cdot k \cdot k$
11. $-6 \cdot y \cdot y \cdot y \cdot y$ **13.** 1   **15.** b **17.** 1   **19.** -7.03
21. 1   **23.** ab **25.** a **27.** 27   **29.** 19   **31.** -81
33. 256   **35.** 93   **37.** 136   **39.** 10; 4   **41.** 3629.84 ft^2
43. $\dfrac{1}{3^2} = \dfrac{1}{9}$ **45.** $\dfrac{1}{10^3} = \dfrac{1}{1000}$ **47.** $\dfrac{1}{a^3}$ **49.** $8^2 = 64$
51. y^4 **53.** $\dfrac{5}{z^4}$ **55.** $\dfrac{x}{y^2}$ **57.** 4^{-3} **59.** x^{-3} **61.** a^{-5}
63. 2^7 **65.** 9^{38} **67.** x^5 **69.** x^{17} **71.** $(3y)^{12}$
73. $(7y)^{17}$ **75.** 3^3 **77.** 1   **79.** $\dfrac{1}{x^{13}}$ **81.** $\dfrac{1}{a^{10}}$ **83.** $x^6 y^{15}$
85. $s^3 t^7$ **87.** 7^3 **89.** y^8 **91.** $\dfrac{1}{16^6}$ **93.** $\dfrac{1}{m^6}$ **95.** $\dfrac{1}{(8x)^4}$
97. x^2 **99.** $\dfrac{1}{z^4}$ **101.** x^3 **103.** 1   **105.** $a^3 b^2$
107. $5^2 = 25; 5^{-2} = \frac{1}{25}; \left(\frac{1}{5}\right)^2 = \frac{1}{25}; \left(\frac{1}{5}\right)^{-2} = 25; -5^2 = -25;$
$(-5)^2 = 25; -\left(\frac{1}{5}\right)^2 = -\frac{1}{25}; \left(-\frac{1}{5}\right)^{-2} = 25$
109. 8 in.; 4 in.   **110.** 51°, 27°, 102°   **111.** 45%; 37.5%; 17.5%
112. Lengths less than 2.5 ft   **113.** $\frac{7}{4}$ **114.** 2   **115.** $\frac{23}{14}$
116. $\frac{11}{10}$ **117.** No   **119.** No   **121.** y^{5x} **123.** a^{4t} **125.** 1
127. $>$ **129.** $<$ **131.** $-\frac{1}{10,000}$ **133.** No; for example, $(3 + 4)^2 = 49$, but $3^2 + 4^2 = 25$.

Calculator Corner, p. 308

1. 1.3545×10^{-4} **2.** 3.2×10^5 **3.** 3×10^{-6} **4.** 8×10^{-26}

Exercise Set 4.2, p. 310

RC1. Multiply   **RC2.** nth   **RC3.** Right   **RC4.** Positive
1. 2^6 **3.** $\dfrac{1}{5^6}$ **5.** x^{12} **7.** $\dfrac{1}{a^{18}}$ **9.** t^{18} **11.** $\dfrac{1}{t^{12}}$ **13.** x^8
15. $a^3 b^3$ **17.** $\dfrac{1}{a^3 b^3}$ **19.** $\dfrac{1}{m^3 n^6}$ **21.** $16x^6$ **23.** $\dfrac{9}{x^8}$
25. $\dfrac{1}{x^{12} y^{15}}$ **27.** $x^{24} y^8$ **29.** $\dfrac{a^{10}}{b^{35}}$ **31.** $\dfrac{25t^6}{r^8}$ **33.** $\dfrac{b^{21}}{a^{15} c^6}$
35. $\dfrac{9x^6}{y^{16} z^6}$ **37.** $\dfrac{16x^6}{y^4}$ **39.** $a^{12} b^8$ **41.** $\dfrac{y^6}{4}$ **43.** $\dfrac{a^8}{b^{12}}$ **45.** $\dfrac{8}{y^6}$
47. $49x^6$ **49.** $\dfrac{x^6 y^3}{z^3}$ **51.** $\dfrac{c^2 d^6}{a^4 b^2}$ **53.** 2.8×10^{10} **55.** 9.07×10^{17}
57. 3.04×10^{-6} **59.** 1.8×10^{-8} **61.** 10^{11} **63.** 4.19854×10^8

65. 6.8×10^{-7} **67.** 87,400,000   **69.** 0.00000005704
71. 10,000,000   **73.** 0.00001   **75.** 6×10^9 **77.** 3.38×10^4
79. 8.1477×10^{-13} **81.** 2.5×10^{13} **83.** 5.0×10^{-4}
85. 3.0×10^{-21} **87.** Approximately 1.325×10^{14} ft³
89. 1×10^{22} **91.** The mass of Jupiter is 3.18×10^2 times the
mass of Earth.   **93.** 1.25×10^9 bytes   **95.** 3.1×10^2 sheets
97. 4.375×10^2 days

99. 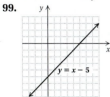   **100.**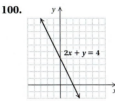

101. 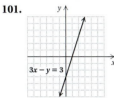   **102.**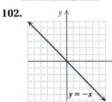

103. 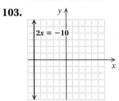   **104.**

105. 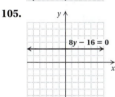   **106.**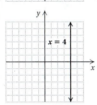

107. 2.478125×10^{-1} **109.** $\frac{1}{5}$ **111.** 3^{11} **113.** 7
115. $\frac{1}{0.4}$, or 2.5   **117.** False   **119.** False   **121.** True

Calculator Corner, p. 318

1. 3; 2.25; −27   **2.** 44; 0; 9.28

Exercise Set 4.3, p. 323

RC1. (b)   **RC2.** (f)   **RC3.** (c)   **RC4.** (d)   **RC5.** (a)
RC6. (e)
1. −18; 7   **3.** 19; 14   **5.** −12; −7   **7.** $\frac{13}{3}$; 5   **9.** 9; 1
11. 56; −2   **13.** 1112 ft   **15.** 55 oranges   **17.** −4, 4, 5, 2.75, 1
19. (a) 5.67 kW; **(b)** left to the student   **21.** 9 words   **23.** 6
25. 15   **27.** 2, −3x, x²   **29.** −2x⁴, $\frac{1}{3}$x³, −x, 3   **31.** Trinomial
33. None of these   **35.** Binomial   **37.** Monomial   **39.** −3, 6
41. 5, $\frac{3}{4}$, 3   **43.** −5, 6, −2.7, 1, −2   **45.** 1, 0; 1   **47.** 2, 1, 0; 2
49. 3, 2, 1, 0; 3   **51.** 2, 1, 6, 4; 6
53.

Term	Coefficient	Degree of the Term	Degree of the Polynomial
$-7x^4$	−7	4	
$6x^3$	6	3	
$-x^2$	−1	2	4
$8x$	8	1	
-2	−2	0	

55. $6x^2$ and $-3x^2$ **57.** $2x^4$ and $-3x^4$; $5x$ and $-7x$ **59.** $3x^5$ and
$14x^5$; $-7x$ and $-2x$; 8 and −9   **61.** −3x   **63.** −8x
65. $11x^3 + 4$ **67.** $x^3 - x$ **69.** $4b^5$ **71.** $\frac{3}{4}x^5 - 2x - 42$
73. x^4 **75.** $\frac{15}{16}x^3 - \frac{7}{6}x^2$ **77.** $x^5 + 6x^3 + 2x^2 + x + 1$
79. $15y^9 + 7y^8 + 5y^3 - y^2 + y$ **81.** $x^6 + x^4$
83. $13x^3 - 9x + 8$ **85.** $-5x^2 + 9x$ **87.** $12x^4 - 2x + \frac{1}{4}$
89. x^2, x **91.** x^3, x^2, x^0 **93.** None missing
95. $x^3 + 0x^2 + 0x - 27$; $x^3 \qquad - 27$
97. $x^4 + 0x^3 + 0x^2 - x + 0x^0$; $x^4 \qquad - x$
99. None missing   **101.** −19   **102.** −1   **103.** −2.25
104. $-\frac{3}{8}$ **105.** −19   **106.** $-\frac{17}{24}$ **107.** $\frac{5}{8}$ **108.** −2.6
109. 18   **110.** $-\frac{1}{3}$ **111.** −0.6   **112.** −24   **113.** 20
114. $-\frac{8}{5}$ **115.** 0   **116.** Not defined   **117.** $10x^6 + 52x^5$
119. $4x^5 - 3x^3 + x^2 - 7x$; answers may vary
121. $x^3 - 2x^2 - 6x + 3$ **123.** −4, 4, 5, 2.75, 1   **125.** 9

Exercise Set 4.4, p. 332

RC1. False   **RC2.** True   **RC3.** False   **RC4.** False
1. $-x + 5$ **3.** $x^2 - \frac{11}{2}x - 1$ **5.** $2x^2$ **7.** $5x^2 + 3x - 30$
9. $-2.2x^3 - 0.2x^2 - 3.8x + 23$ **11.** $6 + 12x^2$
13. $-\frac{1}{2}x^4 + \frac{2}{3}x^3 + x^2$ **15.** $0.01x^5 + x^4 - 0.2x^3 + 0.2x + 0.06$
17. $9x^8 + 8x^7 - 6x^4 + 8x^2 + 4$
19. $1.05x^4 + 0.36x^3 + 14.22x^2 + x + 0.97$
21. $5x$ **23.** $x^2 - \frac{3}{2}x + 2$ **25.** $-12x^4 + 3x^3 - 3$ **27.** $-3x + 7$
29. $-4x^2 + 3x - 2$ **31.** $4x^4 - 6x^2 - \frac{3}{4}x + 8$ **33.** $7x - 1$
35. $-x^2 - 7x + 5$ **37.** −18   **39.** $6x^4 + 3x^3 - 4x^2 + 3x - 4$
41. $4.6x^3 + 9.2x^2 - 3.8x - 23$ **43.** $\frac{3}{4}x^3 - \frac{1}{2}x$
45. $0.06x^3 - 0.05x^2 + 0.01x + 1$ **47.** $3x + 6$
49. $11x^4 + 12x^3 - 9x^2 - 8x - 9$ **51.** $x^4 - x^3 + x^2 - x$
53. $\frac{23}{2}a + 12$ **55.** $5x^2 + 4x$
57. $(r + 11)(r + 9)$; $9r + 99 + 11r + r^2$, or $r^2 + 20r + 99$
59. $(x + 3)(x + 3)$, or $(x + 3)^2$; $x^2 + 3x + 9 + 3x$, or $x^2 + 6x + 9$
61. $\pi r^2 - 25\pi$ **63.** $18z - 64$ **65.** 6   **66.** −19   **67.** $-\frac{7}{22}$
68. 5   **69.** 5   **70.** 1   **71.** $\frac{39}{2}$ **72.** $\frac{37}{2}$ **73.** $\{x | x \geq -10\}$
74. $\{x | x < 0\}$ **75.** $20w + 42$ **77.** $2x^2 + 20x$
79. $y^2 - 4y + 4$ **81.** $12y^2 - 23y + 21$
83. $-3y^4 - y^3 + 5y - 2$

Mid-Chapter Review: Chapter 4, p. 337

1. True   **2.** False   **3.** False   **4.** True
5. $4w^3 + 6w - 8w^3 - 3w = (4 - 8)w^3 + (6 - 3)w =$
$-4w^3 + 3w$ **6.** $(3y^4 - y^2 + 11) - (y^4 - 4y^2 + 5) =$
$3y^4 - y^2 + 11 - y^4 + 4y^2 - 5 = 2y^4 + 3y^2 + 6$
7. z **8.** 1   **9.** −32   **10.** 1   **11.** 5^7 **12.** $(3a)^9$
13. $\frac{1}{x^3}$ **14.** 1   **15.** 7^4 **16.** $\frac{1}{x^2}$ **17.** w^8 **18.** $\frac{1}{y^4}$ **19.** 3^{15}
20. $\frac{x^{18}}{y^{12}}$ **21.** $\frac{a^{24}}{5^6}$ **22.** $\frac{x^2z^4}{4y^6}$ **23.** 2.543×10^7 **24.** 1.2×10^{-4}
25. 0.000036   **26.** 144,000,000   **27.** 6×10^3 **28.** 5×10^{-7}
29. 16; 1   **30.** −16; 9   **31.** $-2x^5 - 5x^2 + 4x + 2$
32. $8x^6 + 2x^3 - 8x^2$ **33.** 3, 1, 0; 3   **34.** 1, 4, 6; 6
35. Binomial   **36.** Trinomial   **37.** $8x^2 + 5$
38. $5x^3 - 2x^2 + 2x - 11$ **39.** $-4x - 10$
40. $-0.4x^2 - 3.4x + 9$ **41.** $3y + 3y^2$ **42.** The area of the
smaller square is x^2, and the area of the larger square is $(3x)^2$, or $9x^2$,
so the area of the larger square is nine times the area of the smaller
square.   **43.** The volume of the smaller cube is x^3, and the volume
of the larger cube is $(2x)^3$, or $8x^3$, so the volume of the larger cube
is eight times the volume of the smaller cube.   **44.** Exponents are
added when powers with like bases are multiplied. Exponents are
multiplied when a power is raised to a power.
45. $3^{-29} = \frac{1}{3^{29}}$ and $2^{-29} = \frac{1}{2^{29}}$. Since $3^{29} > 2^{29}$, we have $\frac{1}{3^{29}} < \frac{1}{2^{29}}$.

46. It is better to evaluate a polynomial after like terms have been collected, because there are fewer terms to evaluate. **47.** Yes; consider the following: $(x^2 + 4) + (4x - 7) = x^2 + 4x - 3$.

Calculator Corner, p. 342

1. Correct **2.** Correct **3.** Not correct **4.** Not correct

Exercise Set 4.5, p. 343

RC1. (a) **RC2.** (a) **RC3.** (d) **RC4.** (e)
1. $40x^2$ **3.** x^3 **5.** $32x^8$ **7.** $0.03x^{11}$ **9.** $\frac{1}{15}x^4$ **11.** 0
13. $-24x^{11}$ **15.** $-2x^2 + 10x$ **17.** $-5x^2 + 5x$ **19.** $x^5 + x^2$
21. $6x^3 - 18x^2 + 3x$ **23.** $-6x^4 - 6x^3$ **25.** $18y^6 + 24y^5$
27. $x^2 + 9x + 18$ **29.** $x^2 + 3x - 10$ **31.** $x^2 + 3x - 4$
33. $x^2 - 7x + 12$ **35.** $x^2 - 9$ **37.** $x^2 - 16$
39. $3x^2 + 11x + 10$ **41.** $25 - 15x + 2x^2$ **43.** $4x^2 + 20x + 25$
45. $x^2 - 6x + 9$ **47.** $x^2 - \frac{21}{10}x - 1$ **49.** $x^2 + 2.4x - 10.81$
51. $(x + 2)(x + 6)$, or $x^2 + 8x + 12$ **53.** $(x + 1)(x + 6)$, or $x^2 + 7x + 6$

55. **57.** 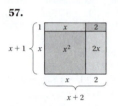 **59.**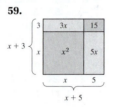

61. $x^3 - 1$ **63.** $4x^3 + 14x^2 + 8x + 1$
65. $3y^4 - 6y^3 - 7y^2 + 18y - 6$ **67.** $x^6 + 2x^5 - x^3$
69. $-10x^5 - 9x^4 + 7x^3 + 2x^2 - x$
71. $-1 - 2x - x^2 + x^4$ **73.** $6t^4 + t^3 - 16t^2 - 7t + 4$
75. $x^9 - x^5 + 2x^3 - x$ **77.** $x^4 + 8x^3 + 12x^2 + 9x + 4$
79. $2x^4 - 5x^3 + 5x^2 - \frac{19}{10}x + \frac{1}{5}$ **81.** $-x + 3$ **82.** 47 **83.** 96
84. 32 **85.** $3(5x - 6y + 4)$ **86.** $4(4x - 6y + 9)$
87. $-3(3x + 15y - 5)$ **88.** $100(x - y + 10a)$ **89.** $75y^2 - 45y$
91. 5 **93.** $(x^3 + 2x^2 - 210)\,\text{m}^3$ **95.** 0 **97.** 0

Visualizing for Success, p. 352

1. E, F **2.** B, O **3.** K, S **4.** G, R **5.** D, M **6.** J, P
7. C, L **8.** N, Q **9.** A, H **10.** I, T

Exercise Set 4.6, p. 353

RC1. Outside; last **RC2.** Descending **RC3.** Difference
RC4. Square; binomial **RC5.** Binomials **RC6.** Difference
1. $x^3 + x^2 + 3x + 3$ **3.** $x^4 + x^3 + 2x + 2$ **5.** $y^2 - y - 6$
7. $9x^2 + 12x + 4$ **9.** $5x^2 + 4x - 12$ **11.** $9t^2 - 1$
13. $4x^2 - 6x + 2$ **15.** $p^2 - \frac{1}{16}$ **17.** $x^2 - 0.01$
19. $2x^3 + 2x^2 + 6x + 6$ **21.** $-2x^2 - 11x + 4$
23. $a^2 + 14a + 49$ **25.** $1 - x - 6x^2$ **27.** $\frac{9}{64}y^2 - \frac{5}{8}y + \frac{25}{36}$
29. $x^5 + 3x^3 - x^2 - 3$ **31.** $3x^6 - 2x^4 - 6x^2 + 4$
33. $13.16x^2 + 18.99x - 13.95$ **35.** $6x^7 + 18x^5 + 4x^2 + 12$
37. $4x^3 - 12x^2 + 3x - 9$ **39.** $4y^6 + 4y^5 + y^4 + y^3$
41. $x^2 - 16$ **43.** $4x^2 - 1$ **45.** $25m^2 - 4$ **47.** $4x^4 - 9$
49. $9x^8 - 16$ **51.** $x^{12} - x^4$ **53.** $x^8 - 9x^2$ **55.** $x^{24} - 9$
57. $4y^{16} - 9$ **59.** $\frac{25}{64}x^2 - 18.49$ **61.** $x^2 + 4x + 4$
63. $9x^4 + 6x^2 + 1$ **65.** $a^2 - a + \frac{1}{4}$ **67.** $9 + 6x + x^2$
69. $x^4 + 2x^2 + 1$ **71.** $4 - 12x^4 + 9x^8$ **73.** $25 + 60t^2 + 36t^4$
75. $x^2 - \frac{5}{4}x + \frac{25}{64}$ **77.** $9 - 12x^3 + 4x^6$ **79.** $4x^3 + 24x^2 - 12x$
81. $4x^4 - 2x^2 + \frac{1}{4}$ **83.** $9p^2 - 1$ **85.** $15t^5 - 3t^4 + 3t^3$
87. $36x^8 + 48x^4 + 16$ **89.** $12x^3 + 8x^2 + 15x + 10$
91. $64 - 96x^4 + 36x^8$ **93.** $t^3 - 1$ **95.** 25; 49 **97.** 56; 16
99. $a^2 + 2a + 1$ **101.** $t^2 + 10t + 24$ **103.** $\frac{28}{27}$ **104.** $-\frac{41}{7}$
105. $\frac{27}{4}$ **106.** $y = \frac{3x - 12}{2}$, or $y = \frac{3}{2}x - 6$ **107.** $b = \frac{C + r}{a}$

108. $a = \frac{5d + 4}{3}$, or $a = \frac{5}{3}d + \frac{4}{3}$ **109.** $30x^3 + 35x^2 - 15x$
111. $a^4 - 50a^2 + 625$ **113.** $81t^{16} - 72t^8 + 16$ **115.** -7
117. First row: 90, -432, -63; second row: 7, -18, -36, -14, 12, -6, -21, -11; third row: 9, -2, -2, 10, -8, -8, -8, -10, 21; fourth row: -19, -6 **119.** Yes **121.** No

Exercise Set 4.7, p. 361

RC1. True **RC2.** False **RC3.** False **RC4.** False
1. -1 **3.** -15 **5.** 240 **7.** -145 **9.** 3.715 L
11. 2322 calories **13.** 44.46 in^2 **15.** 73.005 in^2
17. Coefficients: 1, -2, 3, -5; degrees: 4, 2, 2, 0; 4
19. Coefficients: 17, -3, -7; degrees: 5, 5, 0; 5 **21.** $-a - 2b$
23. $3x^2y - 2xy^2 + x^2$ **25.** $20au + 10av$ **27.** $8u^2v - 5uv^2$
29. $x^2 - 4xy + 3y^2$ **31.** $3r + 7$
33. $-b^2a^3 - 3b^3a^2 + 5ba + 3$ **35.** $ab^2 - a^2b$ **37.** $2ab - 2$
39. $-2a + 10b - 5c + 8d$ **41.** $6z^2 + 7zu - 3u^2$
43. $a^4b^2 - 7a^2b + 10$ **45.** $a^6 - b^2c^2$
47. $y^6x + y^4x + y^4 + 2y^2 + 1$ **49.** $12x^2y^2 + 2xy - 2$
51. $12 - c^2d^2 - c^4d^4$ **53.** $m^3 + m^2n - mn^2 - n^3$
55. $x^9y^9 - x^6y^6 + x^5y^5 - x^2y^2$ **57.** $x^2 + 2xh + h^2$
59. $9a^2 + 12ab + 4b^2$ **61.** $r^6t^4 - 8r^3t^2 + 16$
63. $p^8 + 2m^2n^2p^4 + m^4n^4$ **65.** $3a^3 - 12a^2b + 12ab^2$
67. $m^2 + 2mn + n^2 - 6m - 6n + 9$ **69.** $a^2 - b^2$
71. $4a^2 - b^2$ **73.** $c^4 - d^2$ **75.** $a^2b^2 - c^2d^4$
77. $x^2 + 2xy + y^2 - 9$ **79.** $x^2 - y^2 - 2yz - z^2$
81. $a^2 + 2ab + b^2 - c^2$ **83.** $3x^4 - 7x^2y + 3x^2 - 20y^2 + 22y - 6$
85. IV **86.** III **87.** I **88.** II **89.** 39 **90.** 1.125
91. $<$ **92.** -3 **93.** $4xy - 4y^2$ **95.** $2xy + \pi x^2$
97. $2\pi nh + 2\pi mh + 2\pi n^2 - 2\pi m^2$ **99.** 16 gal
101. $12,351.94

Exercise Set 4.8, p. 370

RC1. Subtract; divide **RC2.** Divide **RC3.** Multiply; subtract
RC4. Multiply; add
1. $3x^4$ **3.** $5x$ **5.** $18x^3$ **7.** $4a^3b$ **9.** $3x^4 - \frac{1}{2}x^3 + \frac{1}{8}x^2 - 2$
11. $1 - 2u - u^4$ **13.** $5t^2 + 8t - 2$ **15.** $-4x^4 + 4x^2 + 1$
17. $6x^2 - 10x + \frac{3}{2}$ **19.** $9x^2 - \frac{5}{2}x + 1$ **21.** $6x^2 + 13x + 4$
23. $3rs + r - 2s$ **25.** $x + 2$ **27.** $x - 5 + \dfrac{-50}{x - 5}$
29. $x - 2 + \dfrac{-2}{x + 6}$ **31.** $x - 3$ **33.** $x^4 - x^3 + x^2 - x + 1$
35. $2x^2 - 7x + 4$ **37.** $x^3 - 6$ **39.** $t^2 + 1$
41. $y^2 - 3y + 1 + \dfrac{-5}{y + 2}$ **43.** $3x^2 + x + 2 + \dfrac{10}{5x + 1}$
45. $6y^2 - 5 + \dfrac{-6}{2y + 7}$ **47.** -1 **48.** $\frac{10}{3}$ **49.** $\frac{23}{19}$
50. $\{r \mid r \leq -15\}$ **51.** 140% **52.** $\{x \mid x \geq 95\}$
53. 25,543.75 ft^2 **54.** 228, 229 **55.** $x^2 + 5$
57. $a + 3 + \dfrac{5}{5a^2 - 7a - 2}$ **59.** $2x^2 + x - 3$
61. $a^5 + a^4b + a^3b^2 + a^2b^3 + ab^4 + b^5$ **63.** -5 **65.** 1

Summary and Review: Chapter 4, p. 373

Vocabulary Reinforcement

1. Exponent **2.** Product **3.** Monomial **4.** Trinomial
5. Quotient **6.** Descending **7.** Degree **8.** Scientific

Concept Reinforcement

1. True **2.** False **3.** False **4.** True

Study Guide

1. z^8 **2.** a^2b^6 **3.** $\dfrac{y^6}{27x^{12}z^9}$ **4.** 7.63×10^5 **5.** 0.0003

6. 6×10^4 **7.** $2x^4 - 4x^2 - 3$ **8.** $3x^4 + x^3 - 2x^2 + 2$

9. $x^6 - 6x^4 + 11x^2 - 6$ **10.** $2y^2 + 11y + 12$ **11.** $x^2 - 25$

12. $9w^2 + 24w + 16$ **13.** $-2a^3b^2 - 5a^2b + ab^2 - 2ab$

14. $y^2 - 4y + \dfrac{8}{5}$ **15.** $x - 9 + \dfrac{48}{x + 5}$

Review Exercises

1. $\dfrac{1}{7^2}$ **2.** y^{11} **3.** $(3x)^{14}$ **4.** t^8 **5.** 4^3 **6.** $\dfrac{1}{a^3}$ **7.** 1

8. $9t^8$ **9.** $36x^8$ **10.** $\dfrac{y^3}{8x^3}$ **11.** t^{-5} **12.** $\dfrac{1}{y^4}$ **13.** 3.28×10^{-5}

14. $8{,}300{,}000$ **15.** 2.09×10^4 **16.** 5.12×10^{-5}

17. 1.564×10^{10} slices **18.** 10 **19.** $-4y^5, 7y^2, -3y, -2$

20. x^2, x^0 **21.** $3, 2, 1, 0; 3$ **22.** Binomial **23.** None of these

24. Monomial **25.** $-2x^2 - 3x + 2$ **26.** $10x^4 - 7x^2 - x - \frac{1}{2}$

27. $x^5 - 2x^4 + 6x^3 + 3x^2 - 9$ **28.** $-2x^5 - 6x^4 - 2x^3 - 2x^2 + 2$

29. $2x^2 - 4x$ **30.** $x^5 - 3x^3 - x^2 + 8$ **31.** Perimeter: $4w + 6$;
area: $w^2 + 3w$ **32.** $(t + 3)(t + 4), t^2 + 7t + 12$

33. $x^2 + \frac{7}{6}x + \frac{1}{3}$ **34.** $49x^2 + 14x + 1$ **35.** $12x^3 - 23x^2 + 13x - 2$

36. $9x^4 - 16$ **37.** $15x^7 - 40x^6 + 50x^5 + 10x^4$ **38.** $x^2 - 3x - 28$

39. $9y^4 - 12y^3 + 4y^2$ **40.** $2t^4 - 11t^2 - 21$ **41.** 49

42. Coefficients: $1, -7, 9, -8$; degrees: $6, 2, 2, 0; 6$

43. $-y + 9w - 5$ **44.** $m^6 - 2m^2n + 2m^2n^2 + 8n^2m - 6m^3$

45. $-9xy - 2y^2$ **46.** $11x^3y^2 - 8x^2y - 6x^2 - 6x + 6$

47. $p^3 - q^3$ **48.** $9a^8 - 2a^4b^3 + \frac{1}{9}b^6$ **49.** $5x^2 - \frac{1}{2}x + 3$

50. $3x^2 - 7x + 4 + \dfrac{1}{2x + 3}$ **51.** $0, 3.75, -3.75, 0$ **52.** B

53. D **54.** $\frac{1}{2}x^2 - \frac{1}{2}y^2$ **55.** $400 - 4a^2$ **56.** $-28x^8$ **57.** $\frac{94}{13}$

58. $x^4 + x^3 + x^2 + x + 1$ **59.** 80 ft by 40 ft

Understanding Through Discussion and Writing

1. 578.6×10^{-7} is not in scientific notation because 578.6 is not a number greater than or equal to 1 and less than 10.

2. When evaluating polynomials, it is essential to know the order in which the operations are to be performed.

3. We label the figure as shown.

Then we see that the area of the figure is $(x + 3)^2$, or $x^2 + 3x + 3x + 9 \neq x^2 + 9$. **4.** Emma did not divide *each* term of the polynomial by the divisor. The first term was divided by $3x$, but the second was not. Multiplying Emma's "quotient" by the divisor $3x$, we get $12x^3 - 18x^2 \neq 12x^3 - 6x$. This should convince her that a mistake has been made. **5.** Yes; for example, $(x^2 + xy + 1) + (3x - xy + 2) = x^2 + 3x + 3$. **6.** Yes; consider $a + b + c + d$. This is a polynomial in 4 variables but it has degree 1.

Test: Chapter 4, p. 379

1. [4.1d, f] $\dfrac{1}{6^5}$ **2.** [4.1d] x^9 **3.** [4.1d] $(4a)^{11}$ **4.** [4.1e] 3^3

5. [4.1e, f] $\dfrac{1}{x^5}$ **6.** [4.1b, e] 1 **7.** [4.2a] x^6 **8.** [4.2a, b] $-27y^6$

9. [4.2a, b] $16a^{12}b^4$ **10.** [4.2b] $\dfrac{a^3b^3}{c^3}$ **11.** [4.1d], [4.2a, b] $-216x^{21}$

12. [4.1d], [4.2a, b] $-24x^{21}$ **13.** [4.1d], [4.2a, b] $162x^{10}$

14. [4.1d], [4.2a, b] $324x^{10}$ **15.** [4.1f] $\dfrac{1}{5^3}$ **16.** [4.1f] y^{-8}

17. [4.2c] 3.9×10^9 **18.** [4.2c] 0.00000005

19. [4.2d] 1.75×10^{17} **20.** [4.2d] 1.296×10^{22}

21. [4.2e] The mass of Saturn is 9.5×10 times the mass of Earth.

22. [4.3a] -43 **23.** [4.3c] $\frac{1}{3}, -1, 7$ **24.** [4.3c] $3, 0, 1, 6; 6$

25. [4.3b] Binomial **26.** [4.3d] $5a^2 - 6$ **27.** [4.3d] $\frac{7}{4}y^2 - 4y$

28. [4.3e] $x^5 + 2x^3 + 4x^2 - 8x + 3$

29. [4.4a] $4x^5 + x^4 + 2x^3 - 8x^2 + 2x - 7$

30. [4.4a] $5x^4 + 5x^2 + x + 5$ **31.** [4.4c] $-4x^4 + x^3 - 8x - 3$

32. [4.4c] $-x^5 + 0.7x^3 - 0.8x^2 - 21$

33. [4.5b] $-12x^4 + 9x^3 + 15x^2$ **34.** [4.6c] $x^2 - \frac{2}{3}x + \frac{1}{9}$

35. [4.6b] $9x^2 - 100$ **36.** [4.6a] $3b^2 - 4b - 15$

37. [4.6a] $x^{14} - 4x^8 + 4x^6 - 16$ **38.** [4.6a] $48 + 34y - 5y^2$

39. [4.5d] $6x^3 - 7x^2 - 11x - 3$ **40.** [4.6c] $25t^2 + 20t + 4$

41. [4.7c] $-5x^3y - y^3 + xy^3 - x^2y^2 + 19$

42. [4.7e] $8a^2b^2 + 6ab - 4b^3 + 6ab^2 + ab^3$

43. [4.7f] $9x^{10} - 16y^{10}$ **44.** [4.8a] $4x^2 + 3x - 5$

45. [4.8b] $2x^2 - 4x - 2 + \dfrac{17}{3x + 2}$

46. [4.3a] $3, 1.5, -3.5, -5, -5.25$

47. [4.4d] $(t + 2)(t + 2), t^2 + 4t + 4$ **48.** [4.4d] B

49. [4.5b], [4.6a] $V = l^3 - 3l^2 + 2l$

50. [2.3b], [4.6b, c] $-\frac{61}{12}$

Cumulative Review: Chapters 1–4, p. 381

1. [1.1a] $\frac{5}{2}$ **2.** [4.3a] -4 **3.** [4.7a] -14 **4.** [1.2e] 4

5. [1.6b] $\frac{1}{5}$ **6.** [1.3a] $-\frac{11}{60}$ **7.** [1.4a] 4.2 **8.** [1.5a] 7.28

9. [1.6c] $-\frac{5}{12}$ **10.** [4.2d] 2.2×10^{22} **11.** [4.2d] 4×10^{-5}

12. [1.7a] -3 **13.** [1.8b] $-2y - 7$ **14.** [1.8c] $5x + 11$

15. [1.8d] -2 **16.** [4.4a] $2x^5 - 2x^4 + 3x^3 + 2$

17. [4.7d] $3x^2 + xy - 2y^2$ **18.** [4.4c] $x^3 + 5x^2 - x - 7$

19. [4.4c] $-\frac{1}{3}x^2 - \frac{3}{4}x$ **20.** [1.7c] $12x - 15y + 21$ **21.** [4.5a] $6x^8$

22. [4.5b] $2x^5 - 4x^4 + 8x^3 - 10x^2$

23. [4.5d] $3y^4 + 5y^3 - 10y - 12$

24. [4.7f] $2p^4 + 3p^3q + 2p^2q^2 - 2p^4q - p^3q^2 - p^2q^3 + pq^3$

25. [4.6a] $6x^2 + 13x + 6$ **26.** [4.6c] $9x^4 + 6x^2 + 1$

27. [4.6b] $t^2 - \frac{1}{4}$ **28.** [4.6b] $4y^4 - 25$

29. [4.6a] $4x^6 + 6x^4 - 6x^2 - 9$ **30.** [4.6c] $t^2 - 4t^3 + 4t^4$

31. [4.7f] $15p^2 - pq - 2q^2$ **32.** [4.8a] $6x^2 + 2x - 3$

33. [4.8b] $3x^2 - 2x - 7$ **34.** [2.1b] -1.2 **35.** [2.2a] -21

36. [2.3a] 9 **37.** [2.2a] $-\frac{20}{3}$ **38.** [2.3b] 2 **39.** [2.1b] $\frac{13}{8}$

40. [2.3c] $-\frac{17}{21}$ **41.** [2.3b] -17 **42.** [2.3b] 2

43. [2.7e] $\{x | x < 16\}$ **44.** [2.7e] $\{x | x \leq -\frac{11}{8}\}$

45. [2.4b] $x = \dfrac{A - P}{Q}$ **46.** [2.5a] \$3.50

47. [4.4d] $(\pi r^2 - 18)$ ft^2 **48.** [2.6a] 18 and 19

49. [2.6a] 20 ft, 24 ft **50.** [2.6a] $10°$ **51.** [4.1d, f] y^4

52. [4.1e, f] $\dfrac{1}{x}$ **53.** [4.2a, b] $-\dfrac{27x^9}{y^6}$ **54.** [4.1d, e, f] x^3

55. [3.3a]

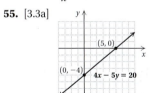

56. [4.1a, f] $3^2 = 9, 3^{-2} = \frac{1}{9}, \left(\frac{1}{3}\right)^2 = \frac{1}{9}, \left(\frac{1}{3}\right)^{-2} = 9, -3^2 = -9,$ $(-3)^2 = 9, \left(-\frac{1}{3}\right)^2 = \frac{1}{9}, \left(-\frac{1}{3}\right)^{-2} = 9$ **57.** [4.4d] $(4x - 4)$ in^2

58. [4.1d], [4.2a, b], [4.4a] $12x^5 - 15x^4 - 27x^3 + 4x^2$

59. [4.4a], [4.6c] $5x^2 - 2x + 10$ **60.** [2.3b], [4.6a, c] $\frac{11}{7}$

61. [2.3b], [4.8b] 1 **62.** [1.2e], [2.3a] $-5, 5$

63. [2.3b], [4.6a], [4.8b] All real numbers except 5

CHAPTER 5

Exercise Set 5.1, p. 389

RC1. (b) **RC2.** (c) **RC3.** (d) **RC4.** (a) **1.** 6 **3.** 24
5. 1 **7.** x **9.** x^2 **11.** 2 **13.** $17xy$ **15.** x **17.** x^2y^2 **19.** $x(x-6)$
21. $2x(x+3)$ **23.** $x^2(x+6)$ **25.** $8x^2(x^2-3)$ **27.** $2(x^2+x-4)$
29. $17xy(x^4y^2+2x^2y+3)$ **31.** $x^2(6x^2-10x+3)$
33. $x^2y^2(x^3y^3+x^2y+xy-1)$ **35.** $2x^3(x^4-x^3-32x^2+2)$
37. $0.8x(2x^3-3x^2+4x+8)$ **39.** $\frac{1}{3}x^3(5x^3+4x^2+x+1)$
41. $(x+3)(x^2+2)$ **43.** $(3z-1)(4z^2+7)$
45. $(3x+2)(2x^2+1)$ **47.** $(2a-7)(5a^3-1)$
49. $(x+3)(x^2+2)$ **51.** $(x+3)(2x^2+1)$
53. $(2x-3)(4x^2+3)$ **55.** $(3p-4)(4p^2+1)$
57. $(x-1)(5x^2-1)$ **59.** $(x+8)(x^2-3)$ **61.** $(x-4)(2x^2-9)$
63. $y^2+12y+35$ **64.** $y^2+14y+49$ **65.** y^2-49
66. $y^2-14y+49$ **67.** $16x^3-48x^2+8x$
68. $28w^2-53w-66$ **69.** $49w^2+84w+36$
70. $16w^2-88w+121$ **71.** $16w^2-121$
72. y^3-3y^2+5y **73.** $6x^2+11xy-35y^2$ **74.** $25x^2-10xt+t^2$
75. $(2x^2+3)(2x^3+3)$ **77.** $(x^5+1)(x^7+1)$
79. Not factorable by grouping

Exercise Set 5.2, p. 397

RC1. True **RC2.** True **RC3.** True **RC4.** False

1.

Pairs of Factors	Sums of Factors
1, 15	16
−1, −15	−16
3, 5	8
−3, −5	−8

$(x+3)(x+5)$

3.

Pairs of Factors	Sums of Factors
1, 12	13
−1, −12	−13
2, 6	8
−2, −6	−8
3, 4	7
−3, −4	−7

$(x+3)(x+4)$

5.

Pairs of Factors	Sums of Factors
1, 9	10
−1, −9	−10
3, 3	6
−3, −3	−6

$(x-3)^2$

7.

Pairs of Factors	Sums of Factors
−1, 14	13
1, −14	−13
−2, 7	5
2, −7	−5

$(x+2)(x-7)$

9.

Pairs of Factors	Sums of Factors
1, 4	5
−1, −4	−5
2, 2	4
−2, −2	−4

$(b+1)(b+4)$

11.

Pairs of Factors	Sums of Factors
−1, 18	17
1, −18	−17
−2, 9	7
2, −9	−7
−3, 6	3
3, −6	−3

$(t-3)(t+6)$

13. $(d-2)(d-5)$ **15.** $(y-1)(y-10)$ **17.** Prime
19. $(x-9)(x+2)$ **21.** $x(x-8)(x+2)$
23. $y(y-9)(y+5)$ **25.** $(x-11)(x+9)$
27. $(c^2+8)(c^2-7)$ **29.** $(a^2+7)(a^2-5)$
31. $(x-6)(x+7)$ **33.** Prime **35.** $(x+10)^2$
37. $2z(z-4)(z+3)$ **39.** $3t^2(t^2+t+1)$
41. $x^2(x-25)(x+4)$ **43.** $(x-24)(x+3)$
45. $(x-9)(x-16)$ **47.** $(a+12)(a-11)$ **49.** $3(t+1)^2$
51. $w^2(w-4)^2$ **53.** $-1(x-10)(x+3)$, or
$(-x+10)(x+3)$, or $(x-10)(-x-3)$
55. $-1(a-2)(a+12)$, or $(-a+2)(a+12)$, or
$(a-2)(-a-12)$ **57.** $(x-15)(x-8)$
59. $-1(x+12)(x-9)$, or $(-x-12)(x-9)$, or
$(x+12)(-x+9)$ **61.** $(y-0.4)(y+0.2)$
63. $(p+5q)(p-2q)$ **65.** $-1(t+14)(t-6)$, or
$(-t-14)(t-6)$, or $(t+14)(-t+6)$ **67.** $(m+4n)(m+n)$
69. $(s+3t)(s-5t)$ **71.** $6a^8(a+2)(a-7)$
73. $-\frac{7}{2}$ **74.** 12 **75.** $\frac{1}{3}$ **76.** −1 **77.** $\frac{5}{4}$
78. No solution **79.** $\{x \mid x > -24\}$ **80.** $\{x \mid x \le \frac{14}{5}\}$

81. $\{x \mid x \ge \frac{77}{17}\}$ **82.** $p = 2A - w$ **83.** $x = \dfrac{y-b}{m}$

84. $c = a + r - 2$ **85.** 73.5 million **86.** $100°, 25°, 55°$
87. $15, -15, 27, -27, 51, -51$ **89.** $\left(x+\frac{1}{4}\right)\left(x-\frac{3}{4}\right)$
91. $\left(x+\frac{1}{3}\right)^2$ **93.** $(x+5)\left(x-\frac{5}{7}\right)$
95. $(b^n+5)(b^n+2)$ **97.** $2x^2(4-\pi)$

Calculator Corner, p. 403

1. Correct **2.** Correct **3.** Not correct **4.** Not correct

Exercise Set 5.3, p. 407

RC1. True **RC2.** True **RC3.** False **RC4.** False
1. $(2x+1)(x-4)$ **3.** $(5x+9)(x-2)$ **5.** $(3x+1)(2x+7)$
7. $(3x+1)(x+1)$ **9.** $(2x-3)(2x+5)$ **11.** $(2x+1)(x-1)$
13. $(3x-2)(3x+8)$ **15.** $(3x+1)(x-2)$
17. $(3x+4)(4x+5)$ **19.** $(7x-1)(2x+3)$
21. $(3x+2)(3x+4)$ **23.** $(3x-7)^2$, or $(7-3x)^2$
25. $(24x-1)(x+2)$ **27.** $(5x-11)(7x+4)$
29. $-2(x-5)(x+2)$, or $2(-x+5)(x+2)$, or $2(x-5)(-x-2)$
31. $4(3x-2)(x+3)$ **33.** $6(5x-9)(x+1)$
35. $2(3y+5)(y-1)$ **37.** $(3x-1)(x-1)$
39. $4(3x+2)(x-3)$ **41.** $(2x+1)(x-1)$

43. $(3x + 2)(3x - 8)$ **45.** $5(3x + 1)(x - 2)$
47. $p(3p + 4)(4p + 5)$ **49.** $-1(3x + 2)(3x - 8)$, or
$(-3x - 2)(3x - 8)$, or $(3x + 2)(-3x + 8)$
51. $-1(5x - 3)(3x - 2)$, or $(-5x + 3)(3x - 2)$, or
$(5x - 3)(-3x + 2)$ **53.** $x^2(7x - 1)(2x + 3)$
55. $3x(8x - 1)(7x - 1)$ **57.** $(5x^2 - 3)(3x^2 - 2)$ **59.** $(5t + 8)^2$
61. $2x(3x + 5)(x - 1)$ **63.** Prime **65.** Prime
67. $(4m + 5n)(3m - 4n)$ **69.** $(2a + 3b)(3a - 5b)$
71. $(3a + 2b)(3a + 4b)$ **73.** $(5p + 2t)(7p + 4t)$
75. $6(3x - 4y)(x + y)$

77.

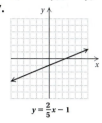

$y = \frac{2}{5}x - 1$

78.

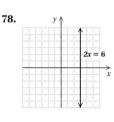

$2x = 6$

79.

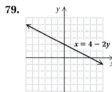

$x = 4 - 2y$

80.

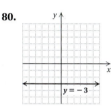

$y = -3$

81.

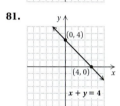

$(0, 4)$
$(4, 0)$
$x + y = 4$

82.

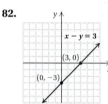

$x - y = 3$
$(3, 0)$
$(0, -3)$

83.

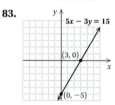

$5x - 3y = 15$
$(3, 0)$
$(0, -5)$

84.

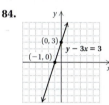

$(0, 3)$
$y - 3x = 3$
$(-1, 0)$

85. $(2x^n + 1)(10x^n + 3)$ **87.** $(x^{3a} - 1)(3x^{3a} + 1)$
89.–93. Left to the student

Exercise Set 5.4, p. 412

RC1. Leading coefficient **RC2.** Product; sum
RC3. Sum **RC4.** Grouping **1.** $(x + 2)(x + 7)$
3. $(x - 4)(x - 1)$ **5.** $(3x + 2)(2x + 3)$ **7.** $(3x - 4)(x - 4)$
9. $(7x - 8)(5x + 3)$ **11.** $(2x + 3)(2x - 3)$
13. $(x^2 + 3)(2x^2 + 5)$ **15.** $(2x + 3)(x + 2)$
17. $(3x + 5)(x - 3)$ **19.** $(5x + 1)(x + 2)$ **21.** $(3x - 1)(x - 1)$
23. $(2x + 7)(3x + 1)$ **25.** $(2x + 3)(2x - 5)$
27. $(5x - 2)(3x + 5)$ **29.** $(3x + 2)(3x - 8)$
31. $(3x - 1)(x + 2)$ **33.** $(3x - 4)(4x - 5)$
35. $(7x + 1)(2x - 3)$ **37.** $(3x - 7)^2$, or $(7 - 3x)^2$
39. $(3x + 2)(3x + 4)$ **41.** $-1(3a - 1)(3a + 5)$, or
$(-3a + 1)(3a + 5)$, or $(3a - 1)(-3a - 5)$
43. $-2(x - 5)(x + 2)$, or $2(-x + 5)(x + 2)$, or $2(x - 5)(-x - 2)$
45. $4(3x - 2)(x + 3)$ **47.** $6(5x - 9)(x + 1)$
49. $2(3y + 5)(y - 1)$ **51.** $(3x - 1)(x - 1)$
53. $4(3x + 2)(x - 3)$ **55.** $(2x + 1)(x - 1)$
57. $(3x - 2)(3x + 8)$ **59.** $5(3x + 1)(x - 2)$
61. $p(3p + 4)(4p + 5)$ **63.** $-1(5x - 4)(x + 1)$, or
$(-5x + 4)(x + 1)$, or $(5x - 4)(-x - 1)$ **65.** $-3(2t - 1)(t - 5)$,
or $3(-2t + 1)(t - 5)$, or $3(2t - 1)(-t + 5)$
67. $x^2(7x - 1)(2x + 3)$ **69.** $3x(8x - 1)(7x - 1)$

71. $(5x^2 - 3)(3x^2 - 2)$ **73.** $(5t + 8)^2$ **75.** $2x(3x + 5)(x - 1)$
77. Prime **79.** Prime **81.** $(4m + 5n)(3m - 4n)$
83. $(2a + 3b)(3a - 5b)$ **85.** $(3a - 2b)(3a - 4b)$
87. $(5p + 2q)(7p + 4q)$ **89.** $6(3x - 4y)(x + y)$
91. $-6x(x - 5)(x + 2)$, or $6x(-x + 5)(x + 2)$, or
$6x(x - 5)(-x - 2)$ **93.** $x^3(5x - 11)(7x + 4)$ **95.** $27x^{12}$
96. $\dfrac{1}{5^{14}}$ **97.** x^5y^6 **98.** a **99.** 3.008×10^{10}
100. 0.000015 **101.** About 6369 km, or 3949 mi **102.** $40°$
103. $(3x^5 - 2)^2$ **105.** $(4x^5 + 1)^2$ **107.–111.** Left to the student

Mid-Chapter Review: Chapter 5, p. 416

1. True **2.** False **3.** True **4.** False
5. $10y^3 - 18y^2 + 12y = 2y \cdot 5y^2 - 2y \cdot 9y + 2y \cdot 6$
$= 2y(5y^2 - 9y + 6)$
6. $a \cdot c = 2 \cdot (-6) = -12$;
$-x = -4x + 3x$;
$2x^2 - x - 6 = 2x^2 - 4x + 3x - 6$
$= 2x(x - 2) + 3(x - 2)$
$= (x - 2)(2x + 3)$
7. x **8.** x^2 **9.** $6x^3$ **10.** 4 **11.** $5x^2y$ **12.** x^2y^2
13. $x(x^2 - 8)$ **14.** $3x(x + 4)$ **15.** $2(y^2 + 4y - 2)$
16. $t^3(3t^3 - 5t - 2)$ **17.** $(x + 1)(x + 3)$ **18.** $(z - 2)^2$
19. $(x + 4)(x^2 + 3)$ **20.** $8y^3(y^2 - 6)$
21. $6xy(x^2 + 4xy - 7y^2)$ **22.** $(4t - 3)(t - 2)$
23. $(z - 1)(z + 5)$ **24.** $(z + 4)(2z^2 + 5)$
25. $(3p - 2)(p^2 - 3)$ **26.** $5x^3(2x^5 - 5x^3 - 3x^2 + 7)$
27. $(2w + 3)(w^2 - 3)$ **28.** $x^2(4x^2 - 5x + 3)$
29. $(6y - 5)(y + 2)$ **30.** $3(x - 3)(x + 2)$
31. $(3x + 2)(2x^2 + 1)$ **32.** $(w - 5)(w - 3)$
33. $(2x + 5)(4x^2 + 1)$ **34.** $(5z + 2)(2z - 5)$
35. $(2x + 1)(3x + 2)$ **36.** $(x - 6y)(x - 4y)$
37. $(2z + 1)(3z^2 + 1)$ **38.** $a^2b^3(ab^4 + a^2b^2 - 1 + a^3b^3)$
39. $(4y + 5z)(y - 3z)$ **40.** $3x(x + 2)(x + 5)$
41. $(x - 3)(x^2 - 2)$ **42.** $(3y + 1)^2$ **43.** $(y + 2)(y + 4)$
44. $3(2y + 5)(y + 3)$ **45.** $(x - 7)(x^2 + 4)$
46. $-1(y - 4)(y + 1)$, or $(-y + 4)(y + 1)$, or $(y - 4)(-y - 1)$
47. $4(2x + 3)(2x - 5)$ **48.** $(5a - 3b)(2a - b)$
49. $(2w - 5)(3w^2 - 5)$ **50.** $y(y + 6)(y + 3)$
51. $(4x + 3y)(x + 2y)$ **52.** $-1(3z - 2)(2z + 3)$, or
$(-3z + 2)(2z + 3)$, or $(3z - 2)(-2z - 3)$
53. $(3t + 2)(4t^2 - 3)$ **54.** $(y - 4z)(y + 5z)$
55. $(3x - 4y)(3x + 2y)$ **56.** $(3z - 1)(z + 3)$
57. $(m - 8n)(m + 2n)$ **58.** $2(w - 3)^2$ **59.** $2t(3t - 2)(3t - 1)$
60. $(z + 3)(5z^2 + 1)$ **61.** $(t - 2)(t + 7)$ **62.** $(2t - 5)^2$
63. $(t - 2)(t + 6)$ **64.** $-1(2z + 3)(z - 4)$, or
$(-2z - 3)(z - 4)$, or $(2z + 3)(-z + 4)$
65. $-1(y - 6)(y + 2)$, or $(-y + 6)(y + 2)$, or $(y - 6)(-y - 2)$
66. Find the product of two binomials. For example,
$(2x^2 + 3)(x - 4) = 2x^3 - 8x^2 + 3x - 12$. **67.** There is a
finite number of pairs of numbers with the correct product, but
there are infinitely many pairs with the correct sum. **68.** Since
both constants are negative, the middle term will be negative so
$(x - 17)(x - 18)$ cannot be a factorization of $x^2 + 35x + 306$.
69. No; both $2x + 6$ and $2x + 8$ contain a factor of 2, so $2 \cdot 2$, or 4,
must be factored out to reach the complete factorization. In other
words, the largest common factor is 4, not 2.

Exercise Set 5.5, p. 423

RC1. False **RC2.** True **RC3.** False **RC4.** False
1. Yes **3.** No **5.** No **7.** Yes **9.** $(x - 7)^2$ **11.** $(x + 8)^2$
13. $(x - 1)^2$ **15.** $(x + 2)^2$ **17.** $(y + 6)^2$ **19.** $(t - 4)^2$
21. $(q^2 - 3)^2$ **23.** $(4y + 7)^2$ **25.** $2(x - 1)^2$
27. $x(x - 9)^2$ **29.** $3(2q - 3)^2$ **31.** $(7 - 3x)^2$, or $(3x - 7)^2$
33. $5(y^2 + 1)^2$ **35.** $(1 + 2x^2)^2$ **37.** $(2p + 3t)^2$ **39.** $(a - 3b)^2$
41. $(9a - b)^2$ **43.** $4(3a + 4b)^2$ **45.** Yes **47.** No **49.** No

51. Yes **53.** $(y + 2)(y - 2)$ **55.** $(p + 1)(p - 1)$
57. $(t + 7)(t - 7)$ **59.** $(a + b)(a - b)$
61. $(5t + m)(5t - m)$ **63.** $(10 + k)(10 - k)$
65. $(4a + 3)(4a - 3)$ **67.** $(2x + 5y)(2x - 5y)$
69. $2(2x + 7)(2x - 7)$ **71.** $x(6 + 7x)(6 - 7x)$
73. $(\frac{1}{4} + 7x^4)(\frac{1}{4} - 7x^4)$ **75.** $(0.3y + 0.02)(0.3y - 0.02)$
77. $(7a^2 + 9)(7a^2 - 9)$ **79.** $(a^2 + 4)(a + 2)(a - 2)$
81. $5(x^2 + 9)(x + 3)(x - 3)$ **83.** $(1 + y^4)(1 + y^2)(1 + y)(1 - y)$
85. $(x^6 + 4)(x^3 + 2)(x^3 - 2)$ **87.** $(y + \frac{1}{4})(y - \frac{1}{4})$
89. $(5 + \frac{1}{7}x)(5 - \frac{1}{7}x)$ **91.** $(4m^2 + t^2)(2m + t)(2m - t)$
93. y-intercept: $(0, 4)$; x-intercept: $(16, 0)$ **94.** y-intercept:
$(0, -5)$; x-intercept: $(6.5, 0)$ **95.** y-intercept: $(0, -5)$;
x-intercept: $(\frac{5}{2}, 0)$

96.
97.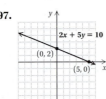

98.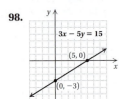

99. $x^2 - 4xy + 4y^2$ **100.** $\frac{1}{2}\pi x^2 + 2xy$ **101.** Prime
103. $(x + 11)^2$ **105.** $2x(3x + 1)^2$
107. $(x^4 + 2^4)(x^2 + 2^2)(x + 2)(x - 2)$
109. $3x^3(x + 2)(x - 2)$ **111.** $2x(3x + \frac{2}{5})(3x - \frac{2}{5})$
113. $p(0.7 + p)(0.7 - p)$ **115.** $(0.8x + 1.1)(0.8x - 1.1)$
117. $x(x + 6)$ **119.** $(x + \frac{1}{x})(x - \frac{1}{x})$
121. $(9 + b^{2k})(3 - b^k)(3 + b^k)$ **123.** $(3b^n + 2)^2$
125. $(y + 4)^2$ **127.** 9 **129.** Not correct **131.** Not correct

Exercise Set 5.6, p. 432

RC1. Common **RC2.** Difference **RC3.** Square
RC4. Grouping **RC5.** Completely **RC6.** Check
1. $3(x + 8)(x - 8)$ **3.** $(a - 5)^2$ **5.** $(2x - 3)(x - 4)$
7. $x(x + 12)^2$ **9.** $(x + 3)(x + 2)(x - 2)$
11. $3(4x + 1)(4x - 1)$ **13.** $3x(3x - 5)(x + 3)$
15. Prime **17.** $x(x^2 + 7)(x - 3)$ **19.** $x^3(x - 7)^2$
21. $-2(x - 2)(x + 5)$, or $2(-x + 2)(x + 5)$, or
$2(x - 2)(-x - 5)$ **23.** Prime **25.** $4(x^2 + 4)(x + 2)(x - 2)$
27. $(1 + y^4)(1 + y^2)(1 + y)(1 - y)$ **29.** $x^3(x - 3)(x - 1)$
31. $\frac{1}{9}(\frac{1}{3}x^3 - 4)^2$ **33.** $m(x^2 + y^2)$ **35.** $9xy(xy - 4)$
37. $2\pi r(h + r)$ **39.** $(a + b)(2x + 1)$ **41.** $(x + 1)(x - 1 - y)$
43. $(n + 2)(n + p)$ **45.** $(2q - 1)(3q + p)$
47. $(2b - a)^2$, or $(a - 2b)^2$ **49.** $(4x + 3y)^2$ **51.** $(7m^2 - 8n)^2$
53. $(y^2 + 5z^2)^2$ **55.** $(\frac{1}{2}a + \frac{1}{3}b)^2$ **57.** $(a + b)(a - 2b)$
59. $(m + 20n)(m - 18n)$ **61.** $(mn - 8)(mn + 4)$
63. $r^3(rs - 2)(rs - 8)$ **65.** $a^3(a - b)(a + 5b)$
67. $(a + \frac{1}{5}b)(a - \frac{1}{5}b)$ **69.** $(x + y)(x - y)$
71. $(4 + p^2q^2)(2 + pq)(2 - pq)$
73. $(1 + 4x^6y^6)(1 + 2x^3y^3)(1 - 2x^3y^3)$
75. $(q + 8)(q + 1)(q - 1)$ **77.** $ab(2ab + 1)(3ab - 2)$
79. $(m + 1)(m - 1)(m + 2)(m - 2)$ **81.** -8.67
82. -3 **83.** $\frac{1}{5}$ **84.** -7.72 **85.** -22
86. -5 **87.** 7 **88.** $>$ **89.** $(t + 1)^2(t - 1)^2$
91. $(x - 5)(x + 2)(x - 2)$ **93.** $(3.5x - 1)^2$
95. $(y - 2)(y + 3)(y - 3)$ **97.** $(y - 1)^3$ **99.** $(y + 4 + x)^2$

Calculator Corner, p. 437

1. Left to the student

Exercise Set 5.7, p. 441

RC1. False **RC2.** False **RC3.** False **RC4.** True
1. $-4, -9$ **3.** $-3, 8$ **5.** $-12, 11$ **7.** $0, -3$ **9.** $0, -18$
11. $-\frac{5}{2}, -4$ **13.** $-\frac{1}{5}, 3$ **15.** $4, \frac{1}{4}$ **17.** $0, \frac{2}{3}$ **19.** $-\frac{1}{10}, \frac{1}{27}$
21. $\frac{1}{3}, -20$ **23.** $0, \frac{2}{3}, \frac{1}{2}$ **25.** $-5, -1$
27. $-9, 2$ **29.** $3, 5$ **31.** $0, 8$ **33.** $0, -18$ **35.** $-4, 4$
37. $-\frac{2}{3}, \frac{2}{3}$ **39.** -3 **41.** 4 **43.** $0, \frac{6}{5}$ **45.** $-1, \frac{5}{3}$
47. $-\frac{1}{4}, \frac{2}{3}$ **49.** $-1, \frac{2}{3}$ **51.** $-\frac{7}{10}, \frac{7}{10}$ **53.** $-2, 9$ **55.** $\frac{4}{5}, \frac{3}{2}$
57. $(-4, 0), (1, 0)$ **59.** $(-\frac{5}{2}, 0), (2, 0)$ **61.** $(-3, 0), (5, 0)$
63. $-1, 4$ **65.** $-1, 3$ **67.** $(a + b)^2$ **68.** $a^2 + b^2$
69. $\{x \mid x < -100\}$ **70.** $\{x \mid x \le 8\}$ **71.** $\{x \mid x < 2\}$
72. $\{x \mid x \ge \frac{20}{3}\}$ **73.** $-5, 4$ **75.** $-3, 9$ **77.** $-\frac{1}{8}, \frac{1}{8}$
79. $-4, 4$ **81.** $2.33, 6.77$ **83.** Answers may vary.
(a) $x^2 - x - 12 = 0$; **(b)** $x^2 + 7x + 12 = 0$; **(c)** $4x^2 - 4x + 1 = 0$;
(d) $x^2 - 25 = 0$; **(e)** $40x^3 - 14x^2 + x = 0$

Translating for Success, p. 449

1. O **2.** M **3.** K **4.** I **5.** G **6.** E **7.** C **8.** A
9. H **10.** B

Exercise Set 5.8, p. 450

RC1. (b) **RC2.** (b) **RC3.** (d) **RC4.** (c)
1. Length: 42 in.; width: 14 in. **3.** Length: 6 cm; width: 4 cm
5. Height: 4 cm; base: 14 cm **7.** Base: 8 m; height: 16 m
9. 182 games **11.** 12 teams **13.** 4950 handshakes
15. 25 people **17.** 14 and 15 **19.** 12 and 14; -12 and -14
21. 15 and 17; -15 and -17 **23.** 32 ft **25.** Hypotenuse: 17 ft;
leg: 15 ft **27.** 300 ft by 400 ft by 500 ft **29.** 24 m, 25 m
31. Dining room: 12 ft by 12 ft; kitchen: 12 ft by 10 ft **33.** 1 sec,
2 sec **35.** 5 and 7 **37.** 4.53 **38.** $-\frac{5}{6}$ **39.** -40
40. -116 **41.** $-\frac{3}{25}$ **42.** -3.4 **43.** $-4y - 13$ **44.** $10x - 30$
45. 5 ft **47.** 30 cm by 15 cm **49.** 11 yd, 60 yd, 61 yd

Summary and Review: Chapter 5, p. 455

Vocabulary Reinforcement

1. Factor **2.** Factor **3.** Factorization **4.** Common
5. Grouping **6.** Binomial **7.** Zero **8.** Difference

Concept Reinforcement

1. False **2.** True **3.** False **4.** True

Study Guide

1. $4xy$ **2.** $9x^2(3x^3 - x + 2)$ **3.** $(z - 3)(z^2 + 4)$
4. $(x + 2)(x + 4)$ **5.** $3(z - 4)(2z + 1)$ **6.** $(3y - 1)(2y + 3)$
7. $(2x + 1)^2$ **8.** $2(3x + 2)(3x - 2)$ **9.** $-5, 1$

Review Exercises

1. $5y^2$ **2.** $12x$ **3.** $5(1 + 2x^3)(1 - 2x^3)$ **4.** $x(x - 3)$
5. $(3x + 2)(3x - 2)$ **6.** $(x + 6)(x - 2)$ **7.** $(x + 7)^2$
8. $3x(2x^2 + 4x + 1)$ **9.** $(x + 1)(x^2 + 3)$ **10.** $(3x - 1)(2x - 1)$
11. $(x^2 + 9)(x + 3)(x - 3)$ **12.** $3x(3x - 5)(x + 3)$
13. $2(x + 5)(x - 5)$ **14.** $(x + 4)(x^3 - 2)$
15. $(4x^2 + 1)(2x + 1)(2x - 1)$ **16.** $4x^4(2x^2 - 8x + 1)$
17. $3(2x + 5)^2$ **18.** Prime **19.** $x(x - 6)(x + 5)$
20. $(2x + 5)(2x - 5)$ **21.** $(3x - 5)^2$
22. $2(3x + 4)(x - 6)$ **23.** $(x - 3)^2$ **24.** $(2x + 1)(x - 4)$
25. $2(3x - 1)^2$ **26.** $3(x + 3)(x - 3)$

27. $(x - 5)(x - 3)$　**28.** $(5x - 2)^2$　**29.** $(7b^5 - 2a^4)^2$
30. $(xy + 4)(xy - 3)$　**31.** $3(2a + 7b)^2$　**32.** $(m + 5)(m + t)$
33. $32(x^2 - 2y^2z^2)(x^2 + 2y^2z^2)$　**34.** $1, -3$　**35.** $-7, 5$
36. $-4, 0$　**37.** $\frac{2}{3}, 1$　**38.** $-8, 8$　**39.** $-2, 8$
40. $(-5, 0), (-4, 0)$　**41.** $\left(-\frac{3}{2}, 0\right), (5, 0)$　**42.** Height: 6 cm;
base: 5 cm　**43.** -18 and -16; 16 and 18　**44.** 842 ft
45. On the ground: 4 ft; on the tree: 3 ft　**46.** 6 km　**47.** B
48. A　**49.** 2.5 cm　**50.** $0, 2$　**51.** Length: 12 in.; width: 6 in.
52. 35 ft　**53.** No solution　**54.** $2, -3, \frac{5}{2}$　**55.** $-2, \frac{5}{4}, 3$

Understanding Through Discussion and Writing

1. Although $x^3 - 8x^2 + 15x$ can be factored as $(x^2 - 5x)(x - 3)$, this is not a complete factorization of the polynomial since $x^2 - 5x = x(x - 5)$. Gwen should always look for a common factor first.　**2.** Josh is correct, because answers can easily be checked by multiplying.
3. For $x = -3$:
$$(x - 4)^2 = (-3 - 4)^2 = (-7)^2 = 49;$$
$$(4 - x)^2 = [4 - (-3)]^2 = 7^2 = 49.$$
For $x = 1$:
$$(x - 4)^2 = (1 - 4)^2 = (-3)^2 = 9;$$
$$(4 - x)^2 = (4 - 1)^2 = 3^2 = 9.$$
In general, $(x - 4)^2 = [-(-x + 4)]^2 = [-(4 - x)]^2 = (-1)^2(4 - x)^2 = (4 - x)^2$.
4. The equation is not in the form $ab = 0$. The correct procedure is
$$(x - 3)(x + 4) = 8$$
$$x^2 + x - 12 = 8$$
$$x^2 + x - 20 = 0$$
$$(x + 5)(x - 4) = 0$$
$$x + 5 = 0 \quad or \quad x - 4 = 0$$
$$x = -5 \quad or \quad x = 4.$$
The solutions are -5 and 4.
5. One solution of the equation is 0. Dividing both sides of the equation by x, leaving the solution $x = 3$, is equivalent to dividing by 0.
6. She could use the measuring sticks to draw a right angle as shown below. Then she could use the 3-ft and 4-ft sticks to extend one leg to 7 ft and the 4-ft and 5-ft sticks to extend the other leg to 9 ft.

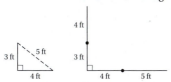

Next, she could draw another right angle with either the 7-ft side or the 9-ft side as a side.

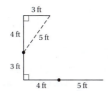

Then she could use the sticks to extend the other side to the appropriate length. Finally, she would draw the remaining side of the rectangle.

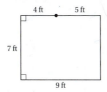

Test: Chapter 5, p. 461

1. [5.1a] $4x^3$　**2.** [5.2a] $(x - 5)(x - 2)$　**3.** [5.5b] $(x - 5)^2$
4. [5.1b] $2y^2(2y^2 - 4y + 3)$　**5.** [5.1c] $(x + 1)(x^2 + 2)$
6. [5.1b] $x(x - 5)$　**7.** [5.2a] $x(x + 3)(x - 1)$
8. [5.3a], [5.4a] $2(5x - 6)(x + 4)$　**9.** [5.5d] $(2x + 3)(2x - 3)$
10. [5.2a] $(x - 4)(x + 3)$
11. [5.3a], [5.4a] $3m(2m + 1)(m + 1)$
12. [5.5d] $3(w + 5)(w - 5)$
13. [5.5b] $5(3x + 2)^2$　**14.** [5.5d] $3(x^2 + 4)(x + 2)(x - 2)$
15. [5.5b] $(7x - 6)^2$　**16.** [5.3a], [5.4a] $(5x - 1)(x - 5)$
17. [5.1c] $(x + 2)(x^3 - 3)$　**18.** [5.5d] $5(4 + x^2)(2 + x)(2 - x)$
19. [5.3a], [5.4a] $3t(2t + 5)(t - 1)$　**20.** [5.7b] $0, 3$
21. [5.7b] $-4, 4$　**22.** [5.7b] $-4, 5$　**23.** [5.7b] $-5, \frac{3}{2}$
24. [5.7b] $-4, 7$　**25.** [5.7b] $(-5, 0), (7, 0)$
26. [5.7b] $\left(\frac{2}{3}, 0\right), (1, 0)$　**27.** [5.8a] Length: 8 m; width: 6 m
28. [5.8a] Height: 4 cm; base: 14 cm　**29.** [5.8a] 5 ft　**30.** [5.5d] A
31. [5.8a] Length: 15 m; width: 3 m　**32.** [5.2a] $(a - 4)(a + 8)$
33. [5.7b] $-\frac{8}{3}, 0, \frac{2}{5}$　**34.** [4.6b], [5.5d] D

Cumulative Review: Chapters 1–5, p. 463

1. [1.2d] $<$　**2.** [1.2d] $>$　**3.** [1.4a] 0.35　**4.** [1.6c] -1.57
5. [1.5a] $-\frac{1}{14}$　**6.** [1.6c] $-\frac{6}{5}$　**7.** [1.8c] $4x + 1$　**8.** [1.8d] -8
9. [4.2a, b] $\frac{8x^6}{y^3}$　**10.** [4.1d, e] $-\frac{1}{6x^3}$
11. [4.4a] $x^4 - 3x^3 - 3x^2 - 4$　**12.** [4.7e] $2x^2y^2 - x^2y - xy$
13. [4.8b] $x^2 + 3x + 2 + \frac{3}{x - 1}$　**14.** [4.6c] $4t^2 - 12t + 9$
15. [4.6b] $x^4 - 9$　**16.** [4.6a] $6x^2 + 4x - 16$
17. [4.5b] $2x^4 + 6x^3 + 8x^2$　**18.** [4.5d] $4y^3 + 4y^2 + 5y - 4$
19. [4.6b] $x^2 - \frac{4}{9}$　**20.** [5.2a] $(x + 4)(x - 2)$
21. [5.5d] $(2x + 5)(2x - 5)$　**22.** [5.1c] $(3x - 4)(x^2 + 1)$
23. [5.5b] $(x - 13)^2$　**24.** [5.5d] $3(5x + 6y)(5x - 6y)$
25. [5.3a], [5.4a] $(3x + 7)(2x - 9)$　**26.** [5.2a] $(x^2 - 3)(x^2 + 1)$
27. [5.6a] $2(2y - 3)(y - 1)(y + 1)$　**28.** [5.3a], [5.4a]
$(3p - q)(2p + q)$　**29.** [5.3a], [5.4a] $2x(5x + 1)(x + 5)$
30. [5.5b] $x(7x - 3)^2$　**31.** [5.3a], [5.4a] Prime
32. [5.1b] $3x(25x^2 + 9)$
33. [5.5d] $3(x^4 + 4y^4)(x^2 + 2y^2)(x^2 - 2y^2)$
34. [5.2a] $14(x + 2)(x + 1)$　**35.** [5.6a] $(x + 1)(x - 1)(2x^3 + 1)$
36. [2.3b] 15　**37.** [2.7e] $\{y | y < 6\}$　**38.** [5.7a] $15, -\frac{1}{4}$
39. [5.7a] $0, -37$　**40.** [5.7b] $5, -5, -1$　**41.** [5.7b] $6, -6$
42. [5.7b] $\frac{1}{3}$　**43.** [5.7b] $-10, -7$　**44.** [5.7b] $0, \frac{3}{2}$
45. [2.3a] 0.2　**46.** [5.7b] $-4, 5$　**47.** [2.7e] $\{x | x \leq 20\}$
48. [2.3c] All real numbers　**49.** [2.4b] $m = \frac{y - b}{x}$
50. [2.6a] 50, 52　**51.** [5.8a] -20 and -18; 18 and 20
52. [5.8a] Length: 6 ft; height: 3 ft　**53.** [2.6a] 150 m by 350 m
54. [2.5a] $6500　**55.** [5.8a] 17 m　**56.** [2.6a] 30 m, 60 m, 10 m
57. [2.5a] $29　**58.** [5.8a] Height: 14 ft; base: 14 ft
59. [3.2a]

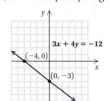

60. [2.7e], [4.6a] $\left\{x | x \geq -\frac{13}{3}\right\}$
61. [2.3b] 22
62. [5.7b] $-6, 4$
63. [5.6a] $(x - 3)(x - 2)(x + 1)$
64. [5.6a] $(2a + 3b + 3)(2a - 3b - 5)$
65. [5.5a] 25　**66.** [5.8a] 2 cm

CHAPTER 6

Exercise Set 6.1, p. 472

RC1. Equivalent **RC2.** Denominator **RC3.** Quotient
RC4. Factors
1. 0 **3.** 8 **5.** $-\frac{5}{2}$ **7.** $-4, 7$ **9.** $-5, 5$ **11.** None
13. $\dfrac{(4x)(3x^2)}{(4x)(5y)}$ **15.** $\dfrac{2x(x-1)}{2x(x+4)}$ **17.** $\dfrac{(3-x)(-1)}{(4-x)(-1)}$
19. $\dfrac{(y+6)(y-7)}{(y+6)(y+2)}$ **21.** $\dfrac{x^2}{4}$ **23.** $\dfrac{8p^2q}{3}$ **25.** $\dfrac{x-3}{x}$
27. $\dfrac{m+1}{2m+3}$ **29.** $\dfrac{a-3}{a+2}$ **31.** $\dfrac{a-3}{a-4}$ **33.** $\dfrac{x+5}{x-5}$ **35.** $a+1$
37. $\dfrac{x^2+1}{x+1}$ **39.** $\frac{3}{2}$ **41.** $\dfrac{6}{t-3}$ **43.** $\dfrac{t+2}{2(t-4)}$ **45.** $\dfrac{t-2}{t+2}$
47. -1 **49.** -1 **51.** -6 **53.** $-x-1$ **55.** $-3t$
57. $\dfrac{56x}{3}$ **59.** $\dfrac{2}{dc^2}$ **61.** 1 **63.** $\dfrac{(a+3)(a-3)}{a(a+4)}$ **65.** $\dfrac{2a}{a-2}$
67. $\dfrac{(t+2)(t-2)}{(t+1)(t-1)}$ **69.** $\dfrac{x+4}{x+2}$ **71.** $\dfrac{5(a+6)}{a-1}$
73. 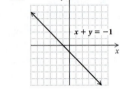 **74.**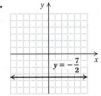
75. $x^3(x-7)(x+5)$ **76.** $(2y^2+1)(y-5)$
77. $(2+t)(2-t)(4+t^2)$ **78.** $10(x+7)(x+1)$
79. $\dfrac{x-y}{x-5y}$ **81.** $\dfrac{1}{x-1}$

Exercise Set 6.2, p. 478

RC1. (d) **RC2.** (e) **RC3.** (a) **RC4.** (f) **RC5.** (b)
RC6. (c)
1. $\dfrac{x}{4}$ **3.** $\dfrac{1}{x^2-y^2}$ **5.** $a+b$ **7.** $\dfrac{x^2-4x+7}{x^2+2x-5}$ **9.** $\frac{3}{10}$
11. $\frac{1}{4}$ **13.** $\dfrac{b}{a}$ **15.** $\dfrac{(a+2)(a+3)}{(a-3)(a-1)}$ **17.** $\dfrac{(x-1)^2}{x}$ **19.** $\frac{1}{2}$
21. $\frac{15}{8}$ **23.** $\frac{15}{4}$ **25.** $\dfrac{a-5}{3(a-1)}$ **27.** $\dfrac{(x+2)^2}{x}$ **29.** $\frac{3}{2}$
31. $\dfrac{c+1}{c-1}$ **33.** $\dfrac{y-3}{2y-1}$ **35.** $\dfrac{x+1}{x-1}$ **37.** $\{x \mid x \geq 77\}$
38. Height: 7 in.; base: 10 in. **39.** $8x^3-11x^2-3x+12$
40. $-2p^2+4pq-4q^2$ **41.** $\dfrac{4y^8}{x^6}$ **42.** $\dfrac{125x^{18}}{y^{12}}$ **43.** $\dfrac{4x^6}{y^{10}}$
44. $\dfrac{1}{a^{15}b^{20}}$ **45.** $-\dfrac{1}{b^2}$ **47.** $\dfrac{a+1}{5ab^2(a^2+4)}$

Exercise Set 6.3, p. 483

RC1. Common **RC2.** Multiple **RC3.** Denominator
RC4. Greatest **RC5.** $2 \cdot 2 \cdot 2 \cdot 2 \cdot 3$
1. 108 **3.** 72 **5.** 126 **7.** 360 **9.** 500 **11.** $\frac{65}{72}$ **13.** $\frac{29}{120}$
15. $\frac{23}{180}$ **17.** $12x^3$ **19.** $18x^2y^2$ **21.** $6(y-3)$
23. $t(t+2)(t-2)$ **25.** $(x+2)(x-2)(x+3)$
27. $t(t+2)^2(t-4)$ **29.** $(a+1)(a-1)^2$ **31.** $(m-3)(m-2)^2$
33. $(2+3x)(2-3x)$ **35.** $10v(v+4)(v+3)$
37. $18x^3(x-2)^2(x+1)$ **39.** $6x^3(x+2)^2(x-2)$ **41.** $120w^6$
43. $120x^4; 8x^3; 960x^7$ **44.** $48ab^3; 4ab; 192a^2b^4$
45. $48x^6; 16x^5; 768x^{11}$ **46.** $120x^3; 2x^2; 240x^5$
47. $20x^2; 10x; 200x^3$ **48.** $a^{15}; a^5; a^{20}$ **49.** 24 min

Exercise Set 6.4, p. 489

RC1. $\frac{2}{2}; \frac{3}{3}$ **RC2.** $\dfrac{x+3}{x+3}; \dfrac{2}{2}$ **RC3.** $\dfrac{x-3}{x-3}; \dfrac{x+2}{x+2}$ **RC4.** $\dfrac{2x}{2x}; \dfrac{3}{3}$
1. 1 **3.** $\dfrac{6}{3+x}$ **5.** $\dfrac{-4x+11}{2x-1}$ **7.** $\dfrac{2x+5}{x^2}$ **9.** $\dfrac{41}{24r}$
11. $\dfrac{2(2x+3y)}{x^2y^2}$ **13.** $\dfrac{4+3t}{18t^3}$ **15.** $\dfrac{x^2+4xy+y^2}{x^2y^2}$
17. $\dfrac{6x}{(x-2)(x+2)}$ **19.** $\dfrac{11x+2}{3x(x+1)}$ **21.** $\dfrac{x(x+6)}{(x+4)(x-4)}$
23. $\dfrac{6}{z+4}$ **25.** $\dfrac{3x-1}{(x-1)^2}$ **27.** $\dfrac{11a}{10(a-2)}$ **29.** $\dfrac{2(x^2+4x+8)}{x(x+4)}$
31. $\dfrac{7a+6}{(a-2)(a+1)(a+3)}$ **33.** $\dfrac{2(x^2-2x+17)}{(x-5)(x+3)}$
35. $\dfrac{3a+2}{(a+1)(a-1)}$ **37.** $\frac{1}{4}$ **39.** $-\dfrac{1}{t}$
41. $\dfrac{-x+7}{x-6}$, or $\dfrac{7-x}{x-6}$, or $\dfrac{x-7}{6-x}$ **43.** $y+3$
45. $\dfrac{2(b-7)}{(b+4)(b-4)}$ **47.** $a+b$ **49.** $\dfrac{5x+2}{x-5}$ **51.** -1
53. $\dfrac{-x^2+9x-14}{(x-3)(x+3)}$ **55.** $\dfrac{2(x+3y)}{(x+y)(x-y)}$ **57.** $\dfrac{a^2+7a+1}{(a+5)(a-5)}$
59. $\dfrac{5t-12}{(t+3)(t-3)(t-2)}$ **61.** x^2-1
62. $13y^3-14y^2+12y-73$ **63.** $\dfrac{1}{x^{12}y^{21}}$ **64.** $\dfrac{25}{x^4y^6}$
65. -8 **66.** $-2, 9$
67. 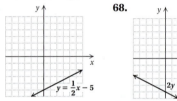 **68.**
69. 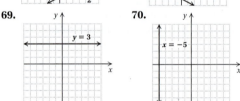 **70.**
71. Perimeter: $\dfrac{16y+28}{15}$; area: $\dfrac{y^2+2y-8}{15}$
73. $\dfrac{(z+6)(2z-3)}{(z+2)(z-2)}$ **75.** $\dfrac{11z^4-22z^2+6}{(z^2+2)(z^2-2)(2z^2-3)}$

Exercise Set 6.5, p. 497

RC1. (a) $3x+5$; (b) $10x-3x-5$; (c) $7x-5$
RC2. (a) $4-9a$; (b) $7-4+9a$; (c) $3+9a$
RC3. (a) $y+1$; (b) $9y-2-y-1$; (c) $8y-3$
1. $\dfrac{4}{x}$ **3.** 1 **5.** $\dfrac{1}{x-1}$ **7.** $\dfrac{-a-4}{10}$ **9.** $\dfrac{7z-12}{12z}$
11. $\dfrac{4x^2-13xt+9t^2}{3x^2t^2}$ **13.** $\dfrac{2(x-20)}{(x+5)(x-5)}$ **15.** $\dfrac{3-5t}{2t(t-1)}$
17. $\dfrac{2s-st-s^2}{(t+s)(t-s)}$ **19.** $\dfrac{y-19}{4y}$ **21.** $\dfrac{-2a^2}{(x+a)(x-a)}$
23. $\frac{8}{3}$ **25.** $\dfrac{13}{a}$ **27.** $\dfrac{8}{y-1}$ **29.** $\dfrac{x-2}{x-7}$ **31.** $\dfrac{4}{(a+5)(a-5)}$

33. $\dfrac{2(x-2)}{x-9}$ **35.** $\dfrac{3(3x+4)}{(x+3)(x-3)}$ **37.** $\dfrac{1}{2}$ **39.** $\dfrac{x-3}{(x+3)(x+1)}$

41. $\dfrac{18x+5}{x-1}$ **43.** 0 **45.** $\dfrac{-9}{2x-3}$ **47.** $\dfrac{20}{2y-1}$ **49.** $\dfrac{2a-3}{2-a}$

51. $\dfrac{z-3}{2z-1}$ **53.** $\dfrac{2}{x+y}$ **55.** $\dfrac{b^{20}}{a^8}$ **56.** $18x^3$ **57.** $30x^{12}$

58. 10 **59.** $-\dfrac{11}{35}$ **60.** 1 **61.** $x^2 - 9x + 18$ **62.** $(4-\pi)r^2$

63. Missing length: $\dfrac{-2a-15}{a-6}$; area: $\dfrac{-2a^3 - 15a^2 + 12a + 90}{2(a-6)^2}$

Mid-Chapter Review: Chapter 6, p. 501

1. False **2.** True **3.** True **4.** False **5.** True

6. $\dfrac{x-1}{x-2} - \dfrac{x+1}{x+2} - \dfrac{x-6}{4-x^2}$

$= \dfrac{x-1}{x-2} - \dfrac{x+1}{x+2} - \dfrac{x-6}{4-x^2} \cdot \dfrac{-1}{-1}$

$= \dfrac{x-1}{x-2} - \dfrac{x+1}{x+2} - \dfrac{6-x}{x^2-4}$

$= \dfrac{x-1}{x-2} - \dfrac{x+1}{x+2} - \dfrac{6-x}{(x-2)(x+2)}$

$= \dfrac{x-1}{x-2} \cdot \dfrac{x+2}{x+2} - \dfrac{x+1}{x+2} \cdot \dfrac{x-2}{x-2} - \dfrac{6-x}{(x-2)(x+2)}$

$= \dfrac{x^2+x-2}{(x-2)(x+2)} - \dfrac{x^2-x-2}{(x-2)(x+2)} - \dfrac{6-x}{(x-2)(x+2)}$

$= \dfrac{x^2+x-2 - x^2+x+2 - 6+x}{(x-2)(x+2)}$

$= \dfrac{3x-6}{(x-2)(x+2)}$

$= \dfrac{3(x-2)}{(x-2)(x+2)} = \dfrac{x-2}{x-2} \cdot \dfrac{3}{x+2}$

$= \dfrac{3}{x+2}$

7. None **8.** 3, 8 **9.** $\dfrac{7}{2}$ **10.** $\dfrac{x-1}{x-3}$ **11.** $\dfrac{2(y+4)}{y-1}$ **12.** -1

13. $\dfrac{1}{-x+3}$, or $\dfrac{1}{3-x}$ **14.** $10x^3(x-10)^2(x+10)$ **15.** $\dfrac{a+1}{a-3}$

16. $\dfrac{y}{(y-2)(y-3)}$ **17.** $x+11$ **18.** $\dfrac{1}{x-y}$ **19.** $\dfrac{a^2+5ab-b^2}{a^2b^2}$

20. $\dfrac{2(3x^2-4x+6)}{x(x+2)(x-2)}$ **21.** E **22.** A **23.** D **24.** B

25. F **26.** C **27.** If the numbers have a common factor, then their product contains that factor more than the greatest number of times it occurs in any one factorization. In this case, their product is not their least common multiple. **28.** Yes; consider the product $\dfrac{a}{b} \cdot \dfrac{c}{d} = \dfrac{ac}{bd}$. The reciprocal of the product is $\dfrac{bd}{ac}$. This is equal to the product of the reciprocals of the two original factors: $\dfrac{bd}{ac} = \dfrac{b}{a} \cdot \dfrac{d}{c}$.

29. Although multiplying the denominators of the expressions being added results in a common denominator, it is often not the *least* common denominator. Using a common denominator other than the LCD makes the expressions more complicated, requires additional simplification after the addition has been performed, and leaves more room for error.

30. Their sum is 0. Another explanation is that $-\left(\dfrac{1}{3-x}\right) = \dfrac{1}{-(3-x)} = \dfrac{1}{x-3}$.

31. $\dfrac{x+3}{x-5}$ is not defined for $x=5$, $\dfrac{x-7}{x+1}$ is not defined for $x=-1$, and $\dfrac{x+1}{x-7}$ (the reciprocal of $\dfrac{x-7}{x+1}$) is not defined for $x=7$.

32. The binomial is a factor of the trinomial.

Exercise Set 6.6, p. 507

RC1. Complex **RC2.** Numerator **RC3.** Least common denominator **RC4.** Reciprocal

1. $\dfrac{25}{4}$ **3.** $\dfrac{1}{3}$ **5.** -6 **7.** $\dfrac{1+3x}{1-5x}$ **9.** $\dfrac{2x+1}{x}$ **11.** 8

13. $x-8$ **15.** $\dfrac{y}{y-1}$ **17.** $-\dfrac{1}{a}$ **19.** $\dfrac{ab}{b-a}$ **21.** $\dfrac{p^2+q^2}{q+p}$

23. $\dfrac{2a(a+2)}{5-3a^2}$ **25.** $\dfrac{15(4-a^3)}{14a^2(9+2a)}$ **27.** $\dfrac{ac}{bd}$ **29.** 1

31. $\dfrac{4x+1}{5x+3}$ **33.** $\{x \mid x \le 96\}$ **34.** $\{b \mid b > \frac{22}{9}\}$

35. $\{x \mid x < -3\}$ **36.** 12 ft, 5 ft **37.** 14 yd **39.** $\dfrac{5x+3}{3x+2}$

Calculator Corner, p. 512

1.–2. Left to the student

Exercise Set 6.7, p. 513

RC1. Rational expression **RC2.** Solutions **RC3.** Rational expression **RC4.** Rational expression **RC5.** Solutions **RC6.** Solutions **RC7.** Rational expression **RC8.** Solutions **1.** $\dfrac{6}{5}$ **3.** $\dfrac{40}{29}$ **5.** $\dfrac{47}{2}$ **7.** -6 **9.** $\dfrac{24}{5}$ **11.** $-4, -1$ **13.** $-4, 4$ **15.** 3 **17.** $\dfrac{14}{3}$ **19.** 5 **21.** 5 **23.** $\dfrac{5}{2}$ **25.** -2 **27.** $-\dfrac{13}{2}$ **29.** $\dfrac{17}{2}$ **31.** No solution **33.** -5 **35.** $\dfrac{5}{3}$ **37.** $\dfrac{1}{2}$ **39.** No solution **41.** No solution **43.** 4 **45.** No solution **47.** $-2, 2$ **49.** 7 **51.** $4x^4 + 3x^3 + 2x - 7$ **52.** 0 **53.** $50(p^2 - 2)$ **54.** $5(p+2)(p-10)$ **55.** 18 and 20; -20 and -18 **56.** 3.125 L **57.** $-\dfrac{1}{6}$

Translating for Success, p. 524

1. K **2.** E **3.** C **4.** N **5.** D **6.** O **7.** F **8.** H **9.** B **10.** A

Exercise Set 6.8, p. 525

RC1. Corresponding; same; proportional **RC2.** Quotient **RC3.** Proportion **RC4.** Rate **RC5.** Distance **1.** Sarah: 30 km/h; Rick: 70 km/h **3.** Ostrich: 40 mph; giraffe: 32 mph **5.** Hank: 14 km/h; Kelly: 19 km/h **7.** 20 mph **9.** Ralph: 5 km/h; Bonnie: 8 km/h **11.** $1\frac{1}{3}$ hr **13.** $22\frac{2}{9}$ min **15.** $25\frac{5}{7}$ min **17.** About 4.7 hr **19.** $3\frac{15}{16}$ hr **21.** $3\frac{3}{4}$ min **23.** $\dfrac{10}{3}$ students/teacher **25.** 2.3 km/h **27.** 66 g **29.** 1.92 g **31.** 8 gal **33.** 287 trout **35.** 200 duds **37.** 4.92 bags **39.** 1.75 lb **41.** $\dfrac{21}{2}$ **43.** $\dfrac{8}{3}$ **45.** $\dfrac{35}{3}$ **47.** About 1700 largemouth bass **49.** 0

50. -2 **51.** x^{11} **52.** x **53.** $\dfrac{1}{x^{11}}$ **54.** $\dfrac{1}{x}$

55.

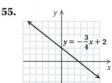

$y = -\frac{3}{4}x + 2$

56.

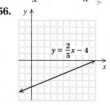

$y = \frac{2}{5}x - 4$

57.

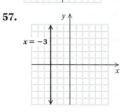

$x = -3$

59. $27\frac{3}{11}$ min

Exercise Set 6.9, p. 536

RC1. (b) **RC2.** (d) **RC3.** (f) **RC4.** (e) **RC5.** (c)
RC6. (a)
1. $y = 4x$; 80 **3.** $y = 1.6x$; 32 **5.** $y = 3.6x$; 72
7. $y = \frac{25}{3}x$; $\frac{500}{3}$ **9. (a)** $P = 12H$; **(b)** \$420
11. (a) $C = 11.25S$; **(b)** \$101.25 **13. (a)** $M = \frac{1}{6}E$; **(b)** $18.\overline{3}$ lb;
(c) 30 lb **15. (a)** $N = 80{,}000S$; **(b)** 16,000,000 instructions/sec
17. $93\frac{1}{3}$ servings **19.** $y = \dfrac{75}{x}$; $\frac{15}{2}$, or 7.5 **21.** $y = \dfrac{80}{x}$; 8
23. $y = \dfrac{1}{x}$; $\frac{1}{10}$ **25.** $y = \dfrac{2100}{x}$; 210 **27.** $y = \dfrac{0.06}{x}$; 0.006
29. (a) Direct; **(b)** $69\frac{3}{8}$ players **31. (a)** Inverse; **(b)** $4\frac{1}{2}$ hr
33. (a) $N = \dfrac{280}{P}$; **(b)** 10 gal **35. (a)** $I = \dfrac{1920}{R}$; **(b)** 32 amperes
37. (a) $m = \dfrac{40}{n}$; **(b)** 10 questions **39.** 8.25 ft **41.** $\frac{8}{5}$ **42.** 11
43. $\frac{1}{3}$ **44.** $\frac{47}{20}$ **45.** -1 **46.** 49 **47.** $P^2 = kt$
49. $P = kV^3$

Summary and Review: Chapter 6, p. 540

Vocabulary Reinforcement

1. Complex **2.** Proportion **3.** Reciprocals **4.** Equivalent
5. Opposites **6.** Similar **7.** Inversely; inverse variation
8. Directly; direct variation

Concept Reinforcement

1. True **2.** False **3.** True

Study Guide

1. 5, 6 **2.** $\dfrac{x - 1}{2(x + 5)}$ **3.** $\dfrac{y + 5}{5(y + 3)}$ **4.** $\dfrac{b + 7}{b + 8}$ **5.** $\frac{71}{180}$
6. $(x + 2)(x - 9)(x + 9)$ **7.** -1 **8.** $\dfrac{x^2 - 4x - 10}{(x + 2)(x + 1)(x - 1)}$
9. $\dfrac{3(2y - 5)}{5(9 - y)}$ **10.** 1 **11.** $y = 150x$; $y = 300$
12. $y = \dfrac{225}{x}$; $y = 22.5$

Review Exercises

1. 0 **2.** 6 **3.** $-6, 6$ **4.** $-6, 5$ **5.** -2 **6.** None
7. $\dfrac{x - 2}{x + 1}$ **8.** $\dfrac{7x + 3}{x - 3}$ **9.** $\dfrac{y - 5}{y + 5}$ **10.** $\dfrac{a - 6}{5}$ **11.** $\dfrac{6}{2t - 1}$
12. $-20t$ **13.** $\dfrac{2x(x - 1)}{x + 1}$ **14.** $30x^2y^2$ **15.** $4(a - 2)$
16. $(y - 2)(y + 2)(y + 1)$ **17.** $\dfrac{-3(x - 6)}{x + 7}$ **18.** -1
19. $\dfrac{2a}{a - 1}$ **20.** $d + c$ **21.** $\dfrac{4}{x - 4}$ **22.** $\dfrac{x + 5}{2x}$ **23.** $\dfrac{2x + 3}{x - 2}$
24. $\dfrac{-x^2 + x + 26}{(x - 5)(x + 5)(x + 1)}$ **25.** $\dfrac{2(x - 2)}{x + 2}$ **26.** $\dfrac{z}{1 - z}$
27. $c - d$ **28.** 8 **29.** $-5, 3$ **30.** $5\frac{1}{7}$ hr **31.** 95 mph, 175 mph
32. 240 km/h, 280 km/h **33.** 160 defective calculators
34. (a) $\frac{12}{13}$ c; **(b)** $4\frac{1}{5}$ c; **(c)** $9\frac{1}{3}$ c **35.** 10,000 blue whales **36.** 6
37. $y = 3x$; 60 **38.** $y = \frac{4}{5}x$; 16 **39.** $y = \dfrac{30}{x}$; 6 **40.** $y = \dfrac{1}{x}$; $\frac{1}{5}$
41. $y = \dfrac{0.65}{x}$; 0.13 **42.** \$288.75 **43.** 1 hr **44.** C **45.** A
46. $\dfrac{5(a + 3)^2}{a}$ **47.** They are equivalent proportions.

Understanding Through Discussion and Writing

1. No; when we are adding, no sign changes are required so the result is the same regardless of use of parentheses. When we are subtracting, however, the sign of each term of the expression being subtracted must be changed and parentheses are needed to make sure this is done. **2.** Graph each side of the equation and determine the number of points of intersection of the graphs. **3.** Canceling removes a factor of 1, allowing us to rewrite $a \cdot 1$ as a. **4.** Inverse variation; the greater the average gain per play, the smaller the number of plays required. **5.** Form a rational expression that has factors of $x + 3$ and $x - 4$ in the denominator. **6.** If we multiply both sides of a rational equation by a variable expression in order to clear fractions, it is possible that the variable expression is equal to 0. Thus an equivalent equation might not be produced.

Test: Chapter 6, p. 547

1. [6.1a] 0 **2.** [6.1a] -8 **3.** [6.1a] $-7, 7$ **4.** [6.1a] 1, 2
5. [6.1a] 1 **6.** [6.1a] None **7.** [6.1c] $\dfrac{3x + 7}{x + 3}$
8. [6.1d] $\dfrac{a + 5}{2}$ **9.** [6.2b] $\dfrac{(5x + 1)(x + 1)}{3x(x + 2)}$
10. [6.3a] $(y - 3)(y + 3)(y + 7)$ **11.** [6.4a] $\dfrac{23 - 3x}{x^3}$
12. [6.5a] $\dfrac{2(4 - t)}{t^2 + 1}$ **13.** [6.4a] $\dfrac{-3}{x - 3}$ **14.** [6.5a] $\dfrac{2x - 5}{x - 3}$
15. [6.4a] $\dfrac{8t - 3}{t(t - 1)}$ **16.** [6.5a] $\dfrac{-x^2 - 7x - 15}{(x + 4)(x - 4)(x + 1)}$
17. [6.5b] $\dfrac{x^2 + 2x - 7}{(x - 1)^2(x + 1)}$ **18.** [6.6a] $\dfrac{3y + 1}{y}$
19. [6.7a] 12 **20.** [6.7a] $-3, 5$
21. [6.9a] $y = 2x$; 50 **22.** [6.9a] $y = 0.5x$; 12.5
23. [6.9c] $y = \dfrac{18}{x}$; $\frac{9}{50}$ **24.** [6.9c] $y = \dfrac{22}{x}$; $\frac{11}{50}$
25. [6.9b] 240 km **26.** [6.9d] $1\frac{1}{5}$ hr **27.** [6.8b] 16 defective spark plugs **28.** [6.8b] 50 zebras **29.** [6.7a] 12 min
30. [6.8a] Craig: 65 km/h; Marilyn: 45 km/h
31. [6.8b] 15 **32.** [6.6a] D
33. [6.8a] Rema: 4 hr; Reggie: 10 hr **34.** [6.6a] $\dfrac{3a + 2}{2a + 1}$

Cumulative Review: Chapters 1–6, p. 549

1. [1.2e] 3.5 **2.** [4.3c] 3, 2, 1, 0; 3
3. [2.5a] About 275 mg **4.** [2.5a] About 7.3%
5. [2.5a] \$2500 **6.** [6.8a] 35 mph, 25 mph
7. [5.8a] 14 ft **8. (a)** [6.9b] $M = 0.4B$;
(b) [6.9b] 76.8 lb **9.** [4.3d] $2x^3 - 3x^2 - 2$ **10.** [1.8c] $\frac{3}{8}x + 1$
11. [4.1e], [4.2a, b] $\dfrac{9}{4x^8}$ **12.** [6.6a] $\dfrac{4(2x - 3)}{17x}$
13. [4.7e] $-2xy^2 - 4x^2y^2 + xy^3$
14. [4.4a] $2x^5 + 6x^4 + 2x^3 - 10x^2 + 3x - 9$
15. [6.1d] $\dfrac{2}{3(y + 2)}$ **16.** [6.2b] 2
17. [6.4a] $x + 4$ **18.** [6.5a] $\dfrac{2(x - 3)}{(x + 2)(x - 2)}$
19. [4.6a] $a^2 - 9$ **20.** [4.6c] $36x^2 - 60x + 25$
21. [4.6b] $4x^6 - 1$ **22.** [5.3a], [5.4a] $(9a - 2)(a + 6)$
23. [5.5b] $(3x - 5y)^2$ **24.** [5.5d] $(7x - 1)(7x + 1)$
25. [2.3c] 3 **26.** [5.7b] $-4, \frac{1}{2}$ **27.** [5.7b] 0, 10
28. [2.7e] $\{x \mid x \geq -26\}$ **29.** [6.7a] 2
30. [2.4b] $a = \dfrac{t}{x + y}$ **31.** [3.3a] Not defined **32.** [3.3a] $-\frac{3}{7}$

33. [3.3a] $-\frac{9}{4}$ **34.** [3.3a] 0

35. [3.2a] y-intercept: $(0, -2)$; x-intercept: $(-6, 0)$

36. [3.2a] y-intercept: $\left(0, -\frac{1}{8}\right)$; x-intercept: $\left(\frac{3}{8}, 0\right)$

37. [3.2a] y-intercept: $(0, 25)$; x-intercept: none

38. [3.2a] y-intercept: none; x-intercept: $\left(-\frac{1}{4}, 0\right)$

39. [3.2b]   **40.** [3.2b]

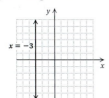

41. [3.2a]   **42.** [3.2a]

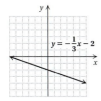   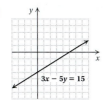

43. [3.1d]

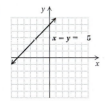

44. [3.1d], [3.2a]   **45.** [6.1a], [6.6a] 0, 3, $\frac{5}{2}$

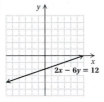

CHAPTER 7

Calculator Corner, p. 556

1. $(-1, 3)$ **2.** $(-4, 1)$ **3.** $(3, 2)$

Exercise Set 7.1, p. 556

RC1. True   **RC2.** True   **RC3.** True   **RC4.** False

1. Yes   **3.** No   **5.** Yes   **7.** Yes   **9.** Yes

11. $(4, 2)$ **13.** $(4, 3)$ **15.** $(-3, -3)$ **17.** No solution

19. $(2, 2)$ **21.** $\left(\frac{1}{2}, 1\right)$ **23.** Infinite number of solutions

25. $(5, -3)$ **27.** $\dfrac{x + 5}{x - 5}$ **28.** $\dfrac{d + 2}{5d^2 - 1}$ **29.** -3 **30.** $\dfrac{2x + 5}{x + 3}$

31. Trinomial   **32.** Binomial   **33.** Monomial

34. None of these   **35.** $A = 2, B = 2$ **37.** $x + 2y = 2$, $x - y = 8$ **39.–41.** Left to the student

Exercise Set 7.2, p. 563

RC1. True   **RC2.** True   **RC3.** False   **RC4.** True

1. $(-2, 1)$ **3.** $(2, -4)$ **5.** $(4, 3)$ **7.** $(2, -3)$ **9.** $(1, 9)$

11. $(4, -3)$ **13.** $(2, -4)$ **15.** $\left(\frac{17}{3}, \frac{16}{3}\right)$ **17.** $(6, 3)$

19. $\left(\frac{25}{8}, -\frac{11}{4}\right)$ **21.** $(-4, 3)$ **23.** $(-3, 0)$

25. Length: $3\frac{1}{2}$ in.; width: $1\frac{1}{2}$ in.   **27.** Length: 365 mi; width: 275 mi   **29.** Length: 110 yd; width: 60 yd   **31.** 16 and 21

33. 12 and 40   **35.** 20 and 8   **37.** Netfix: 10 hr; Hulu: 2.5 hr

39.   **40.**

     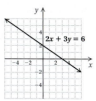

41.   **42.**

     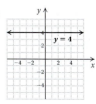

43. $(3x - 2)(2x - 3)$ **44.** $(4p + 3)(p - 1)$

45. Not factorable   **46.** $(3a - 5)(3a + 5)$ **47.** x^3 **48.** x^7

49. $\dfrac{1}{x^7}$ **50.** $\dfrac{b^3}{a^3}$ **51.** $(5.\overline{6}, 0.\overline{6})$ **53.** $(4.38, 4.33)$

55. Baseball: 30 yd; softball: 20 yd

Exercise Set 7.3, p. 571

RC1. False   **RC2.** True   **RC3.** True   **RC4.** False

RC5. False   **RC6.** False

1. $(6, -1)$ **3.** $(3, 5)$ **5.** $(2, 5)$ **7.** $\left(-\frac{1}{2}, 3\right)$ **9.** $\left(-1, \frac{1}{5}\right)$

11. No solution   **13.** $(-1, -6)$ **15.** $(3, 1)$ **17.** $(8, 3)$

19. $(4, 3)$ **21.** $(1, -1)$ **23.** $(-3, -1)$ **25.** $(3, 2)$ **27.** $(50, 18)$

29. Infinite number of solutions   **31.** $(2, -1)$ **33.** $\left(\frac{231}{202}, \frac{117}{202}\right)$

35. $(-38, -22)$ **37.** -5 **38.** $\frac{3}{2}$ **39.** No solution

40. $\left\{y \mid y \geq -\frac{1}{2}\right\}$ **41.** $0, \frac{1}{2}$ **42.** $-4, 5$ **43.** $\frac{13}{3}$

44. $-2, \frac{1}{2}$ **45.–53.** Left to the student   **55.** $(5, 2)$

57. $(0, -1)$ **59.** $(0, 3)$ **61.** $x = \dfrac{c - b}{a - 1}, y = \dfrac{ac - b}{a - 1}$

Mid-Chapter Review: Chapter 7, p. 574

1. False   **2.** False   **3.** True   **4.** True

5.
$$x + x - 3 = -1$$
$$2x - 3 = -1$$
$$2x = -1 + 3$$
$$2x = 2$$
$$x = 1$$
$$y = 1 - 3$$
$$y = -2$$
The solution is $(1, -2)$.

6.
$$2x - 3y = 7$$
$$\underline{x + 3y = -10}$$
$$3x + 0y = -3$$
$$3x = -3$$
$$x = -1$$
$$-1 + 3y = -10$$
$$3y = -9$$
$$y = -3$$
The solution is $(-1, -3)$.

7. Yes   **8.** No   **9.** No   **10.** Yes

11. $(3, -2)$ **12.** $(-2, 3)$ **13.** Infinite number of solutions

14. No solution   **15.** $(5, -3)$ **16.** $(-2, -1)$ **17.** $\left(\frac{5}{3}, -\frac{2}{3}\right)$

18. $\left(\frac{1}{4}, \frac{3}{2}\right)$ **19.** No solution   **20.** $(0, -1)$ **21.** $(-4, 3)$

22. Infinite number of solutions   **23.** Length: 5 ft; width: 4 ft

24. 52 and -34 **25.** 12 and 8   **26.** We know that the first coordinate of the point of intersection is 2. We substitute 2 for x in either $y = 3x - 1$ or $y = 9 - 2x$ and find y, the second coordinate of the point of intersection, 5. Thus the graphs intersect at $(2, 5)$.

27. The coordinates of the point of intersection of the graphs are not integers, so it is difficult to determine the solution from the graph.

28. The equations have the same coefficients of x and y but different constant terms. This means that their graphs have the same slope but different y-intercepts. Thus they have no points in common and the system of equations has no solution.

29. This is not the best approach, in general. If the first equation has x alone on one side, for instance, or if the second equation has a variable alone on one side, solving for y in the first equation is inefficient. This procedure could also introduce fractions in the computations unnecessarily.

Exercise Set 7.4, p. 582

RC1. 10   **RC2.** 15   **RC3.** $0.15y$ **RC4.** 2
1. Discussion post: 15 points; reply: 5 points   **3.** Adults: 90; children: 230   **5.** 4×6 prints: 30; 5×7 prints: 6
7. Two-pointers: 35; three-pointers: 5   **9.** \$50 bonds: 13; \$100 bonds: 6   **11.** Solution A: 40 L; solution B: 60 L   **13.** Hay: 10 lb; grain: 5 lb   **15.** Dimes: 70; quarters: 33   **17.** Brazilian: 200 lb; Turkish: 100 lb   **19.** 70% cashews: 36 lb; 45% cashews: 24 lb
21. Dr. Zeke's: $53\frac{1}{3}$ oz; Vitabrite: $26\frac{2}{3}$ oz   **23.** Type A: 12 questions; type B: 4 questions; 180   **25.** Kuyatts': 32 years; Marconis': 16 years   **27.** Randy: 24; Marie: 6   **29.** $50°, 130°$
31. $28°, 62°$ **33.** 87-octane: 12 gal; 93-octane: 6 gal
35. Large type: 4 pages; small type: 8 pages   **37.** $x^2 + x$
38. $a^2 - \frac{1}{4}$ **39.** $t^4 + 2.4t^2 + 1.44$ **40.** $6x^2 - 11x - 35$
41. $x^3 - 1$ **42.** $3mn - m^2n + mn^2$ **43.** $\dfrac{-2(x - 5)}{(x + 1)(x - 1)}$
44. $\dfrac{-4}{x - 2}$, or $\dfrac{4}{2 - x}$ **45.** $\dfrac{2x^2 - 1}{x^2(x + 1)}$ **46.** $\dfrac{(2a - 1)(a - 4)}{(a - 2)(a - 5)}$
47. $\dfrac{cx}{9}$ **48.** $\dfrac{(t + 2)(t - 2)}{(t + 1)(t - 1)}$

49. 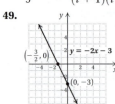   **50.**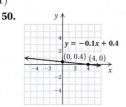

51. 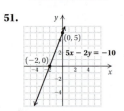   **52.**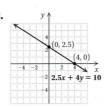

53. 43.75 L   **55.** $4\frac{4}{7}$ L   **57.** 54

Translating for Success, p. 591

1. C   **2.** A   **3.** G   **4.** E   **5.** J   **6.** I
7. B   **8.** L   **9.** O   **10.** F

Exercise Set 7.5, p. 592

RC1. $60t$ **RC2.** $r + 2$ **RC3.** $r - 4$ **RC4.** $125 - r$

1.

Speed	Time
30	t
46	t

4.5 hr

3.

Speed	Time
72	$t + 3$ $\rightarrow d = 72(t + 3)$
120	t $\rightarrow d = 120t$

$4\frac{1}{2}$ hr

5.

Speed	Time
$r + 6$	4 $\rightarrow d = (r + 6)4$
$r - 6$	10 $\rightarrow d = (r - 6)10$

14 km/h
7. 384 km   **9. (a)** 24 mph; **(b)** 90 mi   **11.** $1\frac{23}{43}$ min after the toddler starts running, or $\frac{23}{43}$ min after the mother starts running
13. 15 mi   **15.** $(5x + 9)(5x - 9)$ **16.** $(6a - 1)(2a + 3)$
17. $y(3y - 2)^2$ **18.** $7(x^2 + 2)(x + 1)$ **19.** Approximately 3603 mi

Summary and Review: Chapter 7, p. 594

Vocabulary Reinforcement

1. Pair   **2.** Intersection   **3.** Algebraic   **4.** No solution
5. No solution   **6.** Infinitely many solutions

Concept Reinforcement

1. False   **2.** True   **3.** True   **4.** False

Study Guide

1. Yes   **2.** $(4, -2)$ **3.** $(-2, 1)$ **4.** $(4, -3)$

Review Exercises

1. No   **2.** Yes   **3.** Yes   **4.** No   **5.** $(5, -2)$ **6.** Infinite number of solutions   **7.** No solution   **8.** $(0, 5)$ **9.** $(-3, 9)$
10. $(3, -1)$ **11.** $(1, 4)$ **12.** $(-2, 4)$ **13.** $(1, -2)$ **14.** $(3, 1)$
15. $(1, 4)$ **16.** No solution   **17.** $(-2, 4)$ **18.** $(-2, -6)$
19. $(3, 2)$ **20.** $(2, -4)$ **21.** Infinite number of solutions
22. $(-4, 1)$ **23.** Length: 37.5 cm; width: 10.5 cm
24. Orchestra: 297 seats; balcony: 211 seats   **25.** 40 L of each
26. Asian: 4800 kg; African: 7200 kg   **27.** Peanuts: 8 lb; fancy nuts: 5 lb   **28.** 87-octane: 2.5 gal; 95-octane: 7.5 gal
29. Jeff: 39; his son: 13   **30.** $32°, 58°$ **31.** $77°, 103°$
32. 135 km/h   **33.** 412.5 mi   **34.** A   **35.** A   **36.** $C = 1$, $D = 3$ **37.** $(2, 0)$ **38.** \$960   **39.** $y = -x + 5$, $y = \frac{2}{3}x$
40. $x + y = 4$, $x + y = -3$ **41.** Rabbits: 12; pheasants: 23

Understanding Through Discussion and Writing

1. The second equation can be obtained by multiplying both sides of the first equation by -2. Thus the equations have the same graph, so the system of equations has an infinite number of solutions.
2. The multiplication principle might be used to obtain a pair of terms that are opposites. The addition principle is used to eliminate a variable. Once a variable has been eliminated, the multiplication and addition principles are also used to solve for the remaining variable and, after a substitution, are used again to find the variable that was eliminated.   **3.** Answers will vary.   **4.** A chart allows us to see the given information and the missing information clearly and to see the relationships that yield equations.

Test: Chapter 7, p. 599

1. [7.1a] No   **2.** [7.1b]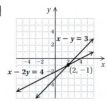

3. [7.2a] $(8, -2)$ **4.** [7.2b] $(-1, 3)$ **5.** [7.2a] $(1, 3)$
6. [7.3a] $(1, -5)$ **7.** [7.3b] No solution   **8.** [7.3b] $\left(\frac{1}{2}, -\frac{2}{3}\right)$
9. [7.3b] $(5, 1)$ **10.** [7.2c] Length: 2108.5 yd; width: 2024.5 yd

11. [7.4a] Solution A: 40 L; solution B: 20 L   **12.** [7.5a] 40 km/h
13. [7.2c] Concessions: $2850; rides: $1425
14. [7.2c] Hay: 415 acres; oats: 235 acres   **15.** [7.2c] 45°, 135°
16. [7.4a] 87-octane: 4 gal; 93-octane: 8 gal   **17.** [7.4a] 20 min
18. [7.5a] 11 hr   **19.** [7.1b], [7.2b], [7.3b] D
20. [7.1a] $C = -\frac{19}{2}; D = \frac{14}{3}$ **21.** [7.4a] 5 people
22. [3.4c], [7.1b] $y = \frac{1}{5}x + \frac{17}{5}$, $y = -\frac{3}{5}x + \frac{9}{5}$
23. [3.4c], [7.1b] $x = 3$, $y = -2$

Cumulative Review: Chapters 1–7, p. 601

1. [1.8d] -6.8 **2.** [4.2d] 3.12×10^{-2} **3.** [1.6c] $-\frac{3}{4}$

4. [4.1d, e] $\frac{1}{8}$ **5.** [6.1c] $\frac{x+3}{2x-1}$ **6.** [6.1c] $\frac{t-4}{t+4}$

7. [6.8a] $\frac{x^2(x+1)}{x+4}$ **8.** [4.6a] $2 - 10x^2 + 12x^4$

9. [4.7f] $4a^4b^2 - 20a^3b^3 + 25a^2b^4$ **10.** [4.6b] $9x^4 - 16y^2$
11. [4.5b] $-2x^3 + 4x^4 - 6x^5$ **12.** [4.5d] $8x^3 + 1$
13. [4.6b] $64 - \frac{1}{9}x^2$ **14.** [4.4c] $-y^3 - 2y^2 - 2y + 7$

15. [4.8b] $x^2 - x - 1 + \frac{-2}{2x-1}$ **16.** [6.4a] $\frac{-5x-28}{5(x-5)}$

17. [6.5a] $\frac{4x-1}{x-2}$ **18.** [6.1d] $\frac{y}{(y-1)^2}$ **19.** [6.2b] $\frac{3(x+1)}{2x}$

20. [5.1b] $3x^2(2x^3 - 12x + 3)$
21. [5.5d] $(4y^2 + 9)(2y + 3)(2y - 3)$
22. [5.3a], [5.4a] $(3x - 2)(x + 4)$ **23.** [5.5b] $(2x^2 - 3y)^2$
24. [5.3a] $3m(m + 5)(m - 3)$ **25.** [5.6a] $(x + 1)^2(x - 1)$
26. [2.3c] -9 **27.** [5.7a] $0, \frac{5}{2}$ **28.** [2.7e] $\{x \mid x \le 20\}$
29. [2.3c] 0.3 **30.** [5.7b] $13, -13$ **31.** [5.7b] $\frac{5}{3}, 3$
32. [6.7a] -1 **33.** [6.7a] No solution   **34.** [7.2a] $(3, -3)$
35. [7.3b] $(-2, 2)$ **36.** [7.3b] Infinite number of solutions
37. [2.4b] $x = \frac{N+t}{r}$ **38.** [2.6a] 35 **39.** [6.8a] $6\frac{2}{3}$ hr
40. [5.8a] Hypotenuse: 13 in.; leg: 5 in.   **41.** [6.8b] 75 chips
42. [5.8a] Height: 9 ft; base: 4 ft   **43.** [6.9d] 72 ft; 360
44. [7.5a] 1.5 hr   **45.** [7.4a] Solution A: 60 mL; solution B: 40 mL
46. [6.9a] $y = 0.2x$ **47.** [3.3a] 0   **48.** [3.3b] $-\frac{2}{3}$, $(0, 2)$
49. [3.4c] $y = -\frac{10}{7}x - \frac{8}{7}$ **50.** [3.4a] $y = 6x - 3$
51. [3.2b]   **52.** [3.2a]

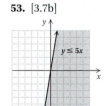

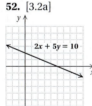

53. [3.7b]   **54.** [3.7b]

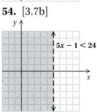

55. [7.1b] $(1, -1)$ **56.** [7.1a] $A = -4, B = -\frac{7}{5}$
57. [2.7e] No solution   **58.** [1.8d], [6.2b], [6.5a] 0
59. [3.6b] $-\frac{3}{10}$

CHAPTER 8

Calculator Corner, p. 605

1. 6.557   **2.** 10.050   **3.** 102.308   **4.** 0.632   **5.** -96.985
6. -0.804

Exercise Set 8.1, p. 608

RC1. True   **RC2.** False   **RC3.** False   **RC4.** True
RC5. True   **RC6.** False   **1.** $2, -2$ **3.** $3, -3$ **5.** $10, -10$
7. $13, -13$ **9.** $16, -16$ **11.** 2   **13.** -3 **15.** -6 **17.** -15
19. 19   **21.** 2.236   **23.** 20.785   **25.** -18.647 **27.** 2.779
29. -168.375 **31.** (a) About 3029 GPM; (b) about 4601 GPM
33. 0.944 sec   **35.** 0.888 sec   **37.** 200   **39.** x **41.** $t^2 + 1$

43. $\frac{3}{x+2}$ **45.** No   **47.** Yes   **49.** No   **51.** c **53.** $3x$

55. $8p$ **57.** ab **59.** $34d$ **61.** $x + 3$ **63.** $a - 5$ **65.** $2a - 5$
67. $11y - 9$ **69.** 61°, 119°   **70.** 38°, 52°   **71.** $10,660

72. $\frac{1}{x+3}$ **73.** 1   **74.** $\frac{(x+2)(x-2)}{(x+1)(x-1)}$ **75.** 1.7, 2.2, 2.6

77. $16, -16$ **79.** $7, -7$

Calculator Corner, p. 615

1. False   **2.** False   **3.** False   **4.** True

Exercise Set 8.2, p. 616

RC1. No   **RC2.** Yes   **RC3.** No   **RC4.** No   **RC5.** Yes
RC6. No   **RC7.** Yes   **RC8.** Yes   **RC9.** No   **1.** $2\sqrt{3}$
3. $5\sqrt{3}$ **5.** $2\sqrt{5}$ **7.** $10\sqrt{6}$ **9.** $9\sqrt{6}$ **11.** $3\sqrt{x}$
13. $4\sqrt{3x}$ **15.** $4\sqrt{a}$ **17.** $8y$ **19.** $x\sqrt{13}$ **21.** $2t\sqrt{2}$
23. $6\sqrt{5}$ **25.** $12\sqrt{2y}$ **27.** $2x\sqrt{7}$ **29.** $x - 3$
31. $\sqrt{2}(2x + 1)$, or $(2x + 1)\sqrt{2}$ **33.** $\sqrt{y}(6 + y)$, or $(6 + y)\sqrt{y}$
35. t^3 **37.** x^6 **39.** $x^2\sqrt{x}$ **41.** $t^9\sqrt{t}$ **43.** $(y - 2)^4$
45. $2(x + 5)^5$ **47.** $6m\sqrt{m}$ **49.** $2a^2\sqrt{2a}$ **51.** $2p^8\sqrt{26p}$
53. $8x^3y\sqrt{7y}$ **55.** $3\sqrt{6}$ **57.** $3\sqrt{10}$ **59.** $6\sqrt{7x}$ **61.** $6\sqrt{xy}$
63. 13   **65.** $5b\sqrt{3}$ **67.** $2t$ **69.** $a\sqrt{bc}$ **71.** $2xy\sqrt{2xy}$
73. 18   **75.** $\sqrt{10x - 5}$ **77.** $x + 2$ **79.** $6xy^3\sqrt{3xy}$
81. $10x^2y^3\sqrt{5xy}$ **83.** $33p^4q^2\sqrt{2pq}$ **85.** $16a^3b^3c^5\sqrt{3abc}$
87. $(-2, 4)$ **88.** $(\frac{1}{8}, \frac{9}{8})$ **89.** $(2, 1)$ **90.** $(10, 3)$
91. 30% insecticide: 80 L; 50% insecticide: 120 L   **92.** 360 ft^2
93. 10 mph   **94.** 211 adults and 171 children   **95.** $\sqrt{5}\sqrt{x - 1}$
97. $\sqrt{x + 6}\sqrt{x - 6}$ **99.** $x\sqrt{x - 2}$ **101.** 0.5   **103.** 2
105. $6(x - 2)^2\sqrt{10}$ **107.** $2^{54}x^{158}\sqrt{2x}$ **109.** $a^2 - 5\sqrt{a}$

Exercise Set 8.3, p. 623

RC1. (e)   **RC2.** (c)   **RC3.** (f)   **RC4.** (d)   **RC5.** (a)
RC6. (f)   **1.** 3   **3.** 6   **5.** $\sqrt{5}$ **7.** $\frac{1}{5}$ **9.** $\frac{2}{5}$ **11.** 2
13. $3y$ **15.** $\frac{4}{7}$ **17.** $\frac{1}{6}$ **19.** $-\frac{4}{9}$ **21.** $\frac{8}{17}$ **23.** $\frac{13}{14}$ **25.** $\frac{5}{x}$

27. $\frac{3a}{25}$ **29.** $\frac{5}{y^5}$ **31.** $\frac{x^9}{7}$ **33.** $\frac{\sqrt{10}}{5}$ **35.** $\frac{\sqrt{14}}{4}$ **37.** $\frac{\sqrt{3}}{6}$

39. $\frac{\sqrt{10}}{6}$ **41.** $\frac{3\sqrt{5}}{5}$ **43.** $\frac{2\sqrt{6}}{3}$ **45.** $\frac{\sqrt{3x}}{x}$ **47.** $\frac{\sqrt{xy}}{y}$

49. $\frac{x\sqrt{5}}{10}$ **51.** $\frac{\sqrt{3}}{3}$ **53.** $\frac{3\sqrt{2}}{4}$ **55.** $\frac{\sqrt{55}}{5}$ **57.** $\sqrt{2}$ **59.** $\frac{\sqrt{55}}{11}$

61. $\frac{\sqrt{21}}{6}$ **63.** $\frac{\sqrt{6}}{2}$ **65.** 5   **67.** $\frac{\sqrt{3x}}{x}$ **69.** $\frac{4y\sqrt{5}}{5}$ **71.** $\frac{a\sqrt{2a}}{4}$

73. $\frac{\sqrt{42x}}{3x}$ **75.** $\frac{3\sqrt{6}}{8c}$ **77.** $\frac{y\sqrt{xy}}{x}$ **79.** $\frac{3n\sqrt{10}}{8}$ **81.** $(-\frac{5}{2}, -\frac{9}{2})$

82. Infinite number of solutions **83.** No solution

84. $\dfrac{x-3}{x-2}$ **85.** $\dfrac{a-5}{2}$ **86.** $\dfrac{(x-2)(x-5)}{(x-3)(x-4)}$

87. 1.57 sec; 3.14 sec; 1.01 sec **89.** $\dfrac{\sqrt{3xy}}{ax^2}$

Mid-Chapter Review: Chapter 8, p. 627

1. True **2.** False **3.** False **4.** True

5. $\sqrt{3x^2 - 48x + 192} = \sqrt{3(x^2 - 16x + 64)}$
$= \sqrt{3(x-8)^2}$
$= \sqrt{3}\sqrt{(x-8)^2}$
$= \sqrt{3}(x-8)$

6. $\sqrt{30}\sqrt{40y} = \sqrt{30 \cdot 40y}$
$= \sqrt{1200y}$
$= \sqrt{100 \cdot 12 \cdot y}$
$= \sqrt{100 \cdot 4 \cdot 3 \cdot y}$
$= \sqrt{100}\sqrt{4}\sqrt{3y}$
$= 10 \cdot 2\sqrt{3y}$
$= 20\sqrt{3y}$

7. $\sqrt{18ab^2}\sqrt{14a^2b^4} = \sqrt{18ab^2 \cdot 14a^2b^4}$
$= \sqrt{2 \cdot 3 \cdot 3 \cdot 2 \cdot 7 \cdot a^3 \cdot b^6}$
$= \sqrt{2^2 \cdot 3^2 \cdot 7 \cdot a^2 \cdot a \cdot b^6}$
$= \sqrt{2^2}\sqrt{3^2}\sqrt{a^2}\sqrt{b^6}\sqrt{7a}$
$= 2 \cdot 3 \cdot a \cdot b^3\sqrt{7a}$
$= 6ab^3\sqrt{7a}$

8. $\sqrt{\dfrac{3y^2}{44}} = \sqrt{\dfrac{3y^2}{2 \cdot 2 \cdot 11}} = \sqrt{\dfrac{3y^2}{2 \cdot 2 \cdot 11} \cdot \dfrac{11}{11}}$
$= \sqrt{\dfrac{33y^2}{2^2 \cdot 11^2}} = \dfrac{y\sqrt{33}}{2 \cdot 11} = \dfrac{y\sqrt{33}}{22}$

9. $-11, 11$ **10.** $\dfrac{x-3}{7}$ **11. (a)** No; **(b)** yes **12.** $8r^3s^3\sqrt{2r}$

13. $5(x-3)$ **14.** $\frac{1}{10}$ **15.** -6 **16.** $-\frac{25}{7}$ **17.** 15 **18.** $10y$

19. $2x - 1$ **20.** $20\sqrt{2x}$ **21.** $\frac{1}{4}$ **22.** $4q^5\sqrt{2q}$ **23.** $\dfrac{9}{z}$

24. 25 **25.** $\frac{3}{7}$ **26.** 8 **27.** $10c^2d^5\sqrt{6c}$ **28.** $12x^3y^5z\sqrt{10yz}$

29. $2\sqrt{15xy}$ **30.** $7a\sqrt{15}$ **31.** $\dfrac{y^{12}}{5}$

32.

$\dfrac{x}{\sqrt{3}}$ $\dfrac{3\sqrt{x}}{x}$

$\sqrt{\dfrac{3}{x}}$ $\dfrac{\sqrt{3x}}{3}$

$\dfrac{3}{\sqrt{x}}$ $\dfrac{x\sqrt{3}}{3}$

$\dfrac{3x}{\sqrt{3}}$ $\sqrt{3}$

$\dfrac{3}{\sqrt{3}}$ $\dfrac{\sqrt{3x}}{x}$

$\sqrt{\dfrac{x}{3}}$ $x\sqrt{3}$

33. The square root of 100 is the principal, or positive, square root, which is 10. **A** square root of 100 could refer to either the positive square root or the negative square root, 10 or -10.

34. It is incorrect to take the square roots of the terms in the numerator individually—that is, $\sqrt{a+b}$ and $\sqrt{a} + \sqrt{b}$ are not equivalent. The following is correct:

$$\sqrt{\dfrac{9+100}{25}} = \dfrac{\sqrt{9+100}}{\sqrt{25}} = \dfrac{\sqrt{109}}{5}.$$

35. In general, $\sqrt{a^2 - b^2} \neq \sqrt{a^2} - \sqrt{b^2}$. In this case, let $x = 13$. Then $\sqrt{x^2 - 25} = \sqrt{13^2 - 25} = \sqrt{169 - 25} = \sqrt{144} = 12$, but $\sqrt{x^2} - \sqrt{25} = \sqrt{13^2} - \sqrt{25} = 13 - 5 = 8$.

36. (1) If necessary, rewrite the expression as $\sqrt{a}/\sqrt{b}$. **(2)** Simplify the numerator and the denominator, if possible, by taking the square roots of perfect square factors. **(3)** Multiply by a form of 1 that produces an expression without a radical in the numerator.

Exercise Set 8.4, p. 633

RC1. Yes **RC2.** No **RC3.** No **RC4.** Yes **RC5.** Yes
RC6. No **RC7.** No **RC8.** No **RC9.** Yes **1.** $16\sqrt{3}$
3. $4\sqrt{5}$ **5.** $13\sqrt{x}$ **7.** $-9\sqrt{d}$ **9.** $25\sqrt{2}$ **11.** $\sqrt{3}$
13. $\sqrt{5}$ **15.** $13\sqrt{2}$ **17.** $3\sqrt{3}$ **19.** $2\sqrt{2}$ **21.** 0
23. $(2 + 9x)\sqrt{x}$, or $\sqrt{x}(2 + 9x)$ **25.** $(3 - 2x)\sqrt{3}$
27. $3\sqrt{2x+2}$ **29.** $(x+3)\sqrt{x^3 - 1}$ **31.** $(4a^2 + a^2b - 5b)\sqrt{b}$
33. $\dfrac{2}{3}\sqrt{3}$, or $\dfrac{2\sqrt{3}}{3}$ **35.** $\dfrac{13}{2}\sqrt{2}$, or $\dfrac{13\sqrt{2}}{2}$ **37.** $\dfrac{1}{6}\sqrt{6}$, or $\dfrac{\sqrt{6}}{6}$
39. $\sqrt{15} - \sqrt{3}$ **41.** $10 - 2\sqrt{7} + 5\sqrt{3} - \sqrt{21}$ **43.** $9 - 4\sqrt{5}$
45. -62 **47.** 1 **49.** $13 + \sqrt{5}$ **51.** $x - 2\sqrt{xy} + y$
53. $-\sqrt{3} - \sqrt{5}$ **55.** $5 - 2\sqrt{6}$ **57.** $\dfrac{4\sqrt{10} - 4}{9}$
59. $5 - 2\sqrt{7}$ **61.** $\dfrac{12 - 3\sqrt{x}}{16 - x}$ **63.** $\dfrac{24 + 3\sqrt{x} + 8\sqrt{2} + \sqrt{2x}}{64 - x}$
65. $\dfrac{2\sqrt{a} - a - 1}{1 - a}$ **67.** $\dfrac{4\sqrt{a} + 4\sqrt{t} + \sqrt{3a} + \sqrt{3t}}{a - t}$ **69.** $\frac{5}{11}$
70. $-\frac{38}{13}$ **71.** $-1, 6$ **72.** $2, 5$ **73.** $\dfrac{x - 3}{4(x + 3)}$
74. $-9, -2, -5, -17, -0.678375$ **75.** 14,270 ft
77. Not correct **79.** $\dfrac{-4\sqrt{6}}{5}$

Calculator Corner, p. 639

1. Left to the student **2.** Left to the student

Exercise Set 8.5, p. 641

RC1. 1 **RC2.** 2 **RC3.** 1 **RC4.** 1 **RC5.** 2 **RC6.** 2
1. 36 **3.** 18.49 **5.** 165 **7.** $\frac{621}{2}$ **9.** 5 **11.** 3 **13.** $\frac{17}{4}$
15. No solution **17.** No solution **19.** 9 **21.** 12 **23.** 1, 5
25. 3 **27.** 5 **29.** No solution **31.** $-\frac{10}{3}$ **33.** 3
35. No solution **37.** 9 **39.** 1 **41.** 8 **43.** 256
45. 62 m/sec **47.** 121 m **49.** 211.25 ft; 281.25 ft
51. $s = 2R - t$ **52.** $\{x | x > -8\}$ **53.** -1728 **54.** y^5w^2
55. $(2x + 1)(x + 5)$ **56.** $(y - 6)(y + 6)$ **57.** $(3t + 4)^2$
58. $(1 + x^4)(1 + x^2)(1 + x)(1 - x)$ **59.** $-2, 2$ **61.** $-\frac{57}{16}$ **63.** 13

Translating for Success, p. 648

1. J **2.** K **3.** N **4.** H **5.** G **6.** E **7.** O **8.** D
9. B **10.** C

Exercise Set 8.6, p. 649

RC1. (d) **RC2.** (c) **1.** 17 **3.** $\sqrt{32} \approx 5.657$ **5.** 12
7. 4 **9.** 26 **11.** 12 **13.** 2 **15.** $\sqrt{2} \approx 1.414$ **17.** 5
19. 3 **21.** $\sqrt{109,444}$ ft ≈ 330.823 ft **23.** $\sqrt{75}$ m ≈ 8.660 m
25. $\sqrt{26,900}$ yd ≈ 164.012 yd **27.** $\left(-\frac{3}{2}, -\frac{1}{16}\right)$ **28.** $\left(\frac{8}{5}, 9\right)$
29. $\left(-\frac{9}{19}, \frac{91}{38}\right)$ **30.** $(-10, 1)$ **31.** $-\frac{1}{3}$ **32.** $\frac{5}{8}$
33. $12 - 2\sqrt{6} \approx 7.101$

Summary and Review: Chapter 8, p. 651

Vocabulary Reinforcement

1. Principal **2.** Radical **3.** Square **4.** Rationalizing
5. Radicands **6.** Conjugates **7.** Hypotenuse, Pythagorean

Concept Reinforcement

1. True **2.** True **3.** True **4.** False

Study Guide

1. $y^2 - 3$ **2. (a)** Yes; **(b)** no **3.** $20y\sqrt{3}$ **4.** $5a^6b^4\sqrt{7b}$

5. $4x^3y^2\sqrt{6x}$ **6.** $b\sqrt{3b}$ **7.** $\frac{5}{9}$ **8.** $\frac{a\sqrt{2}}{5}$ **9.** $(x+6)\sqrt{x-1}$

10. $9 + \sqrt{26}$ **11.** $\frac{47 - 14\sqrt{2}}{79}$ **12.** 6 **13.** 9, 81

14. $a = \sqrt{657} \approx 25.632$

Review Exercises

1. $8, -8$ **2.** $20, -20$ **3.** 6 **4.** -13 **5.** 1.732 **6.** 9.950
7. -17.892 **8.** 0.742 **9.** -2.055 **10.** 394.648 **11.** $x^2 + 4$

12. x **13.** $4 - x$ **14.** $\frac{2}{y-7}$ **15.** Yes **16.** No **17.** No

18. Yes **19.** m **20.** $x - 4$ **21.** $4x$ **22.** $2p - 3$ **23.** $4\sqrt{3}$
24. $4t\sqrt{2}$ **25.** $t - 7$ **26.** $x + 8$ **27.** x^4 **28.** $5a^3\sqrt{3a}$
29. $\sqrt{21}$ **30.** $\sqrt{x^2 - 9}$ **31.** $2\sqrt{15}$ **32.** $2x\sqrt{10}$ **33.** $5xy\sqrt{2}$

34. $10a^2b\sqrt{ab}$ **35.** $\frac{5}{8}$ **36.** $\frac{7}{t}$ **37.** $\frac{c^4}{4}$ **38.** $\frac{\sqrt{2}}{2}$ **39.** $\frac{x\sqrt{15x}}{15}$

40. $\frac{\sqrt{5y}}{y}$ **41.** $\frac{b^3\sqrt{ab}}{a}$ **42.** $\frac{\sqrt{15}}{5}$ **43.** $\frac{x\sqrt{30}}{6}$ **44.** $13\sqrt{5}$

45. $\sqrt{5}$ **46.** $\frac{1}{2}\sqrt{2}$, or $\frac{\sqrt{2}}{2}$ **47.** $7 + 4\sqrt{3}$ **48.** 1

49. $8 - 4\sqrt{3}$ **50.** 52 **51.** No solution **52.** 0, 3 **53.** 9
54. (a) About 63 mph; **(b)** 405 ft **55.** 20 **56.** $\sqrt{3} \approx 1.732$
57. $\sqrt{2,600,000,000}$ ft $\approx 50,990$ ft **58.** 9 ft **59.** B **60.** C
61. $\sqrt{1525}$ mi ≈ 39.051 mi **62.** $b = \pm\sqrt{A^2 - a^2}$ **63.** 6

Understanding Through Discussion and Writing

1. It is necessary for the signs to differ to ensure that the product of the conjugates will be free of radicals. **2.** Since $\sqrt{11 - 2x}$ cannot be negative, the statement $\sqrt{11 - 2x} = -3$ cannot be true for any value of x, including 1. **3.** We often use the rules for manipulating exponents "in reverse" when simplifying radical expressions. For example, we might write x^5 as $x^4 \cdot x$ or y^6 as $(y^3)^2$. **4.** No; consider the clapboard's height above ground level to be one leg of a right triangle. Then the length of the ladder is the hypotenuse of that triangle. Since the length of the hypotenuse must be greater than the length of a leg, a 28-ft ladder cannot be used to repair a clapboard that is 28 ft above ground level. **5.** The square of a number is equal to the square of its opposite. Thus, while squaring both sides of a radical equation allows us to find the solutions of the original equation, this procedure can also introduce numbers that are not solutions of the original equation.
6. (a) $\sqrt{5x^2} = \sqrt{5}\sqrt{x^2} = \sqrt{5} \cdot |x| = |x|\sqrt{5}$.
The given statement is correct. **(b)** Let $b = 3$. Then
$\sqrt{b^2 - 4} = \sqrt{3^2 - 4} = \sqrt{9 - 4} = \sqrt{5}$, but
$b - 2 = 3 - 2 = 1$. The given statement is false. **(c)** Let $x = 3$.
Then $\sqrt{x^2 + 16} = \sqrt{3^2 + 16} = \sqrt{9 + 16} = \sqrt{25} = 5$, but
$x + 4 = 3 + 4 = 7$. The given statement is false.

Test: Chapter 8, p. 657

1. [8.1a] $9, -9$ **2.** [8.1a] 8 **3.** [8.1a] -5 **4.** [8.1b] 10.770
5. [8.1b] -9.349 **6.** [8.1b] 21.909 **7.** [8.1d] $4 - y^3$
8. [8.1e] Yes **9.** [8.1e] No **10.** [8.1f] a **11.** [8.1f] $6y$
12. [8.2c] $\sqrt{30}$ **13.** [8.2c] $\sqrt{x^2 - 64}$ **14.** [8.2a] $3\sqrt{3}$
15. [8.2a] $5\sqrt{x-1}$ **16.** [8.2b] $t^2\sqrt{t}$ **17.** [8.2c] $5\sqrt{2}$

18. [8.2c] $3ab^2\sqrt{2}$ **19.** [8.3b] $\frac{3}{2}$ **20.** [8.3b] $\frac{12}{a}$

21. [8.3c] $\frac{\sqrt{10}}{5}$ **22.** [8.3c] $\frac{\sqrt{2xy}}{y}$ **23.** [8.3a, c] $\frac{3\sqrt{6}}{8}$

24. [8.3a] $\frac{\sqrt{7}}{4y}$ **25.** [8.4a] $-6\sqrt{2}$ **26.** [8.4a] $\frac{6}{5}\sqrt{5}$, or $\frac{6\sqrt{5}}{5}$

27. [8.4b] $21 - 8\sqrt{5}$ **28.** [8.4b] 11 **29.** [8.4c] $\frac{40 + 10\sqrt{5}}{11}$

30. [8.6a] $\sqrt{80} \approx 8.944$ **31.** [8.5a] 48 **32.** [8.5a] $-2, 2$
33. [8.5b] -3 **34.** [8.5c] **(a)** About 237 mi; **(b)** 34,060.5 ft
35. [8.6b] $\sqrt{15,700}$ yd ≈ 125.300 yd **36.** [8.3c] A
37. [8.1a] $\sqrt{5}$ **38.** [8.2b] y^{8n}

Cumulative Review: Chapters 1–8, p. 659

1. [6.1a] $-\frac{1}{2}$ **2.** [8.1e] No **3.** [8.4b] 1 **4.** [8.1a] -14

5. [8.2c] 15 **6.** [8.4b] $3 - 2\sqrt{2}$ **7.** [8.3a, c] $\frac{9\sqrt{10}}{25}$

8. [8.4a] $12\sqrt{5}$ **9.** [4.7f] $9x^8 - 4y^{10}$ **10.** [4.6c] $x^4 + 8x^2 + 16$

11. [4.6a] $8x^2 - \frac{1}{8}$ **12.** [6.5a] $\frac{2(2x + 1)}{2x - 1}$

13. [4.4a, c] $-3x^3 + 8x^2 - 5x$ **14.** [6.1d] $\frac{2(x - 5)}{3(x - 1)}$.

15. [6.2b] $\frac{(x + 1)(x - 3)}{2(x + 3)}$ **16.** [4.8b] $3x^2 + 4x + 9 + \frac{13}{x - 2}$.

17. [8.2a] $\sqrt{2}(x - 1)$, or $(x - 1)\sqrt{2}$ **18.** [4.1d, f] $\frac{1}{x^{12}}$

19. [8.3b] $\frac{5}{x^4}$ **20.** [6.6a] $2(x - 1)$

21. [5.5d] $3(1 + 2x^4)(1 - 2x^4)$ **22.** [5.1b] $4t(3 - t - 12t^3)$
23. [5.3a], [5.4a] $2(3x - 2)(x - 4)$
24. [5.6a] $(x + 1)(2x + 1)(2x - 1)$ **25.** [5.5b] $(4x^2 - 7)^2$
26. [5.2a] $(x + 15)(x - 12)$ **27.** [5.7b] $-17, 0$
28. [2.7e] $\left\{x \mid x \le -\frac{9}{2}\right\}$ **29.** [6.7a] $-\frac{12}{5}$ **30.** [5.7b] $-5, 6$
31. [8.5a] 41 **32.** [2.3b] $\frac{2}{5}$ **33.** [6.7a] $\frac{1}{3}$ **34.** [7.2a] $(4, 1)$

35. [7.3a] $(3, -8)$ **36.** [2.4b] $p = \frac{4A}{r + q}$

37. [3.7b] **38.** [3.2b] **39.** [3.2a]

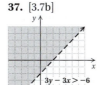
$3y - 3x > -6$

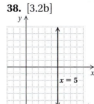
$x = 5$

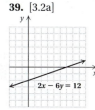
$2x - 6y = 12$

40. [3.4c] $y = \frac{11}{4}x - \frac{19}{4}$ **41.** [3.3b] Slope: $\frac{5}{3}$; y-intercept: $(0, -3)$
42. [4.3a] $-2; 1; -2; -5; -2$ **43.** [7.4a] Hamburger: $2.95; milkshake: $2.50 **44.** [6.9d] 0.4 ft **45.** [2.6a] 38°, 76°, 66°
46. [6.8b] 20 defective resistors **47.** [5.8a] Length: 15 m; width: 12 m **48.** [7.4a] Dimes: 65; quarters: 50
49. [2.5a] $2600 **50.** [6.8a] 60 mph **51.** [7.4a] 300 L
52. [7.3b], [8.5a] $(9, 4)$

CHAPTER 9

Calculator Corner, p. 667

1. 0.6, 1 **2.** $-1.5, 5$ **3.** 3, 8 **4.** 2, 4

Exercise Set 9.1, p. 667

RC1. False **RC2.** True **RC3.** True **RC4.** True
1. $x^2 - 3x + 2 = 0; a = 1, b = -3, c = 2$ **3.** $7x^2 - 4x + 3 = 0$;
$a = 7, b = -4, c = 3$ **5.** $2x^2 - 3x + 5 = 0$;
$a = 2, b = -3, c = 5$ **7.** $0, -5$ **9.** $0, -2$ **11.** $0, \frac{2}{5}$
13. $0, -1$ **15.** $0, 3$ **17.** $0, \frac{1}{5}$ **19.** $0, \frac{3}{14}$ **21.** $0, \frac{81}{2}$
23. $-12, 4$ **25.** $-5, -1$ **27.** $-9, 2$ **29.** 3, 5 **31.** -5
33. 4 **35.** $-\frac{2}{3}, \frac{1}{2}$ **37.** $-\frac{2}{3}, 4$ **39.** $-1, \frac{5}{3}$ **41.** $-5, -1$
43. $-2, 7$ **45.** $-5, 4$ **47.** 4 **49.** $-2, 1$ **51.** $-\frac{2}{5}, 10$
53. $-4, 6$ **55.** 1 **57.** 2, 5 **59.** No solution **61.** $-\frac{5}{2}, 1$

63. 35 diagonals   **65.** 7 sides   **67.** $9(2t - p + 1)$
68. $2x(x - 1)(x - 8)$ **69.** $(x + 7)^2$ **70.** $(6x - 5y)^2$
71. $(t^2 + 9)(t + 3)(t - 3)$ **72.** $(2a + 3)(3a - 5)$
73. $(a + c)(a + 1)$ **74.** $b(5a - 1)(4a + 1)$
75. $(xy + 1)(xy - 1)$ **76.** $\left(c + \frac{1}{2}\right)\left(c - \frac{1}{2}\right)$
77. $3(x + 1)^2(x - 1)^2$ **78.** $\left(\frac{1}{2}x - \frac{1}{4}y\right)^2$
79. $-\frac{1}{3}, 1$ **81.** $0, \dfrac{\sqrt{5}}{5}$ **83.** $-1.7, 4$ **85.** $-1.7, 3$
87. $-2, 3$ **89.** 4

Exercise Set 9.2, p. 676

RC1. False   **RC2.** False   **RC3.** False   **RC4.** True

1. $11, -11$ **3.** $\sqrt{7}, -\sqrt{7}$ **5.** $\dfrac{\sqrt{15}}{5}, -\dfrac{\sqrt{15}}{5}$ **7.** $\frac{5}{2}, -\frac{5}{2}$
9. $\dfrac{7\sqrt{3}}{3}, -\dfrac{7\sqrt{3}}{3}$ **11.** $\sqrt{3}, -\sqrt{3}$ **13.** $\frac{8}{7}, -\frac{8}{7}$ **15.** $-7, 1$
17. $-3 \pm \sqrt{21}$ **19.** $-13 \pm 2\sqrt{2}$ **21.** $7 \pm 2\sqrt{3}$
23. $-9 \pm \sqrt{34}$ **25.** $\dfrac{-3 \pm \sqrt{14}}{2}$ **27.** $-5, 11$ **29.** $-15, 1$
31. $-2, 8$ **33.** $-21, -1$ **35.** $1 \pm \sqrt{6}$ **37.** $11 \pm \sqrt{19}$
39. $-5 \pm \sqrt{29}$ **41.** $\dfrac{7 \pm \sqrt{57}}{2}$ **43.** $-7, 4$ **45.** $\dfrac{-3 \pm \sqrt{17}}{4}$
47. $\dfrac{-3 \pm \sqrt{145}}{4}$ **49.** $\dfrac{-2 \pm \sqrt{7}}{3}$ **51.** $-\frac{1}{2}, 5$ **53.** $-\frac{5}{2}, \frac{2}{3}$
55. About 13.0 sec   **57.** About 9.2 sec
59. $3x^4 + 4x^3 - 2x^2 + 1$ **60.** $3x^4 - 4x^3 + 1$
61. $9x^4 + 6x^3 + x^2$ **62.** $2x^3 - x^2 + x + 1$ **63.** $4x^3 - 5x^2 - \dfrac{x}{3}$
64. $5x + 4 + \dfrac{1}{x - 1}$ **65.** $-12, 12$ **67.** $-16\sqrt{2}, 16\sqrt{2}$
69. $-2\sqrt{c}, 2\sqrt{c}$ **71.** $49.896, -49.896$ **73.** $-9, 9$

Calculator Corner, p. 682

1. The equations $x^2 + x = -1$ and $x^2 + x + 1 = 0$ are equivalent. The graph of $y = x^2 + x + 1$ has no x-intercepts, so the equation $x^2 + x = -1$ has no real-number solutions.

Exercise Set 9.3, p. 683

RC1. (c)   **RC2.** (d)   **RC3.** (a)   **RC4.** (b)   **RC5.** (e)
RC6. (f)
1. $-3, 7$ **3.** 3   **5.** $-\frac{4}{3}, 2$ **7.** $-\frac{5}{2}, \frac{3}{2}$ **9.** $-3, 3$ **11.** $1 \pm \sqrt{3}$
13. $5 \pm \sqrt{3}$ **15.** $-2 \pm \sqrt{7}$ **17.** $\dfrac{-4 \pm \sqrt{10}}{3}$ **19.** $\dfrac{5 \pm \sqrt{33}}{4}$
21. $\dfrac{1 \pm \sqrt{3}}{2}$ **23.** No real-number solutions   **25.** $\dfrac{5 \pm \sqrt{73}}{6}$
27. $\dfrac{3 \pm \sqrt{29}}{2}$ **29.** $-\sqrt{5}, \sqrt{5}$ **31.** $-2 \pm \sqrt{3}$ **33.** $\dfrac{5 \pm \sqrt{37}}{2}$
35. $-1.3, 5.3$ **37.** $-0.2, 6.2$ **39.** $-1.2, 0.2$ **41.** $0.3, 2.4$
43. $\frac{11}{5}$ **44.** $\left\{x \mid x \le -\frac{1}{18}\right\}$ **45.** $\frac{1}{7}$ **46.** $-1, 8$ **47.** $-1, 1$
48. 75   **49.** 5   **50.** $-3, 0$ **51.** $-1, \frac{3}{2}$ **52.** $-4, 4$ **53.** $\frac{1}{2}$
54. $\frac{1}{2}, 4$ **55.** $0, 2$ **57.** $\dfrac{3 \pm \sqrt{5}}{2}$ **59.** $\dfrac{-7 \pm \sqrt{61}}{2}$
61. $\dfrac{-2 \pm \sqrt{10}}{2}$ **63.** Yes   **65.–71.** Left to the student

Mid-Chapter Review: Chapter 9, p. 685

1. True   **2.** False   **3.** True
4. $x^2 - 6x - 2 = 0$
$x^2 - 6x = 2$
$x^2 - 6x + 9 = 2 + 9$
$(x - 3)^2 = 11$
$x - 3 = \pm\sqrt{11}$
$x = 3 \pm \sqrt{11}$

5. $3x^2 = 8x - 2$
$3x^2 - 8x + 2 = 0$ Standard form
$a = 3, \quad b = -8, \quad c = 2$
We substitute for a, b, and c in the quadratic formula:
$x = \dfrac{-b \pm \sqrt{b^2 - 4ac}}{2a}$ Quadratic formula
$x = \dfrac{-(-8) \pm \sqrt{(-8)^2 - 4 \cdot 3 \cdot 2}}{2 \cdot 3}$ Substituting
$x = \dfrac{8 \pm \sqrt{64 - 24}}{6} = \dfrac{8 \pm \sqrt{40}}{6} = \dfrac{8 \pm \sqrt{4 \cdot 10}}{6}$
$x = \dfrac{8 \pm 2\sqrt{10}}{6} = \dfrac{2(4 \pm \sqrt{10})}{2 \cdot 3} = \dfrac{4 \pm \sqrt{10}}{3}$

6. $a = 1; b = -5; c = 10$ **7.** $a = 1; b = 14; c = -4$
8. $a = 3; b = -17; c = 0$ **9.** $0, \frac{1}{3}$ **10.** $-2, 5$ **11.** $0, 1$
12. 7   **13.** $-2, 0$ **14.** $-\frac{4}{3}, \frac{1}{6}$ **15.** $0, 22$ **16.** $2, 3$ **17.** $-\frac{7}{8}, 5$
18. $-3, 1$ **19.** $\dfrac{9 \pm \sqrt{57}}{2}$ **20.** $\dfrac{7 \pm \sqrt{113}}{4}$ **21.** $8, 10$
22. $\dfrac{-3 \pm \sqrt{33}}{4}$ **23.** No real-number solutions   **24.** $-8, 8$
25. No real-number solutions   **26.** $3 \pm \sqrt{2}$ **27.** $-\frac{3}{2}, \frac{1}{2}$
28. $\pm\sqrt{3}$ **29.** $\dfrac{1 \pm \sqrt{5}}{2}$ **30.** $-8, 12$ **31.** No real-number
solutions   **32.** $\dfrac{1 \pm \sqrt{5}}{2}$ **33.** $-5, 5$ **34.** $\dfrac{1 \pm \sqrt{11}}{2}$
35. No real-number solutions   **36.** $-2.4, 3.4$ **37.** $-3.4, -0.1$
38. A   **39.** B   **40.** A   **41.** C   **42.** B   **43.** C   **44.** B
45. Mark does not recognize that the $\pm$ sign yields two solutions, one in which the radical is added to 3 and the other in which the radical is subtracted from 3.   **46.** The addition principle should be used at the outset to get 0 on one side of the equation. Since this was not done in the given procedure, the principle of zero products was not applied correctly.   **47.** The first coordinates of the x-intercepts of the graph of $y = (x - 2)(x + 3)$ are the solutions of the equation $(x - 2)(x + 3) = 0$.   **48.** The quadratic formula would not be the easiest way to solve a quadratic equation when the equation can be solved by factoring or by using the principle of square roots.   **49.** Answers will vary. Any equation of the form $ax^2 + bx + c = 0$, where $b^2 - 4ac < 0$, will do. Then the graph of the equation $y = ax^2 + bx + c$ will not cross the x-axis.
50. If $x = -5$ or $x = 7$, then $x + 5 = 0$ or $x - 7 = 0$. Thus the equation $(x + 5)(x - 7) = 0$, or $x^2 - 2x - 35 = 0$, has solutions -5 and 7.

Exercise Set 9.4, p. 690

RC1. (e)   **RC2.** (a)   **RC3.** (e)   **RC4.** (c)   **RC5.** (d)
RC6. (e)
1. $I = \dfrac{VQ}{q}$ **3.** $m = \dfrac{Sd^2}{kM}$ **5.** $d^2 = \dfrac{kmM}{S}$ **7.** $W = \sqrt{\dfrac{10t}{T}}$
9. $t = \dfrac{A}{a + b}$ **11.** $x = \dfrac{y - c}{a + b}$ **13.** $a = \dfrac{bt}{b - t}$ **15.** $p = \dfrac{qf}{q - f}$
17. $b = \dfrac{2A}{h}$ **19.** $h = \dfrac{S - 2\pi r^2}{2\pi r}$, or $h = \dfrac{S}{2\pi r} - r$
21. $R = \dfrac{r_1 r_2}{r_2 + r_1}$ **23.** $Q = \dfrac{P^2}{289}$ **25.** $E = \dfrac{mv^2}{2g}$ **27.** $r = \dfrac{1}{2}\sqrt{\dfrac{S}{\pi}}$
29. $A = \dfrac{-m + \sqrt{m^2 + 4kP}}{2k}$ **31.** $a = \sqrt{c^2 - b^2}$ **33.** $t = \dfrac{\sqrt{s}}{4}$
35. $r = \dfrac{-\pi h + \sqrt{\pi^2 h^2 + \pi A}}{\pi}$ **37.** $v = 20\sqrt{\dfrac{F}{A}}$
39. $a = \sqrt{c^2 - b^2}$ **41.** $a = \dfrac{2h\sqrt{3}}{3}$
43. $T = \dfrac{2 + \sqrt{4 - a(m - a)}}{a}$ **45.** $T = \dfrac{v^2 \pi m}{8k}$ **47.** $x = \dfrac{d\sqrt{3}}{3}$
49. $n = \dfrac{1 + \sqrt{1 + 8N}}{2}$ **51.** $b = \dfrac{a}{3S - 1}$ **53.** $B = \dfrac{A}{QA + 1}$

55. $n = \dfrac{S + 360}{180}$, or $n = \dfrac{S}{180} + 2$ 57. $t = \dfrac{A - P}{Pr}$

59. $D = \dfrac{BC}{A}$ 61. $\sqrt{65} \approx 8.062$ 62. $\sqrt{75} \approx 8.660$

63. $\sqrt{41} \approx 6.403$ 64. $\sqrt{44} \approx 6.633$ 65. $\sqrt{1084} \approx 32.924$
66. $\sqrt{5} \approx 2.236$ 67. $\sqrt{424}\,\text{ft} \approx 20.591\,\text{ft}$
68. $\sqrt{12{,}500}\,\text{yd} \approx 111.803\,\text{yd}$ 69. $4\sqrt{5}$ 70. $30x^5\sqrt{10}$
71. $3t$ 72. $8x^2\sqrt{3x}$ 73. $3\sqrt{10}$ 74. $2\sqrt{2}$

75. (a) $r = \dfrac{C}{2\pi}$; (b) $A = \dfrac{C^2}{4\pi}$; (c) $C = 2\sqrt{A\pi}$

Translating for Success, p. 696

1. M   2. G   3. F   4. L   5. D   6. N   7. J   8. E
9. B   10. C

Exercise Set 9.5, p. 697

RC1. True   RC2. False   RC3. True   RC4. False
1. Length: 16 yd; width: 5 yd   3. 16 in.; 24 in.   5. Length: 30 ft;
width: 10 ft   7. Length: 20 cm; width: 16 cm   9. 4.6 m; 6.6 m
11. Length: 5.6 in.; width: 3.6 in.   13. Length: 6.4 cm;
width: 3.2 cm   15. 3 cm   17. 7 km/h   19. 2 km/h
21. 0 km/h (no wind) or 40 km/h   23. 8 mph   25. 1 km/h
27. y-intercept: $(0, 4)$; x-intercept: $(\frac{1}{2}, 0)$ 28. y-intercept: $(0, -9)$;
x-intercept: $(15, 0)$

29. 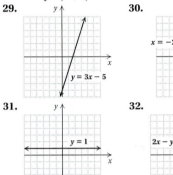   30.

31. 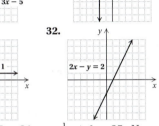   32.

33. $y = -2x - 5$ 34. $y = \frac{1}{2}x + 1$ 35. No   36. Yes
37. $12\sqrt{2}\,\text{in.} \approx 16.97\,\text{in.}$; two 12-in. pizzas

Visualizing for Success, p. 705

1. J   2. F   3. H   4. G   5. B   6. E   7. D   8. I
9. C   10. A

Exercise Set 9.6, p. 706

RC1. Quadratic   RC2. Parabola   RC3. Vertex   RC4. Line
1.   3.   5.

7.   9.

11. 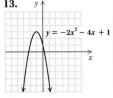   13. 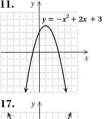   15.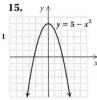

17.   19.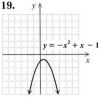

21.   23.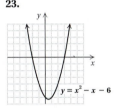

25. $(-\sqrt{2}, 0)$; $(\sqrt{2}, 0)$ 27. $(-5, 0)$; $(0, 0)$
29. $\left(\dfrac{-1 - \sqrt{33}}{2}, 0\right)$; $\left(\dfrac{-1 + \sqrt{33}}{2}, 0\right)$ 31. $(3, 0)$
33. $(-2 - \sqrt{5}, 0)$; $(-2 + \sqrt{5}, 0)$ 35. None   37. $22\sqrt{2}$
38. $25y^2\sqrt{y}$ 39. $y = \dfrac{29.76}{x}$ 40. 35   41. $\frac{47}{60}$ 42. 49.55
43. (a) After 2 sec; after 4 sec; (b) after 3 sec; (c) after 6 sec
45. 16; two real solutions   47. -161.91; no real solutions

Calculator Corner, p. 712

1. -12.8 2. -9.2 3. -2 4. 20

Exercise Set 9.7, p. 715

RC1. Domain; range   RC2. Input   RC3. Vertical
RC4. Linear
1. Yes   3. Yes   5. No   7. Yes   9. Yes   11. Yes
13. A relation but not a function   15. (a) 9; (b) 12; (c) 2; (d) 5;
(e) 7.4; (f) $5\frac{2}{3}$ 17. (a) -21; (b) 15; (c) 42; (d) 0; (e) 2; (f) -162.6
19. (a) 7; (b) -17; (c) 24.1; (d) 4; (e) -26; (f) 6   21. (a) 0; (b) 5;
(c) 2; (d) 170; (e) 65; (f) 230   23. (a) 1; (b) 3; (c) 3; (d) 4; (e) 11;
(f) 23   25. (a) 0; (b) -1; (c) 8; (d) 1000; (e) -125; (f) -1000
27. (a) 34 years; (b) 20 years   29. 2.7 million users; 548.7 million
users; 1082.7 million users   31. $1\frac{20}{33}$ atm; $1\frac{10}{11}$ atm; $4\frac{1}{33}$ atm

33. 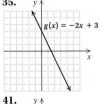   35.   37.

39. 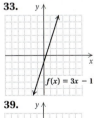   41.   43.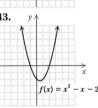

45. Yes   47. Yes   49. No   51. No   53. About 23,800
conventions   55. $\left(\frac{3}{2}, \frac{1}{2}\right)$ 56. $(3, 0)$ 57. $\left(3, -\frac{1}{5}\right)$
58. Infinite number of solutions

59.

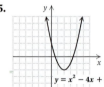

$g(x) = x^3$

61.

$f(x) = |x| + x$

Summary and Review: Chapter 9, p. 720

Vocabulary Reinforcement
1. Quadratic **2.** Square **3.** Discriminant **4.** Vertex
5. Parabola **6.** Quadratic **7.** Vertical-line **8.** Domain

Concept Reinforcement
1. False **2.** True **3.** True **4.** True

Study Guide
1. $-\frac{4}{5}, 2$ **2.** $-\frac{\sqrt{77}}{7}, \frac{\sqrt{77}}{7}$ **3.** $2 \pm \sqrt{3}$ **4.** $\frac{3 \pm \sqrt{21}}{4}$

5.

$y = x^2 - 4x + 2$

6. (a) $h(5) = 9$; (b) $f(0) = -4$

Review Exercises
1. $-\sqrt{3}, \sqrt{3}$ **2.** $-2\sqrt{2}, 2\sqrt{2}$ **3.** $\frac{3}{5}, 1$ **4.** $-2, \frac{1}{3}$
5. $-8 \pm \sqrt{13}$ **6.** 0 **7.** $0, \frac{7}{5}$ **8.** $1 \pm \sqrt{11}$ **9.** $\frac{1 \pm \sqrt{10}}{3}$
10. $-3 \pm 3\sqrt{2}$ **11.** $\frac{2 \pm \sqrt{3}}{2}$ **12.** $\frac{3 \pm \sqrt{33}}{2}$ **13.** No real-
number solutions **14.** $0, \frac{4}{3}$ **15.** $-5, 3$ **16.** 1 **17.** $2 \pm \sqrt{2}$
18. $-1, \frac{5}{3}$ **19.** $0.4, 4.6$ **20.** $-1.9, -0.1$ **21.** $T = L(4V^2 - 1)$
22.

$(0, 2)$ $y = 2 - x^2$

23.

$(0, -2)$ $(2, -6)$ $y = x^2 - 4x - 2$

24. $(-\sqrt{2}, 0); (\sqrt{2}, 0)$ **25.** $(2 - \sqrt{6}, 0); (2 + \sqrt{6}, 0)$
26. 4.7 cm, 1.7 cm **27.** 15 ft **28.** About 8.1 sec
29. $-1, -7, 2$ **30.** $0, 0, 19$ **31.** 2700 calories
32.

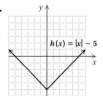

$g(x) = 4 - x$

33.

$f(x) = x^2 - 3$

34.

$h(x) = |x| - 5$

35.

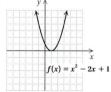

$f(x) = x^2 - 2x + 1$

36. No **37.** Yes **38.** D **39.** A **40.** 31 and 32; -32 and
-31 **41.** $5\sqrt{\pi}$ in., or about 8.9 in. **42.** 25 **43.** $-4, -2$
44. $-5, -1$ **45.** $-6, 0$ **46.** -3

Understanding Through Discussion and Writing
1. The second line should be $x + 6 = \sqrt{16} \ or \ x + 6 = -\sqrt{16}$.
Then we would have
$$x + 6 = 4 \quad or \quad x + 6 = -4$$
$$x = -2 \quad or \quad x = -10.$$
Both numbers check so the solutions are -2 and -10.
2. No; since each input has exactly one output, the number of
outputs cannot exceed the number of inputs. **3.** Find the
average, v, of the x-coordinates of the x-intercepts, $v = \frac{a_1 + a_2}{2}$.
Then the equation of the line of symmetry is $x = v$. The number v
is also the first coordinate of the vertex. We substitute this value for
x in the equation of the parabola to find the y-coordinate of the
vertex. **4.** If $a > 0$, the graph opens up. If $a < 0$, the graph
opens down. **5.** The solutions will be rational numbers
because each is the solution of a linear equation of the form
$mx + b = 0$.

Test: Chapter 9, p. 725
1. [9.2a] $-\sqrt{5}, \sqrt{5}$ **2.** [9.1b] $-\frac{8}{7}, 0$ **3.** [9.1c] $-8, 6$
4. [9.1c] $-\frac{1}{3}, 2$ **5.** [9.2b] $8 \pm \sqrt{13}$ **6.** [9.3a] $\frac{1 \pm \sqrt{13}}{2}$
7. [9.3a] $\frac{3 \pm \sqrt{37}}{2}$ **8.** [9.3a] $-2 \pm \sqrt{14}$ **9.** [9.3a] $\frac{7 \pm \sqrt{37}}{6}$
10. [9.1c] $-1, 2$ **11.** [9.1c] $-4, 2$ **12.** [9.2c] $2 \pm \sqrt{14}$
13. [9.3b] $-1.7, 5.7$ **14.** [9.4a] $n = \frac{-b + \sqrt{b^2 + 4ad}}{2a}$
15. [9.6b] $\left(\frac{1 - \sqrt{21}}{2}, 0\right), \left(\frac{1 + \sqrt{21}}{2}, 0\right)$
16. [9.6a]

$(0, 4)$ $y = 4 - x^2$

17. [9.6a]

$(0, 5)$ $\left(\frac{1}{2}, \frac{21}{4}\right)$ $y = -x^2 + x + 5$

18. [9.7b] $1; 1\frac{1}{2}; 2$ **19.** [9.7b] $1; 3; -3$ **20.** [9.5a] Length: 6.5 m;
width: 2.5 m **21.** [9.5a] 24 km/h **22.** [9.7e] 25.86 min
23. [9.7c]

$h(x) = x - 4$

24. [9.7c]

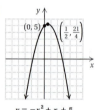

$g(x) = x^2 - 4$

25. [9.7d] Yes **26.** [9.7d] No **27.** [9.7b] D
28. [9.5a] $(5 + 5\sqrt{2})$ ft
29. [7.2b], [9.3a] $(1 + \sqrt{5}, -1 + \sqrt{5}), (1 - \sqrt{5}, -1 - \sqrt{5})$

Cumulative Review: Chapters 1–9, p. 727
1. [4.1a] $x \cdot x \cdot x$ **2.** [4.1c] 54 **3.** [1.2c] $-0.\overline{27}$ **4.** [6.3a] 240
5. [1.2e] 7 **6.** [1.3a] 9 **7.** [1.4a] 15 **8.** [1.6c] $-\frac{3}{20}$
9. [1.8d] 4 **10.** [1.8b] $-2m - 4$ **11.** [2.2a] -8 **12.** [2.3b] -12
13. [2.3c] 7 **14.** [5.7b] $3, 5$ **15.** [7.2a] $(1, 2)$ **16.** [7.3a] $(12, 5)$
17. [7.3b] $(6, 7)$ **18.** [5.7b] $-2, 3$ **19.** [9.3a] $\frac{-3 \pm \sqrt{29}}{2}$
20. [8.5a] 2 **21.** [2.7e] $\{x | x \geq -1\}$ **22.** [2.3b] 8
23. [2.3b] -3 **24.** [2.7d] $\{x | x < -8\}$ **25.** [2.7e] $\{y | y \leq \frac{35}{22}\}$
26. [9.2a] $-\sqrt{10}, \sqrt{10}$ **27.** [9.2b] $3 \pm \sqrt{6}$ **28.** [6.6a] $\frac{2}{9}$
29. [6.7a] -5 **30.** [6.7a], [9.1b] No solution **31.** [8.5a] 12
32. [2.4b] $b = \frac{At}{4}$ **33.** [9.4a] $m = \frac{tn}{t + n}$ **34.** [9.4a] $A = \pi r^2$
35. [9.4a] $x = \frac{b + \sqrt{b^2 + 4ay}}{2a}$ **36.** [4.1d, f] $\frac{1}{x^4}$

37. [4.1e, f] y^7 **38.** [4.2a, b] $4y^{12}$ **39.** [4.3e] $10x^3 + 3x - 3$
40. [4.4a] $7x^3 - 2x^2 + 4x - 17$ **41.** [4.4c] $8x^2 - 4x - 6$
42. [4.5b] $-8y^4 + 6y^3 - 2y^2$ **43.** [4.5d] $6t^3 - 17t^2 + 16t - 6$
44. [4.6b] $t^2 - \frac{1}{16}$ **45.** [4.6c] $9m^2 - 12m + 4$
46. [4.7e] $15x^2y^3 + x^2y^2 + 5xy^2 + 7$ **47.** [4.7f] $x^4 - 0.04y^2$
48. [4.7f] $9p^2 + 24pq^2 + 16q^4$ **49.** [6.1d] $\dfrac{2}{x + 3}$

50. [6.2b] $\dfrac{3a(a - 1)}{2(a + 1)}$ **51.** [6.4a] $\dfrac{27x - 4}{5x(3x - 1)}$

52. [6.5a] $\dfrac{-(x - 2)(x + 1)}{(x + 4)(x - 4)(x - 5)}$ **53.** [5.1b] $4x(2x - 1)$
54. [5.5d] $(5x - 2)(5x + 2)$ **55.** [5.3a], [5.4a] $(3y + 2)(2y - 3)$
56. [5.5b] $(m - 4)^2$ **57.** [5.1c] $(x - 8)(x^2 - 5)$
58. [5.6a] $3(a^2 + 6)(a + 2)(a - 2)$
59. [5.5d] $(4x^2 + 1)(2x + 1)(2x - 1)$
60. [5.5d] $(7ab + 2)(7ab - 2)$ **61.** [5.5b] $(3x + 5y)^2$
62. [5.1c] $(c - 3b)(2a + d)$ **63.** [5.3a], [5.4a] $(5x - 2y)(3x + 4y)$
64. [6.6a] $-\frac{42}{5}$ **65.** [8.1a] 7 **66.** [8.1a] -25 **67.** [8.1f] $8x$
68. [8.2c] $\sqrt{a^2 - b^2}$ **69.** [8.2c] $8a^2b\sqrt{3ab}$ **70.** [8.2a] $5\sqrt{6}$

71. [8.2b] $9xy\sqrt{3x}$ **72.** [8.3b] $\frac{10}{9}$ **73.** [8.3b] $\dfrac{8}{x}$

74. [8.4a] $12\sqrt{3}$ **75.** [8.3a, c] $\dfrac{2\sqrt{10}}{5}$ **76.** [8.6a] 40

77. [3.1d] **78.** [3.2a] **79.** [3.2b]

$y = \frac{1}{3}x - 2$

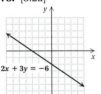
$2x + 3y = -6$

$y = -3$

80. [3.7b] **81.** [3.7b] **82.** [9.6a]

$x \geq -3$

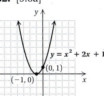
$4x - 3y > 12$

$y = x^2 + 2x + 1$
$(0, 1)$
$(-1, 0)$

83. [9.2c] $\dfrac{2 \pm \sqrt{6}}{3}$ **84.** [9.3b] $-0.2, 1.2$ **85.** [2.5a] 25%
86. [2.5a] 60 **87.** [6.8a] $4\frac{4}{9}$ hr **88.** [9.5a] Length: 12 ft;
width: 8 ft **89.** [9.5a] 2 km/h **90.** [5.8a] 12 m
91. [9.5a] 5 and 7; -7 and -5 **92.** [7.4a] 40 L of A; 20 L of B
93. [9.2d] About 4.4 sec **94.** [6.9b] $451.20; the variation
constant is the amount earned per hour **95.** [2.6a] 6090 cars
96. [7.5a] 140 mph **97.** Nuts: 14 lb; candy: 28 lb **98.** [9.6b] $-3, 2$
99. [9.6b] $(-2 - \sqrt{3}, 0), (-2 + \sqrt{3}, 0)$ **100.** [3.4a] Slope: 2;
y-intercept: $(0, -8)$ **101.** [3.6a, b] Neither **102.** [3.3a] 15

103. [6.9a] $y = 10x$; 640 **104.** [6.9c] $y = \dfrac{1000}{x}$; 8

105. [9.7d] Yes **106.** [9.7d] No
107. [9.7c] **108.** [9.7c]

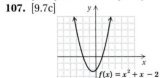
$f(x) = x^2 + x - 2$

$g(x) = |x + 2|$

109. [9.7b] $-4; 0; 0$ **110.** [6.8a], [9.5a] C **111.** [9.3a] D
112. [2.4b] D **113.** [3.2a] C **114.** [1.2e] $-12, 12$

115. [8.1a] $\sqrt{3}$ **116.** [9.2c] $-30, 30$ **117.** [8.6a] $\dfrac{\sqrt{6}}{3}$

118. [4.6b], [5.5d] Yes **119.** [6.1c] No **120.** [4.6c] No
121. [5.5a], [8.2a] No **122.** [8.1f] Yes

APPENDIXES

Exercise Set A, p. 736

1. $(z + 3)(z^2 - 3z + 9)$ **3.** $(x - 1)(x^2 + x + 1)$
5. $(y + 5)(y^2 - 5y + 25)$ **7.** $(2a + 1)(4a^2 - 2a + 1)$
9. $(y - 2)(y^2 + 2y + 4)$ **11.** $(2 - 3b)(4 + 6b + 9b^2)$
13. $(4y + 1)(16y^2 - 4y + 1)$ **15.** $(2x + 3)(4x^2 - 6x + 9)$
17. $(a - b)(a^2 + ab + b^2)$ **19.** $\left(a + \frac{1}{2}\right)\left(a^2 - \frac{1}{2}a + \frac{1}{4}\right)$
21. $2(y - 4)(y^2 + 4y + 16)$ **23.** $3(2a + 1)(4a^2 - 2a + 1)$
25. $r(s + 4)(s^2 - 4s + 16)$ **27.** $5(x - 2z)(x^2 + 2xz + 4z^2)$
29. $(x + 0.1)(x^2 - 0.1x + 0.01)$
31. $8(2x^2 - t^2)(4x^4 + 2x^2t^2 + t^4)$
33. $2y(y - 4)(y^2 + 4y + 16)$
35. $(z - 1)(z^2 + z + 1)(z + 1)(z^2 - z + 1)$
37. $(t^2 + 4y^2)(t^4 - 4t^2y^2 + 16y^4)$ **39.** 1; 19; 19; 7; 1
41. $(x^{2a} + y^b)(x^{4a} - x^{2a}y^b + y^{2b})$
43. $3(x^a + 2y^b)(x^{2a} - 2x^ay^b + 4y^{2b})$
45. $\frac{1}{3}\left(\frac{1}{2}xy + z\right)\left(\frac{1}{4}x^2y^2 - \frac{1}{2}xyz + z^2\right)$ **47.** $y(3x^2 + 3xy + y^2)$
49. $4(3a^2 + 4)$

Exercise Set B, p. 741

1. $y = 4x - 18$ **3.** $y = -2x + 12$ **5.** $y = 3x + 4$
7. $y = -3x - 6$ **9.** $y = 4$ **11.** $y = -\frac{4}{5}x + \frac{23}{5}$
13. $y = x + 3$ **15.** $y = x$ **17.** $y = \frac{5}{3}x - 5$
19. $y = 3x + 5$ **21.** $y = -\frac{7}{4}x$ **23.** $y = \frac{2}{11}x + \frac{91}{66}$
25. $y = 5x + 3$

Exercise Set C, p. 744

1. 5 **3.** -10 **5.** 1 **7.** Not a real number **9.** 6 **11.** 4
13. 10 **15.** -3 **17.** Not a real number **19.** -5 **21.** t
23. $-x$ **25.** 4 **27.** -7 **29.** -5 **31.** 10 **33.** 10
35. -7 **37.** Not a real number **39.** 5 **41.** $3\sqrt[3]{2}$

43. $3\sqrt[4]{4}$ **45.** $\frac{3}{4}$ **47.** $4\sqrt[4]{2}$ **49.** $2\sqrt[5]{4}$ **51.** $\frac{4}{5}$ **53.** $\dfrac{\sqrt[3]{17}}{2}$

55. $5\sqrt[3]{2}$ **57.** $3\sqrt[5]{2}$ **59.** $\dfrac{\sqrt[4]{13}}{3}$ **61.** $\dfrac{\sqrt[4]{7}}{2}$ **63.** $\frac{2}{5}$ **65.** 2
67. 10

Exercise Set D, p. 748

1. $\{3, 4, 5, 6, 7, 8\}$ **3.** $\{41, 43, 45, 47, 49\}$ **5.** $\{-3, 3\}$
7. False **9.** True **11.** True **13.** True **15.** True
17. False **19.** $\{c, d, e\}$ **21.** $\{1, 10\}$ **23.** $\{ \}$, or $\varnothing$
25. $\{a, e, i, o, u, q, c, k\}$ **27.** $\{0, 1, 7, 10, 2, 5\}$
29. $\{a, e, i, o, u, m, n, f, g, h\}$ **31.** $\{x \mid x$ is an integer$\}$
33. $\{x \mid x$ is a real number$\}$ **35.** $\{ \}$, or $\varnothing$
37. (a) A; (b) A; (c) A; (d) $\{ \}$, or $\varnothing$ **39.** True

Exercise Set E, p. 752

1. Mean: 21; median: 18.5; mode: 29
3. Mean: 5.38; median: 5.7; no mode exists
5. Mean: 239.5; median: 234; mode: 234
7. Mean: $36.\overline{5}$; median: 90; mode: 2
9. Mean: $4.19; median: $3.99; mode: $3.99

Exercise Set F, p. 756

1. $(-\infty, 5)$ **3.** $[-3, 3]$ **5.** $(-8, -4)$ **7.** $(-2, 5)$

9. $(-\sqrt{2}, \infty)$

11. $\{x \mid x > -1\}$, or $(-1, \infty)$

13. $\{x \mid x > 3\}$, or $(3, \infty)$

15. $\{x \mid x < -60\}$, or $(-\infty, -60)$

17. $\{a \mid a \leq -22\}$, or $(-\infty, -22]$

19. $\{x \mid x \leq 0.9\}$, or $(-\infty, 0.9]$ **21.** $\left\{x \mid x \leq \frac{5}{6}\right\}$, or $\left(-\infty, \frac{5}{6}\right]$

23. $\{x \mid x < 6\}$, or $(-\infty, 6)$ **25.** $\{y \mid y \leq -3\}$, or $(-\infty, -3]$

27. $\left\{y \mid y > \frac{2}{3}\right\}$, or $\left(\frac{2}{3}, \infty\right)$

Guided Solutions

CHAPTER R

Section R.1

2. Find some factorizations.

$63 = 1 \cdot 63,$
$63 = 3 \cdot 21,$
$63 = 7 \cdot 9$

The factors of 63 are 1, 3, 7, 9, 21, and 63.

7. Using a factor tree gives

The prime factorization is $2 \cdot 5 \cdot 7 \cdot 11$.

8. Find the prime factorization of each number and draw lines between common factors.

$40 = 2 \cdot 2 \cdot 2 \cdot 5$
$100 = 2 \cdot 2 \cdot 5 \cdot 5$

The GCF $= 2 \cdot 2 \cdot 5 = 20$.

16. Find the prime factorizations of each number.

$18 = 2 \cdot 3 \cdot 3$
$24 = 2 \cdot 2 \cdot 2 \cdot 3$
$30 = 2 \cdot 3 \cdot 5$

The LCM is $2 \cdot 2 \cdot 2 \cdot 3 \cdot 3 \cdot 5$, or 360.

Section R.2

4. $\dfrac{18}{45} = \dfrac{2 \cdot 3 \cdot 3}{3 \cdot 3 \cdot 5}$

$= \dfrac{3 \cdot 3}{3 \cdot 3} \cdot \dfrac{2}{5} = \dfrac{2}{5}$

7. $\dfrac{32}{56} = \dfrac{8 \cdot 4}{8 \cdot 7} = \dfrac{4}{7}$

14. $\dfrac{5}{6} + \dfrac{7}{10}$

$= \dfrac{5}{2 \cdot 3} + \dfrac{7}{2 \cdot 5}$

The LCD is $2 \cdot 3 \cdot 5$, or 30.

$= \dfrac{5}{2 \cdot 3} \cdot \dfrac{5}{5} + \dfrac{7}{2 \cdot 5} \cdot \dfrac{3}{3}$

$= \dfrac{5 \cdot 5 + 7 \cdot 3}{2 \cdot 3 \cdot 5}$

$= \dfrac{25 + 21}{2 \cdot 3 \cdot 5} = \dfrac{46}{2 \cdot 3 \cdot 5}$

$= \dfrac{2 \cdot 23}{2 \cdot 3 \cdot 5} = \dfrac{23}{15}$

24. $\dfrac{5}{4} \div \dfrac{3}{2} = \dfrac{5}{4} \cdot \dfrac{2}{3} = \dfrac{5 \cdot 2}{4 \cdot 3}$

$= \dfrac{5 \cdot 2}{2 \cdot 2 \cdot 3} = \dfrac{5}{6}$

Section R.3

1. $0.568.$ $\quad 0.568 = \dfrac{568}{1000}$

3 places

8.
$$\begin{array}{r}
{}^{1}\;{}^{1}\;{}^{1} \\
6\,9.0\;0\;0 \\
1.7\;8\;5 \\
+\;2\,1\,3.6\;7\;0 \\
\hline
2\,8\,4.4\;5\;5
\end{array}$$

Section R.4

8. $0.23\% = 0.23 \cdot \dfrac{1}{100}$

$= \dfrac{0.23}{100} \cdot \dfrac{100}{100}$

$= \dfrac{23}{10,000}$

Section R.5

13. $8 + 2 \times 5^3 - 4 \cdot 20$

$= 8 + 2 \times 125 - 4 \cdot 20$
$= 8 + 250 - 4 \cdot 20$
$= 8 + 250 - 80$
$= 258 - 80$
$= 178$

CHAPTER 1

Section 1.1

5. $A = lw$
$A = (24\,\text{ft})(8\,\text{ft})$
$\quad = (24)(8)(\text{ft})(\text{ft})$
$\quad = 192\,\text{ft}^2$, or
$\quad\quad 192$ square feet

Section 1.3

20. $-\dfrac{1}{5} + \left(-\dfrac{3}{4}\right)$

$= -\dfrac{4}{20} + \left(-\dfrac{15}{20}\right)$

$= -\dfrac{19}{20}$

32. $-x = -(-1.6) = 1.6$;
$-(-x) = -(-(-1.6))$
$= -(1.6) = -1.6$

Section 1.4

11. $2 - 8 = 2 + (-8) = -6$

19. $-12 - (-9) = -12 + 9 = -3$

Section 1.6

21. $\dfrac{4}{7} \div \left(-\dfrac{3}{5}\right) = \dfrac{4}{7} \cdot \left(-\dfrac{5}{3}\right) = -\dfrac{20}{21}$

25. $\dfrac{-5}{6} = \dfrac{5}{-6} = -\dfrac{5}{6}$

Section 1.7

3. $\dfrac{3}{4} = \dfrac{3}{4} \cdot 1 = \dfrac{3}{4} \cdot \dfrac{2}{2} = \dfrac{6}{8}$

4. $\dfrac{3}{4} = \dfrac{3}{4} \cdot 1 = \dfrac{3}{4} \cdot \dfrac{t}{t} = \dfrac{3t}{4t}$

8. $\dfrac{18p}{24pq} = \dfrac{6p \cdot 3}{6p \cdot 4q}$
$= \dfrac{6p}{6p} \cdot \dfrac{3}{4q}$
$= 1 \cdot \dfrac{3}{4q} = \dfrac{3}{4q}$

31. $-2(x - 3)$
$= -2 \cdot x - (-2) \cdot 3$
$= -2x - (-6)$
$= -2x + 6$

44. $16a - 36b + 42$
$= 2 \cdot 8a - 2 \cdot 18b + 2 \cdot 21$
$= 2(8a - 18b + 21)$

52. $3x - 7x - 11 + 8y + 4 - 13y$
$= (3 - 7)x + (8 - 13)y +$
$\quad (-11 + 4)$
$= -4x + (-5)y + (-7)$
$= -4x - 5y - 7$

Section 1.8

13. $5a - 3(7a - 6)$
$= 5a - 21a + 18$
$= -16a + 18$

18. $9 - [10 - (13 + 6)]$
$= 9 - [10 - (19)]$
$= 9 - [-9]$
$= 9 + 9$
$= 18$

25. $-4^3 + 52 \cdot 5 + 5^3 -$
$\quad (4^2 - 48 \div 4)$
$= -64 + 52 \cdot 5 + 125 -$
$\quad (16 - 48 \div 4)$
$= -64 + 52 \cdot 5 + 125 -$
$\quad (16 - 12)$
$= -64 + 52 \cdot 5 + 125 - 4$
$= -64 + 260 + 125 - 4$
$= 196 + 125 - 4$
$= 321 - 4$
$= 317$

CHAPTER 2

Section 2.1

8. $\quad x + 2 = 11$
$x + 2 + (-2) = 11 + (-2)$
$\quad x + 0 = 9$
$\quad x = 9$

Section 2.2

1. $\quad 6x = 90$
$\dfrac{1}{6} \cdot 6x = \dfrac{1}{6} \cdot 90$
$\quad 1 \cdot x = 15$
$\quad x = 15$

Check: $\quad \dfrac{6x = 90}{6x \cdot 15 \ ? \ 90}$
$\quad 90 \mid \quad$ TRUE

2. $\quad 4x = -7$
$\dfrac{4x}{4} = \dfrac{-7}{4}$
$1 \cdot x = -\dfrac{7}{4}$
$x = -\dfrac{7}{4}$

6. $\quad \dfrac{2}{3} = -\dfrac{5}{6}y$
$-\dfrac{6}{5} \cdot \dfrac{2}{3} = -\dfrac{6}{5} \cdot \left(-\dfrac{5}{6}y\right)$
$-\dfrac{12}{15} = 1 \cdot y$
$-\dfrac{4}{5} = y$

Section 2.3

4. $\quad -18 - m = -57$
$18 - 18 - m = 18 - 57$
$\quad -m = -39$
$-1(-m) = -1(-39)$
$\quad m = 39$

11. $\quad 7x - 17 + 2x = 2 - 8x + 15$
$9 \cdot x - 17 = 17 - 8x$
$8x + 9x - 17 = 17 - 8x + 8x$
$17 \cdot x - 17 = 17$
$17x - 17 + 17 = 17 + 17$
$\quad 17x = 34$
$\quad \dfrac{17x}{17} = \dfrac{34}{17}$
$\quad x = 2$

13. $\quad \dfrac{7}{8}x - \dfrac{1}{4} + \dfrac{1}{2}x = \dfrac{3}{4} + x$
$8 \cdot \left(\dfrac{7}{8}x - \dfrac{1}{4} + \dfrac{1}{2}x\right) = 8 \cdot \left(\dfrac{3}{4} + x\right)$
$8 \cdot \dfrac{7}{8}x - 8 \cdot \dfrac{1}{4} + 8 \cdot \dfrac{1}{2}x = 8 \cdot \dfrac{3}{4} + 8 \cdot x$
$7x - 2 + 4x = 6 + 8x$
$11x - 2 = 6 + 8x$
$11x - 2 - 8x = 6 + 8x - 8x$
$3x - 2 = 6$
$3x - 2 + 2 = 6 + 2$
$3x = 8$
$\dfrac{3x}{3} = \dfrac{8}{3}$
$x = \dfrac{8}{3}$

Section 2.4

12.
$$y = mx + b$$
$$y - b = mx + b - b$$
$$y - b = mx$$
$$\frac{y - b}{m} = \frac{mx}{m}$$
$$\frac{y - b}{m} = {}^*x$$

Section 2.5

4. 110% of what number is 30?

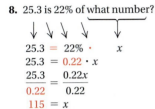

$$110\% \cdot x = 30$$

8. 25.3 is 22% of what number?

$$25.3 = 22\% \cdot x$$
$$25.3 = 0.22 \cdot x$$
$$\frac{25.3}{0.22} = \frac{0.22x}{0.22}$$
$$115 = x$$

Section 2.6

3. Let $x =$ the first marker and $x + 1 =$ the second marker.
Translate and *Solve*:

First marker + Second marker = 627

$$x + (x + 1) = 627$$
$$2x + 1 = 627$$
$$2x + 1 - 1 = 627 - 1$$
$$2x = 626$$
$$\frac{2x}{2} = \frac{626}{2}$$
$$x = 313$$

If $x = 313$, then $x + 1 = 314$. The mile markers are 313 and 314.

8. Let $x =$ the principal. Then the interest earned is $5\%x$.
Translate and *Solve*:

Principal + Interest = Amount

$$x + 5\%x = 2520$$
$$x + 0.05x = 2520$$
$$(1 + 0.05)x = 2520$$
$$1.05x = 2520$$
$$\frac{1.05x}{1.05} = \frac{2520}{1.05}$$
$$x = 2400$$

Section 2.7

10.
$$5y + 2 \le -1 + 4y$$
$$5y + 2 - 4y \le -1 + 4y - 4y$$
$$y + 2 \le -1$$
$$y + 2 - 2 \le -1 - 2$$
$$y \le -3$$
The solution set is $\{y \mid y \le -3\}$.

18.
$$3(7 + 2x) \le 30 + 7(x - 1)$$
$$21 + 6x \le 30 + 7x - 7$$
$$21 + 6x \le 23 + 7x$$
$$21 + 6x - 6x \le 23 + 7x - 6x$$
$$21 \le 23 + x$$
$$21 - 23 \le 23 + x - 23$$
$$-2 \le x, \text{ or}$$
$$x \ge -2$$
The solution set is $\{x \mid x \ge -2\}$.

Section 2.8

9. *Translate* and *Solve:*
$$F < 88$$
$$\frac{9}{5}C + 32 < 88$$
$$\frac{9}{5}C + 32 - 32 < 88 - 32$$
$$\frac{9}{5}C < 56$$
$$\frac{5}{9} \cdot \frac{9}{5}C < \frac{5}{9} \cdot 56$$
$$C < \frac{280}{9}$$
$$C < 31\frac{1}{9}$$

Butter stays solid at Celsius temperatures less than $31\frac{1}{9}^\circ$—that is, $\left\{ C \mid C < 31\frac{1}{9}^\circ \right\}$.

CHAPTER 3

Section 3.1

18. Determine whether $(2, -4)$ is a solution of $4q - 3p = 22$.

$$4q - 3p = 22$$
$$4 \cdot (-4) - 3 \cdot 2 \ ? \ 22$$
$$-16 - 6$$
$$-22 \ \Big| \quad \text{FALSE}$$

Thus, $(2, -4)$ is not a solution.

21. Complete the table and graph $y = -2x$.

x	y	(x, y)
-3	6	$(-3, 6)$
-1	2	$(-1, 2)$
0	0	$(0, 0)$
1	-2	$(1, -2)$
3	-6	$(3, -6)$

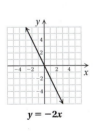

$$y = -2x$$

29. Graph $5y - 3x = -10$ and identify the y-intercept.

x	y
0	-2
5	1
-5	-5

$\leftarrow$ y-intercept: $(0, -2)$

$$5y - 3x = -10$$

Section 3.2

2. For $2x + 3y = 6$, find the intercepts. Then graph the equation using the intercepts.

x	y
3	0
0	2
−3	4

⟵ x-intercept: $(3, 0)$
⟵ y-intercept: $(0, 2)$
⟵ Check point: $(-3, 4)$

6. Graph: $x = 5$.

x	y
5	−4
5	0
5	3

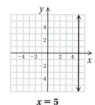

$x = 5$

7. Graph: $y = -2$.

x	y
−1	−2
0	−2
2	−2

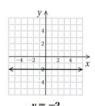

$y = -2$

Section 3.3

1. Graph the line that contains $(-2, 3)$ and $(3, 5)$ and find the slope in two different ways.

$$\frac{5-3}{3-(-2)} = \frac{2}{5}, \quad \text{or}$$

$$\frac{3-5}{-2-3} = \frac{-2}{-5} = \frac{2}{5}$$

8. Find the slope of the line $5x - 4y = 8$.

$$5x = 4y + 8$$
$$5x - 8 = 4y$$
$$\frac{5x - 8}{4} = \frac{4y}{4}$$
$$\frac{5}{4} \cdot x - 2 = y, \quad \text{or}$$
$$y = \frac{5}{4} \cdot x - 2$$

Slope is $\frac{5}{4}$.

Section 3.4

11. Find the equation that contains the point $(3, 5)$ and has slope $m = 6$.

$$y = mx + b$$
$$y = 6x + b$$
$$5 = 6 \cdot 3 + b$$
$$5 = 18 + b$$
$$-13 = b$$

Thus, $y = 6x - 13$.

14. Find the equation of the line that contains $(-1, 2)$ and $(-3, -2)$.
First, determine the slope:

$$m = \frac{-2 - 2}{-3 - (-1)} = \frac{-4}{-2} = 2;$$
$$y = mx + b,$$
$$y = 2x + b.$$

Use either point to determine b. Let's use $(-3, -2)$:

$$-2 = 2 \cdot -3 + b$$
$$-2 = -6 + b$$
$$4 = b.$$

Thus, $y = 2x + 4$.

Section 3.6

1. Determine whether the graphs of
$$3x - y = -5,$$
$$y - 3x = -2$$
are parallel.
Solve each equation for y and then find the slope.

$$3x - y = -5$$
$$-y = -3x - 5$$
$$y = 3x + 5$$

The slope is 3.

$$y - 3x = -2$$
$$y = 3x - 2$$

The slope is 3.
The slope of each line is 3. The y-intercepts, $(0, 5)$ and $(0, -2)$, are different. Thus the lines are parallel.

3. Determine whether the graphs of
$$y = -\frac{3}{4}x + 7,$$
$$y = \frac{4}{3}x - 9$$
are perpendicular.

The slopes of the lines are $-\frac{3}{4}$ and $\frac{4}{3}$.

The product of the slopes is $-\frac{3}{4} \cdot \frac{4}{3} = -1$.

The lines are perpendicular.

Section 3.7

4. Graph: $2x + 4y < 8$.
Related equation: $2x + 4y = 8$
x-intercept: $(4, 0)$
y-intercept: $(0, 2)$
Draw the line dashed.
Test a point—try $(3, -1)$:

$$2 \cdot 3 + 4 \cdot (-1) < 8$$
$$2 < 8 \quad \text{TRUE}$$

The point $(3, -1)$ is a solution. We shade the half-plane that contains $(3, -1)$.

$2x + 4y < 8$

CHAPTER 4

Section 4.1

18. a) $(4t)^2 = [4 \cdot (-3)]^2$
$$= [-12]^2$$
$$= 144$$

b) $4t^2 = 4 \cdot (-3)^2$
$$= 4 \cdot (9)$$
$$= 36$$

c) Since $144 \neq 36$, the expressions are not equivalent.

33. $4p^{-3} = 4\left(\dfrac{1}{p^3}\right) = \dfrac{4}{p^3}$

Section 4.2

10. $(-2x^4)^{-2} = (-2)^{-2}(x^4)^{-2}$
$$= \dfrac{1}{(-2)^2} \cdot x^{-8}$$
$$= \dfrac{1}{4} \cdot \dfrac{1}{x^8}$$
$$= \dfrac{1}{4x^8}$$

15. $\left(\dfrac{x^4}{3}\right)^{-2} = \dfrac{(x^4)^{-2}}{3^{-2}} = \dfrac{x^{-8}}{3^{-2}}$
$$= \dfrac{\dfrac{1}{x^8}}{\dfrac{1}{3^2}} = \dfrac{1}{x^8} \div \dfrac{1}{3^2}$$
$$= \dfrac{1}{x^8} \cdot \dfrac{3^2}{1} = \dfrac{9}{x^8}$$

This can be done a second way.

$\left(\dfrac{x^4}{3}\right)^{-2} = \left(\dfrac{3}{x^4}\right)^2$
$$= \dfrac{3^2}{(x^4)^2} = \dfrac{9}{x^8}$$

Section 4.3

5. $2x^2 + 5x - 4$
$$= 2(-5)^2 + 5(-5) - 4$$
$$= 2(25) + (-25) - 4$$
$$= 50 - 25 - 4$$
$$= 21$$

29. $-2x^4 + 16 + 2x^4 + 9 - 3x^5$
$$= -3x^5 + (-2 + 2)x^4 + (16 + 9)$$
$$= -3x^5 + 0x^4 + 25$$
$$= -3x^5 + 25$$

Section 4.4

14. $(-6x^4 + 3x^2 + 6) - (2x^4 + 5x^3 - 5x^2 + 7)$
$$= -6x^4 + 3x^2 + 6 - 2x^4 - 5x^3 + 5x^2 - 7$$
$$= -8x^4 - 5x^3 + 8x^2 - 1$$

Section 4.5

12. a) $(y + 2)(y + 7)$
$$= y \cdot (y + 7) + 2 \cdot (y + 7)$$
$$= y \cdot y + y \cdot 7 + 2 \cdot y + 2 \cdot 7$$
$$= y^2 + 7y + 2y + 14$$
$$= y^2 + 9y + 14$$

b) The area is $(y + 7)(y + 2)$, or, from part (a), $y^2 + 9y + 14$.

18. $(3y^2 - 7)(2y^3 - 2y + 5)$
$$= 3y^2(2y^3 - 2y + 5) - 7(2y^3 - 2y + 5)$$
$$= 6y^5 - 6y^3 + 15y^2 - 14y^3 + 14y - 35$$
$$= 6y^5 - 20y^3 + 15y^2 + 14y - 35$$

Section 4.6

17. $(6 - 4y)(6 + 4y)$
$$= (6)^2 - (4y)^2$$
$$= 36 - 16y^2$$

27. $(3x^2 - 5)(3x^2 - 5)$
$$= (3x^2)^2 - 2(3x^2)(5) + 5^2$$
$$= 9x^4 - 30x^2 + 25$$

Section 4.7

7. The like terms are $-3pt$ and $8pt$, $-5ptr^3$ and $5ptr^3$, and -12 and 4. Collecting like terms, we have
$$(-3 + 8)pt + (-5 + 5)ptr^3 + (-12 + 4)$$
$$= 5pt - 8.$$

22. $(2a + 5b + c)(2a - 5b - c)$
$$= [2a + (5b + c)][2a - (5b + c)]$$
$$= (2a)^2 - (5b + c)^2$$
$$= 4a^2 - (25b^2 + 10bc + c^2)$$
$$= 4a^2 - 25b^2 - 10bc - c^2$$

Section 4.8

5. $(28x^7 + 32x^5) \div (4x^3)$
$$= \dfrac{28x^7 + 32x^5}{4x^3} = \dfrac{28x^7}{4x^3} + \dfrac{32x^5}{4x^3}$$
$$= \dfrac{28}{4}x^{7-3} + \dfrac{32}{4}x^{5-3}$$
$$= 7x^4 + 8x^2$$

10. $(x^2 + x - 6) \div (x + 3)$

$$\require{enclose}\begin{array}{r} x - 2 \\ x + 3 \enclose{longdiv}{x^2 + x - 6} \\ \underline{x + 3x} \\ -2x - 6 \\ \underline{-2x - 6} \\ 0 \end{array}$$

CHAPTER 5

Section 5.1

7. Find the GCF of $-24m^5n^6$, $12mn^3$, $-16m^2n^2$, and $8m^4n^4$.
The coefficients are $-24, 12, -16$, and 8.
The greatest positive common factor of the coefficients is 4.
The smallest exponent of the variable m is 1.
The smallest exponent of the variable n is 2.
The GCF is $4mn^2$.

19. $x^3 + 7x^2 + 3x + 21$
$$= x^2(x + 7) + 3(x + 7) = (x + 7)(x^2 + 3)$$

Section 5.2

1. Factor: $x^2 + 7x + 12$.
Complete the following table.

Pairs of Factors	Sums of Factors
1, 12	13
−1, −12	−13
2, 6	8
−2, −6	−8
3, 4	7
−3, 4	−7

Because both 7 and 12 are positive, we need consider only the positive factors in the table above.
$$x^2 + 7x + 12 = (x + 3)(x + 4)$$
8. Factor $a^2 - 40 + 3a$.

First, rewrite in descending order:
$$a^2 + 3a - 40.$$

Pairs of Factors	Sums of Factors
$-1, 40$	39
$-2, 30$	28
$-4, 10$	6
$-5, \ 8$	3

The factorization is $(a - 5)(a + 8)$.

Section 5.4

2. Factor $12x^2 - 17x - 5$.
  1) There is no common factor.
  2) Multiply the leading coefficient and the constant:
    $12(-5) = -60$.
  3) Look for a pair of factors of -60 whose sum is -17. Those factors are 3 and -20.
  4) Split the middle term: $-17x = 3x - 20x$.
  5) Factor by grouping:
    $12x^2 + 3x - 20x - 5$
    $= 3x(4x + 1) - 5(4x + 1)$
    $= (4x + 1)(3x - 5)$.
  6) Check: $(4x + 1)(3x - 5) = 12x^2 - 17x - 5$.

Section 5.5

13. Factor $49 - 56y + 16y^2$.
   Write in descending order:
     $16y^2 - 56y + 49$.
   Factor as a trinomial square:
     $(4y)^2 - 2 \cdot 4y \cdot 7 + (7)^2$
     $= (4y - 7)^2$.

27. $a^2 - 25b^2$
   $= a^2 - (5b)^2$
   $= (a + 5b)(a - 5b)$.

Section 5.6

5. Factor $8x^3 - 200x$.
  a) Factor out the largest common factor:
    $8x^3 - 200x = 8x(x^2 - 25)$.
  b) There are two terms inside the parentheses. Factor the difference of squares:
    $8x(x^2 - 25) = 8x(x + 5)(x - 5)$.
  c) We have factored completely.
  d) Check: $8x(x + 5)(x - 5) = 8x(x^2 - 25) = 8x^3 - 200x$.

10. Factor $x^4 + 2x^2y^2 + y^4$.
  a) There is no common factor.
  b) There are three terms. Factor the trinomial square:
    $x^4 + 2x^2y^2 + y^4 = (x^2 + y^2)^2$.
  c) We have factored completely.
  d) Check: $(x^2 + y^2)^2 = (x^2)^2 + 2(x^2)(y^2) + (y^2)^2$
       $= x^4 + 2x^2y^2 + y^4$.

Section 5.7

5. $x^2 - x - 6 = 0$
   $(x + 2)(x - 3) = 0$
   $x + 2 = 0 \quad or \quad x - 3 = 0$
   $x = -2 \quad or \quad\quad x = 3$
   Both numbers check.
   The solutions are -2 and 3.

8. $x^2 - 4x = 0$
   $x(x - 4) = 0$
   $x = 0 \quad or \quad x - 4 = 0$
   $x = 0 \quad or \quad\quad x = 4$
   Both numbers check.
   The solutions are 0 and 4.

CHAPTER 6

Section 6.1

2. $\dfrac{2x - 7}{x^2 + 5x - 24}$
   $x^2 + 5x - 24 = 0$
   $(x + 8)(x - 3) = 0$
   $x + 8 = 0 \quad or \quad x - 3 = 0$
   $x = -8 \quad or \quad\quad x = 3$
   The rational expression is not defined for replacements -8 and 3.

13. $\dfrac{x - 8}{8 - x}$
   $= \dfrac{x - 8}{-(x - 8)}$
   $= \dfrac{1(x - 8)}{-1(x - 8)} = \dfrac{1}{-1} \cdot \dfrac{x - 8}{x - 8}$
   $= -1 \cdot 1 = -1$

16. $\dfrac{a^2 - 4a + 4}{a^2 - 9} \cdot \dfrac{a + 3}{a - 2}$
   $= \dfrac{(a^2 - 4a + 4)(a + 3)}{(a^2 - 9)(a - 2)}$
   $= \dfrac{(a - 2)(a - 2)(a + 3)}{(a + 3)(a - 3)(a - 2)}$
   $= \dfrac{(a - 2)(a - 2)(a + 3)}{(a + 3)(a - 3)(a - 2)}$
   $= \dfrac{a - 2}{a - 3}$

Section 6.2

6. $\dfrac{x}{8} \div \dfrac{x}{5} = \dfrac{x}{8} \cdot \dfrac{5}{x}$
   $= \dfrac{x \cdot 5}{8 \cdot x}$
   $= \dfrac{x \cdot 5}{8 \cdot x}$
   $= \dfrac{5}{8}$

11. $\dfrac{y^2 - 1}{y + 1} \div \dfrac{y^2 - 2y + 1}{y + 1}$

$= \dfrac{y^2 - 1}{y + 1} \cdot \dfrac{y + 1}{y^2 - 2y + 1}$

$= \dfrac{(y^2 - 1)(y + 1)}{(y + 1)(y^2 - 2y + 1)}$

$= \dfrac{(y + 1)(y - 1)(y + 1)}{(y + 1)(y - 1)(y - 1)}$

$= \dfrac{\cancel{(y + 1)}\cancel{(y - 1)}(y + 1)}{\cancel{(y + 1)}\cancel{(y - 1)}(y - 1)}$

$= \dfrac{y + 1}{y - 1}$

Section 6.3

1. $16 = 2 \cdot 2 \cdot 2 \cdot 2$

$18 = 2 \cdot 3 \cdot 3$

$\quad\quad$ LCM $= 2 \cdot 2 \cdot 2 \cdot 2 \cdot 3 \cdot 3$,

$\quad\quad\quad\quad$ or 144

5. $\dfrac{3}{16} + \dfrac{1}{18}$

$= \dfrac{3}{2 \cdot 2 \cdot 2 \cdot 2} + \dfrac{1}{2 \cdot 3 \cdot 3}$

$= \dfrac{3}{2 \cdot 2 \cdot 2 \cdot 2} \cdot \dfrac{3 \cdot 3}{3 \cdot 3} + \dfrac{1}{2 \cdot 3 \cdot 3} \cdot \dfrac{2 \cdot 2 \cdot 2}{2 \cdot 2 \cdot 2}$

$= \dfrac{27 + 8}{2 \cdot 2 \cdot 2 \cdot 2 \cdot 3 \cdot 3}$

$= \dfrac{35}{144}$

Section 6.4

5. $\dfrac{3}{16x} + \dfrac{5}{24x^2}$

$\quad 16x = 2 \cdot 2 \cdot 2 \cdot 2 \cdot x$

$\quad 24x^2 = 2 \cdot 2 \cdot 2 \cdot 3 \cdot x \cdot x$

$\quad\quad$ LCD $= 2 \cdot 2 \cdot 2 \cdot 2 \cdot 3 \cdot x \cdot x$,

$\quad\quad\quad\quad$ or $48x^2$

$\dfrac{3}{16x} \cdot \dfrac{3x}{3x} + \dfrac{5}{24x^2} \cdot \dfrac{2}{2}$

$= \dfrac{9x}{48x^2} + \dfrac{10}{48x^2}$

$= \dfrac{9x + 10}{48x^2}$

10. $\dfrac{2x + 1}{x - 3} + \dfrac{x + 2}{3 - x}$

$= \dfrac{2x + 1}{x - 3} + \dfrac{x + 2}{3 - x} \cdot \dfrac{-1}{-1}$

$= \dfrac{2x + 1}{x - 3} + \dfrac{-x - 2}{x - 3}$

$= \dfrac{(2x + 1) + (-x - 2)}{x - 3}$

$= \dfrac{x - 1}{x - 3}$

Section 6.5

4. $\dfrac{x - 2}{3x} - \dfrac{2x - 1}{5x}$

$\quad$ LCD $= 3 \cdot x \cdot 5 = 15x$

$= \dfrac{x - 2}{3x} \cdot \dfrac{5}{5} - \dfrac{2x - 1}{5x} \cdot \dfrac{3}{3}$

$= \dfrac{5x - 10}{15x} - \dfrac{6x - 3}{15x}$

$= \dfrac{5x - 10 - (6x - 3)}{15x}$

$= \dfrac{5x - 10 - 6x + 3}{15x}$

$= \dfrac{-x - 7}{15x}$

8. $\dfrac{y}{16 - y^2} - \dfrac{7}{y - 4}$

$= \dfrac{y}{16 - y^2} \cdot \dfrac{-1}{-1} - \dfrac{7}{y - 4}$

$= \dfrac{-y}{y^2 - 16} - \dfrac{7}{y - 4}$

$= \dfrac{-y}{(y + 4)(y - 4)} - \dfrac{7}{y - 4} \cdot \dfrac{y + 4}{y + 4}$

$= \dfrac{-y}{(y + 4)(y - 4)} - \dfrac{7y + 28}{(y + 4)(y - 4)}$

$= \dfrac{-y - (7y + 28)}{(y + 4)(y - 4)}$

$= \dfrac{-y - 7y - 28}{(y + 4)(y - 4)}$

$= \dfrac{-8y - 28}{(y + 4)(y - 4)} = \dfrac{-4(2y + 7)}{(y + 4)(y - 4)}$

Section 6.6

2. $\dfrac{\dfrac{x}{2} + \dfrac{2x}{3}}{\dfrac{1}{x} - \dfrac{x}{2}}$

LCM of denominators $= 6x$

$= \dfrac{\dfrac{x}{2} + \dfrac{2x}{3}}{\dfrac{1}{x} - \dfrac{x}{2}} \cdot \dfrac{6x}{6x}$

$= \dfrac{\left(\dfrac{x}{2} + \dfrac{2x}{3}\right) \cdot 6x}{\left(\dfrac{1}{x} - \dfrac{x}{2}\right) \cdot 6x}$

$= \dfrac{3x^2 + 4x^2}{6 - 3x^2} = \dfrac{7x^2}{3(2 - x^2)}$

5.
$$\dfrac{\dfrac{x}{2}+\dfrac{2x}{3}}{\dfrac{1}{x}-\dfrac{x}{2}} \quad \begin{array}{l}\leftarrow \text{LCD}=6\\[4pt]\leftarrow \text{LCD}=2x\end{array}$$

$$=\dfrac{\dfrac{x}{2}\cdot\dfrac{3}{3}+\dfrac{2x}{3}\cdot\dfrac{2}{2}}{\dfrac{1}{x}\cdot\dfrac{2}{2}-\dfrac{x}{2}\cdot\dfrac{x}{x}}$$

$$=\dfrac{\dfrac{3x}{6}+\dfrac{4x}{6}}{\dfrac{2}{2x}-\dfrac{x^2}{2x}}=\dfrac{\dfrac{3x+4x}{6}}{\dfrac{2-x^2}{2x}}$$

$$=\dfrac{7x}{6}\cdot\dfrac{2x}{2-x^2}=\dfrac{7\cdot 2\cdot x\cdot x}{2\cdot 3(2-x^2)}$$

$$=\dfrac{7x^2}{3(2-x^2)}$$

Section 6.7

4. $\dfrac{1}{2x}+\dfrac{1}{x}=-12$

LCM $=2x$

$$2x\left(\dfrac{1}{2x}+\dfrac{1}{x}\right)=2x(-12)$$

$$2x\cdot\dfrac{1}{2x}+2x\cdot\dfrac{1}{x}=2x(-12)$$

$$1+2=-24\cdot x$$

$$3=-24x$$

$$\dfrac{3}{-24}=x$$

$$-\dfrac{1}{8}=x$$

7. $\dfrac{4}{x-2}+\dfrac{1}{x+2}=\dfrac{26}{x^2-4}$

LCM $=(x-2)(x+2)$

$$(x-2)(x+2)\left(\dfrac{4}{x-2}+\dfrac{1}{x+2}\right)=(x-2)(x+2)\cdot\dfrac{26}{x^2-4}$$

$$4(x+2)+1(x-2)=26$$

$$4x+8+x-2=26$$

$$5x+6=26$$

$$5x=20$$

$$x=4$$

Section 6.9

1. $y=kx$

$84=k\cdot 12$

$k=\dfrac{84}{12}=7$

$y=7\cdot x$

Then find y when x is 41.

$y=7x=7\cdot 41$

$\quad=287$

5. $y=\dfrac{k}{x}$

$105=\dfrac{k}{0.6}$

$k=0.6\cdot 105$

$k=63$

$y=\dfrac{63}{x}$

Then find y when x is 20.

$y=\dfrac{63}{x}=\dfrac{63}{20}$

$\quad=3.15$

CHAPTER 7

Section 7.1

2. $(20,40);\ a=\dfrac{1}{2}b,$

$\qquad b-a=60$

Check: $\quad a=\dfrac{1}{2}b \qquad\qquad\qquad b-a=60$

$$\begin{array}{c} \\ \hline 20\ ?\ \dfrac{1}{2}(40)\\[4pt] \Big|\ \ 20 \end{array} \qquad\qquad \begin{array}{c}40-20\ ?\ 60\\[4pt]20\ \Big|\end{array}$$

$(20,40)$ is a solution of $a=\dfrac{1}{2}b$.

$(20,40)$ is not a solution of $b-a=60$.

Therefore, $(20,40)$ is not a solution of the system.

Section 7.2

1. $x+y=5,\qquad$ **(1)**

$\quad x=y+1\qquad$ **(2)**

Substitute $y+1$ for x in equation (1) and solve for y.

$$x+y=5$$
$$(y+1)+y=5$$
$$2y+1=5$$
$$2y=4$$
$$y=2$$

Substitute 2 for y in equation (2) and solve for x.

$$x=y+1$$
$$\ =2+1$$
$$\ =3$$

The numbers check. The solution is $(3,2)$.

3. $x-2y=8,\qquad$ **(1)**

$\quad 2x+y=8\qquad$ **(2)**

Solve for y in equation (2).

$$2x+y=8$$
$$y=8-2x\qquad \textbf{(3)}$$

Substitute $8-2x$ for y in equation (1) and solve for x.

$$x-2y=8$$
$$x-2(8-2x)=8$$
$$x-16+4x=8$$
$$5x-16=8$$
$$5x=24$$
$$x=\dfrac{24}{5}$$

Substitute $\dfrac{24}{5}$ for x in equation (3) and solve for y.

$$y=8-2x$$
$$\ =8-2\left(\dfrac{24}{5}\right)$$
$$\ =\dfrac{40}{5}-\dfrac{48}{5}$$
$$\ =-\dfrac{8}{5}$$

The numbers check. The solution is $\left(\dfrac{24}{5},-\dfrac{8}{5}\right)$.

Section 7.3

6. $3x - 8y = 2,$ **(1)**
    $5x + 2y = -12$ **(2)**
Multiply equation (2) by 4, add, and solve for x.

$$\begin{array}{rcl} 3x - 8y &=& 2 \\ \underline{20x + 8y} &=& \underline{-48} \\ 23x &=& -46 \\ x &=& -2 \end{array}$$

Substitute -2 for x in equation (1) and solve for y.

$$\begin{array}{rcl} 3x - 8y &=& 2 \\ 3(-2) - 8y &=& 2 \\ -6 - 8y &=& 2 \\ -8y &=& 8 \\ y &=& -1 \end{array}$$

The numbers check. The solution is $(-2, -1)$.

9. $5x - 2y = 3,$ **(1)**
   $-15x + 6y = -9$ **(2)**
Multiply equation (1) by 3 and add.

$$\begin{array}{rcl} 15x - 6y &=& 9 \\ \underline{-15x + 6y} &=& \underline{-9} \\ 0 &=& 0 \end{array}$$

The system has an infinite number of solutions.

Section 7.4

2.

$3.10	$1.75	
a	c	166
3.10a	1.75c	$459.25

$a + c = 166,$
$3.10a + 1.75c = 459.25$

3.

x	y	30
50%	70%	55%
50%x	70%y	55% $\times$ 30, or 16.5

$x + y = 30$
$50\%x + 70\%y = 16.5$

4.

$1.40	$1.75	$1.54
x	y	50
1.40x	1.75y	1.54 $\cdot$ 50, or 77

$x + y = 50,$
$1.40x + 1.75y = 77$

Section 7.5

1. Familiarize. Let $t =$ the time for the first car. Then $t - 1 =$ the time for the second car.

2. Translate.

$$d = 56t,$$
$$d = 84(t - 1)$$

3. Solve.

$$\begin{array}{rcl} 84(t - 1) &=& 56t \\ 84t - 84 &=& 56t \\ -84 &=& -28t \\ 3 &=& t \end{array}$$

If $t = 3$ hr, then $d = 56(3) = 168$.

4. Check. The first car travels 168 km in 3 hr, and the second car travels 168 km in 2 hr.

5. State. The second car will catch up with the first car in 168 km.

CHAPTER 8

Section 8.1

31. $\sqrt{x^2 + 8x + 16}$
   $= \sqrt{(x + 4)^2}$
   $= |x + 4|$

37. $\sqrt{x^2 + 8x + 16}$
   $= \sqrt{(x + 4)^2}$
   $= x + 4$

Section 8.2

9. $\sqrt{363q}$
   $= \sqrt{121 \cdot 3 \cdot q}$
   $= \sqrt{121}\sqrt{3q}$
   $= 11\sqrt{3q}$

20. $\sqrt{24x^{11}}$
   $= \sqrt{4 \cdot 6 \cdot x^{10} \cdot x}$
   $= \sqrt{4}\sqrt{x^{10}}\sqrt{6x}$
   $= 2x^5\sqrt{6x}$

Section 8.3

11. $\dfrac{\sqrt{98y}}{\sqrt{2y^{11}}}$

$= \sqrt{\dfrac{98y}{2y^{11}}}$

$= \sqrt{\dfrac{49}{y^{10}}}$

$= \dfrac{\sqrt{49}}{\sqrt{y^{10}}}$

$= \dfrac{7}{y^5}$

15. $\dfrac{10}{\sqrt{3}}$

$= \dfrac{10}{\sqrt{3}} \cdot \dfrac{\sqrt{3}}{\sqrt{3}}$

$= \dfrac{10\sqrt{3}}{\sqrt{9}}$

$= \dfrac{10\sqrt{3}}{3}$

14. $\sqrt{\dfrac{5}{8}}$

$= \sqrt{\dfrac{5}{8} \cdot \dfrac{2}{2}}$

$= \sqrt{\dfrac{10}{16}} = \dfrac{\sqrt{10}}{\sqrt{16}}$

$= \dfrac{\sqrt{10}}{4}$

Section 8.4

4. $\sqrt{24} + \sqrt{54}$
   $= \sqrt{4 \cdot 6} + \sqrt{9 \cdot 6}$
   $= 2\sqrt{6} + 3\sqrt{6}$
   $= (2 + 3)\sqrt{6}$
   $= 5\sqrt{6}$

16. $\dfrac{3}{7 + \sqrt{5}}$

$= \dfrac{3}{7 + \sqrt{5}} \cdot \dfrac{7 - \sqrt{5}}{7 - \sqrt{5}}$

$= \dfrac{3(7 - \sqrt{5})}{7^2 - (\sqrt{5})^2}$

$= \dfrac{21 - 3\sqrt{5}}{49 - 5}$

$= \dfrac{21 - 3\sqrt{5}}{44}$

Section 8.5

2. $\sqrt{3x + 1} = \sqrt{2x + 3}$
$(\sqrt{3x + 1})^2 = (\sqrt{2x + 3})^2$
$3x + 1 = 2x + 3$
$x = 2$

6. $\sqrt{x} - 1 = \sqrt{x - 3}$
$(\sqrt{x} - 1)^2 = (\sqrt{x - 3})^2$
$(\sqrt{x})^2 - 2\sqrt{x} + 1 = x - 3$
$x - 2\sqrt{x} + 1 = x - 3$
$-2\sqrt{x} + 1 = -3$
$-2\sqrt{x} = -4$
$\sqrt{x} = 2$
$(\sqrt{x})^2 = 2^2$
$x = 4$

Section 8.6

2. $a^2 + 11^2 = 14^2$
$a^2 + 121 = 196$
$a^2 = 75$
$a = \sqrt{75}$
$a \approx 8.660$

CHAPTER 9

Section 9.1

8. $2x^2 + 8x = 0$
$2x(x + 4) = 0$
$2x = 0 \quad or \quad x + 4 = 0$
$x = 0 \quad or \qquad x = -4$
Both numbers check. The solutions are 0 and -4.

12.
$$\frac{20}{x + 5} - \frac{1}{x - 4} = 1$$

$$(x + 5)(x - 4) \cdot \left(\frac{20}{x + 5} - \frac{1}{x - 4}\right) = (x + 5)(x - 4) \cdot 1$$

$$(x + 5)(x - 4) \cdot \frac{20}{x + 5} - (x + 5)(x - 4) \cdot \frac{1}{x - 4} = (x + 5)(x - 4)$$

$$20(x - 4) - 1(x + 5) = (x + 5)(x - 4)$$
$$20x - 80 - x - 5 = x^2 + x - 20$$
$$19x - 85 = x^2 + x - 20$$
$$0 = x^2 - 18x + 65$$
$$0 = (x - 5)(x - 13)$$
$$x - 5 = 0 \quad or \quad x - 13 = 0$$
$$x = 5 \quad or \quad x = 13$$

Both numbers check in the original equation. The solutions are 5 and 13.

Section 9.2

3. $2x^2 - 3 = 0$
$2x^2 = 3$
$x^2 = \dfrac{3}{2}$

$x = \sqrt{\dfrac{3}{2}} \quad or \quad x = -\sqrt{\dfrac{3}{2}}$

$x = \sqrt{\dfrac{3}{2} \cdot \dfrac{2}{2}} \quad or \quad x = -\sqrt{\dfrac{3}{2} \cdot \dfrac{2}{2}}$

$x = \dfrac{\sqrt{6}}{2} \quad or \quad x = -\dfrac{\sqrt{6}}{2}$

Both numbers check. The solutions are $\dfrac{\sqrt{6}}{2}$ and $-\dfrac{\sqrt{6}}{2}$.

10. $x^2 - 12x + 23 = 0$
$x^2 - 12x \qquad = -23$
$x^2 - 12x + 36 = -23 + 36$
$(x - 6)^2 = 13$
$x - 6 = \sqrt{13} \quad or \quad x - 6 = -\sqrt{13}$
$x = 6 + \sqrt{13} \quad or \qquad x = 6 - \sqrt{13}$
The solutions are $6 \pm \sqrt{13}$.

Section 9.3

1. Write in standard form:
$2x^2 + 7x - 4 = 0$.
$a = 2, b = 7, c = -4$
$$x = \frac{-b \pm \sqrt{b^2 - 4ac}}{2a}$$
$$x = \frac{-7 \pm \sqrt{7^2 - 4 \cdot 2 \cdot (-4)}}{2 \cdot 2}$$
$$x = \frac{-7 \pm \sqrt{49 + 32}}{4}$$
$$x = \frac{-7 \pm \sqrt{81}}{4} = \frac{-7 \pm 9}{4}$$
$$x = \frac{-7 + 9}{4} \quad or \quad x = \frac{-7 - 9}{4}$$
$$x = \frac{1}{2} \qquad or \quad x = -4$$

3. Write in standard form:
$x^2 + 4x - 7 = 0$.
$a = 1, b = 4, c = -7$
$$x = \frac{-b \pm \sqrt{b^2 - 4ac}}{2a}$$
$$= \frac{-4 \pm \sqrt{4^2 - 4 \cdot 1 \cdot (-7)}}{2 \cdot 1}$$
$$= \frac{-4 \pm \sqrt{16 + 28}}{2} = \frac{-4 \pm \sqrt{44}}{2}$$
$$= \frac{-4 \pm 2\sqrt{11}}{2} = \frac{2(-2 \pm \sqrt{11})}{2 \cdot 1}$$
$$= -2 \pm \sqrt{11}$$

Section 9.4

3. Solve for f: $\dfrac{1}{p} + \dfrac{1}{q} = \dfrac{1}{f}$.

The LCM is pqf.
$$pqf\left(\frac{1}{p} + \frac{1}{q}\right) = pqf\left(\frac{1}{f}\right)$$
$$pqf \cdot \frac{1}{p} + pqf\left(\frac{1}{q}\right) = pqf\left(\frac{1}{f}\right)$$
$$qf + pf = pq$$
$$f(q + p) = pq$$
$$f = \frac{pq}{q + p}$$

7. Solve for r: $A = \pi r^2$.
$$\frac{A}{\pi} = r^2$$
$$\sqrt{\frac{A}{\pi}} = r$$

Section 9.5

1. 1. Familiarize. Let $w = $ the width of the mural. Then the length is $2w + 5$.

2. Translate.

$(2w + 5)(w) = 52$

3. Solve.

$$2w^2 + 5w = 52$$
$$2w^2 + 5w - 52 = 0$$
$$(2w + 13)(w - 4) = 0$$
$$2w + 13 = 0 \quad or \quad w - 4 = 0$$
$$w = -\tfrac{13}{2} \quad or \quad w = 4$$

4. Check. Only 4 checks. When the width is 4 ft, the length is $2(4) + 5 = 13$ ft.

5. State. The width is 4 ft, and the length is 13 ft.

3. 1. Familiarize.

	d	r	t
Upstream	45	$12 - s$	t_1
Downstream	45	$12 + s$	t_2
Total time			8

2. Translate.

$$t_1 = \frac{45}{12 - s}, \; t_2 = \frac{45}{12 + s}$$

$$\frac{45}{12 - s} + \frac{45}{12 + s} = 8$$

3. Solve.

$$(12 - s)(12 + s)\left(\frac{45}{12 - s} + \frac{45}{12 + s}\right) = (12 - s)(12 + s)(8)$$
$$45(12 + s) + 45(12 - s) = (144 - s^2)(8)$$
$$1080 = 1152 - 8s^2$$
$$8s^2 - 72 = 0$$
$$s^2 - 9 = 0$$
$$s + 3 = 0 \quad or \quad s - 3 = 0$$
$$s = -3 \quad or \quad s = 3$$

4. Check. The speed of the stream cannot be negative. A speed of 3 km/h checks.

5. State. The speed of the stream is 3 km/h.

Section 9.6

5.
$$x^2 + 6x + 8 = 0$$
$$(x + 4)(x + 2) = 0$$
$$x + 4 = 0 \quad or \quad x + 2 = 0$$
$$x = -4 \quad or \quad x = -2$$

The x-intercepts are $(-4, 0)$ and $(-2, 0)$.

7. $x^2 + 3 = 0$
$$x^2 = -3$$

Since -3 is negative, the equation has no real-number solutions. There are no x-intercepts.

Section 9.7

9. $p(x) = x^4 - 5x^2 + 8$
$$p(0) = 0^4 - 5 \cdot 0^2 + 8$$
$$= 0 - 0 + 8$$
$$= 8$$

13. $F(r) = \sqrt{r} + 9$
$$F(100) = \sqrt{100} + 9$$
$$= 10 + 9$$
$$= 19$$

Glossary

A

Abscissa   The first coordinate in an ordered pair of numbers

Absolute value   The distance that a number is from 0 on the number line

ac-method   A method for factoring trinomials of the type $ax^2 + bx + c, a \neq 1$, involving the product, ac, of the leading coefficient a and the last term c

Additive identity   The number 0

Additive inverse   A number's opposite; two numbers are additive inverses of each other if their sum is 0

Algebraic expression   An expression consisting of variables, constants, numerals, operation signs, and/or grouping symbols

Area   The number of square units that fill a plane region

Arithmetic numbers   The whole numbers and the positive fractions. All these numbers can be named with fraction notation $\frac{a}{b}$, where a and b are whole numbers and $b \neq 0$.

Ascending order   When a polynomial is written with the terms arranged according to degree from least to greatest, it is said to be in ascending order.

Associative law of addition   The statement that when three numbers are added, regrouping the addends gives the same sum

Associative law of multiplication   The statement that when three numbers are multiplied, regrouping the factors gives the same product

Average   A center point of a set of numbers found by adding the numbers and dividing by the number of items of data; also called the *arithmetic mean* or *mean*

Axes   Two perpendicular number lines used to identify points in a plane

B

Base   In exponential notation, the number being raised to a power

Binomial   A polynomial composed of two terms

C

Circumference   The distance around a circle

Coefficient   The numerical multiplier of a variable

Commutative law of addition   The statement that when two numbers are added, changing the order in which the numbers are added does not affect the sum

Commutative law of multiplication   The statement that when two numbers are multiplied, changing the order in which the numbers are multiplied does not affect the product

Complementary angles   Angles whose sum is 90°

Completing the square   Adding a particular constant to an expression so that the resulting sum is a perfect square

Complex fraction expression   A rational expression that has one or more rational expressions within its numerator and/or denominator

Complex rational expression   A rational expression that has one or more rational expressions within its numerator and/or denominator

Complex-number system   A number system that contains the real-number system and is designed so that negative numbers have defined square roots

Composite number   A natural number, other than 1, that is not prime

Conjugates   Pairs of radical terms, like $\sqrt{a} + \sqrt{b}$ and $\sqrt{a} - \sqrt{b}$ or $c + \sqrt{d}$ and $c - \sqrt{d}$, for which the product does not have a radical term

Consecutive even integers   Even integers that are two units apart

Consecutive integers   Integers that are one unit apart

Consecutive odd integers   Odd integers that are two units apart

Constant   A known number

Constant of proportionality   The constant in an equation of direct variation or inverse variation

Coordinates   The numbers in an ordered pair

Cube root   The number c is called a cube root of a if $c^3 = a$.

D

Decimal notation   A representation of a number containing a decimal point

Degree of a polynomial   The degree of the term of highest degree in a polynomial

Degree of a term   The sum of the exponents of the variables

Denominator   The bottom number in a fraction

Descending order   When a polynomial is written with the terms arranged according to degree from greatest to least, it is said to be in descending order.

Diameter   A segment that passes through the center of a circle and has its endpoints on the circle

Difference of cubes   Any expression that can be written in the form $A^3 - B^3$

Difference of squares   Any expression that can be written in the form $A^2 - B^2$

Direct variation   A situation that translates to an equation described by $y = kx$, with k a positive constant

Discriminant   The radicand, $b^2 - 4ac$, from the quadratic formula

Distributive law of multiplication over addition The statement that multiplying a factor by the sum of two numbers gives the same result as multiplying the factor by each of the two numbers and then adding

Distributive law of multiplication over subtraction The statement that multiplying a factor by the difference of two numbers gives the same result as multiplying the factor by each of the two numbers and then subtracting

Domain The set of all first coordinates of the ordered pairs in a function

E

Elimination method An algebraic method that uses the addition principle to solve a system of equations

Empty set The set without members

Equation A number sentence that says that the expressions on either side of the equals sign, $=$, represent the same number

Equation of direct variation An equation described by $y = kx$, with k a positive constant, used to represent direct variation

Equation of inverse variation An equation described by $y = k/x$, with k a positive constant, used to represent inverse variation

Equivalent equations Equations with the same solutions

Equivalent expressions Expressions that have the same value for all allowable replacements

Equivalent inequalities Inequalities that have the same solution set

Evaluate To substitute a value for each occurrence of a variable in an expression and carry out the operations

Exponent In expressions of the form a^n, the number n is an exponent. For n a natural number, a^n represents n factors of a.

Exponential notation A representation of a number using a base raised to a power

F

Factor *Verb*: To write an equivalent expression that is a product. *Noun*: A multiplier

Factorization of a polynomial An expression that names the polynomial as a product

FOIL To multiply two binomials by multiplying the First terms, the Outside terms, the Inside terms, and then the Last terms

Formula An equation that uses numbers or letters to represent a relationship between two or more quantities

Fraction equation An equation containing one or more rational expressions; also called a *rational equation*

Fraction notation A number written using a numerator and a denominator

Function A correspondence that assigns to each member of a set called the domain *exactly one* member of a set called the range

G

Grade The measure of a road's steepness

Graph A picture or diagram of the data in a table; a line, a curve, or a collection of points that represents all the solutions of an equation or an inequality

Greatest common factor (GCF) The common factor of a polynomial with the largest possible coefficient and the largest possible exponent(s)

H

Hypotenuse In a right triangle, the side opposite the 90° angle

I

Identity Property of 1 The statement that the product of a number and 1 is always the original number

Identity Property of 0 The statement that the sum of a number and 0 is always the original number

Index In the radical $\sqrt[n]{a}$, the number n is called the index.

Inequality A mathematical sentence using $<$, $>$, $\leq$, $\geq$, or $\neq$

Input A member of the domain of a function

Integers The whole numbers and their opposites

Intercept The point at which a graph intersects the x- or the y-axis

Intersection of sets A and B The set of all elements that are common to both A and B

Inverse variation A situation that translates to an equation described by $y = k/x$, with k a positive constant

Irrational number A real number that cannot be named as a ratio of two integers

L

Leading coefficient The coefficient of the term of highest degree in a polynomial

Least common denominator (LCD) The least common multiple of the denominators of two or more fractions

Least common multiple (LCM) The smallest number that is a multiple of two or more numbers

Legs In a right triangle, the two sides that form the right angle

Like radicals Radicals that have the same radicand

Like terms Terms that have exactly the same variable factors

Line of symmetry A line that can be drawn through a graph such that the part of the graph on one side of the line is an exact reflection of the part on the opposite side

Linear equation Any equation that can be written in the form $y = mx + b$ or $Ax + By = C$, where x and y are variables

Linear function A function that can be described by an equation of the form $y = mx + b$, where x and y are variables

Linear inequality An inequality whose related equation is a linear equation

M

Mean A center point of a set of numbers found by adding the numbers and dividing by the number of items of data; also called the *average*

Median In a set of data listed in order from smallest to largest, the middle number if there is an odd number of data items, or the average of the two middle numbers if there is an even number of data items

Mode The number or numbers that occur most often in a set of data

Monomial An expression of the type ax^n, where a is a real number constant and n is a nonnegative integer

Motion problem A problem that deals with distance, speed (or rate), and time

Multiple   A product of a number and some natural number

Multiplication property of 0   The statement that the product of 0 and any real number is 0

Multiplicative identity   The number 1

Multiplicative inverses   Reciprocals; two numbers whose product is 1

N

nth root   The number c is the nth root of a if $c^n = a$.

Natural numbers   The counting numbers: 1, 2, 3, 4, 5, . . .

Negative integers   The integers to the left of zero on the number line

Nonnegative rational numbers   The whole numbers and the positive fractions. All these numbers can be named with fraction notation $\frac{a}{b}$, where a and b are whole numbers and $b \neq 0$.

Numerator   The top number in a fraction

O

Opposite   The opposite, or additive inverse, of a number a is denoted $-a$. Opposites are the same distance from 0 on the number line but on different sides of 0.

Opposite of a polynomial   To find the opposite of a polynomial, replace each term with its opposite—that is, change the sign of every term.

Ordered pair   A pair of numbers of the form (h, k) for which the order in which the numbers are listed is important

Ordinate   The second coordinate in an ordered pair of numbers

Origin   The point on a graph where the two axes intersect

Output   A member of the range of a function

P

Parabola   A graph of a quadratic equation

Parallel lines   Lines in the same plane that never intersect. Two nonvertical lines are parallel if they have the same slope and different y-intercepts.

Parallelogram   A four-sided polygon with two pairs of parallel sides

Percent notation   A representation of a number as parts per 100

Perfect square   A rational number p for which there exists a number a for which $a^2 = p$

Perfect-square trinomial   A trinomial that is the square of a binomial

Perimeter   The distance around a polygon, or the sum of the lengths of its sides

Perpendicular lines   Lines that form a right angle. Two lines are perpendicular if the product of their slopes is -1 or if one line is vertical and the other is horizontal.

Pi (π)   The number that results when the circumference of a circle is divided by its diameter; $\pi \approx 3.14$, or 22/7

Point–slope equation   An equation of the form $y - y_1 = m(x - x_1)$, where m is the slope and (x_1, y_1) is a point on the line

Polygon   A closed geometric figure with three or more sides

Polynomial   A monomial or a combination of sums and/or differences of monomials

Polynomial equation   An equation in which two polynomials are set equal to each other

Positive integers   The natural numbers, or the integers to the right of zero on the number line

Prime factorization   A factorization of a composite number as a product of prime numbers

Prime number   A natural number that has *exactly two different factors*: itself and 1

Prime polynomial   A polynomial that cannot be factored using only integer coefficients

Principal square root   The positive square root of a number

Principle of zero products   The statement that an equation $ab = 0$ is true if and only if $a = 0$ is true or $b = 0$ is true, or both are true

Proportion   An equation stating that two ratios are equal

Proportional numbers   Two pairs of numbers having the same ratio

Pythagorean theorem   In any right triangle, if a and b are the lengths of the legs and c is the length of the hypotenuse, then $a^2 + b^2 = c^2$.

Q

Quadrants   The four regions into which the axes divide a plane

Quadratic equation   An equation equivalent to an equation of the type $ax^2 + bx + c = 0$, where $a \neq 0$

Quadratic formula   The solutions of $ax^2 + bx + c = 0$, $a \neq 0$, are given by the equation $x = \dfrac{-b \pm \sqrt{b^2 - 4ac}}{2a}$.

Quadratic function   A second-degree polynomial function in one variable

R

Radical equation   An equation in which a variable appears in one or more radicands

Radical expression   An algebraic expression written under a radical

Radical symbol   The symbol $\sqrt{}$

Radicand   The expression under the radical

Radius   A segment with one endpoint on the center of a circle and the other endpoint on the circle

Range   The set of all second coordinates of the ordered pairs in a function

Rate   The ratio of two different kinds of measure

Ratio   The quotient of two quantities

Rational equation   An equation containing one or more rational expressions; also called a *fraction equation*

Rational expression   A quotient, or ratio, of two polynomials

Rational number   A number that can be written in the form a/b, where a and b are integers and $b \neq 0$

Rationalizing the denominator   A procedure for finding an equivalent expression without a radical in the denominator

Real numbers   All rational numbers and irrational numbers; the set of all numbers corresponding to points on the number line

Reciprocal   A multiplicative inverse. Two numbers are reciprocals if their product is 1.

Rectangle   A four-sided polygon with four right angles

Relation A correspondence between a first set called the domain, and a second set called the range, such that each member of the domain corresponds to *at least one* member of the range

Repeating decimal A decimal in which a number pattern repeats indefinitely

Right triangle A triangle that includes a 90° angle

Rise The change in the second coordinate between two points on a line

Roster notation A way of naming sets by listing all the elements in the set

Rounding Approximating the value of a number; used when estimating

Run The change in the first coordinate between two points on a line

S

Scientific notation A representation of a number of the form $M \times 10^n$, where n is an integer, $1 \leq M < 10$, and M is expressed in decimal notation

Set A collection of objects

Set-builder notation The naming of a set by describing basic characteristics of the elements in the set

Similar triangles Triangles in which corresponding angles have the same measure and the lengths of corresponding sides are proportional

Simplest fraction notation A fraction written with the smallest numerator and denominator

Simplify To rewrite an expression in an equivalent, abbreviated, form

Slope The ratio of the rise to the run for any two points on a line

Slope–intercept equation An equation of the form $y = mx + b$, where x and y are variables

Solution A replacement for the variable that makes an equation or inequality true

Solution of a system of equations An ordered pair that makes both equations true

Solution set The set of all solutions of an equation, an inequality, or a system of equations or inequalities

Solve To find all solutions of an equation, an inequality, or a system of equations or inequalities; to find the solution(s) of a problem

Square A four-sided polygon with four right angles and all sides of equal length

Square of a number A number multiplied by itself

Square root The number c is a square root of a if $c^2 = a$.

Subsets Sets that are a part of other sets

Substitute To replace a variable with a number

Substitution method A nongraphical method for solving systems of equations

Sum of cubes An expression that can be written in the form $A^3 + B^3$

Sum of squares An expression that can be written in the form $A^2 + B^2$

Supplementary angles Angles whose sum is 180°

System of equations A set of two or more equations that are to be solved simultaneously

T

Term A number, a variable, or a product or a quotient of numbers and/or variables

Terminating decimal A decimal that can be written using a finite number of decimal places

Triangle A three-sided polygon

Trinomial A polynomial that is composed of three terms

Trinomial square The square of a binomial expressed as three terms

U

Union of sets A and B The set of all elements belonging to either A or B

V

Value The numerical result after a number has been substituted into an expression

Variable A letter that represents an unknown number

Variation constant The constant in an equation of direct variation or inverse variation

Vertex The point at which the graph of a quadratic equation crosses its line of symmetry

Vertical-line test The statement that a graph represents a function if it is impossible to draw a vertical line that intersects the graph more than once

Volume The number of cubic units that fill a solid region

W

Whole numbers The natural numbers and 0: 0, 1, 2, 3, . . .

X

x-intercept The point at which a graph crosses the x-axis

Y

y-intercept The point at which a graph crosses the y-axis

Index